Proceedings of the Fifth European Rheology Conference

FIFTH EUROPEAN RHEOLOGY CONFERENCE
Portorož, Slovenia, September 6-11, 1998

ORGANIZING COMMITTEE

Emri I., Chair,	University of Ljubljana, Center for Experimental Mechanics
Cvelbar R.,	University of Ljubljana, Center for Experimental Mechanics
Malavašič T.,	National Institute of Chemistry, Ljubljana
Osredkar U.,	National Institute of Chemistry, Ljubljana
Šebenik Z.,	National Institute of Chemistry, Ljubljana
Zupančič A.,	University of Ljubljana, Faculty of Chemistry and Chemical Technology
Žumer M.,	University of Ljubljana, Faculty of Chemistry and Chemical Technology

European Committee on Rheology, serving as Scientific Committee:

Diogo A.C.,	Instituto Superior Técnico, Lisboa, Portugal
Djabourov M.,	Ecole Superieure de Physique et Chimie Industrielle, Paris, France
Emri I.,	University of Ljubljana, Ljubljana, Slovenia
Ferguson J.,	University of Strathclyde, Glasgow, Great Britain
Gallegos C.,	Universidad de Huelva, Huelva, Spain
Grizzuti N.,	Univerista di Napoli "Fererico II", Napoli, Italy
Ivanov Y.,	Bulgarian Academy of Sciences, Sofia, Bulgaria
Klason C.,	Chalmers University of Technology, Gothenburg, Sweden
Korugic L.,	L-Università tà Malta, Msida, Malta
Kulichikhin V.,	The Russian Academy of Sciences, Moscow, Russia
Lapasin R.,	Universita di Trieste, Trieste, Italy
Marrucci G.,	Univerista di Napoli "Fererico II", Napoli, Italy
Meijer H.E.H.,	Eindhoven University of Technology, Eindhoven, The Netherlands
Mewis J.,	Katholieke Univerisiteit Leuven, Heverlee, Belgium
Mitsoulis E.,	University of Ottawa, Ottawa, Canada
Moldeaners P.,	Katholieke Univerisiteit Leuven, Heverlee, Belgium
Münstedt H.,	Universität Erlangen - Nürnberg, Erlangen, Germany
Öttinger H.C.,	ETH Züerich, Zürich, Switzerland
Piau J.-M.,	Univerisite Grenoble, Grenoble, France
Ribitsch V.,	Univerisität Graz, Graz, Austria
Schausberger A.,	Jahannes-Kepler-Universität Linz, Linz, Austria
Stading M.,	STH/SIK - Biofysik, Gothenburg, Sweden
Šestak J.,	CVUT, Prague, Czech Republic
Wagner M.H.,	Universität Stuttgart, Stuttgard, Germany
Walters K.,	University of Wales, Aberystwyth, Great Britain

Die Deutsche Bibliothek –
CIP-Einheitsaufnahme

Progress and trends in rheology V :
proceedings of the Fifth European
Rheology Conference, Portorož,
Slovenia, September 6 - 11, 1998 /
ed. I. Emri.
Technical co-ed. R. Cvelbar. -
Darmstadt : Steinkopff, 1998

ISBN 978-3-642-51064-9
ISBN 978-3-642-51062-5 (eBook)
DOI 10.1007/978-3-642-51062-5

Originally published by Dr. Dietrich Steinkopff Verlag GmbH & Co. KG, Darmstadt

Chemisty Editor: Dr. Maria Magdalene Nabbe;
Production: Holger Frey

Progress and Trends in Rheology V

Proceedings of the Fifth European Rheology Conference
Portorož, Slovenia, September 6-11, 1998

Editor

I. Emri
Center for Experimental Mechanics
University of Ljubljana, Ljubljana, Slovenia

Technical Co-Editor

R. Cvelbar
Center for Experimental Mechanics
University of Ljubljana, Ljubljana, Slovenia

PREFACE

Global sustainable development of the world economy requires better understanding and utilization of natural recourses. In this endeavor rheology has an indispensable role. The Rheology Conferences are therefore always an important event for science and technology.

The Fifth European Rheology Conference, held from September 6 to 11 in the Portorož, Slovenia, will be the first All-European rheology meeting after the formal constitution of the European Society of Rheology. As such it will be a special historical event.

At this meeting the European Society of Rheology will introduce the *Weissenberg Medal,* to be bestowed every four years to an individual for his/hers contribution to the field of Rheology. The recipient of the first award will be professor G. Marrucci of the Universita degli Studi di Napoli, Italy.

Two mini Symposia will be part of the Conference. The first, on Industrial Rheology, will commemorate the late professor G. Astarita. The second will honor the eightieth birthday of professor N.W. Tschoegl.

This volume comprises extended abstracts of the 15 plenary and keynote lectures and about 300 oral and poster contributions presented at this conference. All contributed papers were reviewed by members of the European Committee on Rheology, assuring the high standard of the Conference.

Besides the scientific program, the Organizing Committee has prepared an extensive social program that will reveal the culture and the natural beauties of Slovenia.

Finally, on the behalf of the Organizing Board, we wish to acknowledge contributions of those who helped to the success of this conference. Important among these is the support of the Slovenian Ministry of Science and Technology.

The Editors

CONTENTS

PLENARY PAPERS

KEYNOTE PAPERS

BIORHEOLOGY

BLENDS

CONSTITUTIVE THEORIES AND EQUATIONS

ELECTRO AND MAGNETORHEOLOGY

FLUID MECHANICS

FOOD RHEOLOGY

INDUSTRIAL RHEOLOGY

LIQUID CRYSTALS

MATERIALS CHARACTERIZATION

MOLECULAR DYNAMICS AND THEORIES

NUMERICAL SIMULATIONS

POLYMER MELTS AND POLYMER PROCESSING

POLYMER SOLUTIONS

RHEOLOGY OF SOLIDS AND COMPOSITES

RHEOMETRY, EXPERIMENTAL METHODS AND INSTRUMENTATION

SURFACTANTS

SUSPENSIONS, EMULSIONS, DISPERSIONS, GELS

TIME DEPENDENT PHENOMENA

TIME-DEPENDENCE IN MATERIALS

N.W. Tschoegl, California Institute of Technology, Pasadena, CA 91125, USA

1. ABSTRACT

The last ten or fifteen years have seen some highly significant changes in what *needs* to, and what *can* be, achieved in the area of the characterization of materials with time-dependent mechanical properties. On the one hand we have seen a tremendous increase in the use and variety of materials with time-dependent properties such as composites and semi-crystalline engineering materials, as well as the development and application of multiphase materials (e.g., block and graft copolymers, and polyblends), requiring new or improved methods of characterization. On the other hand, there have been notable advances in the availability and sophistication of state-of-the-art sensors, and of methods of data acquisition and manipulation. It appears that the need for better characterization has not been universally recognized, and that, consequently, the possibilities for implementing it have not been fully exploited. The lecture examines these issues and suggests some new-old approaches.

The mechanical properties (and other properties: dielectric, optical, etc.) of *ALL* materials are time-dependent. This is simply a necessary consequence of the Second Law of Thermodynamics according to which a part of the energy of deformation is always dissipated by viscous forces even while the rest is stored elastically. Time-dependent behavior may range from virtually ***purely elastic*** to virtually ***purely viscous*** behavior. Polymeric materials are typically viscoelastic and thus fall in-between. As a consequence their mechanical properties must be described by time- (or, equivalently, frequency-) dependent material *response functions*, not elastic or viscous *constants*.

2. SMALL DEFORMATION BEHAVIOR

This presentation will be concerned mainly with *solid* polymeric materials. Their viscoelastic properties are well described by the phenomenological *theory of linear viscoelastic behavior*, as long as the behavior is *linear*, i.e., in the limit of small (theoretically *infinitesimally* small) deformations. Non-linear behavior is quite another matter.

A brief review of the state of the art will highlight some of the salient points of isothermal-isopiestic linear viscoelastic behavior. These may be summarized as follows:

1. Of the response functions most commonly used to characterize this behavior, the time- (or frequency-) dependent shear and bulk moduli and compliances are <u>fundamental</u> as these represent changes in shape or in bulk, respectively. All others are <u>derived</u> and entail changes both in shape *and* in bulk.
2. The response functions potentially allow insight into such structural features as cross-linking, entanglement, long-chain branching, crystallinity, the distribution of molecular mass, long-chain branching, and others more.
3. If their behavior is known over a sufficiently wide (i.e., *effectively* infinite) range of time or frequency in one time regime or mode of deformation, it is possible to predict it adequately in any other including interconverting between relaxation and retardation behavior.
4. Experimentally, an important problem is the necessity of covering a sufficiently wide range of time or frequency spanning anything from 10 to 15 to 20 or so decades of either. In *transient* measurements one can normally span about four and a half decades of time. In *dynamic* measurements another about three and a half decades may be covered.
5. Theoretically, an unsolved problem is the relation, if any, between shear and bulk relaxation or shear and bulk retardation.

The effects of the three cardinal thermodynamic properties: temperature, pressure and composition, cannot, at present, be satisfactorily accounted for in semi-crystalline, in multiphase materials, and in most composites. In these materials the devise, commonly used with amorphous single-phase polymers, for extending the time or frequency scale by shifting several isothermal segments into superposition is not available. The construction of *mastercurves* via superposition requires that *all* relaxation times depend equally on temperature. A full characterization of materials exhibiting more than one distribution of relaxation or retardation times therefore requires two-dimensional *contour plots* in which *isotimics* (lines of equal value, from the Greek *τιμη:* a value) of the portrayed parameter are represented as function of temperature (or pressure, or composition!) as well as of the logarithm of time or frequency. The only device still available for extending the time scale then is the combination of time-dependent and frequency-dependent (and possibly wave propagation) experiments. Measurements on different *specimens* drawn from the same *sample* of a material commonly may vary not insignificantly in their physical properties. I addition, experimental conditions such as temperature and humidity, cannot be guaranteed to be reproducible exactly in different apparatus. Extending the time scale then makes it highly desirable to carry out the measurements *on the same specimen, under identical experimental conditions.* This requires the construction of novel apparatus. One possibility might consist in applying the excitation as a series of A/C signals of varying frequencies imparted to the specimen from a stack of ceramic piezoelectric disks, and following these tests by a D/C signal. Care would have to be taken to avoid charge leakage from the stack during the transient measurement.

Molecular theories (the molecular model theory of Rouse and that of Zimm) are able to describe the dynamic (i.e., frequency-dependent) behavior below a certain frequency limit in infinitesimally dilute solutions, i.e., the behavior of single chainlike molecules. Numerous attempts to extend these theories have not resulted in significant improvement.

3 LARGE DEFORMATION BEHAVIOR

If time-dependence is not an issue, moderately large purely elastic behavior is well accounted for by the celebrated molecular model theory known as the *statistical theory of rubber elasticity*.

The non-time-dependent large (i.e., non-linear) elastic phenomenological behavior of rubberlike materials (elastomers) has been well worked out in continuum mechanics for materials which may be deemed incompressible ($\Delta V = 0$). Time dependence in these materials is often negligible, particularly, if their mechanical properties are determined in mechanical quasi-equilibrium by applying the deformation incrementally, and waiting after each increment until stress relaxation has become effectively negligible.

Continuum mechanicists generally consider any stress-strain relation that does not violate the principles of continuum mechanics a *constitutive equation*. Such equations are routinely derived from fitting data obtained in uniaxial tension tests at constant rate of strain tests. Such equations should be regarded as constitutive only in the *wider sense*, and even then, however, constitutiveness is really a misnomer because the parameters in such equations do not necessarily reflect the constitution of matter. Most often they are not material constants, but merely curve-fitting parameters.

Since this presentation is concerned with time-dependent properties, the state of the art on elastic large deformations will not be reviewed here in detail but it is necessary to point out the continuum mechanical apparatus for describing large elastic behavior. This is governed by the constants appearing in the Helmholtz free energy of deformation (also referred to as strain energy density function, or as elastic potential). In the general case this is

$$W = \sum_{i,j=1}^{\infty} C_{ij}(I_1 - 3)^i (I_2 - 3)^j$$

where the C_{ij} are constants, $I_1 = \lambda_1^2 + \lambda_2^2 + \lambda_3^2$, $I_2 = \lambda_1^2\lambda_2^2 + \lambda_2^2\lambda_3^2 + \lambda_3^2\lambda_1^2$, and (for an incompressible material) $I_3 = \lambda_1\lambda_2\lambda_3 = 1$. The *I's* are the first, second, and third invariants of the deformation tensor, and the λ's are the stretch ratios in the three principal directions.

While mathematically perfectly correct, the continuum mechanics approach is strictly so only if W is expanded to a large (theoretically infinite) number of terms. Truncating the expansion after retaining only a manageably small number (say, five to six) of terms, the equation that fits data in one mode of deformation will not necessarily fit data in another mode with the same numerical values of the parameters and, indeed, might fit the data in another deformation only with another choice of five to six terms. Such equations are therefore simply *curve-fitting* equations—albeit they may be useful if applied with full knowledge of their limited nature.

In the *narrower sense* (the sense of the material scientist, as distinct from the continuum mechanicist), constitutive behavior requires that the material parameters, i.e., the constants or time-dependent functions that appear in a stress-strain relation, reflect the constitution of the matter the behavior of which the equation purports to portray. This, in turn, requires that the *same* parameters describe behavior in *any* mode of deformation. Alas, constitutiveness can never be proven. It is disproven, however, if data obtained in a deformation mode (say, equibiaxial tension) that is different from that from which the equation was derived (say, uniaxial tension), cannot be fitted using the same material parameters. The literature is chockfull of 'constitutive' equations derived from data obtained in uniaxial tension because of experimental convenience, whose material parameters, however, have never been checked out in another deformation.

If the material may be deemed incompressible, the situation is relatively simple. On a plot of the second invariant against the first, all admissible deformations of an incompressible material must lie in a wedge-shaped region bounded by the equibiaxial and the uniaxial tension lines. If, therefore, data in both these deformations can be fitted with the same parameters, these parameters can safely be assumed to be constitutive in the narrower sense since it is then highly unlikely that it should not be possible to fit any other admissible deformation with the same parameters. Some simple elastic potentials that appear to comply with these requirements are available in moderately large elastic deformations.

With respect to *time-dependent* large deformations the situation is not as clear. Molecular model theories are lacking. The completely general continuum mechanics approach requires 15 functions of the three invariants of the deformation tensor, 12 of which are time-dependent. The determination of such a plethora of parameters, although theoretically correct, is nonsense in practice.

Complete generality is, however, not all that often required. It appears that, in moderately large deformations of incompressible materials, i.e., precisely the conditions for which acceptable purely elastic constitutive equations exist, time-shift invariance may be preserved. The material functions of the theory of infinitesimal time-dependent behavior, remain valid in such moderately large deformations when the infinitesimal strain measure (the Cauchy strain) is replaced by a suitable constitutive large deformation measure.

4. ULTIMATE BEHAVIOR

Time-dependence influences also the failure behavior of materials whether that be fracture, tear, rupture, or any other form of failure involving separation into two or more parts, or whether it involves yielding, necking, craze formation, or any other form of failure that does not entail separation. A good illustration of this is the *failure envelope* of Thor Smith. Plotting the stress-at-break against the strain-at-break at various temperatures and strain rates often yields a remarkably reasonable envelope over the endpoints of the stress-strain curves. Although the concept does not seem to hold quite generally, and although no mathematical equation has been found to express the envelope, the concept is useful.

Development of a general theory of failure behavior in constitutive terms appears to be an elusive—if not a downright impossible dream—because failure is an inherently stochastic process.

A CONSTITUTIVE ANALYSIS OF UNIAXIAL, EQUIBIAXIAL AND PLANAR EXTENSION OF LINEAR AND BRANCHED POLYETHYLENE MELTS

M.H. WAGNER[1], H. BASTIAN[1], P. EHRECKE[1], P. HACHMANN[2], J. MEISSNER[3]

[1]Institut für Kunststofftechnologie, Universität Stuttgart, Böblinger Str. 70, D-70199 Stuttgart
[2]Institut für Polymere, ETHZ, CH-8092 Zürich
[3]Swiss F.I.T. Rheocenter, ETHZ, CH-8092 Zürich

1. INTRODUCTION

Nonlinear strain measures and effective damping functions are derived for a linear (HDPE) and a long-chain branched (LDPE) polyethylene melt from constant strain-rate extensional experiments of Hachmann (1996). Experimental results are compared with predictions of two molecular theories, the Doi-Edwards model and the molecular stress function approach of Wagner and Schaeffer. Orientation (but not stretch) of macromolecules at small and large strains is well represented by the Doi-Edwards orientation tensor. Stretch of macromolecules is found to be isotropic as predicted by the molecular stress function model. While the "strain-hardening" behaviour of the linear (HDPE) melt can be accounted for by a tube diameter, which decreases affinely with the average stretch, the LDPE melt shows an enhanced strain-hardening due to the presence of long-chain branches which leads to an overproportional increase of the molecular stress with increasing deformation.

2. EXPERIMENTAL DATA

Experimental data analysed are those for a commercial grade of HDPE (Statoil 870H, M_w=104000g/mol, M_w/M_n=5.5), and a commercial grade of LDPE (Lupolen 1810H from BASF AG, M_w=188000g/mol, M_w/M_n=11.3). Melt rheology was investigated with a new extensional rheometer capable of performing constant strain-rate experiments in uniaxial, equibiaxial and planar extension on commercial polyolefine melts at processing temperatures. Details of the rheometer design and experimental procedures including control of strain rate, correction for surface tension, experimental error and repeatability are given by Hachmann (1996).

Figs.1 and 2 show the uniaxial (μ_u), equibiaxial (μ_e), and planar viscosities. (μ_{p1} is the planar viscosity in the direction of extension, while μ_{p2} is the planar viscosity in the neutral direction.) All viscosities are rescaled to the time-dependent zero-shear viscosity $\eta_0(t)$, which is also shown in Figs.1 and 2.

As seen from Figs.1 and 2, both melts investigated show considerable strain-hardening effects in the elongational viscosity μ_u. From previous elongational experiments with long-chain branched low-density polyethylene (LDPE) melt, it was concluded that long-chain branches lead to strain hardening. However, investigations by Linster and Meissner (1986) on several HDPE melts indicated that also a small amount of very long, linear molecules in the molecular weight distribution of the polymer with no or almost no long-chain branching leads to the same phenomenon, and indeed, for the HDPE melt investigated here, infrared measurements revealed that this material consists of linear molecules, at least within the resolution of present-day IR technique (Kraft, 1996).

3. CONSTITUTIVE EQUATIONS AND NONLINEAR STRAIN MEASURES

Single integral constitutive equations of the type

$$\underline{\underline{\sigma}}(t) = -p\underline{\underline{E}} + \int_{-\infty}^{t} \dot{m}(t-t')\underline{\underline{S}}_t(t')dt' \qquad (1)$$

with the stress tensor $\underline{\underline{\sigma}}$, the isotropic pressure p, the unit tensor $\underline{\underline{E}}$, and a nonlinear strain measure $\underline{\underline{S}}_t$ have been quite useful in analysing the rheology of polymer melts. The memory function $\dot{m}(t-t')$ characterizes the linear-viscoelastic behaviour of the material.

From molecular arguments, Doi and Edwards (1978 and 1979) model intermolecular interaction in concentrated systems of linear polymer chains by the tube concept. Assuming a constant tube diameter a_0, this leads to the nonlinear strain measure $\underline{\underline{S}}_{DE}$,

$$\underline{\underline{S}}_{DE} = 5 \left\langle \frac{\underline{u}'\underline{u}'}{u'^2} \right\rangle \tag{2}$$

if the further assumption is made (the independent alignment assumption I.A.A.) that the orientation of a "primitive step" rotates affinely with the macroscopic deformation. An isotropic distribution of molecular strands normalized with respect to their equilibrium length can be described by an isotropic distribution of unit vectors $\underline{u}$. The inverse deformation gradient $\underline{\underline{F}}^{-1}$ transforms a unit vector $\underline{u}$ into a deformed vector $\underline{u}'$,

$$\underline{u}' = \underline{\underline{F}}^{-1} \cdot \underline{u}$$

Affine rotation of a unit vector can be described by normalizing $\underline{u}'$ with respect to its length u',

$$u' = \sqrt{\underline{u}' \cdot \underline{u}'}$$

<...> denotes an integral over an isotropic distribution of unit vectors before deformation,

$$< ... > = \frac{1}{4\pi} \int ... \, d\Omega$$

where dΩ is an infinitesimal solid angle, and the integration is over the surface of a unit sphere. The strain measure $\underline{\underline{S}}_{DE}$, eq.(2), is thus seen to be the second rank orientation tensor.

The nonlinear strain measure of the molecular stress function theory (Wagner and Schaeffer, 1992 and 1993), $\underline{\underline{S}}_{MSF}$,

$$\underline{\underline{S}}_{MSF} = 5 \, f^2 \left\langle \frac{\underline{u}'\underline{u}'}{u'^2} \right\rangle \tag{3}$$

is the product of the square of the molecular stress function, f^2, and the orientation tensor from eq.(2). It is obtained by relaxing the constant tube diameter assumption of the Doi-Edwards model: The relative tension in the molecular chain (relative to its equilibrium value) is expressed by a molecular stress function f, which is the inverse of the relative tube diameter a/a_0,

$$f = \frac{a_0}{a} \tag{4}$$

Note that in this model, rotation of strands is assumed to be affine, while strand extension is independent of strand orientation, i.e. all strands irrespective of orientation are subjected to the same extension, which is represented by the molecular stress function f.

4. COMPARISON OF EXPERIMENTAL DATA AND MOLECULAR THEORY

The strain measure $\underline{\underline{S}}_{DE}$ (eq.(2)) of the Doi-Edwards theory is unable to predict the nonlinear strain behaviour in uni- and biaxial extensional flows of the polyethylene melts considered. This is demonstrated in Figs.1 and 2, where predictions of the Doi-Edwards model for viscosities in uniaxial elongation, equibiaxial extension, and planar extension are compared with experimental evidence: the strain hardening behaviour of the melts is not described by the model, and both melts are clearly not Doi-Edwards fluids in extensional deformations.

However, as reported recently by Wagner, Ehrecke, Hachmann and Meissner (1998), time and deformation effects are separable for the linear (HDPE) melt, and orientation (but not stretch) of macromolecules at small and large strains is well represented by the Doi-Edwards orientation tensor. Stretch of macromolecules was found to be isotropic as predicted by the molecular stress function model, which is in excellent agreement with the data. From a novel exponential strain energy function w_{EX} (normalized by 3kT),

$$w_{EX} = e^{<\ln(u')>} - 1 \tag{5}$$

where <ln(u')> is the normalized rotational free energy, a molecular stress function was derived with

$$f^2 = \frac{a_0^2}{a^2} = e^{<\ln(u')>} \tag{6}$$

Considering that the average stretch field L/L_0 experienced by the "primitive steps" is given by

$$\frac{L}{L_0} = e^{<\ln(u')>}$$

Fig. 1: Time-dependent extensional viscosities of HDPE melt measured at 150° C and various strain rates $\dot{\varepsilon}_0$. $\eta_0(t)$ is the time-dependent zero-shear viscosity. Broken and solid lines are predictions of the strain measure $\underline{\underline{S}}_{DE}$ (eq.(2)), and $\underline{\underline{S}}_{MSF}$ (eq.(3)) with $f^2=e^{<\ln(u')>}$, respectively.

a) Uniaxial viscosity μ_u:

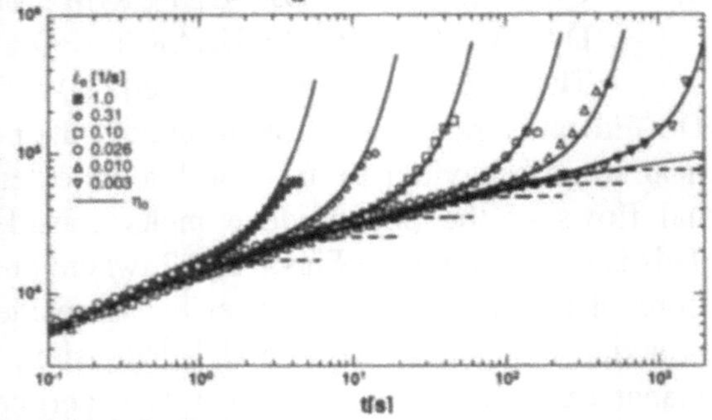

b) Equibiaxial viscosity μ_e:

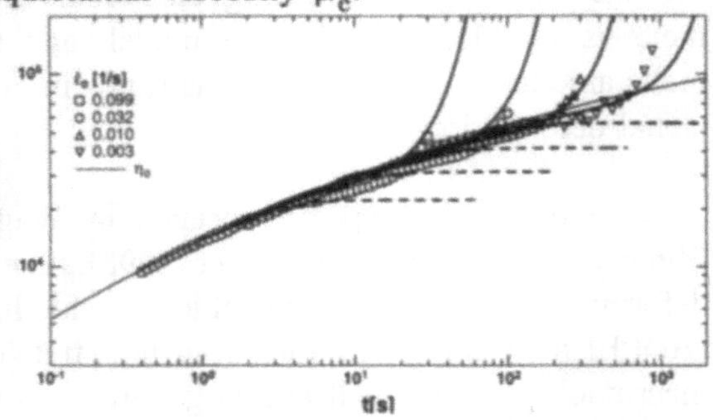

c) Planar viscosity μ_{p1} (direction of extension):

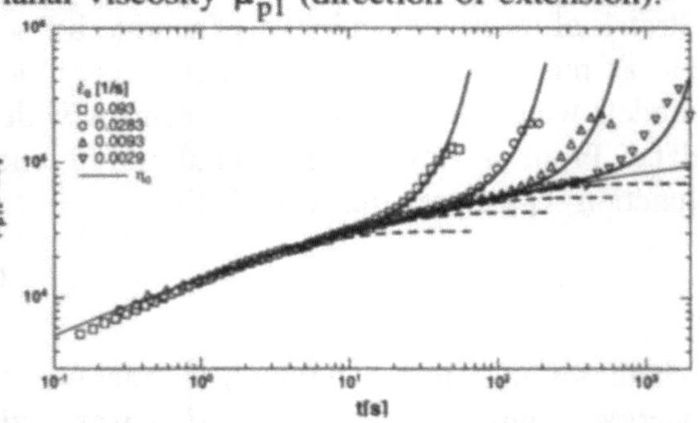

d) Planar viscosity μ_{p2} (neutral direction):

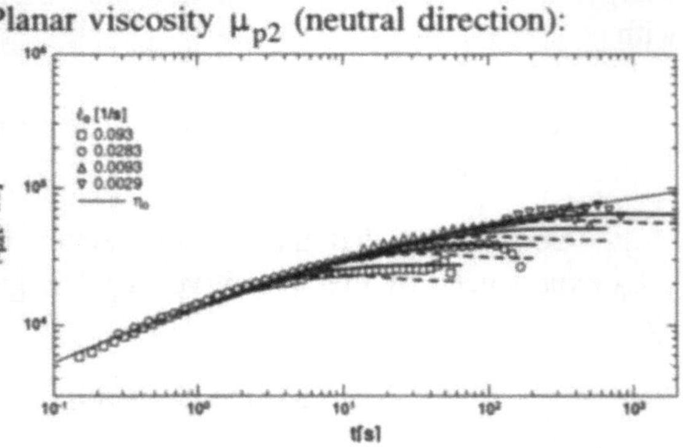

Fig. 2: Time-dependent extensional viscosities of LDPE melt measured at 150° C and various strain rates $\dot{\varepsilon}_0$. $\eta_0(t)$ is the time-dependent zero-shear viscosity. Broken and solid lines are predictions of the strain measure $\underline{\underline{S}}_{DE}$ (eq.(2)), and $\underline{\underline{S}}_{MSF}$ (eq.(3)) with $f^2=e^{<\ln(u')>}$, respectively.

a) Uniaxial viscosity μ_u:

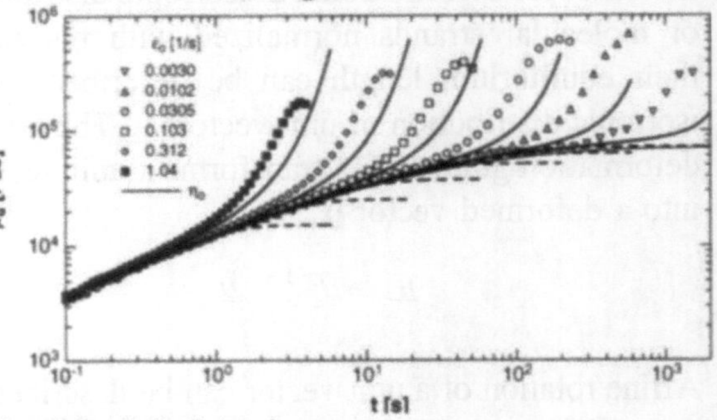

b) Equibiaxial viscosity μ_e:

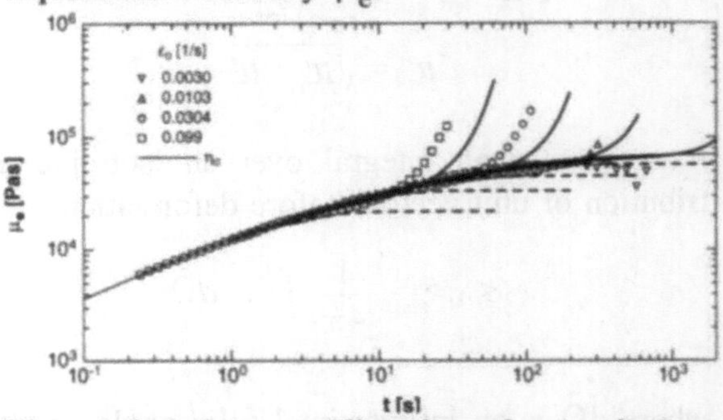

c) Planar viscosity μ_{p1} (direction of extension):

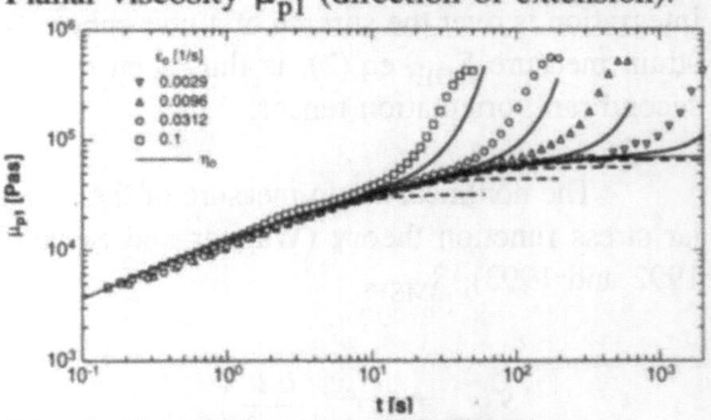

d) Planar viscosity μ_{p2} (neutral direction):

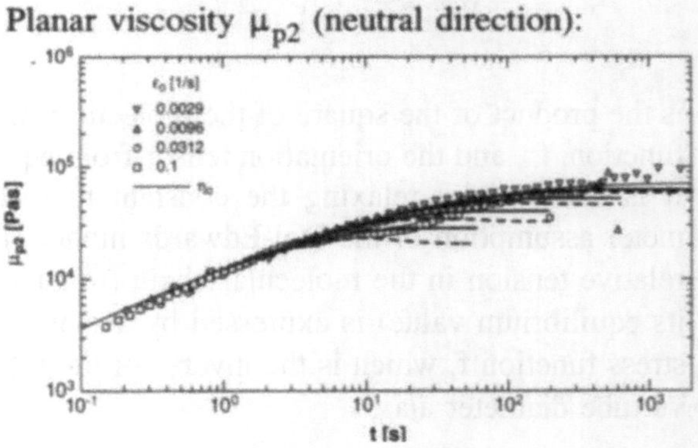

the cross section of the tube (and therefore the tube diameter) is seen to decrease affinely with the average stretch.

If w_{EX} is expressed in terms of the tube diameter a,

$$w_{EX} = \frac{a_0^2}{a^2} - 1 \quad (7)$$

it is obvious that w_{EX} is consistent with the normalized free energy w of a chain with n_s monomer units in a tube of diameter a, which is (Doi and Edwards, 1978)

$$w = \frac{n_s\, b^2}{a^2} - 1 \quad (8)$$

by equating $n_s b^2$ with a_0^2.

In Fig.1, predictions of the strain measure $\underline{\underline{S}}_{MSF}$ (eqs.(3) and (6)) are compared with experimental evidence for the linear (HDPE) melt. For all extensional deformation modes, a consistent description of the strain-hardening effect is found. Therefore up to large extensions, the HDPE melt can be considered as a molecular stress function (MSF) fluid with a tube diameter which is decreasing affinely with the average stretch. At still larger strains, the extensional viscosities are bound. This corresponds to a maximum value of the molecular stress function or a minimum tube diameter. For the linear (HDPE) melt, f^2_{max} is approximately 30, which corresponds to a maximum stretch ratio of the molecular chains of f_{max}= 5.5. The magnitude of f_{max} is intimately related to the strain-hardening effect of the elongational viscosity: a large value of f_{max} indicates a large strain-hardening effect, while non-strain-hardening polymer melts are characterized by small values of f_{max} (Wagner and Schaeffer 1993). As shown by Wagner, Ehrecke, Hachmann and Meissner (1998), f_{max} is the only nonlinear material parameter needed for a quantitative description of all extensional deformation modes of this HDPE melt.

For the long-chain branched (LDPE) melt, however, the extensional viscosities are underpredicted by the strain measure $\underline{\underline{S}}_{MSF}$ (eqs.(3) and (6)). The molecular stress is seen to increase faster with increasing strain than predicted by eq.(6). This can be attributed to the presence of long-chain branches in the melt: at larger deformations, long-chain branches are forced into conformations parallel to the backbone of the macromolecule, thereby increasing the average tension in the chains overproportionally. For the LDPE melt, a maximum value of the molecular stress function in the order of f_{max}=10 is found.

A comparison of the predictions of the strain measure $\underline{\underline{S}}_{MSF}$ (eqs.(3) and (6)) with experimental evidence in extensional deformation can therefore be used to distinguish between the strain-hardening effect caused by long-chain branching versus the strain-hardening effect caused by polydispersity of linear melts.

5. ACKNOWLEDGEMENT

Financial support from the ETH Zürich and the German Science Foundation (DFG) through grant No. Wa 668/3 is gratefully acknowledged.

6. REFERENCES

Doi M, Edwards SF (1978), Faraday Trans.II, 74:1818-1832

Doi M, Edwards SF (1979), Faraday Trans.II, 75:38-54

Hachmann P (1996) Multiaxiale Dehnung von Polymerschmelzen, Dissertation ETH Zürich Nr. 11890

Kraft M (1996) Untersuchungen zur scherinduzierten rheologischen Anisotropie von verschiedenen Polyethylen-Schmelzen, Dissertation ETH Zürich Nr. 11417

Linster JJ, Meissner J (1986), Polymer Bulletin 16: 187-194

Wagner MH, Schaeffer J (1992), J. Rheol. 36:1-26

Wagner MH, Schaeffer J (1993), J. Rheol. 37:643-661

Wagner MH, Ehrecke P, Hachmann P, Meissner J (1998), J. Rheol., 42:621-638

FLOW INSTABILITIES AS STRESS LIMITERS FOR MOLTEN POLYMERS

J. M. DEALY

Department of Chemical Engineering, McGill University, 3610 University Street
Montreal, Canada H3A 2B2

1. INTRODUCTION

Molten polymers and concentrated solutions in which the molecular weight is well above the critical value for entanglements can exhibit quite high shear and tensile stresses in flows that can be easily generated in the laboratory and in processing machinery. Various phenomena occur at these high stresses that have the effect of inhibiting the increase in stress as the rate of deformation is further increased. In the laboratory, these phenomena interfere with efforts to evaluate constitutive equations under conditions of large strain and strain rate, and in plastics forming operations they can produce unacceptable defects in extrudates. The occurrence of such phenomena is often attributed to some inadequacy of the apparatus generating the deformation.

In the computational fluid dynamics of elastics liquids, problems always arise in areas of the flow field where the stress is high. These problems are usually attributed to an inadequacy in the constitutive equation or in the numerical technique.

It seems likely, however, that highly entangled polymeric liquids avoid strain histories in which the stress is very high for a significant period of time. If we impose boundary conditions designed to generate such high stresses, we do not observe a homogeneous, stable flow. The fluid, in fact, has a repertoire of mechanisms for coping with this situation, and the purpose of this paper is to survey these mechanisms. These include hydrodynamic instability, wall slip, and melt fracture.

2. SECONDARY FLOW

We consider here flow fields that result from hydrodynamic instability. Of course Newtonian fluids also exhibit flow instabilities, but these result from inertia, surface tension, gravity or compressibility. We consider here only instabilities arising from the viscoelasticity of the material. We use the term "base flow" to mean the velocity field in the absence of hydrodynamic instabilities of any kind. "Secondary flow" implies a departure from the base flow, but one that is still governed by the equations of motion of continuum mechanics..

2.1 Rotational Flows

In flows in which one or more bounding surfaces rotates, the base flow has curvilinear streamlines. For inelastic fluids, secondary flows arise from inertia, i.e., centripetal acceleration. For example, in the case of the flow between concentric, rotating cylinders, i.e., Couette flow, there is a well known instability that leads to a secondary flow in the form of toroidal, "Taylor" vortices. The secondary flow thus involves flow normal to the streamlines of the base flow.

In the case of viscoelastic fluids, even at vanishing Reynolds number where inertial effects are absent, elasticity interacts with the curved streamlines to produce secondary flow normal to the streamlines of the base flow. Secondary flow of viscoelastic fluids has been observed for Couette flow, cone-plate flow, and parallel disk flow, among others. Larson (1992) has summarized our state of knowledge regarding these instabilities, and McKinley (1996) has proposed a general criterion for the limit of stability in such flows.

2.2 Entrance Flows

The pressure-driven flow from a reservoir into a smaller channel is a converging flow in which there is a strong extensional component of the flow. Although the flow boundaries are stationary, the streamlines of the base flow are curved, and elasticity can generate a secondary flow in a direction across these streamlines.

In the case of an axisymmetric contraction, the instability usually manifests itself as a spiral motion that produces an extrudate in the form of a helix. This has been observed, for example, by Tordella (1956), Berge (1976) and El Kissi and Piau (1990). In the case of flow into a slit, the secondary flow appears as a periodic oscillation of the principal stream surface about the plane of symmetry (Checker et al. 1983).

2.3 Sliding Plate Flow

A sliding plate rheometer is a device intended to generate plane Couette flow. The base flow has no curvature, and there is no secondary flow in the case of Newtonian fluids, although turbulence appears at a sufficiently high Reynolds number. In the elastic fluid, however, normal stress differences generate a pressure gradient that drives the flow inward along the x_1 axis and outward in the x_2 direction. This secondary flow has been observed for molten polyethylene (Koran, 1994) and polystyrene solutions (Larson et al. 1995; Reimers and Dealy 1996) and it can become the dominant feature of the velocity field at high nominal shear rates. It has the effect of limiting the growth of the shear stress as the plate speed is increased.

3. WALL SLIP AND APPARENT SLIP

Slip at a bounding surface is clearly a mechanism for reducing substantially the stress that would result from the motion of a melt along such a surface in the absence of slip. It now seems clear that what has been called "slip" in the flow of molten polymers is actually a group of phenomena, although the detailed nature of these is still not clear. True slip would arise from adhesive failure at the interface, but other phenomena can produce virtually the same effect from a macroscopic point of view.

3.1 True Wall Slip

Adhesive failure at a bounding surface relative to which the melt is flowing would certainly have the effect of reducing the shear stress. It is easy to generate this effect by use of a low-energy surfaces such as one coated by a fluoroelastomer (Hatzikiriakos et al 1993; Wang and Drda 1997), but it is now thought that it occurs rarely at a high energy interface, such as that between a polymer and metal. Long before stresses sufficient for adhesive failure are reached, one or more other mechanisms come into play to reduce the stress. Only after such mechanisms have been exhausted, and the stress is increased much further will adhesive failure occur. It is possible that this is the mechanism responsible for the second "spurt" observed, for example, by Li et al. (1986) for the flow of a polyolefin through a capillary.

3.2 Apparent Wall Slip

There is a growing acceptance of the hypothesis that what is most often referred to as slip in the flow of molten polymers on high energy surfaces is, in fact, the result of a dramatic change in the interaction between the molecules strongly adsorbed at the wall and those in the bulk. This change, which gives rise to the well known "spurt" phenomenon, is often described as "disentanglement" or a "coil stretch transition" (Wang and Drda 1996). Elegant studies involving monodisperse polymers with only one end adsorbed at the wall have provided convincing evidence that such a mechanism does indeed operate in such systems (Leger et al. 1997) but in the case of molten bulk polymers on high energy surfaces, each molecule is adsorbed at multiple points along its chain, forming a "dense pseudobrush" of loops and free ends (Guiselin 1992). While it is not possible to know the precise state of this complex adsorbed layer, empirical evidence based on capillary flow measurements provides strong evidence that true wall slip (adhesive failure) of polymers occurs only on low energy surfaces and that polymers can slip at high energy interfaces by a disentanglement mechanism when the stress and the viscosity are sufficiently high, and this only seems to occur in the case of certain linear polyethylenes (Yang et al. 1998).

4. MELT FRACTURE

Chain scission is the final stress limiter that comes into play when the polymer has exhausted other mechanisms for stress limitation. While it is easy to demonstrate the effect by manually stretching a rod formed from polydimethylsiloxane, it is difficult to subject a polymeric liquid to a tensile test in order to determine its strength. Nguyen and Kausch (1992), however, have shown that chain scission can be produced in a polymer solution in an extensional flow above a critical strain rate.

In a sliding plate rheometer, which subjects a melt to simple shear (plane Couette flow), various types of fracture, including tearing and cavitation, have been observed by Koran (1998). Archer et al. (1997) have observed cavitation in rotational flows, and Joseph (1995) has related the appearance of cavitation to the stress field in the fluid.

4.1 Gross Melt Fracture in Converging Flow

It has been known for many years that some melts exhibit a severe extrudate distortion called "elastic turbulence" or "gross melt fracture" at high flow rates in flow through converging dies Bagley and Schreiber 1961) and it was demonstrated by Howells and Benbow (1962) that this is accompanied by fragmentation of the melt in the die. Gau and Mackley (1994) have shown by use of birefringence that cracking occurs at random intervals of time in the flow of a highly structured "melt" of UHMW polyethylene into a slit die.

5. CONCLUSIONS

The various instabilities that occur in the flow of entangled polymeric liquids at high stresses are not exotic phenomena that occur because of imperfections in the experiment and prevent us from studying the flow of these fluids at high stresses. In fact, these materials do not flow in a homogenous way under such conditions, and the mechanisms that they employ to avoid high stresses provide a challenging field for experimentalists and theoreticians. At the present time, we have rather more observations of bizarre behaviour than we have explanations for them.

6. REFERENCES

Archer LA, Larson RG et al. (1995) J Fl Mech 301:133

Archer LA, Temet, D, Larson RG (1997) Rheol. Acta36:579

Bagley EB and Schreiber HP (1961) Trans Soc Rheol 5:341

Bergem N (1976) Proc. VIIth Intern Cong Rheol 50-54

Checker N, Mackley, MR et al. (1983) Phil Trans R Soc A308:451

El Kissi N and Piau JM (1990) J Non-Newt Fl Mech 37:55

Gau P, Mackley, MR (1994) Polym Mater Sci Eng 71:43

Guiselin O (1992) Europhys Lett 17:225

Hatzikiriakos SG, Stewart CW, Dealy JM (1993) Intern Polym Proc VIII: 30

Howells ER and Benbow JJ (1962) Trans Plast Inst 30:240

Joseph, DD (1995) Phys Rev E 51:1649

Koran K (1994) M.Eng. Dissert. Chem. Eng., McGill University

Koran K (1998) Ph.D. Dissert. Chem. Eng., McGill University

Larson RG (1992) Rheol Acta 31:213-263

Leger L, Hervet H et al. (1997) Trends Polym Sci 5:40

Li H, Hürlimann HP, Meissner J (1986) Polym Bull 15:83

McKinley GH, Pakdel P et al. (1996) J Non-Newt Fl Mech 67:19-47

Nguyen TQ, Kausch HH (1992) Polymer 33: 2611-2621

Reimers MJ and Dealy JM (1996) J Rheol 40:167

Tordella JP (1956) J Appl Phys 27:454

Wang SQ and Drda PA (1996) Macromolecules 29: 2627, 4115

Wang SQ and Drda (1997) Rheol Acta 36:128

Yang X, Ishida H et al. (1998) J Rheol 42:63

Recent Advances in Optical Rheometry

Gerald G. Fuller

Department of Chemical Engineering, Stanford University, Stanford, CA 94305-5025

1.0 Introduction

Optical techniques in rheometry continue to be developed to explore new microstructural dynamics in a variety of complex fluid systems. In this presentation, the application of *dichroism* is emphasized and is used to investigate a wide range of systems ranging from thin polymer films, to miscible and immiscible polymer blends. Dichroism is defined as the polarization-dependent attenuation of light transmitted through a material. Because of structural anisotropies induced by flow, most materials will produce dichroism by either absorption or scattering processes. The use of standard polarization modulation techniques makes it possible to acquire very sensitive, time-dependent measurements of these phenomena. Although the same optical arrangements can be used for either source of dichroism, different light sources are normally required: visible sources for scattering dichroism and often ultraviolet or infrared sources in the case of absorptive dichroism.

This optical phenomena can be used to elucidate the microstructural origins of complex rheology in a variety of systems and this presentation will focus on three types of materials: thin polymer films, miscible polymer blends, and immiscible polymer blends. In the first two examples, absorptive dichroism used to follow the orientation dynamics of polymer chains. It is demonstrated that scattering dichroism is an effective tool to follow the morphology of immiscible blends.

2.0 Absorptive Dichroism Investigations of Polymer Films and Miscible Blends

There is considerable interest in the dynamics of thin polymer films when subjected to flow. Film forming processes such as spin coating and Langmuir-Blodgett deposition direction involve the application of hydrodynamic forces to thin polymer layers and these can induce significant segmental orientation that must be controlled to produce optimal properties of the resulting films. In this work the orientation dynamics of Langmuir films consisting of rigid rod poly(para-phynylene sulfonic-acid) (PPPSH) polymer have been studied. This polymer absorbs strongly in the ultra-violet and for that reason an optical system based on a UV-argon-ion laser has been used to probe the chain orientation. Experiments are presented where these films are subjected to extensional flow processes. It is demonstrated that a nematic order is produced in these systems and evidence of a new "pressure-induced" order-disorder transition is presented.

The same PPPSH rigid rod polymer can be incorporated into bulk matrices of flexible chains to produce a molecular composite. Normally, such mixtures are very difficult to produce and result in phase separation. However, the group of Prof. Eisenbach of the University of Stuttgart has recently succeeded in producing miscible polymer blends of rigid and flexible chains by incorporating specific acid-base interactions between the components. The same UV-dichroism system described above has been used to probe the dynamics of mixtures of poly(para-phynylene sulfonic-acid) (PPPSH) and poly(thylacrylate-co-4-vinyl pyridine) (P(EA-co-4VP). This system is synthesized using a co-precipitation method. It is demonstrated that the dichroism measurements can be used to extract the dynamics of the rods whereas the matrix orientation can be independently monitored using birefringence measurements.

3.0 Scattering Dichroism by Immiscible Blends

Research exploring flow-induced morphology in immiscible blends has considered mixtures of poly(butadiene) dispersed in poly(isobutene). Both small angle light scattering and scattering dichroism have been used to monitor the evolution of drop deformation, break-up, and coalescence using *in situ* measurements. The measurements made at Stanford University are unique since they access the "plane of the flow": the plane containing the flow direction and the velocity gradient direction. In this manner, both the degree of drop deformation and the orientation of the drops relative to the flow axes can be obtained. These measurements provide direct evidence of the process of drop elongation and break up (Vermant *et al.* (1998)). In an extension to this work, we have made extensive measurements comparing both mechanical measurements with scattering dichroism measurements during oscillatory shear flow. This comparison has been used to establish a new "stress-optical" rule where scattering dichroism measurements can be used to identify the interfacial contributions to the stress tensor (Puyvelde *et al.* (1998)).

4.0 Acknowledgments

The work on Langmuir films and miscible rod-coil mixtures is a collaboration with the research group of Prof. Eisenbach of the University of Stuttgart. The work on immiscible blends is a collaboration with the research group of Prof. Mewis of K.U.L, Leuven.

5.0 References

1. P. van Puyvelde, P. Moldenaers, J. Mewis, and G. Fuller, "A Stress-Optical Rule for Sheared Immiscible Polymer Blends", in preparation, 1998.

2. J. Vermant, P. van Puyvelde, P. Moldenaers, J. Mewis, G. Fuller, "Anisotropy and Orientation of the Microstructure in Viscous Emulsions during Shear Flow", Langmuir, in press, 1998.

RHEOLOGY AND MICROSTRUCTURE OF OIL-IN-WATER EMULSIONS

C.GALLEGOS[1], J.M. MADIEDO[1], M.R. MACKLEY[2]

[1] Departamento de Ingeniería Química. Universidad de Huelva. 21819 Palos de la Frontera (Huelva), Spain.
[2] Chemical Engineering Department. Cambridge University. Pembroke Street, Cambridge, CB2 3RA, UK.

1. INTRODUCTION

Most of food emulsions are oil-in-water (o/w) emulsions, typically stabilised by an adsorbed layer of protein at the oil-water interface.

Stability is the most important factor to be considered in emulsion technology. Emulsion stability depends on a number of structural parameters, i.e. droplet size distribution, rheology of the continuous phase, interdroplet interactions, etc (Malik et al., 1988) and all of them affect the bulk rheology of the emulsion. Consequently, emulsion rheology cannot be understood without considering the structural parameters of the emulsion.

The flocculation of oil droplets in oil-in-water emulsions has traditionally been considered as a mechanism of instability, as it can be an intermediate step towards coalescence or creaming processes. However, in concentrated emulsions, a generalised flocculation process may favour emulsion stability because of the formation of a tridimensional network (Dickinson, 1989). This microstructure yields a very complex rheological response.

Several attempts have been made to model the transient and steady state flow of flocculated oil-in-water emulsions by means of non-linear viscoelastic models. Thus, Mackley et al. (1994) have used the Wagner model together with a discrete relaxation spectrum to predict the steady state flow of several materials, including a commercial mayonnaise. Other authors (Gallegos et al., 1995) have used a modified Wagner model where the Wagner damping function was substituted by the Soskey-Winter damping function, to fit the transient flow of different commercial and model emulsions, obtaining relatively good results at low shear rates, although the absolute values of transient stress were hardly predictable.

By assuming that the flow characteristics of oil-in-water flocculated emulsions must be related to the evolution of structure with shear rate, the aim of this research was to study the microstructural modifications induced by shear in flocculated emulsions, using an optical device, as well as the evolution of the steady-state viscosity with shear rate. The overall objective was to clarify the lack of concordance between experimental and predicted values from the non-linear viscoelasticity modelling .

2. NON-LINEAR VISCOELASTICITY MODELLING

The Wagner model derives from the K-BKZ equation (Larson, 1988). Considering simple shear, and assuming that the memory function can be separated into time and strain-dependent components, a form of the K-BKZ equation is:

$$\sigma_{12}(t) = \int_{-\infty}^{t} \frac{dG(t-t')}{dt'} h(\gamma_{12})\gamma_{12}(t,t')dt' \quad (1)$$

where G(t-t') is the linear relaxation modulus and h(γ) the dumping function:

$$h(\gamma_{12}) = e^{-k\gamma(t,t')} \quad (2)$$

k is an empirical parameter which quantifies the level of non-linearity in the material. This function is calculated from the ratio between the non-linear relaxation modulus and the linear relaxation modulus:

$$h(\gamma_{12}) = \frac{G(\gamma, t-t')}{G(t-t')} \quad (3)$$

The linear relaxation modulus is related to the continuous relaxation spectrum by means of the following relationship:

$$G(t-t') = \int_{-\infty}^{\infty} H(\lambda)e^{-(t-t')/\lambda} d\ln\lambda \quad (4)$$

Substitution of eq. (4) into eq. (1) yields:

$$\sigma_{12} = \int_{-\infty}^{t} \int_{-\infty}^{\infty} H(\lambda) e^{-\frac{t-t'}{\lambda}} e^{-k\gamma_{12}(t,t')} \gamma_{12}(t,t') \, d\ln\lambda \, dt' \quad (5)$$

and the resulting steady-state viscosity can be calculated in the following way:

$$\eta(\dot{\gamma}) = \int_{-\infty}^{\infty} \frac{\lambda H(\lambda)}{(1 + k\lambda\dot{\gamma})^2} d\ln\lambda \quad (6)$$

The continuous relaxation spectrum has been calculated using regularization techniques from the experimental values of the linear viscoelasticity dynamic functions, by means of the software RheoSpec for Windows 95 (© 1995-1998 by J.M. Madiedo). The calculation of the damping function and the non-linear viscoelasticity modelling has been also performed with this software package.

3. EXPERIMENTAL

Two emulsions stabilised by different proteins have been studied. Thus, an emulsion containing 65% sunflower oil, 35% water and 4% dried pea protein (Systems Bioindustries, S.A., Barcelona, Spain) was prepared. Besides, a commercial mayonnaise stabilised by egg-yolk proteins was also tested.

The rheological characterisation of both emulsions was performed on a RDSII dynamic spectrometer (Rheometrics Scientific). Dynamic, relaxation and steady-state flow tests were performed.

Experimental observations of the effects of shear rate on the emulsion microstructure were performed on a Linkam optical cell (Linkam Scientific Instruments, Surrey, UK).

4. RESULTS AND DISCUSSION

4.1 Vegetable protein stabilized emulsion

The rheological data of the pea protein stabilized emulsion are shown in figure 1. The dependence of the dynamic moduli with frequency within the linear viscoelasticity range is typical for flocculated emulsions, with G' values higher than G" values in the whole experimental frequency range studied (figure 1a), which indicates a predominantly elastic behaviour in the emulsion. The relaxation spectrum calculated from these moduli shows two different regions (figure 1b): the transition region and the plateau region.

The values of the linear and non-linear relaxation modulus, at different strain values, are shown in figure 1c. It can be noticed that there is an apparent non-linear response even at strains of about 10%. These data have been used to estimate the damping function h(γ) in order to predict the steady-state flow behaviour of this material.

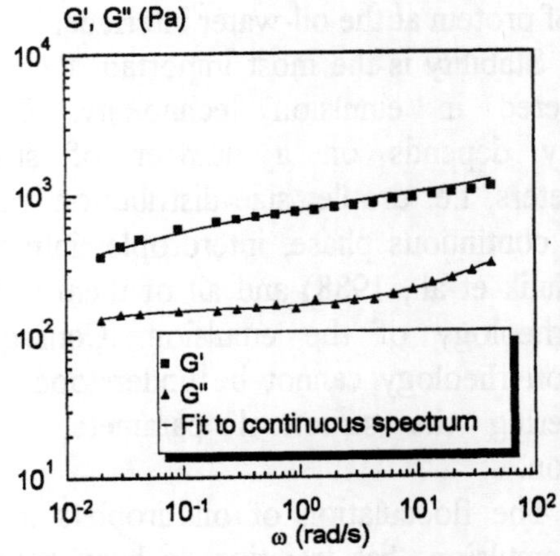

Figure 1a. Frequency sweep (γ=0.4%), with continuous spectrum fit, for the pea protein stabilised emulsion, at 25 °C.

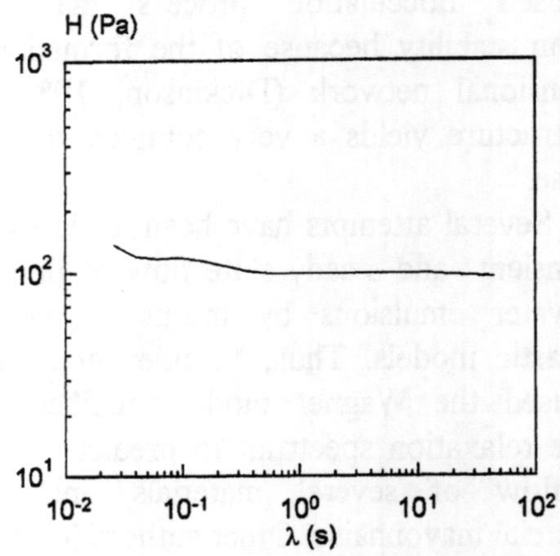

Figure 1b. Continuous relaxation spectrum for the pea protein stabilised emulsion.

Figure 1d shows the experimental flow curve obtained for this emulsion at 25 °C, together with the predicted flow curve from the continuous relaxation spectrum and the Wagner damping function. It can be noticed that experimental and predicted values are in good agreement, although there is an over-estimation of viscosity values at low shear rates.

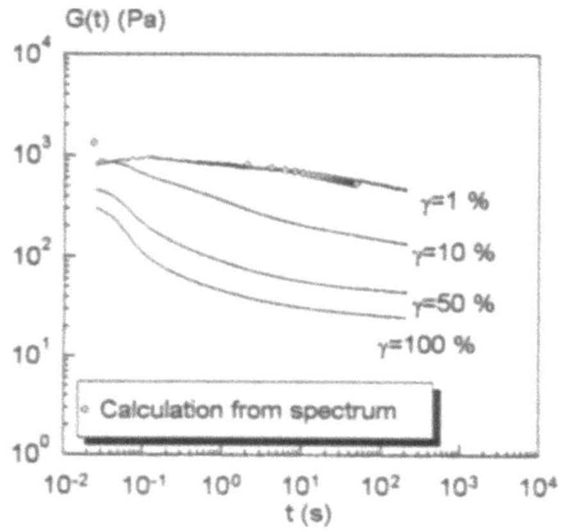

Figure 1c. Relaxation modulus, at different strain values, for the pea protein stabilised emulsion, at 25 °C.

This is related to wall-slip effects developed during the flow up to shear rates of around 2 s^{-1}, as was confirmed by means of microstructure observations, due to the fact that a plate and plate tool with smooth surfaces was used.

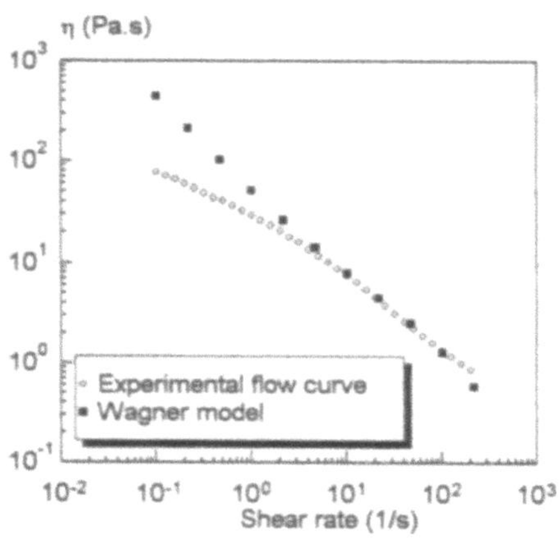

Figure 1d. Experimental and predicted flow curve for the pea protein stabilised emulsion, at 25 °C.

4.2 <u>Commercial mayonnaise</u>

The results of the rheological tests performed on the commercial mayonnaise are shown in figure 2. In general, these data are similar to those previously obtained for the vegetable protein stabilised emulsion. Thus, the frequency sweep shows that these emulsions are mainly elastic, with G' values higher than G" values in the whole experimental frequency range studied (figure 2a). Figure 2b shows the continuous relaxation spectrum for this emulsion. As can be noticed, this spectrum just exhibits the plateau region.

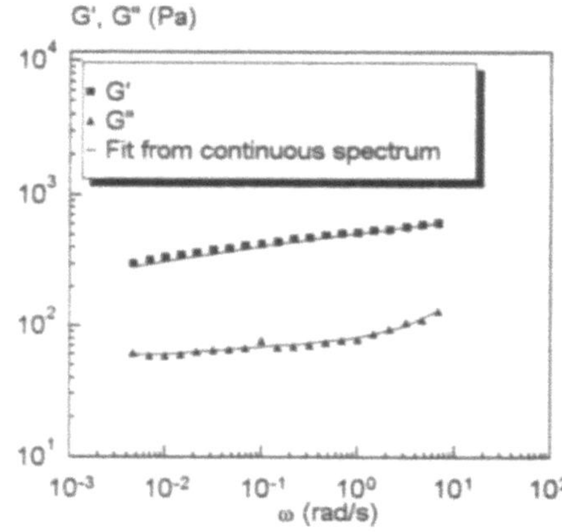

Figure 2a. Frequency sweep (γ=1%), with continuous spectrum fit, for the commercial mayonnaise, at 25 °C.

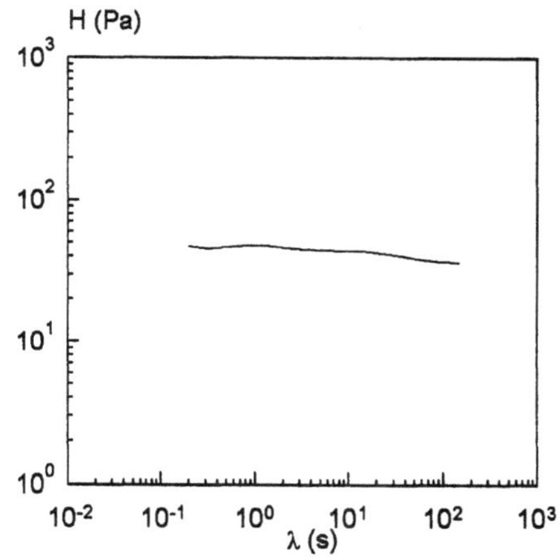

Figure 2b. Continuous relaxation spectrum for the commercial mayonnaise.

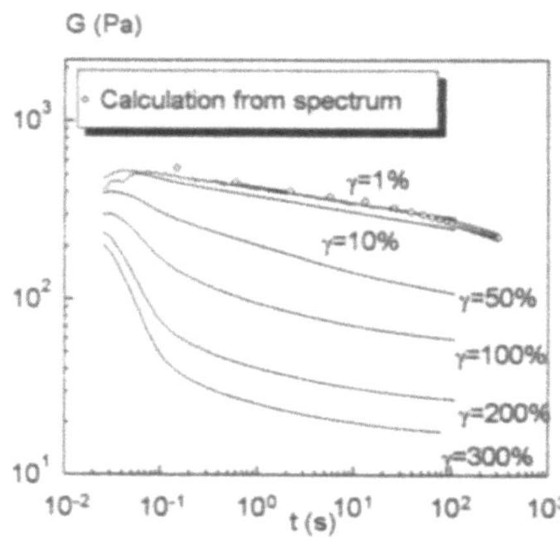

Figure 2c. Relaxation modulus, at different strain values, for the commercial mayonnaise, at 25 °C.

The experimental values of the linear and non-linear relaxation modulus, at different strain values, (figure 2c) show that for this emulsion, the structural network formed by the extensive flocculation of oil droplets is more shear-resistant than the structure of the vegetable protein stabilised emulsion.

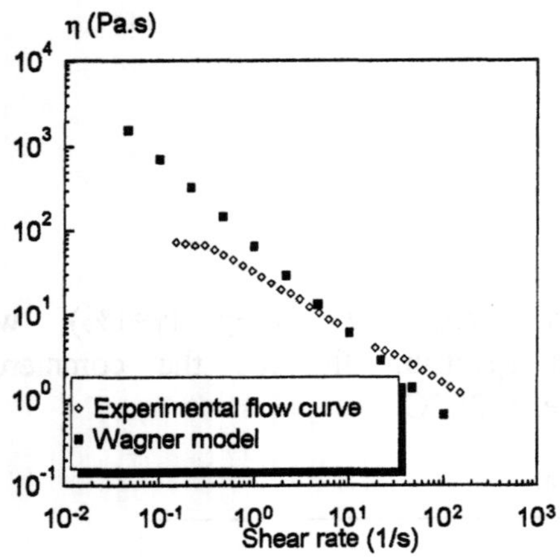

Figure 2d. Experimental and predicted flow curve for the commercial mayonnaise, at 25 °C.

Figure 2d shows the experimental and predicted steady-state flow curves for this emulsion. As can be noticed, the concordance is quite poor in this case, with an overestimation of the apparent viscosity at low shear rate values. Besides, experimental values are higher than predicted ones at high shear rates. These results can be explained by means of the optical tests performed on samples subjected to different shear rates. Thus, at low shear rates the samples exhibited wall-slip effects, which explains the fact that experimental viscosity values are lower than predicted ones. As shear rate increased (100 to 300 s^{-1}), a coalescence of oil droplets was observed. A further increase in shear rate led to a decrease in droplet size, with similar values of droplet size distribution than the corresponding to non-sheared samples. Thus, this seems to indicate that certain shear-induced processes limit the applicability of the Wagner model.

5. CONCLUSIONS

These emulsions appear to fall into a general rheological class of behaviour characterised by a plateau relaxation spectrum and a factorable strain softening damping function. Although the emulsions have high viscoelasticity at small strain, they also strain soften to give an essentially viscous fluid at large strain deformations and in steady shear. The Wagner type integral equation appears to be a reasonable constitutive equation to characterise these emulsions, although it is almost certainly not the unique constitutive equation that can describe the observed behaviour. Wall slip and changes in droplet size distribution may explain the disagreement between experimental and predicted values.

6. ACKNOWLEDGEMENTS

We would like to thank Acciones Integradas, a British Council/Spanish joint collaborative fund, for providing some financial support for this work.

7. REFERENCES

Dickinson, E (1989) Colloids Surf. 42:191-204.

Gallegos C, Franco JM (1995) Les Cahiers de Rhéologie 14:107-116

Larson RG (1988) Constitutive Equations for Polymer Melts and Solutions. Buterworths, Boston.

Mackley et al. (1994) Chem. Eng. Sci. 16:2551-2565

Malik D.H., Fogler H. S. (1988) Encyclopedia of Emulsion Technology, III. Becher, P. (ed.), Marcel Dekker, New York.

CREEP OF SOLIDS - MICROMECHANISMS OF DEFORMATION OF SEMICRYSTALLINE THERMOPLASTICS AND THEIR LONG TERM BEHAVIOUR

H.-H. KAUSCH, P. SCARAMUZZINO, C. J. G. PLUMMER

Laboratoire de Polymères, École Polytechnique Fédérale de Lausanne CH-1015 Lausanne, Switzerland

1. ABSTRACT

A brief review is given of the relative importance of parameters determining the long time behaviour of thermoplastics (chain topology and mobility, sample morphology, physical and chemical aging, creep deformation). An example is then given, namely results from a study of the response of polyoxymethylene (POM) to static and dynamic load conditions at different temperatures. Unstable crack propagation occured in POM at $K_{app} \leq 1.67$ MPa $m^{1/2}$, under static conditions at 100 °C, corresponding to the ductile-brittle transition observed in the un-notched tensile specimens. The slow crack growth behaviour was described by a two stage mechanism involving interlamellar void formation and the appearence of a wedge shaped fibrillar deformation zone ahead of the crack tip (first stage) followed by stable crack propagation via breakdown of the fibrils (second stage). A strong increase in the time to failure with the number average molecular weight M_n was found, which has been accounted for in terms of disentanglement of the molecules in the fibrils.

2. MATERIALS AND METHODS

Polyoxymethylene (POM) is a semi-crystalline engineering thermoplastic often used in load-bearing applications where good strength, toughness and durability are required. Three grades of POM with M_n of 35, 41 and 66 kgmol^{-1} and a degree of crystallinity of around 60 % were supplied as injection moulded plaques (10 mm thick) by DuPont de Nemours. All grades had a highly spherulitic morphology, showing an apparently chaotic arrangement of lamellar bundles about 0.1 μm in thickness (Fig.1). The darker areas in Fig.1a indicate lamellae roughly edge-on to the section, whereas the lighter areas contain flat-on lamellae. The mechanical tests were performed between 60 and 100 °C on a Schenck tensile test machine at a loading rate of 2mm/min, using CT specimens with a/W equal to 0.5. A pre-crack was introduced by sliding a fresh razor blade along the notch tip prior to testing. The applied load was kept constant and chosen to give an initial (t=0) mode I stress intensity factor, K_{app}, in the range of 1.11 to 2.78 MPa $m^{1/2}$.

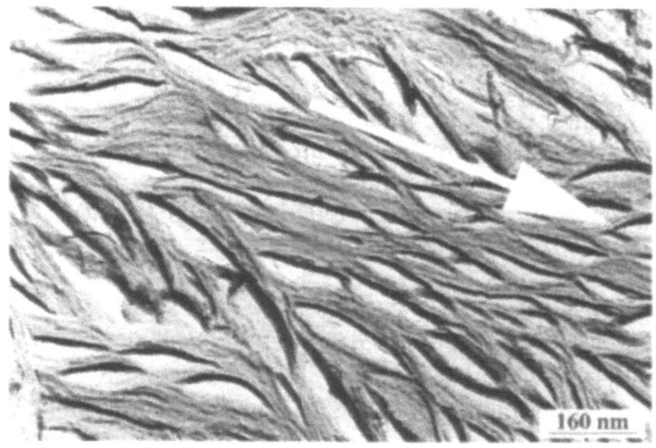

Fig. 1a: Detail of the spherulitic structure of POM M_n = 41 kgmol^{-1}, the arrow marks the growth direction (carbon/platinum replica, from Scaramuzzino 1998 [1])

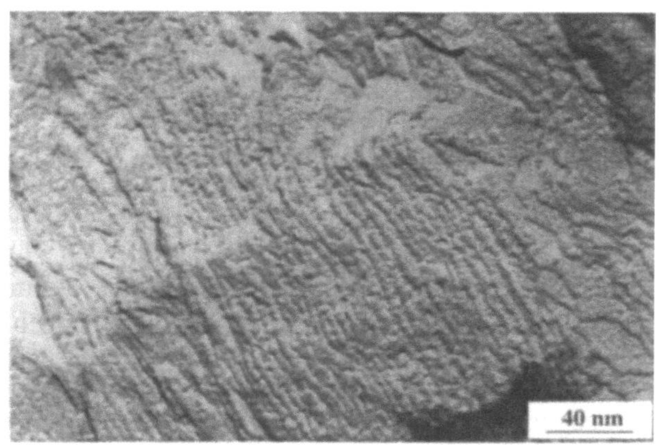

Fig. 1b: Detail of Fig. 1a, showing the piling up of crystal lamellae (carbon/platinum replica, from Scaramuzzino [1])

To assess microdeformation mechanisms ahead of the crack tip, TEM investigations were performed on a Philips EM 300 at 100 kV. Ultrathin sections were microtomed at -120 °C from samples in which a certain amount of stable crack propagation had occurred, and in which the notch faces had been slightly wedged open. Prior to sectioning, the samples were embedded using a low viscosity crosslinkable resin, capable of penetrating and providing mechanical support to the crack tip damage zone. For fixation and to improve the bright field contrast, the sections were post-stained for about 8 hours in RuO_4 vapor (depending on the film thickness).

3. RESULTS AND DISCUSSION

3.1 Steady state tensile deformation

At room temperature and strain rates of the order of 10^{-3} s^{-1}, tensile deformation in POM is characterized by hard visco-elastic behaviour arising from the bending of bundles of lamellar crystals, and substantial internal cavitation. There is little macroscopic ductility in spite of the high strains to failure (up to 70 %), and fracture is qualitatively brittle. At higher temperatures or lower strain rates POM may nevertheless show ductile necking and drawing, and cavitation is reduced. Thus under dead weight loading, samples subject to stresses of about 80 % of the ultimate tensile stress at room temperature deform to at least 200 % strain. However, at high temperatures, a macroscopic ductile-brittle transition has been observed at low loads and indeed this can result in a decrease in the time to failure with decreasing load (in the region of the transition). Under these conditions, sub-critical deformation becomes localized to sparse planar craze-like defects extending perpendicular to the tensile axis. Such behaviour is similar to that giving rise to long term brittle failure in polyethylene (PE) gas pipes, which has prompted many detailed investigations of slow crack growth in PE (Brown et al.[2-4]).

3.2 Slow Crack Growth

Slow crack growth (SCG) behaviour of POM was characterized by measuring the crack opening displacement (COD) in the loading plane of the CT specimens (direct observations confirmed this to be proportional to the *true* COD). The initial value of the applied mode I stress intensity factor K_{app} was 1.67 MPa $m^{1/2}$. Fig.2 shows the evolution of the COD.

The initial anelastic response was identical for all three grades tested. The subsequent creep deformation could be described in terms of a two stage mechanism. The first stage involved crack tip blunting accompanied by local stress whitening and the formation of a wedge-shaped fibrillar deformation zone directly ahead of the crack tip. The maximum thickness of the fibrillar zone increased with time up to a critical value δ_b at time t_b, both of which were molecular weight dependent. This point corresponded to a discontinuity in d(COD)/dt and to the onset of stable crack propagation via breakdown of the fibrils at the crack tip. Finally, at time t_f, the crack propagation became unstable and the specimen failed. In addition to t_b and t_f, the crack growth rate may also be used to characterize resistance to SCG. A measure of the slow crack growth rate is provided by the minimum value of d(COD)/dt [2-4].

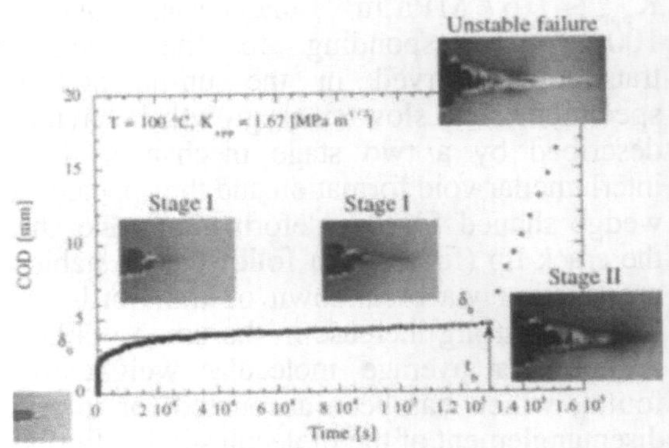

Fig.2: Typical Crack opening displacement (COD) curve versus time of POM tested at 100 °C and K_{app} = 1.67 MPa $m^{1/2}$. The pictures represent the evolution of lateral damage recorded using a CCD camera [1].

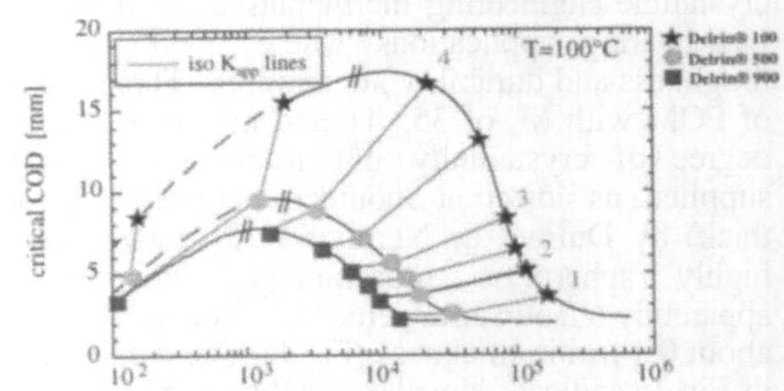

Fig.3: Plot of critical COD showing the influence of the initial applied mode I stress intensity factor K_{app} [1].

The *critical* crack opening at time t_b depends on the initial value of the applied mode I stress intensity factor. Fig. 3 shows this influence for the three grades of POM tested at 100 °C. The ductile-brittle transition is marked by the steep decrease in the critical COD. For a value of K_{app} of 1.67 MPa $m^{1/2}$ the fracture surface morphology at the transition was characterized by the appearence of a fibrillar damage zone. This latter was reminiscent of craze like features observed in un-notched tensile bars, and was thought to be responsible for the fragile creep behaviour of POM. For higher values of initial K_{app}, the specimens displayed fully ductile behaviour.

For lifetime prediction t_b (K_{app}) is clearly a quantity of primary interest. It has been suggested in the literature [2-4] that t_b should depend on the rate of disentanglement of "tie molecules" in the fibrils, although this concept is somewhat ill-defined in the context of drawn material. In rapidly crystallized thin films deformed in tension, the observed draw ratios in localized micronecks are consistent with stabilization by entanglement, as in amorphous glassy polymers. It is therefore reasonable to assume stress transfer to take place at "entanglement points" and to consider crystalline regions to exert a frictional drag on the chains, opposing chain slip. This picture also implies a fringed micelle-type structure, which is admissible for the case of highly stretched chains, and hence total destruction of the chain folded lamellae in the undeformed material. A continuum model for forced disentanglement may therefore be appropriate, in which case one can invoke the simple molecular weight scaling laws derived by Kramer and Berger from models for the kinetics of fibril breakdown in glassy polymers [5]. This accounts at least qualitatively for the increase of t_b with M_n in the present case. By plotting the SCG rate versus 1/T in a semi-log scale (Fig.4), it was also possible to calculate an energy of activation ΔH, assuming a simple Arrhenius temperature dependence. A value of about 100 $kJmol^{-1}$ was determined for all three grades. Similar activation energies have been determined for SCG in PE of different M_n by Brown and Lu [2,7], who suggested a link with the α transition. This in turn implies long range chain mobility in the crystalline regions, consistent with the previous discussion.

Given the presence of a relatively large plastic zone r_p at the crack tip during the different stages of SCG, the small scale yielding criterion for the application of linear elastic fracture mechanics was clearly inappropriate. In an attempt to make the study as quantitative as possible, we therefore calculated a dissipated energy U_d, related to the extent of irreversible processes activated in front of the crack tip as a function of the crack-tip advance Δa, measured by freeze fracture in liquid nitrogen (ESIS [6]). The maximum dissipated energy U_{dmax} prior to unstable fracture showed a systematic increase with M_n, ranging from 5 to as much as 20 kJm^{-2} consistent with the extensive irreversible deformation associated with stable crack advance. It was also shown by confocal microscopic analysis that the roughness of the corresponding regions of the fracture surface was proportional to the measured U_{dmax}, in agreement with observations in PE (Lu et al. 1995[7]).

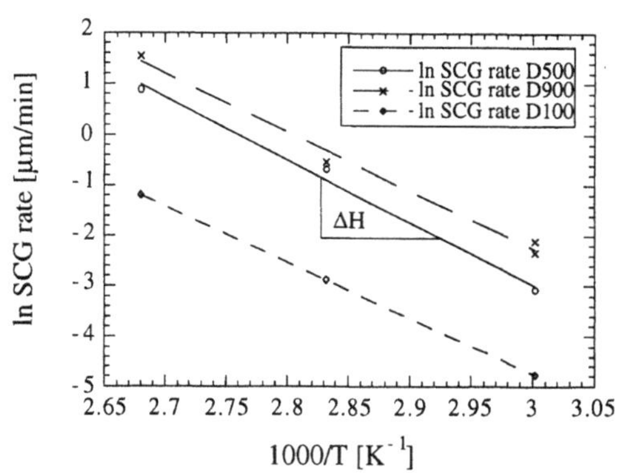

Fig.4: Plot of the SCG rate versus 1/T.

3.3 Dynamic Fatigue

Results from dynamic fatigue (sine wave) tests for M_n = 66 $kgmol^{-1}$ at 100 °C and 0.5 Hz, with ΔK_{init} = 1.67 MPa $m^{1/2}$ showed a similar evolution of the COD as a function of N (number of cycles) (thermal fatigue effects were insubstantial under these conditions). Initially d(COD)/dN leveled off in a manner reminiscent of the static tests (stationary crack growth) and then increased sharply just prior to failure (tertiary growth). However, in contrast with the static behaviour, plastic deformation appeared to be limited to a small zone ahead of the crack tip, and the fibrillar zone was absent. This was reflected by the relatively smooth appearance of the fracture surfaces.

3.4 Microdeformation Mechanisms

Observation of the crack-tip deformation zones in the SCG specimens was complicated by the fact that the dimensions of the damage zone greatly exceeded those appropriate for cryomicrotomy, necessitating considerable retrimming of the sample.

Fig. 5 is a typical thin section, showing a region adjacent to the fibrillar zone in a sample of M_n = 66 kgmol^{-1} deformed under the conditions of Fig. 2. The initial step in the irreversible deformation process appears to be the formation of voids in regions separating lamellar bundles running roughly perpendicular to the principal stress direction. Further extension then leads to craze-like features containing relatively fine fibrils (as shown in Fig. 5). Such features are thought to account for the stress whitening and hard visco-elastic behaviour encountered in room temperature tensile tests). As deformation proceeds, these "crazes" break down, and the intervening regions are forced to rotate towards the principal stress axis. Under conditions favouring shear deformation (high temperature, and in the case of sharp notches, low remote stress levels) further drawing results in the formation of the coarse fibrillar structure characteristic of SCG.

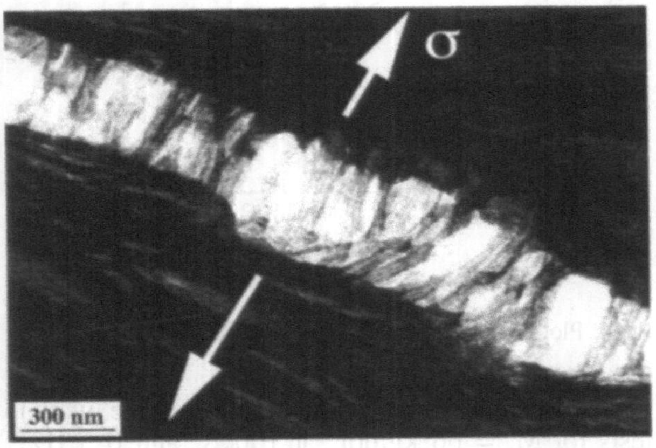

Fig. 5: Micrograph of a craze-like feature observed after creep deformation of POM (from Scaramuzzino 1998[1]).

4. CONCLUSIONS

The initial evolution of SCG in POM has been shown not to depend on the molecular weight. However, fibril extension and breakdown in the crack tip damage zone have been argued to be controlled by the rate of disentanglement of the molecules in the fibrils, consistent with the observed molecular weight dependence of the onset of stable crack propagation. At 100 °C unstable crack propagation occured at $K_{app} \leq 1.67$ MPa m$^{1/2}$, corresponding to the ductile-brittle transition observed in un-notched tensile bars under static loading. For TEM investigations two different techniques were employed; replication and ultramicrotomy at low temperature followed by staining in RuO_4. In both cases the microstructure was observed to be multilamellar, with lamellae organised into bundles of about 100 nm in thickness. Based on the TEM observations, a simple model for the microdeformation mechanisms has been proposed.

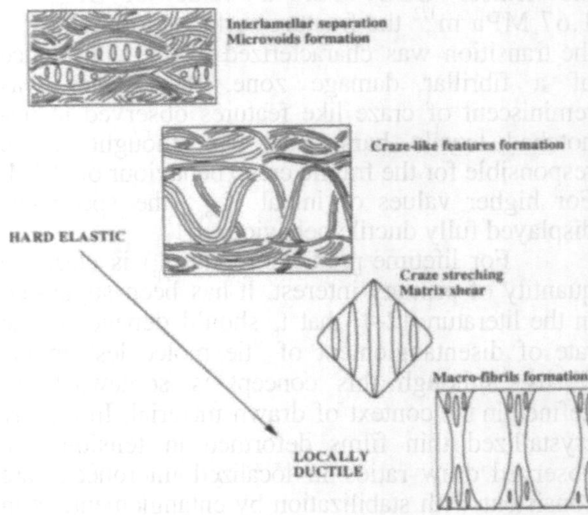

Fig.6: Model of the sequence of microdeformation mechanisms occurring during the creep deformation of POM [1]

According to this model, the main stages in crack-tip deformation are interlamellar separation, microvoid formation, formation of craze-like features and matrix shear leading to the formation of "macrofibrils" whose breakdown controls the slow crack growth under creep conditions.

5. REFERENCES

[1] P. Scaramuzzino, Analyse de la structure et de l'endommagement mécanique dans le polyoxyméthylène, PhD Thesis, EPFL, Lausanne 1998

[2] Brown N, Lu X, Huang Y-L., Qian R (1991) Makromol. Chem., Macromol. Symp. 41: 55 - 67

[3] Lustiger A, Markham RL (1983), Polymer 24: 1647 - 1654

[4] Brown N, Lu. X (1995) Polymer 36: 543 - 548

[5] Kramer EJ, Berger LL (1990) in Advances in Polymer Science 51/52, Kausch HH (ed) Springer-Verlag (Berlin) pp 1 - 56

[6] ESIS (1995), Technical Committee Report *A Testing Protocol for Conducting J-Crack Growth Resistance Curve Tests on Plastics)*

[7] Lu. X, Qian R, Brown N (1995) Polymer 36: 4239 - 4244

VOLUME/SHEAR INTERACTION IN GLASSY POLYMERS

W. G. Knauss, California Institute of Technology,

Pasadena, California 91125

It has been demonstrated in the past that the nonlinear time- or rate-dependent shear behavior of glassy polymers is susceptible to volumetric changes. Following from models based on free volume effects in the glass transition range it has been shown analytically that many effects of nonlinear behavior could possibly be traced to volumetric changes and associated dilatational internal clock mechanisms.

Following studies on the intrinsically nonlinear time-dependent shear behavior of PMMA at various temperatures below the glass transition, it has been found recently that this nonlinear shear behavior is moderated by superposed dilatational response. It turns out that if shear and normal stresses combine to render a constant maximum shear stress, the creep rate is a function of whether there exists a normal stress component.

It had been demonstrated previously that around the glass transition temperature positive dilatation added to shear response accelerated creep, while negative dilatation decelerated it, but to a much lesser degree than the positive added stress. At 80 °C dilatational stresses added to the shear yielded more rapid creep than the shear alone. However, by considering the maximum (or octahedral) shear stress as the driver for the nonlinear behavior and keeping track of the further dilatational influence on the rate of creep, a nearly symmetrical effect is achieved between positive dilatation (accelerated response) and negative dilatation retarded response.

These consequences for testing geometries typically used in material characterizations, whenever inhomogeneous deformations evolve, will be discussed. Moreover, the implications for failure and fracture are imminent,. and the consequences for long term failure by crack propagation will be examined relative to the relaxation or creep properties.

RHEOLOGY OF BIOPOLYMER GEL SYSTEMS

R. LAPASIN, S. PRICL

Department of Chemical, Environmental and Raw Materials Engineering,
University of Trieste, piazzale Europa 1, I 34127 Trieste, Italy

1. INTRODUCTION

Biopolymer rheology is an appealing subject of both academic and practical interest since it deals with materials and properties that are not only industrially important, but are essential for any living systems. In particular, polysaccharides hold a wide range of different functions, from the generation of solid structures or tough fibers in higher plants to the formation of hydrated crosslinked three-dimensional networks in marine seaweeds and to the government of the solution properties of physiological fluids.

The diversity of these functions originates from differences in polymer shapes and structures and reflects in the wide range of rheological behaviors exhibited by biopolymer systems which, in turn, are extensively exploited in a countless number of industrial formulations and applications. Accordingly, the knowledge of the rheology of a given biopolymer system is important when considering its utilization in the formulation of a new industrial product, as well as in the identification of the parameters suitable for process control and optimization and quality control operations of biopolymer-based formulations.

In many aqueous systems, proteins, polisaccharides and their derivatives are the main responsible agents in the formation of physical gel matrices even though, in several other circumstances, surfactants and disperse phase play a fundamental role in the mechanism of gel formation.

The term '*physical gel*' has been coined to distinguish noncovalently crosslinked networks formed from both synthetic and biological macromolecules from *chemically crosslinked materials*, where covalent bonds exist between the polymer chains, and *entanglement networks*, formed by purely topological interaction of the chains. A plethora of physical interactions and mechanisms can be envisaged, by virtue of the diversity in molecular conformations and, therefore, in the interchain spatial arrangements, which eventually lead to the formation of physically cross-linked networks. Non-covalent intermolecular associations, deriving from disparate forces such as Coulombic, van der Waals, dipole-dipole, hydrophobic and hydrogen bonding interactions constitute a rule rather than an exception for biopolymer systems. When interactions act cooperatively - so that extended quasicristalline junction zones are formed between two or more chains or a threedimensional network results from the association of biopolymers in particulate form - then the whole system can manifest a gel-like behavior. The present paper reports an overview of the rheological properties exhibited by biopolymer physical gels in connection with their structural conditions and industrial applications.

2. STRUCTURAL FEATURES

Globular proteins can associate to form particulate networks composed of ordered stringy structures or randomly branched aggregates. This is the case of ovalbumin and β-lactoglobulin forming gels on heating. The final gel structure depends on the polymer concentration, on the environmental and preparation conditions. Small changes in pH or, even more, in the heating rate can produce remarkable changes in the rate and mode of aggregation, so leading to quite different textural and rheological properties of the gel. Low heating rates result into inhomogeneous β-lactoglobulin gels with separate regions of high and low crosslink density and inferior mechanical properties (Stading et al., 1992). The gel compactness may also be a function of its mechanical history and, as it is the case for flocculated colloidal silica gels, the field of applied stresses may induce irreversible structural changes which result in a system hardening (Raghavan and Khan (1995)). As for colloidal dispersions of aggregate particles, the structural and mechanical characteristics of the particulate networks can be analyzed to extract information on the fractal nature of their aggregates, when they are extended enough to exhibit natural fractal behavior. Some discrepancies may be found, however, between the fractal dimensions obtained via scattering studies and those calculated from shear modulus measurements.

The thermoreversible gelation of many polysaccharides and the fibrous protein gelatin is strictly connected with the conformational (coil → helix) transitions induced by temperature variations and/or other changes in the environmental conditions (pH, ionic strength, specific addition of salts or cosolvent). Such transitions can give rise directly to the network through the formation of double helices or ion mediated 'egg box' junctions, or through weak associations of the stiffened molecular structures. In other cases, imperfect matching of chains and interactions involving substantial lateral helix association lead to crosslinking of multiple helices, and then, to more complex structures. Several examples of such associative networks are reported in literature (Clark and Ross-Murphy, 1987; Lapasin and Pricl, 1995).

Even if contradictions and controversies still exist between experimental evidences and theoretical interpretations concerning the specific crosslinking mechanism of some real systems, all these associative networks can be generally treated as composed by more or less extended stiff junction zones which involve two or more chains and by flexible chain segments between the junction zones. The mechanical properties of whole gel system reflect not only the balance between both entropic and enthalpic contributions but also the heterogeneity of the network.

In fact, physical gels are not homogeneous over the entire distance scale, and frequently a specific and intricate structural hierarchy can be distinguished within the system. Various levels of heterogeneity are encountered in noncovalently crosslinked networks, and particularly in polysaccharides gels, depending on the different mechanisms of chain association, the extension of the junction zones and their spatial distribution through the system.

In conclusion, a wide range of structures is exhibited by diverse biopolymer physical gels, by the same biopolymer under different temperature and solvent conditions and also after different thermal and mechanical histories. In any case, all the physical crosslinks are characterized by small but finite energy and/or lifetime, so differing from both covalent bonds and chain entanglements, and that is the reason why the rheological properties of physical gels, even if distributed along a continuum spectrum, stand between those typical of ordinary polymer solutions and those peculiar to chemically crosslinked polymers.

3. RHEOLOGICAL PROPERTIES

In the biopolymers field a disputable but useful distinction between strong and weak gels is usually made on the basis of the peculiar properties at small and, even more, at large deformations.

For all the physical gels the extension of the linear viscoelastic regime is confined at lower strains than for ordinary solutions. However, the viscoelastic moduli of strong gels (as are usually denoted agarose, gelatin calcium alginate) become strain dependent at higher strains ($\geq$ 0.2–0.25) than for weak gels, like xanthan, welan or rhamsan. As for several concentrated particle dispersions, the critical strain derived from strain or stress sweep tests may be sensibly lower than 0.01. Moreover, the transition from linear to nonlinear regime is marked by a redistribution between the viscous and elastic components, so that the loss modulus initially increases with increasing strain. Such a behavior resembles that exhibited by flocculated suspensions, also as its concentration dependence is concerned (Otsubo and Watanabe, 1989), so that similar breakdown processes of the network can be invoked.

Another net discrimination between strong and weak gels derives from their large deformation responses. Under sufficiently high shear stresses, strong gels undergo rupture with the formation of discontinuity surfaces across the sample, whereas homogeneous flow conditions without fracture are achieved for xanthan and other weak gels. Their shear-thinning behavior is characterized by a large power-law region (with $n = 0.1 - 0.2$) which be extended even over 6 shear rate decades. When performed with controlled rate rheometers, steady shear measurements do not generally show a low shear rate Newtonian plateau, which has been experimentally detected only for some weak gels at very low shear rates (10^{-4}–10^{-6} s^{-1}). In several cases the low-shear behavior seem to be associated with an apparent yield stress or, at least, is often described as plastic, and, in this sense, an analogy could be established with the shear dependence of concentrated dispersions, highly filled polymers or liquid crystalline polymers. When steady viscosity data are plotted as a function of shear stress, the flow curves sometimes show an 'hesitation' in the viscosity decrease within the intermediate power-law region as it was often observed for polymeric liquid crystals in the negative N_1 region as well as for colloidal dispersions at very high effective volume fractions. Such a behavior becomes more evident as polymer concentration increases and is associated with net changes in the time-dependent behavior. This could reflect a structural change in the sample during which a substantial breakdown of gel microdomains takes place.

Another peculiar feature of weak gels concerns their time-dependent responses in startup tests, so differing neatly from polymer solutions. Even when low shear rates are applied, significant stress overshoots were attained for a xanthan gel at quite low strain values (0.5-1) (Richardson and Ross-Murphy, 1987). Such initial stress responses are strongly dependent on the previous rheological history so that further information can be derived from the application of on-off procedures. These tests generally show that the timescale of the structural recovery occurring in weak gels at rest after prior shearing is generally much longer than in concentrated polymer solutions.

The thixotropic properties of weak gels can be characterized more suitably and extensively in continuous shear by means of stepwise procedures. The stress transients so obtained closely resemble those exhibited by flocculated dispersions, where structural buildup and breakdown processes regularly result from stepwise shear rate variations of opposite sign (Lapasin and Pricl, 1995).

As for many other structured fluids, the Cox-Merz rule evidently fails for weak gels since no convergence can be noticed between steady and complex viscosities in the $\dot{\gamma}/\omega$ range usually explored. However, it should be remarked that the determination of the asymptotic values (as $\dot{\gamma} \rightarrow 0$ and $\omega \rightarrow 0$) of both these material functions may be uncertain (equivalent problems appear for the limiting behavior of J(t) in creep tests, as $t \rightarrow \infty$).

The elastic character of strong gels can be clearly defined from the profile of dynamic moduli, since the storage modulus is almost independent of ω over a wide frequency range and its value is much greater than that of G'' (generally, much more than one order of magnitude).

The mechanical spectra of weak gels show a slight frequency dependence with no apparent tendency to a Newtonian plateau as $\omega \rightarrow 0$. In most cases the profiles of $G'(\omega)$ and $G''(\omega)$ are almost parallel with G' exceeding G'' in the ω range explored, in a lesser extent than for strong gels. Indeed, such features closely resemble the behavior of chemically crosslinked polymer with strongly unbalanced stoichiometry at the sol/gel transition, since the exponent of the scaling law is far from 0.5.

For these latter systems the efforts of applying the superposition principle lead to poor improvements since, within the field of existence of the weak gel, we do not observe, as a rule, any appreciable variation in the profile of the moduli, both in terms of values and shifting along the frequency axis. Thus, the experimental window cannot be artificially expanded, and any speculation on the terminal zone, and especially on the eventual existence of the equilibrium modulus becomes hazardous. In the case of weak gels, rare are the experiences in which dynamic tests have been performed in combination with relaxation and creep measurements; therefore, for several systems there is a lack of complementary information about the asymptotic behavior. Available data seem to suggest that, in the long time region, the response of such systems in terms of G(t) and J(t) may be approximated by a power law time-dependence, without considerable limiting values of both quantities. This can be taken as a confirmation of the behavior detected under low frequency dynamic tests, and is another rheological feature of distinction between weak and strong gels.

Theoretical models have been proposed to define the dependence of the shear modulus on the network connectivity, as reviewed by Lapasin and Pricl (1995). Starting from the cascade approach, relationships can be established between the shear modulus and biopolymer concentration, molecular weight, and system temperature, and reasonable results have been obtained from data fitting of some thermoreversible gels (Clark and Farrer, 1995). Nevertheless, we must observe that, in the case of physical gels, the model cannot be used for pure prediction, since the functionality parameter (number of potential crosslinking sites per molecule) is not known *a priori* but must be calculated from the data correlation itself. The failure of the cascade model to account for gel time-concentration data can be ascribed also to the uncertainties in these data, which are currently measured from cure curves in various ways, usually performed at 1 Hz. Indeed, very few gel point reported in literature are based on the Winter-Chambon criterium, which implies the collection of cure data at different frequencies. Accordingly, such an approach does not account for the frequency dependence of the viscoelastic moduli and, hence, can be reasonably applied only to strong gels.

The different mechanical spectra observed for strong and weak gels, as well as the sol/gel transitions induced by changes of concentration, temperature and solvent composition can be suitably described by the generalized Maxwell model or, alternatively, by the fractional derivative model suggested by Friedrich and Braun (1992) and its further modifications, as illustrated in a recent paper (Pricl et al., 1997). An interesting convergence can be found between the mechanical spectra described by these novel models and those predicted by the off-lattice simulation model proposed for associative polymer networks by Groot and Agterof (1995). According to the latter approach, it seems possible to predict the general features of the viscoelastic properties of physical gels over very extended frequency/time scales, as well as other aspects of gel behavior including nonlinear responses and microphase separation.

4. CONCLUSIONS

Although many are the literature studied devoted to the experimental characterization of the rheology of biopolymer gel systems, still much more numerous are the open questions about the interrelations between rheological properties and structural features, and the definition of suitable phenomenological and predictive models, able to describe both linear and nonlinear properties of such complex systems.

As concerns the aspects of industrial applications, special attention has been paid to the analysis of the peculiar characteristics of the single systems, to their temperature dependence and to the effect of the solvent environment, since these are key parameters for potential industrial applications. To this respect, an extension of the weak gel analysis under extensional flow conditions seems mandatory. Mixed biopolymer gels deserve much more insight, since the marked synergistic effects they give rise to may result in new, interesting properties to be examined by academia and exploited by industry. Finally, the analysis of even more complex systems, such as suspensions or emulsions in weak gel matrices may reveal attractive aspects for applications in different industrial sectors, and may eventually lead to the identification of new production technologies.

5. REFERENCES

Clark AH, Farrer DB (1995) J. Rheol. 39:1429-1444

Clark AH, Ross-Murphy SB (1987) Adv. Polym. Sci. 83:57-192.

Dickinson E (1995) J Chem Soc Faraday Trans 91:51-58

Friedrich Ch, Braun H (1992) Rheol Acta 31:309-322

Groot RD, Agterof WGM (1995) Macromolecules 28:6284-6295

Lapasin R, Pricl S (1995) Rheology of Industrial Polysaccharides: Theory and Applications, Chapman&Hall, Glasgow.

Otsubo Y, Watanabe K (1989) Colloids and Surfaces 41:303-310

Pricl S, Lapasin R, Grassi M (1997) Les Cahiers de Rheologie XV(4):448-457

Raghavan SR, Khan SA (1995) J Rheol 39:1311-1325

Richardson RK, Ross-Murphy SB (1987) Int J Biol Macromol 9:257-264

Ross-Murphy SB (1995) J. Rheol. 39:1451-1463

Stading M, Langton M, Hermansson AM (1995) J. Rheol. 39:1445-1450

WALL SLIP OF POLYMER MELTS FROM SQUEEZE FLOW AND CAPILLARY RHEOMETRY

H. M. LAUN

BASF Aktiengesellschaft, ZKM - G 201, Polymer Research / Polymer Physics,
D-67056 Ludwigshafen, Germany

1. INTRODUCTION

Partial wall slip of polymer melts is inherent in processes like extrusion and injection moulding. Yet it is a property that is often ignored, with the exception of HDPE melts which exhibit the spurt effect. Partial wall slip reduces the apparent viscosity in processing and prevents long residence times of material adjacent to the die or extruder wall. Another important effect in extrusion is to reduce surface irregularities due to shark skin melt fracture at the die exit lip. Capillary rheometry and squeeze flow are used to effectively quantify wall slip velocities and also to investigate the mechanism of additives.

2. EXPERIMENTAL

A capillary rheometer developed in our laboratory imposes the extrusion pressure by nitrogen gas. The gas pressure is measured by a precision pressure transducer and controlled by a PC (Fig. 1). The flow rate is evaluated on-line by measuring the displacement of a plunger which floats on the meniscus of the melt. The coil of the LVDT is mounted outside and the core inside the pressurized barrel (Laun and Reuther (1999)). The instrument combines high precision of the flow curve measurement with imposed wall shear stress.

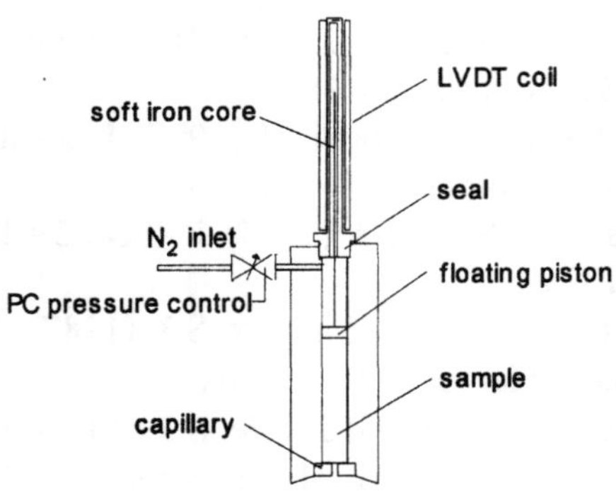

Fig. 1: PC-controlled nitrogen capillary rheometer (schematic)

In the squeeze flow apparatus a disk like sample of initial radius R (R < 40 mm) and thickness $H_o = 4$ mm is squeezed between parallel plates under controlled squeeze force F and the sample thickness H is measured. The tests are performed in a servo-hydraulic testing machine (Fig. 2) with $F_{max} = 10$ kN and 1 µm resolution (Laun (1996)). Creep tests are performed, i.e. the product $FH = F_oH_o$ is kept constant.

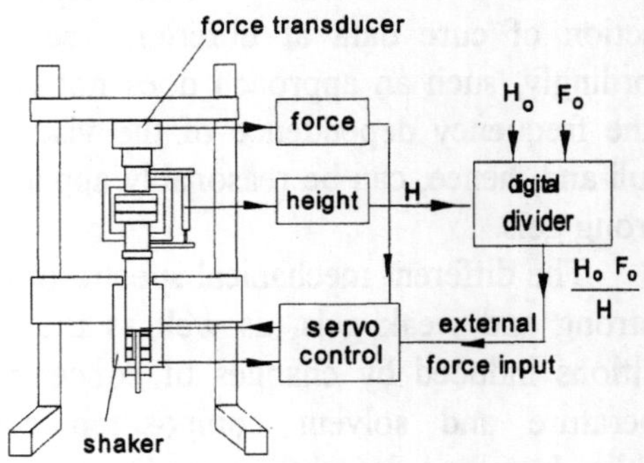

Fig. 2: Creep test on servo-hydraulic testing machine (schematic).

3. THEORY

If v_s is the slip velocity at the wall of the capillary the effective Newtonian wall shear rate due to bulk shear is calculated after subtraction of the plug flow contribution from the total volumetric flow rate:

$$\dot{\gamma}_{N,eff} = \frac{4}{\pi R^3}\left[\dot{V} - \pi R^2 v_s\right] \quad (1)$$

$\dot{V}$ volumetric flow rate, $\dot{\gamma}_{N,eff}$ effective Newtonian wall shear rate, $\dot{\gamma}_a = \frac{4\dot{V}}{\pi R^3}$ apparent Newtonian wall shear rate

Hence,

$$\dot{\gamma}_a = \dot{\gamma}_{N,eff} + \frac{4 v_s}{R}. \quad (2)$$

Following *Mooney*, a plot of the normalized volumetric flow rate (apparent Newtonian wall shear rate) measured at constant wall shear stress (i.e. constant extrusion pressure for a constant L/R of the die) versus 1/R should yield a straight line with slope $4v_s$ the intersect with the ordinate giving the effective Newtonian wall shear rate (Fig. 3). True wall shear rate and stress follow from the *Rabinowitch-Weissenberg* correction.

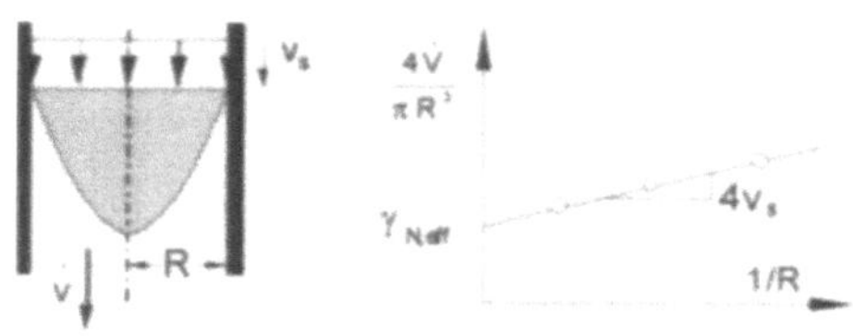

Fig. 3: Capillary flow with slip

In squeeze flow the radial slip velocity at the wall increases up to the rim slip velocity $v_{s,R}$ (Fig. 4). The effective Newtonian rim shear rate at the wall follows from the total volumetric flow rate at the rim by subtraction of the slip contribution:

$$\dot\gamma_{N,eff} = \frac{3}{\pi RH^2}\left[\dot V - 2\pi RHv_{s,R}\right] \tag{3}$$

$\dot\gamma_{N,eff}$ effective Newtonian wall shear rate at the rim

$\dot\gamma_{R,N} = \frac{3\dot V}{\pi RH^2} = \frac{3(-\dot H)R}{H^2}$ apparent Newtonian rim shear rate

Hence,

$$\dot\gamma_{R,N} = \dot\gamma_{N,eff} + \frac{6v_{s,R}}{R} \quad . \tag{4}$$

The rim slip velocity $v_{s,R}$ depends on the wall shear stress at the rim. Therefore, a prerequisite of determining the rim slip velocity is to keep the rim shear stress at the wall constant by keeping FH constant (Laun (1996)):

$$\tau_{N,eff} = \tau_{R,N} = \frac{2HF}{\pi R^3} \tag{5}$$

$\tau_{R,N}$ Newtonian rim shear stress at the wall

Note that (5) is valid for small gaps (negligible contribution of the equibiaxial flow contribution) (Laun et al. (1998)). A plot of the normalized volumetric flow rate (apparent Newtonian rim shear rate) versus 1/H then yields a straight line with slope $6v_{s,R}$, the intersect with the ordinate giving the effective Newtonian rim shear rate. The advantage of the squeeze flow creep test is that the *Mooney*-type plot is obtained from one single run whereas in capillary rheometry separate tests using various die diameters have to be compared.

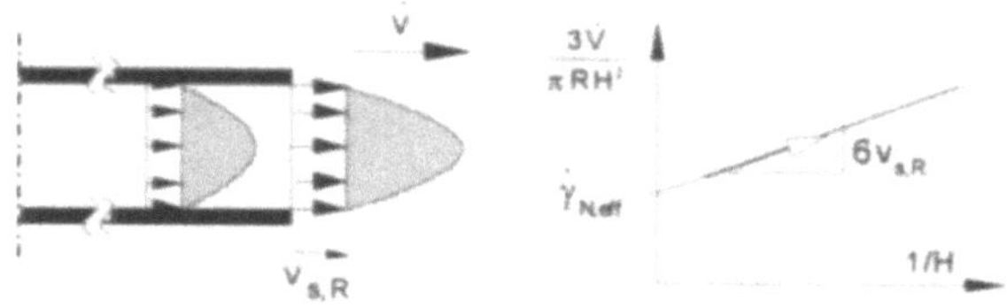

Fig. 4: Squeeze flow with slip

The squeeze force of a power law fluid with wall slip is given for small gaps (shear flow contribution dominant) by the *Rady-Laun* equation (Laun et al. (1996)):

$$F = \frac{(-\dot H)^n}{H^{2n+1}}\left(\frac{2n+1}{n}\right)^n \frac{2\pi kR^{n+3}}{n+3}\left[1 - 2\frac{Hv_{s,R}}{(-\dot H)R}\right]^n \tag{6}$$

n power law exponent

As shown by Laun et al. (1996) equation (5) is recovered also in this case. In other words the determination of $v_{s,R}$ as indicated in Fig. 4 is also valid for the power law fluid. In addition, however, the true rim shear rate and stress have to be determined from the Newtonian values by a correction similar as for capillaries (Laun (1992)):

$$\dot\gamma_R = \frac{2n+1}{3n}\dot\gamma_{N,eff} \tag{7a}$$

and

$$\tau_R = \frac{n+3}{4}\tau_{R,N} \tag{7b}$$

$\dot\gamma_R$, τ_R true wall shear rate and stress at the rim

It is necessary to note that (6) has been derived by assuming that material planes parallel to the wall remain parallel. In this case the slip velocity increases linearly with the radius r. It can be shown, however, that (5) and (6) remain reasonable approximations even for the more general case that the slip velocity exhibits a power law dependence on the radius:

$$v_s(r) = v_{s,R}\left(\frac{r}{R}\right)^m \tag{8}$$

m slip power law exponent

4. RESULTS

Fig. 5 shows Mooney plots for a HDPE melt below (left) and above (right) the spurt effect. The high accuracy of the nitrogen capillary rheometer allows to detect slip velocities also below the critical shear stress τ_c for the onset of spurt!

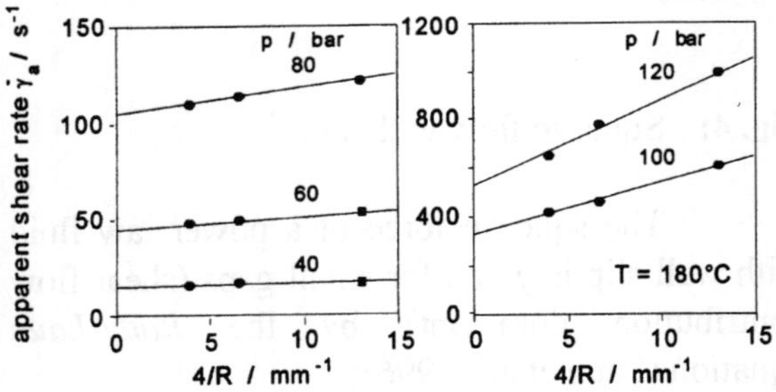

Fig. 5: Mooney plots for an HDPE melt

If the slip velocities are plotted as a function of the wall shear stress (Fig. 6) it becomes obvious that the spurt effect is not a transition from no slip to distinct slip but rather represents a jump in the magnitude of wall slip. As a concequence, two different mechanisms contributing to wall slip of HDPE have to be distringuished.

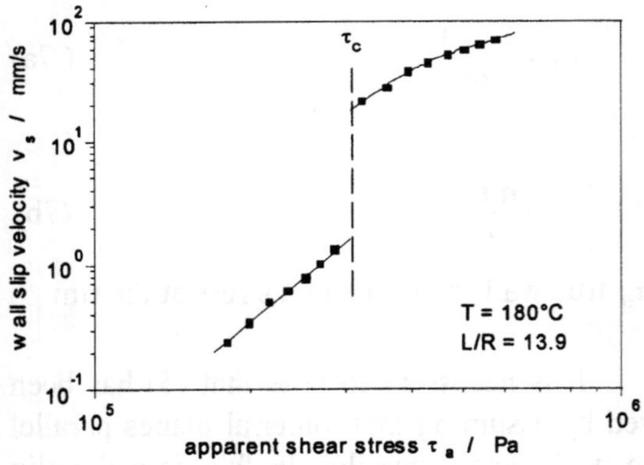

Fig. 6: Jump of magnitude of wall slip velocity at onset of spurt

Mooney plots are also helpful to investigate the mechanism of processing aids. The addition of wax to PP (Fig. 7) increases the bulk shear rate (reduces viscosity) but does not effect the wall slip. A fluoropolymer in LLDPE (Fig. 8) increases the wall slip velocity but does not effect the bulk viscosity.

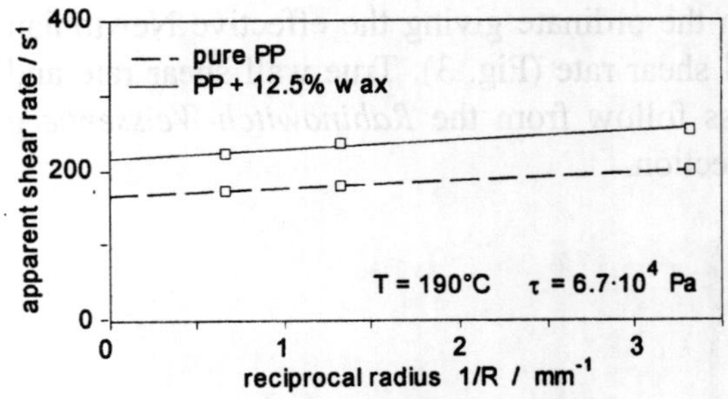

Fig. 7: Effect of wax in PP (capillary)

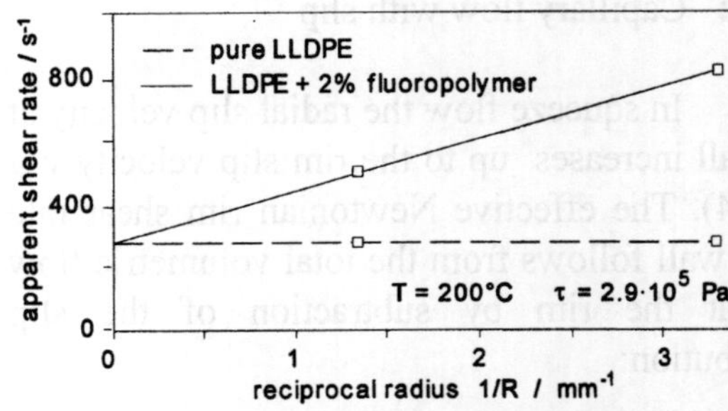

Fig. 8: Effect of fluoropolymer in LLDPE

In squeeze flow the initial part of the test is governed by transient viscoelastic effects before a quasi steady regime is reached (Fig. 9) (Laun et al. (1998)). A comparison of slip velocities from both methods for PP are compared in Fig. 10.

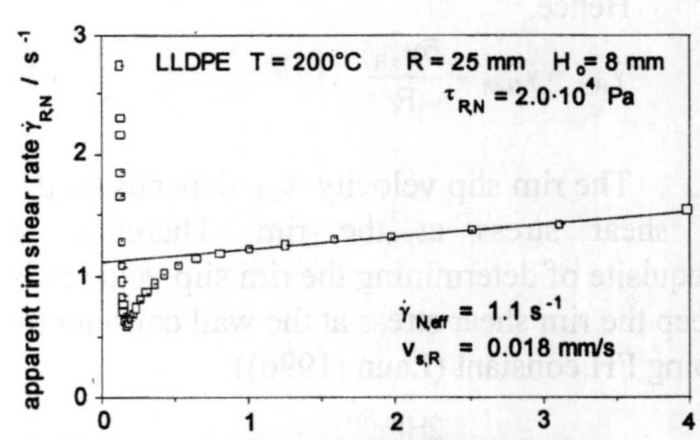

Fig. 9: Wall slip of LDPE (squeeze flow)

Whereas the slope of the slip velocity versus shear stress (true values) is about the same there remains a distinct difference in the absolute values. The discrepancy may be due to different wall materials used in the test and needs further investigation.

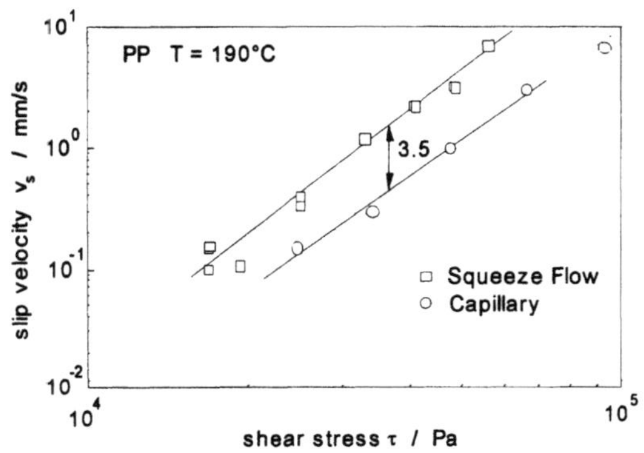

Fig. 10: Comparison of slip velocities

5. REFERENCES

Laun, H. M., Reuther, W., to appear in Rheologica Acta (1999)

Laun, H. M., Proceedings of PPS-12, Sorrento, 31- 34 (1996)

Laun, H. M., Rady, M., Hassager, O., to appear in JNNFM (1998)

Laun, H. M., Makromol. Chem. Makromol. Symp. 56, 55-66 (1992)

SOME PROBLEMS RELATED TO FLOW OF "RHEOKINETIC" LIQUIDS

A.Ya. MALKIN

Research Institute for Plastics, 35, Perovskii proezd, Moscow 111024, Russia

1. INTRODUCTION

"Rheokinetic" liquids are a special branch of time-dependent materials. They change their rheological properties in time due to chemical reactions, and as a general rule these changes are irreversible. The reason for the evolution of the rheological properties of a liquid can be chemical reactions of polymerization, polymer modification, curing and so on. In several cases chemical transformation can result in phase transition of a multicomponent material (for example, in polymer solutions) and it also leads to deep changes in rheological behavior of a matter.

Chemical transformations and phase transitions leading to continuous changes in rheological properties of a liquid. can superimpose on its flow It is rather frequent and trivial situation in particular in numerous technological applications. It is rather evident that changes in rheological properties in time (caused by chemical or physical transformations) must be takes into account in discussing various problems of fluid dynamics of real substances. In this sense the very popular slogan of rheology "everything flows" must be expanded and treated as "everything flows - changing".

2. RHEOKINETICS [1]

Chemical processes lead to enormous changes in viscosity (and other rheological properties). On the one hand there is a correlation between changes in rheological properties and the kinetics of a chemical reaction. Therefore we can follow a chemical reaction by measuring rheological properties of a matter. In the other hand quantitative description of time-dependent rheological properties (no matter directly connected to kinetics of a chemical reaction or not) is the necessary initial information in treating fluid dynamic problems.

Mathematical modeling of a rheokinetic process involves description of the kinetics of a chemical reaction. It can be rather easily done for linear polymerization of any type. In this case viscosity profile is the measure of chemical kinetics by itself. Therefore it is possible to find the direct correspondence between the kinetics of a chemical reaction and rheokinetic properties of a liquid. In this case the final product of a reaction is still a liquid, though the ratio of the final to initial levels of viscosity can reach many decimal orders. Examples illustrating amount of changes in rheological properties of real "rheokinetic" liquids prove that this effect is great.

However this problem becomes rather complicated in three-dimensional reactions of curing, as there is not direct and simple correlations between the degree of chemical transformation caused by curing and changes in rheological properties of a material. Then we need to find an adequate and convenient for applications empirical representation of time-dependent viscosity and other rheological properties of a liquid. Besides, due to curing process a reactive medium becomes a solid. It leads to necessity for special discussing a boundary problem because a solid can move along a channel only sliding.

3. FLOW DYNAMICS OF A "RHEOKINETIC" LIQUID

We need to consider problems related to flow of these liquids as a special branch of fluid dynamics. Indeed results of calculations of velocity profiles, pressure distribution, pressure-vs-output characteristics and so on demonstrate very unusual behavior of "rheokinetic" liquids in flow through different channels.

One of the most characteristic example is flow through a cylindrical channel (capillary). Depending on the flow output we can observe various velocity profiles. The most interesting effects are: appearing of the concave profile, its rearrangement into strongly protruded profile and at last to the break through a capillary along its axis [2].

In application to industrial tubular reactors it means that one must choose the length of a reactor which must be enough for complete transformation according to the kinetics of a chemical reaction. At least in several cases this length is proportional to volume output and pressure appears to be proportional to the squared output.

Due to velocity distribution through a cross section of a reactor the degree of chemical conversion also appears inhomogeneous. Average degree of conversion (determining quality of the end product) depends of rheokinetic properties of the flowing liquid. One of the results of this situation in application to tubular reactors of polymerization is appearance of bimodal "hydrodynamic" MWD of the final product regardless MWD expected from the mechanism of a chemical reaction.

In flow of a curing material other special peculiarities can be observed. They are explained primarily by wall slip. So a hysteresis loop appears at the pressure-vs-output characteristic and this curve have an absolutely unstable branch.

Besides experiments demonstrated that slip velocity can depend on not wall shear stress only but on the radius of a capillary too. This effect might be related not to rheological properties of a material but to its fluid dynamics.

4. FLOW OF "RHEOKINETIC" LIQUIDS THROUGH POROUS LAYERS

Curing oligomers are very typical and important example of rheokinetic liquids, These materials are widely used in technological practice for producing reinforced plastic materials. Therefore problems related to flow of a rheokinetic liquid inside a porous medium are of special interest.

Several typical situations are discussed. Two-dimensional flows are considered. One of the illustrative example is "squeezing" flow: a drop of a rheokinetic liquid is placed on the surface of a porous layer and pressed. A liquid is simultaneouly squeezing along the surface and impregnating a porous space [3].

The next situation is an "inflation" flow which is a model of the RRIM-process. It is also a two-dimensional flow when a curing liquid moves along a porous layer and simultaneously impregnates it. Calculated pressure and velocity profiles were demonstrated for different characteristic parameters of a material and a porous layer [4].

The most important result of model calculations is the demonstration of existence of the limiting profile, or incomplete impregnation. Indeed unlimited growth of viscosity due to curing results in unlimited growth of pressure. In practice pressure is always limited and than it means that flow stops at definite time determined by kinetics of a chemical process and rheokinetic properties of a material. Conditions of incomplete impregnation of a porous (reinforcing) medium were formulated.

4. REFERENCES

1. Malkin AYa, Kulichikhin SG (1996) Rheokinetics, Huthig & Wept, Heidelberg-Oxford, pp 326.
2. Malkin AYa, Zhirkov P.V. (1990) Adv.Polymer Sci.: 95:111-147
3. Malkin AYa,.Kuznetsov VV (1998) Intern.Polymer Processing, accepted for publication
4. Malkin AYa, .Baranov AV, Timofeev SV (1995) J.Polymer Engng 14:141-159

MICROSTRUCTURAL CHANGES IN STRUCTURED COLLOIDAL DISPERSIONS

J. MELLEMA, D. VAN DEN ENDE

Rheology Group, Twente Institute of Mechanics, J.M.Burgerscentrum, Faculty of Applied Physics, University of Twente, P.O.Box 217, 7500 AE Enschede, The Netherlands

1. INTRODUCTION

Globally an interesting program is conducted on the rheology of colloidal dispersions. The objective of this program is to establish the relationship between the macroscopic rheological properties and the micro properties and processes. Micro variables are number density, size and shape (and their distributions) of the particles, the interaction potential between the particles (attractive, repulsive, hard, soft), the thermodynamic forces pertaining to position and orientation distributions and the always present hydrodynamic interactions. Some of them can be varied systematically. Thus the rheological behavior can be mapped out and then it is accessible for theoretical analysis. In this research two quite different regimes can be discerned: near and far from equilibrium. The last years real progress can be observed which is also due to the availability of well defined model dispersions (see e.g. Mellema 1997).

When an interaction potential is present structures of systems of particles are formed. At rest the overall dispersion is not necessarily a one phase system and may phase separate in, say, two phases. Phase separation may take a long time and may not be observed at the time scale of weeks: the time to conduct experiments. In flow the rest structure cannot be maintained and microstructural changes occur. These changes may have the character of an apparent phase separation: one phase with a structure close to the rest structure and one phase in which the rest structure is destroyed by the flow.

The focus of this paper is the concept of this apparent phase separation under flow: far from equilibrium. It will be pursued for two systems, one with repulsive spheres and one with attractive spheres.

2. EXPERIMENTAL METHOD

Most of the rheological experiments are nowadays conducted with commercial apparatuses. These become more and more improved. Advancement in rheometry is however strongly influenced by researchers eager to get more rheological information than contemporary instruments can provide. The home-made apparatus used in the present study superimposes small amplitude harmonic oscillations perpendicular onto a steady shear flow. Thus the obtained linear viscoelastic behavior is informative on the actual structure at a certain rate of shear. The apparatus used (Zeegers et al. 1995) consists of a thin walled cylinder that oscillates in axial direction in an annular holder. The cylinder is suspended from a spring, which in turn is attached to a linearly oscillating exciter. The annular beaker can be rotated, thus providing the steady shear flow. From the measured phase difference and amplitude ratio of the exciter and cylinder displacements the fluid impedance and the complex shear modulus are obtained. The instrument operates in the frequency range from 10^{-3} to 50 Hz and the steady shear rate ranges from 10^{-2} to 100 s^{-1}.

The statement that the actual stress state of a material is a functional of the history of deformation is quite often relevant for colloidal dispersions. The strain history of the fluid has to be defined by a certain treatment before filling and by pre-shear in the apparatus. This procedure is not know a priori and has to be figured out for a chosen fluid (Wolthers et al. 1996, Van der Vorst et al. 1998).

3. FLUIDS

3.1 Depletion flocculating silica spheres

The influence of aggregation was studied on depletion flocculating silica spheres. We synthesized classical stearyl coated silica particles with a radius of 38±3 nm (Wolthers et al. 1996). They were dispersed in cyclohexane and brought to depletion flocculation by adding (126 g/l in cyclohexane) polystyrene polymers with a molar mass of 5970 g/mole. At rest this fluid was in the two phase region far from the phase boundary.

3.2 Ordering latex dispersions

The influence of ordering was studied on classical repulsive latex spheres dispersed in water. For the synthesized particles the radius was 195 nm and the surface charge density -4.6 μC/cm^2 (Van der Vorst et al. 1998). Using an ion exchange resin the final electrolyte concentration was reduced to about 10 μM. At rest the fluids were in the disordered or ordered phase, depending on the volume concentration. The two phase region is for this system negligible small.

4. MODELING

When an ordered or aggregated dispersion is forced into flow the structure cannot be maintained. At low rates of shear one might still expect the structural changes to be small but at higher rates the structure is at least partially destroyed. Before the structure becomes completely destroyed one expects that in a certain shear rate regime a structured phase and a phase with destroyed structure coexist: apparent phase separation.

4.1 Aggregated particles

The aggregates of the structured phase are assumed to have a self similar structure. All the primary particles reside in the aggregates. The dispersion is described as a system of overlapping spherical aggregates which can be interconnected by chains gradually created and broken due to shear and thermal processes. Chains can be rigid and soft. The latter carry negligible stress. The kinetics of transition between rigid and soft is incorporated. The size of the aggregates is determined by a break-up criterion that balances the maximum force which can be resisted by a chain and the force exerted by the flow. Both the hydrodynamics of a concentrated hard spherical aggregate dispersion and the network of the interconnected aggregates contribute to the viscosity (Potanin et al. 1995).

4.2 Ordered latices

The size and the volume concentration of the spherical crystallites are determined by break-up and coalescence processes. The break-up is determined by the condition that the elastic energy stored in the deformed crystallite exceeds the energy to create a new free surface in the crystallite. The deformation is small so that the crystallites can still be considered as spherical. The coalescence is determined by a balance between the shear flow pushing crystallites together and the lubrication force between the crystallites hindering coalescence. In addition modeling of the viscosity of the disordered continuous phase is needed. With this ingredients the flow curve can be calculated (Van der Vorst et al. 1997).

The interpretation of the superimposed complex modulus makes use of the concept of spheres in a continuous phase but rests on a completely different model because the complex modulus probes something different from the steady shear viscosity. Obviously both the crystalline and the continuous phase respond linear viscoelastic to small amplitude harmonic excitations. The phases are described with two different Maxwell models. Then the viscoelastic response can be calculated provided the volume fraction is known (Van der Vorst et al. 1998).

5. RESULTS AND DISCUSSION

5.1 Depletion flocculating dispersions

In a few papers the feasibility of the modeling of the flow curve for aggregated systems has been demonstrated (De Rooij et al. 1993, Potanin et al. 1995). Here we focus on its use to probe the densification of the aggregates during flow (Wolthers et al. 1996). Since the rheological behavior of the dispersion is history dependent the apparent phase separation is more complicated. During shear the aggregates become densified due to the continuous kneading of aggregates by the flow.

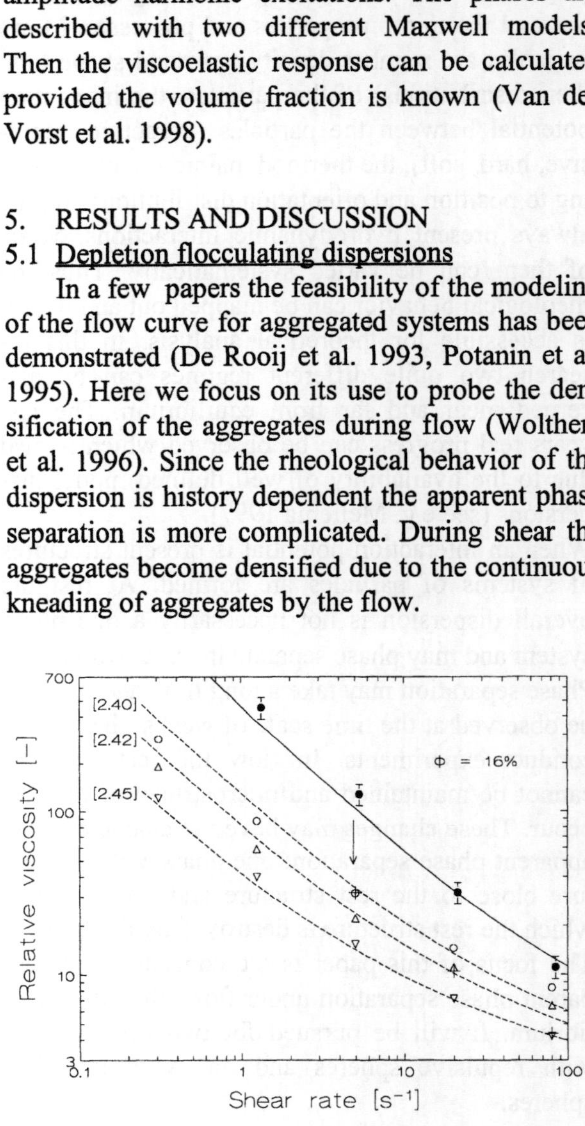

Fig. 1 Flow curves before (closed symbol) and after (open symbols) densification at shear rates of 5-80 s^{-1}. Lines from modeling. Fractal-like dimension given between brackets.

If the effect of this densification in the model is probed by the fractal-like dimension an analysis of the flow curves during shear is possible. Flow curves before and after densification like given in Fig. 1 can be modeled and all lumped together in a scaled representation (Fig. 2).

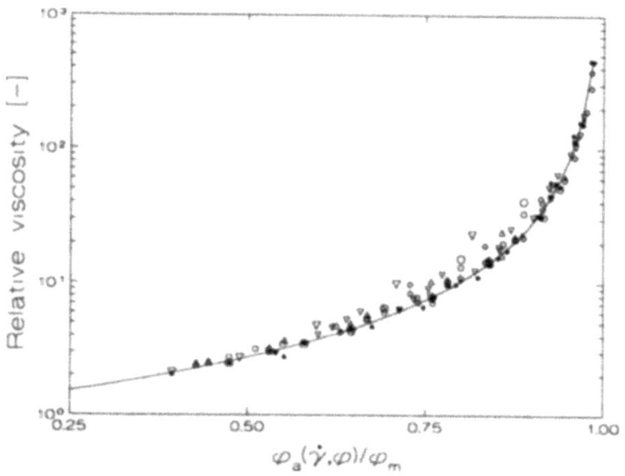

Fig. 2 Relative viscosities of 17 flow curves before and after densification for volume fractions between 8 and 16% lumped together with the adopted model.

5.2 <u>Ordered latices</u>

In Fig. 3 the flow curve given by Van der Vorst et al. (1997) is given to demonstrate the agreement between the flow curve modeling and the measurements for the ordered phases.

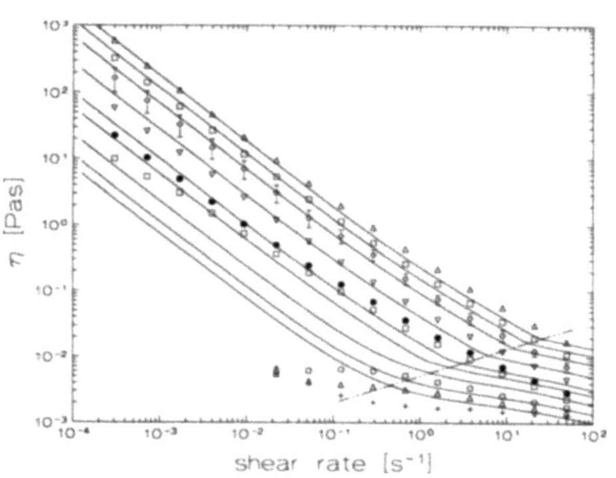

Fig. 3 Flow curves of an ordered latex of volume fractions between 0.05 and 0.20. Lines: theory. Dashed line: transition to completely disordered fluid.

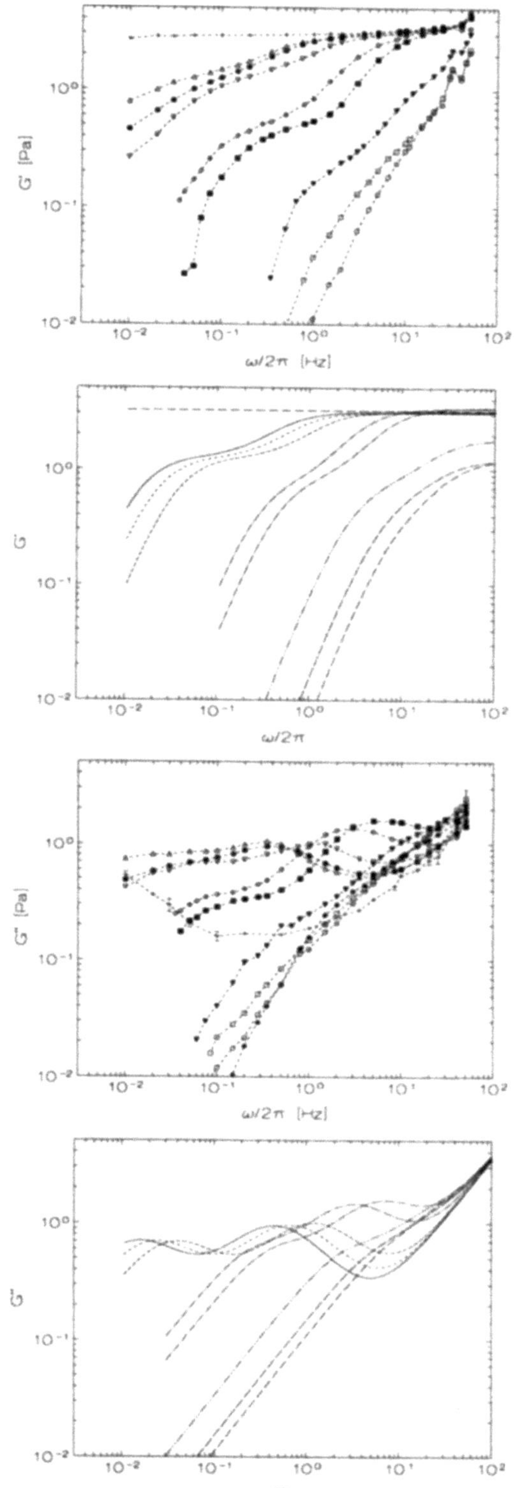

Fig. 4 In the upper two pannels G' (experiment and calculation respectively) and in the lower two G" as a function of frequency at volume fraction 0.2, for shear rates between 0 (+) and 25 (o) s^{-1}.

The complex shear moduli of superimposed oscillations onto the steady shear flow of ordered latices are shown in Fig. 4 (Van der Vorst et al. 1998).

From both the completely different modeling of the flow curves and the complex shear modulus the volume fraction, the viscosity of the continuous phase and the viscosity of the total dispersion can be determined. They are compared in Fig. 5

5.3 Discussion and conclusions

The investigated dispersions are difficult to study experimentally due to history dependence and apparent phase separation in flow but is feasible for skilled experimentalists.

The analysis of such measurements have reached a new stage by a systematic approach based on the 'Ansatz' that flow units of one phase exist in a continuum of another phase. Many of the assumptions in the modeling can be criticized. Nevertheless the overall impression is that the modeling catches essential features of the structural changes. The 'Ansatz' applied to completely different dispersions and structures is promising. The novel use of superimposed oscillation give clear additional information and because the modeling of the complex shear modulus differs completely from that of the flow curves, larger internal consistency can be obtained.

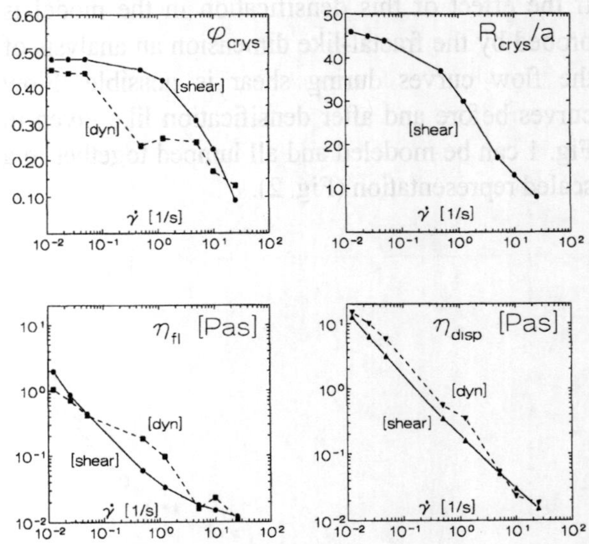

Fig. 5 Volume fraction (ϕ) and mean radius (R) of the crystallites, viscosity of the disordered phase (fl) and the dispersion (disp) at volume fraction 0.2 of primary particles from the flow curves (line) and the complex shear modulus (dashed line).

6. REFERENCES

De Rooij R, Potanin AA, van den Ende D, Mellema J (1993) J Chem Phys 99: 9213-9223

Mellema J (1997) Cur Op Coll & Interf Sc 2: 411-419

Potanin AA, de Rooij R, van den Ende D, Mellema J (1995) J .Chem Phys 102: 5845-5853

Van der Vorst B, van den Ende D, Aelmans NJJ, Mellema J (1997) Phys Rev E 56: 3119-3126

Van der Vorst B, van den Ende D, Tekin NA, Mellema J (1998) Phys Rev E 57: 3115-3122

Wolthers W, Duits MHG, van den Ende D, Mellema J (1996) J Rheo 40: 799-811

Zeegers J, van den Ende D, Blom C, Altena EG, Beukema GJ, Mellema J. (1995) Rheol Acta 34: 606-621

RHEOLOGY AND MORPHOLOGY OF IMMISCIBLE POLYMER BLENDS

P. MOLDENAERS, I. VINCKIER, P. VAN PUYVELDE, H. YANG

Chemical Engineering Dept., K.U. Leuven, de Croylaan 46, B-3001 Heverlee, Belgium

1. INTRODUCTION

Blending immiscible polymers provides an economic alternative to synthesizing new polymers. The properties of a blend depend on the characteristics of the constituent polymers as well as on the microstucture which is generated during the processing. If one wants to optimize the final structure of a blend it is thus essential to understand the interplay between rheology and morphology development during flow. The kinematic conditions during processing can however be complex and morphological changes such as droplet deformation, break-up, retraction and coalescence can all occur simultaneously. Moreover it is common practice to derive the structural information from specimens that have been cooled from the melt after e.g extrusion (Utracki, 1989).

In the present investigation we will use simple, well defined kinematics that allow for the study of the individual morphological phenomena. The time evolution of the structure will be probed in-situ rather than relying on the frozen-in structures. Rheological and rheo-optical techiques are both candidates for the in situ-probing of the morphology development. Rheology has the advantage that, in principle, it can be used for any sample. The rheological response is however difficult to translate directly into structural changes. Rheo-optical data on the other hand provide more direct information on the microstructure development but require samples with low concentrations of the disperse phase and sufficient optical matching between the two phases. It is the aim of this work to interpret the rheological transients by comparing them with the rheo-optical ones. On this basis rheology could be used as an in-situ probing technique for the morphology development.

2. METHODS

2.1 Materials

Model blends have been used that are liquid at ambient temperature. The component polymers are polydimethylsiloxane (PDMS Rhodorsil 47V200000 from Rhône-Poulenc) and polyisobutene (PIB Parapol 1300 from Exxon). They are both Newtonian fluids, the viscosity of PDMS being 2.3 times that of PIB. The elasticity of the PIB sample is negligible whereas that of the PDMS is low but not necessarily negligible. These components have been chosen because they are completely immiscible and their blends are relatively transparent. The interfacial tension is around 3 mN/m at ambient temperature (Sigillo et al., 1997). To ensure a substantial contribution from the interface to the stresses, blends with 10 % disperse phase have been used in the rheological experiments. The rheo-optical experiments could only be performed on very dilute blends, typically with 1% disperse phase.

2.2 Techniques

Rheological experiments have been performed on a Rheometrics Mechanical Spectrometer (RMS 800) using cone and plate geometry. The instrument is equipped with a force rebalanced transducer, which is essential to measure reliable transients. The emphasis will be on normal stress transients. Almost all elasticity in the samples will be due to interface and consequently N_1 can be directly related to the anisotropy of the interface (Onuki, 1987; Doi, 1987).

In the rheo-optical experiments both linear conservative dichroism and SALS have been investigated. Dichroism detects the (flow-induced) anisotropy of the sample. It has been measured on a Rheometrics Optical Analyser (ROA) which uses a polarization modulation technique (Fuller and Mikkelsen, 1989) . The

ROA device has also been modified to perform SALS measurements during flow. A parallel plate geometry has been used on the ROA. Because of the substantial turbidity of the blends the optical path length was kept minimal (typically 0.2 mm).

3. RESULTS AND DISCUSSION

The results of two types of experimental protocols will be discussed in the next sections. First of all, stepwise increases in shear rate have been performed. In these experiments an initial shear rate is applied for a sufficiently long time to ensure well defined initial conditions. This typically requires more than 1000 strain units. The shear rate is then suddenly increased and the resulting stress transients are recorded. For a large increase in shear rate the droplets are expected to deform first, become fibrillar in shape , break-up and eventually coalesce.
In a second series of experiments the previous protocol is repeated but the flow is now stopped before the structure has reached its steady state condition. The evolution of the microstructure is monitored during the subsequent relaxation. The structural mechanism underlying the relaxation can be droplet retraction and/or break-up due to end pinching or Rayleigh instabilities.

3.1 Stepwise increase in shear rate.

Figure 1 displays a typical transient of the first normal stress difference after a step-up in shear rate, in this particular case a fifteenfold increase in shear rate. The evolution of the dichroism after a similar increase in shear rate is shown in figure 2a, whereas figure 2b contains the corresponding SALS patterns. When increasing the shear rate, the droplets are being stretched. This increase in anisotropy reflects itself in growing values of N_1 (figure 1) and of the dichroism (figure 2a). Both curves reach their maximum after approximately 100 strain units. Once the extended fibrils start to break up, the anisotropy of the interface should again decrease as can indeed be seen in both the normal stress and the dichroism. The subsequent, very slow and gradual increase of the anisotropy reflects the coalescence of the small droplets.
Coalescence is likely to occur here as the break-up due to Rayleigh instabilities will produce droplets that are smaller than the equilibrium size at that shear rate.

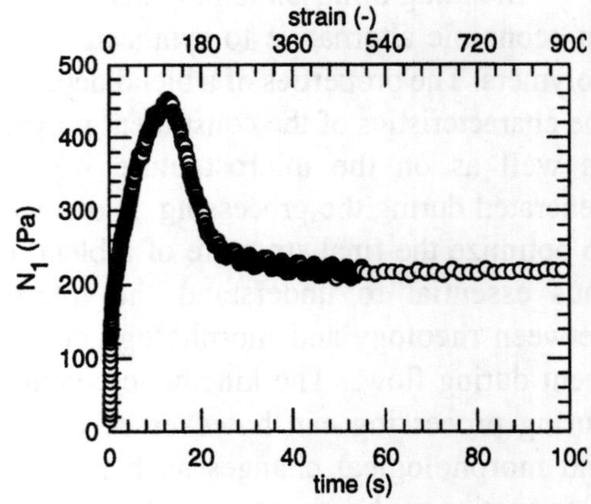

Figure 1: Normal stress transient after a step-up in shear rate from 0.6 to 9 s-1 (10% PDMS in PIB)

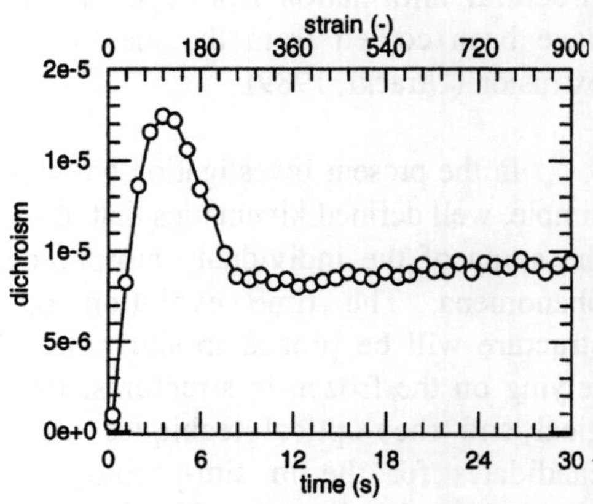

Figure 2a: Dichroism transient after a step-up in shear rate from 2 to 30 s-1 (1% PDMS IN PIB)

The structural interpretation is supported by the evolution of the SALS patterns (figure 2b). In these patterns a bright streak appears perpendicular to the flow direction. This streak is most pronounced when the normal stress difference and the dichroism reach their maximum (figs. 1a and 2). In large step-ups in shear rate the rising part of the curves, either for N1 or dichroism, superimpose for different shear rates when plotted versus strain. This indicates an affine deformation of the droplets. The resulting shape of the interface can be modeled, providing a basis for modeling N_1 and dichroism (Vinckier *et al.*, 1997a)

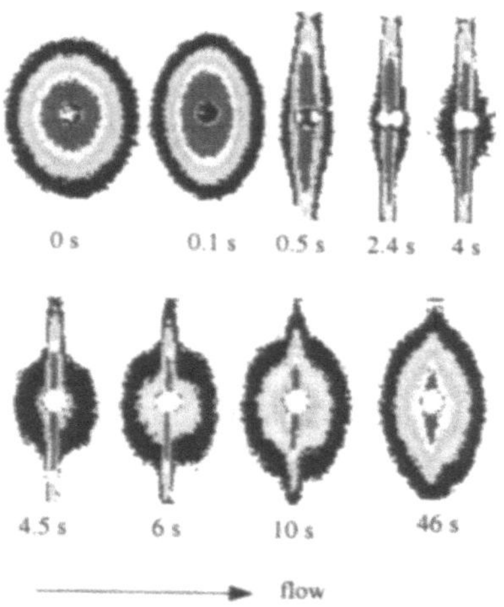

Figure 2b: Time evolution of the SALS patterns after a step-up in shear rate from 2 to 30 s-1 (1% PDMS in PIB), numbers indicate the time that has elapsed since the increase in shear rate.

3.2 Relaxation

If the flow is stopped before the steady state is reached, the resulting transients during relaxation should reflect the mophology existing at the instant the flow was arrested. This is illustrated in figure 3 for N_1 transients and in figure 4 for the relaxation of the dichroism.

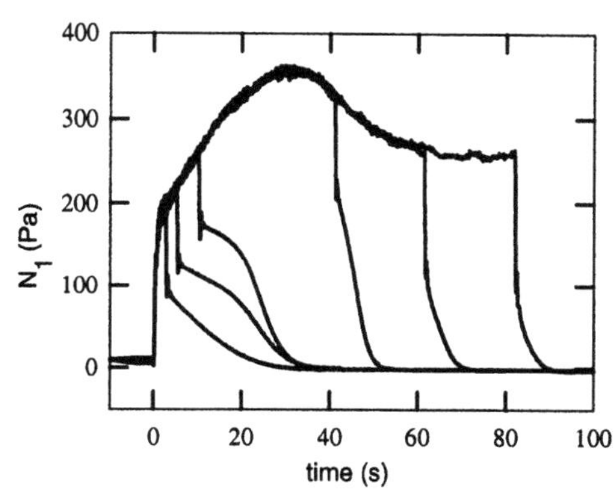

Figure 3: Relaxation of N_1 in an interrupted step-up test from 0.3 to 3 s-1 (10% PIB in PDMS)

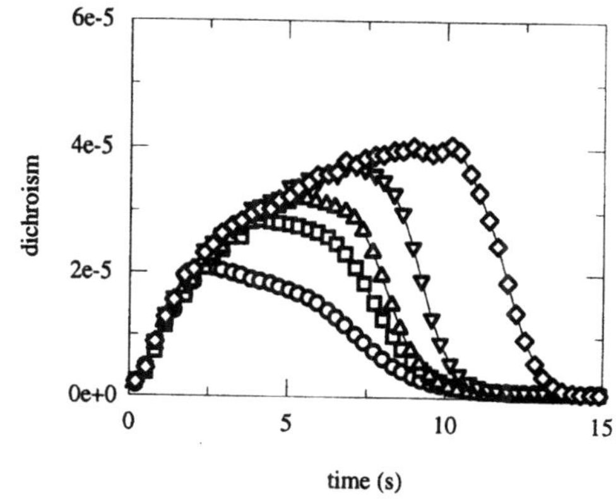

Figure 4: Relaxation of the dichroism in an interrupted step-up test from 2 to 20 s-1 (1% PIB in PDMS)

The initial, almost instantaneous stress decay in figure 3 is due to the relaxation of the component polymers. It does hardly affect the shape of the interface and hence is not visible in the dichroism. The subsequent, slower relaxation is driven by the interfacial tension. When comparing figures 3 and 4, it is clear that after short or long shearing times the relaxation is characterized by a monotonous decay. After intermediate shearing periods a shoulder develops in the relaxation curve. The monotonous decay can be associated with the retraction of the deformed droplets, whereas the curves with a shoulder can be associated with relaxation by droplet break-up due to a combination of end pinching and Rayleigh instabilities (Vinckier *et al.*, 1997b). This picture is consistent with the observations of Stone et al. (1986) on single drops.

It is also confirmed by the SALS results. As an example, figure 5 shows the evolution of the SALS pattern for a relaxation experiment in which a shoulder was observed. After the flow has been stopped (at t = 4s), the streak perpendicular to the flow direction clearly remains visible for a very long time. Nevertheless the dichroism and N1 decay to zero shortly after halting the flow. While this happens two secondary streaks develop in the SALS pattern, parallel to the main streak. It can be shown (Mewis *et al.*, 1998) that this pattern is generated by a string of droplets.

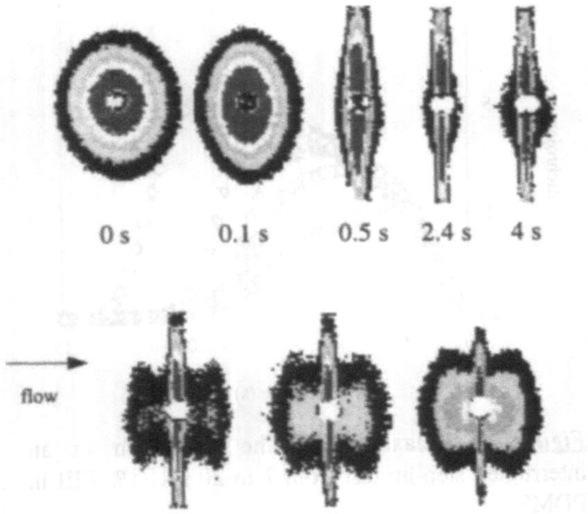

Figure 5: SALS pattern evolution in an interrupted step-up test from 2 to 30 s-1. Flow was interrupted after 4 seconds at 30 s-1 (1% PDMS in PIB)

Break-up of fibrils due to Rayleigh instabilities will result in such string of droplets, they will persist under no flow conditions without entailing a strong contribution to the dichroism.

As in the case of step-up experiments, normal stresses and dichroism during relaxation can be described quantitatively, using models describing the shape of the interface during either droplet retraction or break-up by Rayleigh instability. A corresponding model for end pinching is not yet available. It can easily be taken into account by an empirical factor (Vinckier *et al.*, 1997b, Van Puyvelde *et al.*, submitted). The analysis of the droplet break-up by interfacial instabilities has been used to calculate the interfacial tension of the blends (Van Puyvelde *et al.*, 1998)).

The approach discussed here for stepwise increase in shear rate and for relaxation can be extended to other transient experiments such as recoil. From the model description various scaling relations can also be derived for the transients of the stresses and the dichroism.

4. CONCLUSIONS

When using suitable systems, the flow-induced microstructure in immiscible blends can be investigated by either rheology, dichroism or SALS. These techniques can detect in real time droplet deformation, fibril formation and break-up and coalescence. Time scales for structural changes, during and after stopping flow, can be determined quantitatively. These techniques open possibilities for studying the effect of material and process parameters on the development of the flow-induced microstructure in blends.

5. REFERENCES

Doi M (1987) in Physics of complex and supermolecular fluids, ed. Safran SS, Clark N, Wiley, New York

Fuller GG, Mikkelsen KJ (1989) J. Rheol. 33:761-769

Mewis J, Yang H, Van Puyvelde P, Moldenaers P, Walker LM (1998) Chem. Eng. Sci., in press

Onuki A (1987) Phys. Review A 35: 5149-5155

Sigillo I, di Santo L, Guido S, Grizzuti N (1998) Polym. Eng. Sci. 37:1540-1549

Stone H, Bentley BJ, Leal LG (1986) J. Fluid Mech. 173: 131-158

Utracki LA (1989) in Polymer Alloys and Blends, Hanser, Munich

Van Puyvelde P, Mewis J, Moldenaers P J. Chem. Soc. Faraday Trans, submitted

Van Puyvelde P, Yang H, Mewis J, Moldenaers P, (1998) J. Colloid Interface Sci., 200, 86-94

Vinckier I, Moldenaers P, Mewis J (1997a) J. Rheol. 41: 705-718

Vinckier I, Mewis J, Moldenaers P (1997b) Rheol. Acta 36: 513-523

ELONGATIONAL BEHAVIOUR AND MOLECULAR STRUCTURE OF POLYMER MELTS

H. MÜNSTEDT, S. KURZBECK

Institute of Polymer Materials, Department of Materials Science, University Erlangen-Nürnberg, Martensstr. 7, D-91054 Erlangen, Germany

1. INTRODUCTION

Many processing operations of polymer melts are governed by elongational flow. It is widely accepted today that elongational properties of polymer melts particularly in the non-linear regime of deformation which is dominant in processing cannot be derived from their linear behaviour. Therefore, the importance of rheological measurements in elongation is obvious. On the other hand, the molecular structure of polymers influences their elongational properties in a complex way not completely understood up to now.

2. GENERAL FEATURES OF ELONGATIONAL VICOSITY

One feature of the elongational behaviour of polymer melts in comparison to shear is represented in Fig. 1 (Meißner 1972, Münstedt and Laun 1979) where stressing experiments in shear and elongation are shown on a low density polyethylene (LDPE). Whereas the steady-state value of the shear viscosity becomes smaller with growing shear rate as it is usual for polymer melts in the nonlinear range of deformation, the elongational viscosity defined as $\sigma / \dot{\varepsilon}_0$ increases as a function of time and reaches a steady state which is strongly dependent on the strain rate. σ is the actual tensile stress on the sample and $\dot{\varepsilon}_0$ the elongational rate. At small strain rates the value is $3\,\eta_0$ according to the linear theory of viscoelasticity but increases with rising $\dot{\varepsilon}_0$ and runs through a maximum. The viscosity increase as a function of time or strain, respectively, is called strain hardening. It is a special feature of elongational deformation of polymer melts.

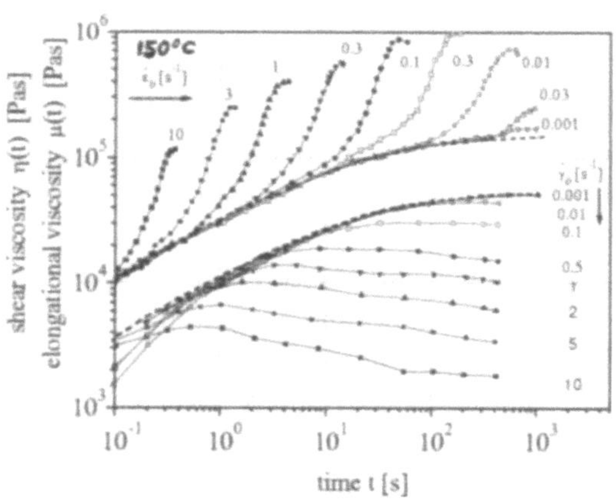

Fig. 1 Comparison of stressing experiments on a low density polyethylene in shear and elongation

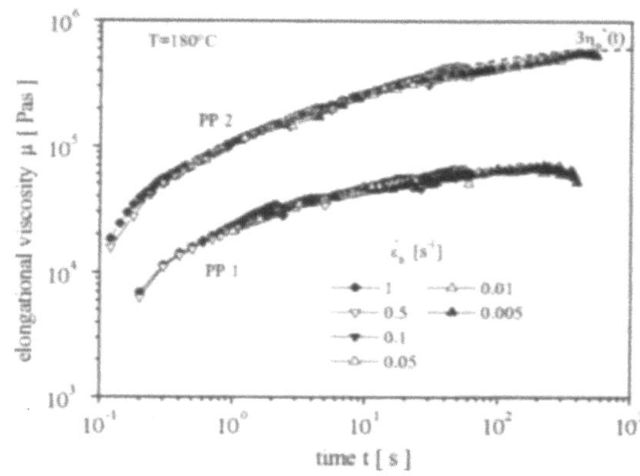

Fig. 2 Time-dependent elongational viscosities of two polypropylenes at different strain rates

The strain hardening is not as general, however, as the viscosity decrease in shear as it becomes obvious from Fig. 2 which shows stressing experiments on two commercial polypropylenes. PP 1 is a homopolymer and PP 2 an ethylene-propylene copolymer. The two materials do not exhibit any indication of strain hardening in the range of strain rates investigated. All curves of each product fall together on one line which corresponds to $3\,\eta_0^+(t)$ the threefold of the linear time-dependent viscosity in shear. That means the two polypropylenes behave linearily in elongation over a range of time and strain rates where the LDPE of Fig. 1 shows a pronounced nonlinear behaviour. Similar results on polypropylenes were found by Trzebiatowski and Wilski (1985).

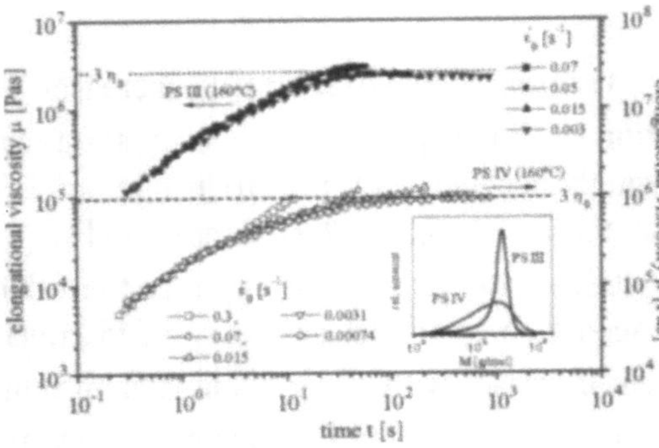

Fig. 3 Time-dependent elongational viscosities of two polystyrenes of different molecular weight distributions

The elongational behaviour of two polystyrenes investigated by Münstedt (1980) is presented in Fig. 3 together with their molecular weight distributions. Whereas for PS III an indication of strain hardening has not been found, PS IV which is a commercial product of a significantly broader molecular weight distribution shows a tendency to strain hardening at high strain rates. The strain hardening is less pronounced, however, than for the LDPE of Fig. 1.

As has been demonstrated on some commercial polymers of various chemical structure two types with respect to strain hardening can be distinguished. If the differences of their molecular structure are regarded it is evident that the polypropylenes and the polystyrenes investigated, which do not exhibit significant strain hardening consist of linear chains whereas the LDPE is a long-chain branched polymer.

Relating long-chain branching and strain hardening to each other as an unambiguous consequence may be too rough a conclusion according to the experimental findings on the two polystyrenes in Fig. 3 that the sample of broader molecular weight distribution shows an indication of strain hardening.

3. INFLUENCE OF MOLECULAR WEIGHT DISTRIBUTION

The influence of the molecular weight distribution and particularly of a high molecular weight component is demonstrated by the results of Fig. 4. PS II shows a much more pronounced strain-hardening behaviour than PS IV the molecular weight distribution of which is distinctly broader, however. From a comparison of the strain hardening of these two polystyrenes with their molecular weight distributions it has to be concluded that the high molecular weight component of PS II clearly distinguished from the main component gives rise to the significant increase of the time-dependent elongational viscosity. A similar result has been reported recently by Koyama et. al. (1997) who found that the strain hardening of a commercial polymethylmetacrylate (PMMA) becomes stronger if a small amount of a PMMA of a ten times higher molecular weight is added.

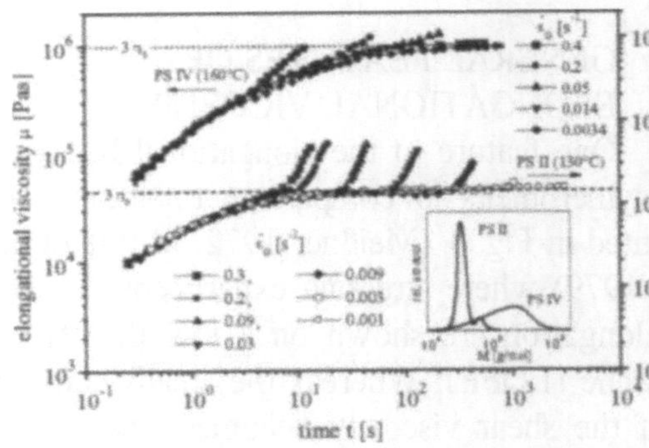

Fig. 4 Influence of a high molecular weight component on the strain hardening of polystyrene

Regarding the molecular weight distributions of the two polystyrenes in Fig. 4 it becomes obvious that not the height of the molecular weight component is decisive for the degree of strain hardening but its difference with respect to the main component. PS IV contains molecules of a much higher molecular weight than PS II but they are part of the broad distribution.

The absence of strain hardening is not a unique feature of polypropylene either, as Fig. 5 shows. Compared to the polypropylenes of Fig. 2, the material PP 3 exhibits a pronounced viscosity increase as a function of time at all strain rates applied. The measurements at different strain rates

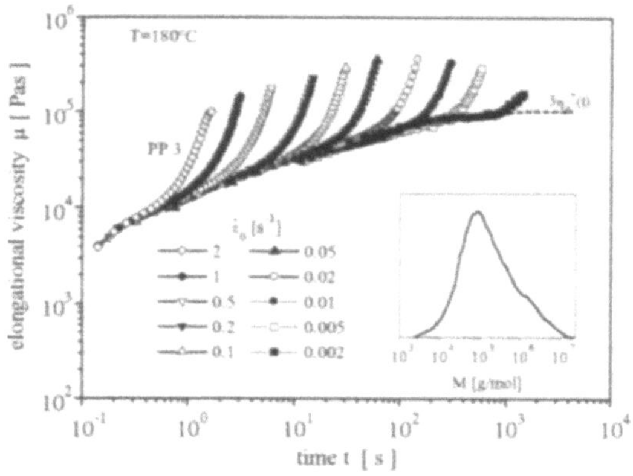

Fig. 5 Strain hardening of a polypropylene modified by electron beam irradiation

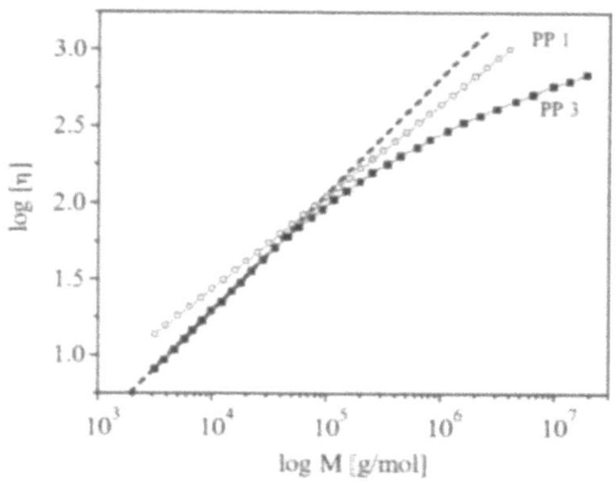

Fig. 6 Relationship between intrinsic viscosity and molecular weight for two different polypropylenes

can be superimposed at small strains to give one curve which coincides in good approximation with $3\,\eta_0^+(t)$ where $\eta_0^+(t)$ is the time-dependent shear viscosity in the linear range of deformation. This agreement which should be valid according to the linear theory of viscoelasticity throws some light on the accuracy of the extensional measurements.

As can seen from the inset of Fig. 5 the polypropylene investigated possesses a rather broad molecular weight distribution which is distinguished by a shoulder at its high molecular weight end. There is at least an indication that some molecules of a higher molecular weight are present in a matrix of lower molecular weight.

It is questionable, however, whether this shoulder alone can be responsible for the observed very strong strain hardening since the high molecular weight component which it indicates is not as pronounced as in the case of PS II .

4. INFLUENCE OF LONG-CHAIN BRANCHING

That long-chain branching has to be regarded as another source for the distinct strain hardening of the polypropylene in Fig. 5 becomes evident from Fig. 6 which shows results by Kurzbeck et. al. (1998). In this figure the intrinsic viscosity $[\eta]$ is presented for several fractions of various molecular weights obtained by gel permeation chromatography (GPC). For linear molecules the Mark-Houwink equation should hold, i.e. a linear relationship should be expected if $[\eta]$ is double-logarithmically plotted as a function of the molecular weight of the fractions. As can be seen from Fig. 6 the Mark-Houwink equation is fulfilled for PP 1 which is assumed to be of linear structure and does not exhibit any strain hardening (cf. Fig. 2). For PP 3, however, a deviation from a straight line is found which can be explained by the assumption of the existence of long-chain branches in PP 3 which give rise to smaller coil dimensions and following from that to lower values of the intrinsic viscosity. Therefore, in the case of PP 3 it has to be concluded that besides a pronounced high molecular weight component long-chain branches as well can be regarded as the molecular origin of the significant strain hardening. To what extent the strain hardening is influenced by the two molecular features cannot be derived from the measurements, however.

5. INFLUENCE OF COMPATIBILITY

The linear low density polyethylene (LLDPE 1) of Fig. 7 shows a strain-hardening behaviour which is in contrast to that presented for other materials. Whereas at the higher strain rates an increase of the elongational viscosity from the linear curve is not found, strain hardening becomes the more pronounced the smaller the strain rate. Investigation of the molecular structure by Gabriel et al. (1998) indicate that the LLDPE 1 contains a distinct linear ethylene homopolymer component of higher molecular weight within the

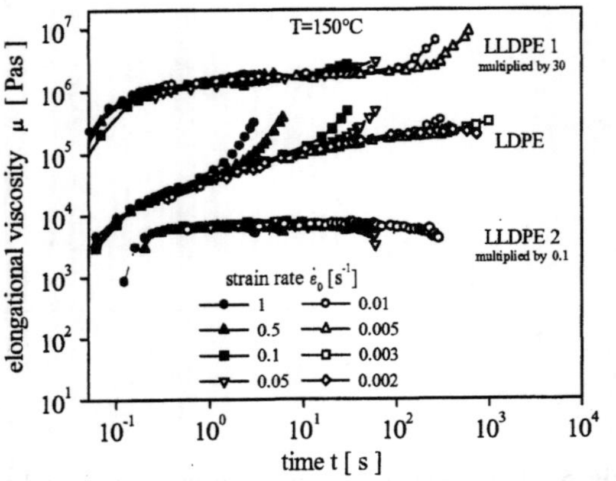

Fig. 7 Time-dependent elongational viscosities of two LLDPE in comparison to a classical LDPE

ethylene-octene copolymer matrix. From other experiments extensively discussed by Gabriel et al. (1998) it can be postulated that the linear molecules are not miscible with the short-chain branched species and form a separate phase which gives rise to the surprising elongational behaviour. The findings of the stressing experiments are supported by results in creep tests (Münstedt et al., 1998) which clearly show two different processes at lower stresses which are not detectable at higher ones.

From these experiments it becomes obvious that in the case of polyethylenes not only properties like long-chain branches and molecular weight distribution influence the elongational behaviour of polymer melts but that the effect of separate phases in the molten state have to be taken into account.

In Fig. 7 another ethylene-α-olefin copolymer (LLDPE 2) is presented which does not exhibit any strain hardening at the strain rates investigated. The decrease of the viscosity or the stress, respectively, at higher elongation is due to a necking of the samples. This behaviour limits the total achievable elongation which is smaller than for the other polyethylenes of Fig. 7. These findings demonstrate the importance of strain hardening for the homogeneity of sample deformation during elongation. The results on th tree polyethylenes may throw some light on the investigations by Schlund and Utracki (1987) who report different features of strain hardening for various LLDPE. This may be explained by the different composition of the samples which are not reported not investigated by the authors, however.

6. REFERENCES

Gabriel C, Münstedt H, Kaschta J (1998) Rheol Acta 37: 7-20

Koyama K, Takahashi T, Nakay Y, Takimoto J (1997) In: Proc. 13th Int. Meeting Polym. Proces. Soc. pp 8B

Kurzbeck S, Oster F, Münstedt H, Nguyen TQ, Gensler R (1998) submitted for publication in J. Rheol

Meissner J (1972) J. Appl. Polym. Sci. 16: 2877

Münstedt H, Laun HM (1979) Rheol Acta 18: 492-504

Münstedt H (1980) J. Rheol. 24: 847-867

Münstedt H, Kurzbeck S, Egersdörfer L (1998) Rheol Acta 37: 21-29

Schlund B, Utracki LA (1987) Polym. Eng. Sci. 27: 359-386

Trzebiatowski T, Wilski H (1985) Rheol Acta 24: 582-587

TURBULENT GAS-LIQUID FLOWS

J.R.A.PEARSON

23 Chaucer Road, Cambridge, CB2 2EB, England

1. INTRODUCTION

Before considering turbulent multi-phase flow in pipes, it is instructive to consider single-phase turbulent flow.

For the familiar Newtonian fluids which are described rheologically by density ρ and viscosity μ, the flow regime (laminar, transitional or turbulent) in a circular pipe of diameter $2R$ is determined by the single dimensionless group

$$\mathrm{Re} = 2\rho RV / \mu = 2\rho Q / \pi R\mu$$

where Q is the mean volume flux, and V is the area-averaged mean velocity. As summarised in Box 1, the turbulent velocity field is unsteady and position dependent; all engineering pressure calculations are based on empirical (experimentally measured) friction factors $f = 2F_N{}^2(\mathrm{Re})$; mean velocity profiles are given in dimensionless form $u = \bar{v}_z(r) / u_*$ in terms of dimensionless distance $\eta = u_*(R - z)\rho / \mu$ from the wall, the observed power law forms having a logarithmic limit first proposed on similarity grounds. Even with the most powerful digital computers, full solutions of the unsteady, incompressible Navier-Stokes equations cannot be obtained except by making assumptions about small-scale Reynolds-stress behaviour. Curiously many of these assumptions introduce the same non-linear features as arise in viscoelastic fluid behaviour (e.g. Gatski & Speziale 1993). What makes turbulent flow difficult to model is the wide range of "eddy" sizes: there are unsteady dynamical processes taking place at all length scales between R and ν/V^{2}; the microstructural length-scales can be regarded as smaller than any flow length-scale.

In practice, friction factors also depend upon a "wall roughness length"; a short development length after the pipe entry is needed before friction factors accurately represent pressure drops. If the pipe diameter varies over a length of the same order as the diameter, the inertial contribution to pressure drop over that length is much larger than the frictional. Thus the response of a Venturi (or orifice plate) flowmeter is determined primarily by Bernoulli's equation applied directly to V, i.e. by a 1D approximation.

For rheologically complex fluids (see Box 2) whose microstructural length scales can be regarded as smaller than any flow length scale, parameters other than ρ and μ are known to influence friction factors and mean velocity profiles. Some progress has been made (Sureshkumar et al.1997) in computational analysis of turbulent pipe flow of viscoelastic liquids, but drag reduction remains a not-fully-explained phenomenon.

2. GAS-LIQUID FLOW

A mixed pipe flow of gas and liquid introduces many complications. The density difference means that gravity, and immiscibility means that interfaces (with interfacial tension σ), have important effects (e.g. the inclination θ of the pipe becomes a fundamental parameter; the compressibility of the gas is more important than in single-phase gas flows; the mean gas fraction will vary with cross-sectional position); there will be 2 mean flow velocities, not necessarily equal. The near universality of the single-phase turbulent flow is replaced by a series of possible regimes (as sketched in Box 3), for all of which a range of length scales associated with continuously-deformable interfaces are introduced.

Clearly, an effective single-phase fluid with a characteristic rheological equation of state (see Box 2 which summarises the now classic Oldroyd approach) is not an appropriate model. Box 3 summarises a mixture-theory approach (for two interpenetrating fluids) which treats each fluid separately; it is most appropriate for vertical bubbly flows where the average bubble diameter is small compared with the pipe diameter, though it has been applied to churn and even slug flows also; the simplest form assumes a (quasi-static) one-dimensional model relating the time- and space-averaged velocities $< \bar{v}_z >_{l,g}$, phase fractions $< \alpha >_{l,g}$ and pressure gradient $d\bar{p}/dz$. For fully developed flow in a uniform pipe, these variables are in principle (empirical) functions of the volumetric flow rates q_l, q_g, the radius R, the pipe inclination θ, the phase densities ρ_l, ρ_g, the phase viscosities μ_l, μ_g and σ. Such evidence as is available suggests that development lengths are large, often larger than the lengths of the pipes used in experiments. Most regime "maps" are coarse summaries of laboratory experiments with only the flow velocities as significant variables; they are restricted to vertical or horizontal geometries. There are no reliable results for the total wall friction F_{wl} + F_{wg} because frictional effects are overwhelmed by uncertainties in estimates of gas fraction $<\alpha>$. The dominant term in the model equations is the phase interaction friction F_i, which has been the subject of much small-scale modelling effort: it has been approximated "constitutively" by a drag term $C_D(<\alpha>)$, relevant in determining (and hence determinable from) the slip velocity in steady-state vertical flow, and an added-mass term, $C_m(<\alpha>)$, that is important in accelerating flow.

Interest in such simple mechanistic models has arisen in connection with Venturi flow meters (based on a gentle decrease of R with z), which in single phase flow can be interpreted to within a few % using no more than Bernoulli's equation over wide ranges of flowrate. Experience of measurements made in gas-liquid flows shows that the assumption of no interphase slip, for flow rates and gas fractions of interest, can be seriously inaccurate. For flows of at least 20% liquid fraction, a 1D model has been developed that is satisfactory for interpreting measurements in terms of mass-flow-rate; a separate measurement of $<\alpha>$ at the throat of the Venturi is needed for this interpretation. However for lower liquid fractions, it is now clear that a core-annular model, which assumes a thin, relatively slowly moving layer of liquid on the pipe walls may be necessary to interpret results; this introduces not only an extra length (h the thickness of the annulus) and an unknown annular layer velocity but also requires 2 "constitutive" relations for mass and momentum exchange between the core and the annulus.

Gas-liquid flows in Venturis exhibit far larger, relatively slow, variations in the measured pressure drop than equivalent (in terms of mass flow rate) single-phase flows. These have led to unsteady models (see Box 3 again) in which the defining variables represent area and short-term time averages; specifically they have been developed to model slug flows and unstable bubble flow (see e.g. Lammers & Bieshuvel 1996).

A different class of models have been used to model horizontal flows, in which stratification occurs and so two-layer models are appropriate.

3. CONCLUSIONS

- Exact modelling of turbulent two-phase flows, taking account of every interface, is not currently practicable.
- "Simple-fluid" rheological models for such flows are not relevant.
- Fluid-flow modelling should be tailored to suit the imposed flow geometry, flow regime and most importantly the issues (engineering or fundamental) of greatest interest in any particular context.
- Experimental verification (validation) of any theory is essential.

REFERENCES

Barenblatt GI, Chorin AJ, Prostokishin VM (1997) *Appl.Mech.Rev.* **50**:413-429

Gatski TB, Speziale CG (1993) *J.Fluid Mech.* **254**:59-78

Lammers JH, Bieshuvel A (1996) *J.Fluid Mech.* **328**:67-93

Oldroyd JG (1950) *Proc.Roy.Soc.*A, **200**:523-41

Sureshkumar R, Beris AN, Handler RA (1997) *Phys.Fluids*, **9**:743-755

BOX 1 **Turbulent single-phase flow in a pipe** $\mathrm{Re} = 2\rho R\langle\bar{v}_z\rangle / \mu > 4000$

Walls at $r = R$ (r,θ,z) cylindrical-polar coordinate system

Velocity field: $\mathbf{v}(r,\theta,z,t) = \bar{\mathbf{v}}(r) + \mathbf{v}'(r,\theta,z,t) = \{0,0,\langle\bar{v}_z\rangle\} + \{0,0,\bar{v}_z^{\,*}(r)\} + \mathbf{v}'(r,\theta,z,t)$

Full Navier-Stokes equations imply predictions for $v_i(\mathbf{r},t)$.
Best direct results from computations using LES or alternative short-scale approximation.

Experimental results available on $v_i(\mathbf{r},t)$ and its correlations in time and space;
the most useful engineering information is on wall friction and the mean axial velocity, i.e.

$$F_N(\mathrm{Re}) = \frac{u_*}{\langle\bar{v}_z\rangle}, \quad u_*^{\,2} = -\left\langle\frac{\partial p}{\partial z}\right\rangle\frac{R}{2\rho} \quad \text{and} \quad \frac{\bar{v}_z(r)}{u_*} \approx C(\mathrm{Re})\left(\frac{u_*(R-r)\rho}{\mu}\right)^{\alpha(\mathrm{Re})}$$

Scaling laws (Barenblatt et al. 1997) explain logarithmic law-of-the-wall as infinite Re limit.

For rough walls $F_N(\mathrm{Re}, k)$ increases with k the roughness length.

For variety of non-Newtonian fluids (dilute and concentrated polymer solutions, suspensions, emulsions) drag reduction, i.e. $F_{n\text{-}N}(\mathrm{Re}) < F_N(\mathrm{Re})$, occurs.

BOX 2 **On the formulation of rheological equations of state** (from J C Oldroyd 1950)

Equations of motion and continuity: $p^{ki}{}_{,k} + \rho F^i = \rho \dot{v}^i, \quad \rho_t + \rho v^i{}_{,i} = 0$ apply to all materials.

It is assumed that the properties of a material element may depend upon the previous rheological history through which that element has passed, but not in any way on the states of neighbouring elements and not on the motion of the element as a whole in space. (This defines a "simple fluid".)

For example, the Newtonian liquid is specified by $e^{(1)}{}_i{}^i = 0, \quad p_{ik} - \frac{1}{3}g_{ik} = 2\mu e^{(1)}{}_{ik}$

In the most general case contemplated, the (rheological) equations of state defining the properties of an element at any instant may involve all the kinematic and dynamic quantities which define the states of the same element during its previous history, ... *moving as part of a continuum.*

Experiments designed to measure the rheological properties of particular materials must, for easy interpretation, relate to simple types of motion, and the resulting simple equations of state will require generalization before they can be applied to more complicated types of motion.

BOX 3 Turbulent gas-liquid flow in pipes

Defining variables/parameters: $\langle \bar{v}_z \rangle_l, \langle \bar{v}_z \rangle_g, \langle\alpha\rangle_l = 1-\langle\alpha\rangle_g, \rho_l, \rho_g, \sigma, \mu_l, \mu_g, \theta, R, \bar{p}$

Regimes (see sketch). Vertical: bubble, slug, churn, annular
Inclined: variants of vertical and/or horizontal
Horizontal: bubble, plug, stratified, slug, annular, mist

Models for vertical flow with varying $A(z)=\pi R(z)^2$, $\langle \bar{v}_z\rangle_{l,g} = u_{l,g}(z), \langle\alpha\rangle_{l,g} = \alpha_{l,g}(z)$

$$\rho_{l,g}\alpha_{l,g}u_{l,g}A = Q_{l,g} \quad \text{(constant)}$$

1-dimensional steady

$$\frac{d(\rho_{l,g}\alpha_{l,g}u_{l,g}A)}{dz} = -\alpha_{l,g}A\frac{d\bar{p}}{dz} - \rho_{l,g}\alpha_{l,g}Ag \pm F_i - F_{wl,g}$$

no-slip $\equiv$ single phase

mixture theory (bubbly flow)

$$F_i = \rho_l A\left\{C_D(\alpha_g)(u_g-u_l)|u_g-u_l| + C_m(\alpha_g)\left[\frac{1}{2}\frac{d(u_g-u_l)^2}{dz} + u_l\frac{du_g}{dz} + u_g\frac{du_l}{dz}\right]\right\}$$

drag added mass

unsteady incompressible

$$\frac{\partial(\alpha_{l,g}A)}{\partial t} + \frac{\partial(\alpha_{l,g}u_{l,g}A)}{\partial z} = 0$$

$$\frac{\partial(\alpha_{l,g}u_{l,g}A)}{\partial t} + \frac{\partial(\alpha_{l,g}u_{l,g}^2A)}{\partial z} = \text{ as for steady case}$$

core-annular core $r \le R-h$, annulus $R-h<r\le R$, $\alpha_{cg} \neq$, $\alpha_{al}=1$

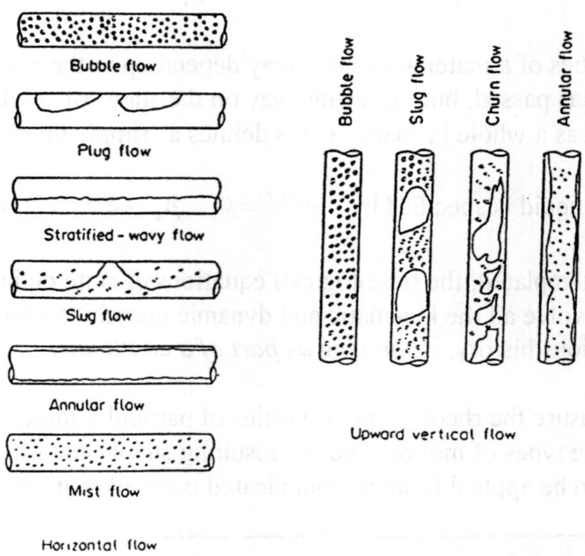

RHEOLOGICAL ASPECTS OF THE DESIGN OF TECHNOLOGICAL PROCESS EQUIPMENT - INTERNAL FLOWS

J. ŠESTÁK

ČVUT - Faculty of Mechanical Engineering, Technická 4, 166 07 Prague 6, Czech Republic,
E-mail: sestak@fsid.cvut.cz

1. INTRODUCTION

Frictional pressure drop for flow of incompressible fluids is traditionally expressed in terms of the Fanning friction factor,

$$f = \frac{D_e \Delta p}{2 \rho u^2 L} \qquad (1)$$

where ρ,u and L stand for the fluid density,mean volumetric velocity and length of the duct. The equivalent diameter D_e=4S/O is defined by the duct cross-sectional area S and the wetted perimeter O. For non-Newtonian fluids, the friction factor for most internal flows may be expressed as f = F(Re_G,ϵ) where Re_G is a suitably defined generalized Reynolds number and ϵ is the relative roughness of the duct. The F-function is usually called the hydraulic characteristic.

2. FLOW OF TIME INDEPENDENT FLUIDS IN STRAIGHT DUCTS

2.1 Hydrodynamically stabilized laminar flow

For most geometries of the cross-section and Newtonian fluids, the hydraulic characteristic may be expressed as, Kozicki, Chou and Tiu (1966), Šesták (1974),

$$fRe/16 = a + b \qquad (2)$$

where Re = $uD_e\rho/\mu$, μ being the dynamic viscosity. In Eq.(2), the parameters a an b depend upon the shape of the cross section only. For a circular cross section a + b = 1.

For flow of purely viscous power-law fluids in non-circular ducts, Kozicki et

$$Re^{*} = \frac{u^{2-n} D_e^{\,n} \rho}{2^{3(n-1)} \left(\frac{a+bn}{n}\right)^n k} \qquad (3)$$

al.(1966) defined a Reynolds number Re*, where k and n are the consistency and flow behavior index. If Re in Eq.(2) is substituted by the Re*, a very useful approximate relation for calculating the hydraulic characteristic in non-circular cross sections is obtained. Miller(1972) reported a different method based upon the idea of defining a mean value of the shear stress along the periphery $\bar{\tau}_s = D_e\,\Delta p/4L$, and of a hypothetic, empirically defined mean value of the shear rate $\bar{\dot{\gamma}} = (8u/D_e)(a+b)$,where again, the fRe product must be obtained from the hydraulic characteristic for a Newtonian fluid. Finally, we have for a power-law fluid,

$$\bar{\tau}_s = k \left(\frac{3n+1}{4n}\right)^n \bar{\dot{\gamma}}^n \qquad (4)$$

Contrary to both the preceding approximate methods, Wichterle and Wein (1980) reported another approximation which does not require the knowledge of the Newtonian hydraulic characteristics.Instead of using the D_e ,the authors suggested to employ a maximum hydraulic radius which is derived from the flow of an ideally plastic material, $n \to 0$. Generally however, the method is tedious for practical purposes since in cross sections with a complicated geometry, a high degree of intuition is needed to reach proper estimates of the slip planes. Šesták et al.(1996) reported numerical and experimental data for flow of power-law fluids in ducts with a L- shaped geometry of the cross section. Comparison of exact and approximate results is given in Fig.1.

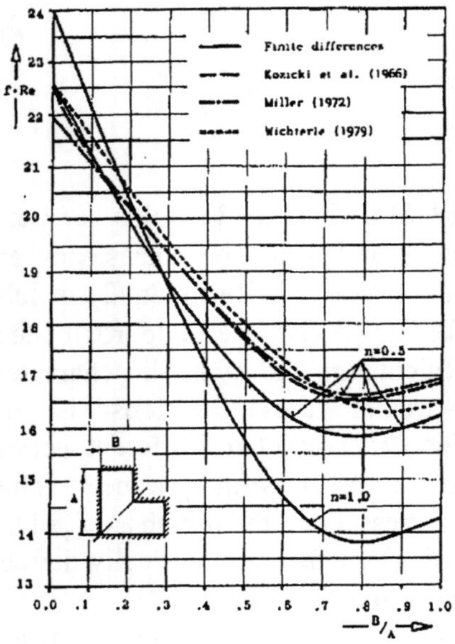

Fig.1.Exact and approximate values of the fRe complex for flow of Newtonian and power-law fluids (n =0.5) in an L-shaped profile

From Fig.1 it is clear that all three approximate methods yield fRe values which are higher than those predicted theoretically. All three methods are associated with an error not exceeding 10%. Wichterle's method is recommended for use for low values of n since for $n \to 0$, the relative error approaches zero.

Since Miller's method employs only one parameter of the geometry, it is, at least for this particular geometry, more advantageous.

2.2. Turbulent flow

In order to estimate the error involved in using the D_e concept in turbulent flow of a Newtonian fluid, the following formula, based on the work of Malák et al.(1975), was found.

$$\Delta p|_{eff} \,/\, \Delta p|_{D_e} = \left[\left[1 + 3\sqrt{(a+b)}\right] /4\right]^{5/4} \qquad (5)$$

Values of the a + b parameters may be found in e.g. Schenkel,(1981).

REFERENCES

Kozicki, W.,Chou, C.H. and Tiu, C.,: Chem.Eng.Sci., **21**, (1966), 665-679.

Malák, J.,Hejna, J. and Schmid, J.: Int.J.Heat Mass Transfer, **18,** (1975), 139-149.

Miller, C.: I.&E.C. Fundamentals, **11**, (1972). 524 - 528.

Schenkel, G.: Fortschritt-Berichte der VDI-Zeitschriften,Reihe: Strömungstechnik, **7**, (1981).

Šesták, J.: Canad. J. Chem.Engng., **52**, (1974), 670-672.

Šesták, J.: Proc.XIIth Int.Congr. on Rheology, Ait-Kadi,A.,Dealy,J.M.,James,D.F. and Williams,M.C.,(Editors),pp. 386, Canadian Rheology Group,Montreal (1966).

Wichterle, K., Wein, O.: Coll Czech. Chem. Comm., **45,** (1980), 2510 - 2515.

ACKNOWLEDGEMENT

The author gratefully acknowledges financial support for the results reported here from the Grant Agency of the Czech Republic under the Project No.: 525/96/0060.

BIOCONVECTION IN THE BLACK SEA UNICELLULAR EUCARIOTIC ALGAE: AGGREGATE'S RHEOLOGY AND QUAZIMULTICELLULAR STRUCTURES FORMATION

M.Y.ALEYEV

Institute of Biology of the Southern Seas, National Academy of Science of Ukraine,
2 Nakhimov avenue, 335011, Sevastopol, Crimea, Ukraine.

1. INTRODUCTION

Algal motility is traditionally treated as mechanism for escaping from nutrient-depleted boundary layer halo: Gavis (1976); or - for reaching nutrient-optimal depths (moving down from euphotic zone and up - to the light - during unicellular phototrph's trophical vertical circadian migrations (TVCM): Taylor(1987)), i.e. motility is an adaptation for cellular metabolism enhancement. Phytoplankton cell need light (in euphotic zone) and nutrients (below), and hence - motility, but not all phytoplankters are actively motile in the same degree. Numerous species are bioconvective-active ones. MAIN objectives of presentation are: 1) - to estimate fluid properties inside bioconvective aggregates (BA) and, 2) - to analise their possible significance for pelagic ecosystems as an adaptation for fast swimming.

2. METHODS

To check the hypothesis of presentation we supposed it reasonable to analyze locomotion of flagellated Black Sea algae as important energy - consumpting function; and to obtain quantitative assay of bioconvection (B) phenomenon in these species. According to Einstein's equation (1) for nonNewton fluids Loyciyanski (1978) it's possible to estimate the viscosity increase (m') inside BA due to hydrodinamic interaction of algae cells boundary layers (according to developed model of aggregate's motility, cell's boundary layers inside BA are overlapped that is the reason of their locomotion speed increase in comparison with single cells one).

3. RESULTS

3.1 Agregates viscosity increase

Total volume of cells forming BA (10 000 - 50 000 specimens in 1 BA) could vary between 1 & 4 % (in various algae species and physico-chemical regimens in culture medium) of total BA volume, a = 0.01 - 0,04.

$$m' = m * (1 + 5 / 2 * a), \quad (1)$$

where m - dynamic cultural medium viscosity without algae cells.

m ' in BA looks to be 3 - 6 % higher than outside it, so it could be treated as a reason of spatio-temporal stability of BA (as well as - probable cellular glycocalyx interaction or gradients of exopolymeric particles & some excreted molecules specific absorption by neighbouring cells: mediator-type regulation inside quazimulticellular structure - QS). Aggregates usually are composed from 10-s thousand of algae cells, sized 1-3 mm in diameter and may exist during several minutes that is enough to cover 10 - 50 cm before aggregate would be damaged by hydrodynamic forces. Fluid properties inside BA differs from outside space, so BA could be treated as QS. Locomotion was shown to be significant energy spending function for pelagic flagellated algae cells energetic metabolism (value of active component is 10-75% of standard one).

3.2 Algae cells respiration

Structure of energetic metabolism in 20 species was obtained. It seemed that standard (total energy expenditures of the cell) metabolic rate allometric equation looks like usual one: Hemmingsen(1960), but energetic metabolism components've revealed marked differencies in allometric growth rates: it may be seen that total degree 0,7 is «composed» from 0,87 and 0,48 (active and basal components that are not usually analysed)

3.3 Bioconvection

Numerous observations of bioconvection patterns: Platt (1961; etc.) - in different representatives of Protista; in algae cultures've shown aggregation being an adaptation of unicells to the real environment: Aleyev (1991, 1992, 1998 - in press). Searching for ecologo-physiological reasons of QS - BA formation for the first time observed in 10 common Black Sea unicellular flagellated eucariotes strains (as Nephrochloris salina, Carteria sp., Exuviaella pusilla, Gymnodinium lanskaya, Olistodiscus luteus, Gymnodinium kowalewskii, Prorocentrum micans, etc.), taking into account their movement-connected O2 consumption, nutrition conditions, energy flows and propulsor's morphology, one may underline that independently from one or another B phenomenon conception accepted by different authors: Kessler et al. (1992; etc.), it's useful adaptation for faster swimming. Mean BA locomotion velocity is 1mm/s instead of 200 microns/s with the same or lower values of respiration rates observed in the same species strains under homogenous cell distribution in culture medium. Analysis of possible ecological advantages for their nutrition & TVCM, originating from aggregation of marine unicellular photoautotrophs of rather wide size range shows that BA may 5 - 7 times faster reach necessary depth with the same energy cost, i.e. - to fulfill TVCM (hence move nutrients in food web) with lower cost. Group locomotion then is energetically reasonable. It means that energy requirements for locomotion per cell in BA are 5 - 7 times lower than in single cell. So B may be treated as unicell's ethological adaptation for optimization of their energy requirements for locomotion by lowering of their significant active energy requirements. Parameters of BA and reasons of energy preservation under intercellular energetic cooperation and exchange (with their boundary layers interaction) leading to optimization of nutrition in nutrient-depleted boundary layer with relatively slow diffusion processes: Bowen et al. (1993) are obtained

4. REFERENCES

Aleyev MY (1991) Ecology of the sea 38: 99 - 107 (in Russian)

Aleyev MY (1992) Locomotion and active metabolism in flagellated sea planktonic algae; Ph.D. Thesis. Sevastopol (in Russian)

Aleyev MY (1998) Algology (Kiev) 8: 20 p. (in press, editorial # 112)

Bowen JD, Stolzenbach KD, Chisholm SW (1993) Limnol. Oceanogr. 38: 36 - 51.

Gavis J (1976) J.Mar.Res. 34: 161-179

Hemmingsen AM (1960) Rep. Steno Mem. Hosp. 9: 7 - 110

Kessler JO, Hill NA, Hader D-P (1992) J. Phycol. 28: 816 - 822.

Loyciyanski LG (1978) Fluid and gas mechanics. Nauka, Moscow

Platt JR (1961) Science: 1766 - 1767.

Taylor FJR (ed.) (1987) Bot. Monogr., v.21.The biology of dinoflagellates. Blackwell sci.publications.

RHEOLOGICAL BEHAVIOUR OF BLOOD UNDER UNSTEADY FLOW CONDITIONS

N. ANTONOVA

Institute of Mechanics and Biomechanics, Bulgarian Academy of Sciences, Akad. G.Bonchev Str., Bl.4, 1113 Sofia, Bulgaria, Email: antrheo@bgcict.acad.bg

1. INTRODUCTION

The alterations of the blood flow with time is typical for viscoelastic and thixotropic fluids. The study aims to examine rheological properties of normal and pathological whole blood and RBC's suspensions under steady and dynamic shear conditions.

2. METHODS

The rotational viscometer with a firmly set amplitude of ± 30 ° " Low Shear 30 sinus " (Contraves, Switzerland) is used. Apparent steady shear viscosity is obtained from 128,5 s^{-1} to 0,0237 s^{-1} , decreasing the shear rates in 11 steps. Complex dynamic viscosity is determined by measuring the loss angle δ and the torque M_{VE} at the frequencies from 0,000222 Hz up to 1,63 Hz, whereas M_V (or viscosity η_V) and M_E (or storage modulus G' respectively) are calculated. The shear stress variations as a function of time for rectangular and triangular change of shear rates are determined.

Normal human blood conserved with glucose-citrate solution L-12 (300ml blood/75 ml L-12); blood of patients with cerebrovascular disease (CVD) as well as RBC's suspensions in physiological solution are investigated. All measurements are carried out at 37°C.

3. RESULTS

The viscous component of the complex viscosity η_V and storage modulus G' for normal human blood conserved with L-12 as a functon of frequency are shown on the Fig.1 The viscous (η_V) component is decreasing exponentially and storage modulus G' is slightly increasing with increasing the frequency from 0 up to 0.025 Hz.

For blood of patients with CVD Antonova (1997) found that the steady shear viscosity and viscous component of the oscillatory viscosity of the group of 20 patients with CVD, tends toward 5.6 mPa.s for $\gamma \approx 128.5\ s^{-1}$, H = 42 ± 5.23 %. The steady and oscillatory shear rate dependence of blood viscosity is non-linear. For $\gamma_{max\ osc} > 10\ s^{-1}$ the steady shear viscosity is larger than the viscous component of the complex viscosity.

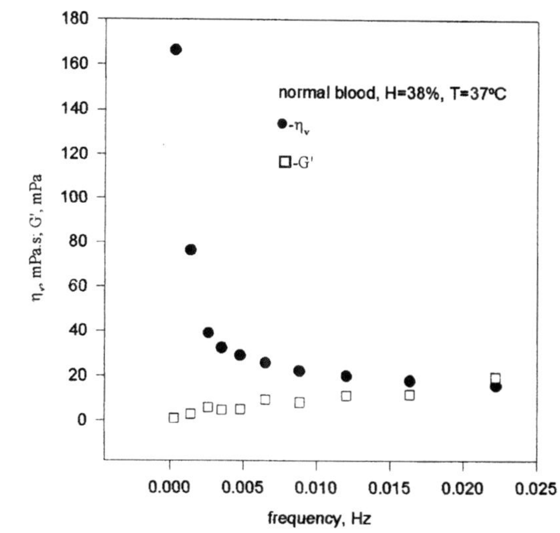

FIG1. Oscillatory frequency dependence of the viscous (η_V) and elastic component (G') component of the viscoelasticity of normal conserved with L-12 human blood;H = 38%,T=37±0.1°C

The character of the steady shear and oscillatory shear rate dependence of viscosity of normal human blood, conserved with L-12 is the same (Fig.2). Both viscosity curves are decreasing, approaching steady viscosity curve between 0.1 and 10 s^{-1}. In this region begins the disaggregation of the erythrocytes lowering both viscous and elastic component. At higher shear rates the only mechanism for storing elastic energy is the deformation of the individual red blood cells. This implies that the elastic energy of the RBC network and aggregates is far more sensitive to a change of their state as the dissipative energy, associated with the viscous component. This signifies a larger energy dissipation probably due to more pronounced cell-cell interactions Thurston (1990).

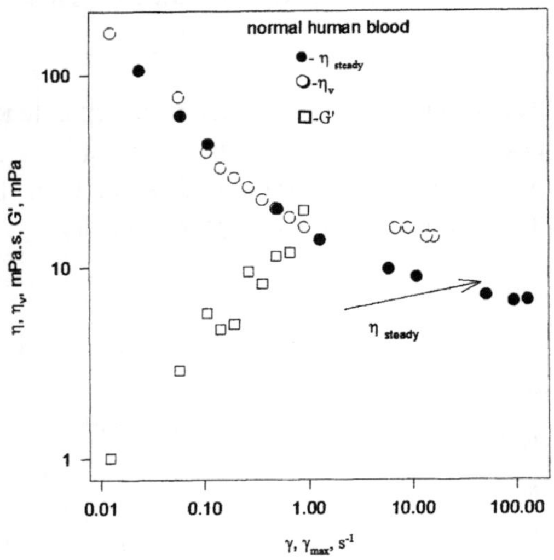

FIG.2.The steady shear and oscillatory shear rate dependence of normal human blood, conserved with L-12, obtained with LS30 sinus Contraves viscometer, H=38%, T=37°C.

The determination of thixotropic properties at rectangular change of shear rates shows the exponential increasing of the torque is found from 0 up to 0.0237 s^{-1}, which gradient is very high in the first 10-15 s. After that very slow increasing of the torque is found, not reaching the steady value during the following 50 s. This process is less pronounced at 00596 s^{-1}. At 1.285 s^{-1} and 5.96 s^{-1} "stress overshoot " is observed in the first 2 to 5 s The steady value is reached after 6 to 26 s from the beginning of flow.

For low shear rates the exponential decay is observed and the full recovery is not reached for 20 s. In the conditions of higher shear rates after cessation of flow the full recovery is reached for 2 to 5 s . The kind of the hysteresis loops (Fig.3) is different for normal blood and blood from patients with CVD.

The viscoelastic properties are significantly affected by the RBC aggregation processes in the conditions of oscillatory flow. In time dependent flow the stored elastic energy changes with time and blood shows properties of thixotropy - Thurston GB (1979). In complex time dependent flows the combined effects of the elastic and viscous energies may control the flow.

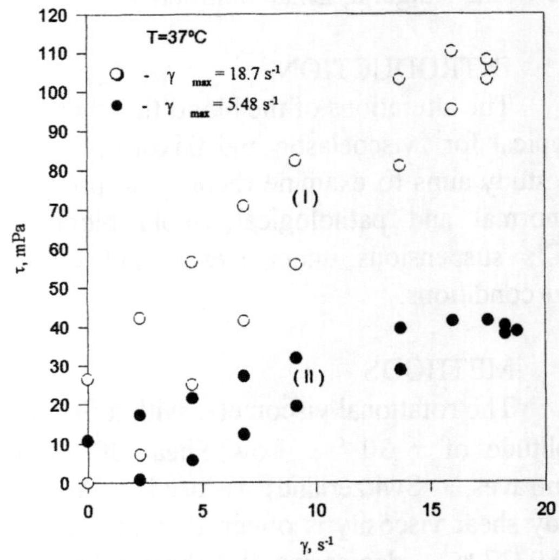

FIG.3. Rheograms at triangular change of shear rates from 0 s^{-1} to 18.7 s^{-1} 0 s^{-1} (I) and from 0 s^{-1} to 5.48 s^{-1} 0 s^{-1} (II) for RBC's suspension in physiological solution H = 42 %, T = 37 °C , t_{max} = 8 s.

4. CONCLUSION

Normal human blood, blood from patients with CVD and RBC's suspensions in physiological solution have non-Newtonian viscosity and nonlinear viscoelasticity, depending on maximum shear rate, frequency, hematocrit and the sample.The effect of oscillatory shear for blood is the decreasing of the viscous component and slightly increasing of the loss modulus G'.

The shear stress variations with time at rectangular and triangular change of shear rates are characterized by exponential increasing of the shear stress, stress overshoot and slow relaxation as well as hysteresis, depending of the applied shear rates and the sample.

4. REFERENCES

Antonova N, (1997) Theoretical and Applied Mechanics (in press)

Thurston G (1990) In: Biofluid Mechanics (Ed. D.Liepsch), Springer-Verlag.

Thurston GB (1979) Biorheology, 16: 149-162

RHEOLOGICAL BEHAVIOUR AND DRAINAGE OF MICROSCOPIC FOAM FILMS FROM INFASURF

N.ANTONOVA[1], R.SEDEV[2], D.EXEROWA[2], YA. IVANOV[3]

[1] Institute of Mechanics and Biomechanics, Bulgarian Academy of Sciences, 1113 Sofia
[2] Institute of Physical Chemistry, Bulgarian Academy of Sciences, 1113 Sofia
[3] Central Laboratory of Physico-Chemical Mechanics, Bulgarian Academy of Sciences, Sofia 1113, Bulgaria

1. INTRODUCTION

Recently foam film formation from therapeutic surfactants was studied by Scarpelli et al. (1997), Exerowa et al. (1997). INFASURF was found the most effective among the three preparations tested - Infasurf, Exosurf and Survanta. Alveolar surfactant is a thin liquid film containing lipids and specific proteins,which lines the alveolar surface and lowers surface tension to prevent collapse at end-expiration. The surfactant is prescribed clinically as surfactant replacement therapy for surfactant-deficient neonatal infants with respiratory distress syndrome (RDS). In this study we report in more details the rheological behaviour of INFASURF (Burroughs Welcome Co., Research Triangle Parc, NC) as well as their links with the morphology and stability of the foam films .

2. METHODS

The bulk (apparent) viscosity has been monitored under steady flow conditions with a Low Shear 30 Contraves viscometer (Switzerland). Measurements have been carried out under steady and unsteady flow at shear rates from 0.2 up to 128.5 s^{-1} and at step rise of shear rates from 0 s^{-1} to 2,37 s^{-1} and from 0 s^{-1} to 11,02 s^{-1} . The temperatures of the experiments were 22 and 37 ° C.

The foam film behaviour has been carried out in a Scheludko-Exerowa cell. The time, τ_{0-1} , from initial film formation to the first black spot appearance has been recorded at capillary pressure P_C = 400, 1200 and 2400 dyn/cm^2 . The lifetime of the black spot τ_{1-2} and the time from the thin film formation to the black film formation τ_{0-2} has also been measured. All experiments were done with fresh aqueous solutions at 22° and 37 °C. The INFASURF concentration was varied from 100 up to 550 µg / mL.

It is assumed that the drainage of foam films is described with Reynolds equation - Sedev et al. (1997). Then

$$\tau_{0-1} = \int_0^{\tau_{0-1}} dt \cong \left(r^2 / h_{CR}\right)\left(\eta / P_C\right) \qquad (1)$$

where η is the bulk viscosity, r is the film radius (~ 100 µm) and h_{CR} is the black spot thickness (~ 30 nm).

3. RESULTS

It is found that τ_{0-1} linearly increases with $1/P_C$ (Fig.1) in agreement with the above equation.

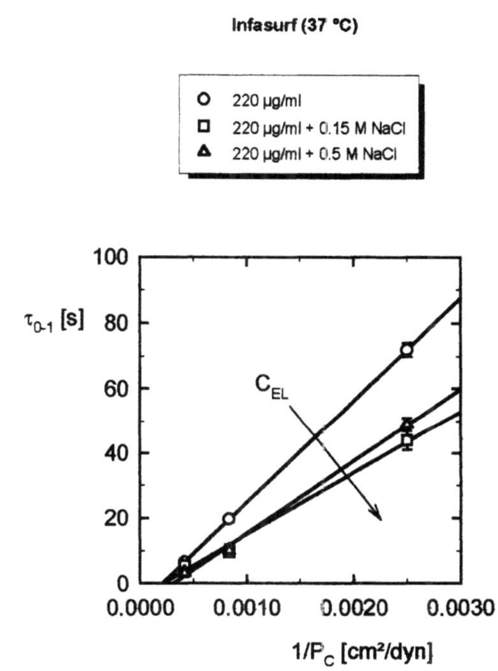

FIG.1. The dependence of the time, τ_{0-1} , from initial film formation to the first black spot

appearance on the reciprocal capillary pressure P_C = 400, 1200 and 2400 dyn/cm^2

Since r and h_{CR} (eq.1) are constant the slope of this linear dependence is directly proportional to η. Thus viscosity can be monitored with film thinning experiments.

Antonova et al. (1995) found that the INFASURF solutions have Newtonian behaviour from 0.2 up to 128.5 s^{-1} within the whole shear rate and concentration range . The mean value of η for Infasurf (35 mg/l) are 1.08 mPa.s (at 22 °C) and 0.88 mPa.s (at 37 °C), i.e. an increase of temperature within 14 °C produced about 19 % decreasing of viscosity.

The concentration - viscosity dependence (Fig.2) showed that the viscosity initially increases and then levels off at a plateau value.The best fit equation is:

$$1/\eta = a + b\,c_s + d\exp(-c_s) \qquad (2)$$

where η is the apparent viscosity, C_s is the concentration of the solution and a,b,d are the characteristic constants of the solution, depending on the temperature, additions and other physico-chemical conditions.

There is a reasonable agreement between the viscosity values from the bulk and film measurements.

The effect of electrolyte concentration (C_{EL}) is significant (Fig.1). Drainage is faster when electrolyte is added. No practical difference between 0.14 M and 0.5 M NaCl is found. The INFASURF concentration was varied from 100 up to 550 μg / mL.

Under unsteady rise of shear rates stress overshoot phenomenon was observed 2 ÷ 3 s after beginning of shear.

4. CONCLUSION

Rheological behaviour of INFASURF solutions under steady shear has been observed without abnomalities, depending on concentration, temperature and electrolyte. At step rise of shear rates stress overshoot has been determined at all concentrations.

It is found that $\tau_{0\text{-}1}$ linearly increases with $1/P_C$ in agreement with the Reynolds' equation. Thus viscosity can be monitored with film thinning experiments. The effect of electrolyte is significant and drainage is faster when the electrolyte is added.

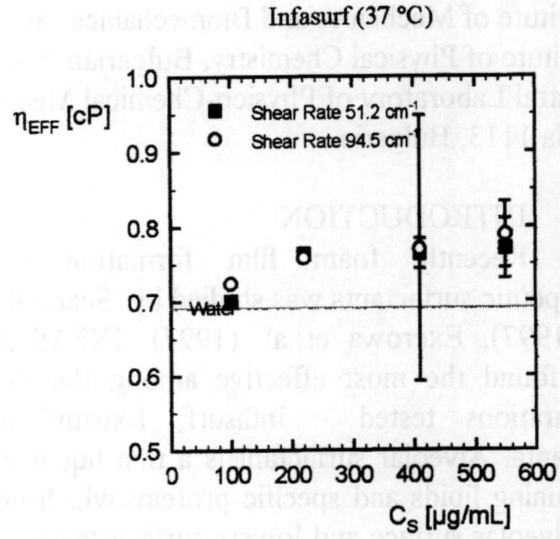

FIG.2. The relationship between apparent viscosity and concentration of INFASURF varied from 100 up to 550 μg / mL.

5. REFERENCES

Scarpelli EM, Mautone AJ, Lalchev Z, Exerowa D (1997) Colloids & Surfaces B 8,133.

Exerowa D, Lalchev Z (1997) In: Toshev BV (ed) Books of abstracts of the 9th International Conference on Surface and Colloid Science, Sofia, Bulgaria, pp 459-460.

Sedev R, Antonova N, Ivanov Ya, Exerowa D (1997) In: Toshev BV (ed) Books of abstracts of the 9th International Conference on Surface and Colloid Science, Sofia, Bulgaria, pp 463-464.

Antonova N, Exerowa D, Ivanov Ya (1995) In: Rachev Al, Toshev Y, Mitutsova L (eds) Proceedings of the 6th International School on Biomechanics, Blagoevgrad, Bulgaria, pp 113-119.

How viscoselastic properties of colloidal biopolymers modify rheological properties of seawater

I JENKINSON[1], T WYATT[2], A MALEJ[3]

[1]ACRO, Lavergne, 19320 La Roche Canillac, France
[2]Instituto de Investigaciones Marinas, CSIC, Eduardo Cabello 6, 36208 Vigo, Spain.
[3]NIB, Marine Biology station Piran, Fornaèe 41, 6330 Piran, Slovenia

At the Third European Rheology Conference, Jenkinson [3] reported that the rheological properties of seawater correspond to those of a lumpy weak gel. Superimposed on the Newtonian viscosity of the aqueous phase, η_W, is a non-Newtonian excess viscosity,

$$\eta_E = k.\dot{\gamma}^{N-1}$$

where η_E is an excess viscosity (Pa.s) due to polymers originating essentially from phytoplankton. When $N = 1$, the polymeric component shows viscosity independent of shear rate, while values of N <0, 0 and >1 correspond, respectively, to gel, Bingham plastic and shear-thickening behaviours.

Table 1 shows that seawater from the Mediterranean at different times of year (no algal blooms were taking place) corresponded to a weak gel, as did North Sea water in June (blooms of *Phaeocystis* and *Noctiluca* were in progress). Values of N were higher in the North Sea blooms, however, perhaps representing less highly annealed polymers recently produced by the algae. Similarly in algal cultures (with algal densities and production rates generally much higher than found in the sea), N values between 0 and +1 showed moderate shear-thinning behaviour characteristic of overlapping, rather than cross-linked polymers. Where measured (not shown) seawater and algal cultures showed elastic moduli G' less than 20% of their excess viscous moduli G''_E.

Northern Adriatic seawater taken as a single sample, then divided into 7 subsamples, showed dramatically different properties, however. Values of G' were an order of magnitude greater than those of G''_E, and the water was shear-thickening (N>1) if only the viscous component of the rheological properties is considered. The phenomenon of large floating organic aggregates, *mare sporco*, characteristic of the Northern Adriatic (see Malej et al, this volume) developed at this site one month later. The rheological properties may have reflected unusual polymeric material or associated fibre-shaped phytoplankton colonies [9] related to this *mare sporco* development.

Intermittency in viscosity (i.e. variance (ln(G''_E))) in was found to be 1.7 in the Mediterranean sampling trips [5]. Higher values found in the North Sea may have been the result of a non-random sampling strategy. Shear thinning and intermittency in G''_E due to algal polymers may be a biological tool to modify turbulent parameters (Jenkinson 1993b)

REFERENCES

1. Carlson DJ (1987) Nature 329:823-825
2. Jenkinson IR (1986) Nature 323:435-437
3. Jenkinson IR (1990) In: Oliver DR (ed.) Third European Rheology Conference. Elsevier pp. 247-251
4. Jenkinson IR (1993a) In: Smayda TJ, Shimizu Y (eds) Toxic Phytoplankton Blooms in the Sea. Elsevier pp. 757-762.
5. Jenkinson IR (1993b) Oceanologia Acta 16:317-334
6. Jenkinson IR, Arzul G (1998) In Reguera B, Blanco J, Fernández ML, Wyatt T (eds) Harmful Microalgae, Xunta de Galicia and IOC of Unesco (in press)
7. Jenkinson IR, Biddanda BA (1995) J. Plankton Res. 17:2251-2274
8. Jenkinson IR, Connors PP (1980) J. Sherkin Isl. 1(1):127-146.
9. Wyatt T, Figueras FG, Rios AF (1993) In: Smayda TJ, Shimizu Y (eds) Toxic Phytoplankton Blooms in the Sea. Elsevier pp. 849-853

Table 1. Rheometric parameters measured in seawater and phytoplankton suspensions. Values represent, or have been calculated from, arithmetic averages.

Material	Shear rate (s^{-1})	k (Pa)	N	# mea -sured	Length scale (mm)	Method, (Reference)
Med (July)	0.0021 to 0.29	2.1×10^{-6}	-0.46	8	0.5	LS30
Med (Nov)	0.0021 to 0.29	10.5×10^{-6}	-0.36	15	0.5	(5)
Med (Feb)	0.0021 to 0.29	4.2×10^{-6}	-0.49	16	0.5	
Med (May)	0.0021 to 0.29	2.0×10^{-6}	-0.22	16	0.5	
N.Sea (June)	0.0021 to 0.97	450×10^{-6} (1)	-0.11	13	0.5	LS30 (7)
N. Adriatic 5 May 1994	0.05 to 2	800×10^{-6}	+1.4	7	0.5	LS40 G''_E Unpubl
7 km off Fano	0.05 to 2	10×10^{-3}	+0.8	7	0.5	G^*_E Unpubl
Surface slicks Atlantic	?	Up to 20% increase in viscosity	?		Macro -mole-cular?	FD (1)
Amphidinium	0.017 to 1.0	8×10^{-3}	+0.92	1	0.5	LS30
$\phi = 7.7 \times 10^{-4}$	6 to 129	30×10^{-3}	+0.49	1	0.5	(2)
Dunaliella $\phi = 1.7 \times 10^{-4}$	0.017 to 129	c. 0.5×10^{-3}	+0.26	1	0.5	LS30 (2)
Dunaliella +	0.017 to 2	1.7×10^{-4}	+0.02	2	0.5	LS30
Noctiluca	2 to 10	1.5×10^{-3}	+0.9	2	0.5	(2)
$\phi = 1.7 \times 10^{-4}$	10 to 129	2×10^{-3}	+0.04	2	0.5	
Noctiluca	0.002 to 0.5	1×10^{-3}	+0.48	2	0.5	LS30
$\phi =$ c. 0.2	0.5 to 8	1×10^{-3}	+0.70	2	0.5	(4)
G. naga.	0.002 to 1	1×10^{-3}	+0.25	3	0.5	LS30
$\phi = 1.4 \times 10^{-5}$	1 to 8	1×10^{-3}	+0.82	3	0.5	(4)
G. naga. $\phi = 7.8 \times 10^{-5}$	0	Yield stress <1 to 60 Pa	?	10	c. 0.05	Flow thro' fish gills (6)
G. naga. $\phi = 1.5 \times 10^{-5}$	If bubble dia 0.1 mm shear rate = 79	0.17	?	1	c. 0.05	Bubbles trapped in a bloom
	If bubble dia 0.5 mm shear rate = 2.8	0.35	?	1	c. 0.25	(8)

Abbreviations: FD - fluorescence depolarisation; G' - elastic modulus; G^*_E - excess complex modulus; LS30 or 40 - Contaves/Mettler Low Shear 30 or 40 rheometer with Couette geometry; Med - Mediterranean (Rade de Villefranche, France); *G. naga* - *Gymnodinium nagasakiense* (syn.: *Gyrodinium aureolum*); ϕ - volume fraction
Notes: (1) For this North Sea sampling trip, $k = 2.0.Chl^{1.3}$, where Chl is chlorophyll content (mg.m^{-3}), an approximate measure of phytoplankton abundance.

MARINE ORGANIC AGGREGATES IN THE NORTHERN ADRIATIC, THEIR PHYSICAL PROPERTIES, MICROSTRUCTURE AND BIOLOGICAL ORIGINS

A. MALEJ[1], I. JENKINSON[2], V. FLANDER[1]

[1]NIB, Marine Biology Station Piran, Fornače 41, 6330 Piran, Slovenia
[2]ACRO, Lavergne, 19320 La Roche Canillac, France

1. INTRODUCTION

Marine organic aggregates of different sizes are common in oceans and coastal seas and usually attain higher abundance and mass in nearshore regions. In the northern Adriatic visible marine snow and large flocks reaching the size of several centimetres can be observed intermittently throughout the year and are recognised as a part of usual plankton dynamics. Extensive mucilage aggregates develop sporadically and can grow to several meters in size and affect large areas (up to 10,000 km^2). This phenomenon appears to be unique to the northern Adriatic and has been known for over 250 years. Analysis of the recurrence pattern indicated that the secular hazard rate of mucilage events did not vary notably over the last 120 years and the calculation of its periodicity revealed an average cycle of 5.74 years (Vollenweider et al. 1995). Since 1872 researchers reported on ten mucilage events spreading over a large part of the northern Adriatic and fourteen more localised ones. In the last decade large mucilage masses appeared in the summers of 1988, 1989, 1991, and 1997 and have caused problems to touristic and fisheries activities. We have followed this phenomenon in the summer of 1997 and compared the results to those obtained in 1991.

2. METHODS

2.1. Field observations and sampling

Observations of the evolution of mucilage phenomenon were carried out weekly from late May to early September 1997 by SCUBA divers in the eastern Gulf of Trieste. In parallel, main water column characteristics (temperature, salinity, fluorescence and oxygen content) were assessed using a CTD fine-scale probe (University of Western Australia, a Sea Tech Inc. fluorometer).

Mucilage aggregates were collected from different depths in the water column by SCUBA divers using a membrane pump and manually from the sea surface. Samples were kept deep frozen (-20 ^{0}C) until analysed.

2.2. Laboratory analyses

After centrifugation of the thawed samples, filtration and rinsing with MQ water, dry weight (dried at 60 ^{0}C), ash-free dry weight (combustion at 480 ^{0}C) and organic content (as the difference between dry and ash-free dry weight) of the mucilage aggregates were determined. Biotic characteristics of the mucilage aggregates were determined using standard optical microscopy and a reverse-phase HPLC for the determination of pigments (chlorophylls, carotenoids, chlorophyll *a* breakdown products) according to Mantoura and Llewellyn and modified by Barlow et al. (1993).

3. RESULTS AND DISCUSSION

3.1. Description of the phenomenon

Large mucilage aggregates in the northern Adriatic were typically observed during the period of water column stratification and the phenomenon dissipated by the time of autumnal vertical mixing. Mucilage aggregates became visible from late May to mid June in a form of transparent opaque mass that reached the size of several meters just bellow the mixed surface layer (depths between 4 to 8 m) and where they remained suspended for several weeks. During the evolution of the phenomenon a fraction of this material became buoyant and floated to the surface. Aged mucilage evolved deeper in the water column into compacted darker clouds and only a minor fraction of suspended mucilage sank to the bottom.

Accumulations of different forms of water column mucilage were related to density discontinuities. However, freshly formed, loose mucilage were observed only in the upper layers. While mucilage aggregates were formed in the upper part of the water column at depths less than 15 m from May to mid-August, aged mucilage

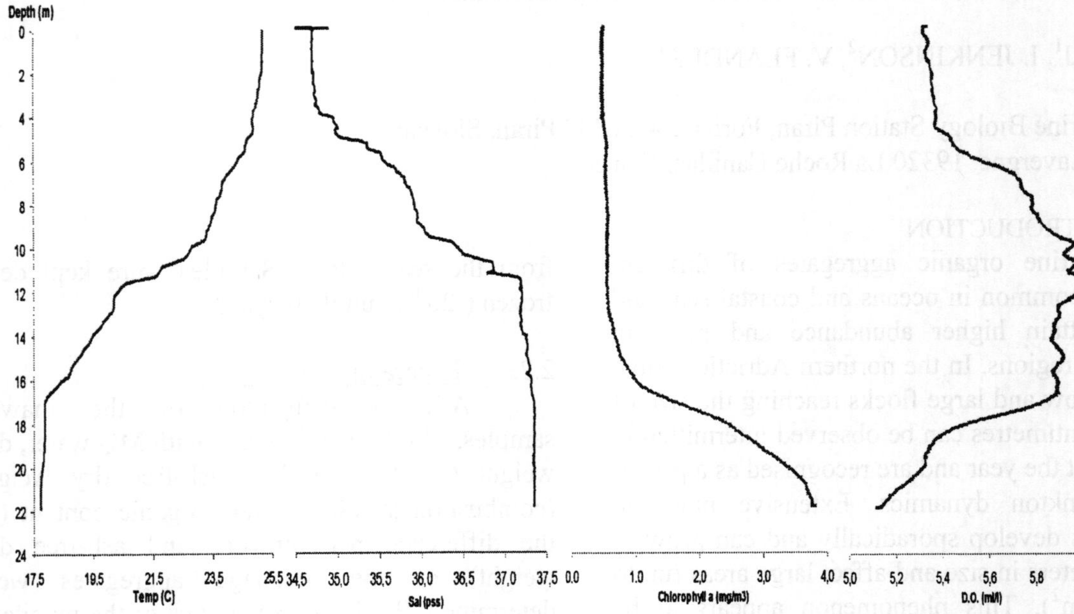

Figure 1. Vertical profiles of temperature, salinity, fluorescence (chl *a*) and oxygen

accumulated in deeper layers from July on. Figure 1 provides some details concerning water column characteristics (temperature, salinity, fluorescence and oxygen) during early August. Characteristically, the bulk of plankton biomass and activity was associated with the aggregates, while surrounding water was very poor in organisms.

Nearshore surface accumulations developed gradually during the day. Surfacing of mucilage usually began at mid morning in open waters. By the afternoon mucilage masses accumulated offshore along surface fronts. Sea breezes and tidal currents pushed them towards the coast where they piled up. Masses of mucilage were either washed ashore or sank from the surface during the night except in sheltered ports where they remained suspended for several consecutive days and underwent extensive transformations that changed their physical appearance. All forms of mucilage aggregates – loose opaque mass, dense compact clouds and surface mucilage could be present at the same time from July to September.

The organic content of the mucilage aggregates collected in the water column was significantly ($p<0.001$) higher (47.71 ± 12.69 % of dry weight) compared to that collected at the surface (32.58 ± 10.94 %). Surface mucilage showed a decreasing organic content from July (47.61 ± 2.60 %) to September (13.69 ± 1.09 %). Moreover, surface mucilage was also enriched in chlorophyll degradation products when compared to mucilage collected in the water column.

3.2. Mechanism(s) of formation

Two mechanisms appeared to be important for the development of mucilage masses in the northern Adriatic, one controlling the production of mucilaginous material and the second controlling mucilage accumulation. Current perception of the phenomenon is that mucilage represents a build-up of organic and inorganic material entrapped in a matrix, generated by exudation of polysaccharides from diatoms and followed by bacterial-mediated further accumulation of polysaccharides. Mucilage accumulation is believed to be related to stability of the water column, longer residence time and low horizontal advection during summer in the northern Adriatic.

Barlow RG, Mantoura RFC, Gough MA, Fileman TW (1993) Deep-Sea Res 40: 459-477

Vollenweider RA, Montanari G, Rinaldi A (1995) Sci Tot Environm 165: 213-224

WHEAT PROTEINS AS POLYMERIC MATERIALS: RHEOLOGICAL PROPERTIES OF PLASTICIZED GLUTEN.

Andreas Redl*, Stéphane Guilbert*, Bruno Vergnes**

*ENSA.M, 2 place Viala, 34060 Montpellier, France; email: redl@ensam.inra.fr
**CEMEF, Ecole de Mines de Paris, BP 207, 06904 Sophia Antipolis, France, email: vergnes@cemef.cma.fr

1. INTRODUCTION

Shaping with extrusion technology is of major importance for succesful implementation of "bioplastics". This requires a good knowledge of the materials viscous behaviour. The aim of the present paper is to characterize the rheological properties of commercial gluten plasticized with glycerol, in order to better understand its structure and flow properties and to derive a phenomenological equation.

2. MATERIALS AND METHODS

Commercial vital wheat gluten (Amylum Aquitaine, Bordeaux, France) with a protein content of 79.8 % (dm) and 6.2 % (dm) humidity and anhydrous glycerol (Prolabo, France) were mixed in a two blade counter-rotating batch mixer turning at a 3:2 differential speed (Rheocord, Haake, Germany). Mixing chamber was not thermostatet and can be considered as moreless adiabatic.

Rheological measurements were carried out using a Dynamic Spectrometer (RMS 800, Rheometrics, USA) with parallel plates (diameter 25 mm) in frequency, strain and time sweep mode.

Molecular size distribution was determined with SE-HPLC after solubilisation in a sodium phosphate buffer containing 1% sodium dodecyl sulphate.

3. RESULTS AND DISCUSSION

Mixing of gluten with glycerol results in a very characteristic evolution of torque and temperature (figure 1). The maximum of torque is reached at a stable level of specific energy (500 - 600 kJ/kg) and coincides with the inflection point of the temperature evolution (50-60°C) regardeless the mixing conditions (rotor speed, filling ratio). Right after t_{peak} the gluten/glycerol dough aspect changes, from a mate, sticky mass to a fairly brilliant and non sticky mass.

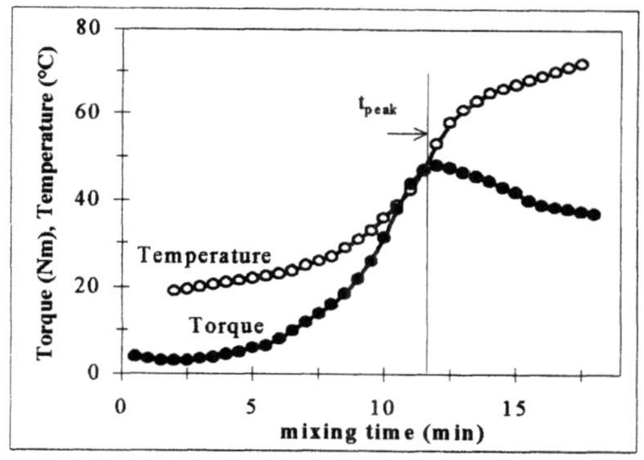

Fig. 1 Torque (●) and temperature (○) evolution during the mixing process in standard conditions (50.2 g, 30 rpm, 45 % glycerol)

The structuration of the material is reflected in small deformations rheological properties in a very different manner. Before t_{peak}, the main change is the reduction of the linear domain (from 5 to 2%) with only a slight change of the dynamic moduli. After t_{peak}, we observe a large in moduli increase, the linear domain remains stable (1.4%). The frequency dependence of moduli is linear on a double logaritmic scale regardeless the mixing time, the slope on a double logaritmic scale decreases with mixing time ($t < t_{peak}$: 0.38 and 0.47, $t = t_{peak}$: 0.32 and 0.43, $t < t_{peak}$: 0.29 and 0.40 for storage and loss modulus, respectively).

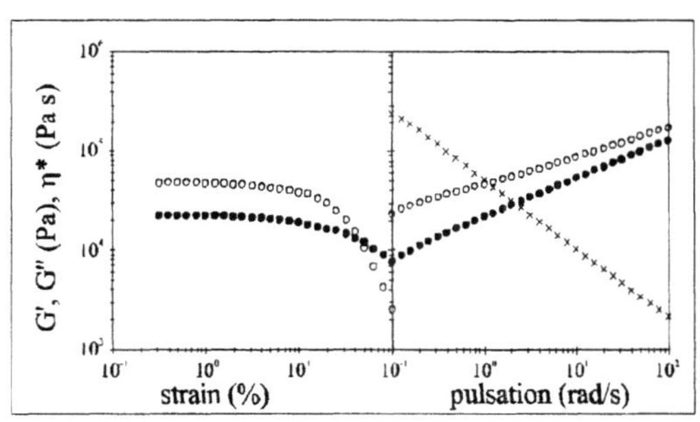

Fig. 5. Evolution of storage G' (●) and loss G'' (○) modulus, and complex viscosity η^* (x) with the strain amplitude at 1rad/s and with the pulsation at 1% strain amplitude (80 °C, 45 % glycerol, 1.5 tpeak)

An example of the strain dependence and the frequency dependence of the dynamic moduli at 80 °C of the gluten/glycerol mixed obtained 1.5 times after t_{peak} is shown in figure 2.

Interpretation of the observed results regarding the structuration of plasticized gluten is not straightforward. The increase of torque is most probably due to a plastification of proteins which enables them to unfold and to form a network, the decrease of the frequency dependency of the dynamic moduli might be a gelation process. But the molecular size distribution remains constant for the whole mixing process, which means that no disulphide bonds or other covalent linkages are involved. The term of partial network systems as proposed by Ross-Murphy (1995), in which polymeric chains are "physically" crosslinked into networks, seems appropriate. However, the origin of the the drastic changes at 50-60°C is not clear yet. Unreversible changes for gluten behaviour have been observed in the same temperature range by other authors (Attenburow et al., 1990; Tsiami et al., 1997, Schofield et al. 1983) but no concluding explanations are proposed.

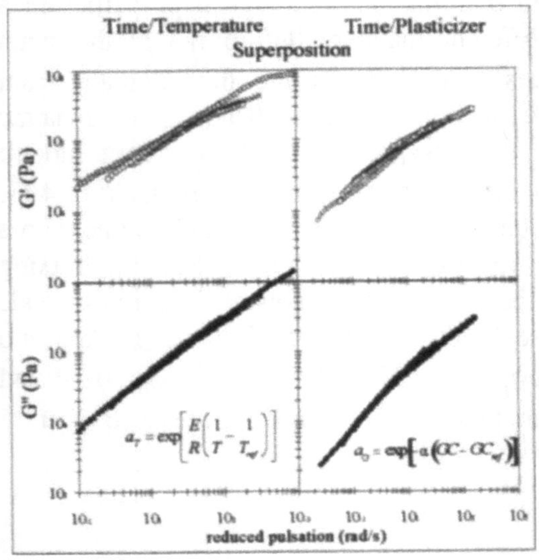

Fig 3. time temperature and time plasticizer superposition for storage (G') and loss modulus (G"). T = 40, 50, 60, 80 °C; GC = 30, 37, 45, 52, 60 %. Reference conditions T_{ref} = 80°C, GC_{ref} = 45%.

The influence of temperature and plasticizer content of a structured material (mixed to 1.5 tpeak) was evaluated with the principle of time temperature and time plasticizer superposition. Figure 3 shows the superimposed curves of storage and loss modulus. Superposition is fairly well for the loss modulus G'', but does not really apply for the storage modulus G'. This indicates that mostly local elastic interactions influence the temperature dependent dynamics. Gluten-glycerol mix does not behave as a thermorheologically simple material, but the effect of temperature and glycerol content can nevertheless be estimated using time-temperature and time plasticizer superposition and a general expression for describing the viscous behaviour of the product can be derived, which is perhaps simplistic for such a complex rheological behaviour, but would remain sufficient for modelling the flow behaviour in a twin screw extruder:

$$\eta^* = K_0 \exp\left[n\left[\frac{E}{R}\left(\frac{1}{T}-\frac{1}{T_{ref}}\right)\alpha(GC - GC_{ref})\right] + \beta(SME - SME_{ref})\right]\omega^{n-1}$$

Following values apply for the range of parameters studied : K_0 = 49800 Pa s^n, n = 0.31, T_{ref} = 353 K, GC_{ref} = 0.45, E/R = 13300 K, α = 17.7 and β = 4.3 10^{-4} kg/kJ. Same kind of rheological behaviour has been previously proposed for molten or plasticized starches (Vergnes and Villemaire, 1987, Della Valle *et al.* 1996, Willet et al., 1995) and the factors for temperature and glycerol dependency are comparable. An important difference for plasticized gluten is that the energy term, β is positive whereas, for thermoplastic starch the energy term is negative. Theroplastic starch is depolymerized by mechanical energy whereas plasticized gluten is rather structured, but the mechanisms involved are not completely understood yet.

4. REFERENCES

Attenburrow G, Barnes DJ, Davies AP, Ingman SJ (1990). J Cereal Sci 12:1-14

Della Valle G, Colonna P, Patria A, Vergnes B (1996) Starches. J Rheol 40:347-362

Ross-Murphy SB (1995). J Rheol 39:1451-1463

Schofield JD, Bottoley RC, Timms MF, Booth MR (1983) J Cereal Sci 1:241-253

Tsiami AA, Bot A, Agterof WGM, Groot RD (1997). J Cereal Sci, 26 (1): 15-29.

Vergnes B, Villemaire JP (1987). Rheol Acta 26:570-576

Willet JL, Jasberg BK, Swanson CL (1995). Polym Eng Sci 35:202-210

RHEOLOGICAL-MICROSTRUCTURE INVESTIGATIONS OF BIOPOLYMER MIXTURES WITH 'EMULSION' CHARACTER

B. WOLF, W. J. FRITH, I. T. NORTON

Unilever Research Colworth, Colworth House, Sharnbrook, Beds. MK 44 1LQ, U. K.

1. INTRODUCTION

Mixtures of two biopolymers in aqueous solution can be designed such that they adopt a de-mixed two-phase state with 'emulsion' character. However, the two emulsion phases are typically not pure solutions of one of the two biopolymer components. The morphology of this 'water-in-water emulsions', like for bioplymer mixtures in general, is strongly dependent on the macromolecular characteristics of the biopolymer components and their concentrations, the physiochemical state of the solvent, as well as the temperature (and pressure).

In this investigation rheological-microstructure relationships are being studied exemplary for a mixture of a protein and a polysaccharide.

2. MATERIALS AND METHODS

2.1 Biopolymer Components and Mixture Preparation

The two-phase biopolymer model mixture under investigation is a mixture of gelatin (250 Bloom ex Extraco, U. K.), and guar gum (Luxara 2180 ex Arthur Bramwell, U. K.). Liquid gelatin/guar mixtures with constant concentration of guar were prepared.

2.2 Experimental Methods

Gelatin/guar mixtures with variable amount of gelatin were 'processed' in constant-stress, as well as constant-strain rate shear flow with a superimposed cooling regime. In 'zero-shear' experiments quiescent cooled mixtures were generated. Cooling at a constant cooling rate was continued beyond the gelation temperature of the gelatin. This allows kinetic trapping of microstructures developed, and, thus, their subsequent visualisation by conventional off-line techniques.

Rheological experiments were carried out in a stress-controlled rheometer (RSR 200 ex Rheometric Scientific) with the software option for strain rate-controlled experiments. A cone-and-plate configuration was chosen in order to expose the biopolymer mixture to a shear field which is at least macroscopically uniform. The material system under investigation does not tend to show significant wall slip effects. The rheological behaviour of 'microstructured' samples was investigated keeping the temperature below the melting point of gelatin.

For their microstructure characterisation the samples were removed from the shear gap and visualised by optical light microscopy. It should be pointed out that the phase volumes of dispersed particles as they appear to be in the micrographs, see part 3, are not representative. Phase volumes were determined by centrifugation methods.

3. RESULTS

Observed microstructures show a widespread spectrum from which a selection is illustrated in the following. Figures 1 and 2 show the morphology of a mixture 'processed' at increasing shear stress, and the influence of gelatin concentration respectively. All systems are gelatin-disperse. The differences in microstructure development are reflected by the viscosity behaviour, which is plotted in Figure 3. The increase in viscosity at 20°C is a result of the gelatin gelation. the description of a viscosity minimum observed for mixtures 'processed' at elevated shear stresses indicates phase inversion which 'flips back' as soon as the gelatin starts gelling. This specific behaviour causes the formation of arbitrary shaped large and small particles. 'Well-behaving' samples develop regularly shaped particles whose degree of anisotropy is a function of the applied shear stress (or strain rate, not shown here).

Further discussion, as well as microstructure and rheology results are provided in the poster presentation.

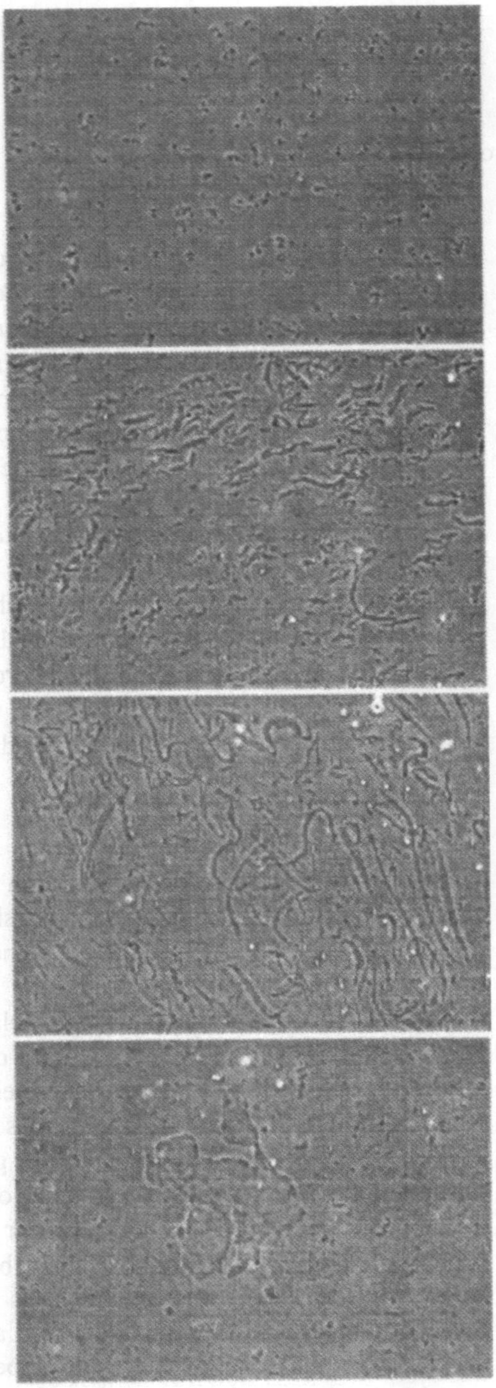

Figure 1: Gelatin/guar mixtures with 0.8 %w/w gelatin (1 %w/w guar)

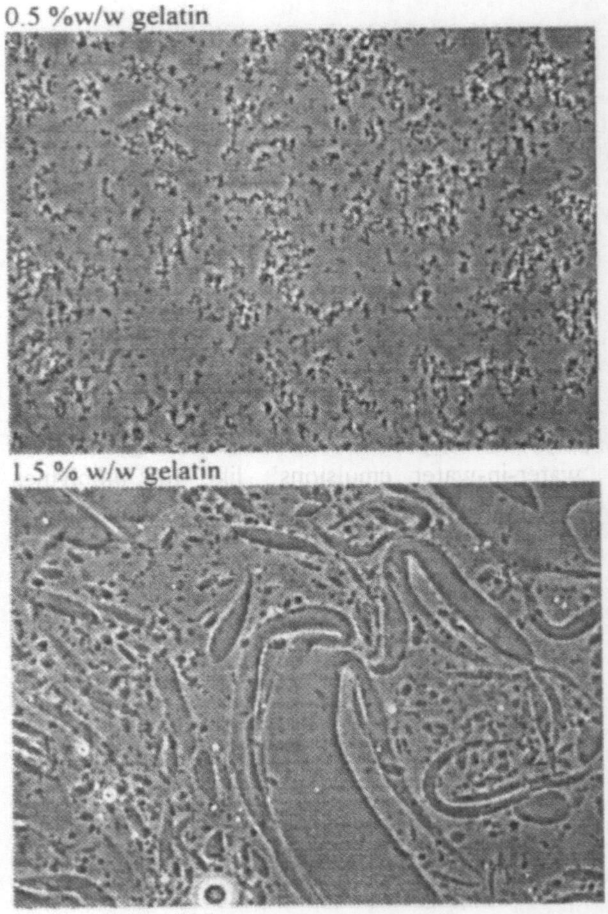

Figure 2: Gelatin/guar mixtures 'processed' at 0.2 Pa

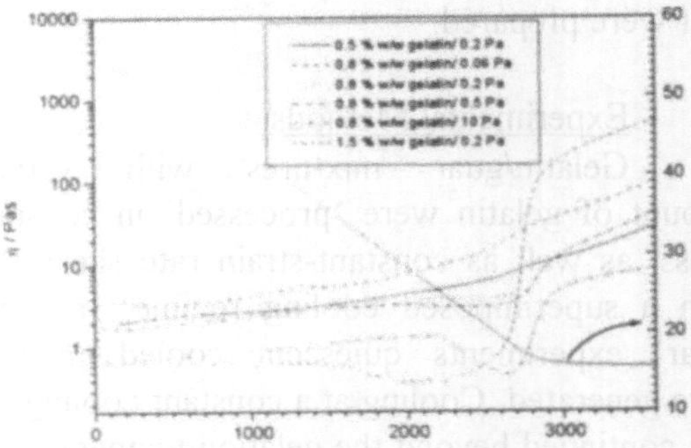

Figure 3: Rheological results

HOW VISCOELASTIC PROPERTIES OF COLLOIDS, TRANSPARENT EXOPOLYMERIC PARTICLES AND MARINE ORGANIC AGGREGATES, MODIFY TURBULENCE AND PLANKTON BIODYNAMICS IN THE SEA.

TIM WYATT[1], I. JENKINSON[2], A. MALEJ[3]

[1]Instituto de Investigaciones Marinas, CSIC, Eduardo Cabello 6, 36208 Vigo, Spain.
[2]ACRO, Lavergne, 19320 La Roche Canillac, France
[3]NIB, Marine Biology station Piran, Fornace 41, 6330 Piran, Slovenia

1. INTRODUCTION

Measurements of viscoelasticity, have already been made using Couette geometry, in the sea itself (See Jenkinson et al, this volume), in $\approx 1\ m^3$ mesocosms, and in phytoplankton suspensions and cultures. These measurements show that the sea is generally a hierarchically lumpy gel, with excess viscous moduli (i.e. in excess of that due to the aqueous phase) and elastic moduli strongly correlated with the quantity of phytoplankton present (measured as chlorophyll content). Formation of these gels starts mainly from colloidal (sub-micrometer) molecules (Gentien, 1995), much of which is probably secreted by phytoplankton.Because they are sticky, they flocculate in deforming water to form Transparent Exopolymeric Particles (TEP - 1 to $\approx 100\ \mu m$) (Mari & Kiørboe, 1996). These TEP are fractal organic aggregates that then flocculate further by sticking to each other and sometimes also to fibres and other biological debris (Mari & Kiørboe, 1996). They are also colonised by bacteria. By these means they can form visible gels (Chin et al., 1998), organic aggregates of size several µm to mm (Kiørboe & Hansen, 1993; Worm & Søndergaard, 1998), or even metres in the northern Adriatic (Malej & Harris, 1993; Malej *et al.*, this volume).

2. TURBULENCE

Fluid strain in plankton-poor, turbulent seas (e.g. near-surface, temperate/boreal high seas in winter) is dominated (except perhaps at exceedingly small scales in and immediately around living cells) by the Newtonian viscosity of the water plus salt, with the non-Newtonian properties of polymeric material being insignificant. In algal blooms under calm conditions, however, strain may be dominated at scales up to several metres by the viscoelastic properties of partly transparent, biopolymeric extracellular material, in addition to TEP and marine organic aggregates.

In marine isotropic turbulence it has classically been assumed that the seawater is Newtonian and that

$$\mathrm{rms}(\dot{\gamma}) = 1.93(\varepsilon/\eta)^{0.5}\ [s^{-1}] \qquad (1)$$

where ε is the rate of energy dissipation ($W.m^{-3}$) and η is the Newtonian viscosity due to water and dissolved salt (Pa.s).

From distributions of measured $\dot{\gamma}$, converted to ε using (1), rms(ε) has been found to have typical values of from 10^{-3} in surface layers in moderate winds to $<10^{-14}$ in tranquil deep waters. ε shows intermittency (i.e. var[ln(ε)]), of 1 to 2 in unstratified surface waters but 3 to 7 in the stratified interior (Sanford, 1997). In Newtonian fluids $\dot{\gamma} \propto \varepsilon^{0.5}$, so intermittency in $\dot{\gamma}$ is always 4 times less than that in ε. In fluids with variable viscosity and in elastic fluids, however, this relationship is decoupled.

3. FEEDING DYNAMICS OF PLANKTON

Modelling of various processes that involve phytoplankton and zooplankton, both in shear flow and in turbulent flow, show that "hunting" type activities, including feeding and sexual encounter show dome-shaped relationships with strain rate (Jenkinson & Wyatt 1992, Wyatt & Jenkinson 1997). We have also shown that the values of strain rate for (1) where these relationships start to become non-linear and (2) for the top of the dome, depend on a sort of biological Deborah number, (swimming speed)*(strain rate)/(radius of perception).

4. FLOCCULATION AND BREAK-UP

Another two physical ways polymer can change feeding dynamics is by physically obstructing (1) swimming and (2) food handling.

While the formation of TEP and marine organic aggregates is described above, in reality, and in purely physical terms, there is a dynamic balance between flocculation and break-up. The break-up is governed by the interaction between the rheological properties of the aggregates and the flow of the water (itself potentially modified by rheological properties), while forces due to Brownian motion and sinking also come into play.

5. ECOLOGICAL IMPLICATIONS

The genes responsible for production of these polymers are subject to Darwinian evolutionary pressures within the phytoplankton or bacterioplankton cells in interaction with their milieu. The polymers must therefore be adaptive, imparting some advantage to the organism in its environment. Qualities like polymer stickiness are thus initially under the genetic control of the species which secrete them. Similarly, if the secretion of a particular polymer is aimed at reducing grazing, for example, the grazer may find it adaptive to have acquired the ability to manufacture enzymes which can neutralise such protective effects. Thus the physical balance between flocculation (Kiørboe & Hansen 1993) and break-up (Jenkinson et al 1991) appears to be paralleled by biological means of shifting the dynamic balance between these opposing trends, while the effectiveness of both flocculation and break-up is regulated by the total energy input from winds, tides, and other sources of turbulence in the system.

6. REFERENCES

Chin W-C, Orellana MV, Verdugo, P. (1998) Nature 391:568-571

Gentien P, Lunven M, Lehaître M, Duvent JL (1995) Deep-Sea Res. I 42:1297-1312

Jenkinson IR, Biddanda BA, Turley CM, Abreu PC, Riebesell U, Smetacek VS (1991) Oceanologia Acta, Sp Vol No 11:101-107

Jenkinson IR, Wyatt T (1992) J. Plankton Res. 14:1697-1721

Kiørboe T, Hansen JLS (1993) J. Plankton Res. 15:993-1018

Malej A, Harris RP (1993) Mar. Ecol. Prog. Ser. 96:33-42

Mari X, Kiørboe T (1996) J. Plankton Res. 18: 969-986

Sanford LP (1997) Mar. Ecol. Prog. Ser. 161: 265-293

Worm J, Søndergaard M (1998) J. Plankton Res. 20:179-186

Wyatt T, Jenkinson IR (1997) J. Plankton Res. 19:551-575

FLOW-INDUCED PHASE INVERSION IN IMMISCIBLE POLYMER BLENDS

M. ASTRUC, P. NAVARD

Ecole des Mines de Paris, Centre de Mise en Forme des Matériaux, UMR CNRS 7635, BP 207, 06904 Sophia-Antipolis, France

1. INTRODUCTION

The morphology of immiscible polymer blends is one of the major factors controlling their final properties. For two-phase immiscible polymer blends, several parameters are important in determining the final morphology: composition of the blend, shear rate, viscosity and elasticity of the both phases, interfacial tension and time of mixing. Paul and Barlow (1980) have proposed an empirical equation to predict the point of dual phase continuity:

$$\frac{\Phi_A \eta_B}{\Phi_B \eta_A} = 1 \qquad (1)$$

where Φ_A and Φ_B are the volume fractions of phases A and B, and η_A and η_B their viscosities. If the ratio in equation (1) is lower than 1, phase A should form the dispersed phase in a continuous matrix of B, whereas B should be dispersed in A for values higher than 1. It has been suggested by Miles and Zurek (1988) that the viscosities should be taken at the shear rate of the actual flow. But it does not always describe satisfactory the inversion for various systems, especially for viscosity ratios diverging from unity. Luciani et al. (1993) formulated another expression for the prediction of dual-phase continuity based on the equilibrium of a bi-fibrillar structure.

The aim of this paper is to study with *in-situ* techniques how a polymer blend of a given starting morphology can invert its morphology under flow by varying the viscosity ratio of the two components.

2. EXPERIMENTAL

The investigated blend is a mixture of poly-dimethylsiloxane (PDMS Rhodorsil 47V200000 from Rhône-Poulenc; M_w =149350) and a solution of 50 wt. % hydroxypropylcellulose in water (HPC Klucel L from Aqualon; M_w =100000), which is liquid crystalline. The concentrations are 67 vol. % of PDMS and 33 vol. % of HPC solution. The HPC solution is birefringent while the PDMS is not. Hence, between crossed polarizers, the two phases can be distinguished: the PDMS appears black and the HPC bright.

The viscosity of the two components was characterized using a Stresstech Rheologica rheometer, using a parallel plate geometry.

Flow-induced morphology characterizations were performed on a Linkam CSS 450 shearing system.

3. RESULTS AND DISCUSSION

3.1 Rheology

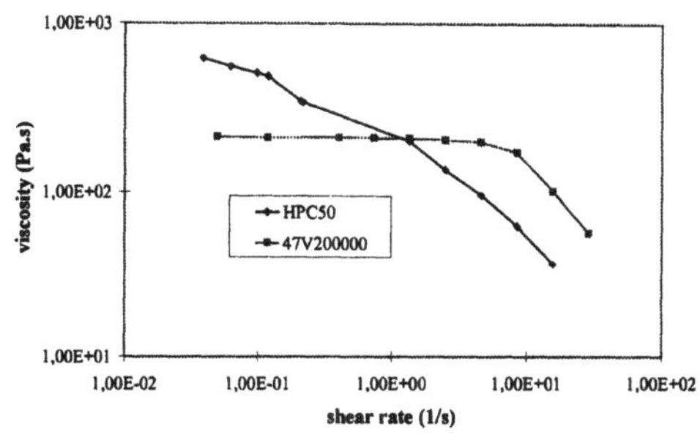

Fig.1: Viscosity data of the PDMS and the solution of HPC at 18°C.

Viscosity data of the two components at 18°C are presented in figure 1 (Fig.1). In the shear rate range under investigation (0.05 s^{-1} to 10 s^{-1}), the PDMS is Newtonian until 8 s^{-1} whereas the solution of HPC is viscoelastic. The viscosity ratio varies, depending on the applied shear rate. Until 2 s^{-1}, the HPC solution is more viscous than the PDMS. Above this shear rate, the PDMS becomes the most viscous. At 0.16 s^{-1} the viscosity ratio is 2 and at 10 s^{-1}, is 0.4.

3.2 Identification of different transient morphologies

A start-up experiment is carried out. We start with the at-rest morphology after mixing and filling the shearing system. The PDMS is the more viscous and constitutes the continuous phase. The

HPC is dispersed in droplets containing smaller droplets of PDMS. Then, a shear rate of 10 s^{-1} is applied. In these shearing conditions, the HPC solution becomes the less viscous component. The different transient morphologies are observed with an optical microscope. After an extensive study on this blend, we are able to distinguish five different morphologies depending on deformation.

In the first stage of the experiment (t_0; t_0+10s), the morphology quickly changes in a bi-fibrillar structure. Then, the fibrils of the two phases break-up.

In the second stage of the experiment (t_0+10s; t_0+60s), the HPC becomes the continuous phase while the PDMS begins to form large domains layered in the flow direction. They contain threads and droplets of HPC (Fig.2a).

In the third stage of the experiment (t_0+60s; t_0+600s), upon further shearing, a "band structure" morphology with local phase inversions is observed. Two kind of bands, layered in the flow direction, can be distinguished. Bands with a matrix of HPC and droplets of PDMS alternate with bands of a PDMS matrix. The bands of PDMS contain droplets and threads of HPC that encapsulate droplets of PDMS. The morphology has just been inverted in the bands where the HPC solution is the continuous phase (Fig.2b).

In the fourth stage of the experiment (t_0+600s; t_0+6000s), two different issues can take place. One of the two kinds of bands enlarges and dominates. Depending on parameters we are not yet able to determine, it leads either to a refined at-rest morphology (encapsulations) (Fig.2c) or to a global phase inversion.

In the latter case, a morphology with simple droplets of PDMS dispersed in the continuous phase of HPC is obtained (Fig.2d). In the whole sample, the simple shear-flow has inverted the morphology of the blend.

4. CONCLUSION

By changing the viscosity ratio, we have been able to inverse the morphology of the blend. But the point of co-continuity is far from the point predicted by Miles and Zurek (1988). However, the phase inversion mechanism has been observed to pass through a bi-fibrillary structure, in the way proposed by Luciani et al. (1993). A point of bifurcation has been found in the last stage of the experiment leading either to a refined morphology or a global phase inversion.

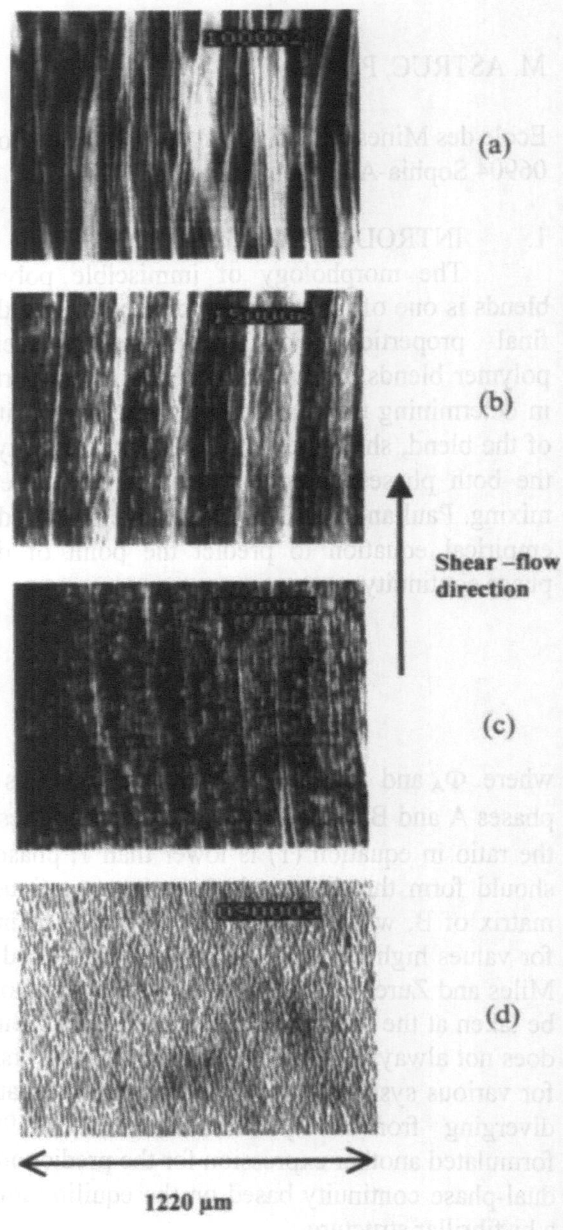

Fig.2: The different transient morphologies of the blend 67/33 vol. % of PDMS/HPC. The PDMS appears black and the HPC bright.

5. REFERENCES

Luciani A, Jarrin J, Dewimille B (1993) J. Phys. IV. France. 3 :1577-1580

Miles I. S, Zurek A (1988) Polym. Eng. Sci. 28 :796-805

Paul D. R, Barlow J (1980) J. Macromol. Sci. Rev. Macromol. Chem. (C) 18 :109

RHEOLOGY/RESISTIVITY CORRELATION IN CARBON BLACK-CONTAINING IMMISCIBLE POLYMER BLENDS

O. BREUER[1], R. TCHOUDAKOV[1], M. NARKIS[1] AND A. SIEGMANN[2]

Departments of [1]Chemical and [2]Materials Engineering, Technion - Israel Institute of Technology, Haifa 32000, Israel.

INTRODUCTION

Immiscible polymer blends are interesting host multiphase systems for the incorporation of low concentrations of fine electrically conductive fillers, such as carbon black (CB). The filler tends to accumulate preferentially in certain regions within the multi-phase matrix, inducing the formation of segregated structures whereupon CB may form a conductive network [Gubbels et al., 1995; Breuer et al., 1997]. Thus, the electrical conductivity is enhanced, and the critical CB content essential for percolation is significantly reduced. The low filler content is of practical advantage due to the tendency of high CB contents to complicate polymer compounding and processing and to diminish mechanical properties.

EXPERIMENTAL

In this study a high impact polystyrene (HIPS, Carmel Olefins Galirene HT-88-5) matrix was blended with other polymers such as various types of polyethylene (UHMWPE, Hoechst Hostalen Gur 412 or LLDPE, Dow Dowlex NG 5056) or ethylene-vinyl acetate copolymers (grades of Exxon Escorene), constituting immiscible systems for incorporation of CB. The CB used was a highly structured electrically conductive grade, Ketjenblack EC-300 (Akzo). All blends were compounded by a Brabender Plastograph, and subsequently underwent either compression molding, injection molding or extrusion. The resistivity of the samples was measured by various methods, depending on the geometry of the samples produced by the specific processing mode. The blend morphology of freeze-fractured samples, gold sputtered prior to observation, was studied by a scanning electron microscope (SEM), Jeol 5400.

A rheological-electrical method was originated, by which extrudates of CB-containing binary immiscible polymer blends were produced by a MCR capillary rheometer mounted on an Instron TT-D machine. The effect of shear level on the extrudates' structure, flow properties and resultant resistivity was determined. The results were compared to CB-containing reference samples based on the individual blend's constituents.

RESULTS AND DISCUSSION

It has been shown that the CB percolation thresholds in the individual polymers vary from one matrix to another, based on effects of surface tension, viscosity and crystallinity [Carmona, 1989]. These parameters determine the ability of the CB to form a network, accordingly affecting the percolation threshold. In immiscible polymer blends, the CB may be situated preferentially within either phase, or at the interface. This induces filler segregation, increases the local CB content and may enhance the quality of the conductive pathways. The double percolation concept, requiring simultaneous continuity of the CB network as well as the polymeric phase in which it is located, has been introduced and found necessary to obtain conductivity in CB-loaded polymer blends. This is depicted in Fig. 1, in which the CB-containing EVA phase is continuous within the HIPS. CB further enhances the networking effect by reducing the dispersed phase size and elongating it, generating fine honeycomb-like structures. These unique segregated structures enable CB-containing blends to attain lower percolation thresholds and lower resistivity values than those depicted for the individual polymers.

Processing has a crucial effect on the electrical properties of CB-containing polymers and blends. Indeed, injection molding severely diminishes conductivity of these compounds, in comparison to those processed by compression molding. [Ezquerra et al., 1988]. Different methods of production, characterized by various levels of shear, may significantly alter resistivity of samples of identical composition, by modifying agglomerate size and distribution, interparticle distances and generating anisotropic structures. This is true for both single or multi-component matrices; however, use of a polymer blend as a matrix introduces additional parameters such as blend structure, dispersed phase shape and dimensions, and CB location, which may be further modified during multiphase flow.

The rheological-electrical method enabled to produce extrudates at specific shear rates, and to determine the electrical, rheological and morphological properties of these samples. Generally, the extrudates based on CB containing polymer blends maintain their conductive features at high shear rates better than the single polymer matrices, dependant on blend components, their composition and CB content, as depicted in Fig. 2. By comparison of capillaries of various lengths, it has been established that capillary entrance effects determine the structure and resultant resistivity of the extrudates. The effect of CB on the flow properties of the CB containing compounds was observed in most of the single phase matrices, even at extremely low filler contents, and may be correlated with its tendency to disperse within the matrix, and the resultant percolation threshold. When a multi-component matrix is employed, the effective filler content within the CB-containing phase should be considered, as the apparent viscosity of this particular phase is increased. This causes the balance of phase viscosities to shift, altering the blend morphology and consequent electrical properties.

CONCLUSIONS

A close interrelation has been observed between the electrical resistivity, morphology and melt flow properties of CB-containing immiscible polymer blends. It is clear that both material-based properties and processing conditions determine the morphology, and accordingly, the electrical performance.

REFERENCES

Breuer, O., Tchoudakov, R., Narkis, M. and Siegmann, A. (1997) J. Appl. Polym. Sci. 64: 1097-1106.

Carmona, F. (1989) Physica A 157: 461-469.

Ezquerra, T.A., Bayer, R.K. and Balta Calleja, F.J. (1988) J. Mater. Sci. 23: 4121- 4126.

Gubbels, F., Blacher, S., Vanlathem, E., Jerome, R., Deltour, R., Brouers, F. and Teyssie, PH.(1995) Macromolecules 28: 1559-1566.

ILLUSTRATIONS

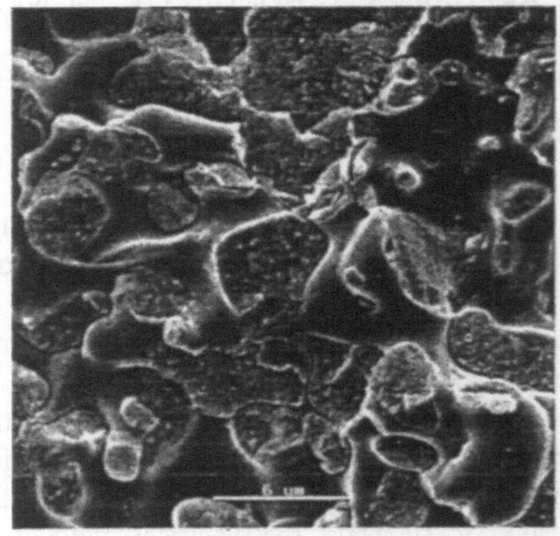

Figure 1: SEM micrographs of 4 phr CB containing 55HIPS/45EVA-9 freeze-fractured blends

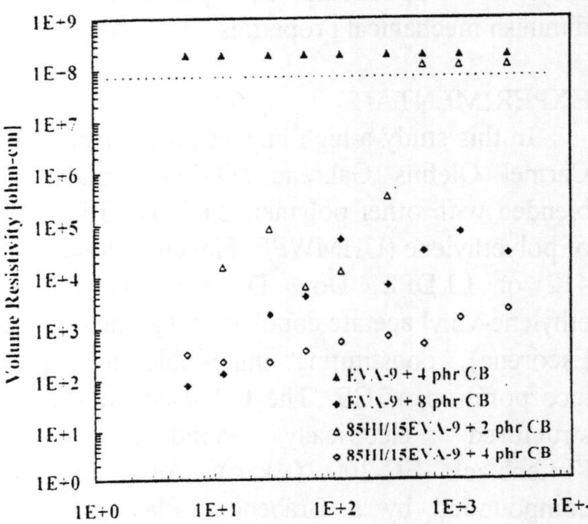

Figure 2: Resistivity versus extrusion shear rate of CB containing EVA-9 and 85HIPS/15EVA-9 blends. Points of resistivity above 1E+8 ohm-cm, exceed the range of the measuring equipment, and are depicted for illustration purposes only.

SHEAR-INDUCED MIXING AND DEMIXING IN POLY(STYRENE-CO-MALEIC ANHYDRIDE)/ POLY(METHYL METHACRYLATE) BLENDS

D. CHOPRA[1], D. VLASSOPOULOS[1], S.G. HATZIKIRIAKOS[2]

[1] FORTH, Institute of Electronic Structure & Laser, P.O. Box 1527, 71110 Heraklion, Crete, Greece
[2] Dept. of Chemical Engineering, The University of British Columbia, Vancouver, Canada V6T 1Z4

1. INTRODUCTION

The response of macromolecular binary mixtures to shear flow is rather complex. Experimental evidence suggests that both shear-induced mixing and shear-induced demixing, observed as suppression or enhancement of concentration fluctuations, respectively, can occur (Hobbie et al, 1994; Larson, 1992; Takebe et al, 1989), depending on the amount of shearing and the molecular characteristics of the polymers (molecular weights, glass transition temperature, T_g). Some of the important open problems on the field are: (i) what happens with *real* industrial blends; (ii) what is the effect of very high shear rates (200 s^{-1} and above); (iii) how does the presence of glassy components and their large viscosity difference interplay with the shear influences on the phase characteristics.

In this paper, we address these issues by investigating the effects of a high shear field on the phase behavior of a polymer blend consisting of a random copolymer of styrene and maleic anhydride and poly(methyl methacrylate). This is a lower critical solution temperature blend with components characteristics shown in Table 1.

2. RESULTS AND DISCUSSION

The phase separation of SMA and PMMA was associated with the thermorheological complexity of the blend, as already known from several relevant investigations (Vlassopoulos, 1996). Due to the strong effects of enhanced concentration fluctuations in the homogeneous pretransitional region (Kapnistos et al, 1996), the rheological phase separation temperature was determined from a combination of the master curves from frequency sweeps and dynamic temperature ramps at low frequencies and very low heating rates. By following this procedure for different compositions, it is possible to obtain the phase diagram of the blend.

The more thoroughly studied blend was the SMA/PMMA 50/50, which demixes at about 215°C. In order to elucidate the interplay between capillary flow and thermodynamics, we examined the extrudates from different capillary extrusion runs (at steady state) corresponding to different apparent shear rates. Extrudates were analyzed by three different means:

(i) visual observations: Samples resulting from shear-induced mixing were transparent, whereas shear-induced demixing yielded opaque extrudates. No yellowness was observed, suggesting no detectable degradation during melt mixing of SMA/PMMA. Additional information was provided by scanning electron microscopy (SEM), which showed that the flow resulted in phase-separated extrudates; there was unambiguous evidence of domain orientation along the flow direction, i.e., the SMA-rich domains elongated and formed large cylinders

(ii) DSC analysis: Extrudates corresponding to the homogeneous regions yielded one T_g; on the other hand, samples well into the two-phase region (corresponding to shear-induced demixing) were characterized by two distinct T_g's. The transitions in the thermographs of blends which were clearly either single-phase or phase-separated ehxibited nearly the same breadth with those of the pure components.

(iii) Shear rheology: To examine the potential degradation, we carried out dynamic frequency sweeps of the "suspect" extrudates in the linear viscoelastic limit, and compared against the results obtained with the "virgin" blend at the same conditions. The extruded quenched samples showed all nearly the same sensitivity to thermodynamics with the virgin ones, independently of the degree of degradation, as demonstrated with temperature ramps of the storage modulus. Based on these investigations, a slightly turbid extrudate with a single T_g, but not broader than the one resulting from the homogeneous virgin blend, and with no drastic change in the linear viscoelastic properties, can be considered as homogeneous.

From the above considerations concerning the characteristics of the extrudates, as well as the dependence of the shear stress on temperature in the capillary studies for different shear rates, and in consistency with the DSC and visual observations, phase diagrams of the SMA/PMMA blend at different shear rates were constructed (Figure 1). The combination of the DSC analysis of the opaque extrudates with the composition dependence of the T_g of the homogeneous blend, yields the coexistence points as follows: each of the two (time-independent) T_g's resulting from the originally 50/50 blend, represents the soft and hard phases, and each of these phases is considered as compositionally homogeneous. Then, for these SMA-rich and PMMA-rich phases we can back-up their composition from the composition dependence of T_g. By following the same procedure for different nominal compositions of SMA/PMMA, it is possible to obtain the coexistence curves of this blend, as shown in Figure 1. The fact that the extracted compositions of the SMA-rich and PMMA-rich phases obey the lever rule of thermodynamics, confirms the validity of this approach. From the above it is evident that the effect of shear on the phase behavior of SMA/PMMA is significant and clearly detected, but on the other hand it is rather complex (Figure 2). As a general trend, at low SMA compositions (typically $\varphi_{SMA}<0.25$), shear-induced mixing takes place, followed by shear-induced demixing at moderate compositions, and finally mixing again at $\varphi_{SMA}>0.65$, depending on the amount of shear.

3. ACKNOWLEDGEMENTS

This research was partially supported by the EU (BRE2.CT94-0610) and the Natural Sciences and Engineering Research Council of Canada.

4. REFERENCES

Hobbie EK, Nakatani AI, Han CC (1994) Modern Phys. Lett. B 8:1143-1161.

Kapnistos M, Hinrichs A, Vlassopoulos D, Anastasiadis SH, Stammer A, Wolf BA (1996a) Macromolecules 29:7155-7163.

Larson RG (1992) Rheol. Acta 31:497-520.

Takebe T, Sawaoka R, Hashimoto T (1989) J. Chem. Phys. 91:4369-4379.

Vlassopoulos D (1996) Rheol. Acta 40:613-622.

Table 1: Molecular characteristics of the homopolymers used for the SMA/PMMA blends.

Polymer	M_w	M_w/M_n	T_g (°C)	η_0 at 205°C (Pas)
SMA	130,000	2-2.5	175	560,000
PMMA	100,000	2	105	33,000

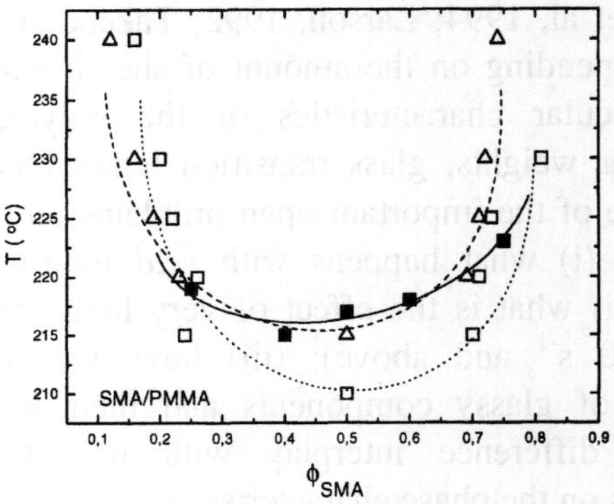

Figure 1: Phase diagrams of SMA/PMMA for various shear rates (■ and solid line: no shear; Δ and dashed line: 10 s^{-1}; □ and dotted line: 100 s^{-1}). Lines are drawn to guide the eye.

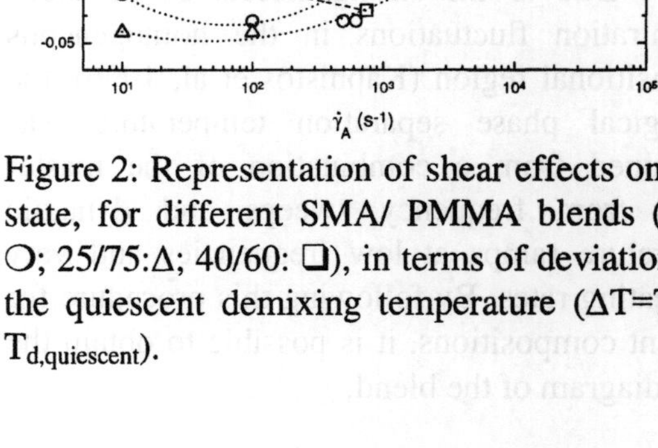

Figure 2: Representation of shear effects on phase state, for different SMA/ PMMA blends (50/50: O; 25/75:Δ; 40/60: □), in terms of deviation from the quiescent demixing temperature ($\Delta T = T_{d,shear} - T_{d,quiescent}$).

EFFECT OF PU-NBR INTERACTIONS ON BLENDS' DYNAMIC PROPERTIES

I. DIMITRIEVSKI, Z. ŠUŠTERIČ, T. MARINOVIĆ

Sava d.d., R & D Institute, Škofjeloška 6, SI 4000 Kranj, Slovenia

1. INTRODUCTION

Butadiene-acrylonitrile rubber (NBR) and polyurethane (PU) are both polar in character and hence expected to exhibit orientational and dispersional interactions. Such interactions yield associations of NBR-NBR, NBR-PU and PU-PU type, acting as secondary molecular linkages and rendering finite values of mechanical dynamic functions, i.e. the storage and loss shear moduli, in uncrosslinked blends or additional linkages in covalently crosslinked blends. These secondary linkages form, together with molecular entanglements, the secondary morphological structure of the blends. By analysis of the blends' dynamic functions as a function of their composition, strain amplitude and temperature, it is possible to confirm the existence of the foregoing associations and to determine the corresponding interactional energies.

NBR/PU blends of different compositions have been made in order to acquire new properties necessary for special applicacions (Dimitrievski et al. (1993)) and the aim of this work is to demonstrate a rheological way to assessment of the blends' morphology.

2. THEORETICAL

For this purpose, a modified version of a model describing secondary structure breakdown with increasing strain amplitude (Šušterič (1989)) has been applied. The model is based on statistical mechanics of chain molecules and on some additional plausible assumptions, the most important among them being the one on gaussian distribution of chain lengths between adjacent linkages.

During the strain induced breakdown of morphological structure, the storage modulus, G', which is proportional to the secondary linkage density, after a short linear range monotonically decreases with increasing strain amplitude all the way towards a small finite value, whereas the loss modulus, G'', proportional to the linkage density change, first increases (due to dissipation of energy caused by structural breakdown), experiences a maximum and then, similarly as G'', decreases, as schematically shown in Fig. 1.

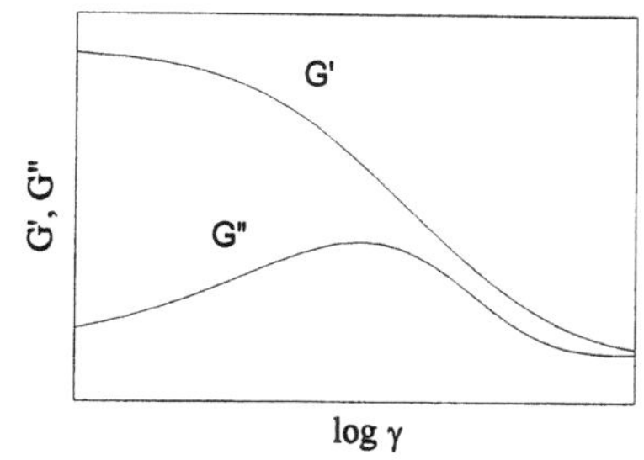

Fig. 1. Storage and loss shear moduli as a function of shear strain amplitude

In accordance with observation, the model correctly describes dependence of dynamic functions on strain amplitude, γ, yielding the following two equations:

$$G'(\gamma) = G'(0)[1 + a(\gamma)\gamma]\exp[-a(\gamma)\gamma], \quad (1)$$

$$G''(\gamma) = G''_{max}\, a(\gamma)\gamma \exp[1 - a(\gamma)\gamma] \quad (2)$$

where $G'(o)$ and G''_{max} are the maximal values of the storage and loss moduli, respectively, and $a(\gamma) \equiv W(\gamma)/3^{1/2}RT$ a function depending on strain amplitude and containing the gas constant, R, the temperature, T, and a characteristic energy for morphological structure breakdown, $W(\gamma)$. This energy is a decreasing function of strain amplitude, meaning that the more destroyed is the structure, less energy is needed for its further destruction. It has the following form:

$$W(\gamma) = W(\infty) + \left\{1/[W(0) - W(\infty)] + \gamma / 3^{3/2}RT\right\}^{-1} \quad (3)$$

where $W(0)$ and $W(\infty)$ are the highest and the lowest characteristic energies, easily determined from Eq. (1) or (2) and measured $G'(\gamma)$ or $G''(\gamma)$, respectively.

Observing the effect of temperature, the secondary structure breakdown appears suggestingly to be a thermally activated process, i.e. the secondary linkage density, n, decreases with increasing temperature as $n \propto \exp(\Delta E / RT)$ where ΔE is a corresponding activation energy. Indeed, taking account of this and the result of the theory of rubber elasticity, $G' = nkT$, where k the Boltzmann constant, the equation

$$G' = CT\exp(\Delta E / RT) \tag{4}$$

(C is a constant) has proved to hold with a high precision. It is hence presumable that the interactional energy, W_{int}, as a decreasing function of temperature, would be of the following form:

$$W_{\text{int}} = C'\exp(\Delta E / RT) \tag{5}$$

where C' is another constant, to be tentatively, though not unreasonably, determined so that W_{int} would be identified with the characteristic energy $W(0)$ at the mean value of 1/RT.

As for composition, PU acts reinforcingly in NBR in all cases, meaning that G' and G'' increase with increasing PU content.

3. EXPERIMENTAL

G' and G'' of NBR/PU blends of various compositions were measured as a function of strain amplitude and temperature by Rubber Process Analyser (RPA 2000, Monsanto) at low frequency of 0.3 Hz, allowing for molecular conformational changes to be prompt enough within the entire strain amplitude range. With the model and so-obtained experimental data, activation energies for secundary structure breakdown and the characteristic energies of molecular interactions were determind, the latter being comparable with those ascertained by other means.

4. RESULTS AND DISCUSSION

As an example, Figs. 2a and 2b present experimentaly obtained (points) and by the model calculated (lines) $G'(\gamma)$ for NBR/PU blends of different compositions at 100°C (a) and for the blend of 50 phr (parts in weight per 100 parts of rubber) PU content at different temperatures (b), together with determined values of ΔE and W_{int}.

(a)

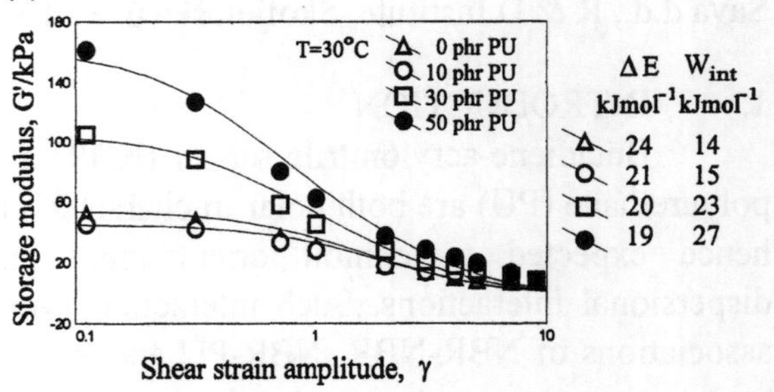

(b)

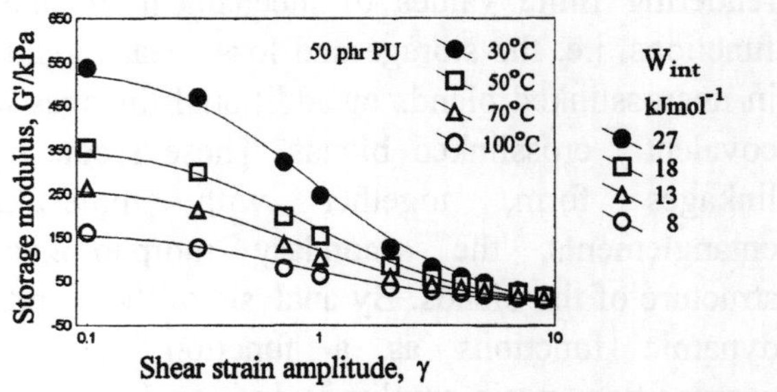

Fig. 2. Strain amplitude dependence of storage modulus (obtained experimentally and calculated from the model) for NBR/PU blends of different compositions (a) and at different temperatures (b)

As can be seen, good agreement of the model with experiment is obtained in all the cases, thus confirming the existence of secondary structure in these blends. Reinforcing effect of PU, at low strain amplitudes clearly perceivable, vanishes at high strain amplitudes (Fig. 2a) due to destruction of the secondary structure. Similar behavior is exhibited at high strains when examining the effect of temperature (Fig. 2b). Moreover, the determined activation energies for secondary structure breakdown and the interactional energies have the order of magnitude characteristic for this type of interactions.

5. REFERENCES

Dimitrievski I, Šušterič Z, Malavašič T (1993) In: K. Finlayson (ed) Advances in Polymer Blends and Alloys Technology, Vol 4, Technomic Publ. pp 19-29.

Šušterič Z (1989) Makromol Chem, Makromol Symp 23: 329-342

RHEOLOGICAL MODELING OF POLYMER BLENDS WITH SALAMI-TYPE MORPHOLOGY

M. Fahrländer and Chr. Friedrich

Institute of Macromolecular Chemistry, University Freiburg, Stefan-Meier-Str. 31, 79104 Freiburg i. Br., Germany

1. INTRODUCTION

Salami-type morphologies are well known from HIPS, a commercial polymer blend of polystyrene (PS) and polybutadiene (PB). Such complicated structures can also be found for polymer blends with PS matrix and poly (methylmethacrylate) PMMA as dispersed phase. We investigate these blends by means of dynamic rheology and develope a constitutive equation wich is based on Palierne's emulsion model (Palierne, 1990).

2. EXPERIMENTAL

The blends of PS (M_w=91 kg/mol) and PMMA (M_w=31 kg/mol) were prepared by solution blending. The volume fraction ϕ of PMMA was varied from 10% to 30%. The morphology was investigated by transmission electron microscopy (TEM) and analyzed by a method developed by Gleinser et al. to obtain the volume average sphere size distribution v(R) and from it the volume averaged radius R_v and the number averaged radius R_n. For the determination of the dynamic moduli $G'(\omega)$ and $G''(\omega)$ the mechanical spectrometer RMS 800 was used.

3. MODEL

The salami-type morphology to be investigated here is presented in figure 1. In the PS matrix PMMA/PS spheres with radius R_2 are shown. Inside of these compound particles, PS spheres with radius R_1 can be found.

To describe the rheological measurements of blends with such a salami-type morphology a modification of Palierne's emulsion model is necessary, since the original model is restricted to simple spherical morphologies. Therefore, we assume that the dispersed phase can be regarded as an emulsion which is discribed by a simple Palierne model. The parameters of this modified model are consequently the two radii R_1 and R_2, the interfacial tension α and the two volume fractions ϕ_1 an ϕ_2, which are related to the bulk volume fraction ϕ of PMMA by equation (1):

$$\phi_2 = \frac{\phi}{1 - \phi_1}. \qquad (1)$$

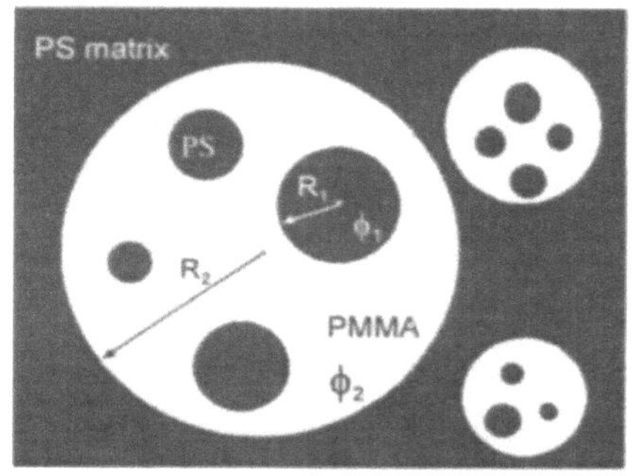

Figure 1: Scheme of the salami-type morphology of PS/PMMA blend.

4. RESULTS

For all blends under investigation salami-type morphologies were found. The volume averaged radius $R_{v,1}$ of PS spheres was found to be near 100 nm, the volume averaged radius $R_{v,2}$ of PMMA/PS spheres is near 1 µm.

The modified Palierne model was found to be suited to describe the measured moduli quantitatively correct in the whole frequency range. From the fits of the model to the experimental data values for the model parameters R_1 and R_2 were obtained and compared to the values obtained from TEM.

It was found that for the PS spheres the number averaged radii R_n obtained from TEM correspond to the radii R_1 from rheology, while for the PMMA/PS spheres the values R_2 from rheology range between the number and volume averages $R_{n,2}$ and $R_{v,2}$) from TEM data.

5. REFERENCES

Gleinser W, Maier D, Schneider M, Weese J, Friedrich Chr, Honerkamp J (1994) J Appl Polym Sci 53:39-50

Palierne JF (1990) Rheol Acta 29:204-214

RHEOLOGICAL PROPERTIES OF GRAFTED PMMA-PARTICLES IN SEVERAL POLYMER MATRICES

M. Fahrländer, M. Bruch, M. Ganter, Chr. Friedrich

Institute of Macromolecular Chemistry, University Freiburg, Stefan-Meier-Str. 31, 79104 Freiburg i. Br., Germany

1. INTRODUCTION

The rheology of particle filled polymers is not only affected by the amount and size of the particles but also by the state of dispersion in the matrix. The influence of the latter on the rheological properties of compounds is not very well understood[1,2]. Therefore, we investigate some model systems to elucidate the importance of this parameter.

2. EXPERIMENTAL

Oxazoline functionalized poly (methyl-methacrylate) (PMMA) particles of 200 nm in diameter were synthesized[3] and chemically modified by grafting carboxy-terminated polystyrene (PS) or carboxy-terminated polybutadiene (PB). Different degrees of grafting up to 14 % (with respect to particle weight) were realized. Samples were prepared by solution blending. Data of the different samples are summarized in table 1. For comparison neat PMMA particles were dispersed in PS matrix using a block copolymer as dispergator.

***Table 1**: Samples under investigation*

system	matrix	wt.-% PMMA particles	grafted polymer	% degree of grafting
A	PCHMA[a)]	0, 8	PS	14
B	PB	0, 8	PB	n. d.
C	PS	0, 8, 16	PS	0, 4, 9,14

[a)] PCHMA = poly(cyclohexylmethacrylate)

The morphology of the compounds was studied before and after the rheological measurements by means of transmission electron microscopy (TEM). For the determination of the dynamic moduli G'(ω) and G''(ω) a mechanical spectrometer RMS 800 was used.

3. RESULTS

It was found that time temperature superposition (TTS) holds for system A. The mastercurve of the compound consisting of PMMA particles grafted with PS in PCHMA matrix only slightly deviates from that of the pure PCHMA matrix. The morphology was found to be homogenous before and after the rheological testing.

A similar result was found for system B where PMMA particles grafted with PB are dispersed in a PB matrix. The isotherms obeyed TTS and the mastercurve showed the typical rheological behavior of the PB matrix with a slight increase in moduli caused by the particles. No shoulder or plateau was found in G' at small frequencies.

The results for system C, where PMMA particles grafted with PS are dispersed in a PS matrix clearly differ from the above results for system A and B.

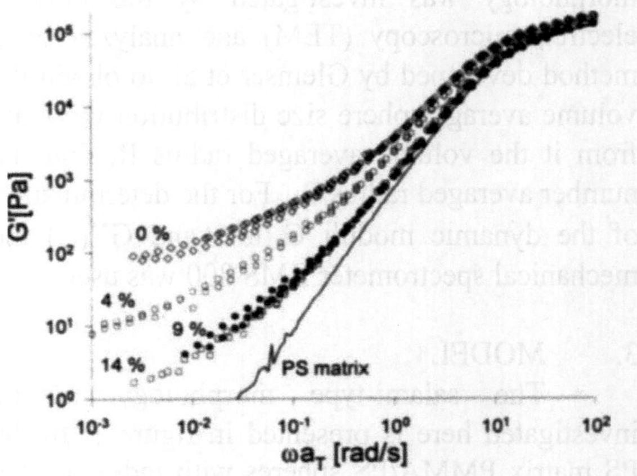

Figure 1: G'-mastercurves of PS matrix containing 8 wt.-% of particles in dependence on the reduced frequency for different degrees of grafting (reference temperature T_o= 190°C).

In figure 1 the dependence of the storage modulus G' on the reduced frequency ωa_T is shown for system C where different degrees of grafting are compared for samples containing 8 wt.-% particles. For the isotherms below 210°C TTS was possible. Above 210°C the isotherms superimpose only in their high frequency parts.

The 'mastercurves' in figure 1 were obtained this way. At frequencies where the matrix reaches its terminal flow region, the G' values of the compounds are shifted to considerably higher values in comparison to the matrix. This effect in G' is the stronger the lower the degree of grafting is. A higher degree of filling leads to higher G' moduli in all cases.

For all samples of system C it was found that the morphology was homogenous before the rheological testing, but agglomerated afterwards as it is shown in figure 2. In general, the degree of agglomeration seems to be the higher the lower the degree of grafting is.

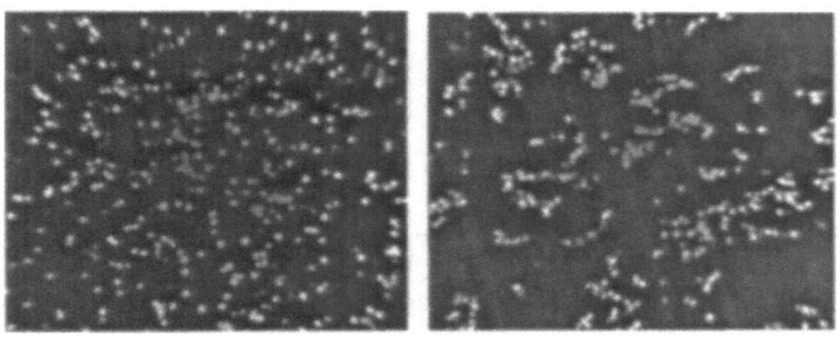

Figure 2: TEM micrographs of a sample containing 8 wt.-% particles with a degree of grafting of 9 %. Left: before, right: after rheological measurements.

Our results are in agreement with the findings of Aoki[2] for a ABS system, the morphological structure of which is comparable to our samples. Superposition of isotherms above 200°C was not possible, due to temperature dependent agglomeration of the particles. The particles in ABS could not form stable morphologies, because they were not covered completely with grafted chains. Since our particles of system C are not completely covered by the grafts they may agglomerate as well. Thus the distribution of particles becomes inhomogeneous. But interactions between the particles or the particles and the matrix should also affect the rheology and are expected to be more pronounced, the lower the degree of grafting is.

For comparison with particles carrying chemically bound polymers on their surface (system A-C) we investigate also neat PMMA particles with physically adherent dispergator molecules.

In figure 3 storage moduli G' of such PMMA particles immersed in a PS matrix are shown in dependence of the degree of filling. For all samples TTS was possible. The G' mastercurve of the sample with 8 wt.-% particles deviates strongly from the G' values of the matrix for frequencies smaller 0.1 rad/s. For higher particle contents the G' moduli at small frequencies become almost ω independent, indicating the formation of a particle network. For all samples the morphology was agglomerated (see figure 3) but stable during the rheological measurements.

Since a high molecular weight block copolymer is used as dispergator, block copolymers from different particles may entangle. Therefore stable agglomerates are formed and TTS is possible.

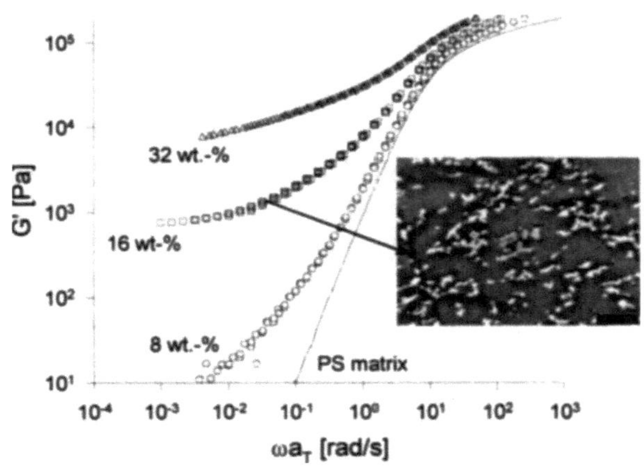

Figure 3: G' mastercurves of PS matrix containing PMMA particles stabilized by a dispergator in dependence of the amount of particles (T_0=190°C).

4. CONCLUSIONS

Our results clearly reveal that the dynamic moduli G' (and G'') depend on the state of dispersion of the particles, which itself strongly depends on the chemistry of particles interface.

5. REFERENCES

1) Bousmina M and Muller R (1996) Rheol Acta 35: 369-381

2) Aoki Y (1987) Macromolecules 20: 2208-2213

3) Hölderle M, Bar G, Mülhaupt R (1997) J Polym Science Polym Chem 35: 2539-2548

RHEOLOGICAL AND MORPHOLOGICAL PROPERTIES OF POLYMER BLENDS WITH COCONTINUOUS STRUCTURE

CHR. FRIEDRICH, C. WEIS, S. STEINMANN, D. MAIER

Freiburg Materials Research Center, University Freiburg, Stefan-Meier-Str. 21, 79104 Freiburg i. Br.

1. INTRODUCTION

Polymer blends with cocontinuous morphology are interesting materials and offer a variety of interesting properties originating from the special arrangement of phases[1]. Our work is focused on the investigation of the phase inversion process and its dependence on structural parameters of the constituent polymers. It is our aim to characterize polymer blends with concentrations around the phase inversion point quantitatively to find out the morphological as well as rheological characteristics of cocontinuity.

2. EXPERIMENTAL

The blends are made from anionically synthesized polystyrene (PS) and polymethylmethacrylate (PMMA). Data of six blends with viscosity ratios $p_0 = \eta_{0PS}/\eta_{0PMMA}$, ranging from 0.0047 to 3.68 are presented in Tab. I. The blends were prepared on a RANDCASTLE mini-extruder at an average shear rate of about 12 s^{-1} and a temperature at the die of 190 °C. The morphology of the melt mixed blends was determined by TEM techniques. Rheological properties like dynamic moduli G', G" and the creep function J(t) were determined using CVO Bohlin stress rheometer.

Table I: Blend viscosity ratios

blend	viscosity ratio p_0 $\dot{\gamma} \to 0$	effective viscosity ratio p_{eff} $\dot{\gamma} = 12{,}5\ s^{-1}$
B1	3.68	0.25
B2	0.5	0.33
B3	0.27	0.18
B4	0.048	2
B5	0.019	0.058
B6	0.0047	0.05

3. RESULTS

3.1 Morphology

A typical example of phase inversion is given for blend B3 in *Figure 1*.

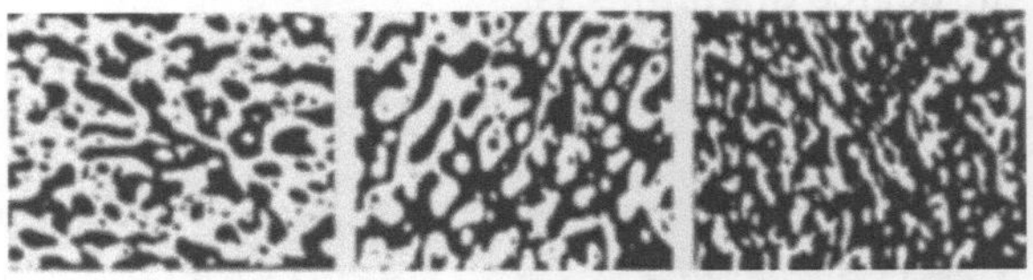

Figure 1: Binarized TEM micrographs of the blend B3 with 40, 50, and 60 wt.% PS(black, from left). The height of the pictures is 10 µm.

We used different image processing techniques to characterize the transition from a particulate morphology to cocontinuity[2]. The simplest concept is based on the form factor[2,3]. It allows to characterize the mentioned transition (see Figure 2) but it does not discriminate between different degrees of cocontinuity originating from compatibilization with different amounts of block copolymer.

The phase inversion concentration was determined from the minimum of the form factor ff_{irr} (see Figure 2), which is the form factor of that fraction of particles which becomes irregular shaped during the increase of minor phase.

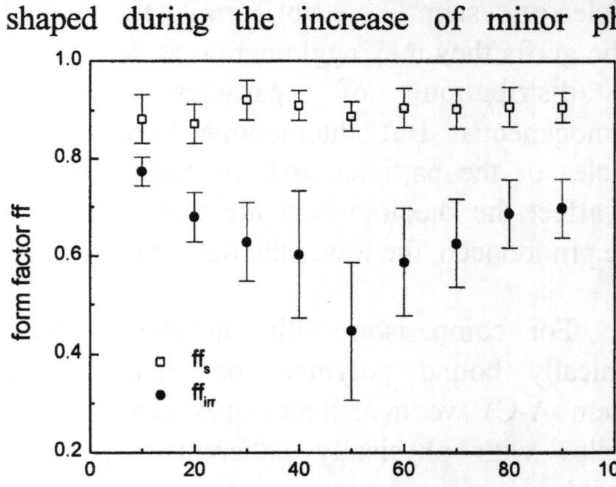

Figure 2: Form factors for blend series B3.

For blend series B3 the phase inversion occurs at composition 50/50.

3.2 Rheology

The plot of the phase inversion concentrations of the PMMA phase, $\Phi_{i\ PMMA}$, in dependence of the viscosity ratio is given for all blends under investigation in Figure 3.

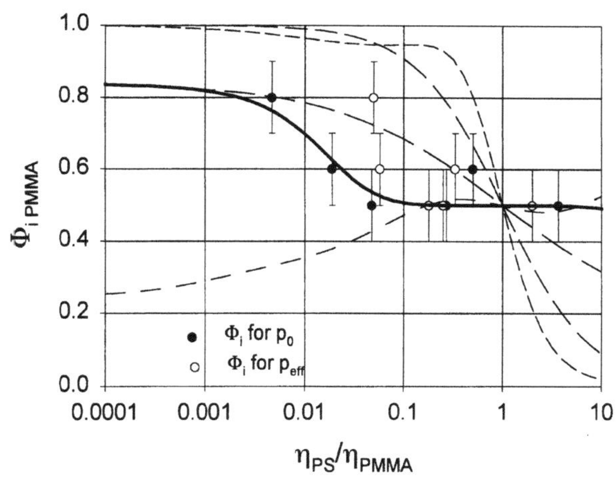

Figure 3: Phase inversion concentration versus viscosity ratio for all blends under investigation.

This figure contains the inversion concentrations (hollow and full circles) and some model predictions (lines). One can see that none of the existing empirical or semiempirical models is able to describe the data well, especially for p_0. The use of the effective viscosity ratio instead of the zero shear rate ratio shifts the data points towards the Utracki-equation[4]. Nevertheless, the data are best presented by the following empirical equation (thick line in Figure 3):

$$\Phi_{i\,PMMA} = \frac{1}{2} - \mathrm{sign}(x)\left(\Phi_m - \frac{1}{2}\right)\frac{\left[\left(\frac{x}{x_0}\right)^2\right]^{\alpha}}{1+\left[\left(\frac{x}{x_0}\right)^2\right]^{\alpha}} \quad (1)$$

In this equation $x = \log(p_0)$ (or $\log(p_{eff})$), and Φ_m, x_0, α are adjustable parameters. Φ_m was chosen the same as in Utracki's equation ($\Phi_m = 0.84$) and the other two parameters where determined by fit: $x_0 = 1.89$ and $\alpha = 4.95$. This model reflects the fact that cocontinuity can be observed for our blends only for zero shear rate viscosity ratios significantly different from 1 (here: $p_0 \approx 10^{1.89}$). From the existing models only the filament stability model[5] describes the very slow change of Φ_i with p_0 in the vicinity of $p_0 \approx 1$ quantitatively correct. This model was first introduced by Metelkin and Blekht[6]. Their solution (line with the highest slope at $p_0 = 1$) is not based on the rigorous solution of the corresponding equations of the filament stability. Metelkin and Blekht introduced some simplifying assumptions leading to the presented strong increase of Φ_i with p_0 for $p_0 = 1$. If we solve the mentioned equations (details will not be given here) the dash-dotted line is observed. This line represents the observed weak dependence of Φ_i on p_0 around $p_0 \approx 1$. The prediction of Φ_i-values for $|x| > 1$ seems to point to the failure of the whole model rather than individual assumptions.

4. CONCLUSIONS

Our experimental data which are based on phase inversion concentrations determined from the analysis of TEM pictures and well defined viscosity ratios indicate that existing theories fail to predict the correct course of the function Φ_i (p_0). We argue that on the one hand the so far used experimental data are not well suited to test the existing theories and on the other hand the theories do not fully reflect the basic processes going on during phase inversion.

5. REFERENCES

1) TRIAX blend series from Bayer Company
2) Weis C et al.(1998) Polym Bull 40:235-241
3) Heeschen WA (1995) Polymer 36:1835
4) Utracki LA (1991) J Rheol 35:1615
5) Tomotika S (1935) Proc Roy Soc A150:322
6) Metelkin VI, Bekht VS (1984) Koll Zh 46:476

THE MORPHOLOGY-DEPENDENT RHEOLOGICAL BEHAVIOUR OF AN IMMISCIBLE MODEL POLYMER BLEND

Nino Grizzuti, Giovanna Buonocore

Dipartimento di Ingegneria Chimica, Università degli Studi di Napoli "Federico II"
Piazzale V. Tecchio, 80 – 80125 Napoli, Italy

1. INTRODUCTION

In this paper, the viscous behaviour of a model immiscible polymer blend has been investigated, using both rotational and capillary rheometry. The blend is composed of a Polyisobutylene (PIB) and a Poly-dimethyilsiloxane (PDMS), both room temperature liquids. PIB is essentially Newtonian over a wide range of shear rates, with a viscosity strongly depending on temperature; PDMS is more viscoelastic and shear thinning. Viscosity measurements on the pure components and on blends of different compositions were performed at temperatures of 23°C and 9°C. In this paper, only results for two blend compositions at 23°C will be presented. The experimental data are compared with viscosity "mixing rules" available in the literature.

2. EXPERIMENTAL RESULTS

Viscosity measurements have been performed on a Rheometric stress controlled rotational rheometer (SR 200, cone and plate configuration) and on a Goettfert capillary rheometer. Fig.1 shows the viscosity as a function of shear rate for the pure components and the two blends, one of 30% PIB volume fraction (PIB is the dispersed phase), the other of 70% PIB (the PDMS is the dispersed phase). An excellent overlapping is obtained between rotational and capillary measurements (the overlapping window is within the 1-10 s^{-1} decade), when the latter are properly corrected to account for entrance losses. This suggests either that the blend morphology in the two different kinematics is similar, or that differences in morphology do not affect significantly the viscous behaviour. At the same time, Fig.1 shows that the viscosity of the blend is non-Newtonian even for shear rates for which a Newtonian behaviour of the pure components is observed.

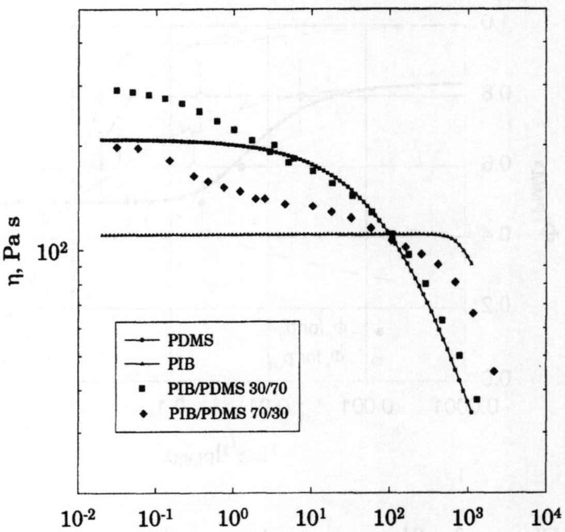

Fig.1

3. MIXING RULES

The results of Fig.1 clearly show that any mixing rule which does not take into account the interactions between the two pahses is not able to explain the observed experimental behaviour. For example, the linear mixing rule:

$$\eta = \eta_A \phi + \eta_B (1-\phi) \qquad (1)$$

underpredicts the viscosity at low rates, and overpredicts that at high shear rates.

Several mixing rules which relate the viscosity of the blend to the pure component properties *and* to the interfacial properties can be found in the literature. Frankel and Acrivos (1970) derived an expression for the viscosity of a Newtonian emulsion of diluted, slightly deformed droplets:

$$\eta = \frac{\eta_m}{1+\Lambda^2\dot{\gamma}^2}\left[1+\frac{5p+2}{2p+2}\phi + \right.$$
$$\left. + \Lambda^2\dot{\gamma}^2\left(1+\frac{5p+2}{2p+2}\phi - \frac{19p+16}{(2p+2)(2p+3)}\right)\phi\right] \qquad (2)$$

where:

$$\Lambda = \frac{(2p+3)(19p+16)}{40(p+1)}\frac{\eta_m R(\dot{\gamma})}{\sigma} \quad (3)$$

In Eqs. (2)-(3) η_m is the matrix viscosity, p is the dispersed to continuous phase viscosity ratio, σ is the interfacial tension, and R is the average droplet radius, which depends in turn on the applied shear rate. Eq. (2) is valid in the limited of very diluted blends. Choi and Schowalter (1975) proposed the following expression for the blend viscosity:

$$\eta = \eta_m\left[1+\frac{5p+2}{2p+2}\phi+\frac{5(5p+2)^2}{8(p+1)^2}\phi^2\right]\frac{1+h_1h_2\dot{\gamma}^2}{1+h_1^2\dot{\gamma}^2} \quad (4)$$

where:

$$h_1 = \frac{(19p+16)(2p+3)}{40(p+1)}\left(1+\frac{3(19p+16)}{4(p+1)(2p+3)}\right)\frac{\eta_m R(\dot{\gamma})}{\sigma}$$
$$h_2 = \frac{(19p+16)(2p+3)}{40(p+1)}\left(1+\frac{5(19p+16)}{4(p+1)(2p+3)}\right)\frac{\eta_m R(\dot{\gamma})}{\sigma} \quad (5)$$

The presence in Eq.(4) of the term proportional to the square of the volume fraction accounts for drop-drop interactions, making in principle Eq.(4) more suitable to describe concentrated blends.

In Eqs.(2)-(4) the effect of the interface on the viscosity behaviour is represented by the term containing the ratio between the drop radius and the interfacial tension between the phases. In order to compare the above models with the experimental data, therefore, independent measurements of these two quantities are necessary. The interfacial tension of the same PIB/PDMS system has been already measured with different methods (Sigillo et al., 1997), and is given by $\sigma=3\cdot10^{-3}$ N/m. The average drop radius as a function of the shear rate has been obtained from oscillatory rheological experiments upon cessation of a steady shear flow, according to the procedure of Vinckier et al. (1996).

Having obtained the morphological information from independent experiments, the comparison between the experimental data and Eqs. (2) and (4) (with no adjustable parameters) is presented in Figs. 2 and 3. The only further assumption is that a shear rate dependent viscosity of the pure components is considered in Eqs (2-5), The model of Frankel-Acrivos is able to predict quantitatively well the viscosity of the blend. On the other hand, that of Choi and Schowalter is found to largely overpredict the experimental behaviour. This result is somewhat surprising, since the model of Choi and Schowalter should be able to describe the concentrated blend behaviour better than that of Frankel and Acrivos.

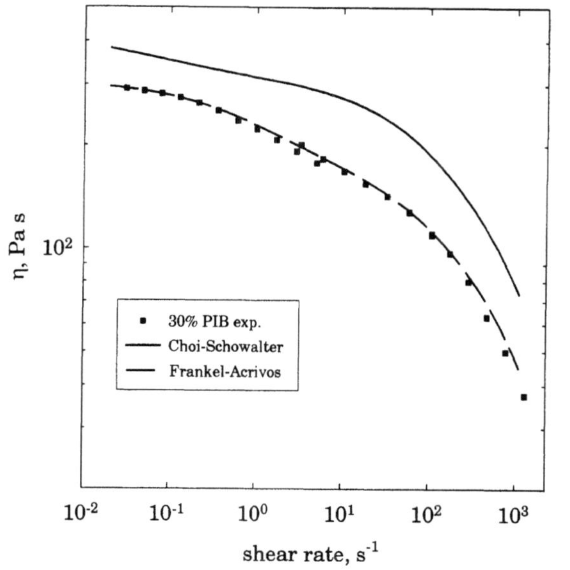

Fig.2

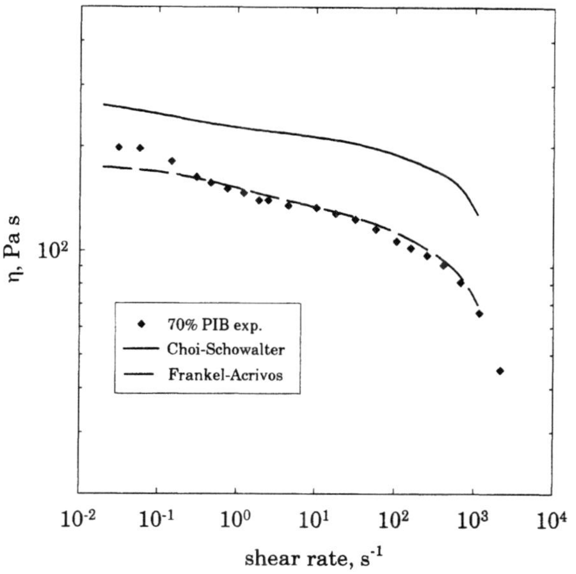

Fig.3

4. REFERENCES

Choi SJ, Schowalter WR (1975) Phys Fluids, 18:420

Frankel NA, Acrivos A (1970) J Fluid Mech, 44: 65

Sigillo I, di Santo L, Guido S, Grizzuti N (1997) Polym Eng Sci, 37:1540-1549

Vinckier I, Moldenaers P, Mewis J (1996) J Rheol 40:613-631

THE EFFECTS OF ADDING POLY(STYRENE)-BLOCK-POLY(BUTADIENE)-BLOCK-POLY(METHYLMETHACRYLATE) TRIBLOCK COPOLYMERS ON THE RHEOLOGICAL PROPERTIES AND MORPHOLOGY OF PPO/SAN BLENDS

THORSTEN KIRSCHNICK

Technical University Hamburg-Harburg, Polymer and Composite Section, Denickestr. 15, 21073 Hamburg, Germany

The idea of adding A-b-B-b-C triblock copolymers to an immiscible polymer blend is to generate a highly dispersed system and simultaniously toughening the blend by elastomeric microphases of the copolymer midblock.

Superior mechanical properties i.e. high elongation at break in combination with only a small reduction of the stiffness has been observed in a PPO/SAN/PS-b-PB-b-PMMA blend prepared by coprecipitation and melt pressing. The excellent mechanical properties are caused by a raspberry like morphology, whichis a result of the strong immiscibility of the midblock to all other blend components forcing the formation of small butadiene spheres at the PPO/PS-SAN/PMMA interface (Fig. 1, 2).

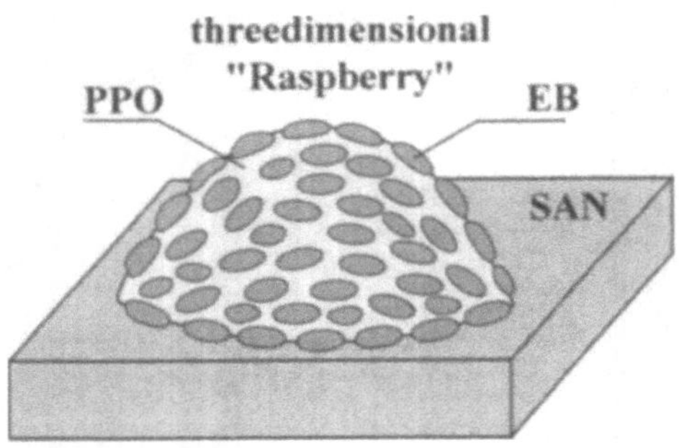

Fig. 1: Schematic representation of SBM/PPO/SAN raspberry; EB-microdomains 30-70 nm in diameter

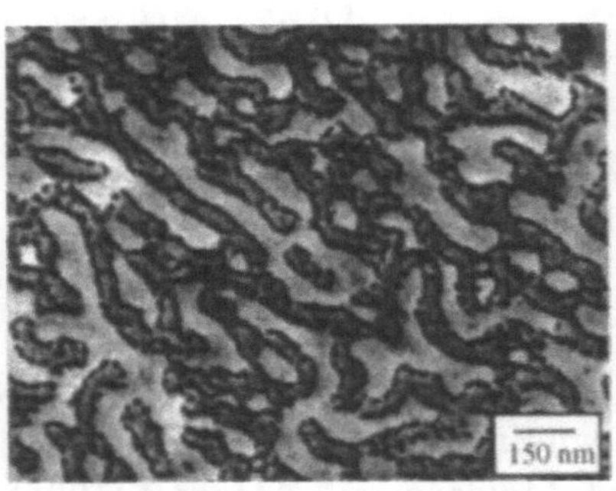

Fig. 2: TEM-micrograph of SBM/PPO/SAN-blend with raspberry-morphology; OsO_4 stained

The subsequent goal was to find out if (and how) this morphology can be created through pure melt processing (extrusion). In this case the macrophases at the beginning have to be highly dispersed into microphases. This is in contrast to coprecipitated blends where the phases develop from a maximum dispersed state.

For the melt preparation of the blend the special rheological behaviour of polymer blends and block copolymers has to be taken into consideration. The rheology of phase separated block copolymers is determined by the chemically and rheologically different blocks. Even if the relaxation times of the single chains are in the vicinity of the deformation frequency they do not want to follow the external force that easily because of the rise in free energy due to the partly forced thermodynamically unfavourable segment-segment interaction. This happens when the sheared block parts come in contact with other block components. Depending on the composition of the triblock copolymers different morphologies occur, leading to a special rheological behaviour and having a considerable influence on the interaction of the blend components in an early stage of compounding. In the case of a symmetric triblock copolymer (PS:PB:PMMA = 33:33:33) a lamellar morphology can be observed i.e. the PS- and the PMMA-surface for interaction with the matrix components are the same size. If one endblock is longer than the other (e.g. PS:PB:PMMA = 70:15:15) spheres or cylinders of the short blocks are surrounded by a matrix of the longest block. This changes the viscosity level and the frequency dependency of the viscosity. As a consequence of the encapsulation of the shorter endblock an interaction with its favourite matrix compound is hindered.

Rheological measurements of triblock copolymers of different block length ratios, of the pure matrix components PPO and SAN and the blends of PPO/PS-b-PB-b-PMMA, SAN/PS-b-PB-b-PMMA and PPO/SAN/PS-b-PB-b-PMMA give fundamental informations about the rheological conditions during the compounding process.

After rheological tests in a rheometer the build up of the blend morphology was studied as a function of time in a melt mixer and a microcompounder, and as a function of processing length in a twin-screw extruder. The phase development was monitored by opening the extruder barrel and taking samples from different extrusion zones (Fig. 3a, b). In this experiment a symmetric PS-b-PB-b-PMMA (Mw 100 kg/mol) was used.

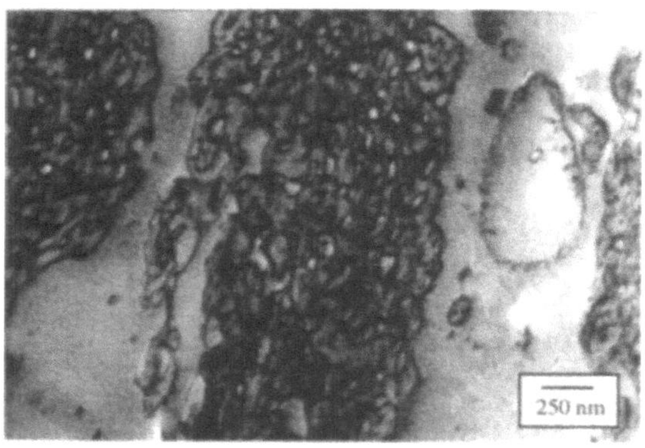

Fig. 3a: TEM-micrograph of SBM/PPO/SAN-blend at early stage of extrusion; OsO_4 stained

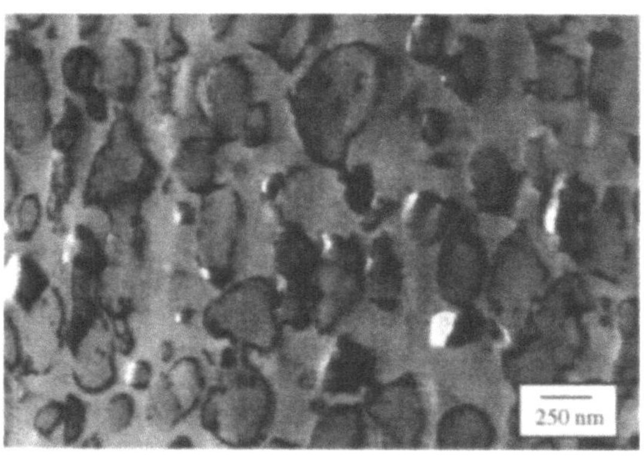

Fig. 3b: TEM-micrograph of SBM/PPO/SAN-blend with raspberry-morphology; final stage of extrusion; OsO_4 stained

After determining several limiting factors for this kind of processing the raspberry morphology was also obtained by an extrusion process.

THE DEFORMATION OF AN ELLIPSOIDAL DROP UNDER VISCOUS FLOW CONDITIONS

P. L. MAFFETTONE, S. GUIDO, M. MINALE

Dipartimento di Ingegneria Chimica, Laboratori Giovanni Astarita
Università di Napoli Federico II, Piazzale Tecchio, 80125 Napoli, Italy

1. INTRODUCTION

Production of articles with suitable mechanical properties is a major issue in the plastic industry. Blending low price immiscible polymers is a good alternative to synthesising new monomers to obtain high performance polymers. In such a case, the mechanical features of the final product depend not only on the properties of the constituent but also on the morphology of the blend (Vinckier et al., 1996). In this work, we consider dilute blends consisting in drops dispersed in a continuous matrix. In these conditions, when the blend is subjected to a viscous flow, both the deformation of the drops and the stress response are only slightly affected by the hydrodynamic interactions between droplets. It so appears that a good description of the deformation of a single drop immersed in a matrix subjected to flow is of value in determining the stress field of the flowing blend.

The problem of drop deformation has been addressed in the literature mainly following two different approaches. The perturbative analysis of the Stokes equations (e.g. Rallison 1984) is a sophisticate analytical method that gives results limited to small deformation. The simulative approach (Kennedy et al. 1994) numerically solves the Stokes equations, and gives detailed predictions even at high deformation, but at a cost of cumbersome calculations.

A phenomenological model has been recently proposed (Maffettone and Minale, 1998) to describe the deformation of an isolated drop. Here we test the model capabilities to predict the drop transient behaviour following the flow start-up and its relaxation upon cessation of flow.

2. THE MODEL

Experimental results (Guido and Villone, 1998) show that the shape of a buoyancy-free Newtonian drop immersed in a Newtonian fluid undergoing a viscous flow is always close to an ellipsoid; this holds true almost up to break-up. Thus, we assume that the drop is always an ellipsoid and we develop a phenomenological evolution equation for such an object.

The drop is represented by a second rank, symmetric, positive definite tensor **S**. The drop deformation results from two antagonist phenomena: The interfacial tension which drives to recover the spherical undistorted shape, and the viscous drag which deforms the drop (see Maffettone and Minale, 1998 for details). The evolution equation in dimensional form is:

$$\frac{d\boldsymbol{S}}{dt} = -\frac{f_1(\lambda)}{\tau}\left[\boldsymbol{S} - g(\boldsymbol{S})\boldsymbol{I}\right] + f_2(\lambda, Ca)(\boldsymbol{E}\cdot\boldsymbol{S} + \boldsymbol{S}\cdot\boldsymbol{E}) + (\boldsymbol{\Omega}\cdot\boldsymbol{S} - \boldsymbol{S}\cdot\boldsymbol{\Omega}) \tag{1}$$

In Eq. 1 λ is the viscosity ratio; τ is the emulsion time given by $\sigma/\mu R$ with σ the interfacial tension, μ the matrix viscosity and R the undistorted drop radius; $\boldsymbol{I}$ is the unit tensor; $Ca=G\tau$ is the Capillary number where G is the intensity of the flow; $\boldsymbol{E}$ and Ω are the rate of deformation tensor and the vorticity tensor, respectively. The function $g(\boldsymbol{S})$ introduced to preserve the drop volume is:

$$g(\boldsymbol{S}) = 3\frac{III_S}{II_S} \tag{2}$$

The functions f_1 and f_2 are chosen to recover the linear analytical limits (Rallison, 1984) and the critical Ca for break-up. They are given by:

$$f_1(\lambda) = \frac{40(\lambda+1)}{(2\lambda+3)(19\lambda+16)}$$
$$f_2(\lambda, Ca) = \frac{5}{2\lambda+3} + \frac{3Ca^2}{2+6Ca^2} \tag{3}$$

For the case of simple shear flow of interest here, the steady state predictions are analytical:

$$
\begin{aligned}
L^2 &= \frac{f_1^2 + Ca^2 + f_2 Ca\sqrt{f_1^2 + Ca^2}}{\left(f_1^2 + Ca^2\right)^{\frac{1}{3}}\left(f_1^2 + Ca^2 - f_2^2 Ca^2\right)^{\frac{2}{3}}} \\
B^2 &= \frac{f_1^2 + Ca^2 - f_2 Ca\sqrt{f_1^2 + Ca^2}}{\left(f_1^2 + Ca^2\right)^{\frac{1}{3}}\left(f_1^2 + Ca^2 - f_2^2 Ca^2\right)^{\frac{2}{3}}} \\
W^2 &= \frac{f_1^2 + Ca^2 - f_2^2 Ca^2}{\left(f_1^2 + Ca^2\right)^{\frac{1}{3}}\left(f_1^2 + Ca^2 - f_2^2 Ca^2\right)^{\frac{2}{3}}} \\
\theta &= \frac{\pi}{2} - \frac{1}{2} Arctan\left(\frac{f_1}{Ca}\right)
\end{aligned}
\tag{4}
$$

where L, B, W are the length, breadth and width of the droplet made dimensionless with R. Notice that L and B are in the shear plane whereas W is along the vorticity axis. The orientation angle θ is the angle between the L-axis and the velocity gradient direction.

3. COMPARISON WITH EXPERIMENTS

The shear device used in this work is a parallel plate apparatus with an optical setup allowing for observations of the drop deformation either along the velocity gradient direction or along the vorticity axis. The fluids used are two immiscible polymers, polyisobutylene (PIB) and polydimethylsiloxane (PDMS), both liquid at room temperature (see Guido and Villone 1998).

The steady state results are reported in Fig. 1 (lines are the predictions and symbols the experiments). The agreement is remarkable even at rather high capillary numbers. Notice that the dimension in the vorticity direction, W, reduces

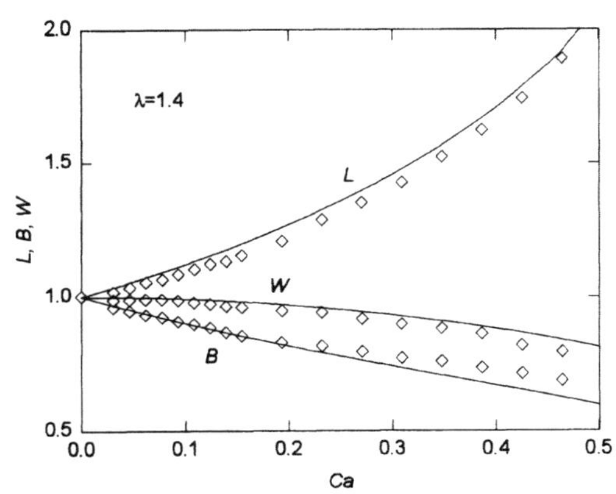

Fig. 1 - The steady state results for $\lambda = 1.4$

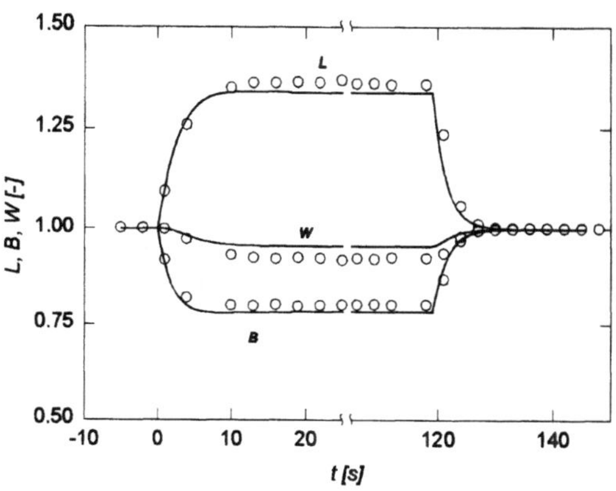

Fig. 2 - The evolution of the axes. ($\lambda = 1.4$, Ca$=0.24$)

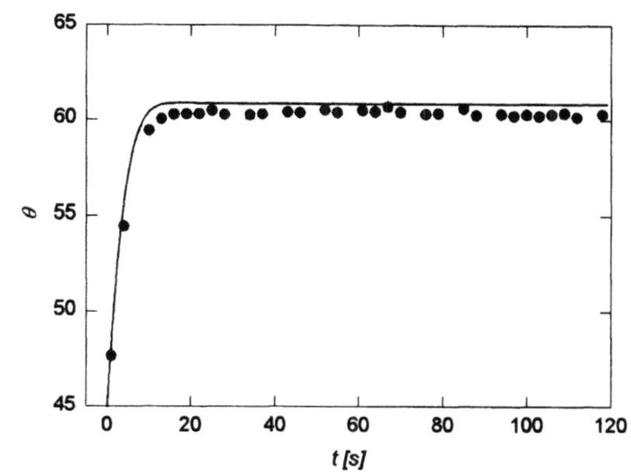

Fig. 3 - The orientation angle dynamics ($\lambda = 1.4$, Ca$=0.24$)

with respect to the undistorted value. This is due to the action of the interfacial tension which tries to keep circular the cross-section.

The transient results are reported in Figs. 2 and 3. In Fig. 2 the evolution of the three axes of the ellipsoid is plotted. The shear flow is started at time zero and is stopped at time 119 (s). The average error is within 4%, i.e it is of the same order of the experimental uncertainties. Both the start-up ramp and the relaxation upon cessation of flow is accurately predicted.

4. REFERENCES

Guido S., Villone M., (1998) J. Rheol, 42:395-415.

Kennedy M. R., Pozrikidis C., Skalak R., (1994) Computers Fluids, 23:251-278.

Maffettone P. L., Minale M., (1998) J. Non-Newtonian Fluid. Mech., in the press.

Rallison J. M., (1984) Ann. Rev. Fluid Mech, 16:45-66.

Vinckier I., Moldenaers P., Mewis J., (1996) J. Rheol., 40:613-631.

DROP DEFORMATION IN IMMISCIBLE POLYMER BLENDS

L. MEYLHEUC, M. BOUQUEY, R. MULLER

Laboratoire de Recherche sur les Procédés et Matériaux Polymères
ECPM, Polymers Department, 4 rue Boussingault, F-67000 STRASBOURG, France

1. INTRODUCTION

Final properties of immiscible polymer blends, such as mechanical or barrier properties, depend strongly on their morphology, which is itself controlled by the flow during melt blending and processing. The difficulty of predicting final morphologies arises in particular from the complexity of temperature and flow fields, the competition between deformation, breakup and coalescence of dipersed domains and the viscoelastic nature of the phases.

In this paper we measured with an original method[1] the deformation of initially spherical inclusions in a dilute blend during uniaxial elongational flow at constant strain rate. The results were compared to Palierne's linear theory of viscoelastic emulsions[2]. The model predicts not only the linear viscoelastic behavior of an emulsion of viscoelastic fluids[3], but also the deformation of an isolated drop in the limit of small strains.

2. EXPERIMENTAL

The method is described in Figure 1: a sample of the matrix polymer is prepared, which contains a small number of inclusions of the second polymer lying all in the same plane so that they can easily be located by optical microscopy. After stretching the specimen in the melt with an extensional rheometer[4], the shape of the droplets is frozen by quenching the specimen to room temperature. The elongation λ_d of each individual droplet can then be measured as a function of its initial diameter, the strain rate and the total strain λ applied to the specimen (which is also the deformation of the matrix far away from the drop).

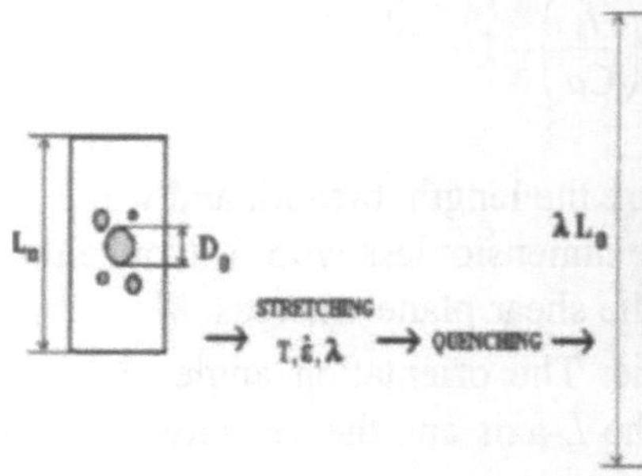

Figure 1: Measurement of drop deformation after elongational flow.

3. RESULTS

For Newtonian fluids[5], the deformation of a dispersed drop is controlled by the Capillary number Ca, which is the ratio of hydrodynamic over interfacial stresses, and the viscosity ratio p. Figure 2 shows the deformation λ_d of drops as a function of λ, for various samples in the case where Ca>>1. The matrix polymer was polystyrene and the dispersed inclusions were either polyethylene or PMMA. The viscosity ratio, taken as the ratio of dynamic viscosities at the strain rate of the experiments was in the range [0.005-13]. The results show that λ_d increases almost linearly with λ, up to λ of the order of 5. As expected, the ratio of drop over matrix deformation decreases with increasing the drop viscosity. On the other hand , for viscosity ratios smaller than 1, the drop deformation is found to be higher than that of the surrounding matrix.

The data in Figure 3 correspond to smaller values of Ca obtained for a blend of viscosity ratio p=0.67 stretched up to a macroscopic elongation of λ=2.5. Since drops of different sizes are present in the sample, a single experiment gives access to values of the capillary number. The data show a linear increase of λ_d with Ca for small droplets (Ca < 0.5) whereas the deformation of larger drops (Ca>1.5) becomes independent of their size.

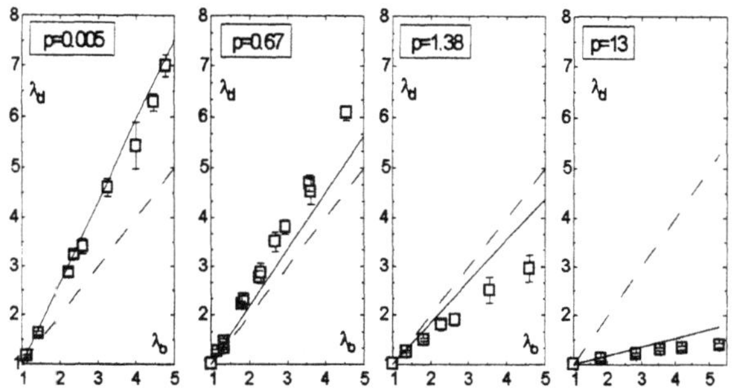

Figure 2: λ_d as a function of λ for different values of p and Ca>>1. Dotted lines: $\lambda_d = \lambda$. Solid lines: Newtonian model (Eq.2).

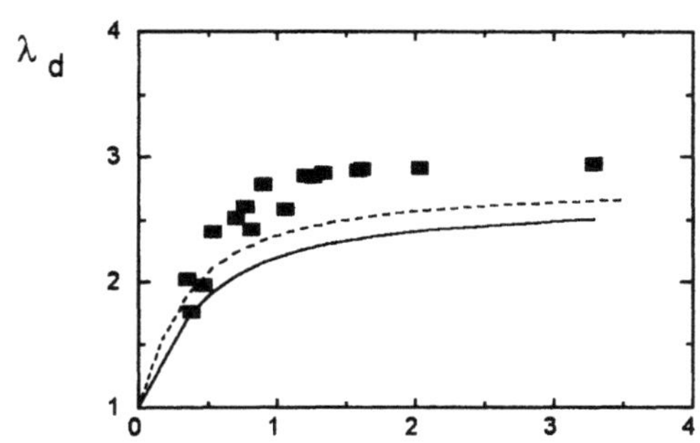

Figure 3: Drop deformation versus Capillary number for a blend with p=0.67 stretched at 150°C and 6×10^{-4} s^{-1}. ■: experimental data; solid line: Newtonian model with p=0.67; dotted line: linear viscoelastic model

4. LINEAR THEORY

In the case of harmonic strain, the following relationship between λ_d and λ is obtained according to the linear theory[1,2]:

$$\lambda_d = 1 + A(\omega)[\lambda - 1],$$

$$A(\omega) = \frac{5G_m^*\left(19G_d^* + 16G_m^*\right)}{\left(2G_d^* + 3G_m^*\right)\left(19G_d^* + 16G_m^*\right) + \frac{40\alpha}{R}\left(G_d^* + G_m^*\right)} \qquad (1)$$

where G^*_d and G^*_m are the complex moduli of inclusions and matrix, α the interfacial tension and R the drop radius. In the particular case of Newtonian phases and negligible interfacial tension (Ca>>1), Eq. 1 simplifies to:

$$\frac{\lambda_d - 1}{\lambda - 1} = \frac{5}{2p + 3} \qquad (2)$$

A good agreement is found between Eq. 2 and the data in Figure 2. More generally, by computing the Fourier transform of $A(\omega)$, the model allows to take into account the viscoelasticity of the components and interfacial tension effects[6]. In the case of Newtonian phases with interfacial tension ($G^*_m = i\omega\eta_m$, $G^*_d = i\omega\eta_d$ and $\alpha \neq 0$), the response to a start-up flow is easily shown to be a single exponential[6]:

$$\lambda_d(t) - 1 = \frac{19p+16}{16p+16}\frac{2\dot{\varepsilon}\eta_m R}{\alpha}\left[1 - e^{-t\times\frac{20\alpha}{R\eta_m}\frac{2p+2}{(19p+16)(2p+3)}}\right]$$

Figure 4 shows the theoretical curves according to the Newtonian and linear viscoelastic models. The Newtonian model underestimates the drop deformation whereas the viscoelastic approach improves the description of the data. The remaining differences may arise from non linear effects which are not taken in account by the model

5. REFERENCES

1. I. Delaby, B. Ernst, Y. Germain and R. Muller, J. Rheol.,**38**, 1705 (1994).
2. J.F. Palierne, Rheol. Acta, **29**, 204 (1990).
3. D.Graebling, R.Muller and J.F.Palierne, Macromolecules, **26**, 320 (1993)
4. R. Muller and D. Froelich, Polymer, **26**, 1477 (1985).
5. H.P. Grace, Chem. Eng. Commun. **14**, 225 (1982).
6. I. Delaby, B. Ernst and R. Muller, Rheol. Acta, **34**, 525 (1995).

MODELING AND SCALING OF DICHROISM RELAXATION IN IMMISCIBLE POLYMER BLENDS

P. VAN PUYVELDE, J. MEWIS, P. MOLDENAERS

Chemical Engineering Dept., K.U. Leuven, de Croylaan 46, B-3001 Heverlee, Belgium

1. INTRODUCTION

The properties of immiscible polymer blends strongly depend on the microstructure which is generated during processing. Therefore, understanding the morphology evolution during flow is important. In this paper, the relaxation behaviour of the droplets in dilute immiscible polymer blends is studied *in-situ* by the combined use of linear conservative dichroism and small angle light scattering (SALS) measurements.

2. METHODS

A model blend consisting of 3% polybutadiene in polyisobutene has been used. These components are Newtonian fluids. Dichroism measurements have been performed on a Rheo-Optical Analyzer (Rheometric Scientific). This instrument uses a polarization modulation technique (Fuller and Mikkelsen, 1989) to measure linear conservative dichroism. In order to obtain additional structural information during flow, the instrument has been modified to perform time-resolved SALS experiments as well (Yang *et al.*, 1998).

3. RESULTS AND DISCUSSION

3.1 Experimental

The study of immiscible polymer blends under transient flow histories is of technological interest. The kinematic conditions considered here consist of stopping the flow, either after steady state shearing or after a limited period of shear. The subsequent relaxation proceeds either by retraction or break-up of the deformed droplets. The different mechanisms can be distinguished with rheo-optical methods (Van Puyvelde *et al.*, 1998). Here we concentrate on relaxation by retraction.

3.2 Scaling of transient dichroism

Interesting scaling relations can be observed for relaxation due to retraction and for droplet break-up by Rayleigh instabilities.

This is illustrated here in figure 1 for the case of relaxation by retraction of the droplets after steady state shearing. The curves for the normalized dichroism (i.e. instantaneous dichroism divided by the value at time zero) can be approximated by a single exponential. The resulting time constant can be used to make the time axis dimensionless. As can be seen on figure 1, the normalized relaxation curves superimpose.

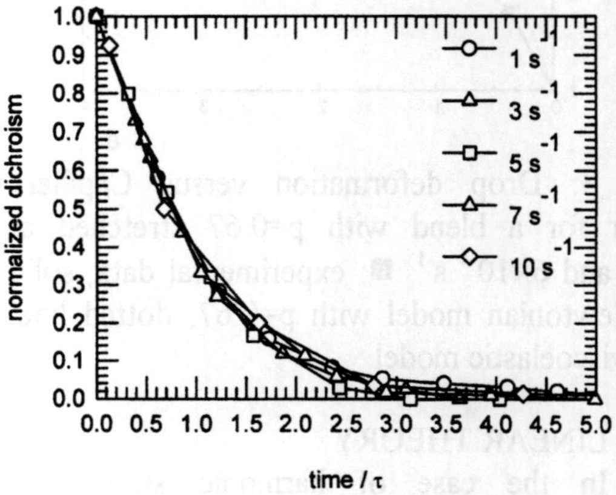

<u>figure1</u>: Scaling of the dichroism evolution during relaxation by retraction after steady shear flow.

The blend morphology created by steady shear flow consists of slightly deformed droplets. In that case the linear model of Palierne (1992) predicts an exponential relaxation behaviour for the shear stress. The experimentally obtained relaxation times can therefore be compared with the theoretical ones. This comparison shows that for low initial aspect ratios of the droplets the relaxation time of the blend approaches that of the linear response whereas for larger aspect ratios relaxation becomes slower than predicted. In this analysis the initial aspect ratio has been calculated by the theory of Choi and Schowalter (1975).

3.3 Modeling of dichroism relaxation

Recently, Marrucci and di Santo (1998) proposed a model to describe the retraction of an ellipsoidal droplet towards its original spherical shape in the absence of flow. Using a suitable scattering theory, their theory can be used to calculate the corresponding evolution of the dichroism. Assuming Rayleigh-Gans-Debye scattering in the limit of small deformations, the following expression for the normalized dichroism can be derived:

$$\frac{\Delta n''(t)}{\Delta n_0''} = \frac{\left[(\rho(t))^{-2} - 1\right]_t \rho(t)^{2/3}}{\left[(\rho)^{-2} - 1\right]_0 \rho_0^{2/3}}$$

where the subscript '0' denotes the value at the moment the flow is interrupted; ρ is defined as b/a with b and a respectively the minor and major axes of the ellipsoid. With the model by Marrucci and di Santo, ρ(t) can be calculated. Under the conditions where retraction is the dominant relaxation mechanism, a good agreement between predicted and experimental dichroism evolution is found. This is demonstrated in figure 2 for the case where a limited shear has been applied to the blend.

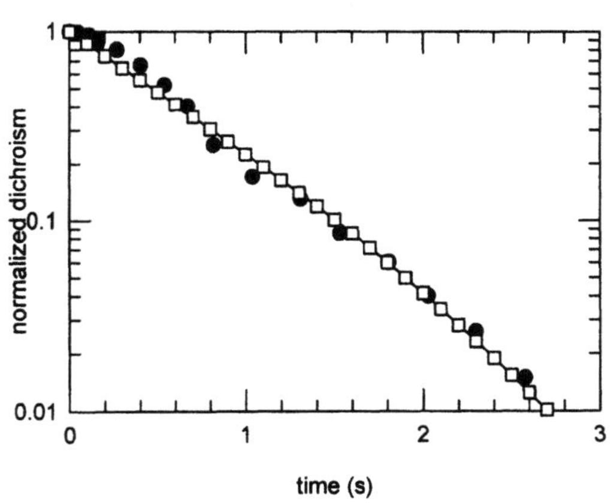

figure 2: Comparison between calculated and measured dichroism during droplet retraction. The droplet are deformed for 3 strain units before relaxation. Closed symbols=measured dichroism, open symbols=calculated dichroism.

The relaxation of the dichroism resulting from droplet break-up by Rayleigh instabilities can be modeled as well. In this case the shape variation of the fibrils is described by the Tomotika theory (1935). It should be pointed out that the evolution of the droplet shape during the various types of relaxation can also be used to compute the corresponding normal stresses during relaxation (Vinckier *et al.*, 1997).

4. CONCLUSIONS

Linear conservative dichroism is a powerful technique to obtain detailed information of the microstrucutre of polymer blends during flow. The technique is capable of determining the relaxation mechanism that is present when flow is stopped. Scaling relations for these mechanisms can be proposed. Attempts have been made to describe the rheo-optical response due to retraction. The results of this kind of measurements can be used to study systematically the different time scales of the different relaxation mechanisms in blends. The similarity between the scaling of the dichroism and the first normal stress difference suggest an underlying stress-optical relation in immiscible polymer blends.

5. REFERENCES

Choi SJ, Schowalter WR (1975) Phys. Fluids 18: 420-427

Fuller GG, Mikkelsen KJ (1989) J. Rheol. 33: 761-769

Marrucci G, di Santo S (1997) Brite Euram project reports, co-n°Breu2.CT920213

Palierne JF (1990) Rheol. Acta 29: 204-214

Tomotika S (1935) Proc. Roy. Soc. Lond. A153: 322-337

Van Puyvelde P, Yang H, Mewis J, Moldenaers P (1998) J. Colloid Int. Sci. 200: 86-94

Vinckier I, Moldenaers P, Mewis J (1997) J. Rheol. 41: 513-531

Yang H, Zhang HJ, Moldenaers P, Mewis J (1998) Polymer, in press

APPLICATION OF PHYSICALLY BASED CREEP DAMAGE EQUATIONS TO LIFE-TIME PREDICTIONS OF KIRCHHOFF- AND V. KÁRMÁN PLATES

J. ALTENBACH[1], H. ALTENBACH[2], K. NAUMENKO[2]

[1]Förderstedter Str. 28, D-39112 Magdeburg, Germany
[2]Technische Mechanik, Martin-Luther-Universität Halle-Wittenberg, D-06099 Halle, Germany

1. INTRODUCTION

The mechanical behaviour of thin-walled structures operating at elevated temperatures is strongly influenced by irreversible time dependent deformations. Creep strains may be cause a significant stress redistribution in a structure leading to creep buckling even by moderate loading, Kraus (1980). Another kind of premature failure is connected with material deterioration processes such as intergranular cavitation in interaction with ageing phenomena corresponding to a microstructural instability of alloys, Lemaitre and Chaboche (1990). Both, the deformation and damage, are depending on the stress levels, the stress states as well as the history of loading, which take place in thin-walled structures at operating conditions, Penny and Marriott (1995).

The appropriate description of the influence of stress levels and stress state on the strain rate is performed using internal state variables, defined by known physical mechanisms of intergranular creep deformation and cavitation. The creep-damage behaviour is assumed to be isotropic.

2. CONSTITUTIVE EQUATIONS

The basis for the general phenomenological modelling of creep behaviour may be a system of constitutive and evolution Eqs (1)

$$\begin{aligned}
\dot{\boldsymbol{\varepsilon}}^{cr} &= \frac{\partial F(\sigma^{eq}, T, H_1, \ldots, H_n, \omega_1, \ldots, \omega_m)}{\partial \boldsymbol{\sigma}}, \\
\dot{H}_i &= \dot{H}_i(\sigma_H^{eq}, T, H_1, \ldots, H_n, \omega_1, \ldots, \omega_m), \\
\dot{\omega}_k &= \dot{\omega}_k(\sigma_\omega^{eq}, T, H_1, \ldots, H_n, \omega_1, \ldots, \omega_m)
\end{aligned} \tag{1}$$

with the creep strain rate tensor $\dot{\boldsymbol{\varepsilon}}^{cr}$, the stress tensor $\boldsymbol{\sigma}$, the creep potential F, the internal state variables H_i, ω_k and the equivalent stresses $\sigma^{eq}, \sigma_H^{eq}, \sigma_\omega^{eq}$ controlling the primary, secondary and tertiary creep. The selection of the state variables and the definition of the creep potential follow from physical mechanisms of the creep process, the available experimental data and the application fields.

Assuming the material behaviour to be isotropic, we consider two special forms (2) and (3) of Eqs (1)

$$\begin{aligned}
&\dot{\boldsymbol{\varepsilon}}^{cr} = \frac{3}{2} A \left(\frac{\sigma^{vM}}{1-\omega}\right)^n t^m \frac{\boldsymbol{s}}{\sigma^{vM}}, 0 \le \omega \le \omega_*, \\
&\dot{\omega} = \frac{B(<\sigma_\omega^{eq}>)^l}{(1-\omega)^k} t^m, \\
&\sigma_\omega^{eq} = \alpha\sigma_1 + 3\beta\sigma_H + (1-\alpha-\beta)\sigma^{vM} \\
&< \sigma_\omega^{eq} > = \left\{ \begin{array}{l} \sigma_\omega^{eq}, \sigma_\omega^{eq} > 0 \\ 0, \sigma_\omega^{eq} \le 0 \end{array} \right., \sigma_H = \frac{1}{3} tr \boldsymbol{\sigma}
\end{aligned} \tag{2}$$

$$\begin{aligned}
&\dot{\boldsymbol{\varepsilon}}^{cr} = \frac{3}{2} \frac{A}{(1-\omega)^n} \frac{\boldsymbol{s}}{\sigma^{vM}} \sinh\left[\frac{B\sigma^{vM}(1-H)}{1-\Phi}\right], \\
&\dot{H} = \frac{h_c}{\sigma^{vM}} \frac{A}{(1-\omega)^n} \sinh\left[\frac{B\sigma^{vM}(1-H)}{1-\Phi}\right]\left(1 - \frac{H}{H_*}\right), \\
&\dot{\Phi} = \frac{K_c}{3}(1-\Phi)^4 \\
&\dot{\omega} = \frac{DA}{(1-\omega)^n}\left(\frac{\sigma_1}{\sigma^{vM}}\right)^\mu N \sinh\left[\frac{B\sigma^{vM}(1-H)}{1-\Phi}\right], \\
&n = \frac{B\sigma^{vM}(1-H)}{1-\Phi} \coth\left[\frac{B\sigma^{vM}(1-H)}{1-\Phi}\right], \\
&N = \left\{ \begin{array}{ll} 1, & \sigma_1 > 0 \\ 0, & \sigma_1 \le 0 \end{array} \right., 0 \le \omega < 0.3, \\
&0 \le \Phi < 1, 0 \le H < H_*
\end{aligned} \tag{3}$$

Eqs (2) are the power law creep constitutive equation generalized by the time hardening function and by Rabotnov's damage parameter ω and the evolution equation for the damage rate. σ^{vM} denote the von Mises equivalent stress and σ_ω^{eq} is the damage criterion proposed by Hayhurst and Leckie (1977). Eqs (3) include the damage variable ω, the hardening variable H and the ageing variable Φ, Perrin and Hayhurst (1996). $A, B, n, m, b, l, k, \alpha, \beta, \omega_*, K_c, h_c, H_*, D, \mu$ in Eqs (2) and (3) are material constants.

3. PLATE EQUATIONS

For the creep–damage modelling of thin plates we use the von Kármán's equations as follows

$$\begin{aligned}
\varepsilon_{ij} &= \frac{1}{2}(u_{i,j} + u_{j,i} + w_{,i}w_{,j}) - x_3 w_{,ij}, \\
\varepsilon_{ij} &= \varepsilon_{ij}^{el} + \varepsilon_{ij}^{th} + \varepsilon_{ij}^{cr},
\end{aligned}$$

$$\begin{aligned}\sigma_{ij} &= C_{ijkl}(\varepsilon_{ij}-\varepsilon^{cr}_{ij})-\delta_{ij}\frac{E\alpha T}{1-\nu},\\ C_{ijkl} &= \frac{E}{2(1-\nu^2)}[(\delta_{ik}\delta_{jl}+\delta_{il}\delta_{jk})(1-\nu)\\ &+ 2\nu\delta_{ij}\delta_{kl}],\end{aligned}$$

$$\begin{aligned}&n_{11,1}+n_{12,2}=0, n_{12,1}+n_{22,2}=0,\\ &m_{11,11}+2m_{12,12}+m_{22,22}\\ &+n_{11}w_{,11}+2n_{12}w_{,12}+n_{22}w_{,22}=-q,\\ &n_{ij}=g^I_{ijkl}e_{kl}-n^{th}\delta_{ij}-n^{cr}_{ij},\\ &m_{ij}=-g^{II}_{ijkl}w_{,kl}-m^{th}\delta_{ij}-m^{cr}_{ij},\\ &g^I_{ijkl}=C_{ijkl}h, g^{II}_{ijkl}=C_{ijkl}\frac{h^3}{12},\\ &e_{ij}=\frac{1}{2}(u_{i,j}+u_{j,i})+w_{,i}w_{,j}, \qquad (4)\\ &n^{th}=\frac{E\alpha}{1-\nu}\int_{-h/2}^{h/2}Tdx_3,\\ &m^{th}=\frac{E\alpha}{1-\nu}\int_{-h/2}^{h/2}Tx_3dx_3,\\ &n^{cr}_{ij}=\int_{-h/2}^{h/2}C_{ijkl}\varepsilon^{cr}_{ij}dx_3,\\ &m^{cr}_{ij}=\int_{-h/2}^{h/2}C_{ijkl}\varepsilon^{cr}_{ij}x_3dx_3,\\ &i,j,k,l=1,2\end{aligned}$$

Together with material models (2) or (3) Eqs (4) define completely the initial-boundary value problem, when the boundary conditions for the plate and the initial conditions (starting solution) are known. The Kirchhoff's plate equations follow from (4) by neglecting the nonlinear terms. Assuming the creep strains and the damage to be known functions at instant time the given set (4) of Eqs can be transformed into a mixed variational problem for the displacement w and the Airy's stress function φ

$$\begin{aligned}I(\varphi,w) &= \frac{1}{2}\int_A[-(g^I_{ijkl})^{-1}N_{ij}(\varphi)N_{kl}(\varphi)\\ &+ g^{II}_{ijkl}w_{,ij}w_{,kl}]dA\\ &- \frac{1}{2}\int_A N_{ij}(\varphi)w_{,ij}wdA \qquad (5)\\ &- \int_A[qw\\ &+ (g^I_{ijkl})^{-1}(n^{cr}_{ij}+n^{th}\delta_{ij})N_{kl}(\varphi)\\ &- (m^{cr}_{ij}+m^{th}\delta_{ij})w_{,ij}]dA\end{aligned}$$

with $N_{11}=\varphi_{,22}, N_{22}=\varphi_{,11}$ and $N_{12}=-\varphi_{,12}$. In Eqs (5) q is the loading component in w-direction and $n^{cr}_{ij}+n^{th}\delta_{ij}$, $m^{cr}_{ij}+m^{th}\delta_{ij}$ are fictitious stress resultants at instant time t.

For Kirchhoff plates a linear variational formulation in w must be used instead of Eqs (5). Details to the solution procedures without the thermal stress resultants are discussed in Altenbach and Naumenko (1997).

4. RESULTS AND CONCLUSIONS

The numerical results enable to a comparative study of different effects influencing the life-time predictions of built-in square plates. Though the material models (2) and (3) approximately describe the uniaxial creep behaviour of an alluminium alloy by qualitativ equal creep functions, significantly different results of the deformations, the stress redistributions and the failure times of square plates can be found. Further the geometrical nonlinear terms must be included in the structural analysis. The neglect of the geometrical nonlinearities leads to significant underestimations of the failure times, Altenbach et al. (1997). Finally the sensitivity of the time dependent plate bending to the material constants α and β in the material model (2) and μ in the material model (3) has been analysed. Neglecting the influence of the principal stress σ_1 in the evolution equations overestimations of the displacement field and the long-time strength can be observed. All here discussed creep-damage equations may be generalized by including load dependent effects and anisotropic creep and damage. Some results to nonclassical creep models are given in Altenbach (1995).

5. REFERENCES

Altenbach H, Altenbach J, Naumenko K (1997) Mech. of Time Dependent Materials 1:181-193

Altenbach H, Naumenko K (1997) Comput. Mech. 19:490-495

Altenbach J (1995) Proc. 1st Int. Conf. on Mechanics of Time Dependent Materials, Ljubljana, 73-78

Hayhurst DR, Leckie FA (1977) Acta Met. 25:1059-1070

Kowalewski ZL, Hayhurst DR, Dyson BF (1994) J. Strain Analysis 29:309-316

Kraus H (1980) Creep Analysis. Wiley, New York

Lemaitre J, Chaboche J-L (1990) Mechanics of Solid Materials. University Press, Cambridge

Lemaitre J (1996) A Course of Damage Mechanics. Springer, Berlin

Penny RK, Marriott DL (1995) Design for Creep. Chapman & Hall, London

Perrin IJ, Hayhurst DR (1996) J. Strain Analysis, 31:299-314

CONSTITUTIVE EQUATIONS FOR TIME DEPENDING BEHAVIOR OF TYPE LOADING SENSITIVE MATERIALS

A. BALTOV, M. DATCHEVA

Institute of Mechanics, Acad. G. Bontchev 4 Street, BG-1113 Sofia, Bulgaria

1. INTRODUCTION

The investigations concerning the theory of creep and fatigue of materials with different behavior in tension and compression have recently significantly increased in importance. The reason is that many structural materials as high- strength steel, sintered materials, cast iron, light alloys, porous materials, polymers, ceramics, concrete, rocks, composites, whose creep behavior depends on the type of loading are extensively used in various fields of the modern technology.

Further the following definition of the load type sensitive materials (LTSM) is used:

Def.1. LTSM is a material with different behavior in tension and compression.

2. MECHANICAL MODELS

Four basic destinations could be found in the theories modeling the effects of the loading type on the materials behavior:

Approach A.1 Direct extension of the Misses type theories introducing by different ways the influence of the hydrostatic pressure $I_1 = \sigma_{ij}\delta_{ij}$ (Bontcheva et al, (1995), Betten (1982)).

$$F(I_1, I_2) = A II_s^2 + B I_1^2 + C II_s + D I_1 \tag{1}$$

where $II_s^2 = \frac{1}{2} s_{ij} s_{ij}$, $I_2 = \sigma_{ij}\sigma_{ij}$, s_{ij} is the deviator of the stress tensor σ_{ij} and *A*, *B*, *C*, *D* are material dependent constants.

Approach A.2 Indirect modification of the Misses type theories (Zolochevsky, (1991), Altenbach and Zolochevsky, (1994)):

The equivalent stress σ_{eq} is introduced as:

Def.2. σ_{eq} is a scalar function of the stress tensor and some internal state parameters $q_1, q_2, \ldots, q_l$. In general:

$$\sigma_{eq} = \sigma_{eq}\left(\sigma_{ij}, q_1, q_2, \ldots, q_l\right) \tag{2}$$

Internal state parameters are used to describe the material properties like damage, hardening, strain sensitivity effects and etc. The equivalent stress enables to compare multiaxial stress states with uniaxial tests data.

The following hypothesis are used to receive the constitutive equations in creep (C) and plasticity (P) respectively:

Hyp.1~C: The behavior of a material can be derived from a potential function: $\Phi = \Phi\left(\sigma_{eq}\right)$ such that:

$$\dot{\varepsilon}_{ij}^{\,c} = \frac{\partial \Phi}{\partial \sigma_{ij}}, \tag{3}$$

where ε_{ij}^{c} is the creep strain tensor.

Hyp.2~C: The dissipated energy rate is positive $\dot{W} = \sigma_{ij}\dot{\varepsilon}_{ij}^{\,c} > 0$.

Hyp.1~P: The yield function depends on the σ_{ij} through σ_{eq} and then for the yield condition we have:

$$F\left(\sigma_{eq}\right) = k^2 \tag{4}$$

Hyp.2~P: (Misses postulate) For the actual stress state the dissipated energy rate $\dot{W} = \sigma_{ij}\dot{\varepsilon}_{ij}^{\,p}$ reaches its maximum. Then we have the associated flow rule to create the constitutive equations:

$$\dot{\varepsilon}_{ij}^{\,p} = \lambda \frac{\partial F}{\partial \sigma_{ij}}, \tag{5}$$

where λ is the Lagrange's multiplier and ε_{ij}^{p} is the plastic strain tensor.

Hyp.3~PC: For $\forall$ $\dot{\varepsilon}_{ij}, \sigma_{ij}$ the $\dot{\varepsilon}_{eq}, \sigma_{eq}$ exist such that

$$\dot{W} = \sigma_{ij}\dot{\varepsilon}_{ij} = \sigma_{eq}\dot{\varepsilon}_{eq} \tag{6}$$

Here ε_{ij} stays for both plastic and creep strains.

Approach A.3 In the stress space special subspaces are determined. In each stress subspace the parameters of the mechanical model are different and they specify the concrete type of loads (Ambarstumyan, (1982), Wesolowski, (1969)). Specially constructed equivalent strain surfaces are used (Baltov, (1994)).

Approach A.4 The type of the stress state is determined introducing special parameters. The constitutive equations directly depend on these parameters (Lomakin and Rabotnov, (1978), Bontcheva et al., (1984)). Another example of usage of internal parameters to describe the load sensitivity of the material is continuum damage theory. The anisotropic damage theories properly interpreted the influence of changes of the principal stress directions on the material response (Datcheva, (1992)).

2.1 Detected Problems

P.1 : It can be seen that in the frame of the *A.2*, from *Def.2. Hyp.1,2,3~P* it follows that:

$$\dot{\varepsilon}_{eq} = \dot{\lambda} \frac{\partial F}{\partial \sigma_{eq}} \quad . \tag{7}$$

This relation is fulfilled only if the yield function F is a homogeneous with respect to the equivalent stress. That condition is not allows satisfied in some of used LTSM theories. In the frame of ***A.1*** the relation (7) is postulated. In this case it follows, that:

$$F(I_1, I_2) = \overline{A}\, II_s^2 + \overline{B} I_1^2 = k^2 - p^2 \quad , \tag{8}$$

where $\overline{A}, \overline{B}, p^2$ are functions of *A, B, C, D*. The equation (8) is an additional condition for I_1, I_2 to satisfy and is in disagreement with (1).

P.2 : Commonly the creep low is introduced as:

$$\dot{\varepsilon}_{ij}^c = \dot{\lambda} \frac{\partial \Phi}{\partial \sigma_{ij}} \tag{9}$$

by an analogy with the plasticity. But in creep theory there is no creep condition and no maximum postulate (*Hyp.2~P*). Therefor (9) is not a correct creep constitutive equation.

In the next paragraphs a model is proposed which avoid the above mentioned difficulties and disagreements.

3. BASIC MODEL

Let us assume that σ_{eq} is a homogeneous function of degree one with respect to the tensor σ_{ij}. Then the equivalent stress can be defined (Altenbach et al., (1995)) as:

$$\sigma_{eq} = \alpha\, \sigma_1 + \sigma_2 + \gamma\, \sigma_3 \text{ , with} \tag{10}$$

$$\sigma_1 = \mu_1 I_1 \quad (11.1)$$

$$\sigma_2^2 = \mu_2 I_1^2 + \mu_3 I_2 \tag{11.2}$$

$$\sigma_3^3 = \mu_4 I_1^3 + \mu_5 I_1 I_2 + \mu_6 I_3 \text{ ,} \tag{11.3}$$

where

$$I_1 = \sigma_{ij}\delta_{ij}\,;\ I_2 = \sigma_{ij}\sigma_{ij}\ ;\ I_3 = \sigma_{ik}\sigma_{kj}\sigma_{ij} \tag{12}$$

The introduced specific internal state parameters functions $\mu_i, i = 1,2,..6$ should be identified by experimental tests, bound estimates and numerical procedures, α and γ are numerical coefficients. If the third order effects are neglected than $\gamma = 0$ and $\alpha = 1$.

4. CREEP

From *Hyp.1~C* the following relation between stresses and strains can be obtained:

$$\dot{\varepsilon}_{ij} = \frac{\partial \Phi}{\partial \sigma_{ij}} = \frac{\partial \Phi}{\partial \sigma_{eq}} \cdot \frac{\partial \sigma_{eq}}{\partial \sigma_{ij}} \quad . \tag{13}$$

From *Def.2* and the Euler's theorem it follows, that:

$$\frac{\partial \sigma_{eq}}{\partial \sigma_{ij}} \sigma_{ij} = \sigma_{eq} \,. \quad (14)$$

From (13) and *Hyp.3~PC* one finds:

$$\dot{\varepsilon}_{eq} = \frac{\partial \Phi}{\partial \sigma_{eq}} = \varphi(\sigma_{eq}) \tag{15}$$

The function $\varphi(\sigma_{eq})$ is defined from the standard creep curves. For the power creep law:

$\varphi(\sigma_{eq}) = L_+ \sigma_{eq}^n$ with L_+, n - material constants.

Therefor:

$$\Phi\left(\sigma_{eq}\right)=\frac{L_+}{n+1}\sigma_{eq}^{n+1} \qquad (16)$$

instead of $\Phi\left(\sigma_{eq}\right)=\sigma_{eq}^{2}$ used by analogy with the plasticity (Altenbach and Zolochevsky, (1994)). Finally:

$$\dot{\varepsilon}_{ij}=\dot{\varepsilon}_{eq}\cdot\frac{\partial\sigma_{eq}}{\partial\sigma_{ij}} \qquad (17)$$

$$\dot{W}=\left(\frac{\dot{\varepsilon}_{eq}^{n+1}}{L_+}\right)^{1/n} \qquad (18)$$

and for the concrete form for the equivalent stress (10):

$$\dot{\varepsilon}_{eq}=\dot{\varepsilon}_s+\dot{\varepsilon}_0 \qquad (19)$$

where:

$$\dot{\varepsilon}_s=-\frac{s}{3}\dot{\varepsilon}_v \text{ and } \dot{\varepsilon}_v=\dot{\varepsilon}_{ij}\delta_{ij}, \text{ and} \qquad (20)$$

$$\dot{\varepsilon}_0=\sqrt{2\left[\frac{\dot{\varepsilon}_{ij}\dot{\varepsilon}_{ij}}{A}+\frac{1}{3}\left(\frac{1}{6B}-\frac{1}{A}\right)\dot{\varepsilon}_v^2\right]} \qquad (21)$$

with constants:

$$A=\frac{2\mu_3\left(3\mu_2+\mu_3-3\mu_1^2\right)}{3\mu_2+\mu_3} \qquad (22a)$$

$$B=\frac{1}{3}\frac{\left(3\mu_2+\mu_3-3\mu_1^2\right)}{3\mu_2+\mu_3} \qquad (22b)$$

$$s=\frac{\mu_1}{3\mu_2+\mu_3-3\mu_1^2} \qquad (22c)$$

5. PLASTICITY

The plastic potential function is given as

$$F=\sigma_{eq}^2=\left(\sigma_1+\sigma_2\right)^2 \qquad (23)$$

which is a modified Misses type potential function. The yield criteria is :

$$F=\sigma_{eq}^2=\left(\sigma_1+\sigma_2\right)^2=\sigma_p^{*2} \qquad (24)$$

where σ_p^* can be calculated as a function of the tension yield limit and the compression yield limit. In the frame of the *A.2* we have:

$$\dot{\varepsilon}_{ij}=\dot{\lambda}\frac{\partial F}{\partial\sigma_{ij}}=\dot{\varepsilon}_{eq}\frac{\partial\sigma_{eq}}{\partial\sigma_{ij}}=\dot{\varepsilon}_{eq}\left(\frac{\partial\sigma_1}{\partial\sigma_{ij}}+\frac{\partial\sigma_2}{\partial\sigma_{ij}}\right)$$

(25)

$$\dot{W}=\dot{\varepsilon}_{eq}\sigma_p^* \qquad (26)$$

and for the concrete form for the equivalent stress (10):

$$\dot{\varepsilon}_{eq}=\dot{\varepsilon}_s+\dot{\varepsilon}_0, \text{ where} \qquad (27)$$

$$\dot{\varepsilon}_s=-\frac{s}{3\sigma_p^*}\dot{\varepsilon}_v \qquad (28)$$

and $\dot{\varepsilon}_0$, *A*, *B* are the same as in rel. (21) and (22a-c);

$$s=\frac{3\sigma_p^*\mu_1}{3\mu_2+\mu_3-3\mu_1^2} \qquad (29)$$

Therefor the creep relations can be received from those in plasticity replacing σ_p^* by $\left(\frac{\dot{\varepsilon}_{eq}}{L_+}\right)^{1/n}$

6. INDUCED ANISOTROPY IN LTSM

Kinematic hardening will be taken into account to describe the induced anisotropy ([13]).
In the frames of the *A.2* that means:

$$q_1=\sigma_{ij}^{\mu} \qquad \text{or}$$

$$\sigma_{eq}=\sigma_{eq}\left(\sigma_{ij}-\sigma_{ij}^{\mu},q_1,\dots,q_l\right) \qquad (30)$$

where σ_{ij}^{μ} is the back stress due to kinematic hardening. $\overline{\sigma}_{ij}=\sigma_{ij}-\sigma_{ij}^{\mu}$ is the actual stress.
There are two famous models for kinematic hardening - one by PRAGER :

$$\dot{\sigma}_{ij}^{\mu}=\tilde{C}\dot{\varepsilon}_{ij},\ \tilde{C} \text{ is a constant;} \qquad (31)$$

and the other by ZIEGLER :

$$\dot{\sigma}_{ij}^{\mu}=\nu\left(\sigma_{ij}-\sigma_{ij}^{\mu}\right),\ \nu \text{ is a constant.} \qquad (32)$$

The invariants $\overline{I}_2=\overline{\sigma}_{ij}\overline{\sigma}_{ij}$, $\overline{I}_1=\overline{\sigma}_{ij}\delta_{ij}$ and the mixed invariants:

$$
\begin{aligned}
I^{\mu} &= \sigma_{ij}^{\mu}\bar{\sigma}_{ij},\ I_1^{\mu} = tr\sigma^{\mu}\cdot\bar{\sigma}^2, \\
I_2^{\mu} &= tr(\sigma^{\mu})^2\cdot\bar{\sigma} \\
I_2^{\mu} &= tr(\sigma^{\mu})^2\cdot\bar{\sigma}^2
\end{aligned}
\tag{33}
$$

form the integrity basis for the representation of σ_{eq}.

In the common case we can form the following

linear $\sigma_1 = L_1\bar{I}_1 + L_2 I^{\mu}$ (34)

quadratic $\sigma_2{}^2 = Q_1\bar{I}_1^2 + Q_2 I^{\mu 2} + Q_3\bar{I}_2 + Q_4\bar{I}_1 I^{\mu}$ (35)

cubic $\sigma_3{}^3 = C_1\bar{I}_1^3 + C_2 I^{\mu 3} + C_3\bar{I}_2\bar{I}_1 + C_4\bar{I}_2 I^{\mu}$ (36)

scalar invariant functions which are used to represent the equivalent stress by analogy with the previous sections.. The scalar functions of the internal state parameters L_1, ..., C_4 have the same meaning as $\mu_i, i = 1,2,..6$ in (11.1 - 11.3). They also can be identified by experiments, bound estimates and numerical procedures.

Then we can find the equivalent strain rate following the algorithm given in the previous section. Finally, if $\gamma = 0$:

$$
\dot{\varepsilon}_{eq} = -\frac{\hat{C}_{11} + \hat{C}_{12}\dot{\varepsilon}_v + \hat{C}_{13}\dot{\varepsilon}_{ij}\sigma_{ij}^{\mu}}{2\hat{C}_0} + \tag{37}
$$

$$
+\sqrt{\hat{C}_0{}^{-1}\left[\dot{\varepsilon}_{ij}\dot{\varepsilon}_{ij} + P_1\dot{\varepsilon}_v{}^2 + P_2\left(\dot{\varepsilon}_{ij}\sigma_{ij}^{\mu}\right)^2 + P_3\dot{\varepsilon}_{ij}\sigma_{ij}^{\mu}\dot{\varepsilon}_v + P_4\dot{\varepsilon}_v + P_5\dot{\varepsilon}_{ij}\sigma_{ij}^{\mu}\right]}
$$

where $\hat{C}_0$, $\hat{C}_{ij}$, P_i are functions of L_i, Q_i, $\sigma_{ij}^{\mu}\delta_{ij}$, $\sigma_{ij}^{\mu}\sigma_{ij}^{\mu}$.

From (37) it is seen, that for the equivalent strain and respectively for the rate of the dissipated energy there are additional terms due to the induced anisotropy and this terms give an additional force when the

7. CONCLUSIONS

The model proposed avoids the existing in the literature disagreement between the used in the models hypothesis.

It is suitable to have a common algorithm to create the constitutive equation for the inelastic behavior.

The proposed constitutive equations are implemented in the FEM code PORO.

8. REFERENCES

Altenbach H, Zolochevsky A (1994) ZAAM 74: 189-199

Altenbach H, Altenbach J, Zolochevsky A (1995) Erweiterte Deformationsmodelle und Ver- sagenskriterien der Werkstoffmechanik, Deutscher Verlag fuer Grundstoffin- dustrie, Stuttgart

Ambarstumyan SA (1982) Different- Modules Elasticity Theory. Nauka, Moscow

Baltov A (1994) Modeling of oriented inelastic deformation and damage in advanced composite materials(*Final report*)Contract NoERB-CIPA-CP-92-2132,

Betten J (1982) in IUTAM Conf. on Deformation and Failure of Granular Materials / Defts / 31 Aug. - 3 Sept. pp 81-89

Bontcheva N, Baltov A, Todorov S (1984) ZAMM 64: 113-115

Bontcheva N, Iankov R, Datcheva M (1996) Mech Res Comm 23: 5: 537-542,

Datcheva M (1992) Jour. of Theoretical and Applied Mechanics 1: 74-82.

Lomakin EV, Rabotnov JN (1978) Isv. AN SSSR MTT 6: 34-38

Wesolowski Z (1969) Archivum Mechaniki Stosowanej 4:**21**: 449-468

Zolochevsky AA (1991) In: Ziczkovski M (ed) Creep in Structures. Springer- Verlag pp 217-220

ON THE USE OF CHARACTERISTIC VARIABLES IN VISCOELASTIC FLOW CALCULATIONS

M. I. GERRITSMA[1] and T. N. PHILLIPS[2]

[1] Laboratory for Aero and Hydrodynamics, Delft University of Technology, 2628 AL Delft, The Netherlands.
[2] Department of Mathematics, University of Wales, Aberystwyth SY23 3BZ, UK.

1. INTRODUCTION

The development of accurate and stable time-dependent approximations to viscoelastic flow problems remains a priority and challenge for researchers in this field. The mathematical statement of viscoelastic flow problems is invariably in terms of velocity, pressure and extra-stress, the so-called mixed formulation. This class of flow problems is characterized by a nonlinear rheological equation of state which relates the extra-stress tensor to the rate of deformation tensor. For high values of the Deborah or Weissenberg numbers it is the discretization of this equation which is the cause of so many numerical difficulties due to the dominating influence of the convective terms.

An investigation of the type of the set of six partial differential equations describing the two-dimensional UCM model reveals that it is of mixed elliptic-hyperbolic type (Joseph et al. (1985)). There are four complex characteristics corresponding to two second-order elliptic operators. These are the Poisson equation for the pressure which ensures that the velocity field is incompressible and the nonlinear elliptic equation for the velocity which ensures that the divergence of the extra-stress tensor is irrotational, in the case of creeping flow. The remaining characteristics are real and correspond to two characteristic variables that are convected along the streamlines. In this paper these characteristic variables are determined.

The hyperbolic nature of the viscoelastic constitutive equations means that inflow conditions have to be prescribed, where appropriate, in order to obtain a well-posed problem. Although there are three convective terms in the constitutive equation there are only two characteristic variables which are transported along streamlines. A consistent numerical scheme should satisfy this property.

2. GOVERNING EQUATIONS

The set of partial differential equations defining the two-dimensional UCM model with a single relaxation time may be written as the following first order system

$$\boldsymbol{A}_0\frac{\partial\vec{\phi}}{\partial t}+\boldsymbol{A}_1\frac{\partial\vec{\phi}}{\partial x}+\boldsymbol{A}_2\frac{\partial\vec{\phi}}{\partial y}+\boldsymbol{S}\vec{\phi}=\vec{0}\,, \qquad (1)$$

where $\vec{\phi}=(p,u,v,\tau_{11},\tau_{22},\tau_{12})^T$. The matrices appearing in this system are defined in Gerritsma (1996).

If one inserts a single Fourier mode of the form $\vec{\phi}(\vec{x},t)=\vec{\phi}_0e^{\imath(\chi t+\vec{\xi}\cdot\vec{x})}$ one obtains

$$(\chi\boldsymbol{A}_0+\xi_1\boldsymbol{A}_1+\xi_2\boldsymbol{A}_2)\,\vec{\phi}=\imath\boldsymbol{S}\vec{\phi}\,.$$

A non-trivial solution exists if for every direction $\vec{\xi}$ a real valued χ exists such that

$$\det B(\chi,\vec{\xi})=0\,.$$

in which the matrix $B(\chi,\vec{\xi})$ is given by

$$B(\chi,\vec{\xi})=(\chi\boldsymbol{A}_0+\xi_1\boldsymbol{A}_1+\xi_2\boldsymbol{A}_2-\imath\boldsymbol{S})\,. \qquad (2)$$

Evaluating this determinant gives

$$\nu^2(\vec{\xi}^T\vec{\xi})(\vec{\xi}^T\boldsymbol{\tau}_A\vec{\xi})=0, \qquad (3)$$

where $\boldsymbol{\tau}_A=\lambda\boldsymbol{\tau}+\eta\boldsymbol{I}$.

3. CHARACTERISTIC VARIABLES

In order to find the characteristic variables one has to compute the left eigenvectors of (2) for $\nu=0$. The equations for these left eigenvectors are given by

$$\xi_1l_1+M_1l_4+M_3l_6=0\,, \qquad (4)$$

and

$$\xi_2 l_1 + M_2 l_5 + M_4 l_6 = 0 \ , \qquad (5)$$

where

$$\begin{aligned} M_1 &= -2\xi_1(\lambda\tau_{11}+\eta) - 2\xi_2\lambda\tau_{12}, \\ M_2 &= -2\xi_2(\lambda\tau_{22}+\eta) - 2\xi_1\lambda\tau_{12}, \\ M_3 &= -\xi_2(\lambda\tau_{22}+\eta) \\ M_4 &= -\xi_1(\lambda\tau_{11}+\eta). \end{aligned}$$

For both left eigenvectors we have $l_2 = l_3 = 0$. These left eigenvectors will in general be a function of x, the point in the computational domain where the characteristics are evaluated, the solution vector $\vec{\phi}$, because the coefficients M_i, $i = 1\ldots4$ depend on the solution, and the direction vector $\vec{\xi}$.

With the left eigenvectors $\vec{l}$ we can compute the characteristic variables, as follows

$$\delta W_i = \vec{l}^{(i)} \boldsymbol{A}_0 \delta\vec{\phi} \quad i = 1, 2 \ ,$$

in which δ denotes a variation in the variable, $\vec{\phi}$, $\boldsymbol{A}_0$ is the singular matrix given in (1) and $\vec{l}$ is the left eigenvector. After a little algebra we obtain the characteristic variables

$$W_1 = \det \boldsymbol{\tau}_A, \ W_2 = \vec{\xi}^T \boldsymbol{\tau}_A \vec{\xi}. \qquad (6)$$

4. NUMERICAL IMPLEMENTATION

Knowledge of the characteristic variables is of vital importance, not only to prescribe the proper boundary conditions in numerical calculations, but also in the discretization in the interior of the computational domain. In weighted residual methods, such as finite or spectral element methods, one has to match the solution between elements. One way of doing this is to assume that all primary variables are continuous between elements. Another way of matching the solution between elements is to treat every spectral or finite element as an independent domain in which the differential equation has to be satisfied. Unique solutions within an element only exist if each element is provided with the correct number of boundary conditions. Continuity of the velocity seems an appropriate choice as far as the elliptic part of the equations is concerned. For the hyperbolic part the characteristic variables need only be prescribed at inflow points. One possible way of prescribing inflow boundary conditions within a finite element or spectral element context is to use the method of Lesaint and Raviart (1974). This is accomplished by prescribing the extra-stress components weakly at the inflow boundary of an element. This approach guarantees that the characteristic variables are continuous.

The approach is demonstrated on the problem in which plane Poiseuille flow is perturbed. In Fig.1 we show the dependence of the maximum error between the exact staionary solution and the time-dependent approximation for both the continuous and discontinuous extra-stress approximations as a function of time. Two spectral elements are used with a time step of 5×10^{-3}.

5. REFERENCES

Gerritsma MI, (1996) PhD thesis, University of Groningen.

Joseph DD, Renardy M and Saut JC, (1985) Arch. Rat. Mech. Anal. 87: 213-251.

Lesaint P and Raviart PA, (1974) In: Mathematical Aspects of Finite Elements in Partial Differential Equations (de Boor C, Ed), 89-123.

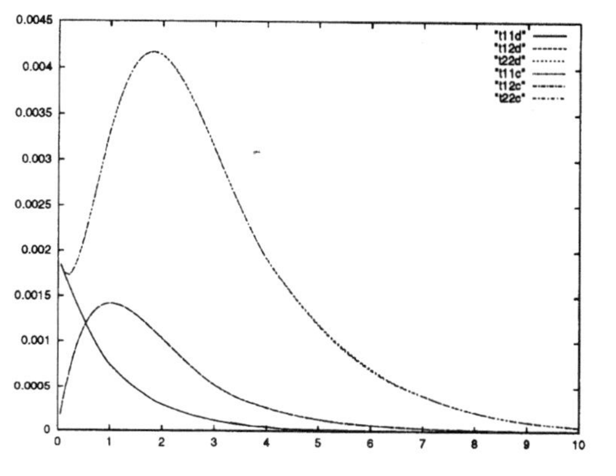

Figure 1: Behaviour of the extra-stress components in time for the continuous and discontinuous approaches

PREDICTION OF SHEAR-BANDING IN FLOWING MICELLAR SYSTEMS

F. GRECO

Chemical Engineering Dept., University of Naples "Federico II", P.le Tecchio 80, 80125 Naples, Italy

1. INTRODUCTION

In recent years, several experimental works on worm-like micelles have shown that a "dynamical phase separation" can arise in these systems when subjected to shear. At steady state, and beyond some critical value of the imposed shear rate, a banded structure can be found, with bands parallel to the velocity direction. Both birefringence measurements (Makhloufi et al., 1995) and direct NMR velocity imaging (Britton and Callaghan, 1997) indicate that bands are actually identified by sharply different values of the local shear rate. The material self-organizes by "splitting" the imposed, nominal shear rate in high-shear and low-shear bands. Details of this process are still largely unknown, however, and explanations and predictions from theory are scarce.

Here, we proceed to explore the consequences of attributing the banding phenomena to a constitutive instability. In fact, the microphysical model for worm-like micellar solutions devised by Cates (Cates, 1990) gives a rheological constitutive equation which is stable at low shear rates only. By adding to Cates' constitutive equation a stabilizing Newtonian contribution which becomes dominant at ultra-high shear rates, we allow (in principle, at least) for inhomogeneous flow fields in an "intermediate" range of shear rates.

It is easily understandable, however, that in order to obtain from the mathematics a dynamical phase separation one is forced to abandon the standard idealization of shear flow between parallel plates. Indeed, were this the case at hand, all points in the sheared slab would behave in the same way, and no separation could occur. What is further needed, then, is a "source" of symmetry breaking, which is readily identified (and always present) in the real geometry of any shearing apparatus (Greco and Ball, 1997).

Here, we examine the case of shear flow in a cone-and-plate geometry, where the shear rate is not strictly uniform, being slightly larger at the cone surface.

2. THE GENERAL EQUATIONS

In isothermal conditions, the motion of any fluid is governed by the equations of mass and momentum balance; for the constant-density, zero-inertia case which is considered here, they are:

$$\nabla \cdot \mathbf{V} = 0 \tag{1}$$

$$\nabla \cdot \mathbf{T} = -\nabla p + \eta \nabla^2 \mathbf{V} + \nabla \cdot \mathbf{S} = \mathbf{0} \tag{2}$$

where **V** is the velocity field, and **T** the total stress tensor. In Eq.(2), the total stress tensor has been broken up into a pressure p, a Newtonian contribution with viscosity η (the stabilizing contribution mentioned in the Introduction), and an "extra" stress term **S** which is the micellar stress. Indicating by G the elastic modulus of the fluid, Cates' constitutive equation for **S** is:

$$\frac{\mathbf{S}}{G} = \int_{-\infty}^{t} B[v(t')] \exp\left[-\int_{t'}^{t} D[v(t'')]dt'' \right] \mathbf{Q}[\mathbf{E}_{t't}]dt' \tag{3}$$

$$v(t') = \frac{\mathbf{S}(t')}{G} : \mathbf{K}(t') \tag{4}$$

where **K** is the velocity gradient, and $\mathbf{E}_{t't}$ is the total deformation applied between t' and t. The explicit expressions for B, D, and tensor **Q** are given in the original reference (Cates, 1990).

A few comments on Eqs.(3) and (4) seem appropriate here. In a micellar system, the present stress consists of a sum of contributions from all earlier times t', when "segments" (i.e., new strands of the micellar physical network) were created at a rate B(t'). However, not all of the strands created at t' still survive at t, because of the death rate D(t''). Thus, a "fading memory" is at work. Notice also that both the B and D functions depend on the flow field through v in Eq.(4), and that the unknown tensor **S** is determined by Eqs.(3) and (4) only implicitly.

The calculations we are going to present

make use of Eqs.(3) and (4) locally, i.e., the quantities **K** and **E** (hence **S**) will be space-dependent. Therefore, because of the intertwining among Eqs.(1)-(4), we must solve for the stress tensor and the velocity at the same time.

3. CONE AND PLATE RESULTS

Let us consider the cone-and-plate geometry. We will assume that the velocity field is always in the form $\mathbf{V} = (0,0,V_\varphi(r,\theta,t))$ in spherical coordinates, with $V_\varphi = r\sin\theta\,\Omega(\theta,t)$. The incompressibility condition, Eq.(1), is then automatically satisfied. The $\varphi\theta$-component of Eq.(2) can be integrated, to give:

$$T_{\varphi\theta} = \eta\dot{\gamma} + S_{\varphi\theta} = \frac{C}{\sin^2\theta} \qquad (5)$$

where $\dot{\gamma} = \sin\theta\,\partial\Omega/\partial\theta$ is the local shear rate. (The quantity C in Eq.(5) is easily related to the torque which must be exerted to impose the chosen angular velocity to the cone.) We can now solve Eqs.(3)-(5) for the unknown fields V_φ and $S_{\varphi\theta}$.

To obtain the dynamical evolution of the system, the gap between the cone and plate is discretized in N "layers" (i.e., coaxial cones), and the resulting set of 2N equations is solved numerically. We present here the results for the case of a shear start-up, when the cone, initially at rest, suddenly starts rotating with a constant angular velocity. In Fig.(1), the histories of the shear rates $\dot{\gamma}_i(t)$ (i is the layer index) are shown, from the inception of motion up to steady state, for a choice of parameters in the equations such that a banded flow is in principle allowed.

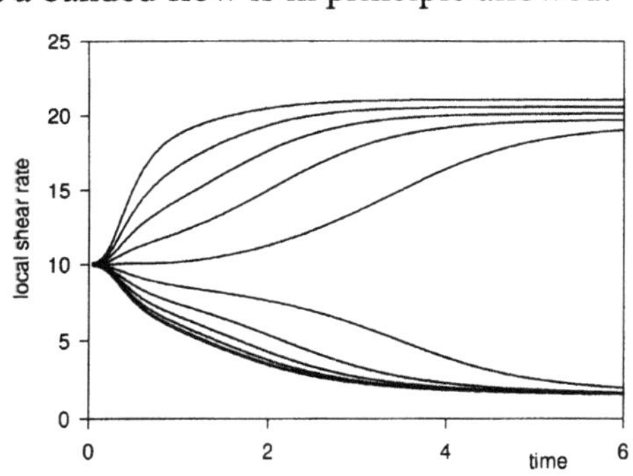

Fig. 1: The "history" of the phase separation

It is clearly seen that, at steady state, two bands at very different shear rates coexist, i.e., *a dynamical phase separation has occurred.* This behaviour persists over the whole unstable region, with the high-shear-rate phase progressively "invading" the gap. For ultra-fast flows, the "phase change" has completely occurred, and the material is again homogeneous.

Another relevant feature of the numerical results is that *the calculated total steady-state stress remains practically constant in the range of shear-banded flow*, in agreement with the available experimental evidence. The physical reason of this selection mechanism is unclear, however; maybe some minimization procedure should be considered (Greco and Ball, 1997).

4. FINAL REMARKS

There is a remarkable qualitative agreement between our simulations and experimental results, namely: i) a stratified flow pattern is found, ii) the "phases" are of a dynamical nature (they are not thermodynamic phases), and iii) there is a plateau in the stress vs. shear rate curve. Nevertheless, no quantitative predictions are possible at this stage. For example, in the transient of the shear start-up it is well known experimentally that huge stress overshoots appear when the final steady state is a banded one (Rehage and Hoffmann, 1991); on the other hand, overshoots calculated with our code are not so severe.

It should be stressed here that our results have been obtained by means of the *ad hoc* introduction of a Newtonian stabilizing viscosity, whose physical origin is left unspecified (Cates et al., 1993). It is the author's opinion that any further advancement in understanding the rheology of these complex micellar fluids will only be possible by deciphering the physics of this "high-shear-rate branch" of the stress.

5. REFERENCES

Britton MM, Callaghan PT (1997) Phys Rev Lett 78: 4930-4933

Cates ME (1990) J Phys Chem 94: 371-375

Cates ME, McLeish TCB, Marrucci G (1993) Europhys Lett 21: 451-456

Greco F, Ball RC (1997) J Non-Newt Fluid Mech 69: 195-206

Makhloufi R, Decruppe JP, Ait-Ali A, Cressely R (1995) Europhys Lett 32: 253-258

Rehage H, Hoffman H (1991) Mol Phys 74:933-938

MODELLING NON-LINEAR AND TIME-DEPENDENT BEHAVIOUR OF VISCOELASTIC MATERIALS USING HIERARCHICAL MODELS

NICOLE HEYMANS

Physique des Matériaux de Synthèse 194/8, Université Libre de Bruxelles, B-1050 Bruxelles, Belgium

1. INTRODUCTION

Hierarchical models have recently been introduced simultaneously by various authors under various formulations, in order to describe viscoelastic behaviour using a restricted number of material parameters. In terms of rheological models, hierarchy leads to the definition of viscoelastic *elements* (or β elements) whose dynamic modulus is

$$E_\beta{}^* = E(i\omega\tau)^\beta \qquad (1)$$

or whose constitutive equation is

$$= E\tau^\beta \frac{d^\beta}{dt^\beta}\varepsilon \qquad (2)$$

where $0 \le \beta \le 1$. For $\beta = 0$ or 1 the element displays purely elastic or purely viscous behaviour respectively.

Models incorporating these viscoelastic elements have been shown to describe linear viscoelastic behaviour extremely well in the dynamic range, and to allow quantitative description of non-linear behaviour such as stress activation, ageing and rejuvenation, in simple loading cases such as creep (under a constant load, equation (2) leads to power law creep) (Hellinckx et al 1994). They also describe multiple transitions successfully in the linear viscoelastic range (Heymans 1996), allowing e.g. different temperature or ageing dependencies of the parameters for each transition.

2. GENERALIZED MODELS

Viscoelastic β elements may be incorporated into classical models as shown on fig. 1. As pointed out by Beris and Edwards, there is necessarily a one-to-one correspondence between any thermodynamically acceptable equation for viscoelastic behaviour and a rheological model made up solely of elastic, viscous and viscoelastic elements. Based on this principle, numerical integration of the complete and somewhat involved constitutive equation may be replaced by integration of equation (2) for each β element; at each time step the new strain for each β element is obtained from compatibility of stresses and strains betweeen the different elements of the model.

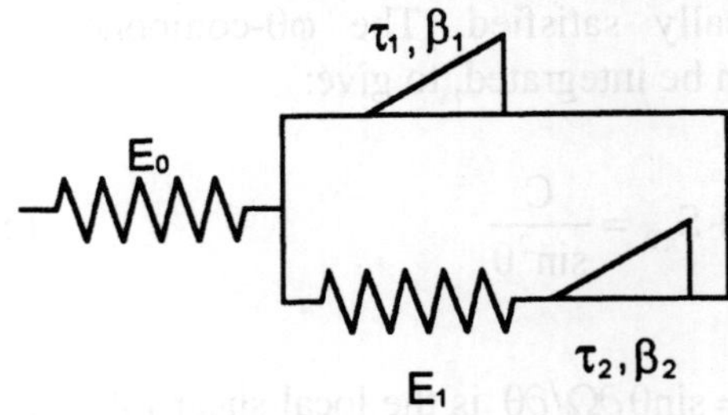

Fig. 1

3. NON-LINEAR BEHAVIOUR

Non-linear behaviour may be accounted for by expressing the stress or strain dependence of the characteristic time τ of the deformation process. For instance, stress activation may be expressed

$$= \tau_0 \frac{(\sigma-\sigma_{th})/\sigma_0}{Sh\left((\sigma-\sigma_{th})/\sigma_0\right)} \qquad (3)$$

where σ_0 is a stress activation parameter and σ_{th} is a threshold stress. Equation (3) is an expression of the Eyring model for stress-activated flow.

Another source of nonlinearity stems from the rubbery-elastic nature of the restraining spring E_1. This is expressed by

$$= \frac{E_1\lambda_M}{9}\left[\mathcal{L}^{-1}\left(\frac{\lambda}{\lambda_M}\right) - \frac{1}{\lambda^{3/2}}\mathcal{L}^{-1}\left(\frac{1}{\lambda_M\sqrt{\lambda}}\right)\right] \qquad (4)$$

For a first approach, no other source of non-linearity is accounted for.

4. VALIDATION FOR POLYETHYLENE

The response of the model is compared with experimental data on HDPE in creep and stress relaxation in the non-linear range. PE was chosen because its behaviour already becomes nonlinear at low strains, and because it displays double yield behaviour requiring incorporation of two β elements in the model; PE therefore allows a more stringent test of the model than materials with simpler viscoelastic behaviour.

4.1 Parameters

Starting parameters were obtained from mechanical spectroscopy at room temperature in the frequency range 10^{-4} to 1 Hz, giving $E_0 = 2000$ MPa, $E_1 = 600$ MPa, $\beta_1 = 0.32$, $\tau_1 = 5$ s. Stress dependence for the high-frequency process was determined from tensile tests, giving a strain-rate sensitivity parameter $\sigma_{01} = 1.38$ MPa. No threshold was taken into account for this process. For the long-term process, $\tau_2=10^7$s, $\beta_2=0.2$, a threshold of 1.5 MPa was introduced, and the strain-rate sensitivity parameter $\sigma_{02} = 0.3$ MPa.

4.2 Discussion

The description of non-linear behaviour of PE in creep and stress relaxation is shown on figs. 2 and 3. The model describes the general features of experimental behaviour extremely well.

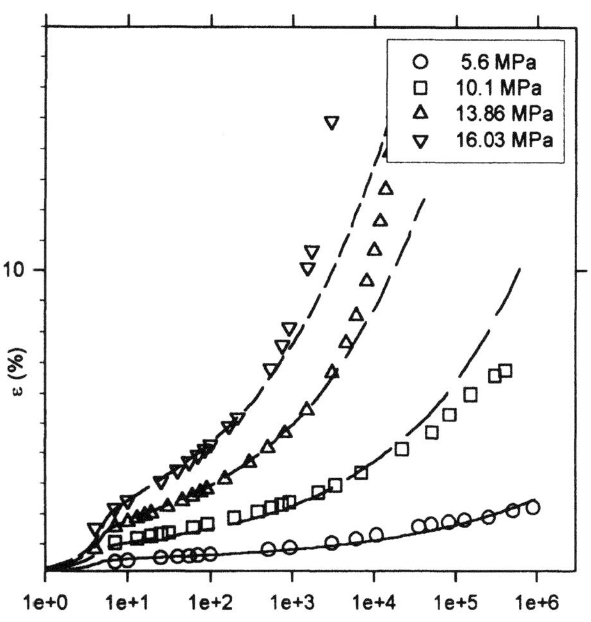

Fig.2

Because of the low strain-rate sensitivity parameter σ_{02}, the upturn of the model creep curves is relatively insensitive to the low-stress time constant for the long-term process, but is far more so to the threshold stress. The threshold stress is a general feature of yield behaviour of semi-crystalline polymers in the low-strain-rate range where tensile curves display a double yield. The construction of the model ensures that the long-term process cannot be activated unless spring E_1 sustains a sufficient stress. This can only occur after sufficient deformation of the fast process. Creep at high stresses is underestimated at long times because strain rejuvenation is not taken into account.

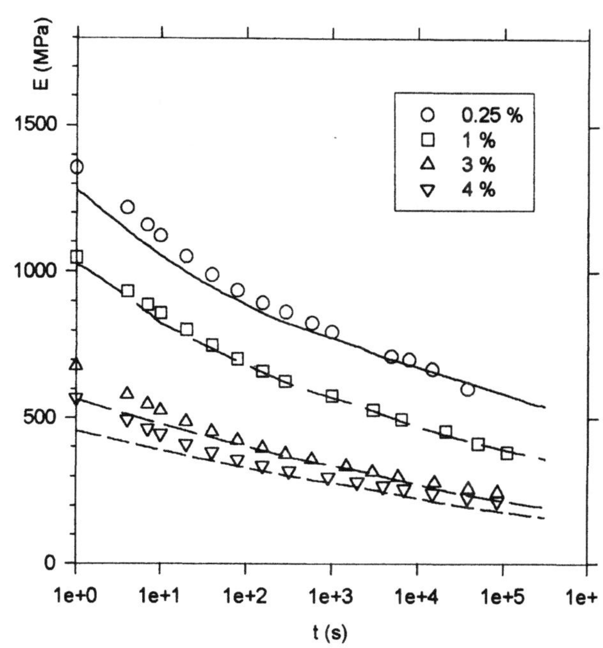

Fig.3

In stress relaxation, at high strains the fast process deforms rapidly at first, so that the stress acting on the slow process increases rapidly. This leads to deformation of the slow process; this deformation becomes irrecoverable when the stress relaxes below the threshold. Thus, the model describes correctly the observation that the relaxed modulus depends on strain, although the stress has relaxed to values which are well within the linear range.

5. REFERENCES

Beris A.N, Edwards B.J (1993) *Rheol Acta* 32:505-510

Heymans N. (1996) *Rheol Acta* 35:508-519

Hellinckx S., Heymans N., Bauwens J.C., *9th International Conference on Deformation, Yield and Fracture of Polymers*, Cambridge, UK, 11-14 April 1994

THE FENE-L AND FENE-LS CLOSURE APPROXIMATIONS FOR THE KINETIC THEORY OF FINITELY EXTENSIBLE DUMBBELLS

G. LIELENS, R. KEUNINGS and V. LEGAT

CESAME, Division of Applied Mechanics, Université catholique de Louvain,
B-1348 Louvain-la-Neuve, Belgium

1. INTRODUCTION

We address the closure problem for the non-linear kinetic model of a dilute polymeric solution known as the Warner Finitely Extensible Non-Linear Elastic (FENE) dumbbell model. This model cannot be translated into an equivalent constitutive equation for the polymer stress without invoking closure approximations which can have significant impact. For instance, the FENE-P macroscopic equation, based on the Peterlin closure approximation, has a dynamical behavior that is markedly different from the FENE model. The main qualitative difference occurs in start-up of elongation followed by relaxation, where the FENE theory exhibits hysteretic behavior for the stress versus average configuration, while the FENE-P does not (see Lielens et al. (1998) and Sizaire et al. (1998)). This hysteresis is observed experimentally by Doyle et al. (1998). The Peterlin approximation hence removes one of the FENE theory essential features. Our aim is to develop improved closure approximations capable of predicting hysteretic behavior and giving a rheological response closer to that of actual polymer solutions.

2. METHODOLOGY

The polymer stress due to FENE dumbbells cannot be computed exactly through constitutive equations, but must be derived, using Kramer's expression, from the probability distribution ψ of the dumbbell connector vector $\mathbf{Q}$:

$$\boldsymbol{\tau} = \int \mathbf{Q}\mathbf{F}^c \,\psi d\mathbf{Q} - \boldsymbol{\delta} = \langle \mathbf{Q}\mathbf{F}^c \rangle - \boldsymbol{\delta},$$

where $\mathbf{F}^c = \mathbf{Q}(1 - Q^2/b)^{-1}$ is the Warner dumbbell connector force , b is the finite extensibility parameter and $\boldsymbol{\delta}$ is the unit tensor (Lielens et al. (1998)). This approach, which involves solving a diffusion equation for the evolution of ψ, is too CPU-intensive for complex geometries. The associated Îto stochastic equation is used instead (Halin et al. (1998)). To further reduce complexity, one develops macroscopic constitutive equations based on scalar or tensorial averaged quantities derived from ψ. For example, the evolution of the second moment of ψ, $\mathbf{A} = \langle \mathbf{Q}\mathbf{Q} \rangle$, reads:

$$\frac{D\mathbf{A}}{Dt} = \boldsymbol{\kappa}\cdot\mathbf{A} + \mathbf{A}\cdot\boldsymbol{\kappa}^T + \boldsymbol{\delta} - \mathbf{A}^c,$$

$$\boldsymbol{\tau} = \mathbf{A}^c - \boldsymbol{\delta},$$

with $\mathbf{A}^c = \langle \mathbf{Q}\mathbf{F}^c \rangle$. To obtain a closed constitutive equation, $\mathbf{A}^c$ must be expressed as a function of $\mathbf{A}$. For a linear connector force, this easily leads to the Oldroyd-B model, but for the Warner non-linear force, closure approximations are needed. As in Lielens et al. (1998), we derive exact closures in *restricted* subspaces of canonical distribution functions. A first step is to use a decoupled distribution function for the length and orientation of the dumbbell:

$$\psi(\mathbf{Q}) = \psi_r(\|\mathbf{Q}\|)\,\psi_\phi\left(\|\mathbf{Q}\|^{-1}\mathbf{Q}\right).$$

In this subspace, $\mathbf{A}^c$ is proportional to $\mathbf{A}$, which makes the closure problem scalar. Decoupling is used for simplicity, even if it loses some features of the FENE theory like the transient second normal stress difference occurring in start-up of shear. In this framework, the Peterlin closure approximation $\mathbf{A}^c = \mathbf{A}(1 - \mathrm{Tr}\,\mathbf{A}/b)^{-1}$ is obtained assuming that all dumbbells have the same length, thus making ψ_r a Dirac distribution. Unfortunately, using only $\mathbf{A}$ as dumbbell state descriptor introduces a one-to-one relation between polymer stress and average polymer configuration which prevents any hysteretic behavior. As pointed out in Lielens et al. (1998) and Sizaire et al. (1998), the key element to reproduce hysteresis is to take into account not only the average dumbbell configuration, but also the dispersion around this average. With a decoupled distribution, this can easily be done by introducing the evolution equation for the scalar parameter $B = \langle (\mathbf{Q} \cdot \mathbf{Q})^2 \rangle$:

$$\frac{D\mathbf{A}}{Dt} = \boldsymbol{\kappa} \cdot \mathbf{A} + \mathbf{A} \cdot \boldsymbol{\kappa}^T + \boldsymbol{\delta} - \alpha \mathbf{A},$$

$$\frac{DB}{Dt} = 4 \frac{B}{\mathrm{Tr}(\mathbf{A})} \boldsymbol{\kappa} : \mathbf{A} + 10\ \mathrm{Tr}(\mathbf{A}) - \beta,$$

$$\boldsymbol{\tau} = \alpha \mathbf{A} - \boldsymbol{\delta},$$

where α and β are function of Tr $\mathbf{A}$ and B which represent the closure problem in the decoupled subspace. These functions are obtained by specifying a two-parameter subspace for the length distribution ψ_r. For example, the Dirac/tail L-shaped distribution used in the unidimensional FENE-L theory (Lielens et al. (1998)) leads to a 3D version of the FENE-L macroscopic equation. Alternatively, we have developed a simplified version of FENE-L, called FENE-LS, by selecting a two-Dirac length distribution of the form

$$\psi_r = (1 - \beta)\delta_{\alpha/K}(\|\mathbf{Q}\|) + \beta\delta_\alpha(\|\mathbf{Q}\|),$$

where δ_x is a Dirac distribution centered on x and K a suitable constant. This new closure is as accurate as FENE-L while being mathematically simpler, which makes it a good candidate for numerical simulations.

3. RESULTS AND CONCLUSIONS

We test in Figure 1 the FENE-L and FENE-LS constitutive equations in start-up of uniaxial elongation followed by relaxation. They reproduce the FENE hysteresis fairly well, and, as a consequence, the FENE stress growth is far better approximated than with the classical FENE-P closure. The stress plateau and relaxation are well reproduced by the three closures, because the plateau FENE distribution function is quite close to a Dirac and is well approximed both in FENE-P, FENE-L and FENE-LS canonical subspaces. The FENE-LS approximation is able to reproduce qualitatively the FENE hysteretic behavior, and is quantitatively closer to this model than FENE-P. Given that this performance is achieved using only one scalar parameter in addition to the FENE-P configuration tensor and that the algebraic closure of the FENE-LS is mathematically almost as simple as the FENE-P, this new closure seems to be a very good candidate for numerical simulation of complex flow where elongation is important.

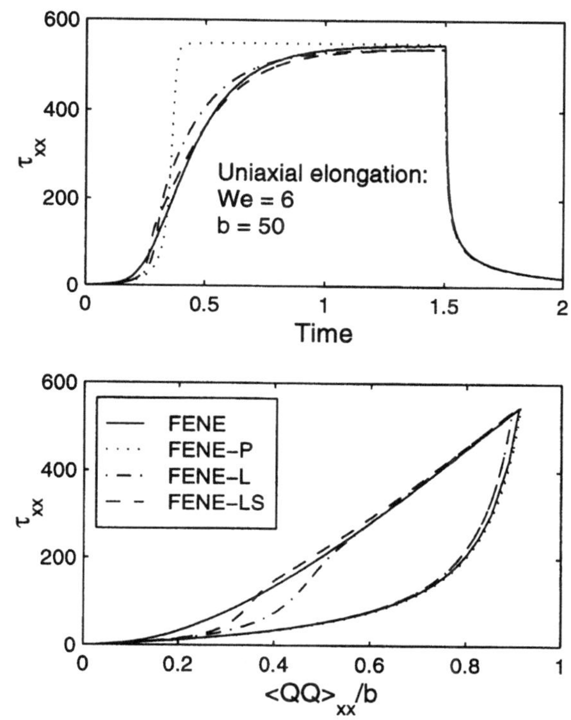

Figure 1: Start-up of uniaxial elongation in x-direction followed by relaxation.

4. ACKNOWLEDGMENTS

This work is supported by the ARC 97/02–210 project, Communauté Française de Belgique. The work of V. Legat is supported by the Belgian Fonds National de la Recherche Scientifique (FNRS).

5. REFERENCES

Doyle PS, Shaqfeh ESG, McKinley GH, Spiegelberg SH (1998) J. Non-Newt. Fluid Mech., 76:79-110.

Halin P, Lielens G, Keunings R, Legat V (1998) J. Non-Newt. Fluid Mech., In press.

Lielens G, Halin P, Jaumain J, Keunings R, Legat V (1998) J. Non-Newt. Fluid Mech., 76:249-279.

Sizaire R, Lielens G, Jaumain J, Keunings R, Legat V (1998) J. Non-Newt. Fluid Mech., Submitted.

EXPERIMENTAL-STATISTICAL MODELLING THE EFFECT OF MULTI-FRACTIONAL FILLER ON RHEOLOGICAL INDICES OF COMPOSITIONS

T. LYASHENKO, I. BARABASH, E. SHINKEVICH, S. SHCHERBINA, V. VOZNESENSKY

Odessa State Building and Architecture Academy, PO Box 76, Odessa 270000, Ukraine

1. INTRODUCTION

Materials science uses mathematical models of different nature and written in different languages. This is why some difficulties might arise when setting computational experiments with such models, in particular, with models describing the influence of mixture proportions on finished product and models of process hydrodynamics and mass exchange. In specific situations some contradictions may be taken off with the help of experimental-statistical models - ESM (Voznesensky et al., 1989, 1998). Structured ESM of the type «mixtures, technology - properties» are of special interest for rheological problems.

2. EXPERIMENT AND MODELLING

Specific features of choosing the solutions on the base of structured ESM are shown for the problem associated with preparing lime suspension (for building application) with multi-fractional fine filler (specific surface 400-500 m^2/kg) in high-speed mixer. Three mixture factors have been the proportions of fillers of different mineral composition: v_1 - proportion of granulated slag, v_2 -quartz, v_3 - limestone, $v_1 + v_2 + v_3 = 1$, water-solids ratio being constant, 1:2, with 1% of superplasticizer. Two independent factors have been the speed of rotor, $n = 40 \pm 20$ s^{-1} - conventionally normalized x_4, and time of mixing $\tau = 20, 60, 120, 300$ s, normalized as $x_5 = \ln\tau = 4.35 \pm 0.35$. Four first factors ($v_i$ and x_4) have been varied according to 16-point optimal design of experiment to be fulfilled at each of four time levels. Rheological curves have been taken from rotational viscosimeter for each of 64 suspensions.

By these data ESM for rheological indices have been built, in the form of incomplete cubic polynomials. Each model is separated into three blocks containing only significant coefficients. The model for effective viscosity η at rate of shear deformations 1 s^{-1} is analysed in the paper.

Block (a) describes the influence of mixture factors on $\ln\eta$ at the centre of "technological" square : $x_4 = x_5 = 0$, i.e., $n = 40$, $\tau = \exp(4.35) = 77$. This part of the model is

$$\ln\eta = 4.3v_1 + 0v_1v_2 + 3.5v_2 + 0v_1v_3 + 4.2v_3 - 3.2v_2v_3 + 0v_1v_2v_3 \quad \text{(a)}$$

$$+ 0v_1v_2x_4 + 0v_1v_2x_5 + 0v_1v_2x_4 + 0v_1v_2x_5 - 0.8v_1v_2x_4 - 0.7v_1v_2x_5 \quad \text{(c)}$$

$$-1.0v_1x_4 - 0.7v_1x_5 + 0.2v_1x_4^2 + 0v_1x_5^2 + 0.3v_1x_4x_5 - 0.8v_2x_4 - 0.8v_2x_5 + 0v_2x_4^2 + 0v_2x_5^2 + 0v_2x_4x_5 - 0.3v_3x_4 - 0v_3x_5 + 0v_3x_4^2 + 0.5v_3x_5^2 + 0.5v_3x_4x_5 \quad \text{(b)}$$

represented by mixture triangle in Fig. 1. This field of viscosity may be generalized by its relative drop $\delta = \eta_{max}/\eta_{min} = 74/20 = 3.7$. However, "anti-gradient" $\nabla_\delta = 3.8$, the ratio of δ to corresponding distance, is more informative as the estimate of system sensitivity to changes in mineral composition. The peculiarity of the system is non-additive decrease in viscosity of suspensions with about equal content of quartz and limestone, the fall being of the same scale as main factor effects, at the same quantity of liquid phase. Analogous plasticization was observed in systems "quartz + marble + clinker" (Voznesensky et al., 1989). One of possible reasons is polymodality of distribution of inter-granular layers thickness, with forming particles associations in non-Newton liquid (Voznesensky and Lyashenko, 1989).

Block (b) describes the variations in viscosity under effects of process factors without

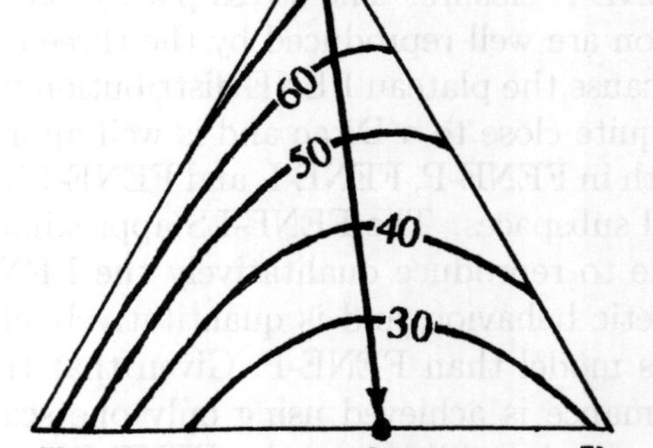

Fig. 1. Three-component diagram "mineralogy – viscosity" at n = 40 and τ = 65

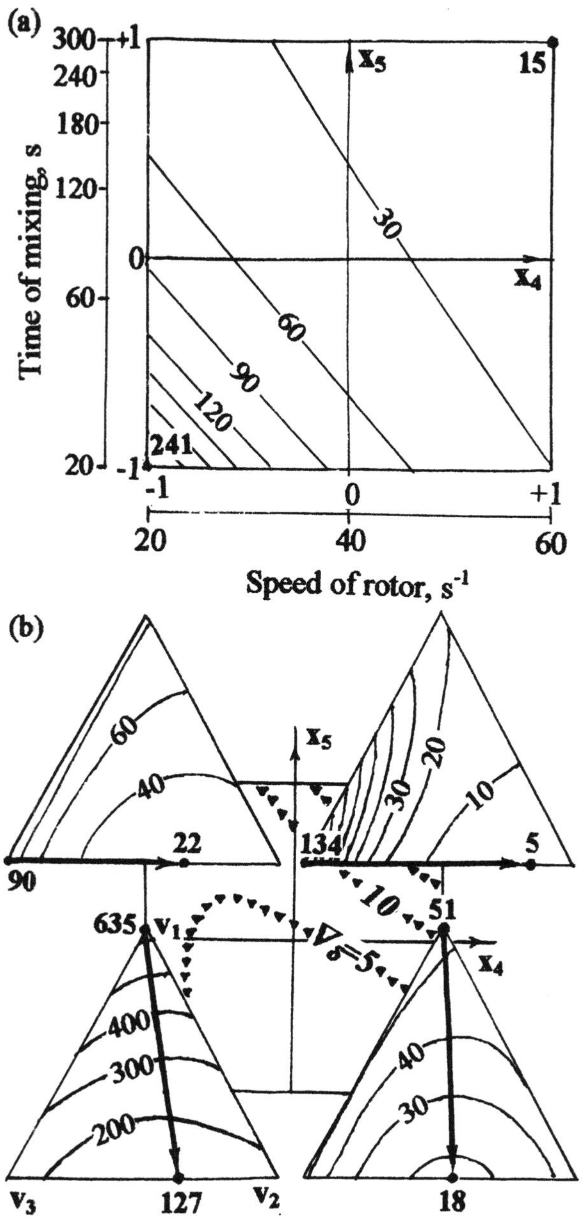

Fig. 2. Influence of n and τ on viscosity of the mixture "v_1=v_2=v_3=1/3" (a) and diagram "triangles on square" (b)

account of synergism in mixture system. Varying non-additivity is evaluated in block (c).

Shown in Fig. 2a is the influence of n and τ on viscosity of composition with equal portions of fillers. Special diagrams "triangles on square" or "squares on triangle" represent the blocks (a-c) as a whole. The diagram in Fig. 2.b is composed of mineralogical diagrams at four combinations of limiting levels of n and τ. The changes in mineralogical field of viscosity are also represented by its generalizing index on carrying square. Some specific features of mixing process may be noted, in particular: 1 – as τ lengthens η_{max} goes from compositions with slag predominating to those filled with limestone, 2 – this gives change in direction of anti-gradient and increase in system sensitivity, 3 – when raising n the effective viscosity of suspensions with limestone begins to increase intensively after some τ.

To evaluate the last peculiarity the model $\ln\eta(v_3,x_5)$, represented in Fig.3, has been got from the full model at v_1=0,v_2=1-v_3, x_4=1. This has allowed varying composition $v_{3.min}(x_5)$, providing maximal plasticization, and the boundary $x5(v3_{.min})$, above which system restructuring begins, to be determined.

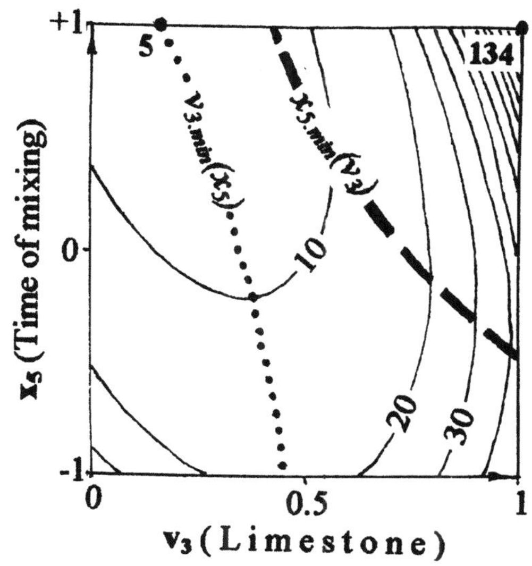

Fig. 3. Influence of mixing time on viscosity of mixture "quartz + limestone"

3. CONCLUSION

The experience in using structured experimental-statistical models when analysing and optimizing the compositions (silicate, polymer and others) shows high efficiency of these means of computational materials science.

4. REFRENCES

Voznesensky V, Lyashenko T (1998) Experimental-Statistical Modelling in Computational Materials Science. Astroprint, Odessa

Voznesensky V, Lyashenko T, Ivanov Ya, Nikolov I (1989) Computers and Optimization of Composite Materials. Budivelnik, Kiev (in Russian)

Voznesensky V, Lyashenko T (1989) In: Concrete Mixes Rheology and Its Technological Problems. Riga. pp 92-94 (in Russian)

A COMPARATIVE STUDY OF THE PERFORMANCE OF SEVERAL CONSTITUTIVE MODELS IN THE MODELLING OF FLUID S1

JOÃO M. MAIA

Department of Polymer Engineering, University of Minho, 4800 Guimarães, Portugal

1. INTRODUCTION

The S1 project is the last in a series of international co-operative programs and is devoted to the study of the Rheology of Polymer Systems, from the Solution to the Melt.

The so-called S1 fluid is a solution of 2.5% w/w of high molecular weight (~1.2x10[6]) Polyisobutylene in a mixed solvent of Polybutene oil and Dekalin. The rheology of a series of PolyIsoButylene solutions based on S1 was measured by a number of groups and the results have been reported (special issue, *J. non-Newtonian Fluid Mech.* (1994)).

In this work, the main concern will be to model theoretically S1 by means of a series of constitutive models and to compare their relative performance and ability to describe the observed behaviour in simple and complex flows.

2. THEORETICAL MODELS

2.1. Simple integral models

Two simple integral models were used in this study: the Wagner model (Wagner (1979)) and the Papanastasiou-Scriven-Macosko -or PSM for short- model (Papanastasiou et al. (1983)), the constitutive equations being of the following type:

$$\underline{\underline{\sigma}} = \int_{-\infty}^{t} m(t-t')h(I_B, II_B)\underline{\underline{B}}dt'. \qquad (1)$$

In this equation, $m(t-t')$ is a memory kernel which is determined by means of a discrete relaxation spectrum, $\underline{\underline{B}}$ is the Finger tensor and $h(I_B, II_B)$ is the so-called damping function:

$$\begin{cases} h(I) = e^{-a\sqrt{I-3}} & (Wagner) \\ h(I) = \dfrac{1}{1+c(I-3)} & (PSM) \end{cases} \qquad (2)$$

where $I = bI_B + (1-b)II_B$. The parameters a and c control the shear behaviour whereas b controls extension.

2.2. Differential models

The used differential models were the Giesekus (Giesekus (1966,1982)) and the Phan-Tien-Tanner -or PTT for short- (Phan-Tien and Tanner (1977)). The constitutive equation of the Giesekus model is $\underline{\underline{\sigma}} = \underline{\underline{\sigma}}_S + \underline{\underline{\sigma}}_P$, where:

$$\begin{cases} \underline{\underline{\sigma}}_s = \eta_s 2\underline{\underline{D}} \\ \underline{\underline{\sigma}}_p + \lambda \overset{\nabla}{\underline{\underline{\sigma}}}_p - \mu \dfrac{\lambda}{\eta_p}\left\{\underline{\underline{\sigma}}_p . \underline{\underline{\sigma}}_p\right\} = -\eta_p 2\underline{\underline{D}} \end{cases} \qquad (3)$$

In equation (3) μ is a parameter related to Brownian motion and anisotropic drag, λ is the relaxation time and $2\underline{\underline{D}}$ is the rate-of-deformation tensor.

For the PTT model it can be written as

$$\exp\left(\frac{\upsilon}{G} tr(\underline{\underline{\sigma}})\right)\underline{\underline{\sigma}} + \lambda\left[\left(1-\frac{\xi}{2}\right)\overset{\nabla}{\underline{\underline{\sigma}}} + \frac{\xi}{2}\overset{\Delta}{\underline{\underline{\sigma}}}\right] = \eta 2\underline{\underline{D}} \qquad (4)$$

where υ is related to the rate of creation of junctions in the transient network and ξ controls the amount of slip between network junctions and the continuum, varying in the following way:

$$\xi(II_{\underline{\underline{2D}}}) = \frac{\sqrt{II_{\underline{\underline{2D}}}}+1}{\sqrt{II_{\underline{\underline{2D}}}}} \qquad (5)$$

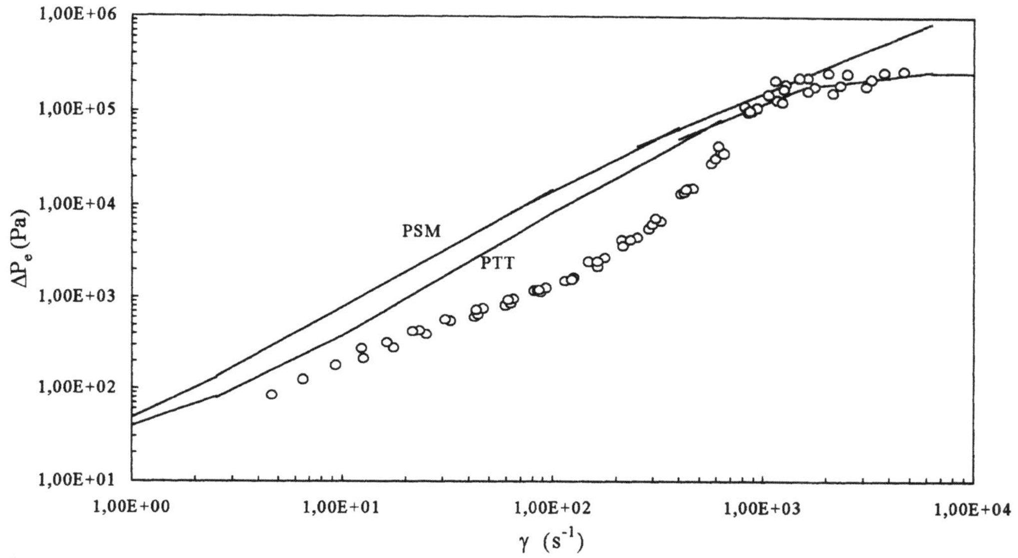

Figure 1 – Entry pressure drop predictions vs. Experimental data (open symbols)

3. RESULTS

The most important function to model is the shear viscosity. In fact, if a model is unable to correctly predict the behaviour of this function, it is almost pointless to pursue the matter further. Next, it was considered that the model's ability to predict the correct level of elasticity, *i.e.*, N_1 (and to a smaller degree, N_2), is also extremely important, as is its ability to predict the correct behaviour of the uniaxial extensional viscosity.

Overall, the PTT yields the best results whereas the integral-type models (and, to a lesser extent the Giesekus model also) under-predict fluid elasticity while the Giesekus and Wagner models do so regarding uniaxial extensional viscosity.

The uniaxial extensional viscosities yielded by the PTT and PSM models were then used, in conjunction with a modification of the Binding analysis for contraction flows (Maia and Binding, 1997), in order to predict the entry pressure drop in abrupt axisymmetric contraction flows. The contraction ratios ranged from 24.4:1 to 124.3:1.

The results (see Figure 1) indicate that the analysis is able to accurately predict the behaviour of the fluid in both the low and high flow rate regimes (which correspond to slow flow and full molecule alignment, respectively). In the intermediate one, which corresponds to molecule unravelling, an overestimation is observed, which is due to deficiencies in the Binding analysis and not the models.

4. CONCLUSIONS

Four different models were used to model the behaviour of the S1 fluid in simple and complex flows. Of these, two were of the integral type (Wagner and PSM) and two of the differential type (Giesekus and (PTT)). The PTT model provides the best description of the behaviour of the fluid in simple flows, when the parameter ξ is allowed to vary with the second invariant of the Finger tensor.

In complex flows (abrupt axysimmetric contraction flows, with high contraction ratios), the PTT model in conjunction with a modification of the Binding analysis is able to capture quantitatively the experimentally observed behaviour for two of the three flow regions and qualitatively in the third.

5. REFERENCES

Special issue, *J. non-Newtonian Fluid Mech.*, 52(1994)

A.C. Papanastasiou, L.E. Scriven and C.W. Macosko, *J. Rheol.*, 27(1983), 387

M.H. Wagner, *Rheol. Acta*, 18(1979), 279

H. Giesekus, *Rheol. Acta*, 5(1966), 29

H. Giesekus, *J. non-Newtonian Fluid Mech.*, 11(1982), 69

N. Phan-Tien and R.I. Tanner, *J. non-Newtonian Fluid Mech.*, 2(1977), 353

J.M. Maia and D.M. Binding, 1997. Accepted for publication in *Rheol. Acta*

Shear viscosity of a molten short fibers composite

Fréderic MESLIN, Arnaud POITOU,
ENS de Cachan, LMT, 94 235 Cachan FRANCE,

1 Introduction

At a molten state, short fibers composites are generally modeled as non Brownian anisotropic viscous fluids [3, 4, 8, 7]. A micromechanical approach enables in simple cases to evaluate the rheological parameters as a function of the fiber's volumic rate and shape factor. However even in these cases where the fibers are in a semi-dilute or in a dilute state, a simple test as a steady shear viscosimeter test is in fact subject nowadays to different interpretations. In particular there is a sort of paradox if we assume for example that the system is described with a slender bodies approximation [8, 4], if one neglects the Brownian motion, a solution for which the fibers are locally aligned along the velocity field is theoretically possible. It can be shown in this case, that the shear viscosity would be less than that of a spherical particles suspension with the same volumic rate of fibers. This result is well known and is usually analyzed in noting that a slight desorientation of the fibers can give rise to a great increase in the shear viscosity. For this reason some authors [6] propose to model the rotation of the fibers with a Brownian diffusion constant which vanishes at zero shear rates and which intends to give a phenomenological modeling of the fibers-fibers interaction. This modeling has shown its efficiency but requires to measure this diffusion constant C_I with respect to the volumic rate and to the fibers aspect ratio. In this presentation we show that this misalignment can be also modeled in considering a large but finite shape factor of the particle. In this case, since early papers of Mason or Hinch, it is known that if their is no interactions between the particles, their is generally no steady solution in shear flows. However in assuming that the time averaging along a trajectory is equal to the local ensemble averaging, we calculate analytically the misalignment and thus the shear viscosity. If we compare this expression to the one computed with the interaction coefficient C_I, we find a way to evaluate C_I.

2 Constitutive relation

The constitutive relation is of the form:

$$\begin{aligned}\underline{\underline{\Sigma}} &= -P\underline{\underline{I_d}} + 2\eta_1\{\underline{\underline{d}}(\mathbf{v}) \\ &+ N_s\left[\underline{\underline{a}}\,\underline{\underline{d}}(\mathbf{v}) + \underline{\underline{d}}(\mathbf{v})\,\underline{\underline{a}}\right] + N_p \underline{\underline{\boldsymbol{a}}}\,\underline{\underline{d}}(\mathbf{v})\}\end{aligned} \tag{1}$$

$$\begin{aligned}\frac{d\underline{\underline{a}}}{dt} &= \underline{\underline{\Omega}}(\mathbf{v})\,\underline{\underline{a}} - \underline{\underline{a}}\,\underline{\underline{\Omega}}(\mathbf{v}) \\ &+ r(\underline{\underline{d}}(\mathbf{v})\,\underline{\underline{a}} + \underline{\underline{a}}\,\underline{\underline{d}}(\mathbf{v})) - 2r\,\underline{\underline{\boldsymbol{a}}}\,\underline{\underline{d}}(\mathbf{v})\end{aligned} \tag{2}$$

$$r = \frac{\lambda^2 - 1}{\lambda^2 + 1} \qquad \underline{\underline{\boldsymbol{a}}} = \underline{\underline{a}}\,\underline{\underline{a}} \qquad \mathrm{div}(\mathbf{v}) = 0 \tag{3}$$

Equation (1) means that the extra stress tensor is a linear anisotropic function of the strain rate tensor $\underline{\underline{d}}(\mathbf{v})$. In this expression, η_1, N_s, N_p are three rheological parameter which depend on the concentration φ and on the shape factor λ of the particles. Analytically, it is shown that N_s is negligible before N_p for realistic values of φ and λ. The tensors $\underline{\underline{a}}$ and $\underline{\underline{\boldsymbol{a}}}$ describe the fibers orientation and are defined in a statistical way as [1]:

$$\underline{\underline{a}} = \int_{\mathbf{p}} \mathbf{p}\mathbf{p}^{\mathbf{T}} \psi(\mathbf{p}) d\mathbf{p} \tag{4}$$

$$\underline{\underline{\boldsymbol{a}}} = \int_{\mathbf{p}} (\mathbf{p}\mathbf{p}^{\mathbf{T}}) \otimes (\mathbf{p}\mathbf{p}^{\mathbf{T}}) \psi(\mathbf{p}) d\mathbf{p} \tag{5}$$

In these expressions, $\mathbf{p}$ is a unit vector aligned along the principal axis of a fiber and ψ is the orientation distribution function. Equation (2) is the evolution equation of $\underline{\underline{a}}$, which depends on the fourth order tensor $\underline{\underline{\boldsymbol{a}}}$. For a complete description of the behavior at this level, it is then needed to use a closure approximation which can be quadratic as written in [2] or even exact as written in [5]. In simple shearing flows, the viscosity is given by

$$\eta_{shear} = \frac{\sigma_{12}}{\dot{\gamma}} = \eta_1\left(1 + N_s + 2N_p < p_1^2 p_2^2 >\right) \tag{6}$$

This indicates that to calculate the viscosity it is only needed to calculate $< p_1^2 p_2^2 >$ where the brackets $< . >$ holds for ensemble averaging and p_1 and p_2 are the first and second component of $\mathbf{p}$.

3 Calculation of $< p_1^2 p_2^2 >$

In the case of a simple shearing motion (Figure 1) from the Jeffery's solution one can write:

$$\mathbf{p} = \frac{\underline{\underline{E}}\mathbf{p_0}}{||\underline{\underline{E}}\mathbf{p_0}||} \tag{7}$$

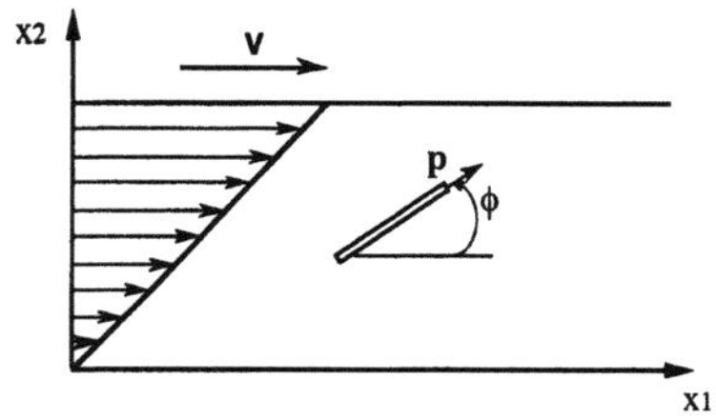

Figure 1: *Simple shearing flow*

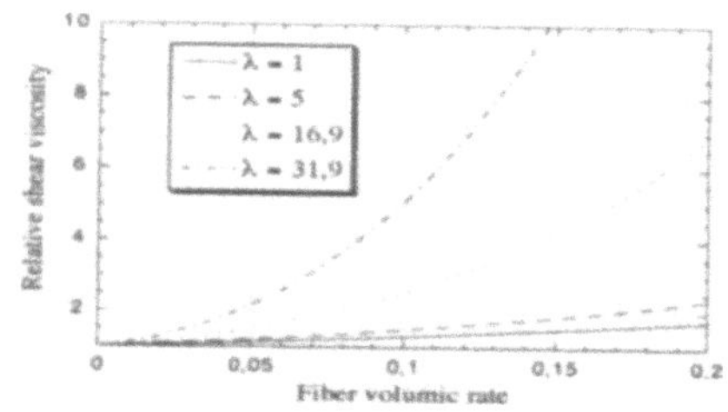

Figure 2: *Shear viscosity from equation (10).*

where $\mathbf{p_0} = x_1$ holds for the initial orientation of the particle at time $t = 0$ and $\underline{\underline{E}}$ is the following matrix:

$$\begin{pmatrix} \cos\left(\sqrt{1-r^2}\frac{\dot\gamma t}{2}\right) & \frac{\sqrt{1-r^2}}{1-r}\sin\left(\sqrt{1-r^2}\frac{\dot\gamma t}{2}\right) & 0 \\ -\frac{\sqrt{1-r^2}}{1+r}\sin\left(\sqrt{1-r^2}\frac{\dot\gamma t}{2}\right) & \cos\left(\sqrt{1-r^2}\frac{\dot\gamma t}{2}\right) & 0 \\ 0 & 0 & 1 \end{pmatrix}$$

We calculate $< p_1^2 p_2^2 >$ with a time averaging approximation instead of the ensemble averaging one. From equation (7), one finds for periodic orbits of period T:

$$< p_1^2 p_2^2 >= \frac{1}{T}\int_0^T p_1^2 p_2^2 dt \tag{8}$$

From this expression it is easy to obtain an analytical expression of both $< p_1^2 p_2^2 >$ and η_{shear} (figure 2).

$$< p_1^2 p_2^2 >= \frac{r^2 - 1 + \sqrt{1-r^2}}{4r^2} \tag{9}$$

$$\eta_{shear} = \eta_1 \left(1 + N_s + \frac{r^2 - 1 + \sqrt{1-r^2}}{2r^2} N_p\right) \tag{10}$$

4 Interaction coefficient

In this model the orientation tensor is governed by:

$$\frac{d\underline{\underline{a}}}{dt} = \underline{\underline{\nabla \mathbf{v}}}\,\underline{\underline{a}} + \underline{\underline{a}}\,\underline{\underline{\nabla \mathbf{v}}}^T - 2\underline{\underline{\mathbf{a}}}\,\underline{\underline{d}} + 2C_I\dot{\bar\gamma}(\underline{\underline{I_d}} - 3\underline{\underline{a}}) \tag{11}$$

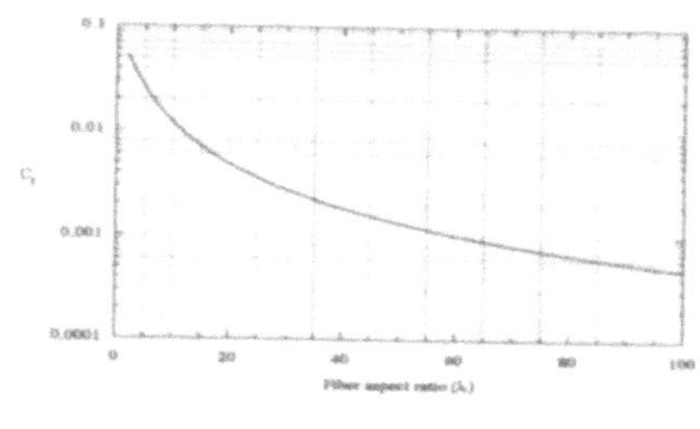

Figure 3: *C_I from equation (13)*

From which we get the shear viscosity

$$\eta_{shear} = \eta_1 \left(1 + N_s + \left\{\frac{C_I(6C_I^2 + \Delta - 3\Delta^2)}{-2C_I^2 + \Delta^2}\right\}^2 N_p\right)$$

or at a first order:

$$\eta_{shear} = \eta_1 \left(1 + N_s + (2C_I^2)^{1/3} N_p\right) \tag{12}$$

From équations 12 and 10, we obtain, at a first order, the expression of C_I as a function of the shape factor of the particle (Figure 3):

$$C_I = \frac{1}{2}\left(\frac{r^2 - 1 + \sqrt{1-r^2}}{2r^2}\right)^{\frac{3}{2}} \tag{13}$$

5 Conclusion

Equation (13) gives the dependance of C_I with respect to the shape factor of the particule for dilute suspensions. If interactions are to be accounted for we have shown that equation (2 is still valid with a coefficient which depends on the volumic rate of fibers. The interaction coefficient is seen here as a consequence of the little time scale associated with the rotation of the particule.

References

[1] S.G. Advani and C.L. Tucker. *J. Rheol.*, 31:751, 1987.

[2] S.G. Advani and C.L. Tucker. *J. Rheol.*, 34:367, 1990.

[3] G.K. Batchelor. *J. Fluid. Mech.*, 44:419, 1970.

[4] S.M. Dinh and R.C. Armstrong. *J. Rheol.*, 28:207, 1984.

[5] F. Dupret and al. ESAFOXM Conference Proceedings, 1997.

[6] F. Folgar and C.L. Tucker. *J. Reinf. Plast. Compos.*, 3:98, 1984.

[7] F. Meslin. PhD thesis, École Normale supérieure de Cachan, LMT Cachan, 1997.

[8] E. Shaqfeh and G. Fredrickson. *Phys. of Fluids*, A2:7, 1990.

CALCULATION OF THE RESIDUAL STRESS DEVELOPMENT IN INJECTION MOULDING USING A NON-LINEAR VISCOELASTIC MODEL

B. A. MLEKUSCH[1], E. A. LEHNER, B. J. SPIEGL

[1]Institut for Designing Plastics and Composite Materials, University Leoben, P.Tunner Str. 27, A-8700 Leoben

1. INTRODUCTION

The residual stress development in amorphous polymers is calculated for the rapid cooling under pressure during processing. This is done preliminary to predict the warpage and shrinkage behaviour of components. A non-linear viscoelastic model based on the free volume concept is used as constitutive equation to describe the polymer behaviour.

2. CONSTITUTIVE EQUATION

The temperature dependence using viscoelastic material laws is in general considered by the assumption of thermorheological simple behaviour and the validity of the Williams-Landel-Ferry (WLF)-equation. Due to Ferry (1980) the WLF-equation is strongly linked with the concept of the free volume. If a viscoelastic expression is used to describe the time dependent free-volume change caused by a temperature or pressure change (Knaus and Emri, 1997), two consequences can be concluded: On the one hand the volume expansion can be described time-dependent from the cooling history, leading to a viscoelastic formulation of the linear thermal expansion coefficients. Therefore in the derived non-linear constitutive equation the strain-stress relation as well as the thermal strain are described viscoelastically:

$$\varepsilon_{ij}(t) = \int_0^t C_{ijkl}(\xi(t) - \xi(u)) \frac{d\sigma_{kl}(u)}{du} du + \int_0^t \alpha_{ij}^{lin}(\xi(t) - \xi(u)) \frac{d\Delta T}{du} du$$

On the other, the time-temperature shift factor and therefore the material time ξ is also dependent on the cooling and pressure history, whereby an iterative solution procedure is needed to determine the reduced times.

3. MODELLING OF THE STRESS-DEVELOPEMENT

The rapid cooling under pressure is regarded as the main cause for the residual stress development, flow induced stresses are neglected. The cooling process in injection moulding can be divided into two stages (Rezayat, (1989): Until the ejection the part is clamped in the mould, while the remaining cooling down to room temperature happens without any constraining. Both steps are calculated viscoelastically in the simulation procedure (Muki, Sternberg, (1961), Lee, Rogers, (1967)).

4. NUMERICAL CALCULATIONES

4.1 Simulations for one material-point

At first stress profiles are determined in one material point of a component using the outlined simulation schema. These results show the possibilities of the non-linear constitutive equation and the influence of different process conditions, such as varying wall temperatures or pressure profiles. Large deviations can be seen, when the viscoelastic free volume concept is used instead of the WLF-equation. For the high cooling rates, which are achieved in the ejection moulding process, the prediction of the temperature-time shift using the WLF-equation seems to fail, the results are physically not reasonable.

4.2 Simulations of a component

To be able to take more complex geometries into account, the non-linear constitutive equation is used in the finite-element program ABAQUS. This is done by the programming of user-defined subroutines. The solution of the coupled thermal-mechanical problem allows the prediction of the residual stresses as well as the warpage and shrinkage behaviour for components with arbitrary geometry.

4.3 Comparison of the simulation results with measurements on components

The calculated warpage and shrinkage values are compared with the behaviour of experimental components, which were specially designed. They posses several geometric-components, most commonly found in real polymeric applications. The simulated behaviour agrees quite well with the behaviour, shown by the experimental components.

5. CONCLUSIONS

The given non-linear constitutive equation is capable of describing the polymer behaviour under processing conditions, which are characterised by high temperature and pressure gradients with respect to time. The equation is based on the free-volume concept and - although this concept is often very disputed - seems to give reasonable results for predicting the residual stresses in the injection moulding process.

6. REFERENCES

Ferry J., (1980), "Viscoelastic Properties of Polymers", John Wiley.

Knaus W.G., Emri I., (1987), "Volume Change and the Nonlinearly Thermo-Viscoelstic Constitution of Polymers", Polymer Engineering and Science, Vol. 27.

Lee E.H., Rogers T.G., (1967), "Solution of Viscoelstic Stress Analysis Problems Using Measured Creep or Relaxation Functions", J.App.Mech.,3.

Muki R., Sternberg E., (1961), "On Transient Thermal Stresses in Viscoelastic Materials with Temperature-Dependent Properties" J.App. Mech., 6.

Rezayat M., (1989), "Numerical Computation of Cooling-Induced Residual Stress and Deformed Shape for Injection-Molded Thermoplastics", Antec 89.

IDENTIFICATION OF VISCOELASTIC RHEOLOGICAL PARAMETERS BY INVERSE ANALYSIS

O. H. NOUATIN, T. COUPEZ, J.-F. AGASSANT
Ecole des Mines de Paris, CEMEF, UMR CNRS 7635, BP:207, 06904 Sophia Antipolis Cedex, France

1 Introduction

This paper deals with a numerical procedure concerning rheological parameters identification of viscoelastic constitutive models by using simulation codes within a loop of optimization. The generic model is a finite element simulation of viscoelastic fluid flows with differential constitutive equations. The inverse problem is solved by minimizing a *cost function* which measures the discrepency between experiment and simulation response. A simplex method is introduced which enables to use any simulation code without any modification.

2 Minimization of a cost function

Different types of experiments data can be expected: velocity, stress field pressure, die swell,... So there are points data (pressure), line-data (velocity, birefringence on the symmetric axis) and field-data (birefringence on the full flow domain). The form of J depends on the experiment data type. Here, we just present the field data case: the cost function J is a L^2 norm.

Let Ω^{exp} denotes the experiment subdomain of the flow domain Ω^{sim}, J is defined by:

$$J(P) = \int_{\Omega} (\sigma^{exp} - \sigma^{sim})^2 d\omega / \int_{\Omega} (\sigma^{exp})^2 d\omega \quad (1)$$

where $\Omega = \Omega^{exp} \bigcap \Omega^{sim}$.

• If we consider that different experiments are carried out simultaneously, the cost function will be a linear combination of the above one. In that case, weights could be used to insist in any particular experiment.

The calculation of (1) can be done exactly by intersecting the mesh of simulation and the mesh of experimental data field if the latter exists. A third mesh is then built which enables the calculation.

The *simplex method*, [Nelder, 65], used in this paper only needs the evaluation of the cost function by direct simulation for different estimated parameters. At each step, the set of parameters giving the highest value of the cost function is eliminated and replaced by a set giving a lower value by using the specific operations: *reflexion, expansion, contraction.* The procedure is repeated until the optimal parameters are reached.

3 Viscoelastic computation

Let's consider the multi-mode Phan-Thien-Tanner (PTT) following system:

$$\left\{ \begin{array}{l} div(-p\underline{Id} + 2\eta s \underline{D} + \sum_{i=1}^{N} \underline{T}_V^i) = 0; \\ \exp(\dfrac{\epsilon\lambda^i}{\eta_V^i} Trace(\underline{T}_V^i))\underline{T}_V^i + \lambda^i \dfrac{\mathbb{D}_a \underline{T}_V^i}{\mathbb{D}t} = 2\eta_V^i \underline{D} \\ \dfrac{\mathbb{D}_a \underline{T_V}}{\mathbb{D}t} = \left(1 - \dfrac{a}{2}\right) \overset{\triangledown}{\underline{T_V}} + \dfrac{a}{2} \overset{\triangle}{\underline{T_V}} \\ div\, u = 0 \quad ; \quad \eta = \eta s + \sum_{i=1}^{N} \eta_V^i \end{array} \right.$$

$$\text{where } \overset{\triangledown}{\underline{T_V}} = \frac{d\underline{T_V}}{dt} - \underline{\nabla u T_V} - \underline{T_V \nabla u}^t,$$

$$\overset{\triangle}{\underline{T_V}} = \frac{d\underline{T_V}}{dt} + \underline{T_V \nabla u} + \underline{\nabla u^t T_V},$$

$$0 \le a \le 2 \text{ and } \epsilon_{min} \le \epsilon \le \epsilon_{max}$$

If we consider $a = 0$, $\epsilon = 0$ in the PTT model, one obtains the Oldroyd-B model.

The viscoelastic computation are performed by using the $P2^+/P1$ (Crouzeix-Raviart element) for the velocity and the pressure fields, and a $P2$ discontinous interpolation of the extra-stress tensor. The numerical solution is performed by using a Lesaint-Raviart technique associated with a quasi-Newton method [Beraudo, 98].

4 Identification results

The calculations are carried out in a 4:1 contraction die, and with a one-mode PTT system.

4.1 Identification of an Oldroyd-B model

The principal stress difference for an Oldroyd-B fluid have been computed with a coarse mesh considered as the experiment one. Then, assuming that $\epsilon = 0$, the aim of the optimization is to find out the parameter $a = 0$ corresponding to the computed field stress.

Starting from the simplex composed of $a = 0.75$ and $a = 0.5$ we can see on figure 1 that the correct value $a = 0$ is reached after 5 iterations. The evolution of the highest and lowest values of the cost function is shown on figure 2. Using a coarser mesh for the experiment, the value of the cost function cannot be zero at the minimun, see figure (2). Nevertheless, and that point is crucial, the identified parameter seems to be idependent of the mesh.

4.2 Identification of a Phan-Thien-Tanner model

In results of table (1), the same experiment and simulation mesh has been used. The PTT model we want to identify is made of $a = 0.5$ and $\epsilon = 0.2$. About 15 simulations have been necessary to find out $(a, \epsilon) = (0.503, 0.19)$. The table (1) shows the initital simplex and the final simplex which determines the identification. The *Point i*, $i = 1, 2, 3$, denote the vertices of a simplex (here a simplex is a triangle because there are two parameters to identify).

5 Conclusion

The simplex method has been implemented to solve a parameters identification of viscoelastic law in complex flow situation. Different types of experiment can be used. The procedure has been experienced with Oldroyd-B and Phan-Thien-Tanner models showing very encouraging results on a good behaviour in terms of convergence. They show that we can characterize a polymer by using complex viscoelastic numerical codes within a loop of optimization. The future prospects consist in identifying parameters with actual experiment data.

Acknowledgements: This work is supported by EEC within the ART Brite Euram project number BE96 3490.

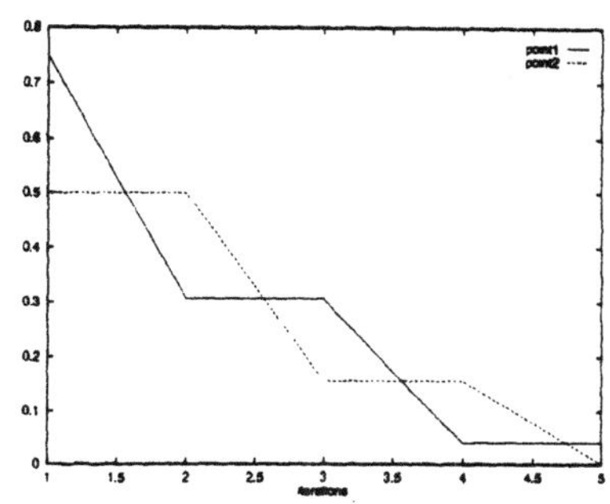

FIG. 1 – *Evolution of a*

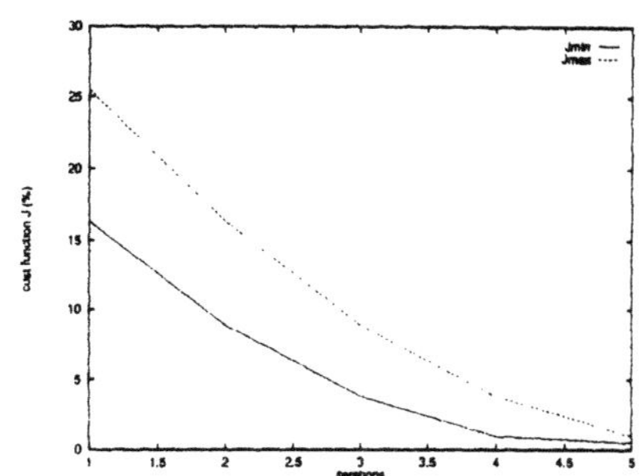

FIG. 2 – *Evolution of Jmax and Jmin*

	Initial simplex		Final simplex	
	a	ϵ	a	ϵ
Point 1	0.35	0.4	0.4937	0.2048
Point 2	0.525	0.4	0.508	0.1999
Point 3	0.35	0.6	0.503	0.1899

TAB. 1 – *The initial and the final simplexes.*

References

[Beraudo, 98] C. BERAUDO, A. FORTIN, Y. DEMAY, B. VERGNES and J.-F. AGASSANT, *A finite element method for computing the flow of multi-mode viscoelastic fluids: comparison with experiments*, J.Non -Newt. Fluid Mech, 75(1998) 1-23.

[Nelder, 65] J. NELDER and R.A. MEAD, *A simplex method for function minimization*, The Computer Journal, 7, 1965.

ENERGY STORAGE AS THE CHARACTERISTIC FEATURE OF INELASTIC DEFORMATION OF SOLID POLYMERS

E. OLEINIK[1], S. RUDNEV[1], O.SALAMATINA[1], M.KRAVCHENKO[1], T. PARAMZINA[1], S. SHENOGIN[1], P. SHANTOROVICH[1], G. HOEHNE[2]

[1]Institute of Chemical Physics Russian Academy of Sciences, Kosygin Str.4, Moscow 117977, Russia-
[2]Section Kalorimetrie, Universitat Ulm, D-89069 Ulm, Germany

1. INTRODUCTION

Deformation behaviour of solids must be characterised not only by conventional mechanical parameters, but also by thermodynamic quantities such as work (W), expended for deformation, heat of deformation (Q), and sample internal energy changes (ΔU). Absolute values and variations of these parameters are connected with the structural alterations in a solid under stress and with physical processes relating to deformation mechanisms operating in them. The aforementioned thermodynamic quantities are well documented for crystalline metals [Bever et al (1973)], rubbers [Godovsky (1993)] and glasses [Oleinik et al (1993), Godovsky (1993)]. However, such data is limited for polymeric crystals, semi-crystalline and LC polymers. The main objective of this paper is to review data of W, Q and ΔU for polymer solids of different structural organization, such as the glassy, LC, crystalline, semi-crystalline and blends. We will than compare the values and their changes with strain for different polymers in order to analyse them in a framework of deformation mechanisms operating in polymers of different structures.

2. METHODS

Measurements of W, Q and ΔU were performed by the isothermal deformation calorimetry method. Samples were deformed at room temperature in a uniaxial compression mode with strain rate (monotonic loading and unloading) $d\varepsilon/dt = 10^{-1}$-$10^{-2}$ min^{-1}. Work was performed working only in the anelastic and plastic strain ranges, between ε=5-80%, Since there is literature for Hookian response [Godovsky (1993)] it was not addressed in this paper.

In addition, we measured the thermally stimulated recovery of residual strain after the complete unloading of sample. DSC measurements were performed for the characterisation of the initial and deformed samples crysialinity, for processes of the stored energy ΔU emmission. Positronium annihilation life time spectroscopy (PALS) was used for the characterisation of free volume changes upon deformation and annealing.

3. RESULTS, DISCUSSIOIN

3.1 Glassy polymers

The study of crystalline metals [Bever et al (1973)] have shown that the main fraction of W is mainly dissipated into deformation heat Q. The ratio of Q/W is 90-95%. The remaining 5% of W is converted to internal energy increase ΔU of metals. The storage is related to the nucleation of dislocations which are the strain carriers in the crystals. The situation is completely different with polymer glasses [Oleinik et al. (1993)]. About 75-95% of W at low strains is converted to ΔU during deformation. At higher strains, the ratio ΔU/W will come to a saturation, but will still stay high, around 40-50%. Such situation reflects very different deformation mechanisms operating in disordered solids and crystals. The energy rising process in glasses is a nucleation of strain carriers, called local shear transformations (LST). However, contrary to the behaviour of dislocations, LST can not expand by gliding due to the structural disorder built in the glass. Thus, accumulation of macroscopic strain in glass is going mainly through concentration increase of LSTs. The number of LSTs per unit of macroscopic strain is much higher than number of dislocations in the crystals. Because each LST in the glass carries energy excess, high concentration of LSTs is the reason for high ΔU storage in plastically deformed polymeric glasses [Oleinik et al (1993)]. We believe that the high energy storage is the characteristic feature of inelastic deformation process in any disordered solid. Different levels of ΔU in different glassy polymers reflect the difference in concentration of the LSTs in them.

The steady level of ΔU depends on the effectivity of the termination processes (carrier's disappearance) in glass. Results have shown that the higher the steady level of ΔU in a sample, the more brittle the polymer glass is. It means that high amount of ΔU during glass deformation may be easily converted to the processes of crack initiation and growth.

When following Q evolution with strain one can trace the processes of carrier's termination. The Q origin is different in a glassy matter in comparison with ordered solids. Q appears in crystals due to lattice friction during the dislocation slippage through the lattice. LSTs can not expand and such origin of Q does not exist in glasses. LSTs are recovered from the excited, deformed state to initial structure (thereby overcoming an energy barrier and Q is emitted in glasses due to such recovery. Thus, the strain of the initial Q emission indicates the moment where the termination (relaxation) of deformation carriers start to operate. The earlier this process starts, the lower ΔU is stored in the polymer upon deformation. When the rates of LSTs nucleation and termination becomes equal the total deformation process reaches steady-state. It is exhibited macroscopically by a constant level of measured ΔU.

It was found that practically all energy storage ΔU in deformed polymer glasses is carried out exclusively by an anelastic strain component which does not contain conformational changes in chains. Conformational changes are appear when given LST is converted to much less excited, but conformationally deformation state. It means that the appearance of new additional strain increment in a sample, but carry an increment which has been produced earlier by the LSTs.

PALS measurements have shown that free volume sometimes increas even at uniaxial compression (e.g. PMMA), but decreases in PC. The means that deformation behaviour in frame of the free volume concept. On the basis of measured thermodynamic characteristics the solid state deformation mechanisms operating in glassy upon their loading and unloading are presented and discussed.

3.2 LC-polymers

We have performed the measurements only with one type of LC polymer, which is polyester Vectra (Hoeghst-Cellanese), where LC order is frozen below T_g (≈100-130°C). Deformation behaviour of Vectra is close to the behaviour of polymeric glasses. There is a high ΔU/W ratio there, steady ΔU level at about ε≈30% which exhibits the steady plastic flow deformation stage. However, the absolute values of ΔU is very low in Vectra. It shows the presence of an effective mechanism of carrier termination upon the straining of such a polymer.

3.3 Semi-crystalline Polymers

The situation is much more complicated in such polymers where amorphous and crystalline components take part both in macroscopic strain and thermodynamics of the process. One should discriminate also between the behaviour of glassy and rubbery states of an amorphous fraction. We have studied PE (different crystalinities), i-PP, PET, PBT, PA-6. The deformation heat is produced mainly by crystallites in PE and PP. The amount of Q per gram of PE crystals is smaller than for polymer glasses. The stored energy ΔU in PE deformation never shows steady-state behaviour. PET demonstrates deformation response of both, the glassy amorphous phase (at room temperature) and the crystallites. Their participation in the deformation process proceeds in an additive manner. The differences in the deformation response of polymer crystals and glasses are discussed.

3.4 Polymer blends

In blends very low DU is quite often. In heterophase blends the presence of inclusions increases the rate of termination mechanisms. Deformation heat starts to appear in blends combines the glassy deformation stage and stress relaxation around inclusions.

4. REFERENCES

Bever M.B., Holt D.L., titchener A.L. (1973) Progr.Mater.Sci.,v. 17, p.1

Godovsky Yu.K. (1993) Thermophysical Properties of polymers, Springer, Berlin.

Oleinik E.F., Salamatina O.B., Rudnev S.N., Shenogin S.V. (1993), Polymer Science v.35. No11. p.11532.

The work supported by: CRDF(RE1-205), RFBR(96-03-32468, 97-03-32762), RFBR-DFG (96-03-00065).

A NUMERICAL/EXPERIMENTAL STUDY ON THE PERFORMANCE OF ENHANCED VISCOELASTIC CONSTITUTIVE EQUATIONS

GERRIT W.M. PETERS, JEROEN F.M. SCHOONEN, FRANK P.T. BAAIJENS and HAN E.H. MEIJER

Materials Technology (http://www.mate.tue.nl), Eindhoven University of Technology,
P.O. Box 513, 5600 MB Eindhoven, The Netherlands, E-mail: gerrit@wfw.wtb.tue.nl

1 INTRODUCTION

A numerical/experimental method is used to test viscoelastic constitutive equations in nominal two dimensional complex flows of a polymer melt (LDPE) with a strong planar elongational component.

Many constitutive equations can fit shear flow data equally well, though they predict different behavior in elongational flows. In complex flows, constitutive models can be tested in both, shear and/or elongation. In this study the performance of the PTT-a (exp. form) and Giesekus model are analyzed and compared with a new class of constitutive models for three different planar flows: 1) 4:1 contraction, 2) flow past a cilinder and 3) cross-slot.

2 METHODS

Experimental: Velocities and stresses are measured using Particle Tracking Velocimetry (PTV) and Fieldwise Birefringence Method (FIB), respectively. The stress optical coefficient is determined using the fully developed inflow profile.

Numerical: Flows are nomally planar and therefore compared with 2D viscoelastic simulations. The simulations are performed with the DEVSS-DG-method as proposed by Baaijens et al. [1]. Stresses are discretized discontinuously over the element interfaces. The method allows for elimination of the stress at element level, offering efficient handling of multimode problems. A second modification is the incorporation of the deformation rate tensor as a global variable, in order to stabilize the method.

3. RESULTS

Figs. 1-3 show, for the three different flow geomtries, the measured and predicted first normal stress difference along the centerlines. For both the models, characterized in shear and uniaxial elongation, the agreement with experimental results is not very good. They can not predict maximum stress levels and stress relaxation in areas with high planar elongational strain.

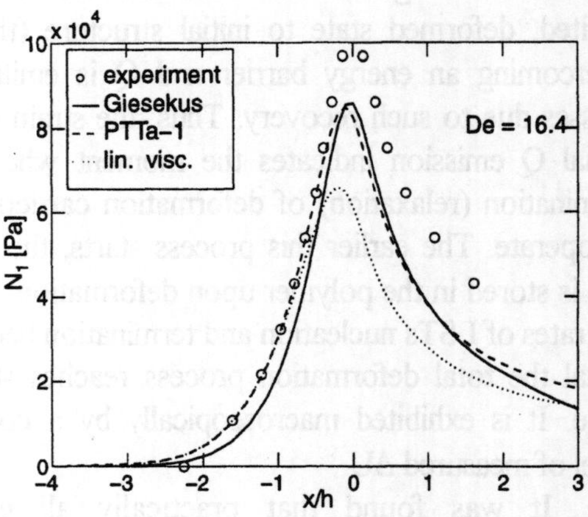

Figure 1: *4:1 Contraction, measured (o) and calculated stresses, on centerline, (-: Giesekus model ($\alpha = 0.29$), - -: PTTa model ($\epsilon, \xi = 0.15, 0.08$), ...: linear viscoelasticity), $T = 170^oC$.*

The enhanced models[2, 3, 4], in which the viscosity function is fixed (Fixed η, *eta: Feta-model*) and thus always predicted equally well, have the flexibility to control independently the shear and elongational properties. Two types of these models (Feta-PTTa = PTT-Ellis and Feta-VD = Variable Drag-Ellis with two parameter sets) were fitted on the cross-slot results and used to predict the flow in the other geometries. The enhanced model based on the concept of variable drag proved to be much better than the PTT and Giesekus model and also better than enhanced PTTa model. A similar approach was applied to the flow of a polymer solution (Pib/C14) in a (3D)

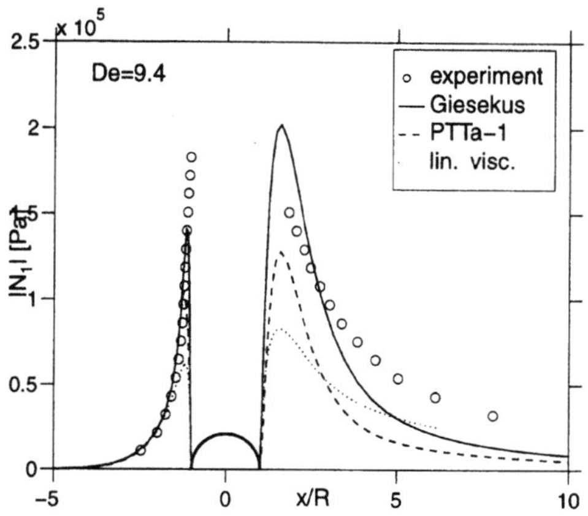

Figure 2: *Flow around a cylinder, stresses along the centerline, (o) from isochromatic pattern, (-) Giesekus model, (- -) PTTa-1 model* (ϵ, ξ = 0.15, 0.08)*,* $T = 170^oC$.

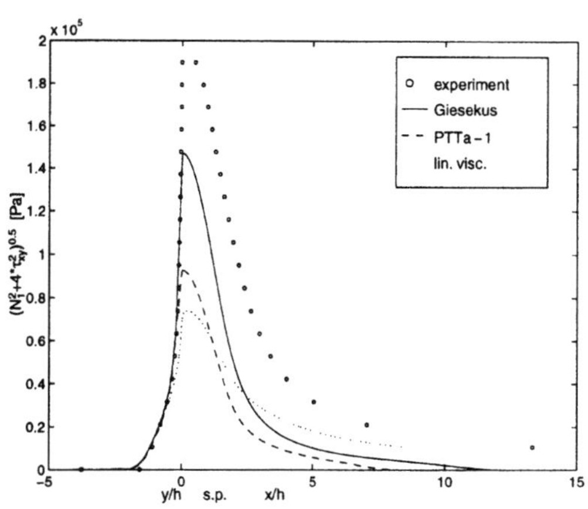

Figure 3: *Cross-slot, stresses along the centerline measured (o) and calculated (Giesekus (-), PTTa-1 (- -)),* $De = 6.0$, $T = 150^oC$.

cross-slot device[3]. The PTTb and Giesekus model were evaluted in that case. Deviations from experimental results were much larger than for the melt. With an enhanced model (Feta-PTTb) it was possible to solve this problem.

4. CONCLUSIONS

The PTT and Giesekus model, characterized in shear and uniaxial elongation, could not predict stresses in areas of complex flows with high planar elongational strain, both, for a polymer melt and solution. Models from a new class of enhanced constitutive equations,

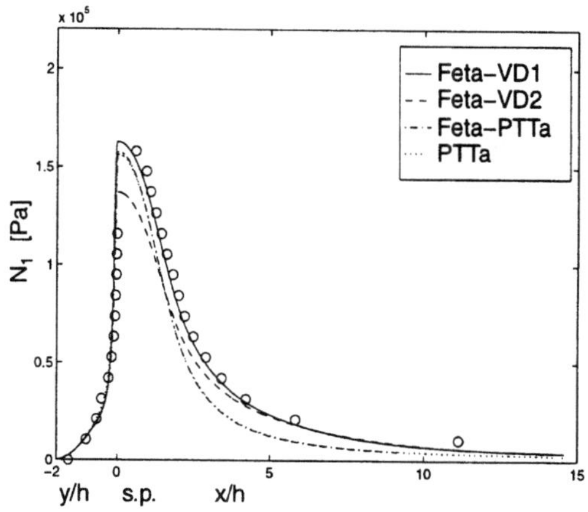

Figure 4: *Stresses along the centerline, cross-slot, Feta-VD1 model* (ϵ, ξ = 0.12, 0.44)*, Feta-VD2* (ϵ, ξ = 0.18, 0.65)*, Feta-PTTa model* (ϵ = 0.045) *and PTTa model* (ϵ, ξ = 0.045, 0.08)*,* $De = 6.0$, $T = 150^oC$.

more flexible in controling independently shear, gave much better results.

References

[1] Baaijens, F.P.T., Selen, S.H.A., Baaijens, H.P.W., Peters, G.W.M., and Meijer, H.E.H. Viscoelastic flow past a confined cylinder of a low density polyethylene melt. *Journal of Non-Newtonian Fluid Mechanics*, 68:173–203, 1997.

[2] Schoonen, J.F.M. *Determination of Rheological Constitutive Equations using Complex Flows.* PhD thesis, Eindhoven University of Technology, the Netherlands, 1998.

[3] Schoonen, J.F.M., Swartjes F.H.M., Peters, G.W.M., Baaijens, F.P.T., and Meijer, H.E.H. A 3D numerical/experimental study on a stagnation flow of a polyisobutylene solution. *Journal of Non-Newtonian Fluid Mechanics*, accepted for publication, 1998.

[4] Peters, G.W.M., Schoonen, J.F.M., , Baaijens, F.P.T., and Meijer, H.E.H. On the performance of enhanced constitutive models for polymer melts in a cross-slot flow. *Journal of Non-Newtonian Fluid Mechanics*, submitted, 1998.

Stress Contribution in the Relaxation of Linear Polymer Melts under Extensional Flow.

P. RUBIO[1], A. ALLAL[1], G MARIN[1], N. ELBOUNIA[2], B. ERNST[3]

[1]Laboratoire de Physique des Matériaux Industriels, E.S.A. CNRS 5067
Université de Pau et des Pays de l'Adour, avenue de l'Université 64000 PAU (France).
[2]Laboratoire de Rhéologie et Mécanique, Elf-Atochem, G.R.L. BP34 64170 Lacq (France).
[3]Laboratoire d'Etude des Matériaux, Elf-Atochem, C.E.R.D.ATO 27470 Serquigny (France).

1. INTODUCTION

The elongational rheological behaviour of polymer melts is of great importance for most of industrial processes. The extensional kinematics may change during the process, like in film blowing and blow moulding. Rheological properties appear to be sensitive on the type of extensional flow (single, biaxial or planar elongation). Predicting both shear and elongational viscosity of polymer melts requires therefore an objective constitutive equation, taking into account the differences between extension modes.

2. CONSTITUTIVE EQUATION FROM MOLECULAR THEORY

Wagner (1976) used a single integral model predicting both linear and non-linear viscoelastic regimes for polymer melts, assuming incompressibility :

$$\sigma(t) = \sigma^s \mathbf{I} + \int_{-\infty}^{t} m(t - t').h(I_1, I_2).\mathbf{C}_t^{-1}(t')\,dt' \quad (1)$$

The time dependant memory function m(t-t') is easily obtained by shear measurements in the domain of small strains. The general objective strain measure $h(I_1 , I_2)C_t^{-1}$ reduces to the Finger Strain tensor $\mathbf{C}_t^{-1}$ at small strain, as the strain invariant function $h(I_1 , I_2)$, called the « damping function », tends to one. If the origin of this function is assumed to derive from an elastic potential, equation (1) can be written as the time-strain separable K-BKZ model as shown by Larson et al. (1984). This approach allows to calculate all the rheological properties for both shear and extension, including the second normal stress difference for polymer melts in shear.

The molecular model of reptation theoretically established by Doï and Edwards (1978b) is of the K-BKZ form and gives an universal expression h_{DE} for the damping function of linear polymer melts.

The expression for the elastic free energy U_{DE} was found by Currie (1982) :

$$U_{DE} = 5.\ln\left(\frac{I_1 + 2\sqrt{I_2 + 13/4} - 1}{7}\right) \quad (2)$$

So that h_{DE} is calculated by the K-BKZ approach:

$$h_{DE}.\mathbf{C}_t^{-1} = \frac{\partial U_{DE}}{\partial I_1}.\mathbf{C}_t^{-1} - \frac{\partial U_{DE}}{\partial I_2}.\mathbf{C}_t \quad (3)$$

The Doï & Edwards model is in good agreement with shear measurements up to 10-20 s^{-1} but underestimates extensional viscosity in the non-linear viscoelastic domain.

3. STRESS CONTRIBUTION UNDER EXTENSIONAL FLOW

We focus on the case of linear polymer melts. According to the Benallal et al. model (1993), the memory function can be determined by a sum of relaxation functions. These contributions describe the reptation and Rouse diffusion processes of the Doï and Edwards model (1978a) for linear entangled polymer. Each process has a characteristic relaxation time which is a function of the elementary time τ_o :

$$\tau_o = \frac{b^2 \xi_o}{\pi^2 k_b T} \quad (4)$$

where b is the length of the Kuhn segment, ξ_o is the monomeric friction coefficient, T is the test temperature and k_b is the Boltzmann constant.

Allal et al. (1996a) wrote an equation for modelling the variations of the Flory characteristic ratio C_∞ with respect to temperature T and polymer mass M. The C_∞ calculation requires the determination of the equilibrium internal energy E_o. Allal et al. (1996b) have shown that molecular parameters like b and ξ_o are therefore functions of E_o. $\tau_o(E_o)$ is then calculated using (4).

We assume that the spherical component σ^s of the full stress tensor increases the free energy of a chain under extension :

$$E(t_o + dt) = E_o + N_a V_a \sigma^s(t_o) \quad (5)$$

where N_a is the Avogadro number and V_a, called the « active volume », is the volume defined by the Kuhn segment b:

$$V_a = \frac{4\pi}{3} b^3 \quad (6)$$

The constitutive equation is of the Doï & Edwards single integral form using (1) - (3), but the time memory function appears now as a function of the stress history in the material under extensional flow, using equations (4) - (6) :

$$\sigma(t) = \sigma^s \mathbf{I} + \int_{-\infty}^{t} m(t - t', \sigma^s) . h_{DE} . C_t^{-1}(t') \, dt' \quad (7)$$

The equilibrium energy E_o = 14310 J/mol was determined for polydisperse PMMA samples ($M_w / M_n \approx 2$) in the range $M_w > 5M_e$ by adjusting the Benallal model with the elastic and storage shear modulus data at 220°C, taking account the tube renewal relaxation process in polydisperse melt sample (Cassagnau et al. 1993). We compare viscosity calculations using our approach (7) with experimental results in biaxial and simple extension flows at 220°C. Uniaxial elongation measurements were performed using a rotary clamp rheometer developed by Meissner (1996) (RME by RHEOMETRIC SCIENTIFIC). Biaxial elongation measurements were achieved under lubricated squeezing flow, following the procedure of Chatraei et al. (1981).

Results for an 245500 g/mol PMMA in single elongation are reported on figure I and calculation for steady state equibiaxial viscosity is plotted on figure II.

4. CONCLUSION

An increase of the free Helmoltz energy for polymer melts under extensional flow, induced by the isotopic stress component growth, modify the memory function by an increase of relaxation times during the elongation. Assuming the reptation relaxation diffusion of the chain, this mechanism can explain the elongational rheological behaviour of linear polymer melts in biaxial flows.

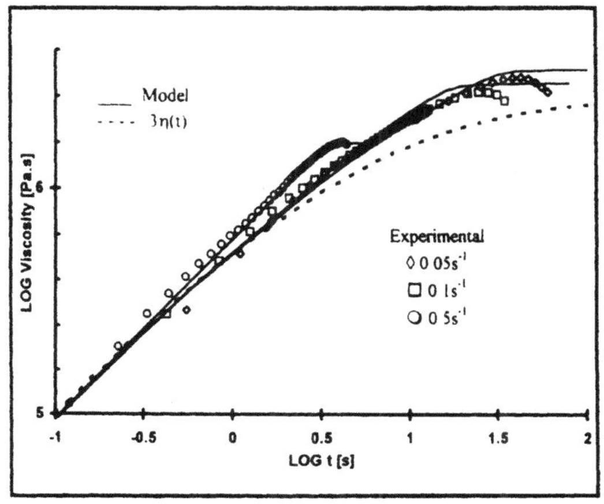

Figure I

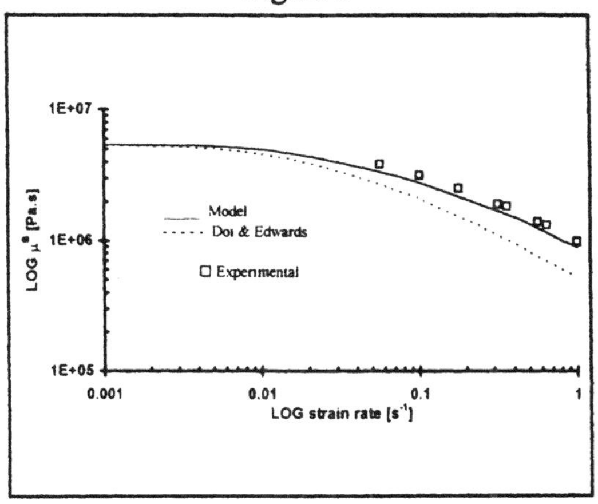

Figure II

5. REFERENCES

M.H.Wagner (1976), Rheol. Acta 15: 136-142

R.G. Larson, K. Monroe (1984), Rheol. Acta 23: 10-13

M. Doï, S.F. Edwards (1978a), J. Chem. Soc., Faraday Trans.2, 74, 1802 - 1817

M. Doï, S.F. Edwards (1978b), J. Chem. Soc., Faraday Trans.2, 74, 1818 - 1832

P.K. Currie (1982), J. Non-Newtonien Fluid Mech., 11, 53-68

A. Benallal, G. Marin, J.P. Montfort, C. Derail (1993), Macromolecules, 26, 7229 - 7235

A. Allal, J.P. Montfort, G. Marin (1996a), Proc XIIth Int. Congr. Of Rheology, 317

A. Allal, G. Marin, J.P. Montfort (1996b), 31ème colloque annuel du GFR, vol. XV, n°1

P. Cassagnau, J.P. Monfort, G. Marin, P. Monge (1993), Rheol.Acta 32:156-167

J. Meissner (1996), Proc XIIth Int. Congr. Of Rheology, 317

S. Chatraei, C.W. Macosko, H.H. Winter (1981), J Rheology, 34, 433 - 443

SAW-TOOTH ARGUMENT TRANSFORMATION METHOD IN THE THEORY OF COMPOSITE MATERIALS

G. STARUSHENKO, N. KRULIK

Ukrainian State Chemical and Technological University, 8 Gagarina av., 320005 Dnepropetrovsk, Ukraine

The analytical method of saw-tooth argument transformation, which can be considered as the argument representation by two non-smooth functions of the special form, is known in the wave oscillation theory. In the present paper saw-tooth transformation technique has developed applicably to the elasticity theory periodic tasks.

From the mathematical view point the suggested method is based on the 4-periodic saw-tooth argument $\tau=\tau(x)$ introduction (fig.1):

$$\tau(x)=\begin{cases} 1/(1+\theta)x, & -(1+\theta)\leq x\leq(1+\theta); \\ -1/(1+\theta)(x-2), & (1+\theta)\leq x\leq(3-\theta); \end{cases} \qquad (1)$$

and any continuous 4-periodic function f(x) identical representation by this argument:

$$f(x)=f_1(\tau) + f_2(\tau)\tau'; \qquad (2)$$

$$f_1(\tau)=0.5\{(1+\theta)f[(1+\theta)\tau]+(1-\theta)f[2-(1-\theta)\tau]\};$$

$$f_2(\tau)=0.5(1-\theta^2)\{(1+\theta)f[(1+\theta)\tau]- -f[(2-(1-\theta)\tau)]\}; \qquad (3)$$

$$\tau' = d\tau/dx.$$

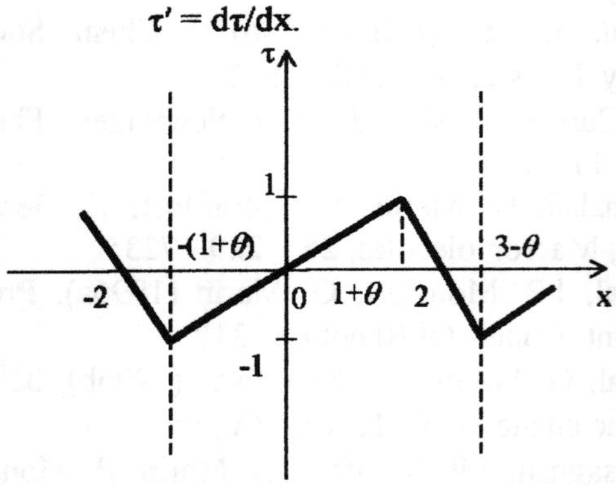

Fig.1

It has been shown that different nature function (continuous, functions that have first kind discontinuity and functions that have specialties of Dirac δ-functions type) can be described with τ-representation help and its derivatives. Such community of approach is essential extended method applications opportunities to the applied tasks solution because of it allows to build mathematically grounded solutions for different functions classes in one method limits. In so doing process of singular terms excludition in the equations is automatizated by the corresponding boundary conditions adoption from the one side and conditions which are provided solution periodic character are got naturally from the other side.

It should be observed that suggested τ-transformation is non-symmetrical with respect to the period quarter (θ is asymmetry parameter, θ=0 being corresponded to the symmetrical case) That gives additional opportunities of any nature periodic processes and structure describing and analysis.

Opportunities of saw-tooth argument transformation method are not restricted by clear periodic solutions building. Combination of τ-transformation method with multi-scaling method and average theory lets to investigate any structure systems, which have both "fast" (periodic) and "slow" (non-periodic) compilers.

Utilization of non-symmetrical saw-tooth transformation technique has been illustrated on the example of elasticity theory periodic task for two-phase composite massif. It is suggested that structure geometric and mechanic characteristics be changed periodically in the one of axis direction. In the τ-method limits elastic constants of composite massif –Lame coefficients and phases geometric sizes $a^{+(-)}$ – can be described by the unanimous analytical expressions for all unit structure:

$$\lambda^*(\tau)=0.5[(1+\theta)\lambda^+ +(1-\theta)\lambda^-]+ +0.5(1-\theta^2)(\lambda^+-\lambda^-)\tau';$$

$$\mu^*(\tau)=\lambda^*(\tau)(\lambda\to\mu); \qquad (4)$$

$$a^*(\tau)=(1+\theta^2)+\theta(1-\theta^2)\tau'; \qquad (5)$$

Elasticity theory task solution in the transference U, V, W are presented by

$$U(\tau)=U^{(1)}(\tau)+U^{(2)}(\tau)\tau';$$

$$V(\tau)=V^{(1}(\tau)+V^{(2)}(\tau)\tau';$$

$$W(\tau)=W^{(1)}(\tau)+W^{(2)}(\tau)\tau'.$$

Further we come to boundary value problem on the structure characteristic period (task on the cell) for U-forming with using τ-transformation algebra after periodic singular terms excludition and equation corresponding parts division under basis elements of the form:

1. $d^2U^{(1)}/d\tau^2 = \Delta^+_{11}X^+_1 + \Delta^-_{11}X^-_1;$

$$d^2U^{(2)}/d\tau^2 = \Delta^+_{12}X^+_1 + \Delta^-_{12}X^-_1;$$

$$U_{12}\,|_{\tau=\pm 1} = 0;$$

$$\{(dU^{(1)}/d\tau)\Delta_{11}+(dU^{(2)}/d\tau)\Delta_{12}\}\,|_{\tau=\pm1}=0;$$

where

$$\Delta^+_{11} = -(1+\theta)^3/2(\lambda^++2\mu^+);$$

$$\Delta^-_{11}=-(1-\theta)^3/2(\lambda^-+2\mu^-);$$

$$\Delta^+_{12}=(1-\theta)\Delta^+_{11};$$

$$\Delta^-_{12}=-(1+\theta)\Delta^-_{11};$$

$$\Delta_{11}=(\lambda^++2\mu^+)/(1+\theta)+(\lambda^-+2\mu^-/(1-\theta);$$

$$\Delta_{12}=(\lambda^++2\mu^+)/(1+\theta)^2-(\lambda^-+2\mu^-)/(1-\theta)^2.$$

2. $d^2V^{(1)}/d\tau^2 = \Delta^+_{21}X^+_2 + \Delta^-_{21}X^-_2;$

$$d^2V^{(2)}/d\tau^2 = \Delta^+_{22}X^+_2 + \Delta^-_{22}X^-_2;$$

$$U_{22}\,|_{\tau=\pm1} = 0;$$

$$\{(dV^{(1)}/d\tau)\Delta_{21}+(dV^{(2)}/d\tau)\Delta_{22}\}_{\tau=\pm1}=0;$$

where

$$\Delta^\pm_{2i} = \Delta^\pm_{1i};$$

$$\Delta_{2i} = \Delta_{1i};$$

$$((\lambda^\pm + 2\mu^\pm)\to\mu^\pm);\ i=1,2.$$

3. $V^{(i)}\to W^{(i)},\ i = 1,2.$

Different composite massif limit states have been considered and corresponding limit crossings have been carried out in the tasks on the cell. Comparative and asymptotic analysis of got results is carried out under interchanging of geometric and cruel composites parameter.

Question about got boundary value problem has been investigated. It has been shown that transference inequality condition was defined by correlation:

$$\int_{-1}^{1} U^{(1)}(\tau)\,d\tau=0 \quad (U\to V\to W)$$

which excludes "slow" (average) part from task periodic solution for self-balanced mass forces on the period.

Solution of elasticity theory flat tasks for different structure composites are got as illustrative example.

STRUCTURE-PHENOMENOLOGICAL RHEOLOGY OF DILUTE SUSPENSIONS IN LIQUID CRYSTALS

E.YU.TARAN, YU.V.PRIDATCHENKO, D.E.TARAN

Faculty of Mechanics and Mathematics, Taras Shevchenko Kyiv University, 64, Vladimirskaya St., 252033, Kyiv, Ukraine

The rheological constitutive equations for dilute suspensions of rigid spherical and axisymmetric elongated particles immersed in an anisotropic dispersive medium are obtained. The effect of anisotropy of dispersive medium on a dynamics of suspended particles and on rheological properties of suspensions in a simple shear flow is investigated.

As a rheological model of suspensions dispersive media, we use the constitutive equations for the anisotropic fluids obtained by Ericksen (1960)

$$t_{ij} = -p\delta_{ij} + 2\mu d_{ij} + \mu_1 n_{ij} + \mu_2 d_{km} n_{kmij} + +2\mu_3(d_{ik}n_{kj} + d_{jk}n_{ki}), \quad (1)$$

$$\dot{n}_i = \omega_{ik}n_k + \lambda(d_{ik}n_k - d_{km}n_{kmi}). \quad (2)$$

In (1), (2), t_{ij} is the stress tensor; d_{ij} is the deformation rate tensor; ω_{ij} is the velocity vortex tensor; n_i is the unit director; $n_{ij} = n_i n_j$, $n_{kmi} = n_k n_m n_i$, $n_{kmij} = n_k n_m n_i n_j$; p is the isotropic pressure; λ and μ's are rheological constants; the dot over n_i denotes a local time derivative; δ_{ij} is the Kronecker delta. The special case $|\lambda| \geq 1$, $\mu_1 = 0$ is considered, where the anisotropic fluid (1), (2) behaves as a liquid crystal (LC).

We assume that suspended particles possess a zero buoyancy and dimensions much more than those of microstructural elements of the anisotropic dispersive medium, so that it interacts with particles as with hydrodynamic bodies. As a hydrodynamic model of suspended particles, a symmetric triaxial dumbbell with axes L_1, L_2, L_3 ($L_1 \geq L_2 = L_3$) is used. We consider the motion of suspended model particles under the influence of hydrodynamic forces only. The frictional coefficient of triaxial dumbbell beads in a slow translational motion in the anisotropic dispersive medium (1), (2) is defined in conformity with the result of Volkov and Vinogradov (1984) by the drug tensor $\xi_{ik} = \zeta_\perp\delta_{ik} + (\zeta_\| - \zeta_\perp)n_{ik}$, where $\zeta_\|$ and $\zeta_\perp$ are the longitudinal and transverse (relative to the director n_i of the Ericksen anisotropic fluid (1), (2)) frictional coefficients of dumbbell beads. Taran and Pridatchenko (1992) demonstrated that a symmetric triaxial dumbbell may be used instead of a sphere (at $L_1 = L_2 = L_3$) or an ellipsoid of revolution (at $L_1 > L_2 = L_3$) as a hydrodynamic model of rigid impenetrable suspended particles in a dilute suspension with the Newtonian (at $\zeta_\| = \zeta_\perp$) dispersive medium.

The constitutive equation for stress tensor T_{ij} in arbitrary gradient flows of dilute suspensions under consideration,

$$T_{ij} = t_{ij} + n_0(L_2^2/2)\{\xi_{ik}(d_{kj} + <\nu_k\mathcal{N}_j> - -\tilde{p}^2 <\nu_j\mathcal{N}_k>) + (\tilde{p}^2 - 1)\xi_{in}d_{nk} <\nu_{kj}> - - <\mathcal{E}_{jnk}\nu_{nmf}\xi_{ki}\mathcal{E}_{lms}\mathcal{N}_l\xi_{sf}(\xi_{\alpha\alpha} - -\xi_{\alpha\beta}\nu_{\alpha\beta})^{-1}> + <\mathcal{E}_{lms}\nu_{ln}d_{mf}\xi_{fs} \times \times\mathcal{E}_{jnk}\xi_{ki}(\xi_{\alpha\alpha} - \xi_{\alpha\beta}\nu_{\alpha\beta})^{-1}>\}, \quad (3)$$

is obtained here with use of the structure-phenomenological approach proposed by Shmakov and Taran (1970) and developed by Taran (1977), Taran and Pridatchenko (1992), Taran et al. (1994). In (3), t_{ij} is the stress tensor in an anisotropic dispersive medium without suspended particles, defined by (1); n_0 is the number of suspended particles per unit volume of suspension; $\tilde{p} = L_1/L_2$; ν_i is the unit vector characterizing the orientation of the main axis L_1 of a triaxial dumbbell in the laboratory coordinate system $Oxyz$; $\nu_{ln} = \nu_l\nu_n$; $\nu_{nmf} = \nu_n\nu_m\nu_f$; $\mathcal{N}_i = \dot{\nu}_i - \omega_{ik}\nu_k$; $\mathcal{E}_{ijk}$ is the Levi-Civita tensor; the angular brackets $< >$ denotes the averaging with use of the distribution function F of angular positions of the vector ν_i, which is the solution of the equation

$$\frac{\partial F}{\partial t} + \frac{\partial}{\partial \nu_i}(F\dot{\nu}_i) = 0, \quad (4)$$

where t is time. A local time derivative $\dot{\nu}_i$ in (3), (4) is defined by the constitutive equation

$$\tilde{p}(\xi_{il}d_{lj}\nu_j - \xi_{ij}\mathcal{N}_j - \xi_{kl}\mathcal{N}_k\nu_{li} - \xi_{jl}d_{jl}\nu_{lki}) -$$

$$-\xi_{jl}\mathcal{N}_i\nu_{jl} - \xi_{jl}d_{ij}\nu_l + \xi_{jl}d_{jk}\nu_{lki} +$$
$$+\frac{\mathcal{E}_{ijk}\xi_{kl}\nu_{jl}}{\xi_{\alpha\alpha} - \xi_{\alpha\beta}\nu_{\alpha\beta}}\mathcal{E}_{nms} \times$$
$$\times(\xi_{sf}\mathcal{N}_n\nu_{mf} - \xi_{ls}d_{ml}\nu_n) = 0. \qquad (5)$$

The obtained constitutive equations (3)-(5) are used for investigation of the effect of anisotropy of the dispersive media of suspensions on a dynamics of suspended particles and on rheological properties of suspensions in a simple shear flow $v_x = v_z = 0$, $v_y = Kx$ (K — const). It is demonstrated that the triaxial dumbbell modelling spherical suspended particles (at $L_1 = L_2 = L_3$) rotates under the action of hydrodynamic forces with a constant angular velocity

$$\omega = K\frac{1 + (\Delta - 1)sin^2\beta}{\Delta + 1}, \qquad (6)$$

where $\Delta = \zeta_{||}/\zeta_{\perp}$; β is the steady angle between the Oxz plane and the director n_i which is defined from (2) for a simple shear flow of the anisotropic dispersive medium (1), (2) by the equation $ctg^2\beta = (\lambda - 1)/(\lambda + 1)$.

An elongated suspended particle (at $L_1 > L_2 = L_3$), whose main axis L_1 lies in shear plane of the simple shear flow, rotates periodically with the angular velocity

$$\frac{d\varphi}{dt} = \left\{\frac{K}{2}(1 + \tilde{\lambda}cos2\varphi) - K(\Delta - 1) \times \right.$$
$$\left. \times \left[\tilde{\lambda}cos\varphi sin\beta sin(\varphi - \beta) - \frac{sin^2\beta}{\tilde{p}^2 + 1}\right]\right\} \times$$
$$\times\left\{1 + \frac{\Delta - 1}{2}[1 + \tilde{\lambda}cos2(\varphi - \beta)]\right\}^{-1}, \qquad (7)$$

if its axial ratio $\tilde{p}$ is less than a certain critical value $\tilde{p}^*$. In (7), φ is the angle between the axis L_1 of the particle and axis Ox; $\tilde{\lambda} = (\tilde{p}^2 - 1)/(\tilde{p}^2 + 1)$.

In the case of the Newtonian dispersive medium ($\zeta_{||} = \zeta_{\perp}$), the angular velocities (6), (7) of model suspended particles coincide with that $\omega = K/2$ and

$$\frac{d\varphi}{dt} = \frac{K}{2}(1 + \tilde{\lambda}cos2\varphi)$$

of a spherical and an ellipsoidal suspended particles obtained by Einstein (1906) and Jeffery (1922), respectively.

With $\tilde{p} \geq \tilde{p}^*$, the rotary motion of elongated suspended particles in a simple shear flow of an anisotropic dispersive medium is stopped. The steady-state orientation in such a situation allows one to predict the possibility of reinforcement of LC polymers of the nematic type by rigid elongated suspended particles in gradient flows of LC polymers.

The apparent viscosity of suspensions and differences of normal stresses in a simple shear flow of suspensions under consideration are invertigated. It is obtained, in particular, that the characteristic viscosity

$$[\mu_a] = \frac{\mu_a - \bar{\mu}_a}{\bar{\mu}_a V} = \frac{5}{2}\left[1 - \left(\frac{\Delta - 1}{\Delta + 1}\right)^2 cos^2 2\beta\right] \qquad (8)$$

of a dilute suspension of spherical particles in liquid crystals under a simple shear flow is less than the corresponding characteristic viscosity of a dilute suspension of spherical particles with the Newtonian dispersive medium $[\mu_a] = 2,5$ obtained by Einstein (1906). In (8), μ_a is the apparent shear viscosity of suspension; $\bar{\mu}_a$ is the apparent shear viscosity of the anisotropic dispersive medium (1), (2), $\bar{\mu}_a = \mu + \mu_2(\lambda^2 - 1)/4\lambda^2 + \mu_3$; V is the volume concentration of suspended particles.

REFERENCES.

Einstein A (1906) Ann Physik 19: 289-306

Ericksen JL (1960) Kolloid Z 173: 117-122

Jeffery GB (1922) Proc Roy Soc A102: 161-179

Shmakov YuI, Taran EYu (1970) Inzh Fiz Zhurn (IFZh) 18: 1019-1024 (in Russian)

Taran EYu (1977) Prikl Mekhanika 13: 110-115 (in Russian)

Taran EYu, Pridatchenko YuV (1992) Inzh Fiz Zhurn (IFZh) 62: 57-65 (in Russian)

Taran EYu, Pridatchenko YuV, Volkov VS (1994) Prikl Mekh Tekhn Fiz (PMTF) 35: 99-107 (in Russian)

Volkov VS, Vinogradov GV (1984) Rheol Acta 23: 231-237

RHEOLOGY AND NON-EQUILIBRIUM THERMODYNAMICS

J. VERHÁS

Technical University of Budapest, Budapest, Hungary, H-1521

Abstract

Non-equilibrium thermodynamics with dynamic degrees of freedom gives the frame of the theories of rheology, liquid crystals, fiber reinforced materials, as well as, fluid turbulence.

1 Introduction

A firm and systematic foundation of rheology stems in modern non-equilibrium thermodynamics ([1] and ref. 90 in it).

Modern thermodynamics (refs. 137, 139, 66, 125/1, 126, 75 in [1]) introduces material equations in the Onsagerian form (refs. 86/1, 173/6 in [1], [2]), starting with the conservation laws and basing on the entropy balance

$$\rho\frac{ds}{dt} + \operatorname{div}\mathbf{J}_s = \sigma_s \geq 0; \qquad (1.1)$$

the entropy function s and flow $\mathbf{J}_s$ have to be modeled (refs. 90, ,135/5, 173/9 in [1]).

2 Local equilibrium

For a local equilibrium material of rheology, the independent variables are the internal energy u and some measure for the deformation[1] $\boldsymbol{d}$, (moreover, a vector or a tensor for the orientation in anisotropic materials). The time derivative of the entropy is arranged into the form

$$\frac{ds}{dt} = \frac{1}{T}\frac{du}{dt} + \boldsymbol{s}_d : \mathring{\boldsymbol{d}}, \qquad (2.1)$$

where $\mathring{\boldsymbol{d}}$ is the symmetric part of the velocity gradient[2]. The material equations are

$$\begin{aligned} \boldsymbol{t} + \rho T \boldsymbol{s}_d &= L_{11}\mathring{\boldsymbol{d}} - L_{12}\frac{1}{T}\operatorname{grad} T \\ \mathbf{J}_q &= L_{21}\mathring{\boldsymbol{d}} - L_{22}\frac{1}{T}\operatorname{grad} T. \end{aligned} \qquad (2.2)$$

giving also the the physical meaning of the tensor $\boldsymbol{s}_d$; $\boldsymbol{t}^e(T, \boldsymbol{d}) = -\rho T \boldsymbol{s}_d$. The coefficients L are higher order tensors reflecting the symmetry of the material[3]. For isotropic materials and linear approximation, they read

$$\begin{aligned} \boldsymbol{t} - \boldsymbol{t}^e &= 2\eta\mathring{\boldsymbol{d}} + \eta_v(\operatorname{div}\mathbf{v})\boldsymbol{\delta} \\ \mathbf{J}_u &= -\lambda \operatorname{grad} T. \end{aligned} \qquad (2.3)$$

The material in local equilibrium is Kelvin-Voigt or Newton body; to obtain other models the local equilibrium is dropped.

Keywords: Deformation, flow, Cauchy's stress, rheology, constitutive modeling, skew-symmetric stress, fluid turbulence, liquid crystals.

This work has been supported by the Hungarian National Scientific Research Funds, OTKA (T-17000) and MKM (FKFP 0287/1997), and the EC (CARNET Contract No: ICOP-DISS-2168-96)

[1] A lot of deformation tensors are known in the literature; equivalent from a general point of view but introduced with different mathematics and suggesting different visualization of the motion and, of course, different ways of thinking.

[2] Nevertheless, several deformation tensors are introduced, the symmetric part of the velocity gradient is widely accepted for a measure of the deformation rate. How it relates to the a deformation tensor is a long story.

[3] In a non-linear theory, they are covariant functions.

3 Rheology

Dynamic degrees of freedom reflect the material structures processes induce ([1], and refs. 79, 90, 125/1 126 in it). With the Morse lemma (ref. 124 in [1]), the entropy function turns into

$$s = s^e(u, \boldsymbol{d}) - \sum_i \xi_i^2 \tag{3.1}$$

In rheology, the term for heat flow is dropped. For the usual models, the dynamic variables are α-type second order tensors [1]. The material equations for volumetric- and deviatoric effects are separate and have the form

$$\begin{aligned} \boldsymbol{t} - \boldsymbol{t}^e &= L_{00}\mathring{\boldsymbol{d}} + \sum_i L_{0i}\boldsymbol{\xi}_i, \\ -\mathring{\boldsymbol{\xi}}_i &= L_{i0}\mathring{\boldsymbol{d}} + L_i\boldsymbol{\xi}_i. \end{aligned} \tag{3.2}$$

The small circle above the symbol ξ means corotational time derivative[4] convenient to define process rate as several tensor component may be present among the dynamic variables[5]. Notice that no cross effect is between the dynamic variables[6].

Nevertheless, the equations are linear in the Onsagerian sense, they are not linear at all in the unknown quantities because of the objective time derivatives.

For fluids, the deformation tensor is replaced by the specific volume $\mathcal{V}$ and the deviatoric part of the equilibrium stress is zero (Pascal's law).

The reduction of the equations to particular motions gives the usual models. For small amplitude oscillations, linear viscoelasticity and any body-model consisting of dashpots and springs can be founded[7].

[4]The use of an objective time derivative is optional and results that the coefficients L are independent of the angular velocity of the material if the Coriolis-force and other "gravitational" effects are negligible.

[5]The symbol $\mathring{\boldsymbol{d}}$ is generally not an objective time derivative but for small deformations and in the case of fluids. For the latter, the deformation tensor $\boldsymbol{d}$ is replaced by the relative deformation gradient, i.e., the deformation with respect to the present configuration.

[6]The reason is that an orthogonal transformation of the dynamic variables leaves the entropy function invariant and a suitable one takes the matrix of the conductivity coefficients into diagonal form. The coefficients L_{ik} form a symmetric matrix (reciprocal relations).

[7]For a unified linear theory of fluids and solids, equation (3.2) is linearized.

The equations give account on the viscometric functions; the shear stress function and the normal stress functions are related by Hilbert transformation. The shear viscosity is the same function on shear rate as the real part of the complex viscosity on the circular frequency; the empirical Cox–Merz rule (ref. 30 in [1]) has been founded. Rheo-optics is based on optical parameters depending on the dynamic variables.

Some parameters result in a decreasing section of the shear stress; the viscometric flow is unstable. For the same material, also the elongational flow is unstable under some load; the displayed effect models the experimental background of the theory of plasticity without a postulated yield condition including also creep and viscoelastic deformations before and after plastic flow (ref. 173/12 in [1]).

The inertial element has also room in the theory with β-type dynamic variables.

Among the non-linear theories one can find all the theories of plasticity. Mieses' theory is founded without dynamic degrees of freedom (ref. 173/3 in [1]). A non-linear theory is able to give account on tixotropy and rheopexy, as well.

The models mentioned up to now fit to the idea of "simple materials", for which the stress is determined by the history of the deformation tensor[8]. The model of "simple materials" does not bear even week non-locality.

4 Some weekly non-local materials

The materials to be mentioned are the melts of fiber reinforced materials, liquid crystals, and turbulent fluids.

4.1 Pseudo tensors

The considerations of section 3 can not do with the skew-symmetric part of Cauchy's stress. The latter does not enter the entropy production but couple stress multiplied with the gradient of the angular velocity does. The gradient

[8]Usually, the history of the deformation gradient tensor is referred to and the "simple fluid" is defined.

of the angular velocity is closely related to the gradient of the deformation rate;

$$\mathrm{Grad}\,\vec{\omega} = (\mathrm{Rot}\,\mathring{\boldsymbol{d}})^T. \tag{4.1}$$

The model needs dynamic variables of second order symmetric or skew-symmetric pseudo tensors. If they are of α-type the Onsager equations are

$$\boldsymbol{\Pi}^{d/a} = L_{00}^{d/a}\frac{1}{T}(\mathrm{Grad}\,\vec{\omega})^{d/a} + \sum_i L_{01}^{d/a}\boldsymbol{\xi}_i^{d/a}$$

$$-\dot{\boldsymbol{\xi}}_i^{d/a} = -L_{i0}^{d/a}\frac{1}{T}(\mathrm{Grad}\,\vec{\omega})^{d/a} + L_i^{d/a}\boldsymbol{\xi}_i^{d/a} \tag{4.2}$$

for the deviatoric and the skew-symmetric parts, respectively. The dynamic degrees of freedom give account on the deformation of fibers.

4.2 Liquid crystals

Liquid crystals are anisotropic fluids; the entropy depends on the internal energy, the specific volume, and a vector or a second order tensor for the orientation (refs. 140, 54 in [1]). Couple stress is assumed and—in some works—internal angular momentum.

Several classes of liquid crystals are known; the material symmetry influences the form of the entropy. Dynamic degrees of freedom have not been introduced. The local-equilibrium theory is reduced (refs. 173/13, 174/2 in [1]) to small gradients of the orientation to obtain the Ericksen–Leslie–Parodi theory (refs. 42, 111, 144 in [1]).

4.3 Fluid turbulence

The thermodynamic theory of fluid turbulence has not been worked out. Dynamic degrees of freedom are surely needed; they concern with the intensity of turbulence. The relation $\mathbf{J}_s = T^{-1}\mathbf{J}_q$ for the entropy flow can not hold as it leads to "simple fluid", which satisfies Reynolds' similarity only for laminar flow [3]. The simplest promising equations (one dynamic degree of freedom with transport and non-linearity) give back the essential features of fluid turbulence (ref. 173/21 in [1]) but the differential equations proved hard and the theory has not been succeeded in quantitative checking.

5 Discussion

Non-equilibrium thermodynamics without the hypothesis of local equilibrium gives a large enough frame to develop a quantitative theory of macroscopic phenomena in a moving continuum. The rich arsenal has not been exhausted.

References

[1] J. Verhás. *Thermodynamics and Rheology.* Akadémiai Kiadó and Kluwer Academic Publishers, Budapest and Dordrecht, 1997.

[2] V. Ciancio and J. Verhás. On the non-linear generalizations of Onsager's reciprocal relations. *J. Non-Equilib. Thermodyn.*, 19:184–194, 1994.

[3] Gy. Béda, I. Kozák, and J. Verhás. *Continuummechanics.* Akadémiai Kiadó, Budapest, 1995.

FOURIER-TRANSFORM RHEOLOGY

M. WILHELM, P. REINHEIMER, D. MARING, H-W. SPIESS

Max-Planck-Institut für Polymerforschung, Postfach 3148, D-55021 Mainz, Germany

1. INTRODUCTION

Viscosity and the related dissipation of energy play an important role in many macroscopic and microscopic processes. As the amplitude of mechanical perturbations increases, non-linear effects become important and induces shear thinning or thickening (Malkin 1995). From the rheological point of view there is an increasing need to describe these phenomena which are important in almost all industrial processes. The aim of our work is to provide a simple theoretical and experimental approach to non-linear rheology in terms of Fourier analysis.

2. BACKGROUND

For a Newtonian fluid the viscosity η has a constant value independent of the applied shear velocity. Non-Newtonian or non-linear behavior occurs when the viscosity η changes as a function of the applied shear rate. In general we expect the viscosity to undergo a transition from a Newtonian behavior to a power law behavior (Macosko 1994) such as Eq.1 which is similar to the Carreau model (Bird 1968):

$$\eta = \eta_0 \Big/ \left(1 + \beta|\dot{\gamma}|^{\alpha}\right) \qquad (1)$$

The application of oscillatory shear at a frequency ω_1 (Eq.2) in the non-linear regime results in the generation of higher harmonic contributions in the response function of the shear force F (Eq.3).

$$\gamma = A_0 \sin \omega_1 t \qquad (2)$$

$$F \propto A \cos \omega_1 t + B \cos 3\omega_1 t + C \cos 5\omega_1 t + \ldots \qquad (3)$$

A magnitude Fourier analysis (Bracewell 1986) for the time evolution of the stress Fig.1a in Eq.3 by a Fourier transformation leads to distinct signals in the Fourier rheology spectrum at exactly ω_1, $3\omega_1$, $5\omega_1$... Fig.1b. These non-linear contributions can be analyzed in Fourier space with respect to frequencies, intensities and phases of the harmonic contributions. This leads to a more detailed description of the material properties (Wilhelm 1998).

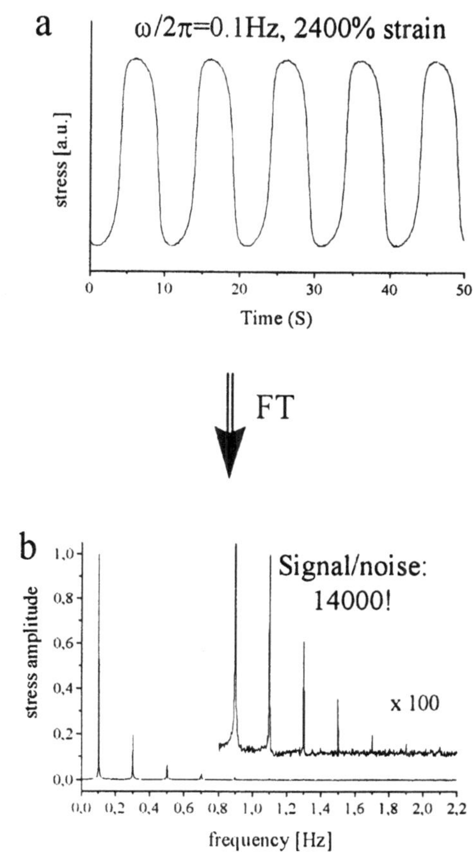

Fig.1 Polyisobutylen, M_n=4,7.10^6 g/mol, T_g=-61°C, 10% in oligomer solution.

3. APPLICATIONS OF FOURIER RHEOLOGY

A straight forward application of Fourier rheology concerns the cross-over from linear to non-linear response which occurs when the shear rate exceeds the inverse of the longest relaxation time of the polymer. The analysis of the odd higher harmonics at $3\omega_1$, $5\omega_1$, $7\omega_1$, etc... allows the elucidation of relaxation phenomena at different non-equilibrium states whereas common bead-spring models can only predict linear properties. Fourier transformation is a reversible mathematical transformation. The information gained by Fourier rheology can be used both in the time and frequency dimension. This allows the reconstruction of the time signal from the FT-rheology spectrum. This representation of time data in terms of analytical functions can be used for the calculation of analytical expressions for derivatives at any point.

4. ANALOGY TO OTHER FOURIER METHODS

Non-linear responses are well known in dielectric and optical spectroscopy. In those cases the intensities and the amount of the higher harmonics are much lower compared with those in rheology. Analogies with FT-IR (Ferraro and Krisham 1990) and FT-NMR (Ernst and Anderson 1966) spectroscopy can also be seen with respect to the so called „multiplex advantage". For instance, in NMR-spectroscopy, the sample is exposed to a range of different frequencies. This is analogous to the application of different shear gradients during a shear cycle. The response function is in both cases a superposition of single responses and the analysis in Fourier-space results in characteristic spectrum.

5. CONCLUSIONS AND PERSPECTIVE

An extended description of the material response functions is possible through explicit consideration of non-linear properties, specially odd multiples of the fundamental frequency under oscillatory shear. Applications of this technique to different polymers will provide a broader understanding of the Fourier transform rheology with respect to different experimental parameters including frequency, shear strain or temperature.

-R.B. Bird and P.J. Carreau, *Chem. Eng. Sci.* **23**, 427 (1968).

-R.N. Bracewell, The Fourier Transform and its Application. *McGraw-Hill.*

-R.R. Ernst and W.A. Anderson, Application of Fourier Transform Spectroscopy to Magnetic Resonance, *Rev. Sci. Instr.* **37**, 93 (1966).

-J. Ferraro and K. Krisham, Practical Fourier Transform Infrared Spectroscopy. *Academic Press.*

-C.W. Macosko, Rheology Principles, Measurements and Applications. *VCH.*

-A.Y. Malkin, Non-linearity in rheology- an essay of classification. *Rheologica Acta* **34**, 27 (1995).

-M. Wilhelm, D. Maring, H.W. Spiess, Fourier-transform Rheology, accepted. *Rheologica Acta.*

THE EFFECT OF PVP POLYMER ON RHEOLOGICAL PROPERTIES OF Na-BENTONITE SLURRY

A. ALEMDAR[1], N. GÜNGÖR[1]

[1]Istanbul Technical University, Faculty Science and Letters, Department of Physics
Maslak Istanbul TURKIYE 80626

Abstract The interactions between electrostatically charged particles in clay-water systems; face-face, edge-edge, face-edge combination models and forming of gel structure are not only important from scientific point of view but also learning of these features are important for technical applications. These interactions can be changed to gain the desired properties with the addition of various electrolytes, polymers, surfactants and organic matters.

The stability of bentonite-water systems can be profoundly affected by the presence of polymers. The adsorption of polymers onto the surfaces of particles can induce either steric stabilization or bridging flocculation which may consequently influence the rheology of the system. There are many cases where the stability is decreased and polymers cause flocculation of the dispersed particles.

Essentially, bridging flocculation occurs because segment of a polymer chain adsorb on different particles, thus linking the particles together. Adsorption is an essential step, and this requires some favorable interaction between polymer segments and particle surface.

The objective of the present study was to determine the influence of PVP(polyvinyl pyrolidone) on the rheological behavior of Na-activated bentonite. PVP ($NC_2H_4C_2OH_{2})$) used in study is pure grade. The molecular weight of PVP is 360.000.

Sample were taken from Enez Bentonite deposits, close to Greece border of Turkey. The sample has a composition (wt %) of Al_2O_3 9.00, SiO_2 62.8, Fe_2O_3 2.20, Na_2O 1.00, CaO 4.60, MgO 1.80, K_2O 0.76, TiO_2 0.45.

Peptization process was done mixing 4 wt% $NaHCO_3$ of clay sample with 35 % humidity and this mixture was left for 7 day period in semi-opened stockroom where it was opened to air circulation in order to obtain better quality drilling day.

The viscosity of Na-bentonite-water system were determined with a Brookfield (model DV-II) viscometer.

PVP polymer, 1-6 % of the dry weight of the sample used, were added to the systems and their viscosities were determined after each addition

PVP resembles a gel structure and it is a linear protein which is made of various pyrolidine rings. It dissolves in water exhibiting various hydrophilic structure and can be curled like a ball. This process involves the desorption of segments.

In PVP adsorption, only the clay surface is involved, not the interlayers. This phenomenon is due to the irregular forms of the polymers. Increasing in PVP concentration increases the viscosity of Na-activated clays. From chemical composition point of view, PVP resembles a gel structure.

PVP macromolecules (non-ionic), consisting of several segments, adopt configurations at the particle-water interface. A number segments are in close contact with the surface leaving and tails dangling in solution. Again adsorption of a nonionic polymer on a charged surface leads to significant modification of the electrical double layer. PVP provide many links between clay particles. Flocs produced by PVP can be considerably stronger than aggregates formed when particles are destabilized with simple salts or by spesifically adsorbed counterions.

PVP adheres on clay surfaces when added into bentonite-water system and absorbs medium water by swelling in slurry and consequently, viscosity of the system increases while fluidity decreases.

In the use of the bentonite-water systems in industry, the addition of PVP polymer is recommended if an increase in viscosity desired.

LINEAR VISCOELASTICITY OF A MODEL MAGNETORHEOLOGICAL FLUID

B.J. DE GANS[1], C. BLOM[1], J. MELLEMA[1], A.P. PHILIPSE[2]

[1] Twente Institute of Mechanics, J.M. Burgers center, Rheology group, Department of Applied Physics, University of Twente, P.O. Box 217, 7500 AE Enschede, The Netherlands
[2] Van 't Hoff Laboratory for Physical and Colloid Chemistry, Debye Institute, Department of Chemistry, Utrecht University, P.O. Box 80.051, 3584 CH Utrecht, The Netherlands

1. INTRODUCTION

Magnetorheological fluids are suspensions of magnetisable particles. Their rheology can be tuned externally: In the absence of a magnetic field they show liquid like behavior, whereas in the presence of a magnetic field solid like behavior is observed. The physical mechanism behind this transition is the formation of chains of dipolarly interacting magnetised particles.

Magnetorheological effects can also be observed in suspensions of non-magnetic particles in a magnetic fluid or ferrofluid. Such a system is commonly referred to as an inverse ferrofluid. The non-magnetic particles are typically several orders of magnitude larger than the nanosized magnetic particles that render the ferrofluid its properties, so that they can be seen as if moving in a medium that is magnetically as well as hydrodynamically continuous. Upon the application of a magnetic field the particles obtain a magnetic dipole moment. One of the first to investigate such a material was Skjeltorp (1983), who studied structures and phase transitions in a two dimensionally confined inverse ferrofluid and observed the formation of particle chains if the magnetic field was directed parallel to the fluid layer.

As non-magnetic particles are available that are very well defined with respect to shape and size, and, consequently, interactions, inverse ferrofluids are very well suited for the fundamental study of magnetorheological effects. Kashevsky et al. (1988) and Popplewell and Rosensweig (1996) have done so while concentrating on the flow properties of their materials.

We present a system of monodisperse silica particles in a ferrofluid of which we have measured the viscoelastic properties as a function of particle volume fraction, frequency and magnetic field strength, using a home made magnetorheometer.

2. EXPERIMENTAL

2.1 The sample

Silica particles with a radius of 184 nanometer were synthesized according to Stöber et al. (1968). Their outer surface was grafted with octadecylalcohol following Van Helden et al. (1981). The ferrofluid consisted of magnetite particles, radius 7 nanometer, stabilized by oleic acid in cyclohexane. Samples were prepared by dispersing a known weight of dry silica in cyclohexane, in which octadecylalcohol coated silica particles are known to behave as hard spheres. After redispersion of the silica with the help of ultrasonication the suspension was mixed with the ferrofluid.

2.2 Apparatus

In order to carry out viscoelasticity measurements on our inverse ferrofluid, we had to modify a rheometer allowing us to apply a magnetic field along the gradient direction of the shear(flow). We used a Bohlin VOR rheometer in combination with a home made cone-plate. The cone had a top angle of 1°. The geometries had a diameter of 6 cm. It was made of a material that was non-magnetic, as it should not disturb the magnetic field, and nonconducting, in order to

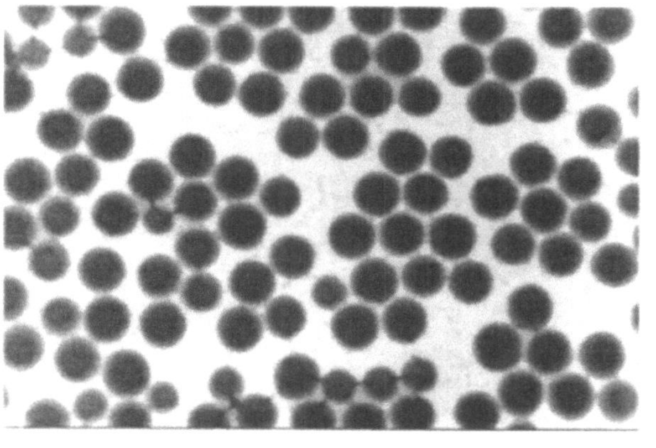

Fig. 1 : Electron micrograph of silica particles. The particles shown have a radius of 125 nm.

Fig.2 : Magnetorheometer

avoid eddy currents due to the rotating motion of the plate. Aluminium oxide was chosen.

The strenght of the field to be applied follows from the magnetisation of the ferrofluid, and the condition of magnetic saturation. A field strength of 0.3-0.4 Tesla suffices. The additional condition for the magnetic field to be fullfilled at each strength is that the field should be as homogeneous as possible over the entire volume of the sample. These requirements can be met by a toroid in combination with a weak iron C-core. The C-core was made from Permenorm 5000H3. The air gap was 6 mm. A photograph of the apparatus is shown in Fig. 2. It is able to create fields with a strength of 0.32 Tesla at maximum.

3. RESULTS

We have measured the complex shear modulus $G^*(\nu)$ as a function of the frequency ν and volume fraction silica particles Φ_v . As G' is at least an order of magnitude larger than G" for all cases studied we mainly concentrate on the behaviour of G'. Typical results are shown in Fig. 3. In the absence of a magnetic field the suspension behaves as a viscous liquid, G' within the experi-

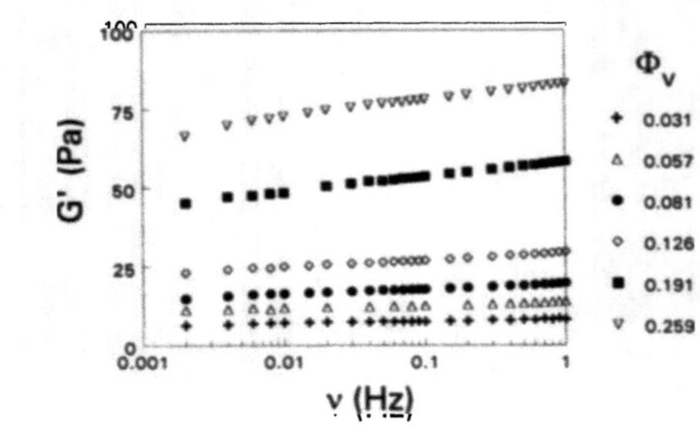

Fig. 3 : Measurement of G'(ν) at 0.308 Tesla for six different volume fractions silica particles.

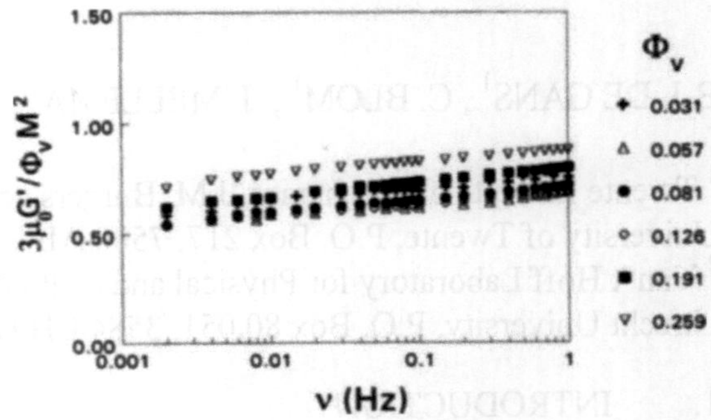

Fig.4 : Scaled G'-data

mental error equal to zero. What can be seen is, that G' does depend only weakly on the frequency. This can be explained assuming a structure of particle chains that are deformed affinely by the applied shear, without any relaxation.

Baxter-Drayton and Brady (1996) have developed a simple theory that predicts a linear scaling with the volume fraction. The data from Fig. 3 have been made dimensionless accordingly, the result being shown in Fig. 4. The agreement is reasonable. Only at the highest volume fraction there seems to be some deviation.

4. CONCLUSIONS

The magnetorheological effect was demonstrated in a suspension of sterically stabilized silica spheres in a ferrofluid. We have measured the behaviour of G' as a function of frequency and volume fraction silica particles using a home made magnetorheometer. G' shows only weak frequency dependance. At low volume fractions it scales linearly with volume fraction.

5. REFERENCES

Baxter-Drayton Y, Brady JF (1996), J Rheol 40, 6: p1027

Van Helden , Jansen JW, Vrij A (1981), J Colloid Interface Sci 81: 354

Kashevsky BE, Kordonsky VI, Prokhorov IV (1988), Magnetohydrodynamics 3: 368

Popplewell J, Rosensweig RE (1996), J Phys D 29: 2297

Stöber W, Fink A, Bohn E (1968), J Colloid Interface Sci 26: 62

Skjeltorp AT (1983), Phys Rev Lett 51, 25: 2306

RHEOLOGY OF MAGNETIZABLE POWDERS RELATED TO THE MECHANICS OF MAGNETICALLY STABILIZED BEDS (gas-solid systems)

J.Y. HRISTOV
Chemical Engineering Department, University of Chemical Technology and Metallurgy, 1756 Sofia, Bulgaria, e-mail: jyh@adm1.uctm.acad.bg

1. INTRODUCTION

The communication focuses on the expansion characteristics of magnetically stabilized beds (MSB) (Rosensweig,1979; Hristov ,1997). The rheology of such beds has been studied by Lee (1981) by the drawing-plate method while a model of magnetizable powder mechanics in compression-shear tests has been reported by Jones et al. (1987). However, MSB do not operate under compression, but under extension due to fluid drag forces (Hristov, 1997; Penchev and Hristov,1990a, 1990b).

The results reported by Penchev and Hristov (1990a, 1990b) indicate that the magnetic interparticle forces allow relative bed expansion $E = \frac{h_i - h_o}{h_o}$ up to 60 %. Moreover, such beds demonstrate (Penchev and Hristov,1990a) rigid-plastic behavior (Brown and Richards, 1970) described by $\Delta P = f(E)$ relationships , where ΔP is the pressure drop .

1.1 Van Wyk equation

Van Wyk (1946) conceived a relationship for description of the compression of woven fabrics.

$$\Delta P = aV^n \quad (1)$$

where V (cm^3/g) is the specific bed volume (inverse of density) . Van Wyk 's law is applied here to describe the deformation of MSB due to the following macroscopic facts:

- fluid driven deformation of a porous body
- quasi one-dimensional extension
- large scale deformations (up to 70 %)

2. EXPERIMENTAL

2.1. Magnetic powders

The experiments reported here concern the deformation characteristics of air-fluidized magnetic particles having wide range of magnetic properties (Table 1). The powders belong to group B of Geldart (1973) where the native cohesion is negligible. Magnetizations at saturation (Ms) were determined earlier by Penchev and Hristov (1990a)

2.2. Experimental set-up

The experiments have been performed in a glass column 65 mm I.D. Penchev and Hristov,1990, 1990b) under the action of two limiting orientations of the field lines- axial and transverse and air as a gas .

Table1. Magnetic powders used

Material	Key	Fraction mm	Density kg/m^3	Ms kA/m
Iron	A1	0.100-0.200	7800	1,700
Fe_3O_4	B1	0.100-0.200	5100	477
	B2	0.200-0.315		
	B3	0.315-0.400		
Metall. dross	C1	0.100-0.200		234
	C2	0.315-0.400		

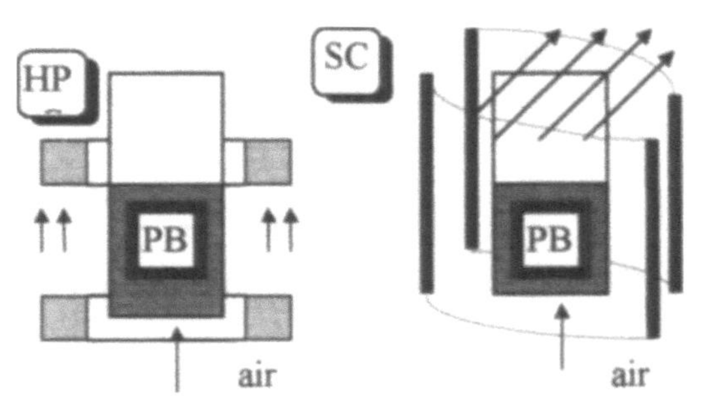

Fig.2 Schematic presentation of the experimental situations . Left - axial field ; Right-transverse field. Arrows indicate the field lines directions. PB - particle bed

The magnetic fields were generated by a Helmholtz pair- (HP) and saddle coils (SC) Penchev and Hristov, 1990a, 1990b). The maximum field intensity was about 50 kA/m . The experimental situations are schematized in Fig.1 .

3. RESULTS

The data reported in the extended abstract are demonstrative and cover a small part of the materials from Table 1.

Pressure drop- relative bed expansion curves are shown in Fig.2 . The data obtained in both fields indicate a rigid -plastic flow behavior , due to the existence of a critical pressure drop required for the onset of bed deformation (expansion). Moreover, during the expansion the pressure drop remains practically constant in axial fields. In the transverse field the pressure drop decreases under the expansion process..

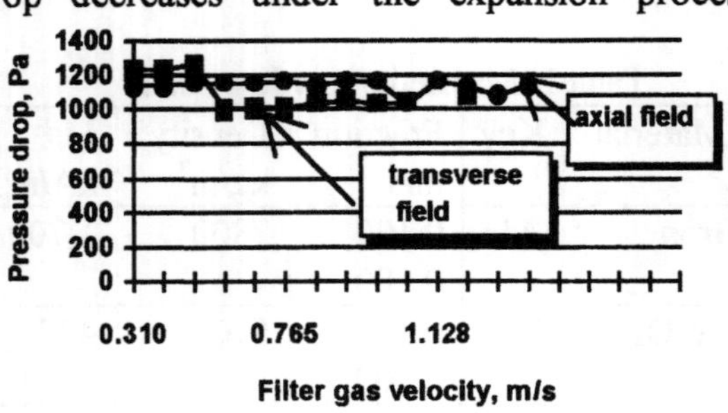

Fig.2a Pressure drop-gas velocity in the state of MSB. Magnetite B3; initial bed height 50 mm. Bed weight , G = 263 g. ; H= 11840 A/m

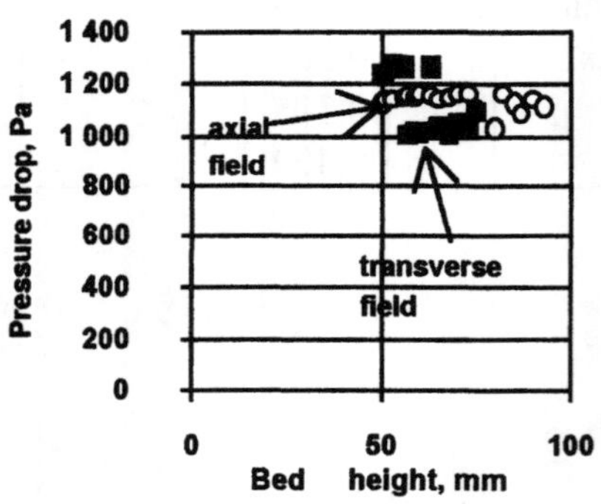

Fig.3 Bed expandability. Conditions like in Fig.2

The data treatment (for magnetite only) with a non-linear regression analysis gave the following form of Eq.(1)

$$\Delta P = 1200 V^{-0.070} \qquad (2)$$

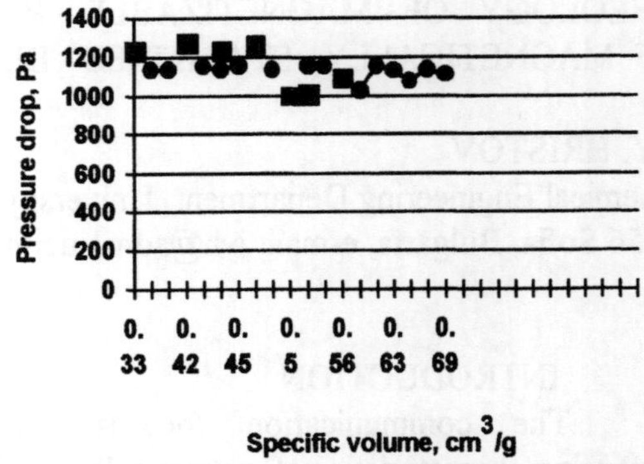

Fig. 4 Expandability by fluid stress of magnetizable particle bed .Conditions and labels like in Fig.2

The coefficient **a** varies in the range of 1200- 1250 (Pa.cm^3/ g) while **n** takes values from 0.070- 0.085 .These values are independent of the field line's orientation and are valid for the deformation (expansion) range of 40 % (relative bed expansion, E). However, the expandability is strongly affected by the magnetic properties of the powders and the field intensity. The materials having greater Ms demonstrate higher ability to expand under the fluid stress.

The field line orientations effect the field intensity required to obtain the desired value of V. In the transverse field the required field intensities are approximately twice that those in the axial field. Further details will be demonstrated in the complete oral presentation.

4. REFERENCES

Brown R.L., Richards, D(1970) Principle of Powder Mechanics, Pergamon Press, Oxford

Geldart D (1973) Powder Technology 7 : 285

Jones, T.B., Whittaker, G.L., Sulenski, T.J. (1987) Powder Technology 49 : 149

Hristov JY, (1997) Powder Technology 87: 59

Lee W-K., (1981) AIChE Symp.Ser 222 79: 87

Penchev IP , Hristov JY (1990a) Powder Technology 61: 103-118.

Penchev IP, Hristov JY (1990b) Powder Technology 62 : 1-11.

Rosensweig RE, (1979) Science 204: 57

Van Wyk CM (1946) Onderstepoort J.Vet.Sci.Anim.Ind. 21 : 99-224

DYNAMICS OF INDUCED CHAIN STRUCTURE IN ELECTRORHEOLOGICAL SUSPENSIONS

P. RIHA[1], H. KIMURA[2], P. VLASAK[1], K. KOYAMA[2]

[1]Institute of Hydrodynamics, Acad.Sci.Czech Rep., Pod Patankou 5, 166 12 Prague 6, Czech Republic
[2]Department of Materials Science and Engineering, Yamagata University, Yonezawa 992, Japan

1. INTRODUCTION

The electrorheological (ER) fluid suspensions display changes in their rheological properties when subjected to an electric field. The changes come from the induced organization of inner structure consisting of polarizable particles. To deform a structure and imposed a flow, the critical adhesive energy between adjacent surfaces of the particles must be expended. When the critical energy is expended and the bridging structure begins to rupture, the structure changes its organization and, consequently, affects the rheological behaviour of the suspension.

2. THEORETICAL

2.1. Description of Stochastic Process

For quantifying the organization of the bridging structure induced by the applied electric field, a novel approach is suggested. It goes beyond the macroscopic description in that it includes a probability density $p_n(t)$ of a stochastic process $\phi = f(e,t)$ which is dependent on a stochastic variable representing a critical adhesion energy e of the bridging structure.

The effect of the structure on the macroscopic behaviour of the ER fluid is assumed to be characterized by the average adhesion function independent of a position and defined as an ensemble average,

$$\Phi(t) = \langle\phi\rangle_t = \sum_n \phi_n p_n(t) \,, \qquad (1)$$

where ϕ_n is a realization of the stochastic process suitably reduced so that when the columns are bridging the electrodes, $\Phi = 1$, and when the columnar structure is destroyed in high intensity flows, $\Phi = 0$. The realization represents a breaking up and formation of the bridging columns, taking discrete value in a state n at time t since its value determines a critical adhesion energy proportional to the number of columns bridging the electrodes.

The evolution of the bridging structure, affected by the reversible process of breaking up and restoring of a temporary adhesion between the particles, is assumed to be of a Markov character and described by the master equation for the transition probability of ϕ_n, Van Kampen (1981).

Since macroscopically observed values are large compared to the possible fluctuations, the probability density $p_n(t)$ for each t is a sharply peaked function. The location of the peak is at a well-defined point in ϕ_n-space. If the width of the peak may be neglected with regard to the small fluctuations, it is possible to consider the peak position in ϕ_n-space as the macroscopic value $\Phi(t) = \langle\phi\rangle_t$, Van Kampen (1981). Thus the time dependence from the probability distribution may be formally transferred onto the macroscopic value Φ in the master equation resulting,

$$\frac{d\Phi(t)}{dt} = -r(\Phi) + g(\Phi) \,, \qquad (2)$$

where the currents on the right hand side are limited to a linear approximation around Φ. The boundary conditions which guarantee that no transition occurs from the state 0 into n-1 and from N to N+1 then give the following approximation,

$$r(\Phi) = k_r \Phi \,, \quad g(\Phi) = k_g (1 - \Phi) \,, \qquad (3)$$

where k_r and k_g are coefficients.

2.2. Rheological Equation

For describing the effect of the bridging of the electrodes on the rheological behaviour of ER fluid suspensions, the Maxwell-like rheological equation is used,

$$f(\Phi)\sigma_{ij}^E + \lambda_E \frac{\partial \sigma_{ij}^E}{\partial t} = \eta_E \dot{\gamma}_{ij} \,, \qquad (4)$$

where $\lambda_E = \eta_E / G_E$ and η_E, G_E are constants. The form of function $f(\Phi)$ must conform to the same conditions as the currents (3). Consequently, the approximated form of the function $f(\Phi)$ when Φ approaches to the boundary value, $\Phi = 1$, may be $f(\Phi) = 1 - \Phi$, giving $f(1) = 0$ and $f(0) = 1$.

3. EXPERIMENTAL

The rheological characteristics have been measured using the coaxial cylinders and parallel disks disposition of the rotational rheometer (MR-300V2, Rheology Co., Ltd., Japan). The disks and cylinders are arranged as electrodes for applying dc electric field. For the observation of the structure organization, a specially designed set-up has been used.

The fluid suspensions used for the experiments are the suspension of polyaniline particles (diameter of 0.4 μm) in silicone oil (see Minagawa et al. 1994), the SR suspension with 30 vol% of suspending particles (average diameter 2.5 μm) composed mainly of carbon (Bridgestone, Ltd.), the EMR suspension of coated iron particles (40 wt%) in liquid silicone rubber (Nippon Oil, Ltd.), and the suspension of coated aluminium particles (average diameter 9.6 μm) suspended (50 wt%) in silicone oil.

4. RESULTS

The observation of formation and breaking up of bridging structure by the specially designed set-up is reflected in the measured rheological characteristics of all tested ER fluid suspensions. The typical result for the case of steady shear flow imposed after the formation of bridging structure in the electric field at rest is shown in Fig.1.

The description of the experimental data presented in Fig.1 resulted: i) the specification of the dependence of function $f(\Phi)$ on the intensity of flow, $f(\Phi)_{\gamma} = 1-(1+k\dot{\gamma})^{-1}$, where $\dot{\gamma}$ is related to the second invariant of the shear rate tensor II_2, $\dot{\gamma} = \sqrt{II_2/2}$, ii) the solution of Eq.(2) taking into account the functions (3), i.e., $1/\Phi = 1 + (k_r/k_g)$, iii) the specification of dependence of $(1/\Phi)$ on the intensity of flow assuming similar form as for the function $f(\Phi)_{\gamma} = 1-(1+k\dot{\gamma})^{-1}$, i.e., $(1/\Phi) = 1+(k_r/k_g) = 1+(k_r^*/k_g)\dot{\gamma}$, satisfying the condition $f(\Phi) = 1-\Phi = 0$ at rest, iv) the dependence of the ratio k_r^*/k_g on the electric field strength, $(k_r^*/k_g) = \alpha/(1+\beta E)$, where E is the strength of the electric field in kV/mm and $\alpha=2.7$, $\beta=-0.08$, v) the dependence of the viscous coefficient η_E on the strength of electric field E, $\eta_E = a_1 + a_2 E^2$, where $a_1 = -12$ and $a_2 = 609$, vi) the dependence of G_E on the strength of electric field E, $G_E = b_1 + b_2 E^{0.5}$, where $b_1 = -686$ and $b_2 = 1130$.

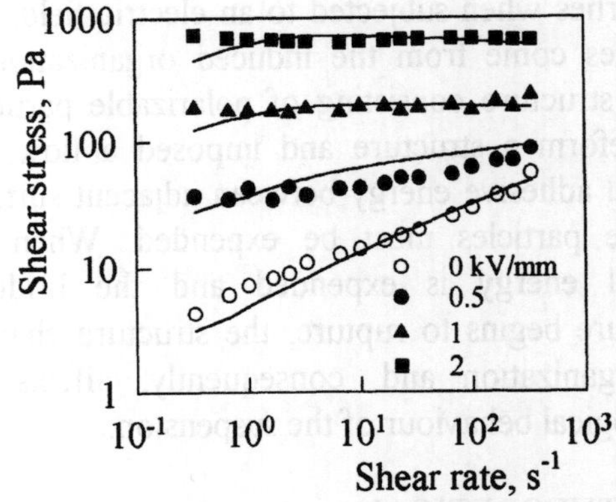

Fig.1 Shear stress vs. shear rate for SR fluid suspension in steady shear flow, 25 °C. Symbols denote the experimental data and the solid curves represent the corresponding solution of Eq.(4).

5. CONCLUSIONS

The induced bridging of the electrodes by suspended particles is a dominant feature of the structure organization of the ER fluid suspensions subjected to an electric field. A stochastic character of this feature may be described and related to the rheological behaviour. For the description of rheological behaviour the Maxwell-like rheological equation may be used.

6. ACKNOWLEDGEMENT

The support of this work by the grant A2060703 from the Grant Agency of the Academy of Sciences of the Czech Republic is gratefully acknowledged.

7. REFERENCES

Kampen Van NG (1981) Stochastic Processes in Physics and Chemistry. North-Holland, Amsterdam, Chpts I, III-VI.

Minagawa K, et al. (1994) J. non-Newtonian Fluid Mech., 52:59-65.

ELECTRORHEOLOGICAL BEHAVIOUR OF POLYANILINE AND CERAMIC POWDER SUSPENSIONS IN SILICONE OIL

J. TRLICA, V. PAVLÍNEK, P. SÁHA

Technical University of Brno, Faculty of Technology in Zlín, Department of Rubber & Plastics Technology, TGM 275, 762 72 Zlín, Czech Republic

2. INTRODUCTION

Electrorheological (ER) fluids are suspensions consisting of dielectric particles dispersed in non-conducting carrier fluid. The fundamental mechanism of ER behaviour is thought to be electric polarisation of particles induced by electric field, which results dipole-dipole interactions, and it causes agglomeration of the particles into chains. The formation of such structure causes an increase in apparent viscosity [1,2].

Experimental studies of ER effect relate mainly to suspensions containing one particle type only. Reports on the behaviour of suspensions where the dispersed phase consists of particles with different polarisabilities are relatively rare [3].

Our investigation focuses in the ER behaviour of the mixture of anhydrous polyaniline (PANI) and ceramic powder based on Al_2O_3 (micro-corundum, MC) dispersed in silicone oil. For the research suspensions with varying composition and concentration were studied.

2. EXPERIMENTAL

2.1 Materials

PANI was prepared by oxidation of aniline. The polymeric powder was treated with alkali and converted into blue, electrically insulating, emeraldine base with the following characteristics: resistivity > 10^8 Ω.cm, density 1.7 g/cm^3, average diameter of particles 27 μm.

The MC was used as delivered. Its density was 2.4 g/cm^3, the average particle diameter was 9.5 μm.

Both MC and PANI were dried in vacuum at 60°C to constant weight.

Silicone oil, Lukoil 200, was used as a carrier fluid. The viscosity at 20°C was 200 mPa.s, the dielectric constant at 100 Hz 2.48.

2.2 Dielectrical and Rheological Measurements

To measure the dielectric constants (DC) of the suspensions a bridge, TESLA BM 595, with a cylindrical measuring cell was used. The frequency was 100 to 20000 Hz.

Rheological measurements were carried out in a concentric rheometer, Rheotest II, with a gap of 0.8 mm and a shear rate range 8 s^{-1} to 650 s^{-1}. The device was modified at the Institute of Macromolecular Chemistry in Prague so that an electric field could be applied across the gap.

2.3 Conditions

Both dielectric and rheological characteristics were determined at the temperature of 25°C. The ER suspensions containing 0.5; 1; 2; 2.5 and 5 vol.% were mechanically stirred before each measurement. The applied electric field varied from 0 to 1.5 kV/mm.

2.4 Results and Discussion

Figure 1 shows the value of dielectric constant for the suspensions used. The DC values are related to the intensity of ER effect. For instance, a suspension containing 5 vol.% MC had a lower DC value and a lower intensity of ER effect than corresponding suspensions based on PANI, indicating a higher polarisability of the PANI particles. The DC value for low concentrations (1 vol.%) of the solid phase was close to that of the pure carrier fluid.

The composition of the suspensions did not influence the shape of the slightly pseudoplastic flow curves when no electric field was applied. This can be seen in Figure 2, which shows the relationship between relative viscosity (η/η_0) and shear rate ($\dot{\gamma}$) at varying strength of the electric field. In the presence of electric field a significant dependence of low-shear viscosity on composition of the suspensions was observed. The decrease of intensity of the ER effect with increasing shear rate is caused by disintegration of the chain structure. The reason for this is a change in the balance between the polarisation and shear forces when the flow rate increases.

3. CONCLUSIONS

It has been found that MC/PANI suspensions show ER effect because PANI is a polymeric semiconductor and as such easily polarised. The intensity of ER effect for such systems depends on the concentration of PANI particles; higher PANI concentrations enhance the intensity of the ER effect.

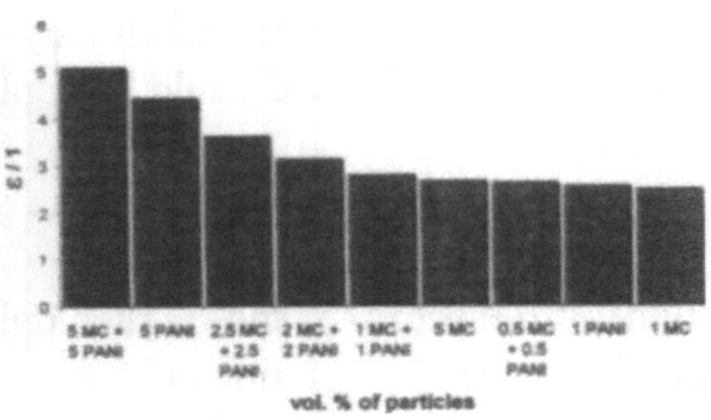

Fig. 1 The value of dielectric constant for the suspensions used

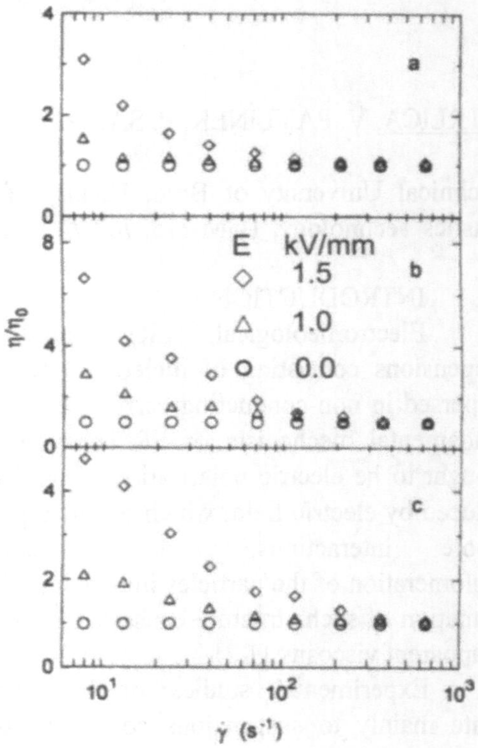

Fig. 2 Relative viscosity η/η_0 vs. shear rate γ for: *(a)* 5 vol. % of MK, *(b)* 5 vol. % of PANI and *(c)* 2.5 vol. % of MK + 2.5 vol. % of PANI

4. REFERENCES

1. Klingenberg, D. J.; Zukoski, C. F. *Langmuir* **1990**, 6, 15
2. Jordan, T., C.; Shaw, T. M. *IEEE Transaction on Electrical Insulation*, **1989**, 24, 849
3. Sakurai, R., See; H., Saito, T. *J. Rheol.* **1996**, 40, 395

MAGNETORHEOLOGY OF MODEL SUSPENSIONS

O.VOLKOVA, S.CUTILLAS, G.BOSSIS,A.MEUNIER

[1] *Laboratoire de physique de la Matière Condensée,CNRS,UMR6622,Parc Valrose,06108 Nice Cedex 2,France*

1. INTRODUCTION

Magnetorheological suspensions (MRS) are suspensions of micronic magnetic particles which exhibit a transformation from a liquid-like structure to a gel like structure when the magnetic field is raised. The yield stress which is necessary to break this structure can be quite important (up to 100 KPa) for concentrated suspensions of particles with a high saturation magnetization "Ginder et al 1996".The interest for applications is to be able to electronically control the yield stress and numerous devices have been proposed which use this effect to control a damping rate or the transmission of a torque or of a force. For these reasons the study of the rheology of MRS has often focused on the yield stress rather than on the dependence between the shear stress and the shear rate. In this paper we focus on the change of structure induced by the flow of a MRS in the presence of a magnetic field.

2. MAGNETIC SUSPENSIONS

Two types of suspensions are studied. The first one is an aqueous suspension of polystyrene particles (average diameter 0.48μm) containing inclusions of magnetite (63% per weight). These inclusions are monodomain but their magnetization can easily jump from one direction of easy magnetization to an other one and the magnetization of the particles follow a Langevin law. The second system is formed of ferromagnetic particles (carbonyl iron) of average diameter 2μm in silicone oil and stabilized against sedimentation by a silica gel. Two kind of experiments have been done: the first one in an oscillatory shear at low Mason number and the second one in a steady shear with a Mason number ranging from zero to a few unities. This Mason number $Ma= 6\pi\mu \dot{\gamma} a^2 / F_m$ ($\dot{\gamma}$ the shear rate, μ the viscosity of the suspending fluid, a the radius of the particles) represents the ratio of the hydrodynamic to the magnetic forces. The two others important dimensionless numbers are the ratio , λ, of the magnetic energy to the thermal energy kT (in this study it is always much larger than unity), and the Peclet number: $Pe=6\pi\mu \dot{\gamma} a^3/kT$

3. OSCILLATING SHEAR FLOW

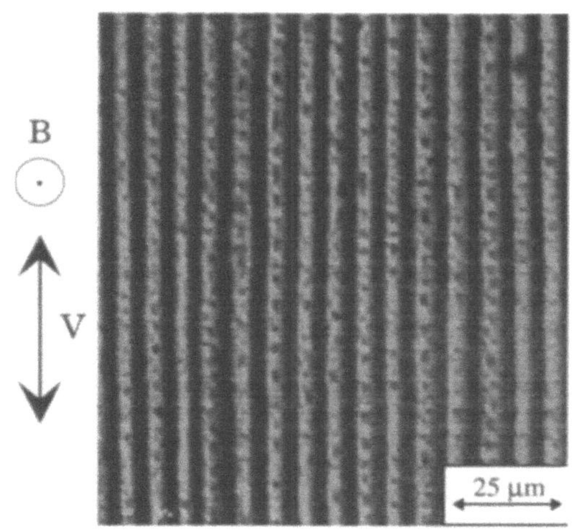

Fig 1 : Parallel stripes formed in magnetic fiel

The experimental cell is made of two disks, one is fixed and the other mounted on a vibrating arm. The fluid is placed between the two disks which are then moved towards each other. The final gap, h, is usually 100μm and the uncertainty on the thickness on the whole surface is less than 1% "Cutillas et al 1998". The magnetic field is perpendicular to the plane of the cell. When the magnetic field is raised above a critical value (λ>5) we have a phase separation with the formation of cylindrical aggregates which arrange on an hexagonal lattice. When we begin to shear the suspension, nothing happens if the strain (γ=x/h) remains

0.15;then , above this value, we observe the formation of stripes which are aligned in the field-velocity plane (cf. fig.1) where the field is perpendicular to the plane of the picture.The period of these stripes is well defined and we have been able to predict its behavior as a function of the field by taking into account the balance between the normal stresses induced by the flow and the osmotic and magnetic pressures.

4.STEADY SHEAR FLOW

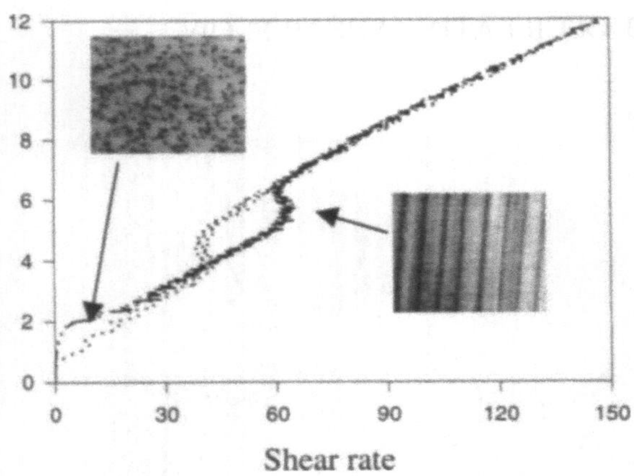

Fig 2. Shear induced phase separation in a steady flow.

When we increase the shear rate in a steady flow and in the presence of a magnetic field we first observe the formation of stripes as in the preceeding case; then ,when we still increase the shear, all the structures disappear. If we reach a high enough shear such that the Mason number is close to unity then we observe a jump in stress (cf. fig.2) which is associated with the onset of a new mesostructure formed of stripes with a typical period which is about 100μm . This new structure can be understood by analogy with the ones which can be found in rotating fields; here the rotational effect comes from the vorticity of the flow. This transition has been found for many different magnetic fluids and is believed to be quite general.

lower than a critical value

It appears that this shear induced transition in a magnetic suspension has many features in common with the one observed on cylindrical micelles. In this kind of system a transition with an increase of stress has also been found recently "P.Boltenhagen et al ,1997" with the existence of a reentrant curve in a stress controlled experiment. We have also found this reentrant behaviour, in a controlled stress experiment. Our system has the advantage compared to micelles to be better characterized and easier to study: the unit is not a surfactant molecule but a spherical particle of micronic size which allows a direct observation of the structures with a microscope. Furthermore the interaction between the particles is simple (dipolar forces) and tunable by the applied field. Also the increase in viscosity is quite easy to understand : the formation of domains with different volume fractions increases the total viscosity due to the non linear behavior of the viscosity with the volume fraction. An other difference with micellar systems is that we have a true hysteresis loop (a waiting time of one hour on the upper part of the loop did not show any tendency for the system to return on the lower branch).So for the same shear rate we have two different equilibrium structures, one corresponding to the homogeneous medium and the other to a phase separated system which is reached by lowering the stress from the domain of high stresses. In this last case the system can remain stable because the stress varies almost linearly with the shear rate in both phases and it must reach the critical stress before it returns to the homogeneous state.

4. REFERENCES

P.Boltenhagen, Y.Hu, E.F.Matthys, D.J.Pine, (1997),Phys.Rev.Lett. **79**,2359

S.Cutillas,G.Bossis,A.Cebers(1998) Phys.Rev E **57**(1) 804-812

J.M.Ginder, L.C.Davis, L.D.Elie(1996) in Electro-Rheological fluids, Magneto-Rheological suspensions and associated technology,Edited by W.A.Bullough (World scientific, suspensions and associated technology, edited Singapore),pp603.

NON-STATIONARY SIMPLE COUETTE FLOW OF THE OLDROYD'S 3 - CONSTANTS MODEL AT SMALL REYNOLDS NUMBERS

C.BALAN

"Politehnica" University, Hydraulics – REOROM, Splaiul Independentei 313, 79590 Bucharest, Romania

1. INTRODUCTION

In the last decade, the experimental and theoretical investigations on the instabilities of viscoelastic fluids put in evidence possible connections between the time oscillations of the flow parameters (in the limit of zero Reynolds numbers), their non-monotone steady flow curves and phenomena as spurt or wall depletion, Larson (1992) and Schowalter (1988). In this context the study of the non-stationary regime in simple shear motions becomes of great scientifical interest. The aim of this study is to model the onset of the motion in the start-up of Couette flow between parallel planes of the Oldroyd's 3 constant model with non-monotone flow curve, respectively to investigate the non-stationary material behaviour in the stressing rheological test. In this case it is important to establish: (i) to what extend the assumption that a unique constant steady strain rate is installed in the gap remains valid for constitutive relations with non-monotone steady flow curve, (ii) to compare the time scale of the equation of motion (i.e. the duration up the velocity distribution becomes steady) with the proper time of the constitutive relation (i.e. the duration of the non-stationary regime under a constant velocity distribution). In this paper are presented the first results of our investigations.

2. THE START-UP COUETTE MOTION OF LINEAR VISCOELASTIC FLUIDS

The start-up Couette motion between infinite plates is a viscometric flow described by the equation of motion,

$$\rho\frac{\partial v}{\partial t}=\frac{\partial \tau}{\partial y} \tag{1}$$

with the boundary and initial conditions,

$v = 0$ for $t \leq 0$ and $0 \leq y \leq h$;

$v = 0$ at $y = 0$ and $v = V$ at $y = h$, for $t > 0$. (2)

In (1) τ in the shear stress and y the co-ordinate normal to the plates and the velocity $v = v(y, t)$; for $h \rightarrow \infty$ this problem is called the Rayleigh's, or the first Stokes's problem. Equation (1) and the conditions (2) describe the propagation of waves of shear in viscoelastic fluid and its study is fundamental in establishing the mechanism of the disturbances propagation in the material and the time scale associated with the equation of motion in the rheological stressing test. For a Newtonian fluid (1) is identically with the heat (diffusion) equation and for the linear Maxwell fluid with the telegraph equation (i.e. a perturbed wave equation) which in the non-dimensional form is given by

$$Wi\frac{\partial^2\bar{v}}{\partial\bar{t}^2}+\frac{\partial\bar{v}}{\partial\bar{t}}=\frac{1}{Re}\frac{\partial^2\bar{v}}{\partial\bar{y}^2} \tag{3}$$

with $\bar{v} = v/V$, $\bar{t} = Vt/h$; $Wi = \lambda_1 V/h$, $Re = \rho Vh/\eta_0$ are respectively the Weissenberg and Reynolds numbers, η_0 is the zero viscosity and λ_1 is the relaxation time of the Maxwell model, see Böhme (1981). The same problem but for the second order fluid, Maxwell and Oldroyd's B model, the Leonov fluid and the Doi – Edwards model has been studied by Ting (1963), Narain (1983) and Vrentas&Vrentas (1995), Braun (1989), respectively Öttinger (1994); for a review on the wave propagation in the linear viscoelastic fluids see Joseph (1990). We have to remark that in the range of the linear viscoelasticity the waves discontinuity propagate into the fluid at rest with the speed $c = \sqrt{G(0)/\rho}$, where $G(0)$ is the present value of the relaxation function (i.e. the corresponding elastic modulus $G = \eta_0/\lambda_1$ in the Maxwell linear model).

3. THE CASE OF MODELS WITH NON-MONOTONE FLOW CURVE

Some differential models which involves the Gordon – Schowalter objective derivatives,

$$\frac{D\bullet}{Dt}=\frac{d\bullet}{dt}-\mathbf{\Omega}\bullet+\bullet\mathbf{\Omega}-a(\mathbf{D}\bullet+\bullet\mathbf{D}) \tag{4}$$

where $d\bullet/dt$ is the material time derivative, $\mathbf{D}$ is the stretching and $\mathbf{\Omega}$ is the spin tensor, as the Johnson& Segalman model or the 3 constant Oldroyd's relation,

$$\lambda_1\frac{D\mathbf{T}}{Dt}+\mathbf{T}=2\eta_0\mathbf{D}+2\eta_0\lambda_2\frac{D\mathbf{D}}{Dt}, \tag{5}$$

disclose non-monotone material functions if the modulus of the "slippage" parameter a is less than one; for $a = \pm 1$ the objective derivatives are convected and the material functions in viscometric motions are independently on the strain rate. In (5) $\mathbf{T}$ is the extra-stress tensor and λ_2 defines the retardation time. The steady flow curve (i.e. $\partial\bar{v}/\partial\bar{y}=1$) of the model,

$$\bar{\tau}=Wi\frac{1+\kappa(1-a^2)Wi^2}{1+(1-a^2)Wi^2} \tag{6}$$

with $\kappa = 0.01$ and a as parameter is shown in Fig. 1. The exact solution of (5) for the viscometric flow is given in Balan&Fosdick (1998), respectively

$$\bar{\tau}(\bar{y},\bar{t}) = \kappa \cos[\sqrt{1-a^2}\gamma(\bar{t})] \frac{\partial \bar{v}(\bar{y},0)}{\partial \bar{y}} e^{-\frac{\bar{t}}{Wi}} + \kappa \frac{\partial \bar{v}(\bar{y},\bar{t})}{\partial \bar{y}} + \frac{1-\kappa}{Wi} \int_0^{\bar{t}} e^{-\frac{\bar{t}-\bar{s}}{Wi}} \cos[\sqrt{1-a^2}(\gamma(\bar{t})-\gamma(\bar{s}))] \frac{\partial \bar{v}(\bar{y},\bar{s})}{\partial \bar{y}} d\bar{s}. \tag{7}$$

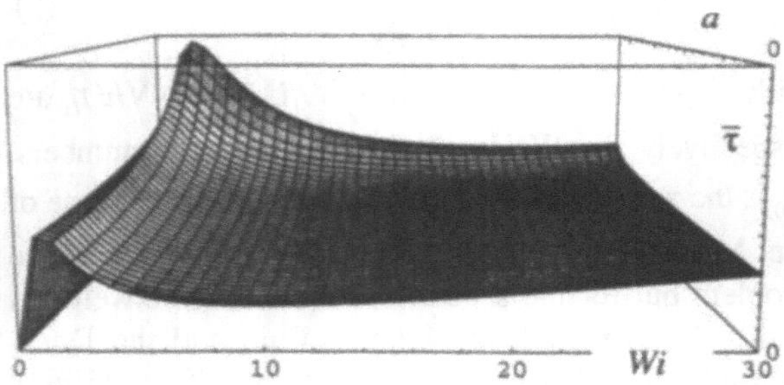

Figure 1 Non-monotone steady flow curve of (5).

The coefficient $\kappa = \lambda_2 / \lambda_1$ establishes the ratio between the infinite and zero shear viscosity and also, for $\kappa < 0.1$, the magnitude of the "jump" at the constant shear stress (for $\kappa \geq 1/9$ the flow curve is strict monotone), see for details Balan&Fosdick (1998), Kolkka et al. (1988). It is important to remark that this type of constitutive models determine coexistent strain rates in the gap, respectively a banding flow structure of the flow, Espanol et al. (1996). To find the solution of (1-2) and (7) is not trivial since the history of the deformation process, the strain $\gamma(\bar{s})$, appears in the kernel of the integral representation of the shear stress. In the case of slow motions, i.e. small Reynolds numbers, is possible to obtain similar solutions with the linear viscoelasticity (3) by solving the equation,

$$Wi \frac{\partial^2 \bar{v}}{\partial \bar{t}^2} + \frac{\partial \bar{v}}{\partial \bar{t}} = \frac{1}{Re}(\bar{\eta}_d \frac{\partial \bar{v}}{\partial \bar{y}} + \bar{\eta}) \frac{\partial^2 \bar{v}}{\partial \bar{y}^2} \tag{8}$$

where $\bar{\eta}_d = \partial \bar{\eta} / \partial(\partial \bar{v} / \partial \bar{y})$ is the differential viscosity and $\bar{\eta} = \eta / \eta_0$ is the non-dimensional viscosity.

We expect that the solutions of (8) for $Re \ll 1$ and moderate Wi numbers to describe at least qualitatively the onset of the Couette motion. The numerical simulations of (3) and (8) with conditions (2) are shown in Fig. 2 for $Re = 0.01$ and $Wi = 0.1$.

4. FINAL REMARKS

An approximate solution of the onset of the Couette flow has been obtained for the Oldroyd's 3 constants model with non-monotone flow curve. The non-linearity of the shear stress introduces "perturbations" in the velocity distribution in the vicinity of the moving plate and damps the waves amplitude. As consequence we have a velocity distribution with "kinks" and different coexisting strain rates in the gap, result which is consistent with the solutions for $Re = 0$. At this level of investigations we did not obtain information about the time scale of the equation of motion.

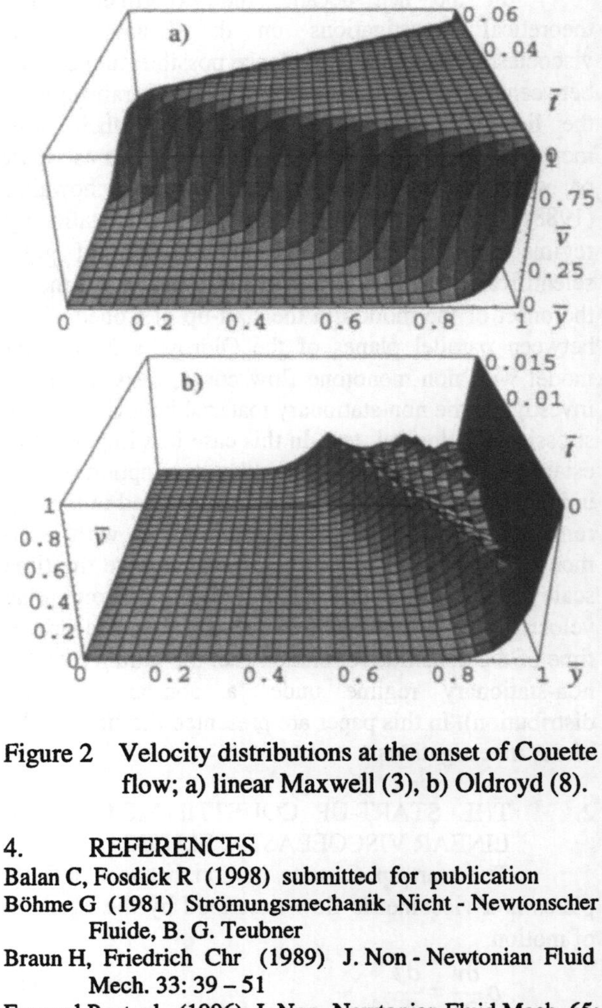

Figure 2 Velocity distributions at the onset of Couette flow; a) linear Maxwell (3), b) Oldroyd (8).

4. REFERENCES

Balan C, Fosdick R (1998) submitted for publication

Böhme G (1981) Strömungsmechanik Nicht - Newtonscher Fluide, B. G. Teubner

Braun H, Friedrich Chr (1989) J. Non - Newtonian Fluid Mech. 33: 39 – 51

Espanol P et al. (1996) J. Non -Newtonian Fluid Mech. 65: 93 - 109

Joseph DD (1990) Fluid Dynamics of Viscoelastic Liquids, Springer Verlag

Kolkka RW et al. (1988) J. Non -Newtonian Fluid Mech. 29: 303 – 335

Larson RG (1992) Rheol. Acta 31: 213 – 263

Narain A (1983) Shearing Deformation of Simple Materials, Univ. Minnesota

Öttinger ChH, Laso M (1994) in: Lopez M, Vanea C (eds) Lectures on Thermodynamics and Statistical Mech. pp. 139 - 153

Schowalter WR (1988) J. Non - Newtonian Fluid Mech. 29: 24 – 36

Ting TW (1963) Arch. Ration. Mech. Anal. 14(1): 1 – 26

Vrentas JS, Vrentas Ch.M (1995) Ind. Eng. Chem. Res. 34: 3203 - 3207

SECONDARY FLOW OF FLUIDS IN PIPES OF NON-CIRCULAR CROSS SECTION

C. BIRD, T. N. PHILLIPS and K. WALTERS

Department of Mathematics, University of Wales, Aberystwyth SY23 3BZ, UK.

1. INTRODUCTION

It has been known for over four decades that rectilinear flow through straight pipes of non-circular cross section is not always possible for elastic liquids (see, for example, Rivlin and Ericksen (1955)). Furthermore, the second normal stress difference N_2 has been clearly identified as the viscometric function of relevance in this connection, since, when $N_2 = 0$, rectilinear flow is always possible in the laminar regime (see, for example, Oldroyd (1965)). Indeed, there were early expectations that the strength and direction of the resulting secondary flow could be used to estimate the sign and magnitude of N_2. However, this was shown to be a false hope, with different signs of N_2 giving rise to the same flow direction and vice versa (see, for example, Dodson et al. (1973), Townsend et al. (1976)).

Notwithstanding this negative conclusion, there has been a lingering fascination with the general problem and a frustration that little or no useful viscometric information can be obtained from the associated experiments. Accordingly, we set about the task of showing unequivocally why the relationship between secondary flow and N_2 is not a simple one. To facilitate this, we felt it prudent to remain within the haven of generality, so that our conclusions could be shown to be sacrosanct.

Complete generality is not possible in such a difficult problem, but we are able to satisfy our stated aim by confining attention to a slow flow regime, where there has already been some useful analysis by Langlois and Rivlin (1963). Our aim is to revisit the problem with a view to expressing the equations in terms of the viscometric functions rather than the plethora of constants involved in the hierachy equations for simple fluids in slow flow (see Coleman and Noll (1961)). This is especially important, since it is already known that, in the problem under discussion, it is necessary to proceed to fourth-order approximation before secondary flows in the cross section of the pipe manifest themselves and the corresponding equation involves fourteen constants.

We may therefore state our problem as follows. Given that it is necessary to consider fourth-order approximation, is it possible to express the resulting differential equations for the flow in terms of the viscometric functions? If it is, then there is every hope that the enduring frustration mentioned earlier can be explained.

2. BASIC EQUATIONS

Langlois and Rivlin (1963) have shown that it is necessary to consider a fourth-order approximation in the speed of flow before secondary flows manifest themselves. The relevant equation can be expressed in the form

$$\sigma_{ik} = -p\delta_{ik} + T_{ik}, \tag{1}$$

$$\begin{aligned} T_{ik} = {} & \alpha_1 A^{(1)}_{ik} + \alpha_2 A^{(2)}_{ik} + \alpha_3 A^{(1)}_{ij} A^{(1)}_{jk} \\ & + \beta_1 A^{(1)}_{jl} A^{(1)}_{lj} A^{(1)}_{ik} + \beta_2 A^{(3)}_{ik} \\ & + \alpha_5 \left(A^{(1)}_{ij} A^{(2)}_{jk} + A^{(2)}_{ij} A^{(1)}_{jk} \right) \\ & + \left(\alpha_6 A^{(1)}_{jl} A^{(2)}_{lj} + \alpha_7 A^{(1)}_{mj} A^{(1)}_{jl} A^{(1)}_{lm} \right) A^{(1)}_{ik} \\ & + \alpha_8 A^{(1)}_{jl} A^{(1)}_{lj} A^{(1)}_{im} A^{(1)}_{mk} + \alpha_9 A^{(1)}_{jl} A^{(1)}_{lj} A^{(2)}_{ik} \\ & + \alpha_{10} \left(A^{(1)}_{ij} A^{(1)}_{jl} A^{(2)}_{lk} + A^{(2)}_{ij} A^{(1)}_{jl} A^{(1)}_{lk} \right) \\ & + \alpha_{11} A^{(2)}_{ij} A^{(2)}_{jk} + \alpha_{12} \left(A^{(1)}_{ij} A^{(3)}_{jk} + A^{(3)}_{ij} A^{(1)}_{jk} \right) \\ & + \alpha_{13} A^{(4)}_{ik}, \end{aligned} \tag{2}$$

where α_1, ..., α_{13}, β_1, β_2 are material constants, $A^{(n)}_{ik}$ is the nth Rivlin-Ericksen (1955) tensor, σ_{ik} is the stress tensor, p is an arbitrary isotropic pressure and T_{ik} is the extra stress tensor.

If we look at the form of the stresses in the case of a steady simple shear we see that the viscosity η and N_2 can be written in terms of the material constants and, for simplicity, we shall define η_0, η_2, $N_2^{(2)}$ and $N_2^{(4)}$ as follows:

$$\begin{aligned}\eta(\dot\gamma) &= \alpha_1 + 2(\beta_1+\alpha_5)\dot\gamma^2 = \eta_0 + \eta_2\dot\gamma^2,\\ N_2(\dot\gamma) &= (2\alpha_2+\alpha_3)\dot\gamma^2\\ &\quad +2(\alpha_8+2\alpha_9+2\alpha_{10}+2\alpha_{11})\dot\gamma^4\\ &= N_2^{(2)}\dot\gamma^2 + N_2^{(4)}\dot\gamma^4. \qquad (3)\end{aligned}$$

It is assumed that secondary flow exists in a pipe whose cross section is given by the surface $f(x,y)=0$. The flow field to fourth order is

$$\begin{aligned}u_1 &= \varepsilon^4 u_1^{iv},\\ u_2 &= \varepsilon^4 u_2^{iv},\\ u_3 &= \varepsilon u_3^i + \varepsilon^3 u_3^{iii} + \varepsilon^4 u_3^{iv}, \qquad (4)\end{aligned}$$

where the Roman numeral superscript indicates the power of ε to which the u_i correspond.

By equating corresponding order terms in the equations of motion, eliminating pressure terms and introducing a stream function so that $u_1^{iv} = \chi_{,2}$ and $u_2^{iv} = -\chi_{,1}$ an equation for $\nabla^4\chi$ is obtained.

As observed by Oldroyd (1965), we see that this resulting equation is, in fact, the fourth order term of

$$\frac{\partial(M^*, w)}{\partial(x,y)},$$

where M^* is defined as

$$M^* \equiv \frac{\partial}{\partial x}\left[\frac{\partial w}{\partial x}\psi_2(\dot\gamma)\right] + \frac{\partial}{\partial y}\left[\frac{\partial w}{\partial y}\psi_2(\dot\gamma)\right]$$

and

$$\begin{aligned}w &= \varepsilon u_3^i + \varepsilon^3 u_3^{iii},\\ \psi_2(\dot\gamma) &= N_2^{(2)} + N_2^{(4)}\dot\gamma^2.\end{aligned}$$

This observation enables us to write the resulting equation for $\nabla^4\chi$ in the form:

$$\nabla^4\chi = -\frac{G^4}{\eta_0^6}(\eta_2 N_2^{(2)} - \eta_0 N_2^{(4)})g(x,y). \qquad (5)$$

From this equation, it is now clear exactly how η and N_2 affect the secondary flow.

3. CONCLUSIONS

We have shown conclusively that it is not possible to associate the sign of N_2 with the direction of the streamlines. The contribution which determines the magnitude and direction of the streamlines is

$$-\frac{G^4}{\eta_0^6}(\eta_2 N_2^{(2)} - \eta_0 N_2^{(4)}). \qquad (6)$$

We know that, for most non-Newtonian liquids, $\eta_0 > 0$, $\eta_2 < 0$, $N_2^{(2)} < 0$, $N_2^{(4)} > 0$, so we cannot say with any assurance whether the function given by (6) will be positive or negative in general.

For completion, we have computed secondary flows using a spectral method for pipes of square and rectangular cross section. Illustrated in Figure 1 is the secondary flow in a rectangular pipe with a 10:9 ratio for the bracketed term negative. This direction is consistent with the experimental results of Dodson et al. (1973) and Townsend et al. (1976).

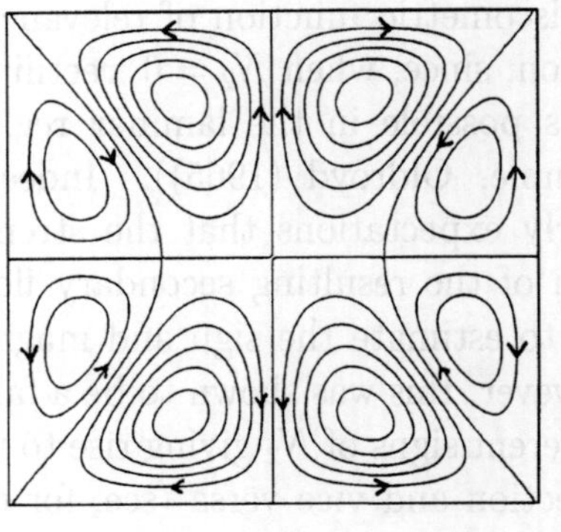

Figure 1: Secondary flow through a rectangular pipe with ratio 10:9 using a spectral method

In conclusion, if we know η_0, η_2, $N_2^{(2)}$ and $N_2^{(4)}$, we will then know the direction of the streamlines, but if we only know the direction of the streamlines, we can say nothing about N_2.

4. REFERENCES

Coleman BD and Noll W (1961) Rev. Mod. Phys. 33: 239–249

Dodson AG, Townsend P and Walters K (1974) Comp. Fluids 2: 317–338

Langlois WE and Rivlin RS (1963) Rendiconti di Mathematica 22: 169–185

Oldroyd JG (1965) Proc. Roy. Soc. A283: 115–133

Rivlin RS and Ericksen JL (1955) J. Ration. Mech. Anal. 4: 323–425

Towsend P, Walters K and Waterhouse WM (1976) J. Non-Newtonian Fluid Mech. 1: 107–123

ANNULAR LINEAL COUETTE-POISEUILLE FLOW OF VISCOPLASTIC FLUIDS

J.DAVID, P.FILIP

Institute of Hydrodynamics, Acad.Sci.Czech Rep., Pod Paťankou 5, 166 12 Prague 6, Czech Republic

SUMMARY

The laminar axial flow of non-Newtonian fluids obeying Robertson-Stiff model in concentric annuli is analyzed. Flow is caused simultaneously by the inner cylinder moving along its axis and by the pressure gradient imposed in the axial direction. Both cases - either pressure gradient assists to the moving cylinder or opposes - are considered. All possible cases with respect to the possible positions of the plug flow regions are uniquely diversified through the derived analytical criteria using the entry (geometrical, kinematical and rheological) parameters. For each possible case there is derived the explicit analytical expression for the volumetric flow rate.

1. INTRODUCTION

A series of papers has been devoted to the description of flow of non-Newtonian fluids in concentric annuli. The problem discussed supposes that the flow is steady, laminar, incompressible, isothermal and axial with negligible end effects of the cylinders. The aim consists in determination of the volumetric flow rate.

Under the assumption of immobile walls the axial flow rate was studied by a series of authors. The literature on the moving inner cylinder (under simultaneous action of pressure gradient) is more scarce. For the case of the inner cylinder moving along its axis Wadhwa (1966) obtained the integral form for the axial velocity profile in the case of Ellis fluid. Lin and Hsu (1980) studied power-law fluids and obtained the integral form for the flow rate. For the same problem Malik and Shenoy (1991) succeeded to derive the analytical form, furthermore they considered both directions of the exerted pressure gradient and provided criteria for diversification among the individual possible cases.

The advantage of analytical expressions for volumetric flow rate over the numerical results is their relative simplicity, determination how the entry parameters participate (functionally) in the solution. However, there still remains one integration parameter that has to be determined numerically (expressing the location where shear stress attains zero value or alternatively expressing the continuity of the velocity profiles).

This contribution deals with axial flow through the concentric annulus. The inner cylinder is moving at a constant velocity along its axis and simultaneously axial pressure gradient is exerted to Robertson-Stiff fluid. The aim is to provide a unique classification of all possible cases (including possible appearance of plug flow regions) and for each case to derive volumetric flow rate in an analytical form.

2. BASIC EQUATIONS

Under the assumptions introduced at the beginning of the Introduction the balance equation is of the form

$$\frac{1}{r}\frac{d(r\tau_{rz})}{dr} = P \tag{1}$$

with the boundary conditions

$$v_z(\kappa R) = V > 0 \quad , \quad v_z(R) = 0 \tag{2}$$

and Robertson-Stiff model (1976) (written in the form useful for the cylindrical coordinates)

$$\tau_{rz} = \left[K^{\frac{1}{n}} \left| \frac{dv_z}{dr} \right|^{\frac{n-1}{n}} + \tau_0^{\frac{1}{n}} \left| \frac{dv_z}{dr} \right|^{-\frac{1}{n}} \right]^n \left(-\frac{dv_z}{dr} \right) \quad \text{for } |\tau_{rz}| \geq \tau_0$$

$$\frac{dv_z}{dr} = 0 \quad \text{for} \quad |\tau_{rz}| \leq \tau_0 \quad . \tag{3}$$

For the sake of simplicity the following transformations converting the problem to the dimensionless form are used

$$\xi = \frac{r}{R}, \phi = \frac{v_z}{V}, T = \frac{2\tau_{rz}}{|P|R}, T_0 = \frac{2\tau_0}{|P|R}, \Lambda = \frac{|P|R}{2K}\left(\frac{R}{V}\right)^n, Q = \frac{q}{2\pi R^2 V} \tag{4}$$

Then, the problem can be reformulated to the form (we put s=1/n)

$$(\operatorname{sgn}(P)) \cdot T = \xi - \frac{\lambda^2}{\xi} \quad , \tag{5}$$

$$\phi(\kappa)=1 \quad , \quad \phi(1)=0 \quad , \tag{6}$$

$$T=\left[\Lambda^{-s}\left|\frac{d\phi}{d\xi}\right|^{1-s}+T_0^{-s}\left|\frac{d\phi}{d\xi}\right|^{-s}\right]^{n}\left(-\frac{d\phi}{d\xi}\right) \quad \text{for} \quad |T|\geq T_0 \tag{7}$$

$$\frac{d\phi}{d\xi}=0 \quad \text{for} \quad |T|\leq T_0 \tag{8}$$

where λ^2 is a dimensionless constant of integration.

If λ_i, λ_o denote the boundary values of the plug flow region, then from eq 5 it follows that

$$\lambda^2=\lambda_i\lambda_o \quad , \quad \lambda_i=\lambda_o-T_o \quad . \tag{9}$$

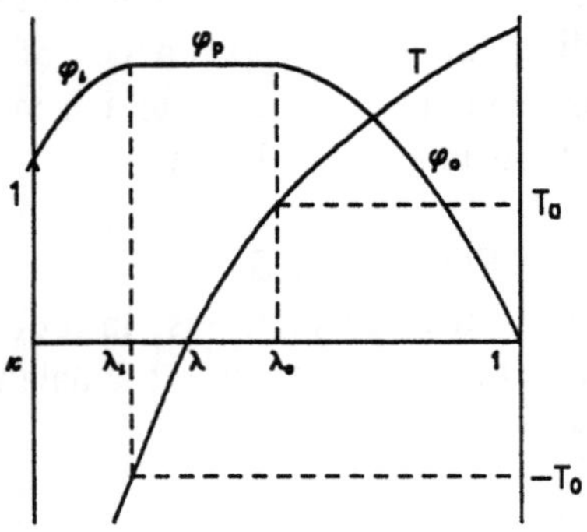

Fig.1 Definition sketch

3. THE INDIVIDUAL FLOW SITUATIONS

It is necessary to analyze the cases P>0 (pressure gradient assists the drag on the fluid caused by the moving cylinder) and P<0 (pressure gradient opposes the drag on the fluid caused by the moving cylinder) separately. In either case there are three possible flow situations. Due to space limitation we restrict the derivation only to one out of three situations for P>0,see Fig.1. Using straightforward calculation, eqs 5-8 imply that

$$\frac{d\phi_i}{d\xi}=\Lambda^s\left[\left(\frac{\lambda^2}{\xi}-\xi\right)^s-T_o^s\right] \quad \text{for}\ \kappa\leq\xi<\lambda_i \quad \left(\text{where}\ \frac{d\phi}{d\xi}>0\right)$$

$$\frac{d\phi_p}{d\xi}=0 \quad \text{for} \quad \lambda_i\leq\xi\leq\lambda_o \quad , \tag{10}$$

$$\frac{d\phi_o}{d\xi}=-\Lambda^s\left[\left(\xi-\frac{\lambda^2}{\xi}\right)^s-T_o^s\right] \quad \text{for}\ \lambda_o<\xi\leq 1 \quad \left(\text{where}\ \frac{d\phi}{d\xi}<0\right)$$

In this case the equation for determination of the parameter λ_o follows from the condition

$$\phi_i(\lambda_i)=\phi_o(\lambda_o) \quad . \tag{11}$$

Using eqs 9-10 we obtain

$$\int_{\kappa}^{\lambda_o-T_o}\Lambda^s\left(\frac{\lambda_o(\lambda_0-T_0)}{\xi}-\xi\right)^s d\xi-\int_{\lambda_o}^{1}\Lambda^s\left(\xi-\frac{\lambda_o(\lambda_0-T_0)}{\xi}\right)^s d\xi-$$
$$-(2\lambda_o-T_o-\kappa-1)\Lambda^sT_o^s+1=0 \tag{12}$$

The limiting situation, when this situation occurs, appears for $\lambda_i \equiv \kappa$. Substituting this λ_i into eq 12 we obtain the critical value of Λ, say $\Lambda_{cr1,2}$, above which there exists an inner plug flow region

$$\Lambda_{cr1,2}=\left[(\kappa+T_0-1)T_o^s+\int_{\kappa+T_o}^{1}\left(\xi-\frac{\kappa(\kappa+T_o)}{\xi}\right)^s d\xi\right]^{-n} \quad . \tag{13}$$

The expression for flow rate is of the form

$$Q=\frac{1}{2}\left(\frac{1-s}{3+s}\lambda^2-\kappa^2\right)+\frac{\Lambda^s}{2(3+s)}\left[\left(1-\lambda^2\right)^{1+s}-\lambda_0^{1-s}\left(\lambda_o^2-\lambda^2\right)^{1+s}\right]+$$
$$+\frac{\Lambda^s}{2(3+s)}\left[\lambda_i^{1-s}\left(\lambda^2-\lambda_i^2\right)^{1+s}-\kappa^{1-s}\left(\lambda^2-\kappa^2\right)^{1+s}\right]-$$
$$-\left[\frac{1+\kappa^3-\lambda_i^3-\lambda_o^3}{6}-\frac{1-s}{2(3+s)}\lambda^2(1+\kappa-\lambda_i-\lambda_o)\right]\Lambda^sT_o^s \tag{14}$$

4. DISCUSSION

Choosing τ_0=0 in Robertson-Stiff model, eq 3, we obtain the power-law model. If we put τ_0=0, eq 14 for volumetric flow rate reduces to the one derived in Malik and Shenoy (1991) concerning power-law fluids. If $\tau_0\neq 0$ and we choose V=0, we get the relation for flow rate given in Gücüyener and Mehmetoglu (1992) that was derived for Robertson-Stiff model with stationary inner cylinder.

ACKNOWLEDGMENTS

The authors wish to acknowledge GA AS CR for the financial support of Grant No.A2060702.

REFERENCES

Gücüyener HI, Mehmetoglu T (1992) AIChE J. 38:1139-1143

Lin SH, Hsu CC (1980) Ind.Eng.Chem.Fundam. 19:421-424

Malik R, Shenoy UV (1991) Ind.Eng.Chem.Res. 30:1950-1954

Robertson RE, Stiff HA (1976) Soc.Pet.Eng.J. Feb:31-36

Wadhwa YD (1966) AIChE J.12:890-893

POWER-LAW FLUID FLOW THROUGH DOUBLY CONNECTED CONCENTRIC ANNULI

J.DAVID, P.FILIP

Institute of Hydrodynamics, Acad.Sci.Czech Rep., Pod Paťankou 5, 166 12 Prague 6, Czech Republic

1. INTRODUCTION

Steady axial Poiseuille flow of power-law fluids in concentric annuli under the assumption of constant pressure gradient was solved by Fredrickson and Bird (1958) and analytically by Hanks and Larsen (1979). Recently attention has been paid to the problem of flow-rate enhancement if we replace a pure concentric annular passage by a combination of a concentric annulus (of the same outer diameter) and a concentrically inserted tube. This combination in sum preserves the cross-sectional area of a pure annulus (see Fig.1). This problem was solved for a laminar Newtonian case by Ranger (1994), and in more detail treated in David and Filip (1996a). It was shown that there exist possibilities of dramatic enhancement of flow rate by a proper choice of geometrical parameters (i.e. of individual radii). Gupta and Chhabra (1996) demonstrated this approach for power-law fluids. They presented results for a few selected geometries. For a determination of flow rate they used the explicit analytical relation derived by Hanks and Larsen (1979). However, this solution contains a parameter λ (representing a location of maximum axial velocity in an annulus) whose determination is not possible to carry out analytically.

The axial flow of power-law fluids in annuli exhibits quasisimilarity behavior, see David and Filip (1994). This means that this type of flow is possible to analyze, to some extent, in a universal form valid in a broad range of entry parameters n, κ. This note uses this quasisimilarity and reveals general characteristics of flow-rate enhancement for power-law fluids. In a numerical form, it offers the application of the approximate analytical result for annular flow rate derived in David and Filip (1996b) that eliminates the parameter λ. The derived resulting relation for flow-rate enhancement differs from the exact one negligibly.

2. ANALYSIS

Using Fubini's theorem, Hanks and Larsen (1979) derived for determination of flow rate in an annulus the following relation

$$Q_{ann,\kappa} = \frac{\pi R^3 n}{1+3n}\left(\frac{PR}{2k}\right)^{\frac{1}{n}}\left[(1-\lambda^2)^{1+\frac{1}{n}} - \kappa^{1-\frac{1}{n}}(\lambda^2-\kappa^2)^{1+\frac{1}{n}}\right] \qquad (1)$$

where R is the radius of the outer cylinder, κ is the aspect ratio of the inner-to-outer radii, and λ denoting the location of maximum axial velocity is the solution of the integral equation

$$\int_{\kappa}^{\lambda}\left(\frac{\lambda^2}{\xi}-\xi\right)^{\frac{1}{n}} d\xi = \int_{\lambda}^{1}\left(\xi-\frac{\lambda^2}{\xi}\right)^{\frac{1}{n}} d\xi \quad . \qquad (2)$$

For $\kappa \to 0$ relation (1) reduces to the classical expression for flow rate through a tube

$$Q_{tube,R} = \frac{\pi R^3 n}{1+3n}\left(\frac{PR}{2k}\right)^{\frac{1}{n}} \qquad (3)$$

Let us denote (for notation see Fig.1)

$$\kappa_i = \frac{r_i}{r_4} \quad , \quad i = 1,2,3 \quad , \qquad (4)$$

$$q = \frac{Q_{ann,\kappa_3} + Q_{tube,r_2}}{Q_{tube,r_4}} \quad , \qquad (5)$$

$$s = \frac{Q_{ann,\kappa_3} + Q_{tube,r_2}}{Q_{ann,\kappa_1}} \quad . \qquad (6)$$

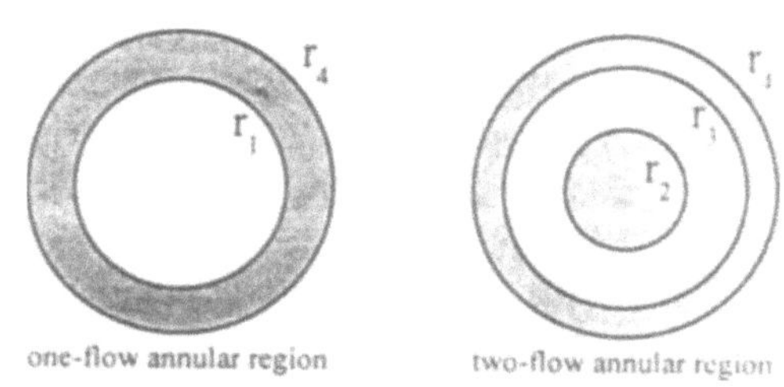

Fig.1 Flow arrangement

To ensure the equality of cross-sectional areas for an annulus (r_1,r_4) with the sum of an annulus (r_3,r_4) and a tube (r_2) the following necessary condition must hold

$$\kappa_2 = \sqrt{\kappa_3^2 - \kappa_1^2} \quad . \tag{7}$$

Then q represents a non-dimensional flow rate and S a flow-rate enhancement as discussed in the introductory section.

Substituting relations (1)-(4),(7) into (5),(6) we obtain

$$q = \left[(1-\lambda_3^2)^{1+\frac{1}{n}} - \kappa_3^{1-\frac{1}{n}} (\lambda_3^2 - \kappa_3^2)^{1+\frac{1}{n}} \right] + \left(\kappa_3^2 - \kappa_1^2\right)^{\frac{1+3n}{2n}} \tag{8}$$

$$S = q \left[(1-\lambda_1^2)^{1+\frac{1}{n}} - \kappa_1^{1-\frac{1}{n}} (\lambda_1^2 - \kappa_1^2)^{1+\frac{1}{n}} \right]^{-1} \quad . \tag{9}$$

The behavior of function $q=q(\kappa_1,\kappa_3,n)$ for an arbitrary n and for $\kappa_1 \in [0,1]$, $\kappa_3 \in [\kappa_1,1]$ is analogous to a Newtonian case n=1 - flow rate attains its maximum for the case of a tube flow, i.e. when $\kappa_3 = 1$ and $\kappa_2 = (1-\kappa_1^2)^{1/2}$.

For a given $n \geq 0.1$ the magnitude q starts from the value q_{ann} for $\kappa_3=\kappa_1$ (and $\kappa_2=0$), then it decreases to q_{min}, increases back to the value q_{ann} and finally for $\kappa_3=1$ (purely tube flow) attains its maximum. As a consequence of a quasisimilarity behavior, the minima of q for given κ_1 and arbitrary n nearly coincide, and the curve of κ_3 for which q attains its minima can be approximated by the relation

$$\kappa_{min} = 0.53 + 0.38\kappa_1^2 + 0.09\kappa_1^3 \quad . \tag{10}$$

Relation

$$\kappa_{ann} = \kappa_1 + \left(1-\kappa_1\right)^2 \tag{11}$$

determines the approximate values of κ_{ann} where q attains again the value q_{ann}.

Consequently, with κ_3 going from κ_1 to 1, the flow-rate enhancement parameter S starts from 1, decreases and minimizes at $\kappa_3=\kappa_{min}$, again goes through 1 at $\kappa_3=\kappa_{ann}$, and finally attains its theoretical maximum for $\kappa_3=1$.

To eliminate the parameter λ from rels.(8),(9) the approximate relation for annular flow rate derived in David and Filip (1996b) is useful

$$Q_{ann,appr} = \frac{\pi R^3 n}{1+3n}\left(\frac{PR}{2k}\right)^{\frac{1}{n}} (1-\kappa)^{2+\frac{1}{n}} (1+\kappa) \cdot Q_{norm}(\kappa,n) \tag{12}$$

where

$$Q_{norm}(\kappa,n) = \frac{1}{1+2n}\left[(1-n)Q_{norm}(\kappa,0) + 3n Q_{norm}(\kappa,1)\right] , \tag{13}$$

$$Q_{norm}(\kappa,0) = \frac{1}{2}\left[1 + (3Q_{norm}(\kappa,1) - 2)^{0.9 \cdot \kappa^{3/8}}\right] , \tag{14}$$

$$Q_{norm}(\kappa,1) = \frac{1+\kappa}{1-\kappa}\left(\frac{1+\kappa^2}{1-\kappa^2} + \frac{1}{\ln\kappa}\right) . \tag{15}$$

Applying these expressions to rel.(8) the obtained approximate result for q

$$q_{appr} = (1-\kappa_3)^{2+\frac{1}{n}} (1+\kappa_3) \cdot Q_{norm}(\kappa_3,n) + \left(\kappa_3^2 - \kappa_1^2\right)^{\frac{1+3n}{2n}} \tag{16}$$

deviates from the exact one for an arbitrary n at most by 2.15% for $\kappa_1 \geq 0.025$, 0.4% for $\kappa_1 \geq 0.5$, and 0.16% for $\kappa_1 \geq 0.6$. The corresponding relative deviation for S

$$S_{appr} = \frac{q_{appr}}{(1-\kappa_1)^{2+\frac{1}{n}} (1+\kappa_1) \cdot Q_{norm}(\kappa_1,n)} \tag{17}$$

does not exceed approximately the same values in the corresponding intervals of κ_1 for an arbitrary n.

ACKNOWLEDGMENTS

The authors wish to acknowledge GA AS CR for the financial support of Grant No.A2060702.

REFERENCES

David J, Filip P (1994) Fluid Dyn.Res. 14: 63-70

David J, Filip P (1996a) Acta Techn.CSAV 41: 51-55

David J, Filip P (1996b) J.Pet.Sci.Eng.16: 03-208

Fredrickson AG, Bird RB (1958) Ind.Eng.Chem. 50: 347-352

Gupta RK, Chhabra RP (1996) AIChE J. 42: 2080-2084

Hanks RW, Larsen KM (1979) Ind.Eng.Chem. Fundam. 18: 33-35

Ranger KB (1994) Phys.Fluids 6: 2224-2226.

Unsteady Free Fall of an Annular, Non-Isothermal, Viscoelastic Film

Kostas Housiadas and John Tsamopoulos

Department of Chemical Engineering, University of Patras, Greece 26500

1. Introduction

Unsteady or free-surface flow arises quite frequently in several processes in the polymer and metallurgical industries. In particular in the extrusion stage of polymer processing operations, such as fiber spinning, film blowing and blow-molding, molten polymer is continuously forced through a die to take the shape of a cylinder or an annulus. It has been observed experimentally in the extrusion stage of a slowly extruded film that the parison thickness and its inner radius vary not only in the axial, but also in the angular direction. The quality of the final product is mainly dependent on the thickness distribution of the film so it is important to examine theoretically this process and to find out which fluid properties, process conditions and equipment characteristics affect these phenomena and in which direction[1]. A schematic of these observations is shown in the figure.

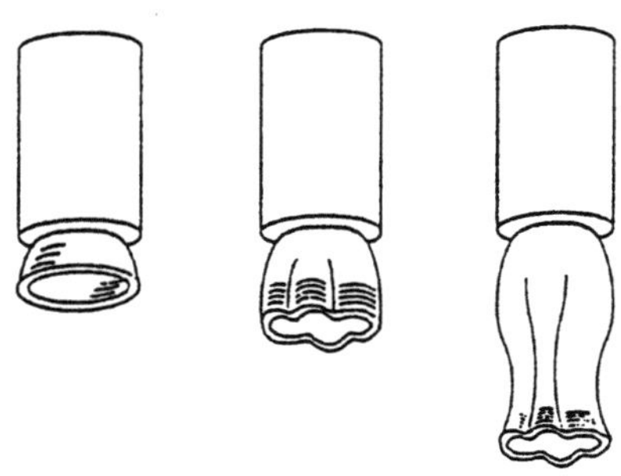

Figure 1

2. Governing Equations and Solution Procedure

In the present study the unsteady extrusion of an annular film is considered. It is composed of an incompressible, viscoelastic and non-isothermal fluid and it enters a passive gaseous atmosphere of temperature T_∞, under the action of gravity. In addition, inertia and forces due to the surface tension of the interfaces are included. The fluid has constant density, ρ, thermal conductivity, k and heat capacity, c_p. The Oldroyd-B fluid model is used. Conservation equations and the constitutive equation for stresses are all written in the time-evolving domain of the film, in dimensionless form:

$$\underline{\nabla} \cdot \underline{v} = 0$$

$$\mathrm{Pe}\left(\frac{\partial T}{\partial t} + \underline{v} \cdot \underline{\nabla T}\right) = -\nabla^2 T$$

$$\mathrm{Re}\left(\frac{\partial \underline{v}}{\partial t} + \underline{v} \cdot \underline{\nabla v}\right) = -\underline{\nabla} p - \nabla \cdot \underline{\underline{\tau}} + \underline{e}_z \mathrm{St}$$

$$\underline{\underline{\tau}} + \mathrm{De}\lambda_1 \frac{\delta \underline{\underline{\tau}}}{\delta t} = -\mu\left(\underline{\underline{\gamma}} + \Lambda \frac{\delta \underline{\underline{\tau}}}{\delta t}\right)$$

where the upper convected derivative is:

$$\frac{\delta(\cdot)}{\delta t} = \frac{\partial(\cdot)}{\partial t} + \underline{v} \cdot \underline{\nabla} - (\cdot) \cdot \underline{\nabla v} - \left((\cdot) \cdot \underline{\nabla v}\right)^+$$

In addition, the Arrhenius equation for the viscosity is used:

$$\frac{\mu(T)}{\mu(T_0)} = \exp\left[-\frac{E_a}{R} \frac{(T-1)}{(T+T_c)(1+T_c)}\right], \quad T_c = \frac{T_\infty}{T_0 - T_\infty}$$

In the above equations, p is the thermodynamic pressure, $\underline{v}$ is the velocity field of the fluid, T is the temperature of the film and $\underline{\underline{\gamma}}$ is the rate-of-strain tensor, $\underline{\underline{\gamma}} = \left(\underline{\nabla v} + \underline{\nabla v}^+\right)$. In the energy equation, viscous dissipation is neglected. The temperature dependence of the fluid's elasticity can be represented by a shift factor, a_T, which relates the relaxation time λ_1 at temperature T to that at the exit of the extruder T_0 (reference temperature), by $\lambda_1 = a_T \lambda_{10}$. When the temperature of the fluid is at least 100 ^{0}C above the glass transition temperature, the shift factor is obtained from the Arrhenius equation. These equations are combined with appropriate boundary and initial conditions: (a) the averaged axial velocity and temperature at z=0 along with the location of the inner and outer interfaces (b) the force balances at the three moving fluid/gas interfaces and (c) their kinematic conditions.

Seven dimensionless groups appear in the equations and the boundary conditions. Their definitions are summarized in table 1. There, g is the gravitational acceleration, R^* the inner radius

Table 1: Dimensionless Numbers

St	$\rho g R^{*2}/(\mu_0 u^*)$
Re	$\rho R^* u^*/\mu_0$
Ca	$\sigma/(\mu_0 u^*)$
De	$\lambda_{10} u^*/R^*$
Λ	$\lambda_{20} u^*/R^*$
Pe	$\rho c_p R^* u^*/k$
Bi	$(h R^*)/k$

of the film at the exit plane, u^* the average velocity at the same plane, μ_0 the viscosity at the reference temperature, λ_{10} the dimensional relaxation time and λ_{20} the dimensional retardation time of the fluid at the same temperature, σ the surface tension of the interfaces and h the heat transfer coefficent. In order to develop a systematic procedure for solving this problem we employ[2]: (a) a double, non-orthogonal mapping of the evolving fluid domain onto a fixed one and (b) a regular perturbation scheme in which every dependent variable except for the film thickness has a zeroth order term. The small parameter is taken to be the ratio of the thickness of the film over its inner radius evaluated at the exit plane. The combination of the above steps reduces the three-dimensional, time-dependent and moving boundary problem to a strongly non-linear partial differential system in the axial and azimuthal direction and in time in a fixed domain[1].

When the magnitude of the dimensionless numbers St and Ca is small, it can be shown that the anulus is moving vertically in a uniform down-flow. This simple solution is subsequently modified to account for the small applied forces. In this way, an analytical solution is obtained. Otherwise, when all forces applied on the film are up to order one, the base state is quite deformed and it needs to be calculated numerically. In the latter case, Galerkin / finite elements are used in the spatial direction and the implicit Euler method is used for time integration. Lagrangian basis functions are employed in order to represent every depedent variable. Hermite basis functions were used too, but it was found that in the case of viscoelastric fuids they could not either: converge, give physical results or achieve the required accuracy in time and space integration. It was also found that the final partial differential system has to be reformulated in quasi-linear form in order to achieve correct numerical integration.

3. Results

First the axisymmetric case is solved numerically. Numerical and analytical results agree qualitatively. Furthermore, under the constraint of small St and Ca number, identical results are obtained.

By examining the influence of each parameter, in turn we have found that:

1. Gravity is the most important force and affects the position of the interfaces deflecting them towards the axis of symmetry of the film, more around the middle of its length and less, if at all, at its two ends.
2. Capillary forces lead to slight increase in the thickness of the film.
3. Inertia leads to more uniform flow of the annulus and, thus, the deviation of the annulus from the vertical position is larger in the inertialess case.
4. The qualitative behavior of the viscoelastic fluid is similar to that of the Newtonian fluid. Quantitative deviations can be observed only if gravity forces become important. In this case, the film is decelerated by elastisity, the interfaces become less curved and the thickness distribution is more uniform than in the Newtonian Fluid.
5. The effect of the retardation time of the model on the interfaces is small.
6. No azimuthal variations can be predicted when gravity and capillary forces are small.
7. Temperature variations influence only the lower part of the film, decelerating the fall of the film.

4. References

1. "Unsteady Flow of an Axisymmetric Annular film Under Gravity" K. Housiadas & J. Tsamopoulos, submitted for publication to *Phys. Fluids* (1997).

2. "Unsteady Flow of a Viscoelastic Annular Film", K. Housiadas & J. Tsamopoulos, to be submitted for publication.

STRAIN-THICKENING IN PLANAR FLOWS

E. MITSOULIS[1,2]

[1]Department of Chemical Engineering, University of Ottawa, Ottawa, K1N 6N5, CANADA
[2]Department of Mining Engineering & Metallurgy, NTUA, Athens 157 80, GREECE

1. INTRODUCTION

The behaviour of polymer solutions and melts under steady-state conditions of planar flow is of particular interest, because it corresponds to such important polymer processing operations as flow through planar sheet dies and in film casting. From the benchmark flow problems used in the comparison between experiments and numerical simulations, the corresponding problem is flow around a cylinder kept between two flat plates. The flow at the centerline in the wake of a cylinder is planar extensional flow and shows long fringes.

While some models, like the PTT and Giesekus models, have the ability to predict strain thickening in planar extension, in the K-BKZ integral constitutive equation the planar extensional viscosity follows the shear viscosity in its shear- (and hence strain-) thinning behaviour, and seems incapable to predict the long fringes behind the cylinder. This defect is corrected by using multiple strain-memory functions in the K-BKZ model with multiple α_k parameters [Mitsoulis (1998)].

2. MATHEMATICAL MODELLING

The flow is governed by the usual conservation equations of mass and momentum while the constitutive equation used is an integral K-BKZ equation proposed by Papanastasiou et al. (1983) and further modified by Luo and Tanner (1988), and is written as:

$$\tau = \frac{1}{1-\theta} \int_{-\infty}^{t} \sum_{k=1}^{N} \frac{a_k}{\lambda_k} \exp\left(-\frac{t-t'}{\lambda_k}\right) \frac{\alpha_k}{(\alpha_k - 3) + \beta_k I_{C^{-1}} + (1-\beta_k) II_C} \left(\mathbf{C}_t^{-1}(t') + \theta \mathbf{C}_t(t')\right) dt' \quad (1)$$

where N is the number of relaxation modes, λ_k and a_k are the relaxation times and relaxation modulus coefficients at a reference temperature, α_k and β_k are material constants, and I_C, $I_{C^{-1}}$ are the first invariants of the Cauchy-Green tensor $\mathbf{C}_t$ and its inverse $\mathbf{C}_t^{-1}$, the Finger strain tensor. The material constant θ is given by $N_2/(N_1+N_2)$, where N_1 and N_2 are the first and second normal stress differences, respectively.

The melts under consideration are LDPE melts used previously in experimental and numerical studies [Baaijens et al. (1997), Hartt and Baird (1996)]. The data used for fitting the parameters of the above K-BKZ constitutive equation are shear viscosities and normal stresses as measured in shear and are given in Table 1. Note that multiple strain-memory functions have been used having different α_k values for each relaxation time, thus ensuring a different planar extensional viscosity as shown in Fig. 1. The steady shear and extensional properties as predicted by Eq. (1) are shown in Fig. 2.

3. RESULTS AND DISCUSSION

The benchmark problem under consideration is flow around a cylinder kept between flat plates. The experiments by Baaijens et al. (1997) showed elongated fringes in the wake of the cylinder, that were difficult to reproduce numerically. Using multi-mode differential models, such as the PTT and Giesekus models, which predict different degrees of strain-thickening behavior in planar extension, rectified the situation. It was found that the degree of strain thickening was crucial in determining the right shape of fringes.

The PSM integral constitutive equation (1) was used in its modified form to accommodate strain thickening in planar extension. The results from the flow simulations are shown in Fig. 3 in terms of the birefringence patterns. The overall shape of the fringes agrees well both with experiments and previous simulations, but subtle differences exist before and after the cylinder, suggesting that the exact behavior in planar extension is of importance [Mitsoulis (1998)]. However, it is still a difficult task to get experimentally planar extension data at high extensional rates, which makes simulations of this kind a valuable tool for determining such rheological information.

4. CONCLUSIONS

The modified PSM integral model has been used for simulations of the benchmark problem of flow around a cylinder kept between parallel plates, which results in planar extension in the wake of the cylinder. The simulations are able to reproduce the long fringes at the centerline, in agreement with experimental observations and with other numerical simulations that employ the multi-mode PTT and Giesekus models, which also show strain thickening in planar extension.

5. ACKNOWLEDGEMENTS

Financial assistance from the Natural Sciences-Engineering Research Council of Canada (NSERC) and the Ontario Centre for Materials Research (OCMR) is gratefully acknowledged.

6. REFERENCES

Baaijens FPT, Selen SHA, Baaijens HPW, Peters GWM, Meijer HEH (1997) J Non-Newt Fluid Mech 68:173-203

Hartt WH, Baird DG (1996) J Non-Newt Fluid Mech 65:247-268

Mitsoulis E (1998) J Non-Newt Fluid Mech in press

Luo XL, Tanner RI (1988) Int J Num Meth Eng 25:9

Papanastasiou AC, Scriven LE, Macosko CW (1983) J Rheol 27:387-410

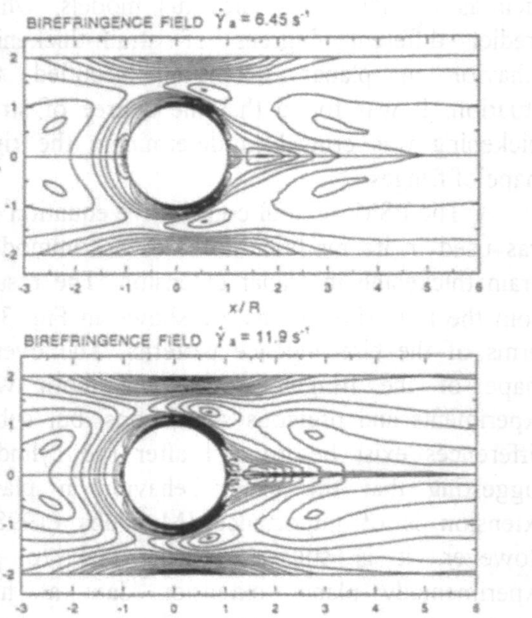

Fig. 3. Simulated birefringence patterns for the LDPE-EUT melt with the modified K-BKZ model given by eq. (1) and the data of Table 1: (a) $\dot{\gamma}_a$ = U/R = 6.45 s^{-1}; (b) $\dot{\gamma}_a$ = 11.9 s^{-1}.

Table 1. Linear viscoelastic spectrum and multiple α_k values for the LDPE–EUT melt at 190°C (β = 0.06, θ = –0.25, $<\lambda>$ = 0.51 s, η_0 = 2,224 Pa·s).

k	λ_k (s)	a_k (Pa)	α_k
1	$1.11x10^{-2}$	$2.43x10^{+4}$	24.62
2	$5.77x10^{-2}$	$9.65x10^{+3}$	5.355
3	$0.79x10^{0}$	$1.66x10^{+3}$	13.41
4	$0.62x10^{0}$	$1.43x10^{+2}$	3390.0

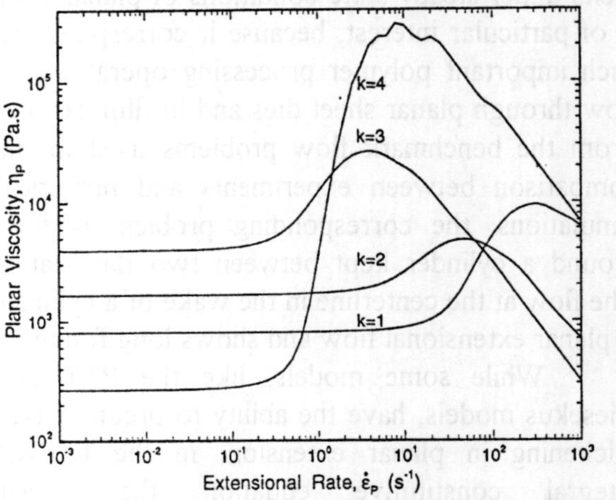

Fig. 1. The contributions of different modes in the planar extensional viscosity η_P for the LDPE-EUT melt (STAMYLAN-L8) using Eq. (1) with data given in Table 1.

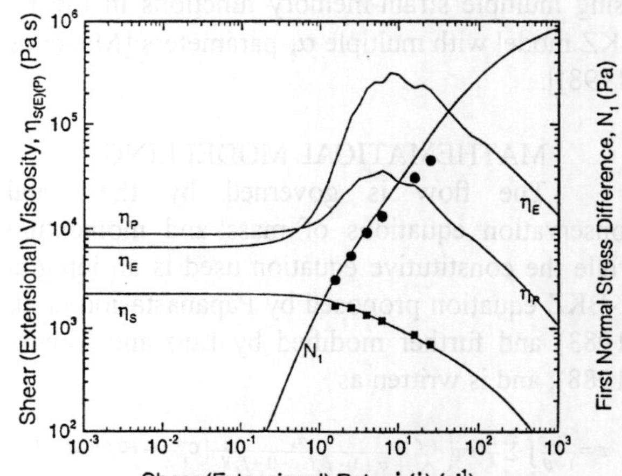

Fig. 2. Model predictions of shear viscosity η_S, first normal stress difference N_1, uniaxial elongational viscosity η_E, and planar extensional viscosity η_P for the LDPE-EUT melt (STAMYLAN-L8) using Eq. (1) with data given in Table 1. Symbols are experimental data by Baaijens et al. (1997).

TOMS EFFECT IN FLOODED JETS

G. NESYN[1], V. MANZHAY[2], I. SHTIN[3], A. MALKIN[4]

[1] Chemical Engineering Dept., Tomsk Polytechnical University, 30, Lenin ave., Tomsk, 634004, Russia
e-mail nesyn@gvn.tomsk.ru
[2] Institute of Petroleum Chemistry RAS, 3, Akademichesky ave., Tomsk, 634021, Russia
[3] Central Siberia Dept. of Transneft Company
[4] Plastics Research Institute, Moscow

Drag reduction in the presence of trace amounts of highpolymers is one of the interesting aspects of rheology. It is not perfectly understandable till nowadays how such small concentrations of macromolecules can cause such a dramatic influence on the flow. Many authors suggest that all events responsible for the Toms effect proceed in elongational flow of wall region. May be it is the very place where extremely deformed macromole- cules realize their interactions. In any case there is some evidence that 'macroobjects' are involved in drag reduction mechanism. For example, the destroying properties of open water jets, including polymer additives are much the same of water - sand mixtures.

Our efforts were concentrated on the practical use of Toms phenomenon, that is drag reduction of crude oil. We have synthesized and tested a number of oil-soluble polymers. Two factors determine the effectiveness of drag reducing agents. They are good solubility and super-high molecular mass. The best results in hydrocarbon liquids showed comb-like polymers of 1-alkenes prepared with Ziegler - Natta polymerization (Nesyn et al., 1997). We managed to produce the polymer with molecular mass exceeding 2 10^7. The field test of such a polymer in the Novorossiysk trunk line (800 mm diameter) gave 23 % drag reduction while polymer concentration was 7 ppm.

But the matter of our special interest was flooded jets, the almost unknown aspect of Toms phenomenon. Introducing of polymer additive gives rise to elongation of compact section of the jet while its diameter diminishes. That is why it would have more hard influence upon the perpendicular barrier than the jet of polymer-free liquid. So, we decided to make an attempt to applicate Toms effect in flooded jets to remove the bottom paraffin sediment from the crude oil tank. Some words about this problem. Storage of crude oil in large tanks is accompanied by precipitation of paraffin, water and particles of soil, which form a bottom residue. Accumulation of sediment makes appreciable part of tank volume useless, stimulate corrosion processes, and causes troubles with water discharge. The large tanks are equipped with wash-out arrangement including washing nozzles as the main elements. These nozzles have a form of a plate. They are situated quite close to the bottom and form fan-shaped jets while tank is filling with oil. The range of the jets overlaps with each other thus preventing the formation of sediment. But when the nozzles become blocked up the dead zones appear, where accumulation of sediment takes place. In some cases the ballast may occupy a considerable part of the tank and its removing is a complicated problem. Neither steam working, nor treatment with surfactants could help it. So, manual labour is used sometimes. Then a hard toil is carried out in toxic and fire dangerous atmosphere. Besides, this operation results in waste of valuble materials as well as in environment pollution.

The tank of Transneft Company with high level of residue was chosen for experiment. Its volume was 20.000 m^3. The level of sediment was 1.95 m. That was about 20 % of actual volume. Washing of the tank was carried out in continual manner, i.e. pouring through the washing nozzles and pumping out were taken at the same time. The level of oil was maintained at the height of 1.5 m over the sediment. The control washing with crude oil was ineffectual. The level of residue hadn't changed, and the water content in the oil was the same in and out of the tank. When the introducing of polymer had begun into admission pipeline the grouth of water content in the oil was registrated at the outcome of the tank. It reached 2.7 % while in the inlet it was not more than 0.2 %. That was the evidence of sediment dispersion. We had the amount of polymer enough for 10 hours of dosage. And during this time the average level of sediment had been reduced from 1.95 m till 0.69 m. It means that about 2000 m^3 of the ballast was removed out of the tank. To tell the truth we couldn't expect such an impressive result.

We suppose that 'cutting' property of flooded jets, containing polymer additives is the effect of dynamic interaction of macromolecules in the elongational flow which results in formation of anisotropic macroobjects. V.G. Pogrebnyak et al. (1992) made an experiment on visualization of pouring out of water solution of poly(ethylene oxide) with the dye through a short capillar under the pressure. It showed the formation of 'flexes' before the outlet, i.e. in the zone of elongational flow. Something similar probably takes place inside the washing nozzle. This may result in reinforcement of the jet with rigid 'fibres' that can disperse the sediment.

Finally, we should point out ecological compatibility of Toms effect application. No heat expenditure, no steam working, no formation of water-oil emulsion, no atmosphere pollution with gaseous hydrocarbons, and no soil pollution with buried residue take place.

REFERENCES

Nesyn GV et al. (1997) Russian patent 2075485

Pogrebnyak VG et al. (1992) Eng.-Phys. J., 2:147-150.

VISUALISATION OF THE FLOW FIELD OF A POLYMER SOLUTION IN AN OPEN CHANNEL

D. NISTORAN, C. BALAN, E. BRUJAN, V. PANAITESCU

University "Politehnica" of Bucharest, Hydraulics Dept., # 313 Spl. Independentei, 77 206, Bucharest, Romania

1. INTRODUCTION

The aim of this study is to investigate how a small percentage of polymer added to a solvent can affect the behaviour of a Newtonian fluid in planar (combined shear and elongational) slow motions in an open channel with abrupt changes in geometry. This research was motivated by some earlier experiments of Walters K. and Webster, M.F. (1982) and Quinzani et. al. (1994) who investigated in depth similar problems in pressure driven planar flows.

An optical technique was used to compare the flow fields of the solvent (mineral oil) and weakly elastic (polyisobutilene) solutions flowing in an open channel around a sharp edged T-profile. The range of Re and De no.s no.s covered is of 0÷60 and 0÷0.01 respectively.

2. EXPERIMENT

2.1. Material characterization

The polymer solutions used in this study consist of a low molecular weight polyisobutilene (PIB; Exxon, Vistanex LM-MH, $M_w \approx 5.5\times10^4$ g/mol) and a Newtonian mineral oil (Tr 30, manufactured in Romania) as a solvent. The mineral oil is a brown, opaque fluid with a density $\rho = 890$ kg/m^3 and a viscosity coefficient $\eta_0 = 0.45$ Pa·s at $\theta_0 = 20$°C. At the same temperature Vistanex is like a soft rubber, with a density $\rho = 910$ kg/m^3 and a viscosity coefficient $\eta_0 = 1.5\cdot10^5$ Pa·. Besides being very studied, other reasons why these constituents were chosen for the present investigation are that the resultant solutions do not evaporate in time and measurements could be undertaken at ambient temperature.

Small pieces (100 g) of PIB were dissolved firstly in a relatively low volume of mineral oil (3 l) by continous mixing at a constant temperature of approximate 70°C for 2-3 days. The resulted very concentrated solution was progressively dilluted by continuous mixing at the same temperature, and than at lower temperatures. Care was taken to maintain a low speed of mixing in order to avoid mechanical degradation (break-up of the polymer chains), as Khalil, K. et al. (1994) mentioned.

An extended rheological characterization has been made for PIB solutions in the range of 1÷10%. Combined effects of concentration and temperature have been studied. Dynamic vistosity master curve is presented in Fig. 1. Rheometry tests showed the value for the overlap concentration c* = 2.55, so for the channel investigations were used only dilute solution of 1 and 2%.

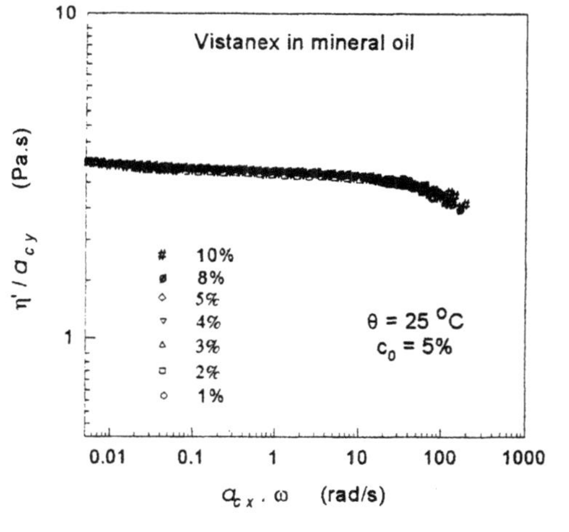

Fig. 1. Dynamic viscosity master curve for the 1÷10% weight PIB in mineral oil at 25°C

Tab. 1. Material constants for the test fluids

Concentration c (%)	Viscosity coefficient η_0 (Pa·s) at 25°C	Relaxation time λ_1
0	0.45	0
1	0.7	$1\cdot10^{-5}$
2	1.05	$1\cdot10^{-4}$

Comparing the results from the oscillatory with the ones from the stationary tests by modelling the viscosity curves with a Maxwell type constitutive relation and taking into account the shear rate range reached in the channel flow, the characteristic relaxation times resulted are listed in Tab. 1.

2.2 Experimental setup

The T-profile is placed in an open channel in which the liquid is recirculated by a vertical impeller. A schematic diagram depicting the visualization area is shown in Fig. 2:

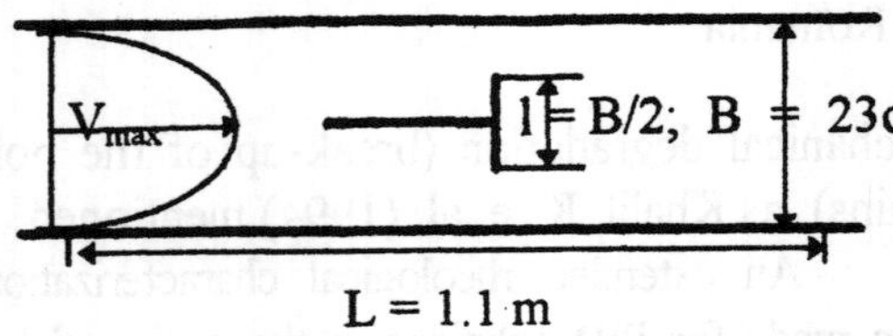

Fig. 2. The visualization zone. The depth h = 18 cm

2.3. The visualisation technique

Velocity field measurements were performed by photographing individiual tracer particles in a definit exposure time. The mean velocity of a section far away upstream the T-profile was taken to be

$$\langle V\rangle = 2\, V_{max} / 3 \qquad (1)$$

where V_{max} is the centerline axial velocity.

The flow pattern around the profile was obtained by adding polystirene particle tracers at the surface of the fluid.

Photographs of the flow field in laminar motion have been taken at different flow rates for the Newtonian solvent and for the 1% and 2% polymer solutions.

Care has been taken to continuously monitor and control the temperature during the photographing sessions. The variations have been taken into account within temperature dependant variables.

The formulas used for computing the shear rate, Re and De no.s, are as follows:

$$Re = \langle V\rangle \cdot B / \nu \qquad (2)$$

$$\gamma_{max} = 12 \cdot \langle V\rangle / B \qquad (3)$$

$$De_{max} = \lambda_1 \cdot \gamma_{max} = 12 \cdot \lambda_1 \cdot \langle V\rangle / B \qquad (4)$$

3. RESULTS AND DISCUSSIONS

A substantial decrease of the wake length has been observed at the non-Newtonian solution compared to the Newtonian solvent as it can be seen in Fig. 3.

Also, an increase with concentration of the local circulation at the edge of the profile has been noticed.

For the non-Newtonian solutions it was found that for all the runs with $Re \geq 35$ the vortex legth remains approximately constant.

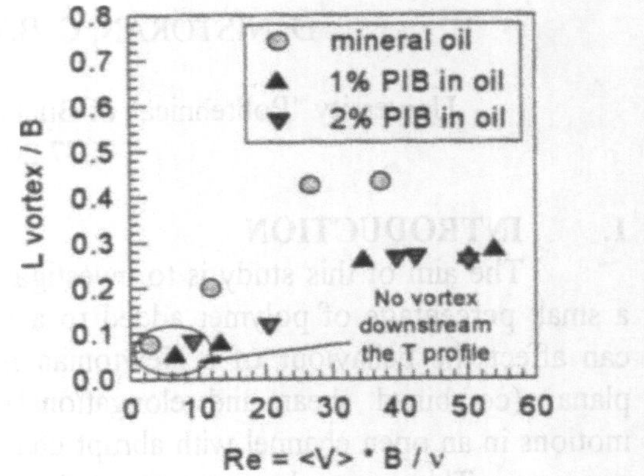

Fig. 3. Wake length versus Re for the three liquids

4. CONCLUSIONS

Although we have not observed a significant influence of liquid rheology on streamlines pattern, a noticeable difference was found concearning the length of the wake behind the T-profile.

We speculate that the extensional viscosity has major influence with this respect since it's value grows very much around the T-profile where the extensional strain rates increase dramatically and the flow is dominantly elongational.

No sistematic attempt has been made to compare the present experimental results with theoretical or numerical analysis.

5. REFERENCES

Walters, K., Webster, M.F. (1982) M.F., Phil. Trans. R. Soc. Lond. A, **308**, 199-218, *On domination elastico-viscous response in some complex flows*

Khalil, K., Tougui, A., Sigli D., (1994), J. Non-Newtonian Fluid Mech., **52**, 375-386, *Relation between some rheological properties of polyisobutylene solutions and their mode of preparation*

Quinzani, L.M., Armstrong, R.C., Brown, R.A., (1994), J. Non-Newtonian Fluid Mech., 52, 1-36, *Birefringence and laser-Doppler velocimetry (LDV) studies of viscoelastic flow through a planar contraction*

6. AKNOWLEDGMENTS

The authors wish to thank Professor Ken Walters and Dr. D. Binding for the helpfull suggestions and for allowing Mrs. D. Nistoran the use of University of Aberystwyth's laboratory equipment.

EXPERIMENTAL AND NUMERICAL STUDY OF THE VISCOUS DAM BREAK PROBLEM

B. NSOM, J.M. PIAU, P. SARAMITO, K. DEBIANE and A. AYADI

Laboratoire de Rhéologie, UJF-INPG-CNRS (UMR 5520),
Domaine Universitaire, BP 53 - 38041 Grenoble Cedex 09, FRANCE

1. INTRODUCTION

When the dam of a water reservoir breaks, a positive wave propagates downstream, with an increase of the depth while a negative wave propagates upstream, with a decrease of the depth. Ritter's inviscid theory (1892) is the simplest description of the inertial flow generated by this phenomenon on a horizontal surface. It was extended to the case where turbulent friction (Dressler, 1952, Whitham, 1955) or laminar friction is present (Piau, 1995). Experimental results available fit reasonnably well with these perturbation theories.

Meanwhile, there exists no study in the litterature reporting the dominating influence of viscosity and that of the reservoir length on the flow development. However a viscous dam break configuration often happens in nature and it is of great interest with regard to the natural risks assessment. After a large quantity of muddy water has been retained behind crosslinked branches and debris (natural dam) or behind a solid wall (artificial dam), the dam may suddenly collapse.

A horizontal channel has been equiped with an upstream obstructing plate (reservoir end) and another one downstream (dam), in order to represent this phenomenon in the laboratory. A high viscosity fluid is poured into the reservoir with given height. At initial time, the dam is suddenly lifted, and the flow generated by the fluid breaking downstream the channel is investigated both experimentally and numerically. Though complex fluids are being investigated, the developments to follow will be focussed on the reference case of the Newtonian model.

2. EXPERIMENTAL SET UP

Water-glucose solutions were chosen to perform this study. Their behavior in the experiment is Newtonian and their density as well as their viscosity can be adjusted. Then, a completely transparent channel was constructed and is used here in its horizontal position. Much care was taken during the experiments in order to avoid the tendency of the very viscous liquids to transform into multiphasic media. Using a Carrimed CSL 100 rheometer, the fluid viscosity was regularly tested. The channel being transparent as well as the fluid itself, we investigated the wave evolution using video-photographic facilities. Immediately after lifting the dam, the flow is very rapid and is inertial. To catch the flow images, we used an ultra fast camera (1000 images/sec) carried by a chart which can move both in the horizontal and in the vertical directions. The images were recorded and then analysed. After the end of the inertial regime, the viscous ones take place. In these regimes, a regular camera can fit (a 25 images/sec camera was used). Furthermore, the photographs of the front flow profile were taken in view to provide realistic boundary conditions to the numerical treatment. The depth was investigated at several stations along the channel using an ultrasonic distance measurement system (Weidmüller LRS 3 Type 262). This probe emits a wave normally to the water surface. The return wave (after reflection) is in direct relation with the flow depth which is obtained with a 0.1 mm accuracy. These measurements were performed both for variable reservoir length and fluid viscosity (glucose concentration).

3. EXPERIMENTAL RESULTS

These flows belong to the large class of open channel flows (Chow, 1959). Piau (1996) proposed a general model for the open channel flow of yield stress fluids. In the special case of Newtonian fluids, shallow water St Venant type equations are recovered. The following set of relations were used to make the results adimensional:

$$x'=\frac{x}{L} \qquad h'=\frac{h}{H} \qquad t'=\frac{\rho g H^3}{12\eta L^2}\, t \qquad (1)$$

where x denotes the abscissa from the dam position, H is the fluid depth in the reservoir at t<0, L the reservoir length, η the fluid viscosity, ρ its density and g the gravity.

Fig. 1 shows the front wave abscissa evolution with time. First of all, in the inertial regime, this evolution follows a Ritter's type law i.e. with unit slope. Then, in the initial part of the viscous regime, the reservoir length and the fluid viscosity have no influence. The law of evolution is:

$$x'_f = \sqrt{t'} \quad \text{if} \quad t' < 0.2 \tag{2}$$

Then, the front wave position obeys the following equation where the reservoir length and the fluid viscosity are implicitely present.

$$x' = 1.86 t'^{1/5} - 0.902 \tag{3}$$

while the fluid depth decreases at the upstream part of the reservoir.

4. THEORY

In the viscous regime, the following differential equation is obtained for kinematic waves on a horizontal plane when the inertial terms have been dropped:

$$\frac{\partial h'}{\partial t'} - \frac{\partial^2}{\partial x'^2}(h'^4) = 0 \tag{4}$$

with boundary conditions

$$h'(t'=0, x') = 1 \text{ when}$$

$$=0 \text{ when } 0 < x' < \infty \tag{5}$$

$$\frac{\partial h'}{\partial x'}(t', -1) = 0 \tag{6}$$

This equation has first been studied analytically in order to derive the scaling laws which govern the phenomenon. A variable separation handling produces the approximate solution:

$$h'^3(x', t') = \frac{3}{40} \frac{[x'_f(t') - x]^2}{t'} \tag{7}$$

which is in very good agreement with the previous experimental data. Then eq. (4) has been solved numerically using the finite difference method. The results of these computations show an excellent concordance with the data.

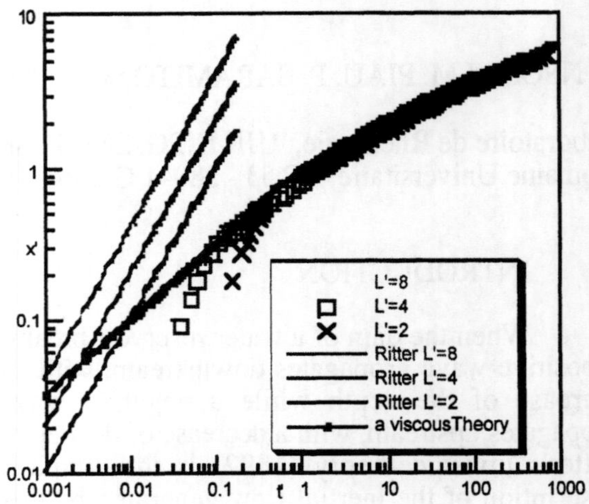

Fig.1 The front wave evolution

5. REFERENCES

Chow, V.T. (1959) "Open channel Hydraulics", eds. Mc Graw Hill, New-York . J.R.

Dressler, R.F. (1952) "Hydraulics resistance effect upon the dam break function", Journal of Research, National Bureau of Standards. **49(3)**, 217-225

Piau, J.M. (1995) unpublished, reproduced in Ayadi, A. (1996): Ph.D. Thesis, Grenoble University

Piau, J.M. (1996) "Flow of a yield stress fluid in a long domain. Application to flow on an inclined plane" J. Rheol., **40(4)**, 711-723

Ritter, A. (1892) "The propagation of water waves", Ver Deutsch Ingenieure Zeischr., Berlin, **36(33)**, 1489-1514

Whitham, G.B. (1955) "The effect of hydraulic resistance in the dam break problem" Proc. Roy. Soc. Lond., **A(227)**, 399-407

Squeeze Flow of Bingham Plastic

Dimitrios Smyrnaios and John Tsamopoulos

Department of Chemical Engineering, University of Patras, Greece 26500

1. Introduction

The axisymmetric squeeze flow of a viscoplastic material is examined. The deformation and flow of such materials is important, since many multicomponent fluids encountered in industrial processes, are viscoplastic. The Bingham plastic constitutive equation is used and the transition from the solid state to the fluid one takes place at the "yield surface". The aspect ratio of the geometry ε (half the distance between the disks to their radius) is taken to be the small parameter of the analysis and a non-standard lubrication approximation is employed.

2. Governing Equations and Solution Procedure

The Bingham plastic constitutive equation follows the von Mises yield condition

$$\underline{\underline{\tau}}^* = -\left(\eta_o + \frac{\tau_y}{\dot{\gamma}^*}\right)\underline{\underline{\dot{\gamma}}}^* \qquad (\tau^* > \tau_y) \tag{1a}$$

$$\underline{\underline{\dot{\gamma}}}^* = \underline{\underline{0}} \qquad (\tau^* < \tau_y) \tag{1b}$$

where $\underline{\underline{\dot{\gamma}}}^* = \left(\underline{\nabla v}^* + (\underline{\nabla v}^*)^T\right)$, η_o and τ_y are the plastic viscosity and yield stress, respectively, and τ^* and $\dot{\gamma}^*$ are the second invariants of the extra stress and rate-of-strain tensors, respectively.

Two distinct regions of flow are thus defined by eqs (1a) and (1b). In the first one, the second invariant of the stress $\underline{\underline{\tau}}$ satisfies $\tau^* > \tau_y$ and the material flows with a viscosity function $\eta(\dot{\gamma}^*) = \eta_o + \tau_y / \dot{\gamma}^*$, whereas in the second one, $\tau^* < \tau_y$ and the material behaves as a rigid solid. The locus of points where the stress is at the yield value is called yield surface.

Three areas that need particular attention because unyielded material may exist there are: The area around the stagnation points of the flow, around the axis of symmetry and around the plane of symmetry. In this problem L and R are taken to be the characteristic lengths in axial and radial directions, respectively, whereas the velocity of each plate, V, is taken as the characteristic velocity in the axial direction. Then, conservation of mass requires that the magnitude of the radial velocity is V/ε, i.e. much larger than the axial velocity. With these scales, the dimensional $\dot{\gamma}^*$ for an axisymmetric flow field is:

$$\dot{\gamma}^* \equiv \frac{V}{L}\frac{1}{\varepsilon}\dot{\gamma} = \frac{V}{L}\frac{1}{\varepsilon}\left\{\left(\frac{\partial v_r}{\partial z}\right)^2 + 4\varepsilon^2\left[\left(\frac{\partial v_r}{\partial r}\right)^2 + \left(\frac{v_r}{r}\right)^2 + \left(\frac{\partial v_r}{\partial r}\right)\frac{v_r}{r}\right] + 2\varepsilon^2\frac{\partial v_r}{\partial z}\frac{\partial v_z}{\partial r} + O(\varepsilon^4)\right\}^{1/2} \tag{2}$$

We have shown that:

i. Unyielded material should exist around the stagnation points of flow at the disk centers, as predicted around stagnation points of various other flows (Beris et al. (1985), O'Donovan and Tanner (1984) and Tsamopoulos et al. (1996))

ii. Upon consideration of both experimental evidence (Chatraei et al. (1981)) and numerical simulations (O'Donovan and Tanner (1984)) we have concluded that at least at the plane of symmetry the material undergoes extensional flow.

Upon inserting the previously defined characteristic velocity and length scales into the radial and axial momentum balances we get

$$\mathrm{Re}\left(\frac{\partial v_r}{\partial t} + v_r\frac{\partial v_r}{\partial r} + v_z\frac{\partial v_r}{\partial z}\right) = -\frac{P_c L^3}{\eta_0 R^2 V}\frac{\partial P}{\partial r} + \left[\varepsilon^2\left(\frac{\partial^2 v_r}{\partial r^2} + \frac{1}{r}\frac{\partial v_r}{\partial r} - \frac{v_r}{r^2}\right) + \frac{\partial^2 v_r}{\partial z^2}\right]\left(1 + \frac{\varepsilon \mathrm{Bn}}{\dot{\gamma}}\right) + 2\varepsilon^3 \mathrm{Bn}\frac{\partial v_r}{\partial r}\frac{\partial}{\partial r}\left(\frac{1}{\dot{\gamma}}\right) + \varepsilon \mathrm{Bn}\left(\frac{\partial v_r}{\partial z} + \varepsilon^2\frac{\partial v_z}{\partial r}\right)\frac{\partial}{\partial z}\left(\frac{1}{\dot{\gamma}}\right) \tag{3}$$

$$\mathrm{Re}\left(\frac{\partial v_z}{\partial t} + v_r\frac{\partial v_z}{\partial r} + v_z\frac{\partial v_z}{\partial z}\right) = -\frac{P_c L}{\eta_0 V}\frac{\partial P}{\partial z} + \left[\varepsilon^2\frac{1}{r}\frac{\partial}{\partial r}\left(r\frac{\partial v_z}{\partial r}\right) + \frac{\partial^2 v_z}{\partial z^2}\right]\left(1 + \frac{\varepsilon \mathrm{Bn}}{\dot{\gamma}}\right) + \varepsilon \mathrm{Bn}\left[\left(\frac{\partial v_r}{\partial z} + \varepsilon^2\frac{\partial v_z}{\partial r}\right)\frac{\partial}{\partial r}\left(\frac{1}{\dot{\gamma}}\right) + 2\frac{\partial v_z}{\partial z}\frac{\partial}{\partial z}\left(\frac{1}{\dot{\gamma}}\right)\right] \tag{4}$$

where the definitions of dimensionless numbers are based on the plastic viscosity, η_o, and the fluid density, ρ:

$$\mathrm{Re} = \frac{\rho V L}{\eta_0}, \quad \mathrm{Bn} = \frac{\tau_y}{\frac{V}{L}\eta_0}, \tag{5}$$

The Reynolds number is typically very small and it is set to zero. The Bingham number measures the ratio of the yield stress to the viscous forces. We will proceed by assuming first that $\mathrm{Bn} = O(\varepsilon^{-3})$

so that viscous and plastic terms are of the same order of magnitude. By considering only the dominant terms and the by taking in account the equation of continuity, we eliminate the axial velocity and the pressure, thus a single equation for $v_r(r,z)$ is obtained:

$$\begin{aligned}
&\frac{\partial^2 v_r}{\partial z^2} \\
&-2\frac{\partial}{\partial z}\left\{\left[\left(\frac{\partial v_r}{\partial r}\right)^2+\left(\frac{v_r}{r}\right)^2+\left(\frac{\partial v_r}{\partial r}\right)\frac{v_r}{r}\right] / \left[\left(\frac{\partial v_r}{\partial z}\right)\left|\frac{\partial v_r}{\partial z}\right|\right]\right\} \\
&\mp\frac{z}{r^2}+2\frac{\partial}{\partial r}\left(\frac{v_r}{r}\right) / \left|\frac{\partial v_r}{\partial z}\right|+4\frac{\partial}{\partial r}\left[\left(\frac{\partial v_r}{\partial r}\right) / \left|\frac{\partial v_r}{\partial z}\right|\right] \\
&+2\frac{\partial}{\partial r}\left[\left(\frac{v_r}{r}\right) / \left|\frac{\partial v_r}{\partial z}\right|\right] \\
&+4A\left[\frac{\partial^2 v_r}{\partial r \partial z} / \left(\frac{\partial v_r}{\partial z}\left|\frac{\partial v_r}{\partial z}\right|\right)\right]_{z=0}=0, \quad z \gtrless 0
\end{aligned} \tag{6}$$

Around the stagnation points of flow we assume that unyielded material extends up to $r=R_1$ $(0 < R_1 \leq 1)$ and that the yield surface is given by the function $h(r)$, $(0<h(r)\leq 1)$ for $0 \leq r \leq R_1$. The function $h(r)$ will be determined by applying on the yield surface the condition

$$\partial v_r / \partial z = 0, \qquad \text{at } z=h(r) \tag{7}$$

On the bounding the flow surfaces the following conditions will be applied:

$v_r=0$ at $r=0$ and at $z=\min(h(r), 1)$ (8)
(axis of symmetry & no slip at fluid/solid interface)

$v_r=Ar$ at $z=0$ (9)
(extensional flow at the plane of symmetry)

$$\sigma_{rr}=0, \quad \text{or} \quad -P+\frac{2}{\dot{\gamma}}\frac{\partial v_r}{\partial r}=0 \quad \text{at } r=1 \tag{10}$$

(balance of the normal force at the liquid/air interface at the edge of the disks)

The final unknown, A, is determined by the overall mass balance

$$\pi R^2 V = 2\pi R \frac{V}{\varepsilon} L \int_0^{h(r=1)} v_r(r=1,z)dz \quad z>0 \tag{11}$$

here $h(r=1)$ is either 1, if the yield surface does not extend all the way to the edge of the disks or less than 1, if it does.

3. Results

The predicted yield surface is shown below. It is seen that at the axis of symmetry the material remains unyielded almost up to the plane of symmetry. The two yield surfaces intersect the axis of symmetry at $z=\pm 0.01$. In spite of its small size, this yielded region is sufficient to permit motion of the disks towards each other. Further away from the axis of symmetry the area of the unyielded material shrinks monotonically but exists all the way to the edge of the two disks.

We are currently using a Finite element code to solve eqs (3) and (4) in their general form (without assuming that L/R is small). Thus we will be able to examine the depencence of the flow field and the yield surface on Bn and L/R.

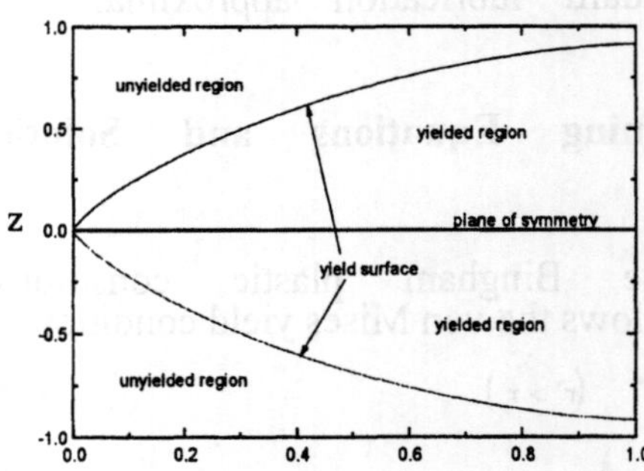

Yield surface seperating the yielded from unyielded material

4. References

1. Beris, A.N., Tsamopoulos, J.A., Armstrong, R.C. and Brown, R.A., J. Fluid Mech. 158, 219-244 (1985).
2. Bird, R.B., Armstrong, R.C. & Hassager, O., Wiley 1, (1977).
3. Chatraei, SR., Macosko, C.W. and Winter H.H.,, J. Rheol. 25(4), 433-443 (1981).
4. Covey, G.H. and Stanmore, B.R., , J. Non-Newt. Fluid Mech. 8, 249-260 (1981).
5. Gartling, D.K. and Phan-Thien, N., J. Non-Newt. Fluid Mech. 14, 347-360 (1984).
6. O'Donovan, E.J. and Tanner, R.I., J. Non-Newt. Fluid Mech. 15, 75-83 (1984).
7. Tsamopoulos, J.A., Chen, M.F. and Borkar, A.V., Rheol. Acta, 35, 597-615 (1996).

THE EFFECT OF WAVEFORMS ON THE VIBRATIONAL PROCESSING OF FRESH CONCRETE PART I - SHEAR FLOW

M.A.O.M. Teixeira, R.J.M. Craik and P.F.G. Banfill

Department of Building Engineering & Surveying, Heriot-Watt University, Edinburgh, EH14 4AS, UK

1. INTRODUCTION

This project addresses the industrial processing problem of compacting fresh concrete by vibration.

Fresh concrete possesses a yield stress but the stress field introduced by a vibrational source is usually large enough to enable the concrete to flow around reinforcement and into the corners of the mould, releasing air which bubbles to the surface.

The flow is analysed here in the case of an oscillatory laminar viscous flow over an infinite flat plate that oscillates harmonically in its own plane, *i.e.* a shear waveform, which is often referred to in the literature as Stokes flow.

Properties of typical concretes and vibrational sources are discussed. A future paper will discuss the compressive waveform.

2. THE FLOW OF INTEREST

Fresh concrete conforms to the Bingham model (Tattersall and Banfill, 1983). For fluid concrete the yield stress τ_y=150 Pa, and the plastic viscosity μ_p=20 Pa.s; and for stiff concrete τ_y=2000 Pa, μ_p=150 Pa.s (Domone et al., 1998).

The compressibility κ, of concrete is given by:

$$\kappa = \frac{1}{c^2 \rho} \tag{1}$$

where c is the velocity of propagation of sound waves in the fluid and ρ its density.

Preliminary tests carried out in the present study with fresh concrete suggest that $c\approx250$ m/s, compared to values of 350 for fresh mortar (Casson et al., 1982) and 1000 for fresh cement paste (Casson and Domone, 1982).

For fresh concrete $\rho\approx2000$ kg/m^3.

In order to establish the frequency and amplitude of a typical vibrational source, measurements were carried out on an internal "poker" vibrator with 45 mm diameter. One accelerometer was fixed at the end of the vibrator and the measured frequency f=300 Hz while the r.m.s. velocity associated with the oscillation u_0=1.6 m/s.

3. SHEAR WAVEFORM

In this section a pure shear waveform is considered and the conditions that justify the neglect of fluid compressibility may be found in Lighthill (1964), and can be expressed as:

$$\frac{u_o}{c} << 1 \tag{2}$$

Considering that for the present study u_0=1.6 m/s and for fresh concrete $c\approx250$ m/s, the condition is satisfied.

In the Stokes flow of a Bingham fluid, the momentum equations reduce to:

$$\frac{\partial u}{\partial t} = \frac{\mu_p}{\rho}\frac{\partial^2 u}{\partial y^2} + \frac{1}{\rho}\frac{\partial \tau_y}{\partial y} \tag{3}$$

Considering that the yield stress τ_y is a material property and has a constant value, the second term on the right-hand side of equation (3) is zero. Assuming that the boundary conditions are:

$$u = u_o \cos 2\pi f \quad \text{at} \quad y=0 \tag{4}$$

$$u \rightarrow 0 \quad \text{as} \quad y \rightarrow \infty \tag{5}$$

The solution for equation (3) is:

$$u = u_o \exp\left(-y\sqrt{\frac{\pi f \rho}{\mu_p}}\right)\cos\left(2\pi f t - y\sqrt{\frac{\pi f \rho}{\mu_p}}\right) \tag{6}$$

see, for instance, Telionis (1981).

Equation (6) shows that there is a well defined region affected by the flat plate oscillation where the Bingham material flows as a Newtonian liquid. But as one moves away from the oscillation source, the shear rate $\dot{\gamma}$ ($\equiv du/dy$) decays according to the exponential function introduced in equation (6), and hence, when the stress level falls below the yield stress, there is no flow. The point at which this occurs is of great practical importance but cannot be defined explicitly using the Bingham model because $\tau = \tau_y$ when $\dot{\gamma} = 0$.

However, if one defines a critical shear rate $\dot{\gamma}_c$ such that when $\dot{\gamma} \leq \dot{\gamma}_c$ there is no flow an exact expression, based on the frequency and velocity of oscillation and on the plastic viscosity and density of the material, may be derived for the length L of the flow region. This may be thought of as the range of effectiveness of a vibrator and is given by:

$$L = -\frac{1}{\sqrt{\frac{\pi f \rho}{\mu_p}}} \ln\left(\frac{\dot{\gamma}_c}{u_o \sqrt{\frac{2\pi f \rho}{\mu_p}}} \right) \quad (7)$$

Figure 1 shows that for both concrete mixes the flow curve is such that there is a value $\dot{\gamma}_c$ below which there is no flow and above which the flow is Newtonian. In the present case $\tau_c = 1.001\tau_y$ and an expression for $\dot{\gamma}_c$ may also be derived as:

$$\dot{\gamma}_c = \frac{0.001\tau_y}{\mu_p} \quad (8)$$

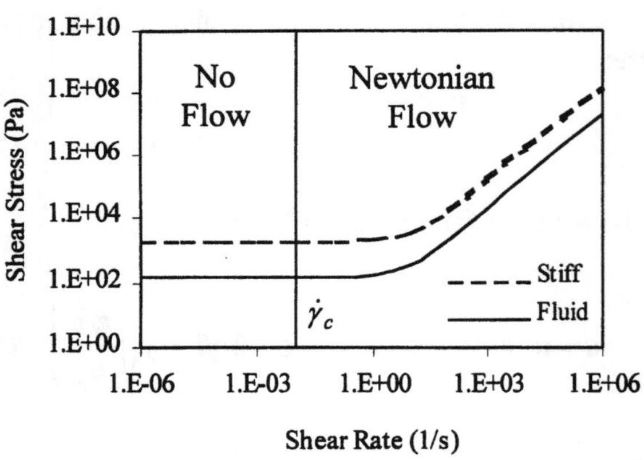

Figure 1. Flow curves for fluid and stiff concrete mixes

By substituting the typical values for fresh concrete in equations (7) and (8), the length of the flow region for a fluid concrete is L=37 mm and for a stiff concrete L=88 mm. These are rather smaller than observed in practice

Additionally, when the effects of frequency and velocity oscillation are analysed with the help of equation (7) for a typical concrete it is clear that the length of the flow region can be increased by reducing the frequency or increasing the velocity of oscillation. Furthermore, the results also indicate that L is more sensitive to a reduction in frequency than an increase in velocity.

4. CONCLUSIONS

In this study, the introduction of a critical shear rate $\dot{\gamma}_c$ for the flow of interest allowed the derivation of an exact expression for the length of the flow region.

The results obtained for the typical vibrational source of an internal poker applied to fluid and stiff concrete mixes, i.e. L=37 mm and L=88 mm respectively, suggest that a pure shear waveform cannot account for the known range of effectiveness of a vibrator. Therefore, an analysis of the compressive waveform is needed to completely understand the dominant mechanism causing the flow around the vibrational source, and this will be considered in a future paper.

5. REFERENCES

Casson RBJ and Domone PLJ (1982) Int. Conf. on Concrete of Early Ages 1: 129-135, Paris.

Casson RBJ, Domone PLJ, Scrivener K, Jennings HM, Gillham CJ and Pratt PL (1982) In: Concrete Rheology, Proc. Symposium M (Materials Research Society) Annual Meeting Jan Skalny (Ed.), pp. 66-75.

Domone PLJ, Xu Y and Banfill PFG (1998) Magazine of Concrete Research, in press.

Lighthill MJ (1964) In: Laminar Boundary Layers, Rosenhead L. (Ed.), Oxford Press, London.

Tattersall GH and Banfill PFG (1983) The Rheology of Fresh Concrete, Pitman Books, London.

Telionis DP (1981) Unsteady Viscous Flows, Springer-Verlag, New York.

STRUCTURE OF THE NEAR-WALL TURBULENCE IN A DRAG-REDUCTING FLOW

J. TIHON[1], C. DESLOUIS[2], B. TRIBOLLET[2]

[1] Institute of Chemical Process Fundamentals, Academy of Sciences, 165 02 Prague 6, Czech Republic
[2] UPR 15 du CNRS 'Phys. des Liquides et Electrochimie', Université P.et M.Curie 752 52 Paris, France

1. INTRODUCTION

The mechanism of frictional drag reduction in turbulent fluid flow due to the addition of minute concentrations of long-chain polymers is still a open topic, *Gyr and Bewersdorff (1995)*. It was shown by *Wells and Spangler (1967)* that this phenomenon is associated with flow changes in the near-wall region. The electrochemical techniques have been proved to be suitable for the study of wall turbulence, *Fortuna and Hanratty (1972)*. In this work, the improved electrochemical sensors have been applied to assess the influence of polymer additives on spatial and temporal correlations of flow fluctuations in the near-wall region.

2. EXPERIMENTAL

The experiments were conducted in a two-dimensional rectangular channel made of Plexiglas (height of 12 mm and width of 12 cm). To minimalize the mechanical polymer degradation the fluid was forced to flow between two tanks (180 l) by imposing an overpressure. The measurement section with the electrochemical electrodes was located 2 m downstream the right tank so as to ensure well developed turbulent flow conditions in the range of the Reynolds numbers investigated (Re=10000÷35000). The pressure taps were used to measure the pressure drop along the channel.

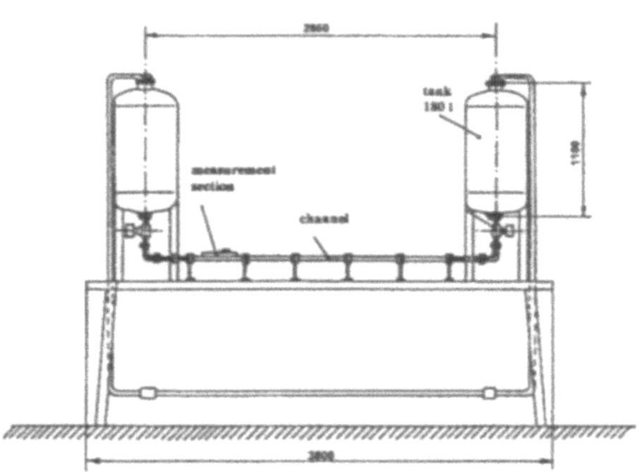

Fig.1: Experimental set-up

The electrochemical method was used to study of flow fluctuations at the wall, see *Hanratty and Campbell (1983)*. The photolithography technique of the probe fabrication (see *Deslouis et al. (1993)*) was improved to turn out very accurate and complicatedly patterned electrodes. Eight two-segment gold electrodes (d=100 μm) located close each other in the channel wall were simultaneously working in the regime of the limiting diffusion current. The current signals from the electrode segments provided information on the two components of the fluctuating velocity gradient at the wall, axial $s_x \sim I_1+I_2$ and transverse $s_z \sim I_1-I_2$.

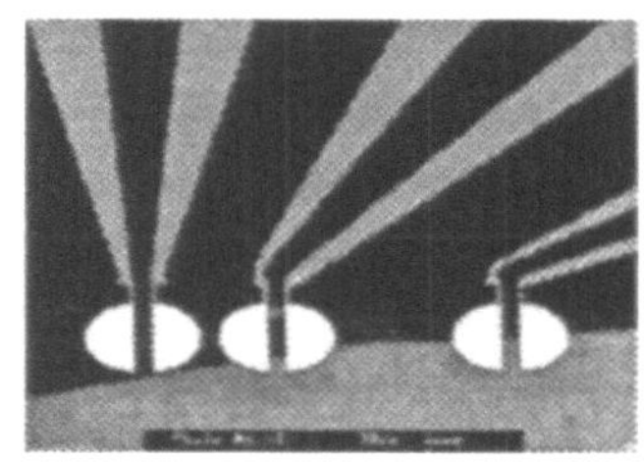

Fig.2: SEM photograph of the double semi-circular probes prepared by photolithography.

Experiments were performed with aqueous solutions (c=20÷100 ppm) of two high molecular weight polymers ($M \sim 5.10^6$): polyethylene oxide (PEO) and polyacrylamide (PAA). The test fluids contained equimolar 0.003 M ferro/ferricyanide and 0.003 M potassium sulphate as a suitable electrochemical redox system.

3. RESULTS

The friction coefficients calculated from the pressure drop measurements are shown in Fig.3. The drag reduction was affected by the concentration and gradual degradation of the polymer on a charge of high shear stresses in the channel and reached the maximum level of DR~75% at high values of Re. The friction data calculated from the electrochemical measurements of the average wall velocity gradient S ($c_f \equiv 2S\mu/\rho u^2$) followed the pressure drop ones ($c_f \equiv (h/L).(\Delta P/\rho u^2)$) very well.

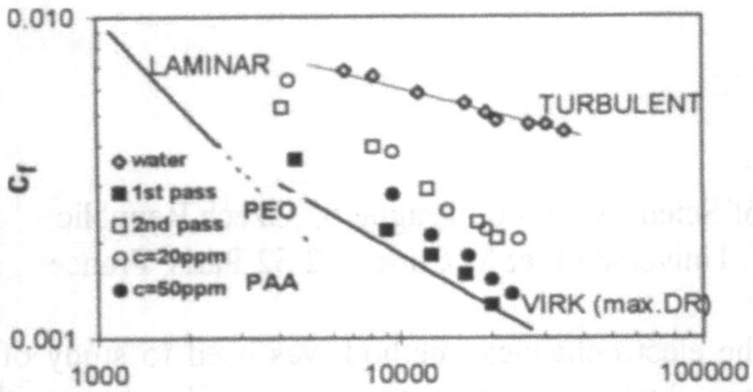

Fig.3: Friction coefficient c_f vs. Re

The effect of drag-reducing polymer on the axial component of the fluctuating wall velocity gradient is illustrated in Fig.4. The comparison of the PSD functions demonstrates reduction and frequency redistribution of flow fluctuations due to the polymer. When the wall velocity gradient and the variance of s_x were used as the scaling parameters for the frequency (f/S) and the PSD function ($W_{sx}/var(s_x)$), the universal, normalized shape of turbulent power spectra was obtained.

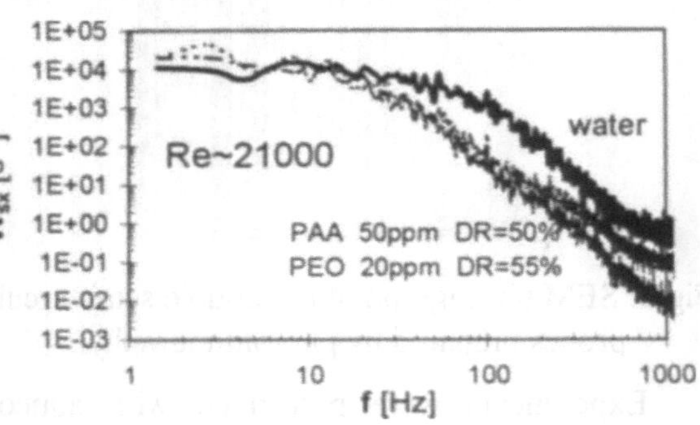

Fig.4: Power spectra of s_x fluctuations

The correlation coefficients of s_x in transverse direction are shown in Fig.5. An increase in the width of the longitudinally oriented eddies with an increase in the drag reduction level is striking. The

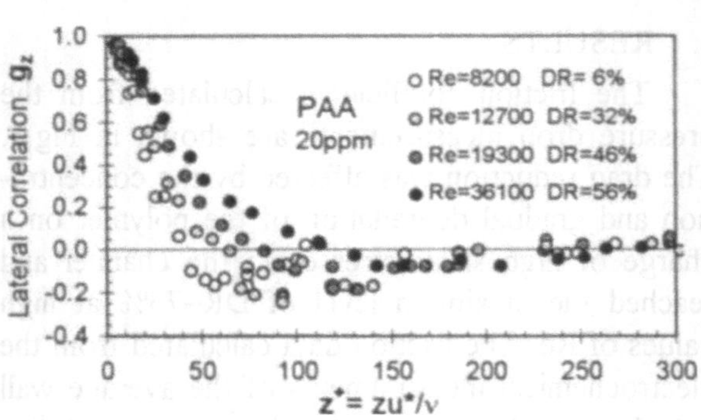

Fig.5: Lateral correlations of s_x fluctuations

average scale λ^+, normalized with respect to the wall parameters and determined from the g_z (twice the distance between the zero points $g_z(z^+)=0$), had the practically constant value of λ^+~100 for water and increased up to the value of λ^+~350 for the maximum drag reduction measured.

Two time traces of current difference signals, corresponding to the s_z fluctuating component for water and the polymer solution at the same value of the wall shear rate ($S \sim I^3$), are shown in Fig.6

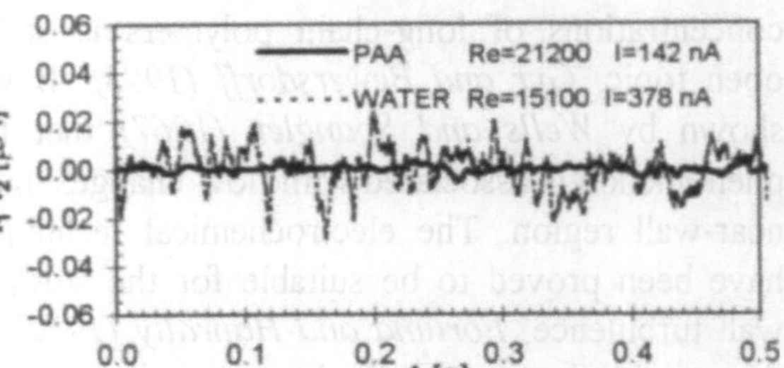

Fig.6: Damping of s_z fluctuations by the polymer

4. CONCLUSIONS

- The frequency and the amplitude of flow fluctuations were greatly reduced by addition of polymer and the corresponding power spectra exhibited a shift towards low frequencies.
- The wall shear rate was found to be a suitable normalization parameter for the power spectra.
- The size of the coherent flow structures was significantly increased due to the drag-reducing additive in both longitudinal, and transverse directions.
- The damping of fluctuations was more pronounced for the transverse velocity component.

5. REFERENCES

Deslouis C, Huet F, Robin S, Tribollet B (1993) Int.J.Heat Mass Transfer. 36:823-829

Fortuna G and Hanratty TJ (1972) J.Fluid Mech. 53:575-586

Gyr A, Bewersdorff HW (1995) Drag reduction of turbulent flows by additives. Kluver, Dordrecht

Hanratty TJ, Campbell JA (1983) in Goldstein JR (ed) Fluid Mechanics Measurements. Hemisphere, Washington, p559

Wells CS, Spangler JG (1967) Phys. Fluids 10:1890

Acknowledgement:

This work was supported by the COST grant under the project OC F2.10.

STABILITY OF THE VISCOELASTIC FILM FLOW DOWN AN INCLINED PLANE

J. TIHON, O. WEIN

Institute of Chemical Process Fundamentals, Academy of Sciences, 165 02 Prague 6, Czech Republic

1. INTRODUCTION

The evolution of travelling surface waves on a flowing liquid film is a classical problem of the hydrodynamic instability. Recent theoretical knowledge of the wavy film flow of Newtonian liquids was reviewed by *Chang (1994)*.

The corresponding problem for a viscoelastic fluid has been much less studied. The linear stability theory predicts destabilizing effect of elasticity in film flows (see *Gupta (1967)*, *Lai (1967)*, and *Ting (1975)*). Two recent theoretical works dealt with an Oldroyd-B fluid. *Shaqfeh et al. (1989)* solved the modified Orr-Sommerfeld equation numerically for a vertical film and they presumed that maximum wave growth rates are probably too low for a purely elastic instability to be observable. *Kang and Chen (1995)* derived a nonlinear long wave evolution equation and investigated the effect of the inclination angle on the bifurcation. The larger equilibrium wave amplitudes were surprisingly found for smaller inclination angles.

In this work we present results of the experimental study of viscoelastic film flows. We demonstrate the effect of elasticity on the wave characteristics and the possibility to observe elastic flow instabilities in the liquid film at small inclinations angles.

2. EXPERIMENTAL

The liquid film flowed down a smooth stainless steel plate, 0.22 m wide and 2 m long. The slope of the plate could be changed from the vertical to the horizontal position. The liquid was fed by a single screw pump through a smoothing vessel and a heat exchanger into a distributor, and then spread by a knife-edge weir formed on the top of the plate. The aqueous solutions of polyacrylamide (SEPARAN AP30) and polyethylene oxide (POLYOX WSR-301) were used as the model viscoelastic liquids. The rheological measurements carried out in the range of the characteristic wall shear rate of the film flow (γ<150 s^{-1}) shown that these solutions exhibited strong elastic behaviour for the concentrations used (c=0.2÷1% b.w.). The measured normal stresses N_1 were greater then the shear stresses τ_{12} and the relaxation times changed in the range λ =0.04÷0.8 s.

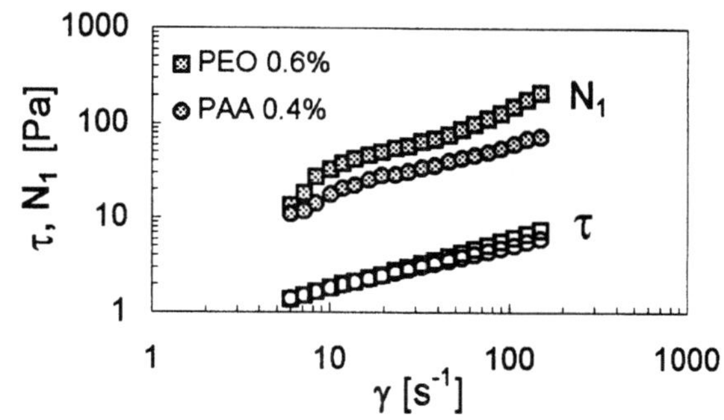

Fig.1: Viscometric data for two polymer solutions

The instantaneous film thickness was measured by the pairs of capacitance sensors aligned 20 mm apart in the flow direction. The wave frequency *f* was estimated from the position of the dominant peak in the power spectral density function. The wave celerity *c* was determined from the time necessary for the waves to travel from the upper capacitance sensor to the lower one. The wave length λ was estimated as the ratio of the wave velocity to the frequency.

3. RESULTS

The typical time trace of the film thickness is shown in Fig.2. The character of the wavy flow differentiated from that exhibited by Newtonian films, e.g. for glycerine *Tihon (1991)*.

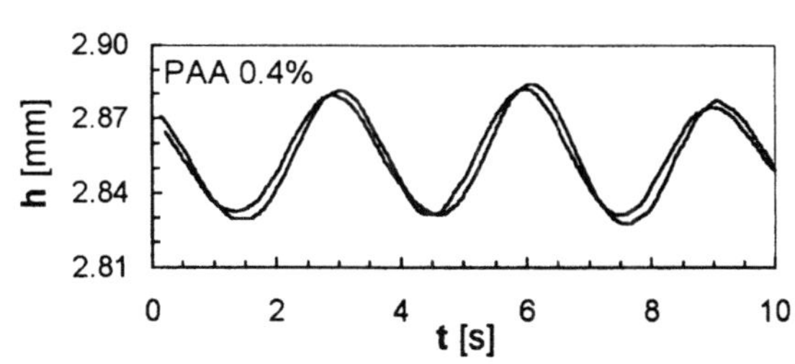

Fig.2: Film thickness measured by two sensors (α=12°, *Re*=4.4, *f*=0.36Hz, *c*=0.127 m/s, λ=35cm)

The dramatic effect of elasticity on wave characteristics is shown in Tab.1, where the experimental results for Glycerine 80%, Polyox 0.6%, and Separan 0.4% are compared. These data were obtained for similar flow conditions (apparent viscosity at the wall μ_a~0.07 Pa.s) at the beginning of the wavy regime.

	Re [-]	h [mm]	u [m.s^{-1}]	f [Hz]	c/u [-]	λ [mm]	a [μm]
PAA	1.1	2.85	0.035	0.33	3.86	409	29
PEO	2.1	2.73	0.050	0.51	4.23	426	23
GLY	6.3	3.40	0.122	4.3	2.81	78	350

Table 1: Effect of elasticity on the wave characteristics (x=1.2m, α=12°)

The elasticity played a destabilising role in the film flow and reduced significantly the value of the critical Reynolds number Re_c below the Newtonian one $Re_c=5/6 \cot(\alpha)$ (see Fig.3).

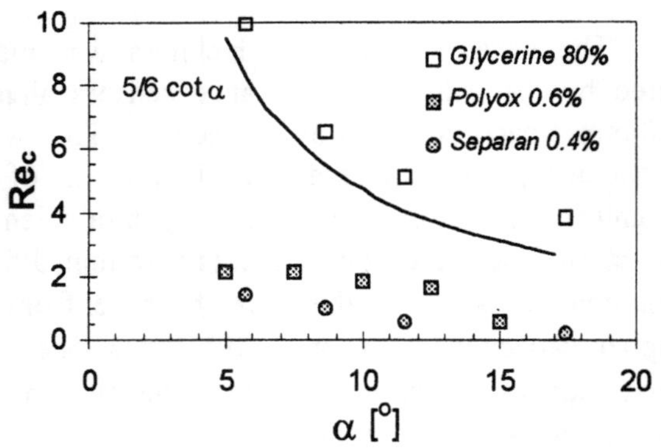

Fig.3: Dependence of the wave inception on the plate inclination angle

The experiments were focused on the region of small inclination angles and low flow rates, where the elastic effects were expected to come to light. Most of experiments were done in the developed wavy flow region (x>1 m). The results were interpreted as dependencies of the wave characteristics (*a, c, λ, f*) on the operation parameters (Re, Wi, α).

The wave amplitudes measured at the fixed downstream position remained small and nearly the same in the range of small Re close to Re_c. The rapid augmentation of the wave amplitude with increasing of the flow rate, typical for Newtonian films, was observed only at higher flow rates, when the inertial forces achieved a sufficient level.

The exponential spatial growth of the amplitude, typical for travelling waves, is demonstrated in Fig.4. The growth rate $r=\delta\ln(a/h)/\delta(x/h)$ obtained for the elastic driven instabilities (r=0.0056 and 0.0112 in Fig.4) were the same order as those for Newtonian films (*Liu and Gollub (1993)*).

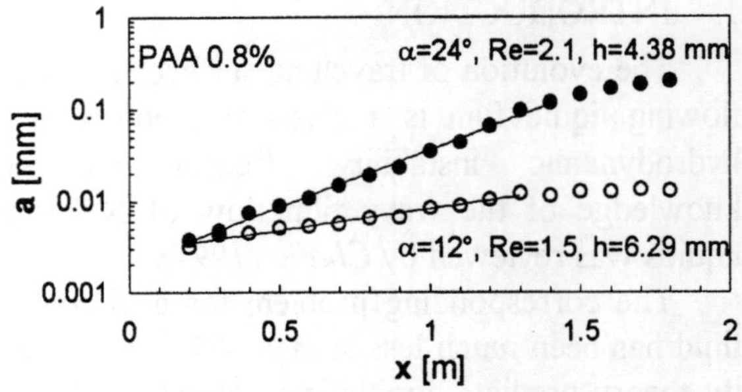

Fig.4: Spatial development of the wave amplitude

4. CONCLUSIONS

- The destabilizing effect of elasticity shifts the wave inception towards lower values of Re.
- The elastic instabilities are observable in falling film flows at small plate inclinations and low flow rates corresponding to the initial region of the wavy regime.
- These elasticity driven waves are very long (λ~10÷60cm), low-frequency (*f*<1Hz) and their equilibrium amplitude remains very small and nearly constant in the certain range of flow rate values.
- This newly observed kind of the waves is assumed to be the result of the competition between elastic and capillary forces in the film.

5. REFERENCES

Chang HC (1994) Annu.Rev.Fluid Mech. 26:103-106

Kang F, Chen KP (1995) J.Non-Newtonian Fluid Mech. 57:243-252

Lai W (1967) Phys.Fluids 10:844-847

Liu J, Paul JD, Gollub JP (1993) J.Fluid Mech. 250:69-101

Gupta AS (1967) J. Fluid Mech. 28:17-28

Shaqfeh ESG, Larson RG, Fredrickson GH (1989) J. Non-Newt. Fluid Mech. 31:87-113

Tihon J (1991) Ph.D. thesis AS CR, Prague

Ting LL (1975) Rheol.Acta 14:503-532

Acknowledgement:

This work was supported by the Grant Agency of the Czech Republic under the contract No. *104/96/0569.*

CURVILINEAR SQUEEZE FILM OF COUPLE-STRESS FLUID WITH REFERENCE TO SYNOVIAL JOINTS

E. WALICKI, A. WALICKA

Technical University of Zielona Góra, Dept. of Mechanics, ul. Szafrana 2, 65-016 Zielona Góra, Poland

1. INTRODUCTION

The presence of small amounts of additives in a lubricant can improve bearing performance by enhancing the lubricant viscosity and thus producing an increase in the load capacity. They also reduce the coefficient of friction and increase the temperature range in which the bearing can operate.

A number of theories of the microcontinuum have been postulated and applied (Lai et al. 1978, Stokes 1966, Walicka 1994). One theory due to Stokes (1966) allows for polar effects such as the presence of couple stresses and body couples.

This theory has been applied to the study of some simple lubrication problems (see e.g. Sinha et al. 1981, Lin 1996). One of the possible application of couple-stress fluid is its use to the study of the mechanismus of lubrication of synovial joints.

The field equations for the motion of an incompressible fluid with couple-stresses are:

$$\operatorname{div} \mathbf{V} = 0, \quad \rho \frac{d\mathbf{V}}{dt} = -\operatorname{grad} p + \mu \nabla^2 \mathbf{V} - \eta \nabla^4 \mathbf{V}$$

where $\mathbf{V}$ - is the velocity vector, p - the pressure, ρ - the density, μ - the shear viscosity, η - the couple-stress viscosity.

In this paper Stokes theory is applied to consider inertia effects on the pressure distribution and the load of synovial joint modelled as a thrust bearing with squeeze film shown in Fig. 1.

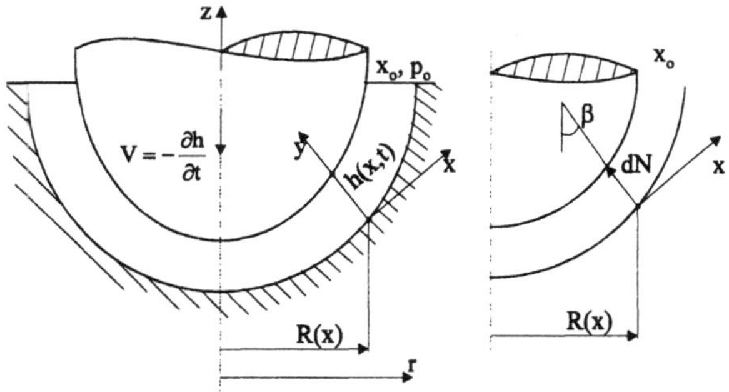

Fig. 1. Coordinate system in the bearing clearance.

2. FLOW OF A COUPLE-STRESS FLUID IN THE BEARING CLEARANCE

Applying the method of averaged inertia we may write (Walicka et al. 1997):

$$\mu \frac{\partial^2 \upsilon_x}{\partial y^2} - \eta \frac{\partial^4 \upsilon_x}{\partial y^4} = f(x) \tag{1}$$

where

$$f(x) = \frac{dp}{dx} + \frac{\rho}{h}\left(\frac{R'}{R} + \frac{\partial}{\partial x}\right)\int_0^h \upsilon_x^2 dy. \tag{2}$$

Here the prime denotes the differentiation with respect to x. Denote the integral of Eq. (1) by $\upsilon_x = F(f, h, y, l)$ we can define the flow rate Q as:

$$Q = 2\pi R \int_0^h \upsilon_x dy \quad \text{or}$$

$$Q \approx -\frac{\pi R f h^3}{6}\left[1 - 12\left(\frac{l}{h}\right)^2\right], \quad l^2 = \frac{\eta}{\mu}. \tag{3}$$

Having Q one can find effectively υ_x and next, after developing of Eq. (2), the formula for pressure distribution:

$$p(x,t) = p_0 + \frac{6\mu}{\pi}\left[S_0 - S(x,t)\right] + \\ + \frac{3\rho}{10\pi^2}\left[T_0 - T(x,t)\right] \tag{4}$$

where

$$S(x,t) = \int \frac{Q}{Rh^3}\left[1 + 12\left(\frac{l}{h}\right)^2\right]dx,$$

$$S_0 = S(x_0, t), \qquad T_0 = T(x_0, t),$$

$$T(x,t) = \int \frac{1}{h}\left(\frac{R'}{R} + \frac{\partial}{\partial x}\right)\frac{Q^2}{R^2 h}\left[1 + 4\left(\frac{l}{h}\right)^2\right]dx. \tag{5}$$

Here the flow rate Q is given by

$$Q=-2\pi\int R\frac{\partial h}{\partial t}dx. \qquad (6)$$

If $T(x,t)=0$ the formula (4) presents the pressure distribution without inertia effects.
The load capacity is given by

$$N_z=2\pi\int_0^{x_0} pR\cos\beta dx. \qquad (7)$$

The sense of the angle β arises from Fig. 1.

3. EXAMPLE OF APPLICATION

Let us consider the simulation of the squeeze-film behaviour of a human joint which geometry is shown in Fig. 2. Using the nondimensional parameters:

$$h^*=\frac{h}{C}=1-\varepsilon\cos\varphi,\quad l^*=\frac{l}{C},\quad \mathrm{Re}=\frac{\rho C^2\dot{\varepsilon}}{\mu}$$

we may present the following nondimensional formulae for the pressure distribution in bearing clearance and the load capacity:

$$p^*=\frac{(p-p_0)C^2}{\mu Rr^2\dot{\varepsilon}}=\frac{3}{\varepsilon}\left\{\frac{1}{(1-\varepsilon\cos\varphi)^2}-1+\right.$$
$$\left.+6l^{*2}\left[\frac{1}{(1-\varepsilon\cos\varphi)^4}-1\right]\right\}+\frac{3}{10}\mathrm{Re}\,F_1(\varphi,l^*,\varepsilon),$$

$$N^*=\frac{N_zC^2}{\mu Rr^4\dot{\varepsilon}}=\frac{6\pi}{\varepsilon^3}\left(\frac{\varepsilon}{1-\varepsilon}-\frac{\varepsilon^2}{2}+\ln(1-\varepsilon)+\right.$$
$$\left.+l^{*2}\left\{2\left[\frac{1}{(1-\varepsilon)^3}-1\right]-3\left[\frac{1}{(1-\varepsilon)^2}-1\right]-3\varepsilon^2\right\}\right)+$$
$$+\frac{3\pi}{5}\mathrm{Re}\,F_2(l^*,\varepsilon)$$

where the given functions F_1 and F_2 represent the inertia effects.

Figs 3 and 4 present the graphs of pressure distributions and load capacity with $(\mathrm{Re}\neq 0)$ and without inertia $(\mathrm{Re}=0)$.

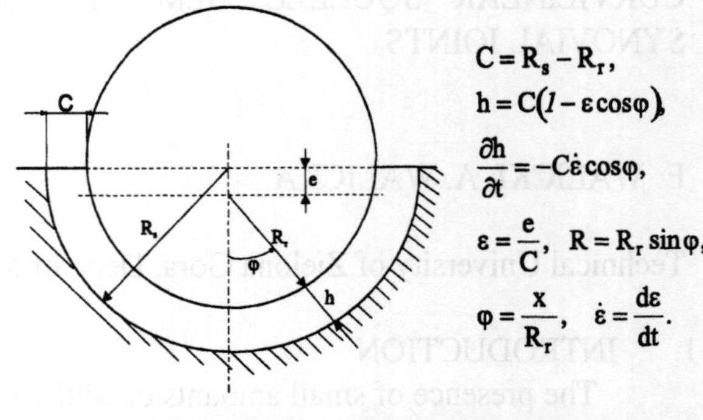

Fig. 2. Hemispherical bearing configuration.

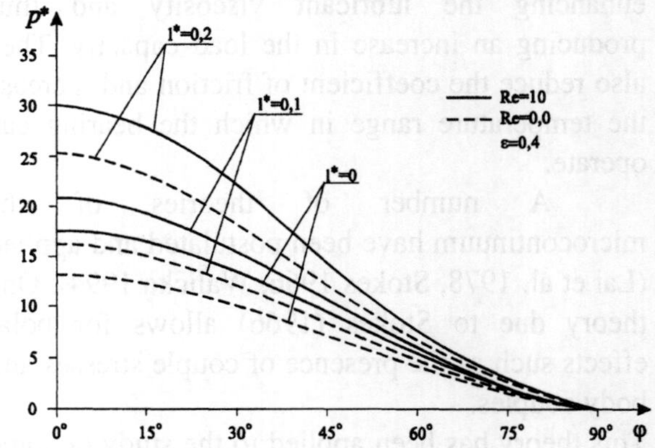

Fig. 3. Pressure p^* vs. φ for different l^*.

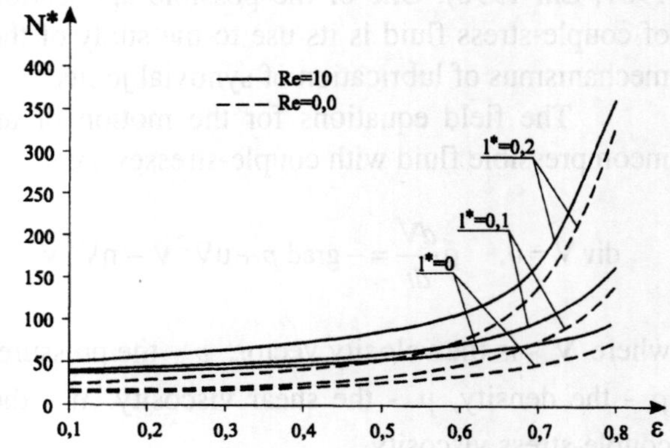

Fig. 4. Load capacity N^* vs. ε for different l^*.

4. REFERENCES

Lai WM, Kuei S.C., Mow VC (1978) J Biomech Engng 100: 169-186

Lin J-R (1996) Appl. Mech Engng 1: 317-332

Sinha P., Singh C, Prasad KR (1981) Wear 67: 15-24

Stokes VK (1966) Phys Fluids 9: 1709-1715

Walicka A (1994) Micropolar flow in a slot between rotating surfaces of revolution. TU Press, Zielona Góra

Walicka A, Walicki E, Raganowicz B (1997) In: Proc 3rd Nat Conf on Non-Conventional Bearing Systems. pp 95-101

PRESSURE DROPS IN THE CONICAL FLOW OF A MOLTEN METAL

E. WALICKI, A. WALICKA, D. MICHALSKI

Technical University of Zielona Góra, Dept. of Mechanics, ul. Szafrana 2, 65-016 Zielona Góra, Poland

1. INTRODUCTION

Conical converging flows occurs frequently in many industrial applications, particularly in manufacturing processes involving metal melts and in those instances the prediction of pressure losses is often of paramount concern.

The manufacturing is often realized in die-casting machines where the molten metal forms a mixture of liquid metal with solidified particles (Hashimoto and Amagai, 1993). The flow of this mixture may be modeled as the flow of viscoplastic fluids (Avenas et al., 1982; Binding, 1988).

To describe the rheological behaviour of such fluids the non-linear model of Shulman (1975) may be used:

$$\tau^{1/M} = \tau_0^{1/M} + (\mu\dot{\gamma})^{1/N}$$

where τ is the shear stress, τ_0 is the yield shear stress, μ is the coefficient of plastic viscosity, M and N are the non-linearity indices, $\dot{\gamma}$ is the shear rate.

By reducing the coefficient in the Shulman equation one can obtain simpler models of the viscoplastic fluid; if $\tau_0 = 0$, $M = N = 1$ we have the classical Newtonian fluid.

Converging flows necessarily involve a component of extensional flow; Cogswell (1972, 1978) was perhaps the first to recognize of extensional viscosity in converging flows. Developed versions of Cogswell's analysis were used by Walicki and Walicka (1994, 1995) in a study of conical converging flow.

Basing on the considerations suggested in the work (Walicki and Walicka, 1995) the authors present in this paper an approximate analysis for converging flow of a molten metal (modeled as viscoplastic fluid) in the conical die.

2. PRESSURE DROPS IN CONICAL DIES

The flow configuration is presented in Fig.1. Corresponding velocity component is υ_z which is parallel to the axis of z. Considering an equilibrum of an element of length dz (Fig.2.) we will obtain from the forces balance the formulae

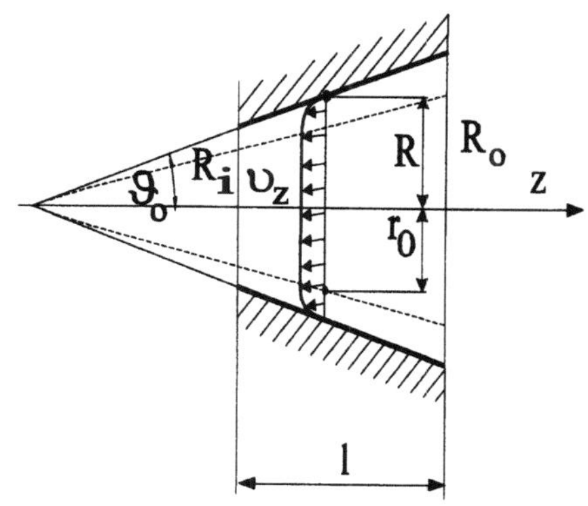

Fig.1. Schematic diagram of conical flow.

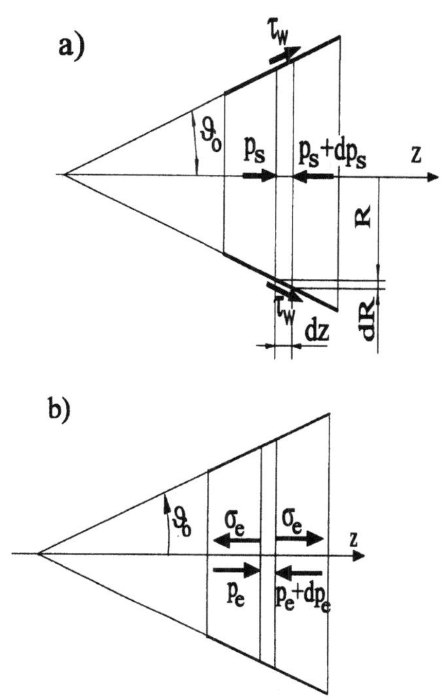

Fig.2. Schematic diagram of forces for the analysis of the pressure drop due to telescopic shear (a) and extensional flow (b).

for the pressure drops (Walicki, Walicka, 1994, 1996). For the telescopic shear there is:

$$\Delta p_s = 2\tau_0 cot\vartheta_0 \int_{R_i}^{R_0} Y dR; \qquad (1)$$

where Y is the solution to the equation:

$$\sum_{i=0}^{N}(-1)^i \frac{M}{3M+N-i} C_N^i \left(Y^{\frac{3M+N-i}{N}} - 1 \right) + \\ - K_{(M)}^{(N)} Y^3 = 0 \qquad (2)$$

and

$$Y = \frac{R}{r_0}, \qquad K_{(M)}^{(N)} = \frac{\mu Q}{\pi R^3 \tau_0^{\frac{N}{M}}},$$

$$Q = 2\pi \int_0^R \upsilon_z r dr, \qquad C_N^i = \frac{N!}{i!(N-i)!}; \qquad (3)$$

here r_0 - is the radius of core flow, Q - is the flow rate.

For the extensional flow there is:

$$\Delta p_e = 2 \int_{R_i}^{R_o} \sigma_e \frac{dR}{R}, \qquad (4)$$

where the averaged extensional stress σ_e is:

$$\sigma_e = \frac{2\tau_0 tan\vartheta_o}{R^4} \int_0^R Y r^3 dr, \qquad (5)$$

3. EXAMPLES OF APPLICATION

Consider for example two models of fluid: power-law fluid ($M = 1, \tau_0 = 0$) and Herschel-Bulkley fluid ($M = 1$).

For the power-law fluid one has:

$$Y_p = F_p R^{-1}, \; \sigma_e = \frac{1}{2} F_p tan\vartheta_o, \; F_p = \left[\frac{\mu(3+N)Q}{\pi R^3} \right]^{\frac{1}{N}},$$

and

$$\Delta p_s = \frac{2NF_p}{3}\left(1 - \beta^{\frac{3}{N}} \right) cot\vartheta_o, \; \Delta p_e = \frac{NF_p}{3}\left(1 - \beta^{\frac{3}{N}} \right) tan\vartheta_o$$

where $\beta = R_i / R_o$.

For the Herschel-Bulkley fluid we have:

- for small core (or small value of τ_0):

$$Y_s = F_s + \frac{4}{3}, \; \sigma_e = \frac{\tau_0 Y_s tan\vartheta_o}{2}, \; F_s = \left[\frac{\mu(3+N)Q}{\pi R^3 \tau_0^N} \right]^{\frac{1}{N}}$$

and

$$\Delta p_s = \frac{2F_{ps} cot\vartheta_o}{3}, \qquad \Delta p_e = \frac{F_{ps} tan\vartheta_o}{3}$$

where

$$F_{ps} = NF_s \left(1 - \beta^{\frac{3}{N}} \right) - 4\tau_0 ln\beta;$$

- for large core (or large value of τ_0):

$$Y_l = F_l + 1, \; \sigma_e = \frac{\tau_0 Y_l tan\vartheta_o}{2}, \; F_l = \left[\frac{\mu(1+N)Q}{\pi R^3 \tau_0^N} \right]^{\frac{1}{1+N}}$$

and

$$\Delta p_s = \frac{2F_{pl} cot\vartheta_o}{3}, \qquad \Delta p_e = \frac{F_{pl} tan\vartheta_o}{3}$$

where

$$F_{pl} = (1+N)F_l \left(1 - \beta^{\frac{3}{1+N}} \right) - 3\tau_0 ln\beta.$$

By the same manner one can obtain the pressure drops for other fluids.

4. REFERENCES

Avenas P, Agassant J-F, Sergent J-Ph (1982) La Mise en Forme des Matiēres Plastiques. Tec-Doc Lavoisier, Paris

Binding DM (1988) J Non-Newt Fluid Mech 27: 173-189

Cogswell FN (1972) Polym Engng Sci 12: 64-73

Cogswell FN (1978) J Non-Newt Fluid Mech 4: 23-38

Hashimoto H, Amagai K (1993) JSME Int J, ser B 36: 584-591

Walicki E, Walicka A (1994) Reports TU Zielona Góra, Fizyka-Chemia 6: 197-217

Walicki E, Walicka A (1995) Solid Metals Alloys 24: 147-154

THROUGHFLOW OF A RIVLIN-ERICKSEN FLUID BETWEEN FIXED SURFACES OF REVOLUTION

E. WALICKI, A. WALICKA

Technical University of Zielona Góra, Dept. of Mechanics, ul. Szafrana 2, 65-016 Zielona Góra, Poland

1. INTRODUCTION

Several constitutive relations are used to model the non-Newtonian characteristics exhibited by some fluids. Integral models, which take the past history of deformation into account, are to be used if memory effects are important. However, for many fluids, only a very short part of the history of the deformations has an influence on the stress. The constitutive relations for these fluids may be expressed as functions of the deformation gradient and its derivatives evaluated at the current time. A material in which the stress depends only a finite number of these time derivatives is called a material of the differential type. The stress in a fluid of differential type of complexity j depends on the velocity gradient and its first $(j-1)$ derivatives. As a non-Newtonian fluid model of differential type, the second grade fluid of Rivlin-Ericksen (1955) is extensively used by both theorists and experimenters. The constitutive relation for this fluid is given as follows:

$$\mathbf{T} = -p\mathbf{I} + \mu\mathbf{A}_1 + \alpha_1\mathbf{A}_2 + \alpha_2\mathbf{A}_1^2$$

where p is the pressure, μ is the viscosity, α_1, α_2 are the material moduli. The kinematic tensors $\mathbf{A}_1$ and $\mathbf{A}_2$ defined through:

$$\mathbf{A}_1 = \mathbf{L} + \mathbf{L}^T,\ \ \mathbf{A}_2 = \dot{\mathbf{A}}_1 + \mathbf{A}_1\mathbf{L} + \mathbf{L}^T\mathbf{A}_1,\ \ \mathbf{L} = \text{grad}\,\mathbf{v};$$

where $\mathbf{v}$ is the velocity vector and $(\dot{\ })$ represents the material derivative with respect to time.

This model of fluid has been applied to the study of some lubrication and hydrodynamic problems (see e.g. Bujurke and Jagadeeswar 1992, Pakdemirli 1994).

The aim of this paper is a study of the pressure distribution in the flow of a Rivlin-Ericksen fluid in a narrow space between fixed surfaces of revolution shown in Fig 1. Using the method of integral approaches one has analysed the influence of inertia terms of the equations of motion on this distribution.

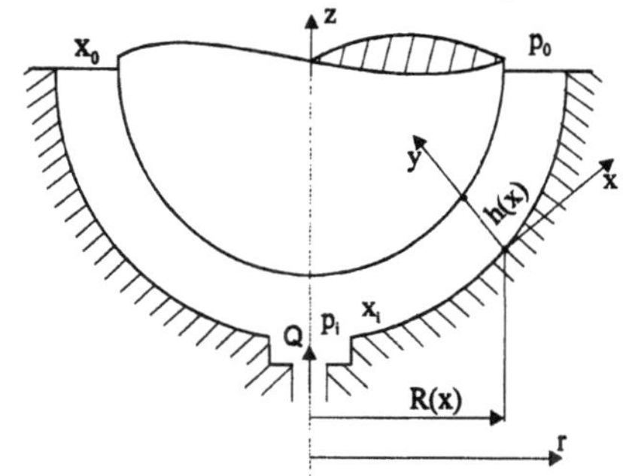

Fig. 1. Clearance between fixed surfaces of revolution.

2. FLOW BETWEEN FIXED SURFACES OF REVOLUTION

Applying the method of integral approaches one obtains the following differential equation for pressure (Walicki 1993, Walicka and Walicki 1997):

$$\frac{dp^*}{dx} = \frac{1}{I_k^{(2)}}\left[\mu I_k^{(3)} + \frac{\alpha R'}{R}\left(I_{k-1}^{(4)} - 2I_{k-3}^{(5)}\right) + \right.$$
$$\left. + \beta I_{k-3}^{(6)} - \frac{\rho}{k+1}\left(\frac{R'}{R} + \frac{\partial}{\partial x}\right) I_{k+2}^{(1)}\right] \qquad (1)$$

where

$$p^*(x) = p(x,y) - (\alpha + 2\beta)\left(\frac{\partial \upsilon_x}{\partial y}\right)^2 \qquad (2)$$

and

$$I_{k+2}^{(1)} = \int_0^h \upsilon_x^{k+2} dy,\ I_k^{(2)} = \int_0^h \upsilon_x^k dy,\ I_{k-2}^{(3)} = \int_0^h \upsilon_x^k \frac{\partial^2 \upsilon_x}{\partial y^2} dy,$$
$$I_{k-1}^{(4)} = \int_0^h \upsilon_x^k \left(\frac{\partial \upsilon_x}{\partial y}\right)^2 dy,\ I_{k-3}^{(5)} = \int_0^h \upsilon_x^k \frac{\partial}{\partial y}\left(\upsilon_x \frac{\partial \upsilon_x}{\partial y}\right)^2 dy,$$
$$I_{k-3}^{(6)} = \int_0^h \upsilon_x^k \left\{\frac{\partial}{\partial y}\left[\upsilon_x \frac{\partial^2 \upsilon_x}{\partial x \partial y} + \upsilon_y \frac{\partial^2 \upsilon_x}{\partial y^2} + 3\frac{\partial \upsilon_x}{\partial y}\frac{\partial \upsilon_x}{\partial x} + \right.\right.$$
$$\left.\left. + \frac{\partial \upsilon_y}{\partial y}\frac{\partial \upsilon_x}{\partial y}\right] - 2\frac{\partial}{\partial x}\left(\frac{\partial \upsilon_x}{\partial y}\right)^2\right\} dy. \qquad (3)$$

Here the primes denotes the derivation with respect to x.

To evaluate effectively the pressure distribution we must calculate the integrals $I_{k+2}^{(1)},\ldots,I_{k-3}^{(6)}$. Taking into account the main values of υ_x and υ_y in the Reynolds approximation ($k=0,\ I_{k+2}^{(1)}=0$) we will obtain after integration of Eq. (1) the formula valid for $k=0$ (momentum integral approach) or $k=1$ (energy integral approach):

$$p(x)=p_0-\frac{6\mu Q}{\pi}\left[A^{(1)}(x)-A_0^{(1)}\right]+$$
$$+\frac{27\alpha Q^2}{5\pi^2}\left[A^{(2)}(x)-A_0^{(2)}\right]-\frac{18\beta Q^2}{5\pi^2}\left[A^{(3)}(x)-A_0^{(3)}\right]+$$
$$+\frac{3(\alpha+2\beta)}{\pi^2}\left(\frac{1}{R^2h^4}-\frac{1}{R_0^2h_0^4}\right)+$$
$$+\frac{C^{(k)}\rho Q^2}{\pi^2}\left(\frac{1}{R^2h^2}-\frac{1}{R_0^2h_0^2}\right)+ \qquad (4)$$

where

$$p(x)=\frac{1}{h}\int_0^h p(x,y)dy$$
$$A^{(1)}(x)=\int\frac{dx}{Rh^3},\quad A^{(2)}(x)=\int\frac{R'}{R^3h^4}dy,$$
$$A^{(3)}(x)=\int\frac{1}{R^2h^5}\left[7h'-4\frac{\left(Rh^3\right)'}{Rh^2}\right]dy,\quad Q=2\pi R\int_0^h\upsilon_x dy,$$
$$A_0^{(i)}=A^{(i)}(x_0),\ C^0=\frac{21}{140},\ C^{(1)}=\frac{27}{140}. \qquad (5)$$

If $C^{(k)}=0$ the formula (4) presents the pressure distribution without inertia effects.

3. EXAMPLES OF APPLICATIONS

Let us consider the pressure distribution in the clearance of constant thickness. To this aim introduce the following dimensionless parameters

$$\tilde{x}=\frac{x}{R_0},\ \tilde{R}=\frac{R}{R_0},\ \tilde{p}=\frac{2\pi(p-p_0)h_0^3}{\mu Q},\ R_\lambda=\frac{\rho Q h_0}{2\pi\mu R_0},$$
$$\tilde{A}^{(1)}(\tilde{x})=A^{(1)}(x)h_0^3,\quad \tilde{A}^{(2)}(\tilde{x})=A^{(2)}(x)R_0^2h_0^4,$$
$$\tilde{A}^{(3)}(\tilde{x})=A^{(3)}(x)R_0^2h_0^4,\ A_\lambda=\frac{\alpha Q}{2\pi R_0^2h_0},\ B_\lambda=\frac{\beta Q}{2\pi R_0^2h_0};$$

here R_λ is the modified Reynolds number and A_λ, B_λ are the parameters of viscoelasticity. Thus the dimensionless pressure is:

$$\tilde{p}=-12\left[\tilde{A}^{(1)}(\tilde{x})-\tilde{A}_0^{(1)}\right]+$$
$$+\frac{6}{5}\left(A_\lambda-4B_\lambda-\frac{10}{3}C^{(k)}R_\lambda\right)\left(\frac{1}{\tilde{R}^2}-1\right).$$

Figs 2 and 3 present the graphs of pressure distributions between parallel disks and between concentric spherical surfaces for different values of $\Delta_\lambda=1.2\left(A_\lambda-4B_\lambda-10/_3C^{(k)}R_\lambda\right)$.

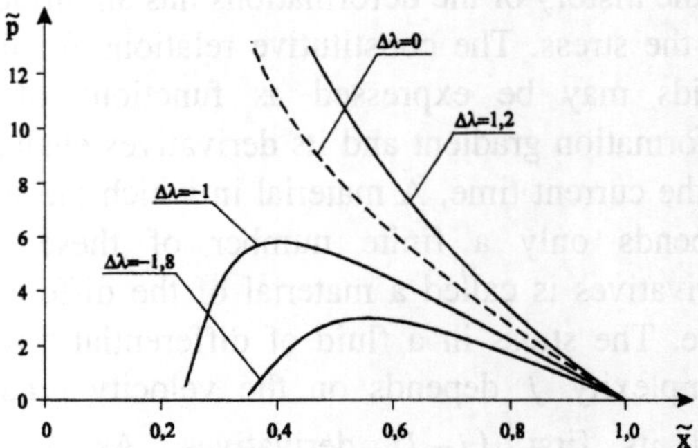

Fig. 2. Dimensionless pressure distribution between parallel disks.

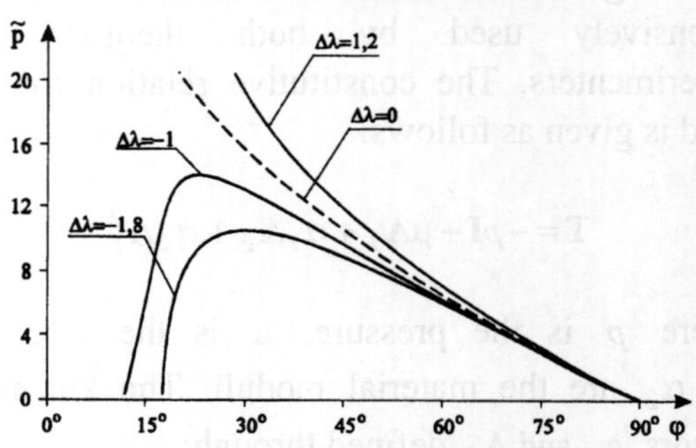

Fig. 3. Dimensionless pressure distribution between concentric spherical surfaces.

4. REFERENCE

Bujurke NM, Jagadeeswar M. (1992) ZAMM 72: 225-228

Pakdemirli M. (1994) Int J Non-Linear Mech 32: 144-154

Rivlin RS, Ericksen JL (1955) J Rat Mech Anal 4: 323-425

Walicka A (1993) Reports TU Zielona Góra 18: 123-135

Walicka A, Walicki E (1997) In: Proc Nat Conf New Tendencies in Tribology and Tribotechnique pp 69-78

INVESTIGATIONS OF FLOW-INSTABILITIES ON HIGH-DENSITY-POLYETHYLENE BY LASER-DOPPLER-VELOCIMETRY

E. WASSNER[1], H. MÜNSTEDT

Institute of Polymer Materials, University Erlangen-Nuremberg, 91058 Erlangen, Martensstr. 7, Germany
[1]Now at: BASF AG, Polymer Research Division, ZKM, 67056 Ludwigshafen, Germany

1. INTRODUCTION

One of the most intriguing effects in the rheology of polymers is the oscillating flow behavior of some melts which is accompanied by pressure fluctuations and a bamboo-like structure of the extrudate.

The mechanism of these instabilities is supposed to be a periodical change between stick and slip of the melt on the die walls.

In order to check this stick-slip-hypothesis direct velocity measurements during instable flow conditions were made on a commercial PE-HD melt in a slit die. As an appropriate measuring method with a high spatial and temporal resolution the laser-Doppler-velocimetry (Durst et al., 1981) was used.

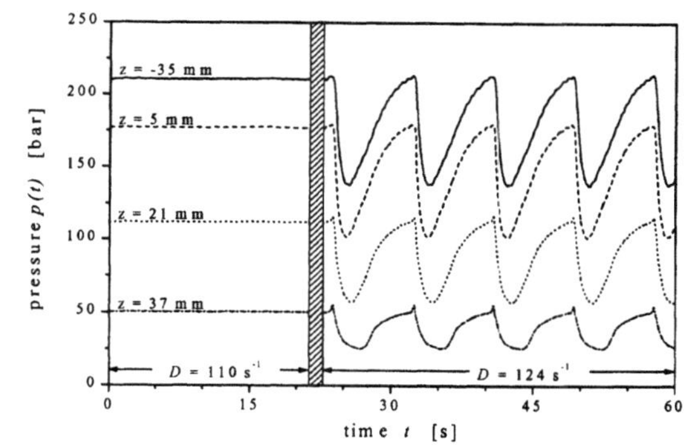

Fig.1: Pressure distribution at 4 positions along the die axis during the transition from stable to instable flow. (z = -35 mm is a position in front of the die.)

2. EXPERIMENTAL

All measurements were performed in a slit die with the height H = 1.03 mm (x-direction), the width W = 14 mm (y-direction) and the length L = 50 mm (z-direction). The origin of the cartesian coordinate system (x, y, z) is located in the center of the die entry plane.

The slit die was fed by a single screw extruder to enable continuous measurements. Along the z-axis of the channel there are 4 pressure transducers to observe the pressure distribution.

During the flow fluctuations the extrudate strand was cut at the die exit in order to calculate an average apparent shear rate $D=6Q/WH^2$.

The laser-Doppler-velocimeter used in this study is a commercial one-dimensional system (Dantec/Invent). The measuring volume is about 30 µm in diameter and 200 µm in length.

3. RESULTS

3.1 Oscillating flow

The flow conditions are stable up to a critical pressure level. When the extruder speed is increased further periodical oscillations set in (cf. fig. 1). The critical shear stress is 0.187 MPa, which is comparable to values given in literature (Hatzikiriakos and Dealy, 1992).

The pressure oscillations are accompanied by large velocity fluctuations in the die (cf. fig. 2, f_p: frequency of the oscillations). The spontaneous pressure drop coincides with a very strong acceleration of the melt which results in an increase in velocity by a factor of about 7-8.

Immediately after the acceleration a strong deceleration sets in. After reaching a minimum there is only a slight increase in velocity till it jumps again to a significantly higher value.

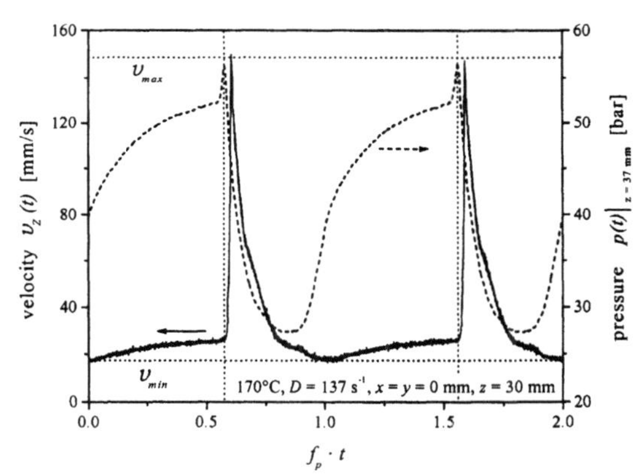

Fig.2: Temporal distribution of the pressure p and the velocity component v_z at a fixed position during oscillating flow.

From the velocity distribution in fig. 2 the minimum velocity υ_{min} and its maximum υ_{max} during one cycle can be determined. In fig. 3 this velocity range is shown as a function of the height of the slit. Firstly, this figure demonstrates the enormous velocity variations during oscillating flow. Secondly, it becomes clear, that even at the moment when the velocity becomes minimum, the extrapolated value at the wall is not zero, i.e. the melt obviously does not stick to the wall.

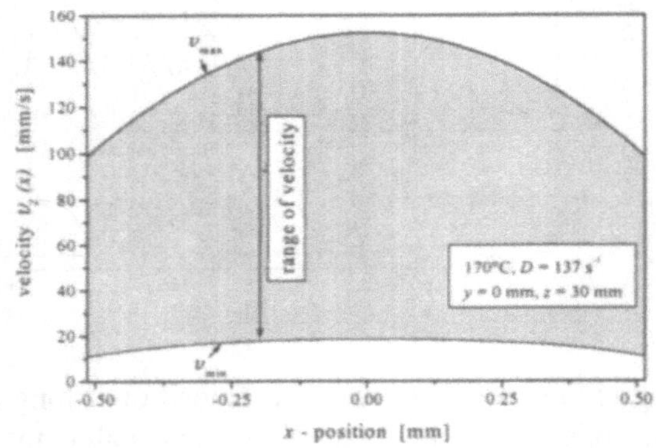

Fig. 3: Range of velocities during fluctuations

3.2 Super-flow

When the output rate exceeds an upper critical value the flow becomes stable again and the pressure oscillations vanish. The structure of the extrudate is nearly as smooth as before the onset of the instabilities. In the following discussion this range is called *super*-flow.

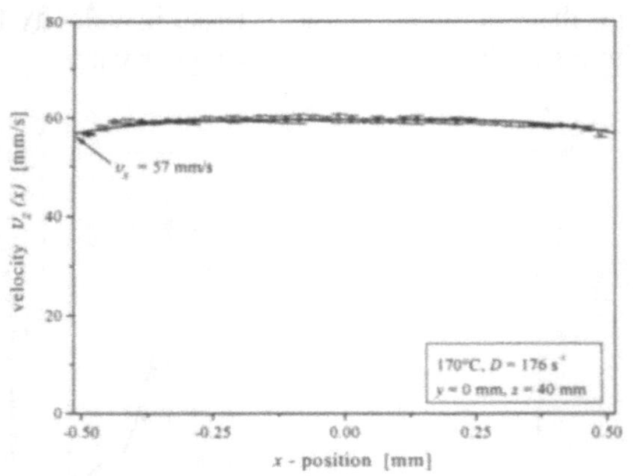

Fig. 4: Velocity profile at *super*-flow conditions

The velocity measurements reveal that at theses high output rates the flow is nearly plug-like. The extrapolated velocity at the die wall (υ_S = 57 mm/s) is almost as high as the velocity in the center (υ_z = 60 mm/s). This indicates that there is nearly ideal slip at the die wall so the melt viscosity only plays a minor role.

4. DISCUSSION

The results suggest that during the whole cycle of the velocity oscillation there is partial slip at the wall. This is due to the orientation of the molecules by the high shear stresses. At a critical stress value the molecules anchored to the wall disentangle from the flowing chains and strong slip sets in (Léger et al, 1996, cf. point 1 in fig. 5). This is accompanied by a output rate which is higher than the output rate of the extruder. This leads to a tearing up of the flow so macroscopic cavities develop. Therefore the pressure cannot be maintained and the melt decelerates immediately (point 2). At point 3 the entanglements are rebuilt and the slip is hindered, i.e. a significant velocity drop can be observed. At point 4 the cavities are filled by the following melt and the velocity increases again.

At *super*-flow the ouput rate of the extruder is high enough so no tearing up of the melt occurs and the strong slip can be maintained.

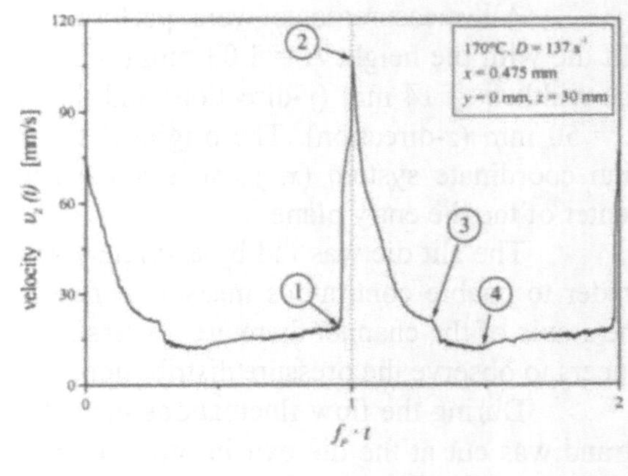

Fig. 5: Characteristics of velocity distribution

5. REFERENCES

Durst F, Melling A, Whitelaw JH (1981) Principles and practise of laser-Doppler-anemometry, Academic press, 2. edition

Hatzikiriakos SG, Dealy JM (1992) Role of slip and fracture in the oscillating flow of HDPE in a capillary. J. Rheol. 36(5):845

Léger L et al. (1996) Slip at the wall. In: Rheology for polymer melt processing. Piau JM, Agassant JF (ed.), Elsevier

Simulation of mixing of viscoelastic fluids in vessels

Olaf Wünsch

Department of Fluid Mechanics, University of the Federal Armed Forces Hamburg, D-22039 Hamburg, Germany

1. Introduction

In many applications of chemical and process engineering the mechanically agitated vessel is the most important equipment to mix liquids. Many fluids used in practice show on the one hand non-Newtonian viscosity properties, on the other hand elastic effects, i.e. they have a memory for their deformation history. These nonlinear properties of the fluid influence the flow in the vessel. For non-Newtonian shearviscosity without elasticity is this shown in literature, e.g. Rubart and Böhme (1991). This paper deals with the numerical simulation of the three-dimensional axissymmetric flow in a cylindrical unbaffed vessel with a rotating disk mounted coaxially on a shaft. The vessel is filled with a viscoelastic fluid and the mixing process works under laminar conditions.

2. Numerical

A single-integral constitutive equation (modified Wagner model) is used to describe real material behavior of viscoelastic fluids,

$$\mathbf{T}(\mathbf{r}) = \int_0^\infty \left[\Psi(s,II)\left(\mathbf{C}_t^{-1}(\mathbf{r},s) - \mathbf{1}\right)\right] ds. \quad (1)$$

The extra stress tensor $\mathbf{T}$ at the position $\mathbf{r}$ is given by the integral of the relative right-Cauchy-Green strain tensor $\mathbf{C}_t$ and the memory function

$$\Psi(s,II) = \left[\sum_{k=1}^{K} \frac{a_k}{\lambda_k} \exp\left(\frac{-s}{\lambda_k}\right)\right] \cdot \exp\left[-n(II-3)^{1/2}\right], \quad (2)$$

which depends on the retarded time s and on the strain history invariant $II = \mathrm{tr}\mathbf{C}_t$. That means, in order to compute the extra stresses at a particular position one have to trace back the position of the material point, to calculate the corresponding relative deformation history and to integrate it along the path line. The free parameter a_k, λ_k and n of the memory function are adapted to the rheological properties of a real viscoelastic fluid checked in rheometer under the kinematics of steady and unsteady shear flow (Table 1).

Table 1: Rheological parameters of a hydrous 1% Separan AP-solution

k	λ_k $[s]$	a_k $[Pa]$
1	$4.00 \cdot 10^1$	$7.38 \cdot 10^{-2}$
2	$8.00 \cdot 10^0$	$9.81 \cdot 10^{-1}$
3	$3.00 \cdot 10^0$	$3.60 \cdot 10^0$
4	$1.00 \cdot 10^0$	$1.85 \cdot 10^1$
5	$3.00 \cdot 10^{-1}$	$3.32 \cdot 10^1$
6	$1.00 \cdot 10^{-1}$	$5.50 \cdot 10^1$
7	$3.00 \cdot 10^{-2}$	$5.52 \cdot 10^1$
8	$1.00 \cdot 10^{-2}$	$1.34 \cdot 10^2$
$n = 0.275$; $\eta_0 = 58.5 Pa\ s$; $\lambda = 4.02 s$		

Note that the relaxation time λ_k differ by a factor of 10^3 which has to be carefully considered in developing an efficient algo:ithm.

Instead of evaluating the improper memory integral (1) we treat an ordinary initial value problem

$$\frac{d\mathbf{H}(\mathbf{r},s)}{ds} = \Psi(s,II)\left(\mathbf{C}_t^{-1}(\mathbf{r},s) - \mathbf{1}\right), \quad (3)$$

$$\mathbf{H}(\mathbf{r},0) = 0, \quad \text{with} \quad \mathbf{T}(\mathbf{r}) = \lim_{s\to\infty} \mathbf{H}(\mathbf{r},s). \quad (4)$$

The continuous problem is discretised by finite elements with 399 Taylor-Hood triangular elements. Because of the axissymmetric of the flow the mesh only extends on one half of the vessel. A new approach to calculate the extra stress tensor in circulating domains can reduce the calculation time enourmously if the memory of the fluid is relatively long. Further details of the numerical procedure is discribed in Broszeit (1997).

3. Quality of mixing

The basic process during the laminar mixing of fluids are the stretching and folding of fluid material by the action of the flow. If the flow in the apparatus are known the right Cauchy-Green strain tensor $\mathbf{C}$ is suitable in order to discribe the deformation of fluid elements. The mean deformation at the time t can be calculated by (Wünsch, 1998)

$$\Lambda(\mathbf{r}_0, t) = \frac{1}{2}\left(\sqrt{\frac{1}{3}\mathrm{tr}\mathbf{C}} + \sqrt{\frac{1}{3}\mathrm{tr}\mathbf{C}^{-1}}\right). \qquad (5)$$

This quantity can be interpretated as a scalar size of the quality of mixing. Fluid elements, which are neighbouring at the beginning of the mixing process, disperse and the elements are deformed strongly if Λ increases with time. So the mixing process is successfull if many fluid elements inside the apparatus reach large values of Λ in a short time. The deformation quantity depends on the position of the starting point $\mathbf{r}_0$ and on the time. Distribution functions of Λ for sufficient start positions show the effects of mixing at fixed times and allows the comparision with other mixing apparatus (e.g. static mixer).

4. Results

Numerical calculations were made for various values of the *De*-number (which discribes the length of the memory of the fluid) where the *Re*-number was fixed. In that case the laminar flows in the vessel depends on the material properties of the fluid. For newtonian fluids the inertia dominates and the secondary flow in the meridian plane moves in two vortices from the disk to the boundary of the vessel (figure 1). With increasing *De*-number the intensity of the vortices becomes smaller and finally the direction of the vortices changes (figure 2). This behavior are not observed for shear-thinning fluids without elasticity. The distribution functions of the mean deformation Λ are similar for the various *De*-number, because the primary velocity components in circumferential direction are comparable in all cases and the circulating velocity of the vortices are much lower than the circumferential velocity component.

5. Conclusion

The quality of the laminar mixing in a vessel depends only a few on the material properties of the fluid. The geometrical parameters (inserts or impellers) are decisive on the mixing process. The use of unbaffed vessels for efficient mixing of Newtonian or non-Newtonian fluids is not recommendable.

6. References

Broszeit, J (1997) J. Non-Newtonian Fluid Mech. 70: 35–58

Rubart L, Böhme G (1991) Theoret. Comput. Fluid Dynamics 3: 95–115

Wünsch, O (1998) Techn. Mechanik, 18: 1–10

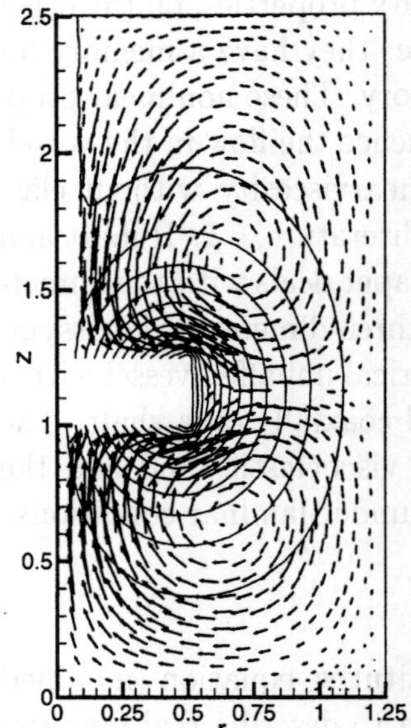

Figure 1: Flow of a Newtonian fluid in a vessel ($De = 0$, $Re = 1.0$). Arrows: Velocity in the meridian plane. Lines: Isolines of the velocity in the circumferential direction.

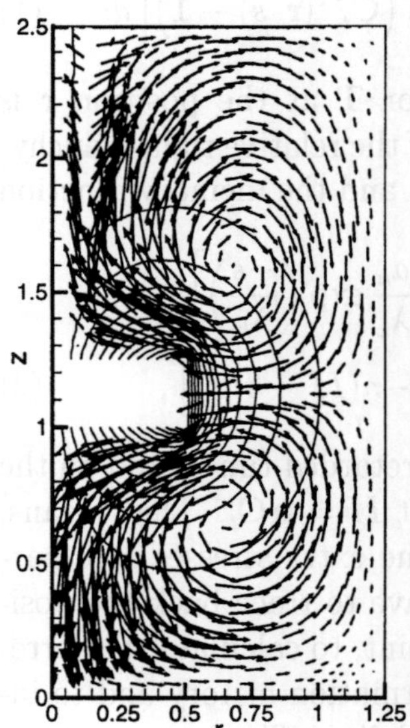

Figure 2: Flow of a viscoelastic fluid in a vessel ($De = 0.25$, $Re = 1.0$). Arrows: Velocity in the meridian plane. Lines: Isolines of the velocity in the circumferential direction.

STOKES MOTIONS OF ASYMMETRICAL FIBROUS PARTICLES IN A STUART VORTEX FLOWFIELD

Youn J. Kim
School of Mechanical Engineering, Sungkyunkwan University, 300 Chunchun-dong, Jangan-gu, Suwon 440-746, KOREA

INTRODUCTION

The motion of non-spherical particles is complex phenomena which is highly influenced by their general lack of spherical symmetry. Analytic formulas are known for some of the classical shapes, such as prolate and oblate spheroids, for which eigenfunction expansion solutions of the Stokes equations are possible.[1] However, the asymmetrical fibrous particles considered in this paper have no symmetries, and all elements of their resistance coefficients are, in general, nonzero. While there are no analytical formulas for them, they can nevertheless be approximated by slender body theory. An especially convenient form of this approximation has been presented by Johnson.[2] Once the resistance tensors have been determined, it is a straight forward matter to integrate the six-degree of freedom equations for the position and orientation of a given particle as a function of time.[3]

THEORY

The linearity of the governing Stokes equations and boundary conditions allows complex flows to be solved by adding the effects of the several terms which describe the flow in which the particle immersed. In order to solve the motion of a particle in a periodic vortical flow field, we may use the following Johnson's integral equation[2] involving a distribution of stokeslets, or point force singularities, along the particle centerline interest (see Fig. 1):

$$V_v(s) = \alpha_v(s)L_v + \int_{S_A}^{2l-S_A} K_v(r_0;\alpha)ds\,,$$

$$K_v = \frac{\alpha_v(s')}{r_0} + \frac{[\alpha(s')^2 r_0]r_{0v}}{r_0^3} - \frac{D_v\alpha_v(s)}{|s-s'|}$$

$, D_n/D_b/D_s = 1/1/2$

In this study the flow field is considered as steady, incompressible, and two-dimensional, and may be specified by a following stream function[4]:

$$\Psi = \frac{\lambda U_\infty}{2\pi}\ln\left[\cosh\left(\frac{2\pi s_2}{\lambda}\right) - k\cos\left(\frac{2\pi s_1}{\lambda}\right)\right]$$

where U_∞ denotes the free stream velocity, λ the characteristics length of each cell. Figure 2 shows the streamlines of two-dimensional Stuart vortex flow field for k=0.25. From the discrete integration of the singularity distribution, we have obtained a set of linearly independent equations relating stokeslet α and velocity for asymmetric fibrous particles, and solved by using the LU-decomposition method.

RESULTS AND DISCUSSION

The equations for the orientations of a fibrous particle are obtained by introducing the Euler angles φ, θ and δ (in this study we call them azimuthal, polar, and roll angles, respectively), and by the two sets of coordinates, one of which is particle-fixed, and the other space-fixed. We have solved the ordinary differential equations using the fourth-order Runge-Kutta scheme. The histories of Euler angles and a particle's center of mass are shown for the particle having the helicoidal geometry $L/D=1, n=2, d/D=0.01$ with various buoyancy parameters. From these figures the terminal orientations can be predicted for given initial orientation. This phenomenon may be explained in full by introducing a phase plane analysis.

The particles settle along the vertical cell boundaries, experiencing successively upflow and downflow regions, and eventually form a closed loop except for the larger **Y**-value cases. The upflow significantly reduces the settling velocity. We may interpret the results for large **Y**-value by considering the motion to consist principally of sedimentation, perturbed by a small shear. As the relative importance of gravity increases, a point is reached where the particle rotation due to sedimentation becomes comparable to that induced by the convective cell flow. At this point, it is possible for a particle to settle without rotation.

CONCLUSIONS

The motions of asymmetrical fibrous particles with the various buoyancy parameters are studied in the presence of the combined effects of a convection cell flow and gravitational sedimentation. More specifically, using Johnson's slender body theory, we were able to find the force and torque required to move a particle in any direction. These results were then expressed in the context of the hydrodynamic resistance coefficients, such as, second- and third-rank tensors, so that all motions can be described at once for a given initial orientation and location.

REFERENCES

1. Chaffey, C. E., Takano, M., and Mason, S. C., *Canadian J. Phys.*, **43**, pp. 1269:1287, 1965.
2. Johnson, R. E, *J. Fluid Mech.*, **99**, pp. 411:431, 1980.
3. Brenner, H., *Chemical Eng. Sci.*, **19**, pp.631:651, 1964.
4. Mallier, R. and Maxey, M., *Physics Fluids*, **A3** **(6)**, pp. 1481:1494, 1991.

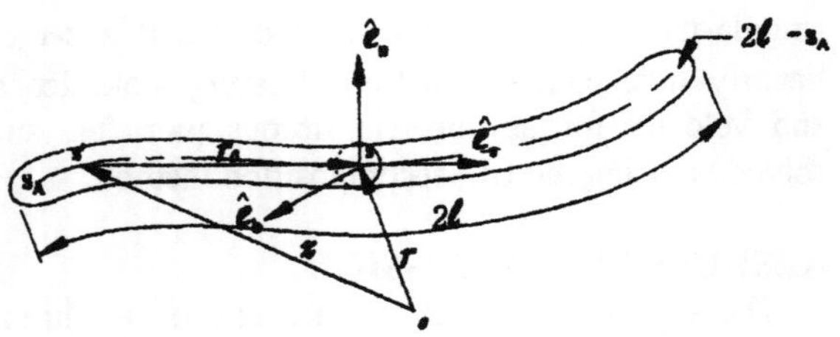

Fig. 1 Configuration of an asymmetrical fibrous particle

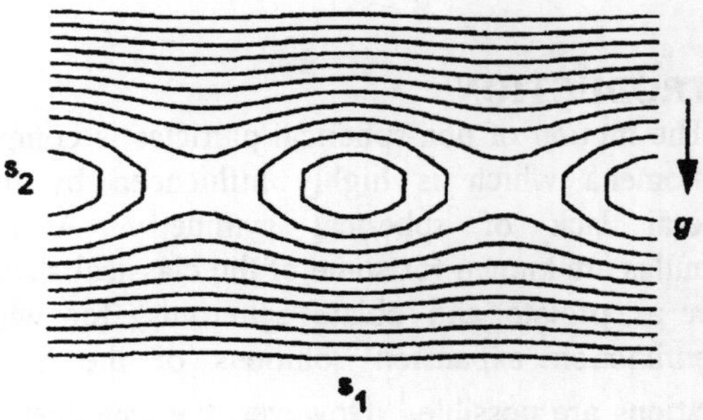

Fig. 2 Stuart vortex flow field for k=0.25

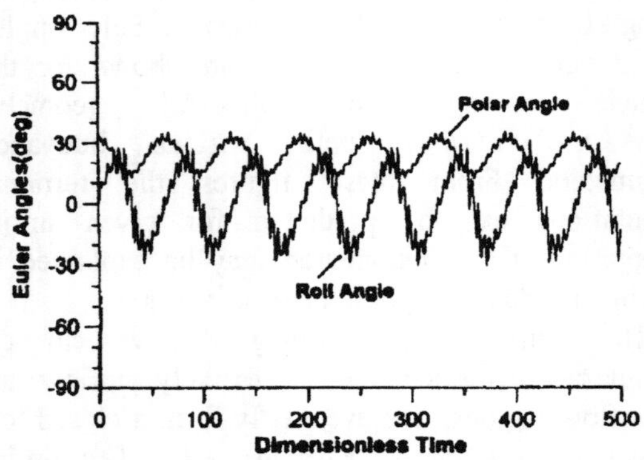

Fig. 3 Euler angles histories for $L/D=1, n=2, d/D=0.01$ **Y**=1 with $\varphi_i/\theta_i/\delta_i$ = 30°/30°/0°

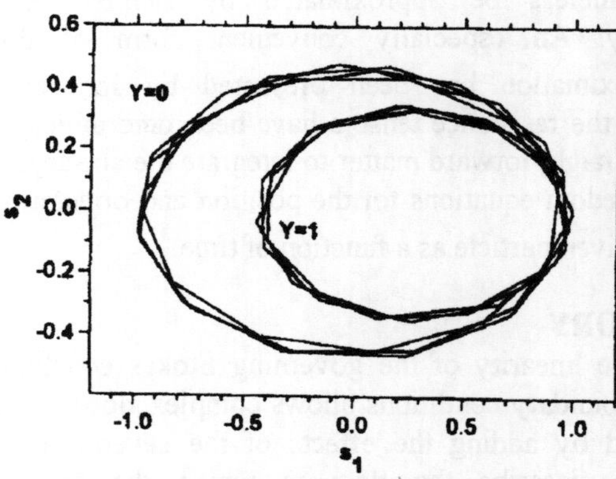

Fig. 4 Trajectories of particles for $L/D=1, n=2, d/D=0.01$ with $\varphi_i/\theta_i/\delta_i$ = 30°/30°/0° and $s_1/s_2/s_3$ = 1/0/1 in the s_1s_2 - plane

CHARACTERISATION OF COMPLEX DOUGHS USING A NOVEL EXTRUSION RHEOMETER

R.A. BAILEY[1], C. BRANCH[2]

[1]University of Newcastle, Newcastle upon Tyne. NE1 7RU. England
[2]AspenTech UK Ltd.,

1. ABSTRACT

The extrusion of doughs and pastes are common operations in the food processing industries. The complex nature and limited understanding of the physical properties of a material and of the processes involved can greatly affect product quality. The rheology of the materials involved is generally the key process variable. If the rheology is accurately characterised it can be used to improve control and increase product output. This paper presents the results of a continuing investigation into the measurement of dough and paste rheology for use as a process control variable and as a measure of product quality. The study uses a novel and versatile parallel plate extrusion rheometer. Dough is extruded in a radial direction between two circular parallel plates, the resulting pressure drop and volumetric flow rate are measured online. From these measurements the dough rheology can be determined by comparison to the integrated equations of motion for a variety of model rheologies. Two different doughs are extruded over a range of shear rates. The shear rate can be altered by varying the flow rate or the gap between the parallel plates. Both doughs used are commercial doughs with ingredients including fats, oils, meat fibres and soya grits. The doughs are shown to be highly non-Newtonian, with shear thinning and viscoelastic behaviour.

2. INTRODUCTION

The initial investigation into this multi-component dough was in two areas. First a Power Law model was produced to provide a simplified and convenient model for use by the dough manufacturer. Second was to investigate if a correlation between dough moisture and it's rheology was close enough to be used as a process control variable. The results of the investigation into these areas are covered in Bailey and Branch (1998). This paper includes an improved model and deals with the further investigation of the complex physical properties of the dough.

2.1 Improved model

The Power Law model constants needed modifying to account for the effect of temperature on the dough's rheology. This change in model requirements was in response to a modification of the production facility.

The dough's rheology can be described under a wide range of conditions by this temperature factor and by considering other models including the full solution to the equations of motion.

2.2 Viscoelasticity

The flow of dough between the parallel plates is assumed to be equal to the volume displaced from the reservoir barrel by the piston. The assumption that the dough is not compressed is valid for this Power Law model. However the dough's recovery after compression can be seen in Figure 2. These recovery curves were collected by monitoring the pressure drop after the piston had been stopped and illustrate a degree of compression does occur.

3. EXPERIMENTAL

3.1 Equipment

The means for producing the extrusion pressure in the rheometer during constant temperature work was from a Testometric 500D Universal Rest rig. For the temperature related and relaxation curve work different equipment was used to provide this pressure. A 2000rpm 0.5Hp motor from Normand Electric Co. Ltd. was linked to a Duff Norton miniature Jactuator - model ME 2625-150 via a 25:1 David Brown Radicon gear box. The motor was controlled by a Zenith E-Trol SCR controller - model ER-50-Z. The temperature control of the dough was achieved by pre-heating or pre-cooling the sample in temperature controlled units. The rheometer's temperature was controlled by the circulation of water from a water bath through copper pipes coiled round the rheometer.

3.2 Operation

The four temperatures, 9°C, 12°C, 18°C, 30°C, chosen so the range investigated was the same as the range experienced in the dough processing. Repeat readings were taken at three flow rates at each temperature.

Eight runs were performed for the recovery curve at one flow rate and at constant temperature, 25°C. Two readings were taken per second for 70 seconds.

4. RESULTS AND DISCUSION

The results have been scaled at the request of the dough manufacturer.

4.1 Temperature effects

Over the small temperature range used in this investigation the difference in the dough's behaviour could be observed. The intercepts and gradients were calculated from a log-log plot of flow rate against pressure drop. Figure 1 illustrates how a linear relationship between the Power Law constants, n and K, and temperature can be assumed under these conditions.

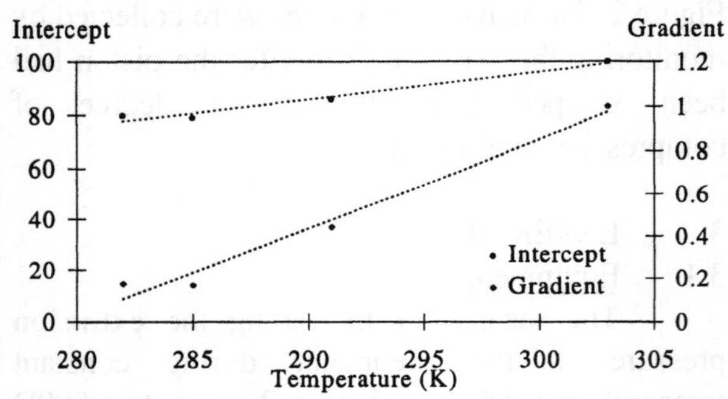

Figure 1: The variation of n and K, as functions of scaled gradient and intercept, with temperature.

4.2 Viscoelasticity

The reduction in the pressure drop across the parallel plates after the piston had been stopped was used to quantify the elastic component of the dough's viscoelastic behaviour.

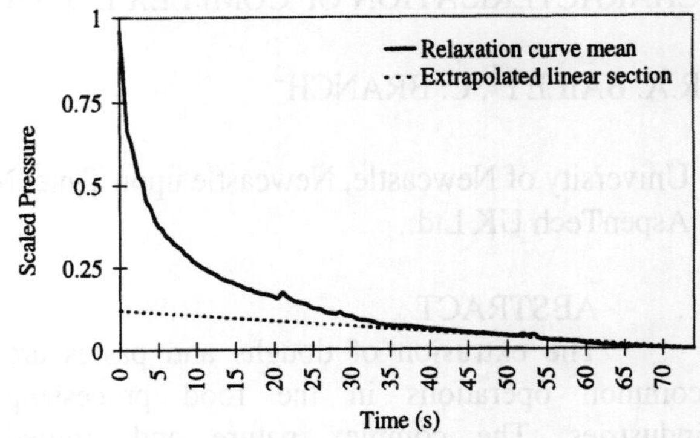

Figure 2: Reduction in pressure drop with time during dough recovery.

5. CONCLUSION

The improvement on the previous Power Law model of the dough's behaviour for use by the manufacturer has provided better information for use in design and operation of the process. The characterisation of the dough's viscoelasticity while illustrating the versatility of the rheometer must be further verified against the extrusion of materials of known rheology.

6. REFERENCES

Bailey, R.A., Branch, C.A. (1998) Proc. Of the IChemE Research Event, Newcastle p 8

RHEOLOGICAL BEHAVIOUR OF FOOD INDUSTRIAL GUMMS

M. BARRACÓ- SERRA, M. A. ADRIÁ- CASAS, R. TORRES- CÁMARA, P. SURIÀ- LLADÓ

Departament de Mecànica de Fluids, Escola Tècnica Superior d'Enginyers Industrials de Barcelona, Universitat Politècnica de Catalunya. Avinguda Diagonal 647, 08028 Barcelona

1. ABSTRACT

The object of this work was to study rheological (rheogrammical and thixogrammical) properties of hydrocolloides solutions and theirs correlations with time and environment conditions.

We have refered our study to the guar gumm and Astragalus gummifer gumm.

2. INTRODUCTION

The gels of hydrocolloids, known as gumms, are frequently use in food industries and pharmaceutical applications.

The main application of gumms is as a viscosizing and gelifing agent. One of the consequences of this effect is its role of emulsing helper. Its reactivity power with proteic macromolecules produce thermodynamic stable complex products that are used as protecting layers.

3. METHODOLOGY

The followed methodology was:

- Water solutions were prepared with 1% to 5% weight/volume concentration.
- Rheological graphs were obtained in order to determinate the rheo- destruction point.

All rheogrammical behaviour results were expressed before getting this break point.

- Thixogrammical characteristics were also obtained by shear stress variation versus time.

However we needed, for this study, the knowledgement of initial viscosity (viscosity that corresponds to shear stress zero). This determination was out of range for our instrumentation.

We had stablished, for this, a different method (following the Green and Weltmann criteria) consisting on analysing the adimensionaly viscosity variation measured on successives rheogramms without stopping phases. Surfaces under rheogramms represent a thixogrammical parameter.

Experimental results were carried out with a cone- plate viscometrical system, type Haake VT 550 (with pk 2.5 and 1° cone).

Temperatures were tested from 20°C to 50°C.

General conclusions were:

- Samples rheological behaviour were plastic or pseudoplastic. Constitutional equation of gumm samples were:

$$\mu_a = K.\ D^{n-1} \qquad (1)$$

Relationships between K (consistency coeficient) and n (non linearity coeficient), which define the characteristical time, were described as follows:

K decreased exponentialy versus increasing temperatures and increased with high concentrations.

n reached the value 1 for high temperatures and low concentrations (samples became newtonians).

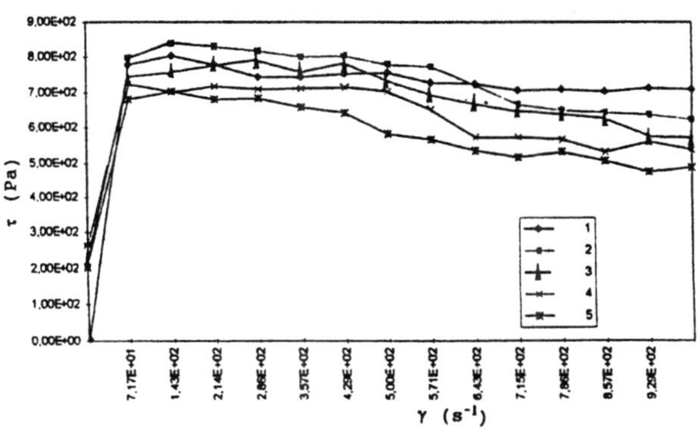

<u>Figure 1</u>

Successives rheogramms for guar gumm (temperature 40°C and concentration 4%)

- Generally, the aparent viscosity (or plastical viscosity) was increasing as concentration was also increasing.

On the other side the aparent viscosity was increasing as temperature was decreasing.

- Each successive rheogrammical cycle showed smaller surfaces. This effect was stronger with higher temperatures.

- Thixogrammical behaviour was detected and the new methodology proposed (surfaces under rheogramms) gave us according values with other refered methods.

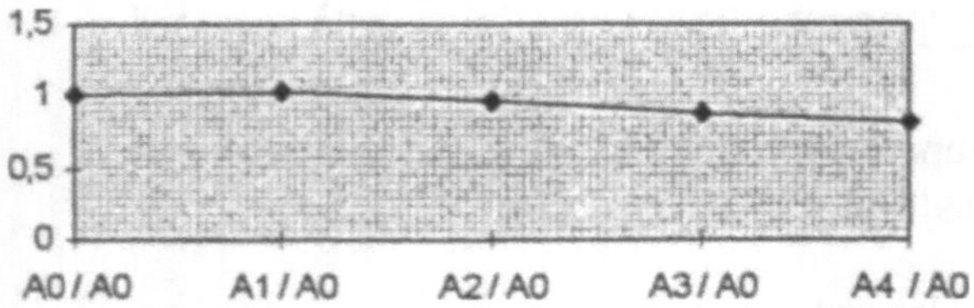

Figure 2

Relationship of thixotropic areas for guar gumm (temperature 40°C and concentration 4%)

The measurement of thixotropic behaviour was made by the "thixotropic area variation speed". This parameter was defined as:

$$A_{sa} = dA_{sa} / dt \quad (2)$$

A_{sa} is decreasing with temperature and is decreasing strongly for high concentrations.
Some structural changes (for any temperature and any concentration) were the exception of the general behaviour.

4. BIBLIOGRAPHY

Bavarian C, Quemada D, Parker A (1996) Proc. XIIth. International Congress on Rheology

Dolz M, Hernández M.J, Pellicer J, (1995) Journal of Pharmaceutical Science 84/6

MODELLING THE SHEAR BEHAVIOUR OF SOFT (CREAM) CHEESE

D. M. BINDING and R. EDWARDS
Dept. Mathematics, University of Wales Aberystwyth, Ceredigion, SY23 3BZ, UK

ABSTRACT

In this paper we briefly report on attempts to rheologically characterise soft cheese. The effect of smooth and roughened plates on both torsional flow and small amplitude oscillatory flow is studied. It will be shown that the flow of soft cheese is, in these situations, dominated by slip at the solid boundaries or by fracture within the bulk of the material. It is suggested that homogeneous shear flow is not possible for such systems.

INTRODUCTION

The most obvious manifestation of slip is that, for a given flow regime, one obtains different values for the viscosity for different sized geometries. For this reason we have employed parallel plate rheometry, which readily lends itself to a study of geometry effects since, in principle, the same kinematics can be obtained using different plate separations. Torsional flow experiments have shown the flow of soft cheese to be dominated by slip effects [Binding and Edwards (1996)]. Typical stress responses for a given nominal shear rate are shown in Fig. 1 where the effect of plate separation is clear and dramatic.

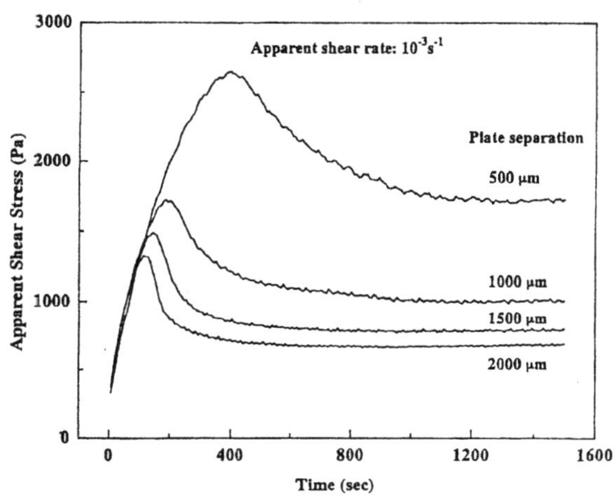

Fig. 1: Apparent shear stress vs time for the torsional flow of soft cheese at a fixed nominal shear rate

Initially the stress is higher for the larger plate separations, which is consistent with the thin depleted layer theory, often applied to explain such slip effects [Barnes (1995)]. However, at longer times the trends are reversed; the stresses peak earlier and fall more rapidly for the larger gaps resulting in lower steady state stresses for the larger plate separations.

EXPERIMENTAL WORK

Clearly, to characterise the bulk rheological properties of complex materials, slip must be eliminated. This has often been achieved by roughening geometry walls [Buscall et al (1993), Magnin and Piau (1987, 1990)]. Slip has been overcome in measurements on ink pastes by roughening the geometry plates with sandpaper [Gregory and Mayers (1993)]. This simple approach has been adopted in this study.

Some researchers have employed small strain oscillatory measurements to examine the viscoelastic properties of soft cheese [Solorza and Bell (1995), Sanchez et al (1996), Sanchez and Hardy (1997)]. In each of these, the authors have failed to examine the possibility of slip effecting their results.

The soft cheese and sandpaper used in this study were locally purchased commercial products. The sandpaper used varied from a very fine paper rated at grit size 800, to a very coarse paper rated at grit size 80. To affix the sandpaper to the plates of a Weissenberg Rheogoniometer a strong double-sided carpet tape was used.

One of the difficulties with materials that exhibit slip is that the experimental procedure employed can influence results. The procedure used for obtaining the data presented here involved using a sharp knife to cut a slice of cheese to a desired width. The sample was then carefully placed onto the lower rheogoniometer plate permitting the upper plate to be lowered into position with a minimum of squeezing. Clearly, however, some squeezing flow was inevitable to ensure proper contact between the sample and the plates.

All experiments were repeated numerous times using fresh samples and sandpaper. Good reproducibility of the data was obtained.

Oscillatory shear flow

For the dynamic data presented here amplitude sweeps were conducted between strains of 0.003 and 0.04 at various frequencies to discover the strain and frequency range over which the soft cheese exhibits a linear viscoelastic response. Subsequently, a strain of 0.008 was employed over a frequency range of 0.5-20 rad/sec to ensure linear viscoelastic behaviour. At higher frequencies no acceptable linear range could be attained with the instrument.

Fig. 2 shows the effect of plate separation on dynamic rigidity and dynamic viscosity measurements using both smooth and roughened plates.

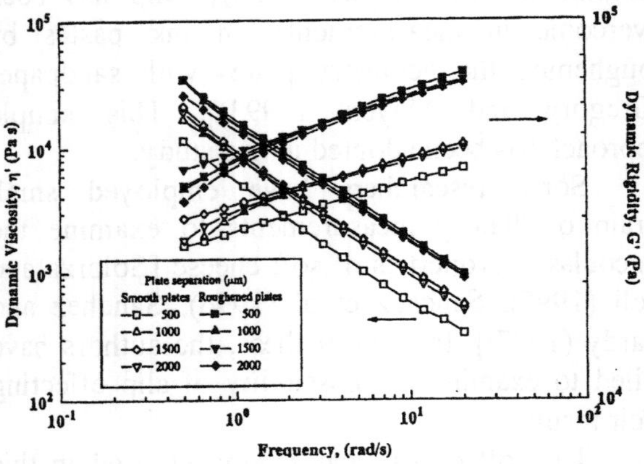

Fig. 2 Dynamic viscosity and rigidity data for soft cheese.

Using smooth plates, gaps between 1000 and 2000 microns exhibit no clear evidence of apparent slip. However, for a separation of 500 microns, a significant decrease in both dynamic rigidity and viscosity is observed. By roughening the geometry plates the measured values of dynamic rigidity and dynamic viscosity not only increase but also the gap effect observed in the smooth plate tests vanishes. Varying the coarseness of the sandpaper had no effect on the outcome of the roughened experiments. It is surprising that the data for smooth plates do not approach the roughened plate data as the plate separation increases.

Torsional flow

In torsional flow, tests have indicated that slip occurs on one plate only and that there is a preference for the lower or upper plate, depending on the nominal shear rate applied. In *all* instances, roughening the plate displaying no apparent slip has only a very small effect on the measured stress response. However, roughening the plate exhibiting slip *always* results in fracture even though one of the plates is smooth. When both plates are roughened slip is *always* eliminated and *always* leads to fracture.

CONCLUSIONS

Slip dominates the flow of soft cheese. The exact nature of the slip processes in torsional flow and oscillatory shear flow are still not fully understood but the effects on rheometric experiments can not be ignored. Even when data appears to be independent of geometry, a significant slip effect may still be present.

When slip is inhibited, the material eventually fractures, so that homogeneous steady shear flow is not possible for these substances. The stress response is therefore determined essentially by the rigidity of the structure and the behaviour of the slip layer.

REFERENCES

Barnes HA, (1995) J. non-Newtonian Fluid Mech., 56: 221-251.

Binding DM and Edwards R (1996), Proc. XIIth Int. Congr. On Rheology, 793-794.

Buscall R, McGowan JI and Morton-Jones AJ (1993), J. Rheology, 37: 621-641.

Gregory T and Mayers S (1993), Surface Coatings Int. (JOCCA), 76: 82-86.

Magnin A and Piau JM (1987), J. non-Newtonian Fluid Mech., 23: 91-106.

Magnin A and Piau JM (1990), J. non-Newtonian Fluid Mech., 36: 85-108.

Sanchez C and Hardy J (1997), Proc. Conf. Food Rheology and Structure, Zurich, 301-308.

Sanchez C, Maurer K and Hardy J (1996), Food Macromolecules and Colloids, Publ. Royal Soc. Chem. 512-515.

Solorza FJ and Bell AE (1995), J. Soc. Dairy Tech., 48: 133-138.

ACKNOWLEDGEMENTS

Support from the Biotechnology and Biological Sciences Research Council (UK) is gratefully acknowledged.

STUDY OF (SSD) BISCUIT BAKING BY DYNAMIC THERMAL ANALYSES (DMTA, DSC, TGA).

S. CHEVALLIER*, G. DELLA VALLE, D. LOURDIN , P. COLONNA

* DANONE, Centre de Recherches Jean Thèves , 91207 ATHIS-MONS Cédex, FRANCE.

INRA, Centre de Recherches Agro-alimentaires , 44316 NANTES Cedex 3, FRANCE

1. INTRODUCTION

The baking process of sheeted short dough (SSD) biscuits involves the transformation from a viscoelastic dough into a cellular solid, with a final texture. This process is induced by the expansion of the material submitted to the vaporization of water and gases from leavening powders. Meanwhile, rheological properties of the matrix, which are modified by water losses, thermal denaturation and melting of dough components, reduce the impact of the expansion phenomenon. The objective of this work is to relate the changes of rheological properties of biscuit dough during thermal treatment to structural modifications of the product from molecular level to macroscopic scale.

2. MATERIAL AND METHODS

2.1 Sample preparation

Ingredients provided by LU (La Haye-Fouassière 44-Nantes) - wheat flour, fats, sugars and leaveners - are mixed at desired levels and sheeted in order to manufacture 19 model doughs. Their respective composition is indicated in the following Table in % of hydrated wheat flour.

dough#	water	sugar	fat	leavener
1	26	-	-	-
2-8	16-26	7-45	-	-
9-11	26	-	6-16	-
12-14	26	-		1
15-18	16-26	45	16	1
19	-	45	16	1

Sample #1 represents the basic dough, sample # 19 the actual composition of the biscuit whereas dough #18 has the same composition as #19 with 16% of added water. Other samples are tested to provide information on the influence of each component (#2-14) and their interaction between one another (#15-17). Water content has sometimes to be adjusted to ensure dough sheetability.

2.2 Rheological and thermal methods

A DMTA Mk III (Rheometric Scientific, Loughborough, UK.) was used in the compression mode at a frequency of 5 Hz with a deformation amplitude of 10 μm. Assuming that the static force, applied (0.045N) to the sample in order to keep contact with the plates, can be neglected, the variation of the sample height was also recorded as a function of temperature. The rise is defined as the ratio of the height variation to the initial height ($\Delta h/h_0$) with h_0 = 2mm.

The weight loss of samples (30 mg) was measured as a function of temperature on a thermogravimetric analyser TGA7 (Perkin Elmer, Norwalk, CT, USA). From these results, a weight loss rate was calculated.

DSC experiments on doughs were performed on a DSC 121 (SETARAM, Caluire, France), using 170 mg samples sealed in stainless steel pans. The reference cell contained 105 mg of alumine. Temperature and enthalpy calibrations were carried out with indium.

Runs were carried out in duplicate from 20 to 160°C at the same heating rate of 3°C/min for all 3 types of experiments and average values are reported.

3. RESULTS AND DISCUSSION

3.1 Basic dough

Figure 1 presents a typical sketch of experimental curves gained with the 3 methods, in the case of dough #1. Events observed on DSC thermogram refer to starch melting in low hydrated medium. Their characteric temperature [70, 101, 122°C] are just above those of wheat starch (Burt and Russel 1983): that may be attributed to the presence of other hydrophilic components, like gluten for instance.

Temperature range corresponding to the rise observed with DMTA ($\Delta h/h_0$~150% on the interval [132°C, 144°C]) and the one corresponding to weight loss obtained from TGA, [135°C, 151°C] overlap. Storage modulus E' is always higher than dissipative modulus and its value at rise onset is 40kPa. For expansion observed in the DMTA oven, air incorporated into the dough during mixing and steam formed by evaporation of dough water are the driving forces of the rise of the basic dough sample.

3.2 Influence and interactions of components

Characteristic values of thermal events noted in Fig.1 are systematically quoted for every dough tested. Detailed results are reported elsewhere (Chevallier et al. 1998).

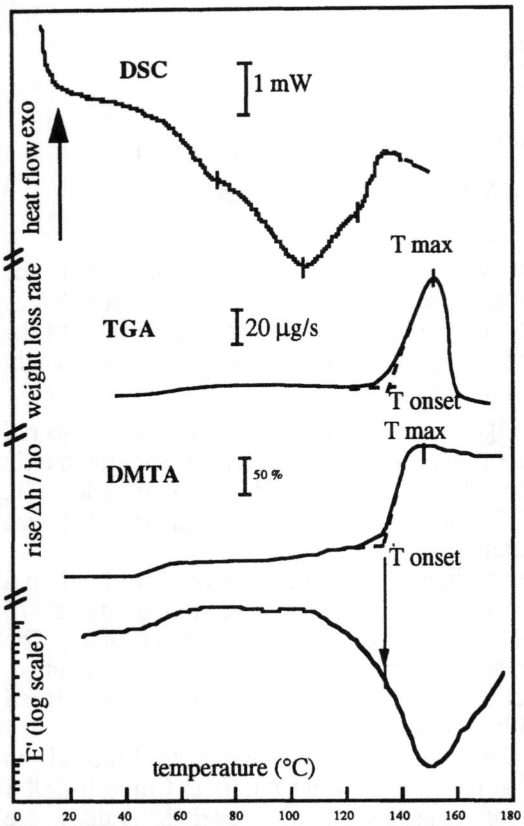

Fig.1: Typical thermogram records for basic dough (#1).

To summarize, sugars increase storage modulus, up to 200 kPa, starch melting temperature, up to 113°C. This last effect, proportional to sugar concentration, may be attributed either to water activity decrease or starch -sucrose interactions (Lelièvre 1976). Fats, by their melting phase transitions, may play a role in the stabilisation of gas cells in doughs (Brooker 1996) and reduce the rise during baking for largest concentration . Phase transitions of starch, even though they have no straightforward influence on the rise, may influence the biscuit texture through the state of cell walls. The only interaction identified with regard to phase transitions is the influence of sugars on thermal transitions of starch. As far as the other ingredients are concerned (fats and chemical leaveners), the system can be considered as additive.

3.3 Biscuit texture development

Rise of all doughs can be represented as function of the elastic modulus E' by DMTA. Indeed, variations of E' over the complete temperature interval of study (20 to 160°C) are very difficult to interpretate mainly because of bubbles expansion. Therefore, values reported for E' correspond to the rise temperature of samples (Fig.2). For doughs with no chemical leaveners (#1 to #10), the lower the rise, the higher the elastic modulus. This variation illustrates that thermomechanical properties of doughs, expressed here by their rigidity increase, preclude expansion of dough samples. Moreover, for a given value of the elastic modulus, rise is more important for doughs with chemical leaveners than doughs with sugars or fats. For the same rigidity, gas produced by chemical leaveners will therefore enhance expansion of dough samples.

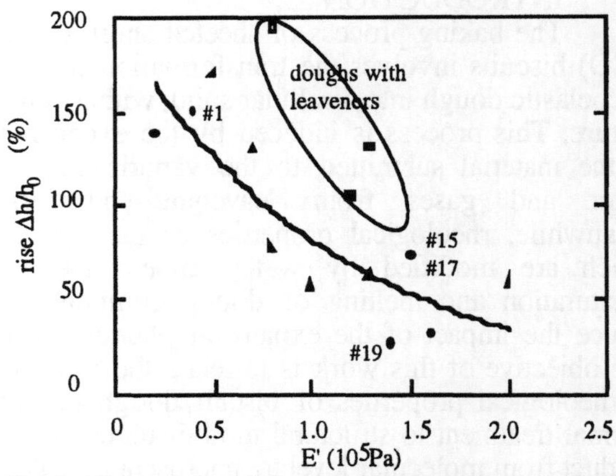

Fig.2: Variations of rise with storage modulus E'

As for extruded foams (Della Valle et al. 1997), final texture of the biscuit is thus governed by two antagonistic phenomena: gas production, function of temperature and water content and matrix viscoelastic properties which are a function of biochemical modification of components.

4. CONCLUSION

The use of DSC, TGA and DMTA shows that thermal transitions, volume and mass changes can be interpreted altogether. This simulation helps the understanding of phenomena involved in SSD biscuit baking process.

However, experiments on such complex materials with reduced water content remain difficult to interprete completely. Roles of flour components, particularly starch and protein, need further studies such as pilot scale baking in order to better define hydrothermal path of the dough and compare the modifications observed to those identified in the present study at the lab scale.

5. REFERENCES

Brooker BE (1996) J. Cereal Sci. 24: 187-198

Burt DJ, and Russel PL (1983) Starch 35: 354-360

Chevallier S, Colonna P, Della Valle G, Lourdin D (1998) J. Cereal Sci. submitted

Della Valle G, Vergnes B, Colonna P, Patria A (1997) J. Food Eng. 31: 277-296

Lelièvre J (1976) Polymer 17: 854-860

INFLUENCE OF THE MIXING CONDITIONS ON THE RHEOLOGICAL BEHAVIOR OF A BISCUIT DOUGH

J.L. DOURNAUX[1], B. VERGNES[1], A.S. CONTAMINE[2]

[1] CEMEF, Ecole des Mines de Paris, UMR CNRS 7635, BP 207, 06904 Sophia-Antipolis Cedex, France
[2] DANONE, Centre de Recherches Jean Thèves, BP 16, 91207 Athis-Mons Cedex, France

1. INTRODUCTION

Biscuits are obtained after rolling, cutting and cooking of a dough preliminary prepared by kneading (Wade, 1995). It has been shown by Contamine et al. (1995) that the viscoelastic properties of the dough are partially responsible for the final quality of the biscuits. Thus, the purpose of the present work is to determine the influence of mixing conditions on the rheological properties of a biscuit dough.

2. MATERIALS AND METHODS

2.1 The dough

The dough formula we studied corresponds to the French "petit beurre" semi-sweet biscuit. On a flour weight basis, it contains 18 % fat, 35 % sugar, 16 % water and 2.4 % leavening agent.

2.2 The kneading process

All the doughs were made using an experimental horizontal type mixer (Loiselet, France), in which 6 kg of dough was prepared. The various ingredients of the dough formula are preliminary homogenized at low speed for two minutes, then mixed at high speed for a fixed time. Process parameters are the rotation speed (85 to 170 rpm), the mixing time (0 to 24 min) and the regulation temperature of the double jacket (4 to 36 °C). After kneading, dough is let at rest during 30 minutes before rheological measurements

2.3 The rheological measurements

They were carried out at 20 °C, using a parallel plate rheometer (Rheometrics RMS 800) in oscillatory mode. Following tests were made :

- frequency sweeps, in the linear domain, between 0.1 and 100 rad/sec,
- deformation loops, to characterise the dough structure. At a fixed frequency (usually 1 rad/sec), oscillatory measurements are made for increasing deformations, until a maximum value, then for decreasing deformations. The non-reversibility of the trajectories is characteristic of the structural changes of the dough.

Details concerning these measurements and the experimental difficulties were given by Charun et al. (1997).

3. RESULTS AND DISCUSSION

3.1 General behavior

Bread doughs have more simple formulations (no sugar, small amount of fat) and have been largely studied. On the contrary, biscuit doughs are more complex and few results are reported in the literature (Contamine et al., 1995; Charun et al., 1997; Baltsavias et al., 1997).

First of all, and as for bread doughs (Lindhal and Eliasson, 1992; Berland and Launay, 1995), the linear viscoelasticity domain is extremely reduced, below 0.02 % deformation. Above this value, G' and G" decrease very rapidly, indicating a high sensitivity to the deformation.

An example of frequency sweep test is presented in Fig. 1. On the whole range of frequencies, the storage modulus G' is higher than the loss modulus G". Moreover, at low frequencies, both moduli tends towards an equilibrium value, independent of frequency. This behaviour is characteristic of a weak physical gel (Ross-Murphy, 1995) and put in evidence the structured nature of the dough.

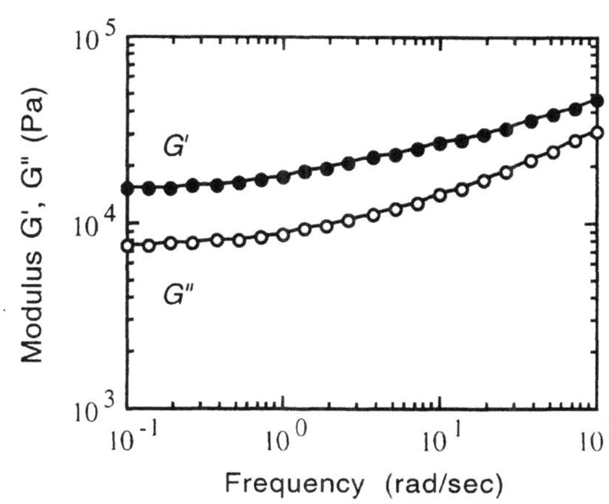

Fig. 1 Variation of storage G' and loss G" modulus with frequency

To characterize this structure, we can perform deformation loops (it means an increase in deformation, followed by a decrease and a return to initial conditions). We observe that there and back trajectories are different, and that the moduli have undergone a drop of approximately 60 % (for a maximum deformation of 80 %). Even for limited deformations, the drop is important (for example, 8 % on G' for 2.5 % deformation), which put in evidence the "weakness" of the structure. If the dough is let at rest a certain time after the return trajectory, we observe a partial recovery of the viscoelastic moduli, which proves that the structural modifications are not totally irreversible (Charun et al., 1997).

These rheological measurements permit one to characterise the viscoelastic behaviour of the dough and its structural properties. We will now apply these techniques for understanding the mixing process and the influence of the main parameters on the dough behaviour.

3.2 Influence of mixing time

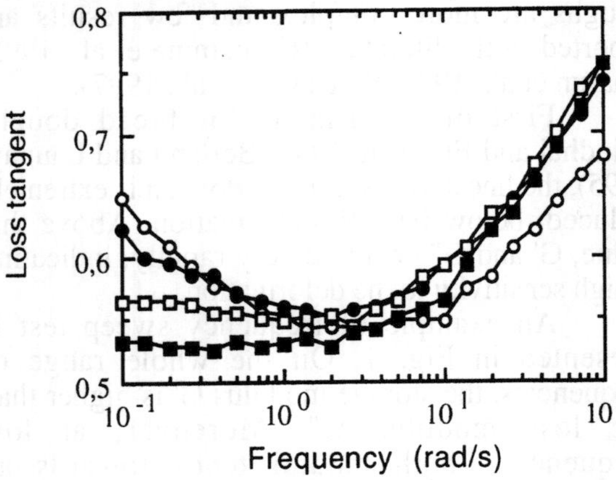

Fig. 2 Influence of mixing time on loss tangent
m : 0 min, l : 1.5 min, p : 6 min, n : 12 min

After a preliminary low speed mixing of the different ingredients during 2 minutes, the dough is mixed at 120 rpm, and samples collected at various time intervals. Fig. 2 shows the evolution of the loss tangent with the frequency. It may be observed that the principal modifications concern the behaviour at low frequency. Just after the pre-mixing (mixing time = 0), we note a strong increase of tg δ below 1 rad/sec. During the mixing, a plateau appears, which indicates the development of the dough structure (Ross-Murphy, 1995). Similar results were reported by Amemiya and Menjivar (1992) on a wheat flour/water dough.

3.3 Influence of regulation temperature

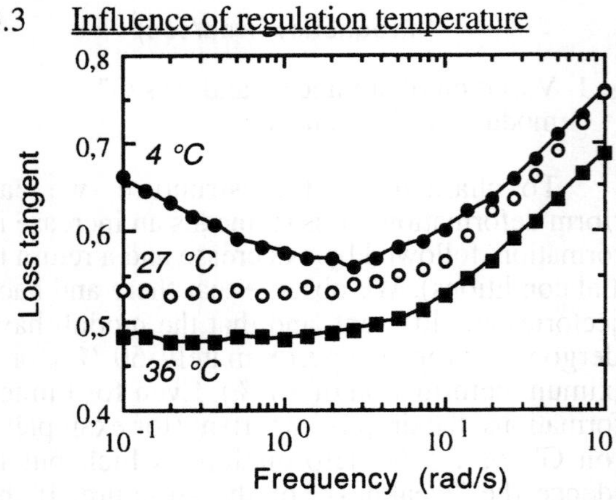

Fig. 3 Influence of regulation temperature

At constant rotation speed (120 rpm) and mixing time (12 min), we observe also a strong influence of the regulation temperature on the rheological properties of the dough. At higher mixing temperature, the dough appears as more consistent, with higher values of the complex modulus G*. If we consider the evolution of the loss tangent with the frequency (Fig 3), the effect of temperature is similar to this of mixing time. The increase of temperature leads to the apparition of a low frequency plateau, indicating a much more elastic behaviour of the dough.

3.4 Influence of rotation speed

At constant regulation temperature (36 °C) and mixing time (12 min), the influence of mixer rotation speed id presented in Fig. 4. As for the preceding parameters, an increase in screw speed leads to a more elastic dough, characterized by a low frequency plateau.

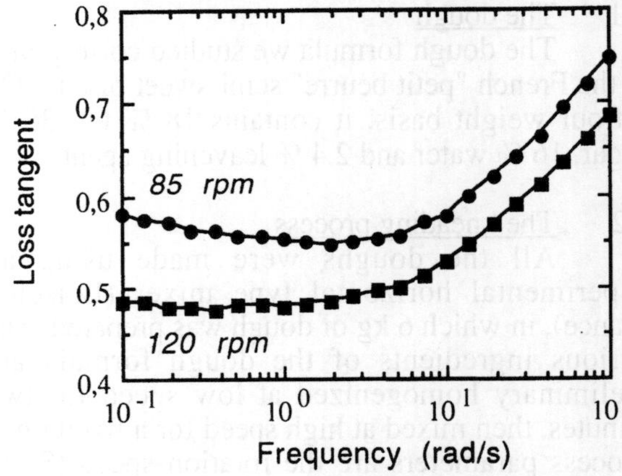

Fig. 4 Influence of rotation speed on loss tangent.

4. CONCLUSION

The biscuit dough behaves as a weak physical gel, with a reduced linear domain. The structure of the dough is very sensitive to the applied deformation. Mixing time, rotation speed and temperature influence the dough structure through the development of a more elastic behavior.

5. REFERENCES

Amemiya JI, Menjivar JA (1992) J Food Eng 16:91
Baltsavias A, Jurgens A, van Vliet T (1997) J Cereal Sci 26: 289-300
Berland S, Launay B (1995) Rheol Acta 34: 622
Charun E, Dournaux JL, Contamine AS, Vergnes B (1997) In: Windhab, Wolf B (eds) Food Rheology and Structure I. Vincentz Verlag pp 193-197
Contamine AS, Abecassis J, Morel MH, Vergnes B, Verel A (1995) Cereal Chem 72: 516
Lindhal L, Eliasson AC (1992) Cereal Chem 69 :30
Ross-Murphy SB (1995) J Rheol 39: 1451
Wade P (1995) Biscuits, Cookies and Crackers. Vol. 1 Chapman & Hall, London

RHEOLOGY OF SPRAY-DRIED EGG YOLK-XANTHAN GUM AQUEOUS DISPERSIONS

J. FERGUSON[1], J. MUÑOZ[2], N. E. HUDSON[1], M.C. ALFARO[2], E. MACDONALD[1] AND G. VELEZ[2]
[1]Department of Pure and Applied Chemistry. University of Strathclyde. 295 Cathedral Street. Glasgow G1 1XL (United Kingdom)
[2]Departamento de Ingeniería Química. Universidad de Sevilla. C/ P. García González, s/n. Edificio de Química. 41012 Sevilla (Spain)

1. INTRODUCTION

Protein-polysaccharide systems have not only scientific interest in the effort to understand the behaviour of complex multicomponents systems, but they also have many technological implications, above all in the development of new food products. They are particularly useful for the formulation of low-fat stuffs and for the enhancement of the funcionality of "natural" food ingredients. In fact, they are the most important structure forming macromolecular ingredients in foods.

The main objective of this work was to explore the protein-polysaccharide interactions in aqueous dispersions of spray-dried egg yolk (SDEY) and xanthan gum (XG) using several rheological techniques.

1. METHODS

2.1 Materials

SDEY-XG aqueous dispersions were prepared in a composition range typical of these biopolymers in formulations of salad cream or reduced calorie (low oil percentage) mayonnaise (Pascual et al.,1998).Commercial spray-dried egg yolk was supplied by Ovosec (Spain), containing 31%wt proteins. Its concentration ranged from 0 to about 3%wt. Xanthan gum (SIGMA) concentration was varied between 0.2 to 1.6%wt. Sodium azide was used as preservative.

2.2 Rheological Techniques

All the samples were characterized at 25°C after allowing an equilibration time of 20 minutes.

A Haake CS Rheostress RS-100 rheometer was used with a cone and plate sensor system (60mm diameter, 4° angle) to carry out multistep steady-state shear flow curve experiments and small amplitude oscillatory shear (SAOS) tests.

Start-up at the inception of shear and flow interrupted experiments were conducted on a Weissenberg rheogionometer using cone and plate geometry (40mm diameter, 4° phase angle).

3. RESULTS AND DISCUSSION

3.1 Optical microscopy of SDEY-XG dispersions

Optical microscopy revealed that the structure of these dispersions is characterized by a wide range of egg yolk granule particle sizes. These particles exhibited different and frequently irregular shapes, probably as a consequence of the atomization process. A marked size reduction of some particles was observed when a rotor-stator homogenizer instead of a magnetic stirrer was used to prepare these dispersions. However, a significant fraction of irregular big particles did still remain unaltered. This higher energy input just provoked a slight decrease in G' and G", without altering the shape of the mechanical spectrum, demonstrating that just a small structure breakdown occurred. In any case, the effect of the higher energy input applied was not more significant than the influence of the batch.

3.2 Multistep steady-state flow curves

Figure 1 shows flow multistep steady-state shear flow curves fitted to the Carreau-Yasuda model. The fitting quality improved as XG concentration decreased. An increase in XG concentration on dispersions containing 2.3% SDEY gave rise to a) a power law increase in the limiting newtonian viscosity at low shear,η_0. In fact, η_0 turned out to be proportional to $XG^{3.6}$, b) the onset of shear thinning behaviour at a steadily lower shear rate and c) a steeper fall in viscosity with shear rate.

3.3 Mechanical spectra

Viscoelastic spectra derived from SAOS experiments gradually changed from those typical of macromolecular solutions to those shown by weak gels as XG concentration increased (Figure 2). Furthermore, G' increased more rapidly than G" with XG concentration and a steadily slower relaxation mechanism occurred as demonstrated by the fact that the onset of the terminal relaxation zone took place at lower and lower frequencies.

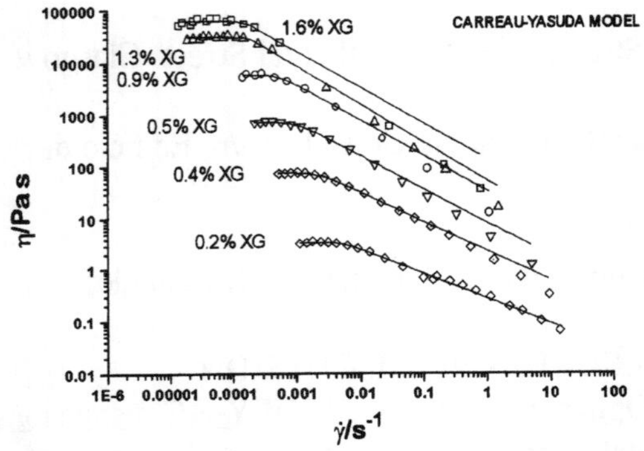

Fig.1. Effect of XG concentration on the flow curves of dispersions containing 2.32%wt SDEY.

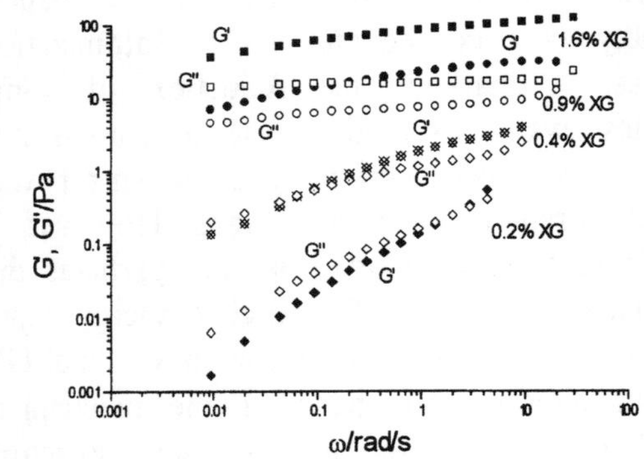

Fig.2 Effect of XG concentration on the mechanical spectra of dispersions containing 2.32%wt SDEY

3.4 <u>Start-up at the inception of shear and flow interrupted experiments</u>

These tests were carried out at fixed shear rates (minimum shear rate reached: $0.2s^{-1}$). At a given shear rate, the shear stress reached a peak value before achieving an equilibrium value as time of shear increased. The strain at which the peak appeared remained essentially constant regardless of rest time. The amount of overshoot (Λ) (Ganani and Powell,1985) was used as a measure of the thixotropic structural breakdown.

$$\Lambda = (\sigma_{peak} - \sigma_{equilibrium}) / \sigma_{equilibrium}$$

The kinetics of structure recovery was followed by plotting the Λ/Λ_0 ratio against rest time (Figure 2), where Λ stands for the amount of overshoot after a certain rest time and Λ_0 stands for the amount of overshoot of the original sample. Fresh samples were used to study the effect of each rest time. Figure 2 demonstrates that the kinetics of structure recovery was quite fast for all the systems studied. In fact, the rate of structure recovery appears to increase with SDEY concentration. The same may be stated concerning the thixotropy of these dispersions as indicated by the increase in the amount of overshoot with SDEY concentration. Finally, the fact that an aqueous dispersion containing only XG showed a recovery kinetics of the same order of magnitude as the SDEY-XG dispersions must be noted.

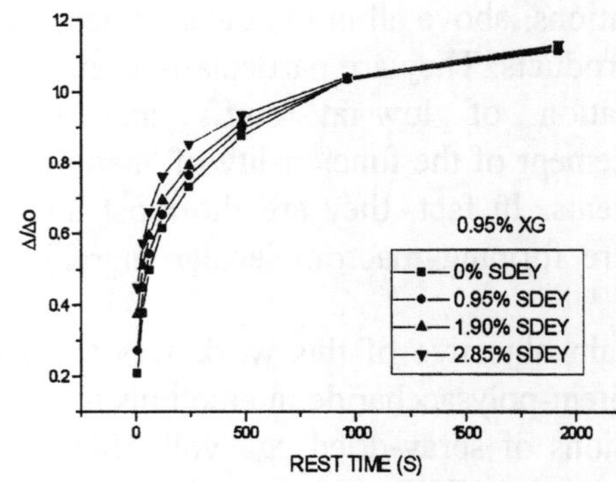

Fig.3 Effect of SDEY concentration on the structural recovery kinetics of dispersions containing 0.95% wt XG after start-up at the inception of shear experiments at $0.4s^{-1}$

4. CONCLUSIONS

A global analysis of the results obtained indicated that the rheology of SDEY-XG dispersion is mainly governed by xanthan gum. Apart from that, no hint of a possible strong complexation between biopolymers was observed. Therefore, the role of SDEY could be considered as that of a mere filler, despite its complex composition.

5. REFERENCES

Ganani, E and Powell, RL (1985) J. Rheol. 29:931-941

Pascual C, Alfaro MC, Muñoz, J (1998) Proceedings of the Food Emulsion and Foams: Interfaces, Interactions and Stability Conference. The Royal Society of Chemistry.

5. ACKNOWLEDGMENT

This contribution is part of a research project (1996-0232) sponsored by United Kingdom-Spain Acciones Integradas Programme. The gratefully acknowledge its financial support.

INFLUENCE OF PROCESSING ON THE RHEOLOGY OF EGG YOLK PRODUCTS

A. GUERRERO[1], J. MIRANDA[1], P. PARTAL[2]

[1]Departamento de Ingeniería Química, Universidad de Sevilla, 41012 Sevilla, Spain.
[2]Departamento de Ingeniería Química, EPS La Rábida, Universidad de Huelva, 21819 Huelva, Spain.

1. INTRODUCTION

The primary industrial application of egg yolk is the manufacture and stabilisation of o/w emulsions such as mayonnaise, salad dressing and sauces. Large lipoprotein complexes provide the maximum functional properties, such as emulsifying capacity, to egg yolk. Processing of egg yolk may impair this capacity by disrupting these complexes.

The main objective of this work is to investigate the influence that spray-drying process (SDP) and lipid-cholesterol extraction (LCE) with pseudocritical CO_2 exert on rheological properties of egg yolk. Variables such as solid concentration (SC), temperature (T) and cholesterol extraction percentage (CEP) or dry basis protein concentration (PC) were investigated.

2. EXPERIMENTAL

The rheological measurements were performed by means of a Control Stress rheometer, RS-100 (Haake, Germany) using cone and plate geometry.

Three types of egg yolk were used: Native (N) yolk containing 10% salt; Low Fat- Low Cholesterol (LFLC) yolk, reconstituted with water up to 45% solids; Spray-Dried (SD) yolk which was reconstituted up to 50% or 45% solids. Futhermore, this last product was used together with spray-dried egg white (reconstituted with water up to 12% solids) in order to prepare mixtures containing 40%, 35%, 30% and 25% solids.

The LFLC yolk was obtained from SD yolk by a pseudocritical extraction process with CO_2 performed in the "Departamento de Ingeniería Química" (Moraga et al., 1996).

3. RESULTS AND DISCUSSIONS.

Figure 1 shows the mechanical spectra of the three types of egg yolk used at a solids concentration of 45%. As may observed, both spray-drying (SD) and lipid cholesterol extraction (LCE) processes provide a dramatic increase in the linear viscoelasticity properties of egg yolk, as well as a shift from fluid-like (N yolk) to gel-like (SD and LFLC yolks) behaviour. This is a consequence of the thermal denaturation process of proteins produced during spray-drying, in spite of the retention of some residual structure. The unfolding of proteins promote the formation of hydrophobic bonds which can participate in the formation of a gel-like network (Rivas and Sherman, 1983). A remarkable increase was also obtained for the flow properties. LFLC yolk shows higher rheological properties because of the highest protein concentration obtained at the same concentration of solids (Bringe and Cheng, 1995).

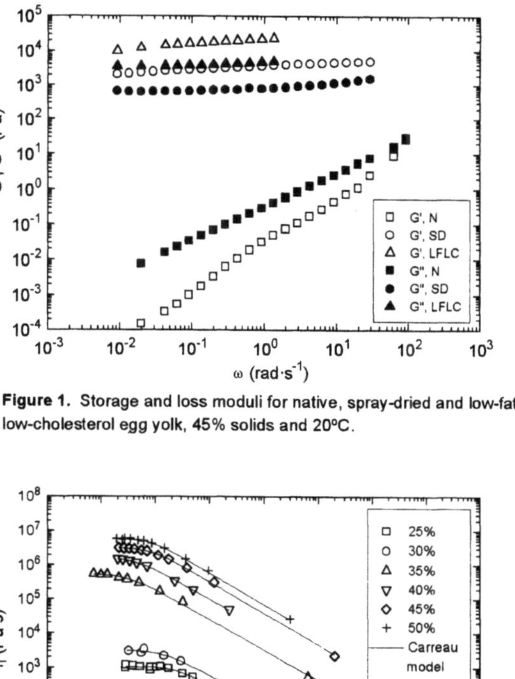

Figure 1. Storage and loss moduli for native, spray-dried and low-fat low-cholesterol egg yolk, 45% solids and 20°C.

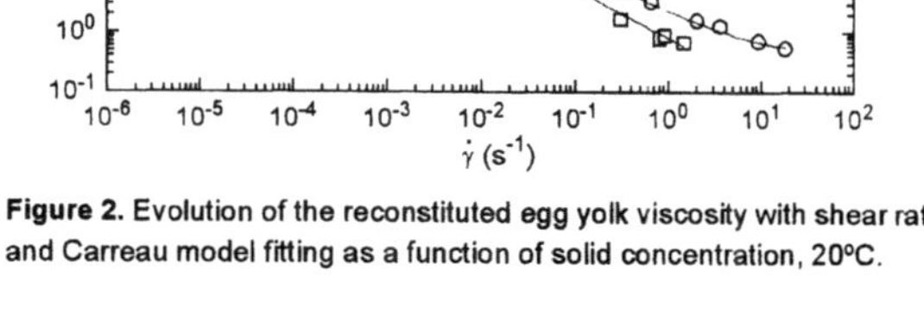

Figure 2. Evolution of the reconstituted egg yolk viscosity with shear rate and Carreau model fitting as a function of solid concentration, 20°C.

Figure 2 shows the flow curves of SD yolk as a function of solids concentration, as well as the experimental fitting to the Carreau Model:

$$\eta = \eta_\infty + \frac{\eta_0 - \eta_\infty}{[1 + (\dot{\gamma} / \dot{\gamma}_c)^2]^s} \qquad (1)$$

An increase in the concentration of solids produces an increase in the flow properties (apparent viscosity, η, and low shear Newtonian viscosity, η_0) related to a higher density of network structures.

The enhancement of the network structures is apparent above 30% since a dramatic change in the evolution of the flow properties takes place. Furthermore, as may be seen in Figure 3, the values of η_0 obtained from creep and flow measurements are rather similar. The same change was obtained for the linear viscoelasticity properties as may be deduced from the values of the plateau modulus, G_N^0. This parameter was approximately calculated as the value of G' corresponding to the frequency at which tg(δ) is minimum (Wu, 1989).

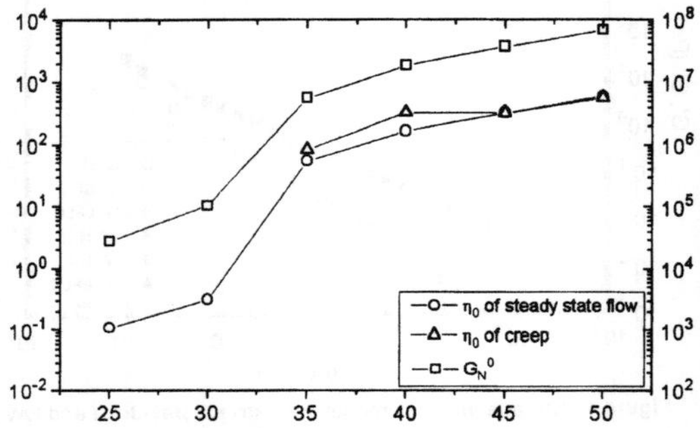

Figure 3. Evolution of the plateau module and the low shear newtonian viscosity with solid concentration of reconstituted spray-dried egg yolk.

Figure 4 shows that both the low shear Newtonian viscosity, η_0 calculated from the Carreau Model, and the plateau modulus, G_N^0, increase with the lipid-cholesterol extraction degree. This is a consequence of an increase in protein concentration that produces an enhancement of the network structures.

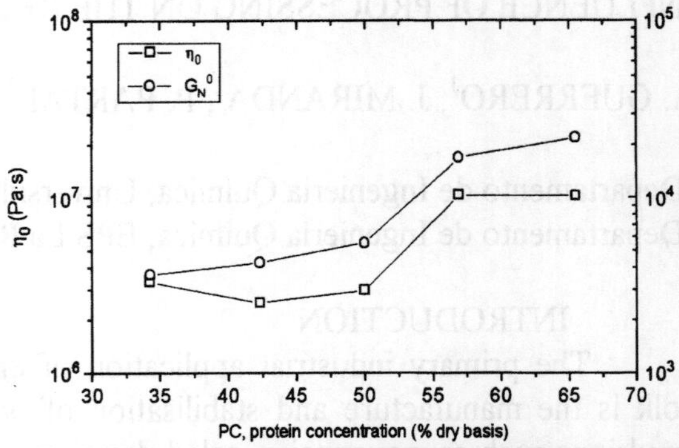

Figure 4. Evolution of the low-fat low-cholesterol egg yolk plateau moduli and the low shear newtonian viscosity with protein concentration, 45% solids and 20°C.

Temperature does not exert a large influence on the rheological properties of SD yolk since only produces a moderate decrease in both moduli, G' and G'' (Figure 5). However, an increase in G'' and above all in G' was obtained from 40°C to 50°C, since a slightly higher degree of protein unfolding could be expected at this temperature.

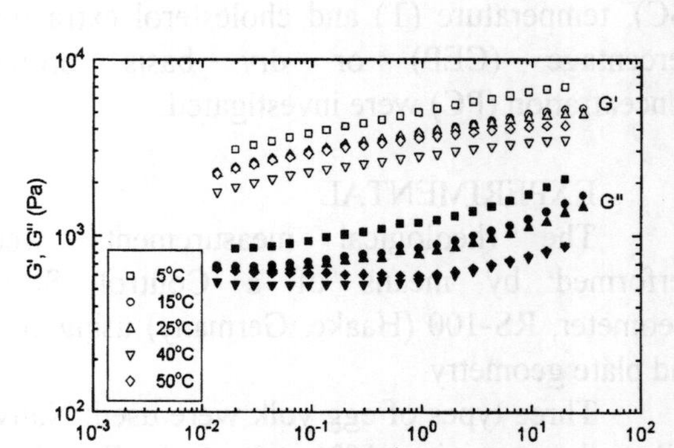

Figure 5. Storage and loss moduli for reconstituted spray-dried egg yolk as a function of temperature, 45% solids.

5. REFERENCES

Seideman WE and Cotterill OJ (1969) Poultry Sc 48:894-900.

Vincent R; Powrie WD; Fennema O (1966) J Food Sci 31: 643-648.

Moraga JM, Cota JA, Roselló A (1996) 7th International Congress of Mediterranean Chemical Engineering (Barcelona), Abstracts: P.3.18, pag. 88.

Wu S (1989) J Polym Sci 27:723-741.

Rivas HJ, Sherman P (1983) J Texture Stud 14:267-275.

Bringe NA, Cheng J (1995) Food Tech 5:94-106.

INFLUENCE OF NaCMC IN VISCOUS SYNERGISM OF LOCUST BEAN GUM + KAPPA CARRAGEENAN MIXTURES

M.J. HERNÁNDEZ[1], J. DOLZ[1], M. DOLZ[1], M. HERRÁEZ[2], J. DELEGIDO[1], J. PELLICER[1]

[1]Dep. Termodinàmica, Fac. Física i Farmàcia, Univ. València. 46100 Burjassot, València. SPAIN
[2]Dep. Farmàcia i Tecnologia Farmacèutica, Fac. Farmàcia. 46100 Burjassot, València. SPAIN

1. INTRODUCTION

Synergistic effects between polysaccharides such as locust bean gum, xanthan gum, guar gum, carrageenan, related to their thickening properties are widely known and used in food industry. (Da Silva and Rao, 1992). These effects are modified due to the existence of salts and other food additives, i.e. cellulose derivatives, in the final products (Therkelsen, 1992).

In this work we have studied the effects produced by the presence of NaCMC, C, in different mixtures of locust bean gum, LB, and kappa carrageenan, κ. To this aim we have compared the synergism produced in the systems LB+κ (Fernandes et al., 1994; Murayama et al., 1995) and the systems LB+κ+C by considering a parameter called viscous synergism index, Iv, (Dolz et al., 1996) which results a function of shear rate and locust bean gum/κ-carragenaan gum concentration ratio, z.

$$I_V = \frac{\eta_{LB+\kappa}}{\eta_{LB}+\eta_{\kappa}} \qquad (1)$$

$$I'_V = \frac{\eta_{LB+\kappa+C}}{\eta_{LB+C}+\eta_{\kappa+C}} \qquad (2)$$

2. MATERIAL AND METHODS

The total concentration in mixtures containing LB and κ is always 1% w/w, being the NaCMC concentration fixed, 0.12% w/w, in all the cases that this component is present. Both in systems with NaCMC and without it, 10 different concentration ratios, z, have been considered, varying from 0.11 to 9. The simple dispersions of LB and κ and those with the fixed proportion of NaCMC, corresponding to the z in mixtures have also been prepared.

The experimental measurements of apparent viscosity have been carried out with a rotatory viscometer Brookfield DVII, at 25°C (shear rate range: 0.063-12.56 s^{-1}).

3. RESULTS AND DISCUSSION

The explicit expressions for I_V (1) and I'_V (2) are obtained from the mathematical expressions calculated for the power law model describing the shear thinning behaviour of all systems considered (Tables 1 and 2).

Table 1. Consistency and power law index as a function of concentration (c) in the ranges indicated. LB: locust bean gum, κ: κ-carrageenan, C: NaCMC.

	K(Pa s^n)	(n-1)	c
LB	$3.07\, c_{LB}^{3.52}$	- 0.118	[0.1,1.0]
κ	$12.1\, c_{\kappa}^{7.30}$	- 0.245	[0.1,0.6]
		$2.07 c_{\kappa}^{2} - 4.33 c_{\kappa} + 1.63$	[0.7,1.0]
LB+C	$8.62\, c_{LB}^{3.42}$	- 0.201	[0.1,1.0]
κ+C	$0.0125\, \exp(2.73\, c_{\kappa} + 103\, c_{\kappa}^{26.9})$	- 0.167	[0.1,1.0]

Table 2. Consistency and power law index as a function of concentration ratio ($z=c_{LB}/c_{\kappa}$) in the ranges indicates. LB: locust bean gum, κ: κ-carrageenan, C: NaCMC.

	K(Pa s^n)	(n-1)	z
LB+κ	$\frac{465(z+0.101)^{4.42}+714}{(z+0.101)^{8.84}+80.1}+3.53$	- 0.623	[0.11,2.3]
		- 0.165	[4.0,9.0]
LB+κ+C	$\frac{3.58\, z^{4.62}+0.154}{z^{9.26}+0.0227}+1.64\, z^{0.5}$	- 0.471	[0.11, 2.3]
		- 0.159	[4.0,9.0]

Figures 1 and 2 show the 3D graphs for I_V. In all cases it is observed that Iv presents a maximum for a determined value of z. The values of these maximums depend on shear rate, while the concentration ratio z at which the maximum is obtained is approximately the same in the whole shear rate range considered (Table 3).

When comparing the values of these maximum synergism produced in the two different kind of mixtures studied, we see that a decrease of about 50% yields when adding NaCMC to the blend LB+κ for the lowest $\dot{\gamma}$, while the values are similar from 0.5 s^{-1}. Furthermore, that maximum synergistic effect takes place at a lower concentration ratio LB/κ when NaCMC is included ($z \approx 1.3$ for systems without NaCMC against $z \approx 0.25$ for systems with NaCMC, Table 3).

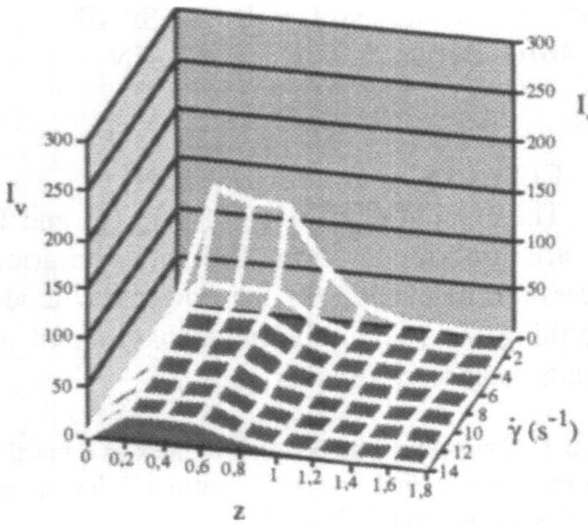

Figure 1. Viscous synergism index vs. concentration ratio and shear rate for locust bean gum + κ-carrageenan gels.

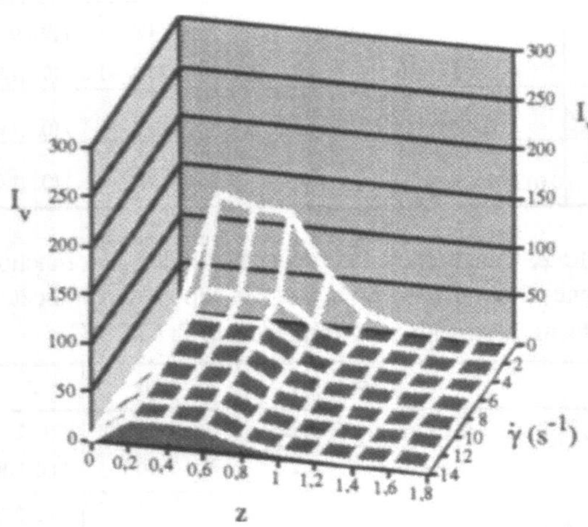

Figure 2. Viscous synergism index vs. concentration ratio and shear rate for locust bean gum + κ-carrageenan + NaCMC gels.

Table 3. Maximum viscous synergism index (I_V^M) and concentration ratio corresponding to those maximums (z^*), for the different shear rates considered.

$\dot{\gamma}$ (s^{-1})	LB + κ		LB + κ + C	
	z^*	I_V^M	z^*	I_V^M
0.063	1.335	282	0.240	128
0.126	1.330	200	0.242	104
0.314	1.324	127	0.244	79.8
0.628	1.319	90.0	0.246	65.2
1.26	1.315	63.6	0.248	53.1
2.51	1.310	45.1	0.250	43.4
6.28	1.304	28.6	0.253	33.2
12.6	1.300	20.2	0.256	27.0

Moreover, we can say that while in the binary blends the synergistic effects are appreciable for $0 < z < 3$, when adding the concentration considered of NaCMC this range is reduced to $0 < z < 1$. In order to observe this better, we show the consistency synergism index, I_c (I_v for $\dot{\gamma} = 1\ s^{-1}$) for both cases (Figure 3).

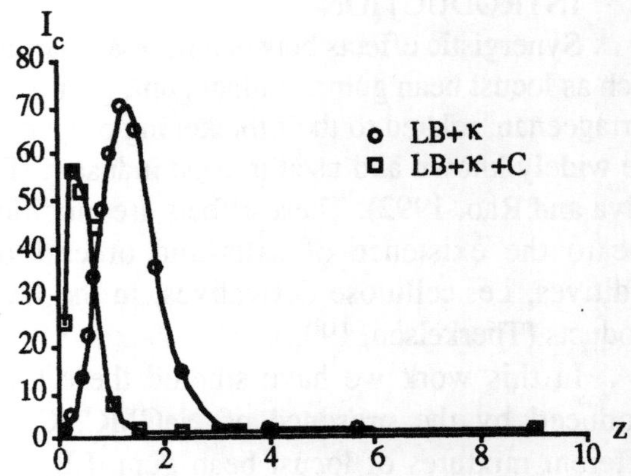

Figure 3. Consistency synergism index as a function of concentration ratio for the systems indicated.

Therefore we can conclude that the addition of this cellulose derivative modifies the synergism expected from the mixture LB+κ, decreasing the final effect for the lowest shear rate and affecting the ratio where the maximum development of the thickening properties of the blend is produced.

4. REFERENCES

Da Silva JAL, Rao MA (1992) In: Rao MA, Steffe JF(eds)Viscoelastic properties of foods. Elsevier Appl Sci. pp 285-315

Dolz J, Dolz M, Hernández MJ, Pellicer J (1996) Proceedings XIIth International Congress on Rheology. Quebec City. p. 807

Fernandes PB, Gonçalves MP, Doublier JL (1994) Food Hydroc 8:345-349

Murayama A, Ichikawa Y, Kawabata A (1995) J Texture Stud 26: 239-254

Therkelsen GH (1992). In:Whistler RL, BeMiller J.N(eds) Industrial Gums. Polysaccharides and Their Derivatives. Academic Press, INC

RHEOLOGICAL BEHAVIOUR OF OIL-IN-WATER EMULSIONS STABILIZED BY A SPRAY-DRIED EGG PRODUCT

J. E. MOROS[1], J. M. FRANCO[2], C. GALLEGOS[2]

[1]Departamento de Ingeniería Química, Universidad de Sevilla, 41012 Sevilla, Spain.
[2]Departamento de Ingeniería Química, EPS La Rábida, Universidad de Huelva, 21819 Huelva, Spain.

1. INTRODUCTION

Most of food emulsions are oil-in-water emulsions, i.e. as mayonnaise, salad dressing, etc. They are traditionally stabilized by egg products derivatives. Native egg yolk is not microbiologically stable and freezing and drying processes are usually required to preserve yolk for food applications. Guerrero and Ball (1994) studied the effect of different egg products as emulsifiers. They found that the rheological behaviour of emulsions stabilized by different egg products depends on the way they were processed.

The overall objective of this work was to study the effect that the spray-drying process exert on the rheological properties of o/w emulsions stabilized by this egg product, in comparison with the rheological behaviour of native egg yolk-stabilized emulsions. Variables such as oil and egg product concentrations were investigated.

2. EXPERIMENTAL

Different o/w emulsions containing 65-77.5% wt sunflower oil, 3.6-7.3% wt reconstituted (1:1.2 solid egg/water weight relation) spray-dried egg product (SDEP) or native egg yolk (NEY), 3% wt vinegar, 0.4% wt salt and the corresponding amount of distilled water were manufactured in a Ultraturrax T-50 rotor-stator turbine homogeneizer, from Ika (Germany). Emulsions were prepared at 6000 rpm during 3.5 minutes and stored at 5°C. Measurements were carried out seven days after their manufacture.

Rheological measurements were performed in a Bohlin CS-50 rheometer (Germany) using a cone-plate sensor system (4°, 60mm) for dynamic shear tests and a serrated plate-plate sensor system (25 mm, 1 mm gap) for steady-state flow measurements.

Droplet size distributions were obtained in a Malvern Mastersizer X analyzer (U.K.). The Sauter diameter was calculated as follows:

$$d_{sv} = \frac{\Sigma n_i d_i^3}{\Sigma n_i d_i^2} \qquad (1)$$

3. RESULTS AND DISCUSSION

Figure 1 shows the evolution of the linear viscoelastic functions, G' and G'' with frequency. As can be observed, both viscoelastic functions increase with oil content, showing a plateau region in G' and a clear minimum in G'' in the frequency range studied. The same behaviour has been found for model and commercial mayonnaises (Gallegos et al., 1992, Franco et al, 1997), which was attributed to an extensive flocculation process.

Figure 2 shows that the steady-state viscosity also increases with the oil concentration. The Carreau model fits the experimental values fairly well.

On the contrary, the mean droplet diameter (Table 1) slightly increases with oil content.

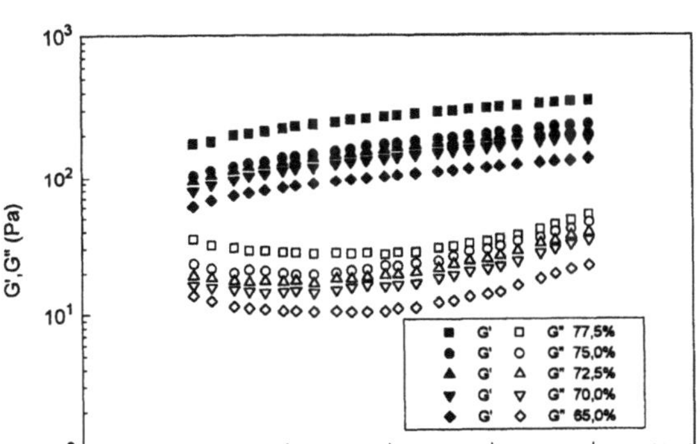

Figure 1. Evolution of the storage and loss moduli with frequency for emulsions containing 6.6% wt processed SDEP and different oil concentration.

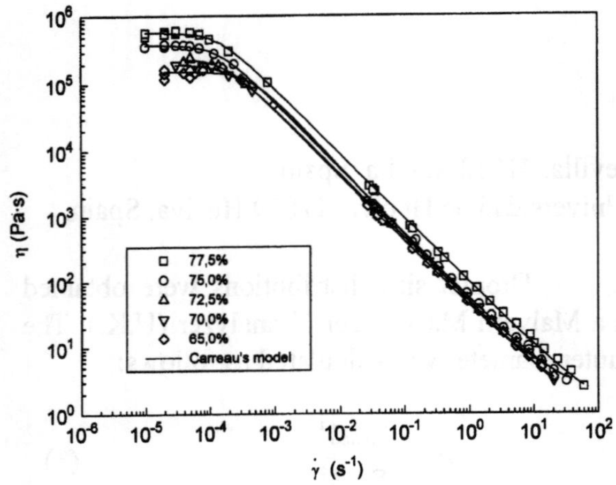

Figure 2. Steady-state viscosity of emulsions containing 6.6% wt processed SDEP and different oil concentrations.

Table 1. Sauter's diameter of emulsions containing 6.6% wt processed SDEP and different oil concentrations.

% oil (wt)	77.5	75	72.5	70	60
d_{sv}	8.04	7.03	6.46	6.33	5.98

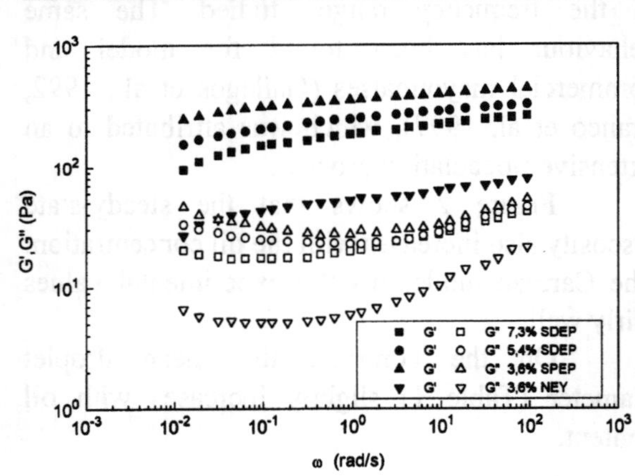

Figure 3. Evolution of the storage and loss moduli with frequency for emulsions containing 72.5% wt oil and different processed egg products.

Figure 3 shows that the values of G' and G'' increase as the spray dried egg content decreases, in spite of the higher mean droplet size of the emulsions (Table 2). These values are also much higher than those found for the native egg yolk stabilized emulsion. This evolution of the viscoelasticity functions with SDEP concentration was previously found by Franco et al. (1997) for NEY-stabilized emulsions. This effect was attributed to the development of the slip planes at the o/w interface, yielding a reduction of the interfacial viscoelasticity (Graham and Phillips, 1980). The same evolution may be observed if the steady-state flow curves of emulsions containing different SDEP concentrations are compared (Figure 4).

Table 2. Sauter's diameter of emulsions containing 72.5% wt oil and different processed egg products.

% EP (wt)	7.3 SDEP	5.4 SDEP	3.6 SDEP	3.6 NEY
d_{sv}	5.90	6.13	7.14	3.36

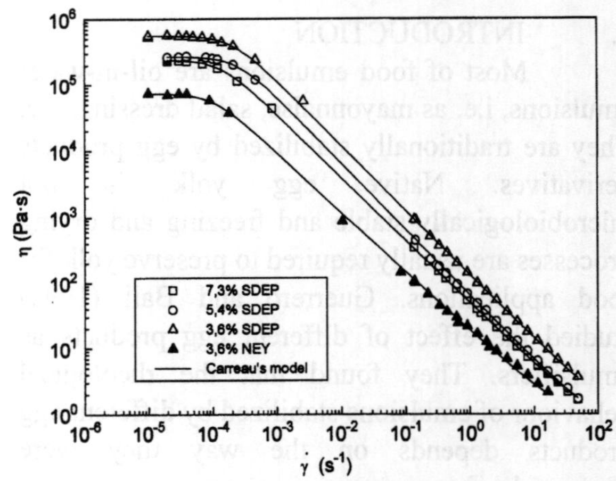

Figure 4. Steady-state viscosity of emulsions containing 72.5% wt oil and different processed egg products.

4. CONCLUDING REMARKS

In comparison with native egg yolk-stabilized emulsions, a spray dried egg product provides emulsions with significantly higher values of the linear viscoelastic functions and steady-state viscosities in spite of a higher mean droplet size. These results must be explained on the basis of protein denaturation due to the spray dry process. In addition to this, linear viscoelastic functions and steady-state viscosity values increase as oil concentration increases or spray dried egg concentration decreases in the range of concentration studied.

5. REFERENCES

Franco JM, Berjano M, Gallegos C (1997) J Agric Food Chem 45:713-719

Gallegos C, Berjano M, Choplin L (1992) J Rheol 36:465-478

Graham DE, Phillips MC (1980) J. Colloid Interface Sci 76:240-250

Guerrero AF, Ball HR (1994) J Texture Stud 25:363-381

ACKNOWLEDGMENTS

This work is part of a research project (ALI96-0892) sponsored by the CICYT, Spain. The authors gratefully acknowledge its financial support.

RHEOLOGICAL BEHAVIOR OF CROSSLINKED WAXY MAIZE STARCH-KAPPA-CARRAGEENAN MIXTURES

A. TECANTE[1] and J.L. DOUBLIER

Institut National de la Recherche Agronomique, Rue de la Géraudière, BP 71627-44316, Nantes, France.
[1]Permanent address: Depto. Alimentos y Biotecnología, FQ-"E", UNAM, México, D.F., 04510, México.

1. INTRODUCTION

The aim of this study is to get a better understanding of the rheological behavior of starch-hydrocolloid mixtures under gelling and non-gelling conditions of the latter. Crosslinked waxy maize starch (CWMS) was used because it is free of amylose and crosslinking avoids granule rupture. Hence, swelled granules are dispersed in a phase containing essentially the hydrocolloid. Application of mixtures with κ-carrageenan (κC) in the food industry provided further motivation to this work.

2. METHODS

2.1 Materials

The CWMS sample (Clearam® CH10) was an adipate/acetate starch obtained from Roquette Frères. κC was provided by SBI (France) and was used without further treatment. Analytical grade KCl (Merck) and deionized water were used throughout.

2.2 Pasting procedure

All starch slurries were pasted in a Brabender viscograph-E at a bowl speed of 75 rpm following the thermal treatment shown in Figure 1. The apparatus was used only as a cooker and to monitor consistency changes during pasting. Dry-weight concentrations were: CWMS 4%, κC 0.5%. In some slurries 20 mM KCl was added to induce κC gelation upon cooling.

2.3 Rheological measurements

Steady shear tests were carried out at 60 °C with a rotational viscometer (Rheomat 120) in a coaxial-cylinders fixture (diameters: bob = 46 mm, cup = 49 mm, height 78 mm). Two consecutive up-down shear scans from 0 to 660 s^{-1} followed by a logarithmic descent from 660 to 0.1 s^{-1} were applied. Low amplitude (0.04) oscillatory shear was carried out at 60 °C and 25 °C in a controlled-stress rheometer (Carrimed CS50) using either the cone/plate (diam. 6 cm, angle 4°) or grooved parallel plates (diam. 4 cm, gap 1 mm) geometries. Gap adjustment considered thermal expansion. The fixture rim was covered with paraffin oil to prevent evaporation.

2.4 Solubility-swelling determination

The degree of granule swelling and starch solubility were determined according to previously described procedures (Doublier 1987).

3. RESULTS AND DISCUSSION

In the following discussion A, B and C designate, respectively, CWMS, CWMS/κC and CWMS/κC/KCl. Figure 1 shows the differences in consistency. The earlier peak in C is due to κC solubilization. Swelling start-up occurred slightly earlier (~ 64 °C) in B than in A and C (~ 66 °C). All pastes displayed a rapid swelling followed by a smooth transition to an essentially constant zone during the holding period which indicates good thermal stability.

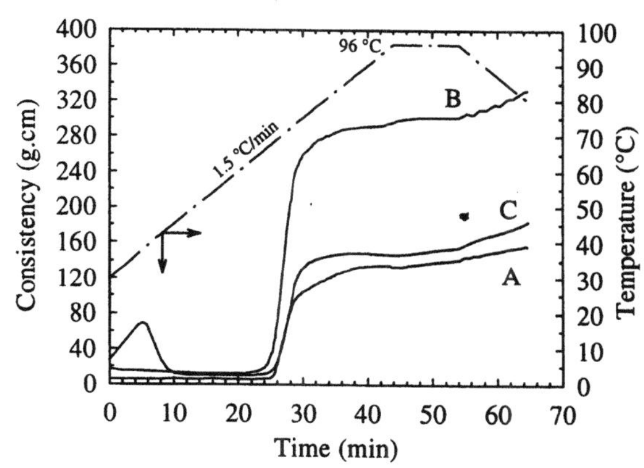

Figure 1. Brabender consistency of starch pastes.

The flow properties of hot pastes are shown in Figure 2. The first up-down scan (continuous line) reveals an anti-thixotropic behavior with C displaying the greatest hysteresis. Time-dependency essentially disappeared after continuous shearing as shown by the second scan (dotted line) which was always superimposed to the descending curve of the first one. Shear-thinning behavior is clear; nevertheless, flow curves could not be described by the power law model over all the shear rate range.

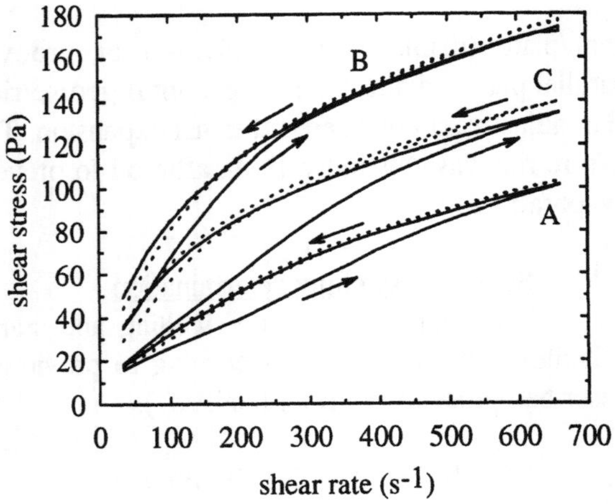

Figure 2. Flow curves of starch pastes at 60 °C.

The viscoelastic properties of hot pastes are shown in Figure 3. None of them was a self-standing gel, yet A and B exhibited apparently a solid-like behavior with G' > G" and

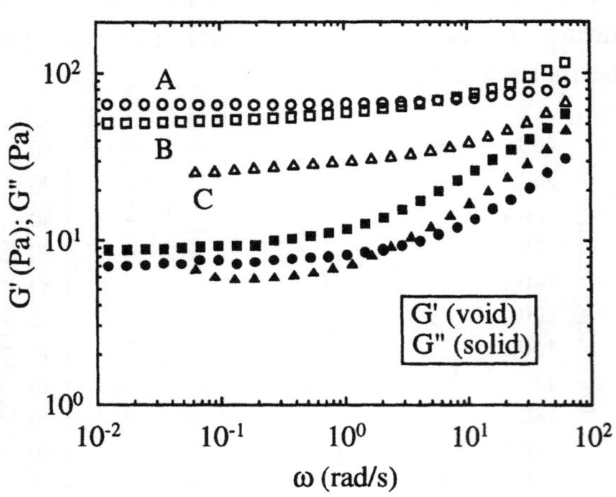

Figure 3. Dynamic moduli of pastes at 60 °C.

both independent of ω below 1 and 0.2 rad/s, respectively. In C, lower and more frequency-dependent moduli were observed. Paste A had the lowest loss angles (tan $\delta \approx 0.1$ at $\omega < 1$ rad/s), while B and C displayed higher and more frequency-dependent loss angles.

Figure 4 shows the mechanical spectra of cooled pastes. As expected, C was a self-standing gel with a typical spectrum; G'>>G" and both independent of ω. In fact, tan $\delta < 0.02$ showing an essentially elastic response. A and B exhibited a behavior similar to that at 60 °C with slightly higher moduli.

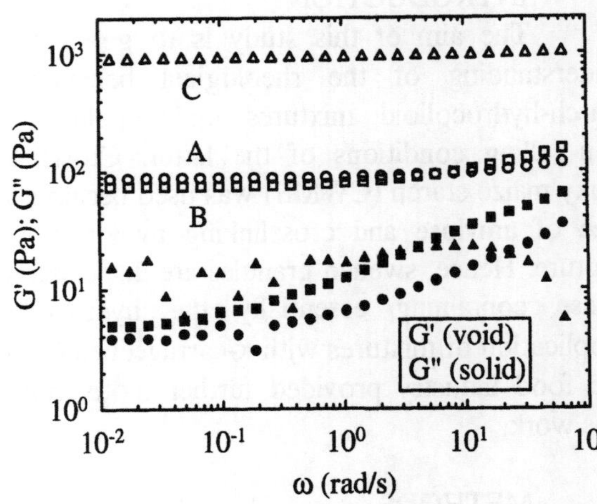

Figure 4. Dynamic moduli of systems at 25 °C.

Starch solubility was low (~ 2%) and swelling was 25.9, 27.3 and 29.3 g/g for A, C and B, respectively, which resulted in starch volume fractions very near to unity. At such value a close-packing of swelled granules can be present (Bagley and Christianson, 1982) leaving a reduced volume to the fluid phase. Therefore, our systems were highly charged and this may explain the initial anti-thixotropic behavior the extent of which depended on the viscosity of the fluid phase. On the other hand, at 60 °C and 25 °C (except C), the increasing dependence of moduli with frequency can be attributed to the response of the continuous phase which is to some extent overwhelmed by that of starch particles. Systems behaved as highly concentrated suspensions of starch particles in water (A at 60 °C and 25 °C) and in a macromolecular solution (B at 60 °C and 25 °C and C at 60 °C), and as a composite gel of swollen particles embedded in a macromolecular network (C at 25 °C).

4. REFERENCES

Bagley EB, Christianson DD (1982) J. Texture Stud., 13: 115 126.
Doublier JL (1987) J. Cereal Sci., 5:247-262.

RHEOLOGICAL BEHAVIOUR OF AQUEOUS CONCENTRATED GLAZE SUSPENSIONS: INFLUENCE OF THE ADDITIVES USED

F.Andreola, G.C. Pellacani, P. Pozzi, M. Romagnoli
Faculty of Engeneering - University of Modena Via Campi 183 - 41100 - MODENA - ITALY
Tel: +39-59-378409 / Fax: +39-59-373543
E-mail:reolab@unimo.it

1. INTRODUCTION

Ceramic glazes are aqueous suspensions with a solid content ranging from 60 to 75 wt% (90-95 wt% frit, 5-7 wt% clayey materials, being additives the remaining 3 wt%). The glassy phase called "frit" is obtained from the fast cooling in water of a fused (around 1450°C) glassy material; the plastic materials are normally kaolin, or bentonite.
The glaze suspensions do not have a Newtonian behavior. "F.Andreola et al, (1994)", thus the relationship between shear stress and shear rate is not linear and time dependent (the structure in these systems is dependent on the shear history and time scale of the experiment). From the industrial point of view it is important for the glaze suspensions to have rheological properties (apparent viscosity, yield stress value, thixotropy, etc.) such as to grant the highest global savings in all processing phases where they are present (wet grinding, mills emptying, stocking in tanks and application tiles). In the phase of wet-grinding additives such as binders, deflocculants and substances to regulate the rheology, wettability, etc. are added.
This study on the effects of additives on the rheological behavior of the frit suspensions has the purpose to deepen the influence of each individual component used in the formulation of glazes to ensure a correct usage and to obtain a defect free glaze coating.

2. EXPERIMENTAL PROCEDURE

2.1 Materials

The frits were prepared from a glass based system $Na_2O.Al_2O_3.SiO_2$ by varying its chemical composition in Na_2O % content (15-9 wt%) and by adding of a constant weight percent (15 wt%) of different metal oxides (CaO, BaO, ZnO, PbO) and borum oxide (B_2O_3). The resulting frits have the chemical average composition of the frits commonly used in the ceramic industrial sector.
A kaolin (BIP 96wt% kaolinite) and a high purity sodium bentonite were used as plastic components. Three sodium carboxymethylcellulose (SIGMA CHEMICAL) at different degree of polymerization were used as binders. Three typologies of deflocculating agents, widely used in the ceramic industry, were selected, different from each other both in their active principle and their deflocculating mechanism "P.Pozzi et al (1987, 1994, 1997). Sodium metasilicate SMS(FLUKA) and sodium tripolyphosphate STP(SIGMA) as inorganic type and among the organic type sodium polyacrylate, SPA(FLUKA).

2.2 Glaze suspensions preparation

For all tests the suspensions were prepared at 65 wt% in distilled water, by grinding for 2 hours to guarantee stability in a fast laboratory ball mill (300 ml) with spherical alumina grinding media "F.Andreola et al., (1997). Two formulations were prepared in order to study the effect of plastic components (95wt% frit +5 wt% kaolin and 98wt% frit + 2wt% bentonite); previous works demonstrated that the 5 wt% of kaolin and 2 wt% of bentonite results as the optimal concentration for obtaining a correct application on the tiles "F.Andreola et al (1996,1997). The percentage of binder used was 0.1- 0.5 wt% with respect to the solid content. To obtain the deflocculation curves were used three defloculating agents in concentrations between 0.1-1 wt%.

2.3 Rheological Measurements

The rheological measurements were conducted on the frits suspensions with and without the additives with a Control Rate rotational viscometer Haake RV20, with concentric cylinders as measuring system. The measurements were thermostated at 20°C. This instrument allows to measure a complete flow curve and determine the rheological parameters yield stress (τ_0) and apparent viscosity (η). For the analysis the data obtained were correlated by Herschel-Bulkley model and used the viscosity values measured at 200 s^{-1}. All samples were tested with a five-step cycle suitable to study this kind of systems "F.Andreola et al (1994). The shear conditions were similar to those occurring in glazing application and optimised to improve the resolution of the flow curves recorded with our rotational viscometer.

3. RESULTS AND DISCUSSION

The rheological behaviour of glaze ceramic suspensions depends on the clayey materials used in the compositions. Their presence causes a change from a dilatant to a pseudo-plastic behaviour at lower shear rates when using bentonite or to a plastic behavior using kaolin.

Fixing the rheological parameters (τ_0, η_γ), a glaze bentonites containing with respect to one kaolin has to be prepared with an inferior solid content and a larger quantity of deflocculant, due to the bentonite gelling effect "C.Galassi, P.Pozzi(1994). Suspensions with bentonites present higher thixotropy and viscosity values. The curves of the suspensions prepared with kaolin and bentonite point out that the suspensions prepared with kaolin express a greater increase of the yield stress (τ_0) and a minor increase of the apparent viscosity (η) than those with bentonite. The latter causes a meaningful increase of the apparent viscosity, and this conditiones in particular manner the quantity to add.

The organic polymers presence (CMC) does not introduce big changes on the rheological behaviour of the suspensions; resulting normally a typical plastic behavior with higher values of apparent viscosity "D.Chiavacci, (1995). The data resulting from the analysis of the curves, demostrate that lower CMC concentrations (0,1wt%) develop a deflocculating action, and furthermore, that a direct relationship among the molecular weight of the CMC added and the resulting viscosity of the suspension exists. Viceversa the effect of the CMC concentration results not to be linear. On the suspensions prepared with bentonite, a synergic effect exists that provokes a remarkable increase of the yield stress as a function of the CMC molecular weight.

The deflocculating agents are substances (polyelectrolites) that added in controlled quanti-ties to a suspension, modify its electrostatic balance, determining a greater fluency and reducing the viscosity, without deminishing the solid content "Q. Malandrino(1976, 1979). From the rheological tests emerges that the STP present a greater deflocculant capability with respect to the others agents.

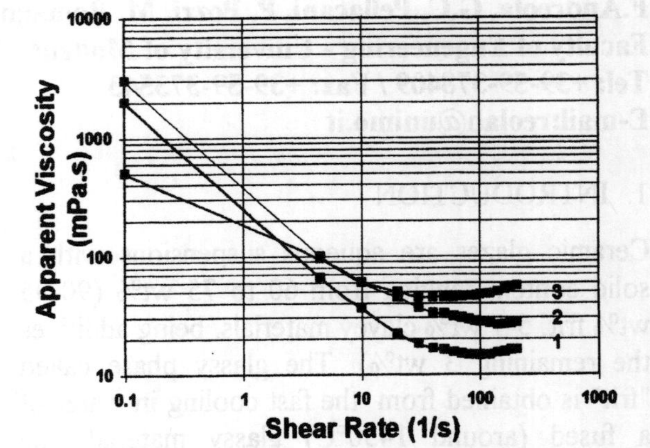

Fig 1 - Flow curves log-log of suspensions at 65 wt% (1) frit; (2) frit/kaolin; (3) frit/bent.

4. CONCLUSIONS

The results obtained in the present study allow to conclude that using the rotational viscometers for controlling industrial glazes is possible to know in advance the effects of the different components and additives on the rheological properties of glaze suspensions. After rationalizing and optimizing them, it is possible to define the composition suitable to obtain the parameters required for each glazing technological application (bell, disk, spraying, etc.).

5. REFERENCES

F.Andreola et al, (1994)Am. Cer.Soc. Bull.73(10), 75-78.

F.Andreola et al., (1997) Mat. Eng.8(2),171-180.

Soc, Cer. Ital.(1991) Reologia Ceramica Applicata, Faenza Ed.

F.Andreola et al., (1996) IV Convegno Nazionale di Reologia Applicata, pp91-97.

C.Galassi, P.Pozzi (1994) in La Reologia dei materiali ceramici, Faenza Ed.pp 67-92,

R.Tanner, (1988) in Engineering Rheology, vol 14, Oxford Eng. Science,pp 11-18.

D.Chiavacci (1995) Cer. Inf., 335(10), 593-599.

Q. Malandrino(1976) Cer. Inf. 124(11).

Q. Malandrino (1979) Cer. Inf. 154(3).

P.Pozzi et al. (1987) Ceramurgia,17 (5), 183-185,

P.Pozzi et al. (1990) Applied Clay Science (5)193-201.

P.Pozzi et al. (1994) J.Am.Ceram.Soc, 77(2), 509-513.

THE MYTH OF YIELD STRESS FLUIDS

Howard A Barnes,

Unilever Research Port Sunlight, UK, L63 3JW.

INTRODUCTION AND OVERVIEW

Originally, it was accepted that the yield stress of a solid was essentially the point at which, when the applied stress was *increased*, the deforming solid first began to show *liquid*-like behaviour, i.e. continual deformation. In the same way, the yield stress of a structured liquid was originally seen as the point at which, when *decreasing* the applied stress, *solid*-like behaviour was first noticed, i.e. no continual deformation. However as time went on, and experimental capabilities increased, it became clear, first for solids and lately for soft solids and structured liquids, that although there is usually a small range of stress over which the mechanical properties change dramatically (an apparent yield stress), these materials nevertheless show slow but continual steady deformation when stressed for a long time *below* this level, having shown an *initial* linear elastic response to the applied stress.

At the lowest stresses, this creep behaviour for solids, soft solids and structured liquids can be described by a Newtonian-plateau viscosity. As the stress is increased the flow behaviour usually changes into a power-law dependence of steady-state shear rate on shear stress. For structured liquids and soft solids, this behaviour generally gives way to Newtonian behaviour at the highest stresses. For structured liquids this transition from very high (creep) viscosity (often $>10^6$ Pa.s) to mobile liquid (often <0.1 Pa.s) can often take place over a *single* order of magnitude of stress. This extreme behaviour, when viewed on a linear basis, gives every reason for believing that the material had a yield stress, and in many cases the flow curve seems to be adequately described by Bingham's simple straight-line-with-intercept equation. However, if viewed on a logarithmic basis, the equally simple Newtonian/power-law/Newtonian description is clearly seen.

'FLOW' OF SOLIDS

When results from creep experiments on solids are collated [1,2], a simple picture for most stressed materials emerges:

(a) after a long-enough time, a steady creep rate *is* achieved;
(b) this creep rate, at low-enough applied stress, usually becomes a linear function of applied stress - in the language of the rheologist, this linearity signifies *Newtonian* behaviour - and is governed by atomic diffusion and,
(c) at higher stresses, the creep rate increases faster than the stress, and usually shows a power-law-like behaviour.

Above this power-law region, other phenomena lead to different kinds of creep behaviour, but among them is a situation where the creep rate (often quite large by now) tends again towards a linear function of applied stress - the so-called drag-controlled plasticity - i.e. in rheological parlance an upper Newtonian behaviour.

The creep results of most solid materials can be fairly well described by the Ellis equation (omitting the usual high-shear-stress asymptotic viscosity).

FLOW OF 'YIELD STRESS' LIQUIDS

As a *qualitative* description of a *general* change in the mechanical response of a material, it is completely satisfactory and unobjectionable to speak about a yield stress, since many materials show a radical transformation in flow (or deformation) behaviour over a relatively narrow range of stress, and have obviously 'yielded'.

The measurement of the response of 'yield-stress' liquids at low stress has changed markedly over the last 20 years or so, since the introduction and improvement of controlled-stress rheometers, which are able to apply small stresses and measure *very* low shear rates. Originally, controlled viscometers (as opposed to static U-tubes and cups etc.) were based on an applied *stress* which was generated by a weights-and-pulleys arrangement. However the limitations of this type of instrument in terms of attaining *low* shear rates due to bearing friction were soon found, as also where the limitations of prolonged shearing at *high* shear rates - the finite length of the string! These methods were then superseded by electrically driven motors and the (what we now call) controlled-strain instruments had arrived, largely replacing the controlled-stress variants. These became more and more sophisticated, and eventually with logarithmic mechanical gear boxes were able to span a shear-rate range typically from ~ 10^{-4} s^{-1} to ~10^{4} s^{-1} (e.g. the Weissenberg Rheogoniometer).

In the mid-1970s, a new generation of controlled-stress rheometers began to appear. Then around 1980, commercial versions of the new generation of electrically driven controlled-stress rheometers began to appear, still based on air bearings that greatly reduced friction, but now using so-called drag-cup electrical motors that allowed controlled stresses to be more easily applied, but still independent of rotation speed. Along with these features came new ways of measuring smaller and smaller rotation and rotation rates. The latest optical-disc technology now means that rotation rates as low as 10^{-8} rad/s (~ 1 revolution in 20 years) can be measured! This has opened up a new range of previously unobtainable flow behaviour for structured liquids which *seemed* to have a yield stress. Access to these ultra-low shear-rate regions is now called creep testing, and in all these tests, creep - very slow flow - is always seen, and hence there is no physical yield stress, see [1].

THE PROPER USE OF 'YIELD STRESS'

Whether or not yield stresses exist as a physical reality, and have a unique value invariant to the method or range of measurement used, there can nevertheless be no arguing with the fact that much useful progress has been made in Applied Rheology by using simple, yield-stress-containing equations that fit steady-state, shear-stress/shear-rate data quite well over a *limited* range of shear rates. In like manner, if these equations are used to predict flow in any other situation, where *the same range of shear rates applies,* this is completely acceptable and uncontroversial, and can indeed be very effective.

The most popular equations that have been used to describe liquids with yield stresses are the Bingham [3], Casson [4] and Herschel-Bulkley [5] (sometimes called the generalised Bingham) models. One interesting concept put forward by Cheng is the possibility in his mind of a hypothetical material that, as the shear rate was decreased, tended clearly to a constant stress, but at much lower shear rates it then decreased again, and eventually showed Newtonian behaviour. This idea is best illustrated by plotting the viscosity against the shear rate on a logarithmic basis, as he did, but translating this to a plot of viscosity/stress is even more informative. This concept, of course, satisfies the idea of having a real yield stress as well as having flow (creep) below this yield stress. Some materials come near to this idealism: the present author [6] quoted results for a Carbopol solution, where the slope of the logarithmic plot of viscosity/shear-rate in the power-law region was - 0.92 instead of Cheng's ideal value of -1; then below this region the viscosity tended towards Cheng's ideal Newtonian plateau [7].

If this concept is cast into the most non-Newtonian form of the Cross model, i.e. $m = 1$, it is easy to show that

$$\frac{\eta-\eta_\infty}{\eta_0-\eta_\infty}=\frac{1}{1+k\dot{\gamma}},$$

which for structured liquids at high shear rates, i.e. $\eta_0 >> \eta_\infty$ and $k\dot{\gamma} >> 1$,

$$\sigma=\frac{\eta_0}{k}+\eta_\infty\dot{\gamma} \quad \text{or} \quad \sigma=\sigma_0+\eta_p\dot{\gamma},$$

where the Bingham-model parameters are the yield stress σ_0, now given by η_0/k and the plastic viscosity η_p now given by η_∞. This idea can be relaxed from its strict use of $m = 1$, to smaller values of m and thereby accommodate many published results [1].

REFERENCES

[1] H.A. Barnes, JNNFM in press.

[2] H.J. Frost and M.F. Ashby, 'Deformation-Mechanism Maps', Pergamon Press, Oxford, 1982.

[3] E.C. Bingham, Bull. U.S. Bur. of Standards, 13 (1916) 309.

[4] C.C. Mills (editor), 'Rheology of Disperse Systems', Pergamon Press, London, 1959.

[5] W.H. Herschel and R. Bulkley, Proc. Amer. Assoc. Test Mater., 26 (1926) 621, and Kollzeitschr., 39 (1926) 291.

[6] H.A. Barnes, 'Theoretical and Applied Rheology', eds. P. Moldenaers and R. Keunings, Proc. XIth Int. Congr. on Rheology, Elsevier, Amsterdam, 1992, pp. 576 - 578.

[7] D.C-H. Cheng, 'Yield stress: a time-dependent property and how to measure it', Report No. LR 540 (MH), Warren Spring Laboratory, Dept. of Industry, UK, August 1985, ISBN 0 85624 395 7.

THE MEASURING OF YIELD STRESS LIQUIDS

C. GALLEGOS, J. M. FRANCO
Departamento de Ingeniería Química. Escuela Politécnica Superior.Universidad de Huelva. 21819 Palos Frontera (Huelva). SPAIN

The yield stress has been defined as the critical stress below which no flow is evidenced. Nevertheless, its value is determined by the experimental technique and the sensitivity of the rheometer employed. Although the existence of yield stress has been questioned by several authors since Barnes and Walter (1985) published their paper "The yield stress myth?", yield stress is still considered as "an engineering reality"(Astarita, 1990), and many methods are nowadays used to estimate its value.

Thus, because yield stress has been related to the concept of structural deformation and breakage of network bonds (Cheng, 1987), it has been characterized as static and as dynamic on a shear rate-shear stress diagram, using controlled-rate rheometers (Schramm, 1994). The yield stress in an undisturbed sample has been associated with the static yield stress, while the dynamic yield stress has been related to sample breakdown. In this last case, the yield stress has been often determined by the extrapolation of the experimental flow curve data, using different well-known rheological models (Herschel-Bulkley, Casson, Bingham, Vocadlo, etc). Another method for determining the yield stress in controlled-rate rheometers is the Vane method (Rao and Steffe, 1997), which uses star-shaped rotors. The yield stress is calculated from the maximum value of torque at low rotational speeds, performing a transient flow experiment.

The development of controlled-stress rheometers has favoured the occurrence of new methods to determine yield values, i. e. by plotting deformation versus shear stress or stress and estimating the yield value from the crossing point of two curve segments (Schramm, 1994).

Nevertheless, the use of these controlled-stress rheometers has also favoured the knowledge of the steady-state flow behaviour of complex fluids in a much wider range of shear rate/stress. This fact has confirmed that many complex materials show a zero-shear rate-limiting viscosity at shear rates lower than 10^{-4} s^{-1} (Barnes and Walters, 1987).

Occasionally, in colloidal systems that show wall slip phenomena, an apparent yield stress may be seen even at very low shear rates. This fact is due to the appearance of an intermediate false pseudo-Newtonian plateau at shear rates/stresses higher than those in which the true zero-shear rate-limiting viscosity is reached (Barnes, 1995). Obviously, this false pseudo-Newtonian plateau disappears by correcting the wall slip effects.

This talk reviews the traditional methods for measuring the yield stress of liquids and gives some insights in the current methods for measuring the steady-state flow curves of "yield stress liquids".

REFERENCES

Astarita, G. (1990) J Rheol, 34, 275-277.
Barnes, HA (1995) J Non-Newtonian Fluid Mech, 56, 221-251.
Barnes, HA, Walter, K (1985) Rheol Acta, 24, 323-326.
Cheng, DCH (1987) Int J Cosm Sci, 9, 151-191.
Rao, M:A., Steffe, J.F. (1997) Food Technol, 51, 50-52.
Schramm, G (1994) A Practical Approach to Rheology and Rheometry, Haake, Karlsruhe.

HIGH RATE TRANSIENT START-UP FLOW WITH GLASS FIBRES FILLED THERMOPLASTICS

Jean L. LEBLANC

University P. & M. Curie, Polymer Rheology and Processing, Paris, France

1. INTRODUCTION

Short glass fibres filled thermoplastics are used for the injection moulding of technical parts. For instance, in the automotive industry, a growing use is made of technical plastics, such as polyamides and polybutylene-terephthalate (PBT) with high levels (above 30%) of short glass fibres (typical length : 200-300 µm ; diameter : 10 µm). By nature however, such reinforced materials are prone to flow induced anisotropy that results in filler orientation during processing operations. This calls for a deeper knowledge of the flow properties of such materials, particularly in the high rate region and under transient flow conditions, typical of injection moulding. This paper describes transient start-up flow experiments conducted at high shear rate with a prototype injection rheometer.

2. EXPERIMENTAL

A prototype injection rheometer that gives access to the 10^4-10^5 s^{-1} shear rate region was specifically developed in our laboratory[1,2]. Essentially it consists in a barrel-and-die system, fixed between the platens of the injection machine (Figure 1), equipped with the suitable transducers for pressure, temperature and screw position, and a fast data acquisition system. Dies with various lengths and diameters are available, as well as several entrance cones.

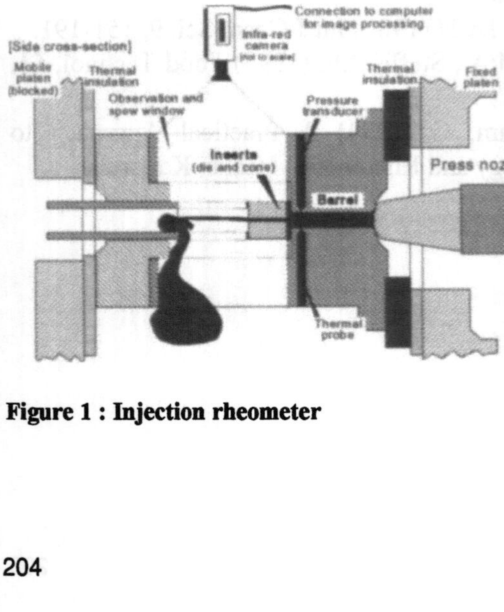

Figure 1 : Injection rheometer

The data acquisition system is designed to provide accurate recording of events with a high frequency, i.e 200 Hz, that is one set of data every 0.005 s. Very fast events can thus be observed, with in principle the possibility to analyse transient start-up flow, in actual injection moulding conditions.

Extensive injection shots experiments were performed with commercial PBT from BASF, either virgin [Ultradur® B4520] or filled with 30% glass fibres [Ultradur® B4300G6], using several dies with two diameters (1.0 and 1.5 mm), various L/D ratios (0 to 20) and different conical entrance angles (30, 60 and 90° , i.e. flat entrance). The injection machine temperature settings were such that the material temperature was 260±2°C at the nozzle.

3. RESULTS

When comparing the start-up flow of virgin and glass filled PBT, a major difference is observed. As illustrated in Figure 2, the virgin material exhibits smooth pressure growth traces until the plateau corresponding to steady flow is reached but the filled polymer systematically gives curves with pressure overshoot.

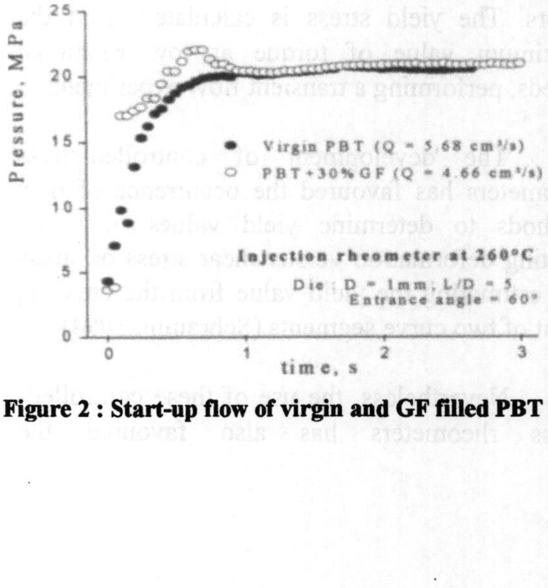

Figure 2 : Start-up flow of virgin and GF filled PBT

Die dimensions and entrance angle affect the intensity and shape of the overshoot, but the modification of the pressure build-up curve is clearly due to fibre presence. At 260°C, virgin PBT has a very high fluidity, does not exhibit a pressure overshoot and therefore it is unlikely that the traces exhibited by GF filled PBT reflect a true non-linear viscoelastic response of the polymer matrix.

4. MODELLING PRESSURE TRACES

With the simple hypothesis that the pressure build-up reflects a dissipation process associated with the ongoing orientation of the fibres along the developing flow lines in the converging path - the latter either imposed by the converging conical entrance, or self developed by the material - then the pressure traces with the glass fibres filled material could be considered as the simple superposition of two processes : a smooth pressure growth associated with the flow development of the virgin polymer, and a pulse pressure consumption as fibres, initially randomly oriented, align themselves along the developing flow paths.

It follows that pressure traces can be modelled by considering two superposed phenomena :

1. a steady growth function, reflecting the response of the polymer ;
2. a pulse function to represent the pressure consumption associated with fibres orientation

A suitable equation is thus :

$$P(t) = P_0 + \alpha \exp\left[-\frac{1}{2}\left(\frac{1}{\beta}\ln\frac{t}{t_m}\right)^2\right] + \left(P_{st} - P_0\right)\left[1 - \exp(-\chi t)\right]$$

where P_0 (MPa) is the initial pressure, α, β and t_m the parameters of the pulse function, P_{st} (MPa) the plateau pressure when steady flow is achieved, χ the rate parameter (s^{-1}) of the growth function and t (s) the time. The parameters of the pulse function have the following significance : α (MPa) corresponds to the maximum extra pressure consumption during the fibres orientation process, t_m (s) is the time to reach a, and b (pure number) the parameter describing the « width » of the pulse function.

The « pulse » pressure function reflects the dissipative process that is associated with fibres orientation when they align themselves along the developing flow paths, in the earlier period of the injection shot. As more and more fibres become oriented, cooperative alignment occurs that reduces the energy dissipation. In other words, the rate of energy dissipation is initially very high but decreases as flow develops in the die entrance region. In terms of local stress (or pressure), this means that a peak value is quickly reached, and is followed by a slow decrease as nearly all fibres are becoming oriented along flow lines.

Figure 3 shows how this model fits recorded start-up pressure traces. The peak pressure is well distinct at low applied output and tends to be masked by the polymer response as the flow rate increases.

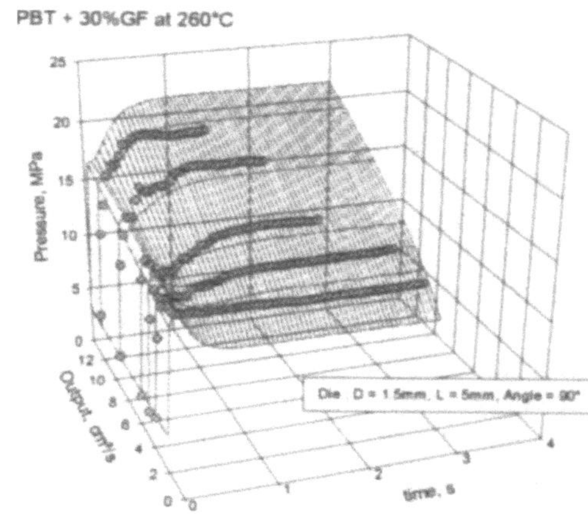

Figure 3 : Modelling start-up pressure traces

The fibres orientation process is thus very fast with peak energy (pressure) consumption occurring within a few 0.01 s after the piston-screw has been activated.

----

[1] Benoit-Cervantès G. (1997) Ph.D. Thesis, Univ. P.& M. Curie (Paris VI), France

[2] Leblanc J.L., Benoit-Cervantès G. (1997) « A few peculiar aspects of the high shear rheological evaluation of short glass fibres filled thermoplastics » - *Eurofillers '97* - Manchester, U.K., 8 -11 Sept. 1997 - proceedings. pp. 187 - 190

RHEOLOGY OF FREE-FLOWING GRANULAR MATERIALS

P. MARCHAL[1], A. BAROIS[2], V. TREGUER[2], D. TONDEUR[2], L. CHOPLIN[1]

[1]Centre de Génie Chimique des Milieux Rhéologiquement Complexes - GEMICO-ENSIC - 1, rue Grandville, BP 451, 54001 Nancy Cedex, France

[2]Laboratoire des Sciences du Génie Chimique - LSGC-CNRS, 1, rue Grandville, BP 451, 54001 Nancy, Cedex, France

1. INTRODUCTION

Rheological properties of granular materials are of paramount interest in many industrial operations involving mixing and handling. The determination of these properties implies suited characterisation methods elaboration. This paper aims at proposing such an experimental approach, based on statistical physic concepts and tested on model powders.

2. MATERIAL and METHOD

Our methodology consists in doing a transposition and an adaptation of existing theories and instrumental techniques on the basis of an analogy between molecular and granular systems.

Rheological properties of a powder are mainly dependent on the spatial arrangement of the grains resulting from sample conditioning. As a result, the measurements show a considerable variability due to the multiplicity of possible configurations.

Therefore, in order to determine significant characteristics of a powder, one need to obtain an average value that takes into account a representative set of spatial configurations.

Under the effect of Brownian motion (thermal energy), a molecular system explores spontaneously these various configurations and a time-average is enough to integrate them during the measurement. Such a property defines the so-called ergodicity of the system.

According to this remark, we have built a measurement cell able to generate the equivalent of a Brownian motion into the powder. On the basis of a Couette geometry, this cell includes a vane device plunging into a cylindrical cup. The bottom of the cylinder has been removed and replaced by an elastic membrane coupled with an electro-magnet, in order to produce vertical sinusoidal vibrations with an amplitude given by :

$$A = A_m \cos 2\pi f t \qquad (1)$$

where f is the frequency of the signal.

These vibrations generate a granular agitation quantifiable by the average kinetic energy of the grains :

$$<\frac{1}{2} m v^2> \propto \left(\frac{dA}{dt}\right)^2 \propto f^2 \qquad (2)$$

According to an analogy with statistical thermodynamics, the square of the vibration frequency can be considered to be proportional to the so-called granular temperature. Hence vibrations source acts as a kind of thermostat in equilibrium with the powder. Two advantages follow from that point. From an experimental point of view, the reproducibility of the measurements is significantly improved and slip and fracture phenomena are minimised. From a theoretical point of view, it is then possible to consider a modelisation of the rheological behaviour of the powder, based on the Maxwell-Boltzmann distribution within the scope of the statistical canonical ensemble.

3. EXPERIMENTAL RESULTS

The experiments have been performed on free-flowing model powders constituted of three series of glass beads with different diameters : 6µm ± 3µm, 60µm ± 15µm and 600µm ± 100µm. The rheological tests were carried out with a Rheometric Scientific SR200 controlled stress rheometer in both "steady-state" and "dynamic" regimes. The influence of the "granular temperature" on the rheological properties of the sample has been investigated by varying the vibration frequency of the membrane.

Figure 1 displays a frequency sweep test that shows the influence of the granular temperature on the viscoelastic properties of the powder. When this temperature is equal to zero (f = 0), the powder behaves like a viscoelastic solid, whereas it behaves like a viscoelastic fluid for a non-zero granular temperature (f ≠ 0). Such an evolution can be viewed as a kind of glass transition from a non-ergodic solid state to an ergodic liquid state.

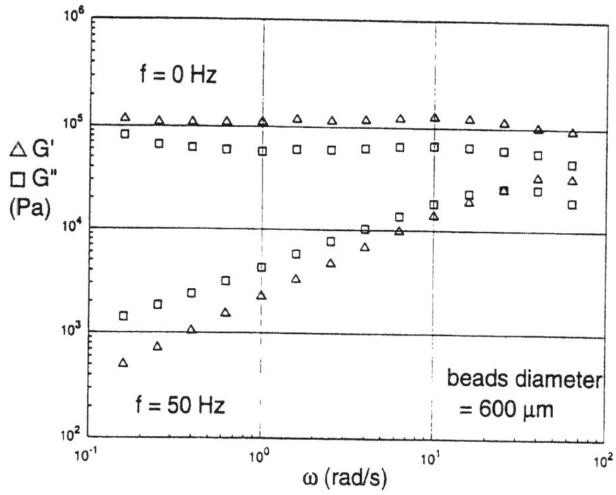

Figure 1. Granular temperature effect on viscoelastic behaviour of glass beads powder

Figure 2 shows the effect of the granular temperature on the steady-state viscosity of the powders for several beads diameter. The behaviour of the samples is typical of condensed fluids inasmuch the viscosity decreased by increasing the temperature. In the theoretical framework exposed in the previous paragraph and in particular on the basis of the "Theory of Rate Processes", it is possible to consider a modelisation of such a dependency and to show that :

$$\eta \propto \frac{1}{f} e^{\frac{A}{Bf^2}} \qquad (3)$$

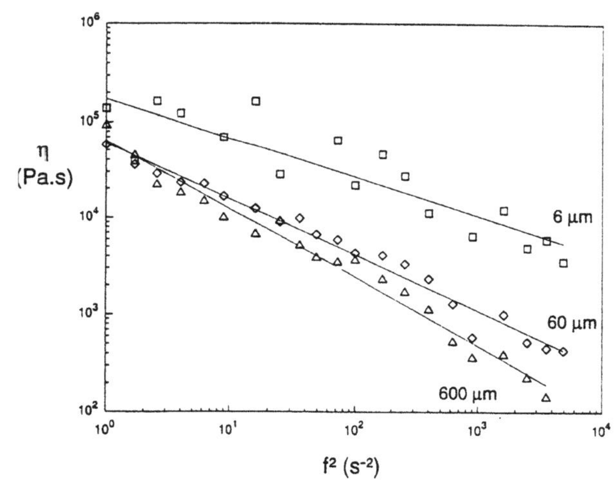

Figure 2. Granular temperature effect on steady- shear viscosity of glass beads powders

4. CONCLUSION

In support to the coherence of the experiments, the analogy between molecular and granular systems seems to be well-founded, at least from a qualitative point of view. Furthermore, at the expense of existing theories adaptation, it is possible to develop, in a first step, a crude model in order to explain and quantify the effect of the granular temperature on the rheological properties of free-flowing powders.

5. ACKNWLEDGEMENTS

The authors wish to acknowledge financial support provided by the "Centre National de la Recherche Scientifique (CNRS)" and the sponsoring companies of the GEMICO-ENSIC.

6. REFERENCES

Astarita G, Ocone R (1994) Ind. Eng. Chem. Res. 33 : 2280-2287
Campbell CS (1190) Annu. Rev. Fluid Mech. 22 : 57-92
Edwards SF (1990) Rheol. Acta. 29 : 493-499
Haff PK (1983) J. Fluid Mech. 134 : 401-430
Jenkins JT, Savage SB (1983) J. Fluid Mech. 130 : 187-202
Laidler J, Eyring H (1941) The Theory of Rate Processes. McGraw-Hill, New-york an London
Laroche C, Douady S, Fauve S (1989) J. Phys. 50 : 699-706
Mehta A, Barker GC (1991) The Am. Phys. Soc. 67(3) : 394-397
Pusey PN, Van Megen W (1987) Phys. Rev. Lett. 59 (18) : 2083-2086
Zik O, Stavans J, Rabin Y (1192) Europhys. Lett. 17(4) : 315-319

THE MIXING OF 'YIELD STRESS' FLUIDS

Dikshika Patel

Unilever Research Port Sunlight, UK

1. INTRODUCTION

Mixing is one of the most fundamental processes in the chemical and biochemical industry. It is necessary to reduce the degree of non-uniformity and to minimise gradients of properties within a system e.g. concentration, viscosity, temperature, pH, etc., and is achieved by material conveyance from one region to another on both a micro and macro scale. In a typical batch process using stirred tanks, the criteria for good bulk mixing is perhaps obvious and is easily achieved for low viscosity liquids. However, in highly viscous shear thinning fluids, bulk flow is necessary to avoid the formation of stagnant or near stagnant regions that would reduce optimal productivity or dramatically increase batch times. The agitation of fluids that are highly shear-thinning and exhibit an apparent yield stress can lead to an important mixing problem; cavern formation. Many investigations [1, 6-7] have visibly shown caverns as the well-mixed region around an impeller where high shear rates predominate. Beyond the cavern boundary, the fluid is assumed to stagnant in zones of very low shear rate and at the cavern boundary, the shear stress is assumed to equal the apparent yield stress of the fluid.

Cavern sizes have been successfully predicted for homogenous single and two-phase 'model' systems from laboratory up to industrial scale [1-2, 5]. The so-called cavern model [1, 6], which uses the sum of torque contributions on the agitator from the yield stress acting on a right circular cylinder shaped cavern, mixing characteristics of the system and important physical properties of the fluid is well known. The following equation [1] is commonly used to predict cavern sizes and agitator speeds required to prevent cavern formation.

$$\left(\frac{D_c}{D}\right)^3 = \left(\frac{1.36\,Po}{\pi^2}\right)\left(\frac{\rho N^2 D^2}{\tau_y}\right)$$ applicable when

$D \leq D_c \leq T$

For this particular application, the yield stress, τ_y, of a fluid is identified as a well defined quantity. During development of the model, τ_y was determined by fitting rheological data to the Casson model over a restricted shear-rate range (below $0.2s^{-1}$), which was considered to represent the residual stress on removal of the shear rate.

2. INFLUENCE OF YIELD STRESS ON CAVERN SIZE PREDICTION

Although the dimensionless cavern diameter in the cavern model is only dependant on the cubed root of the yield stress, studies have shown that for complex two-phase systems where rheological measurements are particularly difficult, the model is sensitive to the yield stress of the fluid.

Mixing experiments to compare experimentally determined and predicted D_c/D were carried out on a filamentous fermentation broth of *Penicillium chrysogenum.* A simple flow visualisation technique was used to detect the speed at which the fluid motion first reached the wall of a standard transparent mixing vessel (T = 0.2m). At this point, $D_c = T$. A standard Rushton turbine (D = 0.1m) was used for agitation, which had a measured Po of 2.7 when the cavern reached the wall. The broth density ρ = 1057g/l.

The complex morphology of the broth prevented the use of 'conventional' rheological techniques. Two methods were used to determine the yield stress. The first was fitting the Casson model to shear-stress/ shear-rate data measured using a calibrated disc turbine geometry [5]. The shear rate range was $0.95\text{-}21s^{-1}$, i.e. not the same as that used previously [1]. The second was a direct method using the rotating- vane technique reported as a superior method for suspensions [3,4]. This is because yielding occurs within the fluid itself avoiding slip at a measuring surface boundary. This method is believed to describe both static stress that must be exceeded for flow to occur, and the dynamic yield stress where flow occurs below the former at relatively high shear rates once the structure has been broken down. Using all the measured parameters and incorporating into the cavern model, Table 1 shows a summary of the results.

Method of τ_y determination	Geometry	τ_y (Pa)	D_c/D	Comments
Casson model (indirect)	disc turbine	1.1	2.9	3 times overprediction of τ_y D_c/D oversized by 45%
rotating vane [4] (indirect)	4 bladed vane	5.7±2.3	1.5-2.0	~ 2 times underprediction of average τ_y, good prediction of D_c/D

Table 1: Experimental D_c/D and τ_y values using two methods for yield-stress determination. For motion at the vessel wall $D_c = T$, $D_c/D = 2.0$ and predicted $\tau_y = 3.3$Pa.

Significantly different yield stress values were obtained using the two methods, the vane technique producing yield stress values five times greater than the curve fitting. With all others parameters being constant, subsequent differences are observed for cavern diameters and yield stress estimations.

3. DISCUSSION

The importance of using a correct definition of yield stress has been highlighted when applying a well established mixing model to a real fermentation broth where rheological characteristics are difficult to define let alone measure. This can have important optimisation and design implications not only in mixing but other processing applications where a yield stress is an engineering reality.

The vane yield stress was significantly greater than that determined using the Casson model. The difference was such that it did have an effect on the cavern diameter; the experimental and predicted *Dc/D* now showing good agreement. This points to the static yield stress being the relevant parameter to use in the cavern model rather than the Casson yield stress which merely provides mathematical constants to the rheological data. The measurement made using the vane technique may be describing an actual broth property corresponding to the yield stress required to overcome a startup stress, which for cavern prediction, is more suitable to indicate the mixing quality. It is important to understand the input made by the system to the 'yielding'. For these types of broths, the cells and their disentanglement may confer this yield stress property, which is overlooked in simple curve fitting.

4. NOMENCLATURE

D_c	cavern diameter (m)
T	tank diameter (m)
ρ	density (kgm^{-3})
D	impeller diameter (m)
Po	Power number (-)
τ_y	yield stress (Pa)
N	impeller speed (rps)

5. REFERENCES

[1] Elson, T. P., Cheesman, D. J., Nienow, A. W. (1986). *Chem. Eng. Sci.* 41. No.10. 2555-2562.

[2] Etchells, A., Ford, W. N. & Short, D. R. G. (1987). Mixing of Bingham plastics on an industrial scale. *Fluid Mixing III. IChemE Symposium series* no.108. Sept 8-10th (Bradford, UK).

[3] Leong-Poi, L., Allen, D. G. (1992). Direct measurement of the yield stress of filamentous fermentation broths with the rotating vane technique. *Biotech. Bioeng.*, 40, 403-412.

[4] Nguyen, Q. D., Boger, D. V. (1983). Direct yield stress measurements with the vane method. *J. Rheol.*, 29, 335-347.

[5] Patel, D. (1997). Mixing and rheological studies of real and simulated fermentation broths. PhD Thesis. The University of Birmingham.

[6] Solomon, J ., Elson, T. P., Nienow, A. W. & Pace, G. W. (1981). Cavern sizes in agitated fluids with a yield stress. *Chem. Eng. Comm.* 11, 143-164.

[7] Wichterle, K., Wein, O. (1979). Threshold of mixing of non-Newtonian fluids. *Inst. Chem. Eng.* 21(1), 116-120.

A NON-LINEAR VISCOELASTIC DESCRIPTION OF THE MOONEY AND THE MOONEY-RELAXATION TEST

PAUL A.M. STEEMAN and JO H.M. PALMEN

DSM Research BV, P.O. Box 18, 6160 MD Geleen, The Netherlands

1. INTRODUCTION

In the rubber industry the standardised Mooney test is used on a routine basis for quantifying the flow behaviour and the Mooney-relaxation test on a *non*-routine basis for characterizing the elasticity of non-vulcanised rubbery materials. These tests involve the measurement of the shear stress, during start-up and after the cessation of steady shear flow, using a viscosimeter equipped with a closed chamber measurement cell which includes a rotor embedded in the material. Because of the very large deformations applied during the Mooney test, typically the deformation amounts to several hundred shear units, the results are highly non-linear viscoelastic. Therefore, the relation between the Mooney results and the linear viscoelastic material functions, known to be a sensitive indicator for details of the molecular structure, is not straightforward.

Alternatively, dynamic mechanical analysis, which gives direct access to the linear viscoelastic material functions, is widely used for the rheological characterisation of polymeric melts and rubbers. It is the purpose of this contribution to elucidate the relations between both approaches.

2. MODELS

It was investigated whether the Mooney tests, both the stress built-up (ML1+4) and the stress relaxation part (MRR), can be described with a non-linear viscoelastic constitutive equation. Wagner modified the linear viscoelastic theory by introducing the so-called *damping function* h(γ), which is a measure for the deformation sensitivity of the material, in the integral kernel of the Boltzmann integral [see e.g. Larson 1988]:

$$\sigma(t)=\int_{-\infty}^{t} m(t-t')\cdot h(\gamma(t)-\gamma(t'))\cdot[\gamma(t)-\gamma(t')]dt'$$

in which m(t) is the so-called *memory function*, which is equal to the time derivative of the stress relaxation modulus G(t).

The memory function can be determined from the stress relaxation after the application of a step strain. Also, oscillatory experiments may be performed from which the relaxation time spectrum can be extracted. With this spectrum the stress relaxation modulus is easily calculated. Spectrum calculations were performed with the program IRIS™. When the memory function is known, the damping function can be extracted from the stress build-up during steady shear flow [Wagner 1978].

3. MATERIALS AND EXPERIMENTS

The (non-) linear viscoelastic response at 125°C of three non-vulcanised EPDM-materials with nearly equal Mooney viscosity, but with varying molecular strucuture, was investigated. First, dynamic mechanical analysis was performed using a Rheometric Scientific DSR 200 controlled stress rheometer, from which the relaxation time spectra and the memory function were calculated. Next, stress-built up experiments were performed with a cone and plate system on a Rheometric Scientific RMS800 controlled strain rheometer, from which the damping function h(γ) could be extracted. Figure 1 shows the results of the dynamic mechanical analysis.

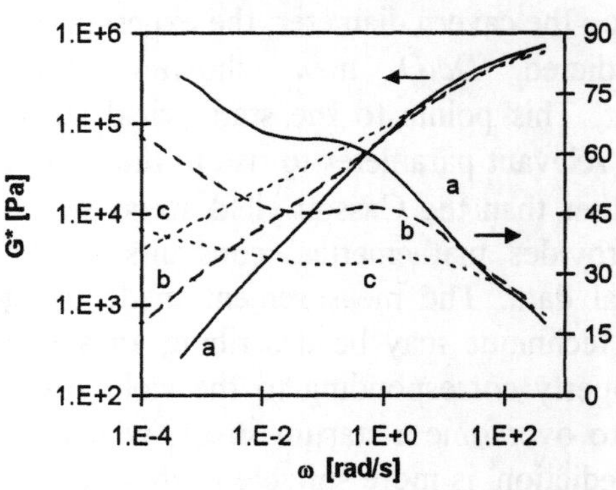

The curves of the phase angle δ clearly display the enormous increase in elasticity of the material from sample *a* to sample *c*, while the Mooney viscosity, estimated from the dynamic modulus G* at ω≈1 rad/s (Cox-Merz rule), is about equal.

Figure 2 shows the damping functions h(γ) for the three materials as determined from the stress built-up experiments. From these results it can be observed that the most elastic material *c*, is also the least deformation sensitive.

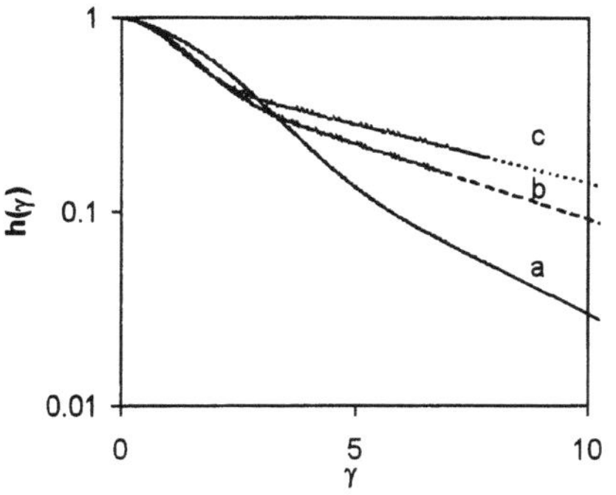

4. CALCULATIONS

The Mooney viscosity (ML1+4 at 125C) and the (logarithmic) Mooney stress relaxation rate (MRR) between 1.2 and 6 seconds after rotor stop were calculated by numerically integrating over the flow field in the Mooney test cell via a non-linear visco-elastic extension of the method used by Nakajima et al. They subdivided the Mooney test cell (Fig.3) into three sections, each of which with a well defined flow field: above and below the rotor (I), the edge of the rotor (II) and the corners (III), the latter regions being disregarded in the calculations. In each section the Wagner constitutive model was used to calculate the stresses during steady shear (stress-built up) and after the cessation of flow (stress relaxation).

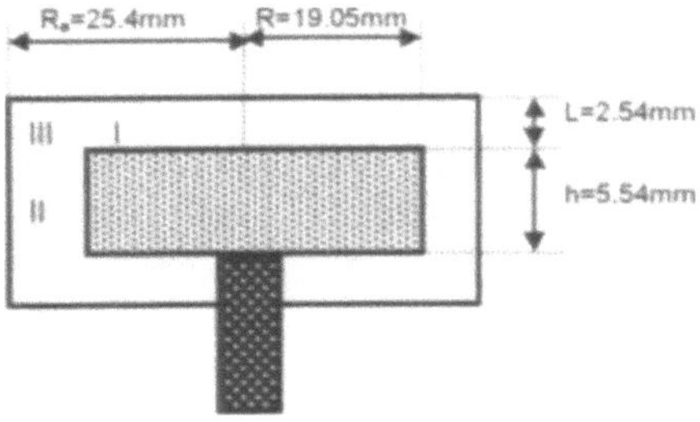

Figure 4 shows the calculated stress built-up curves for the three materials. These curves are qualitatively similar to experimental results. After rotor start the stress increases sharply, passes through a maximum, after which it levels off at a lower plateau value. Table 1 contains some selected results of the calculations in comparison with the experimental data. While the stress relaxation rates (MRR) are in close agreement, the Mooney viscosity is underestimated up to 25%, which is probably due to the simplified approximation of the flow field used, and to possible non-ideal flow effects in the Mooney cell, such as secondary flows.

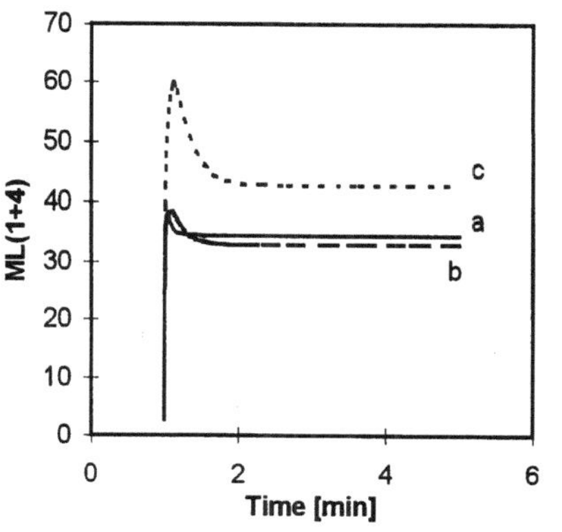

Mat.	ML(1+4) exp	ML(1+4) calc	MRR exp	MRR calc
a	46.7	34.6	0.92	0.95
b	44.[illegible]	33.0	0.64	0.63
c	46.2	42.7	0.46	0.47

5. CONCLUSIONS

Using the material functions determined from dynamic mechanical analysis and from steady shear flow experiments, the Mooney test was successfully modelled with the Wagner constitutive model. The measured and calculated Mooney relaxation rates are in close agreement, while the calculated Mooney viscosity is up to 25% lower than the measured value. These results show that the Mooney test and standard dynamic mechanical analysis are intimately related. However, because of the complicated constitutive equations involved and the need for numerical integration over the Mooney cell, its has become clear that the relations between both results are not simple. Amongst others, they depend on the exact experimental details and conditions during the Mooney measurement such as sample cell geometry and the rotation speed of the rotor.

6. REFERENCES

Larson R.G. (1988), 'Constitutive equations for polymer melts and solutions', Butterworths, New York.

Wagner M.H. (1978), J. Non-Newt. Fluid Mech., **4**, p39-55.

Nakajima N. and Harrell E.R. (1979), Rubber Chem. Technol., **52**:1, p9-19.

INTERDEPENDENCE OF RHEOLOGICAL PARAMETERS AND PRINTING QUALITY IN TEXTILE PRINTING

Dr.Sonja Šostar[1], Dr.Reinhold Schneider[2]

[1]University of Maribor, Faculty of Mechanical Engineering, Institute für Textile Chemistry, Smetanova 17, 2000 Maribor, Slovenia
[2]Institut für Textilchemie der Deutschen Institute für Textil- und Faserforschung Stuttgart, Körschtalstraße 26, 73770 Denkendorf, Germany

1. INTRODUCTION
The use of modified polysaccharides as thickening agents in reactive printing constitutes an interesting alternative to the alginates used up to now. In our research the technical printing properties of these thickeners are presented in succession and categorised according to the rheological and viscoelastic profile of the printing pastes.

2. EXPERIMENTAL
2.1.Substrate
The thickeners used were sodium alginate, non-substituted guar gum and carboxymethylised guar gum .

Table 1: The used thickeners

Thickener	DS	[g thickener/kg printing paste]
Sodium alginate, medium viscosity (ALMV)	0	24
Guar gum, high viscosity (CMG)	1.1	51
Guar gum, low viscosity (CMGN)	0.18	75
Guar gum, high viscosity (CMGH)	0.18	39
Guar gum, low viscosity (DPGN)	0	82,5
Guar gum, high viscosity (DPGH)	0	34,5

All printing pastes were calibrated to the constant viscosity of η=7140±400 mPa s at the shear rate of D =10s^{-1} .

2.2.Methods
The viscometer Rheometrics Fluids Spectrometer RFS II (firm Rheometrics) with a cone-plate measuring system was used to perform the rheological measurements. The sharpness of outline was measured with the picture analysis system Quantimet 970 (firm Cambridge Instruments). Dye penetration and the depth of colour were determined by remission measurements using the colourimetric apparatus Texflach DC 3881 (firm Datacolor).

3.RESULTS AND DISCUSSION
3.1. Rheological properties of thickeners
The shear sensitivity of printing pastes prepared from different thickeners varies depending on the solid content (Fig.1).

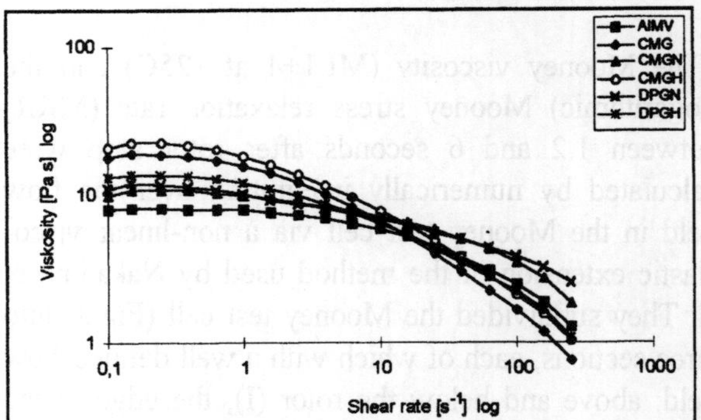

Fig. 1: Viscosity curves of the thickeners

In the first Newtonian range the high viscosity guar thickeners CMG, CMGH and DPGH show a strong decrease of viscosity under the influence of the increasing shear rate in comparison with thickeners with a high solid content and with alginate thickeners. In contrast, the low viscosity guar thickeners (CMGN, DPGN) and the alginate thickener (AlMV) have a Newtonian character, which is evident from their constant viscosity. The transition from the first Newtonian range into the range of shear sensitivity depends on the thickener's solid content. In case of low viscosity thickeners the transition takes place at higher shear rates than it does in case of high viscosity thickeners. The solid content also affects the viscoelastic properties of printing pastes (Fig.2). The loss factor tan δ of high viscosity shear-sensitive thickeners (CMG, CMGH, DPGH) is lower in the entire frequency range than it is in case of all other discussed thickeners. This is the result of higher elasticity caused by the higher molecular mass of the polymers.

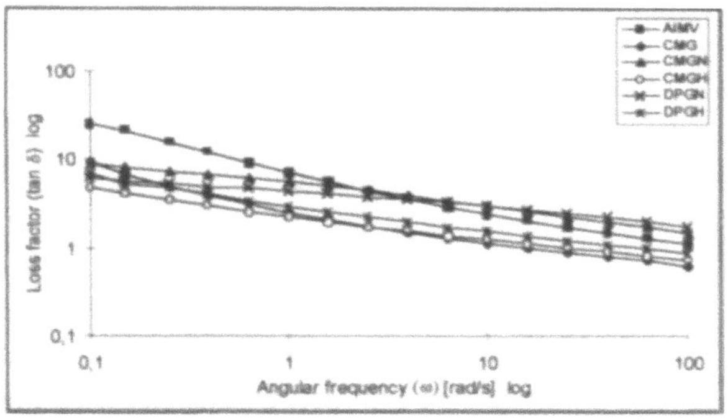

Fig. 2: Loss factor tan δ of the thickener

Irrespective of the thickener used, the loss factor tan δ decreases with the increasing angular frequency, which means that elasticity is increasing. Consequently this thickener shows the lowest elasticity at high shear rates. It is evident from these results that DPGN and CMGN are alternatives to the alginate thickener.

3.2. Sharpness of outline

Printing pastes with low viscosity as well as high paste add-ons prevent the reproduction of very fine patterns, whereas a quick rearrange of high viscosity after shear deformation will support a detailed reproduction of the patterns. As represented in Table 2 the high viscosity and elastic guar gums effect very fine lines, which is due to the low paste add-on.

Table 2: Sharpness of outlines

Thickener	Squeegee ϕ[mm]	Line width [mm] (Original = 0,32 mm)	
		v=5m/min	v=20m/min
AIMV	10	0,582	0,411
	15	0,622	0,532
CMG	10	0,404	0,330
	15	0,413	0,374
CMGN	10	0,520	0,405
	15	0,596	0,468
CMGH	10	0,395	0,078
	15	0,424	0,375
DPGN	10	0,466	0,355
	15	0,572	0,395
DPGH	10	0,409	0,349
	15	0,432	0,345

It is evident from the results that the low viscosity DPGN and CMGN guar gums bring about comparable or better printing results than alginates, whereas high viscosity systems (low solid content) effect the finest lines.

3.3. Penetration and depth of colour

As shown in Figure 3, the penetration of thickeners with a low solid content (CMG, CMGH and DPGH) into the fabric is much weaker than that of thickeners with a high solid content. The reason for the weaker penetration of these systems is the result of the smaller paste transfer and the higher elasticity. Thickeners with a high solid content (AIMV, CMGN, DPGN) yield good penetration, which is caused by the non-elastic behaviour of the printing paste (high loss factor tan δ). Due to the low elasticity and the high solid content of alginates and guar gum a high paste transfer is obtained resulting in high depth of hue and penetration.

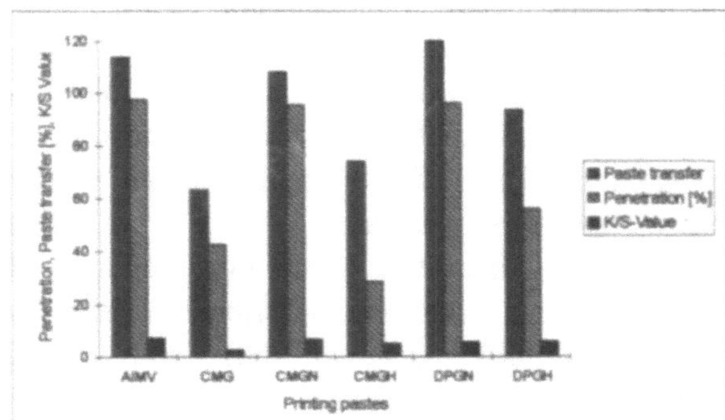

Fig. 3: Penetration, paste add-on and K/S values

4. CONCLUSION

The rheological measurements have shown that the viscosity and the elasticity of thickeners depend on the solid content. Viscosity properties prevail in alginates and guar gum with a high solid content, and elasticity prevails in guar gum with a low solid content. Due to the dominating viscous character, thickeners with a high solid content yield better penetration and depth of colour than thickeners with a low solid content. High molecular mass of the polymers (low solid content) will result in high elasticity which will reduce the quantity of paste add-on as well as penetration. High elasticity will also affect very fine lines in patterns.

REFERENCES

Šostar S., Universität Maribor, Dissertation, 1995.
Šostar S.; Schneider R.: IFVTCC, Wien 1997, 244-246.
Wielinga W.C.: Melliand Textilberichte, 65 (1986) 45-55.
Schneider R. et al. Textilveredlung 30 (1995), 60-64.
Häcker D., Universität Stuttgart, Dissertation, 1995
Šostar S.; Schneider R.: Textilveredlung 33 (1998), 47-51.

DYNAMICS OF POINT DEFECTS IN NEMATIC LIQUID CRYSTALS

J. BAJC and S. ŽUMER

Department of Physics, University of Ljubljana, Jadranska 19, 1000 Ljubljana, Slovenia.

1. INTRODUCTION

The presence of topological defects in liquid crystals is know since the very discovery of this interesting state of matter. However, the number of studies considering the dynamics of defects started to increase only recently [Ericksen (1996), Guidone Peroli (1998)]. One of the most interesting features of liquid crystal dynamics is the coupling between the director reorientation and the mass flow ("back flow" effect). It results from the anisotropy of the liquid crystal molecules and leads to complicated viscous phenomena. Even in the case of the simplest liquid crystal phase (uniaxial nematic), there are five independent viscosities, entering the set of nematodynamic equations, described in detail by Ericksen (1960), Leslie (1966), and Parodi (1970). Recently, dynamics of defects in a cylindrical tube filled with a nematic liquid crystal was studied experimentally for the case of homeotropic anchoring (director prefers perpendicular alignment at the cavity wall) [Hillig (1997)]. Unsolved problems in the theory of defect dynamics and new experiments encouraged us to theoretically examine the annihilation of two point defects with opposite topological charge.

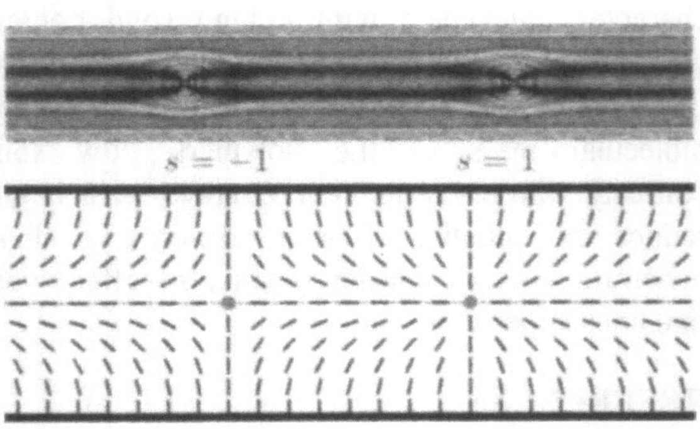

Figure 1 Schematic presentation of the director field (bottom) and simulation of polarizing micro-scope photographs (top). Defects are easy to spot.

We considered a uniaxial nematic liquid crystal confined to a cylinder. Strong homeotropic anchoring is supposed, therefore the most stable structure is the so called escaped radial structure with point defects (Figure 1). In each domain of the structure the director field takes on the escaped radial structure. Equal probability of the escapes "up" or "down" the axis of the tube results in creation of domains with escapes in opposite directions. At the junction of two domains a point defect with a topological charge 1 or -1 is created. Presence of defects increases free energy, so the structures with many domains and point defects are unstable. We focused on two isolated defects of opposite topological charge. It is clear that the defects attract each other as their annihilation lowers the total free energy. The motion of such a pair is studied numerically in the framework of Leslie's nematodynamics. A simplified model with a single effective rotational viscosity γ_1 and equal elastic constant K is used.

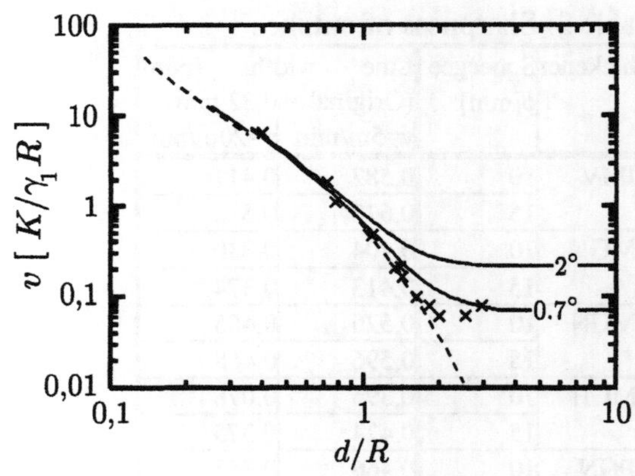

Figure 2 Velocity as a function of distance between defects for small deviations from homeotropic anch-oring. Velocities for tilts $\Delta\Psi$=2° and 0.7° are denoted by full line, for homeotropic anchoring by dashed line, and the experimental ones are denoted by "×" [Hillig and Saupe (1997)]. Standard nematic 5CB in tubes with diameter 300 μm and 400μm was used.

2. RESULTS

In order to incorporate the dynamics of defects into simulation a new numerical method is introduced. The method is based on quasi-stationary director field in the vicinity of a moving defect. Each numerical step consists of two substeps. First explicit scheme is used to calculate the change of the director field after a tiny time step. Each defect moves a bit along with the surrounding structure. Next the structure around the defect is "pulled" back to the original position. The procedure is repeated until quasi-stationary state is reached.

Simulations, obtained with the above procedure, are in reasonable agreement with preliminary experimental results by [Hillig and Saupe (1997)]. In addition, slightly modified boundary conditions result in very good qualitative agreement with the observed annihilation (Figure 1); when defects are more than about $2R$ apart they approach each other with constant velocity and when the distance decreases below $2R$ the velocity of defects dramatically increases. We show that first type of behavior can be ascribed to irregularities on the wall rather than to interaction between defects, while the increasing velocity at smaller distances clearly originates from interaction between the defects.

3. REFERENCES

Guidone Peroli G. (1998) Equilibrium and Motion of Defects in Nematic Liquid Crystals, dissertation.

Ericksen J. L. (1996) Static Theory of Point Defects in Nematic Liquid Crystals, in M. M. Carroll and M. Heyes (eds), Nonlinear Effects in Fluids and Solids (Plenum Press, New York.

Ericksen J. L. (1960), Archs. ration. Mech. Analysis 4: 231.

G. Hillig and A. Saupe, private communication.

Leslie F. M. (1966) Quart. J. Mech. appl. Math. 19: 357; (1968) Arch. ration. Mech. Analysis 28: 265 (1968);

Parodi O (1970) J. Phys. France 31: 581.

THE DIRECTOR TUMBLING-ALIGNMENT TRANSITION IN SHEAR FLOWS OF NEMATIC LIQUID CRYSTAL POLYMERS, MODELLED BY CONTINUUM THEORY

P.A. CAMPOS, A.F. MARTINS

Dept. Ciência dos Materiais and CENIMAT, Faculdade de Ciências e Tecnologia, Universidade Nova de Lisboa, P-2825 Monte da Caparica, Portugal [e-mail: asfm@mail.fct.unl.pt]

1. INTRODUCTION

Constitutive equations for nematic liquid crystalline polymers (LCP's) proposed by Martins (1994, 1996) have been shown to model the observed behaviour of the three flow functions $\eta(\dot\gamma)$, $N_1(\dot\gamma)$ and $N_2(\dot\gamma)$ over the whole range of shear rates $\dot\gamma$ experimentally probed, including the sign changes on the first normal stress difference, observed in some materials (Kiss and Porter, 1980, and Magda et al., 1991).

In this work, the theory of Martins (1994) is applied to the study of the director dynamics of LCP's in simple shear flows. For low molecular mass nematic LC's, alignment or tumbling are predicted by the continuum theory of Leslie (1968), according to whether α_3/α_2 is positive or negative, respectively.

In the case of polymeric LC's, there is evidence that a more complex situation occurs, including a transition from director tumbling to flow alignment at high shear rates (Hongladarom et al., 1991). It will be shown below that the theory of Martins (1994) allows for such a transition and predicts a richer variety of flow behaviours, depending on the Leslie viscosities, the dimensionless quantity $\tau\dot\gamma$ and the parameter ε measuring the degree of non-affinity of the motion.

2. DIRECTOR FIELD EQUATIONS

Representing the velocity, the director and the non equilibrium part of the intrinsic director body force vectors on XY plane by:

$$v = (\dot\gamma y, 0), \quad n = (\cos\theta, \sin\theta) \quad and \quad g' = -\gamma_2\dot\gamma(r_1, r_2)$$

and, neglecting inertia, the director field behaviour in simple shear flow is described by the equations:

$$r_1 \sin\theta - r_2 \cos\theta = \frac{1}{E_K}\frac{\partial^2\theta}{\partial(y/L)^2}$$

$$r_1 + w\frac{\partial r_1}{\partial(\dot\gamma t)} - w\frac{1+\varepsilon}{2}r_2 = (2b+1)\sin\theta\frac{\partial\theta}{\partial(\dot\gamma t)} + (b+1)\sin\theta$$

$$r_2 + w\frac{\partial r_2}{\partial(\dot\gamma t)} + w\frac{1-\varepsilon}{2}r_1 = -(2b+1)\cos\theta\frac{\partial\theta}{\partial(\dot\gamma t)} - b\cos\theta$$

where influent parameters are $\varepsilon \in [-1,1]$, $w = \tau\dot\gamma$, $b = \alpha_3/(-\gamma_2)$ and $E_K = (-\gamma_2)\dot\gamma L^2/K$. Other symbols have their usual meaning (see Martins, 1994).

Leslie equations are recovered with $\tau = 0$. This time constant is negligible for low molecular mass nematics.

Leslie model implies director alignment at a certain angle $\theta_F = (1/2)\arccos(2b+1)$, independent of shear rate, when $b\in[-1,0]$. Martins model gives a flow-alignment angle that decreases with shear rate, approaching the flow direction:

$$\theta_F = \frac{1}{2}\arccos\frac{2b+1}{\sqrt{1+4c^2w^2}} - \frac{1}{2}\arctan 2cw, \qquad c = \frac{1-(2b+1)\varepsilon}{4}$$

and alignment is *possible* not only when $b\in[-1,0]$, but also for b outside this interval, when shear rate exceeds a critical value:

$$w > w_c = \sqrt{b(b+1)}/|c|$$

Relevant director regimes are discussed below in the assumption of $E_K \approx \infty$ (ensuring spatial uniformity: $\nabla\theta = 0$).

3. CASE $b\in[-1,0]$: ALIGNMENT

The evolution $\theta(\dot\gamma t)$ predicted by Leslie approach is determined by the parameter b. Martins model predictions depend upon b, ε and w; high shear rates induce low values of θ_F and promote initial oscillations (Fig.1).

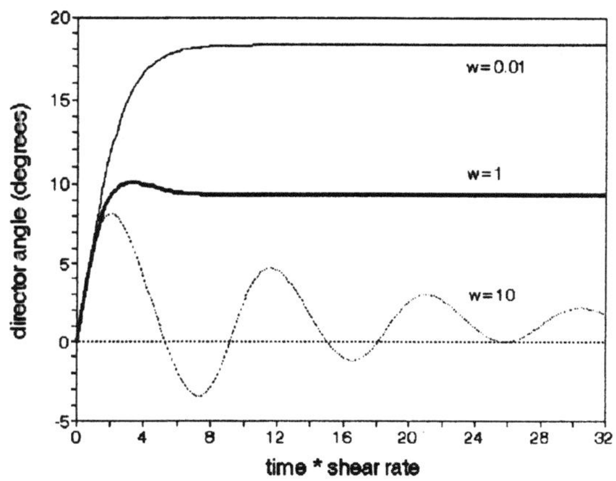

Fig.1 - *Examples of flow alignment regimes, for b = -0.1 and ε=0.1*

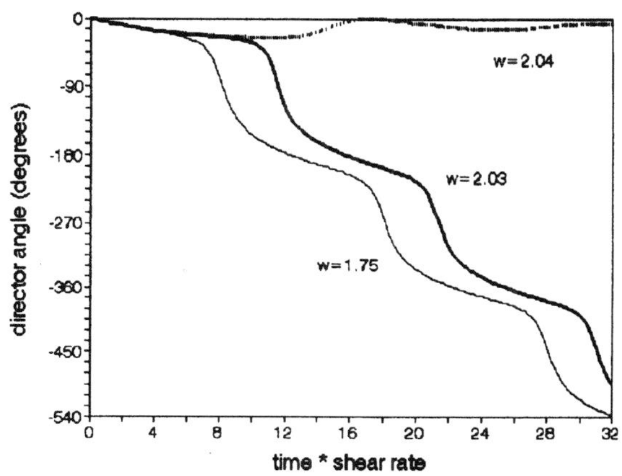

Fig.2 - *Transition tumbling-alignment, when b=+0.1 and ε=0.2; critical value is w_c=1.75 and transition occurs at w_t=2.04*

4. CASE $b \notin [-1,0]$ and $w < w_c$: TUMBLING

Both models predict essentially the same regime, except for fine details of the $\theta(\dot{\gamma} t)$ dependence upon not only b, but also ε and w.

5. CASE $b \notin [-1,0]$ and $w > w_c$: TRANSITION TUMBLING - ALIGNMENT

Numerical simulation on Martins equations has shown that post-critical shear rates can provide flow alignment regimes, typically with intense initial oscillations. When the parameter w exceeds a "transition" value $w_t \geq w_c$, director tumbling is arrested and alignment at the angle θ_F is favoured (Fig.2).

To each pair (b,ε) corresponds a different transition value w_t, and three distinct cases are possible:

I: $w_t = \infty$: *persistent tumbling*
for high values of b and/or ε ; examples:
b = 0.05, ε = +0.8 ... +1
b = 0.1, ε = +0.6 ... +1

II: $w_c < w_t < \infty$: *tumbling* or *alignment*
for moderate b and/or ε; examples:
b = 0.05, ε = +0.3 ... +0.7
b = 0.1, ε = -0.2 ... +0.5

III: $w_t = w_c$: *alignment*
for low values of b and/or ε; examples:
b = 0.01, ε = -1 ... +0.9
b = 0.05, ε = -1 ... +0.3

Numerical simulations have shown that the frequency of the director oscillations preceding its alignment by the flow is roughly proportional to the shear rate $\dot{\gamma}$. For a given b, the (near exponential) decay of the oscillations amplitude is independent of $\dot{\gamma}$. In other words, the time required for the director to stabilize is independent of the shear rate, but the number of oscillations performed during this time increases with $\dot{\gamma}$. In this context, the parameter ε does not play any significant role.

6. REFERENCES

Hongladarom K et al. (1991) *Macromol.* 26, 772, and references therein
Kiss G, Porter RS (1980) *J. Polymer Sci. (Phys.)* 18, 361.
Leslie FM (1968) *Arch. Rat. Mech. Anal.* 28, 265.
Magda JJ et al. (1991) *Macromol.* 24, 4460.
Martins AF (1994) in: *Liquid Crystal. Polymers* ed. C. Carfagna, Pergamon Press, U.K., p.153-161; *ibid* (1996), in: *Proc. XIIth Int. Cong. Rheol.*, ed. A. Ait-Kadi et al., U.Laval, Québec, Canada, p.277-278.

ACKNOWLEDGMENT

Work partly supported by the EU under TMR contract FMRX-CT96-0003, and PRAXIS XXI Programme contract 2/2.1/MAT/380/94.

TEMPERATURE DEPENDENCE OF THE RHEOLOGICAL PROPERTIES OF ACETOXYPROPYLCELLULOSE IN THE THERMOTROPIC CHIRAL NEMATIC PHASE

M.T. CIDADE[1], C.R. LEAL[1], M.H. GODINHO[1] and P. NAVARD[2]

[1]Departamento de Ciência dos Materiais and CENIMAT, Faculdade de Ciências e Tecnologia, Universidade Nova de Lisboa, P-2825 Monte de Caparica, Portugal
[2]Centre de Mise en Forme des Materiaux, École des Mines de Paris, URA CNRS 1374, BP 207, 06904 Sophia-Antipolis, France

1. INTRODUCTION

The rheological behaviour of two thermotropic liquid crystalline (LC) samples of acetoxypropylcellulose (APC) with different molecular weights, at T=120°C, has already been reported by us (Cidade et al., 1995). In this work we present the temperature dependence of the rheological properties of the same two samples.

2. EXPERIMENTAL RESULTS

The synthesis and characterisation of the polymers used in this work have been described elsewhere (Cidade et al., 1995), along with details on the experimental procedure used in the rheological measurements, which were performed in a rotational rheometer (Rheometrics - RMS 800), with a cone and plate geometry.

The observed shear rate dependencies of $\eta(\dot\gamma)$ and $N_1(\dot\gamma)$ for APC 94000 (APC is labelled after its average molecular weight) at 120, 130 and 140 °C, are displayed in figures 1 and 2, along with the theoretical curves given by the following equations:

$$\eta(\dot\gamma) = \eta_0 \exp(-\tau_0 \dot\gamma) + [a_0 (\tau\dot\gamma)^2 + b_0 \tau\dot\gamma + c_0] \cdot [1+(1-\varepsilon^2) \tau^2 \dot\gamma^2]^{-1}$$

$$N_1(\dot\gamma) = \sigma_{11} - \sigma_{22} = N_{10} + \dot\gamma [a_1(\tau\dot\gamma)^2 + b_1\tau\dot\gamma + c_1] \cdot [1+(1-\varepsilon^2) \tau^2 \dot\gamma^2]^{-1}$$

as computer fitted to the experimental data with the MINUIT routine (James and Roos, 1975). These equations above have been derived by Martins (Martins, 1994) for nematic polymers in steady simple shear flow at constant shear rate, with the director lying in the shear plane. τ is the relaxation time for the polymer memory function, τ_0 is related to texture relaxation modes, and ε ($-1 \le \varepsilon \le 1$) is a measure of the non-affine character of the motion. Similar curves were obtained for APC 129000.

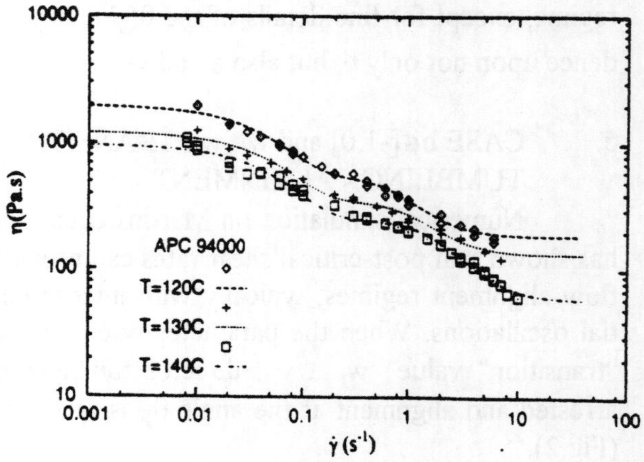

Fig 1. Shear viscosity vs shear rate for APC 94000 as a function of temperature.

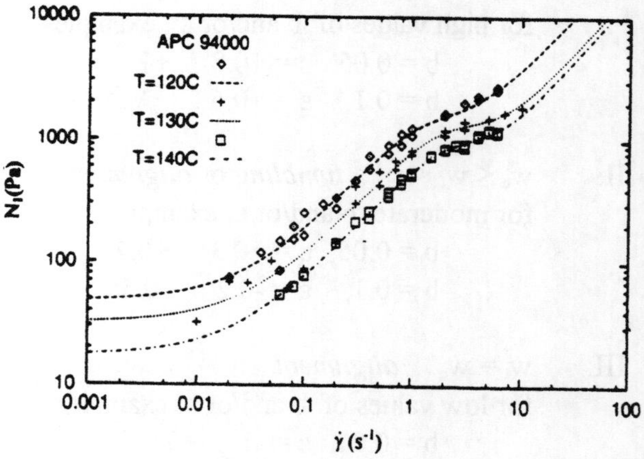

Fig 2. First normal stress difference vs shear rate for APC 94000 as a function of temperature.

DISCUSSION AND CONCLUSIONS

Figure 1 (and the corresponding one for APC 29000, not shown here) shows a decrease of the ıear viscosity with increasing temperature at very experimentally probed shear rate. At a given :mperature the viscosity decreases with ıcreasing shear rate over the whole range of $\dot{\gamma}$, xcept for shear rates between 0.2 and 1 s^{-1} lepending on molecular weight and temperature) 'here a quasi-Newtonian plateau is observed.

Figure 2 shows that the first normal stress ifference decreases with increasing temperature ıd that $N_1(\dot{\gamma})$ is an increasing function of $\dot{\gamma}$ over ıe whole range of shear rates probed. $N_1(\dot{\gamma})$ ıows an inflection for values of $\dot{\gamma}$ slightly above s^{-1}. Actually the experimental data could lead us) believe that a level off of N_1 at the high shear ıte end, or even an onset of a decreasing of N_1 ır higher shear rates was occuring, however the ıflection found by fitting the experimental data to ıe theory is supported by the experimental data n other systems (Baek et al., 1994 and Fried et ., 1994)

Table 1 displays the most important ırameters obtained from the fitting of expressions ›ove to the experimental data, for both molecular eights.

ıble 1. Most important fitting parameters of $\eta(\dot{\gamma})$ and $N_1(\dot{\gamma})$ for APC 94000 and APC 129000 (SI units)

		APC 94 000			APC 129000	
Fitting ırameter		T = 120 °C	T = 130 °C	T = 140 °C	T = 120 °C	T = 140 °C
$(\dot{\gamma})$	τ	1.8	1.3	0.8	3.5	1.7
	τ_0	20.0	17.5	16.6	25.0	21.0
	ε	0.8	0.8	0.8	0.8	0.9
	η_0+c_0	1968.7	1222.3	969.2	2848.0	1764.7
$_1(\dot{\gamma})$	τ	1.9	1.0	1.0	3.3	1.3
	ε	0.9	0.9	0.9	0.9	0.8

For APC 94000, application of an Arrhenius pe law to the temperature dependence of the scosity gives an activation energy, E_η, taking values between 12.8 ($\dot{\gamma}$ =0.01s^{-1}) and 14.8 ($\dot{\gamma}$ =7s^{-1}) kcal/mol, depending on shear rate, and going through a minimum at $\dot{\gamma}$ = 2s^{-1} (E_η, = 8.1 kcal/mol). Comparing these data with similar ones for APC 129000 (e.g. E_η, = 14 kcal/mol for $\dot{\gamma}$ = 2$s^{-1)}$, shows that the activation energy increases with molecular weight, at a given $\dot{\gamma}$.

We have built viscosity "master curves" for the two samples described above, by ploting, in *log scale*, the reduced viscosity $\eta/(\eta_0+c_0)$ vs the reduced shear rate $\dot{\gamma}/\dot{\gamma}_0$, where (η_0+c_0) is the zero shear viscosity and $\dot{\gamma}_0 = 1/\tau_0$, is obtained by fitting the data to the theory. Figure 3 shows the "master curve" so obtained for APC 94000, as an example.

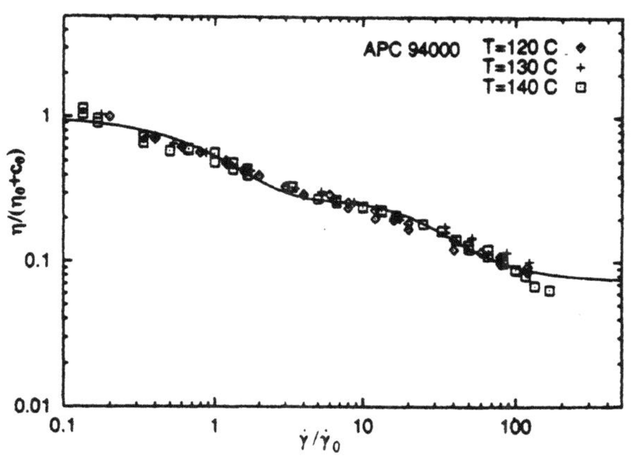

Fig 3. Reduced viscosity master curve for APC 94000.

REFERENCES

Baek S.G., Magda J.J., Larson R.G., Hudson S.D. (1994) J. Rheology 38: 1473

Cidade M.T., Leal C.R., Godinho M.H., Navard P., Martins A.F. (1995) Mol. Cryst. Liq. Cryst. 261: 617

Fried F., Leal C.R., Godinho M.H., Martins A.F. (1994) Polymers for Advanced Technologies 5: 596

James F., Roos M. (1975), MINUIT - Program for Function Minimization and Analysis of the Parameters and Correlations (CERN, Genéve)

Martins A.F. (1994) In: Carfagna (ed) Liquid Crystalline Polymers, Pergamon Press, Oxford p.153

CURE KINETICS BEHAVIOR AND PROPERTIES OF EPOXY-AMINE ADHESIVES

I.Yu. GORBUNOVA, M.L. KERBER, D.A. ALDAKOV, I.N.PONOMAREV

Polymer Processing Department, D.Mendeleev University of Chemical Technology of Russia, Russia, 125190 Moscow, Miusskaya pl., 9

Epoxy resins based adhesives are widely used nowdays. In the present work characteristics of curing processes of epoxy-amine systems were studied.

The cure of epoxy-amine systems was studied by viscometry, dynamic mechanical analysis and differential scanning calorimetry. The temperature intervals both below and above the glass transition temperatures of completely cured composites were studied.

Performance characteristics (strength, deformability, residual stresses) of end products depend on the technology of the cure and are directly related to kinetics of reactions and corresponding the evolution of rheology.

The gelation stage was analyzed by rotational viscometry method. It was found that exponential equation adequately describes the process:

$$\eta = \eta_0 e^{k\tau} \qquad (1),$$

where η_0 is initial viscosity, k is constant and t is current time. Gel-point was found by extrapolation of plots of dependences of reciprocal viscosity versus time to zero. Temperature dependencies of rate constants *k* comply with Arrhenius equation. So they were used to calculate activation energy of gelation process. The modifier and the filler additively reduce the rate of viscosity growth in proportion to their amount that can be explained by partial dilution of the system.

The curing process of epoxy resin near the gel-point can be described by percolation equation:

$$\eta/\eta^* = (1 - \tau/\tau^*)^{-a}, \qquad (2)$$

where τ^* - gel- time. A single generalized curve was found for all the systems studied. Consequently at any temperature the same reaction mechanism takes place which does not change in the presence of neither the filler nor the modifier and which correspons to the percolation equation.

The dependencies of shear modulus G' and loss tangent tg δ versus cure time at different temperatures were obtained by DMA method. The value of shear modulus can characterize the degree of conversion when the system is sufficiently distant from the glass transition point. When this condition doesn't take place the systems vitrify just during the cure process, that changes the mechanism of the process and doesn't correspond with the theories used.

The temperatures of glass transition of fully cured systems were found. Temperature of glass transition decreases with the increasing of cure temperature. Probably it can be explained by the fact with increasing of reaction rate the process slows abruptly at less degrees of conversion that leads to the final under-curing of the system.

The degree of conversion of the process was calculated on the base of shear modulus data: as a ratio of current shear modulus G' to the shear modulus of fully cured polymer in rubbery state. Time dependence of degree of conversion (x) is described satisfactorily by the rheokinetic equation of the first order with self-acceleration:

$$dx/dt = k\,(1 - x)\,(1 + Cx) \qquad (3)$$

To find out the correlation between the glass transition temperature and the adhesion strength of the systems appropriate experiments were carried out. The adhesion strength of unfilled composition grows with the increase of cure temperature. Time dependencies of adhesive strength have maximums. Such character of dependence can be explained by appearing of internal stresses in the system.

Introduction of the modifier and the filler results in the increase of final value of the strength of adhesive join.

Introduction of the filler causes thermal internal stresses to decrease at the expense of decreasing of the linear thermal expansion coefficient (LTEC).

The values of internal stresses were being studied in the course of cure. It was found that increase of temperature leads to the decrease of internal stresses and introduction of the filler decreases them.

It was found out that the less are internal stresses, the more is final strength.

Time dependencies of degree of conversion together with adhesive strength were studied. One can see that the growth of strength begins after the gel-formation and correlates with change of degree of conversion found by DMA.

The achievement of constant value of strength and internal stresses takes place at the moment when $x = 1$. Unambiguous dependence between adhesion, internal stresses and shear modulus is observed.

Cure time dependence of the glass transition temperature in the DiBenedetto equation coordinates was found. The correlation between x and T_g is described satisfactorily by the equation:

$$\frac{T_g - T_{g0}}{T_{g0}} = \frac{(C_1 - C_2)\,x}{1 - (1 - C_2)x} \qquad (4)$$

where T_{g0} is glass transition point of uncured polymer, C_1 and C_2 is relationship between acc. lattice energies and segmental mobilities of cured and uncured polymer.

ELONGATIONAL VISCOSITY OF A THERMOTROPIC LIQUID CRYSTALLINE POLYMER

A.D. GOTSIS, M.A. ODRIOZOLA
Delft University of Technology, dept. of Polymer Technology
Julianalaan 136, 2628 BL Delft, the Netherlands

1. INTRODUCTION

Liquid crystalline polymers (LCP) owe their very good mechanical properties to the high degree of orientation present in their products. The orientation is generated during processing. It is well known that elongational flow is more efficient than shear flow in generating this orientation. The elongational behaviour of LCPs, however, has become the subject of very few studies (e.g. Wilson and Baird, 1992) because an instrument that can measure the elongational viscosity, η_E, accurately at the high temperatures needed for LCPs has become available only recently.

For flexible polymers in uniaxial elongational flow, the elongational viscosity, η_E^+, is a continuously increasing function of the accumulated strain. The steady state, if it exists, is much higher than 3 times the zero-shear viscosity, which is the Newtonian limit. The strain hardening and the high values of the extensional viscosity is usually associated with the molecular entanglements of the flexible chain polymers.

LCPs have rigid-rod like molecules and (presumably) less entanglements, even though they also show elasticity in the form of significant normal forces and G'. This elasticity is the result of the LC character of the material and its domain structure.

The entry flow is another method to obtain an estimate of the extensional viscosity of a fluid. It consists of the flow from a wide reservoir into a narrow capillary. Because of the contraction and the necessary acceleration, a strong elongational flow exists there. By measuring the extra pressure needed for this flow one can calculate an estimate of the elongational viscosity (e.g. using the method of Binding, 1991). This estimate corresponds to the average of the elongational viscosity over strain equal to what should be accumulated in the entry region (Gotsis and Odriozola, 1998).

In this paper we study the response of the thermotropic liquid crystalline copolyester Vectra A950 in uniaxial elongational flow and compare the measured viscosity with the response of the material in the entry flow.

2. EXPERIMENTAL

The stress growth of a thermotropic LC co-polyester (Vectra A950: 73% pHBA / 27% 4,6-HNA) was measured in uniaxial elongational flow (in the Rheometrics extensional rheometer, RME) at 310 °C. The samples were made either by injection molding or by compression of the dried pellets in a hot press. Before starting the measurements, the samples were anealed in the molten state in the rheometer for about 10 min in order for the stresses to relax. Samples were also examined by DSC before and after the measurement for the presence of any melting peaks above the nominal melting point of Vectra 950 at 280 °C.

Entrance pressure measurements were done in a Rosand® capillary rheometer, using capillaries of 0.5, 1 and 2 mm diameter and lengths of 4, 10, 16, 20 and 40 mm. The radial contraction ratios were 24, 12 and 6.

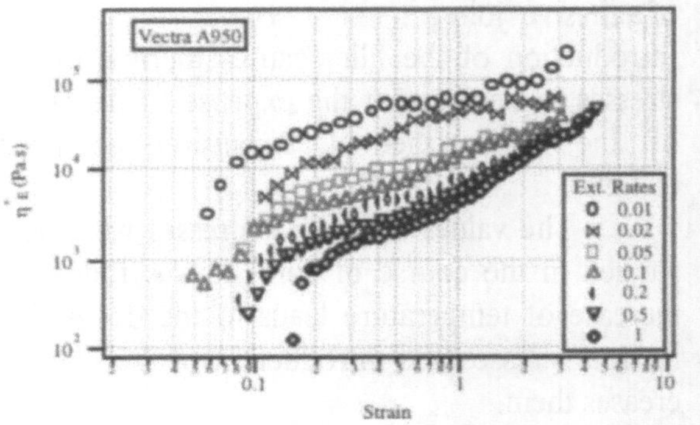

Figure 1. Extensional viscosity growth at 310 °C of compression molded Vectra A950 samples

3. RESULTS AND DISCUSSION

The viscosity growth curves of compression molded samples of Vectra A950 in uniaxial elongation are shown in Fig. 1 as a function of accumulated strain. It can be seen in this graph that the extensional stress increases mildly but continuously with strain and no steady state is reached up to the point where the sample breaks (~3-4 s.u.). In comparison, long branched LDPE shows a much stronger strain hardening, while some linear PP grades of similar MFI show steady state already at 3 to 4 s.u. under the same experimental conditions.

Strain hardening of the extensional viscosity of polymers is usually due to the entanglements in the melt. Rod-like polymers, however, are not expected to have many entanglements. The increase of the viscosity with strain in Fig. 1 has other causes. These may include recrystallization or the effect of the domain structure.

Residual crystallinity or strain induced crystallization could be reasons for a stiff chain polymer to show strain hardening viscosity. By using on-line X-ray diffraction, Nicholson et al. (1992) showed that flow induced crystallization occurred upon shearing at 320 °C. However, measurements in the RME at higher temperatures (up to 330 °C), where these problem would be less important, showed similar increase of the viscosity with strain. Higher melting point crystallites were not found by DSC in the samples before or after the elongation. This indicates that, if such crystallites are formed during the flow, they remelt very fast after the cessation of the flow, or that the crystallinity is too low to be detected by the DSC but enough to be seen in the uniaxial elongation measurements.

On the other hand it has been shown previously (Langelaan and Gotsis, 1996) that the shear stress needs many strain units to come to steady state upon start-up of shear flow (>100 s.u. instead of 10 - 20 for flexible chain polymers). It is possible, then, that 3 - 4 s.u., the strain where the samples break, is not enough to reach steady state. It is remarkable, however, that the viscosity at these strains is already much higher than 3 times the shear viscosity at the same (shear) strain. This was not expected for molecules free of entanglements that, when oriented, can slide past each other.

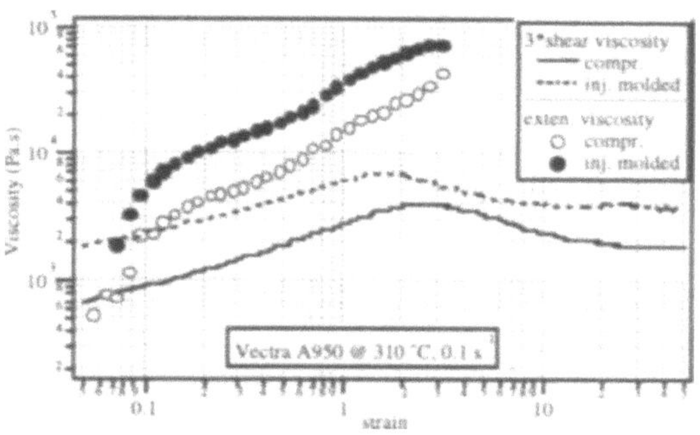

Figure 2 Comparison of η_E^+ of a compressed and an injection molded sample with growth curves of the shear viscosity.

This large difference is probably due to the deformation mechanism of the domain structure of the LCP.

Fig. 2 shows the comparison of the viscosity growth for the strain rate of 0.1 s^{-1} of samples of the same material but prepared either by compression of the pellets in the hot press or by injection molding. The viscosity of the latter is higher than the former. The same is also true for the shear viscosity growth. The domain texture that results from the melting of the injection molded structure is different from the one of the compressed sample. This has a strong influence on the flow properties of the material. On the other hand, the differences in the molecular orientation in these samples seem to be lost during the anealing time in the instrument: The difference between the viscosity measured along the major mold filling flow direction and across were much smaller than the one between the compressed and the injection molded samples.

The variation in the rheology of the samples reflects differences in the domain structure that result from different thermal and flow histories. The measurements shown in Figures 1 and 2 correspond mostly to region I of the (shear) flow curve (Langelaan and Gotsis, 1994), the region where the flow is dominated by the effect of the domain texture (Walker et al., 1995). Whether such regions exist also for the extensional viscosity is still an open question.

One way to compare the values of the extensional viscosity at different strain rates for materials that do not show a steady state is to use the values of η_E^+ at constant accumulated strain. A graph of this vs. the strain rate is shown in Fig. 3. It can be seen there that these curves for Vectra A950 are strain rate thinning.

Fig. 4 shows the results of the entry method for Vectra A950 and for a high viscosity PS of broad MWD, both for a contraction ratio of 12, i.e. for 5 s.u. When the extensional viscosity is averaged correctly, this method gives a good agreement in the case of PS but much lower (extrapolated) values in the case of Vectra A950.

The entrance pressure measurements of Vectra are complicated by the strong non-linearity of its Bagley plots (Langelaan and Gotsis, 1994). A second order polynomial was used here for the extrapolation of the curves to zero L/D, which may account for part of the disagreement. However, it seems that the domain structure is mostly responsible for the higher values of the elongational viscosity that is measured in uniaxial elongation.

4. CONCLUSION

The elongational viscosity of the thermotropic liquid crystalline polymer Vectra A950 is strain hardening up to the strain that the material breaks. The viscosity is much higher

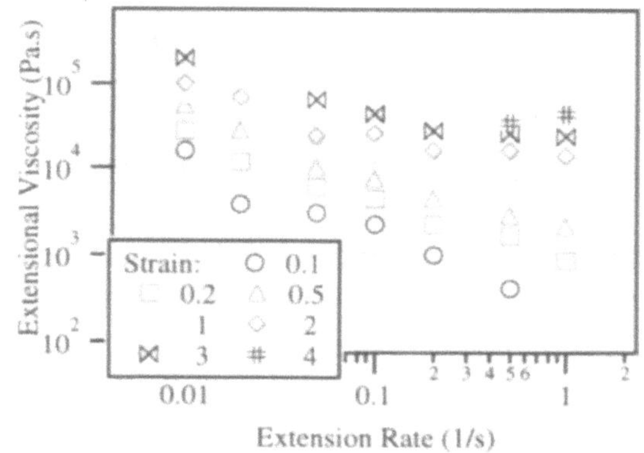

Figure 3. Extensional viscosity of Vectra A950 at constant accumulated strain vs. the extension rate.

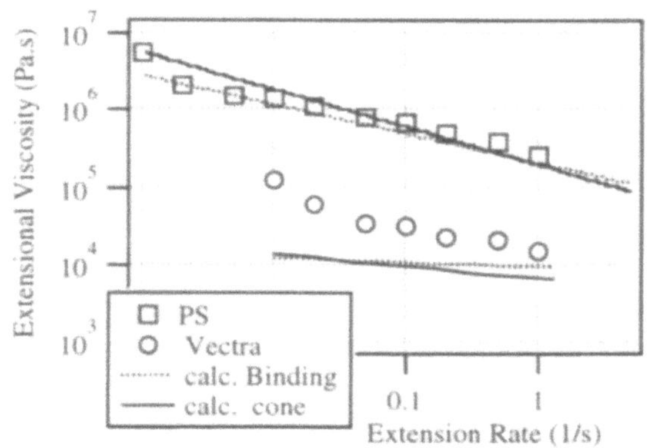

Figure 4. Comparison of the elongational viscosity of Vectra and of PS as calculated from the entrance pressure method and from uniaxial elongation, averaged over the same strain

than 3 times the shear viscosity at equivalent strains. The viscosity depends on the flow and thermal history of the sample because the rheology at these low strain rates is influenced by the domain texture.

5. ACKNOWELEDGEMENT

The present study was financed by the Dutch research organisation "OZS Polymeren PTN".

6. REFERENCES

T.C. Wilson, D.G. Baird (1992), J.Non Newt.Fl.Mech. 44:85-112

D.M. Binding (1991), J.Non Newt.Fl.Mech. 41:27-42

A.D. Gotsis, A. Odriozola, "The relevance of entry flow measurements for the estimation of extens. viscosity of polymer melts", Rheol. Acta, in print, 1998.

T.M. Nicholson, M.R. Mackley and A.H. Windle (1992), Polymer, 33:434-435.

H.C. Langelaan, A.D. Gotsis (1996), J.Rheol. 40:107-129

L. Walker et al.(1995), J.Rheol. 39:925-952

H.C. Langelaan, A.D. Gotsis (1994), J.Rheol. 38:1353-1368

NUMERICAL SIMULATIONS OF LIQUID-CRYSTALLINE POLYMER START-UP FLOW IN AN ECCENTRIC CYLINDER GEOMETRY USING THE DOI THEORY

M. GROSSO*, N. GRIZZUTI**, P. L. MAFFETTONE**
P. HALIN*, R. KEUNINGS*, V. LEGAT*

*CESAME, Unité de Mécanique Appliquée, Université catholique de Louvain,
Avenue G. Lemaitre 4-6, B-1348, Louvain-la- Neuve, Belgium

**Dipartimento di Ingegneria Chimica, Laboratori Giovanni Astarita
Università di Napoli Federico II, Piazzale Tecchio, 80125 Napoli, Italy

1. INTRODUCTION

Liquid Crystal Polymers (LCP) exhibit a complex rheological behaviour which is mainly due to the intrinsic anisotropy of the macromolecules. The Doi (1981) theory is able to capture many of the essential features of LCP rheology. The main limit of this theory is the assumption that the system is spatially homogeneous. This assumption precludes the theory from accounting for the generation and the proliferation of textures.

In order to improve the model in this direction Marrucci and Greco (1991) propose a generalised version of the Doi equation which accounts for spatial distortions. So far, results in flow conditions are yet missing, and the disclination formation is not completely understood.

With a different approach, Feng and Leal (1997) have recently obtained significant results. Indeed, they prove that disclinations can be generated in complex non-homogeneous flows. They consider the start-up flow in an eccentric cylinder geometry by modelling the LCP with the an approximated Doi model. Even though no spatial distortion is accounted for, the model gives a new insight for a possible description of the texture formation. The Doi model is approximated by using a quadratic closure approximation. Furthermore, the system is considered to be symmetric with respect to the flow plane. The results show the formation stable ±1/2-defects. The great significance of this result is undermined by the simplifications introduced. For example, the choice of the closure approximation is not a trivial point. The quadratic closure adopted by Feng and Leal (1997) is known to fail in several homogeneous flow conditions (e.g. in simple shear flow); furthermore, it has not yet been tested in non-homogeneous flow conditions.

Another point is the role played by the symmetry with respect to the flow plane which can severely affect the predictions. In this work, we analyse the problem by removing the symmetry with respect to the flow plane.

2. THE MODEL

The LCP dynamics is modelled with the Doi equation approximated with the quadratic closure:

$$\begin{aligned}\frac{d\langle \mathbf{uu}\rangle}{dt} &= -\frac{1}{Pe}\left(\langle \mathbf{uu}\rangle - \frac{\mathbf{I}}{3}\right) + \\ &\frac{U}{Pe}\left(\langle \mathbf{uu}\rangle \cdot \langle \mathbf{uu}\rangle - \langle \mathbf{uu}\rangle : \langle \mathbf{uu}\rangle\langle \mathbf{uu}\rangle\right) + \\ &\left(\nabla\mathbf{v} \cdot \langle \mathbf{uu}\rangle + \langle \mathbf{uu}\rangle \cdot \nabla\mathbf{v}^T - \nabla\mathbf{v} : \langle \mathbf{uu}\rangle\langle \mathbf{uu}\rangle\right)\end{aligned} \tag{1}$$

where the interactions between the molecules are described with a Maier Saupe potential. Equation (1) along with the equations of motion,

$$\begin{aligned}Re\frac{\partial \mathbf{v}}{\partial t} &= -\nabla p + \nabla^2\mathbf{v} + \frac{c}{Pe}\nabla \cdot \tau^p \\ \nabla \cdot \mathbf{v} &= 0\end{aligned} \tag{2}$$

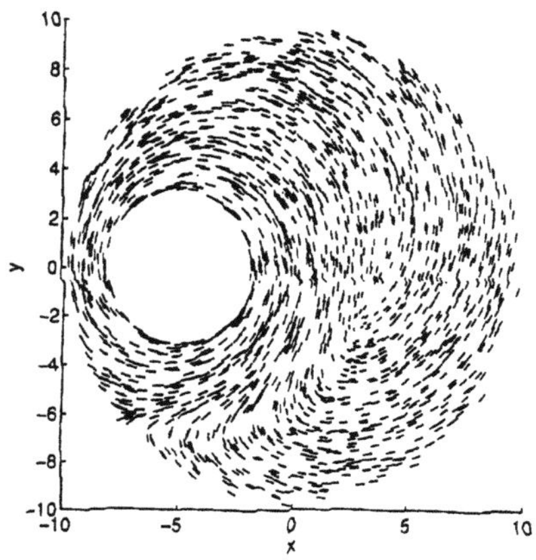

are integrated by using the Lagrangian Particle Method (LPM) (P.Halin et al., 1998) for the simulation of complex flows.

3. RESULTS

The simulations are performed by considering pure kinematics conditions (i.e. the contribution of the polymer stress is neglected). This simplification does not qualitatively affect the results in the parameter range here investigated. Our results are very close to those of Feng & Leal when symmetry is considered as shown in Fig. 1. Thus, it so appears that the generation of the defects is induced only by the flow kinematics.

When the symmetry with respect to the flow plane is removed, the dynamics becomes much more complex, and it shows also out-of-flow plane periodic solutions. In contrast with the results of Feng and Leal, no evidence of the "log-rolling" regions has been detected in such a case.

In the near future we will address the problem by considering the Doi model without the introduction of closure approximations.

4. REFERENCES

Doi M. (1981) J.Poly. Sci. Polim. Phys. Ed. 19:229-243

Feng J. and L.G.Leal, J.Rheol., (1997) 41:1317-1335.

Halin P., R.Keunings and V.Legat, (1998) submitted to J . Non-Newtonian Fluid Mech.

Marrucci G. and F.Greco, (1991) Mol. Cryst. Liq. Cryst. 206:17-30.

INVESTIGATION INTO BAND TEXTURE OCCURRING IN LIQUID CRYSTAL POLYMERS

P. HARRISON[1], P. NAVARD[1], M.T. CIDADE[2]

[1] Ecole des Mines de Paris, CEMEF, UMR CNRS 7635, B.P. 207 - 06904 Sophia-Antipolis Cedex, France
[2]Departamento de Ciência dos Materiais, Faculdade de Ciências e Technologia, Universidade Nova de Lisboa, 20825 Monte de Caparica, Portugal

1. INTRODUCTION

The band texture formation in LCP has been investigated. Rheo-optical techniques have been used to characterise the rate of evolution of the texture with respect to changes in previous shear rate. The effects of the previous shear rate and time of relaxation on the tan(δ) function were subsequently investigated. Temperature effects have also been investigated.

2. METHODS

2.1 Sample Preparation

A lyotropic hydroxypropylcellulose (HPC) 'Klucel EF' water solution was prepared by drying the powder for at least 3 hours in an evacuated oven at a temperature of 70°C and then mixing it with distilled water at a concentration of 50 percent by weight. Two thermotropic acetoxypropylcellulose (APC) polymers of differing molecular weights (i.e. 52000 and 62000 gmol^{-1}) were also prepared by placing them in an evacuated vacuum oven for at least 4 hours at a temperature of about 80°C.

2.2 Rheo-optical Measurements

Optical observation of the band texture was conducted using both a modified 'Instron 3250' rheometer, [see Riti and Navard, 1994] and a 'Linkam CSS 450 Cambridge Shearing System'. The samples were sheared until reaching steady flow conditions (at least 200 strain units) before monitoring their relaxation. All observations were conducted under cross polarized light.

2.3 Rheological Measurements

As with the rheo-optical measurements, samples were sheared until reaching steady flow conditions before testing. Rheometrical oscillatory experiments were conducted using both a controlled shear rate Stresstech rheometer, fitted with a cone and plate geometry measuring 40 mm diameter with a cone angle of 2°, and a controlled stress RMS 800 rheometer, fitted with a cone and plate geometry measuring 50 mm diameter, with a cone angle of 2.3°. The experiments used a frequency of 10 rads^{-1}, and a strain amplitude of up to 5 percent.

3. RESULTS FOR HPC SOLUTION

3.1 Effects of Primary Shear Rate on Optical Evolution of Band Texture

Figure 1 shows the evolution time of the band texture at various stages in its development versus primary shear rate. A critical primary shear rate existed below which the band texture

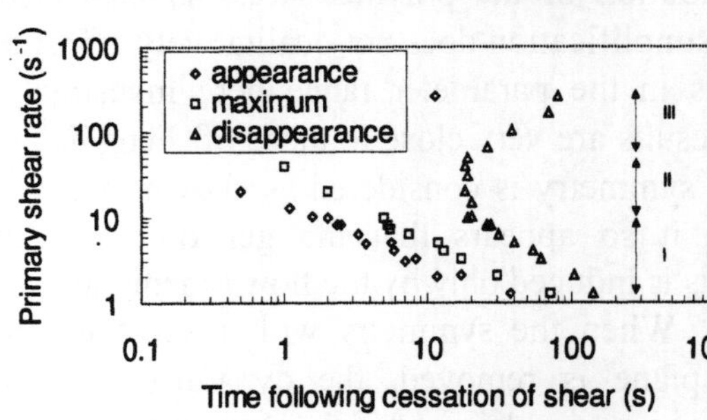

Figure 1. Time to reach various stages in the band texture's evolution versus primary shear rate.

failed to appear following the cessation of shear. This rate was found to be approximately 1.4 s^{-1} for the solution under investigation.

Three different stages in the behaviour of the disappearance time of the band texture can be distinguished according to the primary shear rate. The first stage, region I in figure 1, occurs between about 1 to 10 s^{-1}, here the rate of evolution of the texture increases with increasing primary shear rate. The second stage, region II, occurs between 10 and 40 s^{-1} and the disappearance time of the texture is almost constant. Finally the third stage, region III, occurs above shear rates of 40 s^{-1}, and the disappearance time of the texture increases with increasing primary shear rate.

3.2 Effects of Primary Shear Rate on Evolution of Tan(δ)

Tan(δ) gives the ratio between the viscous and elastic components of flow. The structural organisation of the band texture could be expected to influence these two components and the study of tan(δ) can reveal the relative behaviour of the two functions.

The evolution curves of tan(δ) in figure 2

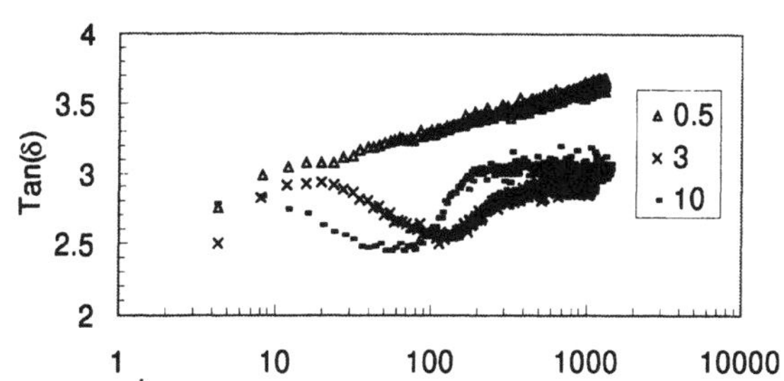

Figure 2. Evolution of tan(δ) verus time following cessation of shear after various primary shear rates.

show various interesting features. In particular, the tan(δ) curve develops a drop soon after the cessation of the primary shear. This drop in tan(δ) only occurs following shear rates of at least 1.4 s^{-1} and therefore coincides with the range of primary shear rates which produce the band texture under

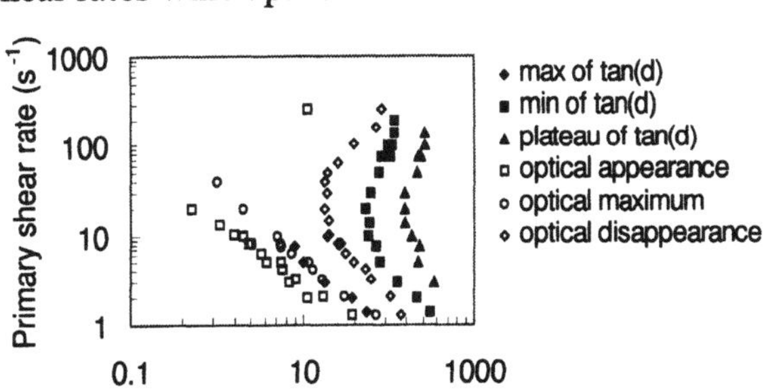

Figure 3. Rate of evolution of drop in tan(δ) versus primary shear rate along with rate of evolution of band texture.

polarized light. Figure 3 shows the time before the tan(δ) versus time curves reach their maxima, minima and plateau values (which signify the beginning, bottom and end of the drop respectively) versus primary shear rate. Also presented are the results shown previously in figure 1, i.e. the rate of optical evolution versus primary shear rate.

The rate of evolution of the drop in tan(δ) shows the same general form as the rate of evolution of the band texture. Furthermore, the maximum visibility of the band texture corresponds to the maxima of the tan(δ) curves. These two facts provide strong evidence to suggest that the drop in tan(δ) is directly attributable to the development of the band texture.

Figure 3 shows that the time to reach the minimum and plateau stages of evolution of tan(δ) are shifted to times greater than the disappearance times of the band texture. This suggests that the structure of the band texture remains even after the optical aspect of the band texture has disappeared.

4. RESULTS FOR APC POLYMER

Experiments conducted over a range of temperatures revealed that the minimum stress required to form the band texture was constant for a given molecular weight and independent of temperature. The size of this minimum stress was found to increase linearly with molecular weight.

Dynamic investigations revealed a drop in the tan(δ) function similar to that found in the HPC solution.

5. CONCLUSIONS

The rate of evolution of the band texture in a HPC solution shows a complicated dependence on the primary shear rate. The rate of evolution of a fall and rise of the tan(δ) function, noted during the relaxation of the sample shows a similar dependence on the primary shear rate. A common cause has been sugested for the two phenomena. Furthermore, the size of the minimum shear stress required to form the band texture in APC polymer increases linearly with molecular weight and is independent of temperature.

6. REFERENCES

Riti and Navard (1994) Synth. Polym. J., 1: 1-22.

TRANSIENT RHEOLOGICAL BEHAVIOUR OF FILLED LYOTROPIC LIQUID CRYSTALLINE POLYMERS

E. HEINRICH, J. VERMANT, J. MEWIS AND P. MOLDENAERS

Chemical Engineering Department., K.U. Leuven, de Croylaan 46, B-3001 Leuven, Belgium

1. INTRODUCTION

Due to their mesoscopic nature, liquid crystalline polymers (LCPs) display unusual rheological properties. They seem suitable for a number of high performance applications, due to their chemical resistance, low thermal expansion and high strength. They can however be strongly anisotropic. This can be overcome by adding filler. Studying the effect of fillers on the anisotropy and the rheology of LCPs is then relevant both from a technical and a scientific point of view. This contribution is devoted mainly to the transient rheological behaviour of filled LCPs.

2. MATERIALS AND METHODS

Only model systems are considered here. The matrix is a main chain lyotropic LCP (hydroxypropylcellulose in water) which exhibits a liquid crystalline phase at high polymer concentrations. As a model filler, monodisperse spherical particles of polystyrene have been synthesized.

The experiments have been performed on both stress and strain controlled rheometers. The former is a Rheometrics DSR (Dynamic Stress Rheometer). The latter device is a Rheometrics Mechanical Spectrometer, RMS 800, equipped with a force rebalanced transducer, which allows for the measurement of reliable normal force transients.

3. RESULTS

3.1 Steady state shear

The effect of adding fillers has been investigated for both isotropic and anisotropic HPC solutions (Moldenaers et al. 1998). For filled "isotropic" polymers, the steady state flow curves are shifted to higher viscosities with respect to those of the pure polymer. Master curves can be obtained by using the internal shear rate concept (Ohl and Gleissle 1993) . The characteristic 3-region viscosity curve is preserved for filled LCPs. Unlike the case of isotropic matrices, the internal shear rate concept does not lead to master curves here. Nevertheless, a superposition of the curves can be achieved by simply reducing the viscosity at constant shear rate with an appropriate factor C.

The effect of the filler on the normal stress is more dramatic than on the viscosity. The negative values of the first normal stress difference, typical for main chain lyotropic LCPs, are no longer observed in the accessible range of shear rates, even for filler volume fractions as low as 1 %. The origin of this disappearance is not completely clear but is certainly due to the complex flow field occurring between the particles (Heinrich et al. 1998).

3.2 Transient Behaviour

3.2.1 Reversal of the flow direction

Transient stresses have been studied in particular upon reversing the flow direction in Region II. The presence of domains in unfilled LCPs gives rise to transient oscillations of the shear and normal stresses that scale with the total deformation (Moldenaers et al. 1994). These oscillations remain present up to volume fractions of the particles of at least 20 %. The presence of the spherical fillers affects the period of the oscillations as well as their damping. In addition, the scaling for these oscillations with strain is still possible for each filler concentration (e.g. the scaling for 10 % of spheres in Fig.1).

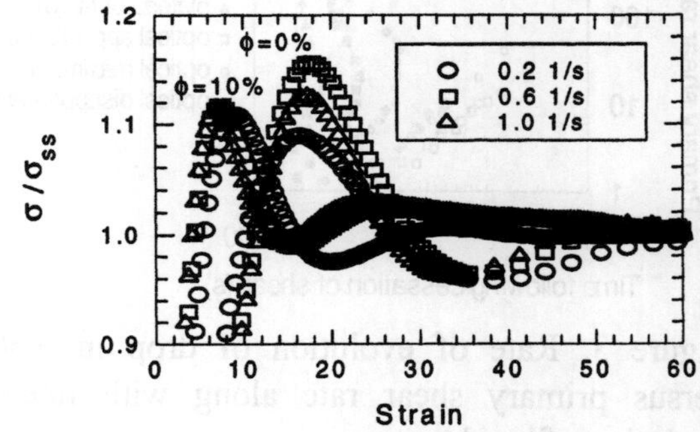

Figure 1: Scaling with strain of the oscillations of reduced shear stress upon reversal of the flow at time zero for the unfilled and 10% filled systems at 3 different shear rates.

The effect of the filler on the time evolution can be scaled by using the shift factor derived for the viscosity in steady state shear flow. The result is shown in Fig. 2, for the shear stress. This scaling applies for the oscillations of the first normal stress differences as well. No physical meaning of this scaling law is available for the time being. It can be observed also that the damping is more pronounced when the filler content increases.

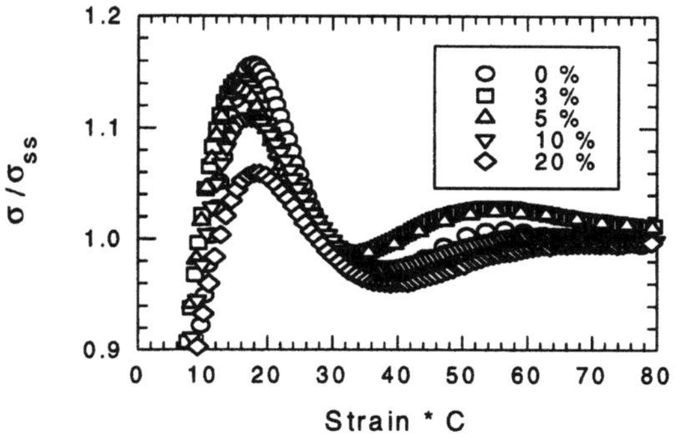

Figure 2: Scaling of the oscillations of reduced shear stress for LCPs with different volume fractions of particles.

3.2.2 Elastic recoil

For many LCPs the elastic strain recovery upon removal of shear stress is a unique function of the time multiplied with the previous shear rate in Region II (Larson and Mead 1989). This still holds for filled LCPs (Heinrich et al. 1998). However, lower final recovered strains are measured for the filled materials. As for the flow reversal experiments, the effect of the dispersed solid phase is drastically larger here than expected from simple volumetric considerations. Hence the distortional energy stored within the matrix is lower, proving that the particles have an important effect on the LCP texture as well.

3.2.3 Evolution of the dynamic moduli after stopping the flow

The time evolution of the dynamic moduli after cessation of flow has also been analysed. It is still characterized by an increase of the moduli with time as for unfilled hydroxypropylcellulose LCPs (Grizzuti et al. 1993). This can be explained by a gradual loss of orientation occurring during the relaxation process. In the case of filled LCPs, the values of the moduli are at all times much higher than in the case of pure LCP solutions. This is especially the case immediately upon cessation of flow. The result suggests that the LC phase is rendered more isotropic by the presence of the filler, both under shear and at rest. Furthermore, the relative change with time in the moduli is smaller in the filled case than in the unfilled one.

4. CONCLUSIONS

The transient rheology of filled LCPs has been investigated systematically. Flow reversal experiments for the filled materials lead to damped oscillatory patterns for both shear and normal stresses. For each filler concentration, these oscillations still scale with strain, as for unfilled LCPs. Recoil experiments show lower values of recovered strain, proving that the textural contribution to the stress is reduced. The recovery curves could also be scaled. Finally the behaviour of dynamic moduli upon cessation of flow has been analysed. As a result, a reduction of anisotropy in filled LC matrices, both at rest and for intermediate shear rates, has been demonstrated.

5. ACKNOWLEDGEMENTS

Financial support for this work has been provided by the European Union, in the framework of the TMR network on "Rheology of liquid crystalline materials" (contract FMRX-CT96-0003). JV is a post-doctoral fellow of the Fund for Scientific Research - Flanders (FWO - Vlaanderen).

6. REFERENCES

Burghardt W, Bedford B, Hongladarom K and Mahoney M (1995) Chap. 21 in Liquid Crystalline Polymers, ACS Symp. Series 597, Eds. Nakatani AI and Dadmun MD

Grizzuti N, Moldenaers P, Mortier M and Mewis J (1993) Rheol. Acta 32:218-226

Heinrich E, Vermant J, Moldenaers P and Mewis J (1998) Proc. of the First Esaform Conference, Sophia-Antipolis (France):405-408

Larson RG and Mead DW (1989) J. Rheol. 33:1251-1281

Moldenaers P, Vermant J, Heinrich E and Mewis J (1998) submitted to Rheol. Acta

Moldenaers P, Mortier M and Mewis J (1994) Chem. Eng. Sci. 49: 699-707

Ohl N and Gleissle W (1993) J. Rheol. 37:381-406

Vermant J, Moldenaers M, Picken SJ and Mewis J (1994) J. Non-Newtonian Fluid Mech. 53:1-23

DETERMINING THE MOLECULAR WEIGHT DISTRIBUTION OF POLYMERS FROM THEIR VISCOSITY FUNCTION USING THE FIRST MIRROR RELATION

B. Hochstein

Institut für Mechanische Verfahrenstechnik und Mechanik, Universität Karlsruhe (TH), D-76128 Karlsruhe

1. Introduction

Efforts to determine the molecular weight distribution of polymers from rheological measurements have centered on the relaxation time spectrum, which has almost always been measured in oscillatory experiments.

2. Relevant Region of the Viscosity Function

From studies of polymer behavior on the molecular scale, it can be shown that only relaxation times within a specific range are relevant for calculating the polymer's molecular weight distribution *(Rouse 1953; Doi u. Edwards 1978; Vinogradov G.V., Malkin A.Ya., 1980)*. In the range of pure viscous behavior, there are no chain entanglements, so that the relaxation time spectrum is due only to chain segment relaxation. In the glassy region, chain entanglements exist as permanent structural features and do not contribute to the relaxation spectrum. In this state, only molecular vibrations exist and the rheological behavior is thus independent of the polymer molecular weight. Rheological measurements in the glassy region therefore do not yield information about the molecular weight. For the viscosity function, this means that the molecular weight distribution of the polymer reflects in the flow region between the beginning of shear thinning behavior and the point in that the viscosity function transitions to a region of constant slope (Fig. 1). Using the first Mirror Relation (Eq. 1) *(Gleißle 1978)*, a discrete relaxation spectrum $(\eta_i, \dot{\gamma}_i)$ can be calculated from the relevant region of the viscosity function.

$$\eta(\dot{\gamma}) = \sum_{i=1}^{N} \eta_i \left(1 - \exp(-\dot{\gamma}_i / \dot{\gamma})\right) \tag{1}$$

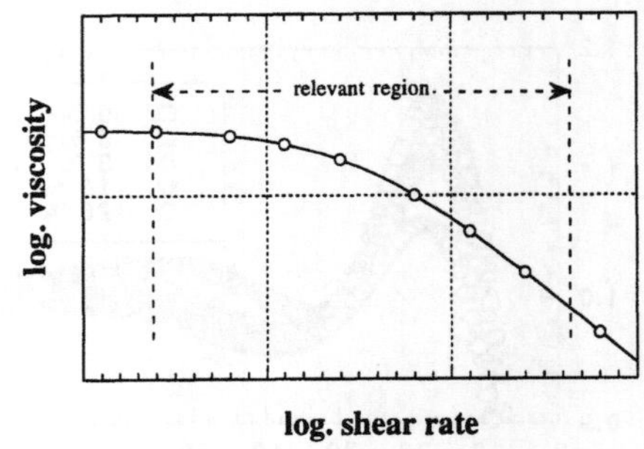

Fig. 1:
Relevant region of the viscosity function for calculatimg the molecular weight distribution

Using the well-known relation from the molecular relaxation theory *(Doi and Edwards 1978)* the molecular weight distribution can be calculated from this relaxation spectrum as given by Eq. 2.

$$W(M_i) = \sum_{I=1}^{i} \eta_I / \sum_{I=1}^{N} \eta_I \tag{2a}$$

$$M_i = \sqrt[\alpha]{\frac{G_N \cdot \pi^2}{12a}} \cdot \sqrt[\alpha]{\frac{1}{\dot{\gamma}_i}} \tag{2b}$$

The quantities "a" and "α" are determined for a given polymer from the limiting viscosity at zero shear rate of various samples with different mean molecular weights *(Vinogradov G.V., et al. 1972; Doi and Edwards 1978)*. Another parameter in Eq. 2 is the plateau modulus, G_N, which is difficult to measure. From dimensional analysis one can also express the molecular weight fraction M_i (Eq. 2b) as follows:

$$M_i \cong \sqrt[\alpha]{\frac{\eta_0 \cdot \omega_c}{a}} \cdot \sqrt[\alpha]{\frac{1}{\dot{\gamma}_i}} \tag{3}$$

Both the limiting viscosity at zero shear rate η_0 and the crossover frequency ω_c (i.e., the crossover of the storage and loss modulus curves) can be calculated from the relaxation spectrum $(\eta_i, \dot{\gamma}_i)$. Thus, all the quantities required to determine the molecular weight distribution of the polymer can be calculated directly from the viscosity function.

3. Results and Discussion

The method proposed here to determine molecular weight distribution was tested on narrowly distributed polystyrene standards. The viscosity functions of the tested polystyrene standards are shown in Fig. 2.

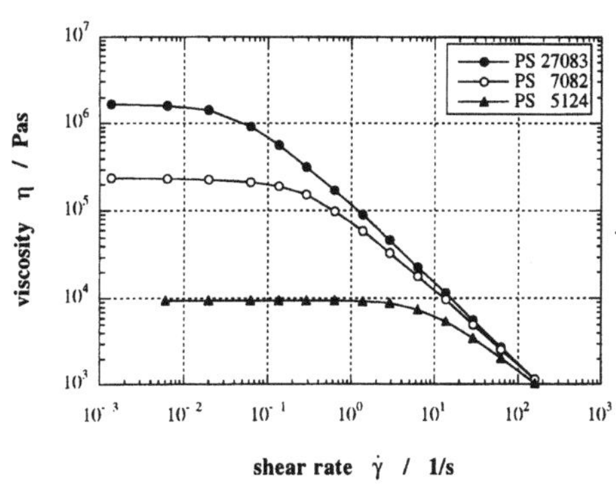

Fig. 2:
Viscosity functions of the polystyrene *standard materials*

Fig. 3 shows the molecular weight distribution determined via GPC analysis compared with that determined using the viscosity function. The mean molecular weight predicted by the two methods is nearly identical, Table 1.

Table 1:
Mean molecular weights of the polystyrene materials predicted by GPC and rheological analysis

M_W g/mol	PS 5124	PS 7082	PS 27083
GPC	125.000	335.000	564.000
Rheologie	132.000	336.000	559.000

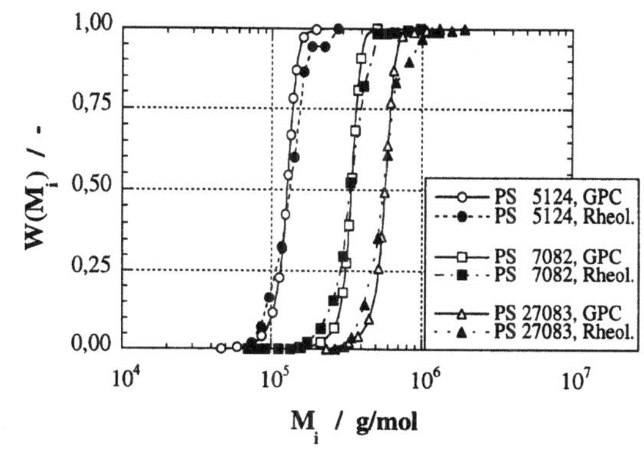

Fig. 3:
Molecular weight distributions of Polystyrene determined via GPC and rheological analysis in comparison

The distribution determined via the viscosity function is somewhat broader. This is partially due to the narrowness of the relevant region. Follow-on experiments showed that agreement between the two methods becomes even better as the distribution width increases.

4. References

Doi M., Edwards S.F., J. Chem. Soc., Faraday Trans. 2, Vol. 74, 1978
Gleißle W.; Dissertation Universität Karlsruhe (TH), 1978
Rouse P.E., J. Chem. Phys, Vol. 21, 1953
Vinogradov G.V. et al., J. Polym. Sci., A-2, 10 1061, 1972
Vinogradov G.V., Malkin A.Ya., Rheology of Polymers, Springer-Verlag, 1980

VISCOMETRIC FUNCTIONS OF LYOTROPIC LCPs AND THEIR THEORETICAL DESCRIPTION

E. Jākobsons, L.A. Faitelson

Institute of Polymer Mechanics, Latvian Academy of Science, 23 Aizkraukles St., Riga LV-1006, Latvia

A brief review of advances in rheology at shear flow of liquid crystalline polymers in a wide concentration range given. It is shown that the dependence of concentration c of Newtonian and quasi-Newtonian viscosity η_0 can be classified in regions according to their kinds of connectivity (KC): within limits of the each KC the relationship $\eta_0 \sim c^{\nu}$ (just as for other physical characteristics) holds its own characteristic power law index and changes up on reaching the critical concentration c_n (see Fig.1, patterned after data of Kiss and Porter 1998).

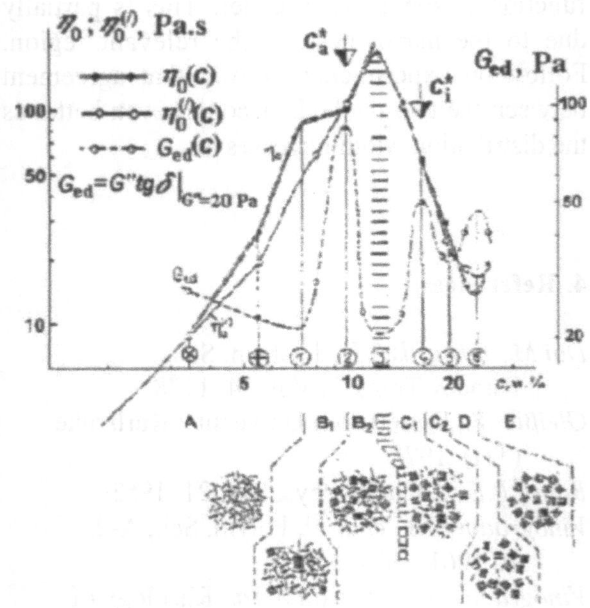

Fig.1. Concentration dependencies of η_0 , $\eta_0^{(l)}$ and G_{ed}. Explanations are given in text and table 1.

At the polymer concentration ① (Fig.1) the inclusions of the anisotropic "islets" form in the isotropic continuous phase; concentration ② corresponds to the percolation threshold of the anisotropic "islets"; the region of polymer concentrations where interpenetrate phases exist is marked by ③; the concentration ④ corresponds to the percolation threshold of the isotropic phase. If the continuous liquid–crystalline phase takes place, the Onogi–Asada flow curve and the yield stress are available (sometimes outside from the achievable measurement region). In this case LCPs forms ultra–weak viscoelastic solids capable to flow at very low values of the shear stress $\sigma_{12} = \sigma_0$. The description of critical concentrations and ranges is also given in more detail in Table 1.

Besides, each KC has a fairly distinctive shape of the viscometric functions such as $\eta(\gamma)$, $N_1(\gamma)$, $\eta'(\omega)$, and $G'(\omega)$. It is significant that beginning with concentrations when in solutions molecular associations of PBG – nucleation of LC "islets" appear, quasi-Newtonian viscosities $\eta_{(0)}$ and $\eta_0^{(l)}$ one can have different values. Here $\eta_0^{(l)}$ is the viscosity and has been obtaned as $\lim_{\omega \to 0} \eta'(\omega)$. Note that the derivative $d\,G_{eo}/d\,c$ change its sign at $c=c_n$ ($J_s^0 = 1/G_{eo}$ is steady–state shear compliance) that allows us to determine c_n better. For practical purposes we can confine ourselves with determination of a dynamic analogue of the J_s^0, that is, $J_{ed} = 1/G_{ed} = 1/G' \cdot tg^2\delta\,)|_{\omega \to 0}$ at the lowest frequency ω used for the measurement of frequency dependence of the complex shear modulus.

The phenomenological Akay–Leslie rheological equation of state for anisotropic viscoelastic liquids (A-L model) was chosen for account of the flow properties of the LCPs. This equation was solved for steady shear flow and small-amplitude (linear) oscillatory deformations. It was revealed that the A-L models describe qualitatively the principal rheological behaviour of the LCP viscometric functions, including phenomena of negative values of the first normal stress difference at some polymer concentration and in some shear rate regions. It is also shown that in the low shear rate region a linear dependence of the shear stress and the first normal stress difference on shear rate and different magnitudes of the viscosity $\eta_{(0)}$ and $\eta_0^{(l)}$ can be obtained by choosing of the value of the model initial ordering parameter. The model must be generalized to the viscoelastic relaxation spectrum for the quantitative description of the rheological behavior of LCPs.

There are a few the possible ways of generalization of the A-L model on the discrete relaxation spectrum. One of the variants of the modifications

ies in the fact that only the equation for stresses is changed. The orientation equation remains unchanged. In this case we can write down the rheological equation of state:

$$\sigma_{ij} = \sum_{p=1}^{N} \sigma_{ij(p)},$$

$$\sigma_{ij(p)} + \lambda_p D\sigma_{ij(p)}/Dt = 2\eta_p \dot{e}_{ij} + \qquad (1)$$
$$+ \left(C_{0(p)} + C_{1(p)}\dot{e}_{kr}n_k n_r\right)n_i n_j - 2\left(\dot{e}_{ik}n_k n_j + n_i n_k \dot{e}_{kj}\right),$$

where $\lambda_p, \eta_p, C_{0(p)}, C_{1(P)}$ and $C_{1(p)}$ denote the stress coefficients; $\dot{e}_{ij} = (dv_i/dx_j + dv_j/dx_i)/2$ is the deformation rate tensor.

Orientation of the fluid microstructure is given by

$$Dn_i/Dt = d_0 n_k n_k n_i + \varepsilon_0 \dot{e}_{ik} n_k + \varepsilon_1 \dot{e}_{jk} n_j n_k n_i +$$
$$+ f_0 \left(D\dot{e}_{ij}/Dt\right)n_j + f_1 \left(D\dot{e}_{jk}/Dt\right)n_j n_k n_i \qquad (2)$$

where □ d_0, ε_0, ε_1, f_0 and f_1 are constants of the material associated with the orientation vector. Here DA/Dt is Jaumann derivative.

For the description of the rheological behavior of microcrystalline cellulose (MCC) dispersions Jākobsons et al (1998) expressed the stress coefficients in equation (1) in form:

$$\lambda_p = \frac{\lambda_1}{(1+p)^{\alpha}}, \quad \eta_p = \frac{\eta_1 p^{\beta}}{(1+p)^{\alpha}},$$

$$C_{0(p)} = C_0 p^{\beta}, \quad C_{1(p)} = \frac{C_1 p^{\beta}}{(1+p)^{\alpha}}, \quad C_{2(p)} = \frac{C_2 p^{\beta}}{(1+p)^{\alpha}}.$$

Such a modification increases the number of parameters in the A–L equation. However, this enables on adequate description of the rheological features of the MMC dispersions established experimentally.

The above remains true for the main-chain TLC copolymers consisting of mesogonic groups and spacers in a certain ratio of t their mole fractions.

Table 1.

Initial states. of LCPs solutions.

Region boundaries in terms of concentration c *)	Explanations	Notes
$c < ⊗$	Transition from dilute to semi-dilute solution.	
$⊗ < c < ⊕$	Semidilute solution.	$\eta' \le \eta$; $(2G'/\omega^2) \le \Psi_1 \vert_{\omega=\dot{\gamma}}$
$⊕ < c <$ ① HS, ICP	Moderate concentrated and concentrated solutions, molecular associations of PBG – nucleation of LC "islets".	
① HS, ICF	Concentrated solution and origiation of "islets" in isotropic matrix of LCP (it is determined in polarized light.)	
① $\le c <$ ② HS, ICP	Increase of number and sizes of "islets" of LCP in isotropic matrix.	$\eta' > \eta$; $(2G'/\omega^2) > \Psi_1 \vert_{\omega=\dot{\gamma}}$
② HS, ICP	Threshold of percolation of LCP "islets".	
② $\le c <$ ③ HS, ICP	Further increase of volume fraction occupied by LCP "islets".	
③ S, ACP	Phase inversion: origination of continuous LCP matrix with "islets" of isotropic phase (it is possible a narrow concentration range of interpenetrated phases)	Complex textures. "Plato" at flow curve. As rule, $\eta' > \eta$; $(2G'/\omega^2) > \Psi_1 \vert_{\omega=\dot{\gamma}}$. Possible negative values of N_1.
③ $\le c <$ ④ HS, ACP	Increase of volume part of anisotropic matrix (reduction of volume occupied by isotropic inclusions).	
④ HS, ACP	Threshold of percolation of isotropic inclusions in anisotropic matrix.	
④ $\le c <$ ⑤ HS, ACP	Increase of anisotropic fraction and reduction of drops fraction of isotropic phase in LCP solution volume.	
⑤	Entire volume represents itself aniso-tropic solution.	
⑤ $\le c <$ ⑥	Concentration increase of completely anisotropic solution.	
⑥	Origination of jelly-like or rigid inclusions.	
$c >$ ⑥	Onset of jelly formation.	

*) HP– heterophase state; ICP– isotropic continuous phase; ACP– anisotropic continuous phase.

REFERENCES

Kiss G, Porter RS (1998) In: Brostow W (ed) Mechanical and thermophysical properties of polymer liquid crystals. Chapman & Hall. pp 342-406.

Jākobsons E, Faitelson LA, Laka MG (1998) Mechanics of Composite Materials. In press.

THE SUSCEPTIBILITY TO ORIENTATIONAL ORDERING OF TWO-DIMENSIONAL POLYMER SYSTEMS DEPENDING ON THE TYPE OF EXTENDING DEFORMATION

A.V. Maximov [1], Yu. Ya. Gotlib [2], O.G. Maximova [1], S.S. Vanjugina [1] and T.V. Korovkina [1]

[1]Institute of Mathematics and Natural Sciences, University of Cherepovets, 8 Soviet Avenue, 162600, Cherepovets, Russia
[2]Institute of Macromolecular Compounds, Russian Academy of Sciences, 31 Bolshoy Avenue, 199004, St.Petersburg, Russia

The effective potential energy of intra-and interchain orientational interactions of two-dimensional ordered both infinite system and finite domain (Fig.1.),like the corresponding Hamiltonian of anisotropic"quasi-lattice" version of the classical plane-rotator Vaks-Larkin (1965) model, has the form

$$V_{ef}\{\varphi_{n,m}\} = -K_l \sum_{n,m} \cos(\varphi_{n,m} - \varphi_{n-1,m}) - K_t \sum_{n,m} \cos(\varphi_{n,m} - \varphi_{n,m-1}) \quad (1)$$

The energetical constant K_l characterises the bending rigidity of chains with the dimensionless persistent length $a=2K_l / k_B$; the corresponding constant K_t describes interchain interactions. The angle $\varphi_{n,m}$ defines the orientation of **n**-th element in **m**-th chain regarding of selected in the plane **(n,m)** direction.

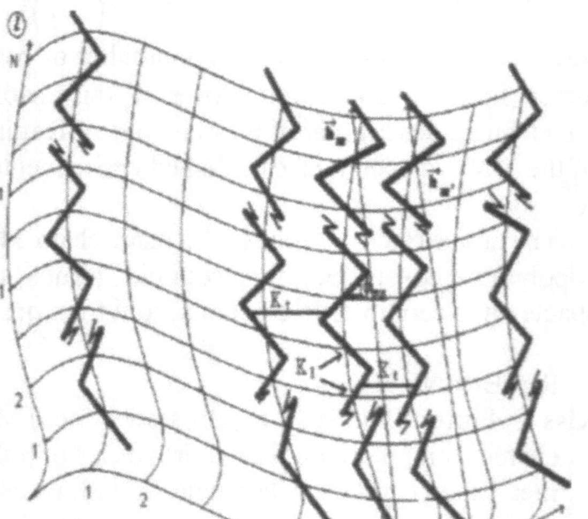

The Fig.1. The two-dimensional domain from of M chains consisting of N rigid elements (rotators).

For the case, when chains are being in almost extended conformations because of strong orientational interactions, Gotlib et al. (1996) used the low-temperature approximation for Eq.(1) :

$$V^*_{ef}\{\varphi_{n,m}\} = (1/2)\cdot K_l \sum_{n,m} (\varphi_{n,m} - \varphi_{n-1,m})^2 + (1/2)\cdot K_t \sum_{n,m} (\varphi_{n,m} - \varphi_{n,m-1})^2 \quad (2)$$

In this approximation the intra-and interchain orientational correlation functions $g(p,q) = <\cos(\varphi_{n+p,m+q} - \varphi_{n,m})>$ decay by power law typical for two-dimensional systems :

$$g(p,q) \cong [1 + \sqrt{(p/p^*)^2 + (q/q^*)^2}\,]^{-\gamma} \quad (3)$$

with the same exponent $\gamma = k_B T / 2\pi \sqrt{K_l K_t}$ determined by the ratio of energies of thermal motion ($k_B T$), intra- and interchain interactions (K_l and K_t, correspondingly). The parameters

$$p^* = (1/\pi)\sqrt{K_l / K_t}\,, \quad q^* = (1/\pi)\sqrt{K_t / K_l} \quad (4)$$

characterise the spatial scale of decreasing of these functions in the corresponding (longitudinal- **l** and transversal- **t**) directions of the system(Fig.1).

The susceptibility of a chain element to orientational ordering caused by the extending deformation of chain as whole by the external mechanical force $\vec{f}$ is defined by Gotlib et al.(1986):

$$\chi = \lim_{f \to 0} \langle \delta h_N \rangle_f / Nf \quad (5)$$

In the cases, when external forces act on one selected chain (1) or on all chains (2) in the two-dimensional system with interchain orientational interactions, the susceptibility is reduced to the following expressions, correspondingly

$$\chi_1 = \langle h_N^2 \rangle / N k_B T$$

$$\chi_2 = \frac{1}{k_3 T} \sum_m \sum_{m'} \langle h_m h_{m'} \rangle \quad (7)$$

where $\vec{h}_m$ и $\vec{h}_{m'}$ - the vectors of length from end to end of m-th and m'-th chains, correspondingly.

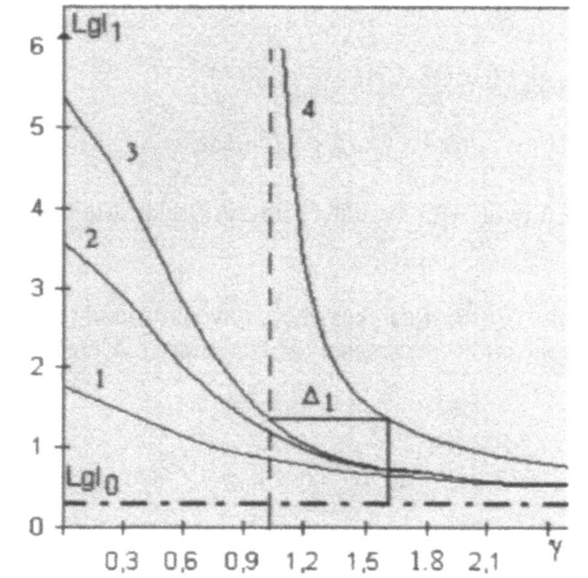

Fig.2.The logarithm of reduced susceptibility $I_1 = \chi_1 \cdot k_B T / l^2$ versus the parameter γ for the chains with the given bending rigidity (a=1) and the number N of chain elements: 1 - $N=10^2$, 2 - $N=10^4$, 3 - $N=10^6$, 4 - $N=\infty$ (the thermodynamic limit).The value Δ_1 characterizes the width of transition (it's shown for the curve 3). The quantity I_0 is the susceptibility of an isolated semirigid chain ($\gamma = \infty$)

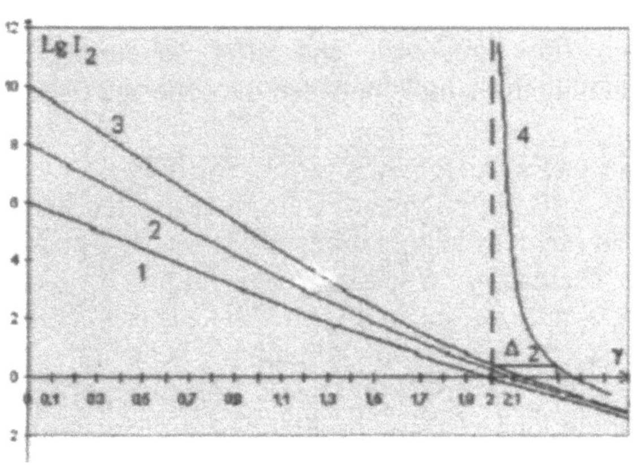

Fig.3.The logarithm of reduced susceptibility $I_2 = \chi_2 \cdot k_B T / l^2$ versus the parameter γ for the system from M chains with the given bending rigidity (a=1) and the number N of chain elements: 1 – N, $M=10^3$, 2 – N, $M=10^4$, 3 - $N=10^5$, 4 – $N,M=\infty$ (the thermodynamic limit).The value Δ_2 characterizes the width of transition (it's shown for the curve 3).

Therefore, the susceptibility is proportional to mean square of the corresponding characteristic vector (of chain length from end to end and dipole moment of all chain elements) and determined by the intra-and interchain correlations (Eq. (3)).

In the thermodynamic limit ($N,M \rightarrow \infty$) the susceptibilities χ_1 and χ_2 are described by the corresponding asymptotic dependences:

$$\chi_{1,\infty}(\gamma) = 2a\gamma/(\gamma - 1)$$

$$\chi_{2,\infty} \cong \frac{4}{\pi^2(1-\gamma)(2-\gamma)}\left(\frac{1}{p^*}+\frac{1}{q^*}\right)^{2-\gamma} \qquad (8)$$

At $\gamma_1 = 1$ and $\gamma_2 = 2$ the values of χ_1 and χ_2 exhibit jump-wise changes (Figs.2, 3), which may be considered as corresponding transitions of "Stanley- Kaplan type" (1966) from isotropic to orientational-ordered state, where the long-range order is absent. For the finite systems (domains, Fig.1) These transitions are "spreaded", i.e. exhibit in the some regions, the width of those (Δ_1 and Δ_2, correspondingly, may be defined from the conditions

$$\chi_1(\gamma = 1, N) = \chi_{1\infty}(1 - \Delta, \infty)$$

$$\chi_2(\gamma = 2, N, M) = \chi_2(2 - \Delta_2, \infty, \infty) \qquad (9)$$

from those the following estimations are obtained:

$$\Delta_1 \cong \gamma / \ln(N/a)$$

$$\Delta_2 \cong \frac{1}{\ln(N/2a + M\pi^2 2a)} \qquad (10)$$

The width of phase transition regions depends not only on the chain length (N) and the number M of chains in the domain,but also on the bending rigidity (a). At $N,M \rightarrow \infty$ the values Δ_1 and $\Delta_2 \rightarrow 0$. Therefore,in the correspondence with Grosberg AYu, Khokhlov AR (1989) these transitions may be determined as the phase ones of second order.

REFERENCES

Gotlib YuYa, Maximov AV,Maximova OG (1996) Polymer Sc A38: 650-656

Gotlib YuYa, Darinsky AA, Svetlov YuE (1986) Physical Kinetic of Macromolecules. Chimiya, Leningrad

Grosberg AYu, Khokhlov AR (1989) Statistical Physics of Macromolecules. Nauka, Moscow

Stanley HE, Kaplan TA (1966) Phys. Rev.Lett.17: 913-915

Vaks VG, Larkin AI (1965) Zh Eksperim i Teor Fiz 49: 975-989

BEHAVIOR UNDER FLOW OF SIDE CHAIN LIQUID CRYSTAL POLYMERS

I. QUIJADA-GARRIDO[1], H. SIEBERT[1], C SCHMIDT[1], C. FRIEDRICH[2]

[1]Institut für Makromolekulare Chemie, Albert-Ludwigs-Universität Freiburg, Stefan-Meier-Str. 31, D-79104, Freiburg, Germany
[2]Freiburger Materialforschungszentrum (FMF), Albert-Ludwigs-Universität Freiburg, Stefan-Meier-Str. 21, D-79104 Freiburg, Germany

1. INTRODUCTION

The Leslie-Ericksen-Parodi theory predicts two classes of nematics depending on their behavior under flow (de Gennes (1993)). The first type of nematics is the *tumbling type*, for which the director is forced by the hydrodynamic torques to continuously rotate in a simple shear flow; for the second type, the *flow-aligning type*, the director aligns at a characteristic angle with respect to the flow direction. In tumbling nematics, the director rotation produces a characteristic oscillating response of the shear stress (σ) and first and second normal stress differences (N_1, N_2). Oscillations in σ and N_1 or optical response have been found for some lyotropic LCPs and thermotropic main-chain LCPs. Low molecular weight nematics belong usually to the flow-aligning type, except when they have a smectic phase at lower temperature, in this case they exhibit tumbling behavior. A recent review of the behavior of nematics under flow was given by Jamieson et al. (1996). Besides these extensive studies, to our knowledge there are no other reports on the transient behavior of σ and N_1 of bulk thermotropic side-chain LCPs.

In this contribution we study the transient stresses of two side-chain LCPs that were previously investigated by rheo-NMR by Grabowski and Schmidt (1994) and by Siebert et al. One of these two polymers is an example of a flow-aligning system, the other one is of the tumbling type. The study of their transient stresses will provide new information on the behavior under shear of side-chain LCPs and the characteristics of tumbling.

2. EXPERIMENTAL SECTION

2.1 Materials

The two side-chain liquid crystalline polymers studied are presented in Scheme 1. The mesogen is connected to the backbone via a flexible spacer of four (PSi4) or six methylene units (PSi6). Both polymers have a broad nematic phase and PSi6 presents also a smectic A phase.

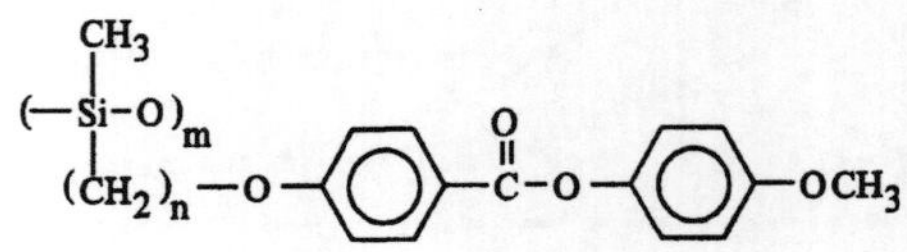

n = 4: g 277 n 367 (K)
n = 6: g 271 s 311 n 376 (K)

Scheme 1

2.2 Rheological Measurements

Rheological measurements were performed on a Rheometrics RSH800 rheometer using cone-and-plate geometry (angle = 0.1 rad, d = 25 mm, gap = 50 µm).

Measurements were performed in the nematic phase after filling the sample in the isotropic state. Transient stresses were measured in first start-up, stepwise increase or decrease of stress, flow reversal, and after squeezing or stretching the sample between the cone and plate.

3. RESULTS AND DISCUSSION

Both polymers were investigated in a broad range of temperatures in the nematic phase. No oscillatory response was found for any temperature or shear rate in PSi4. This polymer shows typical flow-aligning behavior as previously observed by NMR. For the polymer with the longer spacer oscillations in the viscosity and especially in N_1 were found for different temperatures and in the whole range of shear rates studied, in agreement with the rheo-NMR results, which indicated its tumbling character. Oscillations are stronger for low temperatures, close to the nematic-smectic transition.

Transient values of N_1 and η for Psi4 and Psi6 after stepwise increase of the shear rate are presented in Figure 1 at 50 °C. The tumbling pattern is more pronounced in the case of the start-up than in stepwise increase of the shear rate, but in the former experiment strain scaling was not possible.

We attribute the non-reproducibility of the start-up experiments to the squeezing and stretching processes that take place during the setting of the gap and the temperature.

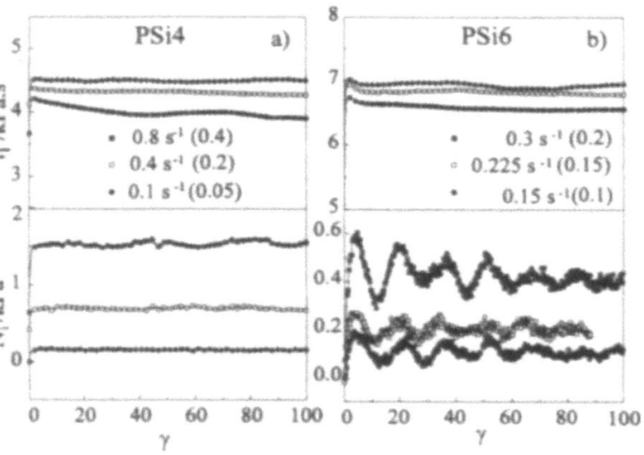

Figure 1. Transient response of η and N_1 to step increase of the shear rate for PSi4 (a) and PSi6 (b) at 50 °C. The sample was presheared at the $\dot{\gamma}$ values indicated in brackets.

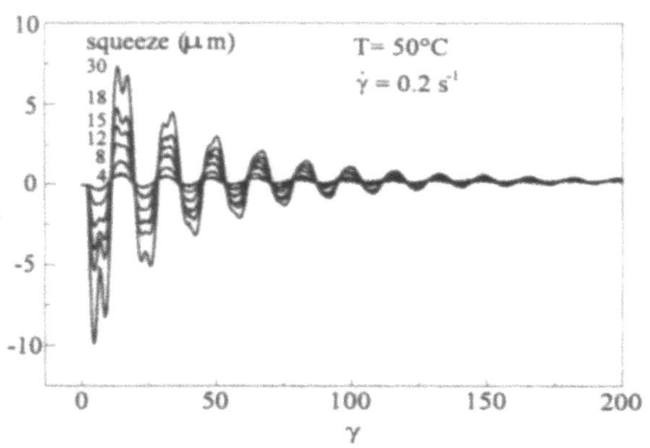

Figure 2. Effect of the degree of squeezing on the transient behavior of N_1 for PSi6 at a shear rate of 0.2 s^{-1} and 50 °C. The samples were sheared prior to squeezing at a shear rate of 0.2 s^{-1}.

Squeezing or stretching of the presheared sample produces very intense oscillations observable up to high strains (Figure 2). This behavior is more similar to low molecular weight systems than to LCPs. The intensity of the oscillations increases as the amplitude of previous squeezing or stretching increases. At the same time double peaks appear. Both squeezing and stretching seem to have two effects on the sample. In presheared samples, they turn the director away from its steady state position and they also disturb the random orientation in the unpresheared sample.

When the direction of the flow is switched after the steady state has been reached, oscillations are not refocused. We can only explain this phenomenon on the basis of an out-of-plane orientation of the director. This out-of-plane orientation has also been observed by Siebert et al.

The viscosity shows in both polymers a slight shear thinning. The steady-state values of N_1 are always positive and they increase with increasing shear rate. In the case of these side-chain LCPs, the behavior of the viscosity and the first normal stress difference is very similar in both systems, tumbling PSi6 and flow-aligning PSi4. This indicates that the flexibility of the polymer backbone plays a more important role in the steady-state value of N_1 than the tumbling character of the polymer.

4. CONCLUSIONS

Damped time-dependent oscillations in the first normal stress difference and shear stress were found in the tumbling PSi6 whereas a rheological response typical of a flow-aligning system was found for PSi4 in agreement with the rheo-NMR results. The intensity of the stress oscillations increases spectacularly by squeezing or stretching the sample by some micrometers between the tools. The tumbling system exhibits an interesting transient rheological response to flow reversal in agreement with the observed *log-rolling* orientation of the mesogen. In the shear-rate range studied, both polymers, the tumbling and the flow-aligning one, show shear thinning and a positive steady-state value of N_1 that increases almost linearly with increasing shear-rate.

Acknowledgement: This work was funded by the European Comission within its TMR program under contract number FMRX-CT96-0003 (DG12-ORGS).

5. REFERENCES

de Gennes PG, Prost J (1993) In: The Physics of Liquid Crystals. 2nd ed. Oxford University Press. New York.

Grabowski D, Schmidt C (1994) Macromolecules 27:2632-2634

Jamieson AM, Gu D, Chen FL, Smith S (1996) Prog. Polym. Sci. 21:981-1033

Siebert H, Becker P, Schmidt C, in preparation

IN SITU NMR STUDIES OF LIQUID CRYSTALS IN SHEAR FLOW

C. SCHMIDT, P. BECKER, A. HASENHINDL, S. MÜLLER, H. SIEBERT

Institut für Makromolekulare Chemie, Universität Freiburg
Stefan-Meier-Str. 31, D-79104 Freiburg, Germany

1. INTRODUCTION

Both thermotropic and lyotropic liquid crystals show a complex flow behavior. When subjected to flow their director can change its orientation. Furthermore, phase transitions may be affected by the flow field. The orientations and phase structures can be investigated on a molecular level by deuteron NMR. We have applied this technique to the study of sheared liquid crystals *in situ*. Two examples from our recent work will be presented.

2. METHODS

In deuteron NMR the coupling between the quadrupole moment of the ^2H nucleus and the electric field gradient at the site of the nucleus is used to probe molecular orientations and rotations. The quadrupole coupling gives rise to a line splitting. In uniaxial liquid crystals, due to motional averaging, the line splitting $\Delta\nu$ and the angle θ between the director and the magnetic field are related by

$$\Delta\nu = K(3\cos^2\theta - 1)\,, \qquad (1)$$

where K is a constant depending on the amount of residual orientational order. Equation (1) can be used to determine the director orientation in a liquid crystalline phase. In an isotropic phase, K becomes zero and a single peak is observed.

Experiments were performed in a home-built rheo-NMR probe developed by Grabowski and Schmidt (1994). A cone-and-plate cell was used, which has its axis aligned parallel to the magnetic field. Thus θ is the angle between the director and the velocity gradient.

3. RESULTS

3.1 Nematic Side-chain Polymers

Based on the Ericksen-Leslie-Parodi theory (de Gennes and Prost 1993), a simple measurement of the director orientation as a function of the shear rate can be used to determine the Leslie coefficients α_2 and α_3 and thus the flow-alignment parameter

$$\lambda = -\frac{\alpha_3 + \alpha_2}{\alpha_3 - \alpha_2}. \qquad (2)$$

Nematics with $|\lambda| > 1$ are flow-aligning; systems with $|\lambda| < 1$ are tumbling. Both types have been observed by Siebert et al. (1998) in the case of side-chain liquid crystal polymers.

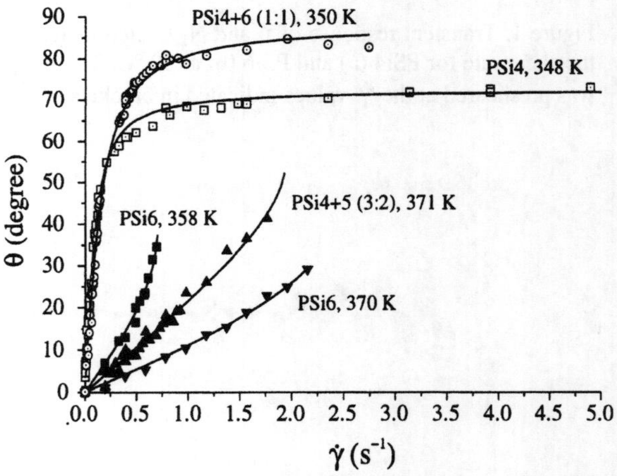

Figure 1: *Director orientation as a function of the shear rate for nematic polysiloxanes PSin (homopolymers) and PSin+m (copolymers) with spacer lengths n and m. For the copolymers the molar ratio of the different spacers is given in parenthesis.*

Figure 1 shows the director orientation determined by NMR as a function of the applied shear rate for several polymers consisting of a selectively deuterated phenylbenzoate mesogen attached to a polysiloxane backbone by spacers of varying length. Homopolymers and copolymers with spacers of four to six methylene units have been used. The lines in Figure 1 are fits based on the Ericksen-Leslie-Parodi theory, from which λ is determined. Two types of curves can be discerned. The ones with open symbols correspond to flow-aligning systems characterized by a stable director orientation approaching the Leslie angle $\pi - \theta$ at high shear rates. The curves with filled symbols were obtained for tumbling systems, for which the director orientation becomes unstable as soon as the shear rate exceeds a critical value and tumbling is no longer suppressed by the magnetic field.

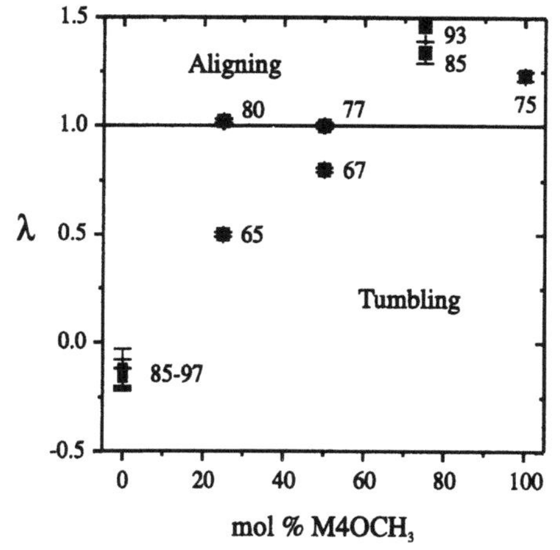

Fig; 2: The flow alignment parameter λ of copolymers PSi4+6 as a function of the molar fraction of comonomers with the shorter spacer. Temperatures in °C are given next to the data points.

Polymers of a very similar chemical structure can show a strikingly different flow behavior. For example, PSi4 is of the flow-aligning type, whereas PSi6 is of the tumbling type. This difference is related to the different phase structure of these polymers; PSi6 has much stronger smectic fluctuations than PSi4 as shown by Siebert et al. (in prep.). In copolymers, the phase structure depends on the polymer composition. As a consequence, the flow properties can be tuned by adjusting the copolymer composition. This is shown in Fig. 3, which displays the λ parameter as a function of polymer composition for copolymers of type PSi4+6.

3.2 The H_1–Isotropic Phase Transition

The phase behavior of lyotropic liquid crystals consisting of surfactant/D_2O mixtures was studied by deuteron NMR. For a polymer surfactant consisting of amphilic side-chains fixed to a polysiloxane backbone a pronounced effect of shear on the transition from the hexagonal (H_1) phase to the isotropic micellar phase was observed. The NMR spectra in Figure 3 illustrate that the normally first-order phase transition (characterized by the superposition of a doublet and an isotropic peak in the biphasic regime) becomes continuouus with increasing shear rate.

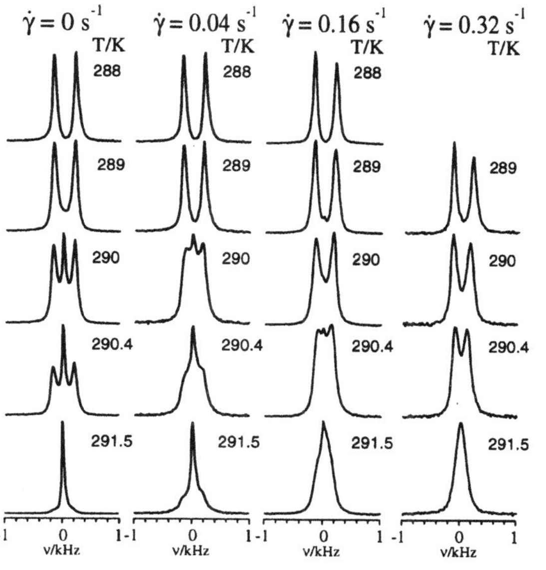

Fig; 3: Deuteron NMR spectra of a polymer surfactant in D_2O as a function of shear rate and temperature.

4. CONCLUSION

NMR can be used to study a large variety of phenomena in sheared liquid crystals. We have shown that the flow behavior of nematic side-chain liquid crystalline polymers can be tuned by inserting chemically similar comonomers that change the phase structure of the system. A shear effect on the hexagonal–isotropic phase transition of a lyotropic liquid crystalline polymer was found.

5. ACKNOWLEDGEMENTS

We thank the Deutsche Forschungsgemeinschaft for funding.

6. REFERENCES

de Gennes PG, Prost J (1993) The Physics of Liquid Crystals. 2nd ed. Oxford University Press, New York

Grabowski DA, Schmidt C (1994) Macromolecules 27:2632–2634

Siebert H, Grabwoski DA, Schmidt C (1998) Rheol. Acta 36:618–627

Siebert H, Becker P, Schmidt C, L. Noirez (in prep.)

NONSYMMETRIC VISCOELASTICITY OF LIQUID CRYSTALLINE POLYMERS

V.S.VOLKOV

Institute of Petrochemical Synthesis, Russian Academy of Sciences, 117912, Moscow, Russia

1. INTRODUCTION

The appearance of LC polymers has aroused interest to the rheology of anisotropic liquids with memory. The popular Leslie-Ericksen theory for small-molecule liquid crystals fails to describe the frequency dependence of dynamic modulus and nonlinear effects observed in common flows of LC polymers. This fact stimulated the studies of modified versions of the Leslie anisotropic liquid model with a single scalar relaxation time [Akay and Leslie (1984); Martins (1994)].

At present it is quite clear that the relaxation anisotropy plays an important role in the rheology LC polymer media. Based on anisotropic character of relaxation processes in LC polymers the theory of symmetric viscoelasticity for nematic polymer melts was proposed earlier [Volkov and Kulichikhin (1990)]. The main feature of the model is that anisotropic relaxation time and viscosity is represented by rank four tensor.

In the present work, the simple theory of nonsymmetric viscoelasticity based on this approach is formulated.

2. CONSTITUTIVE EQUATION

We consider an incompressible viscoelastic liquid with a single director **n**. This unit vector characterizes the preferred orientation of liquid during flow. The total stress tensor of nematic liquid can be represented as the sum of reversible and irreversible parts $\sigma_{ij} = \sigma^e_{ij} + \sigma'_{ij}$. The Ericksen stress tensor σ^e_{ij} describes stresses connected with the spatial inhomogeneity of the director field. To construct the elementary nonsymmetric viscoelastic model we assume that the non-equilibrium part of stress is given by

$$\tau_{ijke}\frac{D\sigma'_{ke}}{Dt} + \sigma'_{ij} = \eta_{ijke}\gamma_{ke} + \beta_{ijk}N_k. \qquad (1)$$

Here γ_{ij} is the strain rate tensor; D/Dt is the Jauman tensor derivative. The stress state in a moving nematic liquid depends on its orientation **n** and the rate of orientation change $N_i = Dn_i/Dt$. For incompressible nematic liquids, the transversely isotropic tensors of viscosity $\underline{\underline{\eta}}(\mathbf{n})$, $\underline{\underline{\beta}}(\mathbf{n})$ and relaxation times $\underline{\underline{\tau}}(\mathbf{n})$ are expressed as

$$\beta_{ijk} = \eta_5 n_i\delta_{jk} + \eta_6 n_j\delta_{ik}, \qquad (2)$$

$$\eta_{ijke} = \eta_1 I_{(ij)(ke)} + \eta_2 n_{ijke} + \eta_3 n_i\delta_{j(k}n_{e)} + \eta_4 n_j\delta_{i(k}n_{e)},$$

$$\tau_{ijke} = \tau_1 I_{ijke} + \tau_2 n_{ijke} + \tau_3 n_i\delta_{jk}n_e + \tau_4 n_j\delta_{ik}n_e + {}$$
$$+ \tau_5 n_i\delta_{je}n_k + \tau_6 n_j\delta_{ie}n_k,$$

were $I_{ijke} = \delta_{ik}\delta_{je}$ is a unit fourth-rank tensor. In these expressions the following notations are introduced

$$n_{ijke} = n_i n_j n_k n_e, \quad 2n_i\delta_{j(k}n_{e)} = n_i\delta_{jk}n_e + n_i\delta_{je}n_k.$$

Here, the symmetrization is made from the indices given in the parentheses. The constitutive equations (1,2) involve six viscosity coefficients $\eta_1 - \eta_6$ and six relaxation times $\tau_1 - \tau_6$. It leads to the set of coupled differential equations for the symmetric and antisymmetric components of the irreversible stress tensor.

3. ORIENTATION DYNAMICS

Using the equation for antisymmetric stress and equation of angular momentum, we obtain the inertialess orientation equation

$$\mu_1 I_{ie}\frac{Dh^\perp_e}{Dt} + \mu_2 I_{ie}n_s\frac{D\sigma'_{(se)}}{Dt} + h^\perp_i =$$
$$= \nu_1(n_e\gamma_{ei} - n_{ien}\gamma_{en}) + \nu_2 N_i \qquad (3)$$

where $\sigma'_{(ei)}$ is the symmetric part of stress tensor and $I_{ie} = \delta_{ie} - n_i n_e$. This equation involves two relaxation times

$$\mu_1 = \tau_1 + (\tau_3 - \tau_4 - \tau_5 + \tau_6)/2,$$

$$\mu_2 = \tau_3 - \tau_4 + \tau_5 - \tau_6$$

and two friction factors

$$\nu_1 = \eta_3 - \eta_4, \; \nu_2 = \eta_5 - \eta_6.$$

The internal elastic forces of a polymer nematic and the external forces acting on the director in a magnetic field **H** can be described by a single molecular field $h_i = h^e_i + h^m_i$:

$$h_i^e = \frac{\partial}{\partial x_j}\left(\frac{\partial F_d}{\partial n_{i,j}}\right) - \frac{\partial F_d}{\partial n_i}, \qquad h_i^m = \chi_a n_e H_e H_i .$$

Here F_d is the Frank free energy density and χ_a is the magnetic anisotropy; $h_i^\perp = h_i - n_i n_e h_e$ is the transverse component of the molecular field. The longitudinal components of **h** has no physical sense.

4. ASYMMETRY OF VISCOSITY

For anisotropic liquids with a non-symmetric stress tensor, two complex viscosities can be defined

$$\eta_1[\omega] = \frac{\sigma_{21}(\omega)}{\dot\gamma(\omega)}, \qquad \eta_2[\omega] = \frac{\sigma_{12}(\omega)}{\dot\gamma(\omega)} \tag{4}$$

related to the oscillating shear stresses $\sigma_{21}(\omega)$ and $\sigma_{12}(\omega)$. Here $\dot\gamma$ is a shear rate. The essential feature of anisotropic liquids is that the viscosities (4) naturally depend on the relative orientation of the director and the flow. Aligning the director by an external field parallel to the direction of oscillating flow (a), parallel to the velocity gradient (b) and perpendicular to both these directions (c), it is possible to determine the three principal complex viscosities

$$\eta_1^a[\omega] = \frac{1}{\tau^+ - \tau^-}\left[\frac{\eta_2^a\tau_4 - \eta_1^a\tau^-}{1 - i\omega\Theta_1} - \frac{\eta_2^a\tau_4 - \eta_1^a\tau^+}{1 - i\omega\Theta_2}\right],$$

$$\eta_1^b[\omega] = \frac{1}{\tau^+ - \tau^-}\left[\frac{\eta_2^b\tau_5 + \eta_1^b\tau^+}{1 - i\omega\Theta_1} - \frac{\eta_2^b\tau_5 + \eta_1^b\tau^-}{1 - i\omega\Theta_2}\right],$$

$$\eta_1^c[\omega] = \frac{\eta_1^c}{1 - i\omega\tau_1} \tag{5}$$

associated with a shear stress $\sigma_{21}(\omega)$ and three principal complex viscosities

$$\eta_2^a[\omega] = \frac{1}{\tau^+ - \tau^-}\left[\frac{\eta_2^a\tau^+ + \eta_1^a\tau_5}{1 - i\omega\Theta_1} - \frac{\eta_2^a\tau^- + \eta_1^a\tau_5}{1 - i\omega\Theta_2}\right],$$

$$\eta_2^b[\omega] = \frac{1}{\tau^+ - \tau^-}\left[\frac{\eta_1^b\tau_4 - \eta_2^b\tau^-}{1 - i\omega\Theta_1} - \frac{\eta_1^b\tau_4 - \eta_2^b\tau^+}{1 - i\omega\Theta_2}\right],$$

$$\eta_2^c[\omega] = \frac{\eta_2^c}{1 - i\omega\tau_1} \tag{6}$$

associated with a shear stress $\sigma_{12}(\omega)$. Here we have introduced the following notations for relaxation times

$$\tau^\pm = \frac{1}{2}\left[\tau_3 - \tau_6 \pm \sqrt{(\tau_3 - \tau_6)^2 + 4\tau_4\tau_5}\right]$$

$$\Theta_1 = \tau_1 + \tau_6 + \tau^+, \quad \Theta_2 = \tau_1 + \tau_6 + \tau^- .$$

The constitutive equation (1,2) leads to the six zero shear viscosities connected with a shear stress σ_{21}

$$\eta_1^a = \frac{1}{2}(\eta_1 + \eta_4 + \eta_6), \quad \eta_1^b = \frac{1}{2}(\eta_1 + \eta_3 - \eta_5),$$

$$\eta_1^c = \frac{1}{2}\eta_1 \tag{7}$$

and shear stress σ_{12}

$$\eta_2^a = \frac{1}{2}(\eta_1 + \eta_3 + \eta_5), \quad \eta_2^b = \frac{1}{2}(\eta_1 + \eta_4 - \eta_6),$$

$$\eta_2^c = \frac{1}{2}\eta_1 . \tag{8}$$

The important feature of viscosities (7,8) is that they are always positive. This condition imposes an additional restriction on viscosities, $\eta_1 - \eta_6$. When deriving the relations (5-8), we have assumed that an applied magnetic field is sufficiently strong to suppress boundary and shear flow aligning effects.

5. CONCLUSION

The relaxation constitutive equations (1,2) describe anisotropy of viscosity as well as the anisotropy of slow relaxation processes in monodisperse polymer nematics. Here, we have shown the effect of "asymmetry" on the shear viscosity of nematics. The measurement of six principal shear viscosities for three basis direction of liquid orientation by nontraditional viscometry is of great interest. There are many open problems which can be solved using this approach.

6. REFERENCES

Akay G, Leslie FM (1984) In: Prog. IX Intl. Congress on Rheology, Mexico, pp 495-502

Martins AF (1994) In: Carfagna C. (ed) Liquid Crystalline Polymers. Pergamon Press UK pp. 153-161.

Volkov VS, Kulichikhin VG (1990) J Rheol 34:281-293

DETERMINATION OF EQUATION FOR A FAMILY OF CREEP CURVES FOR 38 MoCrAl 09 STEEL BASED ON EXPERIMENTAL RESEARCH

N. AMARIEI, C. COMANDAR, D. LEON

Technical University "Gh. Asachi" of Iaşi, Faculty of Mechanics, Bd. Mangeron, 63, 6600 Iaşi, Romania

1. INTRODUCTION

The use of different mathematics models to calculate the residual stresses generated during the nitriding process, Daves and Fischer (1993), Schreiber et al. (1993), Amariei et al. (1997), implies, among others, the knowledge of creep behavior for the nitrided steel. Because the steels for nitriding are not commonly used in conditions of stresses at high temperatures the experimental data base presented until now is insufficient in what concerns the steel creep. The paper presents the functions of the creep families of curves at 500°C and 550 °C for 38 MoCrAl 09 steel. The experimental data were processed using an original methodology, briefly presented in as follows.

2. METHODS

In order to determine the evolution with time of accumulations of plastic strains at constant temperature and stress, Comandar et al. (1997) proposed the following function:

$$\varepsilon^{pl}(t) = A \cdot t + B \cdot \left(1 - e^{-C \cdot t}\right) \quad (1)$$

where: ε^{pl} = plastic strain; t = time; A, B, C = coefficients depending on material, stresses and temperature.

The values of A, B and C coefficients are calculated considering the experimental data, by means of the least squares method.

The data obtained from creep testing for several σ_i stresses are used to determine the following functions:

$$\varepsilon_i^{pl}(t) = A_i \cdot t + B_i \cdot \left(1 - e^{-C_i \cdot t}\right) \quad (2)$$

In order to establish a function for the family of creep curves, the results obtained for A_i, B_i and C_i coefficients are extrapolated. The extrapolation functions considered are polynomial functions. Their coefficients are determined with the least squares method.

For a constant testing temperature the variations of A, B, C coefficients with stress are described by the following functions:

$$A(\sigma) = f_1(\sigma), \; B(\sigma) = f_2(\sigma), \; C(\sigma) = f_3(\sigma) \quad (3)$$

where $f_i(\sigma)$ are polynomial functions:

$$f(\sigma) = a_n \sigma^n + a_{n-1} \sigma^{n-1} + \ldots + a_1 \sigma + a_0 \quad (4)$$

The degree of $f_i(\sigma)$ polynomial functions is chosen so that the best acceptable approximation accuracy is obtained.

Finally, the equation for the family of creep curves, for a constant T temperature is a follows:

$$\varepsilon^{pl}(\sigma, t) = A(\sigma) \cdot t + B(\sigma) \cdot \left(1 - e^{-C(\sigma) \cdot t}\right) \quad (5)$$

Relation (5) represents the equation of a surface generated by the creep curves, for T temperature. By variating the testing temperature it results a family of such surfaces.

3. RESULTS

The results to be presented as follows are referring to the 38 MoCrAl 09 nitriding steel (Romanian standard STAS 791). The chemical composition and mechanical characteristics are presented in Table 1 and 2.

Table 1. Chemical composition, % weight.

C	Si	Mn	S	P	Cr	Mo	Al
0.37	0.36	0.51	0.019	0.021	1.58	0.15	0.98

Table 2. Mechanical characteristics.

Temperature [°C]	Rp_{02} [MPa]	Rm [MPa]	A [%]
20	795	1080	12
500	625	740	20
550	480	530	26

Based on the creep tests and following the methodology above presented, for the $A(\sigma)$, $B(\sigma)$ and $C(\sigma)$ coefficients were determined the following functions:

- for T = 500 °C

$$A(\sigma) = 5.36E\text{-}13 * \sigma^3 - 5.74E\text{-}11 * \sigma^2 + 2.8E\text{-}9 * \sigma + 1.94E\text{-}7 \quad (6)$$

$$B(\sigma) = 2.92E\text{-}13 * \sigma^4 - 1.3E\text{-}10 * \sigma^3 + 3.02E\text{-}8 * \sigma^2 + 1.31E\text{-}6 * \sigma - 5.61E\text{-}7 \quad (7)$$

$$C(\sigma) = 3.65E\text{-}7 * \sigma^2 + 2.56E\text{-}5 * \sigma - 8.5E\text{-}4 \quad (8)$$

-for T = 550 °C

$$A(\sigma) = 1.46E\text{-}13 * \sigma^4 - 3.96E\text{-}11 * \sigma^3 + 3.3E\text{-}9 * \sigma^2 - 5.56E\text{-}8 * \sigma + 1.05E\text{-}7 \quad (9)$$

$$B(\sigma) = 1.44E\text{-}11 * \sigma^4 - 4.05E\text{-}9 * \sigma^3 + 3.65E\text{-}7 * \sigma^2 - 1.71E\text{-}6 * \sigma - 1.03E\text{-}16 \quad (10)$$

$$C(\sigma) = 2.43E\text{-}7 * \sigma^2 + 1.73E\text{-}4 * \sigma + 6.35E\text{-}5 \quad (11)$$

where the stress σ is measured in MPa and the time is expressed in minutes.

Figures 1 and 2 are presenting the surfaces generated by the creep curves functions defined using relations similar to relation (5). The creep curves correspond to the two testing temperatures.

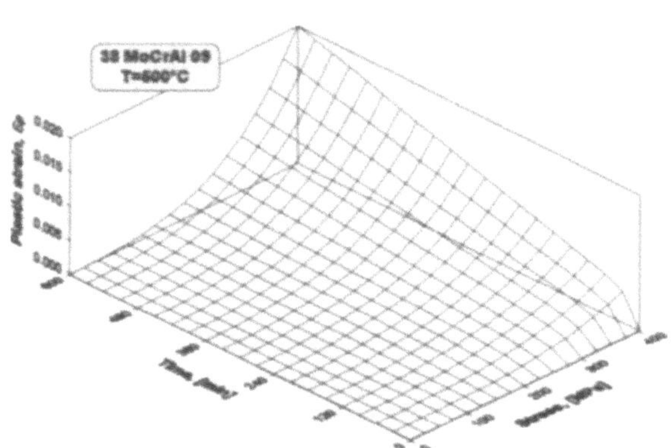

Figure 1. Variation of plastic strain with stress and time (T = 500 °C).

The surfaces illustrated in Figures 1 and 2 suggest that the temperature constitutes a parameter with a strong influence on the growth of plastic strain with time, for a certain creep stress value.

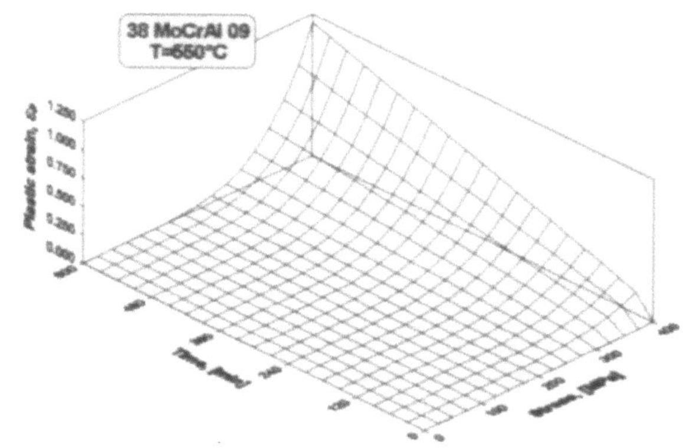

Figure 2. Variation of plastic strain with stress and time (T = 550 °C).

4. CONCLUSIONS

The experimental results can be used as input data in an automatic computation program for the prediction, based on different mathematics models, of the state of residual stresses generated during nitriding.

5. REFERENCES

Amariei N, Leon D, Comandar C (1997) In: Hrabovský M (ed) Proc. of the 35 th International Conference on Experimental Stress Analysis, EAN '97, Olomouc, Czech Republic, 4-6 June 1997, pp. 9-14

Comandar C, Leon D, Amariei N (1997) In: Hrabovský M (ed) Proc. of the 35 th International Conference on Experimental Stress Analysis, EAN '97, Olomouc, Czech Republic, 4-6 June 1997, pp. 73-78

Daves W, Fischer FD (1993) In: Hauk V, Hougardy HP, Macherauch E, Tietz HD (eds) Residual Stresses. DGM Informationsgesellschaft, pp.957-964

Schreiber G, Oettel H, Darbinjan W, Wiedemann R (1993) In: Hauk V, Hougardy HP, Macherauch E, Tietz HD (eds) Residual Stresses. DGM Informationsgesellschaft, pp. 965- 974

IMPREGNATION OF UNWOVEN MATERIALS BY POLYMERS WITH YIELD STRESS

S. ANTONOV[1], V. KULICHIKHIN[2], S. KOTOMIN[2], G. KULICHIKHIN[1], M. KERBER[1]

[1] Mendeleev University of Chemical Technology of Russia, 9 Miusskaya pl,125190 Moscow, Russia
[2] A. V. Topchiev Institute of Petrochemical Synthesis, Russian Academy of Sciences, 29 Leninsky pr, 117912 Moscow Russia

INTRODUCTION

Use of advanced thermoplastics as matrices for polymer composites is restricted mainly by extremely high viscosity of their melts. Since the melts of thermotropic liquid crystal polymers (LCPs) have rather low viscosity their addition to commodity thermoplastics with LCPs can improve processability and mechanical properties of the initial thermoplastics. Viscoplastic rheological behaviour of LCPs, however, can influence on impregnation process. Therefore to determine technological parameters of this process one should study viscoplastic properties of LCPs and blends melts and kinetics of impregnation process. Process of LCP melts filtration in steady-state conditions was investigated by Dreval' et al. (1995) but the real impregnation process is connected with unsteady melt flow through the filament.

METHODS

As LCP we used copolyester of PET and p-hydroxybenzoic acid 20:80 (trade mark Rodrun). Aromatic polysulfone (PSF) obtained from biphenylolpropane and dichlorodiphenylsulfone was chosen as thermoplastic. Blends under consideration contained 5, 10, 25, 50, and 75 % of Rodrun.

Yield stress values were determined using squeezing flow technique and process parameters were determined according to equations reported by Laun (1992). In this method a sample is squeezed between two parallel disks at constant load. The gap between disks is measured. Shear stress decreases with time and when shear stress τ reaches yield stress value τ_f squeezing stops and τ_f can be calculated from the residual gap value.

The capillary viscometer with capillary replaced with metallic net was used for modelling impregnation process. The unwoven fabrics prepared from aramide and carbon fibers were placed on metallic net.

All experiments were carried out at 280, 290, 300, 310°C.

RESULTS AND DISCUSSION

PSF melt is a typical pseudoplastic liquid with weak viscosity anomaly at high shear rates. Melts of Rodrun and all blends show viscoplastic behaviour with yield stress even at small LCP content. The maximal τ_f value was observed for Rodrun/PSF=75/25 blend at 280°C. Yield stress decreases with temperature for Rodrun and all blends. (Fig. 1)

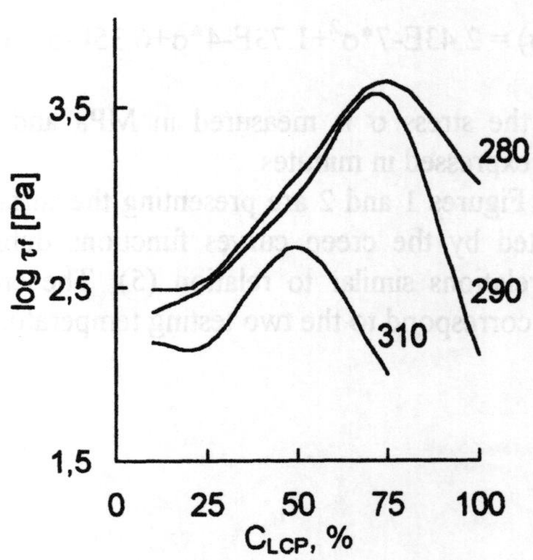

Fig. 1. Concentration dependence of yield stress values at 280, 290 and 310°C.

The fact that polymer blends containing LCP can possess higher yield stress value than that of pure LCP can be explained by essential additional contribution of interphase phenomena to yield stress value.

Capillary viscometer allows us to observe kinetics of impregnation process. Kinetics curves plotted in coordinates $\log(l^2/p)$ - $\log t$ (l - impregnation depth, p - pressure applied, t - time) for systems containing LCP show deviation from linear law inherent for Newtonian liquids.(Fig. 2). For the blends under consideration viscosity increases at approaching τ_f.

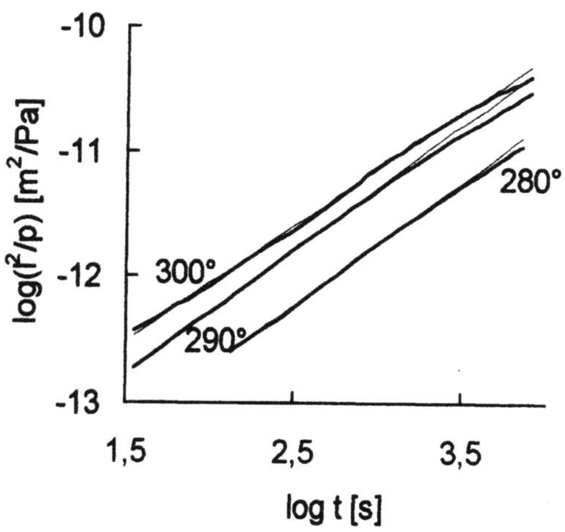

Fig. 2. Impregnation kinetics for PSF/Rodrun=75/25 blend at 280, 290 and 300°C.

Data on impregnation kinetics were used for flow curves plotting. The flow curves obtained by squeezing flow method, rotational viscometry and impregnation showed good coincidence including their shape and location (Fig. 3).

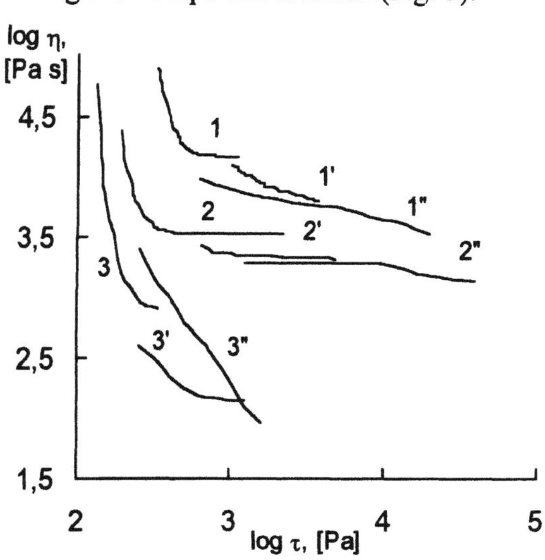

Fig. 3. Flow curves for PSF/Rodrun=90/10 (1, 1', 1"), 75/25 (2, 2', 2"), 0/100 (3, 3' 3") systems obtained by squeezing flow technique (1 - 3), impregnation kinetics (1' - 3') and rotational viscometry (1" - 3") at 290 °C.

ESM method was used for testing the impregnation homogeneity. Morphological study indicate good composite quality for blends containing up to 50 % of LCP.

CONCLUSIONS

Data on yield stress values, which are of greate importance for estimation of proper impregnation process parameters, can be easily obtained by means of squeezing flow technique. Interface phenomena (surface tension, interaction, partial interdiffusion etc.) in heterogenious systems can contribute significantly to yield stress values.

Impregnation kinetics curves for systems containing LCP show deviation from typical for newtonian liquids linear $\log(l^2/p)$ - log t dependencies which is connected with viscosity increase at approaching τ_f.

Data on impregnation kinetics can be replotted to flow curves, which coincide good enough with dependencies obtained by traditional viscometry methods.

Morphological study shows good quality of composites prepared from blends with low and moderate LCP content.

Laun H. M. (1992) Makromol. Chem., Macromol. Symp., 56, 55

Dreval' V. E., Khairetdinov F. N., Litvinov I. A., Kerber M. L., Kulichikhin V. G. Polymer science, 1995, v. 37, N 1, pp 67-76

EMPIRICAL CORRELATIONS BETWEEN FREE VOLUME FRACTIONS FROM PALS DATA AND VARIOUS DYNAMIC QUANTITIES IN ORTHO-TERPHENYL

J.BARTOŠ[1], J.KRIŠTIAK[2]

[1]Polymer Institute of SAS,Dúbravská cesta 9, SK-842 36 Bratislava
[2]Institute of Physics of SAS, Dúbravská cesta 9, SK-842 28 Bratislava, Slovak Republic

1.INTRODUCTION

Phenomenologically, the mobility properties of supercooled liquids such as viscosity ($M = \eta$), relaxation time ($M = \tau$) and diffusion coefficient ($M = 1/D$) are usually described by the Vogel - Fulcher (VF) equation or, equivalently, by the Williams - Landel - Ferry (WLF) equation, Williams et al (1955). Both the equations are often interpreted within the free volume concept at a phenomenological level using the empirical Doolittle equation, Doolittle (1951) or / and at a quasi-molecular one by means of a free volume model of Cohen, Turnbull (1959). The WLF Doolittle formula for the shift factor $a(T) = M(T)/M(T_g)$ can be expressed as follows:

$$\ln a(T) = B_f [\, 1/f(T) - 1/f(T_g) \,] \qquad (1)$$

The use of eq. 1 relies on the two additional assumptions 1) a linearity between the free volume fraction f and the temperature T and 2) the postulate that $B_f = 1$, which have not until now been verified by independent means.

Positron annihilation lifetime spectroscopy (PALS) is a unique method to direct detection of the local regions of lowered density, i.e., free volume holes in disordered materials through annihilation of the ortho - positronium (o-Ps). Two characteristics are measured i) the o-Ps lifetime τ_3 which is connected with the (spherical) hole size V_h via a quantum-mechanical model, Jean (1990) and ii) the o-Ps relative intensity I_3 which is a measure of the number of holes. A quantification of the free volume hole fraction $f_h(T)$ requires to determine the respective coefficients in the following equations:

$$f_h(T) = K_\tau \tau_3(T) I_3(T) \; ; \; f_h(T) = K_{Vh} V_h(T).I_3(T) \quad (2)$$

by means of a suitable model or some calibration procedure.

We present a simple phenomenological model and its applications on some of the dynamic processes in a typical fragile low-molecular glassformer: ortho-terphenyl (OTP).

2.THE MODEL

Our model linking the macroscopic volume from dilatometry with the microscopic free volume data from PALS is based on an assumed linear form of dependence of the total volume V on temperature T in a liquid state above the glass transition temperature $T>T_g$, Krištiak et al (1997):

$$V(T) = V_i + \beta_l (T - T_i) \qquad (3)$$

where β_l is the thermal expansivity of the liquid state and T_i, V_i are free parameters of the model which have to be determined and their physical meanings established. Using the usual definitions of the free volume fractions:

$$f_h(T) = [V(T) - V_i]/V(T) \qquad (4)$$

and of its expansion coefficient:

$$\alpha_{fh} = 1/f_{h,g}[\Delta f_h(T)/\Delta T] \qquad (5)$$

we arrive to a quadratic equation with respect to T_i, whose coefficients obtain both the expansion coefficients of total volume α_V and of free volume α_h,; they differs typically by two order of magnitude. Knowing the initial temperature T_i of the volume change as a solution of this quadratic equation and the macroscopic total volume $V_g(T_g)$, we can calculate the free volume hole fraction f_{hg} at T_g and consequently, the coefficients in eq.2. Finally, all relevant free volume hole quantities, especially the free volume hole fractions f(T) as an additional volume with respect to the initial one V_i, can be determined from PALS data.

3.RESULTS AND DISCUSSION

The V - T data on OTP of Naoki et al (1989) were used to obtain the macroscopic volume expansion coefficient: $\alpha_V = 7.3x10^{-4}K^{-1}$ above $T_g = 243K$. The PALS data on OTP of Malhotra and Pethrick (1982) were analyzed by means of a quantum - mechanical model of o-Ps in the spherical box in order to extract the free volume expansion coefficient: $\alpha_{Vh} = 2.56x10^{-2} K^{-1}$. An application of our model leads to the initial temperatures of volume change: $T_{i,1} = 206$ K and $T_{i,2} = 1702K$. Whereas the second root has evidently no physical meaning, the first value is in an excellent agreement with the Kauzmann temperature $T_K = 206.5K$ from the very recent calorimetric investigation, Murthy et al (1995). Moreover, because of a close relationship between T_K and the Vogel temperature T_0 from the VF equation for the primary mobility, i.e., viscosity η or α relaxation time τ, Angell(1997), $T_{i,1}$ is in an acceptable agreement with this divergence temperature. As for the other model parameter V_i, the packing coefficient V^w/V_i at T_i is equal to 0.722. This value is well above the packing fraction for the closest random packing (0.649) but slightly below that for the closest regular packing of hard spheres 0.7405, Jodrey and Tory (1985). Both these findings seem to support arguments in favor of the underlying phase transition concept and the associated ideal glass idea. The former is hardly accessible due to the known intervening kinetic effects at T_g and the latter notion suggests on the defect-ordered crystal as a possible candidate for the hypothetical ideal glass. Knowing the both parameters, the free volume hole fraction at T_g can be calculated to be 2.7%. Consequently, the coefficients K_τ and K_{Vh} in eq.2 reach the values of $4.17x10^{-2}$ ns^{-1} or $1.096x10^{-3} K^{-1}$, respectively.

Our model has been applied on dynamical data of OTP in the temperature range from T_g up to $1.2T_g$, i.e., the viscosity η and the tracer translational diffusion coefficient D_{tr} in OTP in in order to test the validity of the WLF-Doolittle eq.1 using $f_h(T)$. The viscosity data were taken from Greet and Turnbull (1967) and the tracer translational diffusion coefficients for ACR and TTI photochromic probes from forced Rayleigh scattering (FRS) study by Fujara et al (1992).

Figure 1 demonstrates the case of ACR diffusion in OTP. We can see that the WLF - Doolittle eq. form with the free volume fractions from the PALS data is able to fit the shift factor $a(T) = D(T_g)/D(T)$ over 8 decades of the change. The similar finding holds for the viscosity but with the other $B_h$coefficient.The fact that $B_{h,\eta} = 1.18 > B_{h,Dtr} = 0.84$ indicates the easier course of the translational diffusion process than that responsible for the viscosity phenomenon. This fact is consistent with the known decoupling effect between the diffusion and the viscosity, Fujara et al (1992).

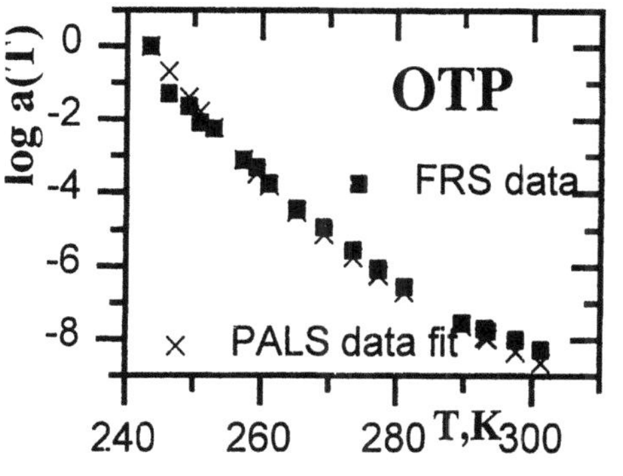

Figure 1. Correlation between the PALS data and the tracer (ACR) translational diffusion coefficient in OTP in terms of the WLF-Doolittle equation

5.REFERENCES

Angell CA(1997) J Res Natl Inst Stands Technol 102: 171-185

Cohen MH, Turnbull D (1959) 31: 1164 - 1169

Doolittle AK (1951) J Appl Phys 22: 1471 - 1475

Fujara F, Geil B, Sillescu H, Fleischer G (1992) Z Phys B88: 195 - 204

Greet RJ, Turnbull D (1967) J Chem Phys 46: 1243 -1251

Jean YC (1990) Microchem J 42: 72 - 120

Jodrey WS, Tory EM (1985) Phys Rev A32: 2341-2351

Krištiak J, Bartoš J, Šauša O, and Bandžuch (1997) Mater.Sci.Forum 255-257: 35 - 39

Malhotra BD, Pethrick RA (1982) J Chem Soc Far Trans 2 78: 297 - 306

Murthy SSN, Paikaray A,,Arya N (1995) J Chem Phys 102:8213 - 8220

Naoki M, Koeda S (1989) J Phys Chem 93: 948-955

Williams ML,Landel RF,Ferry JD (1955) J Amer Chem Soc 77:3701-3706

EXPERIMENTAL VERIFICATION OF THE WLF - DOOLITTLE EQUATION FOR SUPERCOOLED LIQUID DYNAMICS BY PALS METHOD: POLYISOBUTYLENE

J.BARTOŠ[1],J.KRIŠTIAK[2]

[1]Polymer Institute of SAS, Dúbravská cesta 9, 842 36 Bratislava
[2]Institute of Physics of SAS, Dúbravská cesta 9, 842 28 Bratislava, Slovak Republic

1.INTRODUCTION

The WLF equation is one of the most important formulae in polymer physics and for a long time is used to describe the universal non - Arrhenius behavior of viscosity η or of relaxation time τ, Williams et al (1955).Originally, the WLF eq.for the shift factor a (T) was interpreted within the free volume concept using the Doolittle equation, Doolittle (1951):

$$\ln a(T) = \ln \eta(T)/\eta(T_g) = \ln\tau(T)/\tau(T_g) =$$

$$= B_f [1/f(T) - 1/f(T_g)] \qquad (1)$$

However, this explanation is based not only on the use of a functional form of the Doolittle eq., but also on the two additional assumptions concerning the temperature dependence of the free volume fraction f(T) and the B_f coefficient, $B_f = 1$.At present, positron annihilation lifetime spectroscopy(PALS) is, due to the recent methodological progress, able to study the free volume microstructure through measurements of both annihilation characteristics of ortho-positronium (o-Ps) i.e., lifetime τ_3 and relative intensity I_3, Jean (1990). It opens up a possibility to test the validity of the WLF-Doolittle equation directly.

2.SAMPLE AND METHOD

Polyisobutylene(PIB) with M_w=4.2 10^5 and the glass transition temperature T_g = 203K was used.

PALS measurements were performed with the standard lifetime set-up. The annihilation spectra were analyzed with the PATFIT- 88 software package by fitting to the three components. The o - Ps lifetimes τ_3 were transformed into the free volume hole radius R_h and V_h using a semi - empirical quantum-mechanical model,Jean(1990):

$$\tau_3 = 0.5[1 - R_h/R_0 + (1/2\pi)\sin(2\pi R_h/R_0)]^{-1} \qquad (2)$$

where $R_0 = R_h + \Delta R$, ΔR is the fitting parameter from the data on a series of compounds with the known hole sizes. The corresponding expansion coefficients of macroscopic volume α_V and microscopic free volume α_h are defined as usually as:

$$\alpha_X = (1/X_g).(\Delta X/\Delta T) \qquad (3)$$

where X = V or $V_h = (4/3)\pi R_h^3$ respectively, and X_g is the particular X value at T_g.

3.THE MODEL

A phenomenological model linking the macroscopic volume to the microscopic free volume from PALS is presented to explain the order difference between the corresponding expansion coefficients α_V and α_h, respectively, Krištiak et al (1997). Starting assumption is a linear dependence of the macroscopic volume V on temperature T:

$$V(T) = V_i + \beta_l . (T - T_i) \qquad (4)$$

where β_l is the expansivity of liquid state above T_g; V_i and T_i are free parameters of our model whose values and physical meanings need to be determine. By using the usual definitions for the free volume fraction and its expansion coefficient we can arrive to a quadratic equation with respect to T_i. The particular coefficients have rather complicated forms containing both the expansion coefficients α_V and α_h. Subsequently, the knowledge of T_i allows us to determine all free volume characteristics as an expansion volume with respect to V_i. Finally, the semi-empirical coefficients K_τ (or K_{Vh}) in the bridge equation between the free volume fraction $f_h(T)$ and both annihilation characteristics of o-Ps τ_3 (or V_h) and

I_3 :

$$f_h(T) = K_\tau.\tau_3(T).I_3(T) = K_{Vh}.V_h(T).I_3 \qquad (5)$$

can be obtained from the free volume hole fraction at T_g , $f_h(T)$. by means of eq.(4).

4.RESULTS AND DISCUSSION

Typical PALS results, i.e., τ_3 and I_3 vs. T dependences for PIB have already been published Bartoš et al (1997). The τ_3-T plot exhibits dramatic change at around the glass transition temperature T_g which is fully consistent with the known knee effect in the V-T plot. Despite this similarity, the most pronounced fact is the order difference between the corresponding thermal expansion coefficients in the elastic state: $\alpha_V = 6.2\times10^{-4}\ K^{-1}$, Simha, Boyer (1962) and $\alpha_{Vh} = 1.3\times10^{-2}\ K^{-1}$ as determined for the temperature range from T_g up to about 290K.

An application of the above - mentioned model leads to the following findings:
i) the initial temperatures of volume change are $T_{i,1}$ = 129K and $T_{i,2}$ = 2090K respectively. The first value is close to the Vogel temperature T_0 = 122K from the empirical Vogel - Fulcher equation for viscosity, Berry,Fox (1968).
ii) the packing fraction at $T_{i,1}$, defined as the ratio of van der Waals volume of monomer unit V^w to initial volumeV_i, is found to be 0.717. This value lies above the packing fraction for the closest random packing of hard spheres 0.649 but it is close to that for the closest regular packing of hard spheres, 0.7405, Jodrey, Tory (1985). Both the findings seem to support the underlying phase transition hypothesis and the associated ideal glass concept; the former occuring at $T_i \cong T_0$ below the thermodynamic glass transition but being inaccessible due to the kinetic effects and the latter being probably the defect-ordered crystal.

Having both the model parameters we can found the free volume hole fraction at T_g to be $f_{h,g}$ = 4.3 % . Consequently, the corresponding "bridge" coefficients are found to be K_τ = 0.129 ns^{-1} or $K_{Vh} = 4.104\times10^{-3}$ Å^{-3}, respectively.

Figure 1 demonstrates a direct correlation between the free volume hole fractions $f_h(T)$ from PALS data according to our model and the typical dynamic quantity - viscosity data of Plazek et al (1992) in terms of eq.1. We can see that the WLF-Doolittle form fit holds for eight decades of the viscosity variation in PIB between T_g = 203K and about 300K. This finding indicates that the free volume holes as detectable by the PALS method are of relevance for the course of motional process which is responsible for the viscosity phenomenon.

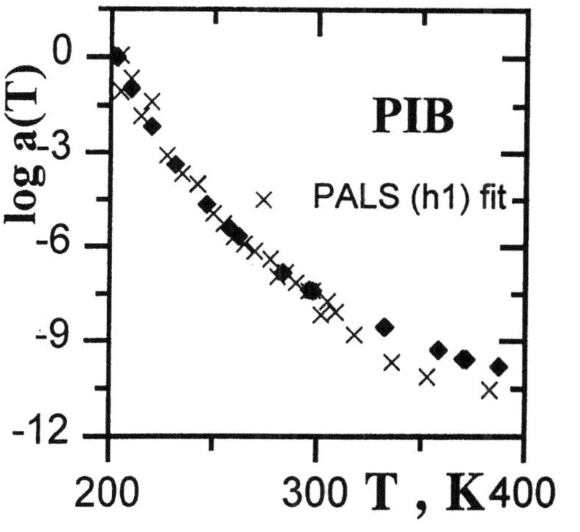

Figure 1. Empirical verification of the WLF - Doolittle equation for the published viscosity data (♦) Plazek et al (1992) on PIB using our PALS data and the presented phenomenological model

5. REFERENCES

Bartoš J, Krištiak J, Kanaya T (1997) Physica B 234 - 236: 425 - 436

Berry GC, Fox TG (1968) Adv Polym Sci 5: 260 - 357

Doolittle AK (1951) J Appl Phys 22:1471 - 1475

Jean YC (1990) Microchem J 42: 72 - 102

Jodrey WS, Tory EM (1985) Phys Rev A32: 2347 - 2351

Krištiak J, Bartoš J, Šauša O, Bandžuch (1997) Mater Sci Forum 255 -257: 35 - 39

Plazek DJ, Zheng XD, Ngai KL (1992) Macromolecules 25:4921-4924

Simha R, Boyer R (1962) J Chem Phys 37: 1003-1007

Williams ML, Landel RF, Ferry JD (1955) J Amer Chem Soc 77: 3701-3706

RHEOLOGY OF A SOLID POLYCARBONATE AT HIGH STRAIN RATE; EXPERIMENTAL STUDY AND NUMERICAL SIMULATION

M. L. BISILLIAT, N. BILLON

Ecole des Mines de Paris, CEMEF, UMR CNRS 7635, BP 207, 06904 Sophia-Antipolis Cedex, France

1. INTRODUCTION

The general purpose of this study is to characterise the mechanical behaviour of thermoplastics polymers over a wide range of strain, including plastic deformation, strain-rates and loading, including impact.

2. EXPERIMENTAL

This paper focuses on a polycarbonate (PC) whose glass transition temperature and average molecular weight are 150°C and 31 027 g per mol, respectively. Cylindrical specimens are machined from extruded rods and characterised using tensile, compression and torsion tests.

Tensile experiments are performed at temperatures ranging from 23 °C to 120 °C, and with strain rates ranging from 10^{-4} to 250 s^{-1}. Hour-glass-shaped samples whose initial smallest diameter is 6 mm are used. *Compression* tests are performed at room temperature. Strain rate ranges from 10^{-4} to 5 10^3 s^{-1}. Samples are 15 mm-diameter samples of height ranging from 3 to 15 mm. Their surfaces are lubricated to minimise friction effects. *Torsion* tests are performed at room temperature up to 25 s^{-1} on non-tubular specimen (diameter 10 mm).

For low strain-rates (≤ 20 s^{-1}) a video camera allows to determine the true stress as a function of the true strain by recording the current diameter of the sample. For strain rates lower than 10^{-3} s^{-1} compression and tensile tests are performed at constant strain rates thanks to a video-controlled system. More rapid deformation are studied at constant cross-head velocities (tension) or using the Split Hopkinson Pressure Bar technique (compression) (Zhao and Gary (1996)). In these cases, local measurements are impossible, the deformation is assumed to be homogeneous in the sample.

This polymer exhibit a typical behaviour depending on the temperature, the strain-rate and the type of loading. However, in any case the stress vs. strain curves is decomposed into four stages (Fig. 1) : *Visco-elastic* stage (strain < 0.1), *softening* stage as plastic deformation is initiated (strain < 0.15), *plastic flow* (necking effect is important in tensile tests), *strain-hardening* process.

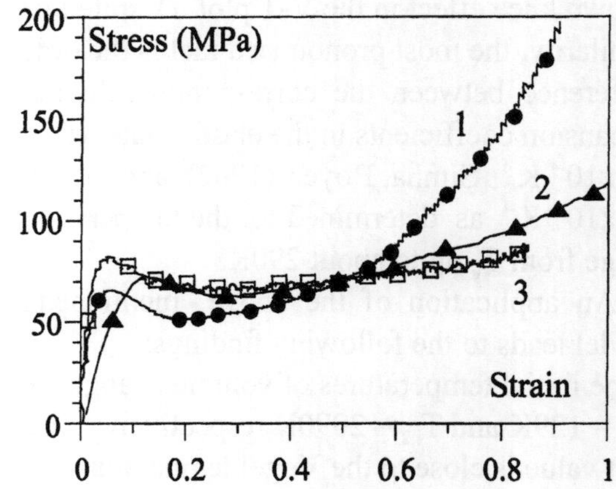

Fig. 1 Comparison between tensile (1), torsion (2) and compression (3) tests (strain rate = 10^{-4} s^{-1}).

Aside from this qualitative agreement significant differences exist: compressive yield stress is higher than others are and strain hardening is noticeably more sensible in tensile conditions. The latter phenomenon is related to an important induced anisotropy in the material.

The higher the strain rates the more important the dependence upon strain rate (Fig. 2). But, as plastic flow induces heat dissipation, it is impossible to unambiguously deduce intrinsic data from rapid tests. Infra-Red measurements show that the temperature of the tensile samples noticeably increases for strain rate as low as 10^{-2} s^{-1} (Tillier (1998)). FEM calculations suggest that adiabatic assumption becomes valid for 1 s^{-1}. Consequently, depending on the strain rate, an important thermal softening is observed once the yield stress has been reached (Fig. 3).

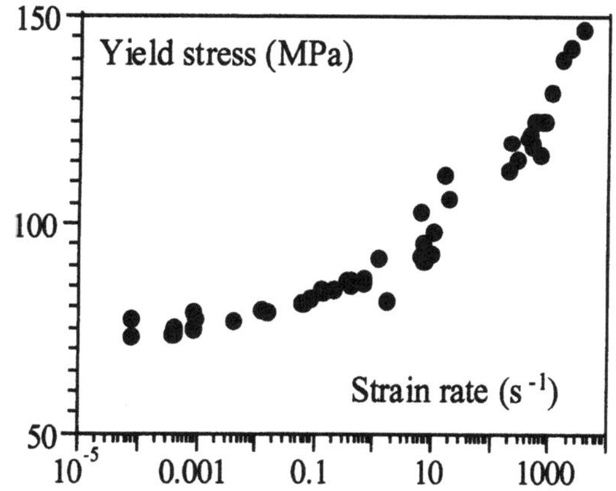

Fig. 2: Compression yield stress vs. strain rate at room temperature.

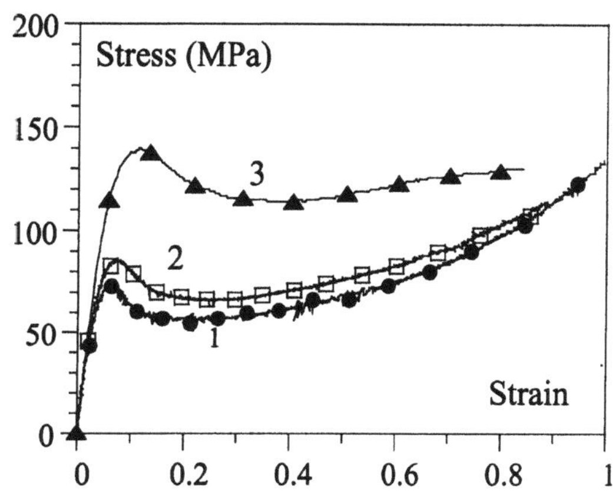

Fig. 3: Compression behaviour of the PC at room temperature: 0.0005 s^{-1} (1), 1 s^{-1} (2) and 250 s^{-1} (3).

3. MODELLING

Constitutive parameters have, then, to be determined using the very low strain rates tests. As a first step, the material is assumed to be homogeneous, incompressible isotropic and to obey the generalised von Mises' flow rule. Within the frame of each test, and revisiting previous phenomenological approaches (e.g., Belcadi (1990), Andrew and Ward. (1970), G'Sell and Jonas (1979)), it is possible to propose an elasto-viscoplastic law (Eq. (1)) to model the polymer:

$$\sigma = K_s\left(1+\vartheta\, \varepsilon_p \exp\left(-w\, \varepsilon_p^2\right)\right)+K \exp\left(\frac{a}{T}\right)\exp\left(h\, \varepsilon_p^2\right)\dot{\varepsilon}^{m} \quad (1)$$

where $\dot{\varepsilon}$ and ε_p are the equivalent plastic strain rate and strain, σ is the equivalent flow stress and T is the temperature, respectively.

This equation allows to model each of the four stages of the deformation, including the softening process, and to reproduces tests up to 250 s^{-1} using thermomechanical FEM calculations and data given table 1.

The effects of hydrostatic stress and strain hardening, which are not taken into account in our approach, are responsible for the discrepancy between data deduced from tensile, compression and torsion tests.

Parameter	Tension	Compression	Torsion
Modulus (MPa)	1300 exp(164/T)	2000	1000
K_s (MPa)	0.9	0.9	1
θ	270	540	272
w	110	140	50
K (MPa s^{-m})	5.3	6.4	5.7
h	1.8	0.52	0.72
m	0.04	0.04	0.02

Table 1: Mechanical characteristics of the PC (T in K). a is equal to 749 K.

4. CONCLUSION

This work represents a complete characterisation of the PC between room temperature and its glass transition. It involves a wide range of strains (up to 1) and strain rates (from 10^{-4} to 5 10^3 s^{-1}) and three kinds of loading: tension, compression and shear. The experimental results allow proposing a constitutive equations valid up to 250 s^{-1}. Inverse method is being developed for analysing more rapid conditions.

In that field, the evaluation of the amount of mechanical work transformed into heat is a problem to be solved since it will govern the temperature in the material.

Nevertheless, the accounting of effects of hydrostatic stress and strain hardening are the main problems to be overcome.

Andrew J M, Ward I M (1970) J Mater Sci 5:411

Belcadi M (1990) Thesis U. de Poitiers, ENSMA

G'Sell C, Jonas JJ (1979) J Mater Sci 14:583

Tillier Y (1998) Thesis EMP (France)

Zhao H, Gary G (1996) Ent. J Solid Structure 33:3363

„DOUBLE REPTATION SPECTRUM" AN ALTERNATIVE MOLAR MASS DISTRIBUTION - MATERIAL FUNCTION CONVERSION

C.J. BRABEC, A. SCHAUSBERGER

Institute of Chemistry, University of Linz, A 4040 Linz, Altenbergerstr. 69

1. INTRODUCTION

The material functions MF of entangled polymer melts depend strongly on the molar mass distribution MMD of the polymer. The idea of double reptation [des Cloizeaux (1988); Tsenoglou (1987)] has brought considerable progress in modelling polydisperse polymers. The blending law

$$\left(\frac{G(t)}{G_N^0}\right)^{1/2} = \sum_{i=1}^{N} w_i F^{1/2}(M_i, t) \qquad (1)$$

combined with the expression for the monodisperse relaxation function F(t, M) of Baumgaertel et al. (1990) or of des Cloizeaux (1990) gives excellent results in calculating the relaxation modulus G(t) and consequently the dynamic moduli $G^*(\omega)$ from MMD as shown by Wasserman and Graessley (1992).

In the reverse calculation - MF to MMD - the same onset creates an ill-posed problem. In order to find practicable solutions this onset is modified e. g. by Mead (1994) and by Wasserman (1995). Unfortunately both,, the modification and the mathematical tool used for solving the ill-posed problem may be the source for less corresponding results. A further disadvantage is that the $G^*(\omega)$ data can not be used directly but must be converted in G(t) data.

The onset presented in this paper incorporates the double reptation idea in the calculation of a spectrum of relaxation times in such a way that the mutual superposition of the relaxation processes of different molecules are resolved („double reptation spectrum"). This is why this spectrum resembles the MMD already and can therefor be transformed into the MMD easily.

2. THEORETICAL BACKGROUND

The relaxation modulus G(t) is related to the number of chain segments still existing at time t. Thus G(t) shows a dependency on the number of entanglements which have survived until the time t. The probability for a chain i to open one of its entanglements by a diffusion process is p(i, t). The probability p(i, j, t) for an entanglement between the chain i and the chain j to open is the product of p(i, t) p(j, t). The number of entanglements between chain i and chain j is proportional to the mass fractions m_i and m_j of the chains with the molar mass M_i and M_j resp. Therefor we write for the relaxation modulus G(t):

$$G(t) = G_N^0 \sum_{i=1}^{N}\sum_{j=1}^{N} m_i m_j p(i,t) p(j,t) \qquad (2)$$

The probabilities p(i,t) show a clear resemblance to the relaxation functions F(M,t).

$$p(i,t) = \sum_{s=1}^{S} a_s \exp\left(-\frac{b_s t}{\tau_i}\right) \quad \text{with} \quad \sum_{s=1}^{S} a_s = 1 \qquad (3)$$

Consequently we introduce a double reptation relaxation strength g which is directly correlated with the mass fraction m. The combination of eqs. (2) and (3) gives the final expression for G(t) (with $\Sigma g_i = G^0_N$)

$$G(t) = \sum_{i=1}^{N}\sum_{j=1}^{N} g_i g_j \left[\sum_{r=1}^{R} a_r \exp\left(-\frac{b_r t}{\tau_i}\right) \cdot \sum_{s=1}^{S} a_s \exp\left(-\frac{b_s t}{\tau_j}\right)\right] \qquad (4)$$

which can be transformed to those for the dynamic modli with $\tau_{i,j,r,s} = (b_r/\tau_i + b_s/\tau_j)^{-1}$

$$G^*(\omega) = \sum_{i=1}^{N}\sum_{j=1}^{N} g_i g_j \left[\sum_{r=1}^{R}\sum_{s=1}^{S} a_r a_s \frac{i\omega\, \tau_{i,j,r,s}}{1 + i\omega\, \tau_{i,j,r,s}}\right] \qquad (5)$$

It is obvious that eqs. (4) and (5) hold for the moduli as long as they ere not masked by the glassy behaviour, i. e. by the motion of chain segments which does not depend on the molar mass. A procedure for detaching the glassy influence from the dynamic moduli has been reported already by Brabec and Schausberger (1996).

The spectrum of the leading relaxation times calculated according eq. (5) (referred to as double reptation spectrum DRS) from „glassy free" dynamic mechanical data can easily be converted into the differential distribution of the molar masses $M > M_c$ via the well-known scaling relation $\tau_i = K(M_i/M_c)^{3.4}$ and $g_i/G^0_N = w(M_i)$. The polymerspecific scaling parameter K depends on the temperature and the volume fraction of the molecules with $M < M_c$ (and of additives).

3. RESULTS

For the calculations of DRS presented there the experimental data, i. e. the dynamic moduli of a nearly monodisperse PS with M_w = 2540 kg/mol and M_w/M_n = 1.13 (standard) and of a polydisperse PS with M_w = 385 kg/mol and M_w/M_n = 2.12 (technical, N 7000, Höchst) are used. In order to record a broad frequency range time - temperature superposition has been applied with T_r = 180°C.

The solution of eq. (5) (with eq. (3) according to Baumgaertel et al. (1990)) has been performed by an iterative algorithm [Brabec and Schausberger (1995)] and by a nonlinear regularization method [Hohnerkamp and Weese (1993)]. Both methods calculate similar results.

The consequence of the DRS is illustrated in Fig. 1 where this spectrum is compared with the usual relaxation time spectrum (spectrum of Maxwel elements) both calculated from the glass free dynamic moduli of the PS standard.

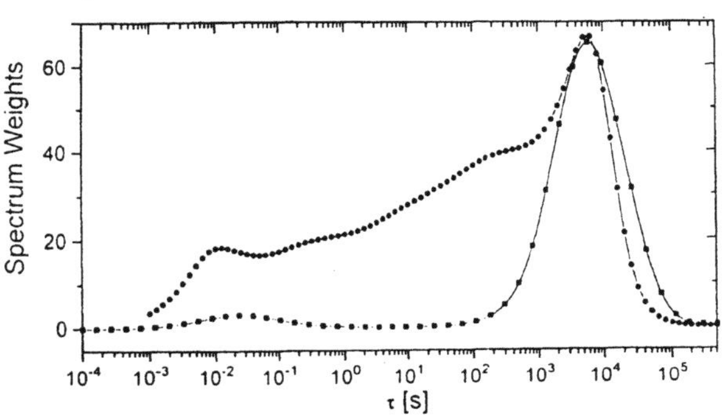

Fig. 1: Relaxation time spectrum (dots) and double reptation spectrum (squares) of the PS standard

The DRS shows that the glassy influence has been detached properly as only molar mass relevant relaxation times show significant strengths and demonstrates the effect of the mixing law (eq. (1)) as the transformation into a MMD as mentioned above gives M_w/M_n = 1.13 whereas the analogous transformation of the molar mass relevant relaxation times of the usual relaxation time spectrum gives M_w/M_n = 1.07.

In Fig. 2 the calculated MMD has been normalized arbitrarily because we do not know the exact value of G^0_N of PS N7000 as it contains not entangled molecules ($M < M_c$). The position on the log M axis is in context with $K = 2 \cdot 10^{-3}$ s.

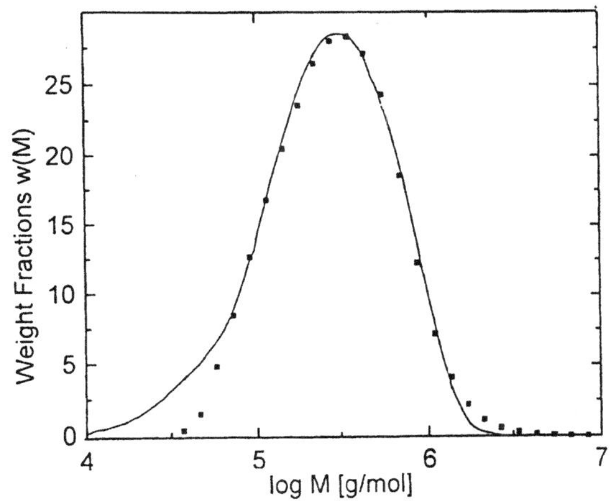

Fig. 2: molar mass distribution of PS N 7000:
measured via SEC (line)
calculated from the experimental glass free dynamic moduli via RDS (squares)

4. CONCLUSION

As is to be expected the DRS as presented in this paper is not able to record the molar masses near and under M_c correctly (see Fig.2, log $M_c \cong$ 4.6). Disregarding this fact the DRS calculation seems to be a proper tool in particular because the superposition of the relaxation processes as expressed by eqs. (4) and (5) reduces the ill-posedness tremendously. But the finding that the dynamic moduli record via DRS calculation larger weight fractions of long molecules is of great interest and has therefor to be proved properly.

5. REFERENCES

Baumgaertel M, Schausberger A, Winter HH (1990) Rheol Acta 29:400-408

Brabec CJ, Schausberger A (1995) Rheol Acta 34: 397-405

Brabec CJ, Schausberger A (1996) Proc XIth Int Congr on Rheology: 51

des Cloizeaux J (1988) J Europhys Lett 5:437-442

des Cloizeaux J (1990) Macromolecules 23: 4678-4687

Hohnerkamp J, Weese J (1993) Rheol Acta 32: 65-73

Mead DW (1994) J Rheol 38(6): 1797-1827

Tsenoglou C (1987) ACS Polym Prepr 28: 185-186

Wasserman SH, Grassley WW (1992) J Rheol 36(4):543-572

Wasserman SH (1995) J Rheol 39(3):601-625

Vibration Transmissivity in Mechanical Systems with Rubber Elements using Viscoelastic Models

Prof.dr.ing. Polidor Bratu[1]
Ing. Aurelia Mihalcea[2]

[1] "Dunărea de Jos" University, Galaţi, Romania, fax 401 255 14 20
[2] Research Institute for Construction Equipment and Technology, 266 Pantelimon Street, Bucharest

The paper presents some results concerning the vibration insulation problem in the case of vibration sensitive equipment. In order to obtain proper vibration insulation solutions, by using rubber based dampers, two rheological models have been considered to describe the rubber behaviour.

Considering a dynamic system with the harmonic disturbing force acting at its base and the Voigt - Kelvin and Hooke - Maxwell rheological models as the means to describe the rubber elements behaviour, the study aimed to establish the necessary conditions for a minimum amplitude of the mass representing the equipment to be protected.

The system transmissivity factor has been determined for each of the two models, in order to evaluate the effectiveness of the vibration insulation solution.

In the case of Voigt - Kelvin rheological model you can obtain:

$$E^* = \frac{k}{\lambda}\left(1 + j\frac{\omega\eta}{k}\right) \equiv E_0(1 + j\delta) \qquad (1)$$

where:

k is the system's equivalent elasticity coefficient;

η - the system's equivalent viscosity factor ;

λ - geometrical multiplication factor, defined by the relation:

$$\lambda = \left(1 + \beta\phi^2\right)\frac{A}{h}; \qquad (2)$$

E_0 - static longitudinal elasticity modulus;

δ - internal loss angle defined by the relation:

$$\delta = \frac{\omega\eta}{k}); \qquad (3)$$

ω - forced circular frequency;

$$\delta_R = \eta / \eta_C$$

where:

$$\eta_C = \frac{2k}{\omega_0} \quad \text{with} \quad \omega_0^2 = \frac{k}{m}$$

$$\delta = 2\,\Omega\,\delta_R$$

$\Omega = \omega / \omega_0$ relative pulsation;
E*- complex elasticity modulus;
j - imaginary unit.

Transmissivity factor is defined by:

$$T = \frac{\left(1 + 4\Omega^2\delta_R^2\right)^{1/2}}{\left[\left(1 - \Omega^2\right)^2 + 4\Omega^2\delta_R^2\right]^{1/2}} \qquad (4)$$

For Hooke - Maxwell rheological model you can obtain:

$$K^* = K + j\,\Delta \qquad (5)$$

where:

$$K = \frac{1}{k_2^2 + \omega^2\eta^2}\left[k_1k_2^2 + (k_1 + k_2)\omega^2\eta^2\right] \qquad (6)$$

$$\Delta = \frac{k_2^2 \eta\omega}{k_1 k_2^2 + (k_1 + k_2)\omega^2 \eta^2} \qquad (7)$$

and for $\frac{d\Delta}{\Delta\omega} = 0$ it results $\omega_M = \frac{k_2}{\eta\sqrt{\alpha}}$

where $\alpha = \frac{k_1 + k_2}{k_1}$

Finally, it results in:

$$\Delta_{max} = \frac{\alpha - 1}{2\sqrt{\alpha}} \qquad (8)$$

$$K = \frac{\alpha k_1 (1 + \upsilon^2)}{\alpha + \upsilon^2} \qquad (9)$$

$$K_0 = \frac{\alpha k_1 (1 + \upsilon_0^2)}{\alpha + \upsilon_0^2} \qquad (10)$$

$$\Delta = 2\Delta_{max} \frac{\upsilon}{1 + \upsilon^2} \qquad (11)$$

$$\Delta = 2\Delta_{max} \frac{\upsilon\upsilon_0}{\upsilon^2 + \upsilon_0^2} \qquad (12)$$

where $\upsilon = \frac{\omega_M}{\omega_0}$ and $\upsilon_0 = \frac{\omega}{\omega_0}$

Taking into account the relation $\frac{G_0}{G} \cong \frac{K_0}{K}$ we obtain:

$$T = \frac{(1 + \Delta^2)^{1/2}}{\left[\left(1 - \Omega^2 \frac{K_0}{K}\right)^2 + \Delta^2\right]^{1/2}} \qquad (13)$$

with

$$\frac{K_0}{K} = \frac{(\upsilon^2 + \alpha\upsilon_0^2)(1 + \upsilon_0^2)}{(\upsilon^2 + \upsilon_0^2)(1 + \alpha\upsilon_0^2)}$$

The values of Δ, G_0 and φ_d for the antivibration rubber produced in Romania are given in the tables.

Rubber hardness ^{0}Sh(A)	30	40	45	50
Δ	0,020	0,035	0,045	0,095
G_0mN/m^2	3,20	4,60	5,40	6,50
φ_d	1,00	1,05	1,15	1,25

Rubber hardness ^{0}Sh(A)	60	65	70	75
Δ	0,15	0,20	0,30	0,40
G_0 mN/m^2	9,40	11,60	16,0	21,0
φ_d	1,50	1,75	2,00	2,50

where:

$$\varphi_d = \frac{G_{din}}{G_0} = \frac{E_{din}}{E_0} \qquad (14)$$

The above tables present the values determined during the tests made on the production batches of antivibration rubber SAB 31, SAB 22, SAB 40 and SAB 9.

CORRELATION BETWEEN VISCOELASTIC SHEAR AND TENSION

O. S. BRUELLER

Dept. of Mechanics and Material Testing, Technical University of Munich, Boltzmannstr. 15, D-85747 Garching, Germany

1. INTRODUCTION

Many papers concerning the viscoelastic behavior of materials under normal stresses, especially under uniaxial tension have been published in the last few decades. The description of the time-dependent behavior under shear stresses has been neglected, in spite of the fact that many structural parts made of viscoelastic materials are loaded in shear, for instance in torsion.

Since long-term shear tests are less frequently performed than tensile tests, it would be helpful to find a direct correlation between the viscoelastic shear behavior of a time dependent material and its behavior under uniaxial loading. For instance, it would be interesting to deduct its viscoelastic tensile response from long term torsional tests. This is the aim of the present paper.

2. THEORY

The nonlinear viscoelastic behavior of a time dependent material is usually investigated by conducting creep experiments (constant stress) or stress relaxation experiments (constant strain) under uniaxial loading. Since constant stress experiments are more suitable, the present work shall be limited to creep; an extension of the procedure to stress relaxation is very easy.

Firstly, one has to remember the relationship between the shear and tensile moduli of an elastic material, given by:

$$G = \frac{E}{2(1+\mu)} \tag{1}$$

where $G = \dfrac{\sigma_{xy}}{2\varepsilon_{xy}}$ and $E = \dfrac{\sigma_x}{\varepsilon_x}$.

Taking into account that Poisson's ratio of viscoelastic materials such as thermoplastics for example, practically do not changes with time or temperature and can be considered as constant, one can write:

$$J_n^* = \frac{J_n}{1+\nu} \tag{2}$$

with J_n^* and J_n as creep compliances in tension and shear, respectively, and ν as Poisson's ratio.

The isothermal response of a time dependent material to a suddenly applied shear stress can be described with good accuracy by the relation:

$$\varepsilon_{xy}(t,\sigma_{xy}) = \sigma_{xy} J_n(t,\sigma_{xy}) \tag{3}$$

where $J_n(t,\sigma_{xy})$ is the nonlinear creep compliance given by:

$$J_n(t,\sigma_{xy}) = g_0(\sigma_{xy})J_0 + g_t(\sigma_{xy})\,\Delta J(t) \tag{4}$$

$g_0(t)$ and $g_t(t)$ are the nonlinearity factors of the instantaneous linear compliance J_0 and the transient linear compliance $\Delta J(t)$, respectively.

The transient linear shear compliance is expressed usually by:

$$\Delta J = \sum_{i=1}^{m} J_i (1 - e^{\frac{-t}{\tau_i}}) \tag{5}$$

Since, as experimentally proved with several polymers, the instantaneous response of these materials is linear, even in the nonlinear viscoelastic range, one can deduct the relation for the nonlinear creep compliance under tension in the form:

$$J_n^*(t,\sigma_x) = \frac{J_0}{1+\nu} + g_t \frac{\Delta J}{1+\nu} \tag{6}$$

or:

$$\varepsilon(t,\sigma_x) = J_0\sigma_x^* + g_t(\sigma_x^*)\Delta J\sigma_x^* \tag{7}$$

with $\sigma_x^* = \dfrac{\sigma_x}{1+\nu}$

3. EXPERIMENTAL

Two kinds of experiments have been conducted:

- Tensile creep tests on thin walled pipes
- Torsion tests on similar specimens

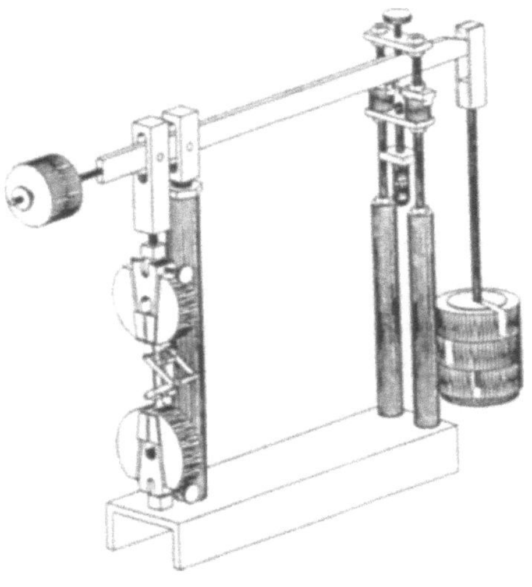

Fig. 1: Tensile creep setup

As an example, the setup for creep in tension is depicted in Fig. 1.

4. RESULTS

Torsion creep experiments have been conducted on thin walled pipes made of PVC under four different shear stresses. Using least squares techniques, the initial shear compliance and the parameters of the exponential series, as well as the values of the transient nonlinearity factor in shear have been computed.

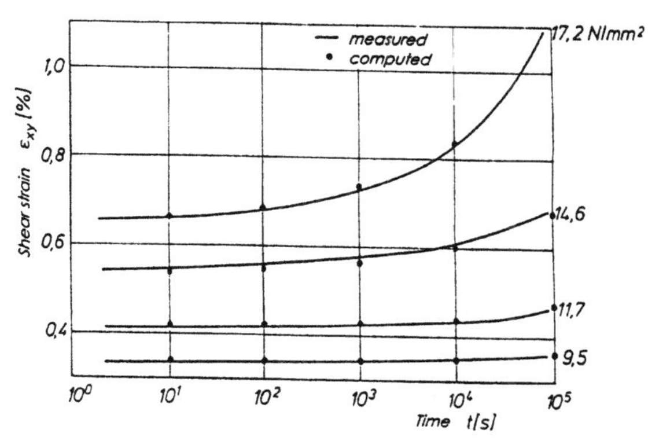

Fig. 2: Creep curves of PVC in torsion

Fig. 2 shows the good agreement between experimental curves and computed values in creep shear of PVC. The course of the nonlinearity factors is depicted in Fig. 3.

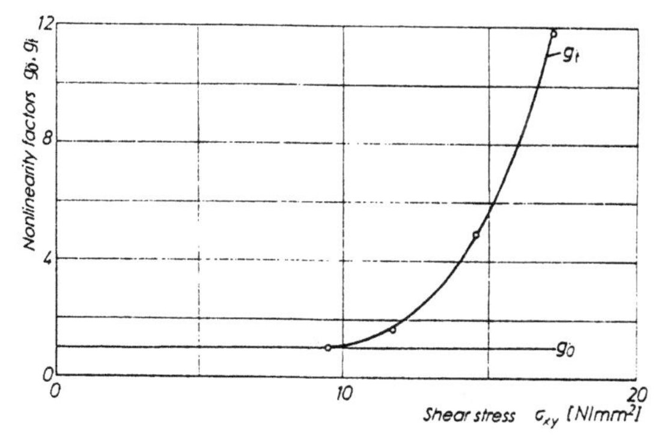

Fig. 3: Nonlinearity factors of PVC in torsion

A comparison between computational results based on torsion and tensile experimental results is presented in Fig. 4. A good agreement can be observed.

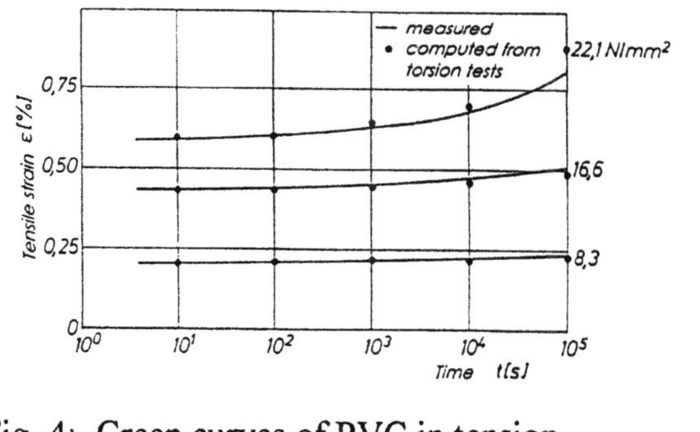

Fig. 4: Creep curves of PVC in tension

DETERMINATION OF EQUATION OF STRESS RELAXATION CURVES BASED ON CREEP EXPERIMENTAL RESEARCH FOR 16 MO 3 STEEL

C. COMANDAR, D. LEON, N. AMARIEI

Technical University "Gh. Asachi" Iaşi, Faculty of Mechanics, Bd. Mangeron nr. 63, 6600 Iasi, Romania

1. INTRODUCTION

The rheological behaviour of metals is illustrated by the phenomenon's of creep and stress relaxation. For alloy steels, these phenomenons are observed over 400°C. The use in high temperatures of metals and their alloys must bring into consideration the time factor. It is necessary to know a function type:

$$f\left(\sigma,\varepsilon_p,t,T,\dot{\varepsilon}_p\right)=0 \qquad (1)$$

where: σ = stress, ε_p = plastic strain, t = time, T = temperature and $\dot{\varepsilon}_p$ = plastic strain rate.

Usually, the number of variables used for analysis is low (3 variables, in general).

The stress relaxation produces a loosening of the pieces fixed with elastic tightening, egg. screw tightening of flange joints for steam pipe - lines, piston rings of internal combustion engines.

The stress relaxation testing consists in maintain constant the initial strain determined by loading. The initial elastic strain decreases with time, part of it turning into plastic strain. The stress needed to maintain the total deformation constant is to decrease with time. The stress relaxation tests are more difficult then the creep tests. For this reason, sometime is easier to work with creep curves, in order to obtain the relaxation curves. The most simple creep testing is done by subjecting the specimen to elongation, by a constant force and at constant temperature. The objective is to determine the residual deformation after a period of time or the rupture time.

2. METHODS

The paper presents the experimental researches carried out by authors on low-alloy steel, type 16 Mo 3, STAS 8184-87. For this, there were performed creep tests, at 500°C and for different stresses. Comandar et al. (1997) par extrapolation of the experimental results are obtained the equation of area represented in figure 1, as follows:

$$\varepsilon(\sigma,t)=A(\sigma)\cdot t+B(\sigma)\cdot\{1-\exp[-C(\sigma)\cdot t]\}+\frac{\sigma_0}{E} \qquad (2)$$

where $A(\sigma)$, $B(\sigma)$ and $C(\sigma)$ function of stress, σ_0 = initial stress and E = Young's modulus, for the testing temperature.

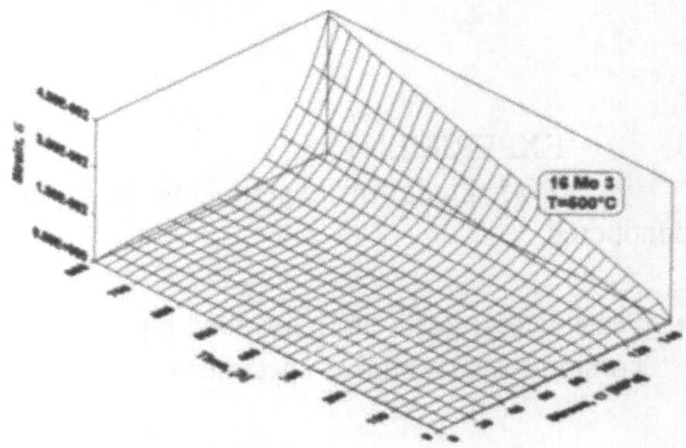

Fig. 1. Variation of strain with time and stress

The intersecting of the area represented by the relation above mentioned with planes ε=const. allows to obtain the stress relation curves.

It is also possible to obtain the equation of the stress relaxation curves by using Katchanov's hypothesis or Belyaev's hypothesis, modified by Malinin, cited by Ponomariov et al. (1963). In the first case (Katchanov's hypothesis) the obtained relation is as follows:

$$\sigma=\sigma_0\cdot\left[1-E\cdot(n-1)\cdot\Omega\cdot\sigma_0^{n-1}\right]^{-\frac{1}{n-1}} \qquad (3)$$

In the second case (Belyaev's hypothesis, modified by Malinin) results the following relation:

$$\sigma=\sigma_0\cdot\left(1-n\cdot E\cdot\Omega\cdot\sigma_0^{n-1}\right)^{-\frac{1}{n}} \qquad (4)$$

where σ = stress at time t, n = coefficient depending on material and temperature, Ω = function of stress and temperature.

3. RESULTS

The experimental results of the creep testing were used to determine the equations of relaxation curves, according to the two hypothesis.

The graphics traced in figure 2, present the variation of the plastic strain rate with time, for a certain stress value, are used to obtain the minimum values of this rate.

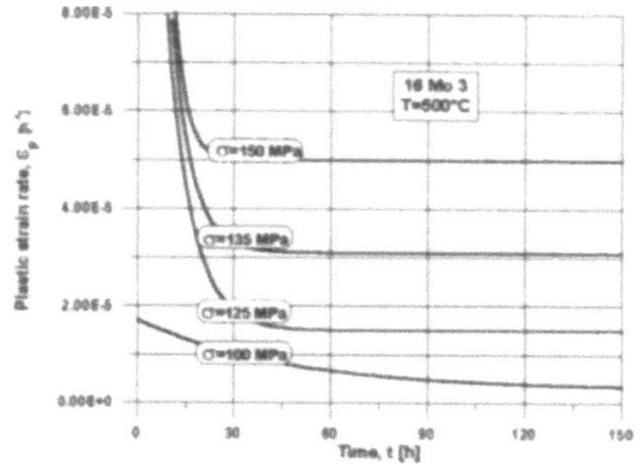

Fig. 2. Variation of plastic strain rate with time

The minimum value of the plastic strain rate is achieved after a relatively short period of time (150 hours). After this time, the phenomenon is considered to take place in the secondary creep domain. The minimum values are used to trace the graphic of stress variation, in logarithmic coordinates. Figure 3 shows the graph.

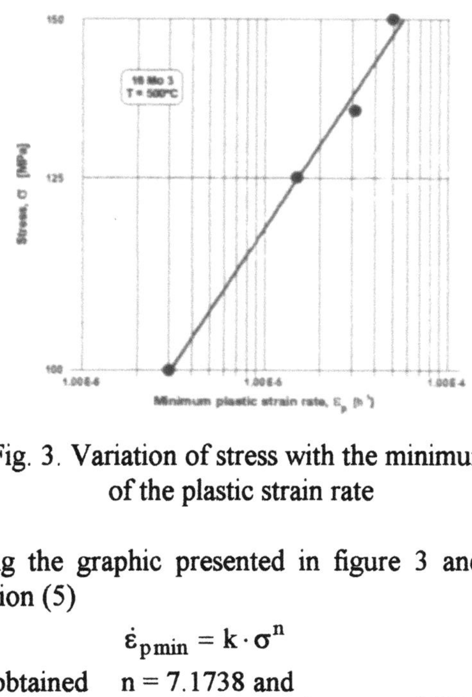

Fig. 3. Variation of stress with the minimum of the plastic strain rate

Using the graphic presented in figure 3 and the relation (5)

$$\dot{\varepsilon}_{p\min} = k \cdot \sigma^n \qquad (5)$$

are obtained $n = 7.1738$ and

$k = 1.3475 \cdot 10^{-3}$ $[(MPa)^{-7.1738} \cdot h^{-1}]$.

The value of coefficient "n" is used to trace the graphics of Ω function and B function, considering:

$$\Omega(t) = \frac{\varepsilon_p}{\sigma^n} \text{ and } B = \frac{d\Omega}{dt} \qquad (6)$$

The graphics are presented in figure 4. The values obtained through calculus are pointed on the graphics.

The least squares method is used for :

$$\Omega = 1.35 \cdot 10^{-20} \cdot t + 1.44 \cdot 10^{-18} \cdot \left(1 - e^{-0.126t}\right) \qquad (7)$$

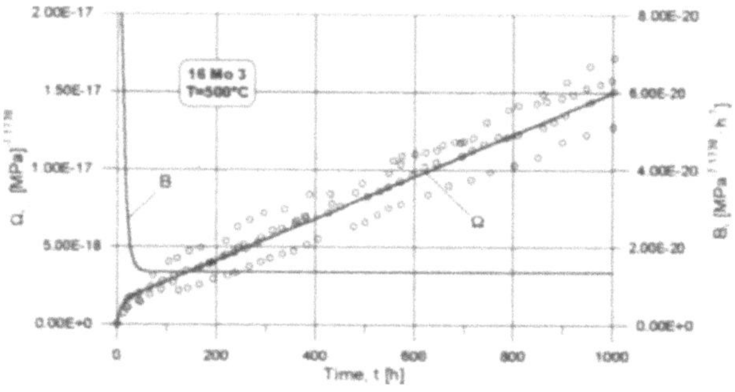

Fig. 4. Graphics of Ω and B functions

The relation for B is resulting from precedent relation:

$$B = 1.35 \cdot 10^{-20} + 1.8144 \cdot 10^{-19} \cdot e^{-0.126 t} \qquad (8)$$

The equation of stress relaxation family curves in Katchanov's hypothesis is obtained by introducing relation (7) in relation (3), as follows:

$$\sigma = \sigma_0 \cdot \left\{1 + 6{,}1738 \cdot E \cdot \left[1{,}35 \cdot 10^{-20} \cdot t + 1{,}44 \cdot 10^{-18} \cdot \left(1 - e^{-0{,}126 t}\right)\right] \cdot \sigma_0^{6{,}1738}\right\}^{-\frac{1}{6{,}1738}} \qquad (9)$$

The Belyaev-Malinin hypothesis conducts to:

$$\sigma = \sigma_0 \cdot \left\{1 + 7{,}1738 \cdot E \cdot \sigma_0^{6{,}1738} \cdot \left[1{,}35 \cdot 10^{-20} \cdot t + 1{,}44 \cdot 10^{-18}\left(1 - e^{-0{,}126 t}\right)\right]\right\}^{-\frac{1}{7{,}1738}} \qquad (10)$$

Relation (10) represents the equation of the stress relaxation's family curves. The surface generated using this equation is illustrated in figure 5.

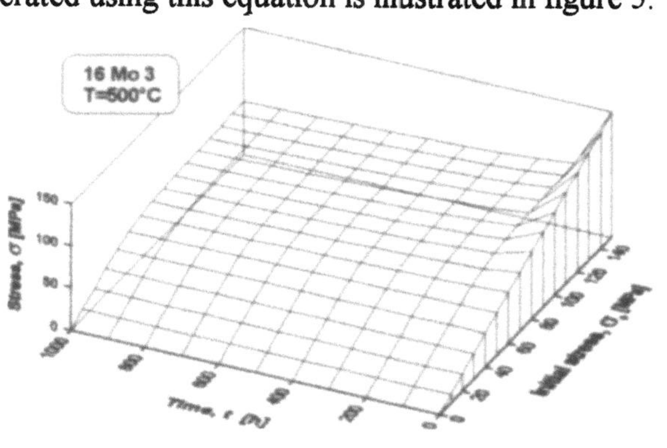

Fig. 5. Stress relaxation with time and initial stress

4. CONCLUSIONS

The interpretation of the creep testing experimental data allows to obtain the equation of stress relaxation curves.

5. REFERENCES

Comandar C, Leon D, Amariei N (1977) In: Hrabovský M (ed) Proc, of EAN '97, Olomouc, Czech Republic, pp. 73-78

Ponomariov SD et al. (1963) Calculul de rezistenţă în construcţia de maşini. Ed. tehnică, Bucureşti

RHEOLOGICAL PROPERTIES OF PLASTICIZED STARCHES

G. DELLA VALLE[1], P. J. CARREAU[2], P. COLONNA[1], B. VERGNES[3]

1 INRA, Centre de Recherches Agro-alimentaires, 44316 NANTES Cedex 3, FRANCE

2 CRASP, Ecole Polytechnique, MONTREAL, QC, H3C 3A7, CANADA.

3 CEMEF, Ecole des Mines de Paris, 06904 SOPHIA-ANTIPOLIS Cedex, FRANCE.

1. INTRODUCTION

Plasticized starches (PLS) are materials of current interest either for food or non-food purposes. Their processability and the control of their end-properties require a good knowledge of their rheological properties measured under physical conditions similar to the ones encountered in industrial processes, like extrusion. In preceding works, stability and linearity conditions and water content have been found for reliable results on PLS viscoelasticity. Semi crystalline structure has been shown tot be partly responsible for their gel-like viscoelastic behavior (Della Valle et al. 1998). Due to this structural feature, rheological properties during processing, i.e. under large deformation, may not be deduced straightforwardly from the linear viscoelastic properties.

Therefore, the aim of this work is to compare preceding results (linear viscoelasticity) to those obtained by capillary rheometry, for two levels of amylose content. Finally, entrance effects are treated to give a first approach of elongational properties in order to adress practical issues on forming and texturizing processes.

2. MATERIAL AND METHODS

2.1 Starch plastication

Wheat and high amylose maize starches were purchased from Roquette (F-62 LESTREM). Their amylose (linear) content is 25 and 70 % respectively. Glycerol (Merck) is used as a plasticizer at a content of 28% on a total weight basis. Starches destructurization and blending with plasticizer were achieved on a twin screw extruder with a specific mechanical energy SME ~1000J/g. For rotational rheometry purpose, extruded pellets were press-molded and then conditioned under relative humidity 0,57 to reach a moisture content MC of 0,14 on a total wet basis. For capillary rheometry study, pellets were stored in sealed plastic bags after equilibrium at MC=0,14.

2.2 Rotational rheometry

A Bohlin CSM (controlled stress) rheometer was used with 25 mm diameter plane/plane geometry, with a rough upper plate. PLS sample outer edge was coated with silicon grease in order to reduce evaporation. Thermogravimetric analysis, performed on sample before and after test, shows that there is no significant loss of plasticizer during rheological test. Frequency sweep tests measurements were performed under constant strain ($\gamma = 10^{-3}$) as recomended previously (Della Valle et al. 1998).

2.3 Capillary rheometry

The melting of PLS requires additional mechanical work. Therefore, a pre-shearing rheometer (Rheoplast) was used, which has already been shown to be adapted to the study of low hydrated starchy materials (Vergnes and Villemaire, 1987). Mechanical treatment is provided by the rotation of the inner piston to the material compacted by an annular piston moving down; measurement of the torque allows to define the specific mechanical energy delivered to the material (SME). Measurements were done for a diameter of 2mm at four values of L/D ratio (0, 4, 8, 16).

3. RESULTS AND DISCUSSION

3.1 Comparison of complex and steady shear viscosities

Frequency sweep tests show that both PLS materials behave like solid more than a fluid since $\eta^*(\omega) >> \eta'(\omega)$ in the range studied (Fig.1).

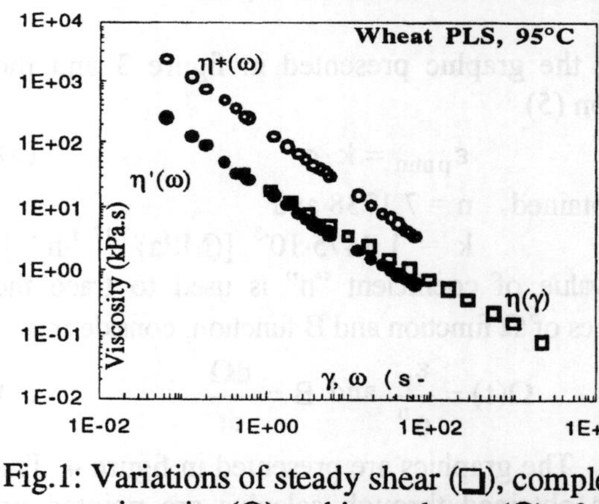

Fig.1: Variations of steady shear (□), complex viscosity (O) and its real part (●) of wheat PLS at 95°C and glycerol /water content (28/14).

This result confirms those obtained for the moduli ($G'(\omega) > G''(\omega)$) of dehydrated materials at

150°C (Della Valle et al. 1998). The present study focused on hydrated PLS, since the presence of water is unavoidable and furthermore eases their processing. The solid-like behavior of PLS at temperatures much higher than Tg value for this starch-glycerol-water ratio has been explained by the existence of a semi-crystalline structure. However, as same level of crystallinity for wheat and amylomaize were found by X-ray diffractometry, suggesting that the larger values of moduli, and of complex viscosities (one decade), in the case of amylomaize, were due to the greater ability of starch linear macromolecules (amylose) to create entanglements (Fig.2).

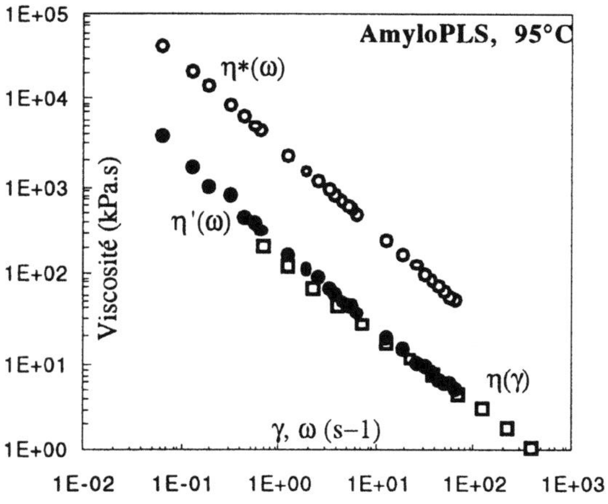

Fig.2: Fig.1: Variations of steady shear (□), complex viscosity (O) and its real part (●) of high amylose maize PLS at 95°C and glycerol /water content (28/14).

Destructuration of PLS is achieved on Rheoplast; SME being 100J/g for wheat PLS and 120J/g for amylomaize PLS. It has been checked that PLS is amorphous just after processing on Rheoplast. After performing Bagley and Rabinowitsch corrections, flow curves can be determined in the range of shear rate ($1\text{-}10^3 s^{-1}$) (Fig 1&2).

Steady shear viscosity follows a power law with flow indices of 0,29 and 0,19 for wheat and amylomaize respectively. No Newtonian plateau has been observed at low shear rate. The same ratio as for complex viscosity is found for shear viscosity when, comparing wheat and amylomaize PLS; this is in agreement with the existence of macromolecular entanglements.

Due to the existence of a structure, Cox-Merz rule is not obeyed and $\eta^*(\omega) > \eta(\gamma)$. However, in both cases, we observe that $\eta(\gamma) \sim \eta'(\omega)$: the dissipative effects measured during frequency sweep tests could only be due to the mobility of the amorphous phase of PLS, which is responsible for the simple shear flow behavior.

3.3 Elongational viscosity

Using Cogswell's analysis (1972), it was possible to derive an apparent elongational viscosity η_E, from the entrance effects (Fig.3).

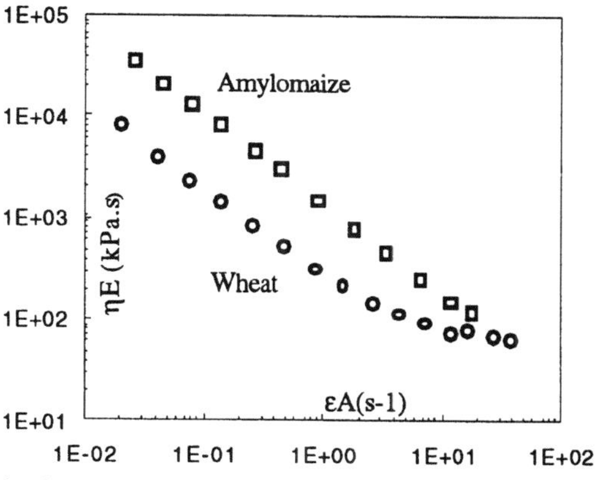

Fig.3: Apparent elongational viscosites of Wheat (O) and amylomaize (□) PLS for same temperature and composition as above.

Elongational viscosity is about ten times larger than shear viscosity. Compared with thermoplastic materials, both PLS exhibit a strong strain rate dependancy and, as expected, wheat PLS has much lower values than amylomaize, chiefly at lower strain rate ($<10s^{-1}$). For same value of shear viscosity (3kPa.s, for instance):

$$\eta_E \text{ (Wheat)} > \eta_E \text{ (Amy)}$$

Elongational viscosity is the property supposed to resist to the bubble growth during foam creation. Therefore, the present result may explain why amylomaize provides higher expansion ratio than wheat in extrusion-cooking of starch (Della Valle et al. 1997).

4. CONCLUSION

Differences of rheological properties of PLS encountered between small and large deformation measurements may be explained by the knowledge of their structure, before and after processing. Predictability of flow properties from linear viscoelastic properties need to be confirmed. Conditions of applicability of Cogswell's analysis have to be examined carefully to confirm the influence of elongational viscosity on starchy product expansion.

5. REFERENCES

Cogswell FN (1972) Polym. Eng. Sci. 12: 64-73

Della Valle G, Buléon A, Carreau PJ, Lavoie P-A, Vergnes B (1998) J. Rheol. 42: 505-527

Della Valle G, Vergnes B, Colonna P, Patria A (1997) J. Food Eng. 31: 277-296

Vergnes B, Villemaire J-P (1987) Rheol. Acta 26: 570-576

NON-LINEAR STRESS RELAXATION IN LIQUID FOAMS.

A.C. DIOGO, E. BAPTISTA

Instituto Superior Técnico (Materials Dep. & unit ICTPOL), Av. Rovisco Pais P-1096 Lisboa Codex, Portugal

1. INTRODUCTION.

Liquid foam rheology has been mainly approached by micromechanical models (see e.g. the review from Kraynik (1988)). Less attention has been paid to a continuum approach, where the rheological behaviour follows from the appropriate constitutive equations, and can be checked by experiments performed at macroscopic scale.

2. EXPERIMENTAL

The non-linear shear relaxation modulus $G(t,\gamma_0)$ of a commercial shaving cream has been computed from the stress response to a step shear strain. The experiments were carried out at ambient temperature, in a RMS-800 mechanical spectrometer (RheometricS), using the parallel plates geometry. Serrated plates were always used, since this was suggested by Diogo (1995) to be the best way to prevent slip. The strain amplitude γ_0 varied from 0.05 to 3. All experimental data points were corrected for strain non-uniformity, according to Soskey and Winter (1984). The corrected data are displayed in Fig.1.

3. DISCUSSION

The data presented in Fig.1 suggest that non-linear behaviour starts at a quite low strain, below 10%. A more precise value for this limit can be found from the measurement of the strain dependence of the complex shear modulus: it is 6-7% strain. These results disagree with the theoretical prediction of Princen (1993) and Prud'homme (1991) for a regular hexagonal foam: $\gamma_c=2/\sqrt{3}=1.155$. The very low linear viscoelastic threshold found, strongly suggests that linear viscoelastic material functions are better computed by taking the limit of the corresponding non-linear ones as the step strain γ_0 tends to zero.

The analysis of the experimental data was performed by assuming that the overall stress relaxation is the sum of the contributions from a few microscopic mechanisms. Each contribution can be described as a non-linear Maxwell element for which time-strain separability holds: for each Maxwell element, the non-linear shear relaxation modulus $G(t,\gamma_0)$ is factorized into a time-dependent linear viscoelastic modulus $G^0(t)$ and a strain dependent "damping function" $h(\gamma)$. So

$$G(t,\gamma)=\sum_{i=1}^{n} g_i . \exp\left(-t/\lambda_i\right) . h_i(\gamma) \quad (1)$$

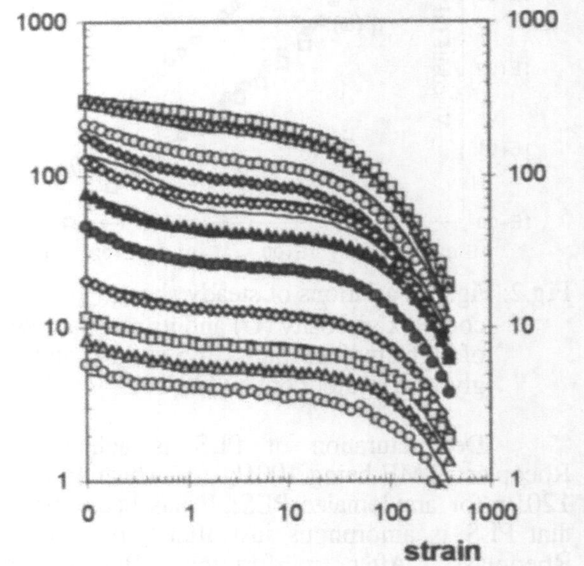

Fig.1 - Non-linear shear relaxation modulus $G(t,\gamma_0)$/Pa of the foam. Points represent the experimental data at γ = 0.05, 0.1, 0.2, 0.4, 0.5, 0.7, 1, 1.5, 2, 2.5, 3; the lines represent the fits according to eqs. (1) and (4).

The following three models were considered for the (simple shear) damping function: Wagner (1976), eq.(2), Papanastasiou-Scriven-Macosko (PSM) (1983), eq.(3), and Wagner-Demarmels (W-D) (1990), eq.(4):

$$h(\gamma) = \exp(-\beta.\gamma) \quad (2)$$

$$h(\gamma) = (1+\alpha.\gamma^2)^{-1} \quad (3)$$

$$h(\gamma) = \left(1 + \theta . \gamma^{m}\right)^{-1} \qquad (4)$$

The analysis of the experimental data was performed by the following procedure: Equation (1) was fitted simultaneously to all data (eleven $G(t,\gamma)$ curves), and the best parameters were computed by a non-linear optimization procedure, using the logarithmic error measure. According to the model chosen for $h(\gamma)$, the number of free parameters per Maxwell element was three (Wagner, eq.2, PSM, eq.3) or four (W-D, eq.4).

In the optimization procedure followed here, also the number n of non-linear Maxwell elements considered was free. However, it was found that, both for the Wagner (eq.2) and PSM (eq.3) equations the overall r.m.s. logarithmic error did not improve significantly for $n \geq 2$. For the W-D model (eq.4), the overall r.m.s. logarithmic error does not improve significantly and remains close to the experimental error for $n \geq 3$ elements. Therefore $4 \times 3 = 12$ parameters were enough to generate the 11 full curves displayed in Fig 1, which are the best fits to the experimental data. The three W-D damping functions generated by these fits are represented in Fig. 2.

The three relaxation times found (0.4 s, 73.6 s, and 268 s) should be put in correspondence to the following three relaxation processes: strain induced draining of water through the Plateau borders (the fastest), hopping of the foam cells and cell coarsening (the slowest). Cell coarsening yields the broader damping function, and hopping the narrower, as shown in Fig. 2. We also found that, within the strain range considered here, there is no need to consider a distribution of relaxation times.

4. CONCLUSIONS.

It was shown that the non-linear relaxation modulus of liquid foams can be analysed as the sum of the contributions from three time-strain factorizable relaxation processes, which correspond to three different mechanisms of stress relaxation in liquid foams: draining, hopping and coarsening. The fastest one (of the order of tenths of second), is related to draining of water through the Plateau borders, induced by the applied strain; the slowest one (of the order of hundreds of seconds) describes the contribution to stress relaxation from foam cell coarsening. The third process (cell hopping) has a relaxation time of the order of some tens of seconds. Therefore, with just four parameters for each of the three relaxation processes above considered, the non-linear relaxation modulus $G(\gamma_0,t)$ can be computed in a step relaxation experiment for $0<\gamma_0<3$, in good agreement with the experimental data.

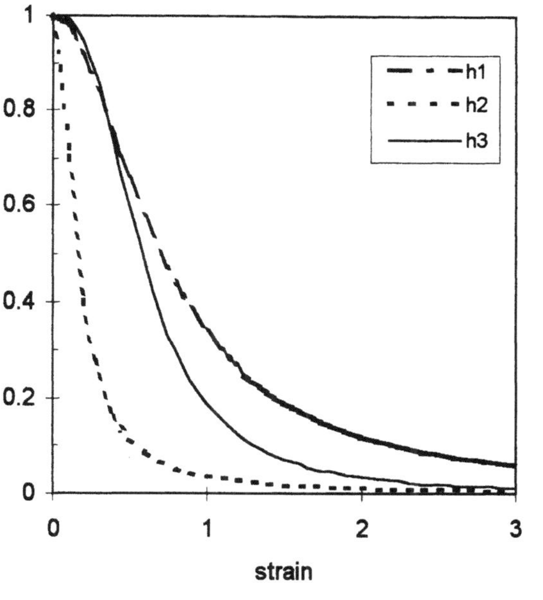

Fig.2 - Damping functions of the foam for cell coarsening (h_1), cell hopping (h_2) and draining (h_3).

5. REFERENCES

Diogo AC (1995) Les Cahiers de Rhéologie, XV-1:231

Kraynik AM (1988) Ann. Rev. Fluid Mech. 20:325

Papanastasiou AC, Scriven LE, Macosko CW (1983) J. Rheol. 27:387

Princen HM (1983) J. Colloid Interface Sci.91:160-75

Prud'homme RK (1981) Proc. Ann. Meet. Soc. Rheology, Louisville

Soskey PR, and Winter HH (1984) J. Rheol. 28: 625

Wagner MH (1976) Rheol. Acta 15:136

Wagner MH, Demarmels A (1990) J. Rheol 34:943

ON SEARCH FOR A GENERALIZED CONSTANT OF POLYMER MELTS

Z. DOBKOWSKI, M. ZIELONKA

Industrial Chemistry Research Institute (ICRI), Dept.of Analysis and Characterisation of Materials, Rydygiera 8, 01-793 Warszawa, Poland.

1. INTRODUCTION

A linear multivariable power function P = f (M, MWD, LCB), representing the dependence of a polymer property P on the polymer molecular characteristics, where M is the molecular weight, MWD - the molecular weight distribution, and LCB - the long chain branching, has recently been applied for investigation of polymer melts, cf. Dobkowski (1982, 1988, 1994, 1998). If the polymer fluidity difference between the Newtonian and non-Newtonian conditions is taken as the polymer property, then the master line can be obtained for polymer melts in reduced log-log coordinates. Thus, the multivariable power function (MVP) for the dependence of η_0 and $\Delta\varphi$ on M, MWD and LCB can be written as

$$\log(\Delta\varphi.\eta_0) = \log B + \log[(\eta_0.\omega)^{b_1}.q^{b_2}.G^{b_3}] \quad (1)$$

where η_0 is the zero shear rate melt viscosity, ω is the frequency, $\Delta\varphi$ is the fluidity difference between the Newtonian and non-Newtonian conditions where, in turn, $\varphi = 1/\eta$ is the fluidity and $\Delta\varphi = 1/\eta - 1/\eta_0$. It should be noted that the frequency can be replaced by the shear rate according to the Cox-Merz rule, cf. Cox and Merz (1958), Dobkowski (1998). The exponents b_1, b_2 and b_3 can be found from respective regression equations using the experimental data from rheological measurements. In particular, the exponent b_1 of the MVP function enables distinguishing linear and branched polymer structures, cf Dobkowski (1982). For linear and monodisperse macromolecules we have

$$\log(\Delta\varphi.\eta_0) = \log B + \log[(\eta_0.\omega)^{b_1}] \quad (2)$$

Eqs.(1) or (2) can be written in a simple form as

$$\log Y = \log X + \log B \quad (3)$$

where X and Y are the respective reduced coordonates.

In this paper, the master lines and the constants B for some selected polymers were obtained. Then, several structural parameters, such as volume or flow parameters, cf Bicerano (1993), Users Guide (1993), Van Krevelen (1990), have been considered, and an attempt to find a correlation between the constant B and some structural parameters has been described.

2. EXPERIMENTAL

Experimental data for various polymers were based on the measurements of melt viscosity vs frequency dependence using the RDS-2 Rheometrics rheospectrometer. Then, the reduced variables $\Delta\varphi.\eta_0$ and $\eta_0.\omega$ (or $\eta_0.\gamma$) were calculated.

Experimental results for PDMS, PP and PC have been considered. Literature data for these polymers, as well as for PIB and PMMA, have also been taken into account for calculations of master line coordinates and for comparison of MVP linear master dependencies.

3. RESULTS AND DISCUSSION

The master lines for investigated polymers are shown in Fig.1, cf. Dobkowski (1998). It is observed that the lines are shifted by a certain value that can be represented by the value of constant B. The shift factors for different polymer molecular structures should depend on some basic structural parameters of the polymer unit.

Several structural parameters were considered, including the Van der Waals volume V_W and the molar viscosity - temperature function.

The Van der Waals volume V_W of a molecule is a space occupied by this molecule which is impenetrable to other molecules at normal temperatures. It is a specific structural parameter

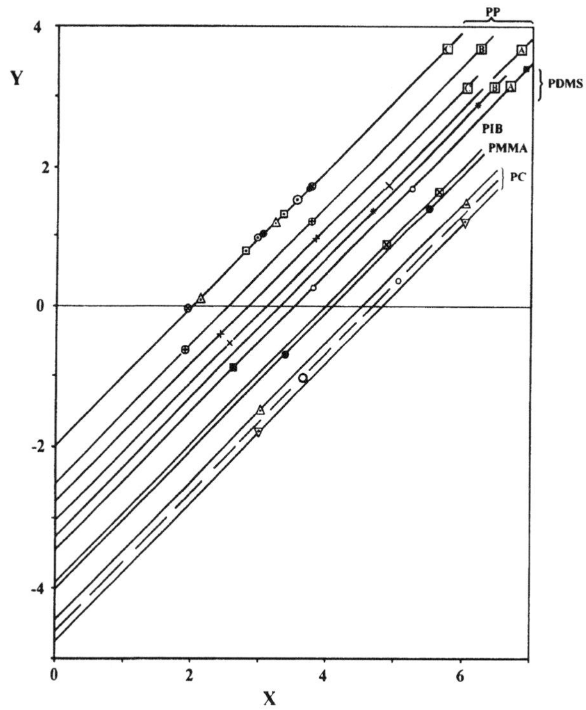

Fig. 1.

Note: $Y = \log(\Lambda\varphi \cdot \eta_0)$; $X = \log(\dot{\gamma} \cdot \eta_0)^{b_1}$ or $X = \log(\omega \cdot \eta_0)^{b_1}$

not influenced by physical changes and transitions.

The molar viscosity - temperature function H_η has been described by Bicerano (1993) and it is correlated with the activation energy of viscous flow E_η

$$E_\eta = (H_\eta / M)^3 \qquad (4)$$

where H_η is expressed in $g.J^{1/3}.mol^{-4/3}$ and E_η in J/mol.

It has been found that the constant B is reasonably correlated with the Van der Waals volume V_W, according to the data given by Users Guide (1993) and Van Krevelen (1990), as well as with the molar viscosity - temperature function, described by Bicerano (1993), see Table 1.

The plots of B vs V_W and B vs H_η, however, are not linear in log-log coordinates. Coefficients of linear regression are $r^2 = 0.80$ and $r^2 = 0.74$, respectively. It should be noted, however, that the linear correlation exists for the dependence of H_η on V_W with $r^2 = 0.985$.

Table 1. The values of constant B and structural parameters for investigated polymers.

Polymer	V_W cm^3/mol	B	H_η $g.J^{1/3}.mol^{-4/3}$
PP	30.7	8.9 10^{-4}	1417
PDMS	39.3	3.5 10^{-4}	1839
PIB	40.9	1.3 10^{-4}	2008
PMMA	56.1	1.0 10^{-4}	3556
BAPC	136.2	3.6 10^{-5}	11119

Note: PP - polypropylene,
PDMS - polydimethylsiloxane,
PIB - polyisobutylene,
PMMA - polymethylmethacrylate,
BAPC - bisphenol A polycarbonate.

4. CONCLUSIONS

It has been found that the shift of master straight lines for polymer melts is reasonably correlated with the Van der Waals volume as a structural parameter of the macromolecules of investigated polymers.

The shift of master straight lines is also corrrelated with the molar viscosity - temperature function, and thus, with the activation energy of the viscous flow.

However, further investigations are needed to find more experimental data for other polymers.

5. REFERENCES

Bicerano J(1993) Prediction of Polymer Properties. Marcel Dekker, New York

Cox WP, Merz EH (1958) J.Polym.Sci. 28: 619

Dobkowski Z (1982) Eur Polym J 18:1051-1059

Dobkowski Z (1988) Polymer Bull 19: 165-169

Dobkowski Z (1994) In: Gallegos C (ed) Progress and Trends in Rheology. IV. Steinkopff Verlag, Darmstadt pp 365-367

Dobkowski Z (1998) Fluid Phase Equilibria, in print.

User Guide (1993) Polymer 6.0. Biosym Technol., San Diego

Van Krevelen (1990) Polymer Properties. Elsevier, Amsterdam

STRUCTURE AND RHEOLOGY OF A MICELLAR CUBIC CRYSTAL

E. EISER[1], F. MOLINO[1], G. PORTE[1], J.F. BERRET[1]

[1]Groupe de Dynamique des Phases Condensées, Université Montpellier II, Place Eugène Bataillon, Montpellier, France

1. SYSTEM DESCRIPTION

Triblock copolymers $(PEO)_n(PPO)_m(PEO)_n$, commercially known as pluronics, have interesting aggregating properties in water: the PPO part becomes increasingly hydrophobic as temperature increases, whereas the PEO remains soluble up to 100°. Starting from a solution of gaussian polymers at 5°, one obtains at room temperature a solution of *micelles*. Their aggregation number (N=50 for m=48, n=127) is quite independant of the polymer concentration. Thus, any increase in the polymer concentration Φ leads to an increase of the concentration in number of micelles. Obviously this results in a geometrical problem of space-filling with (soft) spheres (as long as the basic spherical micellar structure is selected by the system). As a consequence, it is not surprising that one observe in the (Φ,T) phase diagram a region where the system exhibits a *solid-like behavior*, and is *transparent* and *isotropic*. This macroscopic indication of an ordered cubic phase is confirmed by X-ray powder diffraction on a crystallised sample of pluronic F108 (n=127, m=48) at room temperature : the sequence of rings demonstrates the existence of a local *fcc* symmetry, with a mesh size of around 300 Å, which is in good agreement with the aggregation number and the typical micelle size obtained previously. In the same way, *bcc* symmetry could be demonstrated for F68 (n=80, m=30).

The interest of this system is to enable one to study in details the *mechanical* and *structural* response of an ordered phase (having the same structure as aluminum), using standard rheology (compared to a metal, the elastic modulus is typically 10^6 times smaller) and X-ray and neutron diffraction techniques.

2. STRUCTURAL BEHAVIOR

All our structural and mechanical results were obtained in *Couette geometry*. We first investigated the structure of the phase, and our primary goal was to identify the *layer sliding* flow mechanism (see Ackerson 1984, 1996). When it takes place, the original *fcc* polycristalline structure is transformed in an aligned collection of millimetric size compact planes of spheres, of hexagonal symmetry. These planes are parallel to the Couette cell walls and thus perpendicular to the velocity gradient. They slide past each other, driven for example by line defects. Due to this relative motion, these planes are supposed to be uncorrelated in the velocity gradient direction : their stacking order is random. Remember that *fcc* structure corresponds to a fixed stacking order ABCABCAB..., where the letters stand for the regular stacking sites of hexagonal compact planes.

In our X-ray and neutron experiment under shear, we were able to have the beam incident both in the radial direction (i.e. prependicular to the planes), and in the tangential one (thus scanning the stacking order).

Figure 1 represent a typical spectrum in the radial direction for the layer sliding regime.

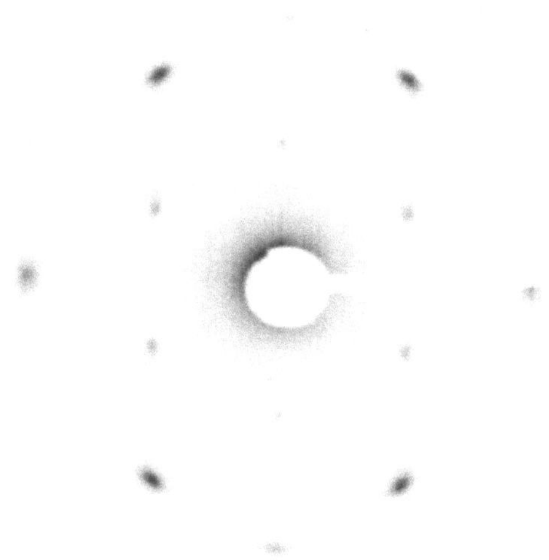

Figure 1 : X-ray spectrum of layer-sliding structure

Both spectra under shear, and at rest after shear on the quenched structure, could be quantitatively analysed in terms of this model. The shear rate range corresponding to this regime goes from 30 s^{-1} to 300 s^{-1}. Above 300 s^{-1}, we observe the *shear-melting* of the sample : the structure factor becomes a liquid ring and all long range order is destroyed. The most interseting regime occurs below 30 s^{-1} : it is the transition from the polycrystal to the layer-sliding regime. For a given shear rate, in the stationnary state, the spectra could be analysed in terms of the superposition of two domains in the sample : one region weakly aligned, but still polycrystalline with local *fcc* order, and the other in layer sliding configuration. Moreover, the ratio of these two phases in the gap could be quantitatively monitored, showing that the layer sliding phase was continuously increasing with the shear rate, before completely filling the gap around 30 s^{-1}. Thus the structural change could be interpreted in terms of an *out of equilibrium phase transition*, with the ratio of the two "phases" governed by the external field of force applied, in a way similar to what has been observed in other systems (see Berret 1997). We call regions I, II and III the shear rate ranges which correspond to the *polycrystal flow*, the *phase transition*, and the *melting* respectively.

3. RHEOLOGICAL RESULTS

We started the mechanical study by the measurment of the *flow curve* (stress versus shear rate) : it exhibits a *weak plateau* on 5 orders of magnitude in shear rate for the F108, and a true plateau in the region II for F68, confirming the analogy with a phase transition system. Moreover the system exhibits a *yield stress* of around 150 Pa (The elastic modulus G_0 is around 2 10^4 Pa).
We investigated in detail the *creep and recovery* behavior under *small stresses*. Under a constant stress (above the yield stress), the system first creeps, after an immediate elastic response, with its $\gamma(t)$ curve never reaching a constant slope, before reaching a turning point in the creep curve, where it begins to flow with a definite slope. The total deformation were this happens is around γ=30 %. The slope versus stress dependence follows the flow curve previously described. If one stops applying stress before the flow takes place, the recovery is divided in two parts : an immediate elastic recovery, followed by a long time recovery due to elasticity being stored in the crystallites (remeber that we are below the flow region : no transition to layer-sliding regime has occured).
These data are coherent with the phase transition picture presented above : the waiting time before the flow takes place can be interpreted as *the nucleation time of the aligned phase*, which we associate with true flow. The *polycrystaline phase* should be responsible for the *creep behavior* (by some mechanism involving mouvement of defects and grain boundaries), and for the *elastic memory* that the recovery behavior imply.

Berret, Porte, Decruppe (1997) Phys. Rev. E **55**
Liu, Weitz, Ackerson (1996) Phys. Rev. E **48**

THE RHEOLOGICAL BEHAVIOUR OF RHAMSAN SYSTEMS

U. FLORJANCIC, A. ZUPANCIC, M. ZUMER, R. LAPASIN[1]

Faculty of Chemistry and Chemical Technology, Department of Chemical Engineering, University of Ljubljana, Askerceva 5, SI 1001 Ljubljana, Slovenia
tel.:++386 61 176 04 23, fax.: ++386 61 126 45 77, e-mail: urska.florjancic@uni-lj.si

[1]Dipartimento di Ingegneria Chimica, dell'Ambiente e delle Materie Prime, Universita di Trieste, piazzale Europa 1, I 34127 Trieste, Italy

1. INTRODUCTION

Microbial polysaccharides are commercially attractive because of their specific properties such as high viscosity and gelling properties, compatibility with a wide variety of salts within a broad range of pH and temperature, high-water solubility, and synergy with other polysaccharides. They were developed for use as thickening, suspending, stabilizing, viscosity-increasing, gel-forming agents in aqueous systems (Paul, 1986).

Rhamsan gum is anionic microbial extracellular polysaccharide produced by an aerobic fermentation using a strain of bacteria Alcaligenes spp. ATCC 31961. Dissolved in water under proper conditions it forms structured systems. Sol-gel transition can be attained even at low polymer concentrations.

The rheological behaviour of aqueous rhamsan systems has been investigated at different concentrations of polymer and the influence of different environmental conditions on the rheological properties of the examined systems has been studied.

2. EXPERIMENTAL

2.1 Material

Rhamsan (Monsanto) was dissolved in different solvents (distilled water, water solutions of ethanol and water solution of ethylene glycol) at polymer concentrations between 0.1 and 1.0 wt.% of rhamsan. The measurments were carried out at three different temperatures between 8 and 45°C.

2.2 Instrument

The rotational controlled stress rheometer HAAKE RS150, equipped with cone and plate (6 cm, 2°) and double cone (6 cm, 4°) sensor systems was used to carry out continuous shear and oscillatory measurments. Temperature was controlled by circulator HAAKE DC5-K20.

2.3 Procedure

Prepared rhamsan systems were investigated under continuous and oscillatory shear conditions.

The shear stress ramps and the stepwise procedures with sequential changes of shear stress (0.03-20 Pa depending on polymer concentration) were applied in order to study shear dependent behaviour under destructive shear conditions. The range of linear viscoelastic response was determined from stress sweep tests under oscillatory shear conditions at frequency of 1 Hz in stress range of 0.005-17 Pa.

The frequency sweep experiments under conditions of linear viscoelastic response were carried out at constant stress or constant strain amplitude in the range of 0.02-10 Hz.

2.4 Analysis of viscoleastic data

The mechanical spectra obtained from frequency sweep tests have been described by the generalized Maxwell model. As a result of viscoelastic data analysis the relaxation spectra have been derived.

3. RESULTS AND DISCUSSION

3.1 Continuous shear conditions

Examined systems exhibited detectable reversible time dependent behaviour of thixotropic type. The appearance of Newtonian plateau in the low shear stress range was observed for all the samples.

All investigated systems were almost insensitive on environmental conditions such as temperature in the temperature range of 8-45°C and different solvents (ethanol, ethylene glycol). The shear dependent behaviour of studied systems was strongly influenced by polymer concentration (Fig. 1).

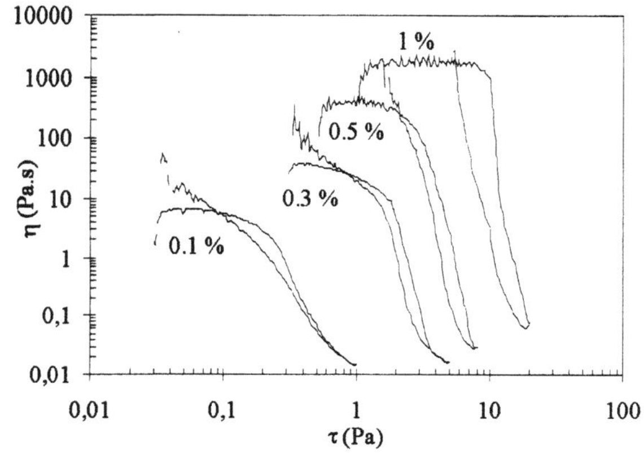

Figure 1. The effect of polymer concentration on viscosity profiles at 25°C

3.2 Oscillatory shear conditions

The results of stress sweep measurements are plotted on Figure 2. The increase of loss modulus followed by sharp decrease was obtained when the system passed from linear to non-linear viscoelastic regime. When polymer concentration increased such overshoot of loss modulus became more pronounced.

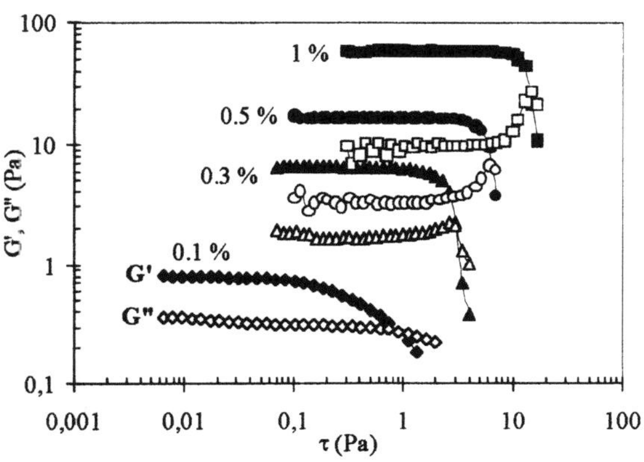

Figure 2. The effect of polymer concentration on both dynamic moduli at frequency of 1 Hz at 25°C

The critical strain characterizing the limit of linear viscoelastic regime was around 10 % and was not strongly dependent on the polymer concentration.

In the frequency range examined the elastic component was predominant for all the systems (Fig 3). With increasing rhamsan concentration both dynamic moduli became less sensitive on the applied frequency. Such viscoelastic behaviour is usually found for structured systems and weak gels.

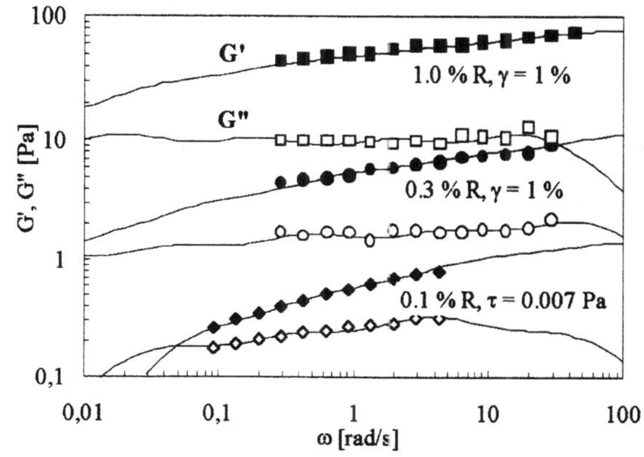

Figure 3. The comparison of experimental data from frequency sweep tests at 25°C with calculated (curves) values by using generalized Maxwell model

As a result of viscoelastic data analysis by the generalized Maxwell model (Lapasin, 1995)

$$G'(\omega) = \sum \frac{g_i \cdot \lambda_i^2 \cdot \omega^2}{\left(1 + \lambda_i^2 \cdot \omega^2\right)},$$

$$G''(\omega) = \sum \frac{g_i \cdot \lambda_i \cdot \omega}{\left(1 + \lambda_i^2 \cdot \omega^2\right)},$$

where λ_i is relaxation time of the ith Maxwell element and g_i is elastic modulus, the discrete relaxation time spectra were evaluated.

The comparison of the oscillatory and continuous shear behaviour showed a failure of Cox-Merz rule. The rheological properties of studied systems differed from those of polymer solutions and were more similar to those of structured systems.

4. CONCLUSIONS

Examined aqueous rhamsan systems were almost insensitive on temperature (8-45°C) and different solvents (ethanol, ethylene glycol).

The structural conditions of examined systems can be described as a dispersion of gel micro domains in polymer solution matrix.

5. REFERENCES

Lapasin R., Pricl S., Rheology of Industrial Polysaccharides: Theory and Applications, Blackie Academic & Professional, An Imprint of Chapman & Hall, Glasgow, 1995

Paul F. et al., Microbial Polysaccharides with Actual Potential Industrial Applications, Biotech. Adv., **4**, 1986, pp. 245-259

CREEP RECOVERY AND MOLECULAR STRUCTURE OF POLYETHYLENE MELTS

C. GABRIEL, J. KASCHTA, H. MÜNSTEDT

Institute of Polymer Materials, Department of Materials Science, University Erlangen-Nürnberg
Martensstr. 7, D - 91058 Erlangen, Germany

1 INTRODUCTION

This contribution deals with the correlation between the molecular structure of polyethylenes and their rheological properties in shear creep recovery experiments. This experimental mode has the advantage that the complete time dependence of the viscoelastic behaviour of polymer melts can be obtained without being masked by the viscous part of deformation.

2 EXPERIMENTAL

The rheometer used was a magnetic bearing torsional creep apparatus (Link and Schwarzl, 1985). With this rheometer measurements in the linear viscoelastic regime at shear stresses between 3 and 20 Pa were performed. The measurements are conducted in nitrogen atmosphere in order to improve the thermal stability of the polymer melts.

3 SAMPLES

Two ommercial polyethylenes were chosen. The LDPE is a stabilized Lupolen® 1810 H from BASF and the LLDPE is sold by the DOW Chemical company as Dowlex® NG 5056 E.

The LDPE is a common high pressure polyethylene polymerized in a tubular reactor. The mean value of 23 branches per 1000 carbon atoms of this LDPE is typical of highly long-chain branched materials. The LDPE has a broad molecular mass distribution characterized by $M_w/M_n = 14$, the mass average molecular weight is 240 kg/mole.

The LLDPE contains octene as the comonomer. It is polymerized in suspension using Ziegler-Natta catalysts. The molecular mass distribution is typical of this kind of polymer with $M_w/M_n = 3$, M_w has a value of 105 kg/mole.

4 RESULTS

The results of the creep and creep recovery experiments are summarized in Fig 1 (Gabriel et al., 1998). The creep and the recoverable compliances of the LDPE overlap in the short time region, this means that the deformation of this long-chain branched polyethylene is mainly elastic up to creep times of 0.2 s. In contrast, the state of rubber-elasticity is not accessible for the LLDPE due to the limited time resolution.

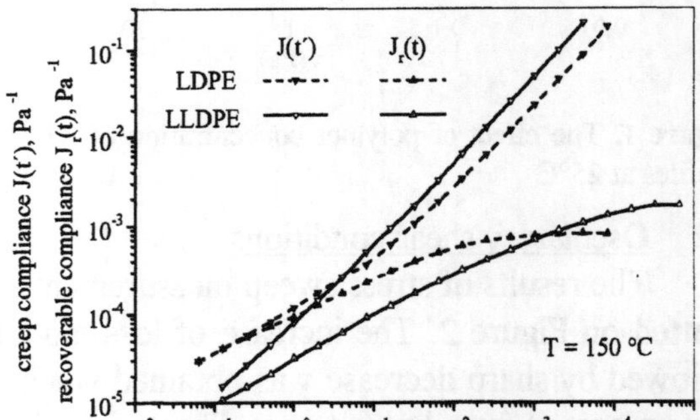

Fig. 1: Creep and recoverable compliances of the two polyethylene melts at T = 150 °C

There are two other obvious differences between the polyethylene melts. At relatively short recovery times, $J_r(t)$ is significantly higher for the LDPE than for the LLDPE but at longer times the curves cross each other and the steady-state recoverable compliance of the LLDPE is twice as high as that of the LDPE. This result is surprising, because it should be expected that the steady-state recoverable compliance is larger for a material with a broader molecular mass distribution like the LDPE.

The second difference is that the steady-state of recovery of the LDPE is reached after recovery times of 1000 s in comparison to about 30 000 s for the LLDPE. From the comparison with the preceding creep times it can be concluded that the behavior of the LDPE is in accordance with the rules of linear viscoelasticity, because also times of about 1000 s are needed to reach a steady-state of creep. In the case of the LLDPE a discrepancy between the preceding creep times needed to reach a steady-state of creep (about 1000 s) and the very long recovery times is found (Fig.2). The rheological behavior of the LLDPE is therefore not in accordance with the theory of linear viscoelasticity. Moreover, the dependence of the recoverable compliances of the LLDPE on the preceding creep times reveals that the retardation behavior of the melt is influenced by two main mechanisms (Abb. 2).

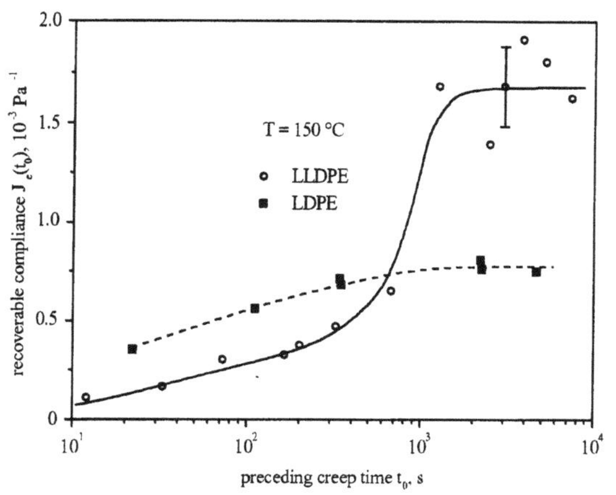

Fig. 2: Influence of the preceding creep time t_0 on the steady-state compliances

Besides that, it has been found that the recoverable compliances of the LLDPE measured at different temperatures cannot be shifted to a mastercurve whereas the time-temperature superposition principle is fulfilled for the LLDPE. These findings lead to the assumption that the LLDPE melt might not be homogeneous.

5 DISCUSSION

Indeed, the TREF-analysis (temperature rising elution fractionation), which was kindly supplied by the DOW Chemical Company, clearly shows that the LLDPE is not a homogeneous material (Fig. 3). It contains a significant component of linear polyethylene together with short-chain branched molecules covering a wide range of branching degrees. From viscosity measurements on the fractions it can be concluded that the molecular mass of the linear component is distinctly higher by a factor of two compared with the branched fraction.

The rheological findings on the LLDPE can be explained by the hypothesis of a phase separation of the linear component and the short-chain branched fraction within the melt. The composition fluctuation leads to the pronounced elasticity of the melt because of the influence of interfacial tensions between the two phases. The different retardation times of the two phases can be observed by performing creep recovery experiments. They are reflected by the bimodal distribution of retardation times obtained from the retardation spectra. The influence of the interface can also be regarded as source for the failure of the time-temperature superposition principle in the case of the recoverable compliances of the LLDPE.

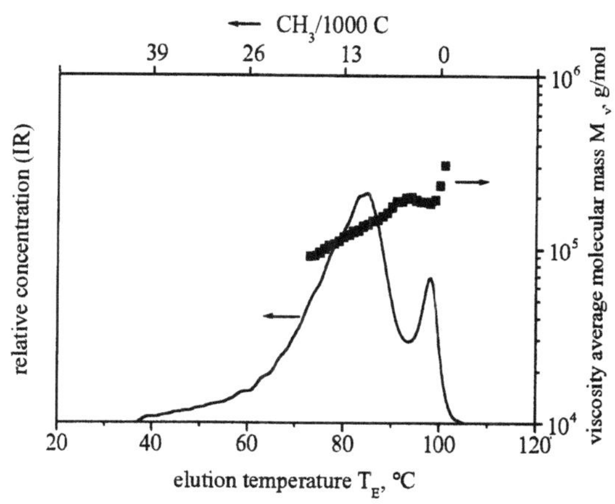

Fig. 3: TREF-analysis of the LLDPE

6 OUTLOOK

In order to confirm the conclusions other Ziegler-Natta catalyzed LLDPE as well as metallocene-catalyzed LLDPE with homogeneous branching distributions have to be investigated. The miscibility of polyolefins with varying branching structures has to be tested with techniques like Imbedded Fiber Retraction method.

7 CONCLUSIONS

The rheological behaviour of the LLDPE, which in certain experimental modes does not obey the theory of linear viscoelasticity and the time-temperature superposition principle, together with the results from TREF-analysis gives strong evidence that the LLDPE can be regarded as an immiscible blend of a linear high molecular mass component and short-chain branched molecules of lower molecular mass.

8 REFERENCES

Gabriel C, Kaschta J, Münstedt H (1998) Rheol Acta 37: 7-20

Link G, Schwarzl FR (1985) Rheol Acta 24: 211-219

OSCILLATORY MEASUREMENTS OF STYRON 457 IN TOLUENE - LOSS AND STORAGE MODULUS MASTER CURVES

Dr. D. Hadjistamov

Ciba Speciality Chemicals Schweizerhalle Inc., Performance Polymers

1. Introduction

It is well known that the complex shear modulus G* and the complex viscosity η^* are obtained in oscillatory shear flow. The complex shear modulus is defined as:

$$G^* = \sqrt{G'^2 + G''^2} \qquad (1)$$

where G' - storage modulus and G" - loss modulus.

The aim of this work is to describe the storage and loss modulus master curves for solutions of the polystyrene styron 457 in toluene from oscillatory measurements.

2. Experimental

Solutions with 30% (⇖), 40% () and 50% (♦) styron 457 in toluene were prepared and rheologicaly investigated. The oscillatory measurements were carried out with a stress controlled viscometer Carri-med CSL 500 with cone-plate (Θ = 4 cm, 2°) at 25 ± 0.2°C. The range for the oscillatory test was 0.5 to 250 rad/s.

3. Results

3.1. Loss modulus master curve for polystyrene solutions

The polystyrene solutions exhibit, in accordance with the shape of the flow curves, structural-viscous flow behavior.

The 30%, 40% and 50% - styrene solutions have zero-complex viscosity of 3.4, 22 and 190 Pa.s respectively. The critical frequency at onset of oscillation (similar to the shear rate at onset of shear thinning) depends on the concentration of the suspentions. The values for the 40% and 50% solutions are 44 and 5.5 rad/s respectively. The critical frequency for the 30% solution will be reached at a higher frequency.

The zero-complex viscosity can be used, to define a loss viscosity master curve (Hadjistamov, 1996)(Fig.1), similar to the viscosity master curve (Vinogradov et al.,1964).

Fig.1 Loss modulus master curve of the polystyrene styron 457 in toluene (Carrimed, cone-plate 2°; 25°C; 30%(⇖), 40%(), 50%(♦))

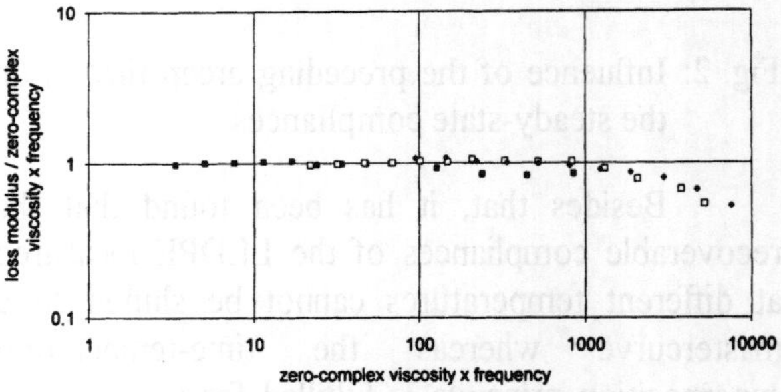

The loss viscosity master curve has a value of 1 up to approx. 970 Pa (Fig.1).

3.2. Storage modulus master curves for polystyrene solutions

The dependence of the storage modulus on frequency leads to straight lines with a slope of n = 2 up to the frequency at onset of oscillation that is similar to that for the loss modulus with 28 and 4.2 rad/s for the 40% and 50% solutions respectively. The values for the storage modulus at onset of oscillation are with 150 Pa about 6 times smaller compared with the values for the loss modulus at onset of oscillation.

It is possible to calculate and represent (Fig.2) a storage modulus master curve (Hadjistamov, 1996) from dynamic measurements, similar to the normal stress master curve from the steady-state measurements.

The polystyrene solutions have up to approx. 950 Pa different straight lines shifted in parallel towards smaller values of the concentration. It is realy surprisingly that the storage modulus master curve has for the 30%

solution higher values compared to the 50% solution. The slope of the straight lines is, as expected (Hadjistamov,1996), n = 1. There is a certain dependence of the storage modulus master curve on the concentration of the suspensions.

Fig.2 Storage modulus master curve of the polystyrene styron 457 in toluene (Carrimed, cone-plate 2°; 25°C; 30%(⧅), 40%(), 50%(♦))

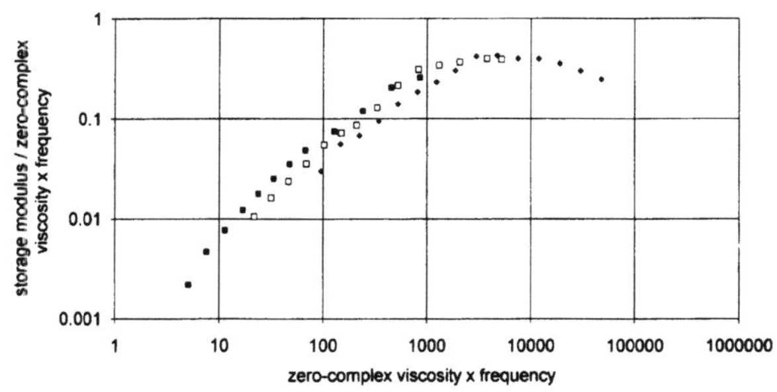

The straight lines of the master curve come together at approx. 1000 Pa, i.e. at the beginning of the shear thinning region. In this region a common master curve is formed that tends toward a maximum with a plateau (Fig.2). The value of the plateau is 0.43. The storage modulus master curve begins to decline after the plateau at $\omega.\eta_o^* \approx 5000$ Pa, i.e. about one decade later than the loss modulus master curve. There is a really storage modulus master curve only in the region of $\omega.\eta_o^* \geq 1900$ Pa. In the region $\omega.\eta_o^* \leq 1900$ Pa the values of $\frac{G'}{\omega.\eta_o^*}$ depend on the concentration of the solutions. Nevertheless, we shall call this curve an oscillatory storage modulus master curve.

3.3. <u>Master curve comparison</u>

A rheological characterization of polystyrene solutions can be done by comparison of the loss and the storage modulus master curves.

The loss modulus and storage modulus master curves have up to approx. 950 Pa slopes of n = 0 and n = 1 respectively .

The relation between the values of the loss modulus and storage modulus master curves (at a certain frequency) is a measurement of the viscoelastic behavior of a system and is equal to the reciprolal value of $\tan\delta$. This ratio shows the relation between the elastic and viscous oscillatory flow behavior. This ratio depends of course on $\omega.\eta_o^*$ (Fig.3).

Fig.3 Viscoelastic master curve of the polystyrene styron 457 in toluene (Carrimed, cone-plate 2°; 25°C; 30%(⧅), 40%(), 50%(♦))

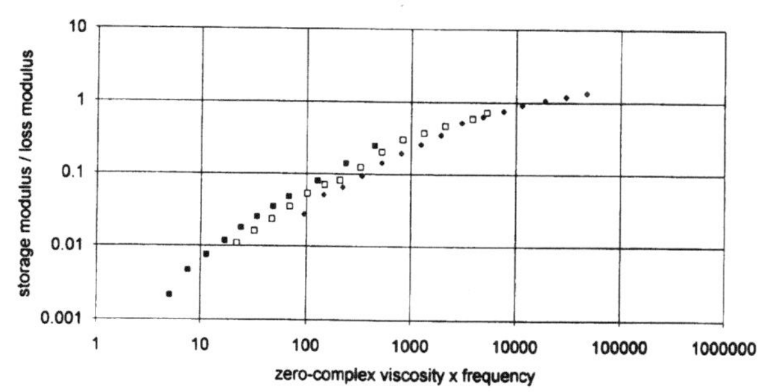

The G'/G''-curve is in the first Newtonian region identical with the storage modulus master curve. The viscous behavior (G'') is better developed in this region. The storage modulus, i.e. the elastic behavior increases with $\omega.\eta_o$.

The viscous and elastic behavior of a system are equal at the cross-over point of the loss modulus and storage modulus master curves:

$$G'/G'' = 1 \quad \text{at } \omega.\eta_o^* \approx 1900 \text{ Pa.}$$

The elastic behavior is better developed compared with the viscous behavior after the cross-over point. The G'/G''-curve seems to tend to a plateau.

4. REFERENCES

Vinogradov GV et al (1964). Doklady Akad.Wiss.Udssr 154:890-893

Hadjistamov D (1996) Rheol. Acta 35:364-368

CHARACTERIZING THE PROPERTIES OF FRESH CONCRETE: BML-VISCOMETER AND BTRHEOM RHEOMETER COMPARATIVE EXPERIMENTS

C. HU [1], G.I. PELOVA [2], J.C. WALRAVEN [2]

[1] LAFARGE Central Research Laboratory, B.P. 15, 38291 St.Quentin Fallavier Cedex, France
[2] Delft University of Technology, P.O.Box 5048, 2600 GA Delft, The Netherlands

1. OVERVIEW OF THE PROBLEM

During the last years, the characterization of the workability of fresh concrete gained increasing interest in concrete technology and building industry. For present-days concretes, made with addition of mineral and chemical admixtures, the classical slump test does not give sufficient information about such an essential property of the concrete mixtures as workability. The development of the two-point workability test by Tattersall (1976) marked the first attempt to quantify the rheological behaviour of fresh concrete, in particular the workability. The implementation of the two-point workability test for concrete is recognition of its Bingham-material characteristics in the case of low shear rates, which usually occur in practice. Wallevik and Gjørv (1990) developed a computerized coaxial cylinders viscometer (BML-Viscometer) and proposed a method to calculate the yield value (τ_0) and the plastic viscosity (μ) from the measured flow resistance (g) and relative viscosity (h) and the dimensions of the apparatus. De Larrard, Hu (1997) and co-workers designed a computerised plane-to-plane rheometer (BTRHEOM), which can also be implemented for field-testing. A drawback of most of the invented two-point tests is that the measured g-value (flow resistance) and h-value (relative viscosity), cannot be related directly (also not easily) to the characteristics of the material (in terms of the Bingham model): yield value of the shear stress and plastic viscosity. The aim of the 'live, in real time" experiments executed with the BML-Viscometer and the BTRHEOM was to compare the rheological characteristics (τ_0 and μ) of different concrete mixes (soft-to-fluid consistency) measured by the two rheometers of different principle with concrete of the same batch.

2. BML-VISCOMETER AND BTRHEOM

The BML-Viscometer is a coaxial cylinders viscometer for coarse particle suspensions that is suitable to measure the rheological properties of mortar and concrete with 80 mm slump or higher. The coaxial system consists of an outer cylinder, an inner cylinder unit and a top ring. The outer cylinder is filled with the concrete mixture to a fixed height and mounted on a rotation disk, which starts to rotate at a certain speed, progressing to the maximum speed (chosen as an input parameter in the software) at which the first torque readings are received. Subsequently, the testing speed Ω is reduced stepwise, and the torque measurements T are made on the downcurve. The total concrete sample volume is about 20.5 litre, 11 litre of which is in the gap between inner and outer cylinder. The rheological parameters are described by the fundamental parameters in the Bingham model, the yield value τ_0 *[Pa]* and the plastic viscosity μ *[Pas]*. If the concrete sample between inner and outer cylinder is overall sheared, τ_0 and μ can be calculated with the Reiner-Rivlin equation:

$$\Omega = \frac{T}{4\pi l \mu}\left(\frac{1}{R_1^2} - \frac{1}{R_2^2}\right) - \frac{\tau_0}{\mu}\ln\left(\frac{R_2}{R_1}\right) \qquad (1)$$

where Ω is the angular velocity *[rad/s]*, l, R_1, R_2 are the height of the inner cylinder in contact with the concrete and the radii of the inner and outer cylinders, respectively.

The BTRHEOM rheometer is a torsion rheometer with a vertical axis. A ring of fresh concrete sample of about 7-litre, is sheared between two horizontal sections, of which the lower one is fixed by a system of blades, and the upper one rotates, driven by another system of blades, linked with the axis of the motor. Vibration can be applied to the plate supporting the container, which allows studying its influence on the rheological behaviour of fresh concrete. In the case of Bingham materials, the relationship between Γ and N in this apparatus is linear. The two Bingham constants τ_0 *[Pa]* and μ *[Pas]*, can be calculated with the following equations:

$$\tau_0 = \frac{3\Gamma_0}{2\pi(R_2^3 - R_1^3)}; \quad \mu = \frac{h\frac{\partial\Gamma}{\partial N}}{\pi^2(R_2^4 - R_1^4)} \qquad (2)$$

where R_1 and R_2 are the inside and outside radii of the sample, h its effective height, Γ_0 the ordinate at the origin of the straight line $\Gamma(N)$ and $\partial\Gamma/\partial N$ its slope. A standard procedure, by which torque values are measured at five different rotation velocities (in decreasing order) can be used to determine the two Bingham constants τ_0 and μ of soft-to-fluid concrete (with or without vibration). By performing a stress-controlled test, the BTHEOM rheometer can measure the peak shear as an important characteristic of the thixotropy of concrete at rest.

3. EXPERIMENTS AND DISCUSSION

Six concrete mixtures of a 50-litre batch were tested simultaneously with both equipments. Moreover, the slump and the DIN flow spread were measured. A comparison of the yield value τ_0 and the plastic viscosity μ received from the BML-Viscometer and the BTRHEOM rheometer showed quite remarkable differences. Furthermore, some results showed a different tendency to increase (mixture 5 and 6, see Fig.1).

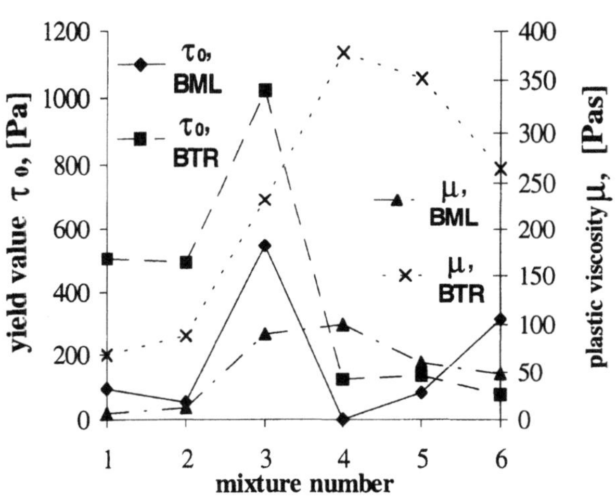

Fig. 1 Comparison of the yield value τ_0 and plastic viscosity μ, measured BML-Viscometer (BML) and the BTRHEOM (BTR).

The results should be regarded only as preliminary, knowing that only three of the six concrete mixtures had suitable stability for the application of the BML-Viscometer and the BTRHEOM. For the BML-Viscometer, the conversion from the pseudo-characteristics (flow resistance (g) and relative viscosity (h)), which are result of a regression analysis of the measured torque at different velocities, to the material characteristics, yield stress (τ_0) and plastic viscosity (μ), needs an assumption of the velocity or the shear profile of the concrete during the test. For certain types of concrete mixtures the viscometer might have a shear zone just outside the inner cylinder and the remainder of the concrete in the gap between inner and outer cylinder might act as a plug.

4. CONCLUSIONS

Remarkable differences in the results obtained with the two different equipments can be observed. However, because of the lack of representative concretes it is difficult to draw definite conclusions. Homogenity of the concrete is important for the determination of the rheological characteristics, in order to avoid the scatter in the measuring values. Further attempts have to be made in order to clarify the question raised and to develop more reliable conversion methods. Otherwise rheological measurements on fresh concrete will be equipment dependent.

5. REFERENCES

De Larrard F, Hu C, Sedran T, Szitkar JC, Joly M, Claux F, Derkx F (1997) A New Rheometer for Soft-to-Fluid Fresh Concrete. ACI Materials Journal 94: 234-243

Tattersall GH (1976) The Workability of Fresh Concrete. Slough, Cement and Concrete Association , Wexham Spring, 138 p

Wallevik O, Gjørv OE (1990) Development of a coaxial cylinder viscometer for fresh concrete, Hanover: Chapman and Hall, pp 213-224

OSCILLATORY GRANULAR FLOWS

J. JAIMES, G. JOSEPH, A. MAGAÑA, & *B. MENA

Instituto de Investigación en Materiales, UNAM, Apdo. Postal 70-360, México, D.F. 04510, MÉXICO
*Instituto Mexicano del Petróleo, Apdo. Postal 14-805, México, D.F. 07730, MÉXICO

1 SYNOPSIS

An experimental device has been used to examine the behavior of granular materials under vertical, horizontal and combined bi-axial vibrations. Different container geometries were used with glass spheres, millet and lentils as granular media. Frontal and lateral visualization of the flow was made possible with high-speed video cameras. Four different phenomena: heaping, arching, small amplitude surface waves and large amplitude surface waves were found. The onset of these depend basically on two parameters: the dimensionless acceleration and the dimensionless depth of the vibrating bed. Comparison was made between experiments and numerical predictions based on molecular dynamics. The numerical solution agrees well with all the experimental observations.

2 INTRODUCTION.

Experiments made with granular material under oscillatory vibrations show an extraordinary behavior. Different phenomena can be observed and the reasons remain unexplained. Under vertical oscillations and low frequencies, the material inside a container flows in a convective pattern and accumulates at a one of the sides of the container forming a "heap". At higher frequencies, small amplitude surface waves (s.a.s.w.) can be observed travelling along one side of the container. These waves appear only at the free surface of the material; the bottom remains plain. In the same way, large amplitude surface waves (l.a.s.w.) are present if the frequency is increased. Within a range of intermediate frequencies between those of s.a.s.w. and l.a.s.w., a fourth behavior, called arching, appears. Arching can be recognized because both the free surface and the bottom of the material present a wave-shaped form. In arching, several convective flows can be present, which depend on the amount of material and the amplitude of vibration. Different materials were used to study these phenomena in order to determine if geometry, density or other properties are related with such a particular behavior. Nevertheless, the phenomenon does not seem to be affected by changes in material properties. Furthermore, changes in container geometry do not affect the observed behavior. Results correlate well with numerical simulations using molecular dynamics.

3 EXPERIMENTS

A simple mechanical device was designed and built in order to supply horizontal and vertical oscillations to containers of various dimensions. A square section container of 12cm and 18cm high, two rectangular containers of 2cm x 20cm x 20cm and 3.6cm x 20cm x 20cm and a cylindrical container of 8cm in diameter and 18cm in height were used in the experiments. The containers were made of Plexiglas to allow for visualization with the help of a high speed video camera (200 frames per second). The granular media used in the experiments were glass spheres of 1.28mm in diameter, millet, with a 2.25mm average diameter and 3.25mm average length; and lentil, with an average diameter of 3.5mm. The oscillating frequency varied from 0 to 35 Hz with amplitudes between 0 and 5.5cm. In order to describe the vibrations, a dimensionless parameter (Γ =aw2/g) was used, where a and w are the amplitude and frequency of vibration and g is the gravity acceleration constant. The dimensionless number of layers (N=H/d) represents the amount of material in the container; where H is the height of material in the container, and d is the particle diameter of the grain. The experimental range was $10<N<40$ and $1<\Gamma<7$.

4 RESULTS

4.1 VERTICAL VIBRATION

Four distinct phenomena are observed when granular material is excited with vertical oscillations. The heap can be present not only on one preferred side of the container, but also at the middle as described by Laroche . Furthermore, for equal values of amplitude, number of layers and frequency, the formation of the heap was observed, in different cases, on both sides of the container or at the mid section of it. It was concluded that the formation of the heap at a changing side of the container is a consequence of small changes of the initial stress conditions inside the container. For values of N<15, arching always appears before l.a.s.w.. The values of Γ at which arching appears are in the range between $5<\Gamma<6.5$ depending on the material and vibration amplitude. For all cases examined with materials other than glass spheres, the value of Γ at which large amplitude waves appear is always greater. For larger values of N, in all cases, l.a.s.w. always appear before arching. When the results of experiments made using different values of vibration amplitude were plotted in a Γ-N graph, this showed that Γ is probably not representing the importance of the oscillation amplitude. It was therefore concluded that new parameters have to be introduced in order to consider not only excitation conditions, but the geometry of containers and material properties such as repose angle, density, restitution and friction coefficients.

4.2 HORIZONTAL VIBRATION

Initially, motion was restricted to the upper layers of particles for $0.5<\Gamma<0.77$. For smaller values of Γ, no motion was observed. For $1.2<\Gamma<7$ particles flow from the center to the sides of the container parallel to the direction of vibration. In a frontal view, two convective flows ascending at the center and descending towards both sides of the container were clearly observed. The speed of the particles increases as they approximate to the lateral walls (the walls perpendicular to the direction of vibration). Using a strobe lamp, stationary waves were also detected along the lateral sides of the container (left and right diameter extremes in the case of the cylindrical container). The frequency of lateral waves is one half the vibration frequency.

5 CONCLUSIONS

For vertical vibrations, the heap could be present on either side of the container or at the mid-section as a consequence of the small changes in the initial stress conditions in the material inside the container. For values of N<15, arching always appeared prior to l.a.s.w.. For larger values of N, l.a.s.w. were always present before arching.

For horizontal vibrations, two convective flows ascending at the center and descending towards the walls were observed; the speed of particles increasing as the lateral walls were approached. As frequency was increased, lateral waves appeared with a frequency of one half the frequency of vibration.

In the case of combined vibrations, a superposition of effects between vertical and horizontal modes was found. Both lateral and vertical waves appeared. For an inclination of 30° the dominating mode was the horizontal while for 60° the corresponding phenomena were dominated by the vertical mode. When a cylindrical geometry was used, there were no regions where particles remained motionless, indicating that the geometry of the container affects the flow of particles in zones close to corners and sharp ends. Finally, comparison was made between experiments and numerical predictions based on molecular dynamics. The numerical solution agreed remarkably well with the experimental observations, particularly for the case of the new results for combined vibrations.

6 Acknowledgements

The authors wish to acknowledge the assistance of the machine shop of the I.I.M. for the construction of the experimental device. Support from grants IN302392 and IN503494 from PAPIIT, UNAM are also gratefully acknowledged. Many fruitful conversations with M.Hunt and C. Brennen and their group at Caltech are also acknowledged.

7 REFERENCES

Hunt, M.L., Weathers, R.C., Brennen, C.E., Lee, A.T. & Wassgren, C.R. " Effects of horizontal vibrations on hopper flows of granular materials", to appear in Phys. of Fluids (1998)

Laroche, Douady & Fauve; "Convective flow of granular masses under vertical vibrations"; Journal of physics 50, No. 7, 699-706 (1990).

Wassgren, C.R., Brennen, C.E. & Hunt, M.L. "Vertical vibration of a deep bed of granular material in a container"; J. of Applied Mechs. 63, 3, 712-719 (1996)

Wassgren, C.R., Hunt, M.L.& Brennen, C.E. "Effects of vertical vibrations on hopper flows of granular materials", submitted to Phys. of Fluids (1998)

THE NEWS METHODS OF RELIABILIY CRITERIA FOR TRANSPORTATION ELEMENTS

M. KOPECKY

Faculty of Industrial Technologies - TrU, 020 32 Puchov, Slovak Republic

1. INTRODUCTION

In some mobile machinery and equipment, or their elements, the problem of fatigue strength is conditioned by a fatigue process and by knowledge of a fatigue curve.

The problem of fatigue strength and service-life, as the most important phenomena of strength reliability under those conditions, is connected more or less with a certain degree of uncertainity.

It is probable that the most significant cause of this unfortunate situation is the fact that so far there has not been a single universal theoretically and experimentally proved fatigue theory which would consider all the factor that have an influence on the phenomen. Apart from this, there is not a generally accepted methodology of fatigue tests.

The methods described in this paper are the ways to reach the solution goals by means of a characteristic curve of fatigue strength and reduced fatigue curve with the maximum use of computer technology.

Also, because of economic and other reasons, it is impossible to carry out sufficient amount of fatigue tests on a finished product in the laboratory in order to get reliable information about the fatigue curve. It is further connected with a choice of a suitable cummulation of a fatigue failure which can also lead to different results.

2. THEORY

If we know the probability density of occurrence of decisive loading quantity, e.q. a moment on the inlet of a mobile machine and we also know the fatigue curve for a given constructional element of unit, we can determine the total strength reliability from the point of view of fatigue strength.

Random operational loading creates a stochastic process of excitation forces. A successful reproduction of the response of this random loading depends on the technical facilities.

With regard to a probability character of the fatigue curve, where its exponents have a character of a random quantity, a total service life is also a random quantity characterized by a probability density.

Let us suppose that the probability density of a moment parameter occurrence on the inlet is $\mathbf{f}_{(M)}$. The following relation can be generally written for stress:

$$\sigma_{(M)} = \varphi_{(M)}.M \tag{1}$$

where $\varphi_{(M)}$ can be generally also a stochastic function, as shown in Fig. 1.

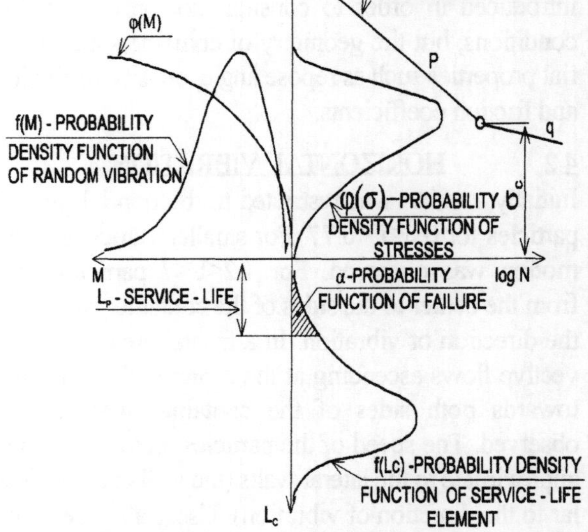

Fig. 1 The dependence of a component service life on loading

To express, by means of failure dependence on damage intensity, reductions in a fatigue curve direction value of the basic material for a constructional element or a complete unit of either a mobile machine or equipment.

Insufficient knowledge of a direction value of a fatigue curve is to be expressed by means of an assessment and reduction according to a failure dependence on a damage intensity.In spite of the fact that a shape of the fatigue curve is idealized to a great extent it was chosen because of its simplicity and possible application of a majority of calculation procedures.

A course of the damage intensity for different loading levels can be a criterion for elimination of nsignificant loading levels and for a proper choice ɪf loading levels.

Determining the total failure intensity we can ;et a basic orientation concerning loading and ɪossible failures in particular elements. The basic cheme is based on an assumed dependence of a ailure on he intensity of damage. We define the ɪasic fatigue curve for a basic material completed vith further information. The method is based on a naximum use of computers.

. THE MEASUREMENT SYSTEM

An analysis of stochastic loading for mobile ɪachines and equipment by means of statistic haracteristics.

To define an equality amplitude and a requency of harmonic excitation to a spectrum of andom loading by means of functions statistic ɪoments of power spectral density.

An identification of dynamic properties in an ɪperational random excitation and an analysis of a ignal with the help of statistical characteristics of tochastic excitation.

For a random stationary and Gaussian istribution time series several parameters escribing the time evolution of the quantity can e expressed by the function of the statistical ɪoments of the power spectral density in elation to the zero frequency axis.

To work out a method of reliability ssessment to transport elements and equipment by ɪeans of a characteristic curve of strength eliability with the maximum use of computer echnology.

An expression of a dependence between a tochastic loading of a given constructional element nd its reliability by means of an extention with a ew magnitude which will express a numerical uarantee in a form of probability.

The dependence between an operating rocess of a constructional component and its ervice life, N, must be expended a variable omponent, $R_{(N)}$, which offers a numerical uarantee in a probability form:

$$R_{(N)} = \exp(-(N - N_{min})/(N_{sig} - N_{min}))^k \qquad (2)$$

ʼhere N_{min} is a minimum of the longevity,
N_{sig} is a modal value of the longevity,
k is a parameter of distribution.

According to the theory of fatigue failure we determine a new fatigue curve, the reduced course of which would correspond to the result of the experimental test. The parameters of the fatigue curve will be reduced. We determined the probability for a given course of the fatigue curve. It is the probability of a failure occurrence for a deviation of a group of parts.

4. APPLICATION

The application of this methology shorten knowledge of the time to failure of mobile machines components and contribute to the safety and economy of mechanical systems.

The results of its application would be presented to mobile facility elements.

5. CONCLUSIONS

Mechanical systems of mobile machinery and equipment differ considerably from the point of view of a possibility of simulation procedures. When assessing properties of a construction model in the stage of its design and prototype test, or innovation and experiment verification, the question of economy is also important.

The presented system of assessment of fatigue strength has been applied in many concrete situations and verified within cooperation with manufacturing enterprises. It enables a simulation of an operating process of loading in laboratory conditions and also a full use of computers. The results gathered so far have shown that the system of assessment of the fatigue strength in mechanical sets of mobile machinery and equipment is good and offers a possibility of generalization.

6. REFERENCES

Kopecky, M., Peslova, F.(1996) Assessment metholdology of elements and constructions reliability criteria for mobile machines and equipment. In: ISTLI special publication 2: Teaching and Education in Fracture and Fatigue, Imprint: E & FN SPON , London, pp.325-330, England.

Kopecky, M. (1996) A view on the ways design of reliability criteria and some applications. In: Preceeding of XXV AIAS National Conf. and Inter. Conf. on Material Engineering, Gallipoli-Lecce, pp.1151-1158, Italy.

Weibull, W. (1951) A statistical distibution function of wide applicability. Journal of Appl. Mechanics, No.3.

MEASUREMENTS OF TORSION RELAXATION UNDER HYDROSTATIC PRESSURE

A. KRALJ[1], I. EMRI[1], N.W. TSCHOEGL[2]

[1]University of Ljubljana, Ljubljana, Slovenia
[2]California Institute of Technology, Pasadena, California, USA

1. INTRODUCTION

The mechanical and other physical properties of polymeric materials significantly depend on time and this can seriously influence their long-term durability. Materials time dependence can be strongly influenced by the temperature and the pressure history to which the material has been subjected during the manufacturing process (as when an injection-molded part emerges from the die).

While the effect of temperature on the mechanical properties is fairly well understood, relatively little information is available on the effect of pressure. It is known that, in principle, pressure has an effect opposite to that of temperature. An increase in pressure slows the relaxation and/or retardation processes. Several years ago these phenomena have been extensively investigated by Fillers & Tschoegl (1977) and by Moonan & Tschoegl (1983). They showed that both temperature and pressure dependence can be described by the FMT-equation which comprises the WLF-equation (Ferry, 1980) as a special case.

2. EXPERIMENTAL APPARATUS

The apparatus we used consists basically of three parts: the high-pressure vessel with its pressurizing equipment, an insert for the shear relaxation measurements (relaxometer) and a stirred thermal bath. The temperature can be kept constant within 0.02 °C. The temperature ranges is from -30 to +120 °C.

The hydrostatic pressure is applied via silicone fluid. A high-pressure hand pump is used for the final compression, which by 10:1 intensifier yields a final pressure up to 650 MPa.

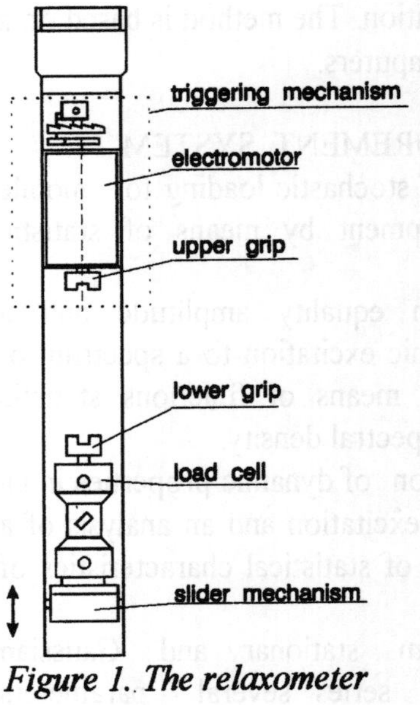

Figure 1. The relaxometer

The relaxometer, shown in Fig. 1, consists of three main parts: a spring triggering mechanism for the specimen twisting, resetting electromotor and a torque load cell. The motor shaft turns the upper grip, which is end-bonded to the specimen with filled epoxy cement. The lower grip is cemented to the specimen in the same manner and is also attached to the torque-measuring cell.

The design of a cell is based on the conventional strain gauge torque cell. The traditional design called for four strain gauges arranged as a Wheatstone bridge and mounted on the flat surface of a sqare shaped, hollow aluminum rod at angles of 45° to the axis of rotation. Its response is nearly insensitive to changes in temperature and pressure.

The another important feature of the relaxometer is the slider mechanism, which allows the torque cell to move vertically as the length of the specimen changes due to pressure and temperature variations.

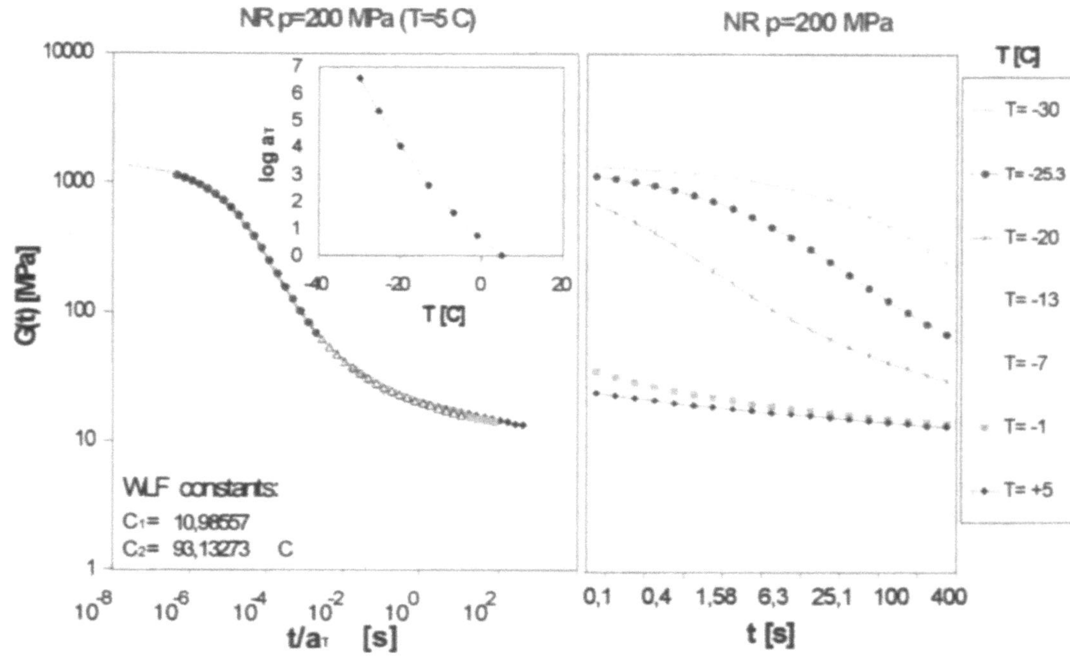

Figure 2. NR relaxation at 200 MPa and mastercurve T=5°C

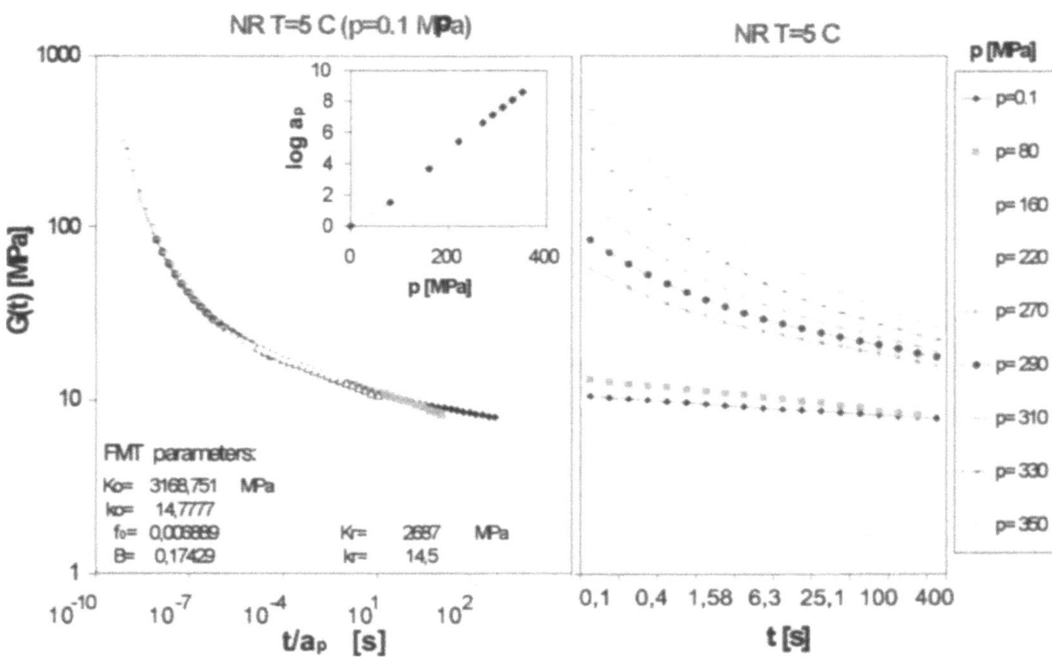

Figure 3. NR relaxation at 5°C and mastercurve for p=200 MPa

3. RESULTS

Three materials were used for measurements: Styrene-Butadiene-Rubber (SBR), Natural Rubber (NR) and Ethylene-Propylene-Copolymer (EPDM). Figure 2 show partial results for NR. The figure 2 shows results of measured at constant pressure of 200 MPa and temperatures indicated on the right scale. Similarly, figure 3 shows results for NR measured at constant temperature of 5°C and pressures indicated on the right scale. Each left graph is a mastercurve with shift factors indicated above. At the lower, left corner there are the corresponding WLF and FMT constants for NR at 200 MPa. Set of measurements on NR taken at constant temperature of 5°C and different pressures was used to find the corresponding FMT constants for NR as an extension of the WLF model.

References:

Ferry, J.D. (1980): Viscoelastic properties of polymers. New York: John Wiley

Fillers, R.W., N.W. Tschoegl (1977) Trans. Soc. Rheol. 21:51-100

Moonan, W.K., N.W. Tschoegl (1983) Macromolecules 16:55-59

DETERMINATION OF THE MOLECULAR WEIGHT DISTRIBUTION OF POLYMERS MELTS

F. LEONARDI[1], A. ALLAL[1], G MARIN[1]
frederic.leonardi@univ-pau.fr
[1]Laboratoire de Physique des Matériaux Industriels, E.S.A. CNRS 5067
Université de Pau et des Pays de l'Adour, Avenue de l'Université 64000 PAU (France).

1. INTRODUCTION

The method presented allows to calculate the molecular weight distribution (MWD) of linear homopolymer melts from the complex shear modulus data measured in a wide frequency domain, (Léonardi et al., 1998). An empirical blending law on complex viscosities is developed. An efficient numerical approach based on a Tikhonov regularization method is used to solve this ill-posed problem; the MWD is hence derived without any assumption on its shape. Then, this algorithm is applied on simulated data to prove its numerical efficiency. The inversion method is applied on complex moduli data of two commercial polymers.

2. A BLENDING LAW ON COMPLEX VISCOSITIES

In order to relate the linear viscoelastic properties to the MWD, we will use a model which is in fact a blending law for complex viscosities (Montfort et al., 1979). For molecular weights above the molecular weight between entanglements M_e, the zero shear viscosity η_o depends strongly on the weight average molecular weight M_w in a power law of exponent "expo". It comes from the definition of M_w:

$$\eta_{o\,\text{polydisperse}}^{\frac{1}{\text{expo}}} = \int_{\ln(M_e)}^{+\infty} \eta_o(M)_{\text{monodisperse}}^{\frac{1}{\text{expo}}} P(\ln M)\, d\ln M \qquad (1)$$

If we associate a single relaxation time to a given molecular weight M, the complex viscosity of a monodisperse polymer may be written as:

$$\eta^*(\omega)_{\text{monodisperse}} = \frac{A(T)\, M^{\text{expo}}}{1 + 1.7 * j\omega A(T) M^{\text{expo}} / G_N^o} \qquad (2)$$

For high molecular weight species, the temperature dependence of the A(T) coefficient is a unique function of the chemical structure of the homopolymer considered.

Thus for a polydisperse polymer using (1):

$$\eta^*(\omega)_{\text{polydisperse}}^{\frac{1}{\text{expo}}} = \int_{\ln(M_e)}^{+\infty} \eta^*(\omega)_{\text{monodisperse}}^{\frac{1}{\text{expo}}} P(\ln M)\, d\ln M \qquad (3)$$

The parameter expo is generally close to 3.4. The use of that numerical method implies to write the blending law on complex viscosity as a linear form which contains vectors and arrays: a normalization plus a decomposition in real and imaginary parts of the complex function leads to two linear integrals.

3. THE TIKHONOV REGULARIZATION

The determination of any distribution by a numerical method is an ill posed problem because the solution is not unique and depends on the parameters introduced to infer it. The Tikhonov regularization method has been widely used. Weese (1992) developed an algorithm based on a QR decomposition and was able to extract degenerated parameters such as the zero shear viscosity in creep experiments. Thus, this procedure selects only positive points because a constraint is added to the solution. We choose this numerical procedure to derive the MWD from complex dynamic moduli data by using a blending law on complex viscosities. The numerical method is first applied on simulated data to check the validity of the method. A bimodal Gaussian distribution simulates the solution vector, the presumed measurement is then evaluated by a linear model (3). After adding a standard noise with a normally distributed random number N(0,1) at a level of 5% to the analytical data, the solution is back-calculated by inversion of the model using a Tikhonov regularization method (see Figure 1).

4. RESULTS

The viscoelastic response of each sample (HDPE, PS) has been determined using a rotary rheometer (DSR) by Rheometrics Scientific®.

The resulting of the MWDs estimated by our method are presented in Figures 2 and 3. We have plotted the MWD measured by SEC (solid line), the Gaussian MWD (dashed line) calculated using the SEC parameters, namely (M_w, I_p) and the MWD estimated from viscoelastic measurements. By a strict conversion (using the viscosity-molecular weight scaling law) of the experimental frequency domain into a molecular weight domain, we can define values of a minimum and maximum molecular weight, which are indicated in all figures by two vertical dashed lines. In these two examples, only the higher part of the MWD lies in this range and the MWD domain must be enlarged to improve stability of the calculated distribution (Friedrich et al., 1996). The transition between the non entangled to entangled domain is close to the molecular weight between entanglement M_e, which has been taken as the minimum molecular weight in the calculation. A good agreement on the M_w calculated from the two distributions of the HDPE sample can be explained by the good fit of the scaling law on viscosity for HDPE samples. The high molecular weights part is also well-predicted (see Figure 2). For amorphous polystyrene samples, an additional problem occurs: the high frequency relaxation domains [Rouse relaxation process in the model of molecular dynamics of Doi and Edwards (1978)] are closed to the terminal relaxation domain. We decided to truncate the data at a frequency corresponding to the minimum of the loss modulus. Figure 3 shows the results obtained for PS sample with truncated viscoelastic data. The higher part of the MWD is reasonably estimated but the cut-off certainly perturbs calculation in the low molecular weights domain.

5. CONCLUSIONS

A linear model based on a mixing law of complex viscosities has been proposed to derive the MWD of linear polymer from mechanical spectroscopy data. Few parameters are needed in the model and a robust numeric method has been used to estimate the MWD of two samples. The use of models derived from molecular dynamics should improve the overall efficiency of the calculations. We have already used this characterization tool to get the MWD of non-soluble or poorly soluble polymers (PVDF).

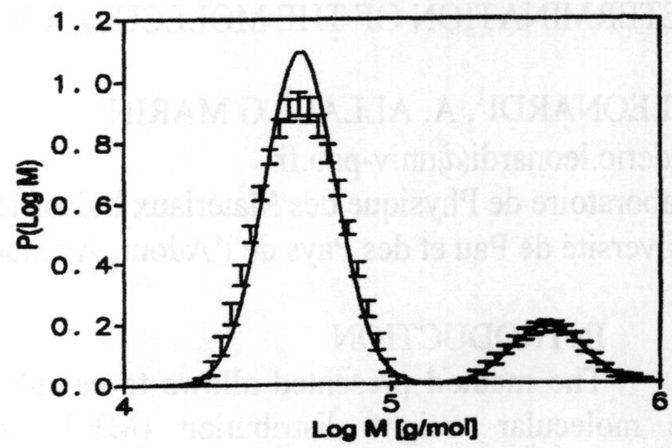

Figure 1

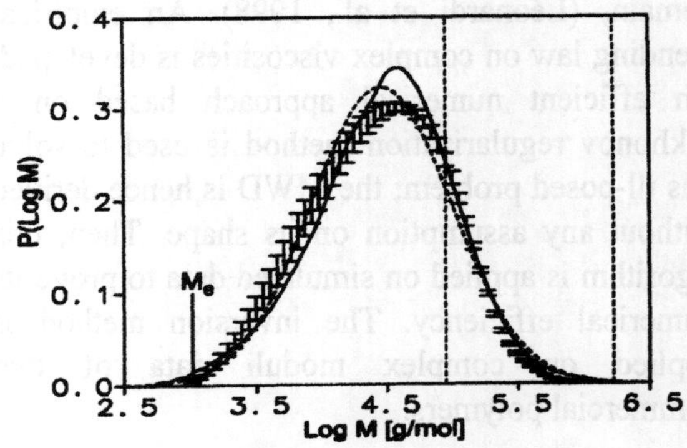

Figure 2

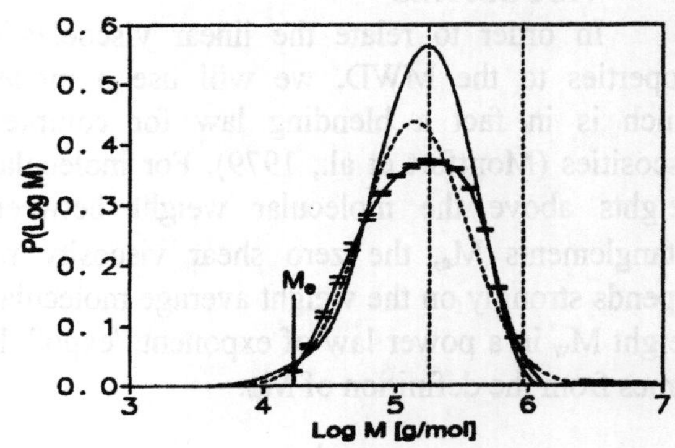

Figure 3

6. REFERENCES

Léonardi F, Allal A, Marin G. Accepted for publication in Rheol Acta (April 1998)

Montfort JP, Marin G, Arman J, Monge P (1979). Rheol Acta 18:623-628

Friedrich C, Honerkamp J, Weese J (1996). Rheol Acta 35:186-193

Weese J (1992). Computer Physics Comm 69:99-111

Doi M, Edward SF (1978). J Chem Soc Faraday Trans II 74:1802-1818

FINITE DEFORMATION ANALYSIS OF ORIGINATION OF NARROW SLOTS IN VISCOELASTIC BODIES

V.A. LEVIN[1], K.M. ZINGERMAN[2]

[1] Composite Mechanics Dept., Faculty of Mechanics and Mathematics, Moscow State University
119899 Moscow, Russia
[2]Computational Mathematics Dept.,Faculty of Applied Mathematics,Tver State University,
170000 Tver, Russia

1. INTRODUCTION

The problem of successive origination of stress concentrators (narrow slots, microdamages) in previously loaded bodies made of hereditary viscoelastic incompressible material is analysed under finite deformations.The mechanical properties of the material are described by integral relationswith the kernel of weak singularity.The problem is considered in a quasi-static tatement. The formulation and the solution of the problem are founded on the theory of repeatedly superimposed large deformations developed by Levin (1988, 1998).

2. PROBLEM FORMULATION

The general statement of the problem is the following. At the initial moment of time large initial strains are brought about in a body by external forces. These strains vary in time due to viscoelastic processes taking place in material. Then, at a prescribed moment of time, a closed surface (giving the boundary of the first cavity) is imagined in the body. Its contents is removed, and the effect of the removed part of the body on the remainder is replaced by forces, distributed over this surface (on the principle of release of braces). Then these forces are reduced to zero "instantly" in the sense of Christensen (1971). As a result, additional finite (at least, at the neighborhood of the originated surface) strains superimposed on the initial ones and correspondent stresses develop in the body. These strains vary in time due to viscoelastic processes. The shape of the boundary surface is changed, too. Note that this shape can be prescribed at a given moment of time.

The following cavities are originated at given moments of time by the same scheme. New finite additional strains superimposing on the existing finite strains develop in the body.

3. CONSTITUTIVE RELATIONS

The following constitutive relations are used:

$$\overset{0}{\Sigma}_{0,n} = \bar{\mu}[I - \frac{1}{3} G_1 G_{0,n}^{-1}] - p_{0,n} G_{0,n}^{-1} , \qquad (1)$$

where

$$\bar{\mu}[\varphi(t)] = \mu_0[\varphi(t) - \int_0^t l(t-\tau)\,\varphi(\tau)\,d\tau ; \qquad (2)$$

$l(t) = \exp(-\alpha t)\, t^{\gamma-1}$; n is the state number; $p_{0,n}$ is a scalar; $\overset{0}{\Sigma}_{0,n}$ is the generalized total for the n-th state stress tensor referring to the base of the initial state, see Levin (1988,1996).It is related with the true total for the n-th state stress tensor $\sigma_{0,n}$ by

$$\sigma_{0,n} = \Psi_{0,n}^T \cdot \overset{0}{\Sigma}_{0,n} \cdot \Psi_{0,n} . \qquad (3)$$

$G_{0,n} = \Psi_{0,n} \cdot \Psi_{0,n}^T$ is the strain measure tensor specifying the strains in transition from the initial to the n-th state (at $n = 1$, $G_{0,1} = G$ is the Green strain measure tensor); $\Psi_{k,n}$ is the deformation gradient specifying the transition from the k-th to the n-th state.

The relations (1)-(2) were approved by Arutyunyan et al. (1986), Drozdov and Kalamkarov (1994) for solution of some problems.

4. COMPUTATIONS AND RESULTS

The particular computations were performed for the two-dimensional case and were concentrated on origination of neighboring narrow elliptical slots of different sizes. The problems are solved using the perturbation technique, Laplace's transform, and Kolosov-Muskhelishvili's tech-nique presented by Muskhelishvili (1951). The "Superposition" program package developed by the authors was used for solution. The specific examples are given for the material parameters corresponding to polydienepoxyurethane.

Some results are presented for the "large slot -- small slot" interaction for the case of uniaxial tensile initial load: $\sigma_{11}^{\infty} = \sigma_{12}^{\infty} = 0$, $\sigma_{22}^{\infty} / \mu_0 = 0.03$. For simplicity, it is assumed that the slots become elliptical at $t = \tau_2$ (τ_1, τ_2 are moments of origination of the slots; the initial load is applied at $t = 0$).

Fig. 1 presents the total true hoop stress $\sigma_{\varphi\varphi}$ at the tips of holes at $t = 3\tau_1$ as a function of a distance between the minor axes of ellipses $(x_2^c - x_1^c)$ for the case $\tau_2 = 2\tau_1$, $a_2 / a_1 = 0.35$, $a_1 / b_1 = a_2 / b_2 = 4$, $(y_2^c - y_1^c) / b_1 = 2.5$, where a_1, b_1, a_2, b_2 are the major and minor semi-axes of the ellipses at $t = \tau_2$; $x_1^c, x_2^c, y_1^c, y_2^c$ are coordi-nates of the centers of the ellipses ($x_1^c = y_1^c = 0$). One can see that the likely microfracturing pattern (Tsukrov and Kachanov, 1997) depend on the mu-tual position of slots.

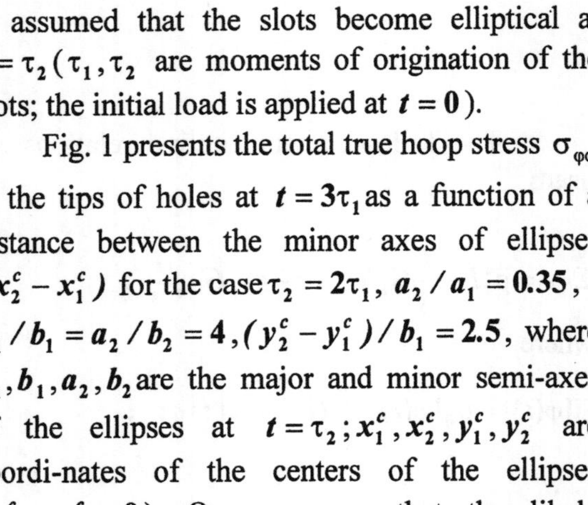

Fig.1

Two collinear holes are considered in fig. 2. Stress concentration at the tip of the smaller hole is plotted as a function of time for different $\tau_2 - \tau_1$ at $a_2 / a_1 = 0.3$, $a_1 / b_1 = a_2 / b_2 = 4$, $x_2^c / b_1 = 6.2$, $x_1^c = y_1^c = y_2^c = 0$.

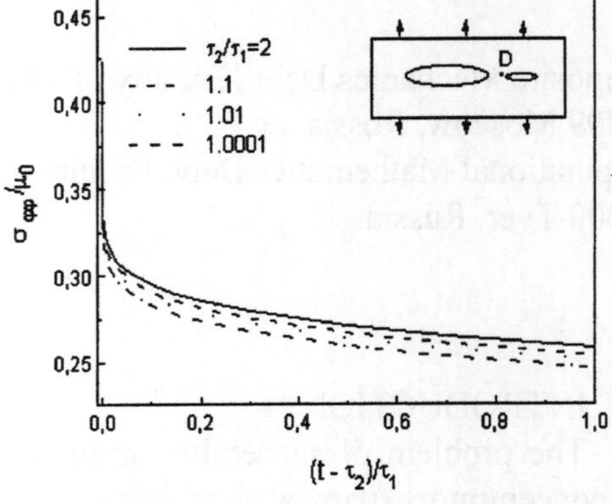

Fig. 2.

REFERENCES

Adamov AA (1982) In: Structural Mechanics of Inhomogeneous Media, Sverdlovsk, pp. 8--11 (in Russian).

Arutyunyan NKh, Drozdov AD, Naumov VE (1987) Mechanics of growing visco-elastoplastic bodies. Nauka, Moscow. (in Russian).

Drozdov AD, Kalamkarov AL (1995) Trans. ASME. J. Appl. Mech. 62:975--982

Christensen RM (1971) Theory of viscoelasticity: an introduction. Academic press, New York, London

Levin VA (1988) Soviet Physics Doklady, 33:296--298

Levin VA (1998) Int. J. Solids and Structures 35:2585-2600

Muskhelishvili NI (1953) Some basic problems of the theory of elasticity. Noordhof, Groningen

Tsukrov I, Kachanov M. Int. J. Solids and Structures 34:2887-2904.

EXTRUSION INSTABILITIES IN POLYMERS AND CHOCOLATE

M R Mackley, H Ovaici, M Ranganathan and R Rutgers

University of Cambridge, Department of Chemical Engineering,
Pembroke Street, Cambridge, CB2 3RA, U.K.

1. INTRODUCTION

This paper reviews experimental data and modelling that we have carried out on the extrusion processing of polyethylene and chocolate. The polyethylene was processed in the molten state whereas the chocolate was processed in the semi-solid state below the material's normal melting point

Extrusion instabilities can occur for a range of situations and reasons. In this paper we make no attempt to classify the full spectrum of possibilities and we will restrict ourselves to four case studies. Two for polyethylene and two for chocolate. These four examples cover a wide but by no means complete range of possibilities.

2. EXTRUSION INSTABILITIES IN POLYETHYLENE

Extrusion instabilities for polyethylene processing can limit commercial production rates and have therefore received considerable attention in literature (see for example Patrie & Denn 1976, Larson 1992). We have recently carried out a series of experiments on linear low density Polyethylene (LLDPE) (Mackley & Rutgers & Gilbert 1998) and these experiments reveal useful information on a small scale extrudate surface instability which is often called sharkskin. Our coupled experimental and numerical simulation results strongly indicate that this instability is linked with the magnitude or the principle stress difference at the exit of the die and the maximum tensile strength of the material.

In contrast to these small scale surface instabilities we have been examining pressure oscillation instabilities that occur with High Density Polyethylene (HDPE) (Ranganathan, Mackley & Spitteler 1998). These instabilities have been observed using a newly developed multi pass rheometer (M P R) (Mackley & Spittler 1996) and the form of the instabilities has been successfully modelled using a relaxation oscillation model which has been used by amongst others (Molenaar & Koopman 1993). This model requires both compressibility effects and a discontinuity in the flow curve to predict the observed pressure oscillation response.

3. EXTRUSION INSTABILITIES IN CHOCOLATE

The cold extrusion of chocolate is a recently reported process for shape forming chocolate below the material's melting point (Becket et al 1994). At low extrusion speeds a stick spurt extrusion instability has been observed and successfully modelled using a similar approach to the HDPE pressure oscillation instability (Ovaici et al 1997). A further unusual and striking instability has been observed which we have named the "Ovaici Necklace" and this formation of this instability can be understood by taking into account that for this particular extrusion condition there is a two component core annular flow.

The four case studies illustrate the delicate sensitivity that, instabilities can have to both extrusion processing and boundary conditions, together with rheological and in some cases compressibility effects.

4. REFERENCES

S.T.Beckett, M.A. Craig, R.J.Gurney, B.S.Ingleby, M.R.Mackley, T.C.L.Parsons The cold extrusion of chocolate, Food and Bio Processing. (Trans I.Chem.E)., 72C, 47-54 (1994)

Larson, R.G., Instabilities in viscoelastic flows. Rheo Acta 31-3, 213-263 (1992)

M.R.Mackley, R.P.G.Rutgers and D.G.Gilbert, Surface instabilities during the extrusion of linear low density polyethylene., J.Non Newtonian Fluid Mechanics. 76 , 281-297 (1998)

M.R.Mackley and P Spittler., Viscoelastic characterisation of polyethylene using a multipass rheometer, Rheol Acta. 35, 202-209 (1996).

Molennaar,J. & Koopmans, R.J. Modelling of polymer melt flow instabilies. J. Rheol.38 - 1, 99-109 (1993)

H Ovaici, M R Mackley, G H McKinley & S Crook. The experimental observation and modelling of an "Ovaici"necklace and stick-spurt instability arising during the cold extrusion of chocolate. J Rheology. 42,1, 125-158 (1998)

C.J.S.Petrie & M.M.Denn, Instabilities in polymer processing., AICHE J. 22, 209-236 (1976)

M Ranganathan, M.R.Mackley and P.H.J.Spittler. Experimental observations and modelling of time dependent capillary flow for a high density polyethylene using a multipass rheometer. Sub J Rheology , June 1998

EVALUATION OF MODELS COMBINING THE MOLECULAR WEIGHT DISTRIBUTION WITH RHEOLOGICAL DATA

D. MAIER, A. ECKSTEIN, C. FRIEDRICH, J. HONERKAMP

Freiburg Materials Research Center, Stefan-Meier-Str. 21, 79104 Freiburg Germany

1. INTRODUCTION

The molecular weight distribution (MWD) affects characteristic properties of polymers. Especially, rheological data depend on the MWD of the investigated polymer. Models combining rheological data with the MWD are described by

$$G(t)=G_N^0\left(\int_{\ln Me}^{\infty} d\ln M w(M)F(t,M)\right)^{\beta}. \qquad (1)$$

G(t) denotes the relaxation modulus, G_N^0 the plateau modulus, *w(M)* the MWD, *F(t, M)* the kernel function, β the mixing parameter, and *Me* the entanglement molecular weight.

The models known from literature are the double reptation model with $\beta=2$ (Tsenoglou,1987), its generalization (Anderssen and Mead,1998) in which β lies between 1 and 2, and the phenomenological model of Maier et al.(1998) who found for PS samples a value of 3.8. Five different kernel functions are known from literature which are presented in detail in Maier et al. (1998). A typical combination used for β and *F* is $\beta=2$ and the *des Cloizeaux-kernel* which will be called the *conventional model.*

In this article we apply the procedure presented in Maier et al. (1998) to poly(methyl metacrylat) (PMMA)-samples in order to fit the parameter β to the data and to optimize *F(t, M).*

2. EXPERIMENTS

Two PMMA samples were generated with M_w = *45 kg/mol* and M_w = *155 kg/mol,* and a polydispersity of *1.10.* These values were obtained using size exclusion chromatography (SEC). Then, 10 mixtures were made with 5% and 10%, 20%, ..., 90% weight fraction *x* of the high molecular polymer. For each sample, the dynamic moduli were measured at different temperatures and shifted into one mastercurve (see Fig. 1).

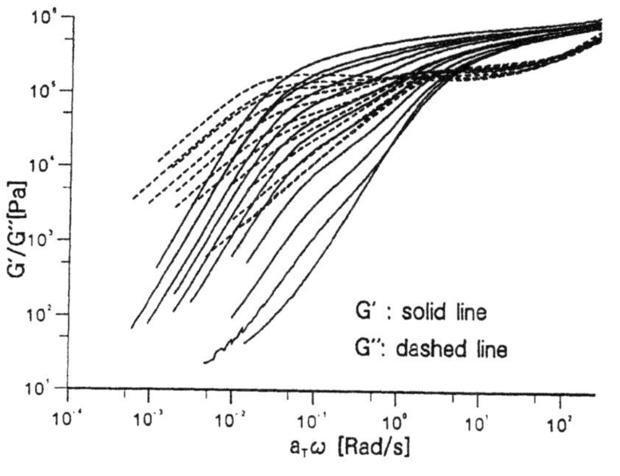

Figure 1: *The dynamic moduli of the PMMA mixtures.*

3. DATA ANALYSIS

The mixtures were analyzed with the procedure described in Maier et al. (1998). Within this procedure the parameter β is varied, so that the term $y(x;t,\beta)=\left(G_x(t)/G_N^0\right)^{1/\beta}$ linearly depend on *x* for every t. The index of $G_x(t)$ makes explicit that the relaxation modulus depend on the weight fraction *x*. β is estimated from the minimum of the distance function $\chi^2(\beta)$ (for details see Maier et al. 1998).

The result for the PMMA data are shown in Fig. 2a. The optimal value of β is **3.3 ± 0.1.** Fig. 2b shows $y(x;t,\beta)$ for $\beta=2$ and $\beta=3.3$ at t = 10s with their corresponding fits. It can be seen, that $\beta=3.3$ linearize better $y(x;t,\beta)$ than $\beta=2.0$.

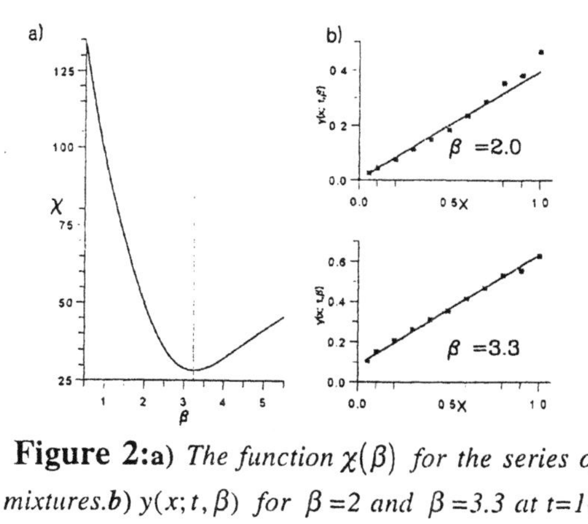

Figure 2:a) *The function $\chi(\beta)$ for the series of PMMA mixtures.* **b)** *$y(x;t,\beta)$ for $\beta=2$ and $\beta=3.3$ at t=10 s with their corresponding fits.*

Furthermore, the time dependence of the fitted parameters are estimates of the kernel function folded with the MWD (see Maier et al. 1998). Consequently, it is possible to optimize the kernel function as well. For the PMMA sample we yield the same kernel function as in our previous paper for the PS samples: the *single exponential kernel*. The combination of the *single exponential kernel* with $\beta=3.3$ will be called the *phenomenological model* for the PMMA sample.

4. COMPARISON OF THE MODELS

For the comparison of the phenomenological with the conventional model, we estimate *w(M)* from *G(t)*. Therefore, the ill-posed inverse problem (see eq.1) is solved with Tikhonov regularization. The two estimates for the different models are compared with the MWD yielded from SEC (see Fig.3). The results demonstrate, that the data are better described with the phenomenological model than with the conventional model.

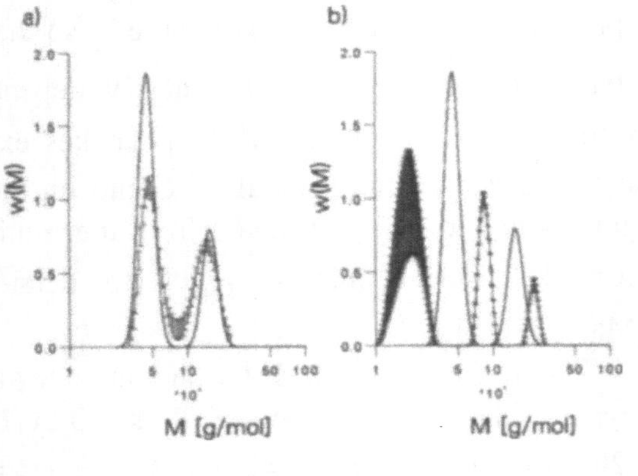

Figure 3: *Estimates of the MWD with* **a)** *the phenomenological,* **b)***the conventional model (error bars) and the MWD obtained using SEC(solid line).The estimates are made from the PMMA sample with x=0.3.*

5. CONCLUSIONS.

For the PMMA samples, we estimate $\beta = \mathbf{3.3 \pm 0.1}$ and the *single exponential kernel*. Therefore, the results of the PMMA and the PS samples are nearly the same. Consequently, we assume, that the model combining rheological data with the MWD can be in general described with a mixing parameter between **3** and **4** far away from 2, and with the single exponential kernel.

REFERENCES .

Anderssen R.S. and Mead D. W. (1998) J. Non-Newton Fluid Mechanics (in press).

Tsenoglou C. ACS Polymer.(1987) Prepr. 28:185-186.

Maier D, Eckstein A, Friedrich C, Honerkamp J (1998) J. Rheol. accepted.

RHEOLOGICAL CHARACTERIZATION OF EPDM'S POSSESSING DIFFERENT STRUCTURES

T. MARINOVIĆ, M. KRALJ NOVAK, Z ŠUŠTERIČ

Sava d.d., R & D Institute, Škofjeloška 6, SI 4000 Kranj, Slovenia

1. INTRODUCTION

The new constrained geometry catalyst, similar to metallocene systems, which is stereoselective for tailor-made polymers and capable of incorporating a variety of co-monomers in a controllable and reproducible way, has led to the rapid development of polymers with "architected" molecular structure, including ethylene-propylene-diene rubber (EPDM). Pillow (1996) explored the effects of variations in polymer composition on both the rubber processing and compounds physical properties.

This work comprises rheological comparison of a newly developed EPDM using a new constrained geometry catalyst, with the "classical" Ziegler-Natta catalyst produced EPDM of similar monomer composition, that is 70 wt.% of ethylene, 25 wt. % of propylene and 5 wt. % of ethylidene norbornene (ENB).

2. RESULTS AND DISCUSSION

Rheological characterization was carried out on two elastomers: (A) Nordel NDX 4770 (produced with new catalyst system by Du Pont Dow) and (B) Dutral TER 4038 (produced with classical system by Enichem).

Ascertained by DSC, the higher glass transition temperature ($T_{g/A}$ = -36^{0}C) exhibited by the sample A in comparison with B ($T_{g/B}$ = -40^{0}C), is in agreement with Reuter and Streit TMA glass transition testing (1997). The higher T_g of the new EPDM points to lower flexibility of its molecules, associated with factors restricting molecular motion in general, such as monomer ratio and distribution. Indeed, the ^{13}NMR analysis data presented in Table 1 indicate higher content of long ethylene block sequences in the sample A, augmenting stiffness of the moving segments and subsequently T_g.

Table 1. The ^{13}NMR analysis' relative peak intensities of ethylene sequences

Samples	A	B
Ethylene block sequen.		
$i(\beta\delta^+)$ / i(PEP)	1,8	1,7
$i(\gamma\delta^+)$ / i(PEP)	2,4	1,8
$i(\delta^+\delta^+)$ / i(PEP)	7,5	5,3

This structure is additionally evidenced by IR spectrum (Figure 1.) of the sample A, displaying relatively higher absorption peak at 721 cm^{-1}, adjacent to ethylene block sequences longer than four monomer units.

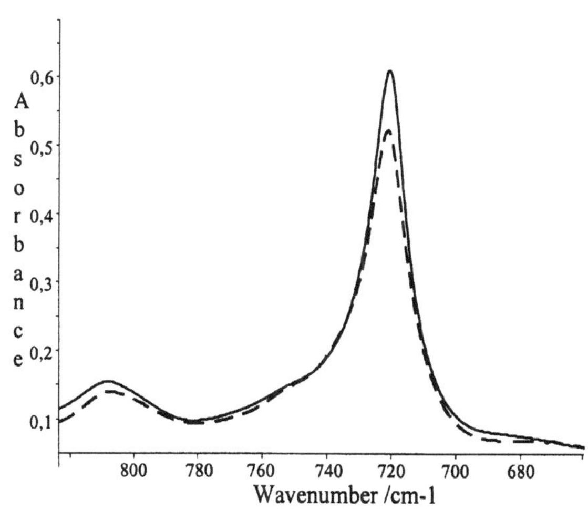

Figure 1. IR spectrum of samples A(solid line) and B(dashed line)

The rubbers' average molecular weights determined by GPC (Table 2) differ too, the new rubber being higher and exhibits bimodal grades of molecular weight distribution.

Table 2. GPC average molecular weights and Mooney viscosity

Samples / Properties	A	B
$M_n \cdot 10^{-5}$(g/mol)	1.08±0.01	1.22±0.01
$M_w \cdot 10^{-5}$(g/mol)	3.67±0.01	3.53±0.01
$M_z \cdot 10^{-5}$(g/mol)	9.66±0.5	6.74±0.01
D	3.41±0.01	2.89±0.02
ML(1+4)/125°C	69±0.5	60±0.5
R/30s(sec)	93	99

Different molecular and morphological structures are expected to yield different rheological properties of the two materials, with subsequent differences in materials' processabilities. The first indication of the foregoing is given by the new rubber's Mooney viscosity, exceeding that of the classical type by 9 units. The differences are further observed through mechanical dynamical measurements. Using dynamic measurement device Rubber Process Analyser (RPA 2000, Monsanto), the two rubbers were rheologically characterized in terms of dynamic functions, i.e. the storage shear modulus, G', the loss modulus, G'', and their common notation in form of complex shear modulus, G*, all as a function of strain amplitude, frequency and temperature. Small but perceivable differences can be noticed in behaviour of the two rubbers. The sample A expectably exhibits higher G' in the plateau frequency zone. Disparity, however, becomes more apparent (Figure 2.), when time-temperature superposition principle is applied as shown previously for polybutadienes by Nakajima (1997).

Whereas G'' of sample A exceeds G' at low frequency (or high temperature), it is just the opposite with B. At the equivalent conditions of low frequencies and elevated temperatures, usual for rubber processing, the new rubber's G' everywhere surpasses G'', meaning that under such conditions elastic response dominates over the viscous one and resulting in quite different processability of this rubber.

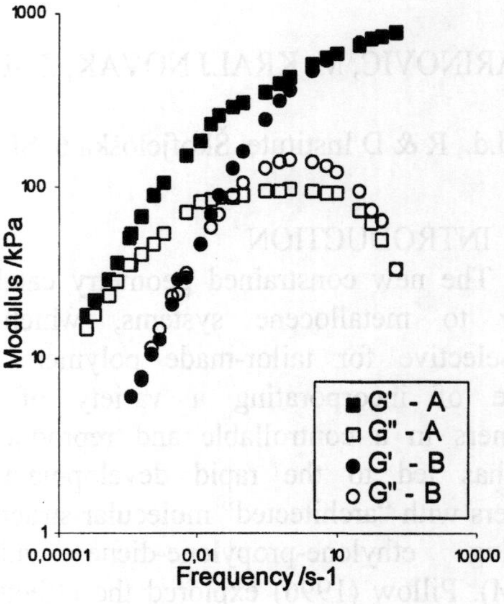

Figure 2. Master curves of storage and loss moduli of EPDM (reference temperature, 40°C)

3. REFERENCES

Nakajima N (1997) Kautsch. Gummi Kunstst. 50:119-125
Pillow JG (1996) New EPDM Materials as Produced by Constrained Geometry Catalyst Chemistry compared with Current Commercial EPDM, International Rubber Conference, Manchester, England
Reuter M, Streit G (1997) Gummi Fasern, Kunstst. 50:201-209

4. ACKNOWLEDGEMENT

This work has been partially supported by the Ministry of Science and Technology of Slovenia grant (Project H2-8728).

RHEOLOGICAL INVESTIGATIONS AND MODELING ON SEMI-SOLID METAL ALLOYS

M. MODIGELL, J. KOKE

Institut für Verfahrenstechnik, RWTH, 52056 Aachen, Germany

1. ABSTRACT

Metal alloys in the semi-solid state with a globular microstructure exhibit time dependent, thixotropic flow properties. Rheological investigations were carried out on a partially solidified Tin-Lead alloy. Based on the experimental results a thixotropic model was developed and implemented in CFD software.

2. INTRODUCTION

If a metal alloy melt is cooled down to the semi- solid state while being sheared, the primary phase solidifies with a globular structure. These kinds of materials behave thixotropically and are used in a new metal forming process with high industrial interest: "Thixoforming", which is the application of classical forming processes as forging and die casting on semi-solid alloys.

The good processability of those alloys is caused by the high mobility of the spherical particles in the liquid matrix. It is generally accepted that the thixotropy is a result of the agglomeration and disagglomeration of the globules. The flow behaviour of these materials is highly dependent on the solid mass fraction, the temperature and the particle size.

Especially the prediction of the filling behaviour during casting requires a good knowledge of the complex rheological properties.

3. EXPERIMENTS

Recently couette rheometer experiments were performed on tin-lead alloys in order to study and model the rheology of semi-solid alloys. At the beginning of each experiment the alloy melt was partially solidified in the couette system while shearing it at a constant shear rate. The solid fraction can be calculated from the temperature by the Scheil equation, neglecting ripening and diffusion in the solid phase.

One set of experiments consisted of hysteresis experiments, where the shear rate was linearly increased and subsequently decreased with the same speed to its initial value. As plotted in Fig.2 the recorded viscosities of the forward and backward curve enclose an area which indicates the degree of thixotropy. The hysteresis area decreases with increasing sweep time, the time interval from the minimum to the maximum shear rate.

While hysteresis experiments rather serve qualitative investigations and are not suitable for modeling, step change of shear rate experiments for quantitative analysis was performed. In this case the shear rate is changed abruptly to a new constant value and the relaxation of the viscosity is recorded until a new steady state is reached. With the help of these experiments the steady state flow curve can be evaluated as well as the kinetics of the viscosity change.

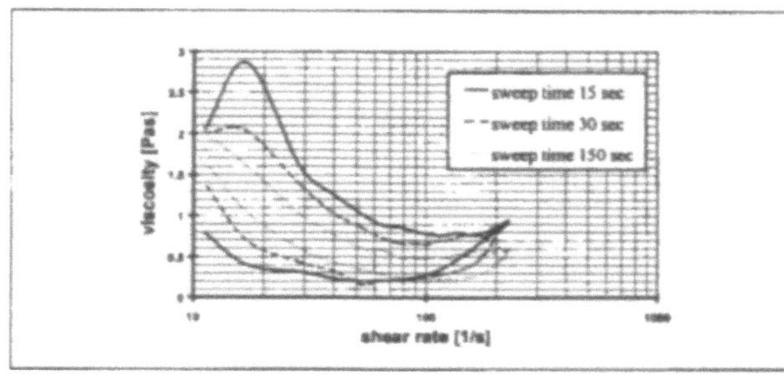

Fig. 2: Hysteresis experiments ar different sweep times (T=198°C, fs=0.45)

4. MODEL

In the suggested model the semisolid alloy is regarded as a homogeneous material with thixotropic behaviour. According to Kemblowski (1979), the viscosity is defined as a function of shear rate and of a structural parameter changing with time, describing the current state of agglomeration.

The shear stress is assumed to grow exponentially with increasing solid fraction. Thus the equation of state is assumed as:

$$\tau = \tau_0(f_s) + \exp(B f_s)\, k\, \kappa\, \dot{\gamma}^{m(f_s)} \qquad (1)$$

with the coefficients B, k, the exponent m, the structural parameter κ, the solid mass fraction f_s. and the yield stress τ_0.

The structural parameter is in the range of zero to infinity: zero corresponds to a fully destroyed structure and infinity to a fully agglomerated structure. The evolution of the structural parameter can be described by a kinetical differential equation which is assumed as a first order reaction kinetics in the following form

$$\frac{\partial\kappa}{\partial t}=c(\dot{\gamma})\,(\kappa_e-\kappa) \qquad (2)$$

where $c(\dot{\gamma})$ describes the rate of approaching the equilibrium value κ_e.

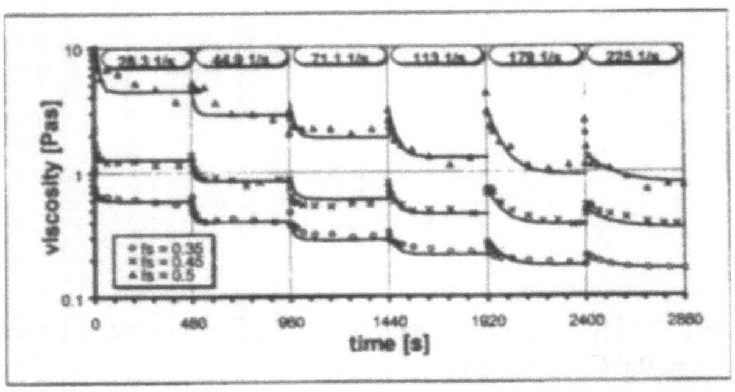

Fig. 3: Step change of shear rate experiments on Sn-15%Pb for different fraction solids (fs = 0.35; 0.45; 0.5)

The structural parameter κ changes in time following the deformation history of the material. In the special case of the step change of shear rate experiment, the parameter will approach the equilibrium value corresponding to the current shear rate.

$$\kappa_e=\frac{1}{\dot{\gamma}^{m(fs)-n}} \qquad (3)$$

Generally the equilibrium parameter is a descending function of the shear rate. If this relation is substituted to the expression for the viscosity Eq.(1), we obtain the equilibrium flow curve, which corresponds to a modified Hershel-Bulkley equation with the exponent n:

$$\tau = \tau_0(f_s) + \exp(Bf_s)\; k\; \dot{\gamma}^n \qquad (4)$$

After parameter evaluation this model matches the experiments quite well. In Fig.3 the experimental data of the step change of shear rate experiments for three different solid fractions, corresponding to different temperatures, is compared to the model.

5. SIMULATION

The presented rheological model was implemented into the CFD software FLOW-3D and a characteristic casting cases was simulated in order to demonstrate the advantage of "Thixocasting" compared to the classical die casting: the filling of a 2-dimensional cavity with a cylindricle obstacle. It has been observed that the filling of the die causes a lot of air entrapment when using fully liquid metal, a low viscosity Newtonian fluid.

The following figure shows three snap shots of the filling process with a semi-solid alloy at a solid fraction of 45%. The pictures demonstrate the advantageous laminar filling behaviour of the thixotropic fluid, which rather smoothly surrounds the cylinder with little air entrapment.

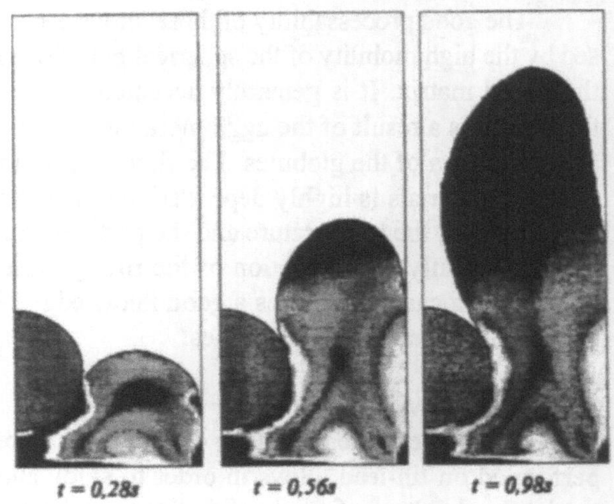

Fig. 4: Snap Shots, Halfviews, f_s=45%, T=198°C

6. REFERENCES

Flemings M.C. (1991) Behavior of Metal Alloys in the Semisolid State, Metallurgical Transac tions, Vol. 22A
Kemblowski Z., Petera J. (1979) Rheological

RHEOLOGICAL STUDY OF THE LAMELLAR, SPONGE AND MICELLAR PHASES OF THE $C_{12}E_4$/CPyCl/BENZYL ALCOHOL/WATER SYSTEM

MONTALVO, G., RODENAS, E., VALIENTE, M.

Dpto. Química Física. Universidad de Alcalá. E28871 Madrid. Spain

INTRODUCTION

The ternary $C_{12}E_4$/benzyl alcohol/water system shows an extensive region of lamellar phase with different subregions of vesicles and extended bilayers structures "Montalvo et al. (1998)". In this system was found a good correlation between the proposed structures and the rheological data. A critical shear rate that induced the transition from extended bilayers to vesicles was also found.

In this communication, the lamellar, sponge and micellar phases of the pseudoternary diluted dodecyltetraethyleneglycol ($C_{12}E_4$)/cetylpiridinium chloride (CPyCl)/benzyl alcohol/water system, at $[C_{12}E_4]$=50 mM and low cationic surfactant and alcohol concentrations are studied by rheological measurements at 25.0±0.1°C. The phase diagram of the system shows a lamellar phase that extends until [CPyCl]= 0.4 mM and a micellar phase that appears from [CPyCl]= 1.2 mM "Montalvo et al (manuscript in preparation)".

METHODS

Phase behavior was determined by visual observation between crossed polarizers and optical microscopy with a Laborlux S LEITZ.

Rheological measurements were carried out at 25.0±0.1°C by a Carri-Med CLS 100 controlled stress rheometer using a cone-plate configuration.

RESULTS AND DISCUSSION

The flow curves for samples with [CPyCl]=0.2 mM and different alcohol conentrations are given in Fig. 1. All the curves belong to the lamellar phase except for the alcohol content of 151 mM corresponding to the L_3 phase.

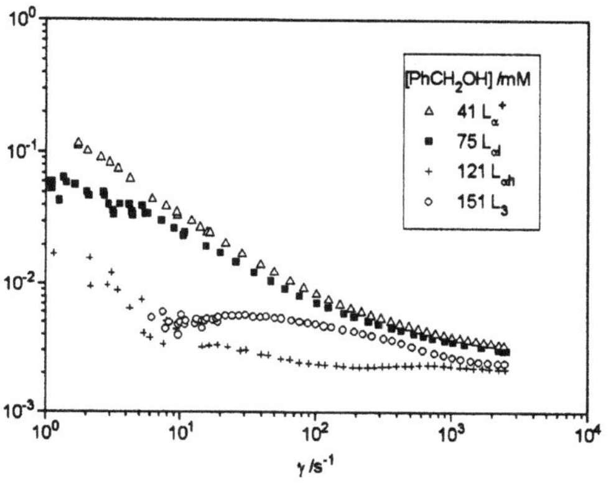

FIG.1. Viscosity as a function of the shear rate for samples with 0.2 mM CPyCl at 25.0±0.1° C.

The viscosity decreases with the shear rate but the dependence of the viscosity at a fixed shear rate, with the alcohol content is more complicated (Fig. 2). We can observe a well defined maximum in the viscosity values with the alcohol concentration. The results let us anticipate another maximum at the highest alcohol concentration.

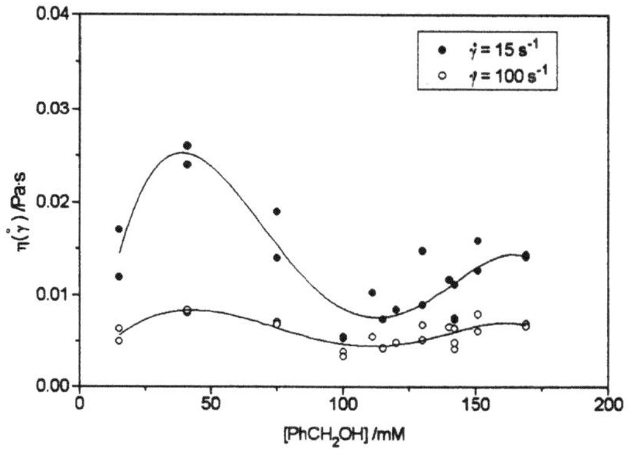

FIG. 2. Viscosity at fixed shear rates as a function of benzyl alcohol concentration at 25.0±0.1° C

The extensive region of the lamellar phase is made up of the subregions with different structures such as ocurred in the ternary system without CPyCl. The maximum and minimum that appears in Fig. 2 is related to the change in the structure of the lamellar phase. The well defined maximum in the viscosity values correspond to the L_α^+ phase, the anomalous lamellar phase " Jonströmer et al. (1992)", and the inflexion point to the transition from the L_α^+ phase to the $L_{\alpha l}$ phase probably composed of the vesicles. The minimum value of the viscosity corresppond to the sample located in the $L_{\alpha h}$ phase built up form flexible and extended bilayers.

The flow curves for the L_1 phase with [CPyCl]=1.4 mM and different alcohol concentrations are given in Fig. 3. At low shear rates the viscosity exhibits a tendency to a constant zero-shear-rate viscosity, η_0, and then it decreases with the shear rate. The non-newtonian behavior of the micelles is observed for the smaller CPyCl concentrations while for higher cationic surfactant contents, the micelles converts to spherical and the behave as newtonian fluids.

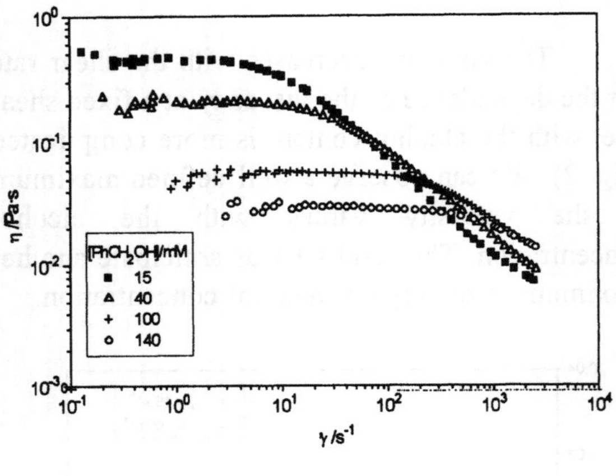

FIG. 3. Viscosity as a function of the shear rate for L_1 samples at 25.0±0.1°C

If we analyze the variation of η_0 values with the alcohol content at fixed surfactant concentrations (Fig. 4), we observe that the viscosity passes through a maximum, then it becomes constant at alcohol contents higher than 100 mM with a viscosity value close to the water viscosity as correspond to spherical micelles.

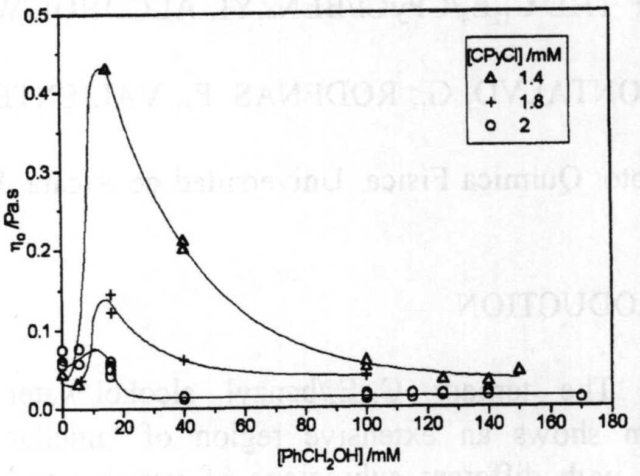

FIG. 4. Zero-shear-rate viscosity as a function of benzyl alcohol at 25.0±0.1° C.

REFERENCES

Jonströmer, M., Strey, R (1992) J. Phys. Chem. 96:5993

Montalvo, G., Rodenas, E., Valiente, M (1998) J. Colloid Interface Sci. 200:000-000

NUMERICAL AND EXPERIMENTAL INVESTIGATION OF THERMAL CONVECTION FOR THERMODEPENDENT HERSCHEL-BULKLEY FLUID IN AN ANNULAR DUCT WITH ROTATING INNER CYLINDER

C. NOUAR

LEMTA (UMR 7563- CNRS-UHP-INPL) , ENSEM, 2 Avenue de la Forêt de Haye, BP 160
54504 Vandoeuvre-lès-Nancy , France.

1. INTRODUCTION

This paper presents an experimental and numerical investigation of the thermal convection for a Herschel-Bulkley fluid, in an annular duct, with rotating inner cylinder. The outer cylinder is heated at constant heat flux density ϕ_p and the inner one is adiabatic. This work has two main motivations. First, it seeks to determine the angular velocity Ω_c for which the plug zone is reduced to zero. Second, to arrive at a better comprehension and better description of the effect of the variation of the consistency K with temperature T on heat transfer. In the case of slot approximation, Bittleston and Hassager (1992) presented for a Bingham fluid an analytical expression of Ω_c .

2. EXPERIMENTS,

2.1 Experimental installation

The experimental tests were conducted on the same flow configuration as described by Nouar *et al.* (1987). Two test section were used: The first one consists of two coaxial cylinders made of Plexiglass. The annular space has a width and a length of 12.5 mm and 160 mm, respectively. The second test section is similar to the first one. The outer cylinder is made of copper and is wrapped by a resistance loop. The annular space has a width and a length of 12.5 mm and 0.369 m.The evolution of the outer cylinder wall temperature is determined by means of thirty one thermocouples. The velocities measurements are performed using a Laser Doppler Velocity system

2.2 Test fluid

The fluid used is an aqueous gel of an acrylic acid polymer (Carbopol 940). The weight concentration of Carbopol is 0.2%. The neutralization of the Carbopol 940 is accompanied by a gelation process. A constant torque rheometer was used in order to study the rheological behavior of the working fluid. The rheograms were determined at different temperatures. The curves fitting indicate that the rheological behavior of the solution can be described by the Herschel-Bulkley model : $\tau=\tau_s+K|\dot{\gamma}|^n$. The results indicate that the dependence of τ_s and n with T are weak, compared to the temperature dependence of K, and thus can be ignored. The relation K-T adopted is $K=a\,exp(-bT)$.

3. PROBLEM FORMULATION AND NUMERICAL SOLUTION

The problem is governed by the conservation of mass, momentum and energy equations. The constitutive equation is given by :

$$\bar{\bar{\tau}} = 2\mu_a \bar{\bar{D}} \quad if \quad (\tau_{II})^{1/2} > \tau_s \qquad \text{(1-a)}$$

and

$$\bar{\bar{D}} = 0 \quad if \quad (\tau_{II})^{1/2} \leq \tau_s \qquad \text{(1-b)}$$

where τ_s the yield stress. Scalars D_{II} and τ_{II} are the second invariant of tensors $\bar{\bar{D}}$ and $\bar{\bar{\tau}}$ respectively. The apparent viscosity μ_a is given by

$$\mu_a = K[4D_{II}]^{(n-1)/2} + \tau_s[4D_{II}]^{-1/2} \qquad (2)$$

The boundary conditions are : (i) at the entrance section, the flow is fully developed and the temperature profile is uniform ; (ii) at the walls of the cylinders, there is no slipping flow. The inner cylinder is adiabatic, and the outer one is heated with constant heat flux density.

It is assumed that the Peclet number is sufficiently high ($Pe > 100$) so as the axial diffusion term can be neglected. To overcome the difficulty due to the singularity in (2) as D_{II} ---> 0, a modified version of the rheological equation is used. Here, we follow, the approach adopted by several authors, notably by Bercovier and Engelman (1980), in replacing the magnitude of the strain rate tensor $[4D_{II}]^{1/2}$ by $[4D_{II}]^{1/2} + \varepsilon$. in (2). Here, ε is a small regularization parameter. The governing equations as well as the associated boundary conditions are solved by means of an linearized implicit finite difference technique. A second order central difference is employed in approximating all partial derivatives in the radial direction and the upwind scheme is used for the convective axial terme

4. RESULTS AND DISCUSSION

4.1 Temperature-independent consistency

The relative dimension ap of the plug zone, is respectively defined by :

$$ap = (r^+ - r^-)/(R_2 - R_1) \qquad (3)$$

where, r^+ and r^- are the external and internal edges of the plug zone. In the case of Poiseuille flow, there is a detached moving plug of unyielded fluid between the two coaxial cylinders. For a given rheological and geometrical parameters parameters, the dimension of the plug flow ap_0, decreases with the axial flow rate (but it is never zero) and increases when the consistency decreases. The rotation of the inner cylinder induces an orthoradial shear rate increasing the second invariant of the rate of deformation tensor. This leads to a reduction of the plug core dimension. The application of π-theorem leads to

$$ap = f(N, n, ap_0, Ro) \qquad (4)$$

Where, Ro is the Rossby number : $Ro = R_1\Omega/U_d$
The critical value Ro_c, from which, the plug zone ap is reduced to zero, is such that :

$$Ro_c = f(N, n, ap_0) \qquad (5)$$

Figure 1 gives in a semi-logarithmic coordinates, Ro_c as a function of ap_0, for $n = 0.5$ and different values of the radius ratio N. A continuous line, represents the curve for the slot approximation.

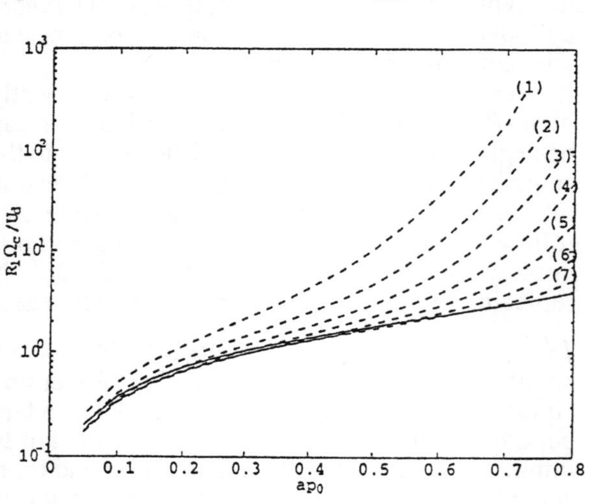

Figure 1 : Ro_c versus ap_0 for $n = 0.5$ and different radius ratio : (1) N=0.2 ; (2) N=0.3 ; (3) N=0.4 ; (4) N=0.5 ; (5) N=0.6 ; (6) N=0.7 ; (7) N=0.8 ; (___) slot approximation.

4.2 Temperature-dependent consistency

The decrease of K close to the heated wall induces a reorganization of the flow, and modifies the heat transfer. The modification of the flow structure, along the heating zone, can be described through the evolution of the plug zone dimension. First of all, recall that for the isothermal situation, when the consistency decreases, ap and Ro_c increases. One can say that the decrease of K favours the development of the plug flow, on the other hand, the rotation of the inner cylinder reduces it. Two cases are distinguished on the evolution of the plug flow, along the heating zone, depending on the difference between Ro and $Ro_{c,e}$. Where $Ro_{c,e}$ is the critical Rossby number evaluated for the entrance conditions :

Case 1 : $Ro<Ro_{c,e}$: Two steps are observed in the evolution of plug core boundaries. In the first step, under the effect of the decrease of K close to the heated wall, the axial velocity gradient increases at the outer cylinder, and decreases at the inner cylinder, du to flow rate conservation. The plug flow is radially displaced towards the heated wall, and its width increases. In the second step, the thermal boundary layer thickness increases and reaches the inner cylinder. This leads to a decrease of K, and therefore, to an increase of the plug flow flow dimension, in the whole annular space.

Case 2 : $Ro>Ro_{c,e}$: At the entry section there is no plug flow. From the inlet section and along the heating zone, the fluid temperature increases and K decreases. The rotating effect of the inner cylinder is dominating until an axial position Z_p. At this position, the fluid temperature increases sufficiently, so that the effect of decrease of K becomes dominating leading to the apparition of the plug flow at a point $P(Z_p)$. Numerical results show, as it was expected, that the point (P) moves downstream and towards the heated cylinder, when Ro increases. From this axial position, the plug flow dimension increases, until practically to occupy all the annular space. Furthermore, the axial pressure gradient decreases with Ro and along the heating zone. If $(\partial P/\partial Z)$ becomes small compared to the yield stress, then numerical results show apparition of plug zone attached to the outer cylinder.

5. REFERENCES

Bercovier M, Engelman M (1980), J. Compt. Physics 36 : 313-326.

Bittleston S. H., Hassager O (1992), J. Non-Newtonian Fluid Mech 42 : 19-36.

Nouar C, Devienne R, Lebouché M (1987) Int. J.Heat.Mass. Transfer 30 : 639-647.

A SEMI-LAGRANGIAN FINITE VOLUME METHOD FOR SOLVING VISCOELASTIC FLOW PROBLEMS

T.N.PHILLIPS and A.J.WILLIAMS

Department of Mathematics, University of Wales, Aberystwyth SY23 3BZ, UK.

1. INTRODUCTION

Recently, there has been considerable interest in solving viscoelastic problems in 3D particularly with the improvement in modern computing power. In many applications the emphasis has been on economical algorithms which can cope with the extra complexity that the third dimension brings. Storage and computer time are of the essence. The advantage of the finite volume formulation is that a large amount of memory space is not required. Iterative methods rather than direct methods can be used to solve the resulting linear systems efficiently.

The SIMPLER algorithm (Patankar (1981)) is used as the basis of the finite volume technique described in this paper together with a semi-Lagrangian treatment of the convective terms. The standard finite volume approach for problems in computational fluid dynamics suffers from restrictions on the size of the time step due to stability considerations. This restriction is not so severe when particle tracking (Lagrangian) methods are used. However, the unrestricted movement of the points in a Lagrangian formulation introduces a new set of difficulties. Semi-Lagrangian methods combine the advantages of fixed grids inherent in Eulerian methods with modifications to treat the convective terms using particle tracking methods.

The hyperbolic nature of the constitutive equation is responsible for many of the difficulties associated with the numerical simulation of viscoelastic flows. Many techniques introduce some form of upwinding in order to obtain numerical solutions at high values of the Deborah number. In some finite volume techniques artificial diffusion has been added to the constitutive equation (Xue et al. (1995)). The Semi-Lagrangian approach described here, which is a generalization of the work of Scroggs and Semazzi (1995)), is a natural way of treating the convective terms in the constitutive equation without resorting to these techniques.

Numerical results are presented for the viscoelastic flow of an Oldroyd B fluid through a planar 4:1 contraction. Particular emphasis is given to the convergence and stability of the method.

2. GOVERNING EQUATIONS

We consider the flow of an Oldroyd B fluid through a planar 4:1 contraction. In dimensionless form the governing equations are given by

$$Re\left(\frac{\partial \mathbf{v}}{\partial t} + \mathbf{v}.\nabla\mathbf{v}\right) = -\nabla p + \nabla.\boldsymbol{\tau} + \beta\nabla^2\mathbf{v}, \tag{1}$$

$$\nabla.\mathbf{v} = 0, \tag{2}$$

$$\boldsymbol{\tau} + We\,\overset{\nabla}{\boldsymbol{\tau}} = 2(1-\beta)\mathbf{d}, \tag{3}$$

where p is an arbitrary isotropic pressure, $\boldsymbol{\tau}$ is the polymeric part of the extra-stress tensor and $\mathbf{v}$ is the velocity vector. The parameter β is defined by

$$\beta = \frac{\lambda_2}{\lambda_1}, \tag{4}$$

i.e., the ratio of the retardation and relaxation times of the fluid. The Reynolds and Weissenberg numbers are defined by

$$Re = \frac{\rho U L}{\eta}, \quad We = \frac{\lambda_1 U}{L}, \tag{5}$$

where the characteristic velocity (U) and length scales (L) are taken as the downstream mean velocity and channel height, respectively.

3. SEMI-LAGRANGIAN APPROACH

Many methods for solving mixed convection-diffusion problems are based on techniques which have been devised for pure diffusion problems. In order to circumvent some of the difficulties encountered when these techniques are applied to convection-diffusion problems some sort of upwinding is necessary. The approach adopted here is to decouple the treatments of convection and diffusion and to use appropriate methods of discretization for each subproblem (Phillips and Williams (1998)).

A staggered grid arrangement is used in which the pressure and normal stresses are located at the cell centres, the shear stress at the cell vertices and the velocity components at the mid-points of the cell edges. The staggered grid ensures that the solution is not polluted by spurious pressure modes and this particular arrangement of the unknowns avoids the need for interpolation of the shear stresses in the momentum equation.

At each time step the convection problems

$$\frac{\partial \mathbf{v}}{\partial t} + \mathbf{v}.\nabla \mathbf{v} = 0, \tag{6}$$

$$\frac{\partial \boldsymbol{\tau}}{\partial t} + \mathbf{v}.\nabla \boldsymbol{\tau} = 0, \tag{7}$$

are solved using a particle tracking method. The idea is to determine the position at the previous time step of the particle paths which pass through the vertices of a control volume at the current time. This is followed by the solution of an unsteady generalized Stokes problem to determine the pressure and the new velocity and the solution of an algebraic problem to update the stresses

$$We\left(\frac{\boldsymbol{\tau} - \boldsymbol{\tau}^{*n}}{\Delta t} - \mathbf{L}\boldsymbol{\tau} - \boldsymbol{\tau}\mathbf{L}^T\right) + \boldsymbol{\tau} = 2(1-\beta)\mathbf{d}, \tag{8}$$

where $\mathbf{L}$ is the velocity deformation gradient.

4. NUMERICAL RESULTS

In this section we show how the size of the salient corner vortex varies with We and Re. In Tables 1 and 2 we give the length (L_1) and width (L_2) of this vortex as a function of We for $Re = 0$ and $Re = 1$, respectively. The algorithm was tested on a sequence of meshes the finest of which has mesh spacings 0.152 and 0.111 in the x and y directions, respectively. The channel is truncated at a distance 16 units from the contraction in the upstream and downstream directions. In Figs. 1 and 2 we display the streamlines for $We = 1$ and $Re = 0$ and $Re = 1$, respectively, to show the influence of inertia.

We	L_1	L_2
0.0	1.452	2.147
0.5	1.467	2.226
1.0	1.513	2.341

Table 1: Dependence of L_1 and L_2 on We for $Re = 0$.

We	L_1	L_2
0.0	1.401	1.863
0.5	1.174	1.544
1.0	1.174	1.632

Table 2: Dependence of L_1 and L_2 on We for $Re = 1$.

5. REFERENCES

Patankar SV (1981) Numerical Heat Transfer and Fluid Flow, Hemisphere, London.

Phillips TN and Williams AJ, (1998) SIAM J. Sci. Comput. submitted.

Scroggs JS and Semazzi FHM (1995) Num. Meth. PDE 11: 445-452.

Xue SC, Phan-Thien N and Tanner RI, (1995) J. Non-Newtonian Fluid Mech. 59: 191-213.

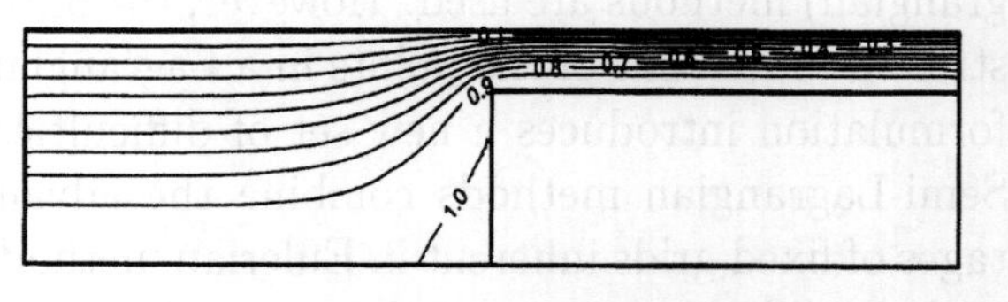

Figure 1: Streamlines for $We = 1$ and $Re = 0$.

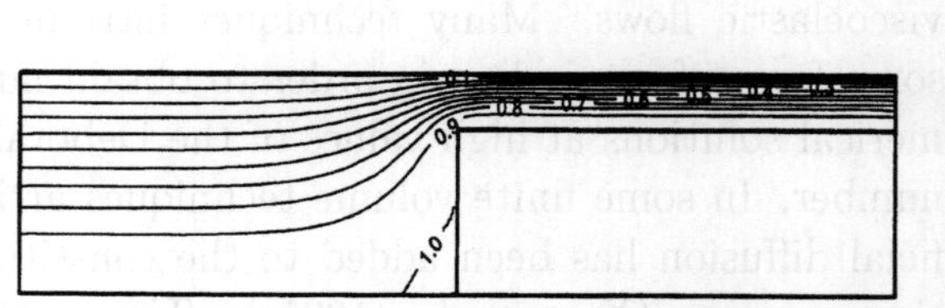

Figure 2: Streamlines for $We = 1$ and $Re = 1$.

THE RELIABILITY AND DURABILITY OF VISCOELASTIC SYSTEMS

V.D.POTAPOV and A.I.MARASANOV
Department of Structural Mechanics, Moscow State University of Means Communication
Obraztsov Str., 15 Moscow 101475, Russia

The problems of the reliability and durability estimation of the stochastic viscoelastic systems are widely investigated in the last decades (Potapov (1985), Potapov and Marasanov (1992), Rusin (1995) and others). The finite element method in combination with the statistical simulation allows to increase considerably the class of the problems that can be solved. The present work is devoted to the application of such methods to the reliability and durability investigation of the viscoelastic systems. For example consider the frame with random material properties subjected to the slowly changing in the course of time random force F(t) (Fig.1) so the quasi-static treatment of the problem is available. The frame consists of the polymeric composite bars with the ring cross sections. The material properties of the bars are assumed to be statistically independent, linearly elastic in axial direction and linearly viscoelastic by interlayer

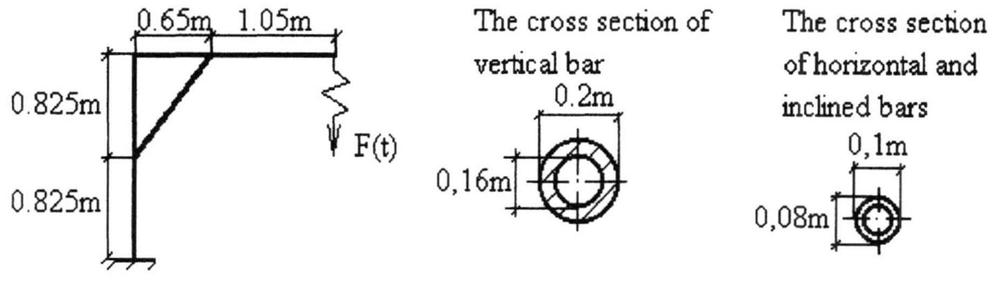

Fig.1 The scheme of the viscoelastic frame.

shear. The kernel of creep is represented in the form:

$$K(t-\eta) = Gba(t-\eta)^{a-1}.$$

Here G is the shear modulus; lnG, lnb and a are the Gaussian random numbers with the following statistical characteristics: $\langle \ln G\rangle = 7{,}95$; $\langle \ln b\rangle = -8{,}44$; $\langle a\rangle = 0{,}23$ are the averages; $S_{\ln G} = 0{,}313$; $S_{\ln b} = 0{,}500$; $S_a = 0{,}059$ are the root-mean-square deviations; $r_{\ln G,\ln b} = 0{,}343$; $r_{\ln G,a} = -0{,}163$; $r_{\ln b,a} = 0{,}083$ are the correlation coefficients. The Young's modulus E is assumed to be the independent Gaussian random value with characteristics: $\langle E\rangle = 5{,}25\cdot 10^4$ and $S_E = 892{,}5$ (G and E are measured in mPa, b is measured in $(\mathrm{mPa}\cdot \mathrm{hour}^a)^{-1}$; a is undimensional parameter). Note that the realizations of the random values lnG, lnb and a can be simulated by the Cholesky method. Assume that the exterior force F(t) is Gaussian stationary random process with the average $\langle F\rangle = 1$ kN and correlation function:

$$R(t_2 - t_1) = Q\cdot \exp(-\theta|t_2 - t_1|),$$

where $Q = 0{,}0025(\mathrm{kN})^2$ is the variance of the process F(t); $\theta = 2{,}083\cdot 10^{-3}\mathrm{hour}^{-1}$ is constant. For the simulation of the process F(t) realizations the autoregressive moving average method is used. Investigate the reduction of the reliability of the presented stochastic viscoelastic system in the course of time. Here the reliability is defined as the probability of no-failure operation. The failure is realized by the first crossing of maximal random displacement y(t) of the given level y_*. The charts of the estimates of the random displacement y(t) root-mean-square deviation $S_y(t)$ and the failure probability P(t) obtained by the analysis of 200 realizations of y(t) are presented in Fig.2-3 ($y_* = 0{,}007$m). Here the dash curves

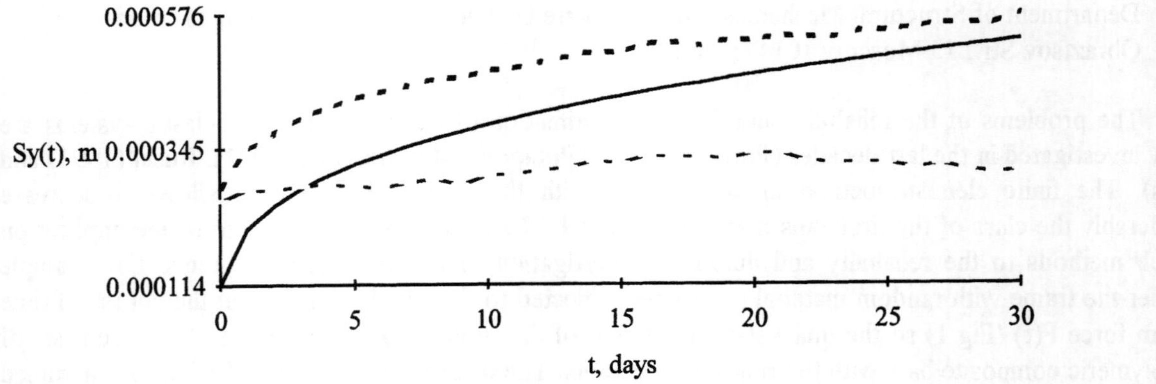

Fig.2 The estimates of the root-mean-square deviations of the random displacement y(t).

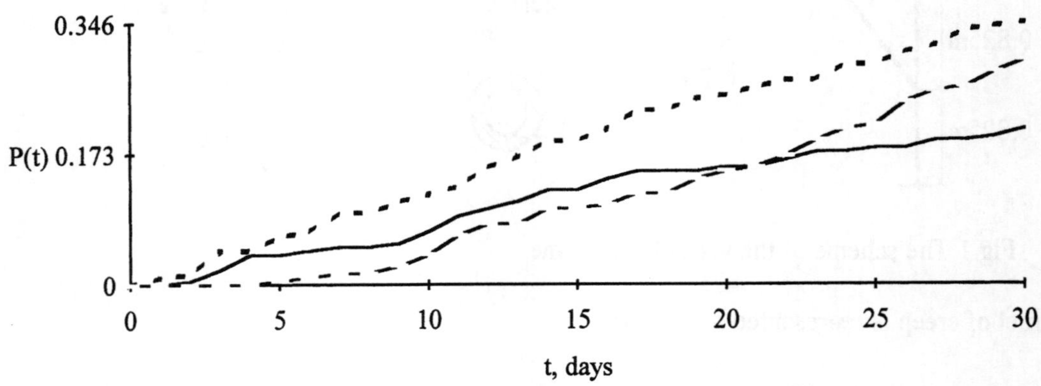

Fig.3 The estimates of the failure probability function.

describe the behaviour of the frame with deterministic material characteristics being equal to mathematical expectations of E, G, b, a and random force F(t). The solid curves pertain to the frame with random material properties and constant force $F(t) \equiv \langle F \rangle$. The dotted curves characterise the system where both the material coefficients and the force are stochastic. Note that the differences between the charts of the mathematical expectation estimates are small so these charts are missed. For the different types of the considered system the histograms of the durability (time t_* passing to the instant of the first crossing by the process y(t) of the given level) can be built. For example the frame with deterministic material properties and random force has the following estimates of the durability expectation and the root-mean-square deviation ($y_* = 0{,}00605$m): $\langle t_* \rangle = 3{,}4$ days and $S_t = 3{,}7$ days.

Acknowledgements - This research was sponsored by Russian Foundation of Fundamental Researches.

References.

Potapov V.D. (1985) Stability of Viscoelastic Elements of Structures. Stroyizdat, Moscow (in Russian).

Potapov V.D. and Marasanov A.I. (1992) Buckling and stability of polymeric composite materials under stochastic excitation. Z. angew. Math. und Mech. 72, 97-100.

Rusin A. (1995) The analysis of the influence of random load, geometry and the material constants on the creep in a bar subjected to bending. Zesz. Nauk. Energ. 123, 77-93 (in Polish).

INVERSE METHOD FOR THE CHARACTERIZATION OF THE RHEOLOGY OF MOLTEN POLYMERS IN BIAXIAL EXTENSION

A.J. Rodriguez-Villa[1], T. Nguyen-Ngoc[2], M. Bellet[2], J.F. Agassant[2], D. Lebouvier[3], R. Jacquemet[3]

[1] TRANSVALOR S.A. – Les Espaces Delta – BP037 – 06901 Sophia Antipolis Cedex - France

[2] ECOLE DES MINES DE PARIS – CEMEF – UMR CNRS 7635 - Sophia Antipolis – 06560 Valbonne – France

[3] ELF-ATOCHEM – CERDATO – 27470 SERQUIGNY - FRANCE

Introduction. The rheology of molten polymers is now well established in shear and even in uniaxial extension. However, several important polymer processes such as blow molding involve biaxial extension for which only poor rheological data is available. We have built a partially open instrumented mold, and recorded with a video camera the free inflation of a parison (i.e. a tube of molten polymer). The true inflation pressure, inside the parison, has been measured. Using, the blow molding software TFORM3®, the rheological parameters of various constitutive equations have been adjusted in order to minimize the residual between computed and experimental radius evolutions.

Experimental setup. A partially transparent mold has been designed (Fig.1).

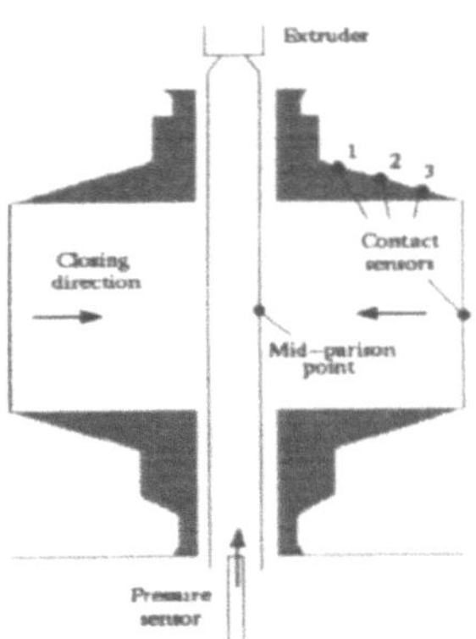

Figure 1. Experimental setup

The free inflation of the parison is recorded through a lateral opening using a video camera (Fig.2). We can measure, for example, the radius at mid height of the parison as a function of time R(t). The internal pressure $P_a(t)$ is also recorded (Fig.3), and is significantly lower than the nominal blowing pressure (100 kPa for Fig.3). Also notice it varies during the blowing stage, which highlights the strong coupling between the thermodynamics of the inflating air and the thermomechanics of the parison.

Figure 2 : Video acquisition during inflation

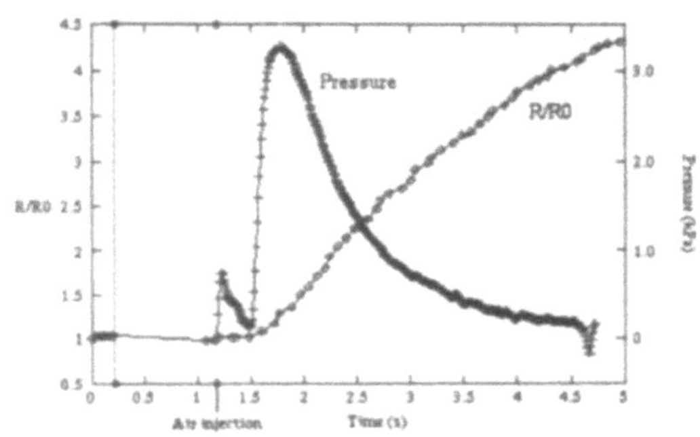

Figure 3 : Measured pressure inside the parison and evolution of the mid parison dimensionless radius R(t)/R₀

TFORM3® FEM membrane model. A membrane formulation with curvilinear coordinates has been chosen. The principle of virtual power expresses mechanical equilibrium:

$$\int_{\Omega} \sigma^{ij} v^{*}{}_{i/j}\; e\; dS - \int_{\Omega_p} P_d\; \mathbf{g}_3 \mathbf{v}^{*} dS = 0$$

in which e is the local thickness, P_d is the differential pressure applied to the region Ω_p of the deformed parison Ω, σ is the Cauchy stress tensor, $\mathbf{g}_3$ is the normal vector to the membrane and $\mathbf{v}^*$ is a virtual velocity field. At each time step, discretization with linear triangles yields a set of non linear equations which is solved by the Newton-Raphson method. The configuration is then updated for the next time step using an implicit time integration scheme. Sticking contact is considered between the polymer and the mold.

Fig.4 compares experimental and computed curves $R(t)/R_0$ for various constitutive equations. Viscoelastic constitutive equations fit reasonably well the experimental radius evolution. The best-fitting rheological parameters have been identified automatically using an inverse method developed hereafter.

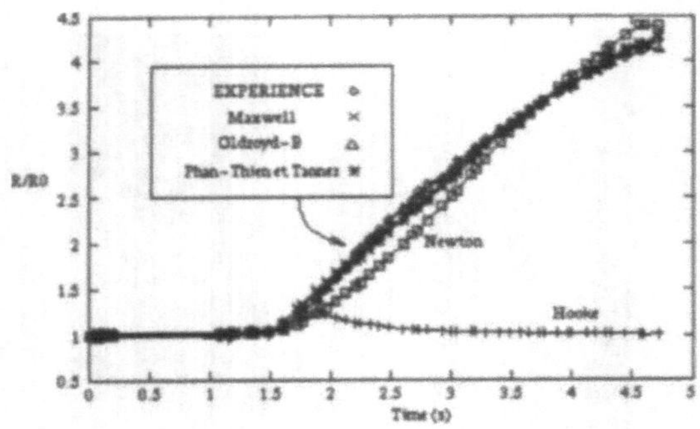

Figure 4 : Experimental vs. computed radius evolutions

Inverse method for the identification of rheological parameters. Our purpose is to find the parameters π of a given constitutive equation which minimize :

$$E(\pi)=\frac{\sum_{t=t_0}^{t=t_N}(R_{cal}-R_{measure})^2}{\sum_{t=t_0}^{t=t_N}R_{measure}^2}$$

where R_{cal} and $R_{measure}$ are the computed and measured radii at mid-height at inflation step t_i. A Gauss-Newton iterative procedure is used to find the parameter increments at each direct simulation of the experiment, solving:

$$G+\frac{\partial G}{\partial \pi}\Delta\pi=0,\text{ with }G=\frac{\partial E}{\partial \pi}$$

Results. Two HDPE (A and B) from Elf-Atochem have been tested. For each polymer, variable process parameters were the thickness and the temperature of the parison as well as the nominal blowing pressure. As the different viscoelastic constitutive equations give the same results (fig. 4), we decided to adjust the two parameters (viscosity η and relation time λ) of a Maxwell constitutive equation. When varying the process parameters, the rheological parameters also vary. In order to compare our results to classical shear and elongational testing, shear viscosity is plotted vs. frequency, uniaxial elongational viscosity vs. elongation rate for a Hencky deformation of 0.66 (i.e. the mean value of our test) and the identified viscosity of the Maxwell model vs. the average rate of strain of our experiments (Fig.5). It is to notice that the uniaxial viscosity is three times the shear viscosity (Trouton behavior). The viscosity determined by an inverse method in biaxial blowing is higher but follows the same dependence to the rate of strain. When comparing HDPE A and HDPE B we can notice that both shear and identified viscosities are higher for the polyethylene devoted to the blowing of big containers (HDPE B), than that used for small containers (HDPE A).

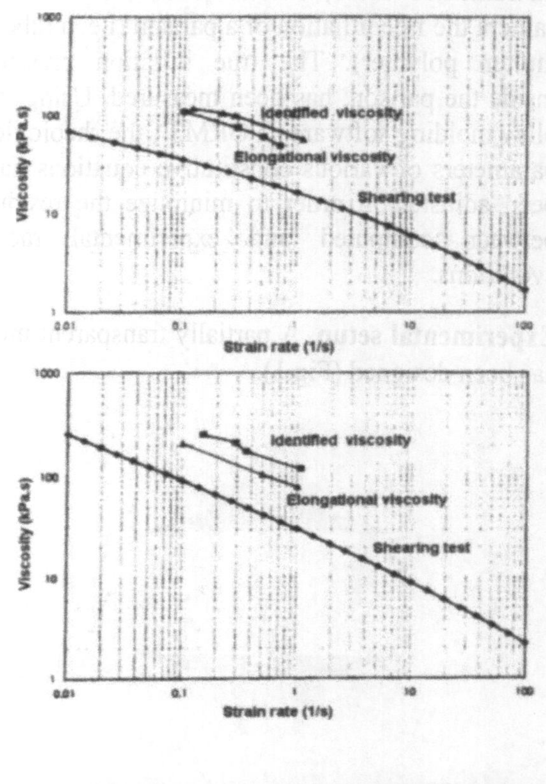

Figure 5 : Comparison of shear viscosity, elongational viscosity and viscosity identified from a blowing experiment by an inverse method (top figure: HDPE A, bottom figure HDPE B)

References

1. **Rodriguez-Villa A., Agassant J.F. and Bellet M.**, "Finite element simulation of the extrusion blow-molding process", Proc. Numiform 95, Balkema, Rotterdam, 1053-1058 (1995)

2. **Rodriguez-Villa A., Bellet M., Agassant J.F. and Lebouvier D.**, "Rheological characterization of molten polyethylene with respect to be blow molding process", Proc. XIIth Int. Cong. on Rheology, Aït Kadi A., Dealy J.M., James D.F., Williams C. edts, Canadian Rheology Group, 695-696 (1996)

3. **Rodriguez-Villa A.**, "Etude théorique et expérimentale de l'extrusion soufflage de corps creux en polymère", Ph.D Thesis, Ecole des Mines de Paris (1997).

4. **Nguyen-Ngoc T.**, "Etude rhéologique de polymères sur machine de soufflage", Mastere Degree, Ecole des Mines de Paris (1997)

DETERMINATION OF RELAXATION TIME SPECTRA FROM RHEOLOGICAL DATA USING AN EDGE PRESERVING REGULARIZATION METHOD

T. ROTHS, D. MAIER, C. FRIEDRICH, J. HONERKAMP
Freiburg Materials Research Center, Stefan-Meier-Str. 21, 79104 Freiburg, Germany

1. INTRODUCTION

In order to describe linear viscoelastic properties of polymers one is interested in the relaxation time spectrum h, as it supplies information about the material properties. Moreover given h it is easy to convert one material function into another one. Additionally many microscopic theories are based on h and provide predictions about its character.

Unfortunately the relaxation time spectrum is not directly accessible. It is only possible to estimate it from measured data, e.g. the dynamic moduli G' and G'', which is an ill-posed inverse problem.

Ill-posed problems are solved with regularization methods: Apart from the constraint from the data additional information about the spectrum independent from the data (*a priori* knowledge) is imposed upon the spectrum.

A typical *a priori* knowledge is the smoothness constraint used in the algorithm proposed by Honerkamp and Weese (1993), which will be denoted the *common method*: The spectrum and its first derivative are assumed to be continuous.

In contrast to this assumption the models due to Rouse (1953), Zimm (1956), Baumgaertel et al. (1990), Chambon and Winter (1987) predict spectra that are piecewise smooth in double logarithmic representation and exhibit edges.

Unfortunately the common method is not able to reconstruct these spectra. Therefore we introduced a new *edge preserving regularization method.* By using the *a priori* knowledge "piecewise smoothness" this method is able to reconstruct even piecewise smooth spectra.

2. MODEL

2.1 Linear viscoelastic model

The dependence of the dynamic moduli G' and G'' on the relaxation time spectrum h is given by the linear viscoelastic model:

$$G'(\omega_i) = \int_{-\infty}^{\infty} \frac{(\omega_i\tau)^2}{1+(\omega_i\tau)^2} e^{\ln h(\tau)} d(\ln\tau) + \varepsilon(0,\sigma_i'),$$
$$G''(\omega_i) = \int_{-\infty}^{\infty} \frac{(\omega_i\tau)}{1+(\omega_i\tau)^2} e^{\ln h(\tau)} d(\ln\tau) + \varepsilon(0,\sigma_i''). \tag{1}$$

$G'(\omega_i)$ and $G''(\omega_i)$ $(i = 1,\ldots,M)$ denote the noisy data. The errors $\varepsilon(0,\sigma_i)$ due to measurement are assumed to be relative.

2.2 Theoretical models for the spectrum h

For polymer solutions Rouse (1953) and Zimm (1956) predicted up to the Rouse time τ_R a linearly decreasing spectrum in double logarithmic scaling with a gradient of -1/2 and -2/3 respectively. The material specific Rouse time is independent of the molecular weight and varies from 10^{-4}s to 10^{-3}s.

The BSW spectrum due to Baumgaertel et al. (1990) predicts another linear increase with a gradient of 0 - ¼ in the time interval from τ_R to the terminal relaxation time $\tau_0 > \tau_R$. In particular the spectrum does not show any peak, i.e. any characteristic relaxation time in this interval. The transition at τ_R may also be smooth.

Finally for relaxation times larger than τ_0 the CW spectrum introduced by Chambon and Winter (1987) predict a linear decay with a gradient much smaller than -1. A strong decrease is necessary in order to obtain a finite zero-shear viscosity, which is correlated with the integral over the relaxation time spectrum.

3. REGULARIZATION

The reconstruction of the spectrum is obtained by minimizing

$$V_{\ln h}(\lambda) = \sum_{i=1}^{M} \frac{\left\{g'(\omega_i) - \int_{-\infty}^{\infty} \frac{(\omega_i\tau)^2}{1+(\omega_i\tau)^2} e^{\ln h(\tau)} d(\ln\tau)\right\}^2}{\sigma_i'^2} + \sum_{i=1}^{M} \frac{\left\{g''(\omega_i) - \int_{-\infty}^{\infty} \frac{(\omega_i\tau)}{1+(\omega_i\tau)^2} e^{\ln h(\tau)} d(\ln\tau)\right\}^2}{\sigma_i''^2} + \lambda R(\ln h) \tag{2}$$

with respect to $\ln h$. The two sums force the spectrum to be compatible with the data. The regularization term $R(\ln h)$ imposes the a priori knowledge upon the spectrum. λ weights between the constraint from the data and the constraint from the *a priori* knowledge. Estimates of $\ln h$ depend on the choice of λ and $R(\ln h)$.

In this article λ is selected using the SC method due to Honerkamp and Weese (1990).
In order to reconstruct even piecewise smooth spectra like the combination of the BSW and the CW spectrum we introduced the following representation of the *a priori* knowledge:

$$R(\ln h) = \int_{-\infty}^{\infty} \frac{\left[\ln h(\tau)''\right]^2}{1+\left[\alpha\left(\ln h(\tau)''\right)\right]^2} d(\ln \tau) . \qquad (3)$$

The parameter α represents an edge preserving measure: The special case $\alpha = 0$ corresponds to the Tikhonov regularization as proposed by Honerkamp and Weese (1993). This common method prefers smooth spectra. Edges, i.e. discontinuities of the second derivative are forbidden.
On the contrary for $\alpha > 0$ edges are allowed. The larger α the better they are reconstructed. However if α is chosen too large artificial edges may be estimated from the measurement errors. In order to reconstruct real edges as well as to prevent artificial edges. $\alpha = 1$ is a proper choice.

3. SIMULATIONS

The following simulations will show the artifacts produced by the common method ($\alpha = 0$) and indicate the improvement due to the new edge preserving regularization method ($\alpha = 1$).
We consider the combined BSW/CW spectrum with $\tau_R = 10^{-3}$, $\tau_0 = 10^1$ and gradients of -2/3, 1/4 and -10 respectively for the linear regions (dashed lines in Fig. 1). Realizations of G' and G'' are calculated according to equation (1). The errors have relative standard deviations of 3%.
Figure 1 shows estimates of $\ln h$ for one typical realization. The common method (Fig. 1a) produces artifacts: Edges are smoothed and thus hard to localize. In particular τ_0 located at the rightmost peak is estimated too small. Even more serious three artificial peaks indicated by arrows are reconstructed. Further simulations prove the common method to have a systematic bias of at least one artificial peak.
In contrast to this the new edge preserving regularization (Fig. 1b) renders very reliable results: The artificial peaks do not appear. Furthermore τ_0 is reconstructed very well. Only the edge at τ_R is smoothed.

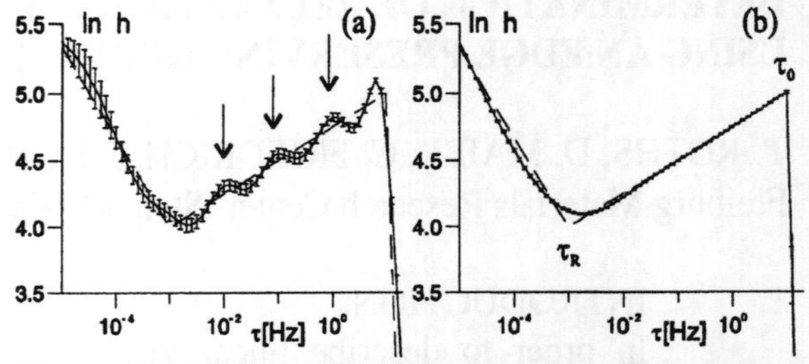

Figure 1: *Simulated BSW/CW-Spectrum (dashed line); spectrum estimated with* (a) *common and* (b) *edge preserving method (solid lines with error bars).*

4. EXPERIMENTAL RESULTS

Both methods were applied to PS data (M_W=1400kg/mol). Figure 2 shows the two estimated relaxation time spectra. The results correspond with the simulations:
Some relaxation times apparent in the commonly reconstructed spectrum (arrows in Fig. 2a) do not appear in the spectrum estimated using the edge preserving method (Fig. 2b).
As seen before the characteristic of these peaks is known to be typical for artifacts of the common method if it is applied to a non-smooth spectrum. Hence the spectrum in Figure 2b seems to be closer to reality than the commonly calculated solution. At least these results demonstrate the theoretical predictions to be compatible with the experimental data.

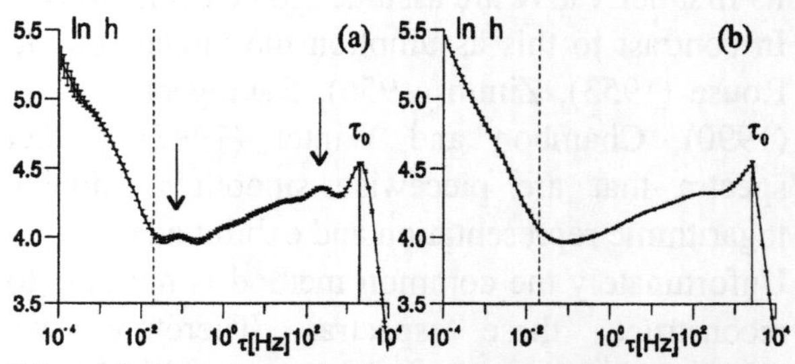

Figure 2: *Experimental PS data: Spectrum estimated with* (a) *common and* (b) *edge preserving method. According to the data set the spectrum right of the dashed line is reliable.*

REFERENCES

Baumgaertel M., Schausberger A., Winter H.H. (1990) Rheol. Acta 29:400
Chambon F., Winter H.H. (1987) J.Rheol. 31:683
Honerkamp J., Weese J. (1993) Rheol. Acta 32:65
Honerkamp J., Weese J. (1990) Continuum Mech. Thermodyn. 2:17
Rouse P.R. (1953) J. Chem. Phys. 21:1272
Zimm B.H. (1956) J. Chem. Phys. 24:269

Disentanglement effects in linear and branched polyolefins

T.Schedenig[1], A. Schausberger[1], M.Gahleitner[2], W.Pirgov[2]

[1] Johannes Kepler University Linz, Department of Chemistry, Altenbergerstraße 69, A-4040 Linz
[2] PCD Polymere AG, St.-Peter Straße 25, A-4020 Linz

1. INTRODUCTION

It is well known throughout the literature (Prichard et Wissbrunn, 1969; Fujiki, 1971) that shear working can appreciably change the rheological behaviour of branched polymers, while the primary molecular parameters like the molecular mass and molecular mass distribution remain unchanged. This effect is called "shear modification". Linear polymers do not show comparable behaviour.

The effect of shear modification on optical properties of blown film, on surface roughness or on extrusion swelling has been entensively studied (Rokudai et al., 1979, 1981). In contrast, the effect of shear working on some basic rheological functions is not well established.

It is the aim of this work to show the effect of shearing in an extruder on rheological functions in shear and extension flow.

2. MATERIALS

Shear modification experiments were carried out for two grades of HDPE and two grades of LDPE, all provided by Fa.PCD Polymere, Linz. The LDPE types were 1840 D and 2410 F. The HDPE types were DH 5973 and BF 5272, which shows a bimodal molecular mass distribution. The densities and MFR's of all materials are given in Table 1.

Table 1: Investigated materials

	MFR [g/10min]	Density [g/cm^3]
1840 D	0,25 (190/2,16)	0,920
2410 F	0,75 (190/2,16)	0,924
DH 5973	0,30 (190/2,16)	0,959
BF 5272	0,25 (190/5)	0,954

3. EXPERIMENTAL

All materials were extruded in a Haake Polylab extruder at an extruder pressure of 30 and 300 bar. The extrudes strands were collected and used for the investigations. Dynamic, creep/recovery and Rheotens measurements were carried out on the granules and the extruded samples.

Dynamic measurements were carried out on a Rheometrics SR 2000 stress controlled rheometer in plate-and-plate geomety.

Creep measurements were carried out on the same instrument in cone-and-plate geomety. Applied stresses varied between 10 and 75 Pa for the different materials. Stress was applied for 1000 seconds, after taking away of the stress the recovery was observed for 2000 seconds.

Rheotens measurements were carried out on a Göttfert Rheotens with a die of 2 mm in diameter and a length of 30 mm. The wheels were accelerated with 120 mm/s^2.

4.RESULTS

a) *Dynamic measurements*

Comparison of the dynamic moduli before and after the shear working showed poor or no changes for both the linear and the branched polyethylenes, respectively. Even in the low-frequency range the measurement curves of the granules and the "300 bar-sample" correspond within the accuracy of measurement.

b) *Creep measurements*

From the stationary part of the creep curves $J_{e,0}$ was determined. The values for $J_{e,0}$ increased strongly after the extrusion for the two LDPE's and showed nearly no change after the extrusion for the HDPE's. Figure 1 shows this fact.

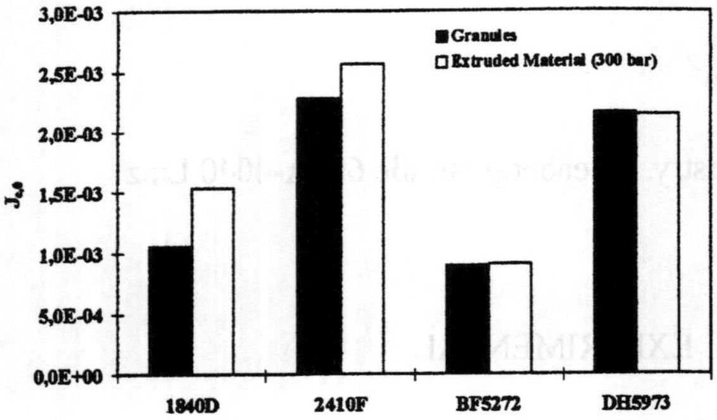

Fig. 1: $J_{e,0}$ before and after extrusion for all materials

c) Rheotens measurement

Significant changes in the extensional behaviour were observed for the LDPE's; no changes for the HDPE's. While the velocity at the break point of the strand did not change remakably with the shear working, the force at the break point decreased strongly for the branched and poorly for the unbranched samples. Figures 2 and 3 show Rheotens curves of the granules and the extruded (300 bar) material for the LDPE 1840 D and the HDPE BF 5272.

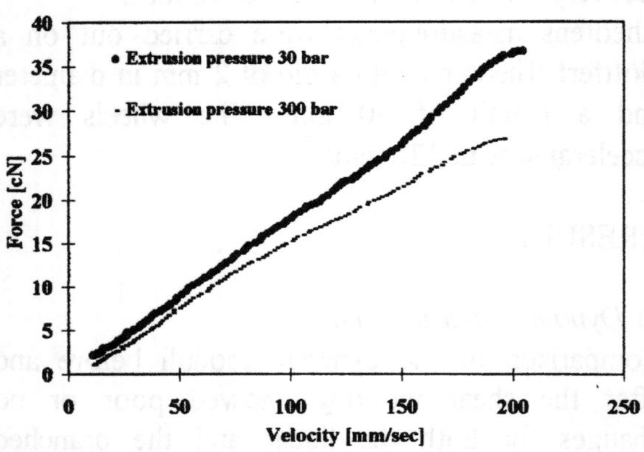

Fig.2: Rheotens-Curves of LDPE 1840 D

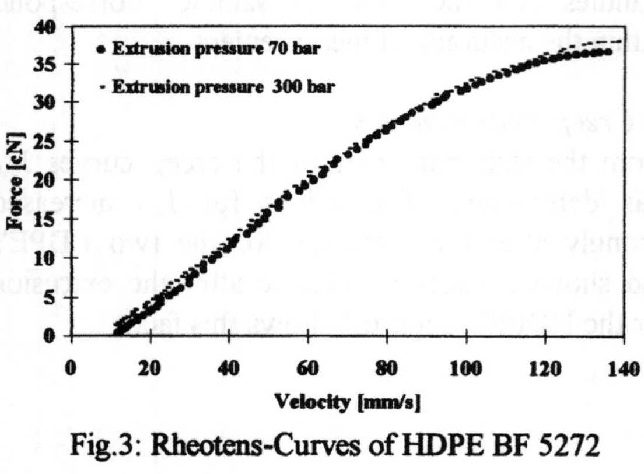

Fig.3: Rheotens-Curves of HDPE BF 5272

d) *Molecular mass distribution*

GPC curves showed not changes in molecular mass distribution after shearing.

5.DISCUSSION

The observed phenomena can be interpreted in terms of different entanglement structures of linear and branched polyethylenes: for long-chain branched polymers, two different types of entanglements are considered. The "normal" entanglement of chains, which also occurs in networks of linear chains can be imagined as looping of the chains. This type of entanglement can easily disentangle in a shear field. The second type of entanglement arises from looping and coiling at the branching points. Because of the steric hindrance, this type of entanglement cannot easily disentangle. For this reason, in a branched polymer "flow units" build up (Busse 1967). These units can be imagined as clusters of some molecules that stick together and behave like a unit. Size and shape of these flow units influece the rheological behaviour of a material strongly. By continuous deformation, the flow units can be sub-divided and therefore the rheological behaviour changes. These changes could be seen in extensional and creep measurements. In oscillatory experiments the total applied deformations are too small to cause an effect of destroying flow units.

6. LITERATURE

Busse WF (1967) J.Polym.Sci.A-2 5: 1261-1281
Fujiki T (1971) J.Appl.Polym.Sci. 15:47-66
Prichard JH and Wissbrunn KF (1971) J.Appl.Polym.Sci. 13: 233-239
Rokudai M et al. (1979, 1981) J.Appl.Polym.Sci.23: 463-471, 3289-3294, 3295-3300, J.Appl.Polym.Sci.26: 1343-1350, 1427-1429

RHEOLOGICAL, MECHANICAL AND ADHESIVE PROPERTIES OF THE FILLED LCP-THERMOPLASTICS BLENDS

V. Shumsky, I. Getmanchuk, V. Babich, Yu. Lipatov

Institute of Macromolecular Chemistry of NAS of Ukraine, 253160 Kiev, Ukraine

Rheological, mechanical and adhesive properties of two-phase polymer blends containing of liquid crystal copolyester of poly(ethylene terephthalate) and p-hydroxybenzoic acid plus three regular thermoplastics (polypropylene or polysulfone or poly(methyl methacrylate)) with varying compositions and concentrations of short fibers (glass fibers, aluminum borate whiskers) have been studied.

Perfect fibrillation of the disperse LC-phase, for example, into the PP-matrix on capillary flow was observed at LCP concentrations > 20 wt.% and temperatures > 488K. This effect leads to a decrease of blend viscosity and a reinforcing of the extrudate's mechanical characteristics. At the same time, more essential reinforcement of thermoplastics is achieved by the simultaneous addition of the reinforcing agents both of the LCP and short fibers. Processing of the thermoplastics is not impaired.

Mechanical properties of filled blends were discussed from standpoint of peculiarities of adhesion of blends to the solid surface.

It was established that concentration dependence of adhesive strength depends on the ratios of the surface tensions of components forming the blend. Two various cases of the structure of the adhesion layer have been considered: the complete wetting of the surface by one of the blend component and incomplete wetting by both components. In the first case the interphase is formed by two layers: a first layer in direct contact with the solid surface (wetting layer) and a second layer between wetting layer and b end matrix. If adhesion between wetting layer and a matrix is lower than adhesion between solid and wetting layer, the failure of adhesion joints will take place in a zone of the contact between two components of the blend. After the phase inversion in the blend, there occurs the transition of the failure zone to the interface with solid. The transition from adhesion failure between the blend components to adhesion failure between the blend matrix and a solid is a specific feature of adhesion of two-phase b ends. In the latter case the adhesion is determined by the adhesion strength of the wetting layer.

FREE ABSOLUTE STRAIN OF LOW DENSITY POLYETHYLENE IN RELATION TO SCREW SPEED IN THE EXTRUSION PROCESS

J. W. SIKORA

Department of Polymer Processing, Technical University of Lublin, 36 Nadbystrzycka St., 20-618 Lublin

1. INTRODUCTION

A test of uniaxial polymer compression is one of the important sources providing information about polymer physical-mechanical properties. Such studies are often used to estimate indirectly other polymer properties as well.

A model dependence of linear dimension of LDPE samples in the state of free absolute strain compressed to 50% of their original height and time after removing the compressing force is presented in figure 1.

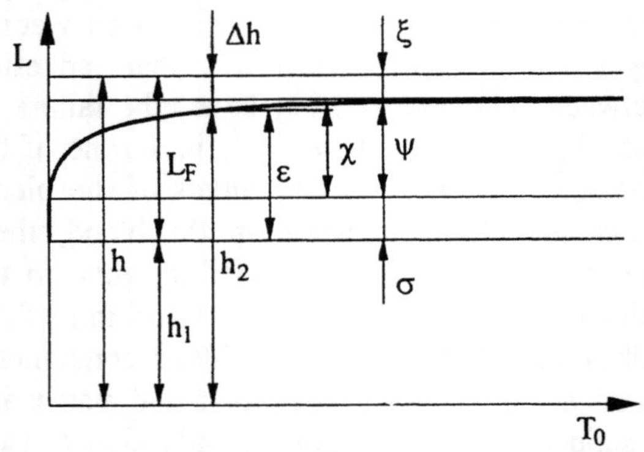

Fig. 1. A model dependence of the linear dimension of samples and time after removing the compressing force with characteristic properties indicated: σ - immediate elastic recovery, χ - delayed elastic recovery, ε - complete elastic recovery, ψ - complete delayed elastic recovery, ξ - permanent set, Δh - free absolute strain, h - height of the samples before compression, h_1 - height of the samples after compression (h_1 = 1/2h), h_2 - height of the samples after the set time since removing them from the testing machine

The strain is defined as the difference of the height of the samples before compression and after the set time given since removing them from the testing machine.

$$\Delta h = h - h_2 \quad (1)$$

The remaining properties given are calculated in the following way:

$$\varepsilon = \sigma + \chi \quad (2)$$

$$h_1 = L_F = 1/2h = \Delta h + \varepsilon = \Delta h + \sigma + \chi \quad (3)$$

$$\sigma + \psi + \xi = L_F \quad (4)$$

2. EXPERIMENTAL

For receiving the samples, low density polyethylene in granulate form of the characteristic granule size (diameter about 3mm) with the addition of antioxidant, antiblocking and lubricant agent of the natural colour was used. The polyethylene is indicated by the symbol: PE FABS 23-D022 and has the trade name Malen E. Basic physical-chemical properties of the polyethylene processed according to the producer's data are set in table 1.

Table 1

Basic physical-chemical properties of LDPE

Properties	Values
Density in the temperature 23°C, kg/m^3	919÷923
Range of the Mass Flow Rate $MFR_{(190;2.16)}$, g/10 min	1.6÷2.5
Numerical value of the swelling factor, % not higher than	160
Tensile stress, MPa, not smaller than	11
Softening point acc. to Vicat, °C, not lower than	90

Samples of low density polyethylene received in the process of autothermal and conventional extrusion at different screw speed of the single screw extruder (4.0; 4.22; 4.38; 4.58 and 4.77 s^{-1}) were compressed axially. The samples were compressed with constant speed 6mm/min to 50% of their original height, and after that the compression was immediately interrupted.

The value of h_2 was measured immediately after removing the samples from the testing machine and then it was assumed that the time of free strain equalled zero ($T_0 \cong 0$). Then h_2 was measured after the time of 0.5; 1.5; 2.5; 7.5; 19.5; 43.5; 67.5; 91.5 and 1000 hours.

The results of the studies of free strain of low density polyethylene samples received by means of conventional and autothermal extrusion at different values of extruder screw speed were presented in the graphic form in figure 2 and in the tabular form in table 2.

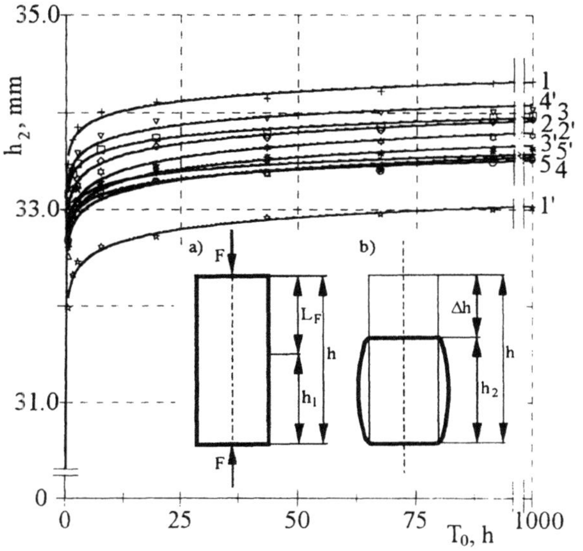

Fig. 2. Influence of time after removing the compressing force on the linear dimension of the samples together with the axial compression scheme (a) and indication of free strain (b): F - axial force compressing the samples; 1 ÷ 5 - conventional extrusion; 1' ÷ 5' - autothermal extrusion; 1 - v = 4.00 s^{-1}, 2 - v = 4.22 s^{-1}, 3 - v = 4.38 s^{-1}, 4 - v = 4.58 s^{-1}, 5 - v = 4.77 s^{-1}

3. RESULTS

The results of the studies of free absolute strain of low density polyethylene samples in the given screw speed range prove that the screw speed almost does not influence free strain, while the value of immediate free absolute strain at compression, immediately after removing from the testing machine the samples received in the process of autothermal extrusion, is slightly bigger than in the process of conventional extrusion. In the scope of delayed free absolute strain its value is slightly bigger in the case of autothermal extrusion than in conventional one.

Table 2

Values in mm characterising free absolute strain

No	Time	Property	K	A
1	0	Δh	9,56	9,33
		σ	10,44	10,67
		χ	–	–
		ε	–	–
2	0 ÷ 20	Δh	6,38	6,62
		σ	–	–
		χ	3,18	2,71
		ε	13,62	13,38
3	0 ÷ 1000	Δh	6,14	6,40
		σ	–	–
		χ	3,42	2,93
		ε	13,86	13,60

Attention: K - indicates conventional extrusion, A - autothermal extrusion

Free absolut strain, complete elastic recovery, complete delayed elastic recovery, immediate elastic recovery, delayed elastic recovery and permanent set of the samples received in conventional and autothermal extrusion and samples compressed to 50% of their original height point out that low density polyethylene behaves like viscoelastic material whose satisfactory rheological model is Burgers model. Immediate elastic recovery is very significant, permanent set is also big which proves a dominant role of Maxwell model as a component element of Burgers model. However, delayed elastic recovery is small which, in turn, proves the significance of Kelvin-Voigt model as a component of Burgers model.

4. REFERENCES

Chang DH (1976) Rheology in Polymer Processing. Academic Press, New York-San Francisco - London

Ferguson J, Kembłowski Z (1991) Applied Fluid Rheology. Elsevier Applied Science, London

Sikora J (1997) Polimery, 42: 111-119

Sikora J (1997) Polimery, 42: 565-571

ELASTOMERIC FRACTURE BY DYNAMIC ARGUMENTS

Z. ŠUŠTERIČ

Sava d.d., R & D Institute, Škofjeloška 6, SI 4000 Kranj, Slovenia

1. INTRODUCTION

Despite extensive studies of elastomeric fracture, including some recent ones (De and Gent (1996), Neumeister and Koenig (1997)), few attempts have been made to quantitatively account for the familiar *stick-slip* and *stable* modes of elastomeric tearing (Stacer et al. (1985)), the usual manner of crack propagation in crosslinked rubbers (Fig.1). This is probably due to problems of nonlinearity and imperfect elasticity of elastomers at very large (critical)

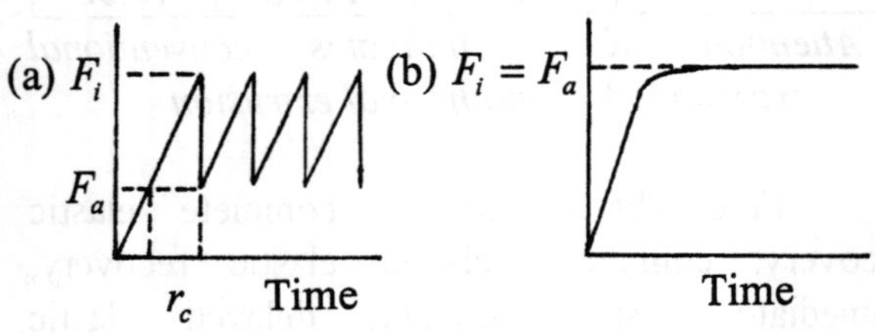

Fig. 1. Stick-slip (a) and stable (b) tearing mode

deformations acquired in the vicinity of the tear tip prior to fracture. In this work an approach circumventing these problems is presented. It considers several theoretical and experimental features, strong energy dissipation at the tear tip in particular, suggesting application of dynamic arguments for quantitative description of elastomeric tearing. With respect to this, it is shown that the stick-slip and stable tearing modes can be treated by a single fracture model.

2. THEORETICAL

The approach assumes elastomeric tearing to be a dynamic process in which, during stressing and deforming the tear tip, stored and dissipated energy density gradients form in the material beyond the tip. The stored energy density gradient is manifested through a stress gradient which, when in critical state, causes the tear tip to rapidly advance, while the force proportional to the stress at the tip and immediate surroundings drops from its *initiation* value, F_i, to the *arrest* value, F_a, an inherent material property (Stacer et al. (1985)) at which the tear tip propagation stops This process, several times repeated, results in the stick-slip tearing, as shown in Fig. 1 (a). The tear initiation and arrest stresses are associated with degrees of molecular orientation, high at tear initiation and low at its arrest. Thermodynamically, stick-slip tearing can thus be qualified as a transition from low entropy and high free energy state to a state of high entropy and low free energy.

The approach using dynamic arguments is also required to agree with Griffith's fracture theory (Griffith (1920), Kausch (1987)).

Considering the foregoing and denoting the stored and disipated energy densities in the critical state of the tear tip by w_s and w_d, the approach yields the following two basic results:

$$F_a / F_i = \exp(-w_d / w_s), \qquad (1)$$

$$r_c = 2w_d G_c / w_s^2 \left[1 - \exp(-2w_d / w_s)\right] \qquad (2)$$

where r_c is a tear growth length in a single stick--slip step and G_c the tearing energy obtained from Griffith' s fracture theory.

Examining extremal cases of Eqs. (1) and (2), correlation of stick-slip and stable tearing modes is immediately apparent. For $w_d = 0$ (ideally elastic material), $F_i = F_a$, meaning a stable tearing. On the other hand, for $w_s = 0$ (viscous fluids), $F_a = 0$ and $G_c = 0$, with r_c being completely uncertain, which is an utmost stick-slip case and obvious outcome for fluids. A perfectly stable tearing mode thus represents an extremal case of the more general stick-slip mode. Since w_s and w_d are functions of rate and temperature, the latter do not appear in the model explicitly. The model is devoid of any geometrical properties of the fractured body or the tear tip. Effects of solid fillers are accounted for implicitly in the values of w_s and w_d, which enables the *knotty* tearing, characteristic of elastomers with active fillers, to be treated as a particular form of stick-slip mode.

3. EXPERIMENTAL

The dynamic tearing model was verified on various crosslinked rubbers, unfilled and filled with active fillers, such as carbon black and silica, and under diverse tearing conditions. The energy densities, w_s and w_d, were determined from tensile hysteresis measurements at critical deformations and G_c from tearing measurements on trouser-type test pieces, which also served to determine F_i and F_a, as well as r_c.

4. RESULTS AND DISCUSSION

As an example, results obtained at tearing rate of 10 cms^{-1} and temperature of 100^oC on unfilled crosslinked rubbers and those filled with *50 phr* (parts in weight per *100* parts of rubber) of N-330 carbon black and some with *50 phr* of silica are presented Fig. 2 in form of plots of calculated (by Eqs. (1) and (2)) vs. measured F_a / F_i ratios and r_c's, respectively.

(a)

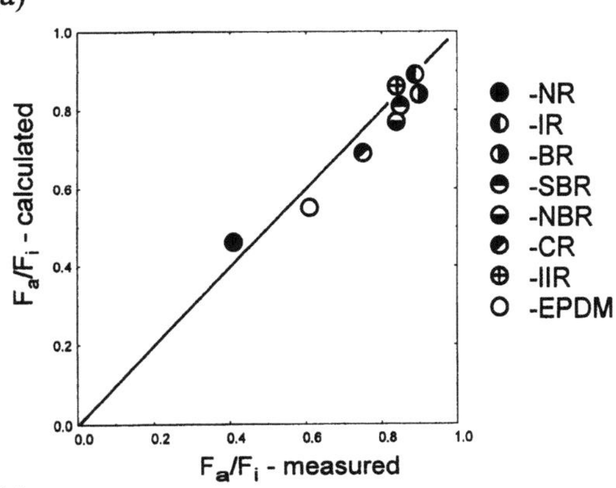

(b)

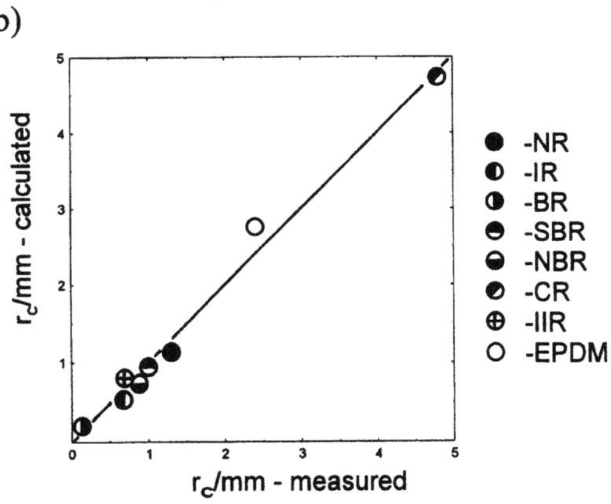

(c)

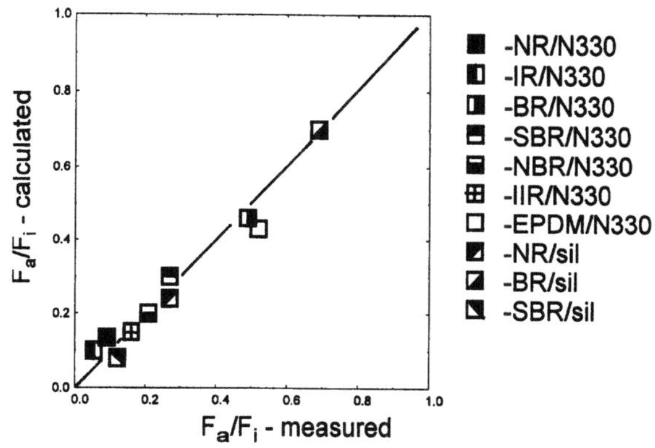

(d)

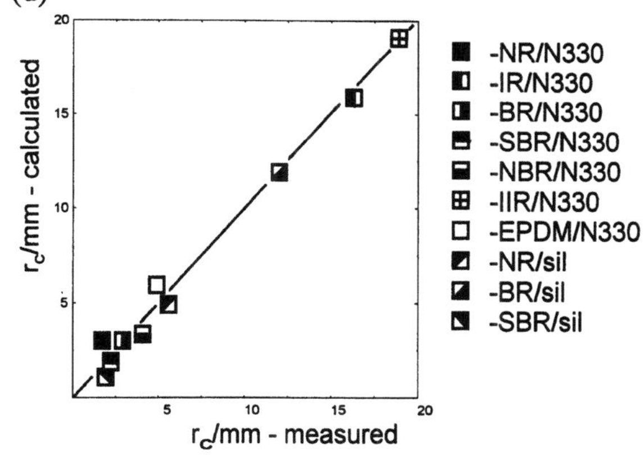

Fig. 2. Calculated vs. measured F_a / F_i ratios and r_c's for unfilled (a) and (b), and with active fillers filled crosslinked rubbers (c) and (d) (solid lines represent hypothetical exact agreement)

Relatively good agreement is obtained in all cases. So vindicated, the approach describing elastomeric tearing by dynamic arguments can now be used to explain the tearing mode transitions, as the additive content or fracture conditions change, as well as predict the nature of elastomeric fracture, i.e. brittle or tough.

5. REFERENCES

De D, Gent AN (1996) Rubber Chem. Technol. 69: 834-850

Griffith AA (1920) Phil. Trans. Roy. Soc. A221: 163-198

Kausch HH (1987) Polymer Fracture, Springer Berlin, Heidelberg, New York, Chap. 9

Neumeister L, Koenig JL (1997) Rubber Chem. Technol. 70:271-282

Stacer RG, Yanio LC, Kelley FN (1985) Rubber Chem. Technol. 58:421-435

REOLOGY OF ROCK MASSIVE AS A FACTOR OF SYNERGETIC MECHANISM OF GEOCATASTROPHE

Prof. V.N. Vilegzhanin

Mining Institute Kamerevo, Kamerevo, Russia

During mining development rock massive saturated by water and gas creates complex of physical and geomechanical process which involves an geocatastrophe such as rock bamp, coal-and-gas outburst and roof fall, quicksand intrusion, movement and landslide of geoterritory's surface. Mechanism of these events is tied with deformation and flowing of rocks such as polyphase anysotropic sphere which have macroreologic qualities: elasticity, crypticy, coal-accomodative rocks, viscosity. However, indignant condition of some areas of rock massive is characterized not only by unbalanced state, but forming a dissipate thermodynamical structure in the area of potential geocatastrophe of fireside. It shows interconnection of synergetic mechanism of geocatastrophe with reological qualities of mining rocks.

THE RELATIONSHIP BETWEEN FOAM STABILITY AND SURFACE RHEOLOGICAL PROPERTIES

K.-D. WANTKE, H. FRUHNER

Max-Planck-Institut für Kolloid- und Grenzflächenforschung, Rudower Chaussee 5, D-12489 Berlin

1. INTRODUCTION

A fluid foam is a multiphase system composed of air bubbles, fluid walls and plateau borders. The air/fluid interface can be considered as an additional phase which properties have an decisive influence on the stability of foams. In particular, the rheological properties of adsorption layers, like surface elasticity, surface dilational and surface shear viscosities are important for investigation of fluid foams. However, this general accepted opinion has only a small influence on the practical foam research because of the difficulties in obtaining reliable data of the surface rheological parameters.

1. SOME REMARKS ON SURFACE RHEOLOGY

The elimination of the influence of the bulk phases is the central problem of all surface rheological measurements. This problem has theoretical and experimental aspects. A fluid interface is considered to be a layer of the thickness of one or several molecules with the structure, composition, and states of energy and tension much different from the analogous properties within the adjacent bulk phases. The choice of the thickness of the interfacial layer is arbitrary, however, the final results of a rheological model of an interface must not be dependent on this thickness value. Due to the molecular exchange between the bulk and interface the fluid interface must be considered as a highly compressible fluid phase of variable composition. Under static conditions, the composition, the state of tension, and the shape of the interface are determined by the laws of equilibrium thermodynamics. This means that any external disturbance of the system, e.g. by stretching or heating, leads not only to a new tension state, but also to relaxation processes as a consequence of the disturbed thermodynamic equilibrium. The tension state is a function of the "history" of disturbance and therefore, the system exhibits typical viscoelastic behavior. In most cases a two-dimensional model of such system independent of bulk phases is an unrealistic idealization. Nevertheless, many authors applied the rheological standard model with storage and loss modulus also in an independent two-dimensional form. However, using such a model the interpretation of the experimental results is difficult because of the influence of the bulk. For example, the dynamic surface tension depends on the bulk diffusion processes of surfactants Besides, the molecular exchange between surface and bulk phase can lead to additional terms in the force balance equations. Therefore, the rheological considerations must be combined with adsorption models.

In the adsorption theory the interface is normally described as a monolayer. For a system in the thermodynamic equilibrium this assumption is justified due to the invariance of the tension state with respect to variations in the position of the dividing plane in the Gibbs model. However, for a rheological model of a fluid interface the knowledge of the real molecular distributions and their dynamics is required. Nevertheless, the most authors used the monolayer model also for dynamic processes. This assumption has an influence on the boundary conditions of the bulk phases (Navier-Stokes and diffusion equations) and on the interpretation of the rheological effects. In particular, the number of interface molecules which have influence on the interface tension state and the dynamic behavior is restricted in the monolayer model.Therefore, the explanation of faster dilational processes becomes difficult in some cases. For sinusoidal homogeneous dilation of a fluid surface (interface) a complex elasticity modulus ε(f, c) can be written as:

$$\Delta\gamma = \varepsilon(f,c)\frac{\Delta A}{A} = |\varepsilon(f,c)|\exp(i\varphi(f,c))\frac{\Delta A}{A} \tag{1}$$

where $\Delta\gamma$ represents the change in the surface tension and $\Delta A/A = |\Delta A/A| \exp(i\omega t)$ the relative change in the surface area. The phase angle φ(f, c) results from the shift between both oscillations. In general the modulus ε(f, c) is a complex function of the frequency f with the bulk concentration c as a parameter. It involves elastic, viscous, and transport properties. An appropriate model leads for surfactant solutions to the following equation

$$\varepsilon(f,c) = \varepsilon_0(c)\frac{1+\zeta+i\zeta}{1+2\zeta+2\zeta^2} + i2\pi f\kappa(1-u(f)) \tag{2}$$

where $\zeta = 1/\sqrt{4\pi f \tau_0}$ is the molecular exchange function, τ_0 the diffusion relaxation time, κ the surface dilational viscosity, and u(f) a weight function. The first term (known as Lucassen/ van den Tempel modulus) describes the influence of the bulk diffusion process on the dynamic surface tension. The second term results from the formal introduction of a intrinsic surface dilational viscosity. This term is related to relaxation processes due to the molecular exchange between surface and subsurface. However, for such an explanation the monolayer model must be replaced by a surface phase model with a greater thickness. The parameters of Eq.(2) (ε_0, τ_0, κ) have a decisive influence on the dynamic processes at surfaces. Therefore the experimental verification of the Eq.(2) is very important. The existing experimental methods for realization such measurements are based on wave and bubble techniques. However, the knowledge about the dynamic processes at fluid surfaces in the important frequency range >10 Hz is very scarce. In the last years the situation has been improved by the development of a new setup enabling bubble oscillations up to 500 Hz. Such apparatus allows the reliable determination of various surface rheological parameters as was demonstrated by Fruhner and Wantke (1996, 1998).

3. FOAM STABILITY

The foams formed from solutions of surface active substances (surfactant or mixture of polymer/surfactant) exhibit significantly different stability. Some solutions, like fatty acid solutions and n-alkanol solutions, form unstable wet foams which stability can be measured using a dynamic method realized by Maŕysa. The rupture thickness of the foam lamella of these solutions is larger of an order of 1 μm and the values of surface viscosity are negligible. A correlation exists between the stability of the wet foams and the "effective" surface elasticity. "effective" means that for this consideration the diffusion kinetics must be taken into account.

Others, like higher concentrated decyl dimethyl phosphine oxide solutions or nonyl dimethyl phosphine oxide phosphine solutions, form stable dry foams with a very small rupture thickness. The concentration range where foam stability behavior changes from wet to dry foam can be easily determined using the method of Gilanyi. The comparison of these results with bubble experiments demonstrates that a measurable intrinsic surface dilational viscosity is necessary for the formation of a dry foam. This means the parasitic waves which cause the lamella ruptures are damped if an influence of the intrinsic surface viscosity exists. For a broader verification of this hypothesis reliable measurements of the dilational properties of fluid surfaces are required.

4. REFERENCES

Fruhner, H., Wantke, K.-D. (1996), Colloids Surface A, 54-59

Wantke, K.-D., Fruhner, H., (1998) In: D. Möbius, Miller, R. (eds), Studies in Interface Science Vol. 6, ELSEVIER, pp 327-365

Maŕysa, K, (1992), Adv. Colloid Intf. Sci., 37

COMPARISON OF THE GENERALIZED MAXWELL AND STASTNA MODEL FOR CHARACTERIZING DYNAMIC MODULI OF MODIFIED ASPHALTS

O.Wein, L.Zanzotto,

University of Calgary, Faculty of Civil Engineering, Bituminous Materials Chair
2500 University Drive N.W., Calgary, Alberta, Canada T2N 1N4

The master curves of dynamic moduli, obtained by using a temperature(θ)-frequency(ω) superposition principle, are commonly available for the reduced frequencies, $\omega' = \omega\, a_T(\theta)$, over 10 to 15 decimal orders of magnitude. For fitting such an extended experimental information, Winter et al. (1993) suggest to use the generalized Maxwell model. As many as 20 parameters are needed for a reasonable fitting,. On a set of experimental master curves for the asphalts modified by addition of styren-butadien copolymers, we demonstrate the ability of the stretched-exponential model, developed by Stastna et al. (1994), to fit experimental data with no more than six adjustable parameters. Such an approach to fitting data on temperature and frequency dependence of the complex moduli makes it possible to rationalize the prevailing manual way of adjusting the temperature shift a_T for the data with some experimental dispersion.

H.H.Winter, M.Baumgärtel, P. Soskey: *"A parsimonious model for viscoelastic liquids and solids"*. in: A.A.Collyer, Ed.:*"Techniques in rheological measurement"*, Chapman and Hall, London UK 1993.

J.Stastna, L.Zanzotto, K.Ho: *"Fractional complex modulus manifested in asphalts"*. Rheol.Acta **33** (1994) 344-354.

MODELING OF RHEOLOGY PROCESSES IN MINING MASSIF OF POTASH SALTS ROCKS WITH LARGE-SCALE UNDERGROUND STRUCTURES

M.A.ZHURAVKOV[1], S.G.DROZD[2]

[1]Mecanical Dept., Faculty of Mechanics-Mathematics, Belarus State University, Skarini ave., 4, Minsk, Belarus, 220050
[2]Mecanical Dept., Belarus State Polytechnical Academy, Skarini ave., 65, Minsk, Belarus, 220027

1. INTRODUCTION

The study of rheology processes in massif of mining rocks of potash salt deposits represents a difficult research problem and has the important practical meaning. Deforming processes of hydrochloric breeds have brightly expressed rheology nature. Therefore important is the construction of deforming models of massif in a vicinity of underground structures in view of creep processes of rocks and relaxation of stresses.

In the report is told about the approaches, used at modeling of rheology processes in mining massif of Soligorsk deposit of potash salts (Belarus).

2. EFFECTIVE CHARACTERISTICS OF A STRUCTURAL INHOMOGENEOUS MASSIF

For building of mechanical properties of multylayered massif is used the theory of quasihomogeneous structured elastic bodies, and as a basic model a model stochastically inhomogeneous mediums is considered "Zhuravkov and Martunenko (1995) ".

General view of the formula for mechanical characteristics of mining massif at presence in it of a system of *m* cavities looks like

$$E(x,t) = \langle E_0(x)\rangle \prod_{j=1}^{m} \left[1 - a_1^{(1-\omega_i^*)} \left(R_j / r_j \right)^{n_1 \omega_i^*} \right] \cdot \eta \left(t + t_m - t_j \right)^{(1-\omega_i^*)} \quad , t \geq t_m \tag{1}$$

Using this formula, it is possible to build functional expressions for effective characteristics of a structural-inhomogeneous massif regions in view of appearances of multylayered, stratification and weak contact of layers, natural change of properties of breeds with magnification of depth of exavation and technological heterogeneity of a massif (in case of production in it of underground mining work). In general case four multiplicands are selected in expression (1) (we shall show on an example of the module of an elasticity):

$$Z(x,t) = Z_1(x) Z_2(x) Z_3(x) Z_4(x,t), \tag{2}$$

where $Z_1(x)$ takes into account multylayered of a massif and local inhomogeneous; $Z_2(x)$ takes into account a stratification of contacting layers; $Z_3(x)$ is introduced for the definition of a modification of properties of breeds with magnification of depth of exavation; $Z_4(x)$ determines a changing of the module of an elasticity by according with technological heterogeneity of massif.

3. METHODS AND SYSTEMS OF DEFINING EQUATIONS

For a construction of mathematical models of investigated geomechanical processes the experimental-analytical approach, based on use of various variants of a Green's quazifunction method for solution of a mechanics of a deformable rigid body problems is used "Zhuravkov and Martunenko (1993, 1995) ". An essential moment that the integral kernels contain an information about the boundary of area at an analytical level. Besides the integrated and integro-differential representations for characteristics of stress-strain state have an equivalent structure at the various approaches to a solution of that or other problem.

The systems of the solving equations for problems of a linear viscoelasticity are created by introduction of concept of fictitious elastic displacements u_i^y and strains ε_{ij}^y . After a solution of an auxiliary elastic problem and definition of fictitious displacements u_i^y the true displacements are

$$u_i(x,t) = u_i^y(x,t) + \frac{1}{I_0} \int_0^t \dot{I}(t-\tau) u_i^y(x,\tau) d\tau \tag{3}$$

It is necessary to underline, that stresses are obtained during a solution of an auxiliary elastic problem are true stresses for initial viscoelasticity problem.

Problems of the linear theory of viscoelasticity for inhomogeneous bodies with the help of the Laplace transformation are reduced to problem solving of the theory of an elasticity of inhomogeneous bodies, which are solution with the help of a Green's quazifunction method.

4 RESEARCH OF DEFORMING PROCESSES IN LAYERED MASSIF OF SALT ROCKS

4.1 Study of a creep processes in a neighbourhood of a counter of underground cavities.

One of basic consequences of deforming processes in layered massif with underground cavities is the process of development of a stratification of layers on contact surfaces. We considered average and large zones of stratifications as the influence of small zones of stratifications to a stability of underground buildings, obviously, is unsignificant. We considered zones of stratifications with rectangular form and located in a contact plane. An initial clearance in all cases we accept equal to zero.

Deformed condition of roof of cavities was studied, as the study of a stratification of layeres in roof of cavity is the problem the most interesting with an engineering point of view, as the roof represents the most vulnerable element in a sense of loss of equilibrium.

In base of analysis of numerical experiments basic positions about development of process of a stratification of layeres in roof of cavities in multylayered massif were formulated.

4.2. Mathematical aspects of a research of effect "unloading camera" for conditions of block-layered massif.

The dangerous strains of a layered massif in a neighbourhood of cavities are determined by mobility on contacts of blocks and structures. One of the important properties "ideal" tier or measures of guards - to suppose a possibility of development of strains owing to reological properties of a massif, as such strains not to keep by any forces. In such cases more effective measures of guards of underground cavities are those, based on introduction of unloading cameras (canity, compensatory slots and etc.). The essence of such measures of guards is based on effect of a unloading of specific zones of a massif, i.e. redistribution of a stress-strain state in a massif.

Influence of some measures of guards of cavities to development and change of a stress-strain state of a massif with a system of underground cavities was studied.

4.3. Study a change with time the deforming state of a multylayered massif in region of underground cavities.

The problem of an evaluation of a condition of a massif in boundary areas near of underground buildings (cavities) is extremely important from a practical point of view, as is connected to a stability of underground buildings. Is developed one of the possible approaches to a construction of an engineering technique of an evaluation of a stability of a condition of underground buildings in a massif of salt rocks. The offered approach is based on use of experimental-analytical methods of Mechanics of Deformable Rigid Bodies (analytical methods and solutions; large volume of laboratory and natural researches).

Is shown, that the behaviour of mining massif in a vicinity of underground structures is satisfactorily described by the theory of linear hereditary creep.The approaches for a choice of a kind of relaxation and creep kernels and authentic determination of the characteristics of these kernel are considered. A procedure of reception of quantitative meanings of the characteristics of creep kernels, based on experimental natural researches and analytical procedures is resulted. Is shown, that the meanings of the characteristics by an essential image depend on a kind of an strees state, in which there are the examined investigated elements of a underground structure.

5. REFERENCES

Zhuravkov M.A., Martunenko M.D. (1993) Method Green's quazifunction in Mechanics of Deformable Rigid Bodies. University Publ. Minsk

Zhuravkov M.A., Martunenko M.D. (1995) Theoretical base of deformed mechanics of block-layered massif of mining rocks of Belarus. University Publ. Minsk

RDS-2 rheospectrometer, Rheometrics. Paralell plates of 25 mm in diameter were applied. Measurement conditions: room temperature, strain 10 %, frequency range 30 - 500 rad/s (about 0.5 - 80 Hz).

3. RESULTS AND DISCUSSION

It has been found that tan δ depends on the type of DVPSO system, i.e. the single- or two-component system (see Table 2).

Table 2. Comparison of DVPSO systems.

Sample symbol	DVPSO, wt. ratio	Crosslink agent, wt.%	Rheological properties *		
			G', kPa	G", kPa	tan δ
8	F1 -	0.78	10.75	6.71	0.62
7	M/F1 10/3.2	0.89	10.54	7.30	0.69

* measured at RT and at frequency of 25 Hz.

For two-component systems, damping properties are influenced by the amount of the specified crosslinking agent (that is compensated by the amount of phenyl groups and by the molecular weight of F component), and by the amount of phenyl groups contained in the investigated material (see Table 3). The value of tan δ is higher for the system containing higher amount of phenyl groups at the same level of crosslinking.

Table 2. Influence of phenyl groups on damping properties of DVPSO two-component system

Sample symbol	DVPSO, wt. ratio	Ph, arbitrary units	Rheological properties *		
			G', kPa	G", kPa	tan δ
3	M/F4 10/5.9	240	10.29	8.64	0.84
4	M/F4 10/3.3	130	10.76	8.63	0.80
7	M/F1 10/3.2	30	10.54	7.30	0.69

* measured at RT and at frequency of 25 Hz; crosslinking agent at the level of 0.9 wt.%.

Michalski (1996) suggested that structures similar to the IPN are formed during the crosslinking process in the case of two-component systems. However, more experimental data are needed in this case.

Moreover, interrelations between shear moduli G' and G" were considered using the relationship given by Han and Kim (1993) for polydisperse polymers

$$\log G' = x \log G'' + (1-x) \log(8G_N^0/\pi^2) \qquad (3)$$

where x values are between 1 and 2, x = 2 being for monodisperse polymers. It has been found that x = 1.7 for PDMS standard material. It is suggested that the approach of Han and Kim (1993) might be useful for characterization of investigated DVPSO systems as damping materials.

4. CONCLUSIONS

It has been found that damping properties are influenced by amounts of the specified crosslinking agent and the amount of phenyl groups contained in the investigated material. The value of tg δ is higher for the system containing higher amount of phenyl groups at the same level of crosslinking.

The results are considered as preliminary ones and a further work on the DVPSO systems is still needed.

5. REFERENCES

Barnes HA, Hutton JF, Walters K (1989) An Introduction to Rheology, Elsevier, Amsterdam

Dobkowski Z, Michalski M, Zielonka M (1997) In: ICRI Annual Report '96, Warszawa, pp 90-92

Han CD, Kim JK (1993) Polymer 34: 2533-2539

Michalski M (1996) ICRI Report, unpublished.

Van Krevelen (1990) Polymer Properties. Elsevier, Amsterdam

RHEOLOGICAL CHARACTERIZATION OF DAMPING MATERIALS BASED ON POLYSILOXANES

M. ZIELONKA , Z. DOBKOWSKI

Industrial Chemistry Research Institute (ICRI), Dept.of Analysis and Characterisation of Materials, Rydygiera 8, 01-793 Warszawa, Poland.

1. INTRODUCTION

Recently, crosslinked polysiloxanes were investigated due to their applications as a component of polymer materials for various vibration damping devices. The elasticity loss factor tan δ , measured at constant temperature and at a specified frequency, is usually applied as a parameter characterizing damping properties of the polymer material. The loss factor tg δ is defined as the ratio of energy dissipated in the material during vibrations to the maximum potential energy stored in the material, i.e. as the ratio of shear moduli measured by dynamic rheological properties, cf Barnes et al. (1989)

$$\tan\delta = G''/G' \qquad (1)$$

where G' is the storage modulus (the elastic component of complex shear modulus G*) and G" is the loss modulus (the non-elastic component). The value of tg δ is correlated with the coefficient of internal friction ψ (or the specific damping capacity) of polymer material, cf Van Krevelen (1990)

$$\mathrm{tg}\,\delta \sim \psi/2\pi \qquad (2)$$

Thus, damping properties of crosslinked polysiloxanes were investigated by Michalski (1996) and Dobkowski et al. (1997) in the Industrial Chemistry Research Institute (ICRI), using the value of tg δ obtained from dynamic rheological measurements. In this paper the parameters influencing the values of tan δ are discussed. Interrelations between shear moduli G' and G" were also considered.

2. EXPERIMENTAL

2.1 Materials

At the first stage, linear polymers - α,ω-divinylpolydimethylsiloxanes (DVPSO) with vinyl and/or phenyl substituents in the main chain - were prepared. Their structure can be written as

```
          Vin      CH3      C6H5
           |        |        |
Vin - O -[-Si-O-]x-[-Si-O-]y-[-Si-O-]z- Vin
           |        |        |
          CH3      CH3      C6H5
```

where Vin denotes the vinyl group $CH_2 = CH-$. Characteristics of linear DVPSO are shown in Table 1.

Table 1. Characteristics of linear DVPSO.

Sample symbol	Substituents of DVPSO	Group contents, mol. % *		Viscosity, η, mPa.s	n_D^{20}
		Vin	Ph		
M	Vin, Me	1	0	10,500	1.403
F1	Ph	0.15	9.2	22,500	1.447
F2	Ph	0.08	9.1	295,000	1.446
F3	Vin,Me,Ph	0.8	23.5	47,000	1.486
F4	Vin,Me,Ph	1.1	40.5	44,500	1.534

* calculated from outcomes (after removing of light fraction of oligomers) and controlled by ^{1}H-NMR, Brucker WP-SY.

At the second stage, DVPSO and their two-component mixtures were crosslinked with methylhydrogensiloxane oil of low viscosity, type M 4640 B of Wacker Co. Pt complex in α,ω-divinylpolydimethylsiloxane of Huls Co. was used as a catalyst for silylation reactions. Reaction conditions, i.e. time and temperature, and amounts of the catalyst, were adjusted for each type of polysiloxane, respectively.

2.2 Measurements of rheological properties

Moduli G*, G' , G" and values of tg δ of crosslinked DVPSO were measured using the

INFLUENCE OF WELAN CONCENTRATION ON RHEOLOGICAL PROPERTIES OF ALUMINA DISPERSIONS IN AQUEOUS WELAN MATRICES

ANDREJA ZUPANČIČ[1], MIHA ŽUMER[1] AND ROMANO LAPASIN[2]

[1]Faculty of Chemistry and Chemical Technology, University of Ljubljana, Aškerčeva 5, 1001 Ljubljana, Slovenia

[2]Department of Chemical, Environmental and Raw Materials Engineering, University of Trieste, piazzale Europa 1, 34127 Trieste, Italy

1. INTRODUCTION

The rheological properties of filled gels are strictly governed by those of the polymeric matrix, even if the influence of particle addition can become very important, depending on the effective volume concentration of the disperse phase and the type and extent of particle-gel matrix interactions. Weak gel matrices are frequently preferred for many practical purposes, particularly when remarkable concentration levels of the disperse phase must be reached in the filled gel, since their viscous and elastic properties are favorably combined to ensure easy manipulation and transportation as well as good stability. Among several polysaccharides giving rise to more or less weakly structured systems when dissolved in water under proper conditions, welan, can be considered a good candidate to form weak gel matrices, since the sol/gel transition can be attained at low polymer concentrations and their gel properties are slightly dependent on environmental conditions (Lapasin R, Pricl S, 1995).

The present paper deals with the problems related to the preparation of concentrated aqueous dispersions of ceramic powders in welan matrices. Highly concentrated aqueous alumina suspensions were used as a source of particle addition in the gel matrices. The specific objective of this work is to examine the influence of welan concentration on the structural conditions of the systems.

2. EXPERIMENTAL

2.1 Materials High purity alumina powder A16 SG (Alcoa) (d_{50} = 0.5 μm), distilled water, three different anionic dispersants (*Duramax D3021*, modified polyacrylic acid, Mw 5000; *Dispex A40,* polyacrylic acid, Mw 10 000; *Dolapix CE64* carbonic acid Mw 230) and Welan K1A96 (Monsanto) were used to prepare alumina suspensions, polysaccharide matrices and alumina dispersions in welan matrices.

2.2 Preparation Alumina suspensions were prepared at the same solid volume fraction of 0.55 by dispersing the alumina powder in aqueous dispersant solutions (through ball milling in polypropylene jars with Si_3N_4 milling balls for 16h). Welan was separately dissolved in distilled water by mechanical stirring at ambient temperature. The two components (alumina suspensions and welan systems) were blended together in proper quantities by hand mixing. For all the systems the final solid volume concentration was 0.38 with welan concentration ranging from 0 to 0.5 wt%.

2.3 Rheological measurements The rheological tests were performed under steady and oscillatory shear conditions at 20 °C using the controlled stress rheometer Haake, RheoStress RS 150, equipped with double cone sensor system (DC 6 cm / 4°).

3. RESULTS AND DISCUSSION

Some preliminary tests were performed to check the influence of environmental conditions on rheological properties of aqueous welan systems. In the concentration range 0.3 - 1 wt% the rheological properties of the systems resulted almost insensible on pH (2-10), temperature (8 - 45 °C) and excess of dispersant in the solution. The rheological behavior of welan systems strongly depends on the polymer concentration. At 0.1 wt% of welan the linear and nonlinear properties are quite similar to those of polymer solutions. Above 0.2 wt% of welan the rheological properties resemble those peculiar of structured liquids or weak gels, i.e. predominant elastic component at small deformations, low sensitivity on the frequency for both moduli and slight time-dependent flow properties.

The stability of aqueous ceramic suspensions depends on the pH and on the addition of dispersants. Three different dispersants were used in order to study their efficacy and the effects produced by the presence of welan in the continuous phase. The rheological characterisation of aqueous alumina suspensions showed that the level of particle stabilisation strongly depends on the dispersant used. A good state of particle dispersion was achieved with Duramax D3021 and Dispex A40, since these suspensions showed relatively low viscosity, shear thinning and negligible time-dependent properties. Conversely, the suspensions prepared with Dolapix CE64 are clearly flocculated, as we can argue from their higher viscosity values and plastic behavior, as well as from their thixotropic and elastic properties in nonlinear and linear conditions, respectively. The mechanical spectra of aqueous alumina suspensions clearly demonstrate the differences in structural conditions of the disperse phase, as shown in Fig. 1.

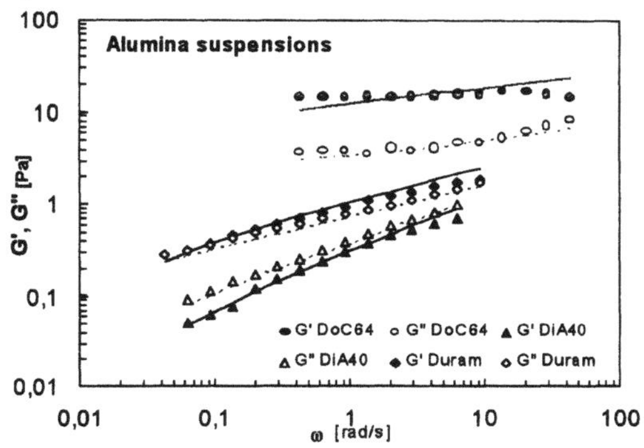

Fig. 1. Mechanical spectra of alumina suspensions formulated with different dispersants at $\Phi = 0.38$

Independently of the dispersant used, the most significant changes in the rheological properties of alumina dispersions were observed for the systems containing 0.1 wt% of welan in the continuous phase. Under linear oscillatory shear conditions the elastic component neatly predominates the viscous one for all alumina-welan systems (see Fig. 2).

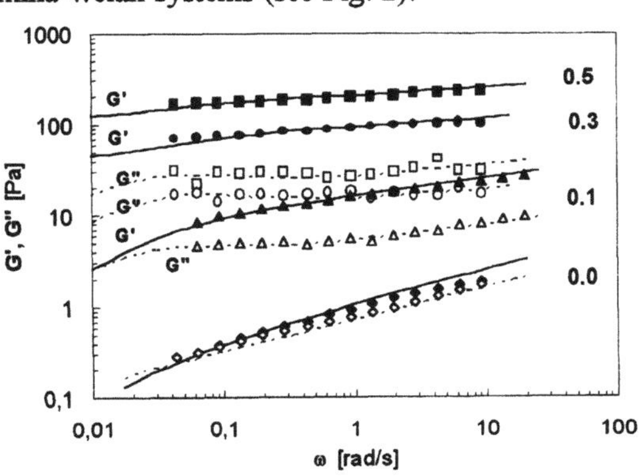

Fig. 2: mechanical spectra for alumina - welan systems, Φ=0.38, prepared with Duramax D3021

At welan concentrations above 0.2 wt%, the dispersions behave like structured liquids or weak gels, since the traces of G'(ω) and G''(ω) become almost independent of frequency and parallel each another, with G' much higher than G''.

The differences observed between stabilized and flocculated alumina suspensions become less pronounced when welan is present in the continuous phase, even at low concentration (0.1 wt%). At higher welan concentrations the linear viscoelastic properties do not depend appreciably on selected dispersant, in spite of quite different level of particle stabilization. Fig. 3 shows that all the systems at 0.3 wt% of welan exhibit similar mechanical spectra of 'weak gel' type. Under such conditions the polymeric gel matrices control the rheological properties of the alumina dispersions, since welan provides an essential contribution to particle stabilization.

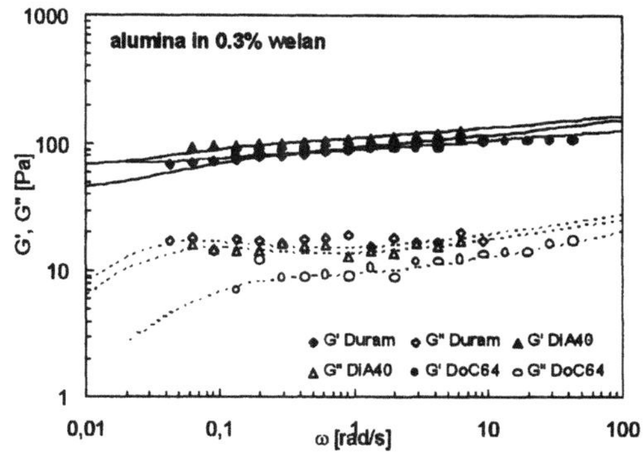

Fig. 3: Mechanical spectra of alumina dispensions in welan matrices, different dispersants at 0.3 wt% of welan.

All the mechanical spectra were compared with the fractional derivative model suggested by Friedrich and Braun (1992) and the generalized Maxwell model. Both models provide satisfactory and almost equivalent fitting results. The curves reported in Figs. 1- 3 represent the profiles of the viscoelastic moduli, calculated from the fractional derivative model.

4. CONCLUSIONS

Aqueous welan systems exhibit 'weak gel' properties even at low polymer concentration (above 0.2 wt%) and their behavior does not vary appreciably when temperature and/or other environmental conditions are changed. Even small additions of welan to the continuous phase produce significant changes in the rheological properties of alumina suspensions and, for welan concentrations above 0.3 wt%, the behavior of the alumina dispersions in welan matrices becomes independent of the dispersant efficiency, assuming typical 'weak gel' features. In such conditions the rheological properties of aqueous alumina/welan systems are governed by welan concentration, since the biopolymer provides an essential contribution to particle stabilisation.

5. REFFERENCES

Lapasin R, Pricl S., "Rheology of industrial polysaccharides: Theory and applications", Chapman&Hall, Glasgow, 1995

Friedrich, Chr. and Braun H.,Rheol. Acta , 1992, 31, 309 - 322

THE MOLECULAR RHEOLOGY OF STAR POLYMER MELTS: BINARY BLENDS

B. Blottière* [1], T.C.B. McLeish[1], A. Hakiki[2], R.N. Young[2], S.T. Milner[3].

[1]IRC in Polymer Science and Technology, Dept. of Physics and Astronomy, University of Leeds, Leeds LS2 9JT, UK.
[2]Dept. of Chemistry, University of Sheffield, Sheffield S3 7HF, UK.
[3]Exxon Research and Engineering Company, Route 22 East, Annandale, New Jersey 08801, USA.
*E-mail: phybb@irc.leeds.ac.uk, Tel:(44) (0)113 233 6648 Fax:(44) (0)113 233 3846

ABSTRACT

Experiments on solution-cast blends of two anionically synthesised monodisperse star-shaped polyisoprene molecules of widely different molecular weight (M_1 = 28000 g/mol and M_2 = 144000 g/mol) exhibit very rich rheological behaviour (see figure below). This work models these new features by extending existing theories for monodisperse melt of star polymers to the blend of two monodisperse star polymers with different molecular weight, keeping the same chemistry. The theory is based on the tube model with constraint release for star polymers in both an approximate and then a more exact level. The latter, with its treatment of non-activated as well as activated breathing modes, is able to account quantitatively for the huge range of blend rheologies. With no extra parameters, it is able to account qualitatively for relaxation functions that vary orders of magnitude on blending. It is also demonstrated that the time-dependent moduli are exponentially dependent on the relative volume fraction of each species. Indeed the inherent time scales of molecular motion of each species in the blend are strongly mutually affected. The relaxation time functions are exponentially dependent on the effective potentials for activated breathing of the arms (free energy functions of the system). It is shown as well that the prefactors have a molecular weight dependence proportional to $M^{3/2}$ and also depend on the volume fraction through the derivatives of the effective potentials. The picture drawn from these calculations shows that when the small stars are less numerous than large ones, they tend to behave like star molecules moving through a fixed network of obstacles built up by the bigger stars. In the reverse case, when the big stars are less numerous, the small stars enhance the constraint release for the big stars, and plasticise their relaxation in a precisely determined way.

FIGURES

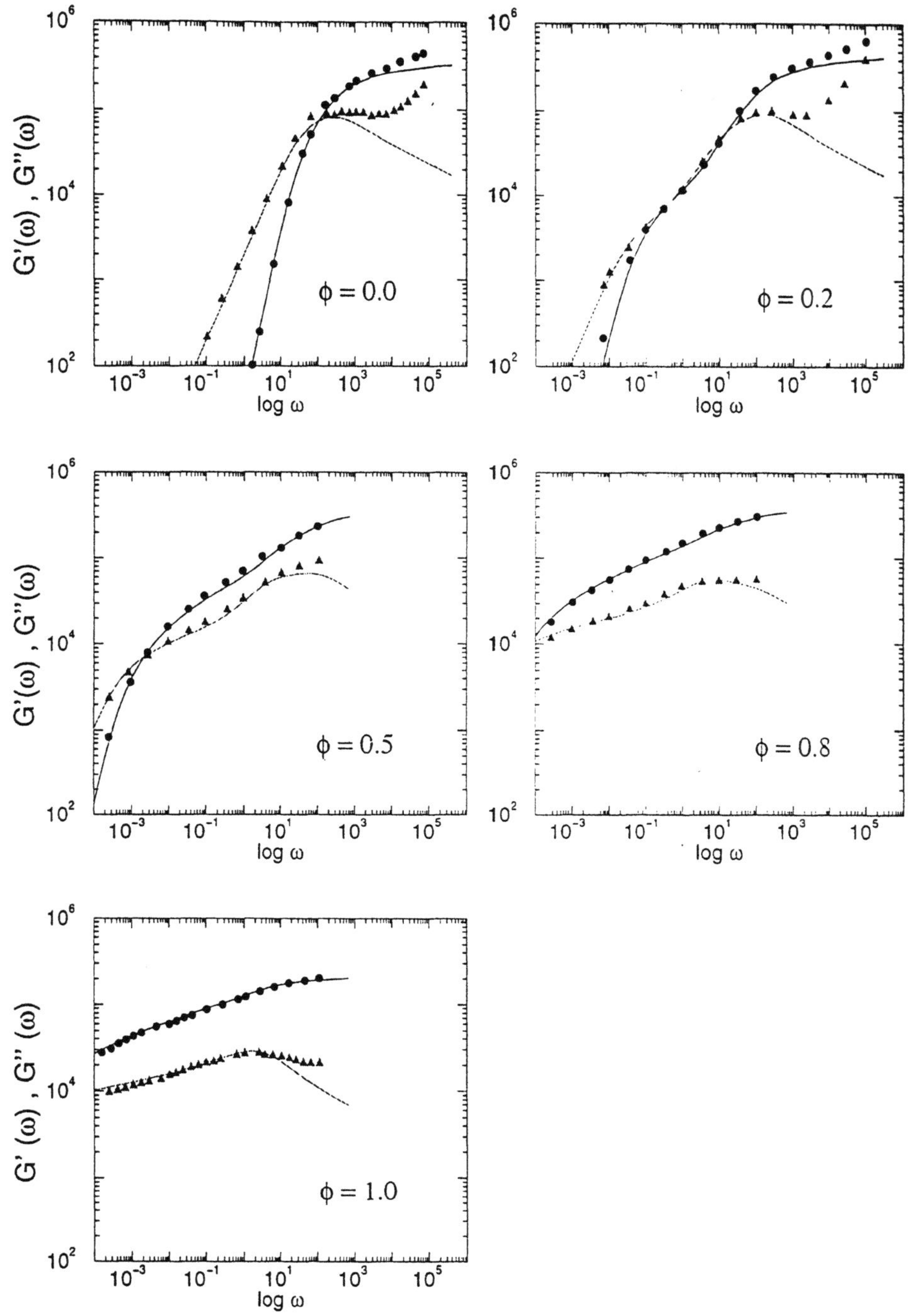

FIG. 1. Full tube blend model (lines) and experiments (circles G' and triangles G"). $M_1 = 28000, M_2 = 144000, M_e = 5500$

PREDICTION OF MOLECULAR WEIGHT DISTRIBUTION FROM RHEOLOGY: POLY(DIMETHYLSILOXANES)S BLENDS.

A. CECCARONI, J. LLORENS, E. RUDÉ, M. R. MARCOS

Chemical Engineering Dept., University of Barcelona. c/ Martí i Franquès 1, 08028 Barcelona, Spain.

1. INTRODUCTION

Several works have attempted to correlate readily measurable viscoelastic properties with polymer molecular characteristics. Among the various techniques to obtain the Molecular Weight Distribution (MWD), W(M), we can classify two main groups: the group that use direct experimental data: Tuminello (1986), Lavallée and Berker (1997), and the group that use integral transforms of experimental data: Llorens (1993), Mead (1996).

2. THEORY

In order to compare different techniques we have chosen a direct experimental technique and an integral transform technique. The used direct experimental technique assumes that the cumulative MWD is closer to $[G'(\omega)/G_N^o]^{1/2}$, G_N^o being the plateau modulus. Consequently, the CMW is :

$$F(M) = 1 - W_u = 1 - \sqrt{\frac{G'(\omega_i)}{G_N^o}} \qquad (1)$$

On the other hand, the used integral transform technique assumes that the characteristic relaxation time of polymer chain, λ, of molecular weight, M, in a polydisperse system is controlled by the surroundings with $\lambda \sim M^2$ and $M \cdot W(M) \sim H(\lambda)$, $H(\lambda)$ being the relaxation spectrum. In order to pass from the experimental frequency-domain to the relaxation spectrum we need the inversion of an integral equation:

$$G'(\omega) + \eta'(\omega) = \int_{-\infty}^{+\infty} H(\lambda) \frac{\omega^2\lambda^2 + \lambda}{1+\omega^2\lambda^2} d\ln\lambda \qquad (2)$$

It is well known that inversion of such integral equation results in an extremely unstable problem (ill-posed problem). In order to solve the integral equation we have used the maximum entropy method with relaxation times equally spaced in log time and constrained with the integral moments given by the zero shear viscosity, $\eta_o = \int_0^{\infty} H(\lambda)\cdot d\lambda$, and the plateau modulus, $G_N^o = \int_{-\infty}^{+\infty} H(\lambda)\cdot d\ln\lambda$.

Both methods have been used to get the shapes of the MWD (unimodal and bimodal) and absolute molecular weight have not been calculated. The definite MWD have been obtained by fitting a particular mean molecular weight obtained by other means ($M_w = (\eta_o/k)^{(1/3.4)}$; $M_v = ([\eta]/K)^{(1/\alpha)}$, $[\eta]$ being the intrinsic viscosity; M_w from light scattering) and the normalization constraint on the weight $\int_0^{\infty} W(M)\cdot dM = 1$.

3. MATERIALS AND METHODS

In this work we have analyzed the prediction of MWD of polydimethylsiloxane blends. Two polydimethylsiloxanes, H and L, which characteristics are shown in Table 1 have been mixed in three different ratios: $(H/L)_i$ = 0.75/0.25, 0.45/0.55 and 0.25/0.75 . The samples H and L have a log-normal distribution.

Table 1. characteristic mean molecular weights of the polydimethylsiloxane blends

Sample	Mn	Mw	Mn	Mw/Mn
H	364,900	631,300	550,000	1.7
0.75 / 0.25	166,600	496,900		3.0
0.45 / 0.55	102,100	341,000		3.3
0.25 / 0.75	79,800	228,100		2.9
L	63,300	93,700	85,200	1.5

4. RESULTS

The dynamic elastic modulus, G', and the dynamic viscosity, η', of the samples were measured on a Haake CS rheometer. Table 2 shows temperatures and frequency sweeps that were made for the different samples. Measurements were made at different strains to ensure we were in the linear viscoelastic region.

Table 2. Tested temperatures and frequency sweeps.

Sample	Temperatures (°C)	Frequencies (radians / s)
H	48.5 -1.0 -36.6	0.0198 - 92.36
$(H/L)_1$	50.3 0.5 -40.0	0.0198 - 92.36
$(H/L)_2$	49.7 0.4 -35.7	0.0198 - 92.36
$(H/L)_3$	50.0 0.2 -38.0	0.0198 - 92.36
L	49.2 -0.1 -49.6	0.0198 - 383.3

The time-temperature superposition principle was employed to reduce G' and η' measured at different temperatures to master curves at 0°C. Master curves for all the samples are displayed in Fig. 1 and 2.

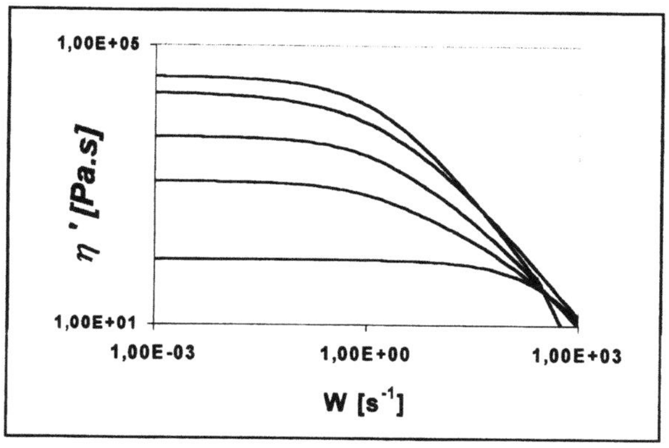

Fig. 1. Dynamic viscosity, η', of the samples.

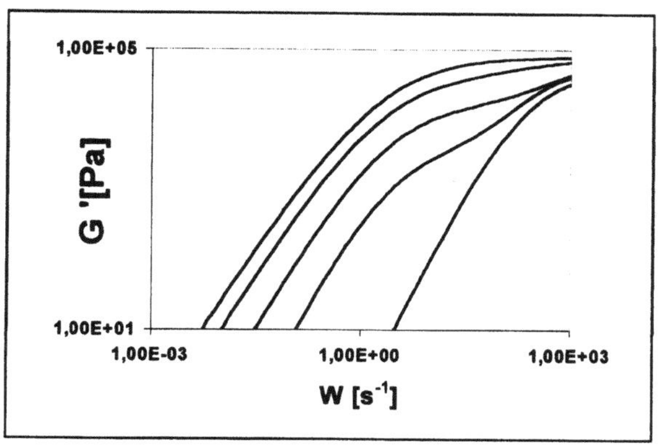

Fig. 2. Dynamic elastic modulus, G', of the samples.

Figures 3, 4 and 5 show the MWD calculated from the rheology with the technique that uses the relaxation spectrum and the exact MWD of the blends.

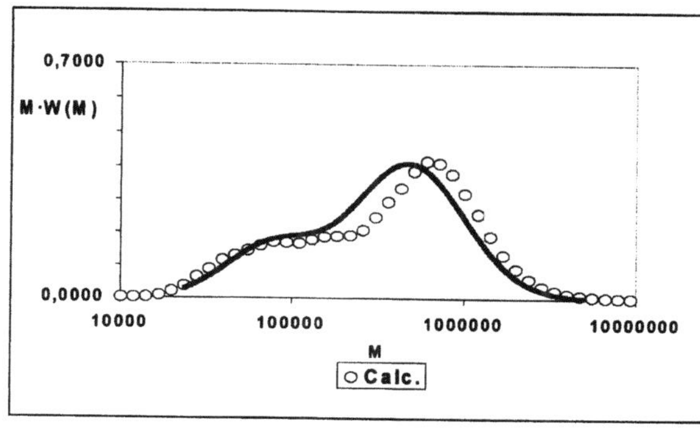

Fig. 3. Calculated and real MWD for $(H/L)_1$

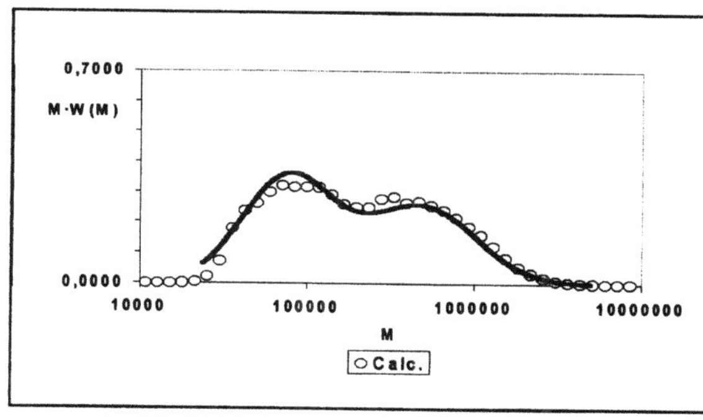

Fig. 4. Calculated and real MWD for $(H/L)_2$

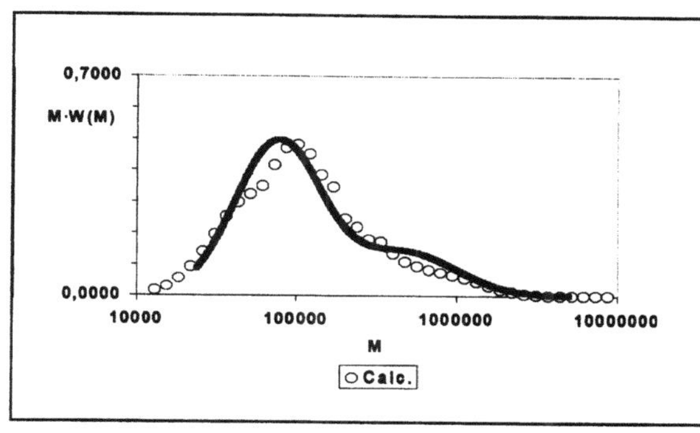

Fig. 5. Calculated and real MWD for $(H/L)_3$

5. CONCLUSIONS

The technique that uses direct experimental data gives approximate MWD for the blends. The technique that uses the relaxation spectrum gives more accurate MWD.

6. REFERENCES

Tuminello WH (1986) Polym. Eng. Sci. 26:1339-1347

Lavallée C, Berker A (1997) J. Rheol. 41:851-871

Llorens J (1993) J. Non Cryst. Solids, 162:188-196

Mead DW (1996) J. Rheol. 40:633-661

VISCOUS MEDIUM CARRYING AWAY BY THE NETWORK

YU. YA. GOTLIB, A. A. GURTOVENKO

Institute of Macromolecular Compounds, Russian Academy of Sciences, Bolshoi Prospect 31, V.O., St. Petersburg, 199004, Russia; Tel. 7-(812)-218-56-01; FAX 7-(812)-218-68-69; E-mail: yygotlib@imc.macro.ru

The previous theories of the relaxation spectra of polymer melts and bulk polymer networks (PN) mostly considered the polymer melts and PN, which move relative to the immobile effective viscous medium (EVM) (or, relative to the EVM with the velocity gradient perturbed by external forces). In the consideration of polymer melts it was suggested that the viscosity of EVM is near to the viscosity of monomer (or, segment) liquid, and is negligible with the calculated macroscopic viscosity of melt.

The perturbation of VM by the chain motions (i.e., the hydrodynamic interactions) were proposed to be negligible. The absence of the correlations between the stochastic motions of different macromolecules leads to the screening of the hydrodynamic interactions between the segments of a given macromolecule and between those of the different chains (see, for example, Doi and Edwards (1986) and Yamakawa (1971)). However, it was mentioned by Doi and Edwards (1986) that residual long-range hydrodynamic interactions exist in the concentrated polymer systems (for example, in polymer melts), and they are governed by the macroscopic viscosity rather than viscosity of monomer liquid. Since the hydrodynamic interactions are proportional to $(1/r\eta)$, the influence of these interactions on the rather high-frequency intrachain motions is small.

In the consideration of polymer networks it was shown by Gotlib and Salikhov (1963), Kloczkowski et al (1990) and the authors (1998) that the intrachain and cooperative interchain relaxation spectra of the network are approximately independent. Therefore, the specific cooperative network relaxation can be described by simplified coarse-grained network model. It may be shown that the long-range perturbation of EVM (or, "residual" hydrodynamic perturbation) can not be neglected for the correlated motions of different chains of the network. The important problem is related to the origin of the viscosity of EVM acting on these network motions. For bulk polymer networks the effective viscoelastic units are the elastically coupled chains between junctions. The macroscopic viscosity η of the network mimics all degrees of freedom which are faster than the network motions, and is near to the macroscopic viscosity of the melt consisting of the network chains between junctions.

In this paper the low-frequency relaxation spectra (RS) of mesh-like polymer networks with mobile EVM (i.e., with long-range hydrodynamic interactions) are considered. The discrete dynamic coarse-grained model of regular polymer network embedded in EVM and continuous viscoelastic models (for network and EVM) are treated. The theory takes into account the incompressibility of the EVM of in bulk polymer. The polymer network must also be incompressible in the average. However, this condition for the rather small-scale motions (normal modes) or for the finite pieces of the polymer network may be not fulfilled.

The theory shows that two types of characteristic network relaxation spectrum exist. (1). The first type of relaxation spectrum arises in the case when remains immobile. The perturbation of the network $\vec{R}(x,y,z)$ has the form $\vec{R} = -\mathrm{grad}\,U$ (i.e., $\mathrm{rot}\vec{R} = 0$). The internal pressure in EVM compensates the force acting on EVM from the polymer network. Then, the relaxation spectrum has the usual form of that of the network embedded in the immobile viscous medium (Gotlib and Gurtovenko (1997)). (2). The second type of relaxation spectrum corresponds to the motions in which the network also behaves as the incompressible medium ($\mathrm{div}\vec{R}(x,y,z)=0$). In this case the relaxation times $\tau(\vec{k})$ of network relaxation (for example, for continuous network model) behaves as

$$\tau^{-1}(\vec{k}) = [k^2 K/\varsigma] + [K/\eta] \qquad (1)$$

where $\vec{k}$ is the three-dimensional wave vector, K - equilibrium modulus of the network, ς - effective friction constant of the unit volume of network system, η - viscosity of EVM. The scale of motion $r(k)$ for given normal mode is given by $r(k) \cong (\pi/k)$.

The first term in expression (1) is the inverse relaxation time $\tau(k)$ of polymer network embedded in the immobile EVM ($\eta \to \infty$); the second that is the inverse relaxation time of the network moving with EVM in phase. In this case the friction relative to EVM does not occur, and the viscosity of EVM is only manifested.

The most important feature of RS (1) is the fact that this spectrum is finite and cut-off for long-range normal modes (if $k \to \infty$ and $r(k) \cong (\pi/k) \to \infty$). The quantity $\tau_{\min}^{-1} \cong (k_{\max} K/\varsigma) + (K/\eta)$ depends on the minimum scale of network interchain motions $r_{\min} \cong (\pi/k_{\max})$. The value of $r_{\min}$ is near to the dimension of mesh-size of the polymer network. The quantity $\tau_{\max}^{-1} \cong (K/\eta)$ does not depend on the macroscopic size of bulk network sample.

The characteristic length $r^* \cong \sqrt{\eta/\varsigma}$ exists so that the value of $r > r^*$ for motions with relaxation times do not practically depend on the scale of normal mode in contradiction to the motion in immobile viscous medium. It means that the polymer networks for the long-range motions behaves as a number of identical Kelvin-Voigt elements connected in series. The most probable assumption is the fact that the viscosity η has the order of magnitude equal to the viscosity of melt of network chains. The quantity r^* appears to be the same order of magnitude of the mesh-size dimension (or, of the mean square dimensions of the chain in the case of random Gaussian network). In general, r^* depends on the prehistory of the network formation, network structure and heterogeneity. The second type of relaxation spectrum for networks with only local heterogeneity is narrow. For heterogeneous network the broadening may exist due to the spread of quantity $\tau_{\min}$. The different types of RS should appear in mechanical and diffusive phenomena, dielectric relaxation, NMR and other dynamic properties of polymer networks. Some special problems arise in tree-like networks where the relaxation spectrum are narrow and finite even without hydrodynamic interactions.

This work was carried out with the financial support of Russian Foundation of Basic Research (grants 96-03-33833 and 96-15-97401).

REFERENCES

Doi M, Edwards SF (1986) The Theory of Polymer Dynamics. Oxford University Press, Oxford

Yamakawa H (1971) Modern Theory of Polymer Solution. Harper and Row, New York

Gotlib YY, Salikhov KM (1963) Akusticheskii zh. (Acoustic J) 9: 301-308

Kloczcowski A, Mark JE, Frisch HL (1990) Macromolecules 23: 3481-3490

Gurtovenko AA, Gotlib YY (1998) Macromolecules (in press)

Gotlib YY, Gurtovenko AA (1997) Macromol. Theory Simul 6: 523-551

INFLUENCE OF MECHANICAL DEFORMATION ON ELECTRO-OPTICAL BIREFRINGENCE IN POLYMER NETWORKS.

Yu. Ya. GOTLIB[1], S.V.LYULIN[1], V.P.TOSHCHEVIKOV[2]

[1]Institute of Macromolecular Compounds of the Russian Academy of Sciences, Bolshoi Prospect 31, V.O., St. Petersburg, 199004, Russia;
[2]Faculty of Physics, St.-Petersburg State University, Ulianovskaya St. 1, Petrodvoretz, St. Petersburg, 198904, Russia

1. INTRODUCTION

A theory of the Kerr effect (electro-optical birefringence) for polymer networks has been developed. The optical birefringence of polymers in the external electric field E is connected with orientation of optically anisotropic chain segments. The orientation of the segment due to induced dipole moment has been considered. The polarizability tensor of segment was assumed to be axial symmetrical with respect to axis of segment. The comparison of electro-optical properties of isotropic polymer networks with those of mechanically deformed networks has been made. The uniaxial mechanical deformed samples (both stretched and compressed) have been considered. Our theory corresponds to an experiment when the boundaries of the deformed sample are fixed so that the volume and the form of the sample remain constant after applying of the electric field (electrostriction effect is not taken into account). The particular case, when external electric field is directed parallel to the direction of mechanical deformation, has been studied.

The field dependence of the birefringence degree Δn is determined by the field dependence of the quadrupole order parameter of chain segments S (see, for example, Volkenshtein (1963) and Tsvetkov (1989)):

$$S=\frac{1}{2}\cdot\left(3<\cos^2\vartheta>-1\right),$$

where $<\cos^2\vartheta>$ is the average (all over the network) square of the cosine of angle between the segment and the electric field. The field dependence $S=S(E^2)$ has been calculated at the strong fields when the birefringence degree Δn (or the quadrupole order parameter of chain segments S) becomes nonlinear in the square of the field strength. The Kerr constant for given network is proportional to the initial slope $[\partial S/\partial(E^2)]_{E^2=0}$.

2. NETWORK MODEL

"Three-chain" network model in detail described by Monnerie (1979) has been used. The elementary cell of this model contains three identical freely jointed chains, end-to-end vectors of which form the rectangular coordinate system.

2.1. Undeformed networks

For the undeformed network the magnitudes of end-to-end vectors in the cell are equal to h_0. The internal structure of the network is characterized by dimensionless parameter h_0/L, where L is the contour length of the chains. The value of this structural parameter depends on the density of the cross-linking, degree of the chain stretching in the network and others structural network parameters.

2.2. Mechanically deformed networks

For the deformed network the condition of constant volume at the process of deformation for elastomers is taken into account. The axis of uniaxial deformation is assumed to coincide with the end-to-end vector of one of the chains in the cell (longitudinal chain). The magnitude of the end-to-end vector of this chain is equal to $h_{\parallel}$. The degree of mechanical deformation is described by the elongation ratio $\lambda=h_{\parallel}/h_0$. The magnitudes of end-to-end vectors of two transversal chains are identical and equal to $h_{\perp}$. The value of $h_{\perp}$ is determined by the condition of constant volume $(h_{\parallel}\cdot h_{\perp}^2=h_0^3)$. The quadrupole order parameter has been obtained as a function of three arguments: $S=S(E^2,h_0/L,\lambda)$.

3. RESULTS

3.1. Birefringence at a Strong Fields

For both isotropic and deformed networks the character of the dependence $S = S(E^2)$ at the strong fields essentially depends on the sign of the polarizability anisotropy of a segment $(a_1^{(0)} - a_2^{(0)})$, where $a_1^{(0)}$ and $a_2^{(0)}$ are the main values of the static polarizability tensor of a segment. The maximum polarizability of segment may be either directed parallel to the segment $(a_1^{(0)} - a_2^{(0)} > 0)$ or lain in the plane perpendicular to the segment $(a_1^{(0)} - a_2^{(0)} < 0)$. In the first case the values of S (at the strong electric fields) are larger than those in the second one at the identical values of h_0 / L and λ.

3.2. Influence of Density of the Cross-Linking

For undeformed networks both initial slope of the curve $S = S(E^2)$ and the limiting value of S decrease if the parameter h_0 / L increases, i.e. when the density of the cross-linking increases.

3.3. Deformed Network in the Weak Field

For deformed networks the initial value of S (at $E = 0$) is not equal to zero. For Gaussian networks with low density of the cross-linking ($h_0 / L << 1$) this value is proportional to a factor $(\lambda^2 - 1/\lambda)$. It corresponds to well-known results of the theory of the stress-optical effect for more general network model considered by Kuhn (1942). The initial slope of the curve $S = S(E^2)$ for deformed networks with given h_0 / L is the complicated function on the λ (see Fig.1):

3.3.1. Stretching. At the stretching $(\lambda > 1)$ for Gaussian networks with low density of the cross-linking ($h_0 / L << 1$) there is a minimum in the dependence of the slope on the λ, for the networks with high density of the cross-linking ($h_0 / L \cong 1$) the initial slope monotonously increases with λ;

3.3.2. Compression. At the compression $(\lambda < 1)$ for networks with $h_0 / L << 1$ there is a maximum in the dependence of the slope on the λ, for the networks with $h_0 / L \cong 1$ there is a minimum in this dependence.

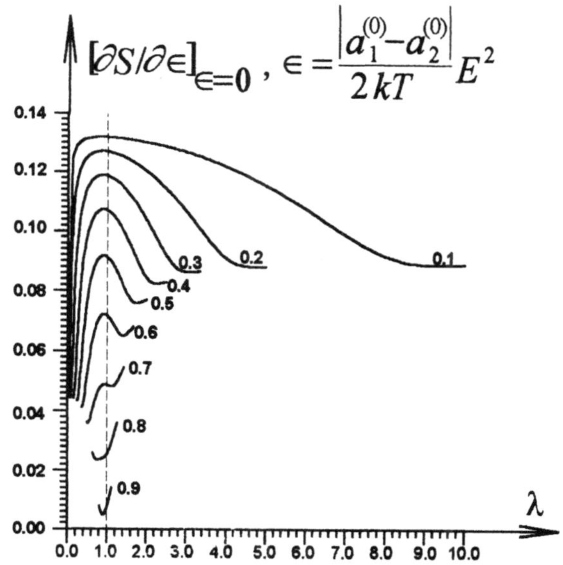

Fig.1. Dependence of the initial slope of the curve $S = S(E^2)$ on the elongation ratio for networks with different values of the density of the cross-linking ($h_0 / L = 0.1 \,..\, 0.9$). Here k is the Boltzmann constant, T is the temperature.

Those complicated dependencies may be explained by the different contributions to S from longitudinal and transversal chains at different values of elongation ratio.

3.4. Deformed Network in the Strong Field

For deformed networks the limiting value of S increases at the stretching and decreases at the compression.

This work is carried out with the financial support of Russian Foundation of Basic Research (grants 96-03-33833 and 96-15-97401).

4. REFERENCES

Kuhn W, Gruen F (1942) Koll. Zeits. 101: 248

Monnerie L. (1979) Macromolecules 12: 316-320

Tsvetkov VN (1989) Rigid-chain polymers. Hydrodynamic and optical properties in solution. N.Y.-London

Volkenshtein MV (1963) Configurational Statistics of Polymer Chains. Interscience Publishers, New York

Simulation of Polymer Solutions in Steady Elongational Flow

J.G. Hernandez Cifre and J. Garcia de la Torre

Departamento de Quimica Fisica, Facultad de Quimica, Universidad de Murcia, 30071 Murcia, Spain (Internet: jgt@fcu.um.es, http://leonardo.fcu.um.es/macromol)

1 Introduction

In this work, we employ Brownian dynamics simulation to calculate steady-state properties of polymer chains in dilute solution when they are subjected to an elongational flow. The well-known coil-stretch transition, (De Gennes, 1974) is studied in terms of the variation of polymer properties on elongational rate, $\dot{\epsilon}$. From this variation we evaluate the critical elongational rate $\dot{\epsilon}_c$, whose dependende on polymer features, such as hydrodynamic interacion and excluded volume is determined. Also, $\dot{\epsilon}_c$ is combined with other solution properties to form dimensionless, length-independent compound quantities.

2 Methods

Polymer molecules are modelled as bead-and-spring chains having $N-1$ bonds that are either Gaussian or of the FENE type. Good-solvent conditions are modelled introducing in the model excluded-volume intramolecular interactions given by either a Lennard-Jones or a soft potential.

The velocity field of the flow is given by $v_x = -\dot{\epsilon}x/2; v_y = -\dot{\epsilon}y/2; v_z = \dot{\epsilon}z$ Trajectories of polymer chains in the steady elongational flow are obtained using the predictor-corrector Brownian dynamics simulation algorithm of Iniesta and Garcia de la Torre (1990). See also Lopez Cascales and Garcia de la Torre (1991). In the simulations we fully include rigorous, fluctuating hydrodynamic interaction given by a Rotne - Prager - Yamakawa interaction tensor between beads in the chain.

3 Results

We have evaluated the dependence of polymer dimensions on $\dot{\epsilon}$, from simulations in various conditions: different spring types, different initial conformation, and with or without excluded volume. interactions. We found that the coil-stretch transition takes place at the very same $\dot{\epsilon}_c$ in all conditions. Particularly relevant is the fact that $\dot{\epsilon}_c$ is the same with and without EV effects. In the literature, there is a long-standing controversy about whether or not the scaling exponent in $\dot{\epsilon}_c \propto N^a \propto M^a$ should be $a = 3/2$ regardless of solvent quality (see, for instance Menasveta and Hoagland (1992) and Nguyen et al (1995)). Our simulation results support the view that a is the same for theta and good-solvent conditions.

From simulations with different number of beads, N=8-80, we evaluated the chain-length dependence of the critical elongational rate, for which we obtained:

$$\dot{\epsilon}_c^* = (14.1 \pm 1.1)N^{-(1.55\pm0.03)}$$

where $\dot{\epsilon}^* = \dot{\epsilon}(\zeta b^2/kT)$, ζ and b^2 being respectively the friction coefficient of the beads and the rms spring length in absence of flow. The exponent $a = -1.55 \pm 0.03$ is in good agreement with the scaling prediction $a = -3/2$.

From the earliest theoretical studies, it is accepted that $\dot{\epsilon}_c$ is of the order of the reciprocal of the longest relaxation time of the polymer chain τ_1, so that the dimensionless combitation

$$\dot{\epsilon}_c \tau_1 \simeq 1$$

is usually formulated. This compound quantitity does not depend on chain length, and it should be a constant for every theta polymer/solvent system. Few theoretical or experimental information is available for the precise value of $\dot{\epsilon}_c \tau_1$. A noteworthy attempt is the calculation of Magda et al (1988), in which the fluctuations in hydrodynamic interaction were somehow averaged out. We have combined the $\dot{\epsilon}_c$ values from the present simulation, with values of τ_1 determined in our previous Brownian dynamics simulation of electric birefringence decay (Navarro et al, 1995), obtaining $\dot{\epsilon}_c \tau_1 = 0.50$

We also consider other dimensionless combinations of $\dot{\epsilon}_c$ with other properties, one of which involves the zero-shear intrinsic viscosity $[\eta]_0$ (Lopez Cascales and Garcia de la Torre, 1991):

$$\nu_c = \frac{M \eta_s [\eta]_0}{N_A k T} \dot{\epsilon}_c$$

Employing previous simulation results for $[\eta]_0$, in which hydrodynamic interaction were included in a rigorous (fluctuating) manner, (Garcia de la Torre, 1984) we obtain $\nu_c = 1.01$.

The experimental determination of the compound quantities $\dot{\epsilon}_c \tau_1$ and ν is problematic due to many reasons: the non-infinite residence time in instrumental setups, the accumulation of errors in the various properties involved in the compound quantities (perhaps coming from different labs), and so on. Nonetheless, our predictions for these quantities are in qualitative agreement with values that we estimated from measurements reported in the literature (see Menasveta and Hoagland (1992) and Nguyen et al (1995)) and may be useful for the analysis of future experiments

4 References

De Gennes PG (1974) J Chem Phys 60:5030-5042.

Garcia de la Torre J, Lopez Martinez M, Tirado MM, Freire JJ (1984) Macromolecules 17:2715-2722.

Iniesta A, Garcia de la Torre J (1990) J Chem Phys 92:2015-2019

Lopez Cascales JJ and Garcia de la Torre J (1991) J Chem Phys 95:9384-9392

Magda J, Larson R, MacKay E (1988) J Chem Phys 89, 2504-2513 (1988)

Menasveta MJ and Hoagland DA (1992) Macromolecules 25: 7060-7062 (1992)

Navarro S, Lopez Martinez MC, Garcia de la Torre, J (1995) J Chem Phys 103:7631-7639.

Nguyen T, Yu G, Kausch HH (1995) Macromolecules 28:4851-4860 (1995)

This work has been supported by grant PB96-1106 from Direccion General de Ensenanza superior. JGHC aknowledges the award of a predoctoral fellowship from Comunidad Autonoma de la Region de Murcia

MOLECULAR MODELS

G. IANNIRUBERTO, G. MARRUCCI

Laboratori Giovanni Astarita, Dipartimento di Ingegneria Chimica
Università di Napoli Federico II, Piazzale Tecchio 80, 80125 Napoli, Italy

1. INTRODUCTION

To understand the flow behaviour of polymeric liquids, it is important to analyse the microscopic origin of the stress. The classical molecular theories relate the stress to the tensile force of entropic origin acting along the chains. The momentum transfer due to collisions between different molecules is not explicitly considered. Though certainly adequate for dilute solutions, that assumption becomes questionable in concentrated solutions and melts

According to the tube model (Doi and Edwards, 1986) entangled chains are enclosed in a sort of tube formed by the topological constraints. Intermolecular interactions can then be seen as a pressure exerted by the confined chain against the tube wall. The question then arises on how to account for these pressure effects in the stress tensor, especially when deformation will destroy the equilibrium axial symmetry of the tube, making the tube cross section "elliptical".

2. STRESS IN ELLIPTICAL TUBES

Let us consider a chain segment in its confining tube oriented along z and with an elliptical cross section having principal directions x and y. In order to compute the stress contributed by that chain segment, we need to account for the interactions of its beads with the surrounding "world". The interaction with the rest of the same chain can be computed in the classical way and gives rise to a traction along the tube axis z. The interactions with the surrounding confining chains, i.e. with the tube wall, are non-classical and require some considerations.

In analogy with the case of circular tubes (Doi and Edwards, 1978) we account for the anisotropic confinement through the elliptic potential V

$$V = \frac{kT}{6}\left(w_x^2 x^2 + w_y^2 y^2\right) \tag{1}$$

which gives rise to a linear force $\mathbf{F} = -\nabla V$ on each bead. In order to determine the single bead contribution to the intermolecular stress we have to calculate the ensemble average of xF_x (and yF_y), and multiply it for the number n of beads in the chain segment.

To this purpose, the Green function G for a chain in an elliptic potential is required: $G(\mathbf{R'},\mathbf{R};n)$ represents the statistical weight of a chain starting from $\mathbf{R'}$ and ending at $\mathbf{R}$ in n steps. By extending the result of Doi and Edwards (1978) to the case of tubes with elliptical cross section, the Green function G is given by (Ianniruberto and Marrucci, 1998):

$$G(\mathbf{R'}, \mathbf{R}; n) \propto n^{-1/2}\, exp\left[-\frac{w_x}{2b}x^2 - \frac{w_y}{2b}y^2 - \frac{3z^2}{2nb^2} - \frac{bn}{6}\left(w_x + w_y\right)\right] \tag{2}$$

where b is the link size, and $\mathbf{R'}$ has been set at the origin. What we really need is the probability of finding at $\mathbf{R}$ the generic k-th bead of the chain, the ends of which have been fixed at the ends of the tube, i.e. at the origin $\mathbf{0}$ and at $\mathbf{L}=(0,0,\ell)$, respectively. In view of the properties of the Green function such a probability is proportional to the product $G(\mathbf{0},\mathbf{R},k)G(\mathbf{R},\mathbf{L},n\text{-}k)$. A further use of the Green function consists is its maximisation with respect to n in order to determine the linear bead density. To a good approximation, this gives:

$$\frac{n}{\ell} = \frac{3}{b\sqrt{bw_x + bw_y}} \tag{3}$$

Once we know n and the appropriate distribution function, we can compute the stress contributed by the test segment. The stress $\mathbf{T}$ in the material is then obtained by averaging over the tube segments. The final result is (Ianniruberto and Marrucci, 1998)

$$\mathbf{T} \propto \frac{\left\langle -\frac{\ell a_y}{2a_x\sqrt{a_x^2+a_y^2}}\hat{\mathbf{x}}\hat{\mathbf{x}} - \frac{\ell a_x}{2a_y\sqrt{a_x^2+a_y^2}}\hat{\mathbf{y}}\hat{\mathbf{y}} + \frac{\ell\sqrt{a_x^2+a_y^2}}{a_x a_y}\hat{\mathbf{z}}\hat{\mathbf{z}} \right\rangle}{\left\langle \frac{\ell a_x a_y}{b^2\sqrt{a_x^2+a_y^2}} \right\rangle} \quad (4)$$

where $a_x^2 = \langle x^2 \rangle = b/2w_x, a_y^2 = \langle y^2 \rangle = b/2w_x$ are the dimensions of the elliptical cross section of the tube. It can be shown that Eq.(4) reduces to the Doi-Edwards expression when $a_x=a_y=a$, where a is a constant. Equation (4) relates the stress to the geometry of the tubes, specified by the principal directions $\hat{\mathbf{x}},\hat{\mathbf{y}},\hat{\mathbf{z}}$ and the principal axes a_x, a_y, ℓ.

3. STRESS-OPTICAL LAW

To check that the stress-optical law remains valid, Eq. (4) has to be compared with the expression for the polarizability tensor **A**. The polarizability arises from the anisotropy of chain links, i.e. it is proportional to $\langle(\mathbf{R}_{k+1}-\mathbf{R}_k)(\mathbf{R}_{k+1}-\mathbf{R}_k)\rangle$. To compute such an average we need the link (rather than the bead) distribution function, which is again proportional to a product of Green functions

$$P(\mathbf{R}_k,\mathbf{R}_{k+1}) \propto G(\mathbf{0},\mathbf{R}_k;k)G(\mathbf{R}_k,\mathbf{R}_{k+1};1)G(\mathbf{R}_{k+1},\mathbf{L};n-k-1) \quad (5)$$

The result of the averaging procedure is:

$$\mathbf{A} = C_1\mathbf{T} + C_2\mathbf{I} \quad (6)$$

where C_1 and C_2 are two constants (Ianniruberto and Marrucci, 1998) and **I** is the unit tensor. Equation (6) is just the statement of the stress-optical law.

4. CONCLUSIONS

This paper emphasises the concept that in entangled polymers the stress results from two different contributions. One of them is due to the traction along the chain (the positive $\hat{\mathbf{z}}\hat{\mathbf{z}}$ term in Eq. 4), while the other arises from the collisions with the confining chains (the negative $\hat{\mathbf{x}}\hat{\mathbf{x}}$ and $\hat{\mathbf{y}}\hat{\mathbf{y}}$ terms in Eq. 4). These pressure effects are relevant in the general case of elliptic confinement, becoming equivalent to a traction along the chain in the limiting case of tubes with circular cross sections.

5. REFERENCES

Doi M, Edwards S (1978) J Chem Soc Farad Trans II 74:1802-1817
Doi M, Edwards S (1986) The Theory of Polymer Dynamics, Clarendon Press, Oxford
Ianniruberto G, Marrucci G (1998) J. Non-Newt. Fluid Mech in press

MODELLING LOW DENSITY POLYETHYLENE RHEOLOGY USING THE POM-POM MODEL

Nathanael Inkson, Tom McLeish and Oliver Harlen

IRC in Polymer Science and Technology , Department of Physics and Astronomy, The University of Leeds Leeds, LS2 9JT, England

A recent constitutive equation derived from molecular considerations (the "pom-pom" equation) captures the qualitative rheological behaviour of LDPE in shear and extension (uniaxial and planar) for the first time. The pom-pom topology of branched polymer has theoretically similar behaviour to LDPE due to its multiple branch points. The idealised pom-pom molecule has a single polymer chain backbone ending in a branch point at each end with multiple arms attached at the branch points. A constitutive equation for the pom-pom class of polymer which involves a set of differential and integral molecular constitutive equations has been recently developed [1]. These are derived from the physics of the polymer molecules based on Doi and Edwards tube model for entangled polymer melts. This approach is coupled with previous work on star arm dynamics[2], of which results have been verified by experiment.

The melt rheology of low density polyethylene (LDPE) is made distinctive by its extreme extension hardening in elongational flow geometries in contrast to its equally strong thinning behaviour in shear flow. Traditional constitutive equations (e.g.K-BKZ) have been unable to describe simultaneously its rheological properties, particularly in planar extensional flow. Polymer melts with the specific pom-pom molecular topology have theoretically similar behaviour to LDPE due to its multiple branch points. We use a hypothetical melt of pom-poms with different numbers of arms to model the behaviour of the long chain branched structure of LDPE. The linear relaxation spectra for various LDPE samples is mapped to the distribution of backbone relaxation times of the pom-pom modes. The longer times are assigned to the modes with the most arms which represent the inner parts of the LDPE molecule. Each linear mode is then "decorated" with the non-linear properties of the pom-pom branched architecture. This involves ascribing a different (shorter) timescale to the stretch relaxation of the active segment in each case, together with a maximum possible value for the stretch. Decoupling of stretch and orientation at the molecular level corresponds to a separation of the dynamics of the scalar and tensorial part of the stress at the macroscopic level. It is this separation that enables this new constitutive class to capture the rheological behaviour of long chain branched polymer melts.

Data from start-up flow in uniaxial extension (from an elongational rheometer) is used to fix the non-linear parameters of each mode. Predictions are then available with no further free parameters in shear and planar extension (see fig. 2). These agree well with data from the literature and from our own measurements on commercial LDPE. The method suggests a concise and functional way of characterising long chain branching in polymer melts.

REFERENCES

[1] McLeish TCB, Larson RC Molecular Constitutive Equations for a Class of Branched Polymers: The Pom-pom Polymer J. Rheology **42(1)** p82, (1998)

[2] Milner ST, McLeish TCB Parameter-Free Theory for Stress Relaxation in Star Polymer Melts. Macromolecules **30**:2159-2166(1997) also
de Gennes PG Reptation of Stars J. de Physique **36**:1199-1203 (1975)

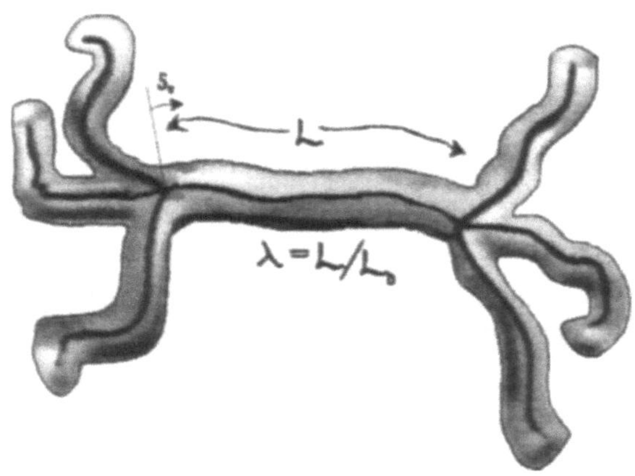

Fig 1) A pom-pom molecule with three arms at each branch point. The primitive path of the polymer chain is surrounded by its tube from the Doi and Edwards tube model.

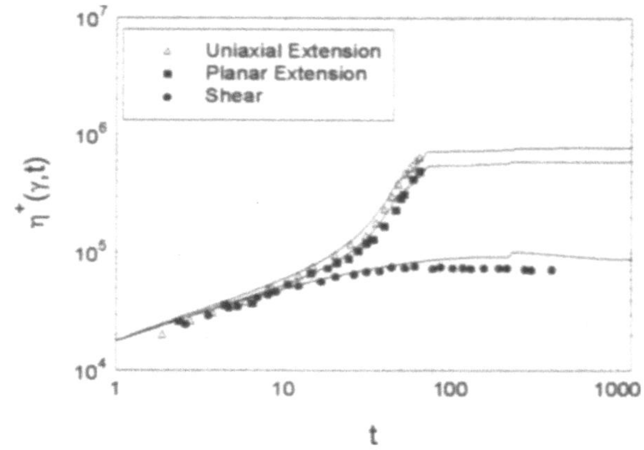

Fig 2) Diagram of multi-modal pom-pom start-up viscosity curves (solid lines) against time at elongation and shear rates of 0.05 s^{-1}. Plotted with the melt viscosity data of a low-density polyethylene IUPAC X (Laun HM, Schuch HJ. Rheol. **33(1)** :119-175 (1989)). The uniaxial extension curve is fitted with 2 parameters per mode, then the planar extension and shear curves can be produced without modifications to the parameters.

DYNAMICS OF ACTIN FILAMENTS

M. KROEGER

Department of Materials, Institute of Polymers, ETH Zurich, Switzerland

INTRODUCTION

Actin filaments, microtubules and intermediate filaments which build up the cytoskeletal network responsible for the mechanical properties of cells are at the relevant length scales (a few microns at most) all semiflexible polymers. In contrast to flexible polymers the persistence length of these polymers is of the same order of magnitude as their total contour length

The statistical mechanics of such macromolecules can not be understood from the conformational entropy alone but depends crucially on the bending stiffness of the filament.

RESEARCH INTEREST

In contrast to the dynamic properties of flexible polymers the dynamics of semiflexible polymers is not yet very well understood.. This is mainly a consequence of the difficulties caused by the constraint of a (virtually) unextensible contour. Models that neglect or relax the constraint like any version of the so called Harris–Hearst–Beals model allow for artificial stretching modes.

METHOD

We introduce a *Semiflexible Finitely Extensible Nonlinear Elastic S–FENE–model* (Fig. 1) which is analyzed by applying Brownian and molecular dynamics computer simulation techniques. Our approach circumvents many previous difficulties. We are able to obtain the relevant equi;librium and nonequilibrium dynamical and structural properties of semiflexible polymers in the dilute, semidilute and concentrated regimes.

The most relevant parameters of the simulation are the *concentration* of actin (mondisperse systems), the *length of the filaments* L and the *bending coefficient* κ. In order to obtain rheological properties, e.g., the shear viscosity, we applied a homogeneous flow field and solved the equations of motion by a NonEquilibrium Molecular Dynamics (NEMD) computer simulation algorithm, as described, e.g., by Kröger (1995).

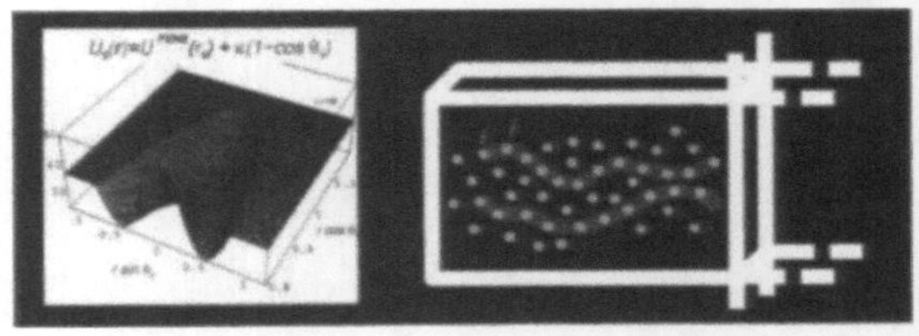

Fig. 1. Multibead S–FENE chain model potential used in the computer simulation between neighboring beads i and j=i+1 along the chains contour.For clarity,the potential has been cut off. Nonbonded interactions are modeled by a Lennard Jones potential.

RESULTS

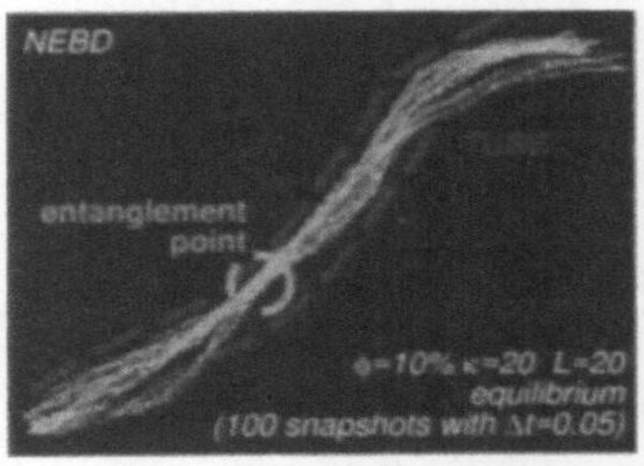

Visualization of a tube around a filament in a solution of 10 %. The picture was obtained by superimposition of 64 traces of the contour taken at time intervals of 0.05 (reduced LJ units) .

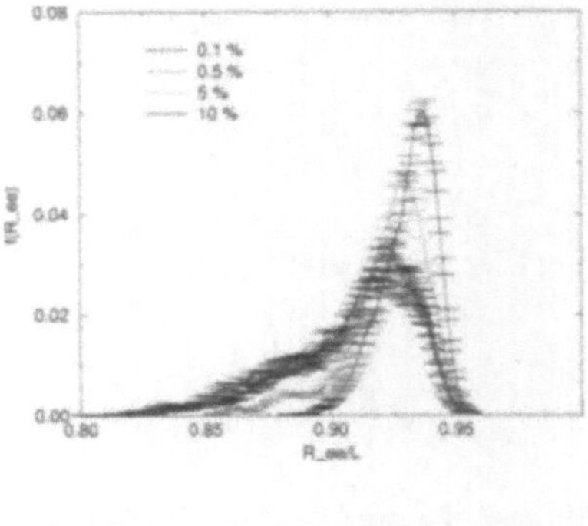

Effect of concentration on the *end–to–end distribution* function of actin filaments (κ=200, L=100).

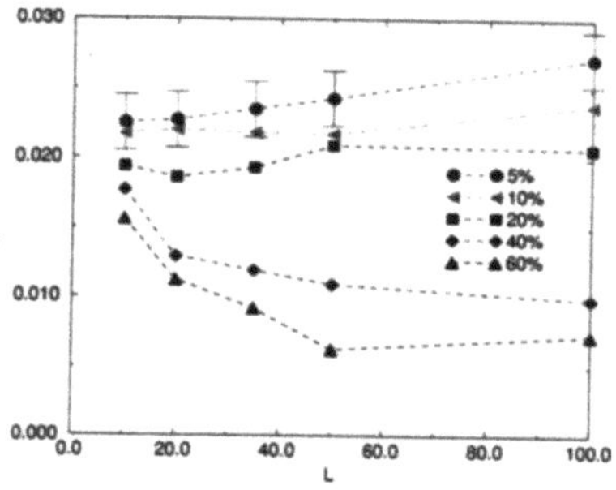

Diffusion coefficient parallel to the tube vs chain length L for various concentrations.

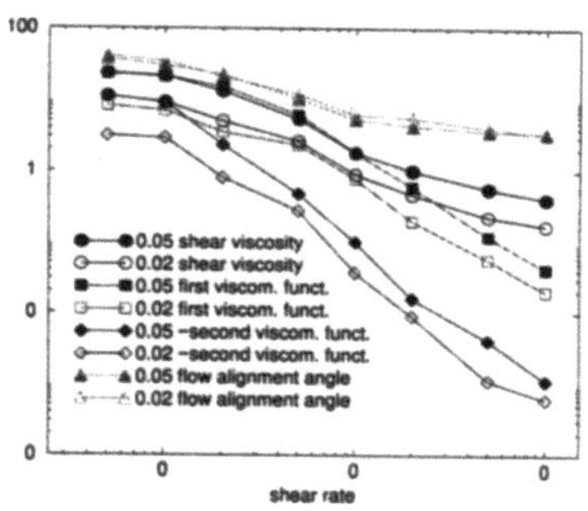

Viscosity coefficients and *flow alignment angle* vs shear rate for both, 2% and 5% solutions of actin (κ=100,L=100).

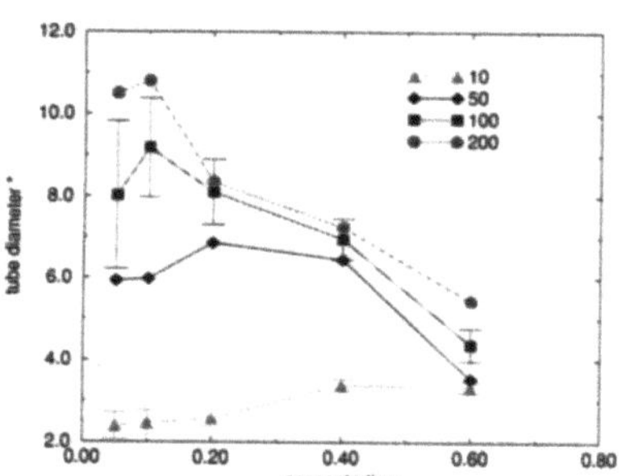

Tube diameter as used, e.g., by Götter et al. (1996), vs concentration for various bending coefficients κ=10,50,100, 200 at fixed L=100.

CONCLUSIONS

The method is capable to study the influence of concentration and bending rigidity on the flexibility and all other configurational properties of actin filaments, single or confined in the bulk. It allows to measure viscoelastic properties through the availability of the stress tensor, as obtained directly from the bead trajectories. The method hence potentially serves as an instrument to measure the relationship between structure and rheology, and allows to test constitutive equations.

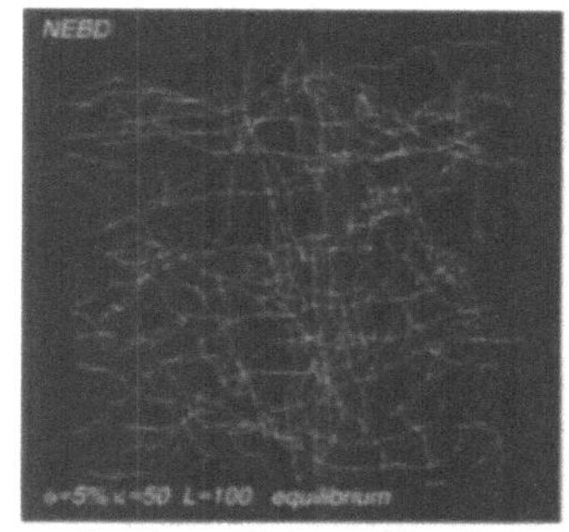

Sample equilibrium configuration

Snapshots of model actin filaments as calculated by a NEBD computer simulation (NonEquilibrium Brownian Dynamics) computer simulation method. In the figures, the concentration ϕ, bending coefficient κ, length L and shear rate are given.

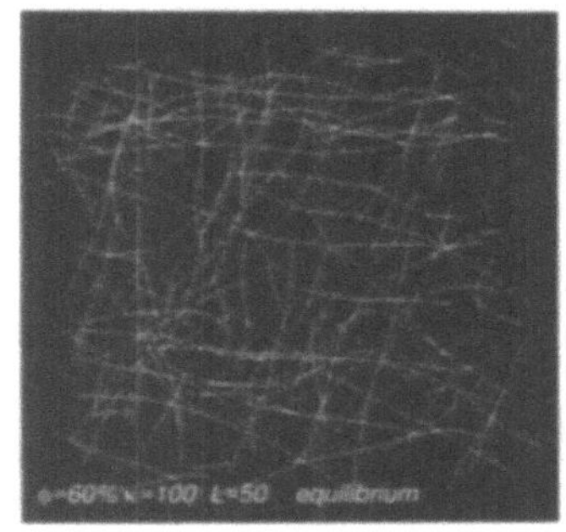

Sample equilibrium configuration

Flow induced alignment

REFERENCES

[1] R. Götter, K. Kroy, E. Frey, M. Bärmann, und E. Sackmann,
Macromolecules 29 (1996) 30

[2] M. Kröger, Rheologie und Struktur von Polymerschmelzen (W&T Publ., Berlin, 1995)

Email: mkroeger@ifp.mat.ethz.ch

MOLECULAR DYNAMICS SIMULATION OF UNSATURATED POLYESTERS MODIFIED BY POLY(ETHYLENE GLYCOL) END GROUPS

I. Neelov[1,2], S. Niemelä[1], F. Sundholm[1], M. Skrifvars[3]

[1] University of Helsinki, Chemistry Department, P.O.Box 55, FIN-00014HY, Helsinki, Finland

[2] Institute of Macromolecular Compounds, 199004, Bolsoi pr.31, St.Petersburg, Russia

[3] Neste Chemicals Research and Technology, P.O.Box 310, FIN-06101 Porvoo, Finland

1. INTRODUCTION

The modification of unsaturated polyesters by poly(ethylene glycol) end groups was described in a recent work by us /1/. The modification was done in order to increase the toughness of the cured unsaturated polyester (UPE), and to lower the shrinkage during curing. This is normally done by physical methods, in which case elastomer particles (low profile additives) are added to the matrix. Chemical modification can also be used, examples are incorporation of long chain diols or unsaturated acids in the polyester chain. In our work the modification was done by an end-group reaction of a carboxyl-terminated unsaturated polyester with poly(ethylene glycol) mono methyl ethers of various molecular weights (range 350 to 2000 g/mol). Analysis revealed that diblock copolymers of the AB type was formed, and that the diblock copolymers had predominantly one poly(ethylene glycol) end group per polyester segment.

2. MOLECULAR DYNAMICS SIMULATION

To understand the structure of the AB diblock copolymer consisting of UPE (B) and PEG (A) blocks on a molecular level we have performed the molecular dynamics simulation of the diblock copolymer. We use the completely extended initial conformation of diblock AB in which all torsion angles are in trans state. This approach was used by Zhan and Mattice /2/ to study the collapse of poly(1,4-trans-butadiene) (PB) to a globule and a thin film. It was found that an isolated PB molecule collapses to a globule and in the case of two-dimensional periodical boundary conditions to a coil, which forms together with its images the polymer film. Similar initial conditions were used by Choi et al. to study miscibility and structure of polyethylene (PE) and polypropylene (PP) mixtures /3/. In this case single extended PE and PP molecules were placed close to each other in such a way that positions of their centers of mass were close and PE and PP chains were oriented perpendicular to each other. It was shown that the final collapsed structure is separated into two (PE and PP) domains.

3. MODEL

Initially we studied the collapse of one molecule of an AB diblock copolymer to a compact structure. The UPE block (B) of the copolymer had a more simple structure than in the synthesised copolymer which was reported in our previous paper /1/. Both blocks had the same molecular weight - 2000 g/mol. We calculated further the collapse and the final structures of UPE and PEG homopolymers (AA and BB). These homopolymers consisted of two blocks of the same molecular weight as in the AB diblock copolymer. The comparison of homo- and block copolymer structures allows to understand how the introduction of a PEG endgroup influences the structure of UPE block. The initial conformations were prepared and molecular dynamics simulations were fulfilled using a molecular modelling software package Discover from MSI /4/. The force field used was CVFF. Molecular dynamics simulations were performed in vacuum at the temperature 298K. For each system (AA, BB and AB) three different initial conditions were used and all results were averaged through these 3 trajectories (1000 ps each).

4. RESULTS AND CONCLUSION

It was shown that there exist two stages of relaxation of initially extended structure to compact final state. During the first stage the fast relaxation of both homopolymers and diblock copolymer occurs with almost the same rate. During the second stage slower relaxation occurred especially for homopolymer AA. For the BB homopolymer the duration and the contribution of the second stage was small. For diblock copolymer AB we have an intermediate behaviour which is between the behaviour in the AA and the BB cases. Such a two stage relaxation of the extended conformation is similar to the results obtained by Zhan and Mattice /2/ for polybutadiene. Final structures of the diblock copolymer AB after 1000ps of simulation were plotted. It can be seen that the UPE block (B) is in compact state (core) inside the copolymer. At the same time the PEG block (A) is surrounding the UPE block as shell. We have thus a separation of the two blocks. Contrary to the case of PE and PP entangled molecules /3/ there is no structure segregation in two parts which are on opposite sides of total structure. In the case of UPE and PEG diblock copolymers the centers of mass of part A and part B are close to each other and to common center of inertia of whole molecule. We therefore get structure like core and shell. The next interesting question is how the sizes of blocks A and B in diblock copolymer AB are affected by each other in comparison with the sizes of the same blocks in homopolymers AA and BB? We calculated the distribution function of gyration radiuses of block A and B both in homopolymers and in diblock copolymer AB. We obtained that the UPE block (B) is essentially more compact in the diblock copolymer (average gyration radius Rg close to 7.7A) than in the BB homopolymer (average value of Rg near 8.5-8.6A). At the same time the size of the PEG block (A) in diblock copolymer is greater than in the homopolymer.

The fluctuation of gyration radii of blocks in diblock copolymer (width of distribution function of Rg) are significantly smaller than in homopolymers. The larger fluctuations for the homopolymers are also due to that both "blocks" of homopolymer can be partly inside and outside of the total structure. For the diblock copolymer the UPE part is always inside and PEG part is outside the total structure.

The relaxation of the extended initial conformation and the final structure of the diblock copolymer containing UPE and PEG blocks were studied by atom-based molecular dynamics simulation. It was shown that relaxation occurs in two stages and that the final structure is like core-shell structure.

5. REFERENCES

1. U.Schulze,M.Skrifvars,N.Reichelt, H.Schmidt. J.Appl.Polym.Sci., in press
2. Y.Zhan, W.Mattice, Macromolecules (1994) **27**, 7056
3. P.Choi, H. Blom, T.Kavassalis, A.Rudin, Macromolecules (1995) **28**, 8247
4. Discover2.9.8/96.0/4.0.0Molecular Simulation Inc., SanDiego (1996)

COMPRESSED POLYMER BRUSHES UNDER THE SHEAR FLOW

I.Neelov[1,2], F.Sundholm[2], K.Binder[3]

[1]Institute of Macromolecular Compounds, Bolsoi pr. 31, St.Petersburg, 199004, Russia
[2] Helsinki University, A.I.Virtasen Aukio 1, PB55, Helsinki, Finland
[3] J.-Gutenberg Universitat, Staudingerweg 7, D-55099, Mainz, Germany

The structure and dynamics of two compressed polymer brushes under the shear deformation were investigated by the methods of stochastic dynamics. The shear was created by the move-ment of grafting planes with the constant velo-cities in opposite directions along the planes.

We varied shear rate,grafting density and the distance between planes. It was shown that both chain elongation and inclination increase with shear rates. As in the case of one brush under shear the dependences of relative chain elogation as function of chain inclination can be fitted by single master curve at all values of parameters. The results were compared both with the theoretical predictions (Rabin, Alexander and Barrat) and with the results of Monte-Carlo (Binder and Lai, Miao) and molecular dynamics (Grest) computer simulations. The dependence of elongation as a function of inclination for two compressed brushes is close to that for one brush under the shear. It is weaker than the dependence predicted by Rabin, Alexander and Barrat theories. At the same time it is close to the curve obtained in MC simulation by Binder and Lai where another method of shear simulaion was applied. It was shown also that up to very large shear values the height of brushes is not affected by shear practically. Only at very large shear rates the chain height begins to decrease with shear rate. This process becomes more and more "step like" if the grafting density increases. Characteristics of translational and rotational motion of chain segments and correlation functions of chain dimensions were also calculated at different shear rates. Interplay between the shear, interpenetration of the brushes and rotational and translational motion are discussed. It was obtained that characteristic times for all correlation functions decrease very slowl with shear up to large shear values. Only at very large shear rates this decrease becomes significant. As in the case of chain height these dependences become more and more "step like" if the grafting density increases. The dependences of dynamical characteristics on shear are similar to that of static characteristics because the main contribution to time decrease is due to the decrease of magnitudes of orientational fluctuations.

1.Neelov I., Binder K. Macromol. Chem., Theory & Simul., (1995) 119-136.

2. Neelov I., Binder K. Macromol. Chem., Theory & Simul. (1995) 1063-1084.

3. Binder K., Baschnagel J., Neelov I. Proceedings of the German-Israeli (BMFTY-MDSA) Meeting on Material Research, Dresden, 1995, 1-22.

4. Neelov I., Binder K. Polym. Sci. (1996) A38, 665-675.

5. Neelov I., Borisov O., Binder K. Macromol. Chem., Theory & Simul. (1998) 141-156

6. Neelov I., Borisov O., Binder K. J. Chem. Phys., 1998, in press

COMPUTER SIMULATION OF DILUTE POLYMER SOLUTIONS IN TRANSIENT ELONGATIONAL FLOWS

M. G. SAPHIANNIKOVA[1], A. A. DARINSKII[1] and N. E. DYAKONOVA[2]

[1]Institute of Macromolecular Compounds of Russian Academy of Sciences, 199004 St.Petersburg, Russia
[2]H. H. Wills Physics Laboratory University of Bristol, UK

The growing interest in the behaviour of polymer solutions in porous media is stimulated by a practical application of polymers for drug reduction, oil recovery etc. Different models of porous media were used to study such a behavior. Particularly, a model device which represents two arrays of glass cylinders was recently suggested by Dyakonova et al [1]. In this flow cell the behaviour of dilute polystyrene solutions in trycresylphosphate flowing through a succession of contractions and expansions was investigated. When the flow rate, Q, exceeded some critical value the accumulation of strain by polymer molecules was observed. It manifested in a slight increase of maximum birefringence value on the symmetry axis of the cell from one pair of cylinders to another along the cell.

This study presents an attempt to reproduce such a behaviour using the Brownian dynamics simulation technique. The advantage of this method is that it does not involve any approximations. The dumbbell with conformation-dependent friction and elastic coefficients was chosen as a model for the polymer chain.

The velocity profile of the flow field was calculated in assumption that the flow modification in dilute polymer solution is negligible. In the vicinity of the cell symmetry axis the flow can be described as an oscillatory elongational planar flow. First we investigated the simplest case of the oscillating flow when the velocity gradient is constant and only changes its sign every half of period. Then we considered the flow with sinusoidal dependence of its velocity gradient on position along the cell. Finally we studied the behaviour of the chain in the flow with the asymmetric position of the gradient maximum. We compare the results of our simulation with the predictions of the theory based on the self-consistent approximation to check the validity of the approximation.

Qualitatively the behaviour of polymer chain does not change if we go from the simplest flow model with constant gradient during the half of the period to more comlicated models where the gradient changes during this interval. The main effect of this variation is difference in the stationary chain deformations in two directions: they are noticeably smaller along the flow than those normally to it for any flow rate considered.

The number of elementary cells which a chain passed until the stationary state is established increases with the increase in Q, but the time of the establishment of this state does not really depend on Q and is equal to $6\tau_0$, where τ_0 is the relaxation time of a chain. The dependencies of maximum and stationary values of polymer extension ratio, $\langle\beta^2\rangle^{1/2}$, on Q show critical behaviour for both values: at small Q they increase with Q but above some Q become independent on it.

When initial state of the polymer chain entering the periodic elongational flow had corresponded to a gaussian coil the birefringence values obtained in the simulation were two orders of magnitude smaller than experimentally observed ones for any flow rate. The polymer deformation followed the oscillations of the flow. After initial increase in the first contraction zone the deformation gradually decreases to the stationary value further in the cell. Moreover, when the initial state had corresponded to almost completely stretched macromolecule the birefringence rather quickly decreased to very small values corresponding to slightly deformed gaussian coil. Simulation did not show also the experimentally observed shift in the position of the maximum birefringence value downstream with respect to the position of a narrowest cross-section. Possible explanation of the disagreement between the simulation results and experimentally observed behaviour could be due to hydrodynamic interactions between the deformed molecules, flow modification and the presence of inertia effects.

Simulation gives the broader peak of $\langle\beta^2\rangle^{1/2}$ dependence on the position along the cell in the converging area and the narrower peak in the diverging area than the theory. This discrepancy is due to the use in the theory of a self-consistent approximation which assumes the δ-character of a conformational distribution function. In contrast, simulations show a rather broad distribution of polymer molecules over the extension ratios.

This work was sponsored by NATO Collaboration Research Grant HTECH. GRG 940365 and by Russian Fond 96-03-33833a.

References

[1] N. E. Dyakonova, J. A. Odell, Yu. V. Brestkin, A. V. Lyulin, A. E. Saez, J.Non-Newtonian Fluid Mech., 67 (1996), 285-310.

RHEOLOGY OF PHYSICAL NETWORKS

S.E.VARYUKHIN, T.F.IRZHAK, V.I.IRZHAK

Institute for Chemical Physics Research Russian Academy of Sciences
Chernogolovka Moscow distr., 142432, Russia

1. INTRODUCTION

The goal of the work is to describe the rheological properties of polymers from viewpoint of physical network taking into account physical bonds of any types: from topological crosslinks (entanglements) to localized strong ones (hydrogen type) including micro phase and mesocrystalline structures (Irzhak et al. (1997)). Such approach is in according to novel theory of chains diffusion developed by Schweizer (1989), Schweizer and Szamel (1995), Tang and Schweizer (1996), namely, mode-coupled theory.

2. MODEL

Model of polymer chain relaxation based on concept physical network was proposed by Irzhak V. et al (1996) and was used successfully by Irzhak T. et al (1997) to interpret the relaxation properties of linear polymers in rubberlike state. The sense of the model is the follow. The conformation of a polymer chain included to physical network can be characterized by set of physical crosslinks: equilibrium ones belong to equilibrium chains and chain fragments with equilibrium conformation, nonequilibrium crosslinks belong to nonequilibrium chains and fragments with nonequilibrium conformation. Naturally, existent crosslinks should hinder the chain dynamics. This phenomenon was found by computer simulation as transfer of diffusion law from $<r^2> \div t$ to $<r^2> \div t^{1/2}$. Evidently, relaxation of the chain conformation must display the same properties, i.e., relaxation time must depend on concentration of nonequilibrium crosslinks x, i.e.,

$$\tau = \tau_0 + \tau_1 x^b = \tau_0(1 + ax^b) \qquad (1)$$

if one supposes the simple power dependence τ on x. τ_0 is the relaxation time conformation of the chain in equilibrium state, i.e., Rouse or "reptation" one.

As far as nonequilibrium (stretched) state of a chain can be connected with applied stress σ in the simplest case one can assume the relation: $x \div \sigma$.

As far as the total concentration of physical crosslinks was shown by Solovyev at al.(1986) to be depended on stress the relaxation equation can be expressed:

$$\frac{d\sigma}{dt} = -\frac{\sigma \cdot \exp\{\beta\sigma - A\}}{1 + a\sigma^b} \qquad (2)$$

A is proportional to $\ln\tau_0$ and to energy of crosslink formation. Value A depends on polymer molecular weight and concentration in the same degree as τ_0 is.

3. RHEOLOGIC REGULARITIES

Comparison of eq.(2) with experimental data on stress relaxation of copolymers of nonylacrylate with acrylamide shows that this equation is applicable with high exactness.

In case of continuous acting stress eq.(2) transforms to:

$$\frac{d\sigma}{dt} = \gamma \cdot \dot{\varepsilon} - \frac{\sigma \cdot \exp\{\beta\sigma - A\}}{1 + a\sigma^b} \qquad (3)$$

where $\dot{\varepsilon}$ is deformation rate.

In case of constant deformation rate $\varepsilon = \dot{\varepsilon} \cdot t$ it is possible to obtain the equation for deformation curve (in differential form):

$$\frac{d\sigma}{d\varepsilon} = \gamma - \frac{\sigma \cdot \exp\{\beta\sigma - A\}}{\dot{\varepsilon} \cdot (1 + a\sigma^b)} \qquad (4)$$

So, as one can see, γ is the initial (σ=0) modulus of the system.

By comparison with experiments on deformation of the above mentioned copolymer eq.(4) was shown to be adequate for the curve σ - ε description.

Eq.(3) describes the qualitative the features of creep and stationary flow of polymer systems including dependencies of viscosity on molecular weight, concentration, phenomenon of flow failure.

4. CONCLUSION.

The proposed model describes different features of relaxation and rheologic behavior of polymer systems. The model is appeared to be suitable for working out the method of determination in block state of molecular weight distribution. So, in spite of phenomenological character it to our mind will find the place among many others empirical approaches.

5. REFERENCES

Irzhak V.I., Korolev G.V., Solov'ev M.E. (1997) Russian Chem.Revs 66:167-186

Irzhak V.I., Varyukhin S.E., Irzhak T.F. (1996) Progress Colloid Polym Sci 102: 4-8

Irzhak T.F., Varyukhin S.E., Ol'khov Yu.A., Baturin S.M., Irzhak V.I. (1997) Polymer Science A 39: 459-463

Schweizer K.S. (1989) J Chem Phys 91:5802-5821, 5822-5839

Schweizer K.S, Szamel G. (1995) J Chem Phys 103: 1934-1945

Solovyev M.E., Raukhvarger A.B., Irzhak V.I., Enikolopyan N.S. (1986) Doklady Akad.Nauk SSSR 290: 293-296

Tang H., Schweizer K.S. (1996) J Chem Phys 105: 779-791

MICRORHEOLOGICAL PROPERTIES IN POLYMERIC HETEROGENEOUS MEDIA BY COLLISION DYNAMICS SIMULATION

Yu.G.Yanovsky, Yu.N.Karnet, N.K.Balabaev

Institute of Applied Mechanics of Russian Academy of Sciences,
Leninsky Prospect 32A, Moscow, 117334, RUSSIA

1. INTRODUCTION

There are several works devoted to computer modeling of bead-rod chains dynamics in a hydrodynamic flow (Liu,1989; Parnas et al.,1991; Pierleoni et al.,1993). In these works polymer-solvent interaction was simulated by the explicit inclusion of the solvent atoms in consideration (Pierleoni et al.,1993) as well as in effective way, using Brownian dynamics method (Liu,1989; Parnas et al.,1991). The behavior of grafted polymer molecules was studied with various theoretical methods, (Bird et al.,1987) including computer simulations (Kremer et al., 1990; Haliloglu et al., 1996). The majority of these simulation studies, however, dealt with the grafted chains under quiescent conditions.

2. MODEL AND SIMULATION DETAILS

New model approach - Collision Molecular Dynamics (CMD) was created and implement for description of the microrheological properties in polymeric heterogeneous media. According to CMD we model the heterogeneous polymer continuum as a "solution" consisted of the polymeric materials macromolecules, an environment "solvent" (virtual solvents atoms) and the repulsive or attractive surfaces (fillers surfaces imitation).

We have studied the response of the chains both end-grafted to a solid wall and freely moving in space to simple shear flow. The main differences between the model used in the present simulations and those in the available BD simulations (Liu,1989; Parnas et al.,1991) are as follows: (i) here the excluded-volume interactions are included in the bead-rod model of polymer chain, and (ii) the interaction between the solvent and the polymer chain is effected by the collisional dynamics (CD) method (Lemak et.al., 1998).

Polymer molecule was simulated using a bead-rod model, which is composed of N identical mass points (beads) of mass m linked by rigid bonds of length l to a linear chain. Hence, the position vectors $\mathbf{r}_k$, which give the location of the beads with respect to some fixed coordinate system, are subjected to the following constraints

$$\left(\mathbf{r}_{k+1}-\mathbf{r}_k\right)^2 = l^2, \quad k=1,\ldots,N-1$$

Contour length of the bead-rod chain is constant and equal to $(N-1)l$.

All pairs of mutually nonbonded beads interact through a shifted, short-range repulsive Lennard-Jones potential

$$U_R(r)=\begin{cases} 4\varepsilon_0[(\sigma/r)^{12}-(\sigma/r)^6+1/4], & r\le 2^{1/6}\sigma \\ 0, & r>2^{1/6}\sigma \end{cases}$$

Here r is the distance between the beads, and ε_0 is the well depth associated with this potential.

For the bead-wall interaction we used the same purely repulsive potential as for the bead-bead interaction, namely, $U_R(z)$, where z is the distance from the bead to the surface.

The effective polymer-solvent interaction was simulated by the collisional dynamics (Lemak et.al.,1998). In this method each bead of the polymer chain suffers collisions with virtual solvent particles. Each stochastic collision is an instantaneous event. Collisions occur in accordance with a Poisson process, which is specified by the only parameter λ, the collisional frequency. Between the stochastic collisions the system evolves in accordance with the equations of motion as in the usual molecular dynamics.

3. RESULTS

3.1. Conformational properties

One of the focal point of our simulations is to elucidate how the presence of the surface affects the chain dimension in shear flow field. The only intramolecular degree of freedom considered in this section is the end-to end distance, $h=\sqrt{\left(\mathbf{r}_N-\mathbf{r}_1\right)^2}$, which is usually used as a measure of the chain dimension.

Figure 1 presents some simulation results. The data for the grafted chains nicely lie on the straight line having a slope equal to 0.570 ± 0.003. Moreover, in the considered range of N ($26 \le N \le 101$) the $<h>$ values of the grafted chains are slightly greater than those of the free chains.

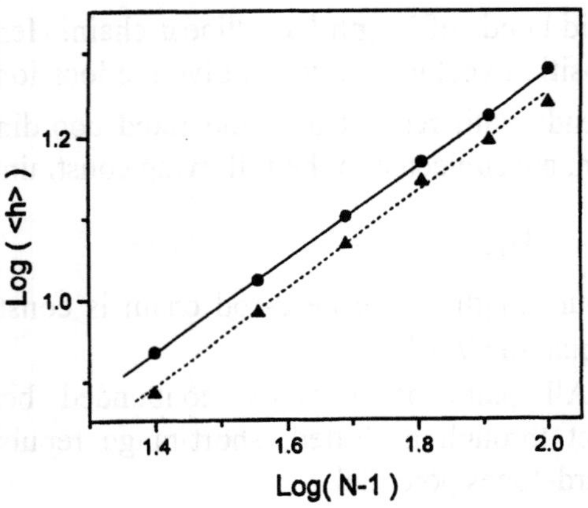

Fig.1. Equilibrium mean end-to-end distance $<h>$ vs. chain length N-1 for both free-space (Δ) and grafted ($\bullet$) chains. The fitted solid and dashed straight lines have slopes of 0.60 and 0.57, respectively.

3.2. Rheological properties

The rheological properties of the polymer media were analyzed with the standard material functions, which are: the shear viscosity η, the first and the second normal stress difference Ψ_1, Ψ_2. The shear rate dependencies of the material functions obtained in our simulations manifest a non-Newtonian behavior typical for polymer solutions. It was shown, that η_0 grows with increasing of the number of beads (N) and its behavior can described by a power law dependence $\eta_0 \propto N^{\nu_\eta}$. For free chains ν_η are equal to 2.37 ± 0.07, whereas grafted chains have ν_η equal to 2.24 ± 0.11. We have found that $\Psi_{1,0}$ vs. N also obeys a power law with the exponent ν_Ψ equal to 4.74 ± 0.23 and 5.08 ± 0.18 for free and grafted chains, respectively.

The main effects of the chain grafting on the rheological properties of polymer solution are a significant increase in the shear viscosity η and the first normal stress difference Ψ_1.

We have compared our results for free-space chains with the results of the Brownian dynamics simulations of Kramers chains. It was found that the inclusion of the excluded volume interactions to the polymer model leads to an increase in the power law exponents describing the dependencies of both the zero-shear-rate properties on the chain length and the material functions vs. the shear rate.

Figure 2 in non-dimensional scale shows the dependencies η vs. shear rate for grafted with different density chains. Curves 1 and 2 are low and very high density of grafted chains respectively.

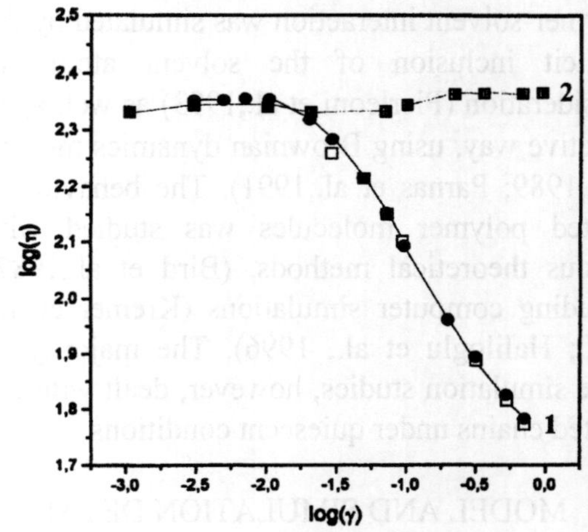

Fig.2.

4. REFERENCES

Liu TW(1989) J.Chem.Phys 90, 5826.

Parnas RS, Cohen Y (1991) Macromolecules 24, 4646.

Pierleoni C, Ryckaert JP (1993) Phys.Rev.Lett. 71, 1724.

Bird RB, Curtiss CF, Armrstrong RC, Hassager O (1987) Dynamics of Polymeric Liquids, Vol.2: Kinetic Theory, 2nd ed. Wiley, New York..

Kremer K, Grest GS (1990) J.Chem.Phys 92, 5057.

Haliloglu T, Bahar I, Erman B (1996) J.Chem.Phys 105, 2919.

Lemak AS, Balabaev NK, Yanovsky YuG, Karnet YuN (1998) J.Chem.Phys. 108, 797.

A LATTICE BOLTZMANN MODEL FOR THE NUMERICAL SIMULATION OF NON-LINEAR VISCOELASTIC FLUIDS

L. GIRAUD[1]

[1]CESAME, Université catholique de Louvain, Division of Applied Mechanics, B-1348 Louvain la Neuve, BELGIUM

1. INTRODUCTION

The Lattice Boltzmann Method (LBM) is a technique based on a discretisation of the Boltzmann equation. LBM for Newtonian fluids has been used as a computational fluid dynamics approach for nearly a decade and numerical experiments on complex flows, such as the suspension of thousands hard spheres or the calculation of transport coefficients in extremely complex porous media, have shown that LBM is quantitatively accurate and computationally efficient. We have recently proposed, Giraud *et al.* (1997), to use the Lattice Boltzmann method as an alternative tool to simulate linear viscoelastic fluid flows. The extension of LBM to the simulation of non-linear viscoelastic flows is presented here.

1.1. The Basic Algorithm

The aim of the original Lattice Boltzmann approach is to sketch the real microworld dynamics with the help of discrete particles populations N_i. The motion of the fluid can be represented by the propagation and collision of the different populations, N_i, $i \in [0, b]$, that live on a regular lattice with b moving discrete velocities, and a zero modulus velocity, that are called $\vec{c}_i$. Figure 1 shows a usual pattern of the lattice with $b = 8$.

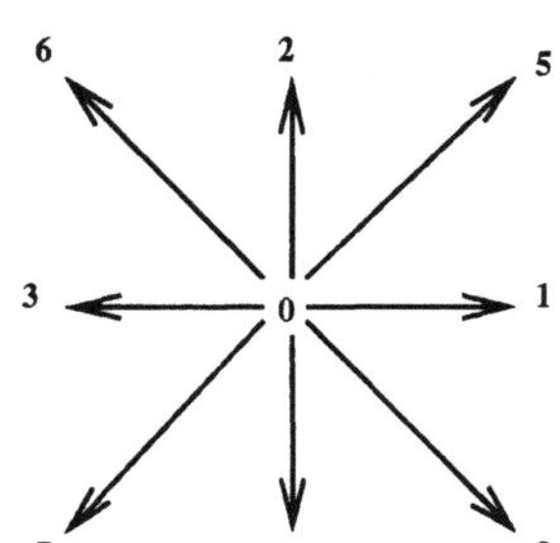

Figure 1: Nine-velocity model in two dimensions

The moments of the particle distributions allow us to define the hydrodynamic modes locally on every lattice site as:

$$\rho = \sum_i N_i \qquad \text{and} \qquad \vec{j} = \rho\vec{u} = \sum_i N_i\,\vec{c}_i,$$

being respectively the density and momentum of the fluid. The Lattice Boltzmann rule that controls the evolution of the populations from the configuration at time t and space position $\vec{x}$ can be written as follow:

$$N_i(\vec{x}+\vec{c}_i, t+1) = N_i(\vec{x}, t) - \sum_{j=0}^{b} \Omega_{ij}\left(N_j - N_j^{(0)}\right),$$

where $\mathbf{N}^{(0)}$ is an equilibrium state that is function of the conserved quantity of the model. The collision matrix Ω is a constant matrix as it does not depend here upon time and space. Ω can be easily determined using the method described in d'Humières (1992). Then, the eigenvalues of Ω, called λ_i, the inverse of a relaxation time, define the hydrodynamic behavior of the fluid. It can be shown, with the help of a Chapman-Enskog expension technique, that the macroscopic behavior of the populations can be described by the Stokes (*resp.* the Navier-Stokes) equations with a suitable linear (*resp.* non-linear) choice of $\mathbf{N}^{(0)}$.

1.2. Introduction of Boundaries

One of the most advocated merit of LBM is its flexibility with respect to complex geometries. This is easily understood by recalling that LBM is basically a particle tracking technique, it allows one to treat complex boundaries in terms of reflexions and bounces at appropriate lattice nodes. The aim of this study is to keep this advantage for viscoelastic fluids.

1.3. The Viscoelastic Model

The key idea to reproduce viscoelastic behaviors with the Lattice Boltzmann Method is to give the viscous modes a very long relaxation time leading to a viscous memory effect. This is simply done by choosing some eigenvalues of Ω related to viscous modes to be small. The viscous modes are simply calculated with the particle distribution by:

$$S_{\alpha\beta} = \sum_i \left(c_{i\alpha} c_{i\beta} - \frac{c_i^2}{2}\delta_{\alpha\beta}\right) N_i,$$

α and β are indices for the two space directions. Based on this idea, a linear Jeffreys-like model has been built by the addition of new degrees of freedom, Giraud *et al.*(1998), in this case two new rest particles were added, but the stress tensor was chosen traceless for a matter of simplification. In order to introduce non-linear effects and a non-zero stress tensor trace, we need to introduce more degrees of freedom, so more lattice directions. It can be shown that the minimal lattice with enough free parameters to have a upper-convective model is the double channel model which is a superposition of two lattices presented in figure 1 excepted for the rest particle leading to a 17 velocity model. A slightly modified Chapman-Enskog expansion, to take viscoelastic effects into acount, gives the following conservation equations:

$$\partial_t \rho + \partial_\alpha j_\alpha = 0,$$

$$\partial_t j_\alpha = \partial_\beta \sigma_{\alpha\beta},$$

with the pressure, viscoelastic and Newtonian contributions to the stress tensor:

$$\sigma_{\alpha\beta} = -P\,\delta_{\alpha\beta} + \sigma^{(V)}_{\alpha\beta} + \sigma^{(N)}_{\alpha\beta},$$

with the respective constitutive equations

$$\sigma^{(N)}_{\alpha\beta} = \nu_\infty \left(\partial_\alpha j_\beta + \partial_\beta j_\alpha - \delta_{\alpha\beta}\partial_\gamma j_\gamma\right), \text{ and}$$

$$\mathbf{A}(\sigma)\cdot\sigma^{(V)}_{\alpha\beta} + \lambda\left(\overset{\nabla}{\sigma}{}^{(V)}_{\alpha\beta} - D_\infty \Delta \sigma^{(V)}_{\alpha\beta}\right) = (\nu_0 - \nu_\infty)\left(\partial_\alpha j_\beta + \partial_\beta j_\alpha\right),$$

the different viscosities, ν_∞, ν_0 and the diffusion coefficient D_∞ are function of free parameters of the model. The relaxation time λ is a function of an eigenvalue of the collision matrix Ω and can vary as a function of the stress; the operator $\mathbf{A}$ can be choosen in order to obtain a Giesekus or Phan-Thien and Tanner model. Note that the model is equivalent to the Oldroyd-B model when we choose $D_\infty \ll 1$ and $\mathbf{A}$ is taken to be the identity operator. One can also notice that the Lattice Boltzmann Model gives rise to a stress diffusion as already pointed out by El-Kareh *et al.* (1989).

2. NUMERICAL RESULTS

Numerical simulations of start-up and steady shear flows are performed to check the validity of the given non-linear constitutive law. We introduce new boundary conditions, based on bounce back, in order to increase the numerical accuracy of the model with moving boundaries. We also performed simulation of flow through simple medium of spatially periodic staggered circular cynlinders. The objective here is to show that LBM behaves well with inertia and other non-linear effects as a function of Reynolds and Weissenberg Numbers.

3. ACKNOWLEDGMENTS

This work is supported by the ARC 97/02-210 project, Communauté Française de Belgique.

4. REFERENCES

EL-KAREH A. W. and LEAL L.G. (1989) *Jour. of Non-Newt. Fluid Mech.*, Vol 33, pp. 257-287.

d'HUMIÈRES, D. (1992) *AIAA Rarefied Gas Dynamics: theory and applications*, Vol. 159 of Progress in Astronautics and Aeronautics pp. 450–458.

GIRAUD L., d'HUMIÈRES, D. and LALLEMAND, P. (1997) *Intl. J. Mod. Phys. C*, Vol 8, No. 4, pp. 806-816.

GIRAUD L., d'HUMIÈRES, D. and LALLEMAND, P. (1998) *to be published in Europhysics Letters*

NUMERICAL SIMULATION OF NONLINEAR CONTACT INTERACTION BETWEEN SHELL STRUCTURES AND SUPPORTS (STAMPS) OF DIFFERENT TYPES

V.S. GUDRAMOVYCH

Institute of Technical Mechanics of the National Academy of Sciences of Ukraine
15, Leshko Popel Str., 320600 Dnepropetrovsk, Ukraine

1. INTRODUCTION

The contact interaction problems for shell structures with different media are of great importance for many branches of engineering. Some problems refer to interaction of complex shell structures with local supports (stamps) of different rigidity. The stamps may be elastic, nonlinearly elastic or rigid.

Methods for the numerical simulation of contact interaction between stiffened shells, such as elastic, and elastoplastic cylindrical, spherical and conical shells and local supports are developed and the numerical simulation is carried out.

2. METHODS

For axially symmetric shell structures the Fourier series method is used. In this case the components of the stressed and strained state are represented as the serias

$$a(\alpha,\beta) = \sum_{m,n}^{\infty}[a_{m,n}\psi_{1m}(\alpha) + \tilde{a}_{mn}\psi_{2m}(\alpha)]\psi_n(\beta), \quad (1)$$

where $a_{mn}, \tilde{a}_{mn}$ are the Fourier coefficients (F.c.) for external loads, contact pressure, forces and displacements, ψ are trigonometric functions of the form $\cos n\varphi$, $\sin n\varphi$.

To determine the F.c. for the contact pressure between the structure and support, an infinite system of linear equations is constructed. The regularity of this system is proved, and the system is solved by the method of reduction.

For complex-shaped highly inhomogeneous structures (the presence of stiffening elements and holes, noncanonical shape of shells) it is appropriate to use the of finite-element method.

For elastoplastic structures known methods of elastic solutions (method additional loads, variable parameters of elasticity method) are used.

3. EQUATIONS

The equation describing the deformation of a structure consisting of shells stiffened with ring ribs under contact interaction has the form

$$DL_1 w = L_2 F, \quad (2)$$

where D is the flexural rigidity, L_1, L_2, w -are differential operators whose form is determined by the theory of shells, w, F-are the deflection and load functions. For elastic supports $F= Q- L_3 w$, (Q is the external load, L_3 is a differential operator determined by the model of the support) for Winkler's model $L_3 = c_{ij}$, for V.Z.Vlasov's model $L_3 = C_{ij} - d_{ij}\,\partial^2/\partial\varphi^2$.

The general solution of the equation is $(\Phi \rightarrow w, F)$

$$\Phi(\alpha,\beta) = \sum_{n}^{\infty}[\Phi_{m,n}\psi_{1m}(\alpha) + \Phi_{mn}\psi_{lm}(\alpha)]\psi_n(\beta) \quad (3)$$

On substitution of (3) into (2), we obtain

$$w_{mn}a_{smn} + \sum_{t,k}^{\infty}(a_{smntk}w_{tk} + a_{imntk}\tilde{w}_{tk})(\alpha) = Q_{smn} \quad (4)$$

$$a_{smn} = \frac{DL_1\psi_{sm}(\alpha)\psi_n(\beta)}{L_2\psi_{sm}(\alpha)\psi_n(\beta)}$$

$$a_{smntk} = \frac{4}{T_\alpha T_\beta}\sum_{i,j}^{N}\iint_{\Omega} C_{ij}\psi_{sm}(\alpha)\psi_{st}(\beta)\psi_k(\beta)d\alpha d\beta$$

$$Q_{smn} = \frac{4}{T_\alpha T_\beta}\int_0^{T_\alpha}\int_0^{T_\beta} Q(\alpha,\beta)\psi_{sm}(\alpha)\psi_n(\beta)d\alpha d\beta,\ s = 1,2,$$

where N is the number of the contact areas, T_α, T_β are the periods of the functions $\psi_{im}(\alpha)$, $\psi_n(\beta)$, Ω is surface of the system, $i = 1,2$.

Using (4) the equations of equilibrium of the system as a solid body

$$\iint_{\Omega} F d\Omega = 0, \quad \iint_{\Omega}[Fxs]d\Omega = 0$$

and the conditions its boundaries

$$(\alpha = \alpha_0, \beta = \beta_0), \quad L_4 w(\alpha_0,\beta_0) = \sigma_i^0$$

we obtain the resolving equations

$$[A]\{w\} = \{Q\}, \quad (5)$$

where $[A]$ is the square system rigidity matrix, $\{w\}$ is the column matrix of the unknowns, $\{Q\}$ is the column matrix of c, F external load.

In (4) a_{mntk}, Q_{mn} are determined for Winkler's model, for other models their form be comes more complicated.

The realizations of the general equation for solving a variety of specific contact problems were made by Gudramovych, Mossakovsky et al (1985, 1988).

For nonlinearly elastic supports the curve Q-w is divided into a number of straight-line segments,the solution is found for each segment and the results are joined. As the load Q increases, the structure can detach from the support. In this case the problem is solved for the case of several contact areas.

The contact pressure having been determined, the stressed-and-strained state of the system is found. The limit load for shell structures my be determined by finding the maximum stresses and setting them equal to the limit stress.

But this approach gives the lower limit of the limit load. Another approach to determining the limit loads involves the theory of limit equilibrium, in particular its kinematic principle. The limit loads are determined by equating the work done by the external loads and the energy dissipations under certain failure mechanismus.

The failure of a structure can also be governed by local loss of stability of the shells. A large body of calculated and experimental data are obtained by Gudramovych et al (1988, 1990, 1997).

4. ILLUSTRATIONS

Shown below are some calculated results.

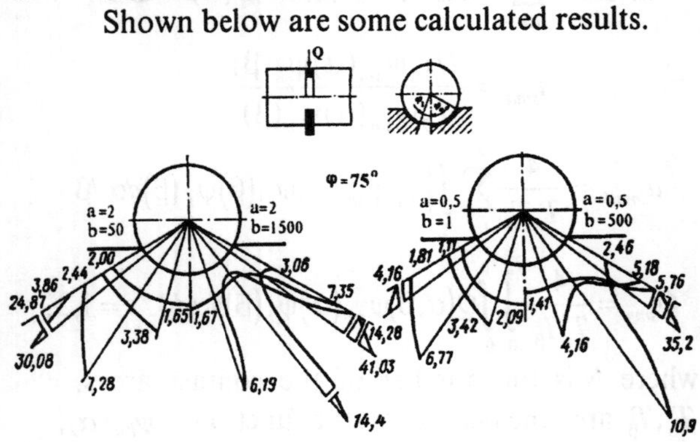

Fig.1 Contact pressure between a stiffened cylindrical shell and support

Shown in fig.1 is the contact pressure distribution (in kgf/cm²) for a cylindrical shell stiffened with a ring which is loaded with supports (stamps) of different rigidity. The rigidity supports are continuous or they have two contact areas. For fig.1a the parameters are as follows:

$$a=\left[3\left(1-\nu^2\right)\right]^{-\frac{3}{4}}\frac{Eh^3R}{E_1J}\sqrt{\frac{R}{h}}=2\,,\; b=\frac{cR^4}{E_1J}=500(\text{right}),$$
$$b=50\ (\text{left}).$$

(R, h- the radius and thickness of the shell, J-the moment of inertia of the ring, E, E_1 - the elasticity modules of material for shell and ring, c- the rigidity of the support, ν - thePoisson coefficient). For fig.1b the parameters are as follows: $a=0{,}5$, $b=500$ (right), $b=1$ (left).

Shown in fig.2 is the contact pressure distribution for conical shell ($\vartheta=15^0$, $a=39$,

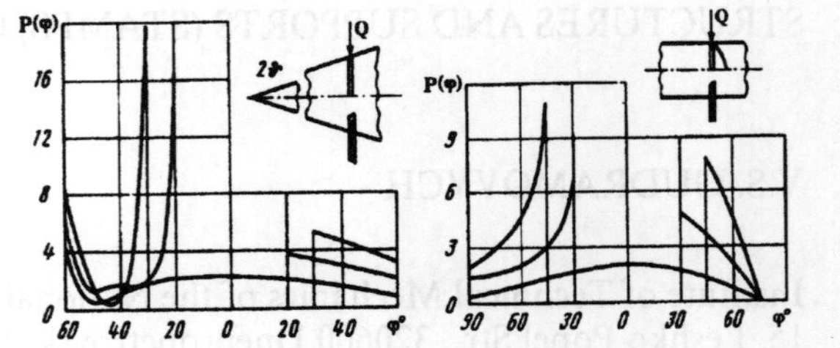

Fig.2 Contact pressure between conical and spherical shells and supports

$R/h=200$, $b=1$ (right), $b=500$ (left)) and spherical shell (the sphere angle 90^0, $b=1$ (right), $b=500$ (left), $\dfrac{EhR^3}{2(1+\nu)E_1J}=600$, $R/h=100$)

As illustrated for a high rigidity support the pressure increases at the edges of the contact areas, and for a low-rigidity support the pressure is close to the cosine distribution. The external load for the examples considered is $Q(\varphi)=q_0\cos\varphi$.

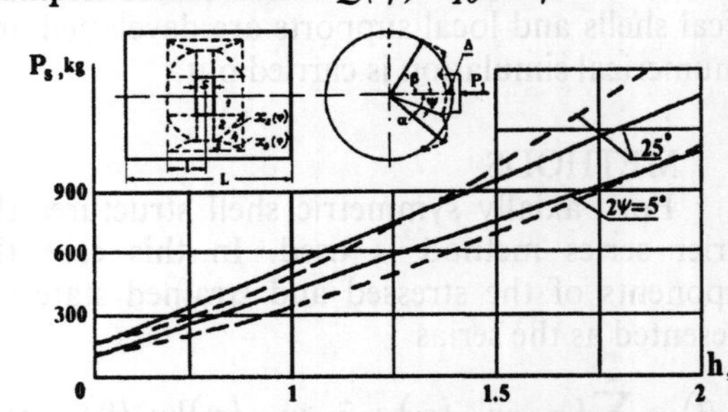

Fig.3 Carrying capacity of a cylindrical shell loaded by a rigid stamp

Shown in fig.3 are calculated and experimental results. The lines are plotted on the basis of experimental data processing (the experimental investigation was performed for 37 shells 200mm in diameter made of an aluminium alloy), and the dashed lines correspond to the theoretical solution (P_s - limit load). Shown at the top of fig.3 are the schemes of loading and failure mechanismus (the dashed lines denote linear plastic hinges).

5. REFERENCES

Gudramovych V.S. (1997) In:Proc. of Intern. Conf. "Carrying capacity of steel shell srtuctures", (Brno, Czech Republ.) pp 145-151

Gudramovych V.S., Demenkov A.F.,Gerasimov V. (1990) Limit analysis of structural elements, Kiev (in Russian)

Gudramovych V.S,Makeev E.M.,Mossakovsky V.I., Nikitin P.I. (1985) In: Problemy prochnosti, 11, pp 108-114 (in Russian)

Mossakovsky V.I.,Gudramovych V.S.,Makeev E.M. (1988) Contact interaction of shell structures, Kiev (in Russian)

THE ADAPTIVE LAGRANGIAN PARTICLE METHOD (ALPM) APPLIED TO THE START-UP FLOW OF FINITELY EXTENSIBLE DUMBBELLS BETWEEN ECCENTRIC CYLINDERS

P. HALIN, X. GALLEZ, G. LIELENS, R. KEUNINGS and V. LEGAT

CESAME, Division of Applied Mechanics, Université catholique de Louvain, B-1348 Louvain-la-Neuve, Belgium

1. INTRODUCTION

The Lagrangian Particle Method (LPM) has been proposed recently (Halin et al., 1998) for the computation of transient viscoelastic flows using either a constitutive equation (macroscopic approach) or a kinetic theory model (micro-macro approach). At the start of the simulation, we drop an ensemble of Lagrangian particles in the flow domain that drift along the fluid trajectories. Each particle acts as a polymer extra-stress calculator. At each time step, the velocity-pressure fields are calculated with a classical finite element method by solving the Navier-Stokes equations slightly modified to account for the polymer stress. To do so, we use the polymer stress values calculated at the previous time step. Next, we track the particles using the new velocity field, and we calculate the polymer stress at each particle location. This is done by solving a macroscopic constitutive equation (e.g. Oldroyd-B, FENE-P, FENE-L (Lielens et al., 1998)) or by simulating polymer dynamics with a kinetic theory model (e.g. the FENE dumbbell theory). In the latter case, each particle contains a large ensemble of dumbbells whose configuration is updated by solving the relevant stochastic differential equation along the fluid trajectories. Finally, the Lagrangian polymer stress field must be mapped onto Eulerian points so as to be accounted for when conservation law equations are solved at the next time step. To this end, we perform a least-squares approximation on an element-by-element basis.

LPM has already been successfully applied to the start-up flow of a viscoelastic fluid between slightly eccentric cylinders (Halin et al., 1998). Results usually obtained with classical finite element methods for the FENE-P model are recovered. More importantly, LPM enables the use of the FENE kinetic theory. In that case, our results are very close to the ones obtained with the Brownian Configuration Field method of van den Brule et al. (1997).

To derive the Eulerian polymer stress field from the Lagrangian data with the help of a least-squares approximation remains an important issue in LPM. Indeed, one must make sure that each finite element contains enough well-located particles. On the other hand, unnecessary particles must be removed from the flow in order to decrease the CPU time of a simulation. This is of upmost importance when LPM is used with a kinetic theory model, where memory needs would simply prevent the simulation from being performed if too many particles are present in the flow. These two requirements are not trivially satisfied for flows in complex geometries. For example, Fig. 1 shows the evolution of the polymer stress at the wall in the start-up flow of a FENE-P fluid in a *large* eccentricity journal bearing. We see that with the conventional implementation of LPM, with a fixed total number of Lagrangian particles, obtaining stable numerical results requires a large number of Lagrangian particles.

Our aim is to develop the Adaptive Lagrangian Particle Method (ALPM) in which the number of particles and their location are adapted when needed, so as to obtain an improved Eulerian representation of the polymer stress while not impairing the high performance of LPM.

2. METHODOLOGY

The main idea behind ALPM is to predict beforehand locations of low particle concentration. New particles can then be added early enough during the simulation to prevent flow regions from being cleared of particles. At each time step, we proceed as follows. Based

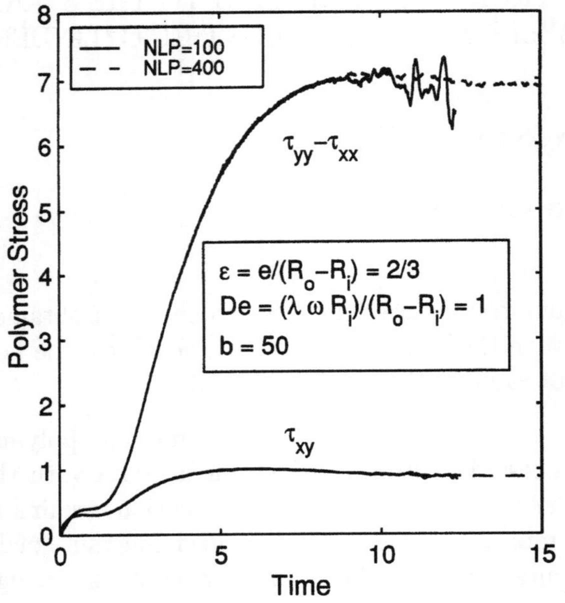

Figure 1: First normal stress difference and shear stress at the inner wall of the large eccentricity journal bearing. The simulation becomes unstable if the mean number of Lagrangian particles (NLP) in each element is not sufficient.

on the current velocity field, the particle ensemble is tracked up to a defined instant of time in order to predict their future location. In each finite element, the predicted locations of the particles are compared to a set of reference locations in that element. If not enough particles are found to be sufficiently close to a given reference location, a new particle is created at that location. The newly created particles are then tracked backwards, down to the current time, so as to determine their current location. Finally, the variables required to calculate the polymer stress contribution need to be initialized for the new particles. These variables, stored in a *configuration vector*, are the connector vector $\mathbf{Q}$ of the dumbbells for a simulation based on the FENE theory, or the configuration tensor $\mathbf{A} = \langle \mathbf{QQ} \rangle$ if the FENE-P approximation is used instead. In each finite element, the configuration vector is initialized from the configuration vectors of the particles currently present in the same element. To do so, we use a least-squares approximation that preserves the polymer stress tensor. Once initialized, the new particles flow along the fluid trajectories, their molecular configuration being updated at each time step. They may or may not, however, participate to the calculation of the polymer stress until the time step at which they have been created is reached. By doing so, we can introduce new particles smoothly, thereby controlling the impact of their presence on the polymer stress contribution. Finally, particles that are too close to one another are merged. This reduces the number of particles, thereby avoiding unnecessary particles that do not bring substantial information for the calculation of the stress tensor. Consequently, computational time is greatly reduced.

3. RESULTS AND CONCLUSIONS

Results are presented for the start-up flow of a FENE fluid between highly eccentric cylinders. We first show how convergent and divergent regions of the flow are saturated with particles or emptied, respectively, in the standard implementation of LPM. Then, we show how ALPM can help us overcome those difficulties. Emphasis is put on the critical parameters of ALPM if one wishes to both modify the number of particles and enforce smooth variations of the calculated stress field.

4. ACKNOWLEDGEMENTS

This work is supported by the ARC 97/02–210 project, Communauté Française de Belgique. The work of V. Legat is supported by the Belgian Fonds National de la Recherche Scientifique (FNRS).

5. REFERENCES

Halin P, Lielens G, Keunings R, Legat V (1998) J. Non-Newtonian Fluid Mech. In press

Lielens G, Halin P, Jaumain I, Keunings R, Legat V (1998) J. Non-Newtonian Fluid Mech., 76:249-278

Hulsen MA, van Heel APG, van den Brule BHAA (1997) J. Non-Newtonian Fluid Mech. 70:79-101

USE OF A LINEAR VISCOELASTIC CONSTITUTIVE LAW TO MODEL THE TIME DEPENDENT BEHAVIOUR OF AN ORTHOTROPIC GLASSFIBRE REINFORCED PLASTIC

R.M. HINTERHOELZL[1], K. MOSER[1]

[1] Institut für Baustatik und verstärkte Kunststoffe, University of Innsbruck, Technikerstr.13, 6020 Innsbruck, Austria

1. INTRODUCTION

One of the aims of our co-operation with an industrial company is to investigate the time dependent behaviour of a centrifugally cast filled glass fibre reinforced unsaturated polyester resin pipe. The pipe consists of several different layers, which exhibit either isotropic or orthotropic behaviour. The fibres have a length of 50 [mm] and calcium carbonate is used as filler.

Polymeric matrix composites show a strong time-dependent behaviour. This has also been observed in long-term tests conducted at our department. For example we applied a constant load at the top of a pipe. This load was maintained constant for 1127 hours. The deformation of the vertical diameter increased in this time by the half of the short-term deformation. (Fig.1)

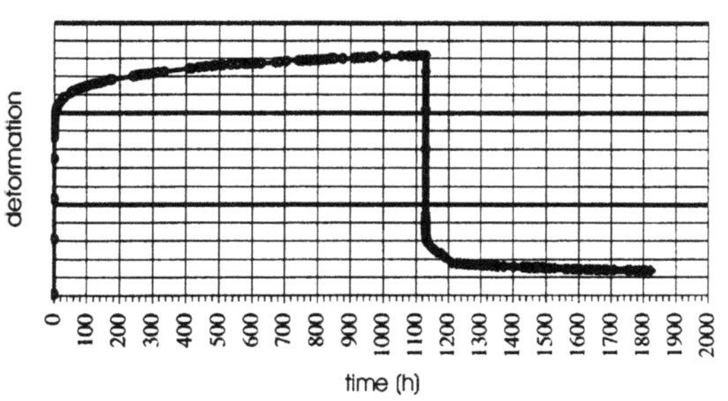

Fig.1

2. THEORY OF VISCOELASTICITY

One of the most powerful tools to model this time-dependent behaviour is the theory of viscoelasticity. For this material linear viscoelastic behaviour is assumed. Different theoretical approaches for linear viscoelasticity are well known: spring-dashpot-models, mathematical models, superposition integrals. For computational strategies, the superposition integrals offer undeniable advantages over that of discrete state variables. The integral formulation is based on the Boltzmann (1874) superposition principle. We derive the superposition integral by considering an arbitrary strain history as consisting of a sequence of steps $d\varepsilon$ at moments τ:

$$\sigma_{ij}(t) = \int_0^t C_{ijkl}(t-\tau)\frac{\partial \varepsilon_{kl}}{\partial \tau}\, d\tau \qquad (1)$$

The term $C_{ijkl}(t)$ represents the fourth order tensor of the relaxation moduli.

3. IMPLEMENTATION OF AN ORTHOTROPIC ISOTHERMIC LINEAR VISCOELASTIC LAW IN A FINITE ELEMENT FORMULATION

The constitutive law for linear viscoelasticity in Eq.1 is used to develop a numerical algorithm for orthotropic media which is implemented in the commercial FE-code of ABAQUS. The constitutive equations are transformed into an incremental algebraic form. This incrementation is accomplished in closed form and results in a recursive relationship. $C_{ijkl}(t)$ can be determined by relaxation tests and the curves can be fitted in the best way by Dirichlet-Prony series corresponding to a generalized Maxwell model:

$$C_{ijkl}(t) = C_{ijkl\infty} + \sum_{m=1}^{M_{ijkl}} C_{ijklm} \cdot e^{\frac{-t}{\rho_{ijklm}}} \qquad (2)$$

$$\rho_{ijklm} = \frac{\eta_{ijklm}}{C_{ijklm}} \qquad (3)$$

C_{ijklm}	spring constants
ρ_{ijklm}	relaxation times
η_{ijklm}	dashpot coefficients

The use of Dirichlet-Prony series leads to the recursive relationship:

$$\sigma_{ij}(t) = C_{ijkl_0} \cdot \left(\underbrace{\varepsilon_{kl}}_{\text{short-term}} - \underbrace{\sum_{m=1}^{M_{ijkl}} \frac{C_{ijkl_m}}{C_{ijkl_0}} \cdot \varepsilon^{ijkl}_{kl_m}}_{\text{long-term}} \right) \qquad (4)$$

4. VERIFICATION

We investigate the ability of the implemented numerical algorithm to predict the response of a long thick-walled cylindrically anisotropic cylinder that is subjected to a internal pressure p=P*H(t) with H(t) as the Heaviside step function. This problem has previously been investigated by Schapery (1967) and Zocher (1995) and an accepted analytical solution is known . In Schapery's analysis, the response of the cylinder was predicted by three different methods: (1) correspondence principle, (2) quasi-elastic method, and (3) collocation. Here we duplicate the first two analyses of Schapery and compare the results to FE prediction (Fig.2). The derivation of a solution by correspondence principle is based on a corresponding elastic solution for general orthotropy by Lekhnitskii (1981). Following Schapery we concern ourselves only with the prediction of the hoop stress, σ_{tang} at the inner wall of the cylinder. The results are satisfying and the FE-prediction fits the analytical solutions very well. A hypothetical material with different time dependent relaxation moduli in radial and circumferential direction was used to calculate the $C_{ijkl}(t)$ which were fitted in form of Dirichlet-Prony series (Fig.3).

$$E_{rad} = E_e \cdot \left[1 + 100 \cdot (t)^{-0.5}\right] \quad (5)$$

$$E_{\tan g} = E_e \cdot \left[1 + 100 \cdot (t)^{-0.1}\right] \quad (6)$$

The FE mesh used in the analysis consisted of 270 plane strain elements (8-node biquadratic). The FE mesh was not equidistant, it was contracted by a factor of 0.6 from the outer to the inner wall. The stress was calculated in the integration points of the elements. Time steps of variable length were used with the magnitude of the time step slowly increasing as time progressed. This variability in time step size was not accomplished adaptively, but assigned in the input file. A formal convergence study of the sensitivity of time step size has not been conducted at this moment.

5. FUTURE GOALS

The main future goal is to combine the study of time dependent behaviour with the analysis of the lifetime durability by means of damage mechanics.

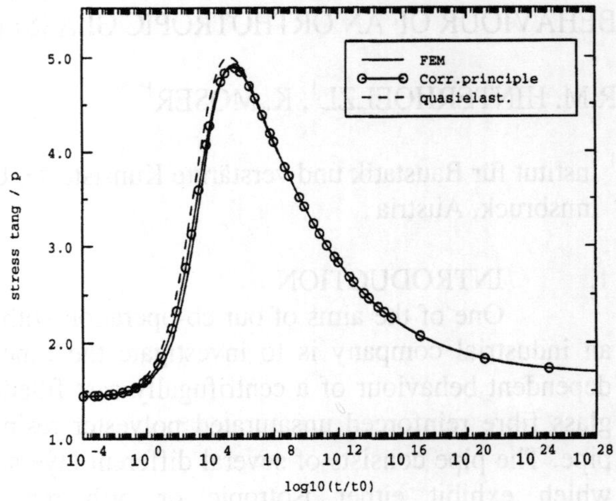

Fig.2

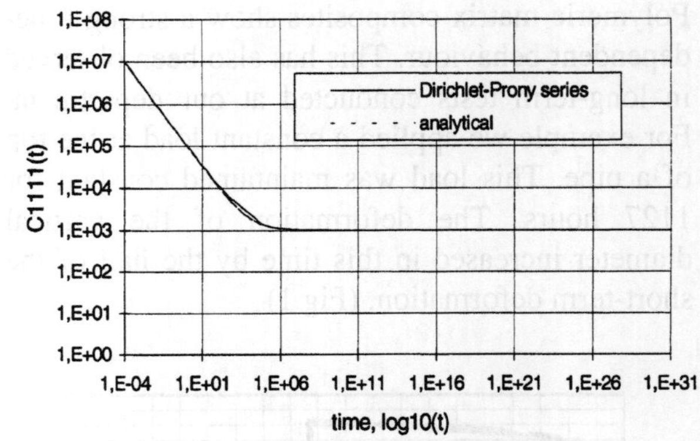

Fig.3

6. REFERENCES

Boltzmann L (1874) Sitzungsber. Kgl. Akad. Wiss. Wien, Math.-Naturw. Classe, (2A) pp. 70-275

Lekhnitskii SG (1981) Theory of Elasticity of an Anisotropic Body. Mir Publishers, Russia, pp. 39-43

Schapery RA (1967) Stress Analysis of Viscoelastic Composite Materials. Journal of Composite Materials, Vol. 1, No. 3, , pp. 228-267

Zocher MA (1995) A Thermoviscoelastic Finite Element Formulation for the Analysis of Composites. Ph.D., Texas A&M University

7. ACKNOWLEDGEMENTS

Hobas Engineering GmbH

GRANULAR DYNAMICS SIMULATION FOR AN HEXAGONAL SILO

G. Joseph, E. Geffroy, B. Mena, O.R. Walton* & R.R. Huilgol**

Instituto de Investigaciones en Materiales ,Universidad Nacional Autónoma de México
Apdo. Postal 70-360, Circuito Exterior, Cd. Universitaria, 04510 México, D.F., MEXICO
*Mechanical Engineering Department, University of Florida
**Department of Mathematics and Statistics, Flinders University of South Australia

1. Synopsis

A new silo design for grain storage is examined using a numerical procedure to model the filling and emptying processes. The simulation uses a molecular dynamics technique which includes contact mechanics for interacting grains. The scheme assumes an elastic-hysteretic model for the *normal* contact forces similar to the one originally proposed by Hertz, and a 3-dimensional model for the tangential contacts which includes simultaneous *rolling* and *sliding* predicting a similar behavior to the model proposed by Mindlin. The new silo design consists of hexagonal cells with the central axis laying horizontally. The bottom walls have an inclination angle slightly higher than the natural repose angle of the granular material, thus providing a large supporting area, where the grains are exposed to minimal stresses during loading and unloading operations. Consequently, the new storage system minimizes grain losses considerably. In this communication, a comparison of the main features of the standard cylindrical silos and the hexagonal silo is presented, together with a detailed account of the contact mechanics for normal and tangential forces involved among grains, the dynamics and kinematics of the granular media, the mechanics of wall-grains interactions for all possible cases, and the predicted behavior during loading and unloading operations.

2. Introduction

Current computational techniques provide two basic approaches to model the behavior of granular materials, each one with specific strengths: finite element methods that analyze the problem as that of a continuum (Haff, 1983; Jackson, 1986), and the granular dynamics (Campbell, 1990; Ristow, 1994; Langston *et al.*, 1995; Walton, 1993a) approach which considers simulation of the bulk behavior via molecular dynamics techniques assuming grain-grain interactions as a problem of contact mechanics (Campbell & Brennen, 1985; Walton, 1993b).

In general, the main results of the simulation presented are based on its predictive capacity to correlate not only the known values on the surfaces, but also the parameters of the flow field in the interior of the flowing materials that otherwise can not be easily known. The principal long-term advantage of the present simulations is the possibility of a detailed knowledge of the operation of the hexagonal silo. In particular, a comparison of the simulation with the main characteristics of the filling and emptying processes, as observed experimentally, are shown, emphasizing the advantages of the new design as compared with the conventional silo. This simulation algorithm gives relevance to a detailed description of the *grain to grain* interactions, and those occurring between a grain and different walls of the silo. The results are based on a grain-dynamics program with fixed boundaries, defining all the walls and the detailed form of the silo.

Once the mechanical properties of the grain interactions are known and the boundaries specified, the simulation continues with a dynamic algorithm in which the grain, prior to a collision, is considered as having a ballistic trajectory due to gravity. It is shown that any arbitrary geometry for the storage silo as well as the characteristics of the grain wall interaction can be taken into account in order to realistically simulate the known behavior of grain flow in an hexagonal silo, with a computationally accessible and efficient algorithm.

3. Conclusions

For the simulation of the filling and discharge of grain silos, it is shown that a compromise is necessary between an elaborate, very accurate model for the particle-particle interactions and a realistic model of such interactions, amenable to computational work with many particles. For these simulations a sufficiently realistic model is desirable for the existing interactions that occur between pairs of grains as well as for the types of walls involved. In particular, the compromise chosen here between the complexity of the model and its efficient calculation shows that a realistic flow is modeled for the hexagonal silo. At this stage, the model for the grain is the most simple —a spherical particle— with particle-particle interactions that include elastic-inelastic contacts with friction. The use of three types of walls makes it possible to simulate the observed behavior of our scaled silo. In these walls, the mechanics of collisions becomes critical, for only with the proper choice of dynamics one may predict the observed behavior. For this silo, it is shown that PCB, or hard mirror walls can predict dynamics that bears a limited resemblance to the real behavior of a granular flow. This is the case for many other results published about granular materials. However, the use of random mirror walls for symmetry planes of the simulation cell allow for an efficient and realistic computation of the observed phenomena. The predicted granular flow resembles closely that observed in a three-ton scaled silo, loaded with corn.

4. Acknowledgements

We thank Ing. Jesús Camacho and personnel of the Machine Shop of the Materials Research Institute (IIM) for their assistance with the construction of the scaled silo. BM would like to thank PAPIIT-DGAPA of UNAM through grants IN302392 and IN503494 for financial support. EG and BM thank CONACyT for funding through grant 0073-A7026.

5. References

Campbell CS (1990), *Rapid granular flows*, Ann. Rev. Fluid Mech. **22**:57-92.

Campbell CS, Brennen CE (1985), *Computer simulation of granular shear flows*, J. Fluid Mech. **151**:167-188.

Haff PK (1983), *Grain flow as a fluid-mechanical phenomenon*, J. Fluid Mech. **134**:401-430.Ahn H, Brennen CE, Sabersky RH (1991), *Measurements of velocity, velocity fluctuation, density and stresses in chute flows of granular materials*, Trans. of the ASME **58**:792-803.

Jackson R (1986), *Some features of the flow of granular materials and aerated granular materials*, J. Rheol., **30**(5)907-930.

Langston PA, Tüzün U, Heyes DM (1995), *Discrete element simulation of granular flow in 2D and 3D hoppers: Dependence of discharge rate and wall stresses on particle interactions.* Chem. Eng. Sci., **50**(6)967-987.

Ristow GH (1994), *Granular dynamics: A review about recent molecular dynamics simulations of granular materials*, Ann. Rev. Comp. Phys. I, Stauffer D (Ed), World Scientific, Singapore, pp. 275-308.

Walton OR (1993b), *Numerical simulation of inclined chute flows of monodisperse, inelastic, frictional spheres*, Mech. of Materials, **16**:239-247.

NUMERICAL ANALYSIS OF THE SORPTIVE BEHAVIOUR IN THE WOOD

P. KOC, M. HOUŠKA

Faculty of Mathematics and Physics, Department of Mathematics and Mechanics,
University of Ljubljana, Slovenia

1 INTRODUCTION

Wood is a natural product of biological origin. It is a low density, cellular polymer composite with quite complex structure. Unlike most materials used in engineering wood possesses strong anisotropy that must be recognised at several levels of material organisation. Particular types of anisotropy can be identified within a layer of the woody cell, in the cell wall as a whole, in the entire cell, in an aggregate of cells, and finally in an aggregation of several types of woody tissue. For deeper understanding of the mechanical behaviour of wood a good knowledge of each constituent in a particular level of material organisation as well as relations between the levels is required. In numerical model, which is based on Finite Element method, two structural levels are used. On the first level a cell wall is taken into consideration, on the second one a typical softwood structure is presented as a assemblage of early- and latewood cells.

2. CELL WALL LAYERING

The cell wall is a composite material in which the stiff cellulose microfibrils act as a fiber reinforcement in a matrix of more compliant hemicellulose and lignin. In a numerical model two basic constituents are used: matrix and fibers. The lignin and hemicellulose may both be described as matrices and each interacts with water to swell and change stiffness, so they are joined into one constituent. The cellulose fibers are supposed not to interact with water. Elastic properties and dilatation coefficient used in the model are given in Bodig and Jayne (1982) and are listed in Table 1.

Table 1. Properties of model constituents

	E [GPa]	ν [/]	β [1/ % MC]
Fibers	132.4	0.1	0.0
Matrix	1.23	0.3	0.0173

The cell wall is a basic element on which the model is established. Regarding the direction of the alignment of cellulose fibers, the cell wall structure consist of the S1, S2 and S3 layers. The microfibrilal angle θ (MFA), measured from the longitudinal axis of the cell, is significant on sort of the wood as well as the cell type. A simplified structure of the cell wall showing the orientation of the microfibrils in each of the wall layers is presented on Fig. 1. Parameters relevant for modelling the earlywood and latewood cell walls of spruce wood are: thickness δ Fengel (1968), ratio k Bodig (1982) between V_{fiber}/V_{matrix} and MFA (Table 2.) The cell wall is represented by a thick shell element with rebars positioned in a proper distance from elements neutral plane.

Table 2. Parameters of cell wall of spruce wood

	Earlywood			Latewood		
Cell wall layer	δ [μm]	k	MFA θ [°]	δ [μm]	k	MFA θ [°]
M+P	0.12	0	/	0.11	0	/
S1	0.23	0.37	±73	0.30	0.37	±73
S2	1.42	1	+22	3.99	1	+9
S3	0.03	1	+76	0.04	1	+76
Total	1.8			4.44		

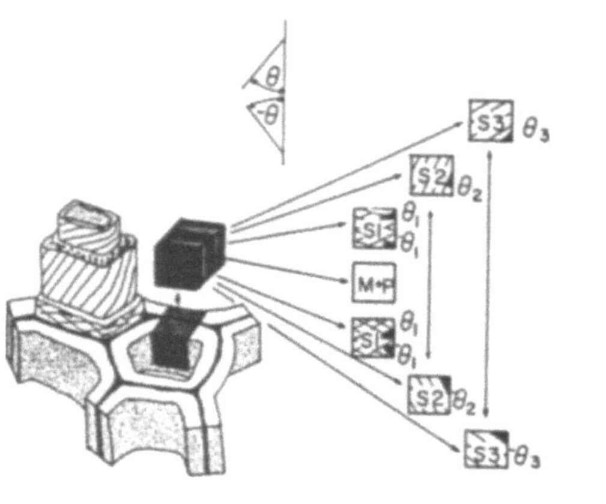

Figure 1. Cell wall layout (after Bodig)

3. GROWTH RING STRUCTURE

Each growth ring consist of two distinct parts: earlywood with large cell cavities and denser part called latewood. The major difference in cell size is due to the difference in radial dimension and in the cell wall thickness. Because of the increased amount of cell wall substance latewood has a great influence on the mechanical properties of wood. The arrangement of the cells strongly influence on the wood behaviour, so the special attention was paid on the FE model layout (Figure 2). The cross-sectional dimensions of the cells and their number matches the earlywood / latewood ratio, which is approximately 5/1. Connections between the cells prevent from the twisting and different longitudinal and tangential dilatation. The constraints, imposed on the numerical model, allow uniform dilatation in each of three principal directions.

4. NUMERICAL VERIFICATION

Since all necessary model data are not found in the literature, some of them are estimated by the numerical analysis of the proposed model. The MFA of the S2 layer fibers and dilatation coefficient of the matrix are computed to obtain an overall wood dilatations measured for spruce wood. The swelling of softwood is described with longitudinal/radial/transversal dilatation ratios, which are approximately 1/33/66, Bodig (1982). The values of those ratios, obtained by our model, using the data from Table 1, 2 and Fig. 1, are 1/42/65. The larger dilatation in a radial direction can be attributed by missing of the ray cells in a numerical model.

Although the S1 and S3 layer are much thinner than the S2 layer, the MFA of these layers have a significant influence on the dilatation coefficient in longitudinal direction. It is found out that small changes of MFA in the S2 layer produce large variations in a dilatation ratios (Fig. 3).

5. SORPTIVE STRESSES ESTIMATION

In the numerical analysis the MC was changed from 8% to 15%. In the case of longitudinal swelling as a consequence of different material properties of basic constituents and their specific structure high internal stresses within the material arise. Their values in particular composite constituents for early- and latewood cells are summarised in Table 3.

Table 3. Sorptive stresses in composite constituents

		Earlywood	Latewood
Matrix	σ_L [MPa]	-205	-200
	σ_T [MPa]	-190	-165
Fibers	σ_{S1} [MPa]	2250	3950
	σ_{S2} [MPa]	390	150
	σ_{S3} [MPa]	2320	4100

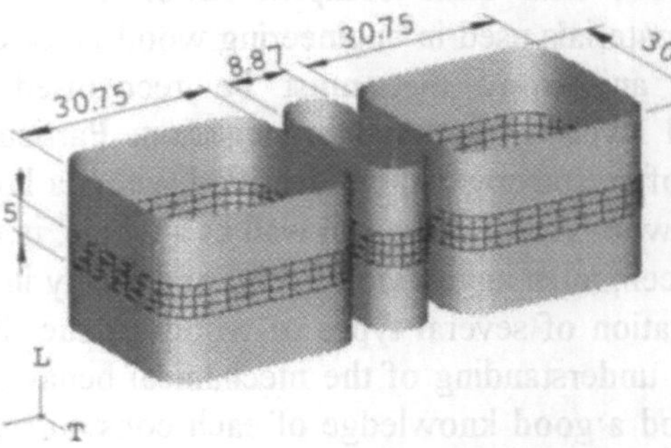

Figure 2. FE model for early- and latewood

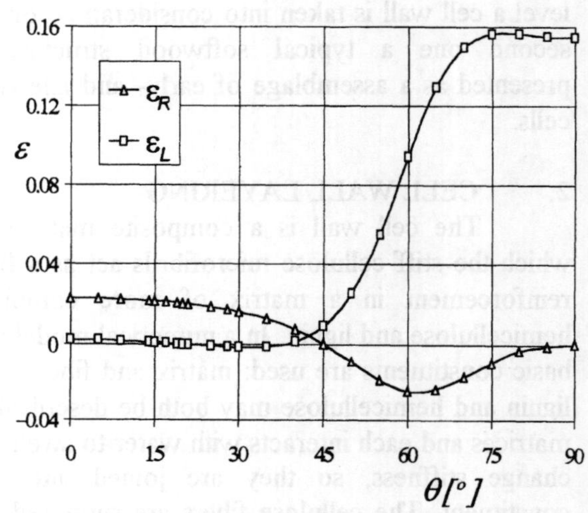

Figure 3. Cell deformations v. MFA of S2 layer

6. REFERENCES

Bodig J, Jayne BA (1982) Mechanics of Wood and Wood Composites. Van Nostrand Reinhold Company Inc. New York,

Fengel D (1968) Wood Sci. Technol. 2:73-83 The ultrastructure of wood.

ABAQUS (1997) Version 5.7, Hibbit, Karlsson & Sorensen, Inc.

M.I. Kolte, P. Szabo and O. Hassager

Danish Polymer Centre, Department of Chemical Engineering,
Technical University of Denmark, Building 229, DK-2800 Lyngby, Denmark

1. INTRODUCTION

The capillary thinning of a polymeric filament is analysed experimentally as well as by means of numerical simulation. The experimental procedure is as follows. Initially a liquid sample is kept between two cylindrical plates. Then the bottom plate is lowered under gravity to yield a given strain. Thereby the sample is stretched into a filament. Provided the filament is sufficiently long, surface tension will induce a thinning of the filament until breakage in finite time. The experimental principle is sketched in Figure 1. This experiment was originally de-

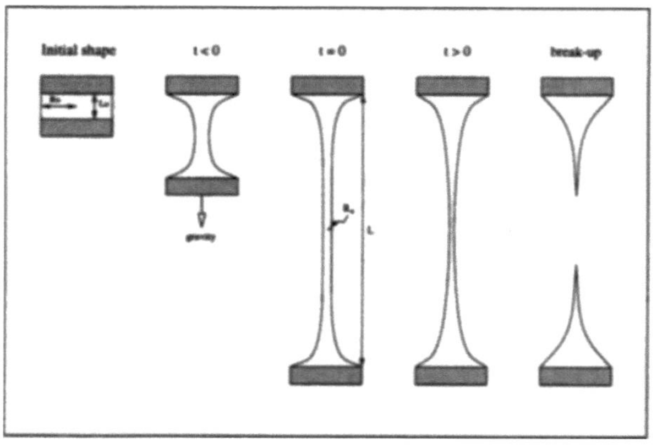

Figure 1: Principles of the different stages in an experiment with the Liquid Filament Rheometer.

viced by Bazilevsky et al.[1]. For the numerical simulations we employ a Lagrangian finite element technique. This method has previously been used in simulations of the transient filament stretching rheometer [2].

2. TEST FLUIDS

2.1 Newtonian fluid

The chosen Newtonian fluid is a polybutene Hyvis 5 from BP Chemicals. At 25°C the viscosity, η_0, and density, ρ, of the fluid were measured to be 8.26 Pas and 850 kg/m^3 respectively. The surface tension coefficient, σ, was estimated to be 0.029 N/m.

2.2 PIB-solution

The non-Newtonian fluid considered here consists of 0.56 % wt high molecular weight polyisobutylene (PIB, Vistanex L-80, $M_w \approx 9 \cdot 10^5$ g/mol) in 94.91 % wt low molecular weight polybutene (BP Chemicals, Hyvis 10, $M_w \approx 950$ g/mol) and 4.53 % wt hexadecane (C16).

The fluid was modelled with an Oldroyd-B model having four modes determined from the linear viscoelastic spectrum. The parameters are shown in Table 1. The density of the PIB-solution was measured to be 850 kg/m^3 at 25°C. The surface tension coefficient was estimated to be 0.029 N/m.

Mode no.	λ_k (s)	η_k (Pa·s)
1	2.14080	0.6523
2	0.31195	2.2221
3	0.04202	1.4241
4	0.00271	0.7204
solvent	···	8.8

Table 1: Linear viscoelastic parameters for the PIB-solution at 25°C.

3. EXPERIMENTS & SIMULATIONS

In the experiments and simulations described here the effect of inertia was found to be negligible. Hence, for a Newtonian fluid, the experiment in non-dimensional time, $\tau = t\sigma/(R_0\eta_0)$, is completely described by the initial aspect ratio, $\alpha = L_0/R_0$, the stretch ratio, $\lambda = L/L_0$, and the Bond number $\mathrm{Bo} = \rho g R_0^2/\sigma$ (the geometrical variable are defined in Figure 1). For the PIB-solution a Deborah number will be important also. We may define the Deborah number based on the largest relaxation time, λ_1 and the apparent strain rate, $\dot{\epsilon} = -2(\mathrm{d}R_s/\mathrm{d}t)/R_s$, near the minimum radius: $\mathrm{De}=\dot{\epsilon}\lambda_1$.

In Figure 2 the minimum radius, R_s/R_0 of the Newtonian filament is depicted as function of the time after the bottom plate has stopped. The effect of two different Bond numbers is illustrated. The lower curve describes conditions with vanishing gravity (Bo=0) whereas the upper curve is produced with the estimated value of the surface tension (Bo=4.62). Apparently, the minimum radius decreases linearly with time until close to break-up.

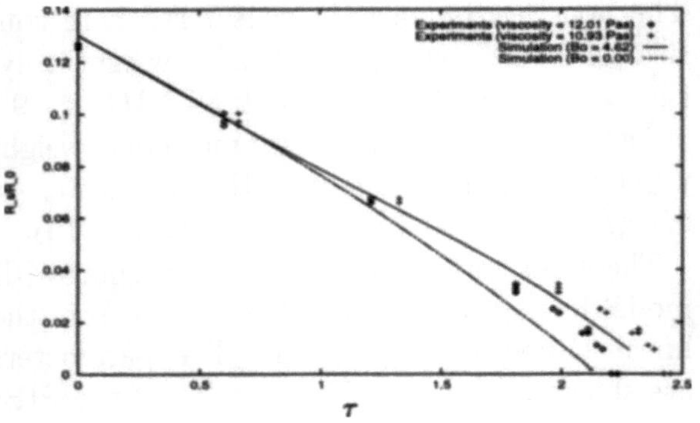

Figure 2: Mininimum radius for the Newtonian fluid: simulations and experiments. λ=13.2, α=0.37 and R_0=4.0 mm.

In Figure 3 the minimum radius of the filament of the PIB-solution is depicted. For this fluid the minimum radius appears to decrease exponentially for the measurements obtained. Hence the Deborah number is constant during the experiment.

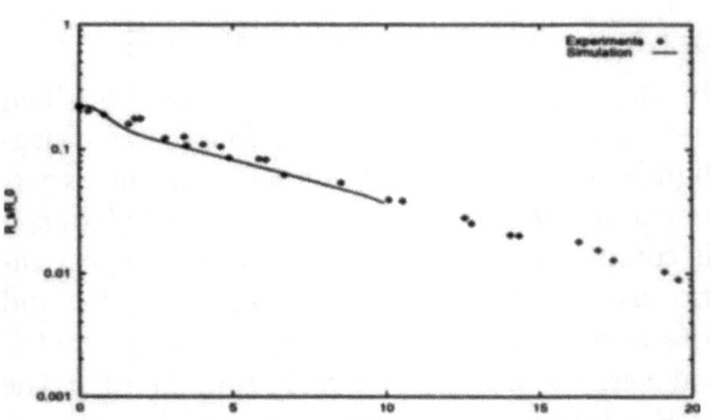

Figure 3: Minimum radius for the PIB-solution: simulations and experiments. De=0.55, Bo=1.80, λ=14.9, α=0.71 and R_0=2.5mm.

In Figure 4 the experimental and simulated profiles of the filament of the PIB-solution are compared.

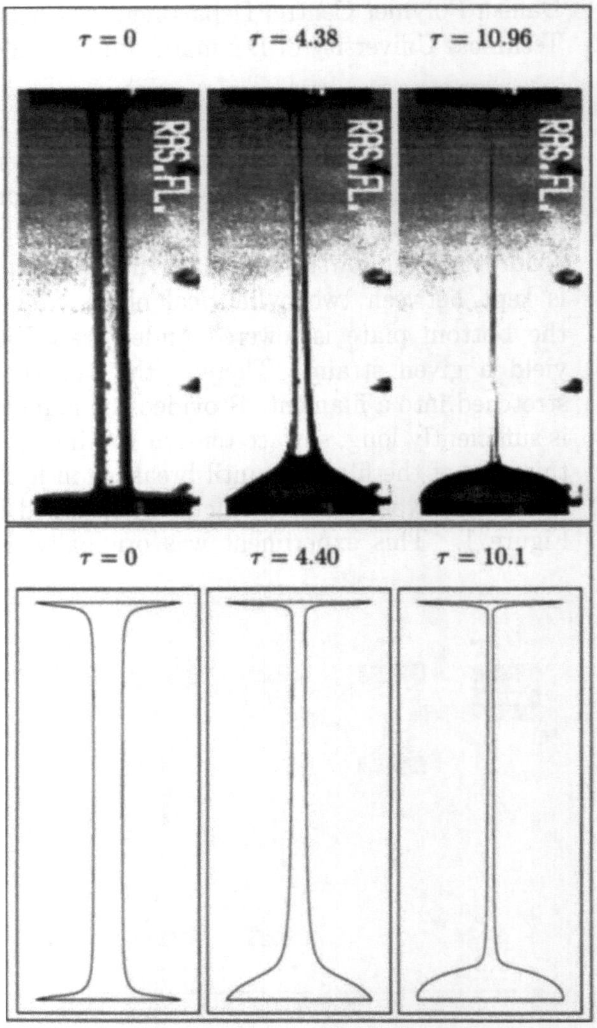

Figure 4: Filament profiles from experiments and simulations with the PIB-solution. The conditions are as in Figure 3.

REFERENCES

1. A.V. Bazilevsky, V.M. Entov & A.N. Rozhkov, Liquid filament microrheometer and some of its applications, Proc. Third European Rheology Conference, editor D.R. Oliver, pages 41-43, Elsevier, London and New York, 1990.

2. M.I. Kolte, H.K. Rasmussen & O. Hassager, Transient filament stretching rheometer, II: Numerical simulation, Rheol. Acta, **36**, 285-302(1997).

The Effect of Lubricant Rheology in Dynamically Loaded Journal Bearings

X. K. Li[1], A. R. Davies[2] and T. N. Phillips[2]
[1]Department of Mathematics, De Montfort University, Leicester LE1 9BH
[2]Department of Mathematics, University of Wales, Aberystwyth SY23 3BZ

The flow between eccentric cylinders is of considerable industrial importance as this is the geometry used to model journal bearings. These bearings are employed in large rotating machinery such as gas turbines, electric power generators and automotive engines. The mathematical modelling aspects of these problems are extremely complex. Consider a lubricant occupying the three dimensional region between two eccentric cylinders, where the inner cylinder (the journal) rotates and is also free to move under a time-dependent applied load, while the outer cylinder (the bearing) is stationary. In dynamically loaded bearings it is important to predict and assess the performance of lubricants within the bearing with respect to friction, wear and load bearing capacity over a wide range of operating conditions.

The classical treatment of the flow in a journal bearing has been via the lubrication approximation introduced by Reynolds. This approximation has served bearing designers well for decades. Nowadays, the use of process fluids as lubricants and the routine treatment of lubricants with additives to impart desirable material characteristics, challenges the validity of the classical lubrication approximation. If the lubrication approximation is not invoked, then there is no option but to solve the full set of coupled equations (kinematic and constitutive) governing the flow of the lubricant and taking proper account of the moving parts of the geometry. Until recently this task has proved too formidable, but with current computing power combined with efficient and accurate numerical methods, the calculation may be attempted.

The present work presents and discusses some results obtained with a very general and physically realistic mathematical model of oil film thickness and hydrodynamic friction in a journal bearing. Particular emphasis is placed upon the automotive connecting rod and main bearings. There are many criteria that need to be taken into account in the modelling of the lubricant in dynamically loaded journal bearings. Some of the most important are oil feed, thermal effects, cavitation and viscoelascity, changes in viscosity with shear-rate, temperature and pressure and three dimensional effects in dynamically loaded journal bearings. Although in the final analysis, it is an understanding of the combined effects of these phenomena which will be of greatest interest to lubricant manufacturers, the investigation of individual effects is essential in the development of this understanding. In this paper we shall investigate the effects of a variable viscosity in the three dimensional dynamically loaded journal bearing. The model has the following features:

- thermal transport with a choice of thermostatted or non-thermostatted boundary conditions on the bearing or journal surfaces.
- non-Newtonian model for the lubricant with a pressure, temperature and shear-rate dependent viscosity law.
- simplified (that is, not two-phase) cavitation model based implicitly on the pressure distribution

In this paper the spectral element method is used to discretize the governing equations in space and the operator-integration-factor splitting technique is used to integrate the energy equation in time. The generalized Stokes problem at each time step is solved using the conjugate gradient method in which the preconditioner is chosen in a dynamic fashion. A computer visualisation technique is used to trace the motion of the dynamic journal bearing system and to demonstrate the 3D variations of temperature and pressure in the oil film around a transiently loaded journal bearing. The effects of these variables together with shear-thinning effects on lubricant viscosity are also shown. The flow of the lubricant in the journal bearing is demonstrated

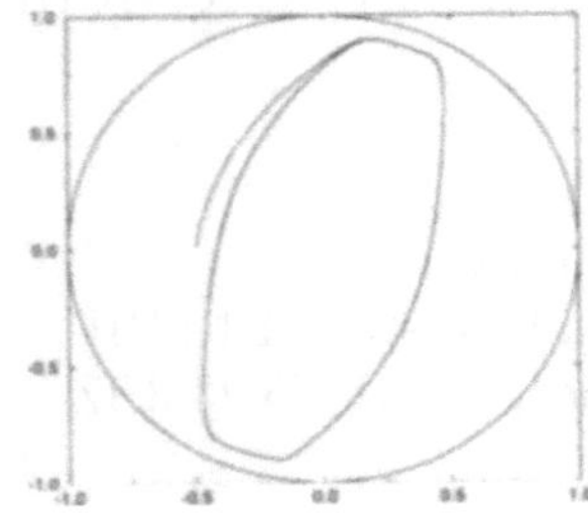

Figure 1: The journal centre orbit.

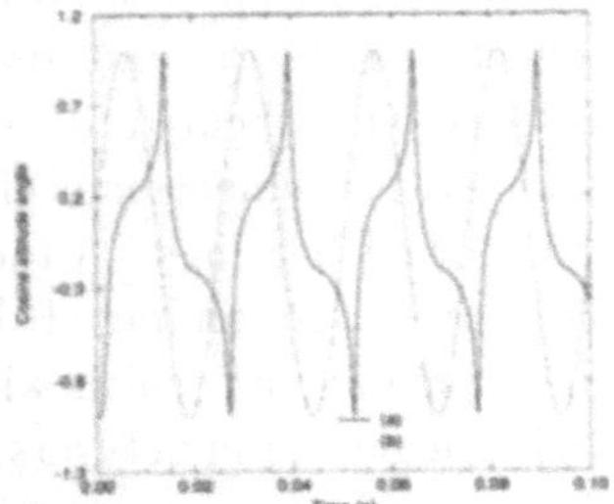

Figure 2: Cosine of the attitude angle.

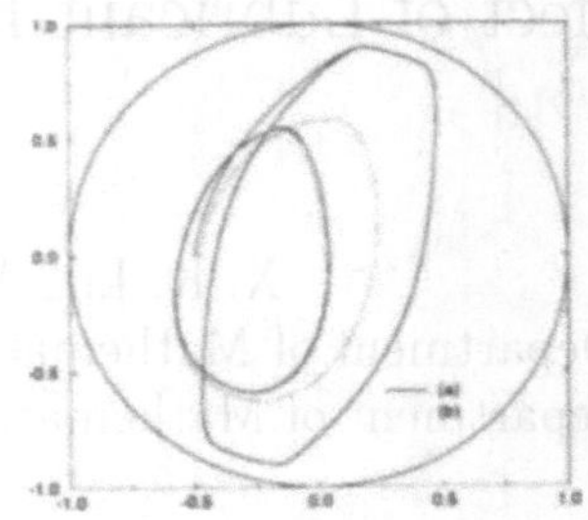

Figure 3: The journal orbits for diffrent loads.

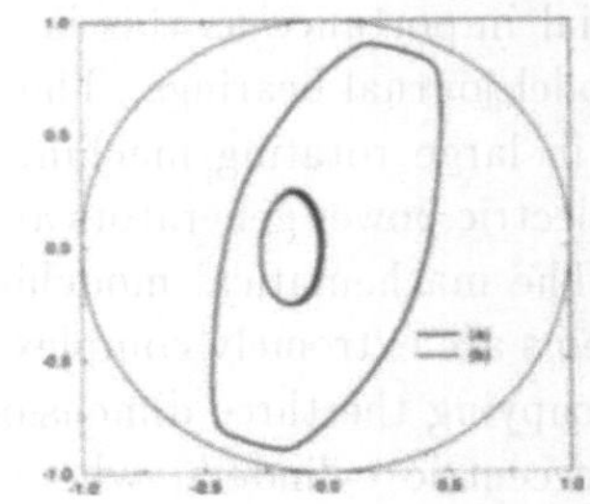

Figure 4: The journal orbits for different lengths.

to be extremely complex, with pressure effects dominating temperature effects in the generation of oil film thickness at the position of closest approach of the journal and bearing surfaces.

Variable applied loads for the dynamically loaded journal bearings have been studied by many investigators using the short or long bearing approximations (see Brindley *et al.* (1983)). To the authors' knowledge, a few have used the full set of governing equations and studied the nonlinear dynamics of the journal bearing under variable applied load conditions (see Gwynllyw *et al.* (1996b)for the 2D case).

An example of the stable closed path is given in Fig.1, where the magnitude of the variable applied load is $F_p = 5 \times 10^4$N. Fig.2 illustrates the dependence of the eccentricity ratio and the attitude angle with time. Referring to Fig.2 and comparing with the plot of $\sin(\omega t)$ clearly shows that the period of the closed path is the same as that of the applied load.

The effect of different magnitudes of variable applied loads is illustrated in Fig. 3. Unsurprisingly, decreasing the magnitude of the applied load results in a larger minimum oil film thickness. There is significant difference as well in the shape of the closed path with the lower variable applied load giving a more rounded path.

An important parameter in 3D journal bearings is the ratio of the length to the diameter of the journal L/D. Changing the length of the journal can have a significant effect on the journal's path as well as journal performance. We support this comment by comparing the two paths in Fig. 4 where the larger path corresponds to a bearing whose length is a tenth of that corresponding to the smaller path. With the same variable applied load F_p, we found that a decrease in the bearing's length always results in larger orbits, or equivalently, smaller minimum oil film thickness. More detailed discussion can be found in Li *et al.* (1997).

This work was supported by Shell Research Ltd., Thornton Research Centre, UK.

References

[1] Brindley, J. & Elliott, L. & McKay, J.T. 1983 The role of cavitation in whirl instability in a rotor bearing. *ASME J. Appl. Mech.* 50, pp.877-890.

[2] Gwynllyw, D.Rh. & Davies. A.R. & Phillips, T.N. 1996b A moving spectral element approach to the dynamically loaded journal bearing problem. *J.Compu. Physics* 123, pp. 478-494.

[3] Li X.K. & Gwynllyw, D.Rh. & Davies. A.R. & Phillips, T.N. 1997 Three dimensional effects in dynamically loaded jpurnal bearings. Research Monograph 7, School of Computing Sciences, DMU. To be appeared in *Int. Journal for Numerical Methods in Fluids.*

DETERMINATION OF RUBBER MELT'S INITIAL VISCOSITY FROM MOLECULAR WEIGHT DISTRIBUTION BY NEURAL NETWORKS

U. LOTRIČ, Z. ŠUŠTERIČ

Sava d.d., R&D Institute, Škofjeloška 6, SI 4000 Kranj, Slovenia

1. INTRODUCTION

Technological properties of rubber melts are related to their physical and chemical properties, in particular to their ability of flowing, characterized by their viscosity. Viscosity of rubber melts, as nonnewtonian liquids, depends on shear rate. The initial viscosity of a melt, i.e. viscosity at zero shear rate, is a material property from which viscosity at arbitrary shear rates may readily be calculated.

An accurate determination of the initial viscosity from measurement is often difficult, since it employs a modified empirical power law relating viscosity and shear rate (Bird et al. (1987)). Although it is well known that the initial viscosity depends on rubber's molecular weight distribution (Vinogradov et al. (1980)) which can be precisely measured, no analytical relation has yet been established between the two.

In this paper, an attempt to relate initial viscosity and molecular weight distribution with a neural network model is presented.

2. THEORETICAL

Neural networks are commonly used to find the relation between quantities when no adequate physical description is available or when such a description requires too many parameters (Bulsari (1995)).

In this work, a threelayered feedforward neural network, presented in Figure 1, is used. The data regarding molecular weight distribution of chosen rubber represent the input to the neural network and the viscosity is obtained on the output.

Each basic computation unit or neuron in the neural network is connected with all neurons from the previous layer in a way that the value on its output is nonlinear function of the weighed sum of outputs from neurons on previous layer (detail in Figure 1). Weights between the neurons represent the knowledge of the neural network. The values of the weights are set in the training process.

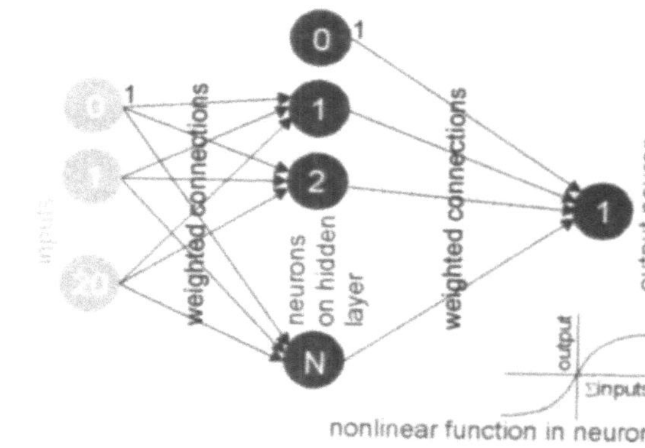

Figure 1: Applied multilayered feedforward neural network.

Neural network models are trained by a set of known input - output samples. All available samples are divided into learning samples and test samples. The first, larger group is used to set up the values of the weights and thus the model, while the second is used to test the model. At the beginning of the learning process, the weights are randomly set. Each molecular weight distribution from the learning group is used as the input to neural network and the calculated viscosity is compared to the measured value. According to the difference between the measured and calculated value, weights of all connections are corrected with use of back-propagation algorithm (Haykin (1994)). This process is repeated as long as the average square difference between the calculated and the measured viscosity is greater than some preset value.

3. EXPERIMENTAL

Molecular weight distribution and viscosity were measured for 36 different rubbers belonging to different structural groups: polybutadiene, styren-butadiene, polychloroprene, butadiene-acrylonitrile, ethylene-propylene-diene and butyl, the measurements on one such group being presented in Figure 2. For each rubber, three samples were used, two for learning and one for testing purposes.

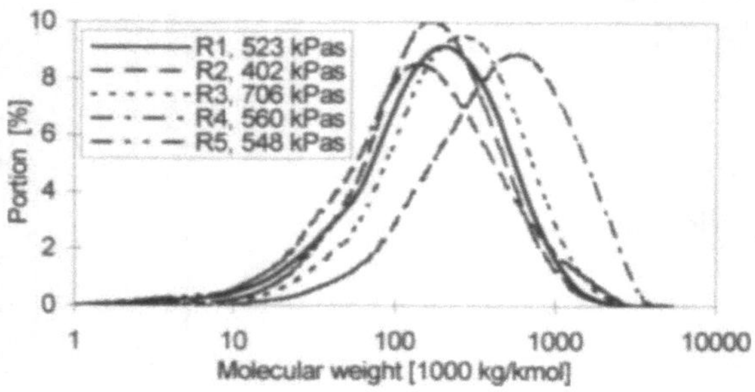

Figure 2: Molecular weight distribution and corresponding viscosity for some rubbers from ethylene-propylene-diene structural group.

A neural network with 20 inputs was used (Lotrič (1996)). The first 14 inputs served to describe the molecular weight distribution. Two inputs held the information about maximal and minimal molecular weight detected. The next three inputs were used for the frequently needed molecular weight averages for rubber characterization, namely the number, the weight and the Z-average. The last input contained information about the oil quantity, having a strong impact on viscosity, in rubber.

Two different types of neural networks were used: one to determine the initial viscosity within one structural group and one to determine it regardless of the structural group. They differ in the number of neurons in the hidden layer.

The best results for a single structural group were achieved with the neural network of only 5 neurons on the hidden layer. As can be seen in the top graph of Figure 3, the calculated initial viscosity for ethylene-propylene-diene rubbers is within the experimental error. Furthermore, the results on the test set (not used in the learning phase) show that the neural network has the ability to generalize the gained knowledge. The results in all the other structural groups are very similar to the foregoing ones.

The general neural network which could handle all structural groups contained 10 hidden neurons. An even higher number could have been used, but the learning time increases dramatically in such case. Results from this model are shown in the bottom graph of Figure 3. Results are not as good as in the first case – the calculated viscosity can be even outside the experimental error, as for rubber 5, for example. Worse results are expected because this network must handle 5 times more samples then the first one. The other reason is that the network should classify the rubber in the correct structural group.

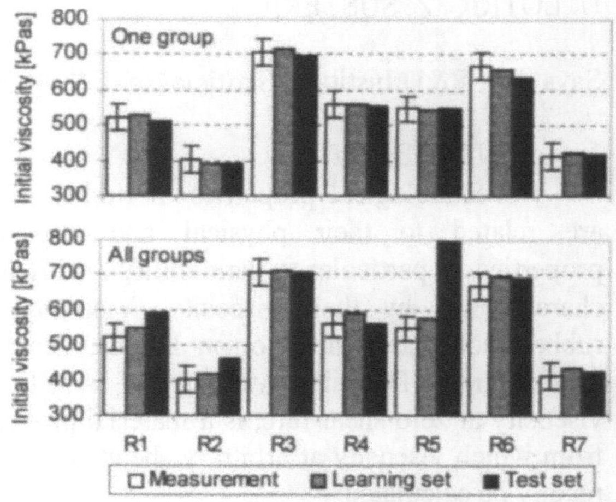

Figure 3: Determination of initial viscosity in ethylene-propylene-diene structural group.

4. CONCLUSIONS

The presented results show that the differences in initial viscosity of rubber melts are mainly due to differences in molecular weight distribution and less due to the structure of rubber molecules themselves. Such result was expected, considering that despite their structural differences, the molecules of different rubbers are all more or less linear and flexible. The latter two properties, namely, make the low glass transition temperature and high elasticity of rubbers possible. Regarding the properties pertaining to rubber technology, i.e. the flowing properties characterized by the melts' viscosity, it was shown that rubber melt morphology prevails in importance over chemical structure of rubber molecules.

Bird RB et al. (1987), Dynamics of Polymeric Liquids. Wiley, New York

Bulsari AB (1995), Neural Networks for Chemical Engineers. Elsevier, Amsterdam

Haykin S (1994), Neural Networks, A Comprehensive Foundation. MacMillan, New York

Lotrič U et al. (1996), Polimeri 17:254-257

Vinogradov GV et al. (1980) Rheology of Polymers. Mir, Moscow

THE ROLE OF MATERIAL IN THE SIMULATION OF THE THERMOFORMING PROCESS

PETR NOVOTNÝ[1], KAREL KOUBA[2], PETR SÁHA[1]

[1] Technical University of Brno, Faculty of Technology Zlín, nám. TGM 275, 76 272 Zlín, Czech Republic
[2] T-SIM CZ Ltd., Tr. T. Bati 299, 764 22 Zlín, Czech Republic

1. INTRODUCTION

Thermoforming is characterized by large deformations of the polymeric material being processed. In order to optimize the process, computer simulation has been found a useful method. However, the simulation is complicated by the fact that an elastic material model cannot principally describe the thermoforming deformation precisely. In this respect non-linear and time dependent viscoelastic behavior of the polymeric material has to be taken into account.

In our research we used K-BKZ model capable to describe viscoelasticity and large deformation. The parameters of this model were determined experimentally. The evaluation was performed using the T-SIM® simulation program.

It has been found that the final shape and the wall thickness distribution shows a significant dependence on the material properties.

2. MATERIAL MODEL

2.1 Elastic and viscoelastic models in thermoforming simulation

For computer simulation of thermoforming, elastic or viscoelastic model can be used to describe a material behavior. However, elastic models give poor results when used for simulations of specific thermoforming situations, such as sagging or plug assisted thermoforming. The reason for that is missing dependence of the stress on the deformation history. To get closely to reality, the only way is to use a viscoelastic, time dependent model.

2.2 K-BKZ model

We have used K-BKZ model proposed by Bernstein et al. (1963) and Kaye (1962):

$$\hat{\sigma}(t) = \int_{-\infty}^{t} \mu(t-t') \cdot h(I_1, I_2) \cdot \hat{B}(t,t') dt' \qquad (1),$$

where $\hat{\sigma}(t)$ is a stress tensor, $\mu(t-t')$ is a material memory function, $h(I_1,I_2)$ is a damping function of the two strain invariants and $\hat{B}(t)$ is the Finger strain tensor. Damping function is one of the functions suggested by Wagner (1976, 1990) or Papanastasiou et al. (1983).

3. MODEL PARAMETERS ESTIMATION

3.1 Measurement

The tests were performed on an Instron machine equipped with a specially developed apparatus enabling deformations similar to those appearing during plug assisted thermoforming. The schema of the apparatus is shown in Fig 1.

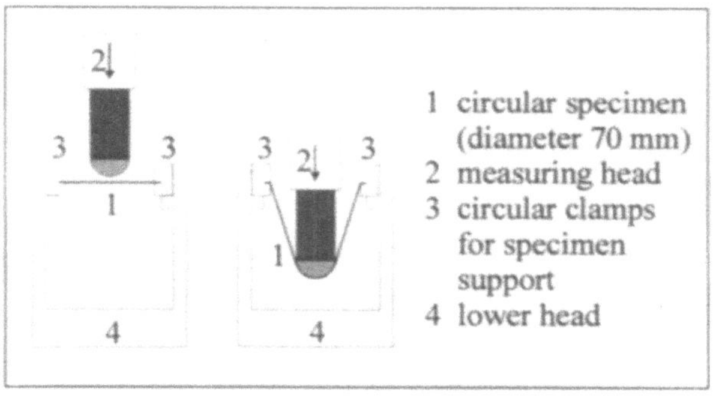

Fig 1: Apparatus used for measurements

During the tests the measuring (upper) head was moving down at a constant speed (100 - 1000 mm/min) and corresponding force was measured and recorded along with the deformation.

Resulting force vs. deformation curves at various temperatures and speeds of deformation has been used for fitting K-BKZ model parameters.

3.2 K-BKZ model parameters fitting

The K-BKZ model cannot be used directly to fit measured force-deformation data. True stresses and strains are unknown. A reverse engineering method is a suitable way to find material constants: A set of materials with variable parameters is used for simulations of the test. Calculated force-deformation curves are compared with the measured data and the best corresponding curve determines fitted material parameters. Fig 2 shows one of the simulation stages.

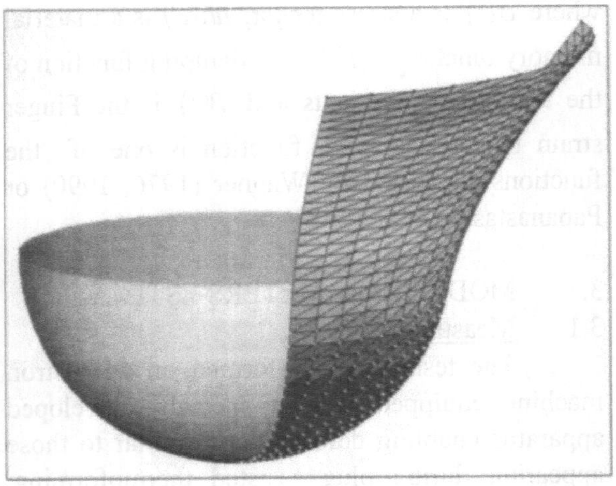

Fig 2: Process simulation

Fig 3 contains a graph showing measured data along with a simulated curve. The speed of the deformation was 1000 mm/min, the temperature 140°C, the material was PS.

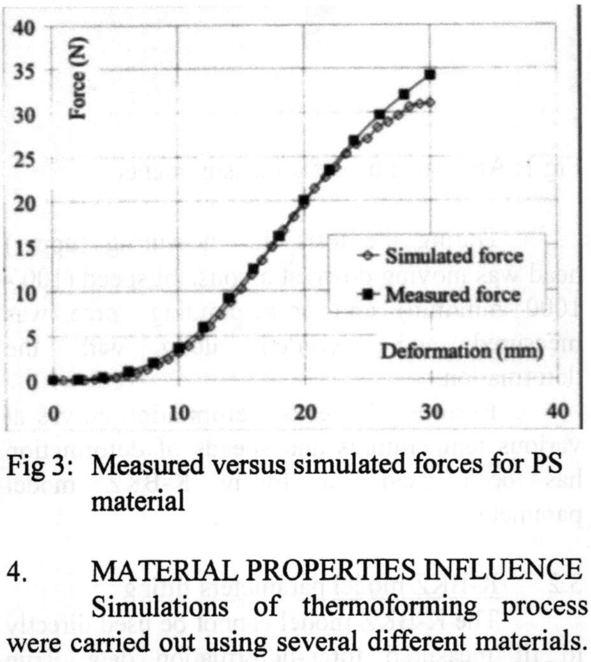

Fig 3: Measured versus simulated forces for PS material

4. MATERIAL PROPERTIES INFLUENCE

Simulations of thermoforming process were carried out using several different materials. The comparison of the resulting thickness profiles confirmed the prediction that the final shape and mainly the final wall thickness distribution are strongly influenced by material properties. Fig 4 shows thickness profile of a cup produced using plug assisted thermoforming. The leftmost part of the graph corresponds to the cup bottom. Material A was more viscous, material B more elastic.

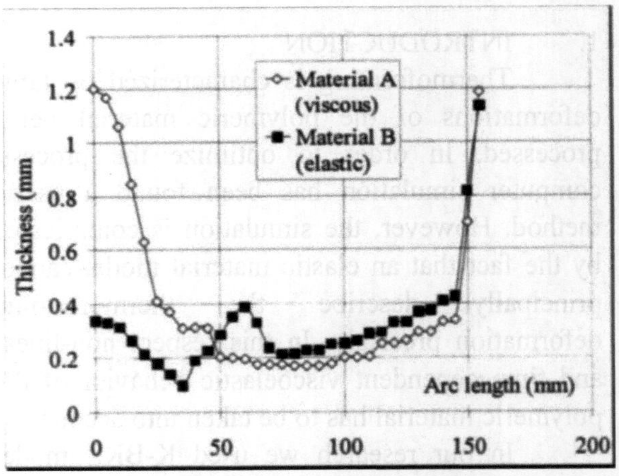

Fig 4: Thickness profile comparison

5. CONCLUSIONS

K-BKZ material model parameters were estimated using a special testing method and consequent computer simulation. The materials were characterized without necessity to perform expensive and complicated biaxial measurements.

The introduced method describes one of many possible ways to get K-BKZ model parameters. These parameters are critical as for the final wall thickness distribution of the thermoformed product.

6. REFERENCES

Bernstein B, Kearsley EA, Zapas LJ (1963) Trans Soc Rheol 7:391-410

Kaye A (1962) "Non-Newtonian Flow in Incompressible Fluids", College of Aeronautics, Cranfield, CoA Note No. 134

Wagner MH (1990) J Rheol 34(6):943-958

Wagner MH (1976) Rheol Acta 15:136-142

Papanastasiou AC, Scriven LE, Macosko CW (1983) J Rheol 27(4):387-410

NUMERICAL PREDICTION OF THE ENCAPSULATION EFFECT FOR POLYMER MELTS

RENAUD SIZAIRE and VINCENT LEGAT

CESAME, Division of Applied Mechanics, Université catholique de Louvain,
B-1348 Louvain-la-Neuve, Belgium

1. INTRODUCTION

In co-extrusion flows, it is well known that a fluid often encapsulates the other one and creeps along the wall while the encapsulated fluid concentrates in the core. This process occurs at two different time scales. Firstly, a fast rearrangement appears at the merging lip. Secondly, a very slow encapsulation effect occurs along the tube. The fast encapsulation effect is explained from the rearrangement of the velocity profiles at the merging lip, but the real challenge is in fact to explain the origin of the slow encapsulation. On one hand, most experimental and theoretical studies suggest that only the viscosity difference between the two fluids leads to encapsulation. The less viscous fluid tends to encapsulate the more viscous one. On the other hand, effect of elasticity in the problem of layer-to-layer encapsulation has also been the subject of a number of investigations. Uwaji (1995) suggested that some slow encapsulation effects explained so far by the viscosity ratio is, in fact, only caused by the second normal stress difference.

In the present paper, we confirm this assumption. In particular, the Reiner-Rivlin model allows us to demonstrate that the slow encapsulation may only be caused by the second normal stress difference. We simulate the experimental device used by Han (1973) with the Giesekus and the Reiner-Rivlin models. A good agreement between the three-dimensional numerical results and the experimental observations is obtained.

2. GOVERNING EQUATIONS

For incompressible fluids, the momentum and mass conservation yields

$$\nabla \cdot \boldsymbol{v} = 0,$$
$$\rho \frac{D\boldsymbol{v}}{Dt} = \boldsymbol{\nabla} \cdot (-p\boldsymbol{I} + \boldsymbol{\tau}) + \boldsymbol{f}, \qquad (1)$$

where $\boldsymbol{v}$ is the velocity, p is the pressure, $\boldsymbol{f}$ denotes the body force per unit volume, and $\boldsymbol{\tau}$ is the extra-stress tensor. The fluid density is denoted by ρ, and the operator D/Dt is the material derivative.

The motion of the interface between the layers is imposed by both the kinematic and the dynamical conditions. Furthermore, an original technique is proposed to predict the motion of the contact line when a stick condition is applied on the die. Our approach can be viewed as the limit of a vanishing slip condition.

In order to close the conservation laws, the viscoelastic contribution $\boldsymbol{\tau}_V$ of the extra-stress tensor has to be related to the deformation of the fluid through a constitutive equation We use the Giesekus model as one relaxation time is enought to obtain a fair fit of the rheometrical data.

$$\boldsymbol{\tau}_V + \lambda \overset{\triangledown}{\boldsymbol{\tau}}_V + \frac{\alpha\lambda}{\eta_V} \boldsymbol{\tau}_V \cdot \boldsymbol{\tau}_V = 2\eta_V \boldsymbol{D}, \qquad (2)$$

where α, λ and η_V are the mobility, the relaxation time and the viscoelastic viscosity, respectively. The rate of deformation tensor is given by $\boldsymbol{D}$ and $\triangledown$ stands for the upper-convective derivative. As an interesting and efficient subsistute, we also use the Reiner-Rivlin constitutive equation

$$\boldsymbol{\tau} = 2\eta \boldsymbol{D} + 4\psi_2 \boldsymbol{D} \cdot \boldsymbol{D}, \qquad (3)$$

where the functions η and ψ_2 are characterized by Bird-Carreau laws in order to reproduce the rheometrical data. Even if this equation is not a popular viscoelastic model and cannot be considered as a realistic approximation of any real material, it is a useful one because it exhibits second normal stress difference, without any first normal stress difference or any memory effect. In other words, this model allows us to focus on the influence of the second normal stress difference. With more realistic models, such as the Giesekus or PTT equations, it would not be so easy to isolate the influence of the second normal stress difference and it would make the interpretation of results more difficult. Finally, the Reiner-Rivlin model does not require to introduce any extra-stress field and then appears to be very attractive for the numerical calculations.

3. DISCUSSION

The numerical technique has been validated for a three-dimensional co-extrusion flow with two Bird-Carreau fluids, but no slow encapuslation effect was observed. With the Reiner-Rivlin and the Giesekus models, we predict the slow encapsulation effect. Calculations with the Giesekus model still remain highly expensive and a lack of convergence is observed even for relatively low Weissenberg numbers. But, those calculations demonstrate that the second normal stress difference can be the cause of the slow encapsulation effect. Indeed, it is always possible to obtain encapsulation only by increasing the second normal stress difference.

At the exit of circular dies, a good agreement is obtained between experimental cross sections as observed by Han (1973) and our numerical calculations with the Reiner-Rivlin model. Shear viscosity and first normal stress difference rheometrical data are used to select the material parameters of the Giesekus model. For the Riener-Rivlin model, we use the second normal stress difference predicted with the Giesekus model as a substitute for the missing measurements of this material parameter. Such a procedure can be considered as a two-step approach. Firstly, we approximate the fluid by a Giesekus equation. Then, we approximate this Giesekus model by the Reiner-Rivlin equation. Even if the use of this Giesekus-Reiner-Rivlin model can be frowned in most cases, it is fully justified here in the analysis of encapsulation in straight tubes, as an interesting compromise between physical reality and computational cost.

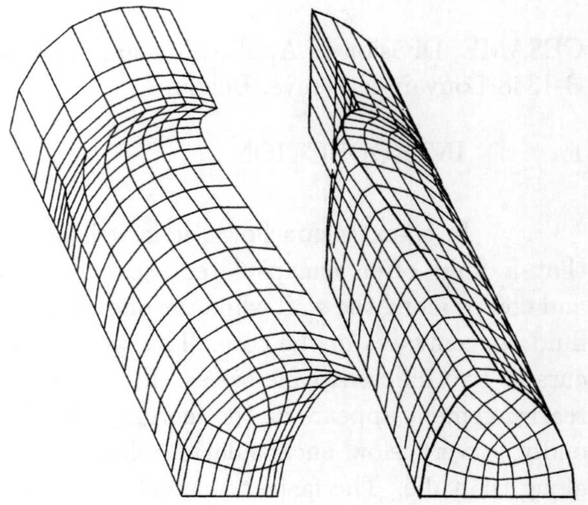

Figure 1: Typical configuration of the layers for the Reiner-Rivlin models for LDPE and PS co-extrusion flow.

4. ACKNOWLEDGEMENT

This work is supported by the ARC 97/02-210 project, *Communauté Française de Belgique*. The work of V. Legat is supported by the Belgian *Fonds National de la Recherche Scientifique* (FNRS).

C.D. Han (1973), Journal of Applied Polymer Science 17:1289-1303

T. Uwaji, (1995), Studies on wrap around phenomena of stratified two phase flow, Master thesis, UCL

R. Sizaire, V. Legat (1998), Journal of Rheology, submitted

R. Sizaire (1998), Numerical study of free surface Newtonian and viscoelastic flows, PhD thesis, UCL

A STUDY OF FINITE VOLUME METHODS FOR VISCOELASTIC FLOWS

P. WAPPEROM, M.F. WEBSTER

Institute of Non-Newtonian Fluid Mechanics, Department of Computer Science University of Wales, Swansea, Singleton Park, Swansea SA2 8PP, U.K.

1 INTRODUCTION

This study investigates the application of a new hybrid finite element/finite volume scheme to the numerical solution of some viscoelastic flows. Such an FE/FV approach (see similarly Sato and Richardson [3]) is contrasted against a purely finite element alternative, that has previously been developed to address highly elastic problems, see Matallah *et al.* [2]. For incompressible viscoelastic flows, with non-trivial source terms, the question arises as to whether a finite element approach may be best suited to solve for the parabolic field equations, concerned with the conservation of mass and momentum, whilst a finite volume approach may be more appropriate for the hyperbolic constitutive law, as originally devised. For example, recent work on advection equations has shown that fluctuation distribution schemes that handle upwinding for fluxes, can yield accurate solutions for problems that manifest steep gradients, see Struijs *et al.* [4]. Efficiency, accuracy and stability are the prime considerations, to be achieved through a localised control volume view, high resolution TVD-type schemes and oscillation-free transient solutions. Computational tractability follows as a consequence.

2 THEORY

The basic equations are those derived from conservation of mass and momentum, and transport of stress. The finite element framework, around which this work is centred and compared to, is a semi-implicit second-order time-stepping Taylor–Galerkin/Pressure-Correction scheme of three fractional stages, that incorporates consistent Petrov–Galerkin streamline upwinding and recovery in the discrete treatment of the constitutive equation (FE/SUPG). In two dimensions, the finite element grid is constructed as a triangular tessellation, with pressure nodes located at the vertices (linear representation) and velocity/stress components at both vertices and mid-side nodes (quadratic representation). In the FE/FV hybrid scheme, a cell vertex approach (Crumpton *et al.* [1]) is adopted in the FV part for the constitutive equation. Lax–Wendroff time-stepping is built into this scheme, as with the FE scheme above. Four linear FV triangular cells are constructed as subcells of each parent quadratic FE triangular cell, connecting the mid-side nodes. With stress variables located at the vertices of the FV cells, no interpolation is required to recover the FE nodal stress values. The governing FV stress matrix is the identity matrix. For the FE method it is sparse, and necessitates the solution of a matrix-vector equation, hereby with an iterative method as for the momentum equation. This step is circumvented using the hybrid FE/FV, with straightforward right-hand side construction, proving a particular advantage for three-dimensional or multimode computations.

Various different alternative treatments are considered for dealing with flux and source terms, capturing such properties as positivity and linearity preservation. This introduces the concept of both linear (LDB and N-scheme) and non-linear (PSI-scheme) fluctuation distribution schemes [4]. Our objective is to compare and contrast the discrete implementation of these various schemes, to indicate their relative strengths and weaknesses. As their initial development was for pure convection problems, here their extension is considered also for those with source terms, as for an Oldroyd-B benchmark flow.

3 PROBLEM DESCRIPTION

Orders of accuracy and efficiency in attaining solutions are established for a two-dimensional Cartesian test problem with analytical solutions on a unit square domain, (x_0, x_1)x(y_0, y_1) with $x_0 = 1$, $x_1 = 2$. This problem may be stated in pure convection form, or for an Oldroyd-B model that exhibits source terms. Structured, uniform, quadrilateral-based, triangular finite element meshes are used to test for accuracy, of 2x2, 4x4, 8x8 and 16x16 elements. For the velocity field, we define $u_x = x$ and $u_y = -y$. Boundary conditions for individual components must be specified, for stress these are required on the inflow boundaries $x = x_0$ and $y = y_1$.

4 RESULTS

For the pure convection problem and comparison of FE versus flux distribution schemes, the difference from the exact solution with mesh refinement, a fixed velocity field, linear representation of stress, and domain 1 ($y_0 = 1$, $y_1 = 2$), are displayed in Table 1. Comparison is made

Table 1: Error norm behaviour for pure convection.

	FE	FV		
	SUPG	N	LDB	PSI
2	$2.0\,10^{-3}$	$1.9\,10^{-2}$	$1.9\,10^{-3}$	$2.4\,10^{-3}$
4	$2.8\,10^{-4}$	$1.2\,10^{-2}$	$4.8\,10^{-4}$	$5.2\,10^{-4}$
8	$3.8\,10^{-5}$	$6.4\,10^{-3}$	$1.2\,10^{-4}$	$1.4\,10^{-4}$
16	$5.0\,10^{-6}$	$3.4\,10^{-3}$	$1.2\,10^{-4}$	$3.5\,10^{-5}$

on maximum error norm behaviour throughout, here between the FE/SUPG, N-scheme (linear, positive), LDB-scheme (linear, linearity preserving) and PSI-scheme (nonlinear, linearity preserving, positive). The linearised advection velocity approach, is assumed initially and results are characterised via component scalar solutions of the form $1 + (xy)^{0.5}$, see Wapperom and Webster [5]. The FE data shows almost $\mathcal{O}(h^3)$ convergence. What is immediately clear for this model problem, is that the linearity preservation property is essential to obtain the higher levels of accuracy. The PSI- and LDB-scheme display $\mathcal{O}(h^2)$ convergence, whilst the N-scheme only achieves $\mathcal{O}(h^{0.8})$. Furthermore, the LDB-scheme has a wider stability range on time step, practically double that of the others. The LDB-scheme renders the fastest convergence rate, that results in 2% of the FE CPU time.

For the Oldroyd-B flow problem that has source terms, we compare in Table 2 the FV method employing linear (FVL) and quadratic integral (FVQ) evaluation with FE/SUPG for the more severe test case of domain 2 ($y_0 = 0.1$, $y_1 = 1.1$), where the velocity is almost parallel to the lower boundary. For this problem and domain 1, FVL- and FVQ-schemes are approximately second-order accurate and FE/SUPG is almost cubic. On domain 2, FE/SUPG achieves

Table 2: Error norm behaviour of τ_{xx} for Oldroyd-B.

	FE/SUPG	FVL	FVQ
2	$1.4\,10^{-3}$	$2.2\,10^{-3}$	$7.3\,10^{-4}$
4	$3.6\,10^{-4}$	$1.2\,10^{-3}$	$1.9\,10^{-4}$
8	$6.5\,10^{-5}$	$5.4\,10^{-4}$	$3.6\,10^{-5}$
16	$9.7\,10^{-6}$	$2.3\,10^{-4}$	$5.7\,10^{-6}$

$\mathcal{O}(h^{2.3})$ accuracy. Table 2 shows the loss of accuracy with linear integral evaluation; this appears in all components. In worst case, we observe $\mathcal{O}(h)$ convergence for τ_{xx}. In contrast, we recover the desired $\mathcal{O}(h^2)$ convergence in FVQ with quadratic integral evaluation.

5 CONCLUSIONS

The introduction of source terms in a consistent manner, does not detract from the accuracy of the FV scheme. The effective coupling of the FE and FV components of the hybrid scheme has been demonstrated elsewhere [5], with accuracy between $\mathcal{O}(h^2)$ and $\mathcal{O}(h^3)$, hence showing no deterioration from the FE discretisation. As the FVQ implementation for stress alone takes 10% of the time for its FE counterpart, solution time is reduced to that of the base Navier–Stokes solver. Hence, a considerable gain in efficiency is anticipated with this hybrid FE/FV scheme for large problems, involving either multi-mode or three-dimensional calculations.

6 REFERENCES

[1] Crumpton PI, Mackenzie JA, Morton KW (1993) J Comput Phys 109: 1-15

[2] Matallah H, Townsend P, Webster MF (1998) J Non-Newtonian Fluid Mech 75: 139-166

[3] Sato T, Richardson SM (1994) J Non-Newtonian Fluid Mech 51: 249-275

[4] Struijs R, Deconinck H, Roe PL (1991) Lecture series 1990-01, Von Karman Institute for Fluid Dynamics

[5] Wapperom P, Webster MF in press J Non-Newtonian Fluid Mech

STARCH MELTING UNDER SHEAR

C. Barron, G. Della Valle, P. Colonna, B. Vergnes*.

INRA, Centre de Recherches Agro-alimentaires, 44316 NANTES Cedex 3, FRANCE

* CEMEF, Ecole des Mines de Paris, 06904 SOPHIA-ANTIPOLIS Cedex, FRANCE.

1. INTRODUCTION

Thermal transitions of starch have been thorougly studied in the last years by differential scanning calorimetry (DSC), X-ray diffractometry and microscopy. Melting temperature T_m may be measured; its variations according to the water content follow the law of Flory: $1/T_m \sim V_1$, V_1 being the volumic fraction of water. However, these experimental conditions are sometimes far from those encountered in industrial processes. During extrusion cooking, for instance, low hydrated starch (MC<40%) is submitted to high heating (≈10 °C/s) and shearing rates and (100 s^{-1}). These conditions lead to the concomittant obtention of an homogeneous molten phase and partial depolymerization of starch. The aim of this work is to better understand the physical mechanisms and to determine the main energetic transfers involved in these phenomena .

2. MATERIAL AND METHODS

2.1 Starch material:

Due to its average amylose content (37%) and very low lipid content, as well as promising mechanical properties (Lourdin et al. 1995) smooth pea starch (Nastar-Cosucra, Belgium) is chosen. It is then hydrated at MC=30% . Melting temperature is previously determined by DSC and Tm (temperature of maximum enthalpy) is found equal at 115°C for MC=30%.

2.2 Starch processing:

Thermomechanical treatment is performed with a pre-shearing rheometer, Rheoplast, which has already shown useful to simulate extrusion-cooking (Vergnes et al. 1987). Material temperature may be increased rapidly by barrel heating or shearing from the inner piston of a Couette geometry, during a well defined duration. this device allows to control processing conditions ($20 < \gamma_w < 80\ s^{-1}$, $10 < t < 30$ s), barrel temperature is set at Tm+/-20°C (95-135°C). Torque measurement allows to define the specific mechanical energy delivered to the material (SME).

Processed material is transfered to the outlet cylindrical chamber of same volume as the Couette zone. Samples (L=8cm, Ø=16mm) can then be cut into slices and vacuum dried (40°C) before further analysis.

2.3 Structural analysis

Starch transformation is evaluated at different scales by intrinsic viscosity measurements (macromolecular degradation), DSC and X-ray diffractometry (cristallinity), and optical and SE-microscopies (granule) according to procedures reported by Planchot et al. (1997).

3. RESULTS AND DISCUSSION

3.1 Destucturation mechanism

Results fom structural analysis are reported in table 1. Most influent processing variable is the barrel temperature. For T>Tm, the whole sample is a plastic translucent homogeneous mass whereas for T<Tm, each slice contains a center part (B) surrounded by powder (A), made of native starch granules easy to recover in water suspension. For T>Tm, optical microscopy shows that granules have lost their structure; only ghosts remain and cristallinity has completely disappeared. In contrast, for T<Tm, semi-crystalline granules are still present and embedded in an amorphous cement. Macromolecular degradation is very low and similar in both cases.

Structural level	T<Tm	T>Tm
macroscopic	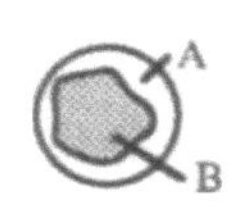	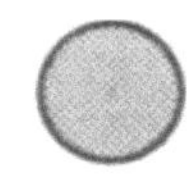
granular	+	+++
crystalline		
loss of crystallinity (X-ray diffraction)	25%	100%
loss of gelatinization enthalpy (DSC)	55-80%	100%
macromolecular		
degradation	15%	15%

Table1: Main structural features on thermo-mechanically processed starch.

These results suggest that:

(i) shear rate promotes starch melting for temperature lower than Tm,

(ii) melting is initiated near the rotating part, i.e. the inner piston,

(iii) loss of cristallinity begins at the hilum, like for starch gelatinization in excess of water, unlike the « onion peeling » phenomenon observed for polymer particles (Todd 1993). Furthermore, no granule deformation or rupture is observed by SEM. Same results are found when processing pea starch at MC=25%, provided that dependency of Tm upon water content is taken into account.

3.2 Pocess- transformation level relationships

Experiments have been carried out with various values of residence time and rotation speed. Mechanical contribution to startch melting can thus be emphasized by comparing the experimental SME to the quantity of powder (part A for T<Tm) recovered by suspending samples in an excess of water: the larger the SME, the smaller the powdery part of starch (Fig.1). For values closed to 100kWh/t (=360J/g) the quantity of powder becomes negligible, in agreement with SME values proposed for the obtention of an homogeneous starchy molten phase in extrusion-cooking (see for instance Guy & Horne 1988).

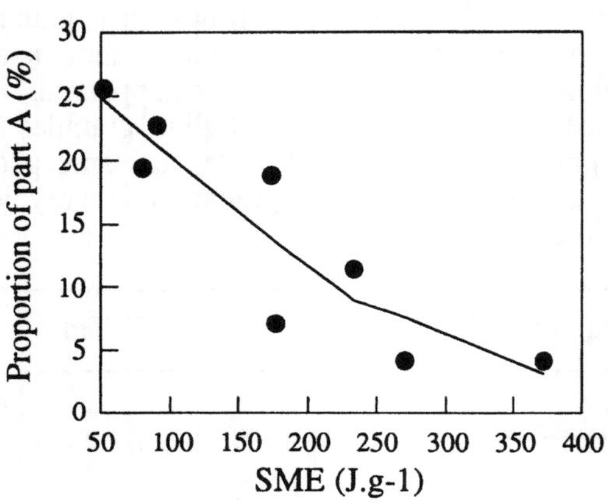

Fig.1: Variations of the quantity of starch powder with the experimental SME during processing (T<Tm).

However, such a correlation cannot be made when comparing SME values to the quantity of residual cristallinity either evaluated by DSC or X-ray diffractometry. This last result underlines the necessary conversion of mechanical energy into heat, in order to reach Tm and achieve granule melting.

4. CONCLUSION: Energetic contributions

Energy transfers in the Couette zone of the Rheoplast can be tackled, in a first approach, by the simple model proposed by Vergnes et al. (1987). However, that can only be employed in our case, for T>Tm, since it is based upon complete melting of starch. Therefore, in agreement with preceding findings, we suggest that this model can be refined by considering an intermediate zone in which flows a concentrated suspension of starch granules in a molten starch phase (zone 2 from Fig.2).

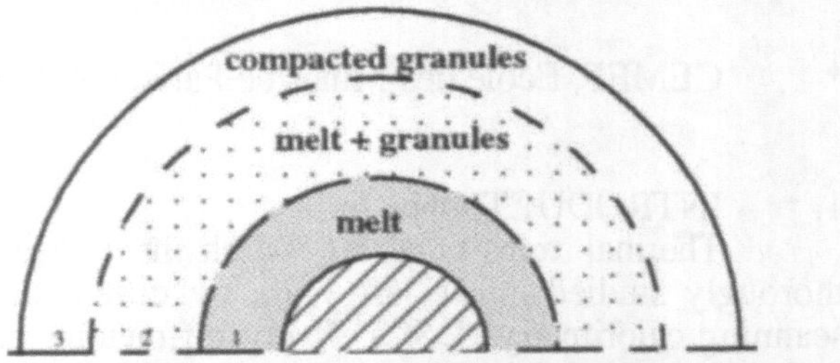

Fig.2: Schematic section of Couette zone of the Rheoplast

The dissipation of mechanical energy (SME) in the three zones may be due to :

(1) melt viscosity η_m

$$SME_1 = \frac{\eta_m \times \gamma^2 \times \Delta t}{\rho}$$

(2) suspension viscosity η_{susp}

$$SME_2 = \frac{\eta_{susp}(\Phi_v) \times \gamma^2 \times \Delta t}{\rho}$$

(3) solid friction

$$SME_3 = \frac{f \times P(\rho) \times \gamma \times \Delta t}{\rho}$$

Δt, γ, ρ being the time, shear rate and density in each zone, repsectively. $P(\rho)$, f compressibility, friction coefficient, respectively, have to be determined by complemetary experiment. Transition zone 1 to 3, i. e. starch melting, is governed by the decrease of Φ_v, the volumic fraction of intact granules in zone 2, in relation to calorimetric transition of starch.Checking that: experimental SME = $SME_1 + SME_2 + SME_3$ can then help to scale-up to extrusion process, and complete existing twin screw extrusion model (Vergnes et al. 1998).

5. REFERENCES

Guy RCE & Horne AW (1988) In: J.M.V. Blanshard JMV, Mitchell JR (eds) Butterworths, London, pp 331-349

Lourdin D, Della Valle G, and Colonna P, (1995) Carbohyd. Polym. 27: 261-270

Planchot V, Colonna P, Saulnier L (1997) In Godon B (ed) Lavoisier Paris pp 215-269

Todd DB (1993) Intern. Polym. Proc. VIII 2:113-118

Vergnes B, Villemaire JP, Colonna P, Tayeb J (1987) J. Cereal Sci 5:189-202

Vergnes B, Della Valle G, Delamare L (1998) Polym. Eng. Sci ,accepted

RHEOTENS EXPERIMENTS AND ELONGATIONAL BEHAVIOUR OF POLYMER MELTS

A. BERNNAT, M.H. WAGNER

Institut für Kunststofftechnologie, Universität Stuttgart, Böblinger Str. 70, D-70199 Stuttgart, Germany

Introduction

Many polymer processes like fibre spinning and blow moulding are strongly influenced by the elongational viscosity of the polymer melt. Therefore the analysis of extensional flows with a complex rheological prehistory, as in the Rheotens test, is of importance.

The Rheotens Test

Meissner [1] developed the Rheotens test (Fig. 1) as a prototype flow for isothermal melt spinning under constant force with shear prehistory. The tensile force needed for elongation of an extruded filament is measured as a function of increasing draw-down speed. Small changes in polymer properties can lead to significant differences in the Rheotens diagrammes.

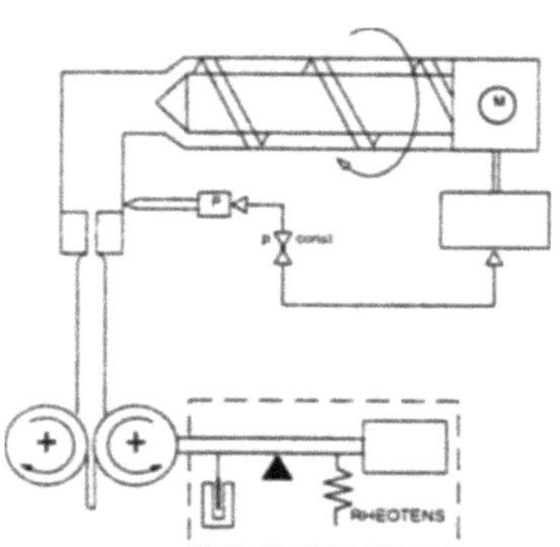

Figure 1: Experimental set-up for Rheotens measurement

Rheotens Mastercurves

For thermo-rheologically simple materials Rheotens curves are invariant with respect to temperature and average molar mass [2]. The concept of so-called Rheotens mastercurves can be extended to invariance with respect to mass flow rates, and die and spinline geometry.

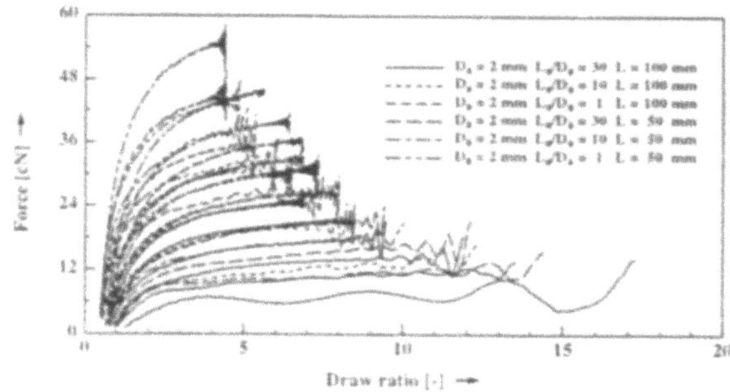

Figure 2: Rheotens curves of HDPE X1010 at different processing conditions, T = 190 °C, extrusion velocities between 15 and 220 mm/s

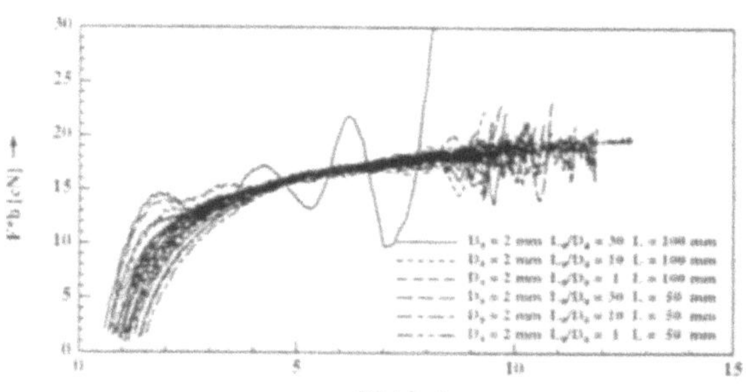

Figure 3: Rheotens Grandmastercurve of HDPE X1010

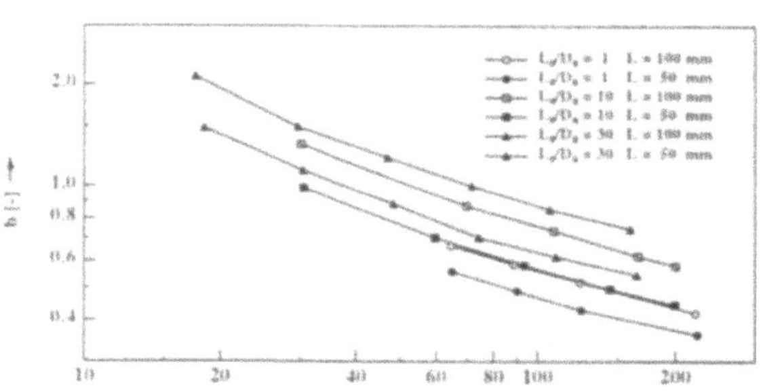

Figure 4: Scaling factor b and its dependence on processing conditions for HDPE X1010

These so-called Rheotens "Grandmastercurves" have been shown to be valid for LDPE [3]. For a HDPE melt (experimental grade X1010) investigated recently (Fig. 2), we found that while agreement of the mastercurves is good for high

values of the draw-ratio (except for the draw resonance effect), deviation of the experimental curves is found around their starting point (Fig. 3). This indicates that the scaling factor b (Fig. 4) which describes the influence of the processing conditions, overpredicts the change of extudate swell for this HDPE for reasons which have to be investigated in more detail.
We also report that for a LDPE melt (Lupolen 3020D), the filaments break at a certain critical stress σ_B independent from mass flow rate and die geometry within experimental error (Fig. 5). However, when melt fracture can be observed on the extruded strand, the critical rupture stress decreases.

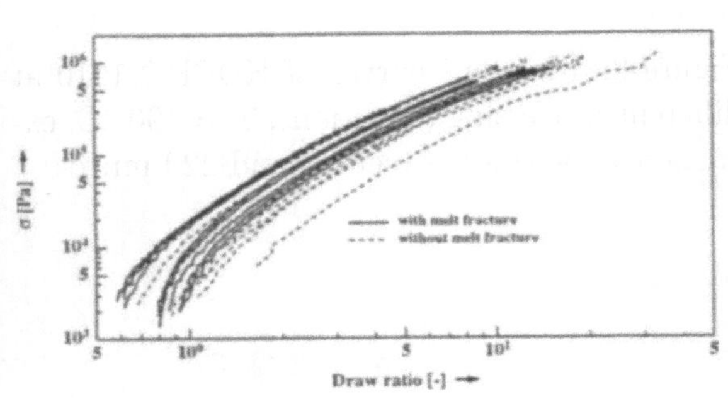

Figure 5: Rheotens stress curves as a function of draw ratio for Lupolen 3020D at various processing conditions. Dependence of critical rupture stress on melt fracture

Elongational Viscosity

The Rheotens diagrammes can be converted to transient elongational viscosities by the analytical model described in [3]. For the LDPE melt the result is given in Fig. 6.
The first part of the elongational viscosity curves corresponds to the viscoelastic start-up until the maximum is reached at a critical tension. At higher extension rates $\dot{\epsilon}$ the elongational viscosity decreases according to a power-law model. With increasing extrusion velocity v_0, the curves are shifted under -45° to smaller viscosities and larger extension rates which shows that the ratio of elongational to shear viscosity does depend on the deformation rate as well as on the rheological prehistory.

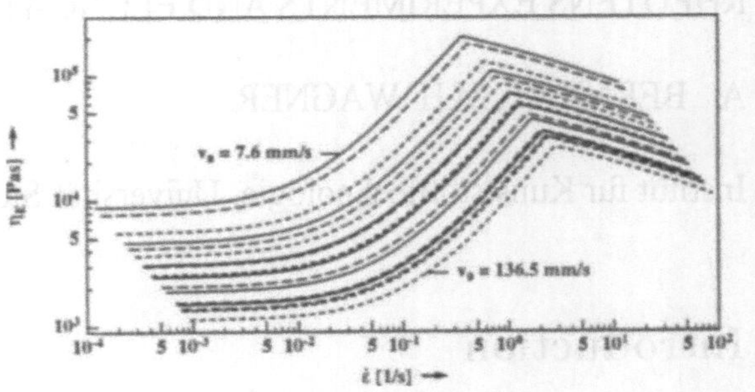

Figure 6: Elongational viscosity $\eta_E(\dot{\epsilon})$ for different extrusion velocities (LDPE 3020D, T = 230 ^{o}C)

Conclusions

The description of the material behaviour in Rheotens tests is simplified by the existence of Rheotens mastercurves. Simple scaling laws are valid for the description of the relationship between tension and deformation in a constant force experiment for LDPE. For a HDPE melt, some deviation is found.
The elongational viscosity can be calculated from the Rheotens curves by an analytical model. It is shown that the elongational viscosity strongly depends on the rheological prehistory of the melt in the extrusion die.

Acknowledgement

Financial support from the Commission of the European Union through the Brite/Euram III project ART (BE96-3490) is gratefully acknowledged.

References

[1] J. Meissner. Dehnungsverhalten von Polyethylen- Schmelzen. *Rheol. Acta*, 10(2):230–240, 1971.

[2] M.H. Wagner, V. Schulze, and A. Goettfert. Rheotens mastercurves and drawability of polymer melts. *Polym. Eng. Sci.*, 36(7):925–935, 1996.

[3] M.H. Wagner, A. Bernnat, and V. Schulze. The rheology of isothermal melt spinning. *Accepted J. Rheol.*, 1998.

PEROXIDE INDUCED DEGRADATION OF POLYPROPYLENE: INFLUENCE OF MOLECULAR STRUCTURE ON THE RHEOLOGICAL PROPERTIES

F. BERZIN[1], B. VERGNES[1], P. DUFOSSE[2], L. DELAMARE[2]

[1] CEMEF, Ecole des Mines de Paris, UMR CNRS 7635, BP 207, 06904 Sophia-Antipolis Cedex, France
[2] Elf Atochem, CERDATO, BP 19, 27470 Serquigny Cedex, France

1. INTRODUCTION

Polypropylenes manufactured using classical industrial polymerisation processes usually have excellent mechanical properties but have a high molecular weight and a broad polydispersity. This makes their processing difficult, in view of their high viscosity and elasticity. In order to adjust the molecular weight distribution, polypropylene resins can be modified in reactive extrusion operations by means of peroxide initiated scission reactions. The process is initiated by free radicals formed from the thermal decomposition of organic peroxide which lead to significant polymer degradation by β-scissions, because of the low stability of tertiary hydrogen atoms of macroradicals (Fritz and Stöhrer, 1986). In this work, we present experimental studies of the rheological behaviour of two polypropylenes obtained by the free-radical initiated molecular weight degradation, in a twin screw extruder.

2. EXPERIMENTS & DISCUSSION

2.1 Sample Preparation

Two kinds of commercial polypropylenes have been studied: the first one was a standard homopolymer (PP 3050 BN 1, Appryl) and the second one was a block copolymer (PP 3060 MN 5, Appryl) which included an addition of a low percentage of polyethylene (less than 10 %). The homo- and copolymer had melt flow indices of 4.5 g/10 min and of 6 g/10 min (2.16 kg/230°C) and initial weight average molecular weights of 301600 g/mol and of 256100 g/mol, respectively. The peroxide used was the 2,5-dimethyl-2,5-di(*t*-butylperoxy) hexane (DHBP, Trigonox 101, Akzo Chemie).

Reactive experiments were made in a co-rotating self-wiping twin screw extruder (model 30/34, Leistritz). By using various amounts of peroxide (0.01 to 0.5 wt.%), we have obtained a series of polymers with different molecular weight distributions. Samples were collected at the die exit, after 15 minutes of steady state extrusion conditions (stable die pressure). Extrudates were quenched in water before being cut into pellets. The samples were dried under vacuum, at 80°C during a 24-hour period, in order to eliminate all residual peroxide.

The molecular weight distributions were analyzed by gel permeation chromatography (GPC). The samples were dissolved at 145°C, at a polymer concentration of 0.8 g/L in the 1,2,4 Trichlorobenzene, which was used as a solvent with the addition of an antioxidant to prevent any degradation. Irrespective of the material, we noted that increasing the peroxide concentration narrowed the molecular weight distribution and decreased the presence of high molecular weight species.

2.2 Rheological Measurements

The viscosity of these different molecular weight polypropylenes (301600 to 62970 g/mol for the homopolymer and 256100 to 74430 g/mol for the copolymer) were measured in a Rheometrics 800 spectrometer using a parallel-plate geometry, in oscillatory mode. The dynamic measurements were carried out for frequencies ranging from 0.1 to 100 rad/s and at five temperatures (185°C to 245°C). Time sweep tests, made at different temperatures and at a frequency of 1 rad/s, showed that the complex viscosity and the storage and loss moduli were constant with time (30 min), whatever the homo- or copolymer.

Fig. 1 shows the result of the time-temperature superposition, for the complex viscosity (reference temperature of 215°C). For the homopolymer, we observe the usual decrease in viscosity with decreasing molecular weight (Fig. 1a) and a pronounced shift towards more Newtonian behaviour (Tzoganakis et al., 1989). For the copolymer (Fig. 1b), these tendencies are verified only for low amounts of peroxide, below 0.1 wt.%. When increasing this amount, and thus decreasing the molecular weight, we observe an increase in the complex viscosity first at low

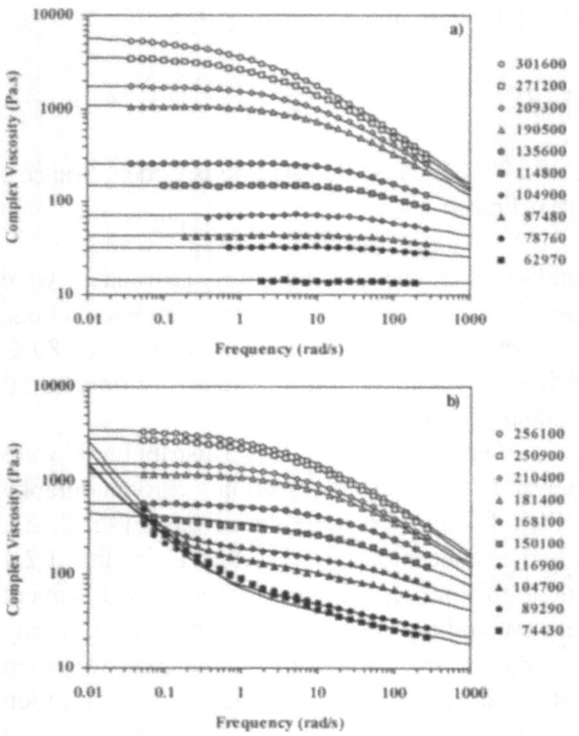

Fig. 1: Viscosity master curves ($T_{ref.}$ = 215°C)

frequencies, and then even for high frequencies. For the more reacted samples, the copolymer has the comportment of a yield stress fluid: the complex viscosity tends to infinite at low frequencies. This behaviour is due to the presence of the polyethylene phase which is dispersed in the polypropylene matrix (Yu et al., 1991). The polypropylene is highly degraded and has nearly a Newtonian comportment (Fig. 1a), whereas the polyethylene is much more viscous and behaves as a "solid" filler.

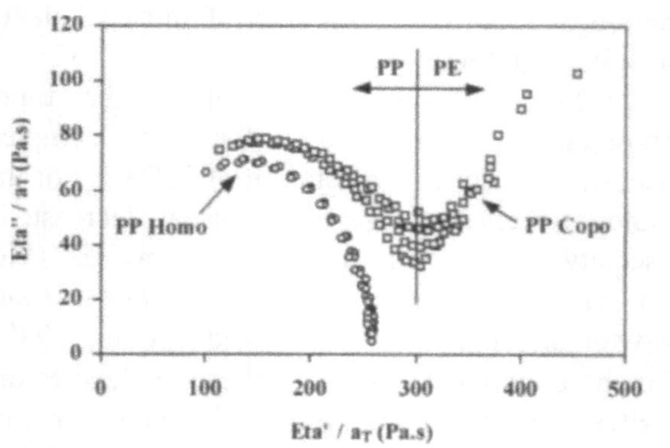

Fig. 2: Cole-Cole plot (master curves: $T_{ref.}$ = 215°C)

The representation of the imaginary part η″, as a function of the real part η′ (dynamic viscosity) of the complex viscosity (Cole-Cole plot) shows clearly the influence of the polyethylene phase as the molecular mass of the samples decreases. Fig. 2 shows the difference in behaviour between the homo- and copolymer, at a peroxide concentration of 0.08 wt.%. The weight average molecular weight is 135600 g/mol and 150100 g/mol for the homo- and copolymer, respectively. From the Cole-Cole plots, we can deduce the average relaxation times of the polypropylene matrix. It can be observed that decreasing the molecular weight decreases the relaxation time (Tzoganakis, 1994), whatever the homo- or copolymer (Fig. 3).

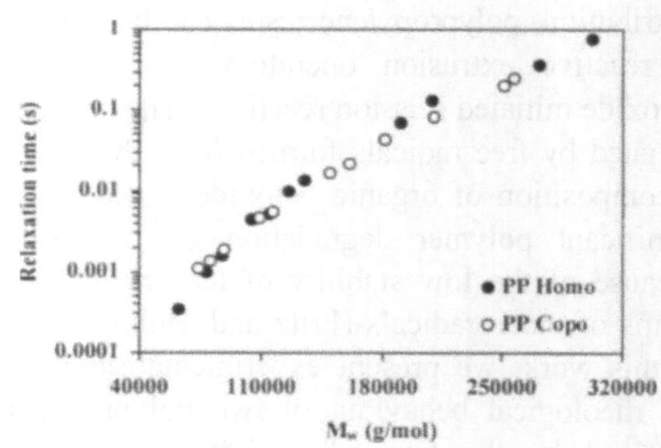

Fig. 3: Evolution of the average relaxation time

3. CONCLUSION

By reactive extrusion process, we have prepared a series of differently degraded samples of both a homo- and a block copolymer. The homopolymer exhibits a classical behaviour, whereas the rheological behaviour of the copolymer is influenced by the presence of the polyethylene phase which greatly increases the viscosity when the polypropylene matrix is highly degraded.

4. REFERENCES

Fritz HG, Stöhrer B (1986) Inter Polym Proc 1: 31-41

Yu DW, Xanthos M, Gogos CG (1991) Adv Polym Tech 10: 163-172

Tzoganakis C, Vlachopoulos J, Hamielec AE (1989) Polym Eng Sci 29: 390-396

Tzoganakis C (1994) Can J Chem Eng 72: 749-754

NUMERICAL SIMULATION OF THE EXTRUSION OF POLYMER MELTS IN THE PRESENCE OF SLIP AT THE WALLS

ERIC BRASSEUR[1], GEORGIOS GEORGIOU[2], ROLAND KEUNINGS[1] and VINCENT LEGAT[1]

[1]CESAME, Division of Applied Mechanics, Université catholique de Louvain, B-1348 Louvain-la-Neuve, Belgium

[2]Department of Mathematics and Statistics, University of Cyprus, Kallipoleos 75, P.O. Box 537, CY-1678 Nicosia, Cyprus

1. INTRODUCTION

Polymer melt flow instabilities in capillary flows have been the subject of research for decades (see Larson (1992)). Beyond a stable regime, three distinct unstable regimes are usually distinguished on the flow curve of polymer melt flow through a capillary. These are generally referred to as the sharkskin, the spurt and the gross-melt fracture regimes. However, not all of them are always encountered, depending on the polymer structure and the nature of the polymer-wall interface. Moreover, some other types of defects may also appear.

The ocurrence of slip and its importance in the mechanisms proposed for explaning the spurt instability have already been emphasized (Ramamurthy (1986), Kalika and Denn (1987), Georgiou and Crochet (1994), Wang (1996)). In a previous work, we have shown that self-sustained oscillations of the mass flow rate and the pressure drop are generated when melt compressibility and non-linear slip are combined in Poiseuille flow. In the present work, we consider the entire flow geometry that includes the barrel and extrudate regions and use a realistic slip equation, i.e. a modified version of the slip model proposed by Hatzikiriakos and Dealy (1992). Our objective is to make comparisons of the numerical simulations with the experiments of Hatzikiriakos and Dealy (1992) with HDPE.

2. MODIFIED SLIP LAW

Hatzikiriakos and Dealy (1992) proposed a two-branch slip equation to fit their experimental flow curve. For the left branch ($\sigma_{c_1} < \sigma_w < \sigma_{c_2}$),

$$u_s = \left\{1 - c_2 \tanh\left[\frac{1}{\mathcal{R}T}\left(E + c_3\frac{\sigma_n}{\sigma_w}\right)\right]\right\} \cdot f_1(T)\left(\frac{\sigma_w}{\sigma_{c1}\, I^{1/4}}\right)^{m_1}, \quad (1)$$

while for the right branch ($\sigma_w \geq \sigma_{min}$),

$$u_s = \alpha' \sigma_w^{m_2} . \quad (2)$$

In the above formulae, u_s is the slip velocity, $\mathcal{R}$ is the gas constant, T is the absolute temperature, f_1 is a temperature-dependent factor, E is the activation energy, σ_w is the wall shear stress, σ_n is the wall normal stress, σ_{c_1} is the critical shear-stress for the onset of slip, I is the polydispersity, and c_2, c_3, α', m_1 and m_2 are empirical constants.

Following our previous work , we build an intermediate branch between $u_{c_2} = u_s(\sigma_{c_2})$ on the left branch and $u_{min} = u_s(\sigma_{min})$ located on the right one, in order to obtain a continuous single-valued expression of the shear-stress as a function of the slip velocity. Below u_{c_2}, we make use of the slip equation, down to $u^* = u_{c_2} \cdot 10^{-2}$, under which we select a polynomial expression, in order to avoid convergence difficulties. The intermediate branch is of negative slope and is thus unstable.

3. NUMERICAL METHOD

We use standard finite-element techniques to solve the compressible flow of a power-law fluid. The field equations to be solved are the momentum equations and the continuity equation:

$$\rho\frac{D\mathbf{v}}{Dt} = \nabla\cdot\underline{\boldsymbol{\sigma}} + \rho\mathbf{g} \quad (3)$$

$$\nabla \cdot (\rho \mathbf{v}) = 0 \quad (4)$$

with

$$\underline{\boldsymbol{\sigma}} = -p\underline{\boldsymbol{I}} + k \left(\sqrt{\frac{1}{2} \underline{\boldsymbol{D}} : \underline{\boldsymbol{D}}} \right)^{n-1} \underline{\boldsymbol{D}} \quad (5)$$

where p is the pressure, $\underline{\boldsymbol{\sigma}}$ is the stress tensor, n is the power-law index, $\underline{\boldsymbol{D}}$ is the rate-of-strain tensor, ρ is the density and k is the zero-shear viscosity. Gravity forces are neglected in the momentum equation and slight compressibility is allowed for by means of the following equation:

$$\rho(p) = \rho_0 (1 + \beta p), \quad (6)$$

where β is the melt compressibility and ρ_0 is the density at atmospheric pressure..

An adaptive time-step predictor-corrector method for time discretization is used for the transient simulations.

4. NUMERICAL RESULTS

The flow curves for various values of the compressibility parameter β are shown in Fig. 1.

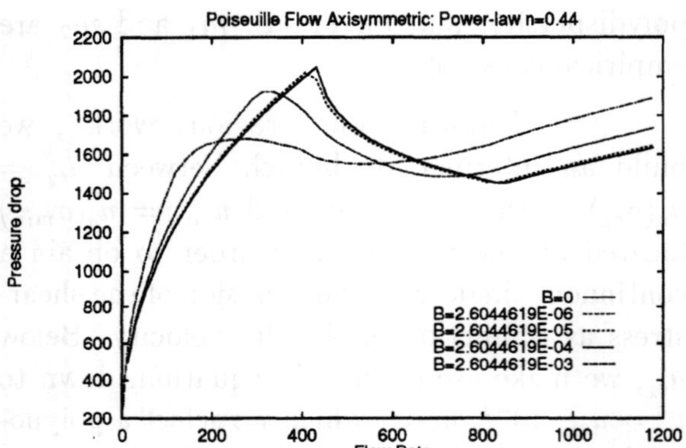

Figure 1: Flow curves for a compressible fluid in the presence of the modified slip law: effect of the fluid compressibility. $n = 0.44$.

In the time-dependent calculations, we start from a steady-state solution located on one of the three branches of the flow curve, and at t=0 we slightly perturb the volumetric flow rate at the inlet to a new fixed value. The flow is found to be unstable only in the negative-slope branch. Simulations of the time-dependent flow of the polymer pushed by a moving piston through the whole geometry of the barrel, the capillary and the extrudate will be presented and the effect of the melt compressibility on the waviness of the extrudate surface will be discussed.

5. CONCLUSIONS

We investigate the stability of capillary flows of molten HDPE in the presence of slip along the walls. We numerically simulate the experiments of Hatzikiriakos and Dealy (1992) through the use of a modified slip equation relating the velocity along the wall to the local wall shear-stress. We make comparisons with experimental observations wherever possible.

6. ACKNOWLEDGMENTS

The work of E. Brasseur is supported by the Belgian F.R.I.A. and the Brite/EURAM project MPFLOW CT96-0145.

7. REFERENCES

Georgiou G and Crochet M, (1994) Journal of Rheology 38:639–654

Hatzikiriakos S and Dealy J, (1992) Journal of Rheology 36:703–741

Kalika D and Denn M, (1987) Journal of Rheology 31:815–834

Larson R, (1992) Rheologica Acta 31:213–263

Ramamurthy A, (1986) Journal of Rheology 30:337–357

Wang SQ, (1996) Journal of Rheology 40:875–898

MELT RHEOLOGY OF REACTIVE POLYMER BLENDS

C. CIRMAN, M. BOUQUEY, J. TERRISSE, R. MULLER

Laboratoire de Recherche sur les Procédés et Matériaux Polymères
ECPM, Polymers Department, 4 rue Boussingault, F-67000 STRASBOURG, France

Interfacial chemical reactions in polymer blends have been the subject of growing interest in fundamental and applied research in recent years. These reactions are very important for compatibilization of polymer blends. As a matter of fact, block and/or graft copolymers or even a crosslinked interphase can be generated depending on the nature and functionnality of the polymers, which leads to a reduction of particle size and/or to a stabilization of morphology by reducing coalescence.

In the present study, our aim was to characterize the interface formed in a reactive blend by using bulk viscoelastic measurements. For this purpose we have synthetized by free bulk radical polymerization two miscible statistical copolymers composed mainly of isobutylmethacrylate (iBMA) and in a weaker proportion (50 functions per chain in average) of glycidylmethacylate (GMA) and for the second copolymer by methacrylic acid (MA). These polymers have glass transition temperatures of the order of 65°C and are thus in the molten state at 125°C.

We have shown that at this temperature the effect of the interfacial reaction between acid and epoxy functions on the bulk viscoelastic properties can be neglected. The morphology is varied by pre-shearing the samples at 125°C whithout beginning the crosslinking reaction. In a second step, the reaction is started by a simple temperature rise (up to about 160°C). This allows to separate the first step where interfaces are produced by shear and where diffusion of the reactive species may take place, from the second step where the interfacial crosslinking reaction occurs.

Experimentally, discs composed of pellets of each copolymer (about 600-800µm in diameter) were compression molded under vacuum and sheared in a cone and plate rheometer at 125°C in order to create a multilayer-type morphology. The applied shear strain allows to control the thickness of the layers and thus the interfacial area. Influence of the crosslinking reaction was studied by measurements of the storage modulus G' at 160°C.

We have checked that the interfacial tension between the two copolymers was low enough to neglect dispersive mixing effects: a calculation gave 0.05mN/m at 20°C; since interfacial tension is a decreasing function of temperature, this value is even lower at 160°C. Calculation of the capillary number confirm that the interfacial effects in the morphology development can be neglected.

Many parameters may influence the bulk viscoelastic properties in this process. We have first examined the influence of applied strain during the pre-shearing step.

Influence of deformation during the pre-shearing step (or of initial interfacial area) on the storage modulus at 160°C after the interfacial reaction has taken place, is shown in Figure 1. We observe in the low frequency range an increase of G' with increasing initial interfacial area.

The limiting case of infinite shear strain would correspond to a homogeneous blend which was obtained by coprecipitation of a solution of both copolymers. The value of the plateau modulus Ge at low frequencies obtained for this blend after a reaction time of 9h at 160°C is 2.9 10^5 Pa (see Figure 1). From the classical relation $Ge=\rho RT/M$ one can estimate the average molecular weight M between crosslinks, and the number of functions per chain which have actually reacted (we find a value close to 8, which is about 6 times below the functionality of the chains determined by chemical analysis).

A simple model allows to account for the order of magnitude of the plateau values in the low frequency range: we have estimated the volumic fraction Φ of the crosslinked interphase by taking into account the interfacial area per unit volume and the thickness on which crosslinking takes place. Without any diffusion of the reactive species, this thickness should be of the order of the radius of gyration of a chain segment of molecular weight Me (the molecular weight between entanglements).

These values of Φ were compared for each applied deformation with the ratio G'(0.01 rad/s)/Ge of the storage modulus taken at the lowest frequency to the plateau modulus of the homogeneous blend. A good agreement between the two sets of values has been found, especially for the blends with the largest applied shear strains.

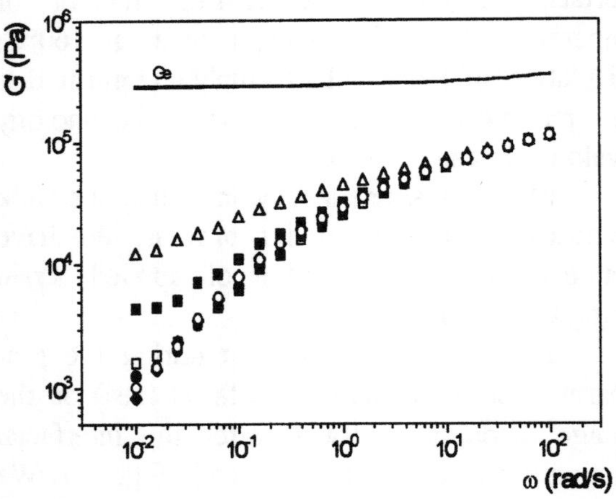

Figure 1 : Storage modulus at 160°C as a function of frequency for heterogeneous blends with different applied shear strains: ◆: γ=0, ○: γ=200, ●: γ=450, □: γ=790, ■: γ=1110, Δ: γ=2170. Solid line: homogeneous blend.

The second parameter which has been studied was the annealing (or diffusion) time at 125°C. We have applied to a series of samples the same deformation (γ=450); these samples were then annealed at the same temperature during different times (between 0 and 10h). During this time, the chains of the two copolymers can diffuse and the final thickness of the crosslinked interphase is expected to increase.

This was expected to show up in the final viscoelastic properties, once the interfacial reaction has taken place. Indeed we have found an increase in the low frequency range of G' with the diffusion time (see Figure 2). Moreover, G' varies almost proportionnaly with the square root of diffusion time.

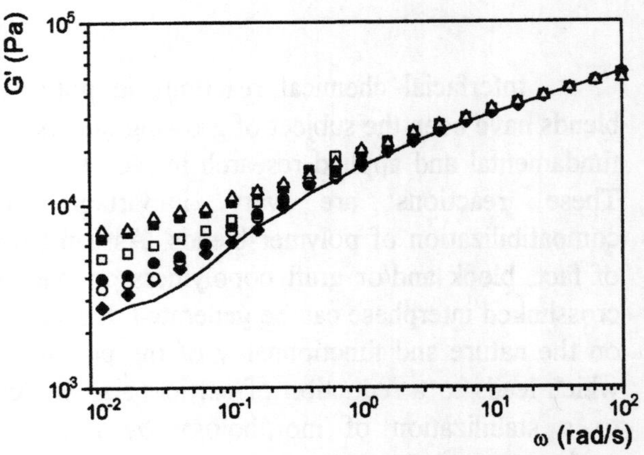

Figure 2 : Storage modulus at 160°C as a function of frequency for blends with different diffusion times at 125°C: (—): 0h, ◆: 15min, ○: 30 min, ●: 3h, □: 5h, ■: 7h, Δ: 10h.

A simple model was proposed to verify the agreement between experience and theory. We have estimated the volumic fraction of the crosslinked interphase from the interfacial area and a diffusion length which depends on the molecular weight of the chains of copolymer.

In the future, we will focus on the influence of other parameters like temperature of reaction and functionality of the copolymer chains. Then the immiscible case will be studied with copolymer MATRIFE-MA (where the main comonomer will be trifluoroethylmethacrylate) and iBMA-GMA. In the same time we will characterize the morphology by SEM or by optical microscopy (measurement of the thickness of the crosslinking interphase).

A NEW INSTRUMENT FOR MEASURING OF RHEOLOGICAL PROPERTIES OF GAS-CONTAINING POLYMER MELTS

S. DJOUMALIISKY, N. TOULESHKOV

Institute of Metal sci., Bulgarian Academy of Sciences, 67 Shipchensky prohod str., 1574 Sofia, E-mail str.djum@ims.acad.bg

1. INTRODUCTION

In the structural foam processes (extrusion, injection moulding) an understanding of the rheology of gas-containing polymer melts is extremely important. Their complicated flow characteristics are connected mainly with phase change in the flow channel. Experimental studies of foam extrusion reported by Han et al (1975), Big et al (1976), Oyanagy and White (1979), shown that the measurements of rheological properties is difficult task due to the absence of a reliable rheological instrument. According to Han and Villamizar (1978) the usual plunger type rheometers can not provide a constant pressure profile and are not suitable for measurements of rheological properties of gas-containing polymer melts.

The present study is an attempt to investigate the flow properties of gas-containing polymer melts by experimental technique based upon known injection moulding gas-counterpressure process.

2. EXPERIMENTAL

2.1. Equipment and method

An experimental apparatus consisting of a Brabender extrusiograph and a distributive device has been developed. The apparatus has been described in detail previously, Djoumaliisky and Touleschkov (1996), so we need to summarize briefly it here. The gas-containing polymer melt from the measuring extruder 30/25D can be diverted by means of pneumatic distributor in two separate flow channels which end-up with identical capillaries. A set of capillaries having constant diameter (d = 1,5 mm) and various length to diameter ratio (l/d = 4; 8; 12; 24) were used. Melt pressure transducers were installed at the entrance of the capillaries. A pressurized chamber was attached to the exit of one of the capillaries. The apparatus permits a differential determination of the flow properties by free extrusion and by extrusion into the chamber with counterpressure. This method is analogous to the injection moulding of structural foams by gas-counterpressure. The basic idea is to suppress the bubble nucleation in the entrance region and to prevent foaming within the capillary by means of counterpressure, thus assuring accurate viscosity measurements.

2.2. Materials

The apparatus was tested with isotactic polypropylene "Buplen" 6631, Bulgarian product, (density ρ = 0,901 g/cm3 and MFI (230/2,16) = 2,12 g/10min)into which chemical blowing agent azodicarbonamide "Genitron" EPA was added in proportions 0,5; 1,0; 2,0 wt%. The counterpressure was varied in the range from 0,1 to 1,2 MPa at constant melt temperature (230°C).

3. RESULTS AND DISCUSSION

Figure 1 shows axial pressure profile of polypropylene containing 0,5 wt% of Genitron EPA at 230°C. It is clearly seen the curved pressure profiles near the exit of the capillary by free extrusion at investigated range of shear rates (curves **a** ,**d**). Similar results confirmed as well

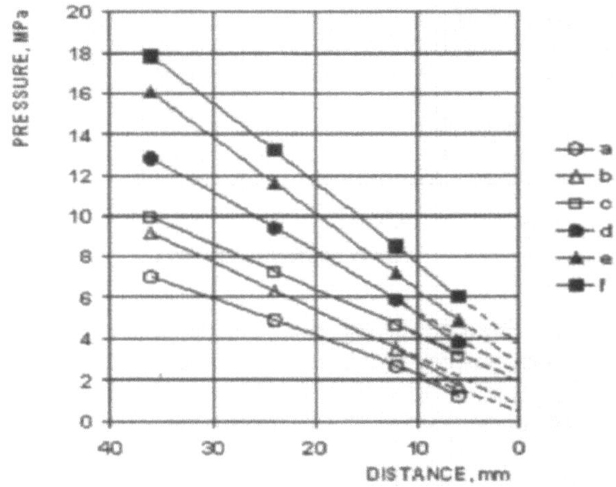

Fig.1. Axial pressure profile of gas-containing polypropylen melt at different shear rates: **a,b,c** - 70 s^{-1}; **d,e,f** - 250 s^{-1} and counterpressure: **a,d** - 0,1 MPa; b,e - 0,6 MPa; **c,e** - 1,2 MPa.

from visual observation of bubble growth in exit region has been reported in work of Han and Villamizar (1978). The other curves (**b**,**c**) show that the deviation from a linear pressure profile decreases with the increase of counterpressure at low shear rates, while the effect of counterpressure is slightly expressed at higher shear rates (**e**,**f**). We need higher values of counterpressure to suppress the foam formation in the capillary at high level of blowing agent concentration (2,0 wt%), because an increase in blowing agent concentration leads to increase in the volume fraction of gas bubbles. The axial pressure profiles correlate with the density of gas-containing polymer melt. The density of gas-containing polymer melt can be calculated following the work of Thron (1976) and the density of the extrudate in pressurized chamber can be experimentally determined. The results plotted in Figure 2 show that the extrudate density increases to a constant value with the increase of counterpressure. It is of interest to note that when the shear rate is low the action of counterpressure on the extrudate density is more efficient, while it is comparatively limited at high shear rate. There is a strong connection between cell morphology, shear rate and counterpressure.

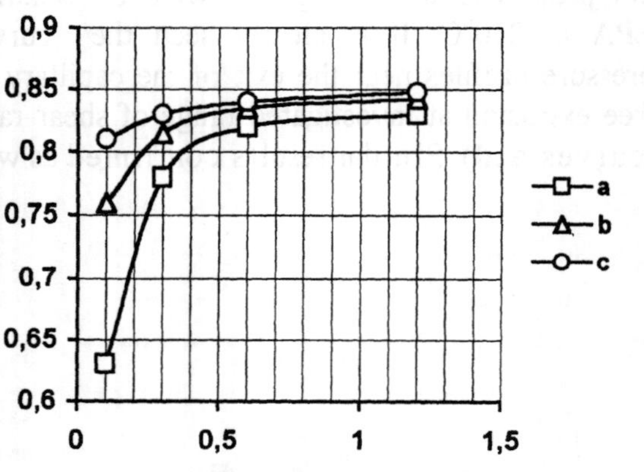

Fig. 2. Extrudate density versus counterpressure for gas-containing polypropylene melt et various shear rates **a** - 70 s^{-1}; **b** - 185 s^{-1}; **c** - 450 s^{-1} (blowing agent concentration - 0,5 wt%; melt temperature - 230°C).

Figure 3 compares the differential obtained flow curves of polypropylene melt containing 0,5 wt% Genitron EPA. The flow curves were constructed after corrections including entrance

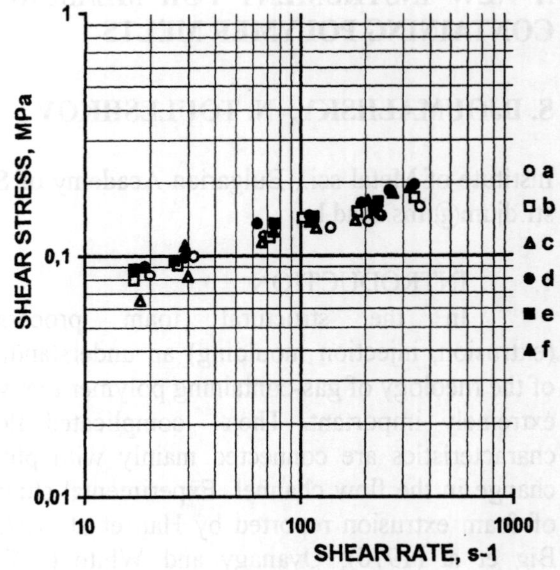

Fig.3. Shear stress versus shear rate for gas-containing polypropylene melt at 230°C and different capillary dimensions (l/d): **a,d** - 4; **b,e** - 8; **c,f** - 16 (empty symbols- free extrusion; full symbols extrusion with counterpressure, 0,6 MPa)

effects and apparent shear rate at the wall of the capillary. The slight separation in the flow data by free extrusion shows that the flow curve are dependent of capillary dimensions at low shear rates. These results are in agreement with the attempt of Bigg et al (1976) to correlate extrusion data in capillary and slit dimensions. However the flow curves obtained by extrusion with counterpressure became independent of capillary geometry.

4. CONCLUSION

The results presented demonstrate the applicability of the method. The applied counterpressure can guarantee a linear pressure profile, absence of foaming in the flow channel at low shear rates and optimum cell morphology of the extrudate.

5. REFERENCES

Bigg DM, Preston JB, Brenner D (1976) Polym Eng Sci 16:706-712

Djumaliisky S, Tuleschkov N(1996) Kunststoffber 9:39-42

Han CD, Kim YW, Malhotra KD (1976) J Appl Polym sci 20:1583-1595

Han CD, Villamizar CA (1978) Polym Eng Sci 18:687-698

Oyanagi Y, White JL (1979) J Appl Polym Sci 23: 1013-1026

Thron JL (1976) J Cell Plast 12:161-176

MELT RHEOLOGY OF RANDOMLY BRANCHED POLYSTYRENES

Dino Ferri, Paolo Lomellini

ENICHEM Research Center, Via Taliercio 14, 46100 Mantova, Italy

1. INTRODUCTION

Branching effect on shear and elongational flow properties is an important topic for polymer industry as well as for polymer science.

Randomly branched polystyrenes (RBPS) were synthesized by copolymerization of styrene and divinylbenzene (DVB), and the melt rheological properties of different homologous series were investigated both in the linear and nonlinear viscoelastic response region.

2. EXPERIMENTAL

A Rheometrics Mechanical Spectrometer (RMS 800) was used for sinusoidal and steady state measurements. The linear viscoelastic functions were measured over a wide temperature and frequency range using the parallel plate geometry. The steady state functions were measured with the cone and plate geometry.

For the capillary measurements a Göttfert Rheograph 2002 was used.

The elongational properties were investigated using a constant Hencky strain rate rheometer of the type developed by Meissner (1987,1994).

Samples from different homologues series were investigated, whose designations and main properties are reported in Table 1.

Sample	Topology	DVBppm	$M_w \cdot 10^{-5}$	B(K)
A1	linear	-	1.85	1371
A2	branched	380	2.17	1408
A3	branched	760	8.68	1515
B1	linear	-	2.42	1360
B2	branched	300	4.47	1424
B3	branched	600	16.4	1533
C1	linear	-	3.11	1388
C2	branched	170	5.11	1386
C3	branched	340	13.1	1544
D1	linear	-	2.11	-
D2	branched	457	-	-

Table 1: Sample designation, DVBppm used for branching, M_w by laser light scattering and the values of the B parameter (see Equation 1).

3. RESULTS AND DISCUSSION

In Figure 1 the molecular weight dependence of the zero-shear melt viscosity at 533 K for the series A, B, and C is reported. The dashed line represents the 3.4 slope curve, and the M_w values are those determined by LLS. The smaller value of the viscosity for branched samples with respect to the linear ones is due to a decrease of the radius of gyration determined by the branching.

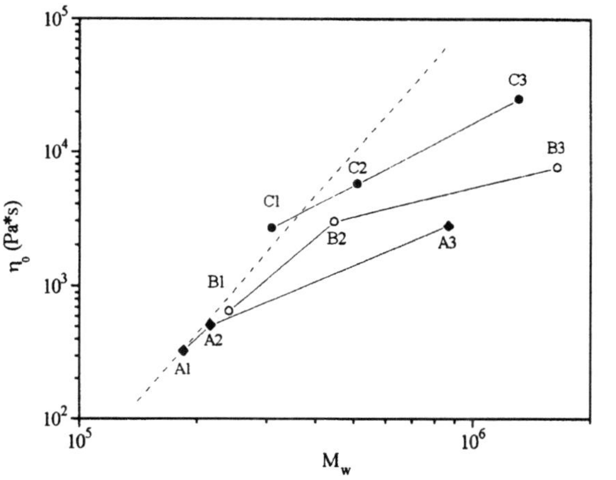

Figure 1: Molecular weight dependence of the zero-shear melt viscosity at 533 K for the series A, B and C.

The temperature dependence of the shift factors a_T, calculated building master curves of the experimental linear viscoelastic functions, in all cases followed the Vogel-Tammann-Fulcher (VTF) equation:

$$a_T \propto \exp[B/(T - T_\infty)]. \qquad (1)$$

Though the values of the Vogel temperature T_∞ coincided for all samples, the value of the pseudo activation energy B slightly increased with increasing the degree of branching (see Table 1).

It is possible to account for this in the frame of free volume theory, in which B is inversely proportional to the free volume expansion coefficient α_f. Greater B values for RBPS indicate that branching inhibits the free volume expansion. Another theory able to explain the stronger temperature dependence of the viscosity of branched polymers is the coupling model recently proposed by Ngai (1979). The topological constraints experienced by a single macromolecule are described in this model through a coupling parameter, n. An increase of the B value accounts for an increase of coupling parameter, thus revealing more severe entanglements constraints in the case of branched polymers.

Nonlinear behavior was also investigated with capillary rheometer measurements. Entrance pressure effects were enhanced by branching, highlighting a stronger melt elasticity of RBPS. The shear rate dependence of the steady state viscosity $\eta(\dot{\gamma})$, corrected for both non-Newtonian and end effects, was compared to the frequency dependence of the dynamic complex viscosity $|\eta^*(\omega)|$ to test the validity of the Cox-Merz rule. In the non newtonian flow region, linear PS showed $\eta(\dot{\gamma})$ values smaller than the $|\eta^*(\omega)|$ ones. On the other hand, branched polystyrenes exhibited, in the same region, an unusual failure of the Cox-Merz rule with $\eta(\dot{\gamma})$ values greater than the $|\eta^*(\omega)|$ ones. Any interpretation of this appealing behavior, to be further investigated, is not straightforward and can be possibly related to the elongational flow properties of RBPS. In particular, correlations have been proposed by Utracky and Gendron (1984) between the validity to the Cox-Merz rule and the existence of strain hardening.

Elongational flow of RBPS was also tested at different temperatures and strain rates. Wagner type single integral constitutive equation:

$$\sigma(t_0, T_0, \dot{\varepsilon}_0) = \dot{\varepsilon}_0 \int_0^{t_0} h(\varepsilon_0 t) G(t, T_0) \Gamma(\dot{\varepsilon}_0, t) dt \quad (2)$$

where $\Gamma(\dot{\varepsilon}_0, t) = (2e^{2\dot{\varepsilon}_0 t} + e^{-\dot{\varepsilon}_0 t})$ is the appropriate component of the Finger tensor, suggests the possibility to apply the time-temperature superposition principle (TTS) in a somewhat extended way also in the field of nonlinear viscoelasticity. In fact, the non linearity function h depends on the dimensionless quantity $\dot{\varepsilon}_0 t$ only. Putting $\dot{\varepsilon}_1 = \dot{\varepsilon}_0 / a_T$ and $t_1 = t_0 a_T$ in Equation 2 one obtains:

$$\sigma(t_0, T_0, \dot{\varepsilon}_0) = \frac{\rho_0 T_0}{\rho_1 T_1} \sigma(t_1, T_1, \dot{\varepsilon}_1) \quad (3)$$

where a_T is an appropriate temperature dependent shift factor. Eq. 3 states that elongational stress curves measured at different temperatures should match the same master curve if both time and strain rate are reduced with the correct shift factor. Using the same a_T values calculated from linear shear experiments it was possible to build elongational stress master curves as that in Figure 2 for RBPS, thus demonstrating the validity of the TTS principle also for elongational tests. Muller and Froelich (1985) demonstrated in a similar fashion the applicability of TTS to the elongational stress curves of linear polystyrene melts, but their data do not extend to high Hencky strain values. In our case the data reach the strain hardening region (Figure 2) and the validity of the extended TTS principle is observed in a wide range of strains. In addition the elongational shift factors obey the VTF equation, since they are identical to those obtained from linear shear viscoelasticity.

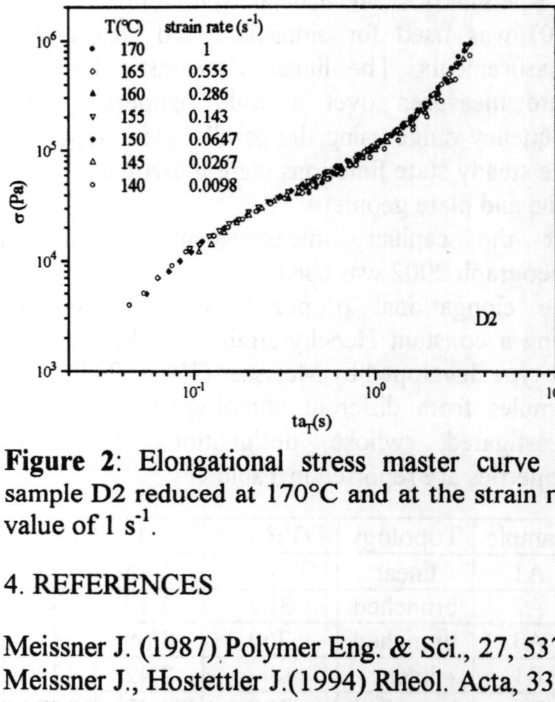

Figure 2: Elongational stress master curve for sample D2 reduced at 170°C and at the strain rate value of 1 s^{-1}.

4. REFERENCES

Meissner J. (1987) Polymer Eng. & Sci., 27, 537.
Meissner J., Hostettler J.(1994) Rheol. Acta, 33,1.
Muller R., Froelich D. (1985) Polymer, 26, 1477.
Ngai K.L. (1979) Comm. Solid State Phys., 9, 147.
Utracky L.A., Gendron R. (1984) J. Rheol., 28, 601.

DETERMINATION OF PLATEAU MODULI AND ENTANGLEMENT MOLECULAR WEIGHTS OF POLYPROPYLENES OF DIFFERENT TACTICITY

CHR. FRIEDRICH[#], A. ECKSTEIN[#], J. SUHM[#], R.-D. MAIER[#], J. SASSMANNSHAUSEN[+], M. BOCHMANN[+], and R. MÜLHAUPT[#]

[#]Institute of Macromolecular Chemistry, University Freiburg, Stefan-Meier-Str. 31, 79104 Freiburg i. Br., Germany
[+]School of Chemistry, University of Leeds, Leeds, LS2 9JT, UK

1. INTRODUCTION

The plateau modulus G_N^0 is an important material parameter reflecting structural aspects of molecular interaction in polymers such as entanglement molecular weight M_e. We determine these parameters for highly stereo-regular polypropylenes (PP) which are synthesized with the aid of metallocene catalysts.

2. METHODS

We use and compare two methods for the determination of G_N^0. Firstly, there is the empirical tangent-delta-minimum method which derives the plateau modulus from that frequency ω of the storage modulus, G'(ω), for which the ratio G'/G'' (= tan δ) is minimal:

$$G_N^0 = G'(\omega)\big|_{\tan\delta \longrightarrow \min} . \qquad (1)$$

The second method is based on the dissipation-fluctuation theorem and combines the plateau modulus with the integral over that part of the loss modulus, $G_{FT}''(\omega)$, belonging to the flow transition (FT, it includes the terminal and the plateau zone):

$$G_N^0 = \frac{2}{\pi} \int_{-\infty}^{\infty} G''_{FT}(\omega)\, d\ln\omega . \qquad (2)$$

3. EXPERIMENTAL

The molecular characteristics of the used polymers are presented in Table I. For rheological mesurements a mechanical spectrometer, RMS 800, from Rheometrics was used. The isotherms of the corresponding polymers were shifted to master curves.

***Table 1**: Molecular characteristics: Molecular weight, polydispersity and percentage of mm, mr and rr triads.*

Polymer	M_w kg/mol	M_w/M_n	mm %	mr %	rr %
iPP9	871	2.38	98.7	0.7	0.6
sPP5	483	2.09	0.7	6.3	93.0
aPP36	3670	1.94	30.2	52.8	17.0

4. RESULTS

Only for the atactic PP (aPP) it was possible to determine the plateau modulus by both methods. The crystallization of isotactic PP (iPP) and syndiotactic PP (sPP) and limitations in the available frequencies did not allow to detect the minimum in tan δ. But it was possible to use the integration technique in all cases. The plateau moduli and the corresponding entanglement molecular weights are given in Table II.

***Table II**: Plateau moduli, entanglement molecular weights and number of entanglements per chain for the polymers.*

Polymer	G_N^0, kPa eq. (1)	G_N^0, kPa eq. (2)	M_e g/mol	M_w/M_e
iPP9	427	-	6900	126
sPP5	1350	-	2170	222
aPP36	418	415	7050	520

The agreement between both methods is good in case of the aPP. The most important result is the tacticity dependence of the entanglement molecular weight. In comparison to aPP and iPP for which the backbone coils in the melt, for sPP zig-zag conformations persisting into the melt state serve as a first hypothesis to explain the observed effect.

EFFECT OF FLOW-INDUCED PHASE TRANSITIONS IN TLCPS ON THE VISCOSITY OF PE/TLCP BLENDS

P. Gao

Department of Chemical Engineering, Hong Kong University of Science and Technology, Clear Water Bay, Kowloon, Hong Kong

The incorporation of thermotropic liquid crystalline polymers (TLCPs) into engineering thermoplastics has long been known to have viscosity, η_{ap}, reducing effects[1,2]. For a η_{ap} reduction to occur the TLCP must have a nematic transition in the same temperature range as that required for processing the thermoplastic[3]. For extrusion grade high density polyethylene (HDPE) this is 170-190 °C and for the more intractable high molecular mass polyethylene (HMMPE) it is 220-230 °C. Recently B.P. Chemicals Ltd has synthesized some TLCPs based on hydroxybenzoic acid, hydroquinone and sebacic acid which have nematic transitions in these ranges[4]. Notably, these TLCPs produce η_{ap} reductions in excess of 90% at very low concentrations in a polyethylene matrix, *i.e.* 0.2-2 wt%. For the two polymer blend systems, HDPE/TLCP1 and HMMPE/TLCP3, there are different η_{ap} reducing modes. For HDPE/TLCP1 blend, with the TLCP1 in the fully nematic state at 185C there is a clear TLCP1 concentration dependent, critical shear stress, 44-100 kPa, beyond which η_{ap} reduction takes place[5]. The contribution of wall slip to the reduction was found to be insignificant.

It is difficult to operate with TLCP3 in the fully nematic phase as this occurs in the narrow temperature range 179.6-182°C. Therefore, capillary rheometry measurements were conducted at 190°C, with the biphasic TLCP3 predominantly nematic, and 230°C, when the biphasic TLCP3 is predominantly isotropic. At 190°C η_{ap} reductions >90% with the 1 wt% TLCP3/HMMPE blend were observed, Figure 1, making TLCP3 an excellent processing aid for HMMPE. Melt flow stability also improved, with no indications of pressure oscillation up to apparent shear rates of 1000 s^{-1}. In comparison, the pure HMMPE showed melt fracture and oscillatory behaviour at shear rates as low as 30 s^{-1}. At 230°C the η_{ap} of the 1 wt% TLCP3/HMMPE blend is higher than at 190 °C until a critical wall shear stress of 230 kPa. This critical wall shear stress is TLCP3 concentration dependent.

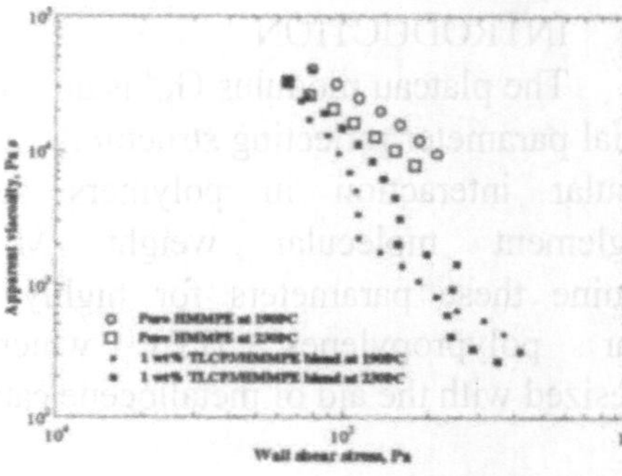

Figure 1 Flow curves for both pure HMMPE and a 1 wt% TLCP3/HMMPE blend, both at 190°C and 230°C (capillary die with 180 entrance angle, 1 mm diam. and 30 mm length)

These rheological characteristics are linked with the TLCP3's optical textures using hot-stage microscopy. The isotropic domains in the biphasic TLCP3 undergo a phase transition to the nematic phase at a critical oscillatory shear rate (Figure 2).

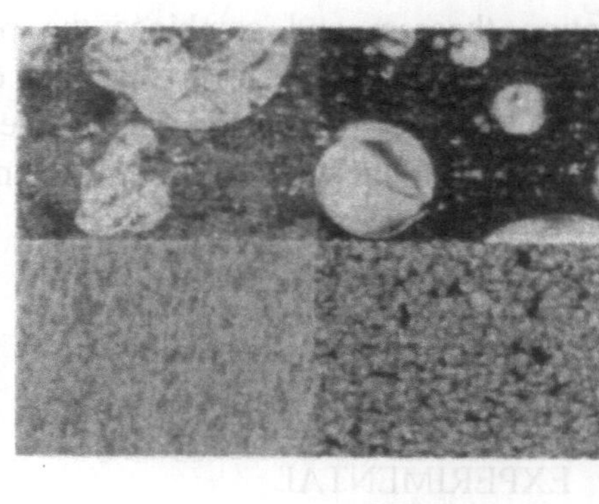

Figure 2 Photomicrographs of TLCP3 melt structure at 230°C. Top, l to r: quiescent melt ; 1 hr after pre-shear of 0.16 s^{-1}. Bottom, l to r: 15 min after steady texture achieved under shear of 30 s^{-1}; 1 hr after cessation of 30 s^{-1} shear

Speculations, based on these experimental observations, about the η_{ap} reducing mechanisms are made. Initial η_{ap} reductions are caused by TLCP3 structure effects, fibrillation and phase change, giving rise to order-induced molecular orientation in the neighbouring HMMPE chains. This is only effective in the presence of nematic structures. Above the critical wall shear stress value the lubrication effect due to TLCP3 migration is dominant.

ACKNOWLEDGMENTS

This project was funded by grant number HKUST 664/96P from the Research Grants Council of Hong Kong.

REFERENCES

1. S.G. James, A.M. Donald and W.A. Macdonald, *Molec. Cryst. Liq. Cryst.*, 1987, **153**, 491
2. W. Brostow, T. Sterzynski and S. Triouleyre, *Polymer*, 1996, **37**, 1561
3. F.N. Cogswell, B.P. Griffin and J.B. Rose, U.S. Patents, 4,386,174 (1983) and 4,433,083 (1984)
4. P. Gao, X.H. Lu and C.K. Chai, *Polym. Eng. Sci.*, 1996, **36**, 2771
5. C. Whitehouse, X.H. Lu, P. Gao and C.K. Chai, *Polym. Eng. Sci.*, 1997, **37**, 1944

EFFECT OF EXTRUSIONAL SHEAR FLOW ON LAMELLAR POLYSTYRENE-POLYISOPRENE DIBLOCK COPOLYMERS

K. GEIGER[1], H. LEIST[2], and U. WIESNER[2]

[1]Institut für Kunststofftechnologie, Universität Stuttgart, Germany
[2]MPI für Polymerforschung, Ackermannweg 10, 55129 Mainz, Germany

Block copolymers are made of sequences of chemically distinct repeat units connected covalently into chains. They are widely used as adhesives, compatibilizers, additives and a variety of automobile parts. Due to the interplay between enthalpic and entropic driving forces block copolymers self organize into periodic microstructures on a length scale of 10-100 nm and exhibit a rich equilibrium phase behavior which has been elucidated with the help of numerous experimental and theoretical studies in the past[1,2]. The ordered phases can be controlled through the block copolymer molecular weight and composition. At a specific temperature the microphases undergo a phase transition, the so-called order-disorder transition (ODT). The associated temperature is referred to as the order-disorder transition temperature (T_{ODT}). Above T_{ODT} block copolymers exist in a spatially homogenous state. Figure 1a depicts schematically what happens at the ODT for a lamellar diblock copolymer. Initially block copolymers below T_{ODT} are macroscopically isotrop exhibiting a grain structure on a µm length scale. Under the influence of shear the microstructure can be macroscopically aligned (Fig. 1b).

The effect of shear fields to macroscopically align the microstructures of block copolymers has received significant attention in the last years. This area of research was initiated by the work of Keller et al. [3] in 1970. They applied extrusional flow fields to orient a block copolymer and demonstrated dramatic success in inducing global orientation in extruded plugs of a hexagonally packed poly(styrene-butadiene-styrene) (SBS) triblock copolymer. The molten material was squeezed through an ortifice into a heated glass capillary, before beeing cooled to room temperature. They found an alignment of the polystyrene cylinders along the flow direction during extrusion.

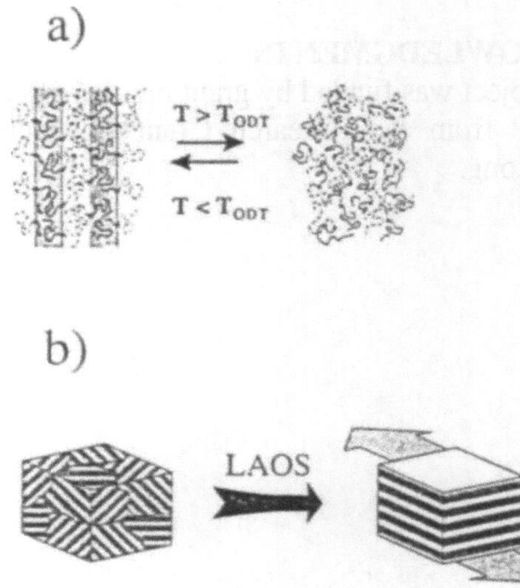

Fig. 1:
a) Schematic representation of the order-disorder transition of a lamellar diblock copolymer
b) Schematic drawing of the effect of shear on the microstructure of a lamellar diblock copolymer

In 1979 Hadziioannou et al.[4] showed that large amplitude oscillatory shearing of lamellar poly(styrene-isoprene) (PS-PI) leads to a parallel orientation state with unit normal of the lamellae parallel to the velocity gradient direction. In 1992 Koppi et al.[5] showed that large amplitude oscillatory shearing of a lamellar poly(ethylene-propylene)-poly(ethylene-ethylene) (PEP-PEE) diblock copolymer leads either to a parallel orientation or a perpendicular orientation (with unit normal of the lamealle perpendicular to both the flow and velocity gradient direction) depending on the shear conditions. Numerous studies of the effect of steady or oscillatory shear flow on cylindrical and lamellar-ordered phases have since elucidated conditions for producing single crystal type morphologies[6,7,8]

Although the conditions to obtain parallel as well as perpendicular lamellar orientations are well known for large amplitude oscillatory shear flow experiments it is not yet understood whether the same orientational states can be observed for extrusional flow fields. The latter should be particulary interesting since changes of the macroscopic orientation through varying parameters of extrusion processes should open up new applications based on the anisotropic properties of the oriented microstructure. In the present study we therefore want to focus on the effect of extrusional shear flow on a lamellar PS-PI diblock copolymer (Fig.2).

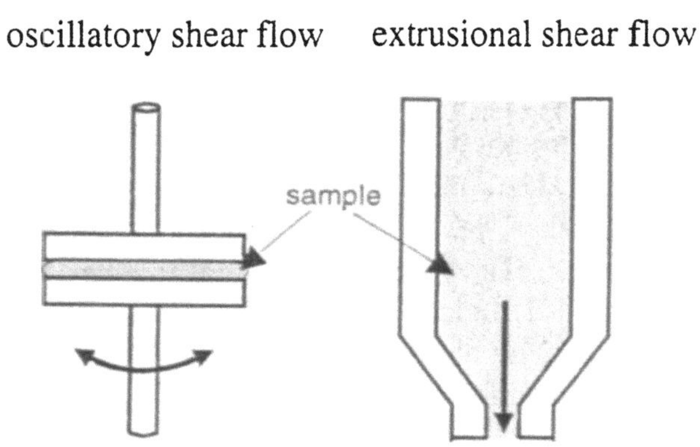

Fig. 2: Schematic comparison of oscillatory and extrusional shear flow

For the extrusion experiments a miniaturized high pressure capillary rheometer was applied. The extruder works with apparent shear rates in the range of 1-1000 s^{-1} and extrusion temperatures between room temperature and 200 °C can be applied. The morphology of the extruded plugs was studied with 2 dimensional small angle x-ray scattering experiments and correlated to the viscoelastic properties of the melt.

Literature:

[1] F.S. Bates, *Science*, **1991**, *251*, 898

[2] K. Almdal, K.A. Koppi, F.S. Bates, K. Mortensen, *Macromolecules*, **1992**, *25*, 1743

[3] A. Keller, E. Pedemonte, F.M. Willmouth, *Nature*, **1970** , *255*, 538

[4] G. Hadziioannou, A. Mathis, A. Skoulis, *Colloid Polym. Sci.*, **1979**, *257*, 136

[5] K.A. Koppi, M. Tirrell, F.S. Bates, K. Almdall, R. H. Colby, *J. Phys. 2 France*, **1992**, 1941

[6] C. C. Honeker, E.L. Thomas, *Chem. Mater.*, **1996**, *8*, 1091

[7] U. Wiesner, *Macromol. Chem. Phys.*, **1997**, *198*, 3319

[8] Z.R. Chen, J. A. Kornfield, S.S. Smith, J.T. Grothaus, M.M. Satowski, *Science*, **1997**, *277*, 124

SHEAR FLOW OF HIGHLY CONCENTRATED CARBIDE POWDER SUSPENSIONS

B. HAUSNEROVÁ[1], P. SÁHA[1], J. KUBÁT[1], T. KITANO[2], J. BECKER[3]

[1]Technical University Brno, Faculty of Technology Zlín , 762 72 Zlín, Czech Republic
[2]National Institute of Materials and Chemical Research, Tsukuba, Japan
[3]Dept. of Polymeric Materials, Chalmers University of Technology, Gothenburg, Sweden

1. INTRODUCTION

Rheological behaviour plays a key role in powder processing by powder injection moulding and extrusion. The flow behaviour of concentrated suspensions consisting of polymeric binder and metal powder is to a large extent determined by the characteristics of the binder system. This has to provide an acceptable flow during the forming process, proper interaction with the powder, and an easy extraction after moulding. Low molecular weight polymers and waxes are often the basis for polymeric binders, mainly because of easier mixing and removal of short polymer chains. Due to their complexity, considerable difficulty is encountered in interpreting the flow behaviour of highly filled polymer systems. The literature provides only a limited amount of data relating to such systems, especially when the polymeric vehicle consists of several components.

In this paper we present the results of an experimental study relating to highly filled metal powder suspensions in capillary flow, where the polymeric matrix has three components and the metal powder consists mainly of tungsten carbide and cobalt particles with a relatively broad size distribution. The role of binder composition, mixing intensity and metal powder loading is discussed with specific regard to unstable flow occurring at a critical powder concentration.

2 EXPERIMENTAL

2.1 Materials

The powder used in the experiments described below was mainly composed of tungsten carbide and cobalt. The average density of the carbide mixture was 13.2 $g.cm^{-3}$. The powder was obtained from Sandvik Coromant AB Sweden.

The binder components were LDPE Lacqtene (Atochem), EBA (27% BA) OE 6427 (Borealis), EVA (28% VA) OE 5028 (Borealis), EAA (5% AA) Ex 225 (Du Pont), and paraffin wax. The binder composition contained: 53 w% polyethylene, 21 w% paraffin wax, and 26 w% copolymer (EAA, EBA, or EVA).

2.2 Methods

The compounds were prepared in a Brabender mixer PL-2000-6, mixer type W 50E, at a temperature of 180°C. The mixing was continued until a constant torque value was reached. The rheological behaviour of the suspensions was studied with a capillary rheometer (Davenport, model ER 406/30), capillary L/D ratio 20 and 10. The pressure was measured with a Dynisco transducer (PT 420A-10M) fitted close to the die.

3. RESULTS AND DISSCUSSION

The maximum powder loading was calculated from the Maron-Pierce relationship. The layer of polymer adsorbed on the particle surface was estimated by comparing measured and calculated values of relative viscosity. As expected, the layer thickness decreased at higher powder concentrations. The adsorption of a polymer layer on the particle surface was confirmed by SEM photographs (Figs.1,2).

The suspensions exhibited two different kinds of shear flow. The flow of suspensions containing 10 to 40% (vol.) of metal powder showed pseudoplastic behaviour without any instabilities within the whole range of the processing conditions used. With increasing particle concentration the portion of the matrix was reduced and the flow changed from pseudoplastic to plug flow and became unstable. The instabilities, appearing at metal concentrations of around 50 % (vol.) and higher, were of the pressure oscillation type with the flow curve split into two branches between which the pressure oscillates. At 50 % (vol.) the gap between the branches diminished when the extrusion was repeated with the same material. At a concentration of 55% (vol.) the pressure oscillations were even more pronounced, remaining unaffected by repeated extrusions.

Fig.1 SEM photograph of hard-metal carbide powder.

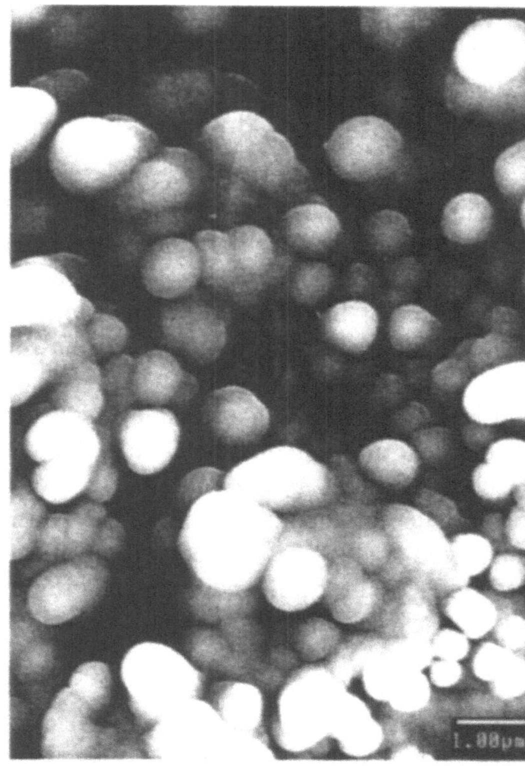

Fig.2 SEM photograph of hard-metal carbide powder covered by polymeric binder. 40 vol. % carbide powder suspension.

A CLOSE LOOK AT THE "SHARKSKIN" MELT FRACTURE OF POLYBUTADIENE

YONG WOO INN[1], MATTHEW G. DALY[2], ROBERT J. FISHER[2] AND MONTGOMERY T. SHAW[1,2]

[1]Polymer Program and [2]Department of Chemical Engineering, Institute of Materials Science, U-136, University of Connecticut, Storrs, CT 06269-3136 USA

1. INTRODUCTION

Sharkskin melt fracture (SSF) refers to a fine surface distortion found on extrudates of certain polymers at shear stress levels around 0.1 MPa. SSF can limit the productivity of extrusion lines. Two schools of thought concerning the mechanism of SSF have developed: one assigns the initiation of fracture to slip in the die land [e.g., Ramamurthy (1986)] whereas the other attributes the fracture to excessive stress at the exit as the flow profile redistributes [e.g., Howells and Benbow (1962), Kurtz (1984)]. Both may contribute to the problem. The former is supported strongly by the seeming invariant shear stress required for SSF and by the reported kink in the flow curve at this stress.

2. EXPERIMENTAL

We chose polybutadiene (PBD) as the melt because it exhibits large SSF. The PBD was Firestone Diene 35AC10, with reported M_w = 182,000 and M_w/M_n = 1.87. Most of the extrusion used 12.7-mm-diameter circular dies with L/D's of 4 and 6 attached to a 19-mm Brabender single-screw extruder operated at 50 °C. The exit of the die was tapered to a sharp angle to provide a clear view. A CCD camera (Elmo SE360) fitted with macro lenses provided output to S-VHS VCR (Panasonic AG1310) for recording the SSF development at the exit. For digitization a frame grabber (Snappy) captured frames of 640×480 pixels with an 8-bit gray scale.

3. RESULTS AND DISCUSSION

The flow curve of PBD at 50 °C is shown in Fig. 1. PBD exhibits both SSF and stick-slip spurt flow, as does LLDPE.

On extrusion, the extrudate surface was initially smooth and transparent. At the wall stress of approximately 80 kPa, cracks began to appear on the extrudate surface. The surface cracks increased in amplitude with stress and evolved to periodic waves of the SSF. An extrudate having the spurt instability from the Instron capillary rheometer at stress of 670 kPa had a far different appearance.

In Fig. 2, the wavelength λ and the time interval τ_{ss} of the SSF waves are plotted vs. stress. For PBD, τ_{ss} of was about two orders of magnitude larger than that of LLDPE ($\tau_{ss} \approx 0.01$ s) [Kurtz (1992), Wang *et al.* (1996) and Venet and Vergnes (1997)]. The large τ_{ss} of PBD makes detailed observation of SSF development possible.

The pictures of SSF shown in Fig. 3 were taken every 1 s at an approximate wall stress of 150 kPa. It was observed that a ridge grew and rolled outward with a tearing of the outside layer from the central core of the extrudate. Because the layer of polymer melt close to the die wall was barely moving, it stuck to the rim of the die exit as it was pushed out with the die swell. After the ridge had grown to a certain size, it was released from the die exit and moved forward with the whole extrudate; immediately a new ridge started at the rim of the die exit. This is a similar mechanism to the ring formation of chocolate extrusion [Ovaici *et al.* (1998)].

A soap solution was applied with a cotton swap gently touched to the top, outside rim of the die exit while PBD was extruding. (Because the soap solution spread easily on the metallic high-energy surface of the die exit, a migration of the soap solution into the die land can not excluded.) As shown in Fig. 4, when the polymer melt lost adhesion to the die exit, the SSF was eliminated.

Our observations agree with a previous hypothesis that the SSF starts at the die exit with tearing and peeling, as suggested by several researchers [Howell and Benbow (1962), Cogswell (1977), Kurtz (1984), El Kissi and Piau (1990, 1994)]. The origin of the periodicity can be qualitatively modeled by using the idea of the ring formation in chocolate extrusion suggested by Ovaici et al. (1998). More details can be found in Inn et al. (19XX).

4. REFERENCES

Cogswell FN (1977) J. Non-Newtonian Fluid Mech. 2: 37-47

El Kissi N, Piau JM (1994)" J. Rheol. 38: 1447-1463

Howells ER, Benbow JJ (1962) Trans. J. Plastics Inst. 30: 240-253

Inn Y-W, Shaw MT, Fisher RJ (19XX) Rheol. Acta (submitted).

Kurtz SJ (1984) In: Proc. IX Intl. Congress on Rheology, pp. 399-407

Kurtz SJ (1992) In: Proc. XI Intl. Congress on Rheology, pp 377-379

Ovaici H, Mackley MR, McKinley GH, Crook SJ (1998) J. Rheol. 42: 125-157

Ramamurthy AV (1986) J. Rheol. 30: 337-357

Venet C, Vergnes B (1997) J. Rheol. 41: 873-893

Wang SQ, Drda PA, Inn Y-W (1996) J. Rheol. 40: 875-898

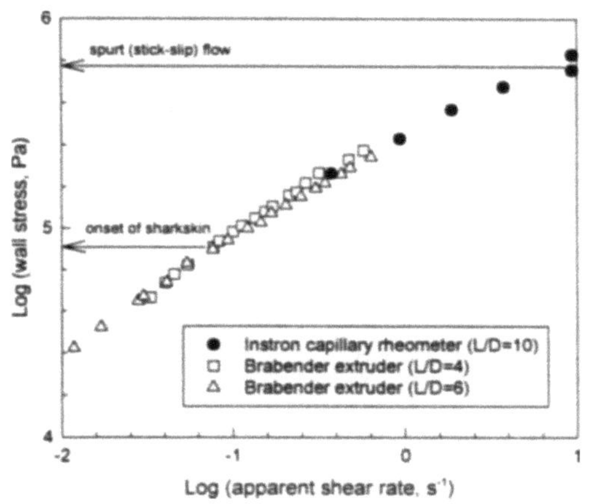

Fig. 1. Flow curve of PBD at 50 °C.

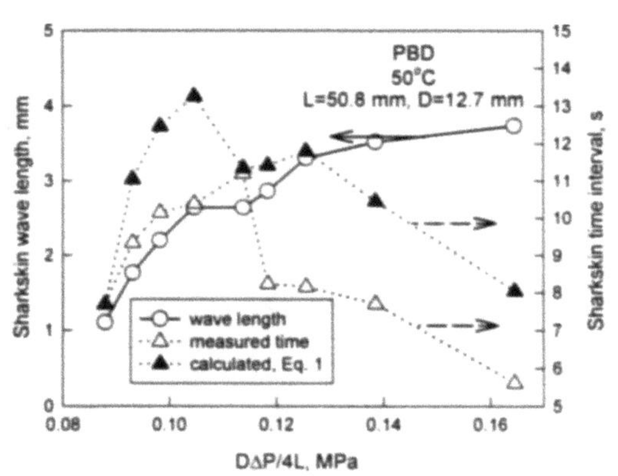

Fig. 2. SSF wavelength λ and time interval τ_{ss} as a function of shear stress σ_w.

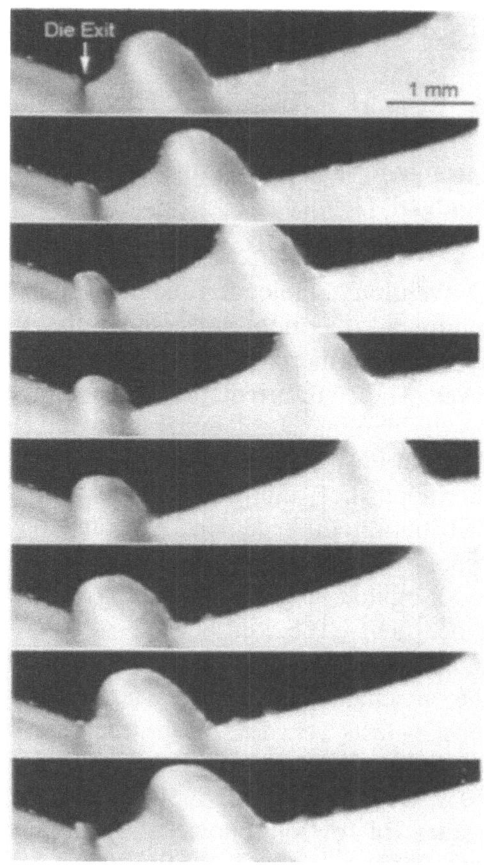

Fig. 3. SSF taken every 1 s at σ_w = 150 kPa.

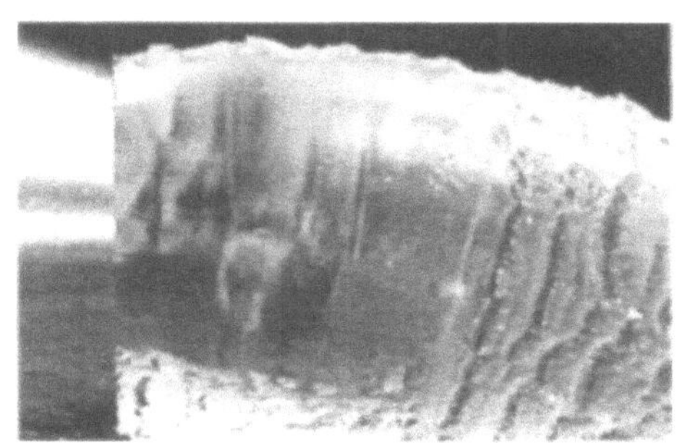

Fig. 4. Extrudate after soap solution coating near top of die.

THE INFLUENCE OF PROCESSING CONDITIONS ON MECHANICAL PROPERTIES OF UNREINFORCED AND DISCONTINUOUS GLASS FIBRE REINFORCED THERMOPLASTICS

A. Kech, H.-C. Ludwig, U. Fritz, U. Müller, B. Möginger, P. Eyerer

Institute for Polymer Testing and Polymer Science, University of Stuttgart, Pfaffenwaldring 32, D-70569 Stuttgart, Germany

1. ABSTRACT

Conventionally injection moulded short glass fibre reinforced thermoplastics show the well-known symmetrical morphology of skin-, shell- and core-layer. The fibre orientation in each layer depends on the dominance of either shear or elongational flow. In Push-Pull processed glass fibre reinforced thermoplastics, morphology and fibre orientation distribution differ distinctly from conventionally processed ones. This processing technique yields to a much higher anisotropy of the elastic properties. The influence of the modified morphology and fibre orientation distribution on the mechanical properties are investigated in tensile tests, tensile creep tests and DMA for isotactic polypropylene (VESTOLEN P 7000, P 7032 G, DAPLEN K2 XMOD) and two different grades of discontinuous glass fibre reinforced polyamide 66 (VERTON RF 70010, ULTRAMID A3EG10).

2. INTRODUCTION

Morphology and fibre orientation influence the elastic properties distinctly. Unreinforced isotactic polypropylene (i-PP) shows a special morphology when using Push-Pull processing. Three different grades of polypropylene, two unreinforced and one fibre reinforced grade, were processed. Orientation of the molecules results in advanced material properties parallel and perpendicular to the flow direction. This behaviour can be observed for static as well as for dynamic loading. The material behaviour of polymers depends on time and temperature. Measurements using a DMA temperature and frequency sweeps are performed.

Fibres align parallel to the flow direction under shear forces. Fibre orientation measurements on polished cross sections of samples produced with Push-Pull illustrate that the core-layer, characterized by fibres aligned perpendicular to the flow direction, is reduced, even for long discontinuous glass fibre reinforced grades. A higher rate of anisotropy of the mechanical properties can be obtained for short and long term properties.

3. EXPERIMENTS

Tensile tests were performed on a universal testing machine to determine static stiffness and tensile strength properties. The cross head speed was 5 mm/min (constant strain-rate) at 25 and 75 °C. The temperature and frequency dependent behaviour was investigated using a DMA in the temperature and frequency sweep mode and torsional deformation. The long term properties were investigated on tensile creep tests.

4. RESULTS

4.1 DMA

Figure 1 shows the results of i-PP for storage shear modulus G' and the damping factor tan δ at a frequency of 0.1 Hz and Figure 2 at 50 Hz.

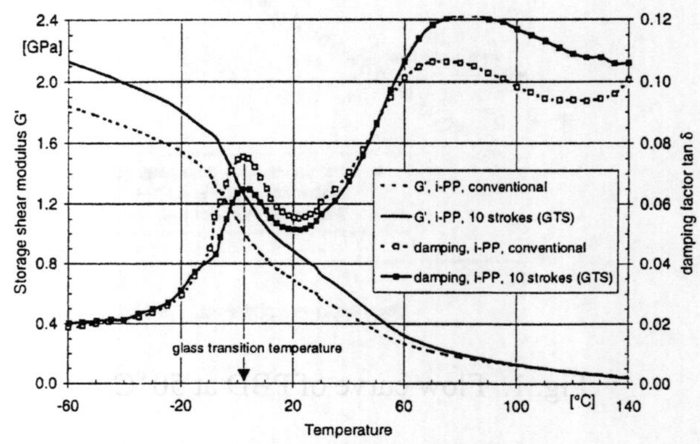

Figure 1: Temperature sweep at 0.1 Hz

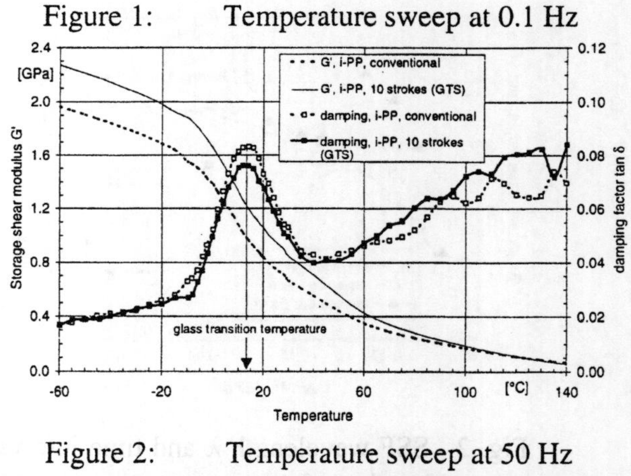

Figure 2: Temperature sweep at 50 Hz

The temperature range was -60 to 140° C. Below the glass transition temperature up to a temperature of 60° C a significantly higher storage modulus for the samples produced with Push-Pull can be determined. The glass transition temperature does not dependent on the processing technique.

4.2 Tensile Creep

After dynamic loading the long term properties under constant tensile load ($\sigma_m = ½\, \sigma_{max,\ static}$) were investigated for glass fibre reinforced PA 66 grades.

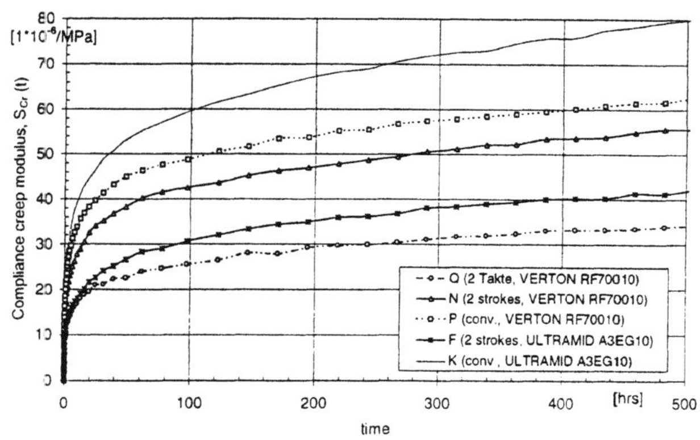

Figure 3: Tensile Creep behaviour

4.3 Tensile Test

Tensile tests on unreinforced and fibre reinforced polypropylene illustrate that the dominance of shear forces during the processing results in enhanced mechanical properties. Table 1 and 2 show results for injection moulded tensile bars processed conventionally and with push-pull.

Materials	Processing	σ_{max} [MPa]	E_l [MPa]
VESTOLEN P 7000	conv.	30	1436
VESTOLEN P 7000	PPP	41	2160
DAPLEN K2 XMOD	conv.	37	2145
DAPLEN K2 XMOD	PPP	44	2835
VESTOLEN P 7032 G	conv.	44	4619
VESTOLEN P 7032 G	PPP	50	5365

Table 1: Results for tensile tests (23 °C)

The effect of push-pull on mechanical properties is even visible at elevated temperatures (75 °C).

Materials	Processing	σ_{max} [MPa]	E_l [MPa]
VESTOLEN P 7000	conv.	15	670
VESTOLEN P 7000	PPP	25	972
DAPLEN K2 XMOD	conv.	22	1140
DAPLEN K2 XMOD	PPP	30	1365
VESTOLEN P 7032 G	conv.	29	3151
VESTOLEN P 7032 G	PPP	33	3602

Table 2: Results for tensile tests (75 °C)

5. SUMMARY AND CONCLUSION

The investigations show that the dominance of shear forces during processing result in advanced mechanical properties. Push-pull processing is one technology to produce parts with a rate of orientation and the corresponding mechanical properties. Not all materials are suitable for an orientation of the molecules. Polypropylene is one materials which provides a high potential of orientation in injection moulding.

6. REFERENCES

(1) A. Kech et al., Push-Pull Processing of Short Fiber Reinforced Thermoplastics, AIChE/PPS meeting, Chicago, Il, November 1996

(2) A. Kech et al., Push-Pull processing of unreinforced and short fibre reinforced thermoplastics and LCP; Processing and mechanical properties, PPS Meeting, Gothenburg, August 1997

(3) H.-C. Ludwig et al., Push-Pull injection moulding - A new processing method for achieving advanced mechanical properties of short fibre reinforced and virgin Polypropylene (i-PP), European Conference on Macromolecular Physics, Lausanne, June 1997

(4) A. Kech et al., Push-Pull processing of unreinforced and short fibre reinforced thermoplastics, ECCM-8, Naples, June 1998

SHEAR AND ELONGATIONAL FLOW OF SMECTITE / PMMA HYBRIDS AND GLASS FIBER / PMMA COMPOSITES

R. KOTSILKOVA[1], M. OKAMOTO[2], H. TAGUCHI[2], N. SATO[2], H. SATO[2], T. KOTAKA[2]

[1]Central Lab Physico-Chemical Mechanics, Bulgarian Academy of Science, Acad. G. Bonchev Str. Block 1, 1113 Sofia, Bulgaria
[2]Advanced Polymeric Materials Engineering, Toyota Technological Institute, Hisakata 2-12-1, Tempaku, Nagoya 468, Japan

1. INTRODUCTION

The intercalated polymer / silicate hybrids (nanocomposites) are relatively new material with a nanosize layered silicate filler. Moreover, the reinforcing is mainly governed by the interfacial effects. A presence of long-range order and domain structure with grains in nanocomposites and similarities with block copolymers and smectic liquid crystals is discussed [1-2].

From processing point of view it is important to relate the rheological properties of both the hybrid composites and the convential polymer composites for an optimizing the processing conditions [3]. In the present study we examine the oscillatory shear and elongational flow behavior of intercalated hybrids and compare with another anisotropic material - glass fiber reinforced polymer (GFRP).

2. EXPERIMENTAL

The silicate filler used for preparation of the Smectite / PMMA Hybrids was *SPN* - a lipophilized with quaternized ammonium salt $[(C_2H_5)_2(CH_3)N^+(O\text{-}iPr)_{25}]Cl^-$ smectic montmorilonite. The intercalated hybrids are synthesized by polymerization of MMA monomer in the presence of SPN, becoming a structure with multilayers of about 200 nm length and 20 nm thickness [2]. The GFRP ($a_r \approx 10$) is prepared by an extruder mixing of short fibers with the synthesized PMMA.

The linear viscoelastic behavior was studied in a small-strain oscillatory shear by *Rheometric Scientific RDA II* and was described by means of the relaxation time spectrum. Simultaneous measurements of stress and birefringence as a function of time in a constant strain rate elongational experiment were carried out by an *elongational flow opto-rheometer (EFOR).* The elongational viscosity and stress-optical behavior in the context of both the effect of intercalation on the relaxation processes and the flow-induced filler orientation during elongation were evaluated.

3. RESULTS AND DISCUSSION

3.1. Linear Viscoelastic Behavior

The linear viscoelastic response was measured by the storage (*G'*) and loss moduli (*G"*), using the lowest possible strain amplitudes. Data were acquired at several temperatures and shifted using the time-temperature superposition principle to construct the master-curves (Tref=180°C).

A generalized Maxwell-type model was used to fit the dynamic moduli - G'(ω) and G"(ω) and to describe the linear viscoelastic behavior by means of relaxation time spectrum [Ferry (1980)]. The both moduli are related to the relaxation spectrum coefficients (G_i, τ_i), by the following equations:

$$G'(\omega) = \sum_i G_i\,(\omega\tau_i)^2 / (1+\omega^2\tau_i^2) \qquad (1)$$

$$G''(\omega) = \sum_i G_i\,\omega\tau_i / (1+\omega^2\tau_i^2) \qquad (2)$$

The discrete relaxation time spectrum in wide range of relaxation times was determined from the master-curves of storage *G'*(*ω*) and loss *G"*(*ω*) dynamic moduli. The evaluation of the discrete spectrum (G_i,

τ_i) was carried out by the Rheometric software. In Fig. 1 we present discrete spectra (G_i, τ_i),

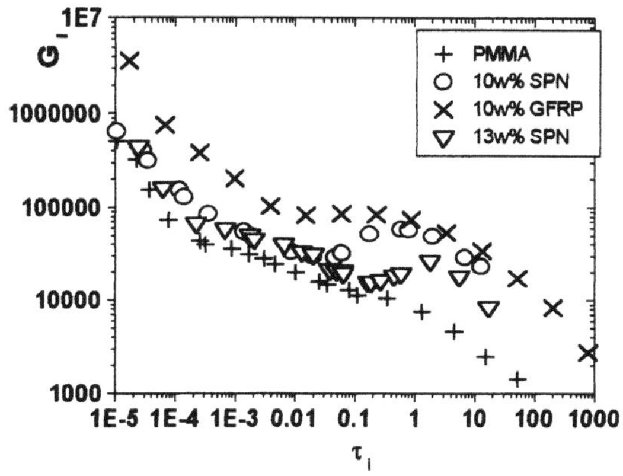

Fig. 1 Relaxation-time spectra

of PMMA, GFRP and Smectite/PMMA hybrids at T_{ref}=180°C. The presence of filler shifted the spectra curves toward the long time scales. The tail of the spectra profiles of the Smectite/PMMA hybrids shows an additional peak, which is not reported for the fiber composites and could be attributed to the relaxation of the interface. The time at which the maximum occurs in the spectra is determined as a *characteristic time of the interfacial relaxation* (τ^*_{int}), as well as the magnitude of the peak (G^*_{int}) could be related to the amount of interfacial area.

3.2. Elongational Flow and Birefringence

In an elongational flow field, the hybrid composites display a nonlinear *strain-hardening* behavior, which is not manifested by PMMA and GFRP. Fig. 2a,b compare the time variation of (*a*) elongational viscosity $\eta_E(t)$ and (*b*) birefringence $\Delta n(t)$ of 10% SPN/PMMA hybrids, measured at various constant strain rates at 170 °C. At low strain rates, the $\eta_E(t)$ for intercalated hybrids show linear viscoelastic behavior, but at high strain rates thay exhibit tendency of a rapid increase (*strain hardening*) above a critical *up-rising* time $t_{\eta E}$. . The birefringence was not detected for PMMA and GFRP, but for Smectite/ PMMA hybrids $\Delta n(t)$ curves tend to deviate upwards, above the same critical time $t_{\Delta n} = t_{\eta E}$ depending on strain rate.

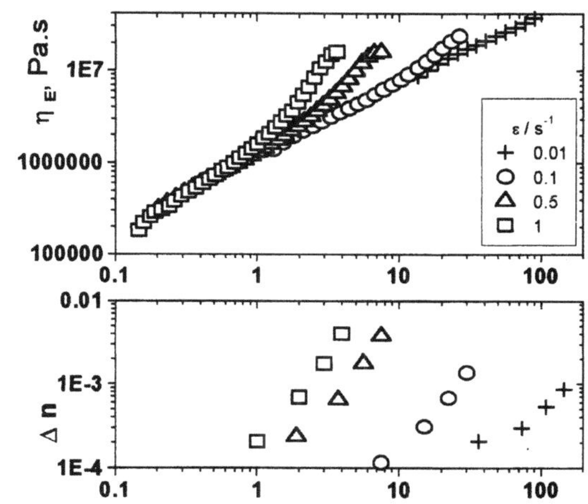

Fig.2a,b. The elongational viscosity and birefringence of 10v% SPN/PMMA, T=170°C

The birefringence results could be contributed to the flow-induced orientation and alignment of the silicate layers in the stretch direction. It seems to produce nonlinear effects on the elongational viscosity, displaying by the *strain hardening*.

Both, the increasing of the silicate content and the decreasing of the temperature is found to increase the *strain hardening* effect. It is correlated with the interfacial effects and the retardation of the relaxation processes.

Acknowledgments: The study is supported by the grants from Japan Society for Promotion of Science and the Science Fund of Bulgaria (Project X- 645).

REFERENCES

1. Krishnamoorti, R., R. Vaia, E. Giannelis. *Chem. Mater*, **8**, (1996) 1728-1734
2. Sato, H., H. Taguchi, N. Sato, M. Okamoto and T. Kotaka (1997) In: *Proc. SPSJ 46th Ann Meeting*, Part I and II, Tokyo, Japan, E144-145
3. Kotsilkova, R., M. Okamoto, N. Sato, H. Sato and T. Kotaka, In: *Proc.44th Japan Rheol. Soc. Symp.*, Miyazaki, Japan, B 03, (1997)

RELATIONSHIPS BETWEEN PROCESSING CONDITIONS AND RHEOLOGICAL BEHAVIOUR OF POLYETHYLENES

V.KULICHIKHIN[1], E.BORISENKOVA[1], E.PLOTNIKOVA[1], G.VASIL'EV[1], J.MERRICK-MACK[2], N.O'REILLY[2]

[1]Institute of Petrochemical Synthesis, Russian Academy of Sciences, 29 Leninsky Pr., Moscow 117912, Russia

[2]Lyondell Technology Center, P.O.Box 2917, Alvin, Texas 77512, USA

1. INTRODUCTION

In spite of a lot of papers concerning the rheological properties of polyethylenes (PEs) (see, for example, Sebastian & Dearborn (1983), Linster & Meissner (1986), the interest in this problem remains, especially with respect to processing-properties and molecular structure-property relationships. The main portion of these papers is devoted to attempts to separate the influence of MMD and branching on the most sensitive rheological characteristics. As a rule, the investigations were carried out on specially synthesised samples with regular change of molecular characteristics.

But until this time we can not separate reliably the role of either MMD, or branching in rheological response. We will try to do this, based on a series of PEs which have been treated at processing in a definite manner. The treatment involves melt compounding a high density, linear PE's resin produced in a Zeigler-Natta slurry process, at high shear rates and temperatures in the presence of oxidizers, such as oxygen (series O), or peroxides (series P), or both (series PO). The samples inside series were varying by intensity of treatment (concentration of oxidizer and compounding temperature). The compounding process is controlled in such a way that the major effect on the resin is to generate long branches. The only effect of MMD may be a slight changes as consequence of the presence of larger macromolecules resulting from branching.

2. METHODS

2.1 Rheological measurements

A set of rheological methods has been used: rotational and capillary viscometers, squeezing flow and uniaxial extension. The experiments were conducted at temperatures of 150 and 190°C, corresponding to temperature range of the blow moulding regimes. For identification of time-temperature prehistory all samples tested at shear deformation were "homogenized" at shear rate of 0.36 s^{-1} for 5 min, then after a defined rest time (see below) the flow curves were measured in two ways: from low γ to high and in reverse.

The rheological behaviour at extension deformation was investigated in a quasi-stationary regime of constant force. The main results are presented as dependencies of extensional deformation according to Hencky $\varepsilon=\ln(l_t/l_o)$ on time t, where ε is the ratio of current length of sample l_t to its initial dimension l_o. By the angle coefficient of linear sections of $\varepsilon(t)$ dependencies the extension rate ε was calculated. Assuming the constancy of the sample volume during extension, the current cross-section and tensile stress were determined, and used for calculation of the apparent extensional viscosity λ.

The molecular characteristics of neat and treated samples were determined by GPC and IR methods. IR measurements of PE's powders were conducted to found any change of absorption bands as a result of treatment.

3. RESULTS AND DISCUSSION

3.1 Start-up shear experiments and steady-state shear flow

All samples under investigation were rheologically stable at 150°C and unstable in more or lesser degree at 190°C. It means that shear stress, measured at low shear rates, monotonously rises in time, but the rate depends on prehistory of each sample. This effect leads to non-reproducibility of low shear rates sections of flow curves in direct and reverse directions.

In a single case it is possible to obtain three different branches on the flow curves "a", "b" and "c". Just after loading and thermostating (20 min) the pseudo-Newtonian branch "a" takes place. After homogenization at γ=0.36 s^{-1} and rest during 20 min the structural branch "b" is realized. If after determination of flow curves from low γ to high ones, we will repeat the experiments in an opposite direction the new branch "c" with higher viscosity is obtained. As the level of oxidation or peroxidation was increased, region of divergence is extended to higher shear stress and deflection between "b" and "c" branches is increased.

In addition to irreversible changes created by treatment and reflected on shear viscosity increase, an additional network is forming during measurements, but is destroyed at high shear rates. The use of the word "network" was chosen because of tendency to exhibit increased yield behaviour with higher residence times at increased temperatures. This hypothesis was confirmed by stress relaxation experiments. The residual level of stresses is increased with intensity of treatment for all PEs, but especially prominent for PO series.

3.2 Extensional rheology

The shape of the $\varepsilon(t)$ curve was found to be quite different for all treated samples versus the neat PE. This change consists of the appearance of two sections of the deformation curves with various slopes, for all values of initial tensile stress. After achieving of a definite deformation, the extension is decelerating (the slope is decreasing), and as a result, the process continues for a much longer time before break than for untreated sample. An accumulation of characteristic changes of deformation curves increases with intensity of treatment inside any series, but the most drastic changes takes place for PO series.

3.3 GPC and IR data.

We did not observe a significant change of MMD for neat and treated samples from GPC data. But there was an important feature of dilute solutions (0.04 %) in o-chlorobenzene of treated PEs at repeated elutions (temperature145^{o}C), namely significant decrease of Mw at nearly the same value of this parameter for solutions of untreated PEs. The main reason of such destruction process (in spite of addition of Irganox) is the presence of reactive sites in the treated samples.

IR spectra of neat and treated samples are very similar, and only detail analysis of them allowed us to observe the difference of absorption bands at 1380 (CH_3-groups) and 1450 cm^{-1} (C=O-groups), intensity of which increases with the treatment depth. The first band can play role of indicator of branching. The second band (1450 cm^{-1}) indicates on oxidative processes, mainly on presence of carbonyl groups.

An accumulation of methyl (presumably, long branches) leads to increase of entanglements density, increase of steady shear and extension viscosities and growth of melt resistance to extension deformation. Presence of C=O-groups reflects the existence of oxidized defect links, probably capable to "live" in melt state and transform to branches and partially even to crosslinks. This effect results in instability of transient processes at 190°C and increase of shear viscosity with residence time.

4. CONCLUSION

4.1. During mechanochemical treatment with addition of oxygen or peroxide the reactive sites along the PE's chains were created. Part of them initiates the rare, long-chain branches formation just before the blow-moulding procedure, another part is conserving.

4.2. Presence of long branches leads to increase of absolute values of shear and extension viscosities, presumably, due to increase of molecular coil dimensions and entanglements density. The most prominent consequence is increasing the resistance to extension deformation with essential growth of time before break.

4.3. During prolonged residence at T=190°C the remaining sites can transform to additional branches or even weak crosslinks. This leads to instability of transient processes and non-reproducibility of the low shear rate sections of flow curves.

5. REFERENCES

Sebastian D.H., Dearborn J.R. (1983) Polym. Eng. Sci 23:572

Linster J.J., Meissner J. (1986) Polym. Bull. 16:187

ELONGATIONAL PROPERTIES OF POLYOLEFIN MELTS AND CORRELATIONS WITH THEIR PROCESSING BEHAVIOUR IN FILM BLOWING AND THERMOFORMING

S. KURZBECK, H. MÜNSTEDT

Institute of Polymer Materials, Department of Materials Science, University Erlangen-Nürnberg, Martensstr. 7, D-91054 Erlangen, Germany

1. INTRODUCTION

Elongational flow plays a dominant role in many processing operations such as film blowing and thermoforming. An essential elongational property to assess these processing operations is the transient elongational viscosity or stress growth function, especially in the nonlinear transient range of deformation or deformation rate. The molecular structure of polymers has a strong influence on the nonlinear elongational behaviour characterized by the so-called strain hardening which cannot be deduced from the nonlinear flow in shear (Meissner 1972, Münstedt and Laun 1979).

2. ELONGATIONAL VISCOSITIES

2.1 Low-density Polyethylenes

The elongational behaviour was investigated by an elongational rheometer described by Münstedt et al. (1998).

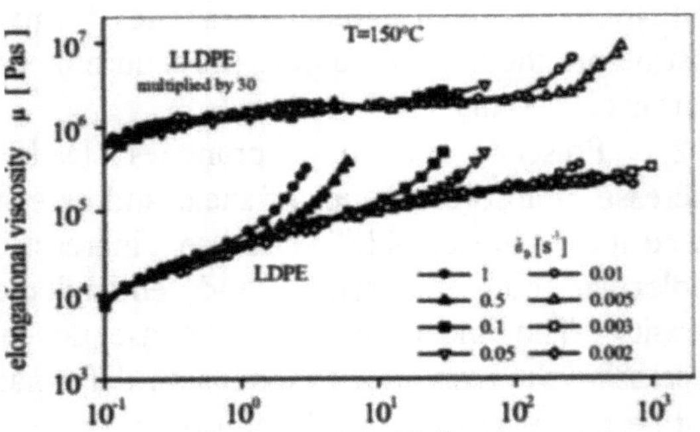

Fig. 1 Time-dependent elongational viscosities of two low-density polyethylenes

For film blowing application a conventional low density polyethylene (LDPE) which contains long-chain branches and a linear low-density polyethylene (LLDPE) which is short-chain branched were chosen. As Fig. 1 shows the LDPE exibits a strain hardening which is well known for long-chain branched polyethylene (Laun and Münstedt 1978, Koyama and Ishizuka 1983). It is pronounced at higher strain rates and gets weaker with decreasing strain rate. The LLDPE shows a strain hardening which is contrary to that of the LDPE. No strain hardening can be observed at high strain rates whereas strain hardening increases steadily from strain rates of $\dot{\varepsilon}_0 = 0.1\,s^{-1}$ to $\dot{\varepsilon}_0 = 0.005\,s^{-1}$. This behaviour cannot be related to the short-chain branched structure but to the existence of an ethylene homopolymer fraction in this resin which seems to be immiscible with the copolymer matrix (Gabriel et al. 1998, Münstedt et al.1998).

2.2 Polypropylenes

With respect to thermoforming two polypropylenes have been chosen which are different in their molecular structure. One resin (PP 1) is an ethylene-propylene block copolymer suitable for thermoforming operations. It consists of linear molecules. The second resin (PP 2) is a homopolymer which contains long-chain branches which were induced by electron beam irradiation.

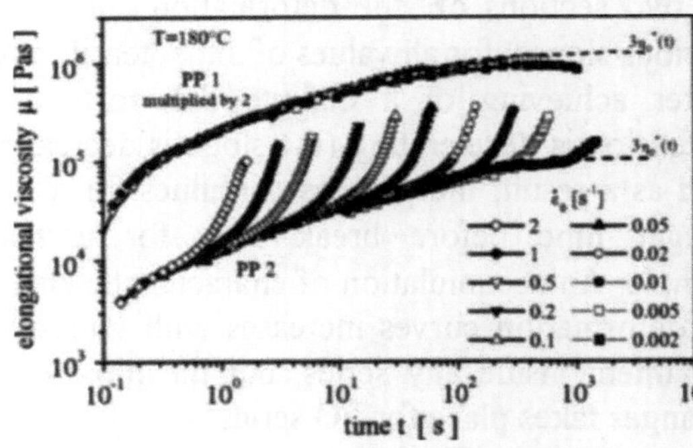

Fig. 2 Time-dependent elongational viscosities of two polypropylenes

Fig. 2 shows the elongational viscosities of the two polypropylenes for a wide range of strain rates. For PP 1 all curves for the transient elongational viscosity superimpose on the linear viscoelastic curve which corresponds very well with the threefold value of the linear start-up viscosity $3\eta_0^+(t)$ in shear. This can be expected for linear polymers as shown by Münstedt and Laun (1981) for an HDPE, for example. PP 2, however, shows an elongational behaviour which is very similar to that of long chain-branched polyethylenes (cf. Fig. 1). The difference to LDPE is a more pronounced strain hardening which is significant even at very small strain rates (Kurzbeck et al., 1998).

It is obvious from simple considerations that strain hardening should lead to a kind of self-healing effect beneficial for the homogeneity of deformation. This has been tested on two processes, namely film blowing and thermoforming.

3. BLOWN POLYETHYLENE FILMS

The film blowing experiments were performed on a laboratory equipment. The produced films were investigated with respect to the homogeneity of the film thickness.

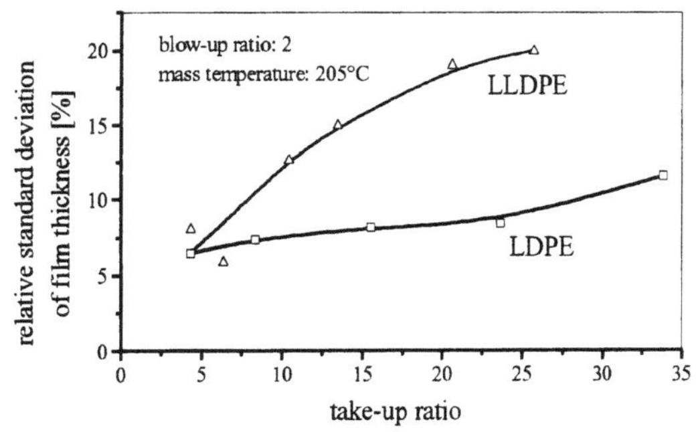

Fig. 3 Homogeneity of film thickness of blown polyethylene films

In Fig. 3 the relative standard deviation related to the mean value of film thickness is plotted as a function of the take-up ratio. At low take-up ratios both polyethylenes exhibit a similar homogeneity. The LDPE shows only a slight increase whereas the inhomogeneity of LLDPE films increases more pronounced with the take-up ratio. This corresponds very well with the strain hardening found in the rheological experiments in uniaxial elongation where the LDPE shows increasing strain hardening with increasing strain rate while the LLDPE shows a contrary behaviour.

4. THERMOFORMED POLYPROPYLENE

The thermoforming experiments were performed on a small size GEISS thermoforming equipment using the drape forming technique at a sheet temperature of 180°C. The formed parts are beakers with a geometry given in Fig. 4. They were investigated with respect to the homogeneity of the wall thickness. Fig. 4 shows the wall thickness as a function of the position on the side wall. The wall thickness is more evenly distributed along the mold height for PP 2 than for PP 1 which exhibits a minimum in the wall thickness in the upper part of the formed beaker. This behaviour can also be correlated with the observation of strain hardening in uniaxial elongation for PP 2.

These two examples demonstrate that strain hardening behaviour found in uniaxial elongation is of advantage for the homogeneity of the multiaxial elongational deformations which are predominant in thermoforming and the film blowing process as well.

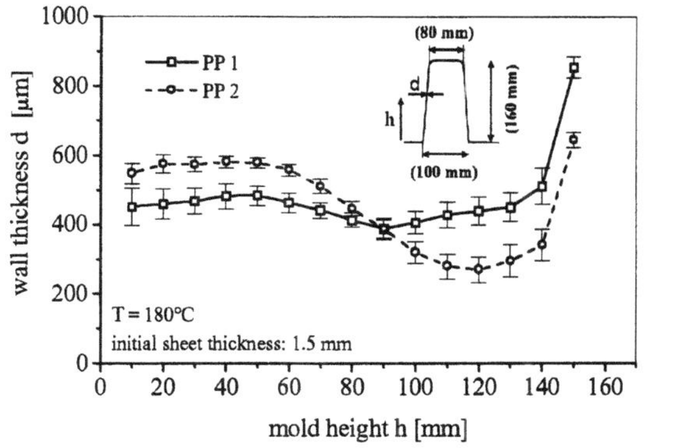

Fig. 4 Wall thickness of thermoformed polypropylene parts

5. REFERENCES

Gabriel C, Münstedt H, Kaschta J (1998) Rheol. Acta 37: 7-20

Koyama K, Ishizuka O (1983) Polym. Proc. Eng. 1: 55-70

Kurzbeck S, Oster F, Münstedt H, Nguyen TQ, Gensler R (1998) submitted for publication in J. Rheol

Laun HM, Münstedt H (1978) Rheol. Acta 17: 415-425

Meissner J. (1972) J. Appl. Polym. Sci. 16: 2877

Münstedt H., Laun HM (1979) Rheol. Acta 18: 492-504

Münstedt H., Laun HM (1981) Rheol. Acta 20: 211-221

Münstedt H, Kurzbeck S, Egersdörfer L (1998) Rheol Acta 37: 21-29

RHEOLOGICAL CHARACTERIZATION OF POLYETHYLENE MELTS WITH LONG RELAXATION TIMES

Joachim MEISSNER and Markus KRAFT[1]

Institute of Polymers and Swiss F.I.T. Rheocenter, ETH Zürich, Switzerland

Abstract

Rheometry of melts of linear polyethylenes (HDPE) in not easy because of the necessary long time measurements due to the very broad retardation and relaxation spectra. Only by a careful stabilization of the material under test these spectra can be determined using a combination of linear viscoelastic tests, viz. shear oscillations and creep followed by creep recovery. For the latter test, an incomplete creep period is followed by long time recovery. This procedure combines experimental flexibility such that the strain is always within the linear viscoelastic limit and still has a good resolution, and a convenient determination of the equilibrium recoverable shear compliance and the zero shear viscosity.

With an appropriate stability of the rheometer, creep/creep recovery tests are to be performed up to 30 hours at a temperature of 170° C. In fact, the weighted relaxation spectrum shows a maximum at 10 000 s of relaxation time. This maximum is not measurable with traditional oscillatory testing. The shape of the spectrum is very different from that of branched polyethylene melts. In addition, the recoverable shear compliance of the HDPE melt turns out to be very high. This long time relaxation is the reason that the limiting values of shear viscosity and first normal stress difference (for zero shear rate and t ➔ □) are never reached when measured in conventional start-up experiments. To complete the characterization, also results of the second normal stress difference are given.

EXTRUSION MODEL FOR NATURAL RUBBER COMPOUNDS

A. Mesec[1], Z. Šušterič[2] and M. Žumer[3]

[1]Goodyear Engineered Products Europe, Škofjeloška 6, SI-4502 Kranj
[2]Sava d.d., R&D Institute, Škofjeloška 6, SI-4502 Kranj
[1]Faculty of Chemistry and Chemical Technology, University of Ljubljana, Aškerčeva 5, SI-1000

1. INTRODUCTION

Rheological and some other mechanical properties of rubber compounds are important for the process of extrusion where compounds get the shape appropriate for their further use in the tire production line. The aim of this work is to develop a natural rubber based compounds with properties enabling determination of extrusion conditions, such that would optimize the extrusion output.

To optimize the compounds, where mastication time of raw rubber and concentration of two different process additives have been varied, the non-Newtonian isothermal model of extrusion is proposed. The result of the model is the operating point (volume flow rate, Q, and pressure drop, ΔP) of the process of extrusion.

2. THEORETICAL

In order to determine conditions yielding the optimal extrusion output, an extrusion model have been developed by solution of extruder's screw dynamic equation by Middleman (1980) and of the extruder's matrix by Michaeli (1991). For rubber compounds, it has been found that the relationship between the volume flow rate, Q, and pressure drop, ΔP, in the extruder matrix in not linear, as for Newtonian liquids (Poiseuille's law).

If we assume power law behaviour Bird, Armstrong and Hassager (1987) of the rubber compounds in the circular matrix, the solution of the equation of motion for two directions is as follows:
where R is the matrix radius and L the matrix length. Parameter K is material constant and n power law exponent or flow index $(0<n<1)$. At fixed geometry of the screw extruder the volume flow rate depends explicitly on the rheological properties of the rubber compounds.

$$Q = \frac{\pi n R^3}{(3n+1)}\left(\frac{\Delta P}{2KL}\right)^{1/n}, \qquad (1)$$

Solving the dynamic equation of motion for screw it is possible by neglecting the transverse flow in the screw. For uniform screw channel down the helical axis, z, the dynamic equation of motion can be written as one-dimensional ordinary differential equation in the following form:
where Z is the helical length and u_z velocity down helical axis, z. As the velocity gradient du_z/dy changes sign in the region $0 \leq y \leq B$, the numerical solution is carried out, resulting in the graphical presentation of screw characteristics for different flow indices, Middleman (1980).

$$0 = -\frac{\Delta P}{Z} + K\left(\left|\frac{du_z}{dy}\right|^{n-1}\frac{du_z}{dy}\right), \qquad (2)$$

Characteristic equations of extruder's screw and matrix are non-linear and have no analytical solution. The volume flow rate and the pressure drop are obtained as an intersection of graphical plot of both characteristic equations. The operating point of the process is a function of screw speed and can be approximated by the following power relationship:

$$Q = A\Delta P^{1/n}, \qquad (3)$$

where A is an extruder geometry dependent constant, implicitly encompassing viscosity. Q thus depends on extruder's screw speed explicitly and implicitly through the rate of dependence viscosity.

3. EXPERIMENTAL

The basic material investigated in this work is natural rubber ($M_w=1.25\times10^6$ gmol^{-1}) based compounds with fixed content of carbon black and variable content of two different plasticizers (mineral oils). The plasticizers differ in acid number, which is much higher for plasticizer B as for plasticizer A.

Rheological measurements at steady shear conditions were performed by capillary rheometer (Gotffert 2001) and at oscillatory shear conditions by instrument for dynamic rheological testing of rubbers (RPA 2000, Monsanto). Extrusion process of the compounds was carried out bay laboratory extruder with Garvey type matrix (Colmec, Italy). The volume flow rate and pressure drop were measured at five fixed screw speeds; 10, 20, 30, 40 and 50 rpm.

4. RESULTS AND DISSCUSION

Figure 1 shows the calculated and experimentally obtained operating points for three different compounds, which differ in rubber mastication time at constant concentration of both plasticizers. The model agrees well with experimental results for the compounds with the longest mastication time, but predicts much higher volume flow rate at modestly higher pressure drop due to neglecting the transverse flow in the screw. Dash line presents the fit of experimental operating points with the power relationship from the Equation (3). The power exponent is in a good agreement with the inverse value of the power law exponent.

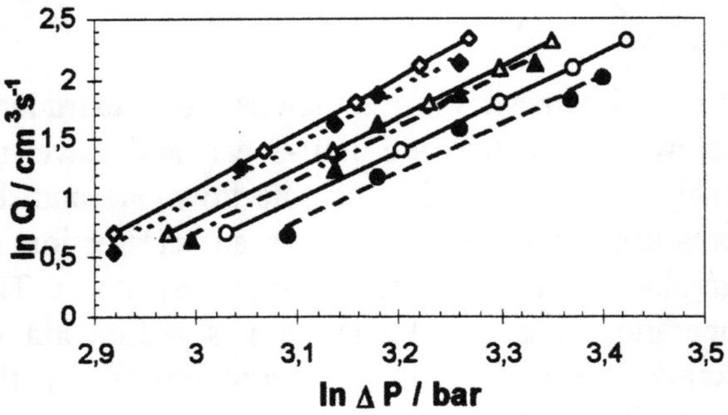

Figure 1. Volume flow rate versus pressure drop at constant concentration of softeners A and B. Mastication times are 0s (●), 120s (▲) and 240s (◆). Full lines with open symbols represent the results of the model and dash line approximation power relationship.

Figure 2 shows Q(ΔP) curves, where concentration of plasticizer A has been varied. Plasticizer A is a neutral process additive (low acid number) and enables better homogenisation of the compounds. The result is better agreement between the experimental operating points, model and approximation power relationship.

Figure 3 shows the Q(ΔP) curves, where concentration of plasticizer B is varied. All curves line in a thin range of pressure drops due to higher acid number of plasticizer B. Model shows worse agreement between experimental operating points and approximation power relationship.

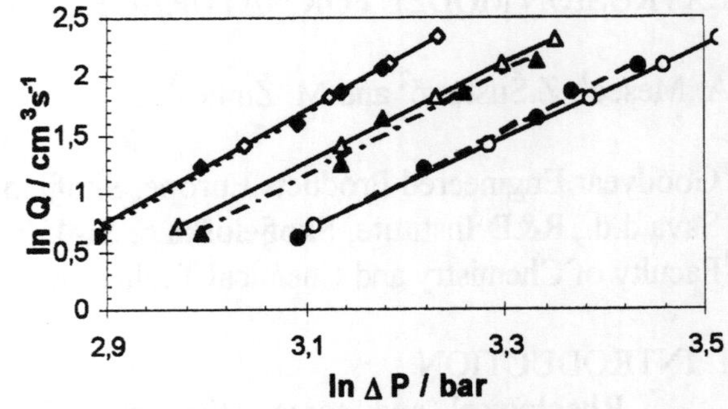

Figure 2. Volume flow rate versus pressure drop at constant concentration of softeners B and mastication time. Concentrations of plasticizer A are 0 phr (●), 2 phr (▲) and 4 phr (◆). Full lines with open symbols represent the results of the model and dash line approximation power relationship.

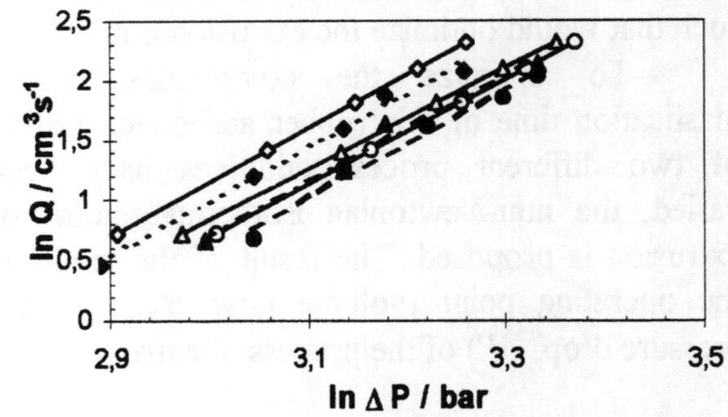

Figure 3. Volume flow rate versus pressure drop at constant concentration of plasticizer A and mastication time. Concentrations of plasticizer B are 0 phr (●), 2 phr (▲) and 4 phr (◆). Full lines with open symbols represent the results of the model and dash line approximation power relationship.

5. CONCLUSION

Model fit the experimental results very good, but it depends on rheological properties of compounds and the change of rheological properties during the process of extrusion. By knowing the rheological properties we are able to quantitatively calculate the volume flow rate and pressure drop for the process of extrusion. Model enables to optimize the compounds for the highest possible volume flow rate.

6. REFERENCES

Bird RB, Armstrong RC, Hassager O (1987) Dynamics of polymeric Liquids, Wiley&Sons, New York

Middleman S (1980) Fundamentals of Polymer Processing, Wiley&Sons, New York

Michaeli W (1991) Extrusion Dies for Plastics and Rubbers, Haser Publishers, Munich

EFFECT OF FILLER CONCENTRATION ON THE TRANSIENT VISCOSITY OF FILLED POLYPROPYLENE MELTS

J. MIKEŠOVÁ, I. FORTELNÝ
Institute of Macromolecular Chemistry, Academy of Science of the Czech Republic
Heyrovský Sq. 2, 162 06 Praque 6, Czech Republic

1. INTRODUCTION

Flow properties of filled polymer melts are fully characterized by both linear and nonlinear viscoelastic data and the stress growth experiment is an elegant way to investigate both types of behaviour in a single measurement. The time dependence of the shear stress of a viscoelastic fluid consists of an initial linear viscoelastic portion followed by a nonlinear one – the steady state flow. In the case of thixotropic behaviour of highly filled systems, the initial linear part of the stress curve indicates the elastic solid-state behaviour of the physical network structure in the system, the stress overshoot shows the kinetics of the breakdown of the structure and the peak in the stress is a measure of the static yield stress.

In a few previous studies concerning the time dependent viscosity of filled polymers, contradictory conclusions were published about the influence of the filler content on the attainment of the steady flow. Bretas and Powell (1985) found that PVC melt filled with glass beads reached the steady state flow faster than the matrix alone. Suspensions of limestone in silicone oil and neat oil attained the steady state flow at approximately the same time, as presented by Gleissle and Ohl (1985). On the other hand, various authors observed an increase in the relaxation time of suspensions with increasing concentration of particles, recently, Aral and Kalyon (1997) for the glass beads in silicone oil.

2. EXPERIMENTS

Two polypropylene matrices with different molecular weight averages and distributions, filled with calcium carbonate were chosen for the study. The first matrix (PP) was a commercial polymer (Mosten 52 422, Chemopetrol, Czech Republic) with the molecular weights $M_w=3.6\times10^5$, $M_n=7\times10^4$. The second (DPP) was a laboratory sample with $M_w=2\times10^5$ and $M_n=6.4\times10^4$, prepared from the first matrix by the dynamic degradation with dicumyl peroxide. The preparation was carried out in a Brabender plastograph at 210 °C. The degradation process not only decreased the zero-shear-rate limiting values of the steady viscosity and the first normal stress coefficient of the second matrix, but it also significantly broadened the region of the independence of both material functions on the shear rate.

The filler (Durcal 2, Omya, Austria) was without a surface treatment and its average particle size and the specific surface area were 3.6 μm and 3.1 m^2g^{-1}, respectively. The components were mixed at a constant temperature 190 °C using a Brabender Plasti-Corder. The samples for rheological measurements were prepared by compression moulding at 190 °C.

Rheological measurements were performed on a rotational rheometer Rheometrics SYS 4 in cone and plate geometry (radius - 12.5, 25 mm; cone angles - 0.101, 0.0401 rad). The gap between cone and plate was 14 times greater than the average size of particles. All measurements were carried out at 190 °C. The transducer of the rheometer was kept at the constant temperature of 45 °C. The samples exhibiting the yielding behaviour were heated at 200 °C after the gap setting to shorten the relaxation of a resistive normal force.

3. RESULTS

The time dependent viscosity ($\eta^+(t)$) of melts, filled with particles of volume fractions $\varphi=0.05 - 0.35$, was measured in the range of shear rates $10^{-3} - 10^{0}$ s^{-1}. In comparison with the pure matrices, all filled melts showed the damped transient behaviour at the same shear rate. An example of the data $\eta^+(t,\varphi)$ is provided in Fig.1. The effect of filler concentration on the transient viscosity was quantified by parameter t_s – the time at which the steady state viscosity was reached. Special attention was paid to the investigation of the filler effect on the envelope curves - the low-shear-rate limits - of the transient viscosity. These linear viscoelastic data were only obtained for melts with filler contents $\varphi<0.2$. The increase in the time t_s of the envelope curves with φ was found nearly the same for both studied matrices and can be described by the empirical function: $t_s(\varphi) = t_s(0)\exp(7.5\varphi)$, where $t_s(0)$ is the parameter t_s for the pure matrix.

In highly filled melts ($\varphi>0.25$), thixotropic behaviour was observed in both matrices. The typical peak on the first shear stress growth curve $\tau_{12}(t)$ occurred only in the system with the DPP matrix. The static yield stresses τ_S were determined. The evaluated stresses τ_S were higher than the values of the steady shear stresses extrapolated from the plateau at low shear rates τ_E (e.g., τ_S = 380 Pa and τ_E = 250 Pa were obtained for the melt with φ = 0.35). The maximum on the first curve $\tau_{12}(t)$ did not appear in the melts with a commercial PP matrix (see Fig. 2). In these very elastic melts, the effect of destruction of the particle network structure (a decrease in $\tau_{12}(t)$) was eliminated with the effect of stretching the entanglement structure of long molecules in PP matrix.

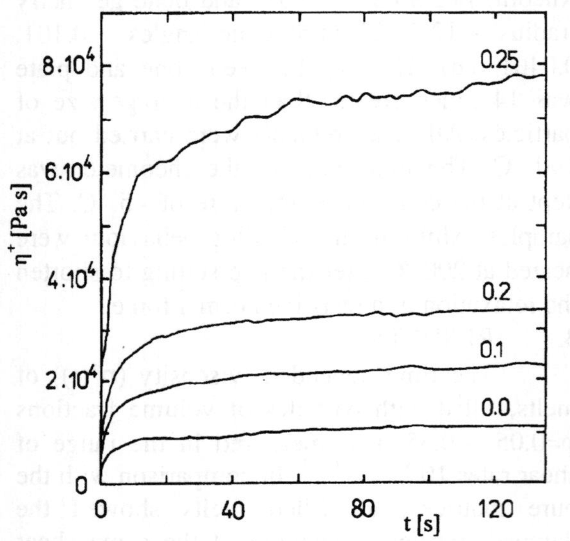

Fig.1. The time dependence of the transient viscosity of the commercial PP melts filled with φ = 0.0, 0.1, 0.2, 0.25 measured at the shear rate 0.001 s^{-1}. The average values of $\eta^+(t)$, calculated from the data obtained in the clockwise and counter-clockwise direction, are plotted

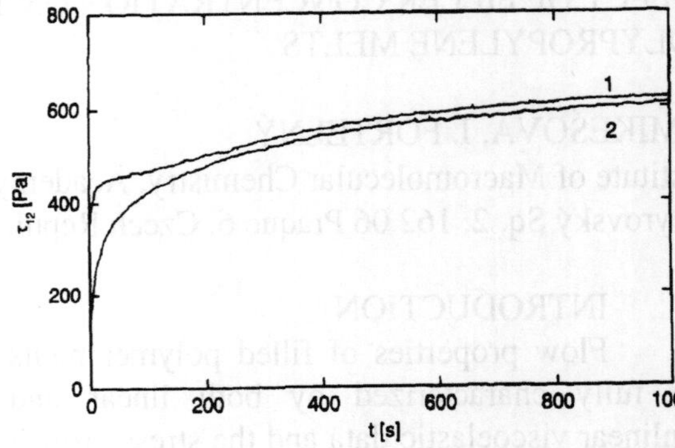

Fig. 2. The growth of the shear stress of the commercial PP melt filled with φ = 0.35 during two following experiments carried out in the clockwise direction at the shear rate 0.001 s^{-1}. The next experiments followed the curve 2.

4. CONCLUSIONS

Critical concentrations of the filler ($\varphi_{crit}\approx0.2$) were found for polypropylene matrices with different molecular weights and distributions in time dependent behaviour. This concentration limit - φ_{crit} - coincides with the formation of a physical network structure in melts. In considering absence of the longest molecules in the degraded polypropylene, no influence of these molecules on the foundation of the physical network was found.

In the region $\varphi<\varphi_{crit}$, the filler effect was quantified for the envelope curves. The found increase in the time required for the attainment of the steady viscosity with increasing filler concentration was approximately the same in both matrices and was described by an exponential function.

In highly filled melts ($\varphi>0.25$), a significant difference in the elasticity of the studied matrices appeared in the shape of the first growth curves of the shear stress.

5. ACKNOWLEDGEMENT

The authors wish to thank to the Grant Agency of the Czech Republic for support (Grant No. 106/98/0274).

6. REFERENCES

Aral BK, Kalyon DM (1997) J Rheol 41: 599-620

Bretas RES, Powell RL (1985) Rheol Acta 24:69-74

Gleissle W, Ohl N (1987) Proceedings of CHISA Meeting, Prague

DEPENDENCE OF RHEOLOGICAL PROPERTIES OF POLYHYDROXYETHYLENE UPON THE CONTENT OF ADDITIVES OF OLIGOHYDROXYETHYLENE.

Satyanarayan Panigrahi, T.A.Kostinova

State Acedemy of Lighy Industry of Ukraine, 25015 Kiev, Ukraine

The investigation of rheological properties of polyhydroxyethylene (PHE) is of practical interest firstly for the production of fibers out of polymer melt. However, the application of this method meets difficulties due to the great viscosity of liquid PHE, valuable properties of which appear only when the molecular mass is very high.

The preliminary experimental data have shown that the viscosity of the melt of PHE with $M=1x\ 10^6$ at shear load of $5.8x10^4$ Pa amounts to $4.48x10^5$ Pa x s, which is greater by 2 or 3 orders than viscosity at which fibers are usually formed. Let us note, that it is impossible to decrease the viscosity essentially by the increase of the temperature, because the quick destruction of the polymer occurs in that case.

The increase of the viscosity of PHE melts is possible if use the plasticizing effect of low-molecular additives. Their selection for PHE is to be determined by not only efficiency of plasticizing effect but also the influence of added substance upon the structure formation and properties of the fiber formed, and also upon the flocculating capacity and hydrodynamic activity, properties important in practice for these solutions. These requirements are most completely satisfied by low-molecular and oligomer compounds of homologous series of ethylene oxide, the composition of which is ultimately close to the composition characteristic for polyhydroxyethylene. We investigated the plasticizing effect of diethyleneglycol and oligohydroxyethylenes ("OOE") with molecular masses of 600, 1500 and 3000 upon polyhydroxyethylene, they will be further called OOE-134, OOE-600, OOE-1500 and OOE-3000 correspondingly. The compounds were obtained by mutual solving of components in ethanol at 60°C, followed by evaporation of the solvent and drying films obtained in vacuum at 35°C up to the constant weight. The films milled then are pressed in the form of tablets at the room temperature and at pressure of $152x10^5$ Pa. The rheological properties of the melts have been studied with the use of microviscosimeter MB-2 in the wide range of shear stresses. For the elimination of the influence of losses of the pressure at the input upon the conditions of the appearance of the non-stable flow, the relation (L/D) of length by diameter of a capillary is assumed to be equal to 21. The curves of the flow of the original PHE as well as its compounds with OOE in the shear stress range examined follow the exponential law $D = k\tau^n$. The extent of the reliability of the calculation of exponent n at the Student's factor of 0.95 was 0.2.

The addition of oligohydroxyethylenes to polyhydroxyethylene results in the change of the flow conditions. From the dependence n=f(c) (Fig. 1) it is seen that the curves can be divided into two sectors: the first is up to 30% of the additive introduced, and the second is from 30 to 60. At the first sector the influence of the molecular mass of the additive upon the flow conditions is proved to be small for all OOE (n=3÷4). However, at the second sector when the composition of the additive is 30%, this difference becomes essential.

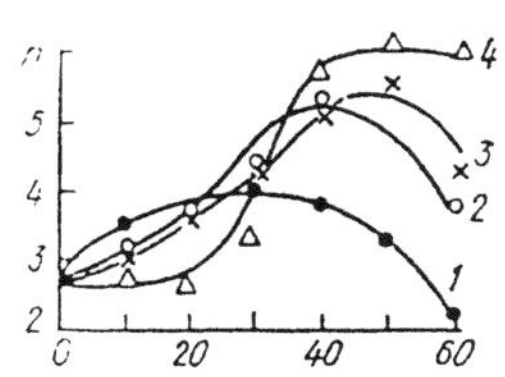

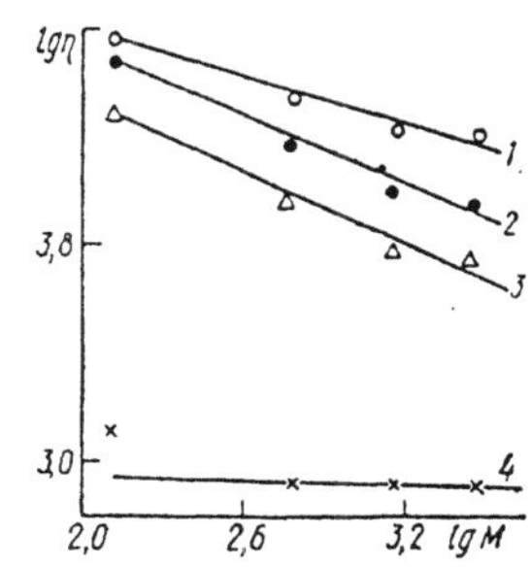

Fig. 1. Exponent for PHE melt versus the composition of OOE at the temperature of 90°C and shear stress of $5.8x10^4$ Pa: 1 - PHE+OOE -134; 2 - PHE + OOE - 600; 3 - PHE + OOE - 1500; 4 - PHE -+ OOE – 3000.

Fig. 2. Logarithmic dependence of the viscosity of PHE melt upon molecular mass of OOE at temperature of 90°C and shear stress of $5.8x10^4$ Pa: 1 - PHE + 10% OOE; 2 - PHE + 20% OOE; 3 - - PHE + 30% OOE; 4 - - PHE + 40% OOE.

The tendency is observed that the greater the molecular mass of oligohydroxyethylene, the greater distortions of the conditions of Newton fluid until the specific additive concentration C_{cr}. For example, for OOE=134 it amounts to 30%, OOE-600 - 40%, OOE-1500 - 50%, OOE-300 - (50-60) %. As the molecular mass of an oligomer increases, the concentration of the additive shows the tendency to the reduction, especially at $M \leq 1500$. All oligomers studied decrease the viscosity of polyhydroxyethylene melt (Fig.2), the most active in this relation being the oligomer OOE 600. The dependence of the viscosity of mixture of PHE and OOE upon the molecular mass of OOE is satisfactorily described by the equation for polymer composites:

$$\eta = \eta_o \cdot M_{OOE}^{\alpha} \quad (1)$$

or in logarithmic coordinates:

$$\lg \eta = \lg \eta_o + \alpha \lg M_{OOE} \quad (2)$$

The constant η_o, can be defined as the viscosity of the mixture of PHE with atomic hydrogen (at $M_{OOE} = 1$, that is, $\lg M_{OOE} = 0$): It gives the picture of the character of the effect of additive, and it, together with constant α indicating the intensity of the plasticizing effect of OOE, allows to describe effects observed quantitatively.

The value of η_o remains almost unchanged at C_{pl} =30% (Fig. 3 a), but dramatically declines at $C_{pl} \geq 30\%$. The plasticizing effect of OOE grows as the concentration increases (Fig. 3 b), and reaches the extremum at 30%, then it dramatically declines. The viscosity of the PHE melt decrease as the volume portion of the polymer decreases in the mixture of PHE and OOE. In logarithmic coordinates, the experimental points lie on a common line (Fig. 4) with the exception of $\varphi = 0.5$ (that is, when the OOE was added in the amount of 50%). Therefore, the dependence of the viscosity of the melt upon the volume polymer portion φp is described by the equation

$$\eta = \eta_{PHE} \cdot \varphi^{z} \quad (3)$$

or in logarithmic coordinates:

$$\lg \eta = \lg \eta_{PHE} + z \lg \varphi \quad (4)$$

In this equation constant η_{PHE} does not need an explanation, and constant Z characterizes destructive effect of the additive.

Proceeding from rheological data obtained, we have determined additives and their doses which allow to count on obtaining satisfactory fibers based on polyhydroxyethylene.

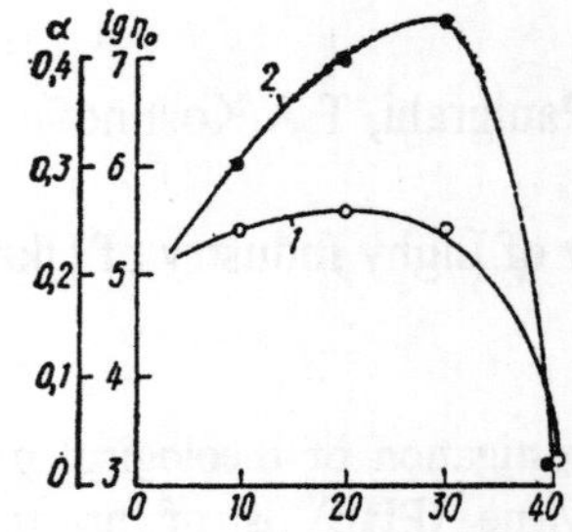

Fig. 3. The influence of the composition of OOE on the viscosity of the PHE melt (1) and on the constant of the intensity of plasticizing effect (2) at temperature of 90°C and at shear stress of 5.8x10 Pa.

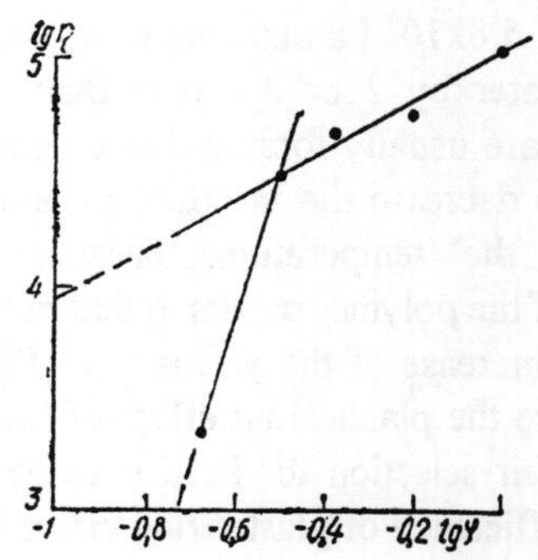

Fig. 4. The logarithmic dependence of the viscosity of PHE melt upon the volume polymer portion in the mixture with OOE-1500 at temperature of 90°C and shear stress of 5.8×10^4 Pa.

It is not expedient for the concentration of oligomer in the melt to be increased over 30% because in the 30% area an acute changes in the mechanism of the destructive effect of the additive takes place (Fig. 3 a, b; 4). The prospective use of oligomers is at M$\leq$1500, when their plasticizing effect on the PHE stops improving as the molecular mass of OOE increases (See Fig. 1). In whole, on the basis of the data of rheological investigation of compositions of PHE with OOE, it is possible to affirm that oligohydroxyethylenes are acceptable for the reduction of the viscosity of polyhydroxyethylene melt viscosity.

RHEOLOGICAL PROPERTIES OF POLYPROPYLENE - CO-POLYAMIDE MIXTURE MELTS AT PHASE INVERSION

M. V. TSEBRENKO, N. M. REZANOVA, I. A. TSEBRENKO

State Light Industry Academy of Ukraine, 2 Nemirovich-Danchenko Street, Kiev-11, 252011, Ukraine

Two polymer mixture melts presents, most often, a bi-phase system in which one of components (disperse phase) is distributed in certain order within the second one (disperse medium), i.e. in terms of its structure the molten mixture of incompatible polymers is a colloid system. In any colloid system, when component ratio is changed, phase inversion occur: under the influence of external factors the disperse medium and disperse phase change their roles. Such external factor of influence can be the change of either composition or temperature or introducing new substance. Within low molecular emulsions the phase inversion occurs unevenly at component ratio close to one. Within polymer mixtures the area of phase change is usually expanded over the wide composition range, when both phases are continuous.

It is known from the works of Tsebrenko et al. (1976, 1983) that disperse phase polymer forms, under certain flow conditions, in another (matrix) polymer a lot of microfibres, that are strictly oriented towards flow direction. Such phenomenon was called the specific fibre-formation. It is realised in the most distinct way at disperse phase polymer content as much as 20 to 30 mass %.

Micro- and macrorheological processes during polypropylene - co-polyamide (PP/CPA) mixture melts' flow have been investigated in this work. The component ratios (PP/CPA) in mixtures are as follows: 40/60; 50/50; 60/40. For mixtures of the above-mentioned compositions the specific fibre-formation phenomenon is not being realised, as far as the phase inversion occur. In this work it has been shown that such microrheological properties as deformation of the disperse phase

polymer's drops into liquid streams, coalescence of liquid streams being formed, in longitudinal direction, migration of drops, fracture of liquid streams into droplets, can be regulated by introduction of the third component that is capable to display the specific interactions with CPA macromolecules. The copolymer of ehtylene and vinylacetate (CEVA) was chosen as such third component. Infra-red spectra have shown that dipole-dipole interactions take place between CPA and CEVA macromolecules at the interphase. On the other hand, CEVA displays affinity to PP macromolecules at the expense of non-polar part of macromolecules. This is the reason for improvement of miscibility between PP and CPA, when CEVA is added. The phase separation in such ternary mixtures is shown, and phase inversion does not occur at the above-mentioned component ratios. The amount of CEVA being introduced was equal to 5; 10; 20; 25 mass % of the total PP mass within the mixture. Blending the components was carried out within the melt using the combined screw-disk type extruder. Viscous properties were evaluated on capillary constant pressure viscometer. Flow regime "n" was evaluated in terms of tangent value in certain point of the flow curve. Elasticity of the mixture melts were evaluated indirectly in terms of swelling parameter "B" of annealed extrudates. Microrheological processes were determined by results of optical and electronic microscopy.

Results (see Table) evidence that polymers being blended demonstrate close viscosity values (η) of melt. Sharp decrease of the mixture melt viscosity takes place as compared to η value of initial polymers. Such regularity is typical for polymer mixture melts, and it can be explained by the processes of disperse phase polymer's structure formation.

Table. Viscoelastic properties of molten initial polymers and their mixtures

Mixture Composition, mass %	η*, Pa x s	n	B
PP 100	520	1.8	1.7
CPA 100	540	1.1	1.4
CEVA 100	400	1.4	1.5
PP/CPA 40/60	220	1.8	4.5
PP/CEVA/CPA 40/5/60	400	1.8	6.0
PP/CEVA/CPA 40/10/60	410	1.7	6.2
PP/CEVA/CPA 40/20/60	430	1.8	5.8
PP/CEVA/CPA 40/25/60	380	1.8	5.6

* at T = 185 °C and shear stress 5.69 x 10^4 Pa

When 5 to 25% of CEVA is introduced, the sharp increase of viscosity of mixture melt is observed, that confirms indirectly the manifestation of specific interactions between CEVA and CPA macromolecules.

Microscopic analysis has shown that the predominant structure type for extrudates of PP/CPA mixtures with composition 40/60, 50/50 are films, coarse fibres, at that, extraction of matrix polymer goes very slowly, and disperse phase polymer residues are impossible to be dispersed after dissolving the matrix in order to perform the qualitative microscopic analysis. When CEVA is introduced, the structure formation is changed sharply: the phenomenon of specific fibre-formation is realised distinctively. So, for example, within the extrudate of PP/CEVA/CPA mixture of the composition 40/10/60 the predominant (up to 80%) structure type appear to be microfibres with average diameter 3.2 μm. The trend revealed is confirmed distinctively as well by the values of mixture extrudate swelling (see Table).

In the works of Tsebrenko *et al.* (1979, 1983) it was demonstrated that the value of swelling of the polymer mixture extrudate serves as an indirect characteristic of the specific fibre-formation phenomenon: the higher will be the number of microfibres within the extrudate and the finer will be those, so, the greater will be the "B" parameter. Actually, introducing CEVA enables the PP microfibre formation within the CPA matrix. Due to this reason, extrudate swelling increases sharply for ternary mixtures.

So, the investigation carried out has shown for the first time the possibility for slowing processes of polymer phase separation and realising specific fibre-formation within the mixtures which are characterised under normal conditions by occurrence of interlacing morphology or phase inversion.

REFERENCES

Tsebrenko M. V., Yudin A.V., Ablazova T.I., Vinogradov G.V. (1976) Polymer 17: 831-834.

Tsebrenko M. V. (1983) Intern. J. Polymeric Mater. 10: 83-119.

Tsebrenko M. V., Rezanova N. M., Yudin A.V.(1979) Khimicheskiye Volokna. 6: 36-38.

Tsebrenko M. V.(1983) Khimicheskiye Volokna 3: 28-31.

TIME EVOLUTION OF NORMAL STRESSES IN STRONG SHEAR FLOWS OF POLYETHYLENE MELTS.

A.M. VELHO[1], A. C. DIOGO[1], G. MARIN[2]

[1] Instituto Superior Técnico, Av. Rovisco Pais, P-1096 Lisboa Codex, Portugal
[2] Labo. Physique des Mat. Industriels, IURS, Univ. Pau-Adour, Av. de l'Université, F- 64000 Pau, France

1. INTRODUCTION

The development of (transient) tangential and normal stresses has been widely studied in steady shear flows. In strong shear flows, one can find some measurements of shear stresses as the data presented by Doshi and Dealy (1987) or Züllle et al. (1987), but not of normal stresses.

Defining strong flows as those for which the largest principal extension ratio, λ, increases exponentially with time, we have considered the strong shear flow where the deformation γ is given by the equation

$$\gamma = A.\exp(\alpha.t) + (1/A).\exp(-\alpha.t) \qquad (1)$$

The parameter α controls the growth rate.

The aim of this work is to study the time evolution of normal as well as tangential stresses in strong shear flows and compare them with the corresponding values, at same shear rate, in simple shear. Several constitutive models, specially Wagner's model, will be compared with the experimental results.

2. EXPERIMENTAL

The time evolution of both the normal and shear stresses, was measured in strong shear flows, generated according to equation (1), for different values of the growth parameter α, and during the following period of rest. Tests in steady shear situations have been performed too.

All measurements were performed on a RMS-800 mechanical spectrometer, for two different polymer melts, a high density polyethylene (HDPE), and a low density polyethylene (LDPE) at 140° C, using parallel plates as well as cone-plate geometries.

3. DISCUSSION

The values of normal forces obtained in steady shear flow, both with cone-plate and plate-plate geometries, for PEBD, are in a good agreement if one takes N_2=-.25N_1. For exponential shear flows, the measurements in the two geometries can show values of (N_1-N_2) which are about 3 times N_1, as in the curves shown in Fig.-1. Also, the results obtained with the cone-plate geometry show an increase of the force during the rest period, after cessation of deformation, that is much stronger than what is observed in plate-plate geometry .

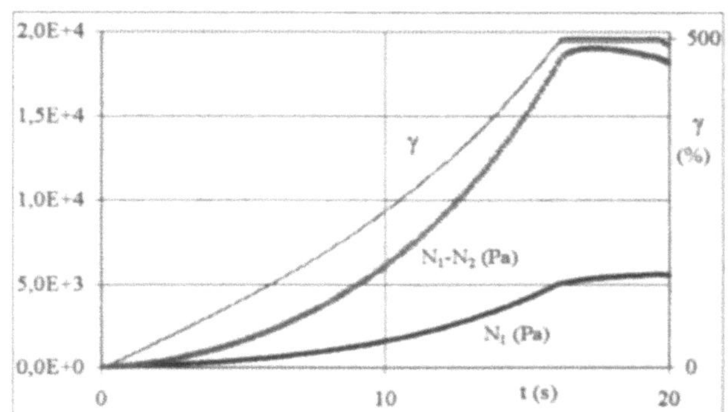

Fig.1. Time evolution of N_1 (o) measured with cone-plate geometry and N_1-N_2 (x) measured with plate-plate geometry, during the deformation, γ (eq. (1)) with α=.1 and the rest period, for LDPE at 140° C.

These differences are related with the fact that in strong shear flows, normal stresses development is strongly sensitive to the growth rate parameter α. This is clearly visible for values of α where the attained rates of deformation are greater than the inverse of the longest time constant of the material: normal forces became smaller compared with steady shear for the same values of the rate of deformation.

In plate-plate geometry, non-homogeneity of the flow between the plates makes this flow "less strong" than if it were in cone-plate geometry. So, the transient normal stresses reach larger values, but not so high as it would have been in steady shear.

In particular, the (transient) normal stress differences can present quite smaller values as compared with the steady values at the same shear rate. In Fig-2 this effect is clear for LDPE, when $\alpha \geq .01$. For HDPE the difference is not so strong and it starts with higher values of α (=.1), because the longest time constant of this material is much smaller.

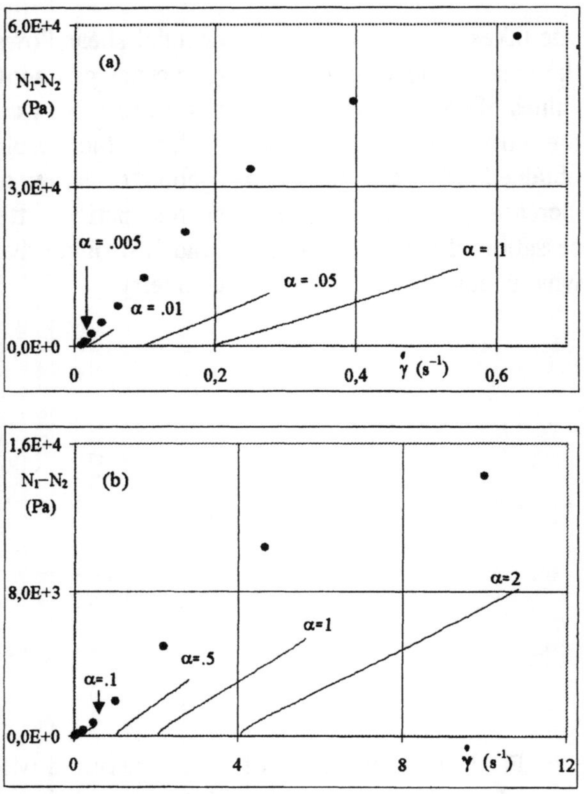

Fig.2 – Evolution of N_1-N_2 vs $\dot{\gamma}$, for LDPE (a), and for HDPE, (b), both in steady shear (o), and in strong shear (-) with different values for the growth parameter, α. Measurements were done in plate-plate geometry, both at 140° C

Tangential stresses development in strong flows is also sensitive to the values of α, but the dependency is not so strong. The resultant form of the curves shown in Fig.3, reflect the "time delay" in the development of normal stresses, which occurs in steady shear flow. This "normal stress time lag", is now accentuated by the strong shear flow in a way that depends on the relation between the rate of change of $\dot{\gamma}$, (or on α) and the (largest) times of the spectrum, responsible for the growth of normal stresses. This occurs in both materials but for the same values of α it is much stronger in LDPE, (Fig. 3), because the largest times of this material are bigger.

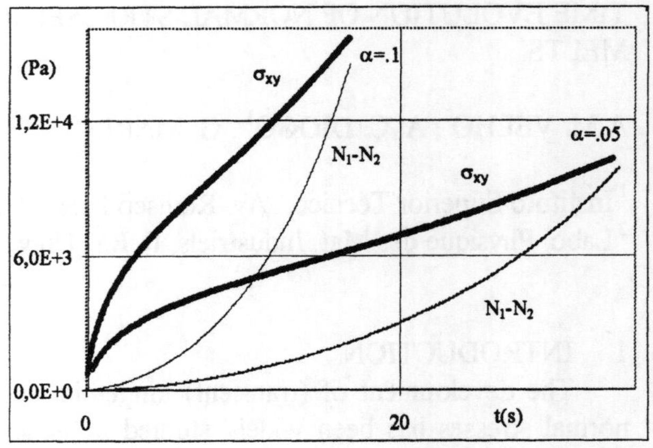

Fig.3. Time evolution of σ_{xy} (o) and N_1-N_2 (-) with time, in exponential shear flow for LDPE, at 140°C.

The curves obtained with the models of Lodge, Doi-Edwards or Wagner and Schaeffer (1993), couldn't reproduce in a coherent form both tangential and normal stresses in this flow. Even not describing the curve, all the models preview that normal forces are smaller in strong shear than in steady shear at the same values of $\dot{\gamma}$. They permit also to show that if the non-homogeneity of the flow in plate-plate geometry is considered, one will have higher values of the normal force in plate-plate compared with those obtained in cone-plate geometry for the conditions of the flow considered in Fig.1.

4. CONCLUSIONS

The measured values of normal or tangential stresses, in a strong shear flow have smaller values, compared with those obtained in steady shear at the same values of $\dot{\gamma}$. The differences found are particularly great for the normal stresses.

The development of normal stresses in strong shear flows is strongly sensitive to the growth parameter, α.

The tested constitutive equations can't preview the strongest part of these variations, specially for the normal stresses.

5. REFERENCES

Doshi, S.,R. and Dealy, J., M., J. Rheol. 31, 563 (1987)

Wagner M.H. and Schaeffer, J.J. Rheol. 37, 643 (1993)

Zülle, J. Linster, J. Meissner, H. Hürlimann, J. Rheol. 31, 583 (1987)

PARTICLE COALESCENCE AND BUBBLE DISSOLUTION IN ROTATIONAL MOLDING

J. VLACHOPOULOS[1], M. KONTOPOULOU[1], C.T. BELLEHUMEUR[2]

[1]Centre for Advanced Polymer Processing and Design, Dept. of Chemical Engineering, McMaster University Hamilton, ON, Canada L8S 4L7

[2]Dept. of Chemical and Petroleum Engineering, University of Calgary, Calgary, AB, Canada, T2N 1N4

1. INTRODUCTION

The production of a hollow plastic part from polymer powder particles during rotational molding requires the formation of a homogeneous melt. The latter involves the coalescence of particles, often referred to as sintering, followed by the formation of bubbles and the densification of the melt through bubble dissolution.

The identification of material properties which influence the two above processes is crucial in the fundamental understanding of rotational molding. Previous work by Vlachopoulos et al. (1996) and Kontopoulou et al. (1997) has shown that, although viscosity is the most important parameter used to characterize resin rotomoldability, melt elasticity should also be taken into account. The present study intends to clarify the contribution of material properties, by presenting a thorough investigation of all the aspects involved in particle coalescence, bubble formation and densification during rotational molding.

2. METHODS

The present research involved various experimental studies of particle coalescence, bubble formation and bubble dissolution, using a hot stage microscopy unit and an imaging system. Several rotational molding grade polyolefin resins were used, including polyethylene (PE), polypropylene/polyethylene (PP/PE) copolymers and PE copolymers. The experimental studies were complemented by the development of models describing particle coalescence and bubble removal.

3. RESULTS

3.1 Particle coalescence and bubble formation

Particle coalescence (Bellehumeur et al. (1996)) and bubble formation experiments revealed the significance of viscosity in both processes. The higher the viscosity, the more and/or larger bubbles are formed inside the polymer melt and longer times are needed for particles to coalesce, lose their identity and form a homogeneous melt. **Figure 1** shows the size distribution of bubbles initially formed inside the polymer melt for three LLDPE resins with different MFIs. The variation of viscosity with temperature is of great importance as well. Especially in the case of amorphous polymers, viscosity decreases gradually starting from very high values, resulting in the early entrapment of bubbles, which subsequently take long times to dissolve.

It was also found that in the case of PP/PE copolymers with similar viscosities and different melt elasticities, significantly higher amounts of bubbles were formed when the resins were more elastic and the onset of the formation of a homogeneous melt showed a considerable delay. Observations of coalescence also suggested that melt elasticity slows down the process. Based on these findings, a viscoelastic model for the coalescence of two particles was developed. The model is based on the approach first presented by Frenkel (1945) and modifications proposed by Pokluda et al. (1997). The convected Maxwell constitutive equation was used to introduce viscoelastic behavior into the description of the sintering process. Details of the model development are presented by Bellehumeur et al. (1998).

The current model for the coalescence of two particles predicts that the sintering rate decreases as the material relaxation time increases. However the coalescence rate becomes nearly independent of the material relaxation time as either the viscosity or the particle radius become very large. The model predicts quite satisfactorily trends obtained experimentally with

PP/PE copolymers, which purely Newtonian models failed to describe **(Figure 2)**.

3.2 Bubble dissolution

The dissolution of a single spherical bubble in an infinite polymer melt under isothermal conditions was modeled by solving the coupled continuity, conservation of momentum and diffusion equations. Viscosity was included in the model by substituting the Newtonian constitutive equation into the momentum equation. The concentration of dissolved air inside the polymer melt and at the bubble/melt interface was determined from Henry's law.

The results obtained by this model confirmed previous reports by Xu and Crawford (1993) that bubble dissolution is mainly diffusion controlled. It was found that dissolution depends significantly on initial bubble size, surface tension and air concentration in the polymer melt, but is not influenced significantly by melt viscosity, when the latter lies within ranges typical for rotational molding resins. Experimental bubble dissolution curves for LLDPEs with different MFIs confirmed this result. The above seem to contradict the general belief that increased viscosity results in slower bubble dissolution. However, when melt viscosity is higher, more and larger bubbles form initially, as shown in the previous section. This can slow down the dissolution process significantly, as shown in **Figure 3**.

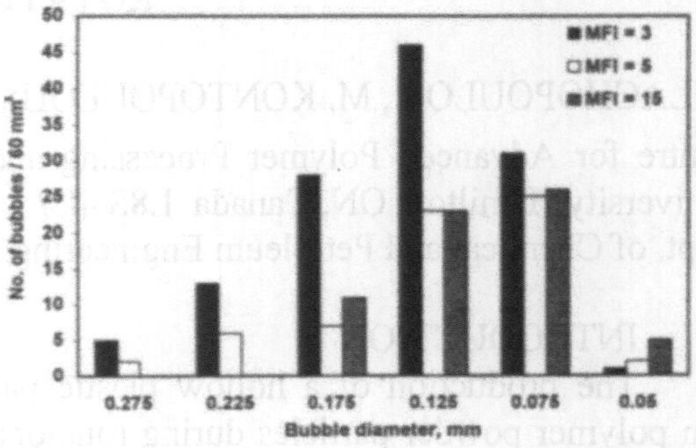

Figure 1: Bubble content distribution in LLDPE melts with different MFIs

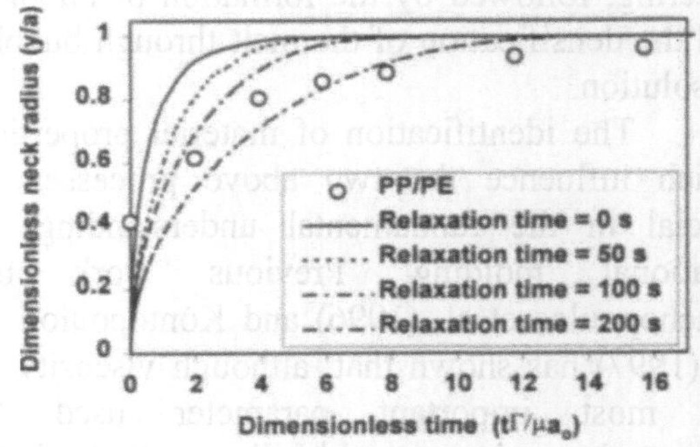

Figure 2: Comparison of viscoelastic sintering model predictions with experimental data for PP/PE (y, a, t, Γ, and η, are the neck radius formed between the particles, particle radius, time, surface tension and viscosity, respectively)

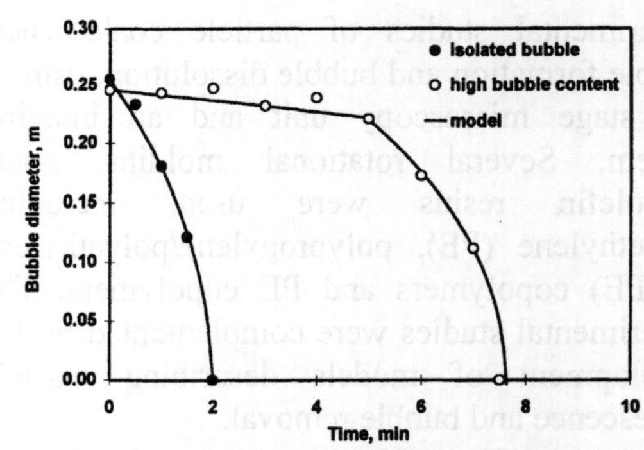

Figure 3: Effect of high bubble content on dissolution curves in an LLDPE melt

4. REFERENCES

Bellehumeur CT, Kontopoulou M and Vlachopoulos J (1998) Rheol Acta (accepted for publication)

Bellehumeur CT, Bisaria M and Vlachopoulos J (1996) Polym Eng Sci 36: 2198-2207

Frenkel J (1945) J Phys 9:385-391

Kontopoulou M, Bisaria M and Vlachopoulos J (1997) Int Polym Process 12:165-173

Pokluda O, Bellehumeur CT and Vlachopoulos J (1997) AIChE J 43: 3253-3256

Vlachopoulos J, Bellehumeur CT and Kontopoulou M (1996) Proc XIIth Int Congr Rheol, Quebec City, 693-694

Xu L and Crawford RJ (1996) J Mat Sci 28: 2067-2074

SIMULATION OF POLYMERS THROUGH DIES OF FORMING DEVICES

E. WALICKI, A. WALICKA

Technical University of Zielona Góra, Dept. of Mechanics, ul. Szafrana 2, 65-016 Zielona Góra, Poland

1. INTRODUCTION

Forming of plastic products is performed generally by the dies (dies, sizing dies, moulds and other forming devices). The die, which plays a role of major importance for a line of extrusion equipment, is to give various, desired cross sectional shape to a melt continuously transferred by the screw.

Physical analysis of the polymer flow in the die indicates that the following three factors must be taken into consideration at designing the die:

- rheological properties of polymer melt inside the die;
- problems concerning the die structure;
- problems of temperature control and adjustment.

To describe the rheological behaviour of polymer melt in complex geometries the non-linear model of Shulman (1975) has been successfully applied: $\tau^{1/M} = \tau_0^{1/M} + (\mu\dot{\gamma})^{1/N}$ where τ is the shear stress, τ_0 is the yield shear stress, μ is the coefficient of plastic viscosity, N and M are non-linear indices.

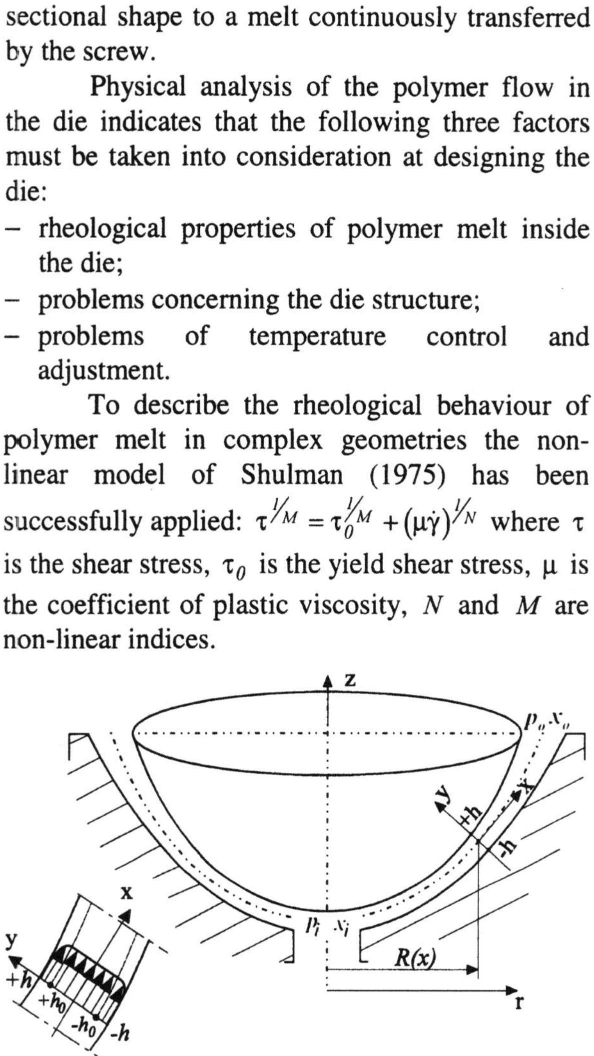

Fig. 1. Geometry of the die.

The die structure may be presented generally as the narrow space between two surfaces of revolution, having a common axis of symmetry, as shown in Fig. 1.

The aim of this work is a study of the pressure distribution in the die in function of two first factors mentioned above. Using the method of averaged inertia, one has analysed the influence of inertia terms of the equations of motion on this distribution.

2. EQUATIONS OF MOTION OF A POLYMER IN THE DIE CLEARANCE

Applying the method of averaged inertia we may write the following equation (Walicki and Walicka 1994, 1996):

$$\frac{\partial \tau_{yx}}{\partial y} = f(x), \tag{1}$$

where the non-zero component of stress tensor is:

$$\tau_{yx} = \left[\tau_0^{\frac{1}{M}} + \left(\mu\left|\frac{\partial \upsilon_x}{\partial y}\right|\right)^{\frac{1}{N}}\right]^M \tag{2}$$

and

$$f(x) = \frac{dp}{dx} + \frac{\rho}{h}\left(\frac{R'}{R} + \frac{\partial}{\partial x}\right)\int_0^{+h} \upsilon_x^2 dy. \tag{3}$$

Here the prime denotes the differentiation with respect to x. Denoting the integral of Eq. (1) by $\upsilon_x = F(f,h,y)$ we can define the flow rate Q as:

$$Q = 2\pi R\int_{-h}^{+h} \upsilon_x dy = 4\pi R\left(\int_0^{h_0} \upsilon_{xc} dy + \int_{h_0}^{h} \upsilon_{xs} dy\right) \tag{4}$$

here υ_{xc} is the velocity in core flow and υ_{xs} - in shear flow.

Note that from Eq. (1) we have the following relationship for core flow: $h_0(-f) = \tau_0$; putting it to Eq. (4) one obtains:

$$\sum_{i=1}^{N}(-1)^i \frac{M}{2M+N-i} C_N^i \left(X^{\frac{2M+N-i}{M}} - 1\right) + \\ - K_{(M)}^{(N)} X^2 = 0 \tag{5}$$

where $$C_N^i = \frac{N!}{i!(N-i)!}$$

$$X = \left(-\frac{h}{\tau_0} f\right), \quad K_{(M)}^{(N)} = \frac{\mu Q \tau_0^{-\frac{N}{M}}}{4\pi R h^2}, \tag{6}$$

Denoting the final solution of Eq. (5) by f_s one has the formulae for pressure distribution:

$$p = p_o + \left[S(x) - S_o\right] + \left[T(x) - T_o\right] \tag{7}$$

where

$$S(x) = \int f_s dx, \quad S_o = S(x_o); \tag{8}$$

$$T(x) = -\rho \int \left[\frac{1}{h}\left(\frac{R'}{R} + \frac{\partial}{\partial x}\right) \int_0^{+h} \upsilon_x^2 dy\right] dx,$$
$$T_o = T(x_o) \tag{9}$$

If $T(x) = 0$ the formula (7) presents the pressure distribution without inertia effects.

3. EXAMPLES OF APPLICATIONS

Equation (5) corresponds to the viscoplastic flow of the Shulman fluid. By reducing M and N one can obtain simpler models of fluid. Taking into account the results given by Adams et al. (1994), Covey and Stanmore (1981), Walicki and Walicka (1994, 1996) we will present the pressure distribution in clearances of constant thickness; to this aim introduce the following dimensionless parameters:

$$\tilde{x} = \frac{x}{R_o}, \quad K_{(0M)}^{(0N)} = \frac{\mu Q \tau_0^{-\frac{N}{M}}}{4\pi R_o h^2},$$

$$R_\lambda = \frac{\rho Q^{2-\frac{M}{N}}}{(2\pi)^{2-\frac{2M}{N}} \mu^{\frac{M}{N}} R_o^{3-\frac{M}{N}} (2h)^{1-\frac{2M}{N}}},$$

$$\tilde{p} = \frac{2^{\frac{4M}{N}} (p - p_o) \rho h^{\frac{4M}{N}} Q^{2-\frac{2M}{N}}}{(2\pi)^{2-\frac{2M}{N}} \mu^{\frac{2M}{N}} R_o^{4-\frac{2M}{N}} R_\lambda};$$

here R_λ is the modified Reynolds number and $K_{(0M)}^{(0N)}$ is the plasticity number.

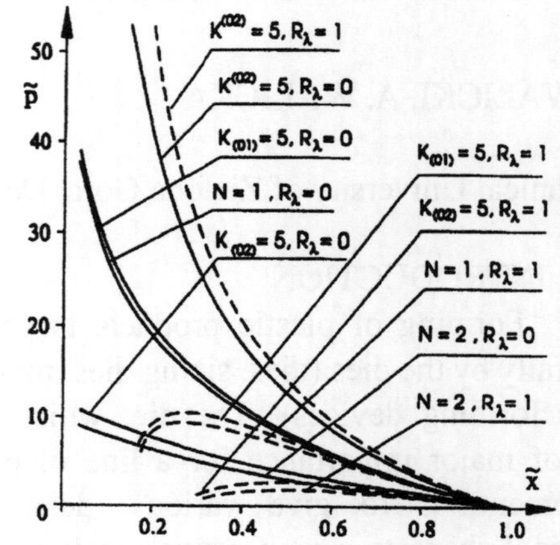

Fig. 2. Dimensionless pressure distribution $\tilde{p}$ for small pressure drop in the clearance.

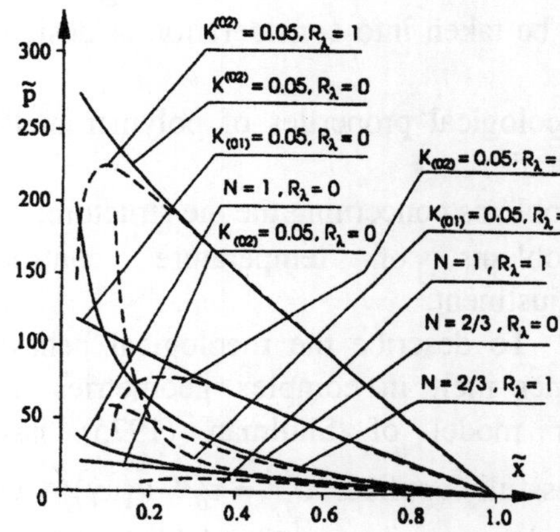

Fig. 3. Dimensionless pressure distribution $\tilde{p}$ for large pressure drop in the clearance.

Figs 2 and 3 present the graphs of pressure distributions between parallel disks with $(R_\lambda \neq 0)$ and without inertia $(R_\lambda = 0)$ for two cases of flow with small core (large K) and with large core (small K).

4. REFERENCES

Adams MJ, Edmondson B, Caughey DG, Yahya R (1994) J Non-Newt Fluid Mech 51: 61-78

Covey HG, Stanmore BR (1981) J Non-Newt Fluid Mech 8: 249-260

Shulman ZP (1975) Convective Heat Transfer of Rheologically Complex Fluids (in Russian). Energy, Moscow

Walicki E, Walicka A (1994) Reports TU Zielona Góra, Fizyka - Chemia 8: 219-242

Walicki E, Walicka A, Rupiński D (1996) In: Atti di IV Conv AIMETA di Tribologia, S. Margherita Ligure pp 83-90

COMPUTATION OF THE MORPHOLOGY CHANGES OF A POLYMER BLEND ALONG A SINGLE-SCREW EXTRUDER

K. WILCZYŃSKI, B. ŁĄCZYŃSKI, A. TYSZKIEWICZ

Warsaw University of Technology, Production Engineering Department, Polymer Processing Institute
02-524 Warsaw, Narbutta 85, Poland

1. INTRODUCTION

A computer model for a single-screw plasticating extrusion SSEM developed by Wilczyński (1990, 1996) has been modified to model polyblend extrusion. The model has been applied for description of the nascent morphology of an immiscible binary polyblend (LDPE/PS) compounded in the single screw extruder. The SSEM model describes five zones of the extruder (hopper, solids conveying, delay zone, melting zone, melt conveying) and the die, and simulates an operation of the extruder-die system. Given the material and rheological properties of the polymer; the screw, hopper and die geometry; and the extruder operating conditions (screw speed and barrel temperature profile), the model predicts: flow rate of the polymer, pressure and temperature profiles along the extruder screw channel and in the die, solid bed profile, and power consumption.

2. EXPERIMENTAL

The computer model was used to simulate a 45-mm single-screw plasticating extruder with an L/D (length/diameter) ratio of 27:1, equipped with a rod cylindrical die. Two different screws (conventional and non-convetional with a mixing rings section), and different operating conditions (within a range of screw speeds N=20 up to N=110 rpm) were used. Experiments were performed on the LDPE/PS blends (Lupolen 3020 / Polystyrol 158 K, BASF) of different composition (1/0, 95/5, 85/15). The rheological parameters were computed from data obtained on an Instron capillary rheometer within a processing range of temperature and shear rate - by curve-fitting Klein equation to the viscosity and elasticity data (for example: the viscosity and elasticity ratio data are presented in Fig.1). The predictions of the model were in acceptable agreement with the experimental runs.

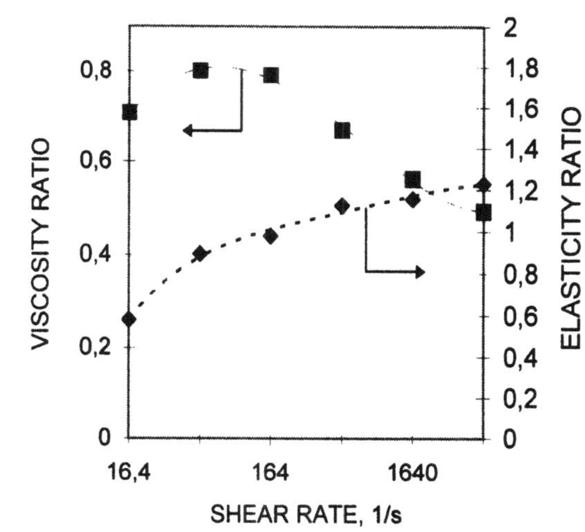

Fig.1. Effect of the shear rate on viscosity and elasticity ratio of PS/LDPE, at the temperature T=220 °C.

3. MORPHOLOGY EVOLUTION MODEL

The SSEM program was assumed as a basis for modeling of morphology development in the single-screw extruder. The main assumptions of the morphology evolution model were taken after Utracki and Shi (1992), Shi et al. (1992), Bordereau et al. (1992), Huneault et al. (1995), Wilczyński and Tyszkiewicz (1997):

* Mixing takes place in the shear field, and in the melt conveying zone of the screw channel only.

* The concentration of the dispersed phase is low enough so the blend morphology can be described considering drop deformation and breakup only; coalescence is not considered.

* Two mechanisms contribute to dispersion:
- Fibrillation and disintegration, as well as

- Droplet deformation and breakup (split-up mechanism).

* The occurrence of one or other breakup mechanism depends on the value of the reduced capillary number κ^*, i.e. the ratio of the capillary number κ to the critical capillary number κ_{cr}.

The capillary number is defined as the ratio of the viscous stress σ to the interfacial stress D/ν, which is the ratio of the drop diameter D to the interfacial tension, $\kappa = \sigma \cdot D/\nu$. The κ_{cr}-value depends on the viscosity ratio, i.e. the ratio of the viscosity of the dispersed phase to the viscosity of the matrix, and is calculated from De Bruijn's empirical equation for shearing flows.

* It is assumed that:

- If $\kappa^*<1$, the morphology does not change.
- If $1<\kappa^*<4$, the drop split-up mechanism is valid.

The droplets deform and split into two identical droplets if the local residence time is greater than the break-up time t_b which is computed from the dimensionless time t_b^*. The dimensionless time t_b^* is defined as the ratio of the break-up strain $t_b \cdot \gamma$, where γ is the shear rate, to the capillary number κ, $t_b^* = t_b \cdot \gamma/\kappa$. The t_b^*-value is calculated from the experimental data of Grace (extrapolated to κ^*) and those of Elemans.

- If $\kappa^*>4$, the fibrillation / disintegration mechanism occurs.

Fibrillated drops disintegrate when the fiber diameter becomes smaller than its critical value (assumed D_f=0.5 mm), or when the capillary number is reduced below the critical value, i.e. for $\kappa^*<4$.

In the proposed model calculations are made in the elemental volumes of the screw channel over the melt conveying zone, defined by the SSEM. First, the stress field is computed. Then, the micro-rheological parameters: capillary number, viscosity ratio, strain, residence time, dimensionless time for break-up, and critical capillary number are calculated. And next, the dispersion mechanism is determined, and finally, the filaments or drop dimensions are computed.

Calculations were made for different data. An example is shown in Fig.2. It was observed that the fibrillation/disintegration breakup mechanism prevails in the (conventional) single-screw extruder, and the filaments break when their diameter falls below the critical value ($D<D_f$) rather than the reduced capillary number decreases below four, $\kappa^*<4$. The validity of the evolution model should be verified experimentally. The model must be refined, and the coalescence should be taken into account. The elasticity effects are currently being tested.

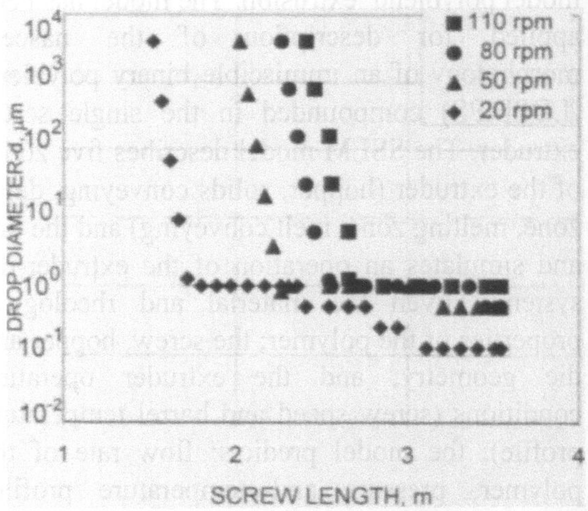

Fig.2. Dispersed phase diameter d as a function screw length for LDPE/PS blend (95/5) for conventional screw, at various screw speeds.

4. REFERENCES

Bordereau V et al. (1992) Polymer Eng. Sci., **32**, 1846

Huneault MA et al. (1995) Polymer Eng.Sci., **35**, 115

Shi ZH et al. (1992) Polymer Eng. Sci., **32**, 1834

Utracki LA, Shi ZH (1992) Polymer Eng. Sci., **32**, 1824

Wilczyński K (1990) "Theory of Polymer Single-Screw Extrusion", Warsaw University of Technol. Publ., Warsaw (in Polish)

Wilczyński K (1996) Polym.-Plast. Technol. Eng., **35**, 449

Wilczyński K, Tyszkiewicz A (1997) 2-nd Pacific Rim Conference on Rheology, Melbourne, 301

COMPARISON OF MEASURED SHEAR AND ELONGATIONAL VISCOSITIES WITH SOME RHEOLOGICAL MODELS

M.ZATLOUKAL [1], J.VLČEK [2], T. DOBBIE [3], P. SÁHA [1]

[1]Technical University of Brno, Faculty of Technology Zlín, 762 72 Zlín, Czech Republic
[2]Compuplast International, Inc., Tř. T.Bati 299, Zlín, Czech Republic
[3]Rosand Precision Ltd., Balds Lane, Lye, Stourbridge DY9 8SH, UK

1. INTRODUCTION

The aim of the presented work is to compare measured shear and elongational viscosities with several rheological models. The elongational viscosity was determined by using the Cogswell method [2,4]. Obtained material viscosities were fitted to the models which are able to describe the measured properties. The used rheological models are as follows: Phan Thien model [6], modified Leonov [1] and Larson [3,5] models .

2. EXPERIMENTAL

2.1 Materials

The measurement was performed for two materials: LDPE DOW 150 (film grade) and PP BASF BOVELLEN (film grade).

2.2 Methods

Both viscosities were measured on a RH7-2 capillary rheometer produced by Rosand Precision Ltd. The measurements were carried out at 190°C in shear rate region $10^{-0.5}$ - $10^{3.2}$ s^{-1} and the calculated elongational rates were 1-10^{3} s^{-1}. The obtained elongational viscosity curves are shown in Fig.1 and 2.

3. MODEL FITTING AND RESULTS

In our model fitting we have assumed that the material behavior is described by a relaxation spectrum. In such a spectrum we choose a set of relaxation times λ_i and we find corresponding moduli G_i. The sets of relaxation constants for both materials are shown in Tab. I. The moduli G_i were determined in the following way: For each model we assume that the shear stress can be defined as a sum of stresses corresponding to each couple λ_i, G_i (eq. 1). The shear viscosity is defined as (eq. 2), where $\dot{\gamma}$ is the shear rate for which the shear stresses are calculated. Each model provides a set of equations for calculation of the stresses in shear flow. These stresses are dependent on the values of λ_i and G_i, on the shear rate $\dot{\gamma}$ and on specific model parameters. The values of G_i and model parameters are obtained by fitting as precisely as possible the shear viscosity curve and the elongational curve. We did not use the classical approach for obtaining relaxation spectrum from oscillatory measurement because we only wanted to use the capillary rheometer.

Our fitting approach takes away the physical meaning of G_i and makes them fitting parameters only. On the other hand we found that for all three models the values of G_i were almost identical, which indicates that G_i are more than fitting parameters only. They keep their meaning as material characteristics because they are model independent.

The detailed model behavior depends on the model parameters. The ability of the model to fit well the studied material behavior is given by the role of the parameters in the model. We have found that all three models can describe the shear viscosity very well. Therefore, we do not show the shear viscosity curve fitting. We have found some differences in fitting the elongational viscosity.

The fitting of the elongational viscosity for both materials is shown in Fig. 1 and 2. It can be seen that Larson model describes this viscosity most accurately. A little less conformity was found for Phan Thien model. Both models can be used for both linear and branched polymers. Leonov model describes linear polymers well (elongational viscosity always decreases with elongational rate).

As a matter of fact, we performed the measurements for five commercial materials (LDPE, PP, HDPE, LLDPE and PS all of them will be presented) but for this paper we have used two of them only. In general it can be said that the behavior and conclusions were the same for all the five materials.

3.1 Tables

Tab. I The relaxation constants for LDPE and PP at 190°C

i	LDPE λ_i / s	LDPE G_i / Pa	PP λ_i / s	PP G_i / Pa
1	10^{-4}	49735	10^{-4}	107995
2	10^{-3}	48362	10^{-3}	79797
3	10^{-2}	36623	10^{-2}	64924
4	10^{-1}	19066	10^{-1}	32316
5	1	4716	1	12399
6	10	99	10	70

4. EQUATIONS

$$\tau_{x,y} = \sum_{i=1}^{k} \tau_{x,y_i} \tag{1}$$

$$\eta_S = \frac{\tau_{x,y}}{\dot{\gamma}} \tag{2}$$

5. ILLUSTRATION

5.1 Figures

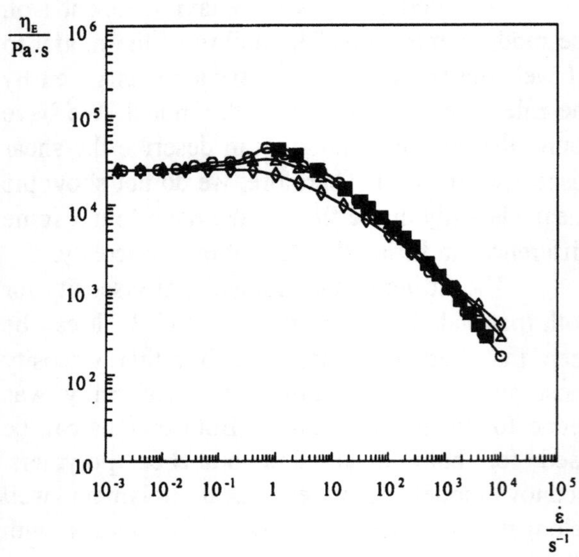

Fig. 1 Comparison of Phan Thien ($\Delta, \varepsilon = 0.22$, $\xi = 0.1$), modified Leonov (◇, n = -0.1, m= -2.5) and Larson (○, $\alpha = 0.25$) models with experimental data (■) for LDPE in steady elongation flow at 190°C

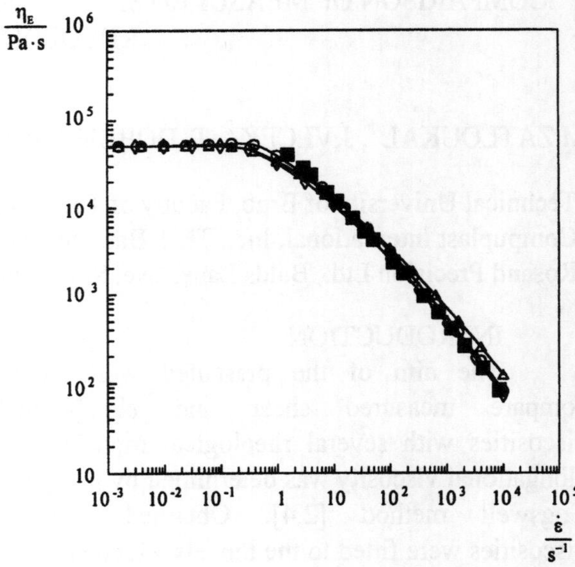

Fig. 2 Comparison of Phan Thien ($\Delta, \varepsilon = 1.2$, $\xi = 0.5$), modified Leonov (◇, n = -0.86, m = -10) and Larson (○, $\alpha = 0.98$) models with experimental data (■) for PP in steady elongation flow at 190°C

6. REFERENCES

1. Leonov AI (1995) Rheol Acta 34:259-273
2. Cogswell FN (1994) Polymer Melt Rheology, England
3. Isaki T, Takahashi M, Takigawa T, Masuda T (1991) Rheol Acta 30:530-539
4. Collyer AA, Clegg DW (1988) Rheological Measurement. London, New York
5. Khan SA, Larson RG (1987) Journal of Rheology 31(3):207-234
6. Phan Thien N, Tanner RI (1977) J. Non-Newtonian Fluid Mechanics 2:353-365

THE INFLUENCE OF SIDE-CHAIN MODIFICATION ON SHEAR THICKENING IN SOLUTIONS OF SOME HIGH MOLECULAR WEIGHT COPOLYMERS

J. FERGUSON[1], N. E. HUDSON[1], J. R. EBDON[2], N. J. FLINT[2], D. J. HOURSTON[3], B. C. H. WARREN[4]

[1]Department of Pure and Applied Chemistry, University of Strathclyde, Glasgow G1 1XL, UK
[2]Polymer Centre, School of Physics and Materials, Lancaster University, Lancaster LA1 4YA, UK
[3]Institute of Polymer Technology & Materials Engineering, Loughborough University of Technology, Loughborough LE11 3TU, Leics, UK
[4]Defence Evaluation and Research Agency, CBD, Porton Down, Salisbury, Wilts SP4 0JQ, UK

1. INTRODUCTION

An investigation has been carried out on the influence of chemical structure on the rheological properties of polymer solutions over a range of shear rates. Previous research has shown that the size of droplets produced when a fluid is dispersed in air is related not only to the structure and concentration of any dissolved polymers, but also depends upon the extensional strain rates imposed during the break-up process. Under both strong shear and elongational flows, complexation of the polymer molecules can be induced, leading, with the correct solvent, to a change of phase. The polymer molecules separate into a gelled or crystalline region surrounded by a solvent rich layer. These two phases have different break-up characteristics, thereby indicating the importance of the type of polymer, and its interactions, in controlling drop size. At high rates of shear, such polymers would be expected to show shear thickening, which can be attributed to the transition from intra- to stronger intermolecular associations made possible by elongation under shear flow. The viscosity of such a solution will initially decrease with shear rate, then at intermediate values of shear rate will start to increase with shear rate, as the number of inter-chain junctions is increased. Ultimately, at very high shear rates, the inter-chain junctions are broken more rapidly than they can reform, and shear thinning occurs once again.

2. MATERIALS AND METHODS

Poly(methyl methacrylate-co-n-butyl acrylate-co-ethyl acrylate) has been shown to be shear thickening at moderate to high shear rates, in solutions of relatively low concentrations (3-5% w/w). Polymer/polymer interactions can be maximised by, e.g., copolymerising interacting groups, and/or attaching side chains. The base polymer (coded as Polymer 1) was therefore modified by one of three different titanium alkoxide mediated trans-esterifications of the aliphatic acrylate side-chains (Polymers 2, 3 & 4). The lengthening of these side-chains altered the shear thickening behaviour of the modified polymer in polar solvents, by promoting the formation of clusters of the aliphatic groups in such solvents. Rheological measurements were carried out at shear rates between 10^{-3} and $2 \times 10^5\ s^{-1}$ on solutions in triethyl phosphate (TEP), which has a solubility parameter matched to that of poly(methyl methacrylate). TEP is hygroscopic. In order to saturate the solvent, 5 g of water per 95 g of TEP was added to each prepared solution. As a starting point, all polymers were dissolved in TEP to give solutions with w/w concentrations of 0.5 to 5%. Later, a new solution of each polymer was prepared so that the zero shear viscosity (η_0) of each was equal, enabling direct comparisons of flow behaviour to be made. Samples were tested on a TA Instruments Controlled Stress Rheometer (CSR) in flow, creep and oscillation modes; on a TA Instruments Weissenburg Rheogoniometer (WR) in flow mode; on a R-16 Weissenburg Rheogoniometer in shear and first normal stress difference modes; and on a Rheometrics Capillary Rheometer (ACER) in steady shear mode. The rheology laboratory is maintained at 20°C and 65% humidity. All experiments were performed at 20°C.

3. RESULTS

Figure 1 shows results for Polymer 1 from both rotational and capillary rheometers, covering a shear rate range from 10^{-3} to over $10^5 s^{-1}$. The most important feature of these results is the sudden shear thickening which occurs at a critical shear rate, different for each solution, above a weight concentration of 2%. Two to three-fold increases in viscosity are seen, followed by renewed shear thinning. At the lower concentrations, there is even a second occurrence of this phenomenon. It must be emphasised at this point that this is not a manifestation of the flow instability recently reported for a number of different polymer solutions. Three different rotational instruments were used, with a number of different geometries, and the shear thickening occurred at approximately the same value of shear rate, not the same value of angular velocity. Hence the effect is genuine, and is due to the chemical structure of the fluid.

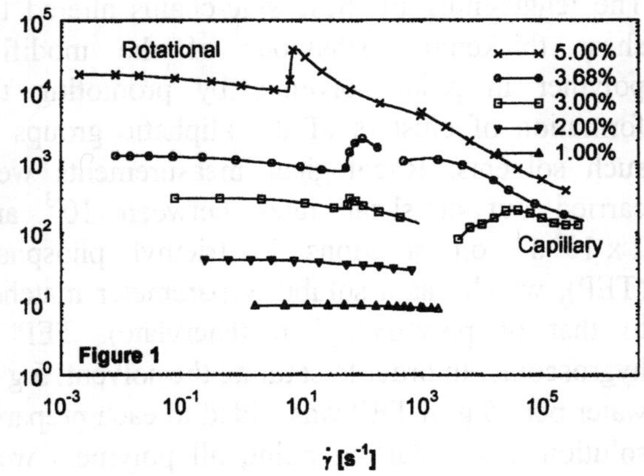

Flow curves for several solutions of Polymer 2 are shown in Figure 2. As can be seen, as concentration of polymer increases, the stress required to induce shear thickening also increases, but due to the levels of viscosity of the solutions, the rate at which it occurs is decreasing.

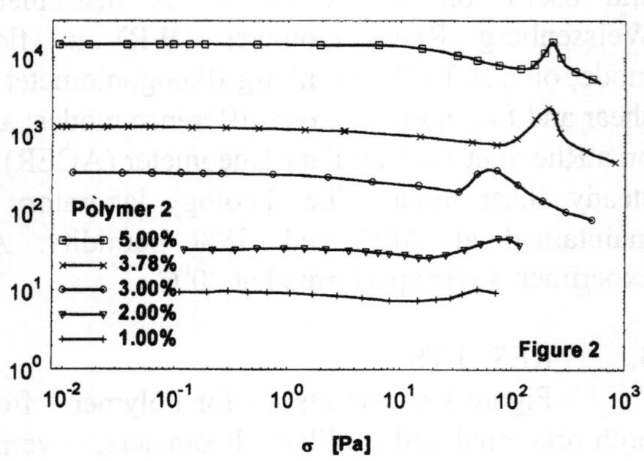

Figure 3 shows results from the selected solutions of constant zero shear viscosity of each of the four polymers in both rotational and capillary rheometers. At low values of the shear rate, the viscosities are equal, and at very high rates of shear (about $10^5\,s^{-1}$) it would appear that the viscosities are once again nearly equal. However, the behaviour in between is certainly different. There appear to be two regions in which shear thickening is present. The values of the shear rates at which shear thickening occurs, and the magnitude of the shear thickening, varies between polymers.

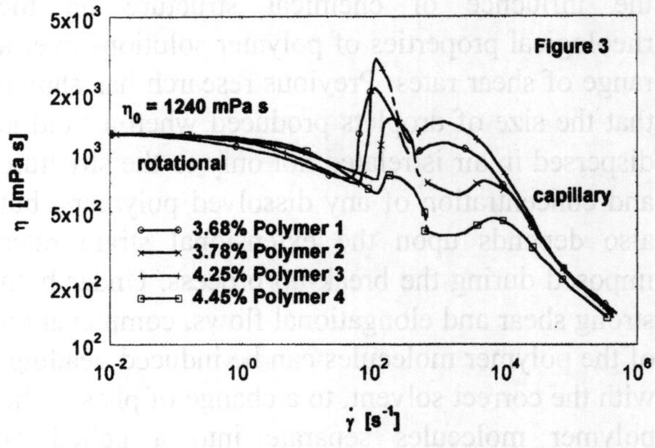

4. DISCUSSION

The differences in shear thickening in solutions of the trans-esterified polymers are attributed to association of the aliphatic chains in polar solvents to form clusters. These clusters will increase the number of entanglement junctions and hence increase the viscosity at an equivalent shear rate, or cause shear thickening at higher shear rates. In the Newtonian region, fairly large changes in viscosity have been induced by the substitution of very small mole fractions of the acrylate units by hexyl, decyl and benzyl groups. Pronounced shear thickening was observed in all solutions at a shear rate dependent upon concentration. There were also indications of a second strain hardening region at high shear rate in the capillary rheometer. The origin of such effects is still a matter of discussion. The possibilities are that it is due to either (a) the high Deborah number, which induces solid-like behaviour with subsequent breakdown of laminar flow, or (b) structure formation being induced by shear.

RHEOOPTIC PROPERTIES OF HIGHLY DILUTE POLYMER SOLUTIONS - EXPERIMENT AND NUMERICAL SIMULATION OF THE FLOW FIELD

B. GAMPERT, T. EICH, H. HAHN

Applied Mechanics, University of Essen, 45127 Essen, Germany

1. INTRODUCTION

While adding a small amount of polymers to a Newtonian solvent different flow phenomena like a change in the turbulence structure in internal flow may occur. This is accompanied by a sometimes drastic loss in frictional drag in comparison to the solvent at the same flow rate.

However, these observations could not be explained by the pseudoplastic shear viscosity functions of these fluids. Even highly dilute polymer solutions with a shear rate independent viscosity slightly increased compared to the values of the solvent exhibit these flow effects. This gives rise to the idea that the described behavior has to be related to the elastic fluid properties.

2. POLYMER SOLVENT SYSTEM

In this work two different types of polyacrylamide were solved in tap water, the technical polymer Praestol 2350 from Stockhausen with an average molecular weight of 5-10x10^6 g/mol and a non-ionic additive polymerized under laboratory conditions with a molecular weight of 5x10^6 g/mol. The polymer concentrations were 50 wppm in the ionic and 250 wppm in the non-ionic case. These concentrations are lower than the individual critical concentration c^* of 70 wppm and 950 wppm respectively where transition from particle to network molecular solution structure occurs .

Rheological measurements were carried out in stationary and oscillatory shear flow in different rotational rheometers. The applied fluids do no show any significant first normal stress difference nor complex viscosity within a shear rate range up to 100 1/s. It was virtually not possible to determine elastic material properties. A possible reason could lie in the limited shear rate range of the applied cone-plate rheometer.

3. RHEOOPTIC EXPERIMENTS

The results obtained by rotational rheometry may indicate that elastic properties become apparent only if a critical strain is exceeded. To investigate this an opposed jet system with a rheooptic measuring device was developed. Figure 1 shows the flow layout of the test setup.

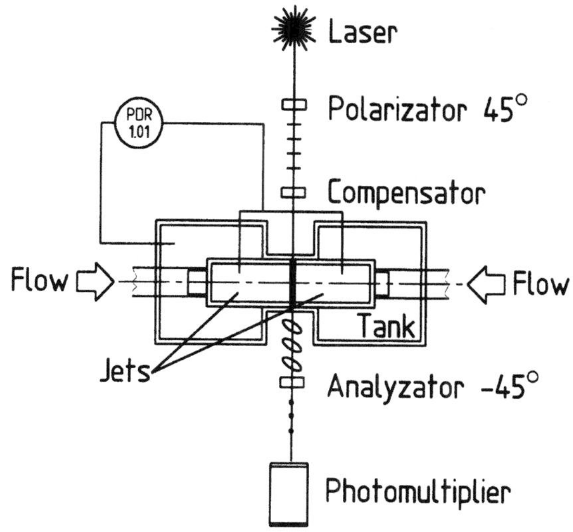

Figure 1: Flow layout.

A planar elongational flow field was realized by means of two jets with a slit width of 20 mm and a slit height of 1 mm. The jets with a distance d of 2 mm are mounted in a storage tank. However, measurements of the elongational viscosity are affected by the applied jet geometry. So the jet design was based on two dimensional numerical simulations of the flow field with a commercial FEM program.

The flow rate $\dot{Q}$ was controlled by a computer regulated injection unit. The average elongational rate $\dot{\varepsilon}$ is then given by the jet cross section A and the distance d:

$$\dot{\varepsilon} = \frac{\dot{Q}}{dA}. \qquad (1)$$

From the differential pressure between the jet inflow and the fluid in the storage tank the elongational viscosity can be computed:

$$\Delta p = \Delta p_1 + \Delta p_2 = \eta_E \dot{\varepsilon} + \frac{1}{2}\zeta \rho d^2 \dot{\varepsilon}, \qquad (2)$$

when omitting the normal stress σ_{yy} in the equation for the elongational viscosity

$$\eta_E = \frac{\sigma_{xx} - \sigma_{yy}}{\dot{\varepsilon}} \qquad (3)$$

as shown by Petrie (1979). The second term in equation 2 is the dissipative pressure component (see Tatham, 1993), in which ζ is the friction factor and ρ is the fluid density. This influence was eliminated by calibration measurements with the Newtonian solvent. The contribution of the shear viscosity to the pressure was neglected. So, an apparent elongational viscosity η_{AE} could be computed from the pressure difference $\Delta p - \Delta p_2$.

Polymers show flow birefringence when orientated (see Janeschitz-Kriegel, 1969). This leads to an elliptical polarization of linear polarized laser light passing the flow field as shown in figure 1. As the molecules are orientated parallel to the flow axis the polarization angles are 45° and -45°. The maximum change in light intensity can be detected by a photomultiplier.

The Stokes-vector $\vec{P}$ is composed of the Müller-matrices M_p of the polarizator and M_b of the birefringent fluid and the Stokes-vector of linear polarized light $\vec{P}_\alpha$:

$$\vec{P} = M_p M_b(\delta) \vec{P}_{\alpha=-45°} . \qquad (4)$$

The theoretical light intensity depends on the retardation angle δ:

$$I_{\alpha=45°} = 0.5E^2(1-\cos\delta) . \qquad (4)$$

The field intensity E can be determined by measuring the intensity under an analyzator angle of 45° using a non-birefringent fluid.

5. RESULTS

In figure 2 the results from the differential pressure measurements are shown. Exceeding critical elongational rates of 1200 1/s for the ionic solution and 1550 1/s for the non-ionic case a significant increase of the apparent elongational viscosity could be noticed.

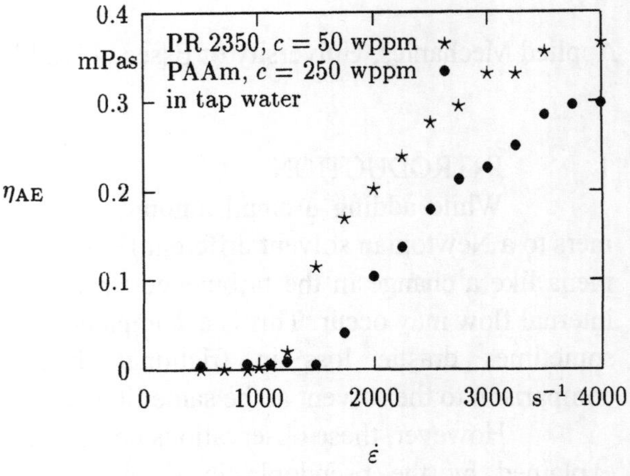

Figure 2: Pressure measurements.

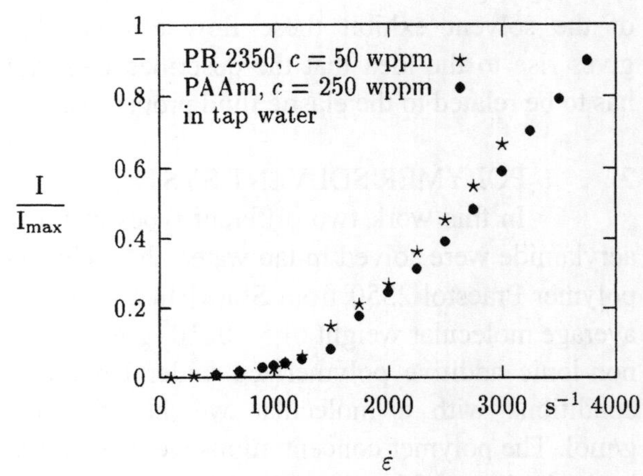

Figure 3: Flow birefringence.

The optical measurements confirm these results. The light intensity ratio increases for the same critical rates of strain. The initial increase is due to the elliptical deformation of the polymers. The elongational thicking and therefore elasticity was observed while the macromolecules were orientated in the flow and therefore elongated.

6. REFERENCES

Janeschitz-Kriegel, H. (1969) Ad. Pol. Sc. 76 (55): 170-318

Petrie, C.J.S. (1979) Elongational Flows, Pitmen

Tatham, J.P. (1993) Extensional flow dynamics of macromolecules of different flexibility in solution, Ph.D. thesis, Univ. of Bristol

RHEOLOGICAL PROPERTIES OF POLYELECTROLYTE SOLUTIONS

B. Gampert, C. Wilkes, T. Eich

Applied Mechanics, University of Essen, Schützenbahn 70, D-45127 Essen
e-mail: christoph.wilkes@uni-essen.de

1 INTRODUCTION

For turbulent channel flow of highly dilute polymer solutions, a rather drastic reduction in friction loss could be observed in comparison to the Newtonian solvent in the turbulent flow regime. Are there changes in drag reduction depending on the concentration of the polyelectrolyte if the shear viscosity of these solutions remains the same? Within highly dilute polyelectrolyte solutions in comparison to non-ionic solutions of polymers there are different structure regimes. For instance, for very highly diluted polyelectrolyte solutions a maximum value for the reduced viscosity could be observed. This maximum is shifted to higher polyelectrolyte concentrations in the presence of a low salt content. Due to rheometric difficulties especially for polymers of a molecular weight greater than 10^6 g/mol, no experimental data of this maximum exists in literature as far as the authors are aware. In this paper, an attempt has been made in order to characterize the structure of highly diluted solutions of polyelectrolytes with a molecular weight greater than that of 10^6 g/mol.

2 THEORY

The significant behaviour of the internal turbulent flow of highly dilute polymer solutions is due to the structure of the polymer in the solution. The value of the reduced viscosity is one possibility for characterizing these polymer solutions. If the dissolved polymers move independent of each other in the solution, the reduced viscosity is proportional to the concentration of the dissolved particle. This is only possible if the solution is highly dilute. The reduced viscosity extrapolated to zero concentration and zero shear rate yields the intrinsic viscosity as:

$$[\eta] = \lim_{\substack{c \to 0 \\ \dot{\gamma} \to 0}} \eta_{\text{red}}. \qquad (1)$$

The intrinsic viscosity is also a measure of the space required of one molecule in the solution and can be used in order to determine the molecular weight and the shape of the polymer in the solution. For neutral polymer solutions the polymer concentrations can be classified into two regions. Above the critical concentration $c_p^\star$ a network could be observed. The polymer chains in this region either make contact with one another or form a network. Under this critical concentration $c_p^\star$ the solution is said to be a particle solution. In the case of these polymer solutions, the intrinsic viscosity is simple to define. The curve obtained when plotting the reduced viscosity against the concentration is almost linear, so that the graph could be extrapolated in accordance with equation 1, Huggins (1942).

The polyelectrolyte solutions can be divided into two distinct regions - dilute or semidilute. The semidilute region is further separated into three regimes, called isotropic, transition and lattice in the order of decreasing concentration, while the dilute region is divided into two regimes, known as order and disorder. The isotropic and lattice regimes are depicted by de Gennes et al. (1976) and the transition regime is described by Odijk (1979). The order and disorder regimes are distinguished from each other by whether an intermolecular single correlation due to electrostatic repulsive forces exists or not.

The existence of a maximum of the reduced viscosity as a fuction of the polymer concentration of ionic polymer solutions in terms of the intermolecular and intramolecular interactions was derived theoretically by De Gennes et al. (1976) and Wolff (1978). In their work, Hess and Klein (1983) defined this aforesaid maximum by means of an approximation of the hydrodynamics of the charged Brownian spheres. As yet, the number of published experimental results concerning this topic have numbered but a few for example

Cohen and Priel (1994) as well as Vink (1992). Polyelectrolytes with relatively small molecular weights of the order $M_w < 1 \cdot 10^6$ g/mol were investigated during the course of these dissertations.

3 RESULTS

The experiments undertaken here, were carried out using a commercial, anionic polyacrylamide with a hydrolysis factor of 70% and a molecular weight of $M_w \approx 1.4 \cdot 10^7$ g/mol. 11.7 g/l or respectively 23.4 g/l sodium chloride was also added to the solution. Polymers with high molecular weight values which are influential to turbulent flow behaviour with the aim of a technical signification of the drag reduction were investigated here.

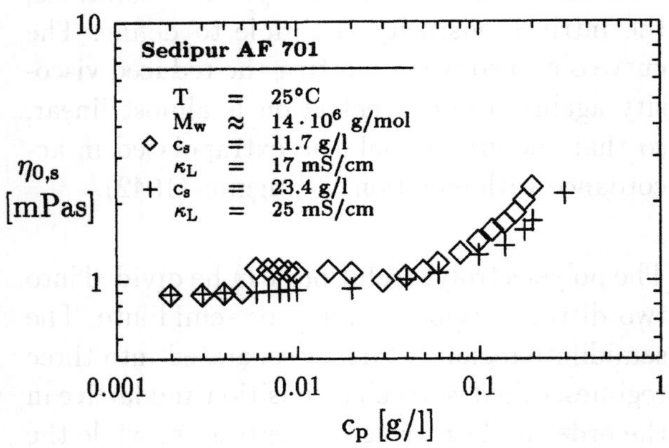

Fig. 1: The shear viscosity $\eta_{0,s}$ as a function of polymer concentration for different salt concentrations

Fig. 1 shows the zero shear viscosity, $\eta_{0,s}$ as a function of the polymer concentration. A fluctuation of the trend line can be seen at a polymer concentration of $c_p = 0.005$ g/l and salt concentration of $c_s = 11.7$ g/l. The value of the reduced viscosity in fig. 2 at this concentration reaches a maximum. Wolff (1978) defines the maximum values as the border between the disordered and ordered structure of the polymer in the particle solution.

It can be clearly seen, on the basis of these results, that the determination of the intrinsic viscosity for ionic polymer solutions is extraordinarily difficult. This is due to the fact that measurements are required when the polymer concentration is very small and the extrapolation of equation 1 is very difficult to achieve due to the association of large uncertainties during the measurement phase. Furthermore, a maximum cannot always be detected because there are instances when the maximum could move to areas of even smaller concentrations which becomes apparent when dealing with solutions of especially high anisotropy.

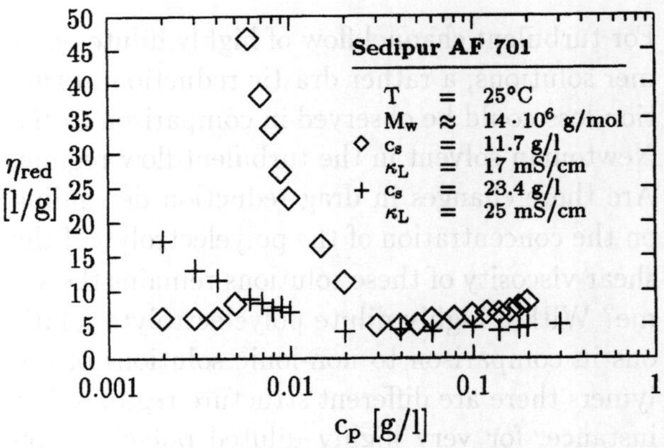

Fig. 2: The reduced viscosity η_{red} as a function of polymer concentration for different salt concentrations

Moreover, within highly dilute solutions of Sedipur AF 701 with the concentration of 0.006 g/l and 0.02 g/l having nearly the same shear viscosity, the drag reduction of both fluids is completely different. The drag reduction of the 0.02 g/l concentrated solution is higher than that of 0.006 g/l. Complementary measurements in an elongational flow field confirm this observation. The experimental arrangement is described in detail in Gampert and Eich (1998). The extensional viscosity for elongational rates greater than that of 1500 1/s for the 0.02 g/l solution is higher. This points to the elastic behaviour of the single macromolecules.

4 REFERENCES

Cohen J, Priel Z (1994) Trends in Macromol. Res. 1:201-213

De Gennes PG et al. (1976) J. Phys. 37:1461-1473

Gampert B, Eich T (1998) Proc. of the GALA Meeting Essen 1998

Hess W, Klein R (1983) Adv. Phys. 32:173-283

Huggins MI (1942) J. Chem. Soc. 64:2716-2718

Odijk T (1979) Macromol. 12:688-693

Vink H (1992) Polymer 33:3711-3716

Wolff C (1978) J. Phys., Coll. C2 39:169-174

DYNAMICS AND RHEOLOGY OF HAIRY-ROD POLYMERS

G. PETEKIDIS[1], D. VLASSOPOULOS[1], G. FYTAS[1], R. RÜLKENS[2], G. WEGNER[2]

[1] FORTH, Institute of Electronic Structure and Laser, P.O. Box 1527, 711 10 Heraklion, Crete, Greece
[2] Max-Planck Institut für Polymerforschung, P.O. Box 3148, 55021 Mainz, Germany

1. INTRODUCTION

The dynamic behavior of stiff-chain molecules, affected by their highly anisotropic shape, the formation of mesophases, anisotropic diffusion and the translation-rotation coupling, and its interplay with their rheological response, remains virtually unexplored and represents a great scientific and technological challenge. Hairy-rod poly(p-phenylenes), PPPS, substituted by bulky aromatic sulfonate esters, represent good model systems for addressing this issue (Vanhee et al, 1996). These macromolecules are soluble at very high concentrations (typically 60% by weight), while in the isotropic regime, and this allows a systematic investigation of their orientational dynamics.

2. MATERIALS

The PPPS's used in this work are depicted in chart 1 (Vanhee et al, 1996). The molecular characteristics are summarized in Table 1 (the overlap concentrations for semidilute and concentrated regions are c^* and c^{**}, respectively; L_w is the weight average contour length, M_n the number average molecular weight, and L_w/L_n the polydispersity). These molecules were chosen because of their large inherent optical anisotropy and enhanced solubility in common solvents.

3. RESULTS AND DISCUSSION

Figure 1 shows the experimental time orientational correlation function C(q,t) for two PPPS solutions in two different solvents at 25°C and scattering angle 30°. A two-step relaxation is clearly evident from the C(q,t) of S1, as well as the distribution relaxation functions L(lnτ) of the two solutions, irrespectively of M_w and solvent used (toluene, chloroform and TCE). For all samples but S10 above about 6%, in three different solvents, C(q,t) conforms to this bimodal relaxation and is distinctly different from the isotropic light scattering (from concentration fluctuations) over the same time range. These pertinent findings support the molecular origin of this bimodal orientation relaxation, observed also in other PPP's (Petekidis et al, 1998).

To establish the phenomenology of this unusual orientation relaxation behavior, the variation of C(q,t) with the wavelength (q^{-1}) of the fluctuations and solute concentration was thoroughly examined for five PPPS samples. The variation of the rate Γ_f and the intensity I_f associated with the fast process, with q, suggests spatial correlation in the random orientation fluctuations of the isotropic solutions; the coherence length, ξ, of the orientational fluctuation correlations is O(40 nm). In contrast, the q-dependence of the slow process is unusual. It increases with q for all PPPS samples.

From the phenomenological evidence above, and in view of the unusual q-dependence of the slow mode, we suggest the following qualitative picture of the orientation dynamics of hairy-rod polymers: The fast process is assigned to the collective random rotational motion of segments of the PPPS chains, typically of the size of the persistence length. On the other hand, in addition to their collective rotational relaxation leading to randomly correlated in space orientational fluctuations, pairs of such Kuhn segments can exhibit orientational correlations, resembling local ordering; the latter would give rise to a peak in the scattering intensity at finite q's. This situation, which essentially represents two "dynamic populations", namely orientationally correlated single Kuhn segments and pairs of Kuhn segments, is schematically illustrated in Figure 2. The figure attempts to depict a snapshot of a concentrated solution of a hairy-rod polymer. The Kuhn segments are relaxing their orientation correlations through a "classical" rotational diffusion-like mechanism. The space correlation between orientations of neighboring Kuhn segments can persist over a typical coherence length which defines the size ξ of (hypothetical) swarms (de Gennes and Prost,

1993). The latter can be visualized as orientationally correlated regions, randomly correlated in space, with local orientational ordering.

We complement this study, with information on the dynamic shear moduli for two PPPS samples at different concentrations. As seen in Fig. 3, the linear viscoelastic data of a concentrated entangled 8% S1 solution in toluene, at 25°C, exhibit the standard terminal regime scaling behavior, which allows the estimation of the longest chain relaxation time τ and the zero-shear viscosity η_0. The measured viscosities also exhibit a much stronger than c^2 dependence, in qualitative agreement with the sharp concentration dependence of both Γ_f and Γ_s. The longest viscoelastic relaxation rates are much slower (by nearly 2 decades) than the slowest orientational relaxation rates. This can be explained by the fact that the former are actually the disentanglement times (chain orientation) in the transient physical network formed at the high concentrations investigated, whereas on the latter relate to orientations of Kuhn segments, i.e., length scales shorter than the chain length.

4. REFERENCES

de Gennes PG, Prost J (1993) *The Physics of Liquid Crystals;* 2nd Ed. Clarendon Press: Oxford.

Petekidis G, Vlassopoulos D, Fytas G, Fleischer G (1998) Macromolecules 31:1406.

Vanhee S, Rulkens R, Lehmann U, Rosenauer C, Schulze M, Kohler W, Wegner G (1996) Macromolecules 29:5136.

Table 1
Molecular characteristics of PPPS samples

Sample	M_n (g/mol)	L_n (nm)	L_w/L_n	c^* (mg/ml)	c^{**} (mg/ml)
S1	99900	133.8	1.91	0.4	17.8
S3	55400	73	2.09	0.73	21.4
S7	22000	29.7	1.82	2.39	34.9
S9	16500	22.3	1.64	3.74	42.7
S10	8150	11	1.42	12.6	75.8

Overlap concentration, c^*, and c^{**} correspond to toluene solutions.

Chart 1: PPPS molecular structure

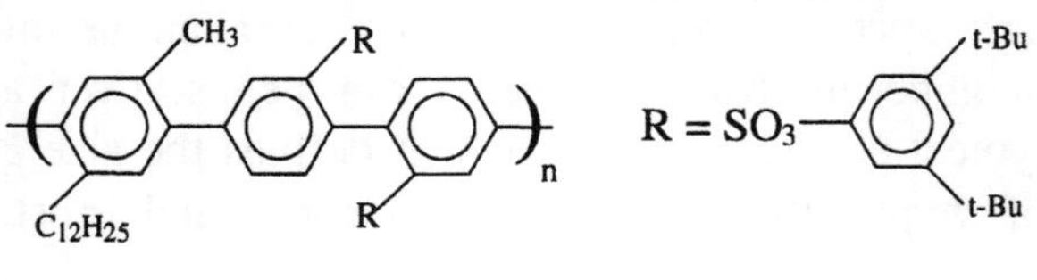

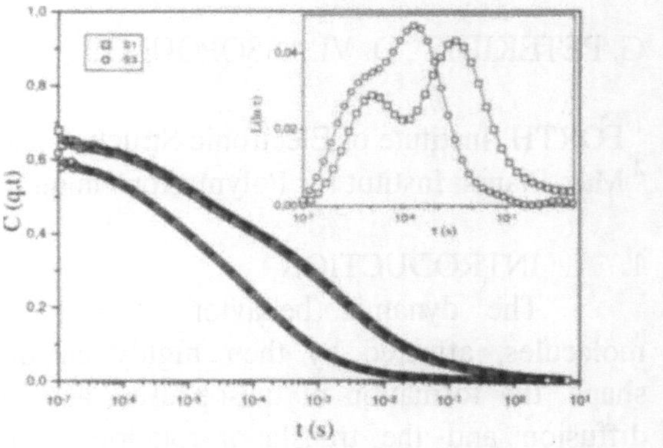

Figure 1: Experimental orientational relaxation functions at $q \approx 9 \times 10^{-3}$ nm^{-1} for PPPS S1 (❑, 65.6 wt%) and S3 (O, 17.5 wt%) in toluene and chloroform, respectively. The two distributions $L(\ln\tau)$ of relaxation times are shown in the inset.

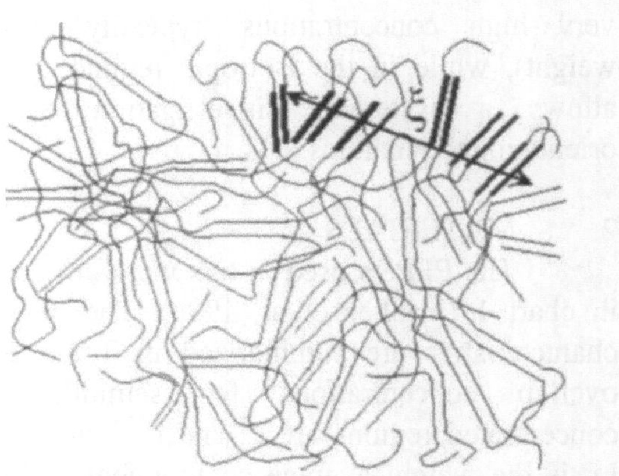

Figure 2: Schematic representation of the local orientational order in entangled semistiff chains.

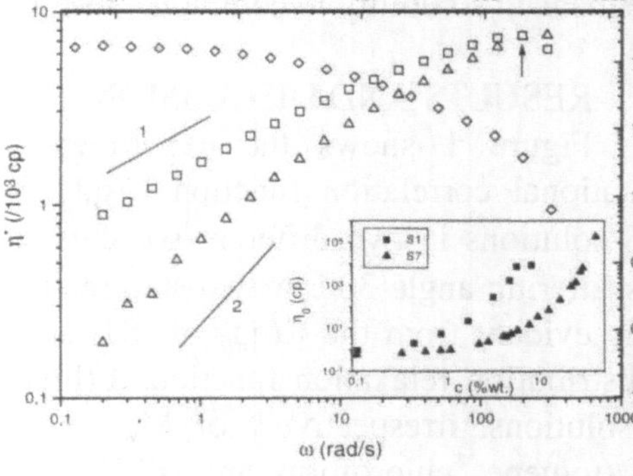

Figure 3: Frequency dependence of the dynamic shear moduli (G′, Δ; G″, ❑) for a typical semidilute S1 solution in toluene (8%) at 25°C. Inset: Concentration dependence of η for S1 (■) and S7 (▲) in toluene.

A NEW TRANSIENT NETWORK MODEL APPLIED TO GUAR GUM

R.H.W. WIENTJES, M.H.G. DUITS, R.J.J. JONGSCHAAP, J. MELLEMA

Twente Institute of Mechanics & J.M. Burgers Centre, Rheology group, Department of Applied Physics, University of Twente, P.O.Box 217, 7500 AE Enschede, The Netherlands.

1. INTRODUCTION

Guar gum is a linear polymer with associative sites which can connect and disconnect reversibly. Experimental visco-elastic moduli showed two storage modulus plateaus, a broad relaxation spectrum and partial time temperature superposition.

We developed a new model for linear viscoelastic behaviour in which the connectivity of the polymer chain is taken into account by assuming several sticky points on the polymer chain at which the chain can connect to other polymers. Thus a network of polymer chains is formed.

2. GENERAL MODEL CONCEPT

In our model (Wientjes et al, 1998) we focus upon the state of a whole chain instead of individual segments. We distinguish two types of chain segments:

1. Active segments belonging to the network.
2. Inactive segments which are part of either a loose end or a free chain.

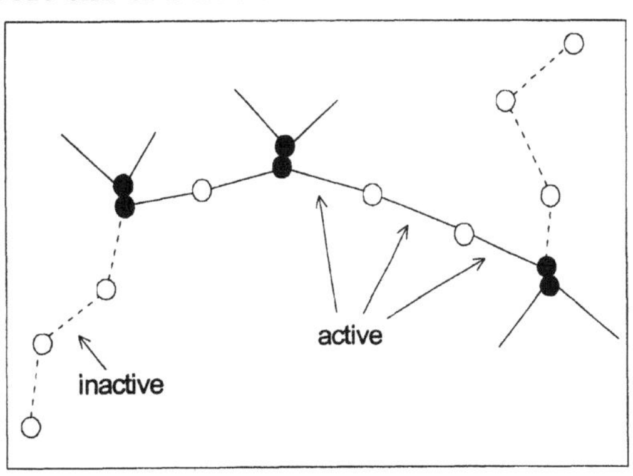

Figure 1: Active and inactive segments

We have based our model on the following assumptions:

- Only active segments deform affinely and contribute to the stress $\underline{T}$ (Green and Tobolsky, 1946):

$$(1)\quad \underline{T} = \sum_{s} \sum_{i \in A_s} N_s \kappa \langle \underline{q}_i \underline{q}_i \rangle_s$$

$\underline{q}_i$ is the segment vector of segment i which is in chain state s, κ is the segment spring constant and N_s the number-density of chains in state s. A_s contains all active segments.

- Transitions between chain-states are possible via various processes involving one or more associative sites.
- Dynamic exchange between active and inactive states described by kinetic equations which do not depend on deformation.
- Arrhenius equations relate the rates of the various chain state transitions to the activation energies involved.
- 'Single chain' model
- Active segments are confined in a tube as shown in figure 2, similar to the reptation model (Doi and Edwards, 1986).

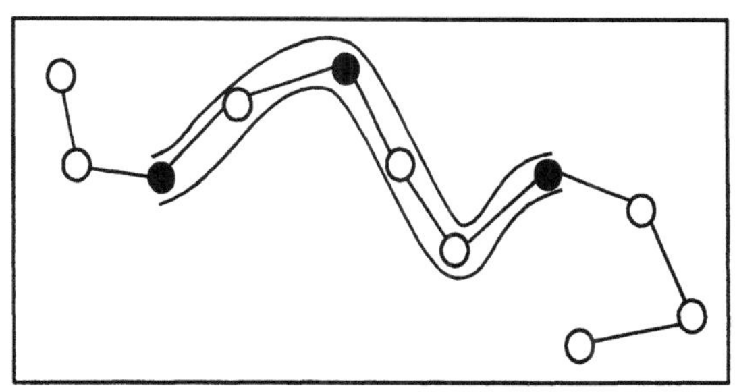

Figure 2: Confinement in a tube

To obtain an expression for the second moment $\langle \underline{q}_i \underline{q}_i \rangle$ we start with the balance equation that governs the time dependence of the distribution function Ψ_s for chains in state s:

$$(2)\quad \frac{\partial \Psi_s}{\partial t} = \sum_{k=1}^{N-1} \frac{\partial}{\partial \underline{q}_k} \cdot \Psi_s \underline{q}_k - \sum_{s \neq s'} f_{s,s'} \Psi_s + \sum_{s' \neq s} f_{s',s} \Psi_{s'}$$

The first term is a convection term and the second term represents the annihilation rate of chains in state s to all other possible chain states noted as s'.

$f_{s,s'}$ is the transition rate constant belonging to the transition from s to s'.

$$(3)\quad f_{s,s'} = \alpha \exp(-E_{s,s'} / kT)$$

with α^{-1} a reference timescale and $E_{s,s'}$ resulting from a superposition of activation energies of the individual associative sites involved in the transition. The third term is the creation term.

Multiplying with $\underline{q}_i\underline{q}_i$, and integrating over the configuration space leads then to a number of Maxwell like equations:

$$(4)\quad \frac{\delta}{\delta t}\langle \underline{q}_i \underline{q}_i \rangle_s = -A_{si}\langle \underline{q}_i \underline{q}_i \rangle_s + B_{si}\underline{1}$$

with A_{si} and B_{si} resulting from summations over constants.

3. TYPICAL MODEL RESULT

In figure 3 the dynamic moduli are shown for a multi-segment chain. The number of sticky points per chain was 8 and we took the activation energies of breaking and forming of bonds equal to 6 kT.

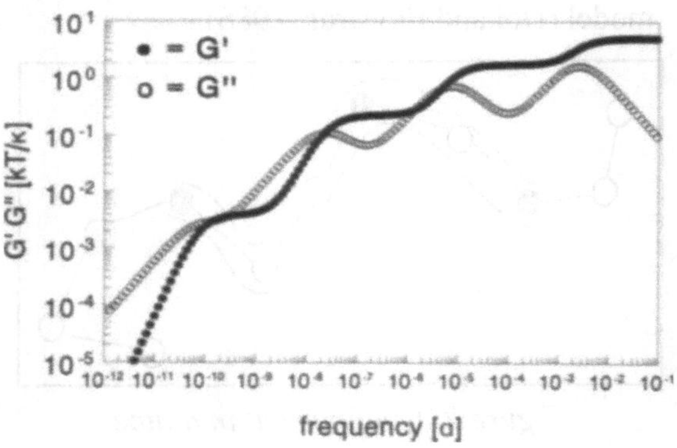

Figure 3: Dynamic moduli.

The essential features of this relaxation behaviour i.e. the number of storage modulus plateaus, the stress-levels of these plateaus and the dominant relaxation times can be predicted in terms of activation energies of forming and breakup of bonds and the number of sticky points per chain.

Also our model does not necessarily satisfy time temperature superposition as can be seen in figure 4 where the storage modulus is shown for two different temperatures. Only the high frequency part can be superimposed by scaling.

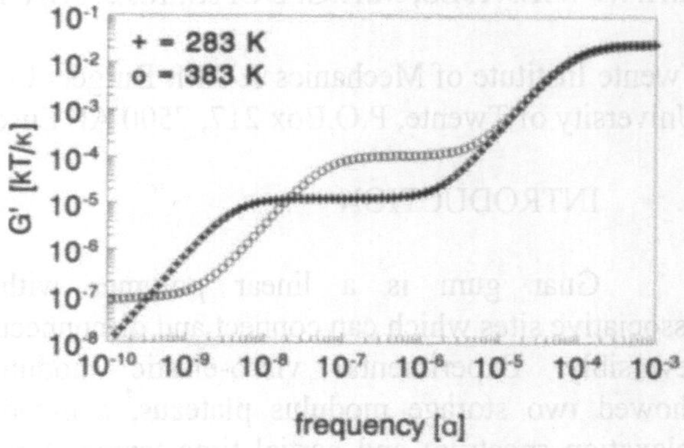

Figure 4: Partial time temperature superposition

4. EXPERIMENTAL RESULTS

In figure 5 temperature mastercurves of an aqueous 1 % guar gum solution are shown.

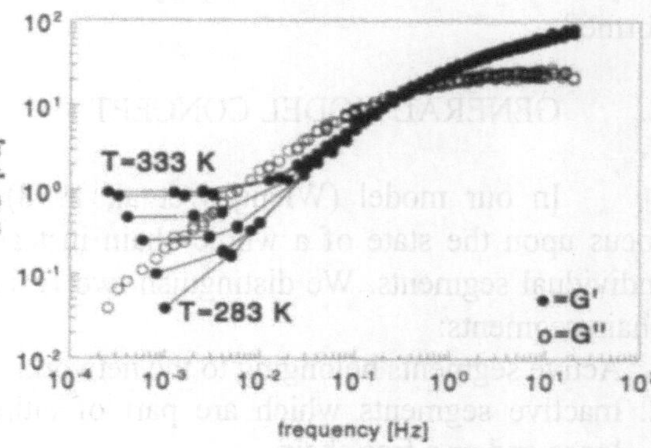

Figure 5: Mastercurves

Time temperature superposition is only possible for high frequencies, yielding an activation energy for breakup of 10 kT. At low frequencies there is a second G' plateau.

5. CONCLUSIONS

Various new phenomena can be predicted with the new model and therefore it seems to be a promising model for transient networks.

Doi M, Edwards SF (1986) The theory of Polymer Dynamics. Clarendon, Oxford

Green MS, Tobolsky AV (1946) J. Chem. Phys. 14: 80-89

Wientjes RHW, Jongschaap RJJ, Duits MHG, Mellema J (1998) submitted

TEMPERATURE AND TIME RELAXATION OF THE COMPLEX YOUNG'S MODULUS FOR ISOTROPIC AND UNIAXIALLY DRAWN POLYMERS

S.V. BRONNIKOV

Institute of Macromolecular Compounds, Russian Academy of Sciences, 31 Bolshoi Prospekt, 199004 St. Petersburg, Russia

1. INTRODUCTION

The mechanical behavior of an idealized linear polymer has been described in the term of the generalized Maxwell relaxation theory by Tobolsky and Eyring (1943). They have resulted the following equations for both storage (E') and loss (E'') elastic moduli as functions of frequency ω

$$E'(\omega) = \int_{-\infty}^{+\infty} H(\tau)\frac{\omega^2\tau^2}{1+\omega^2\tau^2}d(\ln\tau) \qquad (1)$$

and

$$E''(\omega) = \int_{-\infty}^{+\infty} H(\tau)\frac{\omega\tau}{1+\omega^2\tau^2}d(\ln\tau) \qquad (2)$$

where $H(\tau)$ is the distribution function for the relaxation times τ. The methods of the $H(\tau)$ evaluation have been elaborated and described in many monographs.

To describe temperature dependence of the components of modulus, the Boltzmann-Arrhenius equation

$$\tau = \tau_0 \exp\frac{U}{kT} \qquad (3)$$

should be taken into account (τ_0 is a constant, U is the activation energy, k is the Boltzmann constant and T is the absolute temperature).

Such a description is suitable especially for isotropic polymers which exhibit distinguished relaxation peaks in their relaxation spectra.

An uniaxial drawing of polymers is known to cause a broadening of relaxation peaks and an overlapping of their wings. In the region of superimposed wings, the relaxators 'lose' their individuality and become indistinguishable. In this case, the relaxation formulae (Eqs. (1) and (2)) should be modified.

2. EXPERIMENTAL

2.1. Objects

The objects under investigation were isotropic and uniaxially drawn polymers of different chemical structures. In this paper, experimental data are presented for a polyimide of the following structure

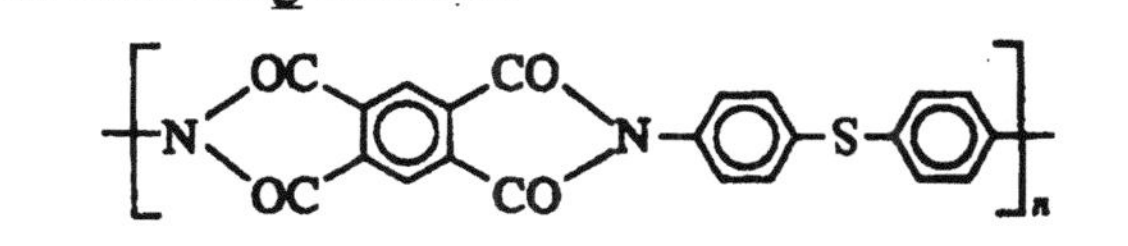

2.2. Methods

The components of the complex Young's modulus have been investigated in the time windows from 10^{-6} to 10^{6} s in the temperature range from 120 to 700 K.

3. RESULTS

Figure 1 shows temperature dependences of both storage and loss dynamic Young's modulus for the polymer of different draw ratios λ varying from 1 to 4. For isotropic polymer ($\lambda = 1$), two relaxation processes (β and α) may be seen distinctly. In this case, both E'(T) and E''(T) dependences may be described by Eqs. (1) and (2).

The highly oriented ($\lambda = 4$) polymer demonstrates superimposed wide distribution of β and α processes. In this case, the approximation of a rectangular ('box') distribution of the relaxation times should be taken into account in the region indicated by dotted lines (Fig. 1,b).

Following Andrews et al. (1948) the rectangular distribution may be written in the time window $\tau_{min} < \tau < \tau_{max}$ (τ_{min} and τ_{max} are the minimum and maximum relaxation times) as follows

$$H(\tau) = H_0 = \text{const} \qquad (4)$$

A substitution of Eq. (4) into Eqs. (1) and (2)

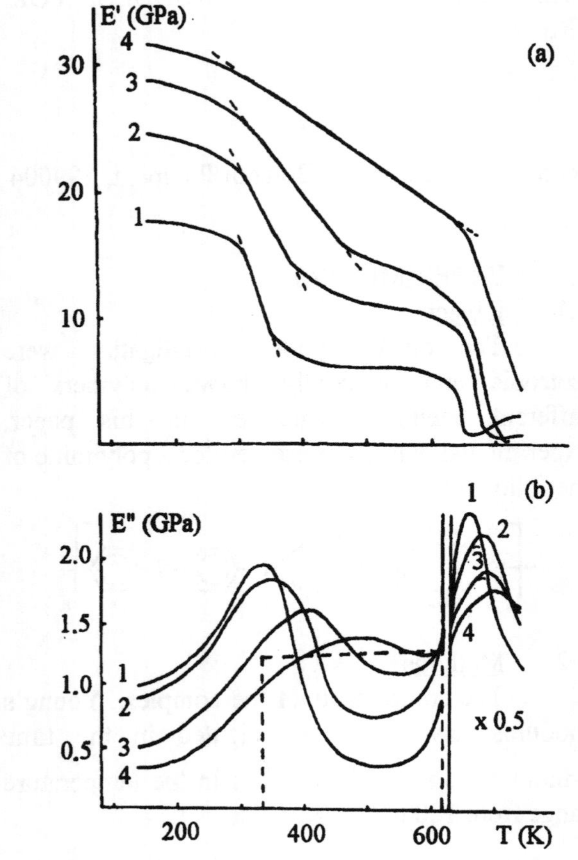

Figure 1. Temperature dependences of (a) storage modulus and (b) loss modulus of the polymer with different draw ratios (see the figures by the curves)

results

$$E'(\omega) = \frac{H_0}{2} \ln \frac{1+\omega^2 \tau_{max}^2}{1+\omega^2 \tau_{min}^2} \qquad (5)$$

and

$$E''(\omega) = H_0 \left(\tan^{-1} \omega\tau_{max} - \tan^{-1} \omega\tau_{min} \right) \qquad (6)$$

Figure 2 presents both experimental and calculated frequency/time dispersion of the Young's modulus at different temperatures for the highly oriented polymer. For calculation, Eqs. (5) and (6) have been used.

For an adequate description of the presented data, a validity of Eq. (3) for the relaxators with minimum and maximum relaxation times and some peculiarities of the $E'(T,\omega)$ and $E''(T,\omega)$ dependences have been taken into consideration. The final equation for

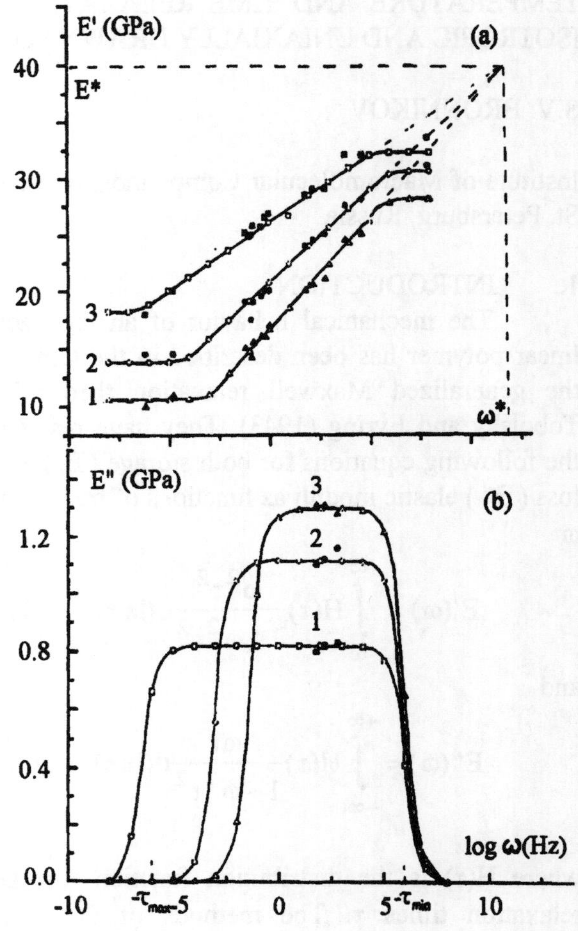

Figure 2. Frequency dispersion of (a) storage modulus and (b) loss modulus of the oriented polymer (λ = 4) at temperatures (1) 298, (2) 423 and (3) 523 K

the temperature and frequency/time relaxation of the storage modulus for a highly oriented polymer has been derived in the form

$$E'(T,\omega) \cong E^* \left(1 - \frac{kT}{U_{max}} \ln \frac{\omega^*}{\omega} \right)$$

where U_{max} is the activation energy for relaxators with τ_{max}, $\omega^* \cong 10^{12}$ Hz and E^* is the extrapolated E' value at $\omega = \omega^*$.

4. REFERENCES

Andrews RD, Hofman-Bang N, Tobolsky AV (1948) J Polym Sci 3:669

Tobolsky AV, Eyring H (1943) J Chem Phys 11:125

SOME MODELS FOR THE TRANSVERSE CREEP AND DAMAGE BEHAVIOUR OF A UNIDIRECTIONAL REINFORCED POLYMER MATRIX COMPOSITE

A.H. CARDON, C. VAN VOSSOLE

Composite Systems and Adhesion Research Group of the University Brussels (COSARGUB), Faculty of Applied Sciences and Engineering, Free University Brussels (VUB), Pleinlaan 2, B1050 Brussels, Belgium.

1. INTRODUCTION

The transverse behaviour of polymer matrix composites is matrix dominated and exhibit time dependent effects. For the use of such composites in structural components, it is necessary to have a good knowledge of those time dependent characteristics and some predictions of the behaviour under an estimated complex mechanical loading history in interaction with environmental variations.

It is important to develop simple models, compatible with more generalised thermodynamic based models, in order to analyse this time dependent behaviour including the initiation and development of damage.

2. GENERAL MODEL BASED ON IRREVERSIBLE THERMODYNAMICS

We consider as basic model the nonlinear viscoelastic equations proposed by R. Schapery (1966). This model can be considered as a nonlinear version of the classical 3-parameter, Burgers type, model for solid viscoelasticity introducing nonlinearising functions for the initial elastic response, for the relation between the instantaneous response and the viscoelastic behaviour, and for the viscoelastic evolution including a reduced time as introduced in the WLF-principle.

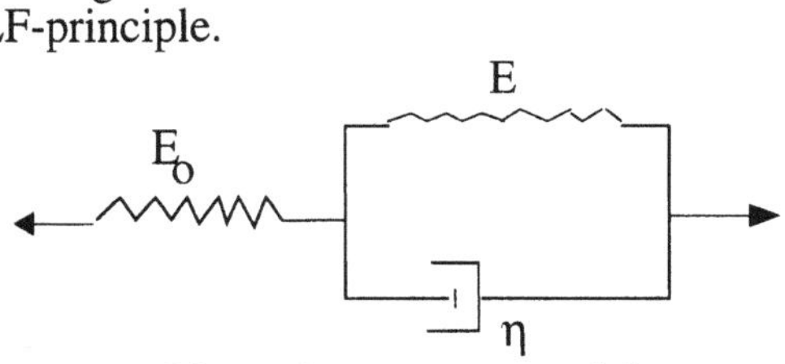

Linear 3-parameter model

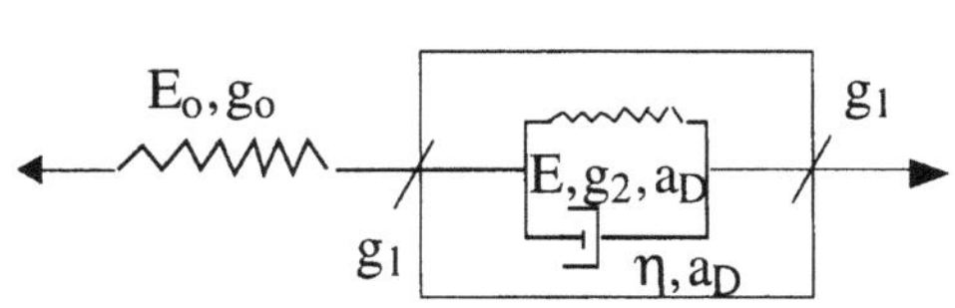

Nonlinear Schapery model

The nonlinearizing functions : g_o, g_1, g_2 and $a_\sigma = \frac{a_D}{a_G}$ are dependent from some invariant related to the stress state, from temperature and other environmental conditions.

3. RESIDUAL TIME INDEPENDENT STRAIN

If after recovery a permanent residual strain is observed, following creep under a given stress level, independent of time, it is necessary to complete one of the above models by a viscoplastic component.

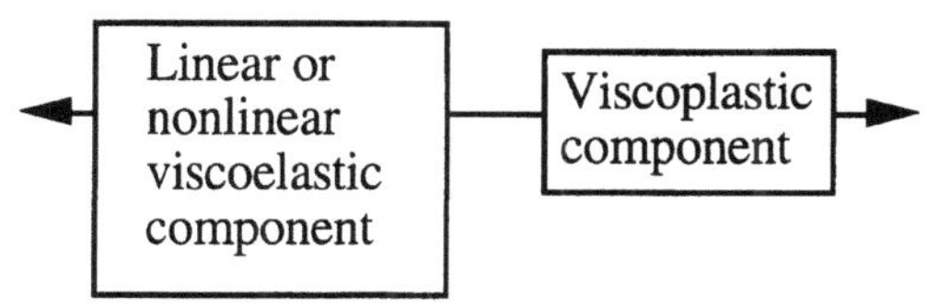

Viscoelastic-viscoplastic model

As was shown by Yan Qin (1996), using a neuro-fuzzy inference approach, the application of a Zapas and Crissman type of functional gives good results for some graphite-epoxy composite systems.

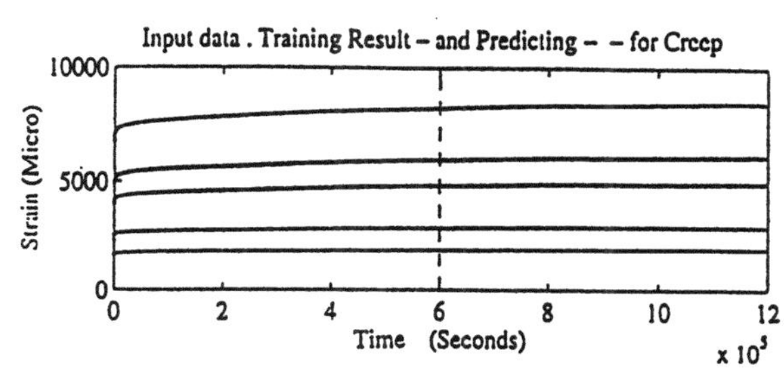

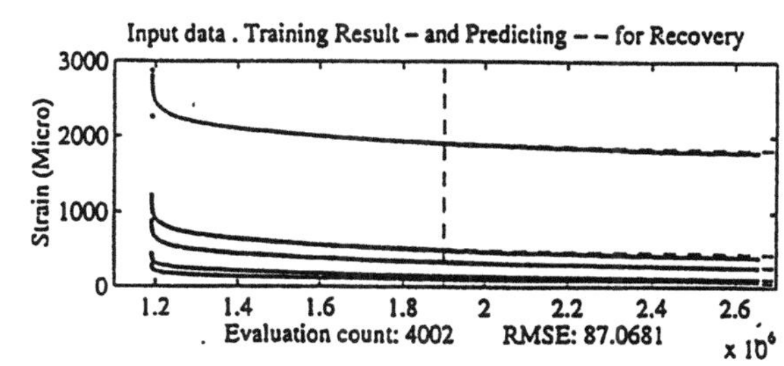

4. INTERACTIONS

If satisfactory results are obtained with a viscoelastic-viscoplastic model for creep and creep-recovery this model is unable to describe the behaviour of graphite epoxy composite systems under combined static and dynamic loadings. This was shown by Yan Qin et al. (1997-1998), and is illustrated by the following figure showing the difference between the responses of pure creep behaviour, with a very good corresponding model output, and the creep behaviour after a first 20 cycle dynamic loading sequence.

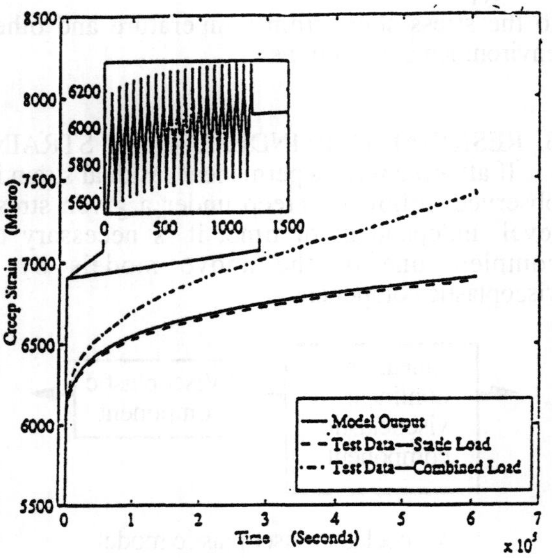

5. DAMAGE MODEL

To model this accelerated creep behaviour after a starting, low frequency, dynamic loading, it is necessary to introduce a damage component in the model.

This damage model can be based on a viscoplastic model where the yield stress is a decreasing function of the number of cycles of the dynamic loading component, of the environmental temperature-moisture cycles, of the time and also of some internal time, if ageing has to be considered.

A combined VE-VP-Damage model, linear or nonlinear, in relation to the specific material system, including the fibre-matrix interaction, and the stress level, can predict the observed stiffness evolution under combined loading and the final failure as consequence of the stiffness degradation.

6. CONCLUSIONS

The transverse behaviour, in stiffness and strength, of a unidirectional fibre reinforced polymer matrix composite can be predicted on the basis of a combined VE-VP-Damage model. This model has to be applied to other fibre reinforced composite systems for validation and extended to laminates before it can be applied to a complete structural composite element.

7. REFERENCES

Schapery RA (1966) A theory of nonlinear thermovscoelasticity based on irreversible thermodynamics. Proc. 5th U.S. National Congress on Applied Mechanics, ASME, pp. 511-530

Qin Yan (1996) Nonlinear Viscoelastic-Viscoplastic characterization of a polymeric matrix composite. PhD thesis, University Brussels, VUB

Qin Yan, Guedes R, Cardon AH (1998) Characterisation and prediction of nonlinear VE-VP behaviour of CRP under cyclic loading by using a neuro-fuzzy model, in Advances in Durability Analysis of Composite Systems (Ed. K. Reifsnider), Balkema.

CREEP BEHAVIOUR OF SHORT GLASS FIBER REINFORCED AND VIRGIN POLYBUTYLENE TEREPHTHALATE (PBT) - A COMPARISON OF EXPERIMENTAL DATA TO MODELED DATA USING THE ELEMENTARY VOLUME CONCEPT (EVC)

B. MÖGINGER; H.-C. LUDWIG; A. KECH, P. EYERER
Institute of Polymer Science and Polymer Testing (IKP), University of Stuttgart, Pfaffenwaldring 32, D-70569 Stuttgart, Germany,

1. INTRODUCTION

The mechanical behaviour of short fiber reinforced composites depends on the mechanical properties of both matrix and fiber (e.g. Young's moduli), the fiber volume content, the fiber length distribution or at least the mean fiber length, the fiber matrix adhesion, and the fiber orientation distribution in a quite complicated way. In the Elementary Volume Concept (EVC) it is assumed that the fundamental mechanical properties of a composite are completely determined by the mechanical properties of a square column shaped Elementary Volume (EV) containing a single fiber of length ⟨L⟩ in its center, Fig. 1.

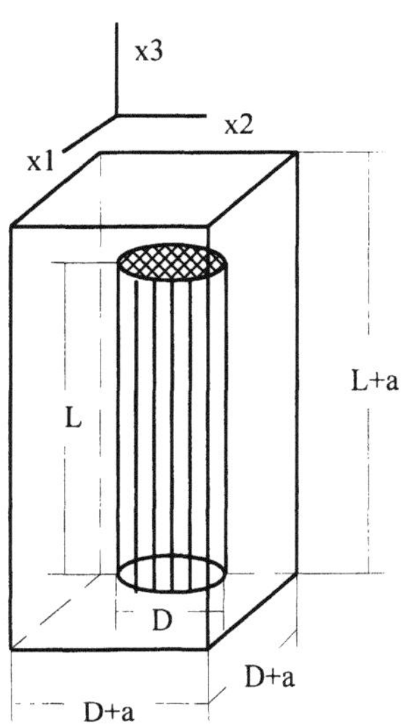

Fig. 1: Elemenary Volume

with L as fiber length, D as fiber diameter and a as half fiber distance.

For ideal fiber matrix adhesion it is possible to calculate the Young's moduli parallel and perpendicular to the fiber axis by separating the EV in a composite and a matrix part [Möginger et al., 1995]. This yields for the corresponding strains

$$\varepsilon_{\parallel} = \frac{r}{r+d}\varepsilon_F + \frac{d}{r+d}\varepsilon_M \qquad (1)$$

and

$$\varepsilon_{\perp} = \frac{n_Z}{1+d}\varepsilon_F + \frac{1+d-n_Z}{1+d}\varepsilon_M \qquad (2)$$

with $\varepsilon_{\parallel}$, $\varepsilon_{\perp}$ ε_M, ε_C as strain parallel and perpendicular to the fiber axis of the EV and strain of matrix and composite part, respectively, r = L/D as aspect ratio, d = a/D as normalized fiber distance and n_Z as Zdrazilova number [Möginger et al., 1998].

2. APPLICATION OF THE EVC TO CREEP

If the creep behaviour is considered in the EVC, one has to derive expressions for the time and stress dependent creep $\varepsilon(t,\sigma_0)$ and compliance $C(t, \sigma_0)$ parallel and perpendicular to the fiber axis, respectively, under the boundary conditions

$$\sigma_0 \quad = \sigma_M \quad = \sigma_C \qquad (3)$$

$$\varepsilon(t) \quad = \varepsilon_{EV}(t) \quad = \varepsilon_{EV}(t, \varepsilon_M, \varepsilon_C) \qquad (4)$$

with σ_0, σ_M, σ_C as applied stress, stress of the matrix part and the composite part.

If one assumes purely elastic behaviour for the composite part the calculation yields a creep compliance parallel to the fiber axis

$$C_{EV}^{\parallel}(t,\sigma_0) = \frac{d}{r+d}C^M(t;\sigma) + \frac{r}{r+d}\frac{1}{E_F A_F^{\parallel} + E_M(1-A_F^{\parallel})} \qquad (5)$$

$$\text{with} \quad A_F^{\parallel} = \frac{\pi}{4}\frac{1}{(1+d)^2}$$

and a creep compliance perpendicular to the fiber axis

$$C_{EV}^{\perp}(t,\sigma_0) = \frac{1+d-n_Z}{1+d} C^M(t,\sigma_0) + \tag{6}$$

$$\frac{n_Z}{1+d} \frac{1}{E_F A_F^{\perp} + E_M (1 - A_F^{\perp})}$$

$$\text{with} \quad A_F^{\perp} = \frac{n_Z r}{(r+d)(1+d)}, n_Z = 0.886.$$

This means that the time dependent creep behaviour of the EV parallel and perpendicular to fiber axis is mainly determined by the creep compliance $C^M(t)$ of the matrix part. The consequence is that the time dependent creep compliance of a short fiber reinforced composite can be determined - or at least estimated - if only the time dependent creep compliance of the matrix is known.

3. EXPERIMENTAL

In order to check this consequence highly oriented short fiber reinforced composites close to the unidirectional limit are necessary. Therefore PBT and glass fiber reinforced PBT with a fiber weight content of 20 and 30 %, respectively, were processed using push-pull technique [Ludwig et. al., 1995]. This yields plates with an fiber orientation factor between 0.8 and 0.9 in flow direction. Tensile test bars were taken out from these plates and investigated in an automized creep stand at room temperature under initial stresses σ_0 of 12.5 MPa, 25 MPa, 37.5 Mpa, and 42 MPa over 2 weeks.

4. EVALUATION AND RESULTS

In order to compare measured creep compliances with calculated ones the creep compliances have to be weighed with an orientation factor of f=0.85. For calculating the compliance the following values have been used: E_M=2,700 MPa, E_F=74,000 Mpa, r=20, d=1.38 and v_F=0.13. The model yield a calculated creep compliances in the order of the measured ones, Fig. 2.

5. REFERENCES

Ludwig et al. (1995) Comp. Sci. Techn.53: 235-240

Möginger et al. (1995) PPS Proc. European Meeting, Stuttgart, contr. 5.25

Möginger et al. (1998), Kautschuk Gummi Kunststoff, submitted for publication

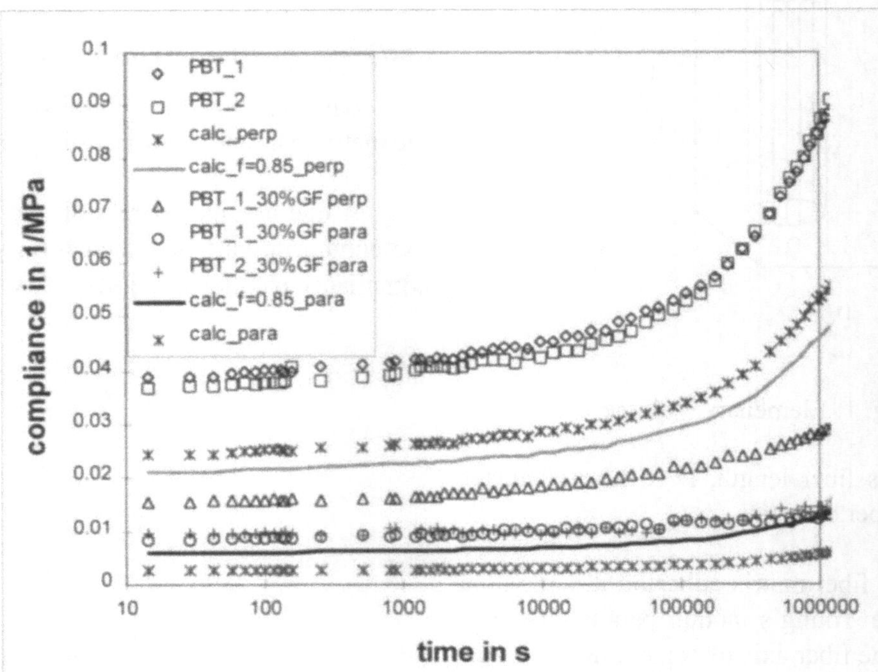

Fig.2: Comparison between measured and calculated creep compliances

AVERAGING OF ELASTIC AND SHRINKAGE PROPERTIES FOR VISCOELASTIC COMPOSITES

J. Orlik

Institut of Technical and Industrial Mathematics (ITWM), Erwin-Schroedinger-Str., 67663 Kaiserslautern, Germany, e-mail: orlik@itwm.uni-kl.de

1. INTRODUCTION

A multi-phase composite with periodically distributed smooth inclusions is considered. The composite components are linear viscoelastic and aging (of the non-convolution integral Volterra type) and are subjected to isotropic shrinkage. The shrinkage deformation can be associated with a fictitious thermoelastic deformation in the constitutive law. An averaging procedure is presented enabling to obtain effective (homogenized) viscoelastic and shrinkage (thermoelastic) properties of the composite and the homogenized stress - field from known properties of the components. Known homogenization results given by [1], [2] have dealt with viscoelasticity of a rather particular differential form.

The procedure proposed in this paper consists in an extension of the asymptotical homogenization technique, known for pure elastic non-homogeneous bodies, to the general linear non-homogeneous aging viscoelasticity, described by a Volterra integral operator of the second kind, with a thermoelastic term. The homogenization procedure is based on the construction of an asymptotic solution with respect to a period of the composite structure. This reduces the original problem to several auxiliary boundary value problems of viscoelasticity on the periodicity cell. The existence and uniqueness results for such problems were obtained for viscoelastic kernels belonging to appropriate function classes. This is done by extension of the classical Volterra integral operator theory to partial Volterra operators with respect to time, which kernels are linear operators with respect to the space coordinates. The effective homogenized characteristics are calculated by averaging of the cell-problems solutions over the cell. Convergence estimates of the averaged and exact solutions as the periodicity cell decreases are obtained. A numerical example for two-phase composite was calculated under the assumption of isotropy of all composite components. The inclusion material is supposed to be pure elastic and the matrix one – viscoelastic. The unit-cell problems are solved by the finite element method.

2. MAIN RESULTS

We consider boundary value problem (BVP) of the viscoelasticity with additional shrink age term, mixed homogen boundary conditions and homogene initial conditions.

$$(A_{ox}^{hk}(t) + A_x^{hk}\star)\,u(x,t) = f_0(x,t) + \frac{\partial f_h(x,t)}{\partial x_h} \qquad (1)$$

where u is a displacement vector, $A_o^{hk}(x,t)$ and $A^{hk}(x,t,\tau)$ are the tensors of elastic and viscoelastic coefficients, f_0 is the vector of external forces, f_h is the tensor of the so-called shrinkage stresses,

$$A_{[o]x}[\star]u(x,t) := \frac{\partial}{\partial x_h}\left(A_{[o]}^{hk}(x,t)\,[\star]\frac{\partial u(x,t)}{\partial x_k}\right) \qquad (2)$$

$$A^{hk}(x)\star e_k := \int_o^t A^{hk}(x,t,\tau)\cdot e_k(x,\tau)d\tau. \qquad (3)$$

We are looking for asymptotics of the solution to this problem of the form

$$u(x,t) = \sum_{p=0}^{\infty}\varepsilon^p u_p(x,\xi,t), \quad \xi \in]0,1[^3, x\in\Omega \qquad (4)$$

where ξ is a space coordinate in the 1-periodic cell, $\xi = \frac{x}{\varepsilon}$, ε is a period of structure (small parameter), t is a time variable, $u_p(x,\xi,t)$ are vector-functions 1-periodic with respect to ξ. The following substitution is proposed: $\star$

$$u_1(x,\xi,t) := (N_{0p}(\xi,t) - N_p(\xi)\star)\frac{\partial u_0(x,t)}{\partial x_p} + y(x,\xi,t), \quad (5),$$

where $N_{0p}(\xi)$, $N_p(\xi)$, $y(x,\xi,t)$ are auxiliary 1-periodic in ξ functions. The first one is determined from the BVP of elasticity for each t, the last ones – from the BVPs of viscoelasticity on the unit periodic cell. The following existence and uniqueness theorem is proposed:

Theorem 1 *Let Ω be a domain with a periodic structure and let the tensor of elastic coefficients $A_0^{hk}(t)$ satisfies positivity and continuity condition. Furthermore, let $f_h \in B^\infty([0,T];L^2(\Omega))$, $h = 0,\ldots,n$. Then, if operator $A_x(t)$ belongs to $V(B^\infty;[0,T];L(H_0^1(\Omega),H^{-1}(\Omega)))$ there exists a unique solution of the problem* (1) *in* $B^\infty([0,T];(H_0^1(\Omega))$.

The following notations are used here: $L(X,Y)$ is the space of contineous linear operators from X to Y with the norm $\|T\|_{L(X,Y)} := \sup_{\|x\|X\le 1}\|Tx\|_Y$,

$$B^\infty(J,Y) := \{K_x : \|K_x\|_{B^\infty(J,Y)} = \sup_{t\in J}\|K_x(t,\tau)\|_Y, \|K_x\|_{B^\infty(J,Y)} < \infty\},$$

$$V(B^\infty;J;Y) := \{K_x : \|K_x\|_{V(B^\infty;J;Y)} := \sup_{t\in J}\int_0^t \|K_x(t,\tau)\|_Y, \|K_x\|_{V(B^\infty;J;Y)} < \infty\}$$

3. NUMERICAL EXAMPLE

We performed calculations for cubic shape inclusion, under assumption of the cell symmetry.

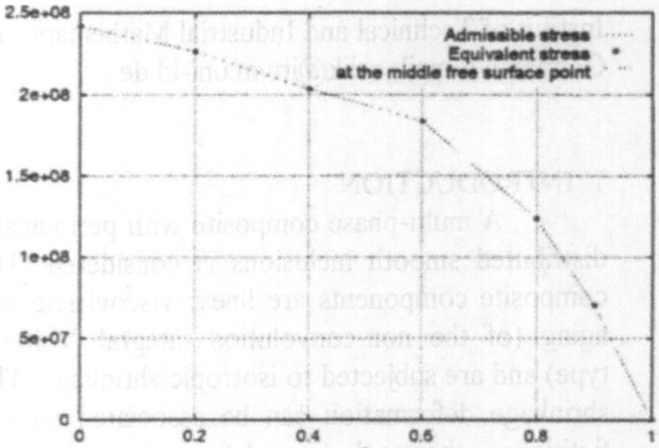

Figure 1: Von Mises and admissible homogenised stress depending on the volume fraction of the inclusions.

4. REFERENCES

Sanchez-Palencia E. (1980) *Non Homogeneous Media and Vibration Theory,* Springer-Verlag Berlin/York.

Francfort G.A. et al. (1986) *Homogenization and Mechanical Dissipation in Thermoviscoelasticity,* Arch. for Rat. Mech. and An., vo1.96, no.3. p.265-93.

S.E.Mikhailov (1998) *On some weighted Hardy type classes ...,* J. of Math. An. and Appl.. vol.222, p.374-396.

PHYSICAL AGING EFFECTS IN *a*-PMMA AND PEO/*a*-PMMA BLENDS

R.W. RYCHWALSKI[1], J. VERNEL[1], V. PELÍŠEK[2], P. SÁHA[2], A.V. VANNIKOV[3]

[1]Department of Polymeric Materials, Chalmers University of Technology, Göteborg, Sweden
[2]Faculty of Technology Zlín, Technical University Brno, Czech Republic
[3]A.N. Frumkin Institute of Electrochemistry, Russian Academy of Sciences, Moscow, Russia

1. INTRODUCTION

Evolution of physical properties originating from non-equilibrium of glasses is referred to in the literature after Struik (1978) as physical aging, and can take place in homopolymers, polymer-polymer blends and polymer matrix composites. Important results have been published for many homopolymers (see, for example, McKenna (1994a, 1994b), Hutchinson (1997)). Physical aging of polymer blends and composites is much less analysed, although for example, recently there is a great interest in this behaviour in composites towards aerospace applications (Gates et al. (1997)). Thus, there is a need to further understand physical aging behaviour, in particular, in more complex systems (McKenna and Hutchinson (1997)).

PEO/PMMA blends have been in the mainstream of research for several years but there still remain issues which are unresolved or disputable, for example the level of miscibility / structure related properties. Physical aging behaviour of *atactic*-poly(methyl methacrylate)(*a*-PMMA) and poly(ethylene oxide) (PEO)/(*a*-PMMA) blends at low content of PEO is studied up to 18 vol. % of PEO. Differential scanning calorimetry (DSC), mercury-in-glass dilatometry, dynamic mechanical analysis (DMA) and second harmonic generation (SHG) are used.

2. DSC

From the dependence of the fictive temperature, T_f, measured from the equal area construction (Moynihan (1976)), the peak temperature, T_p and the excess enthalpy, ΔH, on the cooling rate and aging time, the non-linearity parameters are calculated, namely the apparent activation energy, Δh^*, and the Narayanaswamy parameter, x (Narayanaswamy (1971), Godard et al. (1996)). Above calculated parameters are next used to describe the time evolution (physical aging) in thermal properties, on the basis of the stretched exponential formula. The prediction is compared with experimental results.

3. MERCURY-IN-GLASS DILATOMETRY

Volumetric isothermal contractions in the T_g-region, following temperature programmes comparable to the ones used for DSC, are measured and analysed.

4. DMA AND STRESS RELAXATION

Time scales and aging rates are studied by means of DMA and by stress relaxation, in shear, from the time-aging time superposition principle (Struik (1978)), where a purely horizontal shift is used by us

$$\mu = -\frac{d\log(\omega/\omega_r)}{d\log(t_e/t_{er})} \qquad (1)$$

where ω and t_e denote frequency and aging time, and r denotes reference values.

It is found that the aging rate measured at the same negative temperature distance to T_g, monotonically decreases with increasing content of PEO in the materials, and a non-synergistic variation of the aging rate as a function of composition is observed.

5. SHG

The decay of the second harmonic intensity with increasing time in the present materials mixed with 4-(dimethylamino)-4′-nitrostilbene chromophore (a guest-host system) is measured in spin cast films, following a determined temperature history. The SHG set-up is based on a pulsed laser.

It is found that the decay is governed by the surface potential (electric field) generated by corona poling, used at an earlier stage to align the low molecular weight chromophores (dipoles) (Vannikov et al. (1998)). The kinetics are discussed in terms of free volume depending on composition and physical aging. It is suggested that the control of SHG by electric field and blend composition can be employed in photonic devices. The non-linear optical method of SHG is an interesting way to characterise the time dependence of chromophore rotational diffusion in polymer-polymer blends.

6. ACKNOWLEDGEMENT

The financial supports of the Swedish Board of Engineering Sciences (TFR), the Royal Swedish Academy of Sciences, and the Russian Foundation for Basic Research, is gratefully acknowledged.

7. REFERENCES

Moynihan CT, Easteal AJ, DeBolt MA, Tucker J (1976) J Am Ceram Soc 59: 12-16

Narayanaswamy OS (1971), J Am Ceram Soc 54: 491-498

Godard ME, Saiter JM, Burel F, Bunel C, Cortes P, Montserrat S, Hutchinson JM (1996) Polym Eng Sci 36: 2978-2985

Struik LCE (1978) Physical aging in amorphous polymers and other materials. Elsevier, Amsterdam

McKenna GB (1994a) J Non-Crystalline Solids 172-174: 756-764

McKenna GB (1994b) J Res Natl Inst Stand Technol 99: 169-189

Hutchinson JM (1997) In: Haward RN, Young RJ (eds) The physics of glassy polymers. Chapman and Hall pp 88-153

McKenna GB and Hutchinson JM (discussion leaders) (1997) Measurement and modelling of glassy phenomena: future needs and perspectives, 3rd Int Discussion Meeting on Relaxations in Complex Systems, 30 June-11 July, Vigo, Spain

Vannikov AV, Vernel J, Mal´tsev EI, Savel´ev VV, Rychwalski RW (1998) Polym Eng Sci, accepted

Gates TS, Veazie DR, Brinson LC (1997) J Comp Mater 31/24: 2478-2505

KEVLAR FIBRE - ANHYDRIDE EPOXY RESIN COMPOSITES THERMIC AND MECHANICAL CHARACTERISATION

J.J. Suay Antón[1], E. Gimenez[1], M. Monleón[2], J.L. Gómez[2].

[1] Area de Ciencia de los Materiales e I. M. .Dep. Tecnología. U. Jaume I. Campus Penyeta Roja. Castellón 12.071

[2] Departamento de Termodinámica Aplicada. Escuela Superior Ingenieros Industriales. Universidad Politécnica de Valencia. Valencia 46071.

1. INTRODUCTION.

Epoxy resins are widely used as the matrix material for high performance composites. Epoxy anhydride systems are preferred for electrical and electronic applications or when chemical safety has to be taken into account [1]. Various chemical reactions take place during cure and determine the resin morphology which, in turn, determines the properties of the cure formulation. At present, the uncatalyzed cure of epoxy-anhydride system is quite well understood [2], but in the case of using a catalyst to accelerate the cure, the mechanism is more complex [3]. Over last years, differential scanning calorimetry (DSC) has been a common analytical technique in the attempt to study the cure kinetics [4]. Mechanical measurements, have frequently been utilised to facilitate calculation of viscoelastic properties in heterogeneous polymers systems and to elucidate if any matrix modifications has been made because reinforcing addition.

2. EXPERIMENTAL.

The formulations used in this study were composed of Ciba-Geigy's Araldite F epoxy resin, which contains basically DEGEBA molecules, HY-905 hardener (which is basically metil-tetrahydroftalic anhydride acid MTHPA), 50% in weight of each component, and continuos Kevlar 49 fibre as reinforcement used in different contents. Kevlar 49 fibre was supplied by Dupont in a filament bobbin and commercial glass fibre by Cristaleria Española S.A. In order to determine Kevlar reinforcing content for all formulations a Dupont Thermogravimetric Analyser 951 was used. T_g samples determination was done with a Dynamic Mechanical Analyser (Polymer Laboratories).

3. RESULTS AND DISCUSSION.

Fibre influence on T_g.

Glass transition temperature of a epoxy system increases with crosslinking density increasing as epoxy curing reaction progress [5][6], so controlling T_g is a good method to know at which curing reaction stage is the system. It is important to get maximum T_g for neat and reinforced formulations in order to compare dynamic-mechanical results of total cured samples, with similar crosslinking density, so Kevlar fibre reinforcing effect be obvious.

Two different curing cycles were used in our epoxy-anhydride mixtures: (cycle 1) isothermal curing at T_c=150°C during 118 hours and (cycle 2) which is cycle 1 plus a post-curing consisting in one ramp 45°C to 200°C at 1°C/min. A dynamic DSC scan 45°C to 200°C at 10 °C/min was done to each sample. Figure 1 shows dynamic DSC results for neat and reinforcing formulations and the two curing cycles applied. A significant decrease in T_g can be observed as Kevlar reinforcing fibre content increase (until 22°C with curing cycle 1 between neat and 80% reinforcing fibre formulations). Post-curing reduce differences in T_g between neat and reinforcing formulations but it did not equlate values (even it had more influence increasing T_g in those reinforced formulations with more Kevlar content).

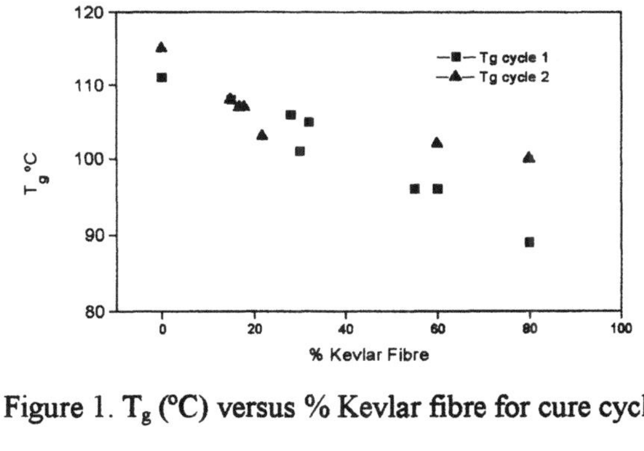

Figure 1. T_g (°C) versus % Kevlar fibre for cure cycles.

A trapping effect by the fibre could reduce the mobility of the system, decreasing the encounter probability between reactive species, proportional to the increase in viscosity of the medium. So there could be regions in the bulk material with epoxi and anhydride groups without reacting which have a plastificant effect and reduce T_g [7]. Others authors have pointed out in other epoxy systems that catalyst addition accelerate curing reactions but produce significant decrease in T_g because catalysed formulations have different network with shorter chain segment [8].

Dynamic Mechanical Properties.
We choose cycle 2 as curing treatment, trying to minimise differences in T_g between different systems. Isothermal dynamic-mechanical scans were done from 30° to 200° at frequencies 0.1 to 30 Hz. Loss tangent and storage modulus master curves versus frequency were obtained from isothermal scans. These curves would represented E′ and Tang δ relaxation spectra in a wide frequency range at the reference temperature (105°C in our case). Figures 2 and 3 show loss tangent and storage modules master curves versus frequencies for different neat and reinforced formulations.

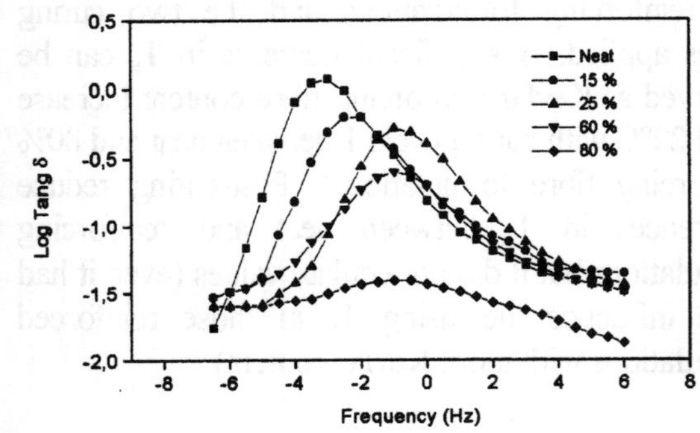

Figure 2. Loss tangent δ versus frequency for different % Kevlar content.

Alpha relaxation (which is associated with polymeric glass transition temperature) and its evolution versus frequency with fibre addition was object of our study. With Kevlar content increase there is a decrease of maximum loss tangent peak high, and a shift to high frequency or lower temperatures.

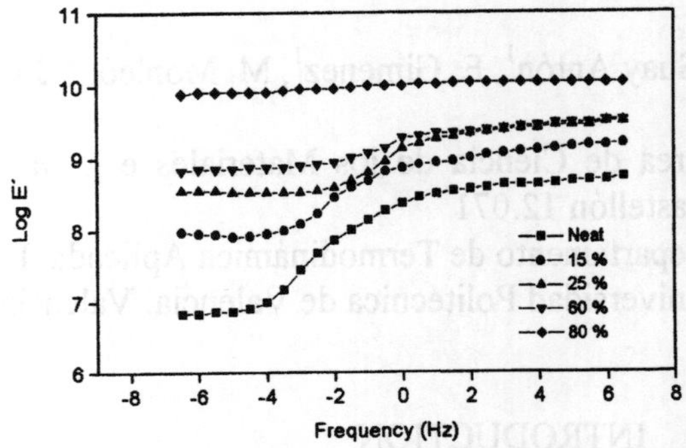

Figure 3. Log E′ versus frequency for different % Kevlar content.

Maximum loss tangent shift to high frequency (or low temperatures) was in the same magnitude order than differences in T_g found between neat and reinforcing formulations (until 15° C by DSc and applying curing cycle 2). All these factors can be representative of an increase of heterogeneity in the chain morphologies and a decrease in crosslinking density as Kevlar fibre content in formulations is bigger. Reinforcement effect on E′ is the one expected: the modulus increases in the glassy region as well as in the plateau region, but this last increase in considerably greater, and thus the increment of the modulus in the transition is lower the higher the fibre content.

REFERENCES.

[1] P. Guerrero et al, Polymer, 37, 11, 1996.
[2] B. Steinmann, J. of Applied Polym. Sci., 37, 1753 (1989).
[3] J. M. Barton, Epoxy Resins and Composites I, K. Dusek, Ed., Advances in Polymer Sci., Vol 72, Springer Verlag Berlin, 111-154, (1985).
[4] R. B. Prime, in Thermal Characterization of Polymeric Materials, E.A. Turi, Ed. Academic Press, New York, 1981, Chap 5, 435-569.
[5] J. Mijovic, j. of Applied Polym Sci., 31, 1177-1187 (1986)
[6] D.J. Plazek, J. of Applied Polym. Sci., 24, 1303 (1986).
[7] M. F. Grenier-Loustatot, P. Grenier, Polymer, 33, 6 (1992).
[8] K. Fryauf et al, Polymer 34, 2 (1993).

EXPERIMENTAL STUDY OF ANYSOTROPIC VISCOELASTICITY OF HOMO AND HETEROPHASE POLYMER SYSTEMS.

A. TERESHIN, V. KULICHIKHIN

A.V. Topchiev Institute of Petrochemical Synthesis,
29 Leninsky pr., 117912 Moscow, Russia.

1. INTRODUCTION

There are numerious number of anisotropic polymeric systems. As a rule, these are complex liquids, such as LCPs, emulsions, two-phase blends, polymers filled by fibrous fillers, etc. For part of them an anisotropy appears as a result of deformation action (elongation of disperse phase droplets), for another ones an anisotropy is thermodynamically inherent property (LCPs). Unfortunately, until this time there no exists the special equipment for the experimental study of anisotropic viscoelasticity, though some theoretical predictions were already appeared (see, for example, Volkov and Kulichikhin (1990)). Our first attempts to investigate the anisotropic dynamic moduli on an example of LCP's melts and their blends with commodity thermoplastics (Kulichikhin, Tereshin (1997)) showed us some advantages and disadvantages of proposed experimental procedure. In this paper except the results on well-known thermotropic LCPs and their blends at high temperature the new ones will be shown on mixtures of two immixible solutions (one of them can form the lyotropic LC phase) at room temperature.

2. OBJECTS AND METHODS

2.1. Equipment

The new vibrorheometer (VR-04) with operating unit cylinder-cylinder was designed in such a way that three kinds of movement (rotation, longitudinal and transversal oscillations) are coming from one engine to the outer cylinder. A superposition of rotation and one mode of oscillation is also possible. The measuring system for tangential and normal forces connects with the inner cylinder, suspended on air bearing. For measurements of torque and normal force the sets of torsions or elastic rings with different rigidities are using, respectively. The upper shaft on which the measuring system is mounted is fitted by traper roller air bearing. Two kinds of cylindrical operating units were used: the traditional one and the Pochettino type (practically, without bottom effects). The total frequency range is of 0.001-100 Hz, the angular displacement - of 3'-1°; the axial displacement - of 10-280 μ, the deformation amplitude - of 0.001-0.2, the shear rate – of 0.001-0.35 s^{-1}, the measured dynamic moduli - of 1-10^6 Pa.

2.2 Objects

We used two groups of model objects:

a) thermotropic liquid crystal copolyester of PET and p-hydroxybenzoic acid (HBA) 40:60 and its blends with poly-4-methylpentene-1 (PMP) in which two kinds of anisotropy are existing: due to anisotropic structure of LCP melt and due to probable elongation of disperse phase droplets at steady state shear. A composition of LCP/PMP blends were 25/75 and 75/25. All measurements were performed at 250°C.

b) lyotropic LCP — the 50% aqueous solution of hydroxypropylcellulose (HPC) and its two-phase mixtures (emulsions) with 50% solution of polyisobutylene (PIB) in decaline. Compositions of mixtures were the same as in the case of PET/HBA–PMP system. Measurements were performed at room temperature.

3. RESULTS AND DISCUSSION

For PMP melt the both kinds of superposition of steady shear and longitudinal or transversal oscillations lead to the same effect, namely to decrease of dynamic moduli in comparison with pure oscillation modes. The scale of moduli decreasing is approximately the same for both kinds of superposition and more pronounced at low frequencies.

For LCP a non-coincidence of storage and loss moduli measured even by pure oscillations was registrated: both of moduli are a little higher at angular than at axial oscillations. The reason of such effect is preferential orientation of LCP at loading along to cylinders axis and very slow relaxation in constraint conditions (gap width is ~ 1 mm). At superposition the moduli are changing in different directions depending on oscillation mode.

At longitudinal superposition the storage modulus decreases compared with pure oscillation mode, but it increases at transversal superposition. The final values of the elastic moduli at any kind of superposition are equal to initial values at pure oscillation modes. Thus, at longitudinal superposition G' achieves the values inherent for transversal oscillations (without superposition of them on steady shear flow) and vice versa. It means, that steady flow performs the same orientation along the shear direction as at loading of samples along the cylinders axis. The similar situation takes place for loss moduli, but at low frequency range only. The scale of moduli change reaches [0.3] order for G' and [0.5-0.6] order for G".

At addition of LCP to PMP the scale of moduli change increases for longitudinal superposition and achieves maximal values for blend containing 75% of LCP (-0.8 order for G' and -0.65 order for G"). In orthogonal direction of superposition a scale of moduli fall decreases, reaches zero for the same blend and then starts to increase. So, when PMP is a matrix the elongation of LCP's droplets takes place and this process influences moduli change (especially in longitudinal direction) more pronouncely, than orientation of LC phase inside droplets.

When LCP becomes a matrix (its content >50%), a phase inversion proceeds and capability of PMP's droplets to orient in direction of steady shear is lesser than LCP's droplets in the complementary composition. It means that anisotropy of a nematic matrix predominates at longitudinal superposition. As to transversal one, the presence of non-elongated PMP's droplets compensates the moduli increase due to corresponding response of LC phase. So, the change of PMP relaxation spectrum inside almost spherical droplets at transversal superposition should be negative, the LCP response in a matrix should be positive, and a result is practical constansy of both of moduli in pure axial oscillation mode and its superposition on steady shear flow for blend with 75% of LCP.

Generally, behaviour of HPC solution and its blends with solution of PIB is similar to behaviour of thermotropic LCP and blends in melt state. For LC solution we also observed decrease of dynamic moduli along the steady shear and increase of moduli in orthogonal direction. It is interesting that HPC solution exhibits distinct asymmetry of these effects: increase of moduli at orthogonal superposition (+0.8 order at 0.1 Hz) is much higher than their decrease at longitudinal superposition (-0.3 order). We believe that molecular orientation is not parallel to the shear direction, i.e. long axes of molecules form some non-zero angle with the vector of steady shear.

The most pronounced anysotropic effects was obtained for the mixture of solutions with LC matrix (content of HPC solution — 75%). Decrease of moduli in longitudinal direction exceeds one decimal order. It means that some kind of synergism takes place: elongation of PIB droplets acts in the same direction as molecular orientation of LC matrix, but overall result is much higher than one can expect for sum of two effects. Results obtained in orthogonal direction for this mixture are closely resembled that for blend melts (PET/HBA-PMP 75/25): counteraction of moduli increase due to orientation of LC phase and decrease because of droplets elongation leads to practically unchanged values of G' and G''.

4. CONCLUSIONS

1) The expected effect of viscoelastic anisotropy was obtained for LC melt and LC solution: the dynamic moduli decrease at longitudinal superposition ans increase at transversal superposition, compared with pure oscillation modes.

2) If LCP is a matrix in blend composition, its behaviour is similar to the neat LCP behaviour, but more pronounced due to heterophase morphology of blend melts. In the opposite case the change of PMP relaxation properties predetermines the direction and scale of moduli change.

5. REFERENCES

Volkov V.S., Kulichikhin V.G. (1990) J.Rheol. 34: 281

Kulichikhin V.G., Tereshin A.K. (1997) In:Proc. PPS-13 Meeting, Ghotenbugr, Sweden, August 19-21, p4:16

BULK MODULI FROM PHYSICAL AGING AND STRESS RELAXATION DATA

J. VERNEL[1], M.J. KUBÁT[2], R.W. RYCHWALSKI[1], J. KUBAT[1]

[1]Department of Polymeric Materials, Chalmers University of Technology, Göteborg, Sweden
[2]Department of Polymer Technology, Royal Institute of Technology, Stockholm, Sweden

1. INTRODUCTION

During physical aging of an amorphous polymer below its glass transition temperature, T_g, thermodynamic properties such as enthalpy, H, and volume, V, will decrease with time towards their equilibrium value. This results, among others, in an overall increase of stiffness.

It is shown here that by correlating the excess enthalpy with the excess volume for polystyrene, PS, and poly(vinyl acetate), PVAc, the bulk modulus can be calculated.

Moreover, it is shown, using polyethylene, PE, as an example, that time dependent volume changes recorded during stress relaxation can be used to calculate bulk moduli.

Further details of the present analysis are given in Kubát et al. (1997).

2. RESULTS

2.1 Excess enthalpy vs. excess volume. Bulk modulus calculation.

Using standard thermodynamics, the variation of enthalpy, H, with volume, V, at constant temperature can be expressed as

$$\left(\frac{\partial H}{\partial V}\right)_T = -\frac{1-\alpha_P \cdot T}{\beta_T}, \qquad (1)$$

where the isobaric cubic coefficient of thermal expansion, α_P, is defined as $(\partial V/\partial T)_P/V$ and the isothermal compressibility, β_T, is defined as $-(\partial V/\partial P)_T/V$. Inserting the bulk modulus, K, instead of the compressibility, we can rewrite eq. (1) as

$$\left(\frac{\partial H}{\partial V}\right)_T = K \cdot (1-\alpha_P \cdot T). \qquad (2)$$

Since at normal conditions the product $\alpha_P{\cdot}T$ is much less than unity, the derivative of the enthalpy with respect to the volume at constant temperature is approximately equal to K.

Based on data for polystyrene, PS, extracted from the work of Adachi and Kotaka (1982), Hourston et al. (1996) and Greiner and Schwarzl (1989), and according to eq. (2), $(\partial H/\partial V)$, or the slope of the excess enthalpy, δ_H, vs. specific volume, v, should be equal to K when the $\alpha_P{\cdot}T$ correction is neglected. From the plots, K is found to be between 3.3 and 4.3 GPa for an aging temperature of 100.6°C and around 2.7 GPa for an aging temperature of 72°C. Taking $\alpha_P{\cdot}T$ into account brings a correction of about 20%.

Considering the fact that the graphs were obtained by combining results from different investigations, K values are in fair agreement with p-V-T values for PS reported to be between 2.3 and 3.3 GPa by Zoller and Walsh (1995) and Quach and Simha (1971).

Based on our calorimetric and volume measurements during aging of poly(vinyl acetate), PVAc, at 35°C, the relaxed enthalpy, ΔH, vs. specific volume, v, was plotted. In this case the K value of about 1 GPa at early aging increases with aging time, approaching a value of about 3.0 GPa at equilibrium. The increase of K is in agreement with the well-known stiffening effect following temperature down-jumps. From p-V-T measurements, Zoller and Walsh (1995) obtained for PVAc a K value of 3.6 GPa at 37.1°C and an applied pressure of 20 MPa.

2.2 The Grüneisen parameter

The termodynamic Grüneisen parameter, γ_G, links thermal expansion, α, to the bulk modulus, K, and the heat capacity at constant volume, C_V, in the following way

$$\gamma_G = \frac{\alpha \cdot K \cdot V}{C_V}. \qquad (3)$$

For solids, values between 2 and 4 are reported. Using the following relation between C_V and C_P

$$C_P = C_V \cdot (1+\gamma_G \cdot \alpha \cdot T) \qquad (4)$$

it is possible to show that the Grüneisen parameter links the quantities ΔH and ΔV in the following way

$$\frac{\Delta H}{\Delta V} = \frac{K}{\gamma_G} \cdot (1 + \gamma_G \cdot \alpha \cdot T). \tag{5}$$

As can be seen, the Grüneisen approach yields slightly higher values than straightforward application of the thermodynamic formulae.

3. APPARENT BULK MODULUS FROM STRESS RELAXATION

According to standard solid mechanics the volume increase of an isotropic elastic solid subjected to a uniaxial stress can be related to a negative pressure equal to one third of the stress. The volume strain is thus given by

$$\frac{\Delta V}{V} = \frac{\sigma}{3 \cdot K}. \tag{6}$$

When applying the same reasoning to stress relaxation data measured by Delin et al. (1995), where both the volume strain and the stress decrease with increasing time, we obtained K values amounting to 0.9 ± 0.05 GPa for LDPE and 1.37 ± 0.19 GPa for the LLDPE, where LD and LLD stands for low and linear low density, respectively. Hydrostatic bulk moduli are usually much higher, as reported by Zoller and Walsh (1995): 3.15 GPa for LDPE at 23.5°C and 3.71 GPa for at 29.7°C. In both cases, the applied pressure was 20 MPa.

4. CONCLUSIONS

It is shown that, based on standard equilibrium thermodynamics, an apparent bulk modulus, K, of reasonable magnitude can be calculated from the excess enthalpy and the excess volume measured during physical aging of PS and PVAc. An alternative calculation of K based on the Grüneisen parameter γ_G is also shown.

Bulk moduli extracted from the linear volume-stress relations measured during stress relaxation of polyethylene were significantly lower than reported K values. When discussing possible reasons for this, the differences in time scale between conventional K measurements and the extraction of K data from a stress relaxation experiment must be taken into account.

5. REFERENCES

Adachi K, Kotaka T (1982) Polymer J 14: 959-970

Delin M, Rychwalski RW, Kubát MJ, Kubát J (1995) Rheol Acta 34: 182-195

Greiner R, Schwarzl FR (1989) Colloid Polym Sci 267: 39-47

Hourston DJ, Song M, Hammiche A, Pollock HM, Reading M (1996) Polymer 37: 243-247

Kubát MJ, Vernel J, Rychwalski RW, Kubát J (1997) Polym Eng Sci, in press

Quach A, Simha R (1971) J Appl Phys 42: 4592-4606

Zoller P, Walsh DJ (1995) Standard Pressure-Volume-Temperature Data for Polymers. Technomic Publ. Co.

Quantitative Analysis of Mixer-Type Rheometers

A. Ait-Kadi[1], P. Marchal[2], A.S. Chrissemant[2], M. Bousmina[1] and L. Choplin[2]

[1]Chemical Engineering Department and CERSIM, Laval University, Quebec, Canada
and [2]ENSIC/GEMICO, Nancy, France

I. Introduction

Rheological characterization of complex fluids may not be an obvious task mainly when one deals with materials having constituents with charcteristic dimensions of the same order of magnitude as the space available for flow in the measuring device. This is the case for example of solid suspensions and some structured food materials. Mixer-type rheometry consisting of a mixing device with a more or less complex geometry rotating in a fluid contained in a cylindrical tank provides an alternative solution to such rheological characterization problems (Choplin and Marchal, 1996). Mixing devices with large local spacing can indeed be used to handle fluids with relatively large heterogeneity. Choplin and Marchal (1996) found that monitoring torque and rotational speed during the mixing process can provide a « rheological signature » of the material under consideration. However, a more detailed analysis is required to translate this information into a complete rheological characterization such as the one obtained from conventional rheometers. This work reports on recent developments to quantify speed/torque data of different types of mixing devices to extract the rheological parameters of interest. The analysis is based on the Couette analogy.

II. Theory and results

The quantitative analysis of torque-rotor speed data of mixer-type rheometers is based on the Couette analogy (Bousmina et al., 1998). In this approach, the complex geometry of the mixing device is replaced by a simple bob and cup geometry which will produce the same torque, Γ, at the same rotational speed, N, of the mixer or the bob. Figure 1 schematically illustrates the approach.

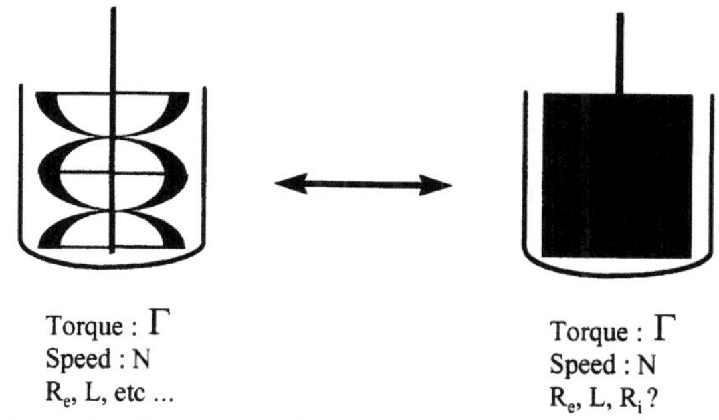

Figure 1 - Couette analogy

Hydrodynamic equations were solved in the particular case of a generalized Newtonian fluid obeying a power law model (Bird et al., 1987) and the following formula for the equivalent internal radius of the mixing device was obtained (Bousmina et al., 1998) :

$$R_i = \frac{R_e}{\left(1 + \frac{4\pi N}{n}\left(2\pi ML\, R_e^2 / \Gamma\right)^{\frac{1}{n}}\right)^{\frac{n}{2}}} \qquad (1)$$

where R_i and R_e are respectively the equivalent radius of the mixing element and the radius of the cylindrical mixing chamber, L the length of the mixing chamber, Γ the measured torque, N the imposed rotational speed, and M and n the consistency and the flow index of the power law model respectively. All parameters in Eq. (1) are in consistent dimensions.

Bousmina et al. (1998) found that the equivalent internal radius R_i given by Eq.1 varies only slightly with the power law index n. The authors concluded that a calibration procedure can be performed using for example a Newtonian fluid or any other well characterized power law fluid.

The approach is validate here with a relatively simple geometry namely a double walled Couette device (see Fig. 2) for which an analytical expression of the equivalent internal radius is given by :

$$R_i = \frac{R_4}{\left[1 + R_4^{2/n}\left(\frac{R_2^2}{\left[(R_2/R_1)^{2/n}-1\right]^n} + \frac{R_3^2}{\left[1-(R_3/R_4)^{2/n}\right]^n}\right)^{(-1/n)}\right]^{n/2}} \qquad (2)$$

where $R_k, k \in \{1,4\}$, are defined on Fig.2.

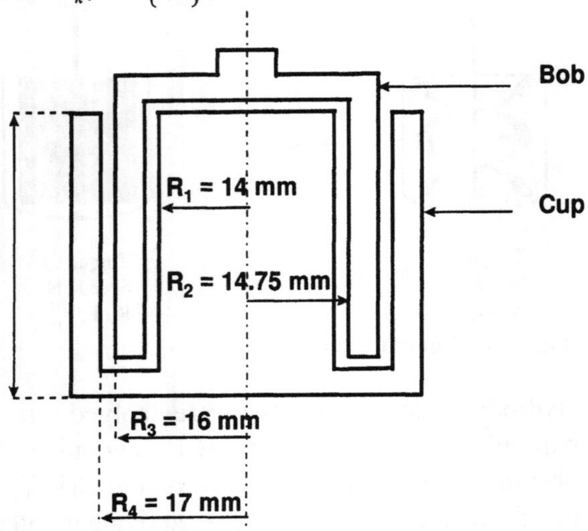

Figure 2 - Double walled Couette

In this case, the external radius of the cup in the Couette analogy is $R_e = R_4$. For the particular dimensions given on Fig.2, the internal equivalent radius R_i varies from 16.5 mm to 16.9 mm for a power law index ranging from 1 to 0.3 which are typical values for polymer solutions and melts (see Fig 3).

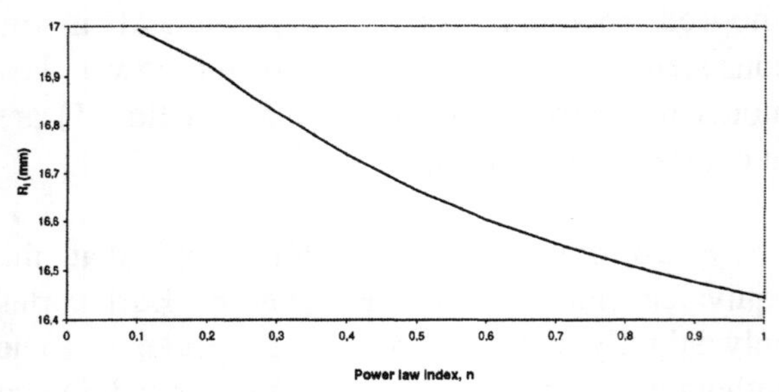

Figure 3 - Effect of power law index on R_i for the double walled Couette geometry of Fig.1.

These variations are relatively small and if one takes an arithmetic mean value for the internal equivalent radius, the error in the evaluation of the viscosity (see below) is smaller than 2.5 %, which is below the maximum error tolerated for the measurement of viscosity using conventional rheometers (below 5%). Rheological data obtained for several model and complex fluids with conventional rheometers (cone-and-plate geometry) and mixer type rheometers (double walled Couette and several other geometries) are compared. Some of these data are shown on Fig. 4 for a water/caboxymethyl cellulose (CMC) solutions. As can be seen, the results extracted from torque/speed data with mixer type rheometers (helical ribbon, RH-2, and anchor geometries) are in excellent agreement with those obtained with a conventional cone-and-plate (CP) rheometer

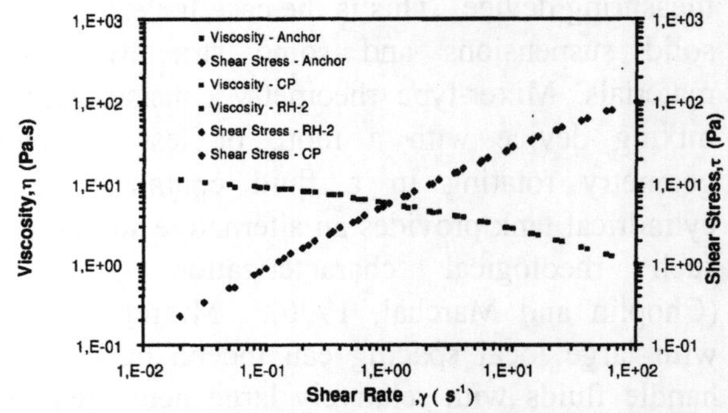

Figure 4 - Viscosity and shear stress data for a 2% CMC in water solution.

Other specific features related to Couette analogy and mixer type rheometry will be discussed.

III. Conclusions

The Couette analogy used here to quantify torque/rotor speed for mixer type rheometers is found to be very effective. The approach was successfully used with several geometries and several model and complex fluids.

References

Bird, R.B., R. Armstrong and O. Hassager, *Dynamics of Polymeric Liquids - Vol. 1, Fluid Mechanics*, 2nd Edition, John Wiley and Sons, New York (1987).

Bousmina, M. A. Ait-Kadi and J.B. Faisant, Submitted to AICHE, (1998).

Choplin, L. and P. Marchal, Proc. XIIth Int. Congr. on Rheololgy, edited by Ait-Kadi et al. , 665-666 (1996).

RHEOLOGICAL MEASUREMENT DURING THE SETTING OF HYDRAULIC CONCRETE BY ACOUSTIC TEST

L. ARNAUD, S. THINET and C. BOUTIN

Ecole Nationale des Travaux Publics de l'Etat, Labo. Géomatériaux, rue Maurice Audin, 69518 Vaulx-en-Velin Cedex France
E.mail : laurent.arnaud@entpe.fr ; sandrine.thinet@entpe.fr ; claude.boutin@entpe.fr

1. Experimental device and tested material

The control of the behaviour of fresh hydraulic concrete is an actual challenge with interest for both scientists and industrials, e.g. for optimising the time of casing or the use of new additives.

In this paper, setting and hardening of hydraulic concrete is studied using of non destructive vibratory tests. The main characteristic is the frequency range. Small wavelength as compared to heterogeneity size are diffracted and so strongly attenuated. Then, either the sample must be desaerated [Sayers, Dahlin (1993)] or the generated signal must present a large frequency range [Reinhardt, Grosse (1996)]. We avoid this difficulty by using long wavelength achieved with low frequency (0.1-10kHz) [Boutin, Arnaud (1995)]. Transient plane compressive waves are emitted in a sample of material. Pressure transducers (P3,P2) plunged into the paste detect the signals (fig.1). The large variation of the compressibility when the material evolves from a fluid to a solid state is followed up. At the beginning, due to the gas bubbles trapped during the mixing, the fluid paste is very compressible. With the hydration of cement, the paste turns to a solid state : the compressibility strongly decreases. Celerity C and damping coefficient of the wave are obtained from the time arrivals and the amplitudes of the pressure signals P3/P2 [Villain (1997)].

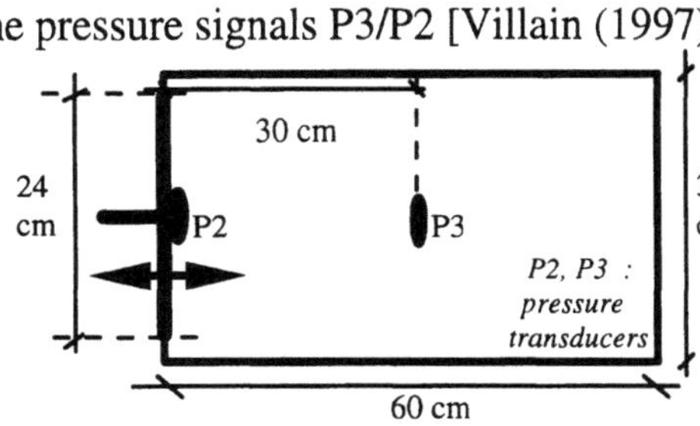

Figure 1. Experimental device

The studied concrete formulation is usual. The cement is a CPA-CEM I 52,5 and the granulates are 5/20 and rolled. The water/cement ratio is 0,41 and the granulates/sand ratio is 1,5.

2. Behaviour law and inverse analysis

The test is non homogeneous, so the rheological parameters are obtained from the measurements by an inverse analysis. A theoretical study based on the homogenization method leads to the linear behaviour law of the macroscopic equivalent medium. The concrete is considered as a suspension of air bubbles in a viscoelastic paste (water, cement and granulates). The behaviour law of the homogenized material describes a compressible viscoelastic medium [Villain (1997)] :

$$\sigma = (K + aM^*).\mathrm{div}(u)\, I + 2bM^*.\, \varepsilon(u)$$

where σ represents the macroscopic stress tensor, ε the deformation tensor, u the macroscopic displacement, M^* the viscoelastic modulus of the interbubble paste, K the elastic compressibility of the air bubbles in concentration ϕ in the paste. In adiabatic condition, $K = \gamma P/\phi$ (γ=1,4 and P is the atmospheric pressure). a and b are real shape factors. It is noticeable that even under compressive waves, the ***shear*** properties of the interbubble paste ($\mathrm{Im}(M^*)$) are tested. From the measurements, we obtained the oedometric modulus $K+(a+2b)M^*$ and the viscoelastic modulus $(a + 2b)M^*$.

The hypothesis of the linear behaviour is experimentally verified. For proportional solicitations (Ai/Ao), normalised pressure responses (Pi/Po) are plotted (fig 2).

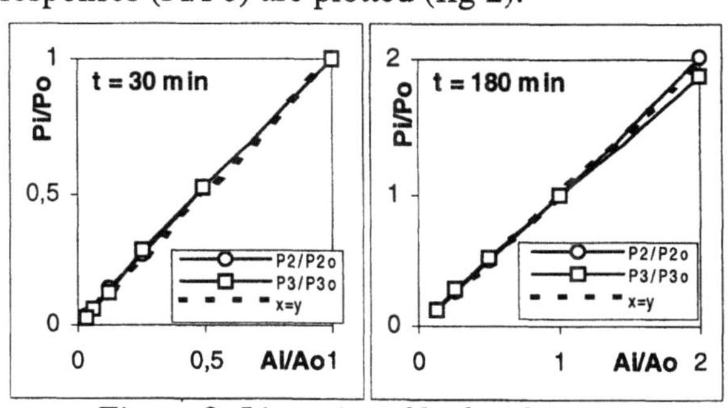

Figure 2. Linearity of hydraulic concrete

For both fresh and hardened concrete, the curves are aligned on the x = y line. The material behaves linearly and so, the test is non destructive.

Two experiments have been done with the same initials conditions. The evolution of celerity shows clearly that the test is repeatable for this material (fig 3).

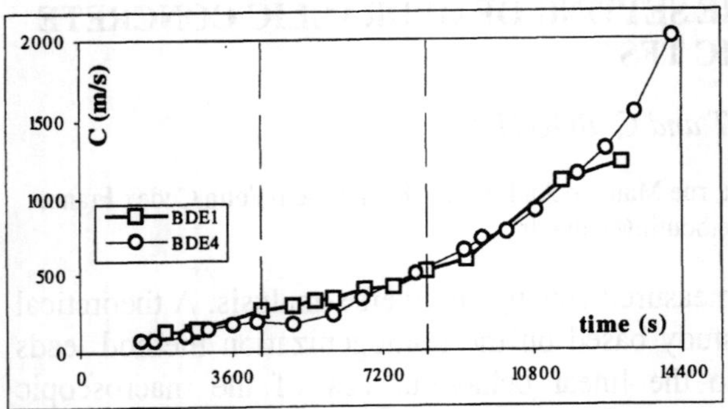

Figure 3. Evolution of celerity versus time

The typical evolution of celerity shows an increase from 100 to 2000 m/s (x 20) and presents two slopes. During the two first stages, the increase is slower. After the celerity increases strongly for 1.5 hours.

The ratio P3/P2 decreases at the beginning, so the damping increases. After 1.5 hour, P3/P2 goes up strongly. The increasing of P3/P2 is slower during the third stage and the ratio reaches 0.75 at the end of the test (fig 4).

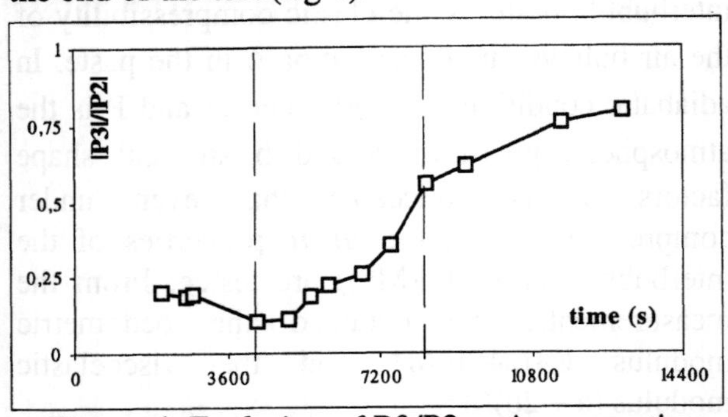

Figure 4. Evolution of P3/P2 ratio versus time

The evolution of the viscoelastic modulus (modulus (a), phase φ (b) versus time) is presented (fig 5.). The norm increases slowly during the first period, a little faster during the second one, and stronger after (x 5). The phase increases rapidly from 45 to 75° during the first hour and then decreases fastly. It remains constant (≈30°) during the last period.

At the beginning, we observe C=100m/s . From the inverses analysis, C is given by $(|K+(a+2b)M^*|/\rho)^{1/2}$. In case of K>>M*, $C=(K/\rho)^{1/2}$=80m/s for ϕ =1%, which is closed to the measured value. The elasticity of the material is due to the compressibility of the air (K). During the first stage, the hydration of cement leads to the creation of hydrates in the interbubble paste. The hydrates are not yet connected. They lead to increase the viscosity of the interbubble paste, so we observe an increase of φ. |(a+2b)M*| remains weak and so C does not increase strongly. During the second stage, |(a+2b)M*| increases a little and φ decreases hardly. K≈M* : both the air and the paste contribute to the elasticity of the material. The transition between the two first phases can be related to the percolation threshold that can define the setting time of concrete. In the third stage, the phase φ decreases by the way that |(a+2b)M*| increases dramatically. The interbubble paste is becoming solid (M*>>K). A solid skeleton is building. It is the hardening of concrete.

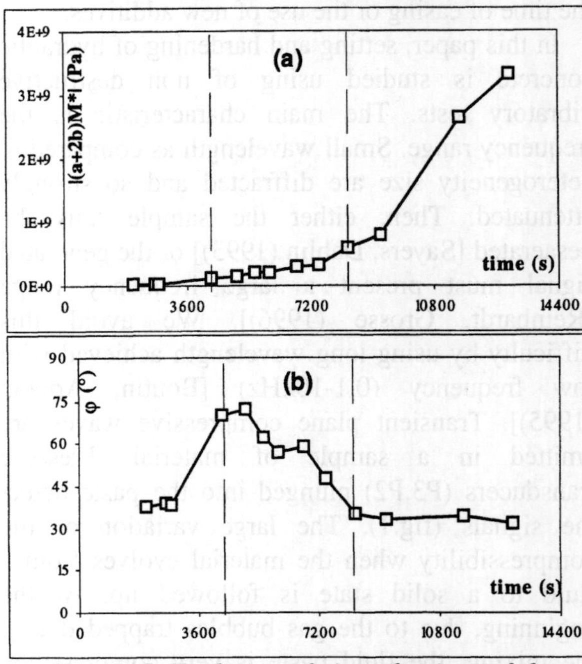

Figure 5. Evolution of viscoelastic modulus

6. Conclusion

The vibroscope allows to perform rheological measurement during the setting of hydraulic concrete in laboratory. We can obtain on the curves the setting time of concrete. We followed up the mechanical characteristics which are related to the chemical reactions in the material and then to the microstructure evolution. Influence on the setting of some parameters as the water/cement ratio, the casting temperature, admixtures...can be studied too. This experimental device can be adjusted to test some others time-dependant heterogeneous materials during their mechanical evolutions in civil engineering or in order fields.

REFERENCES

BOUTIN C., ARNAUD L., (1995*), Eur. J. Mech., A/Solids, 14 , n°4, pp 633-656.*

REINHARDT H.W., GROSSE C., (1996), *Proc. of the Int. RILEM Conference, Paisley, Scotland, pp 415-425.*

SAYERS C.M., DALHIN A., (1993), *Adv. Cement Bas. Mat., 1, pp 12-21.*

VILLAIN G., (1997), *Doctorat Thesis, INSA-ENTPE Lyon.*

SHEAR RHEOMETRY OF UNCURED RUBBER COMPOUNDS USING A SLIDING CYLINDER RHEOMETER

Claire BARRÈS, Jean L. LEBLANC

University P. & M. Curie, Polymer Rheology and Processing, Paris, France

1. INTRODUCTION

Rubber compounds are heterophasic materials that exhibit peculiar rheological properties such as yield stress, wall slip and flow induced anisotropy effects. Strong interactions between the elastomer and the filler, for instance carbon black, result in a complex morphology that obviously affects the flow properties. Current rheometric methods used with these materials, namely capillary rheometry and dynamic flow tests, deal either with relatively high shear rates (10^1 - 10^3 s^{-1}) that modify the rubber-filler morphology, or with low strain harmonic conditions that do not allow an accurate approach of phenomena like yield stress or anisotropy. The low shear region is of particular interest with such polymer systems because, in shear flow, their tridimensional soft morphology is expected to exhibit a specific response. In this respect, plate (or sandwich) rheometers have been used for experiments on filled rubber compounds in the low rate region[1,2,3] but plates parallelism is difficult to achieve and edge effects may become detrimental.

The purpose of this paper is to describe a sliding cylinder rheometer (SCR) that has been designed and constructed to test filled rubber compounds in the very low shear rate region.

2. EXPERIMENTAL

As shown in Figure 1, the sliding cylinder rheometer consists in two symmetric shells that determine a 26 mm diameter outer cylindrical cavity. An inner cylinder of 24 mm diameter allows a 1mm annular testing gap to be obtained. All parts are made out of hardened steel. At both ends of the inner cylinder, two rods are screwed. When the system is operated in creep mode, the upper rod moves a linear displacement transducer (LDT) and the weights necessary to induce the displacement are hung on a hook attached on the lower rod. To operate the system in constant speed mode, a tensile machine is used with the appropriate grips.

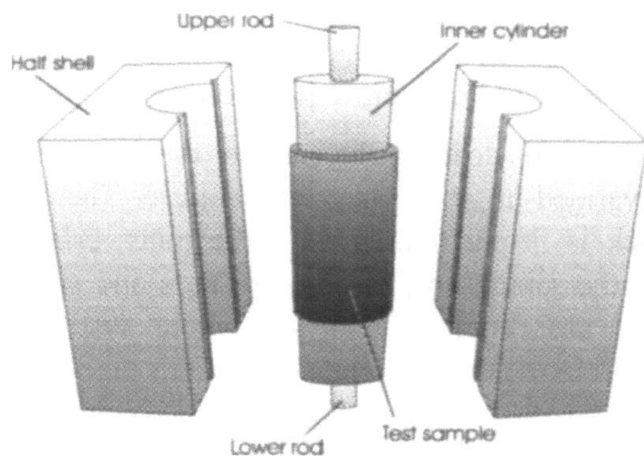

Figure 1 : Sliding cylinder rheometer

Test samples are small sheets of material, sheeted out on open mill and compression moulded to a thickness slightly higher than the annular gap so that the latter can be perfectly filled. When loading the system, the sheet is wrapped around the inner cylinder and, since it is cut as a parallelogram, the «welding line» is not aligned with the shear direction. The length L of the cylindrical sample is well controlled by means of a system of rings (see Figure 2) which simultaneously ensure the constancy of the gap and the axial alignment of the inner cylinder. Once the specimen is wrapped around the cylinder, the two rings (thickness 1 mm, length 65 mm, threaded over 25 mm) are positioned at top and bottom and the whole system is placed into the cavity of a half-block. The spacing between the rings (that will be the length L of sample after complete tightening) is adjusted by using annular spacers of calibrated height, so-called « struts ». Then nuts are screwed on top of each of those spacers until the desired spacing is obtained, and thus firmly maintain the struts.

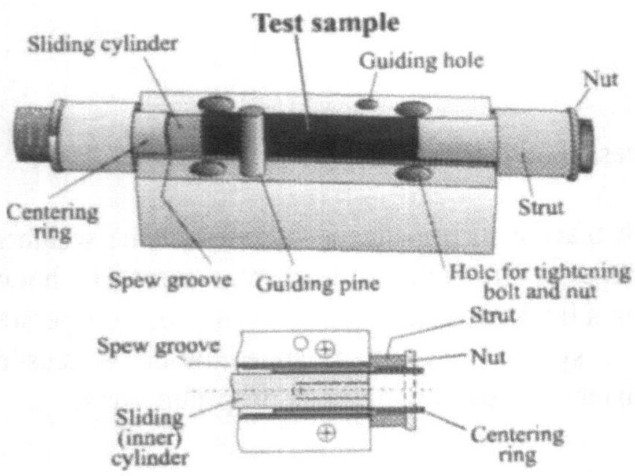

Figure 2 : SCR; sample positioning

After loading the sample, the full system is warmed-up at the test temperature. Then the block is firmly tighten with the inner cylinder and the sample in position, the nuts and spacer rings are removed and the whole system is placed in the oven on the appropriate supporting rack. At this moment, the specimen is perfectly in position and its length L is well controlled. The upper and lower rods are then screwed at both ends of the sliding cylinder. Note that the whole device is completely symmetric and can be operated either downwards or upwards.

Working equations for shear viscosity calculation are fairly simple and have been developed by Segel[4], i.e. :

$$\eta = \frac{K.F}{V} \qquad (1)$$

where V is the longitudinal speed of the inner cylinder under the application of a force F.
K a shape factor given by :

$$K = \frac{\ln\left(\frac{b}{a}\right)}{2\pi L} \qquad (2)$$

where L is the effective length of the sample in the annular gap, a and b respectively the diameters of the inner and outer cylinders.

Simple rubber formulations based on commercial high *cis*-1,4 polybutadiene (NEOCIS BR40, EniChem), N330 carbon black, and the usual quantities of zinc oxide, stearic acid, processing oil (Dutrex 729) and antidegradants (IPPD and TMQ) were tested in order (i) to assess the capabilities of the device as a low shear rate rheometer, (ii) to evaluate its versatility and (iii) to check the experimental reproducibility.

3. RESULTS

Experiments were performed at 100°C with the SCR, operated either in creep mode or in constant rate mode. For comparison, tests were also made at the same temperature with a Mooney viscometer and a capillary rheometer. With the latter, a series of dies with different L/D were used in order to apply the Bagley correction for entrance pressure loss. The Rabinowitsch correction for the non-Newtonian character was made in the usual manner. Figure 3 shows how SCR data compare well with results from other rheometrical techniques, allowing the viscosity function to be obtained over 9 decades of shear rate. As can be seen, no Newtonian flow region is observed with the tested material and the power law fits well the data over the whole shear rate range investigated.

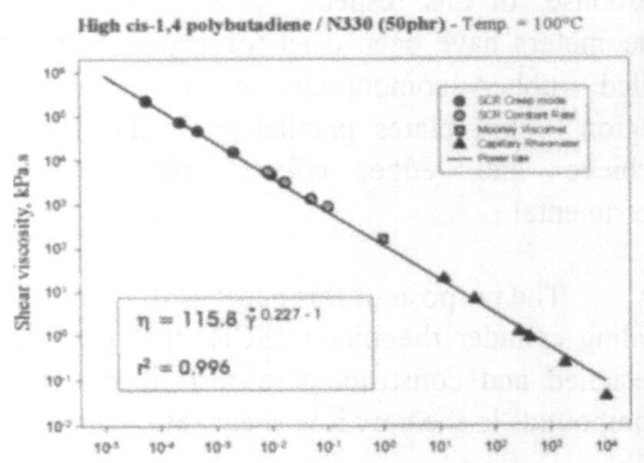

Figure 3 : Shear viscosity function

---

[1] ZAKHARENKO N.V., TOLSTUKHINA F.S and BARTENEV G.M. *Rubb.Chem.Technol.*, **35**(1), 26 (1962)
[2] TOKI S., WHITE J.L. *J. Appl. Polym. Sci.*, **27**, 3171 (1982)
[3] SHIN K.C., WHITE J.L. and NAKAJIMA N. *J. Non Newt. Fluid. Mech.*, **37**, 95 (1990)
[4] SEGEL M. *Physikalische Zeitschrift.*, **18**, 492 (1903)

A TECHNIQUE FOR MEASURING THE UNIAXIAL EXTENSIONAL VISCOSITY OF ELASTIC LIQUIDS

M. BARROW[1] AND P.R. WILLIAMS[1]

[1]Dept. Chemical Engineering, University of Wales Swansea, Singleton Park, Swansea SA2 8PP UK

1. INTRODUCTION

Extensional flows may be generated within a flow field that experiences a sudden change in geometry, such as contractions or orifice plates. The need for appropriate rheological data pertaining to such flows represents a major obstacle to the development of improved process simulation, monitoring and control, and requires not only a knowledge of fluid shear viscosity, η_S, but also an appropriate measure of its extensional viscosity, η_E. The problematical effects of extensional viscosity in process engineering become evident when one considers that η_E may be several *orders of magnitude* higher than the corresponding shear viscosity in elastic liquids.

In marked contrast to measurements of shear rheological properties, such as apparent viscosity in steady shear, or of complex viscosity in small amplitude oscillatory shear, extensional viscosity measurements are far from straight forward. This is particularly so in the case of mobile elastic liquids whose rheology mitigates against the generation of well defined extensional flow fields. This paper reports current work which is intended to realise the basic rheometrical principles devised by Jones et al. (1986) which involve the generation of a uniaxial extensional flow by the imposition of a centrifugal force upon a sample that has experienced little or no prestraining and in which no direct measurement of extensional stresses is required.

2. EXPERIMENTAL

The test fluid consisted of 1000ppm of polyacrylamide dissolved in a mixture of glucose syrup and water, producing a "dilute" solution of polymer in the solvent. About 0.5cm^3 of liquid was placed on a cylindrical drum of diameter and length 2 cm driven by a variable speed electrical motor, so that it rotated about its axis which was horizontal. Photographs were taken of the fluid in its initial state and as it stretched into a filament as the drum rotated (Fig.1). Five different speeds of revolution of the drum were used: 6, 9, 12, 13, and 35 revolutions per second. The camera employed was a Kodak Ektapro system running at 2000 frames per second.

The behaviour of the solvent, compared with that of the glucose/water mixture solution was also investigated. Although no filament formed in this case, a portion of the fluid became detached, pulling a thin 'neck' of fluid (which was still attached to the drum) with it. This rapidly thinned and broke.

Three forces act on the filament as it rotates: (1) the centrifugal force (2) air drag (3) gravitational force (weight). Surface tension may be neglected. The centrifugal force always acts along the radial filament with air drag at right angles to it. The weight acts along the filament when the filament is vertical and acts perpendicular to the filament (cantilever) when the filament is horizontal.

Consider a small element dx of filament at x from the centre of rotation then the centrifugal force on it is $\rho A\omega^2 x dx$ where ρ is the density of the fluid, A the area of cross-section of filament, and ω the angular speed of rotation. The gravitational force is $\rho A g dx$. Integrating these forces over the filament whilst assuming A is constant along the filament at any instant and then taking the ratio gives

$$\frac{\text{centrifugal force}}{\text{weight}} = \frac{\omega^2 L}{2g}$$

where L is the length of the filament. The observed filaments grow from ca. 10 mm to 200 mm so the centrifugal force predominates over the weight in determining L as a function of time and of position.

The "radius" R is measured from the centre of the drum to the base of the filament from where the filament is of uniform cross section. It has been shown (Jones W., 1986) that the gravitational force on the filament is negligible compared with the centrifugal force and the air drag is always just about at right angles to the centrifugal force hence the centrifugal force is the sole force stretching the filament. The centrifugal force increases outwards along the length of the filament and integrating it over L gives the effective stretching force

$$F = \frac{M\omega^2}{2}(L+R)$$

where M is the mass of the filament. Hence at any instant

$$\eta_E(t,\varepsilon,\dot{\varepsilon}) = \frac{F}{A}\cdot\left(\frac{dL}{dt}\cdot\frac{1}{L}\right)^{-1}$$

where A is the instantaneous cross-sectional area of the filament and η_E is an apparent elongational viscosity dependent on time (history), total strain ε, and rate of strain $\dot{\varepsilon}$. Substituting $A = M / L\rho$ gives

$$\frac{\omega^2\rho}{2}\cdot dt = \eta_E \frac{dL}{L^2(L+R)}$$

Integrating yields

$$\frac{\omega^2\rho Rt}{2} = -\eta_E\cdot\left[\frac{1}{L}-\frac{1}{R}\cdot\ln\left(\frac{L+R}{L}\right)\right]...$$

A plot of (1/L)-(1/R)ln[(L+R)/L] as a function of time would be a straight line if η_E were constant, but, as expected, curves are obtained, the slopes of which at any time give the values of the elongational viscosity at that time.

For the filaments studied here the maximum Trouton ratio η_E/η_S, where η_S is shear viscosity, is about 200. The Trouton ratio (η_E/η_S) obtained for the solvent is in good agreement with the expected value (3) for a Newtonian fluid as is that for the elastic liquid in the limit of zero ε, and rate of strain, $\dot{\varepsilon}$ (see Fig.2). The advantage of this experiment over the usual fibre-spinning experiments is that a single sample is being stretched throughout the range of stretching, and the relation between the apparent extensional viscosity and the total strain and between apparent extensional viscosity and time can be estimated. As $\dot{\varepsilon}$ is nearly constant throughout the growth of the filament, with any increases in strain rate being towards the end of the experiment, there is some justification therefore in quoting results in terms of a single $\dot{\varepsilon}$ for each experiment.

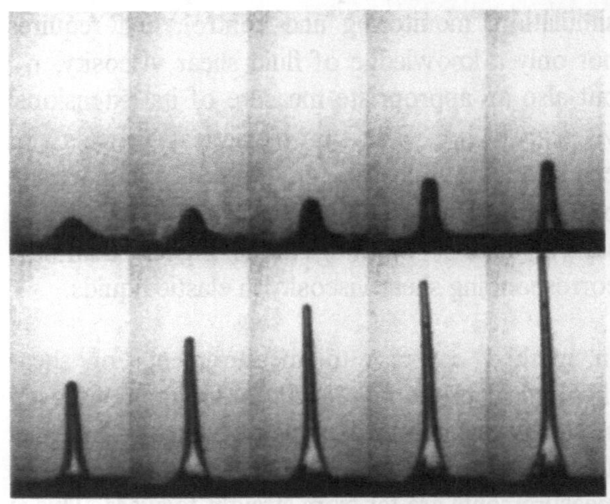

Fig.1 : Stages in the radial growth of a fluid filament upto 45 mm

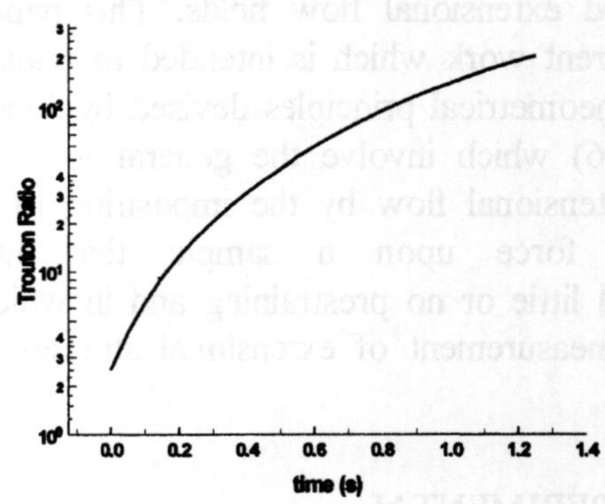

Fig.2 : η_E/η_S as a function of time

3. REFERENCES

Jones WM, Williams PR and Virdi TS (1986) J. Non-Newtonian Fluid Mech. 21: 51-64

QUALITY CONTROL BASED ON RHEOLOGICAL MOLECULAR WEIGHT DISTRIBUTION FINGERPRINTS

Hartmut Braun, Hans-Michael Petri and José-Maria Madjedo*
Gebrüder HAAKE GmbH, Dieselstraße 4, 76227 Karlsruhe, GERMANY
*Escuela Politecnica Superior, 21819 Palos de la Frontera, Huelva, SPAIN

INTRODUCTION

Conventional gel permeation chromatography (GPC) is a well-established method to measure the molecular weight distribution (MWD), but it is also known for its weaknesses, which are e.g. solubility of the polymer, expensive intruments, time-consuming procedures and less sensitivity for high molecular components.

To get the MWD of linear polymers using rheology seems to be a very interesting alternative. In principle it becomes realistic, because a mechanical excitation of polymers leads to a characteristic response of the material depending on its macromolecular structure. Based on the current state in this field, the application of a rheological method for MWD determination in quality control could be a next consequent step.

The aim of this paper consists in (i) comprehensive tests of a fast rheological experiment and (ii) in an analysis of the method with respect to the sensitivity limits to detect small MWD deviations.

THEORETICAL BACKGROUND

Following the idea of the so-called double reptation [1] which is based on binary chain contacts, one gets Eq. (1) with n=2 for the relaxation function G(t) of a mixture of monodisperse linear polymers with the MWD P(M). In accordance with our previous work [2] we prefere the Doi term for the relaxation behavior F(M,t) of the monodisperse components as well as an exponent n>2 in Eq. (1).

$$\frac{G(t)}{G_o}=\left[\int_{-\infty}^{\infty} P(M)F(t,M)^{1/n}\, d\ln M\right]^n \quad \text{with} \quad F_{Doi}(M,t)=\frac{8}{\pi^2}\sum_{\substack{p\\ odd}}\frac{1}{p^2}\exp\left(-\frac{t}{\lambda(M)/p^2}\right) \qquad (1)$$

In order to determine the desired MWD function P(M) in Eq. (1), one needs the necessary data for G(t) and a transformation equation $\lambda=aM^b$ between relaxation times λ and molecular weight M. We solved the ill-posed problem for P(M) using a non-linear regularization method [3]. From P(M) we further calculated the number and weight averaged molecular weight M_n and M_w. Both numbers are the basis to quantify the polydispersity M_w/M_n and all together give the so-called MWD fingerprint of the sample.

RHEOLOGICAL METHOD - MULTIPLE WAVE DYNAMIC EXPERIMENTS

For QC tests we need a less time-consuming method which, however, gives all material information necessary. Therefore, we used the multiple wave dynamic experiment (10 harmonics overlayed; 0.01 ... 10 Hz = 3 decades in frequency) on a HAAKE RS150 rheometer. This method is also called Fourier-transform spectroscopy, because the response of the overlayed dynamic input signals is analysed by the Fourier analysis. In this way, we obtained G' and G" *curves at once* in a minimum time period of less than 2 minutes.

In all cases we used parallel plate fixtures with 20 mm diameter. Measuring conditions are 185°C with nitrogen environment for PMMA and 25 °C for polyisobutylene.

The relaxation function G(t) of the samples, which is used in Eq.(1), was not separately measured but calculated according to Eq.(2) from the measured dynamic data G' and G" via the relaxation time spectrum H(λ). The spectrum was determined by PolySoft® - HAAKE's software package dedicated to polymer rheology.

$$G(t)=\int_{-\infty}^{+\infty} H(\lambda)e^{-\frac{t}{\lambda}}\,d\ln\lambda \qquad (2)$$

POLYMER SAMPLES

The materials under consideration are all narrowly distributed: polymethylmethacrylate (PMMA, M_w ranging from 20.000 to 150.000 g/mol and polydispersities M_w/M_n from 1.27 to 1.11) and polyisobutylene (PIB, M_w ranging from 42.000 to 93.000 g/mol and polydispersities M_w/M_n from 1.17 to 1.26). In addition to the pure components we also investigated some blends of both types.

RESULTS AND DISCUSSION

The dynamic data obtained in the very fast multiwave experiment are given in Fig.1. Both curves, G' and G", consist each of 10 values and the numerical method is able to get a reasonably smooth MWD curve given in Fig.2.

We find two main aspects that have to be taken into account in order to get satisfactory results: (i) Three frequency decades in the multiwave experiment correspond to a molecular weight 'measuring window' of only about 1 decade. Therefore, the maximum polydispersity which can be detected within this range is limited. (ii) On the other hand, it is necessary to measure just the complete transition range (flow to plateau region) in the dynamic experiment to get the MWD just in this 'observation range'.

The molecular weight sensitivity of this method is limited both, by the remaining small oscillations of the numerical solution as well as by the relative difference of the molecular weights between the main and a second component.

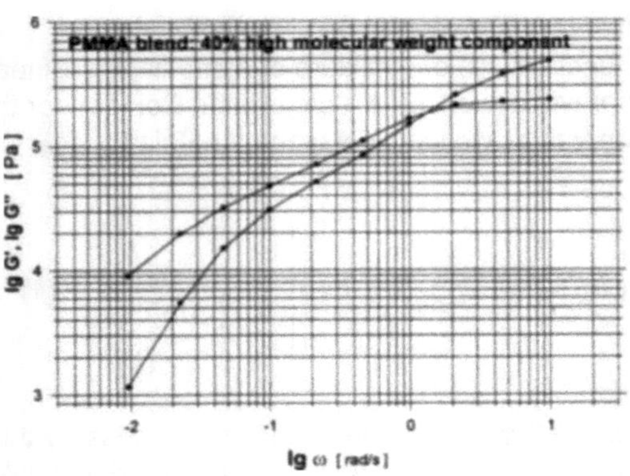

Fig. 1: dynamic multiwave data G' and G"

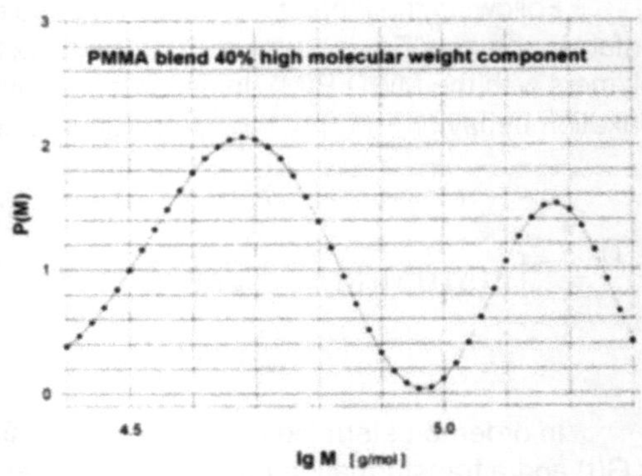

Fig.2: MWD P(M) 'observation window'

CONCLUSIONS AND OUTLOOK

The rheological method based on Tsenoglou's advanced theory of mixing behavior of linear polymers, including Doi's generalized relaxation functions in the integral kernel, gives encouraging results also for quality control applications in the MWD field. Some disadvantages of the GPC method can be avoided by the rheological MWD characterization. It should be emphasized that the determination of the high molecular weight tail is totally unmatched by GPC.

The multiple wave dynamic experiment is well suited for fast QC information. Together with the theoretical method it can really be used to get a quick MWD fingerprint. However, some aspects have to taken into account: The method works only for linear, non-cristallized polymers and a completely measured flow and plateau region in G' and G". In addition to that, there are few sensitivity limits to distinguish between two similar samples to make a QC assessment.

Summarizing the first results one can see that quantitative QC-predictions for the MWD function can be obtained from short-term rheological experiments. This is one way to increasingly make rheology to an analytical physical-chemical quality-control tool in addition to its conventional use.

REFERENCES

1. Tsenoglou C., ACS Polym. Preprints, **28** (1987) 185
2. Braun H., Friedrich Chr., Maier D., Proc. XIIth Int. Congr. on Rheology, Quebec 1996, p.75
3. Weese J., Comput. Phys. Commun., **69** (1992) 69

FLOW BIREFRINGENCE OF SEMIDILUTE POLYMER SOLUTIONS

S. COSTEUX, M. DJABOUROV, J. C. CHARMET

Laboratoire de Physique et Mécanique des Milieux Hétérogènes, UMR 7636,
ESPCI, 10 rue Vauquelin - 75231 Paris Cedex 05 - France

1. INTRODUCTION

Shear-induced orientation of viscoelastic materials usually leads to an anisotropy of the refractive index of the sample. For a polymeric solution, the extent of orientation depends on its nature —chain rigidity, presence of charges for polyelectrolytes— and on its structure —entanglements, conformation etc. Flow birefringence then reveals a suitable technique to study these phenomena.

2. EXPERIMENTAL DEVICE

A polarimetry laser chain was adapted on a HAAKE RS100 constant stress rheometer. It consists of a modulation stage —polariser and rotating half-wave plate— and an analysis stage — circular polariser and detector. The modulated laser beam enters the 1 mm gap between the concentric cylinders of the Couette cell through the 57 mm optical path. Amplitude and phase of the detected signal give access to the macroscopic birefringence Δn and to the average orientation angle χ of the sample. The rheogram (stress τ or viscosity η versus shear rate $\dot{\gamma}$) is measured simultaneously.

3. SAMPLES

Three different polymer samples were investigated, two synthetic polymers (PEO and PSS) and one biopolymer (xanthan gum).

Poly(styrene sulfonate) (PSS Mw=10^6g/mol), from *National Starch and Chemical*, was dissolved at two concentrations (5% and 10% w/w), close to the entangled regime [1].

Poly(ethylene oxide) (PEO Mw=5.10^6 g/mol) was obtained from *Polysciences*, and was studied at 20°C at two concentrations (0.5% and 1% w/w) corresponding to the entangled regime [2].

These samples were dissolved into demineralised water with no added salt, and the solutions were filtered on Millipore 1 μ before the experiments.

The xanthan gum was obtained from *SKW* (Mw=4.10^6 g/mol) and dissolved into 5.10^{-3} M NaCl at 0.1% (w/w). This concentration was assumed to be in the semidilute regime [3].

4. RESULTS

4.1 PSS and PEO

Experiments on the two PSS samples were carried out at 20°C and 40°C : the extinction angle χ was measured versus shear rate $\dot{\gamma}$ and shows a decrease from 45° at $\dot{\gamma}=0$, corresponding to random orientations, to 37° for $\dot{\gamma}=10^3\ s^{-1}$, which corresponds to a very weak orientation effect. The theory of flow birefringence states that [4]

$$\lim_{\dot{\gamma}\to 0}\frac{d\chi}{d\dot{\gamma}}=-\frac{1}{12D_r},$$

where the rotational diffusion coefficient D_r can be determined for each concentration and temperature. Extinction angles can then be plotted using normalised shear rates $\dot{\gamma}/D_r$, leading to a superimposition of the four curves (Fig. 1).

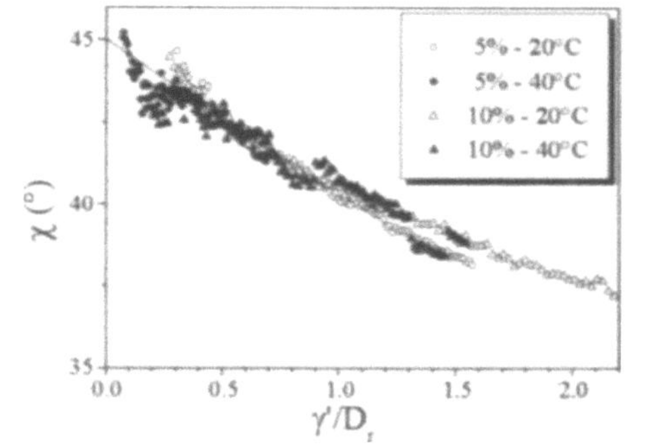

Fig. 1: Extinction angle versus normalised shear rate for PSS

Another classical representation of birefringence data is the stress-optical rule: [5]

$$\Delta n.\sin(2\chi)=2C\tau$$

We found that all curves superimpose (Fig. 2), therefore the value of the stress-optical coefficient (C=0.82.10^{-8} Pa^{-1}) appears to be weakly dependent on concentration and temperature. According to the molecular theory in dilute solution, it is expected to depend only on the anisotropy of electronic polarisability of the monomers and on the chain conformation in solution.

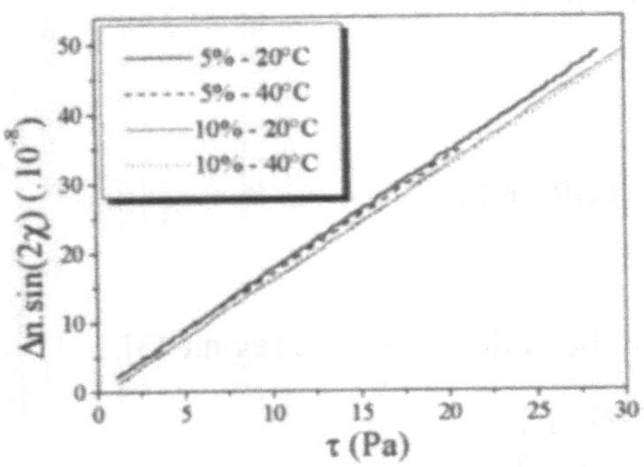

Fig. 2 : Stress optical rule for PSS

The same experiment was carried out for both PEO solutions at 20°C. The extinction angle displays a very steep decrease with shear rate, indicating a large deformation and orientation of the chains under shear. The normalisation of the shear rate with the diffusion coefficient still leads to superimposition (Fig. 3), as for the stress-optical rule, which gives C=0.10.10^{-8} Pa^{-1} (namely an extremely small value).

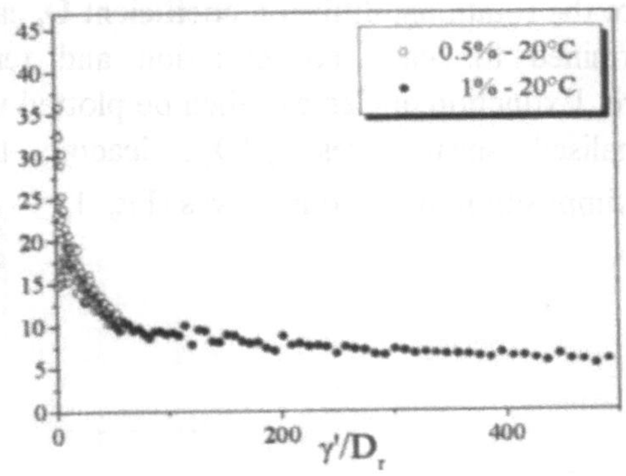

Fig. 3 : Extinction angle versus normalised shear rate for PEO 0.5% and 1% at 20°C

4.2 Xanthan gum

The same experiments have been carried out for 0.1% xanthan solutions, and the ionic force was adjusted in order to keep the polymer in its helical structure at 20°C and up to 40°C, in a semi-rigid rod conformation. The viscosity displays a very large shear-thinning behaviour accompanied by a strong shear-induced orientation, and a huge birefringence (C=130.10^{-8} Pa^{-1}). The extinction angle was then plotted for three temperatures versus normalised shear rate (Fig 4a). It can be noted that using the diffusion coefficients derived from $\chi(\dot\gamma)$, the plots $\eta/\eta_0 = f(\dot\gamma/D_r)$ also superimpose (Fig. 4b), which shows the correlation between molecular orientation and rheological behaviour under flow.

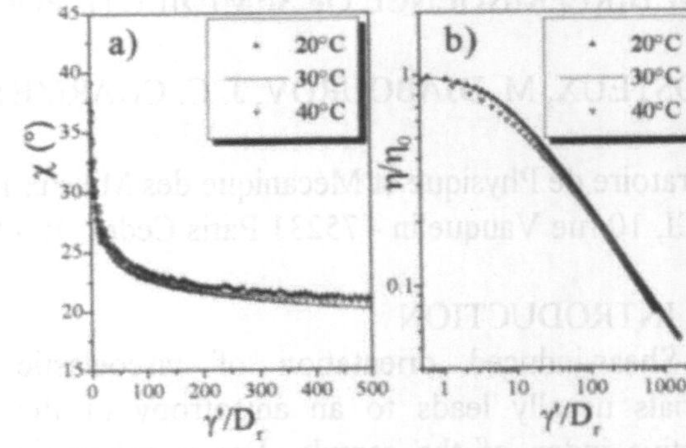

Fig. 4 : Extinction angle and viscosity versus normalised shear rate for xanthan 0.1% (w/w) in 5.10^{-3} M NaCl at 20, 30 and 40°C.

5. DISCUSSION

The flow birefringence experiments for PSS and PEO bring information upon the organisation of polymer chains in the semidilute regime.

For PSS, which is a strong polyelectrolyte, the orientation effect is very weak and the zero-shear viscosity is very low (75 mPa.s for 10% w/w) : if the chains were elongated due to electrostatic repulsion, we would expect a strong orientation and a large birefringence under flow, and a much larger viscosity. Therefore the PSS chains appear as collapsed coils avoiding entanglements, maybe due to the presence of unwanted salt in the sample. The situation is different for PEO which is a very flexible neutral polymer ; for 0.5% and 1% solutions, the chains are entangled and are readily oriented under flow. The birefringence remains very small due to the very weak optical anisotropy of the monomers.

Xanthan rods are very well oriented by shear, which explains why the high shear viscosity approaches water. Besides, C decreases slightly with temperature (from 130.10^{-8} Pa^{-1} at 20°C to 100.10^{-8} Pa^{-1} at 40°C, data not shown) due to progressive increase of flexibility of the helical rods.

6. REFERENCES

[1] V. Dobrynin, R.H. Colby, M. Rubinstein, *Macromolecules*, **28**, 1859 (1995)

[2] B. Cabane, R. Duplessis, *J. Physique*, **43**, 1529 (1982)

[3] I. Capron, G. Brigand, G. Muller, *Polymer*, **38**(21), 5289 (1997)

[4] R. Cerf, H.A. Scheraga, *Chem. Revs.*, **51**, 185 (1952)

[5] H. Janeschitz-Kriegl, *Adv. Polym. Sci.*, **6**, 170 (1969)

ELONGATIONAL RHEOMETRY WITH PRE-SHEARING HISTORY

[1]J. CRÚZ-MENA, [1]F.SERRANÍA, [1]B. MENA.

[1]Instituto de Investigación en Materiales, UNAM, Apdo. Postal 70-360, México, D.F. 04510, MÉXICO
[2]Instituto Mexicano del Petróleo, Apdo. Postal 14-805, México, D.F. 07730, MÉXICO

1 SYNOPSIS

A shear-elongational rheometer has been developed to relate shear properties of complex fluids in a viscometric flow situation to simultaneous measurements on elongational flow. The viscometric flow imposes a pre-shearing history on the fluid, allowing for the systematic study of its influence on the extensional properties of various materials. Measurements are presented for Newtonian oils, viscoelastic polymeric solutions and a constant shear viscosity, elastic fluid of the Boger type.

2 INTRODUCTION

Progress in Rheometry has been as much a success, in providing experimental rheologists with shear viscometers of ever increasing accuracy and ease of operation, as a disappointment, as Walters (1992) has pointed out, with regard to the seemingly insurmountable difficulties of measuring extensional viscosities of complex fluids. The field is plagued by discrepancies in measurements of extensional properties of Non-Newtonian fluids, and perhaps most of them may be attributed to the deformation histories to which these materials are subjected on their way to the test sections, almost always involving some shearing.

3 EXPERIMENTS

The elongational viscometer described by Gama et al. (1992) aims at this very problem by subjecting the fluid to a Couette simple shearing regime as it enters the gap between two concentric cylinders and flows in a descending spiral. This provides the controlled pre-shearing motion, after which the liquid exits through a small hole at the bottom of a cone and plate arrangement.

A vacuum pump applies suction to the jet through a second, vertically aligned hole, and the tensile stress driving the elongation is measured by means of pressure transducers and a vacuum meter. As the filament deforms, its shape is recorded by a video camera and the image is digitised for later analysis, yielding the axial variation of the jet diameter. The rate of elongation of the fluid can then be calculated from knowledge of the volumetric flow rate, measured upstream of the entrance to the viscometric section.

Data obtained with polyacrylamide aqueous solutions under various pre-shearing histories yield reasonably constant rates of extension in the test section, ignoring end effects.

3.1 Numerical simulations

To provide further confirmation of the nature of the deformation regime in the test section, we performed numerical simulations of the whole process (including the upstream pre-shearing portion) by means of a commercially available finite element software package (Polyflow), specifically designed to deal with non-Newtonian flows. Space integration along selected streamlines yield axial velocity profiles from which the rate of extension follows readily.

Qualitative agreement is apparent, inasmuch as axial velocity profiles are evidently linear in the test section.

3.2 Elongational viscosity measurements

A well characterised Newtonian oil (Mobil SAE-30), with a shear viscosity of 3.6 poise, was employed for calibration. The elongational measurements reproduced the correct value of Trouton's ratio, which is remarkable for such a low value of shear viscosity. Next, we tested various aqueous solutions of polyacrylamide (Separan AP30), a viscoelastic fluid with well known properties, and a Boger fluid, Roshfran's 30, a transmission oil which, although not too elastic, does maintain a fairly constant value of shear viscosity within our range of deformation rates.

In order to be able to assess the degree of influence of the shearing history on the way the liquid responded to extension, each series of runs commenced with no shearing at all, and then step increments of the inner cylinder's rotational speed set up subsequent runs. In all cases, the tensile stress --calculated from a momentum balance on the filament-- increased with pre-shearing, as a function of axial distance. Similar results were obtained for theBoger fluid.

4 THEORETICAL ANALYSIS

Some theoretical relations between simple shearing and simple extension have been examined by Crúz-Mena et al. (1995), on the basis of a continuum analysis. By calculating the relative strain tensor and the deformation gradient tensor for each deformation regime, and by considering a simple integral constitutive equation for the extra stress, they have showed that the effect of shearing the fluid prior to elongation is to increase the tensile stress needed to perform the extension. A similar result follows from a KBKZ model. This is in agreement with the experimental results presented here.

5 RESULTS AND DISCUSSION

Experimental readings, as well as numerical simulations, indicate that this apparatus is indeed capable of generating a uniaxial extensional flow in the test section, along with varying pre-shearing histories. It thus provides a means of systematically studying the influence of the latter deformation regime on the extensional properties observed during the former.

The elongational viscosity consistently turns out to be greater for those liquids which were sheared prior to the elongation. An interpretation of such behaviour may follow from consideration of the hydrodynamics at the molecular level. As they travel through the seemingly eternal pre-shearing portion of the rheometer, the polymer molecules are dragged in the direction of the flow, uncoiling as a consequence of the velocity gradient. The stronger the flow is, the more pronounced the uncoiling of the molecules. By the time they reach the exit orifice, those which have been more strongly sheared are less liable to yield to the extension than those which still have some uncoiling capacity left in them.
These results are reminiscent of those reported by James et al. (1984), and more recently by Schümmer and Thelen (1988) and Gama et al. (1992).

6 ACKNOWLEDGEMENTS

B. Mena wishes to acknowledge the financial support of PAPIID project IN301189 of the National University of México and the financial aid from CONACYT. J. Crúz-Mena acknowledges the invaluable advice from Prof. B. Caswell, at Brown University, and Prof. R. Huilgol, at Flinders. The authors wish to acknowledge that a patent application on the system by the Universidad Nacional Autónoma de México is pending.

7 REFERENCES

Walters K (1992) Proc. XIth. Int. Congr. on Rheology, 1:16-23

Gama J., Nuñez F., Zenit R., von-Ziegler A. and Mena B. (1992), Proc. XIth. Intl. Cong. Rheol., 2, 917

Crúz-Mena J., Mena B. and Huilgol R. (1995), Proc. III Caribbean Congress on Fl. Dynamics, 2

James DF, Saringer JH, Welch AB, MacLean B (1984) Proceedings of the 3rd Int. Conf. on Drag Reduction, A.2-1-2

Schümmer P and Thelen HG (1988) Progress and Trends in Rheology, 192-195

VISCOELASTIC MEASUREMENT AT VERY LOW TEMPERATURE ON LUBRICANT SYSTEMS

P. D'ANTONA[1], A. VETTOR[1], L.MATTEI[2], A. BELLONI[1], M. VANELLI[1] AND B. DE CINDIO[3]

[1] Eniricerche S.p.A., via Maritano 26, 20097 San Donato Milanese (MI), Italy
[2] Agip Petroli- Euron, via Maritano 26, 20097 San Donato Milanese (MI), Italy
[3] Chem. Eng.ing & Mat.s Dept., University of Calabria, 87036 Arcavacata di Rende, (CS), Italy

1. INTRODUCTION

Rheological behavior of lubricant base oils is very important especially at low temperature because it can be very helpful in predicting the onset of possible motor failure. It is well known that oils that have passed official standard requirements may nevertheless show failures, depending on some particular conditions to which the engine may be exposed.

It was observed that during flow at low temperatures, according to Deysarkar and Clampitt, (1990) two different phenomena occur depending on the shear rate level: cracking at high shear rates and air-binding at low shear rates. The first phenomenon is due to the fact that oils show a flow-limited mechanism if, at the considered temperature, they are too viscous to flow at the rate needed to satisfy pump demand after engine start-up. On the contrary the air binding is a failure mechanism that occurs at low shear rates as a consequence of the air drag into the oil pump due to a partial or complete gelification of the oil.

The flow limited failures may be related to both oil viscosity characteristics and engine pumping system design, while air-binding strongly depends on the formed gel structure and therefore its strength and extension. From this point of view the thermal history to which the oil is subjected becomes a very important factor. Then to properly design lubricating oils as well as to test special additives, a careful rheological characterization is needed to predict and prevent gel formation. In this work the results of an oscillatory time cure test at 1 Hz are discussed as a suitable tool to characterize lubricants in terms of low temperature performance.

2. MATERIALS AND METHODS

The rheological tests were performed by means of a controlled rate rheometer (Rheometrics RMS 800, USA) equipped with a test cell consisting of two serrated parallel plates (ϕ =25 mm and ϕ =8 mm) to avoid surface slippage when the material solidifies. Samples were priory subjected to a strain sweep test to determine the linear range behavior, that resulted to be strongly influenced by the considered materials and by its aggregation state. Then a frequency sweep test was performed in the linear range at different temperatures by varying angular speed (ω) between 0.1 and 100 rad/s. The results are reported in terms of storage (G') and loss (G") moduli versus ω.

All the materials were also subjected to a time cure test. Samples were cooled down from room to glass transition temperature, where they were maintained for half an hour; then the sample was heated at a rate of 1°C / min. At three fixed frequencies, i.e. 0.1, 1 and 10 Hz, G' and G" moduli were measured as a function of temperature.

A series of three different classes of commercial base oils were tested: mineral, synthetic and non-conventional.

3. RESULTS AND DISCUSSION

All the tested samples showed two typical behaviors during the timecure measurements, which will be thereafter denoted as Case 1 and Case 2. In case 1, during the timecure test, the moduli G' and G'' showed a sharp step change at a well defined temperature (Cross-Over Temperature, Tco), where the storage modulus G' crosses the loss modulus G''. It is possible to consider that at a temperature higher than Tco the lubricant is "liquid-like" (G''>G'), while at a temperature lower than Tco the lubricant behaves as "solid-like" (G'>G''). Thus the observed sudden increase of the moduli at Tco, is associated with a liquid-solid transition during the cooling process, that may be ascribed to the crystallization of some paraffinic components. In fig.1 it is shown (for simplicity) only the value of the modulus G''.

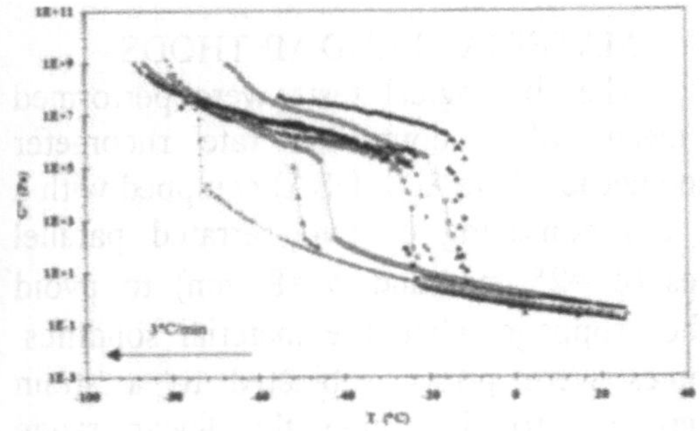

Fig.1: Time cure measurement for Case I.

The case II refers to a situation where the increase of the moduli G' and G'' as the temperature decreases is rather smooth, and the sharp change of the two moduli disappears. In this case we can consider that the lubricant becomes a gel as the temperature decreases, without a sharp liquid-solid transition; however, the definition of the Tco temperature is still possible, even if its value is very low. In Fig.2 it is reported G" for this case.

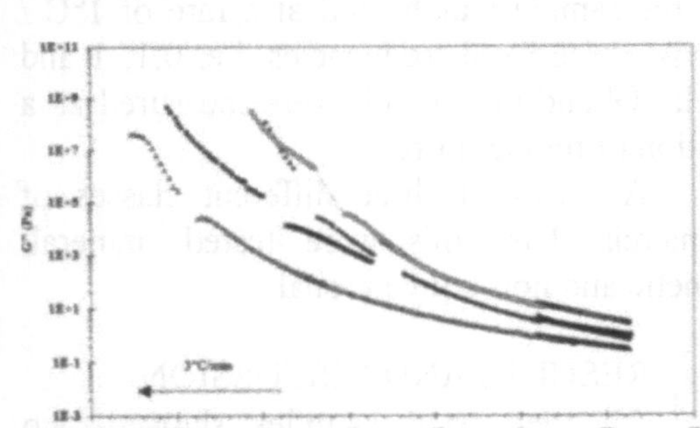

Fig.2: Timecure measurement of Case II.

All the studied base oils showed a rheological behavior that follows case I or case II. When the lubricant shows a sudden change of G" at Tco, its flow properties will also suddenly change near Tco, in spite to what showed by lubricants holding to case II. This result doesn't mean that low temperature performances of case I lubricants are worst than case II lubricants, because the G' and G'' moduli of these latter might be greater than the other, even if it is still liquid.

To better understand and quantify the flow properties associated to the structure at low temperature, D'Antona and de Cindio (1998) performed frequency sweep measurements around the Tco.

The time cure test may be used as a screening tool for the characterization of lubricants that were additivated with polymer based temperature depressants. In fig 3 a typical example is reported, where it appears that it is possible to modulate the amount of additive, by choosing the desired effect, in terms of decreasing Tco.

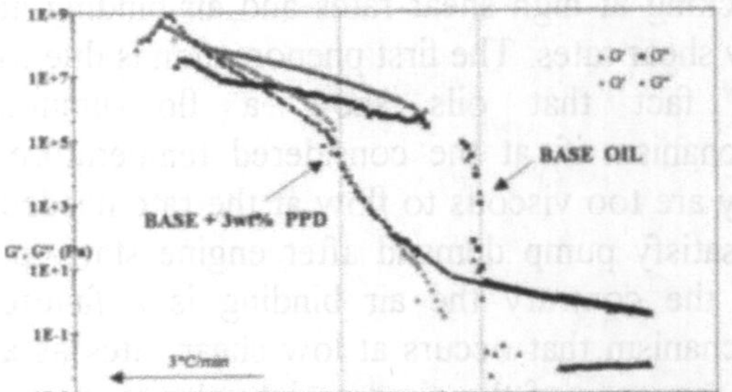

Fig.3: Effect of polymer additive on a base oil.

4. REFERENCES

D'Antona P. and de Cindio B. (1998), Rheol. Acta, submitted.

Deysarkar A.K. and Clampitt B.H. (1990), Lubric. Eng.ing 46:502-507.

HIGH PRESSURE VISCOSITY OF LUBRICANTS: EFFECT OF POLYMERIC ADDITIVES

P. D'ANTONA[1], A. VETTOR[1], L.MATTEI[2], A. BELLONI[1], M. VANELLI[1] AND B. DE CINDIO[3]

[1] Eniricerche S.p.A., via Maritano 26, 20097 San Donato Milanese (MI), Italy
[2] Agip Petroli- Euron, via Maritano 26, 20097 San Donato Milanese (MI), Italy
[3] Chem. Eng.ing & Mat.s Dept., University of Calabria, 87036 Arcavacata di Rende, (CS), Italy

1. INTRODUCTION

Lubricants in the automotive applications are very often subjected to very high pressures. Thus in order to properly design the lubricant, it is necessary to measure rheological and tribological properties at these extreme conditions, that are not usually investigated. For instance the values of the operative conditions in an elastohydrodynamic regime (EHD), are approximately T>100°C, pressure up to 1 GPa and shear rates ranging between 10^5 to 10^8 s^{-1}. During this regime, the lubricant properties will determine the film thickness, the load carrying capacity, the friction, etc., therefore its performance is essentially based on the the capability to reduce friction, viscous dissipation.

Thera are few instruments able to perform measurements at very high pressure with an acceptable reproducibility, so the experimental data are rather poor. One of the main obstacle is that when there are rotating or moving mechanism, the measurement becomes very difficult and often not possible to be performed.

2. MATERIALS AND METHODS

The viscosity of mineral oil lubricants was measured up to 1.4 GPa by means of a new quartz viscometer produced by Flucon Fluid Control GmbH (Germany).

Its functioning principle is based on the determination of the material response to a pure shear wave generated by the piezoelectric oscillation of a quartz crystal at high frequency (KHz). Of course the resulting shear rate is limited to a single value that was 125000 s^{-1}. The viscosity of the fluid is determined by measuring the damped resonance frequency of the crystal completely immerged in it.

The technique was applied to a typical mineral oil and to a mixtures obtained by adding polymeric additives used as Viscosity Index Improver (VII).

3. RESULTS AND DISCUSSION

A first comparison was done between mineral oil and a pure system like n-dodecane. (Fig. 1). It appears that in the case of the pure system a sharp change occurs at a well defined

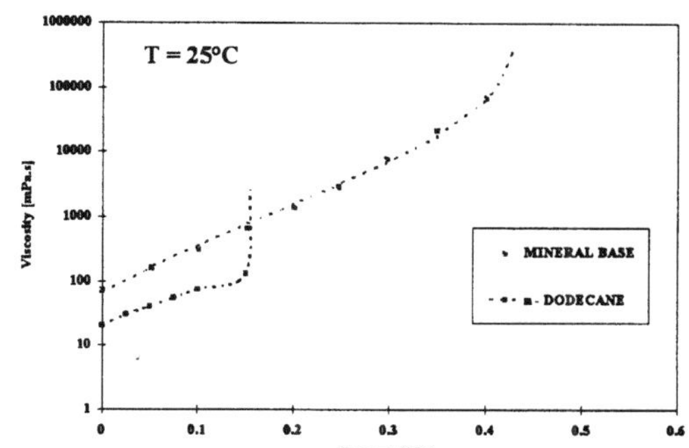

Fig. 1: Viscosity-Pressure plot.

pressure, that corresponds to a glass transition (Sargent 1982). When mixing VII additives the glass transition is shifted at higher temperature depending on the amount and kind of additive (Fig.2)

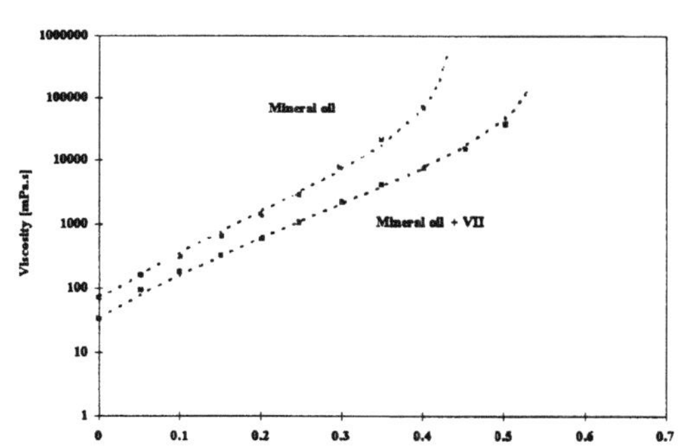

Fig.2 VII effect on glass transition.

The difference between several additives is enhanced only at rather high pressure (see fig.3). If we magnificate the high pressure

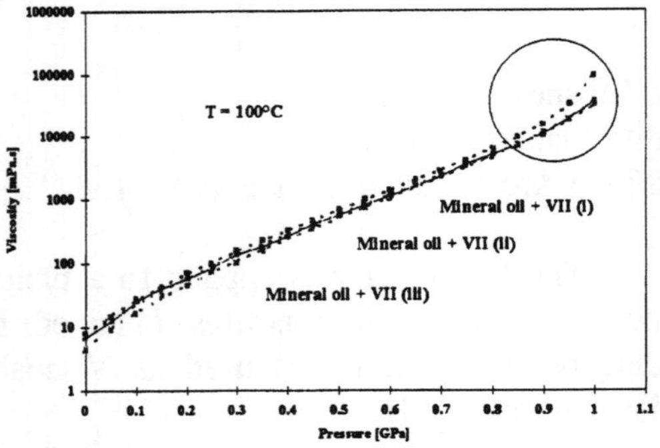

Fig. 3 Viscosity of differentVII-oil systems

region the effect of the the polymer additives becomes rather evident (fig.4).

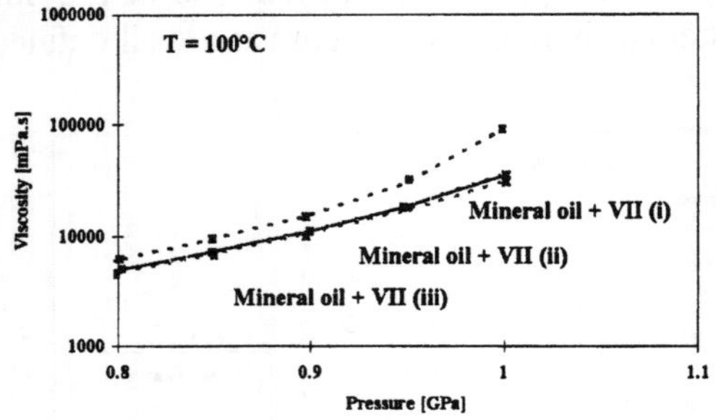

Fig. 4 High pressure zone

As concerning with the the fitting equations relating viscosity to pressure and temperature, it appears from the literature that the isothermal state equations do not predict the glass transition and hold only for low pressure. Difference between different models stands only on the amount of model parameters. Other equations like those proposed by Rodermund or Bair and Winer are capable to predict a well defined the glass transition as found in pure solvent. Therefore all these seem to be not valid for real systems as those investigated during this work. The only equation that is capable to predict a smooth behaviour is due to to Bode (1991) who proposed to include contemporary two effect glass transition and crystallisation. Our experimental data were in a good agreement with Bode model.

4. REFERENCES

Sargent L. B. (1982), ASLE Transactions, 26, 1: 1-10.

Bode B. (1984), Dissertation, TU Clausthal.

INVESTIGATION OF THE STRESS - INDUCED GEL - SOL TRANSITION IN POLYMERS USING PARALLEL SUPERPOSITION AND NORMAL STRESS MEASUREMENTS

B.A. de L. Costello

TA Instruments Ltd., Cleeve Road, Leatherhead, Surrey KT22 7UQ, U.K.

1. INTRODUCTION

Gelatin and other gelling agents have extensive uses as industrial rheology modifiers and food stabilisers. Their rheological properties have been much investigated over the years and are described in detail by Clark and Ross-Murphy. But it is conventional to discuss the thermal properties of these materials in the terminology of physical chemistry (phase diagrams), and their rheological properties in the terminology of rheology (shear-thinning, elastic). In this paper, shear-thinning is described as a stress induced sol - gel transition. The techniques of parallel superposition and normal stress measurement are used to provide qualitative and quantitative support for this notion, and to establish the non-equilibrium phase boundaries.

2. EXPERIMENTAL

2.1 Parallel superposition

The rheological techniques of steady shear and oscillation are well known. The former is used to provide quantitative viscometric information on the properties of materials, but does not usually allow qualitative interpretation. The latter can be used in a more analytical way, for example to identify thermally induced phase transitions, but its operation is usually restricted to the regime of linear viscoelasticity. In the technique of parallel superposition, the motions of steady shear and oscillation are combined. An expression for the shear stress at low amplitudes is provided by Moldenaers and Mewis:

$$\sigma_{xy}(\dot{\gamma}, \gamma_0, t) = \eta(\dot{\gamma})\cdot\dot{\gamma} + G_{\parallel}^{*}(\omega, \dot{\gamma}, \gamma_0)\,\gamma_0 \sin\omega t \qquad (1)$$

where γ_0 is the shear strain amplitude, $\dot{\gamma}$ is the shear rate, ω is the angular frequency of the superimposed oscillation, and t is time.

The complex parallel superposition modulus, $G_{\parallel}^{*}$, can be decomposed into in and out of phase components, $G_{\parallel}'$ and $G_{\parallel}''$ respectively. If the amplitude of oscillation is sufficiently small, then one motion will not interfere with the response of the other, and the oscillation can be regarded as a linear perturbation of the steady shearing motion.

Although data interpretation is not straightforward, this technique provides a way of identifying phase transitions which occur in materials undergoing shear.

2.2 Experimental details

In this study a series of gelatin gels was used. Data are presented here for a commercial hair gel. Parallel superposition experiments were conducted one each sample at temperatures from 20°C to 60°C inclusive, at 5°C intervals. Preliminary investigations established that the condition of linearity was observed for an oscillatory shear stress of 1 Pa, at all temperatures and values of superimposed steady shear. Frequency ramps were therefore conducted at this stress amplitude at each temperature, with superimposed stress of zero, 1 Pa, 2 Pa, and subsequent doubling until the instrument ran over speed.

The steady shear viscosity and superposition moduli are directly obtained from parallel superposition experiments. The normal stress is also reported; no correction has been made for the parallel plate geometry, but to a first approximation, the reported normal stress will be equal to the first normal stress difference, N_1, as shown by Barnes et al.

3 RESULTS AND DISCUSSION

Figure 1 shows $G'_{\parallel}$ against angular frequency on logarithmic axes, at 35°C over a range of superimposed shear rates. Data at the other temperatures are similar.

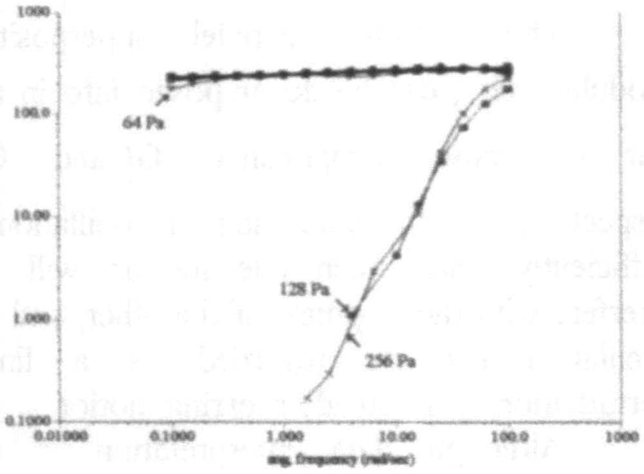

Figure 1: $G'_{\parallel}$ against angular frequency at 35°C

At superimposed shear rates of less than Pa, the shape of the curve is typical of a soft gel, as reported by Clark and Ross-Murphy. At 64 Pa, a slight divergence begins to occur at low frequencies, and at 128 Pa, complete phase transition has occurred. The shape of curve is now typical of a polymer gel.

Figure 2 shows $G''_{\parallel}$ against angular frequency for the same data set: once again the sharp stress induced phase transition is observed to occur at a shear stress between 64 Pa and 128 Pa.

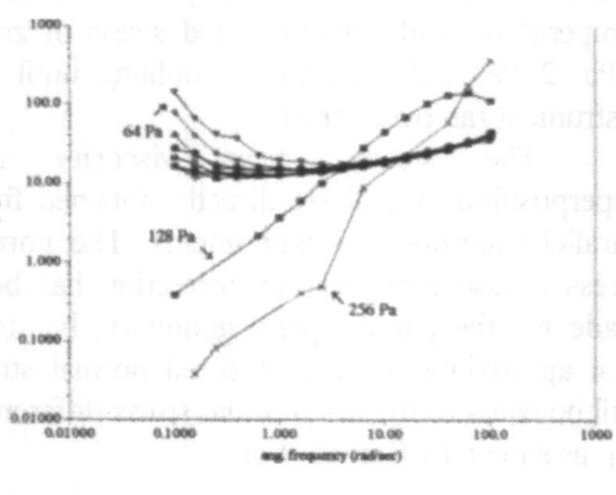

Figure 2: $G''_{\parallel}$ against angular frequency at 35°C

Figure 3 shows viscosity $\eta(\sigma)$ against shear stress (σ) on logarithmic axes: the scatter in the data arises from the viscosity being measured at each frequency. The onset of the transition at a shear stress of about 80 Pa can clearly be seen.

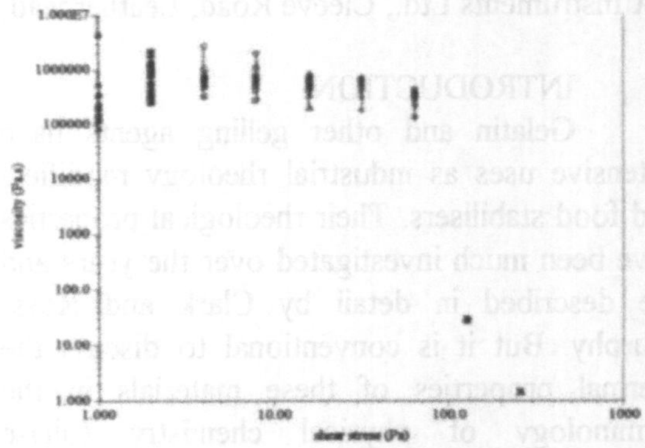

Figure 3: viscosity, $\eta(\sigma)$ against shear stress, σ

Figure 4 shows the same data plotted as viscosity, $\eta(\dot{\gamma})$ against shear rate, $\dot{\gamma}$. The material appears as a typical power law fluid with low shear Newtonian plateau.

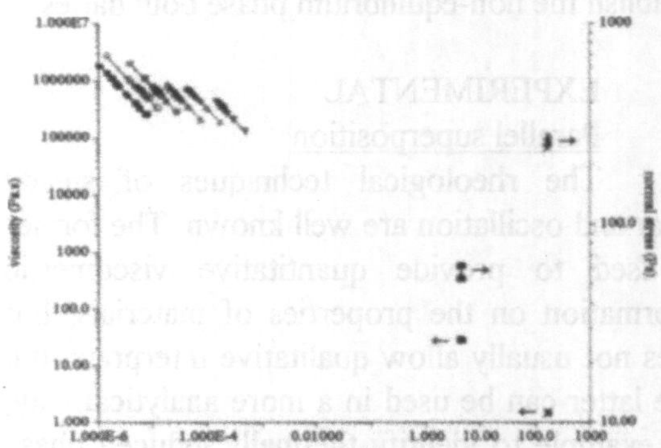

Figure 4: viscosity, $\eta(\dot{\gamma})$ against shear rate, $\dot{\gamma}$

4. REFERENCES

Barnes HA, Hutton J, Walters K (1996) An Introduction to Rheology, Elsevier, Amsterdam

Clark AH, Ross-Murphy SB (1987) Adv. Polym. Sci. 83:57-193.

Kao N, Battacharya SN, Shanks R, Coopes IH (1996) Colloid Journal 58:448-459

Moldenaers P, Mewis J (1993) J. Rheology 37:565-584.

SPECIAL CASE OF OSCILLATORY DIFFERENTIAL EQUATION FOR CAPILLARY

M. FIALKA, S. VAŠUT
Technical University Brno, Faculty of Technology in Zlín, nám. TGM 275, CZ - 762 72 Zlín, Czech Republic

1. INTRODUCTION

The solution of problems dealing with oscillation phenomena presents a substantial part of engineering applications[1-10].

The aim of the paper is to study the case, when a nonlinear[3] differential equation for reservoir-capillary system is brought into linear differential equation. Moreover, we require the linear equation to be oscillatory, i.e. any of its solutions is oscillating.

To analyze the problems we apply some results from the theory of differential equations. The above mentioned analysis can be an initial approximation to further studies of linear oscillation phenomenon when the influence of more parameters is considered.

The onset of the oscillatory flow corresponds to the flow conditions when the critical shear stresses or the critical shear rates appear. In certain polymers oscillatory flow can be observed which is characterized either I. by periodic oscillation of flow rate[1] when a critical value of pressure drop in the capillary rheometer is reached, or II. by periodic pulsation of pressure drop[9] when a critical mean velocity is reached. These cases appear in capillary rheometers which can be operated either ad I. by constant pressure drop in capillary, or ad II. in the constant piston speed mode. The remaining part of the paper deals especially with analysis of the case I.

2. PRELIMINARIES

We consider a second order nonlinear differential equation as a possible model for a sample class of molten polymer flows in capillary with reservoir

$$\alpha\,\ddot{x} + F(\dot{x}) + \gamma\, x = 0, \tag{1}$$

where constants hold

$$\alpha = \frac{1}{2} R\rho, \quad \gamma = \frac{3}{2}\frac{G_0 R}{L \cdot L_R}. \tag{2}$$

$F(\dot{x})$ is a nonlinear function of derivative of dependent variable x with respect to time t, $t \in J = \langle 0, T\rangle$, called system characteristic, R and $D = 2R$ are the radius and diameter of the capillary, respectively, L its length, L_R height of molten polymer in reservoir, ρ is density of molten polymer and G_0 its elastic modulus.
In the paper (cit.[3]) there were formulated conditions for all solutions of (1) to be oscillating.

Considering that L_R is a function of time t, i.e. $L_R = L_R(t)$, we obtain more general nonlinear differential equation

$$\alpha\,\ddot{x} + F(\dot{x}) + \gamma(t)\, x = 0, \tag{3}$$

where $t \in J$, $\gamma(t) > 0$ on J. Thus, the equation (3), general enough, can also include the case when $G_0 = G_0(t)$ is the relaxation modulus.

Let $L_0 = L_R(0)$ be the initial condition, and let $L_T = L_R(T)$, $T \in J$ be the final condition at time T when the observed oscillatory flow has been finished. Therefore the instant height of polymer melt in reservoir holds the weak inequality

$$0 \le L_T \le L_R(T) \le L_0 .$$

Suppose the characteristic $F(\dot{x})$ is a linear function of $\dot{x}$ in the next part of the paper, i.e.

$$F(\dot{x}) = \beta \cdot \dot{x}, \ \beta = const \neq 0, \tag{4}$$

β is proportionality constant. In this very special case we get the second order linear differential equation

$$\alpha \ddot{x} + \beta \dot{x} + \gamma(t)\, x = 0, \qquad (5)$$

where

$$\gamma(t) = \frac{3}{2} \frac{G_0 R}{L \cdot L_R(t)} \, . \qquad (6)$$

3. MAIN RESULTS

Let us briefly notice that trinomial equation (5) defined on sufficiently large time-interval $J = \langle 0, T \rangle$, and describing a linear polymer melt flow in reservoir-capillary system, can be transformed into the Jacobian form

$$\ddot{y} - Q(t)\, y = 0, \qquad (7)$$

where support

$$Q(t) = \left[\beta / (\rho R)\right]^2 - 3G_0 / \left[\rho L \cdot L_R(t)\right]. \qquad (8)$$

Relationship between a solution $x(t)$ of equation (5) and the solution $y(t)$ of equation (7) corresponding to $x(t)$ is given by formula

$$x = h(t)\, y,$$

where the function $h(t)$ is defined and positive on J, $h(t)$ can be specified.

Due to the above mentioned formula, the zero points of corresponding solutions of both equations (5) and (7) are coincident. The amplitudes of these solutions $x(t)$ and $y(t)$ are different, of course.

4. CONCLUSION

We derived the special second order linear differential equation of Jacobian type (7) with support $Q(t)$, keeping the oscillatory property of (7) that $|Q(t)|$ increases when the height of certain polymer melt $L_R(t)$ decreases. Therefore the frequency of linear oscillatory flow should increase with time and also depend on the velocity of the piston in reservoir at constant pressure drop. Both properties of frequency in our linearized case are also well known in a more general nonlinear case during measurements in capillary rheometers, if oscillation occurs.

Acknowledgement:

This paper was partially supported by grant FUZ no. 70002 of The Technical University Brno Fund for Sciences and Arts.

5. REFERENCES

1. Bartoš O, Holomek J (1971) Polym. Eng. Sci. 11,324-334

2. Feireisl E, Herrmann L (1992) Appl. Math. 37, 469-478

3. Fialka M, Vašut S (1997) Extended Abstracts, Eur. PPS Meeting, Göteborg, August 19-21, Sweden

4. Hatzikiriakos SG, Delay JM (1992) J. Rheol. 36, 703-741

5. Herrmann L (1980) Czech. Math. J. 30, 177-206

6. Herrmann L (1992) Appl. Math. 37, 369-382

7. Laitochová J (1982) Acta Univ. Palacki. Olomuc., Fac. Rer. Nat., Math. 21, 63-67

8. Larson RG (1992) Rheol. Acta 31, 213-263

9. Pavlínek V, Sáha P et al. (1996) Proc. 12th PPS Ann. Meeting, Sorrento, May 27-31, Italy

10. Švábík J, Maňas M et al. (1997) Extended Abstracts, Eur. PPS Meeting, Göteborg, August 19-21, Sweden

STRUCTURAL INFORMATION FROM RHEOLOGICAL DATA WITH A NEW SOFTWARE EVALUATION PACKAGE

A.J.P. FRANCK[1], T. VAN DYKE[2], D. MEAD[2]

[1]Rheometric Scientific, Schwanheimer Str. 144a, D-64625 Bensheim
[2]Mechanical Engineering and Applied Mechanics, University of Michigan, Ann Arbor, Michigan, USA

Analytical rheology is the determination of the polymer structure from rheological measurements. The determination of MWD from linear viscoelastic data is of major interest and economic importance.

Since there currently exists many viable methods of determining the molecular weight distribution of flexible polymers such as gel permeation chromatography, light scattering, intrinsic viscosity etc., it is important to appreciate what specific advantages, a MWD determination method based on rheological data provides. Virtually all conventional MWD methods are predicated on the ability to readily dissolve the polymer in a solvent at ambient temperatures. However, many polymers of commercial importance such as Teflon, polyethylene and polypropylene are only slightly soluble if at all in common solvents at ambient temperatures. Rheological methods do not require this time consuming solvating step.

Even when conventional MWD methods can be made to work for these polymers, the sensitivity and resolution of the analytic techniques is poor especially for the high molecular weight tail. As a matter of fact, characterization of the high molecular weight tail is particularly important for characterizing the processability of a commercial polymer. Rheological methods of determining the MWD are intrinsically sensitive to the high end of the molecular weight distribution.

1. MIXING RULES FOR POLYMER MELTS

The quantitative relationship that relates the observed mechanical properties of a polydisperse melt and the underlying polymer-structure is called a mixing rule. The mixing rule is a relatively simple mathematical approximation to a more complex molecular theory of polydispersity. The double reptation mixing rule (Tsenoglou, Des Cloizeau), used in the ^{RSI}*Orchestrator* has the following mathematical structure:

$$G(t) = G_N \left[\int_0^{\infty} F^{\frac{1}{2}}(M,t) W(M) dM \right]^2 \qquad (1)$$

The function $F^{1/2}(M,t)$ is the monodisperse relaxation function which represents the time dependent fractional stress relaxation of a monodisperse polymer. $W(M)$ is the weight based molecular weight distribution. The magnitude of each component's contribution to the stress will depend on the details of the interaction with the other molecules in the molecular weight distribution.

2. MATERIAL DEPENDENT INPUT PARAMETERS

To calculate a molecular weight distribution for a given material using (1) it will be necessary to supply material dependent data to the application. Specifically, the plateau modulus G_N and the form of the monodisperse relaxation function $F^{1/2}(M,t)$ must be supplied. The plateau modulus is tabulated in several references. The monodisperse relaxation function can take several forms. Generally we shall use a single exponential form,

$$F^{\frac{1}{2}}(M,t) = \exp\left\{\frac{-t}{2\lambda(M)}\right\} \qquad \text{where}$$

$$\lambda(M) = K(T) M^x \qquad (2)$$

Here, λ (M) is the characteristic relaxation time for the monodisperse system and $K(T)$ is a coefficient that depends on temperature. The exponent x is typically ~3.4 for flexible polymers.

3. CALCULATION OF THE RELAXATION MODULUS (Mead)

In determining the relaxation modulus to calculate a molecular weight distribution there are two distinct experimental situations possible: 1) Complete dynamic moduli data and 2) incomplete dynamic moduli data. A data set is "complete" if the data span a dynamic time/frequency range from fully terminal behavior through transition to the glassy modes. In practical situations involving routine characterization of commercial polymers with broad molecular weight distribution, a far more common occurrence is to have an "incomplete" data set. In this case, either fully terminal behavior is not achieved because of the exceptionally long relaxation times of the high molecular weight tail or the high frequency regime could not be fully.

4. COMPLETE DATA SETS

When dynamic moduli data are available from the terminal region through the plateau region, all rheological information required to invert the mixing rule is present. The molecular weight distribution is calculated using regularization methods applied to moments of the MWD calculated from the experimental data.

5. INCOMPLETE DATA SETS

When the experimental data does not span the entire frequency range from the terminal to the plateau region, the data set is incomplete and the double reptation model cannot be rigorously inverted. The method we adopt is to assume a priori knowledge about the shape of the molecular weight distribution.

Effectively the numerical method fits the predicted dynamc moduli curve from model molecular weight distributions to the measured experimental data in an optimal manner. A variety of model molecular weight distributions is allowed as well as binary combinations thereof.

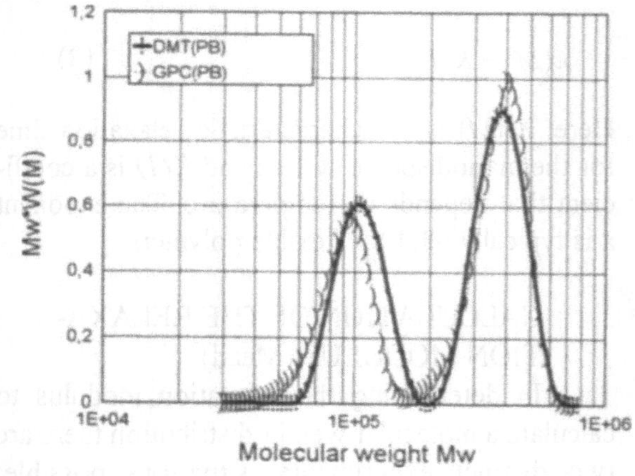

Figure 1: MWD of a bidisperse polybutadiene blend

The [RSI]*Orchestrator* software provides for automated fitting or alternatively, a user defined manual fit to the data. Two types of model molecular weight distributions are built into the [RSI]*Orchestrator* software. The first is the Wesslau or log normal molecular weight distribution, that is typical of addition polymers or polymers produced by Ziegler-Natta catalyst systems. The second is the Schultz distribution which is a generalized most probable molecular weight distribution, more typical of polymers produced via condensation reactions or with metallocene catalysts.

6. REPRESENTATIVE APPLICATIONS

Figure 1 shows the calculated MWD for a bidisperse system of polybutadiene. The dynamic moduli data set was complete and the MWD calculated from the complete data set method is shown juxtaposed with the MWD measured from GPC.

Figure 2 shows two metallocene catalyzed binary polyethylene blends. The dynamic moduli data sets are not complete either in the terminal region or the plateau region. The molecular weight distributions were obtained using the incomplete data set method with a combination of two Schultz models.

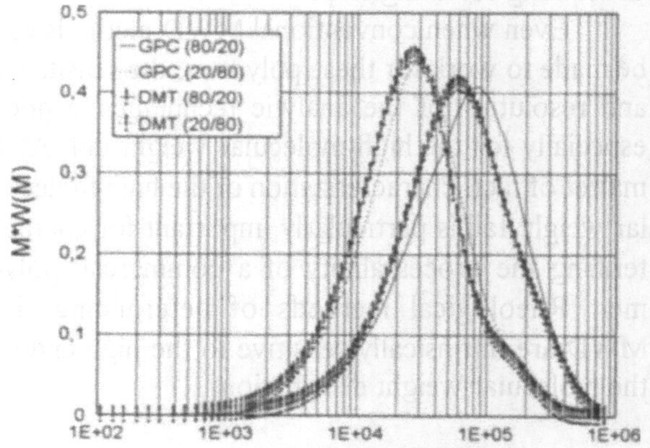

Figure 2: MWD of a binary polyethylene blend

7. REFERENCE

Tsenoglou, C; ACS Polymer Preprints 28, 185 (1987)

Des Cloizeau, J; Macromolecules 23, 4678 (1990)

Mead, D; J.Rheol. 38,1797 (1994)

RHEOLOGY AS A QUALITY CONTROL INSTRUMENT

M. GAHLEITNER, C. PAULIK, W. NEIßL
PCD Polymere GmbH, St. Peterstraße 25, A-4021 Linz, Austria
e-mail: markus.gahleitner@pcd.at

1. INTRODUCTION

In the last few decades, rheology has attained a status in polymer research which makes it an indispensable tool in the development of polymer materials. Regardless of whether the problems in question concern the analysis and characterization of new or unknown materials, production or quality control or the design of molding processes, the determination of rheological parameters is a significant element.

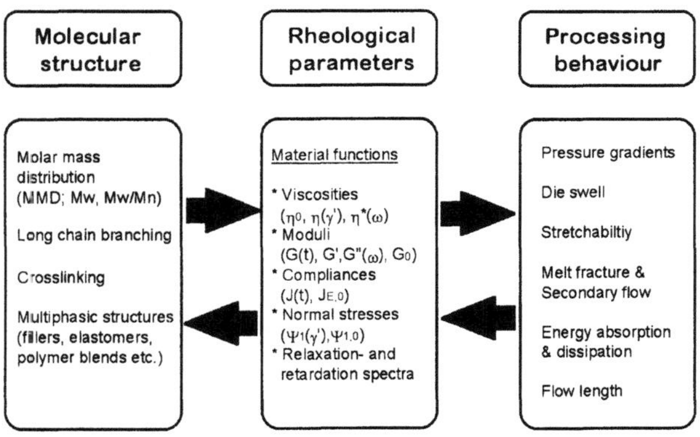

Figure 1 - Rheological parameters acting as a „link" between molecular structure and final properties of a polymer

2. PARAMETER DETERMINATION

Rheological parameters of polymers have a central role in polymer research, as they both reflect the molecular structure and determine the processing and application behaviour (see figure 1). For investigation reasons, the determination of the linear viscoelastic material behaviour in the oscillatory test is often preferred, yielding the storage and loss modulus of the material as a function of frequency (G′, G′′(ω)). The advantages of this method lie in its relative simplicity as well as in the information content of the result, which separates viscous and elastic components of the material behaviour. Moreover, the conversion to other linear viscoelastic parameters as well as a connection to the molar mass distribution is possible via a spectral analysis of the data (Shroff & Mavridis 1995, Carrot *et al.* 1996)

3. QUALITY CONTROL

Several examples for the importance of rheology as a quality control tool can be given here:

- **HDPE pipe grades**: Problems in the production process can significantly change the polydispersity of the molar mass distribution (MMD) while the MFR remains nearly unchanged (see figure 2). As the mechanics of these materials is largely determined by the fraction of longest molecules acting as ties between crystalline segments (Shroff et al. 1996), only a control of the flow curve can assure constant quality.

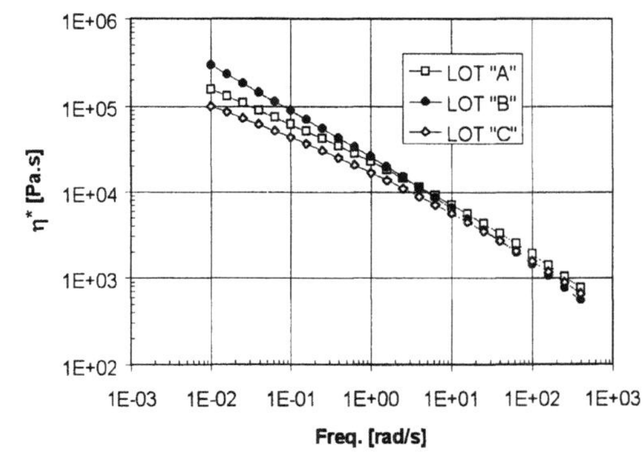

Figure 2 - Fluctuations of MMD in actual production of HDPE pipe grades reflected in the viscosity curve; deviations at low frequencies are relevant for long-term mechanical properties, while not influencing the MFR measured at higher shear rates

- **Homogeneity of bimodal grades**: One of the most important developments in the field of polymer design in recent years was the production of Materials with a bimodal MMD. The combination of high molar masses providing mechanical strength and lower molar masses supporting processability has opened some completely new fields for application and facilitated applicability in others. However, one of the preconditions for application of such grades in critical fields is the homogenous distribution of the high molar mass component. Rheology has proven to be a very sensitive tool for checking homogeneity of bimodal PE- and PP-grades, as in case of a non-homogeneous distribution the rheological behaviour is dominated by the low molar mass component.
- **Degradation and crosslinking effects**: In PP-homopolymers mechanics is determined by the MMD and crystallinty with strong interaction between these two factors. Degradation reduces the polydispersity of the MMD leading to a decrease in impact strength because of the reduced number of tie-molecules and in stiffness as well because of the reduced number of crystallization nuclei (Gahleitner *et al.* 1996). With PP-impact-copolymers (HECO's), the situation is more complex - mechanics are mainly determined by

their morphology (particle size and distribution). Through the different reaction of the matrix (PP) and the elastomer phase (EPR/PE) to radical reactions, the ratio between the molar mass of the matrix and the disperse phase will change leading to aggregation effects and loss in toughness (Martuscelli 1995). This effect is strongly reflected in the rheology, where a secondary plateau in G' is developed at low frequencies reflecting a contribution of long relaxation times of the elastomer phase (see figure 3).

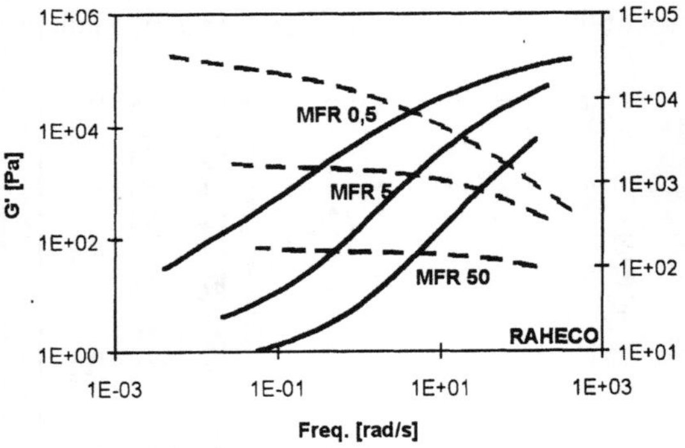

Figure 3 - Evolution of rheological behaviour in the degradation series of a heterophasic EP-copolymer (impact type); increase of elasticity at long relaxation times with increasing MFR

- **Mineral-filled compounds**: Compounds of thermoplastics with a high volume content of mineral fillers show a viscoelastic behaviour which is significantly different from homogeneous systems. A strong nonlinearity is combined with the development of a second plateau region of the dynamic moduli at lower frequencies. These effects are determined by particle size, loading and dispersion quality, all of which also define mechanical properties . The determination of the linear viscoelastic properties of a mineral filled thermoplastic material in the melt can provide a quick and reliable way to investigate filler properties and dispersion quality (Suetsugu & White 1983, Gahleitner *et al.* 1994), even if severe limitations exist and a certain reliability can only be attained based on experience.

4. PRECISION

In a round-robin test organised by PCD (Gahleitner *et al.* 1995) the primary scope was to check the comparability and reliability of rheological measurements on polymer melts. Eleven laboratories in four european states took part in the test; three polymer systems were included: Polystyrol PS 168N (BASF AG), Daplen FSC 1012 (heterophasic EP-copolymer) and Daplen FSC 65 T 30 (talc-compound on copolymer-base), the latter two being products of PCD Polymere GmbH. The tests to be carried out were in the dynamic mode (including a strain sweep - G′, G′′ (γ) at ω = 1 and 100 rad/s - to determine the linearity region (γ_{crit}) of deformation) and a steady shear measurement of $\eta(\gamma')$ at the same temperature; all of these in cone/plate- or plate/plate-geometry as well as under nitrogen to avoid degradation at 230°C for PP and 200°C for PS. Each measurement was to be carried out twice and, in case the difference in the two results exceeded 5%, a third time.

The results (see figure 4) show a good coincidence between the different laboratories' results for the two unfilled systems at least. The standard deviation varies with both frequency and temperature, being highest (i.e. 15-25%) for low and lowest (i.e. 5-8%) for high frequencies; temperature deviations are observed for the PS-sample only. The variation was significantly stronger for the compound sample, maybe because of the inability to reach a truely linear viscoelastic regime in this case. Similar results have been obtained in other studies (Bafna 1996, Rides & Allen 1996), so the reliability of results obtained in dynamic mode can be assumed to be rather high with deviations in the <10%-range.

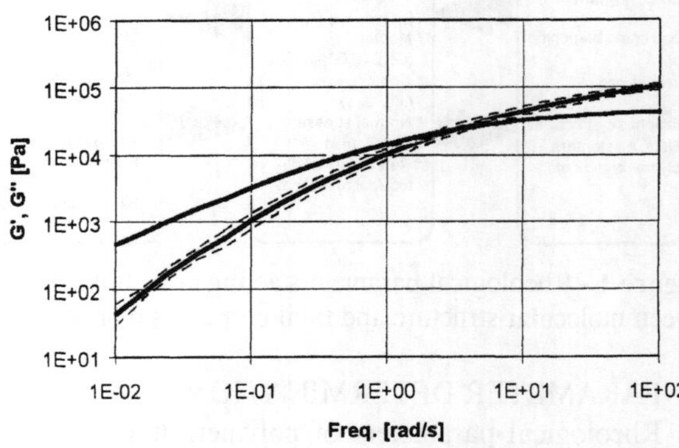

Figure 4 - Results of round-robin test for PS 168 N at 200°C; average values and 95% confidence interval

5. REFERENCES

Bafna SS (1996) Polym.Eng.Sci. **36**:90

Carrot C, Revenu P, Guillet J (1996) J.Appl.Polym.Sci. **61**:1887

Gahleitner M, Bernreitner K, Neißl W (1994) J. Appl. Pol. Sci. **53**:283

Gahleitner M, Wolfschwenger J, Bernreitner K, Neißl W, Bachner C (1996) J.Appl.Polym.Sci. **61**:649

Gahleitner M, Knogler B, Pirgov W (1995) Kunststoffe/PlastEurope **85**:31

Martuscelli E, in: Karger-Kocsis J (1995) *Polypropylene Vol.2*, Chapman & Hall, London, 1995

Rides M, Allen CRG (1996) Round robin for parallel plate oscillatory rheometry using polyethylene and polypropylene melts, NPL report CMMT(A)11

Shroff R, Mavridis H (1995) J.Appl.Polym.Sci. **57**:1605

Shroff R, Prasad A, Lee C (1996) J.Polym.Sci.B:Polym. Phys. **34**:2317

Suetsugu Y, White JL (1983) J. Appl. Pol. Sci. **28**:1481

WHAT'S NEW IN CAPILLARY RHEOMETRY ?

A.GÖTTFERT, E.-O. REHER
Göttfert GmbH, Siemensstr. 2, 74722 Buchen , Germany

BROAD SPECTRUM RHEOLOGY

Traditionally, the linear flow behavior of polymer melts at low shear rates has been characterized with rotational rheometers, while the non-linear shear thinning region at high shear rates is measured with capillary rheometers.

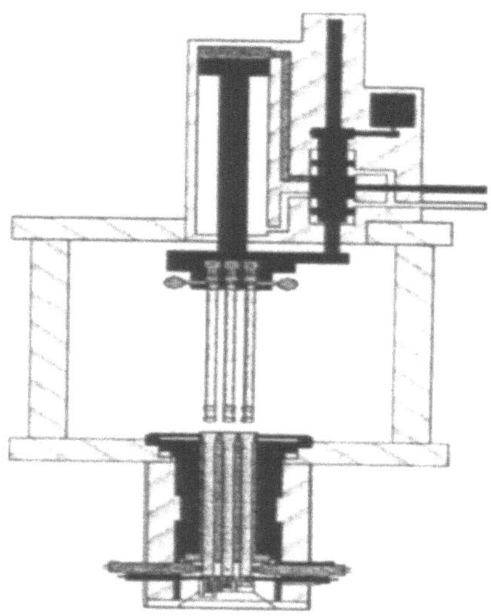

Fig.1: Schematic of a servo-hydraulic drive controlled capillary rheometer with triple die-single barrel, manufacturer : GOETTFERT

It has long been a desire to broaden the speed range of capillary rheometers to cover a significant part of the low shear rate range. This is possible by using a servo-hydraulic drive system. As shown schematically in Figure 1, this servo-system combines a high pressure cylinder with a low pressure cycle with a high precision control. In the rheometer concept described here, it is a stepper motor with microprocessor supported speed control which activates the high load hydraulic system by means of a permanently controlled valve. Minimum piston speeds of .0001 mm/sec can be obtained. With a maximum speed of 40 mm/sec, a dynamic speed range of 1 : 400 000 can be achieved.

The development of a new triple bore system in a single barrel expands the operation range even further. Parallel operation of two dies with .5 mm and 3 mm diameter gives a shear rate ratio larger than 1:500. With this configuration a total shear rate range of 8 decades can be achieved.

Although extremely low shear rates in the region of 10^{-2} sec^{-1} are possible under the described conditions, the precision and accuracy of the pressure transducer defines the practical limits of this operation range.

MODELING THE DEPENDENCE OF VISCOSITY ON SHEAR RATE

To overcome rheometer/sensor limitations as discussed and reduce time consuming experiments, modeling of the viscosity function followed by extrapolation into the Newtonian plateau can be very useful. This approach, however, requires a good agreement between experimental data and the model being selected. In the following, the 3-parameter models Winter-Carreau and Sabia will be analyzed in comparison to the 4-parameter model according Yasuda.

EXPERIMENTAL

The experiments have been performed on several commercial LDPE grades from BASF which are quite different in the molecular structure. The viscosity functions measured with a capillary rheometer have been corrected according to Rabinowitsch and Bagley. In the low shear rate range as well as for reference purposes the viscosity functions have been measured also with a commercial cone&plate rheometer in the oscillatory mode, assuming the applicability of Cox-Merz . Only the capillary data have been used for curve fitting and extrapolation based on the models described below.

GENERALIZED POWER LAWS AND EXP. FUNCTIONS

In representing the shear rate dependence of the viscosity function of polymer melts it is convenient to employ a generalized power law of the simplified form[1] according to Yasuda (1981):

$$\eta = \frac{\eta_0}{\left[1+\left(\lambda\gamma\right)^a\right]^{\frac{1-n}{a}}} \tag{1}$$

This functional form describes the viscosity at both asymptotes as well as the transition in between and is being reviewed elsewhere, Plumley (1994) The curvature, or the breath of transition, is described by the coefficient a. With a=1 the Winter-Carreau (Winter, 1980) can be derived :

$$\eta = \frac{\eta_0}{\left[1+\left(\lambda_C\gamma\right)\right]^{1-n}} \tag{2}$$

While λ from (1) is a relaxation time, λ_C is precisely the relaxation time of the crossover of both asymptotes. Consequently 3-parameter form (2) provides less flexibility to fit the shape of the viscosity function in the transition region compared to the 4-parameter Yasuda model. An other 3-parameter model is described by Sabia (1963), according :

$$\ln\left(\eta/\eta_0\right) = \left[\eta/\eta_0 - \alpha\right]\ln\left[1-\left(\lambda_S\gamma\right)^{1/3}\right] \tag{3}$$

While λ_S is a relaxation time, the coefficient α has a similar function like Yasuda's coefficient "a" which makes the Sabia model quiet attractive to work with.

RESULTS

All three models have been applied to the data from the capillary rheometer. The extrapolation capabilities to approximate the Newtonian plateau, have been compared with the experimental data from the rotational rheometer. It has been found that the model coefficients and correlation coefficients indicate that Sabia and Yasuda approximate the experimental data better than Winter-Carreau does. This can be seen more clearly from the modeled viscosity function for LDPE 1800 S (Fig.2). Sabia predicts well the expected plateau region, while the Yasuda data show pronounced curvature as does even stronger the Winter-Carreau data. As a first conclusion, approximation methods based on the above evaluated models are a useful tool to expand the application range of capillary rheometery. The described models are available as subroutines in the rheometer program Göttfert (1997).

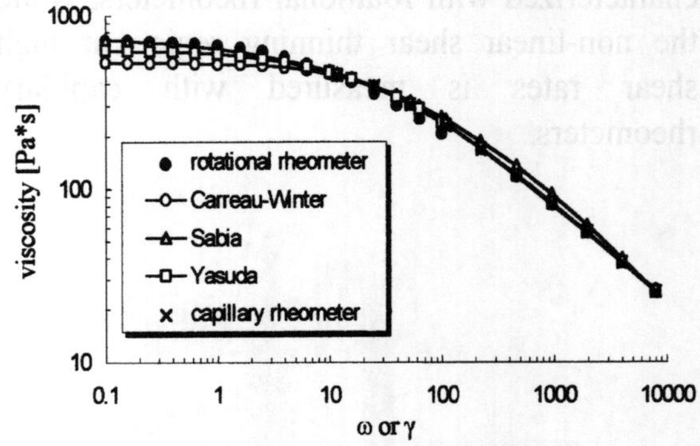

Fig.2: modeled viscosity function ftom LDPE 1800 S based on capillary rheometer data

CONCLUSION

Wide speed coverage of modern capillary rheometers with servo-hydraulic drive systems in combination with user friendly software tools for appropriate modeling of the viscosity function significantly extend the application of these rheometers for broad shear rate rheology. It is also shown that models from Sabia and Yasuda predict the Newtonian plateau based on capillary data alone.

REFERENCES

Göttfert, (1997), "WinRHEO", software for Göttfert capillary rheometers

Göttfert, unpublished investigations,

Plumley T.A, (1994), SPE proc., ANTEC,

Yasuda K., Armstrong R.C., Cohen R.E., (1981), Rheologica Acta 20, 163-178,

Winter H.H., (1980), Polymer Eng. Sci. 20,

Sabia R. , (1963), J. Appl. Polym 7, 347,

[1] In the original form according [1], η_∞ is incorporated, which is however negligible assuming experimental accuracy.

3D DEFORMATION OF A DROP IN SIMPLE SHEAR FLOW BY VIDEO MICROSCOPY

S. GUIDO, M. SIMEONE, M. VILLONE

Dipartimento di Ingegneria chimica, Università degli Studi di Napoli "Federico II", 80125 Napoli, ITALY

1. INTRODUCTION

The deformation of an isolated drop immersed in an immiscible liquid phase undergoing shear flow has been the subject of a number of studies, both theoretical and experimental, as reviewed by Rallison (1984) and by Stone (1994). The dynamics of an isolated sheared drop can be regarded as a sort of elementary event, which can provide some insight into the complex rheological behavior of a flowing dispersion of drops, such as the one encountered in emulsions, polymer blends and food industry.

Quantitative data on 3D drop shape in shear flow, which would allow a rigorous test of theories and numerical simulations, are lacking. In fact, most of the experimental work on shear flow has been focussed towards the measurement of the deformation parameter $D = (a - b)/(a + b)$, where a and b are , respectively, the major and minor axis of the drop, and the angle θ that a makes with respect to the velocity gradient. Such measurements were performed by observing the drop just along the vorticity direction and can not provide a full 3D picture of drop shape. In this work, a methodology to study the 3D deformation of an isolated drop under shear flow has been developed.

2. METHODS

Shear flow was generated by a parallel plate apparatus, which allowed one to observe the deformed drop either along the velocity gradient direction or along the vorticity axis. A 2-axis motorized stage was used to displace one plate with respect to the other. A set of micrometric rotary and tilting stages was used to adjust parallelism between the two glass plates by exploiting the reflections of a laser beam from the surfaces confining the sample. The fluids selected for the experiments were polyisobutylene (PIB) and polydimethylsiloxane (PDMS), both transparent and liquid at room temperature. In the shear rate range of the experiments, the two fluids were essentially Newtonian. Furthermore, the small density difference bewteen PIB and PDMS coupled to their high viscosity ensured that buoyancy effects were negligible. The sample was observed through a transmitted light microscope, equipped with a B/W CCD video camera and a motorized focus system. The microscope itself was mounted on a motorized translating stage, which was used to keep the deformed drop within the field of view during shear flow. Once the fluid selected as the continuous phase was loaded between the glass plates, a few drops of the other fluid were injected by using a tiny glass capillary. Drop diameter was always at least 10 times smaller than the gap, in order to minimize wall effects. Quantitative information on drop shape was obtained by an automated procedure based on image analysis techniques. The two axes a and b were calculated as the two axes of an ellipse having the same area and the same first and second moments of area as the observed drop projection in the plane of observation. Analogously, the angle θ was calculated as the angle that the major axis of such equivalent ellipse makes with respect to flow direction.

3. RESULTS

Let's call λ the viscosity ratioμ_d/μ_c, where μ_d and μ_c are, respectively, the viscosities of the dispersed and continuous phase, and Ca the capillary number, which is equal to $\mu_c\ \dot{\gamma}\ R_0\ /\sigma$, where $\dot{\gamma}$ is the shear rate, R_0 is the radius of the undeformed drop and σ the interfacial tension. In Fig. 1 a digital image of the steady state shape observed along the vorticity direction at $Ca = 0.24$ and $\lambda = 1.4$ is compared to the identified drop contour (continuous line) and the equivalent ellipse (dashed line) in the principal frame of reference. The x and y coordinates in Fig. 1 are made nondimensional with respect to the undeformed radius. The equivalent ellipse is visibly coincident with the experimental contour, and the result is even more remarkable if one takes into account that Ca is not small. An analogous result was obtained for the view along the velocity gradient direction.

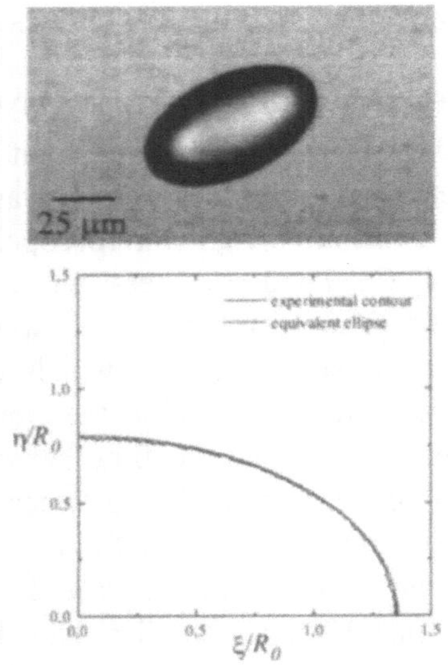

Figure 1

The three axes of the deformed drop at steady state, normalised with respect to the diameter at rest, are presented in Fig. 2 as a function of *Ca* at values of λ around 1 and 2.

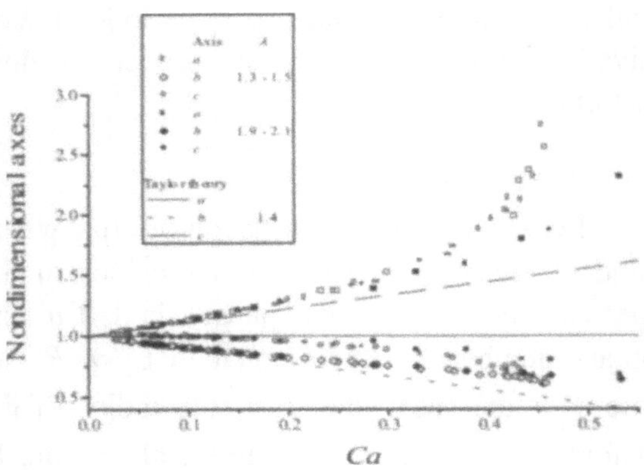

Figure 2

The dashed, dotted and continuous lines correspond, respectively, to *a*, *b* and *c* (nondimensionalised with respect to the diameter at rest) calculated from Taylor theory at $\lambda = 1.4$. Deviations between theory and experiments become significant at *Ca* around 0.1. It can be noticed that the *c* axis is always larger than the *b* axis. A comparison with the numerical simulations of Uijttewaal and Nijhof (1995) was made in terms of a deformation parameter D_p perpendicular to the longest axis of deformation, which can be calculated as $(c-b)/(c+b)$. In Fig. 3 the deformation parameter D_p calculated from the experimental data of *b* and *c* presented in Fig. 2, is plotted as a function of *Ca*. In the same figure, the continuous and dashed lines correspond to the numerical simulations of Uijttewaal and Nijhof for $\lambda = 1$ and 2, respectively. The agreement between experimental and numerical data is good up to *Ca* around 0.3. At higher *Ca* the experimental values of D_p go to 0, corresponding to the presence of a central cylindrical region which can be observed at high deformations. Such trend is not reproduced by the numerical simulations, which do not capture the elongated, sigmoidal shapes in proximity of the critical value of *Ca*.

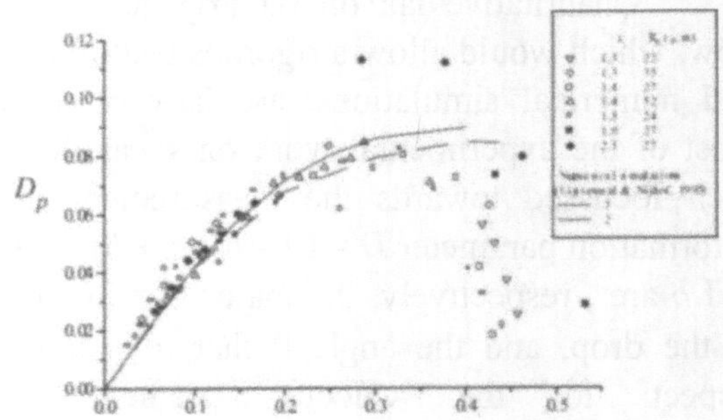

Figure 3

4. CONCLUSIONS

From the analysis of steady state data so far presented, it can be concluded that, up to moderate values of deformations, drop shape is well represented by an ellipsoid with three different axes. The main evidence supporting such conclusion is that (i) both drop projections along two perpendicular views look apparently as an ellipse and (ii) direct measurements of *c* are in good agreement with the values calculated from volume conservation. At higher values of *Ca* the two projections start to deviate from an elliptical shape.

4. REFERENCES

Rallison JM. (1981) J. Fluid Mech. 109: 465-482

Stone HA. (1994) Ann. Rev. Fluid Mech. 26: 65-102.

Uijttewaal WSJ. and Nijhof EJ (1995) J. Fluid Mech. 302: 45-63.

SPECKLE-PHOTOGRAPHY IN RHEOLOGICAL MEASUREMENTS

Ya. IVANOV[1], V. KAVARDJIKOV[2], R. KOTSILKOVA[1], D. PASHKOULEVA[2], V. STOYANOV[1]

[1]Central Laboratory of Physico-Chemical Mechanics - Bulgarian Academy of Sciences, Acad. G. Bonchev str., Block 1, 1113 Sofia, BULGARIA, e-mail: yaiv@bgcict.acad.bg
[2]Institute of Mechanics - BAS, Acad. G. Bonchev str., Block 4, 1113 Sofia, BULGARIA

1. INTRODUCTION

The non-contact techniques provide complementary information for conventional rheological measurements (e.g. Fuller GG et al., 1994). The visualization of the flow is founded on the optical methods. The investigations are carried out with pure liquids, emulsions and very dilute suspensions. The transparency of the sample is a general condition to use these techniques. The transparency of the walls, which confine the samples, is an extra necessity. These circumstances determine the visualization of flow as a self-dependent branch of the rheology.

In this paper, the potentialities of speckle-photography to provide a full field velocity information in tube and entrance flow regimes are described.

2. SPECKLE-PHOTOGRAPHY

Small deformations are registered and measured by speckle-photography technique. When we investigate the process of deformation, the velocity in each point of section of the sample can be determined.

The basis of speckle-photography is the simple dependence between the point velocity and a fringe spacing at the same "point" region. This distance between two neighboring interference fringes is obtained by scanning a specklegram. The random diffraction over a screen is used. The specklegram is a registration of the speckle-structure, which is formed by introducing the spheroidal solid particles (markers) in the sample (e.g. Kavardjikov V et al., 1998). For optical view-point the particles need to reflect the light (with good reflective index) and for rheological view-point is necessary that the particles do not act on the rheological characteristic of the sample.

For concrete example of use the speckle-photography the optimization of the volume fraction and the diameter of the markers, the optical system for recording speckles and the time interval between two neighboring light pulses is accomplished. These parameters of speckle-photography technique depend on flow velocity. This means, that if the velocity field is not homogeneous and the vectors of the velocity differ in a marked degree in this region the velocity can not be measured. This fact is the main restriction for the use of this technique.

We studied the flow of glucose (ρ=1420 kg/m^3, η=12.5 Pa.s) with volume flow rate Q$\approx$200 mm^3/s. The mercury lamp was a suitable light source. When we studied the gravitational flow (Q$\approx$2 mm^3/s) of epoxy - dibutilphtalate mixture (ρ=1140 kg/m^3, σ_y=4.04 Pa, η_{pl}=56.9 Pa.s) a krypton-ionic laser was a light source.

3. RESULTS

The used technique allows to measure the velocity and the slope of the stream-lines. In this way the speckle-photography is able to determine two components of velocity at each point of studied plane of flow field.

3.1 Tube Flow

Fig.1 shows data of the velocities, which are obtained for 24 points from a section, which is perpendicular to the tube axis. The angle between the stream-lines and the direction of flow is approximately equal to zero. The velocity profile is determined by approximation procedure on point values. The parabola index corresponds to the power law index of Ostwald-de Waele model. Two bounder conditions are used - fluid with wall adhesion ($v(r=R)=0$) and fluid with wall slip effects ($v(r=R)=v_s\neq 0$).

Direct measurements of the thickness of slip layer are yet unavailable - due to the difficulty to determine the distance from the wall where the change of the velocity occurred. By used technique the slip velocity v_s can be determined by extrapolation.

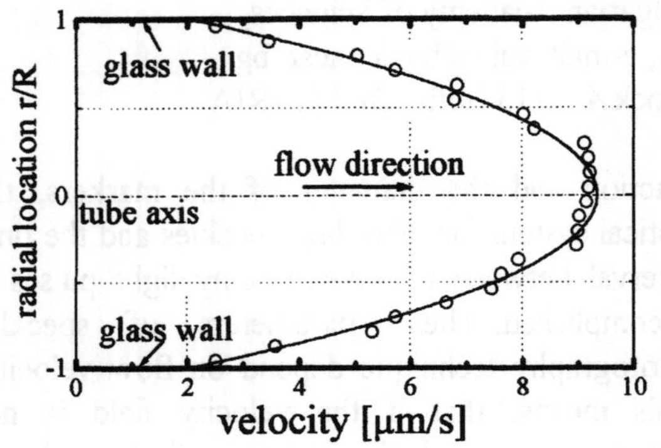

Figure 1. Axial velocity profile as a function of radial location [tube radius R=10.65 mm; fluid: epoxy - dibutilphtalate mixture with ρ=1140 kg/m^3, σ_y=4.04 MPa, η_{pl}=56.9 Pa.s] (∘ experimental data; — calculated)

The data on Fig.1 show that in the central region of the tube a plug flow is observed. The measurement in the core region shows the great possibilities of the speckle-photography. The exact rheological behavior of the sample can be determined by comparison between the values of the measured and calculated data.

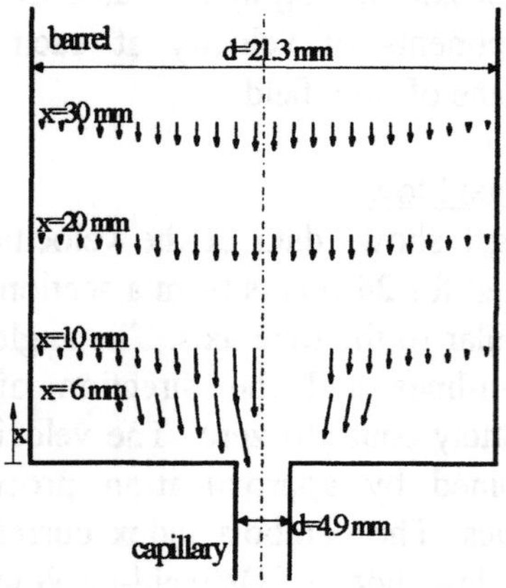

Figure 2. Velocity vector as a function of location [sample: fluid - epoxy - dibutilphtalate mixture with ρ=1140 kg/m^3, σ_y=4.04 MPa, η_{pl}=56.9 Pa.s; phase - glass fiber (ρ=2250 kg/m^3; l=150 μm; d=10 μm; concentration of 0.1 wt%);]

3.1 Entrance Flow

Fig.2 shows an entrance flow pattern with stable symmetrical circulating flows in the vicinity of the capillary entrance for low volume rate (shear rates receptively). The point velocity data and the slope of the streamline are shown as vectors. Four sections, which are perpendicular to the tube axis, are presented.

The velocity profile is obtained using the equation given in Ivanov Y et al., 1997. In this case the point velocity vector has two components - along the flow axis and radial one. Using the equation of continuity, the boundary of the vortex, formed in entry region, is determined. The last step is to obtain with a fair approach of accuracy the vortex detachment length L. The L value allows to calculate the magnitude X used by Cable and Boger (1978) to describe the characteristics of the circulating flow. X is defined as $X=L/d$ where d is the diameter of the barrel as shown in Fig.2.

4. CONCLUSIONS

The speckle-photography allows us to study the flow behavior of the distinct type of fluids. It can be used for measurements of the velocity profile, the slip velocity, the "power law" index and the vortex detachment.

5. REFERENCES

Cable PJ, Boger DV (1978) AIChE J 24:869

Gerald GG, Zawada JA, Atcher LA, Friedenberg MC (1994) Progress and Trends in Rheology. Proc IV-th Eur Congr. Rheology. pp 6-8

Ivanov Y, Kavardjikov V, Kotsilkova K, Pashkouleva D, Stoyanov V (1997) J. Th Appl Mech, Sofia, Year XXVII, (in press)

Kavardjikov V, Nikolov I, Manolova A, Illieva P (1998), Opt Quantum Electr 30:33-38

ACKNOWLEDGMENTS

The authors thank the National Fund "Scientific Investigations" for the financial support.

EXPERIMENTAL MODELING OF POLYMER BRUSHES RUBBING

S.V.KOTOMIN

Institute of Petrochemical Synthesis, 29 Leninskii Prospect, 117912 Moscow, Russia

1. INTRODUCTION

In spite of numerous publications on polymer brushes friction on molecular level, many of them lack essential experimental confirmation of theoretical speculations.

Witten et al. (1990) showed that interpenetration of polymer brushes during friction was quite possible. Joanny (1992) predicted a critical velocity for sharp decrease of the sliding viscosity. As noted by Semenov (1995), the structure of the interpenetrating layer should be non-dependent on the velocity of displacement. The limited experiments with polymer brushes were fulfilled mainly on block-copolymer polystyrene-polyvinylpyridine (PS-PVP).

In view of above a certain interest may be attributed to experimental modeling of the friction process on macrobrushes. We suggested for this purpose very simple samples - ordinary toothbrushes.

2. EXPERIMENTAL

Two types of toothbrushes used for experiments: "medium" and "hard" type brushes were obtained from the "Kruidvat" store in the Netherlands. We call the first as a mild, and the second as a rigid brush. Experiments were performed on pair of brushes with various mutual penetrations and speed of displacement on Instron 4301 testing machine (Fig.1). A force-displacement response during cyclic shift in two directions was studied.

The dependence of the axial force response on the brushes displacement from the start is shown on Fig.2. For all samples the force growth is nonlinear but rather smooth up to critical displacement, after that a noticeable drop is observed. The onset of the unsteady force changes connected with filament ends separation. This disconnection may happen in two ways:

1. The sliding of bended filament ends along each other up to complete pulling off the brushes ends, we call it "slipping-pulling" mechanism
2. The declination of the filament axis leading to slipping across common axis, we call it "rake-on-rake" mechanism.

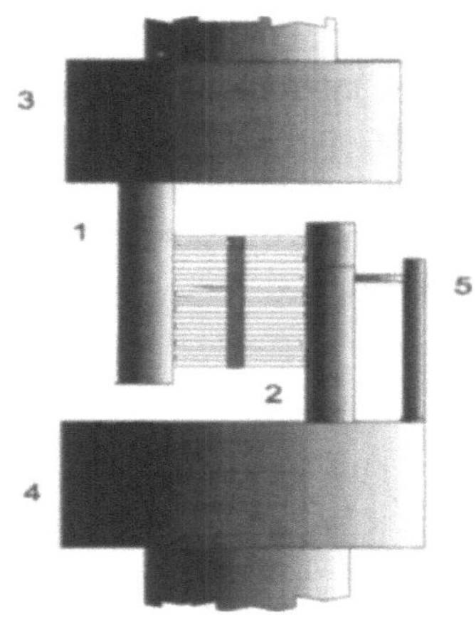

Fig.1 Schematic image of the samples fixed in holders: 1,2- brushes;3 -moving holder; 4- unmovable holder; 5 -normal displacement sensor

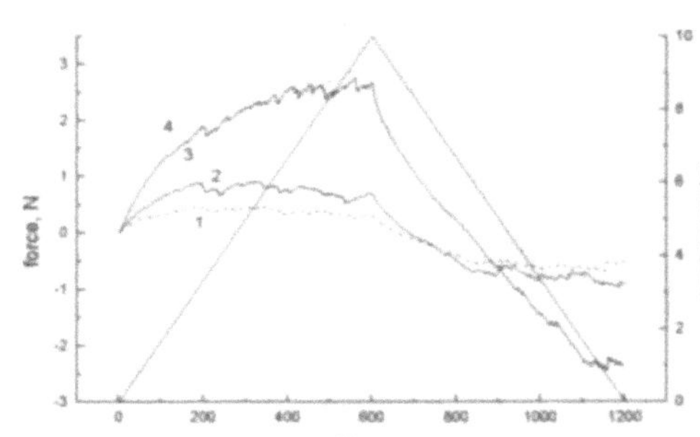

Fig.2 Force response for mild (1,3) and rigid (2,4) brushes for 1 mm (1,2) and 3 mm (3,4) interpenetration for velocity 1 mm/min

On the basis of parabolic model of disconnecting consoles, for the critical displacement s_C, distance between brushes basements l we can estimate the effective penetration ξ using the following simple equation:

$$\xi = \frac{6}{5}\frac{s_c}{l} \qquad (1)$$

For both mild and hard brushes from Fig.1 we have in the experiment l=11 mm, $s_C \approx 3$ mm and thus from (1) ξ =0.98 mm , that is very close to the real value of the initial penetration (1 mm).

Comparison of dynamic response diagram for rubbing brushes studied on Instron machine (Fig.2) with experimental results obtained on surface force instrument (SFA) for real macromolecular polymer brushes (Fig. 3) reveal striking resemblance in spite of principal differences in sample types and in measuring technique.

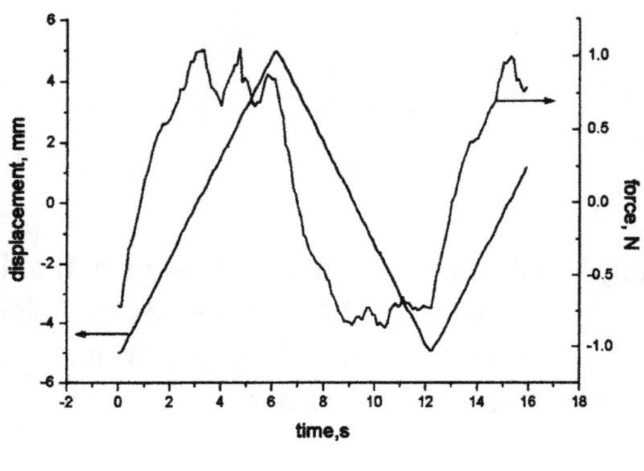

Fig.3 Force response in cyclic displacement for rigid brushes at 500 mm/min

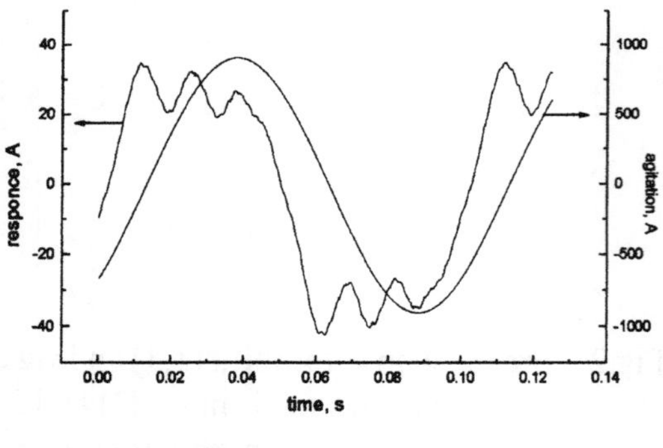

Fig.4 Displacement response in sine agitation on SFA for polymer brush at 10 Hz

The negative normal displacement at the beginning of the cycle corresponds to "attraction" of rubbing brushes while moving. It is a consequence of friction force reaction along chain axis, the normal component of which acts as an attraction force. When this force becomes equal to increasing opposite normal reaction of the chain due to bending, the sign of the resulting force changes to opposite and the latter becomes repulsive.

Although the amplitude of shear-plane force for the pair of rigid brushes was twice higher then that for the mild brushes, the normal repulsive force in case of mild brushes is on half higher than for rigid one. This can be explained from different mechanisms of moving for mild brushes and for rigid ones.

We believe that thermodynamic factors and relaxation phenomena in polymer brushes friction may also be simulated further using the same experimental model. The detailed description of the experiment may be found later on in our paper (1998).

4. CONCLUSIONS

For experimental study of polymer brushes friction macromodel approach using ordinary tooth brushes proved to be useful. Experimental results revealed two mechanisms of friction, so-called "slipping-pulling" and "rake-on-rake". The "attraction" effect was observed at the beginning of the moving cycle, connected with the friction of chain ends along axis. Controversial influence of friction force between constituent filaments of opposite brushes and the elastic reaction on bending during mutual displacement was shown. Contrasting with the shear force (along crosshead axis), the normal repulsive reaction for mild brushes is stronger than that for rigid brushes due to presence of different rubbing mechanisms. An existence of critical velocity for drop of friction force was proved and possible link of critical displacement with interpenetration level was estimated.

5. REFERENCES

Joanny J-F (1992) Langmuir 8: 989-992
Kotomin SV (1998) Wear (to be published)
Milner ST (1991) Science 251: 905-907
Semenov AN (1995) Langmuir 11: 3560-3562
Witten TA, Leibler L, Pincus PA (1990) Macromolecules 23: 824-826

MICROSQUEEZING TECHNIQUE FOR RHEOLOGICAL STUDIES

S.V.KOTOMIN, V.G.KULICHKHIN

Institute of Petrochemical Synthesis, 29 Leninskii Prospect, 117912 Moscow, Russia

1. INTRODUCTION

Squeezing flow is of certain interest for rheological studies due to simplicity of the procedure and developed mathematical analysis of the experimental data. Application of high-precision linear dilatometer based on the thermomecanical analyzer was demonstrated by us (1996) for rheological studies by micro-squeezing flow technique.

The possibility of using the squeezing flow technique for evaluation of the yield stress for liquid-crystalline thermoplastic melts was shown by Langelaan and Gotsis (1996) and the study of temperature dependency of the yield stress was fulfilled by us (1996).

To the best of our knowledge, no other experimental technique may provide the direct determination of the yield stress.

In this communication we consider the application of this device for a complex study of viscosity, relaxation modulus and adhesion durability.

2. METHOD

For experiments we use the high-precision linear dilatometer on the basis of DTMD termomechanical analyzer, developed in the Institute of Petrochemical Synthesis, Moscow.

The basic measuring cell of the device is shown on Fig 1. The polymer sample is placed between flat silica surfaces formed by loading rod and support.

At the present thanks to Gordon and Shaw (1994) the sophisticated software is available for computer analysis of the data obtained from squeezing flow technique for plate-plate geometry.

On the basis of equations developed for surface force apparatus (SFA) by Monfort and Hadziioannou (1988) we further extended this software applicable for sphere-plate geometry. The basic equation for maximal shear stress τ and shear rate $\dot{\gamma}$ during squeezing in the gap for this case are:

$$\dot{\gamma}_{max} = 1.3778 \frac{R^{0.5}}{D^{1.5}} \frac{dD}{dt} \tag{1}$$

$$\tau = 0.22963 \frac{F}{\pi R^{1.5} D^{0.5}} \tag{2}$$

Where R -sphere radius, D - distance between plate and the sphere, F - applied force.

In DTDM device the load force may be applied in both direction corresponding to compression or extension of the polymer sample. The absence of load when $F=0$ takes place during balance procedure and recovery after squeezing. The full test procedure includes three stages: squeezing, relaxation recovery and normal extension of stuck plates (Fig.2).

From the data obtained on the stage I the flow curves may be constructed and also the yield stress can be estimated.

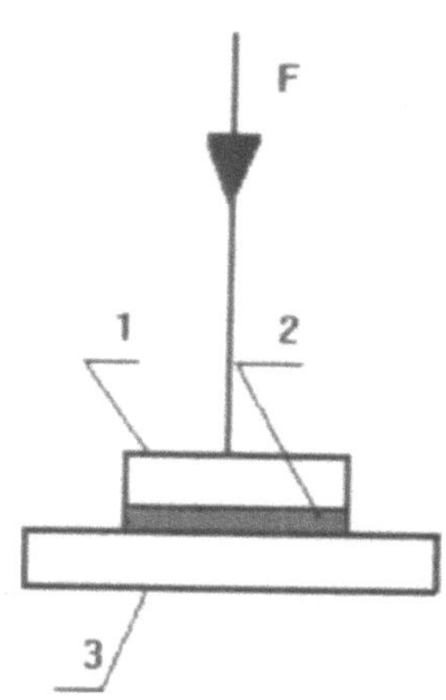

Fig.1 Main test scheme: upper plate (1); polymer sample (2); bottom plate (3).

durability of the bonding differs in three orders.

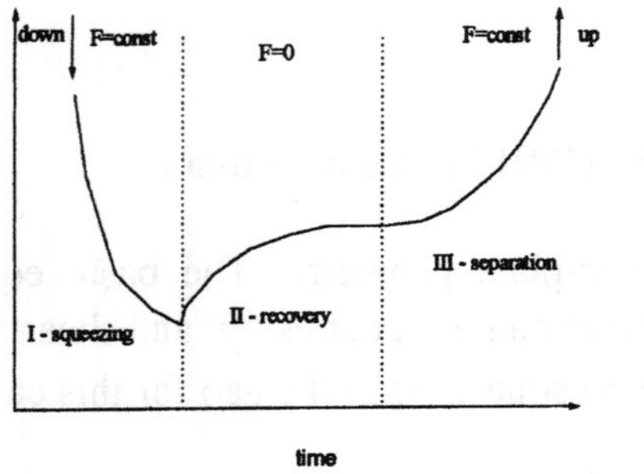

Fig.2 Test procedure stages - squeezing (1); recovery (2) and plates separation (3)

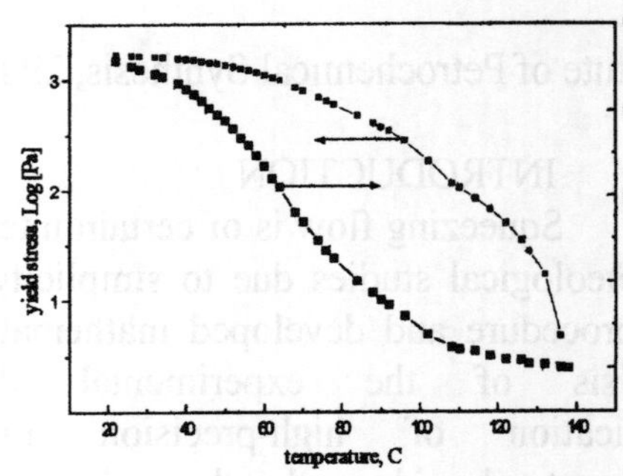

Fig.3 Gap and yield stress vs. temperature for the blend PVP:PEG 2:1 for 1N load

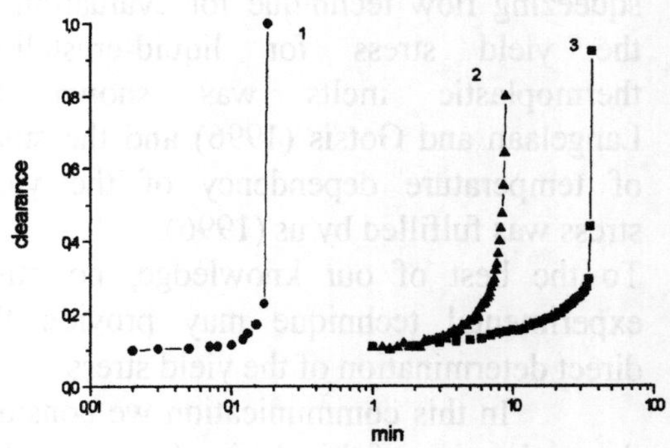

Fig.4 Clearance vs. time during plates separation for the blend PEG:PVP at composition 2:1 (1); 1:1 (2); and 1:2 (3) and extension force 0.3 N (1,2) and 1 N (3).

Using data from the stage II we performed long-term viscoelastic response analyses based on Chasset-Thirion equation . For simplification we considered recovery as the process of stress relaxation after compression, when the initial stress corresponds to the compression stress of the applied load, and the final zero -stress state - after full recovery. For calculation we also used the programes presented by Gordon and Shaw (1994) in the same book.

The last stage is used for estimation of the strength of adhesive joint in the terms of durability (time for adhesive bond rupture). In this case the durability of the adhesion bond is the moment of full plates separation or separation on definite distance. The work of debonding A in this case is calculated as

$$A = \int_{h_0}^{h_t} (F - m\frac{d^2h}{dt^2})dh \qquad (3)$$

where F- extension force, m - mass of moving part (the rod), h_o, h_t - initial and final distance between plates

3. EXPERIMENTAL

Below some experimental results for the polymer blend of poly (N-vinyl pyrrolidone) (PVP) and poly(ethylene glycol) (PEG) are presented. This system is a hydrogel with a network structure, which exists up to 140 ^{o} C. The disappearance of the yield stress shown in Fig.3 corresponds to disruption of this network. Next figure shows the development of stuck plates separation for the same adhesive PVP-PEG with different composition. As is seen, the

4. REFERENCES

Gordon GV, Shaw MT (1994) Computer Programs for Rheologists Hanser Publ. pp 48-55, pp 126-137

Kotomin SV, Kulichikhin VG, (1996) Polymer Science, Ser.B, 38,11-12:2079-2083

Kotomin SV (1997) Software Adjustment for Analysis of Squeezing Flow on SFA. Report Dept of Polymer Chemistry, University of Groningen, The Netherlands.

Langelaan HC, Gotsis AD, (1996) J Rheology, 8:107

Monfort JP, Hadziioannou G, (1988) J Chem Phys, 88:7187

A STUDY OF RUPTURE CONDITIONS OF POLYMER MELTS IN EXTENSIONAL UNIAXIAL FLOW

J. M. MAIA, J. A. COVAS, V. C. BARROSO

Department of Polymer Engineering, University of Minho, 4800 Guimarães, Portugal

1. INTRODUCTION

Knowledge of the extensional behaviour of polymer melts is extremely important due to the industrial relevance of extensional flows in common processing techniques and sequences such as blow moulding, film blowing, fibre spinning, melt flow through extrusion dies and injection mould filling.

In this communication, a modification to a commercially available controlled rate rotational rheometer is proposed in order to produce a cheap, easy to set-up, flexible extensional rheometer. The feasibility of the modification is evaluated. The results also include a study of types of rupture as well as rupture conditions. Different attainable flow regimes and corresponding types of rupture are discussed.

2. MATERIALS AND EXPERIMENTAL TECHNIQUE

The extensional rheometer used in this work is described elsewhere (Maia *et al.*, 1997), being essentially a modification to a commercially available rotational rheometer (TA Instruments Weissenberg Rheogoniometer).

Several types of experiments in controlled-rate uniaxial extension are possible (Maia *et al.*, 1997). For instance, information about the viscoelastic response of materials in extension can be obtained by performing stress relaxation in extension. An example is shown in Figure 1 for the case of a polyisobutylene (BASF Oppanol B15), PIB, for imposed Hencky strains ranging between 0.028 and 1.149.

The materials used in this study were the above mentioned PIB, which was supplied by BASF, Portugal and a low-density polyethylene (LUPOLEN 1840H), supplied by BASF, Germany. This is known as IUPAC-X and is a different batch of the well-known IUPAC-A.

3. THEORETICAL FRAMEWORK

Rupture in extension is often neglected by experimentalists. However, since one can not continue a given extensional experiment *ad infinitum*, this implies that the understanding of the set of conditions under which the polymer looses cohesion (which may be a crucial physical limitation to the attainable strains/strain rates) is important for extensional rheometer design and

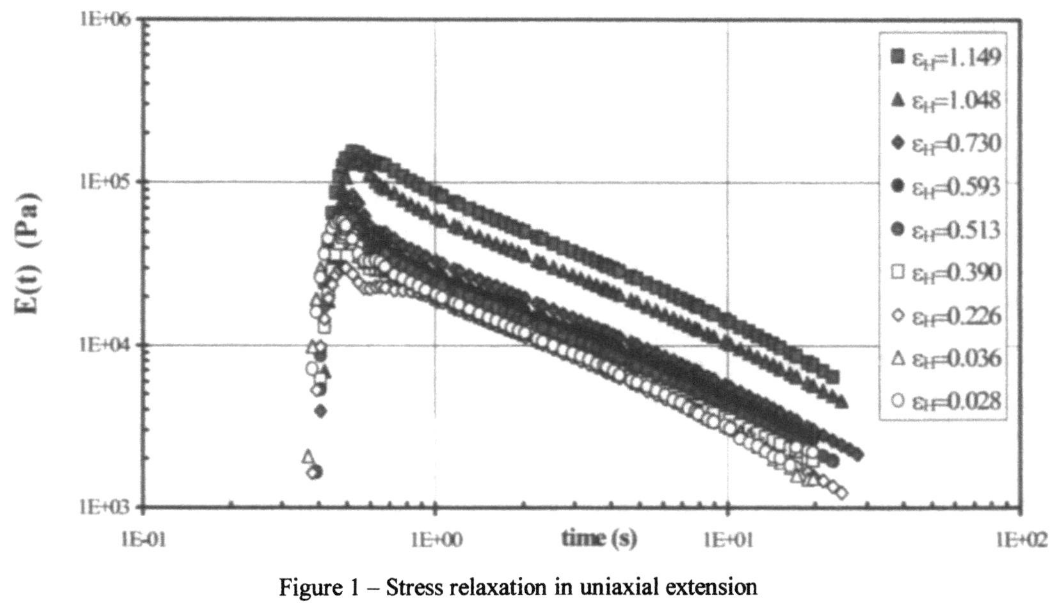

Figure 1 – Stress relaxation in uniaxial extension

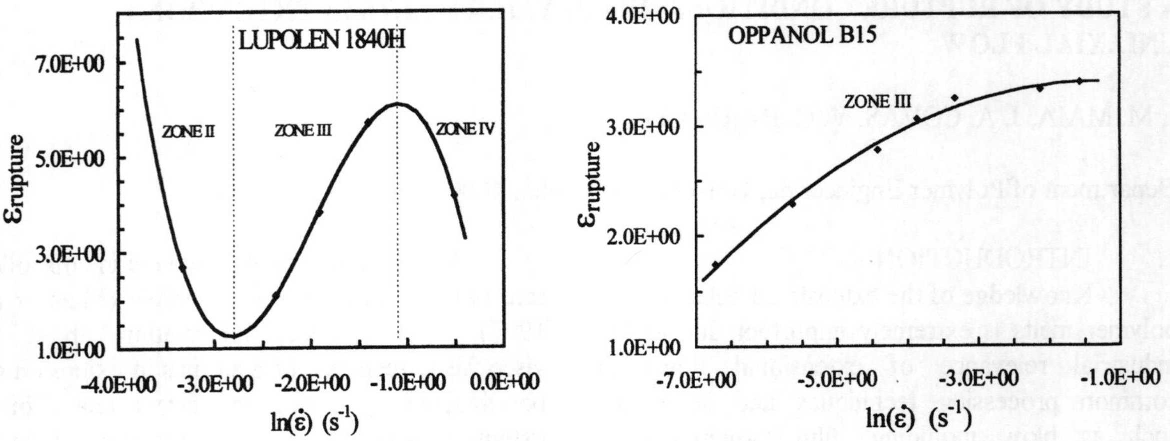

Figure 2 – Hencky strain at rupture

replication of practical processing conditions.

One of the major factors influencing the rupture of polymer melts is the transition from the fluid to the rubbery state, at high enough strain rates (Vinogradov *et al.*, 1975 and Malkin and Petrie, 1997). Malkin and Petrie (1997) distinguish four zones, which correspond to as many different types of behaviour, when one studies the dependence of the Hencky strain at rupture on strain rate.

Zone I corresponds to low strain rates. In this region steady elongational flow is possible because extension due to real viscous flow can continue without limit. In zone II superposition of elastic and viscous elongational flows occurs. Steady flow is impossible due to sample rupture. In zone III the deformation is almost entirely elastic and there exists virtually no viscous flow. Again, steady-state is not achievable. Finally, zone IV is the *glass-like zone*. In regions II and III rupture is ductile, whereas in region IV it is brittle.

4. RESULTS

The experimental results are shown in Figure 2, where a simple scaling of the strain rate (the usual time-temperature superposition principle works for these flows) allows results at different temperatures to be analysed together.

It is apparent that the experiments spanned all the zones for LDPE except that corresponding to steady flow. This is relevant since it means that if one wants to achieve steady state for this polymer (which has been used, to some extent, as a "standard"), it becomes necessary to go into the extremely low strain rates region (at least lower than 0.03 s^{-1}). This is probably the explanation for the fact that steady state was never reported for LDPE. It is also pertinent to consider whether it is worth, in practical terms, to try to achieve steady state for LDPE. Probably not, since the required low strain rates would hinder the practical usefulness of the exercise.

For PIB all experiments were performed in the region corresponding to elastic flow since the results fall within the rubbery behaviour region. This is not surprising since PIB is elastomeric at room temperature. The maximum attainable strains are lower than those for IUPAC-X, presumably due to the almost complete inexistence of viscous, irreversible deformation.

5. CONCLUSIONS

Studies of sample rupture conditions were carried out in two polymer melts. These studies showed that, for IUPAC-X, it will be very difficult (if not nearly impossible) to achieve steady-state and that even if that was accomplished the results would be of little practical relevance. The same applies to PIB since the deformation is almost entirely elastic, there existing little or no viscous deformation. This, again, makes it extremely difficult to achieve steady-state.

6. REFERENCES

J.M. Maia, J.A. Covas, J.M. Nóbrega, T.F. Dias and F.E. Alves, accepted for publication in *J. Non-Newtonian Fluid Mech.*, 1997

G.V. Vinogradov, A. Ya. Malkin and V. V. Volosevitch, *Appl. Polym. Symp.*, 27(1975), pp. 47-59.

A.Ya. Malkin and C. J. S. Petrie, *J. Rheol.*, 41(1997), pp. 1-25.

COMPARISON OF CREEP AND STRESSING EXPERIMENTS ON POLYOLEFIN MELTS IN ELONGATION

H. MÜNSTEDT, S. KURZBECK

Institute of Polymer Materials, Department of Materials Science, University Erlangen-Nürnberg, Martensstr. 7, D-91054 Erlangen, Germany

1. INTRODUCTION

The advantage of creep experiments in characterizing polymer melts is based on the fact that a steady state is reached after a shorter period of time than in stressing experiments at a constant deformation rate. The steady state is of interest as some material properties can quantitatively be correlated with molecular parameters like the average molecular mass. Moreover, creep experiments are of importance in all the cases where the measuring time is limited due to problems with the thermal stability of a polymer or because the maximum deformation is restricted as a consequence of the given equipment or the accuracy of the experimental setup.

2. EXPERIMENTAL

There exist two different principles for investigating elongational properties of polymer melts. The principle of rotating belts by Meissner and Hostettler (1994) is the base of an apparatus commercially available, the other principle goes back to a design by Münstedt (1979) which is schematically presented in Fig. 1. The instrumental challenge with respect to creep experiments stems from the fact that during elongation by

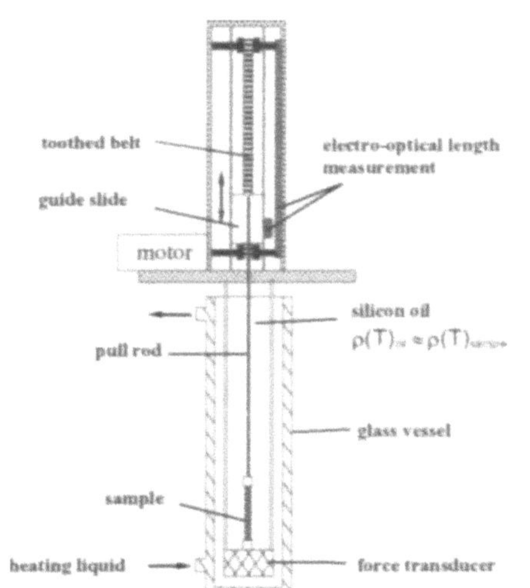

Fig. 1 Schematic drawing of the extensional rheometer

factors up to 50 a significant reduction of the sample cross section occurs. Performing well defined experiments requires a force control in such a way that the quotient of force and cross section is kept constant. This mode has been realized for the apparatus of Fig. 1 using a rotary current servo motor together with a stored program control unit (Münstedt et al. 1998).

3. COMPARISON OF CREEP AND STRESSING EXPERIMENTS

3.1 Time Dependence

Fig. 2 shows a typical creep experiment on a polypropylene. The upper diagramme demonstrates how constant the stress on the sample can be kept as a function of time. Furthermore, it becomes obvious from Fig. 2 that the rise time of the

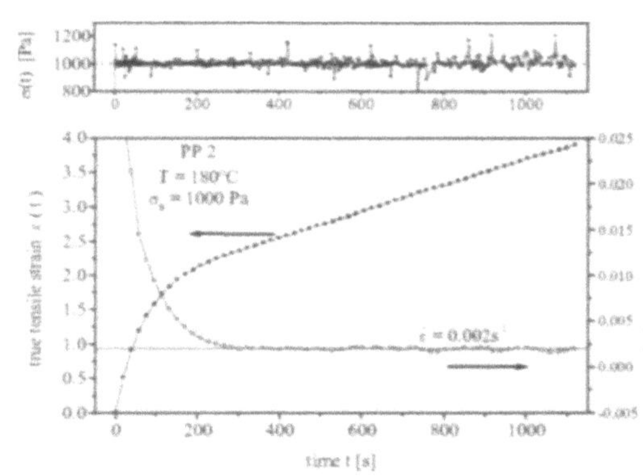

Fig. 2 Creep test on the polypropylene PP 2

applied stress is smaller than 1 second. The resulting elongation as a function of time for a polypropylene is plotted in the lower part of Fig. 2. The molecular data of the polypropylene are given by Kurzbeck et al. (1998). It shows the typical creep behaviour. After some time the steady state is reached. From the time dependence of the strain rate which is also given in Fig. 2 it can be seen that a constant strain rate of $\dot{\varepsilon}_0 = 0.002\ s^{-1}$ is reached after a creep time of about 300 s.

In the stressing experiment at the constant strain rate of $\dot{\varepsilon}_0 = 0.002\ \mathrm{s}^{-1}$ (cf. Fig. 3) a steady state has not be attained even at times up to 2000 s corresponding to a Hencky strain of $\varepsilon = 4$. The agreement of the measurements for the two different initial sample lengths is satisfactory and the good coincidence between $\mu_0^+(t)$ and $3\eta_0^+(t)$ points to the reliability of the experiments. Taking the value of the steady-state viscosity of

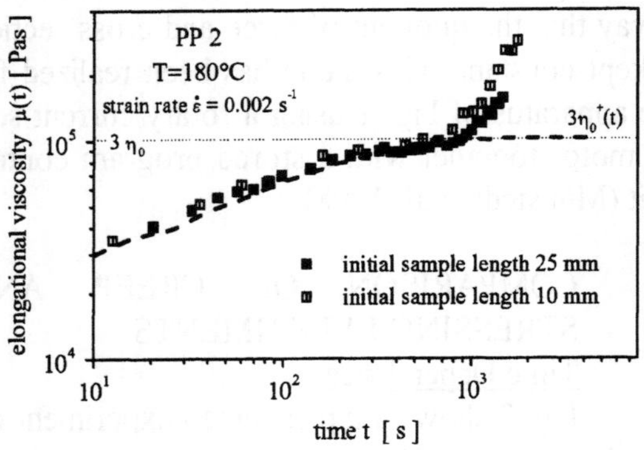

Fig. 3 Stressing experiments on PP 2

$\mu_s = 5 \cdot 10^5$ Pas from the creep experiment in Fig. 2 into consideration much higher strains than $\varepsilon = 4$ would be expected for the attainment of a steady state in the stressing experiment of Fig. 3. Measurements of elongations of this order are not reliably to perform by the experimental methods available. Creep experiments are very suitable to

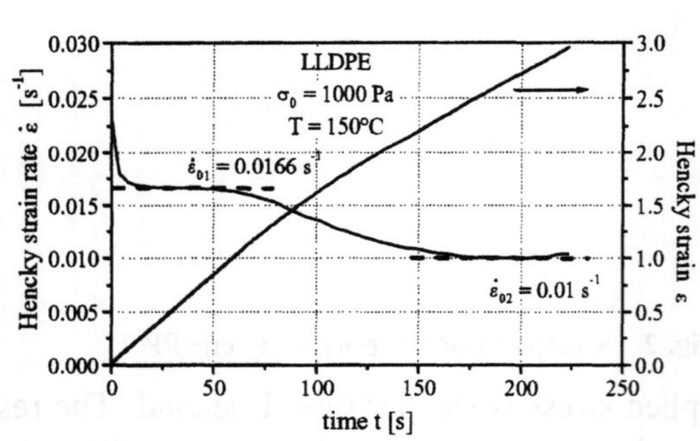

Fig. 4 Creep behaviour of a linear low density polyethylene

reveal different molecular processes as Fig. 4 demonstrates for a linear low density polyethylene (LLDPE). A first plateau of the strain rate is reached after a short creep time and a second at longer times. The importance of these two plateaus and their relevance with respect to stressing experiments are discussed by Münstedt et al. (1998).

3.2 Steady State

The reliability of elongational experiments on polymer melts can be assessed by the fulfillment of two requirements. Firstly, in the steady state results from different modes of deformation have to be in agreement and secondly, the predictions of the theory of linear viscoelasticity ha-

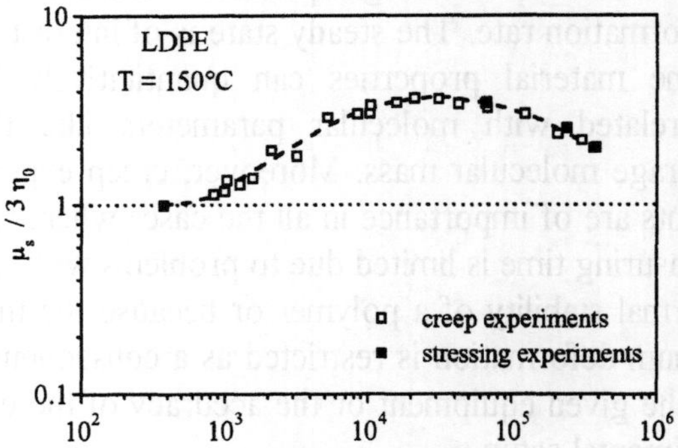

Fig. 5 Comparison of the steady-state elongational viscosities obtained in creep and stressing experiments

ve to be found valid, that means for example $\mu_0/3\eta_0 = 1$, where μ_0 and η_0 are the steady-state linear viscosities in elongation and shear, respectively. As can be seen from Fig. 5 for a low density polyethylene (LDPE) this relationship is very well fulfilled. On the other hand, it is obvious from Fig. 5 that the steady-state viscosities obtained in creep and stressing experiments are in good agreement. This result indicates the excellent suitability of the elongational rheometer for creep tests.

4. REFERENCES

Kurzbeck S, Oster F, Münstedt H, Nguyen TQ, Gensler R (1998) submitted for publication in J. Rheol.

Meissner J, Hostettler J (1994) Rheol. Acta 33: 1-21

Münstedt H (1979) J. Rheol. 23: 421-436

Münstedt H, Kurzbeck S, Egersdörfer L (1998) Rheol Acta 37: 21-29

IN-LINE RHEOLOGICAL MEASUREMENTS AND FLOW VISUALISATION USING DOPPLER ULTRASOUND METHOD.

B. URIEV[1], E. J. WINDHAB[2]

[1,2]Swiss Federal Institute of Technology Zurich, Department of Food Science, Laboratory of Food Engineering, ETH Zentrum, LFO-E18, CH-8092 Zurich, Switzerland.

1. INTRODUCTION

In this work a new in-line non-destructive rheological measuring technique was developed and tested for use in pilot plant and industrial scale applications. This method is based on using an Ultrasonic Pulsed Echo Doppler device in combination with a pressure difference method. The measuring system was installed as a section of the transporting pipe in a fat crystallization process. The rheological flow behaviour was derived from simultaneous recording and analysing of the velocity profile and shear stress calculations based on pressure difference measurements along the flow channel.

2. EXPERIMENTAL PROCEDURES

2.1 Materials

Fat crystal suspensions with crystal fractions of 1.6 % to 10.7 % were produced using a batch crystallisation process. The shear viscosity of the fat suspension changes in the range between 0.1 to 0.3 Pa*s.

2.2 Experimental Equipment

The principle of the ultrasonic method has been described extensively by previous workers (e.g., Satomura, 1957; Baker, 1970; Takeda, 1986), and thus only a brief description will be given here.

This method is based on the Doppler phenomenon. Considering a simple case of one moving target, the frequency shift Δf between transmitted f_0 and received frequency f_0-Δf of scattered ultrasonic wave can be observed. A Doppler shift frequency Δf and related to it velocity of the moving target can be derived. Since a sample volume contains a large number of moving particles, the echo signal received contains a spectrum of frequencies although the transmitted signal would be of a single basic frequency.

A pulsed Doppler technique was used to obtain a distribution of local velocity vectors along the tube diameter. This technique is based on transmitting wideband bursts of 8 MHz basic acoustic frequency. The measuring distance between volume samples was varied in the range between 0.5 mm to 0.9 mm.

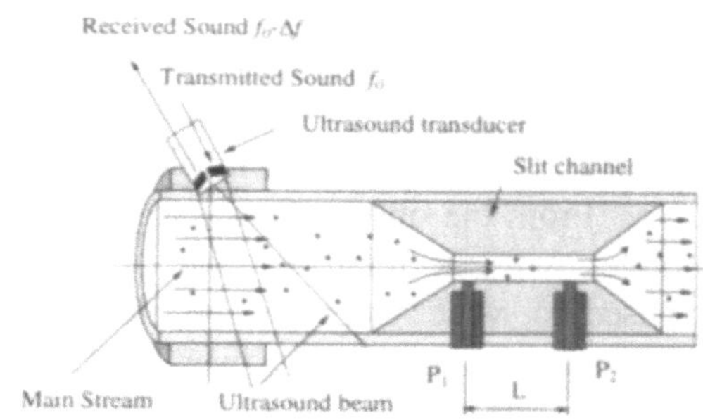

Figure 1. Flow velocity measurement by Doppler ultrasound in a cylindrical pipe with the attached slit channel at the end. The pressure difference is measured using two pressure sensors (P1 and P2) at fixed distance L apart.

Simultaneously with recording the velocity profile a shear stress measurement was performed. The micro-pressure transducers were mounted flush to the wall of the slit channel. The slit channel was attached at the end tube channel of 8 mm of inner diameter. The slit dimensions are 5 x 0.8 mm (see Fig. 1).

3. WORKING EQUATIONS

The velocity profile generated by a power law fluid flowing through a cylindrical pipe section with radius R can be described according to Wilkinson (1960), using the following expression:

$$v = v_m\left(\frac{3n+1}{n+1}\right)\left[1-\left(\frac{r}{R}\right)^{(n+1)/n}\right] \quad (1),$$

where v is a local velocity component, v_m stands for average flow velocity, n is a power law exponent and r is a radial co-ordinate.

Based on non-slip, steady flow conditions a volume flow rate can be determined as follows:

$$Q = \int_0^R 2\pi r v(r) dr \quad (2),$$

Consequently we can calculate the apparent shear viscosity from the measured pressure difference in the slit channel and volumetric flow rate derived from the velocity profiles (see Fig. 1).

4. RESULTS

In-line measurements of the velocity profiles of a fat suspension with 6.68% of solids content (see Fig. 2) are shown below.

From the observation it follows that the samples with the relatively high solids content exposed to wide range of the flow velocities showed a reduction in the power law exponent n with an decrease of the flow velocity (Fig. 2). This effect could not be observed for lower crystal amount suspensions (Fig. 3).

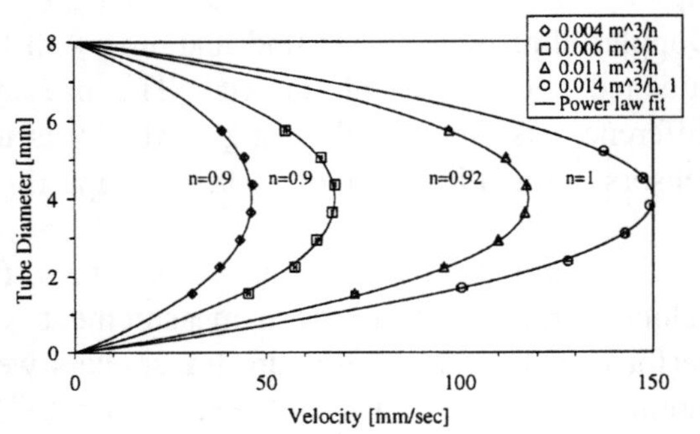

Figure 2. A power law, shown as a line, can describe the measured flow profiles for different flow rates, shown as symbols. Each velocity profile represents a set of averaged 1024 profiles.

The dependence of the n-factor on the fat suspension structure, i. e., the solid content and on the maximum flow velocity suggests the use of the n-factor as a process indicator (see Fig. 3).

To verify the reliability of the ultrasonic flow visualisation, the mass flow rate and density were measured simultaneously. The comparison of the two mentioned methods is shown on Figure 4.

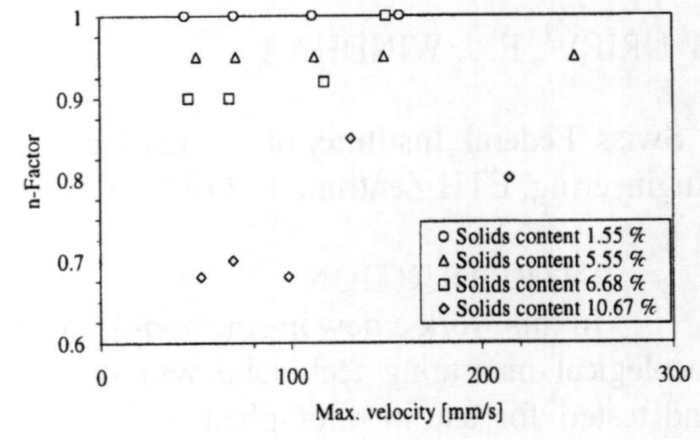

Figure 3. The power law exponent derived from the power law fit for velocity profiles versus maximum flow velocity.

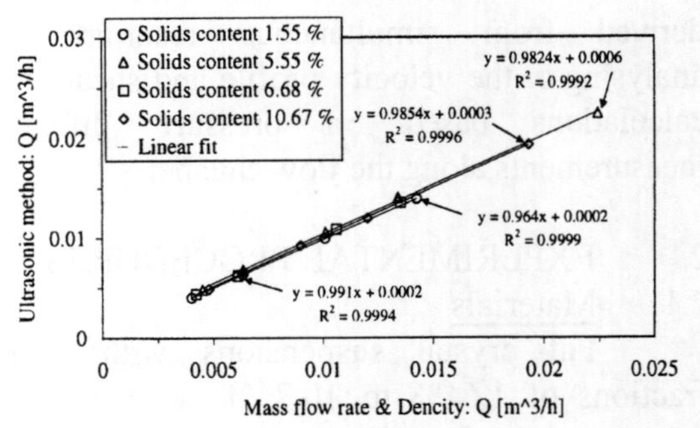

Figure 4. Comparison between volumetric flow rate Q derived from velocity profiles and Q derived from mass flow rate and density measurements.

6. ACKNOWLEDGEMENT

The authors wish to thank Dr. Birgit Breitschuh for providing this work with the sample, Dr. Yasushi Takeda for his help in optimising of the data acquisition software.

7. REFERENCES

Y. Takeda, (1986) Velocity Profile Measurements by Ultrasound Doppler Shift Method, Int. J. Heat Fluid Flow.

W. L. Wilkinson, (1960) Non-Newtonian Fluids

E. Windhab, B. Ouriev, T.Wagner, M. Drost (1996) Rheological study of non-Newtonian fluids Proc. of the 1st Inter. Symp. on Ultrasonic Doppler Methods for Fluid Mechanics and Fluid Engineering, Villigen PSI, Switzerland.

STUDIES OF SHEAR INDUCED MICROSTRUCTURAL CHANGES IN VISCOELASTIC PARTICLE GELS BY WAVE DISPERSION MEASUREMENTS.

S.R.REN, P.R.WILLIAMS

[1]Chemical Engineering Dept., University of Wales, Swansea, Singleton Park, Swansea SA2 8PP UK

1. INTRODUCTION

Particle-gels are an interesting type of gel that involve the nonbonding forces between an assembly of solid units. These forces are principally van der Waals or net-charge Coulombic in origin. Such systems are sufficiently well-defined at the basic unit level to enable a fundamental understanding of their structure to be made. We report how high-frequency shear waves have been used to investigate the time dependent viscoelastic properties of particle-gels formed from aqueous dispersions of a synthetic clay colloid (Laponite XLG, Laporte Industrie) and their modification under applied shear fields.

2. METHOD

In plane shear wave propagation in a linear viscoelastic medium of density ρ, the complex shear modulus G^* (= $G' + iG''$) is given by

$$G^* = -\rho\omega^2 / \Gamma^2, \qquad (1)$$

where Γ (= $[1/x_o + i2\pi/\lambda]$) is a constant in which x_o is the exponential damping length and λ is the shear wave length.The phase velocity, v, is given by ω/k, where k and ω are the angular wavenumber ($2\pi/\lambda$) and frequency respectively. G' and G" are the dynamic rigidity and loss modulus respectively. Viscoelastic systems produce a frequency dependence of v and x_o giving rise to *anomalous* dispersion in which v is exceeded by the group-velocity, U (= $d\omega/dk$), where,

$$U/v = 1/[1-(\omega/v)\{dv/d\omega\}] \qquad (2)$$

The rheometer employs a dual concentric-cylinder arrangement of shear wave generating and detecting surfaces, mounted on an R18 Rheogonoimeter. The geometry consists of 3 surfaces A, B, and C. A undergoes HF (>100 Hz) harmonic displacement, generating shear waves which are reflected at B and C, situated at distances x_1 and x_2 respectively from A, thereby creating the 'virtual gap' Δx (= $x_1 - x_2$). B and C are connected to piezo-crystals which produce voltages in response to the motion of A (see Williams and Ren, 1996). A undergoes HF oscillation driven by 2 PZT-5A actuators (Vernitron, UK) through a silicon collar; the latter also connects A to the R18 drive mechanism to provide steady rotation. This produces two sheared regions, of identical radius ratio (but unequal gap size), as A undergoes continuous rotation with superimposed HF oscillation. The actuators are controlled through a 5MHz D/A, voltages from piezocrystals (BSM 641 Bimorph) at B and C being acquired by 1MHz A/D's. U and v are determined by cross-correlation of the nodal and near-nodal regions of received waveforms with the corresponding phase content of the driving voltage waveforms.

Aqueous dispersions of a synthetic clay colloid, Laponite RD (Laporte Ind. UK) formed the test systems. Of particular relevance here, its thixotropic behaviour under shear has been studied (Ramsay and Lindner, 1993). Samples were prepared to give the requisite solids concentration and pH (> 9) as described by Ramsay (1986).

Measurements of U and v were made on 0.036 g/ml samples using a carrier frequency of 400 Hz, in the presence, and absence, of imposed shear fields (at shear rates, γ of 0 s^{-1}, 5 s^{-1}, 20 s^{-1} and 50 s^{-1}). The results are shown in Figs.1 and 2.

3. RESULTS AND CONCLUSIONS

In accordance with the highly elastic solid-like nature of the system, the results recorded yield very low values of the wave dispersion number, (U-v)/v. The lowest value is recorded in the absence of imposed shear but increases progressively as shear rate increases. When the shear field is removed, the system exhibits a gradual recovery of structure, (U-v)/v returning to its equilibrium value over *ca.* 1400 seconds. The repulsion of the electrical double layers about the particles gives rise to viscoelastic behaviour, of which the ability to sustain shear wave propagation is characteristic. The results obtained reinforce our previous findings, i.e. that the equilibrium structure consists of virtually immobile particles whose translational diffusion is constrained by interactions between overlapping double layers.

4. REFERENCES

Ramsay, J.D.F., J. Colloid Interface Sci., 109 (1986) 441

Ramsay, J.D.F. and Lindner, P., J.Chem.Soc.Faraday Trans., 89 (1993) 4207

Williams, P.R. and Ren, S.R., Journal of Physics D: Applied Physics, 29 (1996) 2718

Fig. 1

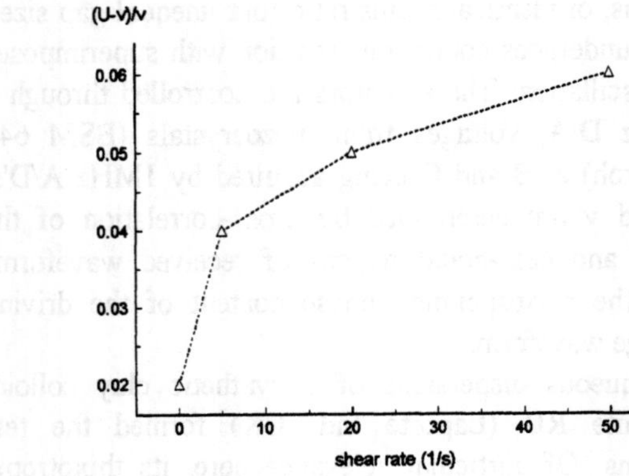

Fig. 2

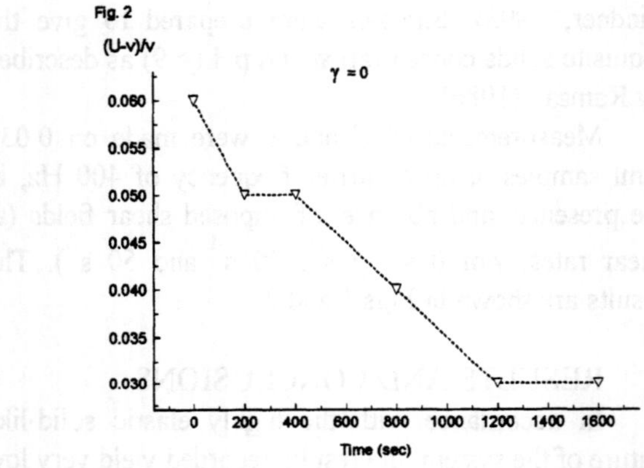

APPARATUS FOR MEASURING THE TIME-DEPENDENT POISSON'S RATIO IN UNIAXIAL TENSION

M. SAMARIN[1], I. EMRI[1] and N.W. TSCHOEGL[2]

[1]Center for Experimental Mechanics, University of Ljubljana, Cesta na Brdo 49, SI-1125 Ljubljana, Slovenia
[2]California Institute of Technology, Pasadena, CA 91125, USA

1. INTRODUCTION

Polymers and composite materials are widely used in applications requiring lightweight and high damping capacity. Since these materials show viscoelastic behavior under static and dynamic loading, the stress analysis involving such materials has recently become an important subject. The time-dependent Poisson's ratio as well as the tensile relaxation modulus are the primary input properties for finite element methods (FEM), and are also required in the constitutive modeling of a material's mechanical behavior (Emri et al., 1997). Previous attempts at measuring the time-dependent Poisson's ratio have shown convincingly that it is indeed difficult to obtain experimentally. Because the total range of response is compressed effectively between 0.3333 and 0.5 the measurements require high accuracy. This is made more complicated because they are strongly influenced by number of parameters such as e.g. temperature, pressure, humidity, etc., and the reproducibility of the material properties of the specimen. We report here on the development of an apparatus designed to enable us to determine the time-dependent. Poisson's ratio, $\nu(t)$, with the required accuracy while simultaneously determining the tensile relaxation modulus, $E(t)$. Using the E-T algorithm (Emri and Tschoegl, 1992-1997) we hope to be able to obtain the fundamental moduli, the shear relaxation modulus, $G(t)$, and the difficult-to-determine bulk relaxation modulus, $K(t)$, by computer calculation.

2. THEORETICAL BACKGROUND

The simultaneous measurement of two material functions on the same specimen under identical experimental conditions is indispensable for the successful calculation of any other material functions. Several authors who obtained, $\nu(t)$, by calculation from two non-simultaneously measured material functions found maxima and minima in contradiction to the prediction of linear viscoelastic theory (Thomson Ker., 1966, Crowson et al., 1979, Heydemann et al.,1962, 1964).

The behavior of $E(t)$ and $\nu(t)$ in a constant strain deformation is shown schematically in Fig. 1.

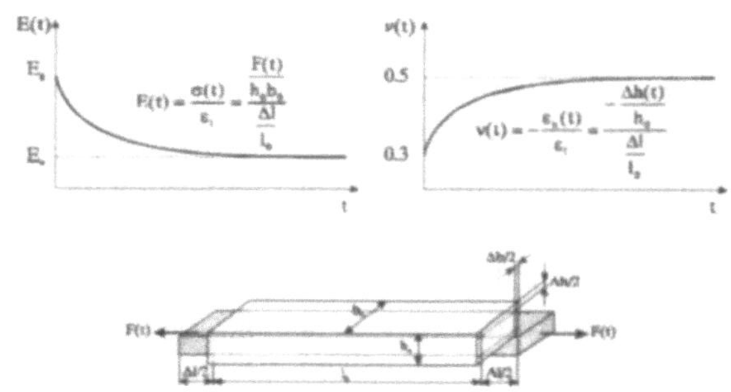

Fig. 1: Relaxation of a specimen in a constant strain deformation

3. EXPERIMENTAL EQUIPMENT

The design of an apparatus for the direct measurements of Poisson's ratio and the tensile relaxation modulus is governed by the basic definition of both these material functions. By their definitions one must measure the change with time of the transverse strain to determine the Poisson's ratio and change with time of the normal force to obtain the relaxation modulus. The difficulty lies in determining thickness changes of the order of 0.005% (50με) particularly in the glassy and equilibrium regions. Today's sophisticated non-contact optical methods for measuring displacement can achieve the required level of accuracy.

Equally obvious is the need for precise temperature control. The temperature must be kept constant within at least ±0.01°C. The apparatus, shown in Fig. 2, principally consists of the loading device, the environmental chamber, and the measuring unit.

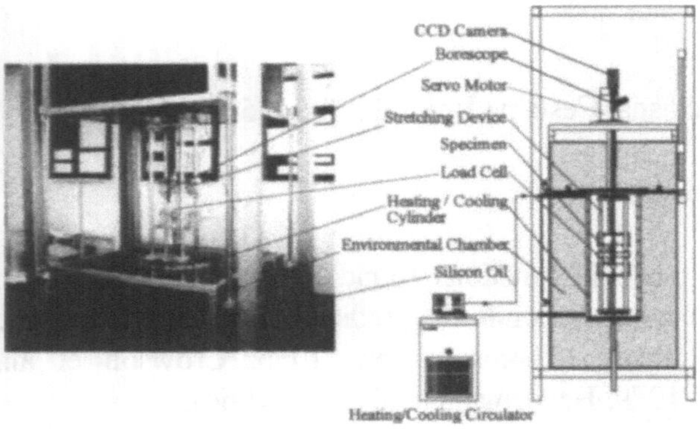

Fig. 2: Apparatus assembly

The stretching device is shown in the perspective drawing of Fig. 3. It is actuated by an electric servomotor (not shown). The measuring equipment consists of a load cell for measuring the applied load, the optical extensometer for measuring the transverse as well as longitudinal strain, and a unit for measuring and controlling the temperature. The environmental chamber is an well-insulated double walled vessel stirred by a circulator. The temperature range extends from -40ºC to 150ºC.

The apparatus allows simultaneous measurements of the time-dependent Poisson's ratio and the tensile relaxation modulus in a dry air or nitrogen atmosphere. Alternatively, the stretching device can be placed in liquid baths. The electric servomotor is equipped with feed back control and thus makes it possible to perform measurements at constant rates of strain as well as in tensile creep.

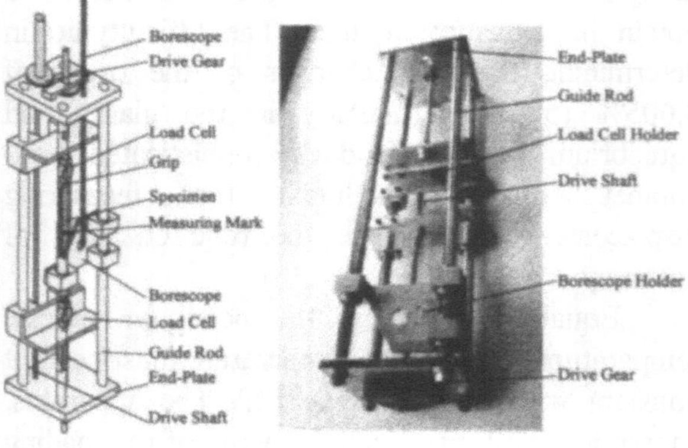

Fig. 3: Tensile stretching device

4. REFERENCES

Emri I, Samarin M, Tschoegl NW, Conference on ATEM'97, (1997), 481-486.

Emri I, Tschoegl NW, Rheol. Acta., 32, (1993), 311-321, 32, (1993), 322-327.

Emri I, Tschoegl NW, Intern. J. Polymeric Mater., 18, (1992), 117-127,

Tschoegl NW, Samarin M, Emri I, to be submitted to Mechanics of Time Dependent Materials, (1997).

Thomson Ker.C, (1966), J. Appl. Polym. Sci., 10: 1133-1136.

Crowson RJ, Arridge RGC, (1979), Polymer, 20: 737-746.

Crowson RJ and Arridge RGC, (1979), Polymer, 20: 747-754.

Heydemann P, Zosel A, (1962), Acustica, 12: 360-365.

Heydemann P, Nagarel H, (1964), Acustica, 14: 70-75.

INVESTIGATIONS OF THE CONVERGING FLOWS OF LOW-DENSITY-POLYETHYLENES BY LASER-DOPPLER-VELOCIMETRY (LDV)

M. SCHMIDT, E. WAßNER AND H. MÜNSTEDT

Institute of Polymer Materials, University Erlangen-Nürnberg, Martensstraße 7, D-91058 Erlangen, Germany

1. INTRODUCTION

Due to a high spatial (20μm) and temporal resolution (2μs) the used Laser-Doppler-Velocimeter is an appropriate instrument for local velocity measurements in translucent fluids without disturbing the flow itself. The measured Doppler-frequency is directly proportional to the particle velocity. Detailed knowledge of die entry flows is important for the optimization of processing operations of thermoplastic polymers. Details of the LDV-method are reported elsewhere (Durst et al., 1981).

2. FLOW CHANNEL

The flow behaviour of a commercially available branched LDPE (*Lupolen 1840 H*) is investigated at 150°C in a specifically developed flow channel. The die entry geometry consists of a flow-channel with a square cross-section (*reservoir*) of 14x14 mm² and an inserted *slit die* with a flat inlet (180° entry angle). The coordinate system is shown in *Fig.* 1.

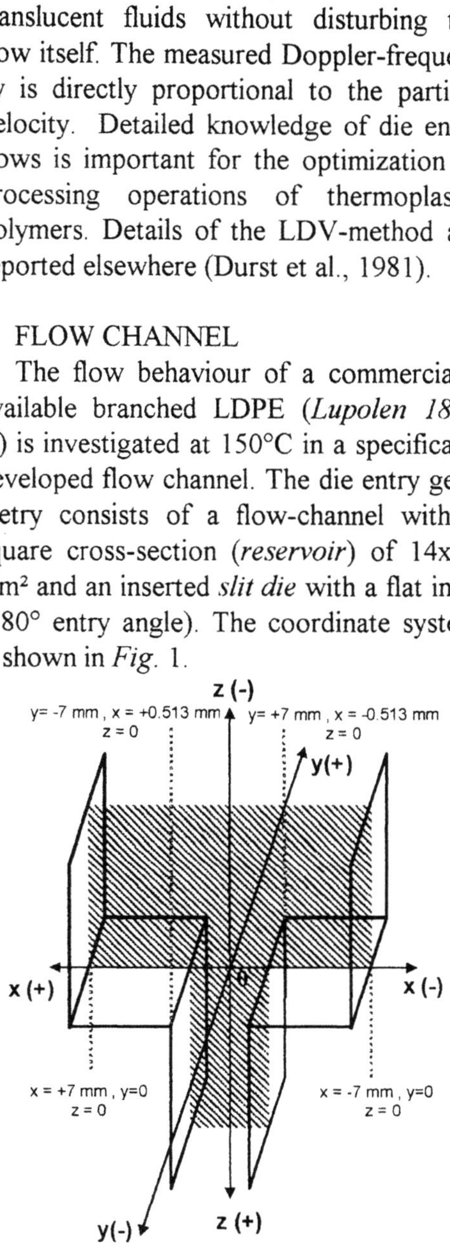

Figure 1: Determination of the coordinate system (x,y,z)

The dimensions of the flow channel are as follows (in millimeters):$-100 \leq z \leq 0$, $-7 \leq y \leq 7$, and $-7 \leq x \leq 7$ in the upstream channel (*reservoir*), and $0 \leq z \leq 50$, $-7 \leq y \leq 7$, $-0.513 \leq x \leq 0.513$ in the downstream channel (*slit die*). The relation of depth to width (contraction ratio) of the *slit die* is 14:1.026. The flow channel is fixed by a cross head to extruder which provides the melt.

3. DIE ENTRY FLOW IN THE SYMMETRY PLANE y=0

Using streak-photography *Bagley* and *Birks* (1959) illustrated the existence of large circulating vortices in the corner of abrupt *tubular* entry flows of melts of *branched* polyethylenes. The aim of our work is to quantitatively determine the flow fields connected with vortices.

The flow in the symmetry plane y=0 which is indicated by the shaded region in *Fig.1* can be assumed to be two-dimensional. Hence, the die entry flow in the y=0-plane can be recomposed by measuring the velocity components in the extrusion direction z (i.e. the v_z-component) and normal to the extrusion direction (i.e. the v_x-component). For a complete characterisation of the die entry flow of a branched polymer melt in a slit die geometry the velocity profiles $v_z(x)$ and $v_x(x)$ were measured at designated displacements z from the die entry plane z=0 at a fixed flow rate of $D=118.93\ s^{-1}$. The resulting velocity profiles $v_z(x,z)$ and $v_x(x,z)$ are illustrated in *Fig. 2* and *3*. Approaching the die entry plane (z=0) the flow starts to converge at positions $z \geq -10$ mm and the velocity profiles $v_z(x,z)$ change respectively. A higher resolution of the marginal ranges ($-7<x<-1$) of the velocity profiles $v_z(x,z)$ in *Fig. 2* shows that approaching the die entry plane very low negative velocity components v_z are present at positions $z \geq -5$ mm.

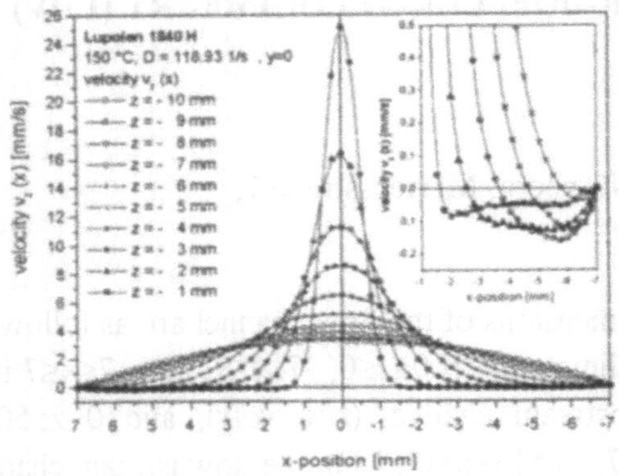

Figure 2: Velocity profiles $v_z(x,z)$, y=0

These negative velocity components v_z are caused by the circulating vortices and are also present in the marginal ranges of the left half (7>x>0) of the flow channel. In consequence of the circulating vortices a flow reversal takes place in the vicinity of the die wall and hence negative velocity components v_z (i.e. contrary to the extrusion direction) are detected. The velocity components v_z within the circulating vortices are very low (150 μm/s) and their order of magnitude amount to only 0.2-1.5% of the maximum velocities $v_z(x=0,y=0,z)$ of a profile $v_z(x,z)$. As the flow starts to converge at positions z ≥-10 mm first velocity profiles perpendicular to the extrusion direction $v_x(x)$ can be measured at this z-position. *Figure 3* shows the corresponding velocity profiles $v_x(x,z)$.

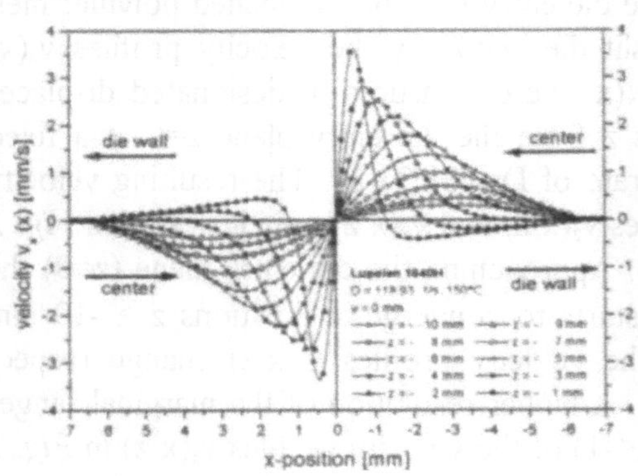

Figure 3: Velocity profiles $v_x(x,z)$, y=0

Considering the right half of the flow channel (-7<x<0) it can be seen that within the converging flow the polymer melt flows towards the center of the flow channel (in planes z=constant) and while approaching the die entry plane also negative velocity components v_x are detected at positions z ≥-2 mm (i.e displacement direction of the polymer melt is towards the die wall). These negative components v_x result from a flow reversal within the circulating vortices. Flow behaviour is analogous in the left half of the flow channel (7>x>0) considering reversed velocity signs. By recomposing the measured velocity profiles $v_z(x,z)$ and $v_x(x,z)$ within the y=0-plane (*Fig. 2* and *3*) a vector plot $\underline{v}(x,z)$ is obtained. The resulting vector plot $\underline{v}(x,z)$ shows a characteristic circulating vortex which forms on both sides of the converging flow (see *Fig. 4*).

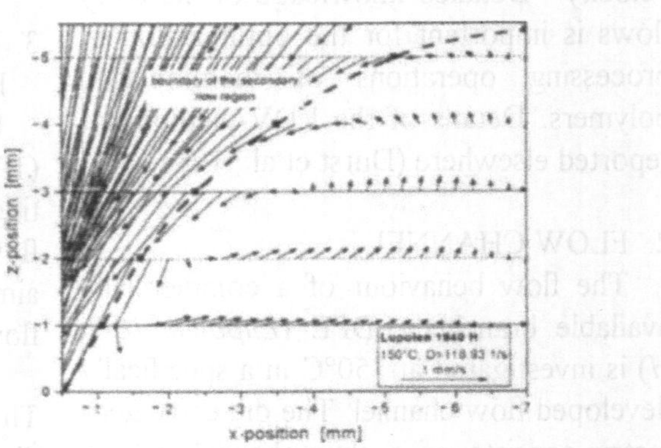

Figure 4: Vector plot $\underline{v}(x,z)$, y=0

4. CONCLUSION

The accuracy of the Laser-Doppler-Velocimeter (i.e. detection of very low velocities of 150 μm/s within the secondary flow region) has been demonstrated for steady state flow behaviour of a *branched* polyethylene melt in a slit die entry geometry. The converging flow as well as the secondary flow region (i.e. the vortices) has been recorded (detection of the velocity magnitudes and signs) for the first time quantitatively.

5. REFERENCES

Bagley E, Birks M, 1959. *Flow of PE into a capillary*. J.Appl.Phys. 31(3): 556-561

Durst F, Melling A, Whitelaw JH, 1981. *Principles & practice of LDV*. Acad. Pr., 2nd ed.

Schmidt M, 1997. *Investigation of polymer melt flows by means of LDV, Diploma-thesis*, Univ. Erlangen-Nürnberg

Waßner E, 1998. *Investigation of polymer melt flow behaviour by means of LDV*, Ph.D.thesis, Univ. Erlangen-Nuernberg

THE SINGLE POINT EVALUATION METHOD FOR THE FLOW CURVE OF VISCOELASTIC FLUIDS AND RELATED PROBLEMS

P. SCHÜMMER

Institut für Verfahrenstechnik, RWTH, 52056 Aachen, Germany

1. INTRODUCTION

To get the flow curve of a non Newtonian fluid one has to evaluate the material measurement relations related with special viscometric flow geometries. The mathematical procedure usually is not trivial. Following the lines of power law - finally without explicit use - an elementary single point evaluation method will be presented. Single point evaluation means obtaining the pair of flow curve variables ($\eta = \frac{\tau}{\dot\gamma}$ and $\dot\gamma$) from one pair of the respective measured relation.

2. VISCOMETRIC FLOWS

We will deal with the following viscometric flows: 1. pipe flow (c/s), slit flow (c/s), cylindrical Couette flow (c/s), axial annular flow (c/c), 2. torsional flow (c/s), small gap rotating concentrical sphere-sphere flow (c/c), one side wetted rotating disc flow (c/c).

Within the first group ideal viscometric flows are listed while in the second group there are viscometric flows under creeping flow conditions or other approximative flow conditions. The abbreviations c or s indicate if the mathematical evaluation of one of the two viscometric variables is complicated or simple.

3. REPRESENTATIVE VISCOSITY

For all c/s-cases mentioned it has been shown that under rather general conditions there exists one "representative" place r_R within the flow field for which the variable viscosity is given by the related value for Newtonian liquids. To get the point of the flow curve from one point of the measured curve we only have to find or to estimate this r_R.

E. g. for pipe flow we get η_R from Hagen-Poiseuille-relation $\eta_R = \frac{p'd^2}{32\,\bar v}$ and $\dot\gamma_R$ as $\dot\gamma_R = \xi\frac{\bar v}{d}$ wherein the factor ξ is given by the representative place r_R: $\xi = 8\frac{r_R}{R}$.

For the Rabinowitsch approximation of the flow curve (or related polynomial functions) we know the constant value of ξ. We call it ξ_3. For the more general case we may introduce a power law $|\tau| = k|\dot\gamma|^m$ to calculate ξ as a function of m. (cf. table 1)

For the physically interesting regime m = 0.2...1.0 we get a monotonous variation of ξ between 6.7 and 6.2 . If we choose for ξ a mean value $\bar\xi = 6.53$ in the whole m-regime mentioned, we get a variation of η_R only between 1.02 and 1.00.

Now for the c/c-cases we are able to use the same procedure of the power law calculation. The example of annular flow shall demonstrate it.

From the integration of the velocity profile we find the mean velocity

$$\bar v = \left(\frac{p'R}{2k}\right)^{1/m}\frac{R}{(1-\kappa^2)(3+\frac{1}{m})}\left((1-\lambda^2)^{\frac{1}{m}+1}-\kappa^2\left(\frac{\lambda^2}{\kappa}-\kappa\right)^{\frac{1}{m}+1}\right) \quad (1)$$

where λ follows from the equation given by Frederickson and Bird (1958) as a function of m and κ. (cf. table 1)

As for the pipe flow we now write

$$\dot\gamma_R = \xi\frac{\bar v}{\tilde d},\quad \eta_R = \frac{\tau_R}{\dot\gamma_R} = k\dot\gamma_R^{\,m-1}$$

introducing

$$\tilde d = d\left(1+\kappa^2-2\,\frac{1-\kappa^2}{\ln 1/\kappa^2}\right)^{1/2} \quad (2)$$

so that we can transform into a Hagen-Poiseuille-form:

$$\Lambda(m,\kappa,\xi)\cdot\eta_R = \frac{p'\tilde d^2}{32\bar v} \quad (3)$$

From $\Lambda \equiv 1$ we get ξ as a function of m and κ which now allows the same procedure as for the pipe flow. For a mean value of ξ, say $\bar\xi$, we calculate the variation of Λ to find it sufficiently small. That means: $\eta_R = \frac{p'\tilde{d}^2}{32\bar{v}}$ and $\dot{\gamma}_R = \bar{\xi}\frac{\bar{v}}{\tilde{d}}$ describe one point of the flow curve.

4. RESULTS

The table shows the results for all types of viscometric flows listed above.

Table 1: Table of results

	η_R	$\dot{\gamma}_R$	$\bar{\xi}$	$\xi(m)$	remarks
pipe flow	$\frac{p'd^2}{32\bar{v}}$	$\xi\frac{\bar{v}}{d}$	6.53	$8\left[\frac{4}{3+\frac{1}{m}}\right]^{\frac{m}{1-m}}$	$\xi_3 = \xi\left(\frac{1}{3}\right)$
slit flow	$\frac{p'\tilde{d}^2}{32\bar{v}}$	$\xi\frac{\bar{v}}{\tilde{d}}$	7.36	$4\sqrt{6}\left[\frac{3}{2+\frac{1}{m}}\right]^{\frac{m}{1-m}}$	$\tilde{d} = \left(\frac{8}{3}\right)^{1/2} s;\ \xi_3 = \xi\left(\frac{1}{3}\right)$
Couette flow	$\frac{M(1/\beta-1)}{8\pi^2hR^2v}$	$\xi\ v$	$\bar{\xi}_1 = \frac{12.6}{\ln 1/\beta}$ $\bar{\xi}_2 = \frac{11.9}{1-\beta}$	$\frac{4\pi}{1-\beta}\left[\frac{m(1-\beta^{1/m})}{1-\beta}\right]^{\frac{m}{1-m}}$	$\left(\frac{R_1}{R}\right)^2 = \beta$ $\bar{\xi}_1$ for $\beta \geq 0.5$; $\bar{\xi}_2$ for $\beta \geq 0.8$; $\xi_3 = \xi\left(\frac{1}{3}\right)$
torsional flow	$\frac{Mh}{\pi^2R^4v}$	$\xi\ v$	$4.93\frac{R}{h}$	$\frac{R}{h}\ 2\pi\left(\frac{m+3}{4}\right)^{\frac{1}{1-m}}$	creeping flow; $\xi_3 = \xi(3)$
axial annular flow	$\frac{p'\tilde{d}^2}{32\bar{v}}$	$\xi\frac{\bar{v}}{\tilde{d}}$	7.36	$8\left[\frac{4}{3+\frac{1}{m}}\right]^{\frac{m}{1-m}} \cdot \left[\frac{(1-\lambda^2)^{1+\frac{1}{m}} - \kappa^2\left(\frac{\lambda^2}{\kappa}-\kappa\right)^{1+\frac{1}{m}}}{\left(1+\kappa^2-2\frac{1-\kappa^2}{\ln 1/\kappa^2}\right)^{\frac{1}{2}\left(1+\frac{1}{m}\right)}(1-\kappa^2)}\right]^{\frac{m}{1-m}}$	$\tilde{d} = d\left(1+\kappa^2-2\frac{1-\kappa^2}{\ln 1/\kappa^2}\right)^{1/2};\ \kappa = \frac{R_1}{R};\ \alpha = \frac{1}{m}$ $\int_{\kappa/\lambda}^{1}\left(\frac{1}{\xi}-\xi\right)^{\alpha} d\xi = \int_{1}^{1/\lambda}\left(\xi-\frac{1}{\xi}\right)^{\alpha} d\xi \Rightarrow \lambda(\alpha, x)$
rotational concentric sphere-sphere-flow	$\frac{M(1/\delta^3-1)}{16\pi^2R^3v}$	$\xi\ v$	$\frac{5.2}{1-\delta}$	$\left[\frac{4(m+2)\Gamma\left(1+\frac{m}{2}\right)\left(\frac{m}{6}\right)^m(1-\delta^{3/m})^m}{\pi^{m-1/2}\Gamma\left(1+\frac{m+1}{2}\right)(1-\delta^3)}\right]^{\frac{1}{1-m}}$	$\delta = \frac{R_1}{R} > 0{,}8$ $\Gamma(x)$: Gammafunction
rotating disc flow (one side wetted)	$\frac{3\ M}{4\pi d^3 v}$	$\xi\ v$	10.1	numerical calculation	creeping flow

From the deviation between ξ and $\bar{\xi}$

$$\Delta_\xi = \left|1-\frac{\bar{\xi}}{\xi}\right|$$

we can calculate the deviation between $\Lambda(\xi) = 1$ and $\Lambda(\bar{\xi})$ as

$$\Delta_\Lambda = |1-\Lambda(\bar{\xi})| = \left|1-\left(\frac{\bar{\xi}}{\xi}\right)^{1-m}\right|$$

which is **smaller than** Δ_ξ. It vanishes identically for the Newtonian case (m=1).

5. Summary

As we have shown it is possible to get one point of the flow curve by elementary evaluation from one point of the measured relation. The factor ξ_3 in c/s-cases is known exactly. Usually ξ can be calculated as $\xi(m)$ by an estimation of the double-logarithmic slope of the measured curve or by using the $\bar{\xi}$-value for which between m = 0.25 and unity the deviation of η_R is smaller than 5%, normally much smaller. That means only for high shear thinning behaviour the real value $\xi(m)$ is needed. The method has also been tested successfully for a lot of real fluids.

6. REFERENCES

Frederickson A G, Bird R B, (1958) Ind. Eng. Chem. 50: 347

Schümmer P (1969) Chem. Ing. Techn. 41: 1020

Chmiel H, Schümmer P (1972) Rheol. Acta 11: 1, (1975) Rheol. Acta 14: 666

von Brachel H, Schümmer P (1975) Rheol. Acta 14: 219

Hoffmann L, Schümmer P, Schwerdt H (1975) Rheol. Acta 14: 626

Schümmer P, Worthoff R H (1975) Rheol. Acta 14: 919

A NEW METHOD OF STANDARDIZATION OF UNSTEADY MEASUREMENTS APPLIED TO A SHEAR RATE CONTROLLED RHEOMETER

I. TALBI-BOUCENNA and D. QUEMADA

Laboratoire de Biorhéologie et d'Hydrodynamique Physico-chimique. Université PARIS VII - 2 Pl. Jussieu. 75251 Paris-cedex 05, FRANCE.

INTRODUCTION

Rheological measurements in response to unsteady solicitations exhibit effect of coupling the mechanical characteristics of the rheometer and the unsteady behavior of the fluid (thixotropic and viscoelastic). This coupling occurs if the « thixotropic » and instrumental times lie in the same range. Hysteresis curves obtained for some standard oils submitted to a rapid solicitation are anomalous responses due to such a coupling. The aim of this study is the development of a new standardization method for unsteady measurements performed in shear rate controlled rheometers.

METHOD

The method consists in applying shear rate steps to standard silicon oil in order to determine corrections taking account of electromechanical characteristics of the instrument.

The rheometer used for this study is a Couette rotational rheometer The measuring bob (the inner cylinder) is coupled to a deflection system in the measuring head. When the deflection (due to the torque exerted by the fluid on the measuring bob) is transmitted to the system, it is scanned optically and compensated in order to keep the measuring bob fixed under steady conditions.

We write the equation of motion for the measuring bob.

$$I\frac{d^2\theta}{dt^2} - M_m + M_r = 0 \qquad (1)$$

where M_r is the reaction torque and M_m the measured torque of the inner cylinder. I is the moment of inertia and θ the rotation angle (deflection) of the measuring bob.

Using the ratio of the shear rate and shear stress constants ($\beta = F_\sigma / F_{\dot{\gamma}}$) equation (1) can be written in terms of deformation γ and derivatives, and $\dot{\gamma}_o$ (applied shear rate in the fluid):

$$\beta I \ddot{\gamma}(t) + \eta(\dot{\gamma}(t) - \dot{\gamma}_o) + \beta k \gamma(t) = 0 \qquad (2)$$

Where k is the constant of the torsion wire, and η the viscosity of the fluid.

In order to correct measurements for the effect of electromechanical characteristics of the instrument, for a given sensitivity (S = 1, 2, 3, 4, 5), we assume that the instantaneous measured $\gamma(t)$ which corresponds to the intensity i(t), is governed by the following equation :

$$\alpha \ddot{\gamma}(t) + D(\dot{\gamma}(t) - \dot{\gamma}_o) + E\gamma(t) = 0 \qquad (3)$$

For the steady conditions, the limit of the equation (3) gives $D = \eta_{oil}$, then the data fitting by the theoretical model will permit us to determine α and E, which cumulate both mechanical and electrical characteristics including sensitivities.

RESULTS AND DISCUSSION

A low shear rate rheometer (Low Shear 40, coaxial cylinders geometry) was used. Measurements are expressed in terms of shear stress in the fluid, $\sigma(t) = \eta_{oil}\dot{\gamma}(t)$, for a given sensitivity (1, 2, 3, 4 and 5), in response to shear rate steps (15 steps from 0,1 sec^{-1} to 5 sec^{-1}).

Figures 1, 2, 3 and 4 show the variation of the shear stress versus time for two silicon oils (10 CSt and 50 CSt). For simplicity only results for two shear rates are shown on the figures.

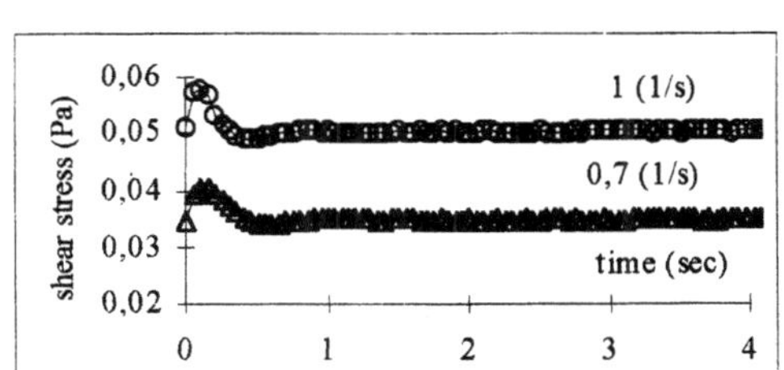

Figure 1 : Unsteady shear stress response of silicon oil (50 CSt) at $\dot{\gamma}$ =0,7 sec^{-1} and $\dot{\gamma}$ =1 sec^{-1} (sensitivity = 3).

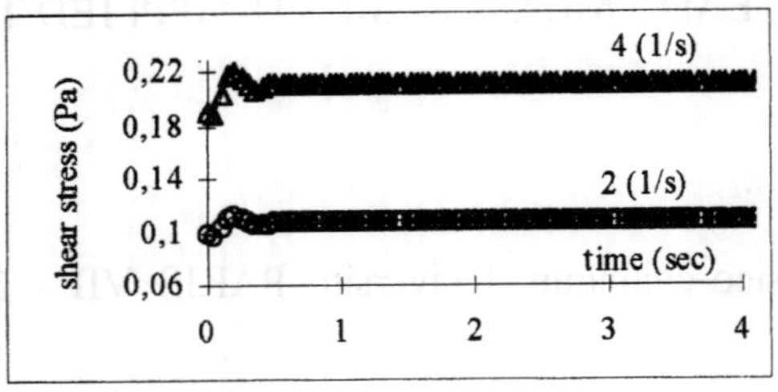

Figure 2 : Unsteady shear stress response of silicon oil 50 CSt) at $\dot{\gamma}$ =2 sec^{-1} and $\dot{\gamma}$ =4 sec^{-1} (sensitivity = 4).

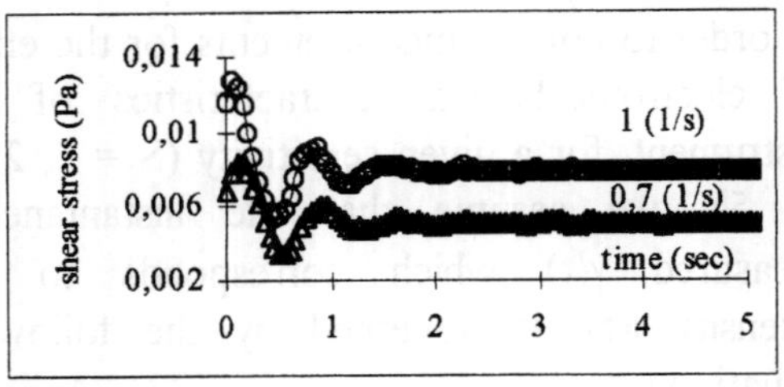

Figure 3 : Unsteady shear stress response of silicon oil (10 CSt) at $\dot{\gamma}$ =0,7 sec^{-1} and $\dot{\gamma}$ =1 sec^{-1} (sensitivity = 3).

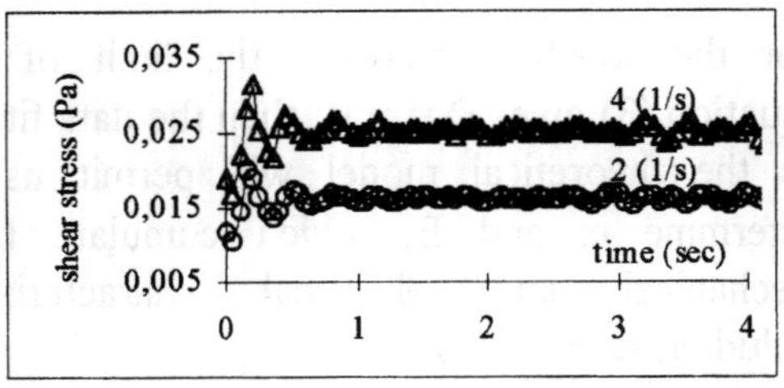

Figure 4 : Unsteady shear stress response of silicon oil (10 CSt) at $\dot{\gamma}$ =2 sec^{-1} and $\dot{\gamma}$ =4 sec^{-1} (sensitivity = 4).

Responses on figures 1, 2, 3 and 4 depend on time, oil viscosity, and sensitivity.

Results of data fitting by a theoretical model, based on equation (3) are represented on figures 5 and 6, which show very good agreement between modeling and experiment.

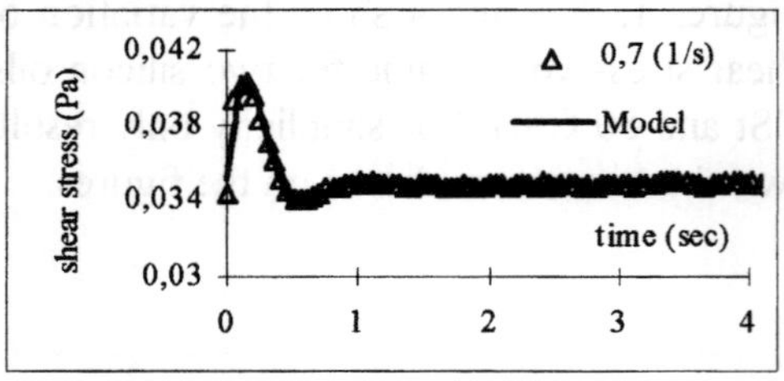

Figure 5 : Fitted plot of figure 1.

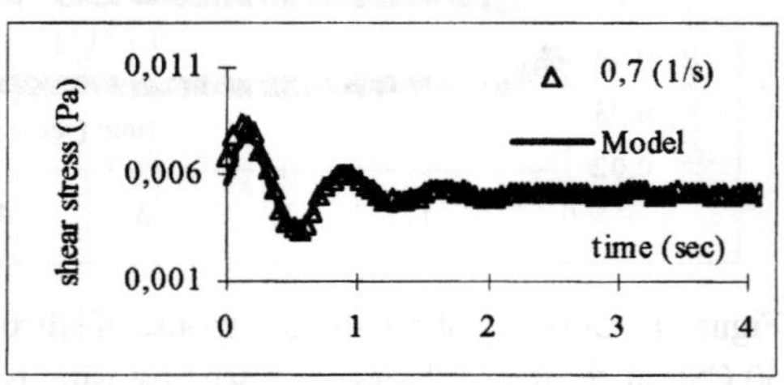

Figure 6 : Fitted plot of figure 3.

All the data fittings by the equation (3) for two silicon oils (for two sensitivities and different shear rates) and 15 shear rate steps show us that α and E are independent of fluid viscosity and shear rate but vary with sensitivity, such as (IS units) :

S = 3 :
$\alpha = 7\ 10^{-3}$ (kg.m^{-1})
E = 0,55 (kg.m^{-1}s^{-2})

S = 4 :
$\alpha = 9\ 10^{-3}$ (kg.m^{-1})
E = 4,36 (kg.m^{-1}s^{-2})

Once the parameters of the theoretical equation are defined, we can model the unsteady response of red blood suspensions using a « thixo - elastic » model for these fluids. Such a behavior is shown on figure 7.

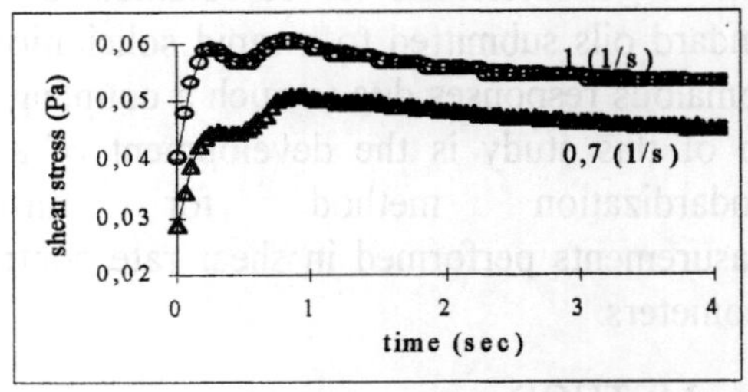

Figure 7 : Unsteady shear stress response of 40% red blood cell suspended in Dextran (m. w.= 72000) at $\dot{\gamma}$ =0,7 sec^{-1} and $\dot{\gamma}$ =1 sec^{-1} (sensitivity = 3).

CONCLUSION

In the case of shear rate controlled rotating rheometers, a phenomenological differential equation was used for the bob's movement. A « dynamic standardization » of shear stress measurements (which includes all electromechanical effects of the fluid - bob coupling) is determined from a series of measurements on standard silicon oils. Predictions of shear stress vs. time variation in blood samples, using a « thixo-elastic » model, taking this standardization into account will be discussed elsewhere.

REFERENCES

Operating Instructions : Low Shear 40, technical report, M E - 709208, Mettler - Toledo AG1991

J.C. Healy and M. Joly. *Biorheology*, **12**, 335-340 (1975).

U. Pollice, G. Viggiano, B. Accarino, and R. Santamaria. *Rheol. Acta* **21**, 201-206 (1982).

RHEOLOGY OF CEMENT PASTES WITH ULTRASONIC SHEAR WAVE METHOD

M.I. VALIČ[1], J. STEPIŠNIK[2]

[1]Faculty of maritime and transport studies, University of Ljubljana, Portorož, Slovenia
[2]Department of physics, FMF, University of Ljubljana, Ljubljana, Slovenia

1. INTRODUCTION

Presented in this contribution are some applications of a novel nondestructive method based on the ultrasonic shear waves reflection (USWR) for studies of rheological properties of cement pastes in unsheared conditions, made with a new apparatus.

2. EXPERIMENTAL

Materials; The cements used, CEM I 52.5R and PC 30dz 45s, are commercial, generally available products produced by Cementarna d.o.o., Salonit, Anhovo, Slovenia.

Apparatus; The measurements were made with a home-built *'Hardening meter'*. The apparatus comprises a measuring head, ultrasonic transmitter/receiver electronics and a PC computer with proprietary software (Valič, Stepišnik, 1998).

Method; When an ultrasonic wave hits an interface it is partially reflected and partially refracted. In the case of a normal incidence, the amplitude ratio of the reflected (A_2) and of the incident (A_1) wave is the reflection coefficient *r*,

$$r = A_1/A_2 = (Z_2-Z_1)/(Z_2+Z_1). \quad (1)$$

Z_1, Z_2 are the acoustic impedances determined by the viscoelastic properties of each of the two interface forming media. For the case of an incident shear wave

$$Z = [\rho(G+i\eta\omega)]^{1/2}. \quad (2)$$

G is the dynamic shear modulus, η the dynamic viscosity, ρ the mass density and ω the angular sonic frequency. Thus, by knowing Z, the viscoelastic properties of a medium can be evaluated. One way of determining Z of a medium is by measuring *r* from an interface formed with the medium of known Z. It has been shown experimentally (Stepišnik et. all, 1981), for a quartz-cement paste interface, that Z_2 is real and

$$G_2 = (1/\rho_2)[Z_1(1-r)/(1+r)]^2. \quad (3)$$

Since the Δr changes during the early hydration of PC cements are small, eq. (3) can be approximated with

$$G_2 = (2/\rho_2)[Z_1(\Delta r)]^2 \quad (4)$$

The shear modulus is thus $\propto (\Delta r)^2$.

Measurements of *r* in the manner suggested by eq. (1) are not possible with enough accuracy. Instead, the first and second back echoes are used in the apparatus, their amplitudes being related to *r* in a straightforward manner (Valič, Stepišnik, 1998).

3. RESULTS

In Fig. 1 the Δr change with time during the hydration (first 8 h period) for PC 30dz 45s cement pastes with w/c ratios (0.26, 0.30 and 0.40) and at T = 22 °C is shown.

In Fig.2 the Δr time dependence due to the hydration (first 8 h period) for CEM I 52.5R paste (w/c =0.40) at different temperatures (20, 26, 28, 35, 43 and 58 °C, respectively, is plotted; (the bottom curve corresponding to the lowest, the top to the highest temperature).

In Fig. 3 the Δr time dependence of CEM I 52.5R paste with w/c = 0.27 close to (26,8 %) required for the normal consistency, is shown. The setting times (initial, final) for this paste (110 and 155) min, determined by Vicat needle test, are indicated by two vertical lines.

3. DISCUSSION

The results of Fig.1 show that the reflection coefficient r of cement pastes decreases with continuing hydration consistently with the increase of its rigidity. The rate of Δr change decreases progressively with the increase of the amount of water.The forms of hydration curves seem to be sigmoidal with an induction period the length of which depends on the amount of water.

In Fig. 4 an Avrami type fit of the curve with w/c = 0.26 of Fig. 1 is shown. The fit is remarkably good in view of the complexity of the multicomponent, polysize PC cement system not being encountered for by Avrami equation. From this result it follows, that $-\Delta r(t)$ is proportional to the amount of hydrated phase $\alpha(t)$. On the other hand, it is well known, from the cement rheology data, that the compressive strength σ of a cement pastes is $\propto \alpha$. Therefore, the growth $\sigma(t)$ is thus $\propto -\Delta r(t)$. This is not surprising since σ is $\propto E^{1/2}$ and E is $\propto$ G via Poisson relation. In the diagrams presented, $-(\Delta r)$ vs. t is plotted. In this way they resemble to the classical diagrams of compressive strength time development during hydration.

The result of Fig. 2 demonstrate a very strong dependence of the reflection coefficient r on temperature. This is consistent with the high activation energy for hydration process obtained also from the measurements with other techniques. The shape of the dependence is influenced, in particular at longer hydration times (after setting), by the humidity and other prevailing conditions in the variable temperature unit used. A detailed analyses of a larger amount of measurements made will be published elsewhere.

The $\Delta r(t)$ dependence in Fig. 3 indicates that the setting of this particularly dense paste commences immediately after smearing the paste on the measuring head. In contrast with Vicat, the USWR method gives a continuous information of the setting and hardening process with time. Obviously, there are no abrupt changes in the Δr curve at Vicat setting times. However, their magnitudes at these times (0.34 and 0.62 db, respectively) can be used to determine a correlation between the two methods. Calibrating the USWR apparatus for a given cement type in this way, the setting times of production samples can be effectively measured. Repeated measurements on identically made pastes have demonstrated the reproducibility and the sensitivity for different cement types.

4. CONCLUSION

The measurements of the reflection coefficients of cement pastes with USWR method render very useful rheological information of the material. The prototype 'Hardening meter' developed for this purpose performs well offering several industrial applications, one of which is the setting time determinations in cement production as a complementary method to the standard one used.

5. REFERENCES

Valič MI, Stepišnik J (submitted to J. Metals, Alloys and Technologies)

Stepišnik J, Lukač M, Kocuvan I (1981) Cer. Bull. 60: 481-483

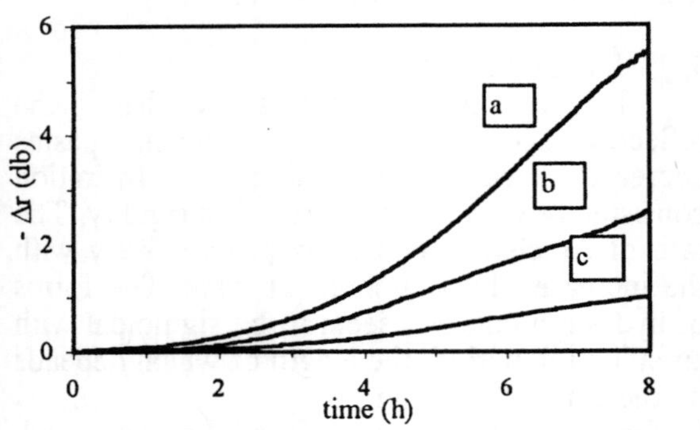

Fig.1: Early hydration of PC 30dz 45s pastes with w/c ratios; a) 0.26, b) 0.30, c) 0.40

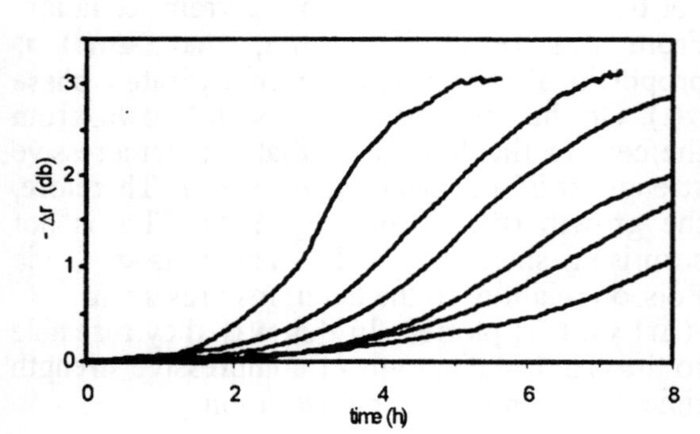

Fig.2: Hydration of CEM I 52.5R paste (w/c =0.40) at different temperatures 20. 26. 28., 35. 43 and 58 °C (the lower curve corresponding to the lower T)

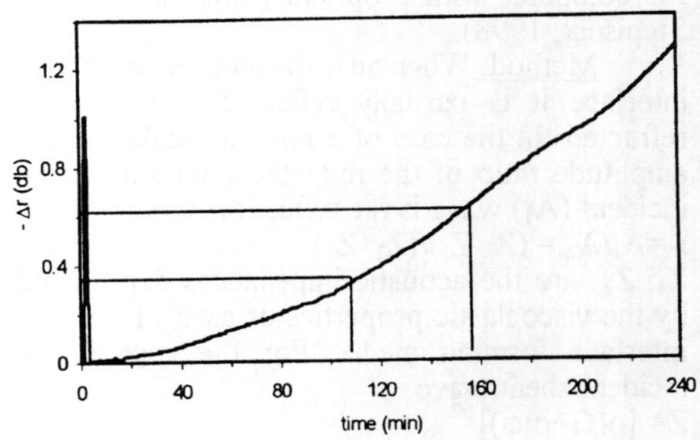

Fig.3: Early hydration of CEM I 52.5R paste with w/c of normal consistency

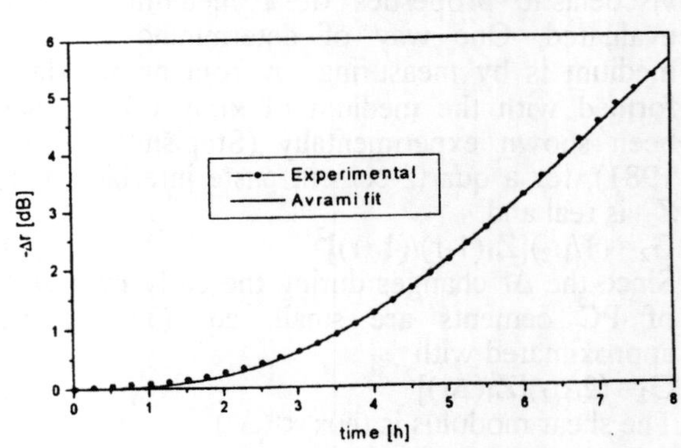

Fig.4: Fit of the experimental $-\Delta r(t)$ cement hydration curve (w/c = 0.26, Fig. 1) with the Avrami equation

PARALLEL VERSUS ORTHOGONAL SUPERPOSITION RHEOMETRY

J. VERMANT, P. MOLDENAERS AND J. MEWIS

Chemical Engineering Department., K.U. Leuven, de Croylaan 46, B-3001 Leuven, Belgium

1. INTRODUCTION

In superposition rheometry, the nonlinear condition, resulting from a steady-state shear flow, is analysed by applying a linear perturbation at various frequencies. In this manner, a mechanical spectroscopy of complex fluids under flow can be performed. Two different superposition methods are available, depending on the direction of the superimposed oscillatory deformations. In *parallel* superposition measurements the oscillations are parallel to the main flow direction, whereas in *orthogonal* superposition they are perpendicular to it.

Superposition measurements have received some attention, both theoretically and experimentally, in the 60's and 70's (see e.g. Booij 1966, Tanner 1968, Leonov et al. 1976, Mewis and Schoukens 1978). Recently, superposition measurements regained interest as parallel superposition became available on a number of commercial instruments, data interpretation is however not straightforward. Instruments for orthogonal superposition measurements have been described in the literature (see e.g. Simmons 1966). It has been shown recently that a simple modification of a rheometer equipped with a force rebalanced transducer makes it also possible to perform this type of measurements (Vermant et al. 1997).

Here we compare, both experimentally and theoretically parallel and orthogonal superposition moduli. Some applications on complex fluids will be adressed as well.

2. ANALYSIS OF SUPERPOSITION FLOWS

For superposition measurements, the experimental data are often discussed in terms of the analysis by Yamamoto (1972), in which a perturbation spectrum is associated with each shear rate. It can be shown that in *orthogonal* superposition, the moduli depend on the corresponding spectra in the same way as derived in the theory of linear viscoelasticity. Hence they seem to retain their physical meaning as a storage ($G'_\perp$) and a loss ($G''_\perp$) modulus:

$$G'_\perp = \int_{-\infty}^{\infty} H(\tau,\dot\gamma^2)\cdot\frac{(\omega\tau)^2}{1+(\omega\tau)^2}\,d\ln\tau \qquad (1)$$

$$G''_\perp = \int_{-\infty}^{\infty} H(\tau,\dot\gamma^2)\cdot\frac{\omega\tau}{1+(\omega\tau)^2}\,d\ln\tau \qquad (2)$$

Also, the Kramers-Kronig relationship between the moduli is preserved. It can be noted that :

$$\lim_{\omega\to 0}\left(\frac{G''_\perp(\dot\gamma)}{\omega}\right) = \eta(\dot\gamma) \qquad (3)$$

These results seem to make orthogonal superposition a suitable method to analyze effects of flow on complex fluids. The expressions for the moduli obtained for the case of *parallel* superposition reveals an intrinsic disadvantage. They do not only contain the perturbation spectra, but also the derivatives with respect to shear rate squared (see e.g. Yamamoto 1972). This constitutes a practical difficulty while using this method. The Kramers-Kronig relation is no longer valid for parallel superposition moduli. The zero limit of the parallel dynamic viscosity yields:

$$\lim_{\omega\to 0}\left(\frac{G''_{//}(\dot\gamma)}{\omega}\right) \neq \eta(\dot\gamma) \qquad (4)$$

It can be concluded that the coupling of the main and superimposed flows causes the parallel superposition moduli to loose the physical meaning of a storage and a loss modulus and they should not be interpreted as such.

However, an issue that needs to be resolved is whether the approach using perturbation spectra adequately describes nonlinear effects. To assess this, the superposition moduli were calculated for a Wagner I model. The orthogonal superposition moduli now become:

$$G'_{\perp} = \int_{-\infty}^{\infty} H(\tau) \cdot \frac{(\omega\tau)^2}{(n\dot{\gamma}\tau + 1) \cdot \left[(n\dot{\gamma}\tau + 1)^2 + (\omega\tau)^2\right]} \, d\ln\tau \qquad (5)$$

$$G''_{\perp} = \int_{-\infty}^{\infty} H(\tau, \dot{\gamma}^2) \cdot \frac{\omega\tau}{\left[(n\dot{\gamma}\tau + 1)^2 + (\omega\tau)^2\right]} \, d\ln\tau \qquad (6)$$

It can be seen from eqns. (5) and (6) that the Wagner I model leads to different predictions for the superposition spectra, calculated from either $G'_{\perp}$ or $G''_{\perp}$. This result suggests that care should be taken in interpreting orhogonal data in terms of a shear rate dependent spectrum. However eqn. (3) is still satisfied by the Wagner model.

3. MATERIALS AND METHODS

As a model viscoelastic fluid, a 4% (w/v) solution of a high molecular weight poly-(isobutene) (PIB - BASF Oppanol B200) in decalin (Fluka Chemical) has been prepared.

A rheometer equipped with a force rebalanced transducer (FRT) has been modified to perform orthogonal superposition measurements (Vermant et al. 1997). Care was taken in selecting the flow geometry, i.e. a hollow double wall Couette cell was used. The parallel superposition moduli during flow were measured using a Rheometrics DSR equipped with a cone and plate geometry.

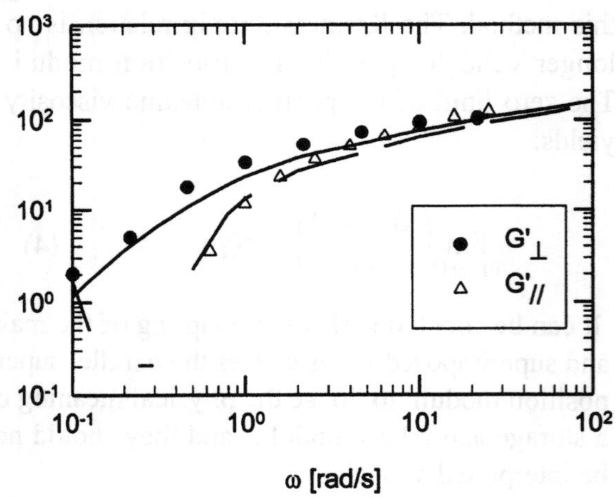

Fig.1. $G'_{\perp}$ and $G'_{//}$ for a PIB solution at 0.8 s^{-1}. The lines given the predictions for the Wagner model (— $G'_{\perp}$ and — — $G'_{//}$)

4. RESULTS

Figure 1 shows typical results for the PIB solution under shear. It can be seen that the parallel moduli are indeed lower than the orthogonal ones, a trend which seems to be qualitatively predicted by most models. The difference is mainly caused by the difference in coupling between the two types of superposition flow with the main steady state flow. The qualitative agreement between the predictions of the Wagner model and the experimental values is fairly good.

5. CONCLUSIONS

Superposition measurements have been discussed by using a perturbation approach to the study of nonlinear viscoelasticity. Both orthogonal and parallel superposition are considered. The interrelations between the various functions, as known from viscoelastic theory, do not hold anymore for the linear superposition moduli. Predictions for the Wagner I model are presented to demonstrate this point.

Experimental results of both orthogonal and parallel superposition moduli on a model viscoelastic fluid confirm the general features of the theoretical analysis.

6. ACKNOWKLEDGEMENTS

JV is a post-doctoral fellow of the Fund for Scientific Research - Flanders (Belgium).

7. REFERENCES

Tanner RI (1968) Trans. Soc. Rheol. 12:155.

Yamamoto M (1972) Trans. Soc. Rheol.15:331.

Booij H.C. (1966) Rheol. Acta, 5:215.

Simmons J (1966), J. Sci. Instrum. 43:887.

Leonov AI, Lipkina EH, Paskhin ED, Prokunin; AN (1976) Rheol. Acta 15:411.

Mewis J, Schoukens G, Farad. Disc. Chem. Soc., (1978), 65:58.

Vermant J, Moldenaers P, Mewis J., Ellis M, Garritano R (1997) Rev. Sci. Instr., 68:4090.

Vermant J, Walker LM, Moldenaers P, Mewis J (1998) J.Non-Newt. Fluid. Mech., in press.

APPARENT WALL SLIP IN ROTATIONAL VISCOMETRY

ONDŘEJ WEIN
Institute of Chemical Process Fundamentals, Academy of Sciences, 165 02 Prague, Czech Republic

1. INTRODUCTION

The concept of *apparent wall slip* (AWS) was used *ad hoc* for the viscometric flows with constant wall shear stress, see Oldroyd (1955). Effect of AWS in parallel-plate viscometer (PPV), a situation with varying wall shear stress, was considered by Yoshimura (1988) and by Wein (1994) in a narrow-gap approximation (NGA) without stating correctly the related 2D problem. We formulate here correct AWS boundary conditions for viscometric flows. We confirm that the NG approximation is accurate enough for treating primary data from the PPV.

2. AWS BOUNDARY CONDITION

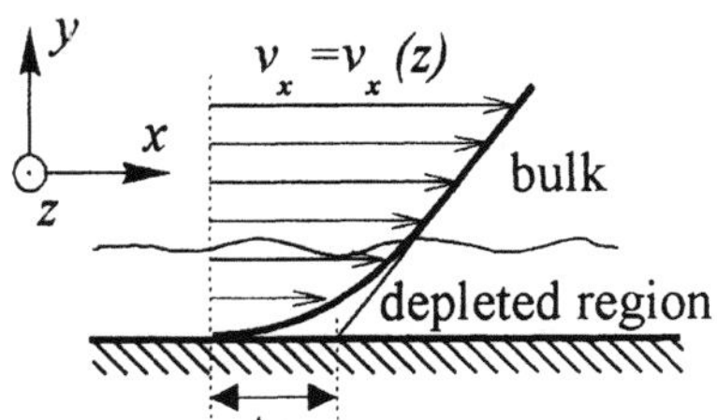

Fig 1a
Microscopic view on velocity profile within the depleted region at wall

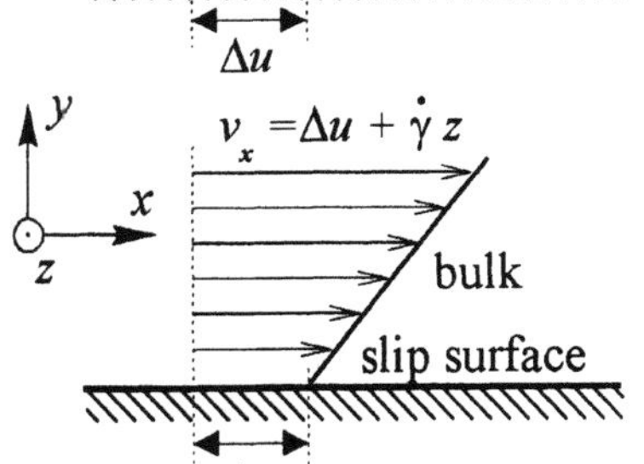

Fig 1b
Concept of apparent slip velocity as a fluid/wall interfacial property

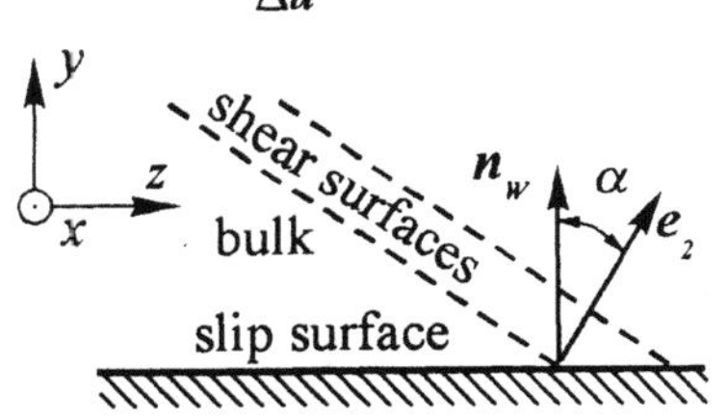

Fig 1c
Kinematics of the AWS effect for a general viscometric flow with non-parallel slip/shear surfaces

The AWS manifests interactions between a smooth solid wall and a flowing non-Newtonian microdisperse liquid. The simplest way of explaining AWS on a microscopic level is based on the idea of a partially depleted layer close to the wall. Within the depletion layer, Fig 1a, the local apparent fluidity $\varphi = \varphi(\tau; y)$ may be substantially higher than in the bulk flow, where it depends merely on the shear stress τ. From macroscopic view, a thin depletion layer at wall is reduced to the *slip surface* between the wall and a bulk material, Fig.1b. The anomalous behavior can be represented by the apparent slip velocity Δu. In a general viscometric flow, the wall slip surfaces do not coincide with viscometric shear surfaces, Fig 1c. A frame-indifferent formulation of the AWS boundary condition at wall should be formulated with respect to the local viscometric base ($\boldsymbol{e}_1$-flow direction, $\boldsymbol{e}_2$ -shear direction) which can differ from the wall-embedded base ($\boldsymbol{e}_z = \boldsymbol{n}_w$ is normal to the wall):

$$v_{1,\text{liquid at wall}} - \Delta u = v_{1,\text{wall itself}}\,,\quad \Delta u = u\cos(\alpha). \qquad (1)$$

The material function $u[\tau]$ of the scalar argument $\tau = |\tau_{12}|$ expresses a property of the interface between a smooth wall and a microdisperse liquid. In general case, $\alpha \neq 0$, the wall shear stress component τ_{xy} differs from τ_{12}, and the same holds for the corresponding shear rate and slip velocity components: $\boldsymbol{e}_2 \cdot \boldsymbol{n}_W = \cos(\alpha) = \tau_{xy}/\tau = \dot{\gamma}_{xy}/\dot{\gamma} = \Delta u/u$.

3. AWS IN A STEADY ROTATIONAL FLOW

For a given bulk fluidity $\phi[\tau] = \dot{\gamma}/\tau$ and slip coefficient $\chi[\tau] = u/\tau$, any steady rotational flow is completely characterized by the field of angular velocities $\omega(r,z)$ in polar cylindrical coordinates $\{\phi,r,z\}$: $v_\phi = r\omega(r,z)$, $v_r = 0$, $v_z = 0$; $\dot{\gamma}_{\phi z} = r\partial_z\omega$, $\dot{\gamma}_{\phi r} = r\partial_r\omega$, $\tau_{\phi z} = \phi^{-1}\dot{\gamma}_{\phi z}$, $\tau_{\phi r} = \phi^{-1}\dot{\gamma}_{\phi r}$. The local momentum balance $\partial_z(r^2\tau_{\phi z}) + \partial_r(r^2\tau_{\phi r}) = 0$, can be written as

$$\partial_z(r^3\phi^{-1}\partial_z\omega) + \partial_r(r^3\phi^{-1}\partial_r\omega) = 0, \qquad (2)$$

and should be solved with the AWS boundary condition (1) at axisymmetric walls,

$$\{\omega - \chi\phi^{-1}\boldsymbol{n}_W\cdot\nabla\omega\}_{\text{liquid}} = \omega_{\text{wall}}, \qquad (3)$$

where $\boldsymbol{n}_W\cdot\nabla\omega = (n_{W,r}\,\partial_r\omega + n_{W,z}\,\partial_z\omega)$ stands for the derivate of ω normal to the wall and $\omega_{\text{wall}} = \Omega$ (or $=0$) for the rotating spindle (or fixed vessel). In addition, there are homogeneous boundary conditions at the axis and free surface.

With a known solution to this problem, one can calculate the macroscopic torque on the rotating bob, $M = \int_{bob} \tau \cos(\alpha)\, 2\pi r^2\, dl$. A useful normalizing of the torque to an average shear stress over the bob surface, proposed by Mitschka et al. (1987), can be generalized for the flows under AWS effect in the following way

$$T \equiv \int_{bob} \tau \cos(\alpha)\, 2\pi r^2\, dl \Big/ \int_{bob} 2\pi r^2\, dl. \quad (4)$$

4. NARROW-GAP APPROXIMATION

Well-known NGA to the above problem can be applied for a class of general torsional flows if the gap width $h = h(r)$ between the plate and rotating bob is small and smoothly varying with r. The resulting equation for radial stress profiles under presence of AWS,

$$G_r \equiv \Omega r/h(r) = \dot{\gamma}[\tau] + 2u[\tau]/h(r), \quad (5)$$

was used tacitly by Yoshimura (1988) and explicitly by Wein (1994) to obtain a working equation for treating the PPV primary data ($h(r)$ = const) :

$$T = \tau_R - \int_0^{\tau_R} (G_r/G)^3\, d\tau(G_r, h). \quad (6)$$

Here, the kinematic consistency variable G and maximum shear stress τ_R are related through the condition at the disk rim, $r = R$:

$$G \equiv \Omega R/h(R) = \dot{\gamma}[\tau_R] + 2u[\tau_R]/h(R). \quad (7)$$

5. SOLUTION TO COMPLETE 2D PROBLEM

Numerical solutions by Wein (1997) to the Eqs (2),(3) for the PPV indicate that the NGA to the torque according to Eq.(6) is good enough for treating the PPV primary data, which consist of individual points $\{T, \Omega, R, h\}$ for at least two levels of h. For known material functions $\dot{\gamma}[\tau]$, $u[\tau]$, the theory provides the results in rather compact way, using the normalized gap width $H = h/R$ and normalized slip velocity, $A = u_G/(\Omega R)$, where $u_G = u[\tau_G]$, $\tau_G = \tau[G]$. The local material behavior at the proper stress level, $\tau = T$, is characterized by the corresponding logarithmic derivatives,
$m \equiv d \log \dot{\gamma}[\tau] / d \log \tau$, $w = d \log u[\tau] / d \log \tau$.

From such a viewpoint, the complete theory involves four independent parameters $\{A, H, m, w\}$. A result of the correct theory, shown below, confirms that the NGA prediction of the torque is very close to the exact values (the deviations below 1% even for H close to unity):

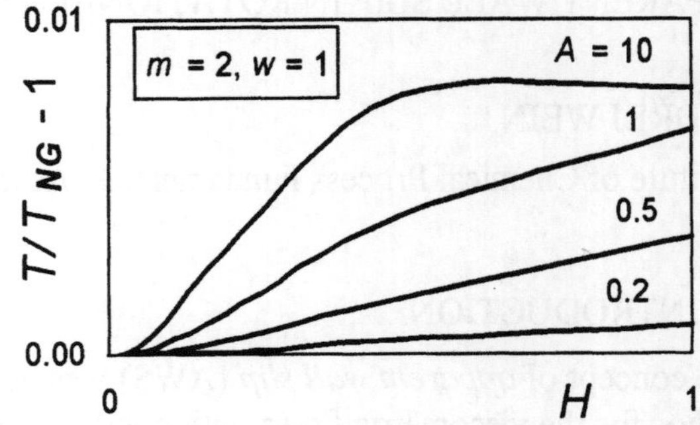

Fig.2 Comparing the NGA and correct torque predictions

The following picture shows a map of the shear surfaces, $\omega(r,z)$ = const. Thin lines corresponds to the NGA Eq.(5), thick lines to the correct theory. Notice that NGA starts with the assumption of ideal torsion flow, α =0, i.e. $\partial_r\omega$=0 (ω=const should then lay on the coaxial surfaces z = const).

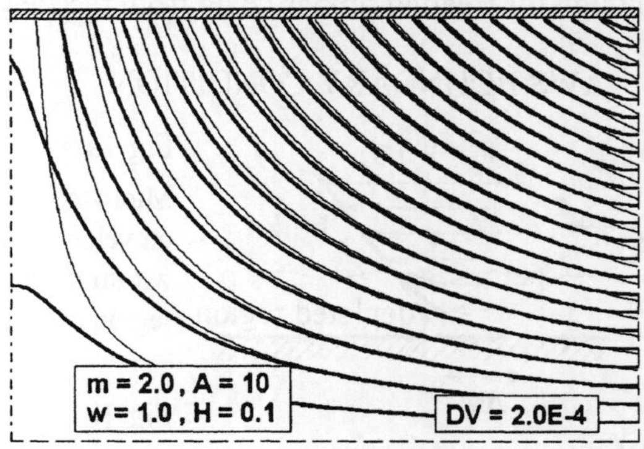

Fig 3. Mapping the shear surfaces, ω= const, for PPV

6. REFERENCES

Mitschka P, Wein O, Wichterle K (1987) In: Mashelkar RA (ed) Adv. Transport Processes 5. Wiley Eastern, New Delhi. pp 37-116

Oldroyd JG (1956) In: Eirich FR (ed) Rheology 1. Academic Press, New York. pp 653-681

Wein O (1994) In: Gallegos C (ed) Progress and Trends in Rheology 4. Steinkopff Verlag, Berlin. pp 528-530

Wein O (1997) Rheol Acta, submitted

Yoshimura A, Prud'homme RK (1988) J.Rheol 32:53-67

7. ACKNOWLEDGEMENTS

The Grant Agency of the Czech Republic, contract 104/94/1028, supported the project.

HIGH FREQUENCY VISCOELASTICITY OF CONCENTRATED COLLOIDAL DISPERSIONS

N. Willenbacher[1], J. Bergenholtz[2], F.M. Horn[3], W. Richtering[3], N.J. Wagner[4]

[1]BASF AG, Polymer Research Division, D-67056 Ludwigshafen, Germany
[2]Institut für Physik, Universität Konstanz, Germany
[3]Inst. f. Makromolekulare Chemie, Albert-Ludwigs-Universität Freiburg, Germany
[4]Department of Chemical Engineering, University of Delaware, Newark, DE, USA

1. INTRODUCTION

The high-frequency viscoelastic properties of concentrated colloidal dispersions are discussed. 'High-frequency' here means, that times probed by the experiment are much shorter than the timescale of diffusive colloidal motion, but still longer than momentum relaxation and vorticity propagation times. For typical liquid-like, industrial dispersions (particle radius a = 100 nm) this high frequency limit is around 10 kHz. The viscosity η'_∞ is mainly a hydrodynamic quantity, while direct interactions among the colloidal particles determine the modulus G'_∞. The concept to extract information about particle interactions from shear modulus measurements is not new, but here we present a method to determine the effective colloidal charge Q in concentrated (ϕ > 0.3), but liquid dispersions [1]. For electrostatically stabilized dispersions η'_∞ is investigated for the first time here and is compared to short time self diffusion coefficients D_s from fiberoptical quasi-elastic light scattering (FOQELS) [2].

2. METHODS

2.1 Measurements

A commercial torsional resonator (Physica Rheoswing) (resonance frequency in air $\omega_0^{air}/2\pi$= 8858.4 Hz, width of resonance curve $d^{air}/2\pi$ = 2.3 Hz) was used to determine the high frequency values of G' and $\eta' = G''/\omega$. Upon immersion into a liquid the resonance frequency ω_0 decreases and the damping d increases. For aqueous dispersions the penetration depth of the shear wave propagating into the liquid ranges roughly from 5 to 50 μm, thus probing bulk properties of the surrounding fluid. Since δ is much smaller than the radius of the oscillating rod (6 mm), the plane wave approximation applies [3] and accordingly, the moduli G' and G'' of the surrounding fluid are related to the changes of the resonance characteristics as follows:

$$\rho G' = K_1^2 [(\Delta d)^2 - (\Delta\omega)^2] \qquad (1a)$$

$$\rho G'' = K_2^2 \Delta d \, \Delta\omega \qquad (1b)$$

with $\Delta d = d^{fluid} - d^{air}$, $\Delta\omega = \omega_0^{air} - \omega_0^{fluid}$ and ρ = fluid density. The calibration constants K_1 = 74.2 kg/m² and K_2 = 128.6 kg/m² were obtained using a series of Newtonian liquids covering the viscosity range $1 < \eta < 80$ mPas. An ideal resonator would have $K_2 = 2K_1$.

FOQELS measurements were conducted using a self-built instrument [4] modified to yield two different wavelengths λ = 633 nm and 458 nm. Light is detected in the backscattering mode (q = 0.0268 nm^{-1} and 0.0366 nm^{-1}). For large enough particles these q-values are larger than that corresponding to the maximum of the structure factor, thus providing a probe of selfdiffusion.

2.2 Samples

Monodisperse poly(n-butyl-metacrylate) (PBMA) and poly(styrene) (PS) dispersions were prepared by emulsion polymerization. Sodium and potassium persulfate, resp., were used as initiator introducing strong (-SO_3^-) acid groups. Some of the PBMA dispersions were copolymerized with acrylic acid (PBMA/AA) giving rise to additional weak (-COO^-) acid groups. Styrenesulfonate was used as comonomer for the PS latexes, imparting strong acid sulfonate groups on the particle surface. All samples were dialyzed against de-ionized water, followed by ion-exchange and a final dialysis against salt solutions of known ionic strength and pH.

3. RESULTS AND DISCUSSION

3.1 High frequency modulus / surface charge

G'_∞ is related to the interaction potential Ψ and hence to the surface charge of electrostatically stabilized latices by the well known Zwanzig and Mountain formula [5], which for liquid-like dispersions with high ionic strength (κa >> 1) simplifies to

$$G'_\infty \frac{a^3}{kT} = \frac{3\phi}{4\pi} - \frac{6\phi^2}{5\pi} a_{eff}^4 g^{HS}(2;\phi) \frac{d(\frac{\Psi(r)}{kT})}{dr}\Big|_{r=2a_{eff}} \qquad (2)$$

if the pair correlation function g(r) is approximated by the hard sphere correlation function $g^{HS}(r/a_{eff})$ and if a short range interaction potential is assumed. The effective particle radius a_{eff} > a accounts for the extra excluded volume among particles due to their charged nature and is obtained experimentally from the divergence of the zero shear viscosity η_0 at ϕ_m^{exp} according to

$a_{eff} = \sqrt[3]{\phi_m^{HS} / \phi_m^{exp}}$, with $\phi_m^{HS} = 0.61$ being the hard sphere maximum packing fraction.

The well-known Carnahan-Starling expression is used for the contact value of $g^{HS}(\phi)$. The classical DLVO interaction potential is taken as Ψ, accordingly the corresponding surface charge Q_{eff} is an effective quantity (charge renormalization [6]). The counterions from the dissociated surface chemical groups and the reduced volume available for the ions due to the presence of the latex particles have been considered when calculating the ionic strength of the concentrated dispersions and are self-consistently included in the calculation of Q_{eff} from Eq. 2 .

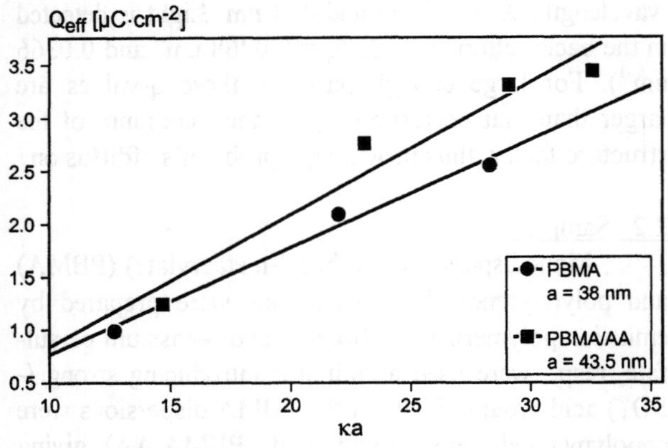

Fig. 1: Q_{eff} vs κa for PBMA and PBMA/AA dispersions

Q_{eff} increases linearly with increasing ionic strength for different PBMA dispersions with and without weak acid groups (Fig. 1). For particles with strong acid groups only, this increase is a consequence of the charge renormalization [6]. If weak acid groups are present an additional shift of the dissociation equilibrium has to be considered. From the slopes in Fig. 1 we get surface potential values Ψ_s = -56 mV and -78 mV for the PBMA and the PBMA/AA dispersions, respectively. The later value is in good agreement with ζ = -68 mV obtained from electrophoresis.

3.2 High frequency viscosity / short-time selfdiffusion

In Fig. 2 we compare η_0 with η'_∞ as functions of volume fraction for a series of ionic strengths. Both quantities are scaled with the solvent viscosity μ_0. As seen, $\eta_{0,r} = \eta_0/\mu_0$ strongly depends on ionic strength and diverges well below ϕ_m^{HS} showing that these dispersions are far from hard sphere like. In contrast, $\eta'_{\infty,r} = \eta'_\infty/\mu_0$ shows essentially no dependence on ionic strength and particle size, its increase with increasing ϕ is only modest. This demonstrates the hydrodynamic character of η'_∞. Moreover, the data for our charge-stabilized dispersions are in good agreement with Beenakker's hard sphere theory [7]. The short-time self diffusion coefficient D_s is also unaffected by changes in particle size and ionic strength, i.e. changes in the interaction potential. In Fig. 3 $(D_s/D_0)^{-1}$ is compared to η'_∞/μ_0. As seen,

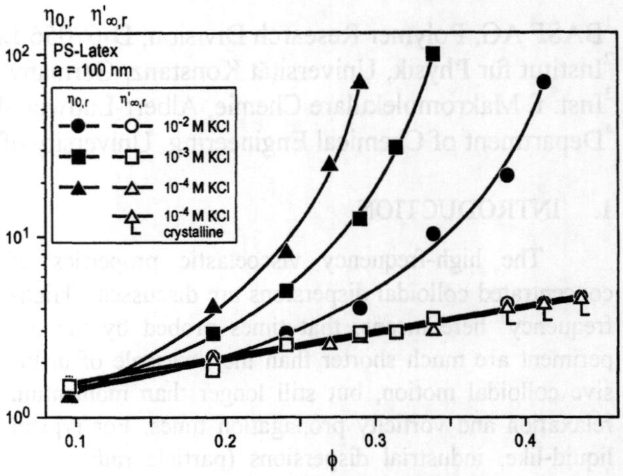

Fig. 2: η_0 and η'_∞ for PS dispersions of different particle size and ionic strength

$D_s(\phi)$ and $\eta'_\infty(\phi)$ are not simply related by a generalized Stokes-Einstein equation $D_s(\phi) = kT/6\pi a\eta'_\infty(\phi)$. Instead, $(D_s/D_0)^{-1}$ lies below η'_∞/μ_0 for all volume fractions as predicted theoretically by Benakker and Mazur. Our data agree better with their virial expansion [8] than with their many-body hard sphere theory [9] and more recent results from computer simulations [10].

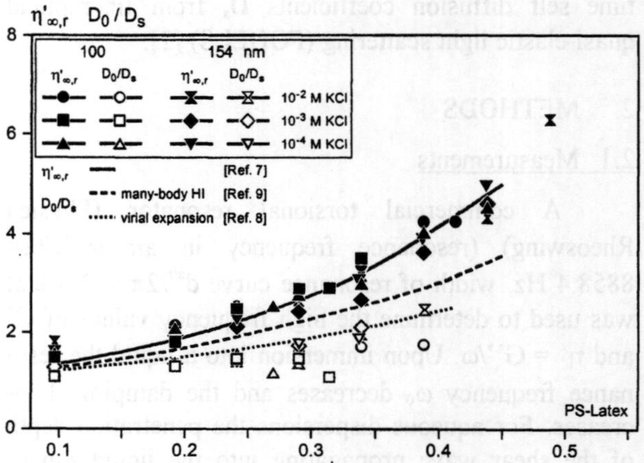

Fig. 3: D_s and η'_∞ compared to hard sphere theories

4. REFERENCES

[1] J. Bergenholtz et al., *Phys. Rev. Lett.*, submitted 1998
[2] J. Bergenholtz et al., *J. Colloid Interface Sci.*, in press 1998
[3] W.P. Mason, *Trans. ASME* 69, 359 (1947)
[4] H. Wiese and D. Horn, *J. Chem. Phys.* **94**, 6429 (1991)
[5] R. Zwanzig, R.D. Mountain, *J. Chem. Phys.* **43**, 4464 (1965)
[6] Alexander et al., *J. Chem. Phys.* **80**, 5776 (1984)
[7] C.W.J. Beenakker, *Physica A* **128**, 48 (1984)
[8] C.W.J. Beenakker and P. Mazur, *Physica A* **120**, 388 (1983)
[9] C.W.J. Beenakker and P. Mazur, *Physica A* **126**, 349 (1984)
[10] A.J.C. Ladd, *J. Chem. Phys.* **93**, 3484 (1990)

SHEAR WAVE DISPERSION IN GEL NETWORKS.

P.R.Williams and R.L.Williams,
University of Wales Swansea,Swansea, UK.

INTRODUCTION

At the gel point (GP), a material changes from one where there is short range connectivity to one where there is sample spanning structural self similarity. The linear stress relaxation modulus, $G(t)$, is characterised at the GP by the *Gel Equation*; $G(t) = St^{-\alpha}$ where α is the viscoelastic stress relaxation exponent (in the range $0 \leq \alpha \leq 1$) and S is the strength of the gel structure. The problems associated with GP determination by oscillatory shear measurements (eg. strain sensitivity and non-equilibrium nature of 'critical gels') can be overcome by high-frequency wave propagation measurements. Here the conjunction of viscoelasticity and wave propagation results in wave dispersion; Williams and Williams (1997). Wave dispersion can be used to characterise the gel-like properties of systems ($0 \leq \alpha \leq 1$) prior to the GP, in terms of the gel equation, by measurements at any *single* frequency. This is illustrated using self-similar networks known as 'ladder models' for various α; Williams and Williams (1998a).

THEORY

In plane shear wave propagation in a linear viscoelastic medium of density ρ, the complex shear modulus $G^*(= G' + iG'')$ is given by

$$G^* = -\rho\omega^2/\Gamma^2 \qquad (1)$$

where $\Gamma(= 1/x_o + i2\pi/\lambda)$ is a constant in which x_o is the exponential damping length and λ is the shear wave length. The phase velocity, v, is given by ω / k (where $k=2\pi/\lambda$ and ω is the frequency). G' and G'' are the dynamic rigidity and loss modulus respectively. In viscoelastic systems the v, is exceeded by the group-velocity, U, where

$$U/v = 1/[1-(\omega/v)\{dv/d\omega\}] \qquad (2)$$

G' and G'' are related, via the gel equation, as

$$G'(\omega) = G''(\omega)/(\tan \alpha\pi / 2) (3).$$

Equations (1),(2) and (3) have been used, Williams and Williams (1998b), to determine $(U-v)/v$ for all critical gels, $0 \leq \alpha \leq 1$. The wave dispersion of critical-gels, which is bounded by the elastic solid ($(U-v)/v=0$, $\alpha=0$) and the Newtonian fluid($(U-v)/v=1$, $\alpha=1$), is frequency independent at the GP.

Ladder models in which all nodal elements are *identical* simulate critical gels; Schiessel and Blumen (1995). Such 'self similar' systems are equivalent to series-parallel models embodying discrete relaxation time spectra (DRS) derived from the Gross-Marvin models, Tschoegl N. (1989). For an infinite number of nodes, n, this system simulates a critical-gel with $\alpha = ½$. Use of modified Gross-Marvin models allows simulation of critical gels for α in the range $0 \leq \alpha \leq 1$. Herein the pre-GP wave dispersion characteristics of these systems (i.e. networks with $n < \infty$) are examined.

RESULTS

Network evolution is depicted in Figure 1, in terms of tanδ for α=0.2, 0.5, 0.67 and 0.9 respectively, where the broken line indicates the frequency independent loss tangent predicted by the gel equation for an infinite network at each value of α (ω is made dimensionless by the inverse of the smallest relaxation time in the infinite network). For all α, no frequency independent tanδ is observed for $n < 10^4$ nodes, that is, in terms of the *Gel Equation* the system displays no 'gel-like' characteristics. Figure 2 depicts network evolution in terms of the wave dispersion number ($(U-v)/v$) as a function of frequency (broken line being the frequency independent wave dispersion number) for a range of α. It can be seen that in all systems the dispersion number tends to zero at the extremes of high frequency (the elastic limit) whilst approaching unity in the low frequency limit, as expected for networks of finite size. The

proximity of a system undergoing gelation to an equivalent 'critical gel' (CG) state at any single frequency can be assesed using the gel equation by comparison of the system wave dispersion (system dispersion) with that of the equivalent CG state wave dispersion (CG dispersion). Figure 3 compares system dispersion and CG dispersion (broken line) during network evolution for $\alpha=0.2$. For $n<75$ the curves are clearly separate apart from their viscous and elastic limits but tend to merge as n increases. For $n=100$ the curves are coincident over a small range of frequency indicating that the system dispersion is characteristic of a critical gel. It should be noted that a frequency independent loss tangent is not observed in this system for $n<10^6$ (see Figure 1).
Figure 4 shows the wave dispersion results for an evolving network for $\alpha=0.67$. Whilst the coincidence of system dispersion and CG dispersion does not occur until the network is far more extensive than for $\alpha=0.2$, it is still observed before the establishment of a frequency independent loss tangent.

CONCLUSIONS

Use of a modified Gross-Marvin model allows the simulation of network growth for gels in the range $0 \leq \alpha \leq 1$. Comparison of system wave dispersion with CG wave dispersion indicates that 'gel-like' behaviour can be observed during network growth and prior to any development of a frequency independent loss tangent.

REFERENCES

Schiessel H and Blumen A (1995) Macromol. 28: 4013-4019

Tschoegl N W 1989 The Phenomenological Theory of Linear Viscoelastic Behaviour (Berlin: Springer).

Williams P.R. and Williams R.L. (1997) J. Non-Newtonian Fluid Mech. 68: 311-322

Williams P.R. and Williams R.L. (1998a) J. Non-Newtonian Fluid Mech. *in press*

Williams P.R. and Williams R.L. (1998b) J. Phys.D 31:1083-1091

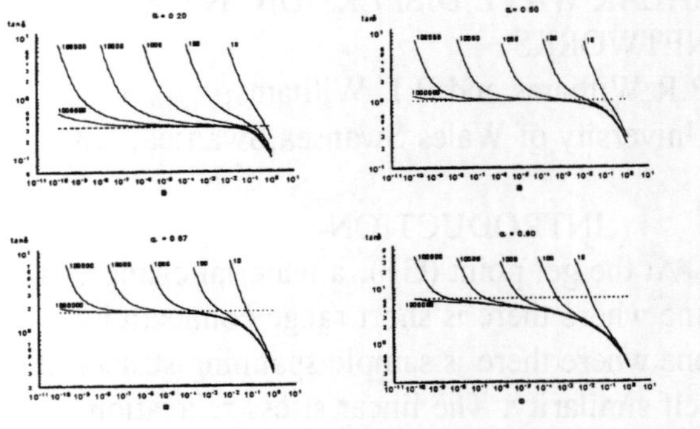

Figure 1 $\tan\delta(\alpha)$ as a function of network growth

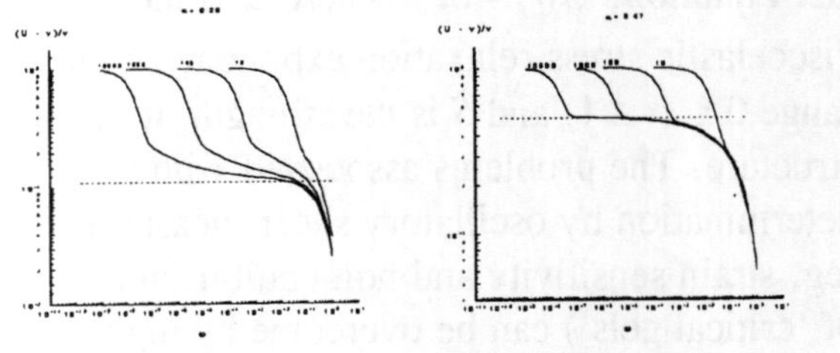

Figure 2 Frequency dependence of $(U-v)/v$ as a function of network growth and α

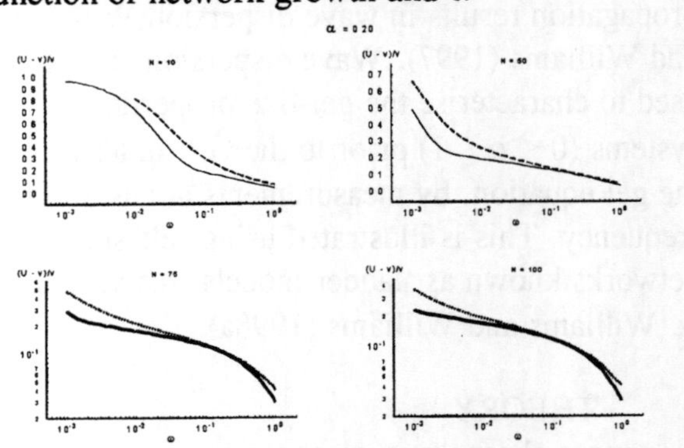

Figure 3 System dispersion and CG dispersion for $\alpha=0.2$ as a function of network growth.

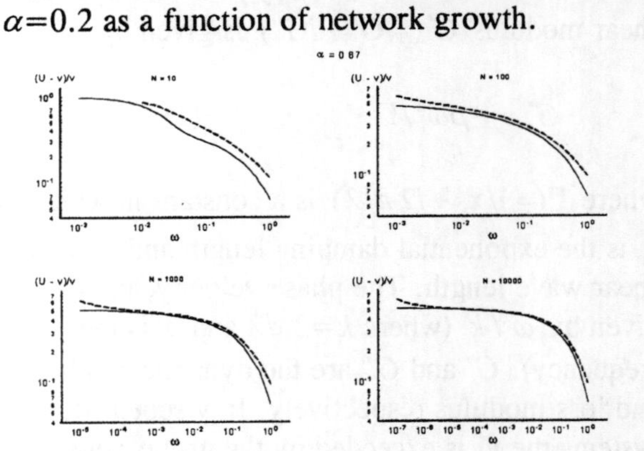

Figure 4 System dispersion and CG dispersion for $\alpha=0.67$ as a function of network growth.

ELASTIC EFFECTS IN LIQUID JETS FORMED BY BUBBLE COLLAPSE

P.R.WILLIAMS[1], S.W.J.BROWN[1], P.M.WILLIAMS[1]

[1]Chemical Engineering Dept., University of Wales, Swansea, Singleton Park, Swansea SA2 8PP UK

1. INTRODUCTION

Using a newly developed technique (Williams *et al.* 1997), we recently reported studies of liquid jets formed by bubble collapse under shockwaves and identified a mitigating effect of viscoelasticity on liquid jet formation. In that work, liquid jet production in commercial multigrade motor oils and their Newtonian counterparts was studied. The results showed that, in the former, liquid jet formation is markedly suppressed, with up to an order-of-magnitude reduction in length being recorded. This paper reports experiments on systems involving Newtonian solvents consisting of glycerol/water solutions, and their equivalents (in terms of shear viscosity) made elastic by the addition of polyacrylamide ('Magnafloc' E10) as in the `Worthington Jet` experiments reported by Cheny and Walters (1996).

2. METHOD

The technique used herein facilitates a type of dynamic stressing in which a liquid is subjected to a tension pulse. The apparatus has recently been described in detail (Williams et al 1997) but its basic aspects, and those of the associated technique, are now briefly described. The apparatus involves a cylindrical tube (length 100cm., i.d. 2.1 cm.) incorporating a syringe in its base through which a small gas bubble is introduced into a liquid column within the tube. Having risen through the liquid, the bubble rests beneath the free surface. The tube used in the experiments reported here is forced upwards and brought suddenly to rest, following which the liquid continues its upward motion, thereby generating a tension pulse, *ab initio*.

In order to study the formation of the liquid jets a Kodak EM high-speed video camera was used, with illumination being provided by a pair of Xenon flash lamps.

3. RESULTS

Fig.1 shows the video camera recording of events involving the jet formation in a column of 0.3 Pa.s Newtonian silicone oil, whereas Fig. 2 shows the comparative sequence of events in a column of 0.3 Pa.s elastic liquid containing 75ppm polyacryamide polymer, in which both sequences run sequentially from top-to-bottom of the page. The first two frames of each sequence record the growth of the bubble as the liquid goes into tension, the second frame recording the bubble close to maximum diameter immediately prior to the arrival of the shock wave. This causes the rapid collapse of the bubble towards the free surface of the liquid (frame 3) and is immediately followed by the formation of a liquid jet which rapidly grows in length (frames 4-6).

4. CONCLUSION

It is reasonable to conclude from the results of the present study, as in related work (Cheny and Walters 1996), that the suppression of the liquid jets is due to an increased resistance to elongational flow conferred upon the liquid by the polymer additive. The additional significance of the results, in the context of cavitation damage, is that this effect is found in studies involving jets formed by bubble collapse under shock waves. Insofar as jet production is either markedly reduced or, in some cases entirely suppressed, the findings reported here certainly identify a 'mitigating effect of viscoelasticity on cavitation' as mooted by Berker *et al.* (1995)

5. REFERENCES

A. Berker, M.G. Bouldin, S.J. Kleis and W.E. VanArnsdale (1995) J. Non-Newt. Fluid Mech., 56; 333.

P.R. Williams, P.M. Williams and S.W.J. Brown (1997) J. Non-Newt. Fluid Mech. 72:1 101.

Cheny, J M and Walters, K (1996), J. Non-Newtonian Fluid Mech., 67 125.

Figure 1 - Jet formation from bubble collapse in 0.3Pa.s Newtonian Glycerol/water liquid

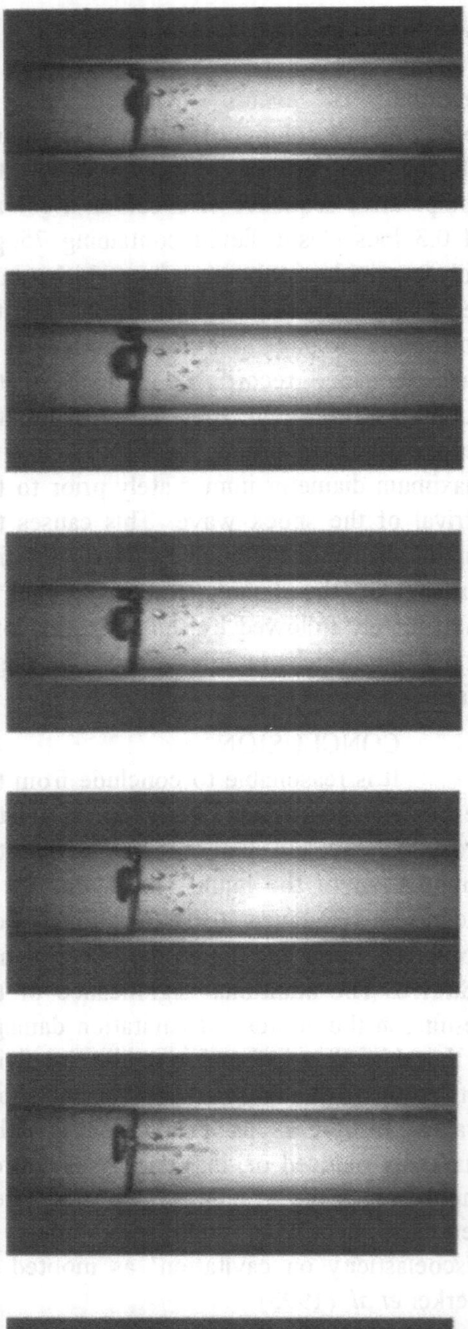

Figure 2 - Jet Formation in Elastic liquid (contains 75ppm polyacrylamide- same 0.3Pa.s shear viscosity)

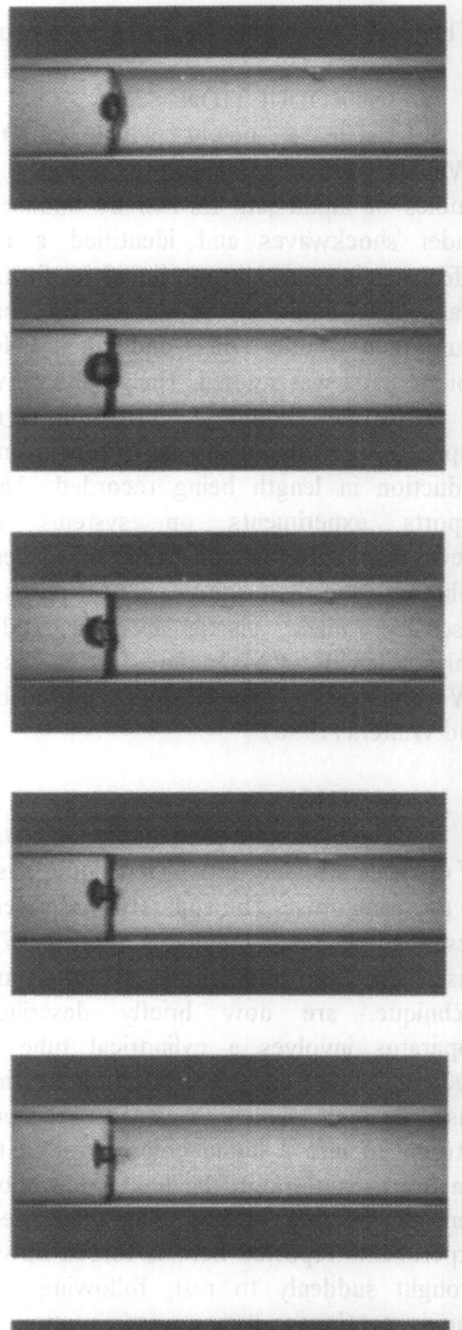

MEASUREMENTS OF THE TENSILE STRENGTH OF LIQUIDS BY THE FREE-SURFACE REFLECTION OF COMPRESSIONAL PULSES.

P.R.WILLIAMS[1], S.W.J.BROWN[1], P.M.WILLIAMS[1]

[1]Chemical Engineering Dept., University of Wales, Swansea, Singleton Park, Swansea SA2 8PP UK

1. ABSTRACT

Measurements of the velocity of tension pulses arising from the reflection of compressional waves from the free surface of water confirm predictions concerning the acoustic properties of a metastable transition layer (Temperley and Trevena 1979, 1987, 1994). This liquid-vapour layer, which leads to a reduction in the effective speed of sound, constitutes a source of cavitation nuclei. Accordingly, values of the tensile strength of water, F_c, derived from tension pulse velocities are commensurate with the lowest values in the range of theoretical estimates (~ 600 bar).

2. INTRODUCTION

A puzzling feature of research into the tensile strength of liquids over the past fifty years has been the discrepancy between theory and measurements involving the reflection of compressional waves as tension (Temperley and Trevena 1994). This is disquieting as the tensile strength of water, F_c, is estimated to be *ca.* 1300 bar (Fisher 1948) while some pulse reflection measurements yield only *ca.* 20 bar - a low value given that some biomedical applications of ultrasound might involve tensions greater than 40 bar (Marston and Unger 1986).

Attempts to explain this discrepancy invoke the concept of a metastable transition layer between liquid and vapour (Temperley and Trevena 1994). This involves the idea that, at a free surface, the density falls off with height with a characteristic distance of the order of a few intermolecular distances. Assuming that the equation of state of the system repesented by the metastable, vapour-liquid transition layer resembles the bulk equation of state then, at a certain height, the isotherm becomes flat and the effective velocity of sound is much reduced, becoming small compared with that in the bulk liquid. Temperley and Trevena's theory implies, *inter alia*, that this layer would constitute a source of cavitation nuclei and that, due to the dispersion and attenuation of pulses engendered within it, pressure transducers would be unable to measure F_c directly in pulse reflection experiments.

In order to test this theory we have eschewed the direct measurement of tension by pressure transducers, using them instead to measure the *velocity* of tension pulses generated by reflection of pressure from a free surface. The values of F_c derived therefrom exceed 600 bar and are the largest values hitherto reported for measurements which do not involve extrapolation using equations of state for water determined in the range of positive pressure, e.g. Zheng *et al.* 1991.

3. EXPERIMENTAL

The apparatus consists of a vertically mounted cylindrical tube containing deionised water: the tube is pulled downwards against springs and released. After the tube's arrest the liquid continues its upward motion generating a tension pulse, *ab initio*. Pressure transducers in the tube wall record the tension pulse which initates cavitation, resulting in the production of a shockwave which travels upwards through the liquid column to be reflected at the free surface as tension (Williams *et al* 1997).

The reflected tension pulse velocity, v_r , is measured by timing the passage of the pulse between two pressure transducers mounted in the lower section of the tube. From a knowledge of the distance, z, between the transducers (10 cm.) and the pulse transit time, t, the velocity v_r ($= z / t$) was computed.

An Imacon 468 CCD camera (Hadland Photonics) recorded cavitation in the liquid as the reflected pulse passed its optical axis. Typical results are shown in Fig.1, in which frame 2, recorded 0.6 ms after the passage of the pulse, shows a group of approximately spherical cavities distributed in the bulk liquid. This cavitation was evident only during the period immediately following the arrival of the tension pulse.

4. CALCULATION OF F_C

The potential tensile strength of a pure liquid may be illustrated by considering two molecules separated by a variable distance, r. Equilibrium occurs at the separation, r_o, typically of the order of 10^{-10} m. The attractive force, F, between the molecules is equal to df/dr where f is the potential energy associated with the intermolecular forces. F is a maximum at some distance, r_1, where typically r_1/r_o is of the order of 1.1 or 1.2. In a bulk liquid this corresponds to a fractional volumetric expansion, DV/V_o ,of about 1/3 (Frenkel 1955). Consequently the application of a constant tensile stress equal to that corresponding to r_1 would completely rupture the liquid since, for $r > r_1$, F is insufficient to counteract that tensile force and the velocity of sound is given by

$$c^2 = K / \rho$$

where K is the bulk modulus of the liquid, i.e. the reciprocal of its compressibility (the ratio of volume strain to stress); then

$$K = \rho c^2$$

At peak pressures of the order of hundreds of bar (as considered here) there is no serious error in using linear acoustic theory for water. The *linear* model of the relation between stress, F, and volumetric strain is

$$F = K \, . \, \Delta V / V_o$$

which becomes

$$F = \rho c^2 \, . \, \Delta V / V_o$$

using $\rho = \rho_o \, (r_o / r_1)^3$ and $\Delta V / V_o = (r_1 / r_o)^3 - 1$ then

$$F = \rho_o c^2 \, . \, \{1 - (r_o / r_1)^3\}$$

Williams *et al.* (1997) have reported that the velocity of the reflected tension pulses accompanying cavitation of deionised water in the 'tube-arrest' experiment corresponds to that in water-*vapour*, some 494 ms^{-1}. For ρ_o = 1000 kg/m^3, and c = 494 ms^{-1} this gives F_c = 607, 1023 bar for r_1 / r_o = 1.1, 1.2 respectively. This is in good agreement with the lower values in the range of theoretical estimates, some 500 – 1000 bar (Temperley 1947).

5. CONCLUSIONS

We conclude that the estimate of F_c reported herein is in order-of-magnitude agreement with the predictions of nucleation theory, but emphasise that the values obtained greatly exceed those reported in previous studies involving the reflection of low frequency ultrasound pulses at the free surface of a column of water. The present results are sufficient to support the claim of Temperley and Trevena (1994) that, in terms of such reflection measurements, no *serious* discrepancy in fact exists between theory and experiment. This suggests that the correspondence of those (anomalously) low values of F_c from previous work with estimates of the tension generated by some forms of ultrasound is unlikely to have significance in terms of assessing its safety in biomedical applications.

6. REFERENCES

Fisher J C 1948 *J. Appl. Phys.* **19** 1062

Frenkel J 1955 *Kinetic Theory of Liquids* (New York: Dover)

Marston P L and Unger B T 1986 *Shock Waves in Condensed Matter* (New York: Plenum)

Temperley H N V 1947 *Proc. Phys. Soc.* **59** 199

Temperley HNV and Trevena D H 1994 *J. Stat. Phys.* **77** 501

Williams P R, Williams P M and Brown S W J 1997 *J. Phys. D: Appl. Phys.* **30** 1197

Zheng Q, Durben D J, Wolf G H and Angell C A 1991 *Science* **254** 829.

NON-LINEAR VISCOELASTICITY OF MICELLAR SYSTEMS

F.BAUTISTA, J.F.A.SOLTERO, J.E.PUIG, O.MANERO*

Dept. de Ingeniería Química, Universidad de Guadalajara, Guadalajara, Jal. 44430.
*Instituto de Investigaciones en Materiales, UNAM. AP 70-360, México, D.F. 04510.

1. ABSTRACT

The non-linear viscoelastic behaviour of the Cetyltrimethylammonium tosilate (CTAT)-water system is investigated in steady and unsteady flow. This system forms elongated micelles at low and intermediate concentrations. A simple model consisting of the Upper-Convected Maxwell constitutive equation and a kinetic equation for construction and destruction of the structure is used here to reproduce the complex rheological behaviour of these elongated micellar solutions. Predictions show qualitative agreement with experiments that exhibit the presence of "banded flow".

2. INTRODUCTION

Systems of elongated or "wormlike" micelles exhibit a very rich and interesting rheological behavior, due to the fact that the dynamics of these systems are influenced by the presence of reversibly breakable polymer-like chains. The prediction of the non-linear viscoelastic behaviour has been attempted by the so-called reptation-reaction model, developed by Cates (1990) using the constitutive equations of polymer dynamics, Doi and Edwards (1986), adapted to the kinetics of breakage-reformation of the micellar units.

It is the aim of the present work to propose an alternative model based on a phenomenological description of the kinetic process mentioned above coupled to a non-linear viscoelastic model. The motivation to choose the Upper-Convected Maxwell equation is based on the fact that the linear viscoleastic behavior of wormlike micellar systems is near-Maxwellian.

The suggested model qualitatively predicts the correct behaviour in steady state and the very long relaxation times for attainment of steady state conditions when a critical stress is reached in a controlled stress experiment. Furthermore, it may well predict a plateau-like region of the stress versus shear rate curve, a relaxation function with two characteristic times, overshoot at the inception of shear flow and oscillations in the stress past the overshoot time scale. Hysteresis in exponential shear flow can also be accounted for. The model, therefore, is capable to simulate the relevant features of the "shear-banded" flow in which the flow field is inhomogeneous.

3. THE MODEL.

The model used here consists of the upper convected Maxwell equation:

$$\tau + \frac{1}{G_0\varphi}\overset{\nabla}{\tau} = \frac{2D}{\varphi} \qquad (1)$$

and a kinetic equation for construction and destruction first proposed by Fredrickson:

$$\frac{d\varphi}{dt} = \frac{(\varphi_0 - \varphi)}{\lambda} + k(\varphi_\infty - \varphi)\tau : D \qquad (2)$$

Here $\boldsymbol{\tau}$ is the stress tensor, $\overset{\nabla}{\boldsymbol{\tau}}$ is the convective derivative of the stress tensor, $\mathbf{D}$ is the strain rate tensor, φ is the fluidity or the reciprocal of the shear viscosity η, G_0 is the shear modulus, φ_o and φ_∞ are the fluidities at zero and infinite shear rate, λ is a structure relaxation time, and k can be interpreted as a kinetic constant for structure breakdown.

4. RESULTS.

Experimental data of CTAT solutions is similar to those reported for other wormlike micellar systems, for which the shear banding behavior is clearly established and has been confirmed by birefringence data . The presence of a plateau-like region in the stress-shear rate curve indicates that for a given shear stress equal or larger than a critical stress τc, the corresponding mean strain rate lying in this region may split into "shear bands" of low and high strain rates. The plateau has been observed for the CTAT system in the controlled stress mode as soon as the stress reaches τc. Moreover, around this characteristic stress, there is a clear evidence of slow transient behaviour in which the system takes a long time to attain steady state after the inception of shear flow. The presence of slow relaxation processes may be interpreted as the time taken for macroscopic flow inhomogeneities to establish. Fig. (1) shows experimental data on the inception of shear flow exhibiting the time required for the solution to achieve steady state as a function of the normalized shear rate. For shear rates larger than $\dot{\gamma}c$, overshoot peaks shift to shorter times as the shear rate increases, but the attainment of steady state takes progressively longer times. At a given value of $\dot{\gamma}$, a second overshoot in the stress is recorded. For very high shear rates, the attainment of steady state takes progressively shorter times, as shown by the decreasing steady state branch in figure (1).

The model qualitatively predicts the long times required for the establishment of steady state, as shown in figure (2). Furthermore, these long times can also be predicted by the model in a controlled stress situation or in a step change in the stress, as discussed elsewhere by Bautista et al.(1998).

The presence of thixotropic loops when the sample is subjected to transient stress cycles can also be simulated by eqs. (1) and (2). Such behaviour has been observed with surfactant-based viscoelastic micelar solutions, dispersions and liquid crystalline phases. The role of the characteristic time for structure buildup, λ, in the extent and shape of the thixotropic loops has been demonstrated in Bautista et al.(1998).

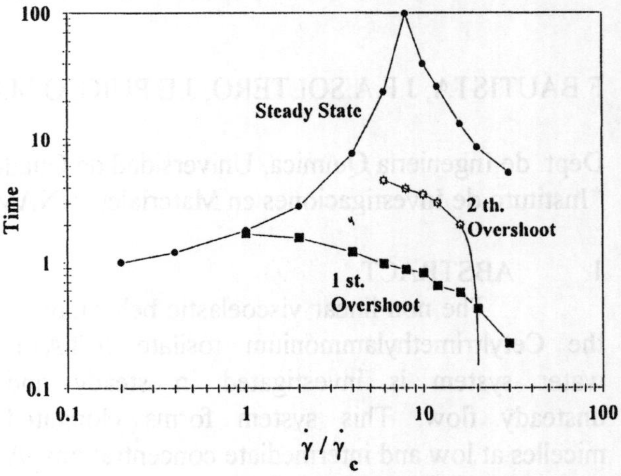

Figure (1). Time for attainment of steady state that follows the inception of shear flow, as a function of normalized shear rate. The time scales of the first and second overshoot peaks are also shown.

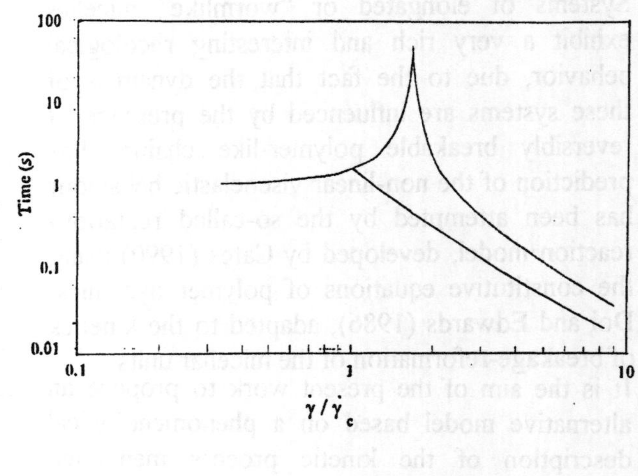

Figure (2). Model predictions of the curves shown in figure (1).

REFERENCES.

Bautista F, de Santos JM, Puig JE, Manero O (1998). Jour. non-Newt. Fluid Mech. (to appear).

Cates ME, (1990) J.Phys.Chem. 94:371.

Doi M and Edwards SF (1986) The Theory of Polymer Dynamics.Clarendon Press, Oxford.

Fredrickson, AG (1970) AICHE J. 16:436.

THE INFLUENCE OF AGEING OF A CATIONIC SURFACTANT ON FLOW CHARACTERISTICS

Z.CHÁRA, P.VLASÁK, P.ŠTERN

Institute of Hydrodynamics, Acad. Sci.Czech Rep., Pod Paťankou 5, 166 12 Prague 6, Czech Republic

In the past several years a number of papers concerning the investigation of different types of surfactants from many points of view has been published. It was proved that this type of drag reducing additives is very suitable for recirculating flows as, for example, central heating or cooling systems. In the study presented here we have focused on the problem how a process of ageing of additive (or a time of storage) can influenced the rheological parameters as well as turbulent flow characteristics. For this reason we have investigated several samples of the cationic surfactant Habon-G (main components- hexadecyl dimethyl hydroxyethyl ammonium, 3-hydroxy-2-naphtoate, industrial product of Hoechst AG-FRG) – *sample 1* - four years old product, *sample 2* - two years old and *sample 3* – fresh product. All samples were dissolved in water, mass concentration of active tenside was 0.095%. A very good drag reducing efficiency of this type of surfactant was proved by many laboratory as well as field tests. Zakin et all. (1996), Pollert et al. (1994).

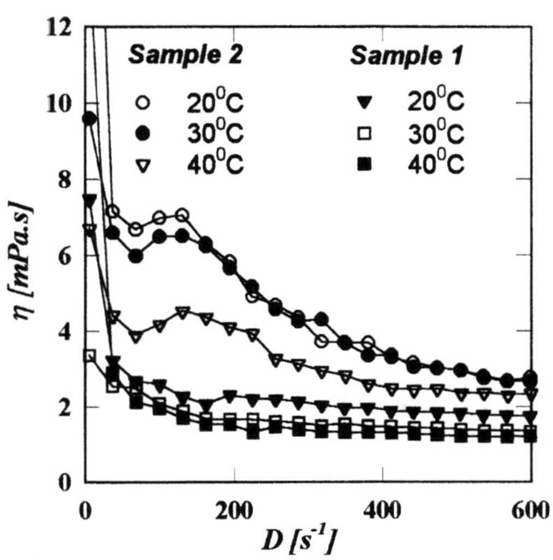

Fig. 1

The rheological measurements (apparent viscosity and the first normal stress differences) were performed with the Haake Rheometer Rotovisco RV 20 / CV 20N with a 2^0 cone-plate system and also with a coaxial cylinder system. The solutions were pre-sheared before the measurements by shearing in a Couette flow cell for 2 hours at shear rate 1170 s^{-1}. It was found no significant differences between the *sample 3* (fresh sample) and the *sample 2* (two years old), but the behaviour of the *sample 1* (the oldest sample) was significantly changed. Fig. 1 and Fig. 2 show the effect of shear rate and temperature on the apparent viscosity and on the first normal stress differences. The *sample 2* shows a remarkable sign of SIS (Shear Induced Structure) formation in the region 100-200 s^{-1} (compared to the *sample 1*) and also the viscosity of the *sample 2* is higher. (Fig.1). The values of the first stress differences of

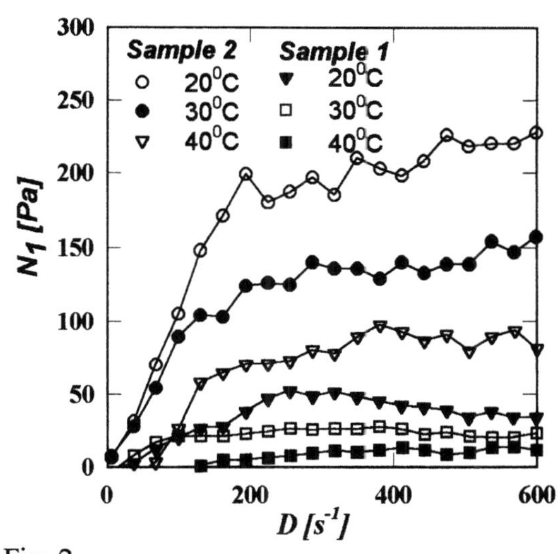

Fig. 2

the oldest sample are about 10 times less than the values of *sample 2* (Fig.2).

To show the influence of ageing of the surfactant on flow turbulent characteristics a set of experiments was carried out. The experiments were performed in a recirculating laboratory loop provided with glass circular pipes of inner diameter 39.4 mm. The total length of the pipe loop was 14.2 m. Differential pressure transducer Hottinger-Baldvin (PD1) was used for measurements of pressure drop along the 2 m section of straight pipe. The flow rates were controlled by a ball valve located in a distance 5 m upstream the measuring section. To determine the turbulent flow characteristics the LDA method was applied. LDA components were assembled from Dantec optics including the BSA processor to evaluate Doppler frequencies.

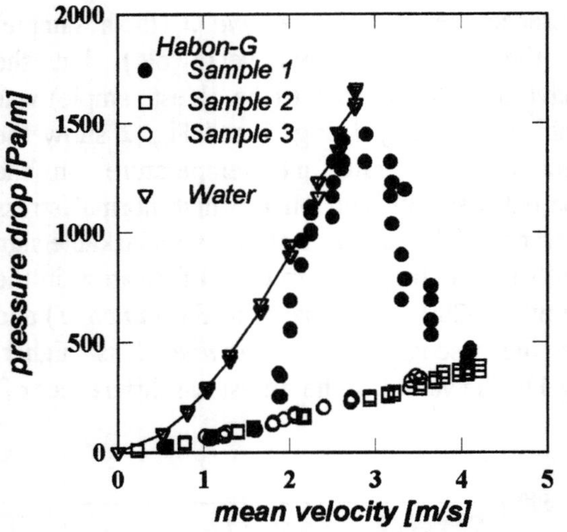

Fig.3

The dependence of pressure drops on mean velocities is shown in Fig.3. The drag reducing effectiveness of the *sample 2* and *sample 3* was very high (about 80%) in the whole range of attainable velocities ($v < v_{max} = 4.3\ m/s$). The drag reduction effectiveness of the *sample 1* was losing in a regime of relatively low mean velocities (*1.7 m/s*), but the effectiveness of this oldest sample was surprisingly fully regained if the velocities reached values just below the maximal velocity (corresponding to the full opening of the ball valve). In the transition zone, where the effectiveness increased from no drag reduction to maximal drag reduction, a very interesting behaviour was observed. Fig. 4 shows time series of instantaneous velocities measured in the centre of the pipe together with pressure gradients for mean velocity $v=3.4\ m/s$. As can be seen time series of velocities possess low frequency waves with relatively high amplitude, where parts of high and low velocity fluctuation alternate. An increasing of velocity is accompanied with a decreasing of the velocity fluctuations and a decreasing of the pressure gradient. When

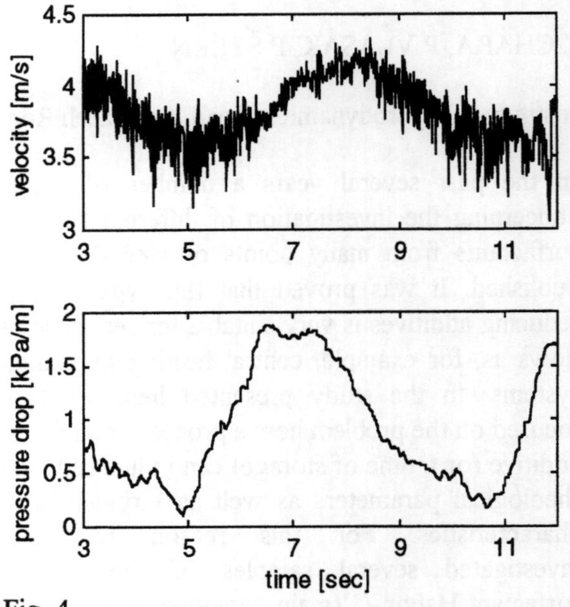

Fig. 4

the local maximum velocity is achieved drag reduction effect is vanishing, fluctuation is increasing and local velocity is falling down. This behaviour was probably caused by the ball valve located upstream the test section. When the valve was not completely opened a sharp edge inside the valve destroyed a micelle structure of the oldest surfactant and there was not sufficient space to rebuild the inner structure. When the valve was fully opened, the micelle structure was not destroyed and drag reduction effectiveness could be restored.

The investigation leads to the conclusion, that the lifetime of the drag reduction effectiveness (of this type of surfactant) can be limited, because during the process of ageing the resulting forces, which drive the balance inside the micelle network, become weaker and inner structure could be much more sensitive to any disturbance.

Acknowledgements

This research was supported in part under Grant No.A2060804 of the Grant Agency of the ASCR.

REFERENCES

Zakin JL, Myška J, Chára Z (1996) AIChE Journal 12, 3544-3546

Pollert J, Zakin JL, Myška J, Kratochvíl K (1994) 85th Conf. Int. District Heating and Cooling, Seattle , 18-21 June, 141-156

RHEOLOGICAL BEHAVIOUR OF THE HEXAGONAL LIQUID CRYSTALLINE PHASE FOR A HEPTANE/NONIONIC SURFACTANT/WATER SYSTEM

F. CORDOBÉS[1], C. GALLEGOS[2]

[1]Departamento de Ingeniería Química, Universidad de Sevilla, 41012 Sevilla, Spain.
[2]Departamento de Ingeniería Química, EPS La Rábida, Universidad de Huelva, 21819 Huelva, Spain.

1. INTRODUCTION

Lyotropic surfactant liquid crystals are typically formed by surfactants, water, and/or an organic compound. They can be considered as ordered networks of giant micelles, and find interesting applications in different fields: detergents, cosmetics, pharmaceuticals, emulsions, etc. The structure of hexagonal liquid cristalline phases consists of long cylindrical micelles which are parallel to each other within a hexagonal array, although this liquid crystal is polycrystalline, being the cylinder aggregates orientated within compact packs of microscopic domains. This microstructure yields a very complex rheological response.

The aim of this work was to model the linear viscoelastic behaviour of systems exhibiting a direct hexagonal liquid-crystalline structure using small amplitude oscillatory shear experiments. Binary aqueous systems of a nonionic surfactant and ternary systems, solubilizing a linear chain hydrocarbon, were characterized. The effects of composition and temperature were analyzed.

2. EXPERIMENTAL

A poly(ethylene glycol)nonylphenylether (Triton N-101, NPFE-10) from Sigma (St. Louis, U.S.A) was used as received. Hexagonal systems were prepared by mixing this nonionic surfactant with single distilled, deionized water and n-heptane of analytical grade, obtained from Panreac (Barcelone, Spain).

Sample concentrations ranged from 40 to 50% (w/w) surfactant and 0 to 20% (w/w) heptane. A binary system (45 % (w/w) surfactant) and a ternary system (45% (w/w) surfactant and 20% (w/w) heptane) were selected to study the influence of temperature, between 5 and 25°C.

Oscillatory shear measurements were performed in the linear viscoelasticity region, using a Rheostress RS-100 rheometer from Haake (Karlsruhe, Germany) with a cone-and-plate sensor system (20 mm, 4°).

3. RESULTS AND DISCUSSIONS

The continuous relaxation spectrum, H(λ), is related to the linear viscoelasticity functions by means of Fredholm equations of the first kind:

$$G'(\omega)=G_e+\int_{-\infty}^{\infty} H(\lambda)\frac{\omega^2\lambda^2}{1+\omega^2\lambda^2}d(\ln(\lambda)) \quad (1)$$

$$G''(\omega)=\int_{-\infty}^{\infty} H(\lambda)\frac{\omega\lambda}{1+\omega^2\lambda^2}d(\ln(\lambda)) \quad (2)$$

Tikhonov's regularization method has been proposed as a powerful tool for solving the difficulties that arise with Fredholm equations of the first kind (Tikhonov and Ibinskji, 1980).

Figure 1 shows the values of the storage and loss moduli recalculated from equations (1) and (2) using Tikhonov's regularization method to obtain H(λ) (Madiedo, 1996). As can be observed, there is a good agreement between experimental and calculated values.

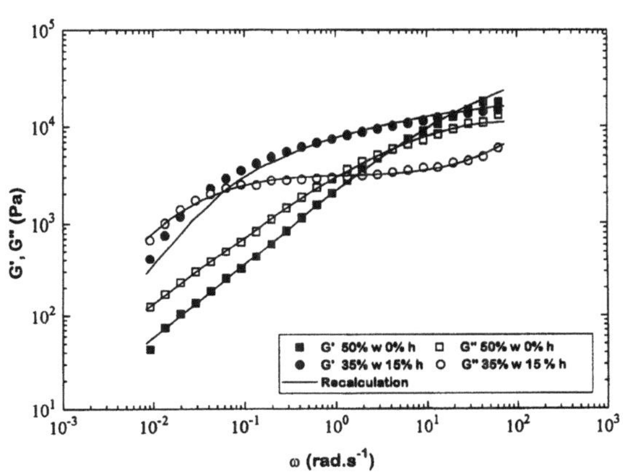

Figure 1. Selected storage and loss moduli curves recalculated from the relaxation spectrum of the hexagonal phase

Recently, De Rosa and Winter (1994) have proposed an empirical model, derived from BSW-spectrum (Baumgaertel et al. 1990) and CW-spectrum (Chambon and Winter 1985), which may describe the hexagonal phase relaxation spectrum in the experimental frequency range studied, comprising the transition, plateau, and terminal regions,

$$H(\lambda)=A\left[\left(\frac{\lambda}{\lambda_c}\right)^m+\left(\frac{\lambda}{\lambda_p}\right)^n\right] \qquad \text{for } 0<\lambda<\lambda_p \quad (3)$$

$$H(\lambda)=A\left(\frac{\lambda}{\lambda_p}\right)^c \qquad \text{for } \lambda_p<\lambda<\infty \quad (4)$$

where λ_c and λ_p are the characteristic relaxation times for the onset of the plateau and terminal regions, respectively, m, n and c are the power law exponents for the three different regions and A is an empirical constant.

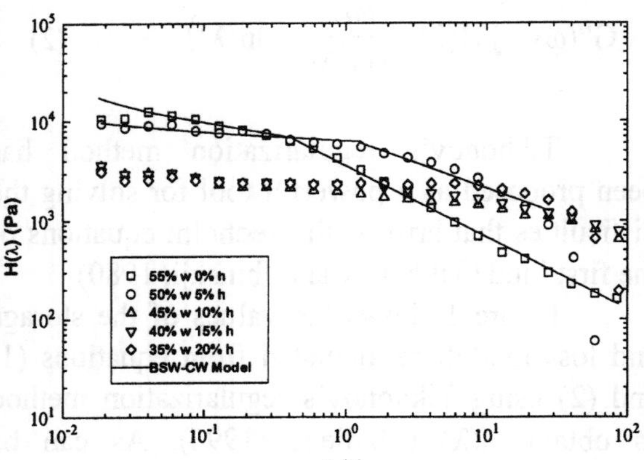

Figure 2. Influence of composition on the relaxation spectrum of the hexagonal phase (45% (w/w) surfactant

Figure 2 shows the relaxation spectra for binary and ternary systems containing 45% (w/w) surfactant, as a function of heptane concentration. Heptane solubilization into H_1 enhances the formation of an elastic network, as demonstrated by the occurrence of a plateau region in the continuous relaxation spectrum. The same effect is achieved by a decrease in temperature for the binary system (Figure 3). As can be observed in Figure 2, the BSW-CW model describes the behaviour of the binary system fairly well, but it frequently fails with ternary systems. This is due to the fact that only for binary system do the value of λ_p (that indicates the beginning of the terminal region) coincide approximately with the reciprocal of the crossover frequency, ω_c, in the frequency sweep tests. For ternary systems, λ_p is related to a pseudo-terminal region, which appears at relaxation times significantly lower than that corresponding to the crossover frequency. In relation to this parameter, an increase in heptane concentration and a decrease in temperature yield higher values of the terminal (or pseudoterminal) relaxation time, λ_p. Steady-state viscosity, η_o, obtained from the relaxation spectrum, decreases as temperature increases. An Arrhenius-like relation describes this dependence for the binary system fairly well.

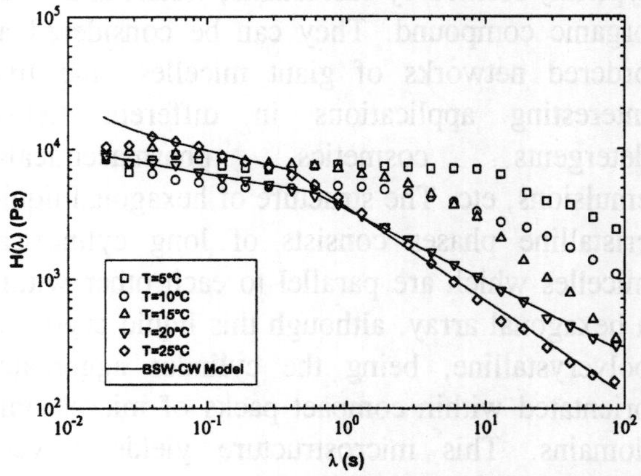

Figure 3. Influence of temperature on the relaxation spectrum of the hexagonal phase (binary system)

4. CONCLUDING REMARKS

Heptane solubilization into H_1 enhances the formation of an elastic network, as demonstrated by the occurrence of a plateau region in the continuous relaxation spectrum. The same effect is achieved by a decrease in temperature for the binary system. The BSW-CW model does not describe, in most cases, the continuous relaxation spectrum of the hexagonal phase.

6. REFERENCES

Baumgaertel M, Schausberger A, Winter HH (1990) Rheol Acta 29: 400-408

Chambon F, Winter HH (1985) Polym Bull 13: 499-503

De Rosa ME, Winter HH (1994) Rheol Acta 33:220-237

Madiedo JM (1996) PhD Thesis Univ. of Seville

Tikhonov AN, Ibinskji AS (1980) Probleme der Numerischen Mathemathick. Verlag der Moskauer Universität, Moscow

ACKNOWLEDGMENT

This work is part of a research project sponsored by the CICYT, Spain (PB92-0664). The authors gratefully acknowledge its financial support.

Stress Optical Coefficient of Viscoelastic Solutions

J.P. Decruppe and C. Humbert

June 9, 1998

1 Introduction

The rheo-optical properties of different viscoelastic solutions of surfactants are investigated in order to gather experimental data which will allow for the study of the stress optical law and the computation of the stress optical coefficient C [1, 2]. This law when it applies, states the proportionality of the stress tensor and the refractive index tensor; it is expressed mathematically through two simple equations relating the extinction angle χ and the birefringence Δn of the medium to the components of the stress tensor. The proportionality coefficient is the so called " stress optical coefficient C " which is found to vary with both the concentrations C_D of the surfactant and C_S of the salt. The knowledge of its accurate value is of primary importance when one wants to deduce the first normal stress difference N_1 from optical experiments.

2 Theoretical part

In the case of simple shear flow which is the case in the Couette cell at moderate shear rates, the only non-diagonal term of the stress tensor which is different from zero is σ_{yx}. Let N_1 be the first normal stress difference, the stress optical law reduces simply to the two equations:

$$\Delta n \sin(2\chi) = 2C\sigma_{yx} \tag{1}$$

$$\Delta n \cos(2\chi) = C(\sigma_{xx} - \sigma_{yy}) = CN_1 \tag{2}$$

In the frame of Kühn and Grün theory, the coefficient C is given by [3] :

$$C = \frac{2\pi}{45\, k_B\, T} \frac{(n^2+2)^2}{n} \Delta\alpha \tag{3}$$

n is the average index of refraction of the solution, $\Delta\alpha$ the polarisability anisotropy of the micelles. T is the absolute temperature and k_B Boltzman's constant.

The index n varies only little with the concentrations C_D or C_S; the main factors will be $\Delta\alpha$ which is related to the persistence length q of the particle and the temperature T which is an important parameter for the length of the micelles.

Equation (1) allows for the computation of C once the shear stress σ_{yx} is known; the second equation (2) gives then the first normal stress difference N_1 and this is of particular interest since measuring N_1 precisely is not always an easy task to do.

Laun's law [4] forms a good way to check the accuracy of the different values of C; it states that :

$$\left(\frac{G'}{\omega^2}\right)_{\omega\to 0} = \left(\frac{N_1}{2\dot{\gamma}^2}\right)_{\dot{\gamma}\to 0} \tag{4}$$

G', the storage modulus being measured independently as a function of the angular frequency of an oscillating cone-plane device.

3 Experimental results

The surfactant used in this work is the well-known cetyltrimethylammonium bromide (CTAB). It is mixed with potassium bromide (KBr) in distilled water. The rheological studies are performed with a cone-plane or Couette device while the optical experiments are all realized in a conventionnal Couette cell with the inner cylinder rotating.

Three different sets of results shall be presented and discussed; they refer to the measurements of χ and Δn.

3.1 Optical results: flow birefringence

3.1.1 Equimolar solution of CTAB and KBr

The solution is a single solution of CTAB and KBr at the same molar concentration ($0.3M$); the temperature is varied from 30 to 38°C. The overall behaviour of χ and Δn is consistent with what is usually observed: $\chi(\dot{\gamma})$ decreases while $\Delta n(\dot{\gamma})$ increases with the shear rate as a result of a better orientation of the micelles. At small values of the shear rate, both curves show a linear part the slope of which (absolute value) increases with T; this is consistent with the fact that the average length of the particules diminishes with T, the influence of the Brownian motion which tends to disorganize the particles is thus strengthened and the birefringence gets smaller. The values of χ and Δn are used to check the validity of the stress optical law. For each value of the temperature T, the points are well distribute near a straight line the slope of which is twice the coefficient C. The average value of C is $0.4\ 10^{-7} Pa^{-1}$ which is not in agreement with Shikata's [5] value. However their measurements have been made on a different system: CTAB/NaSal.

3.1.2 Constant surfactant concentration, increasing salinity

C_D is kept constant while C_S takes four different values (0.1, 0.2, 0.3 and $0.4M$). The temperature is kept constant to 30°C. The general behaviour of χ and Δn is qualitatively the same as the one described in the previous section. The stress optical law holds over a wide range of shear stresses and half the slope of the different straight lines gives C which lies between 0.25 and $0.62\ 10^{-7} Pa^{-1}$.

3.1.3 Constant ionic force but increasing surfactant concentration

$C_S = 0.1M$ and $C_D = 0.3$ to $0.6M$, the last value being the highest we can use without the solution forming a liquid crystal even under the lowest value of the shear rate. Again the stress optical law holds for these solutions and C can be obtained by fitting the experimental results with a straight line.

3.2 Rheology: dynamical measurements

The same solutions are subjected to an oscillating shear stress in a cone-plane device to record G' and G''. The qualitative aspect of the curves $G'(\omega)$ or $G''(\omega)$ is the same for all the solutions. The values of G' are used to compute $\frac{G'(\omega}{\omega^2}$, these data are then compared to $\frac{N_1}{2\dot{\gamma}^2}$ according to the fourth equation. The first normal stress difference N_1 computed from the birefringence measurements using the stress optical coefficient C agree well with the values obtained from the dynamical measurements of G'

4 Conclusion

These experimental results show that the stress optical law is valid in a wide range of shear stresses for various solutions of a surfactant mixed with KBr. The optical coefficient allows for the computation of the first normal stresses difference N_1. The main result is that C is not constant but depends significantly on the concentration of the surfactant and of the salt.

References

[1] Lodge AS (1956) Trans Faraday Soc 52:120-130

[2] Doi M , Edward SF(1978) J Chem Soc Faraday Trans 2 74:418-432

[3] Kuhn W, Grün F, Kolloïd Z, 28:248-271 (1942)

[4] Laun H.M. (1986) J Rheol 30:459-501

[5] Shikata T, Dahman SJ, Pearson DS (1994) Langmuir 10:3470-3476

STRUCTURE AND RHEOLOGY OF FOAM AND EMULSION FILMS OF GELATIN-SURFACTANT

S.R.DERKATCH[1], L.A.PETROVA[1], V.N.IZMAILOVA[2], B.N.TARASEVITCH[2], N.G.VORONJKO[1]

[1]Murmansk State Technical University, 13 Sportivnaya Street, 183010 Murmansk, Russia
[2]Moscow State University, Vorobjevu goru, 119899 Moscow, Russia

1. INTRODUCTION

The natural polypeptide gelatin has the clearly defined surface activity, emulsifying, stabilizing properties and film-forming ability. To change its properties the gelatin macromolecules can be modified by the low-molecular-mass surfactants.

When surfactants are added to aqueous gelatin solutions, binding of surfactants by gelatin with complex formation on consequence of electrostatic and hydrophobic interactions takes place. The properties of these complexes are not additive to individual components ones. They are determined by a binding degree of surfactant by gelatin. The maximum amount of cetylpyridinium chloride (CPC) and sodium dodecylsulphate (SDS) (A_{max}) bound by 1 g of gelatin is $1.7 \cdot 10^{-3}$ mol (pH 6.0) and $2.1 \cdot 10^{-3}$ mol (pH 6.0) respectively.

At the certain ratio of components these complexes are surface-active more than macromolecules of gelatin and molecules of surfactant singly. For instance, for SDS without gelatin the surface activity $G = \lim_{C \to 0} (d\sigma/dC)$ is about $0{,}9 \cdot 10^5$ $N \cdot m^2/mol$, for SDS in presence of gelatin ($C_G = 0{,}1\%$) G is about $1{,}2 \cdot 10^5$ $N \cdot m^2/mol$.

The work has been aimed to study stabilizing ability of the gelatin-surfactant complexes on the model systems: thin bilateral foam and emulsion films (elements of foam and emulsion).

2. MATERIAL AND METHODS

The photographic gelatin was a sample decomposed in alkaline solution. The isoelectric point was 4.9. The hydrocarbon low-molecular-mass surfactants were: cation-active – cetylpyridinium chloride (CPC) with the critical micelle concentration c.m.c. = $1.3 \cdot 10^{-3}$ mol/dm^3; anion-active – sodium dodecylsulphate (SDS) with c.m.c. = $8 \cdot 10^{-3}$ mol/dm^3.

The rheological parameters of the emulsion films (viscosity η_0^f, yield shearing stress P_r^f) and the interfacial adsorbed layers (surface creep viscosity η_0^s, yield shearing stress P_r^s) were measured on a surface elastoviscometer. A method of progressively increased load was used, the load τ_s applied on a system was 0,28 mN/m every 15 s. The thickness of the interface adsorbed layers were measured by a method of multiple attenuated total reflection spectroscopy in infrared area of spectrum (MATR IR spectroscopy). The interface tension (σ) was measured using a Wilgelme method.

3. RESULTS AND DISCUSSION

The stable foam and emulsion films are not formed from gelatin solution. The range of surfactant-gelatin ratio (complex composition) and pH values were found when the stable foam and emulsion films are formed.

It was shown that the thin bilateral films stabilized by gelatin-surfactant complex give synergism of the rheological parameters when the surfactant concentration is less than the maximum amount bound by gelatin. At the specific ratio of components $Z = 5{,}9 \cdot 10^{-4}$ mol CPC/g gel and $5{,}0 \cdot 10^{-4}$ mol SDS/g gel at $C_G = 0{,}1\%$; $Z = 6{,}7 \cdot 10^{-5}$ mol SDS/g gel at $C_G = 0{,}3\%$ the bilateral films are characterized by high values of the rheological parameters.

For example, for the foam films in case of CPC ($C_G = 0{,}1\%$) η_0^f is about 160 mN·s/m, P_r^f is about 3,1 mN/m, in case of SDS ($C_G = 0{,}1\%$) η_0^f is 0,9 mN·s/m, P_r^f is 2,2 mN/m. For the emulsion films at the interface with decane in case of CPC ($C_G = 0.1\%$) η_0^f is about 0,4 mN·s/m, P_r^f is about 0,6 mN/m; in case of SDS ($C_G = 0{,}3\%$) η_0^f is 0,2 mN·s/m, P_r^f is 0,4 mN/m.

Decrease of rheological parameters of the emulsion films at the surface with organic liquid (decane, heptane) in comparison with foam films at the surface with air has been explained by solubilization of hydrocarbon adjoining with aqueous phase by hydrophobic zones of the gelatin-surfactant complex. Volume of gydrophobic zones of gelatin (1 mol) bound into complex with SDS (Z= $1{,}7 \cdot 10^{-4}$ mol SDS/g gel, $C_G = 0{,}35\%$) is about $14{,}5 \cdot 10^4$ A^3.

Simultaneously there is a plateau in the surface and interface tension isotherms made by Wilgelme method, which indicates the saturation of the surface by the surface-active gelatin-surfactant complexes. At the interface with air in case of CPC ($C_G = 0{,}1\%$) σ is about 46,5 mN/m, at the interface with decane in case of SDS ($C_G = 0{,}3\%$) σ is 6,1 mN/m in the area of plateaus.

Under such conditions interface adsorption layers (element of thin films) of gelatin and its mixtures with surfactant at the solution/air and solution/organic liquid interfaces were investigated. Rheological streaming curves of adsorbed layers detect yield stress and are characterized by elastic and viscous parameters.

Increasing strength of stabilizing layers formed by complexes in comparison with gelatin ones is accompanied by decrease of the layer thickness (Table 1). This is connected with conformational changes (increase of parallel chains) of gelatin macromolecules interesting in complex with surfactant in adsorption layer.

Table 1. Rheological parameters (P_r^s, $P_v = P_r^s/d$) and thickness (d) of the interface adsorption layers of gelatin and complexes with SDS at the interface with air at $C_G = 0{,}3\%$, T = 295 K, pH 5,8-6,1

C_{SDS}, mol/dm^3	P_r^s, mN/m	d, nm	$P_v \cdot 10^{-6}$, mN/m^2
0	0,16	42,6	3,8
10^{-5}	0,24	56,7	4,3
$5 \cdot 10^{-5}$	0,52	51,0	10,2
10^{-4}	0,83	25,9	32,0
$1{,}5 \cdot 10^{-4}$	0,69	35,1	29,7
10^{-3}	0,20	37,0	5,4
10^{-2}	0,03	43,5	0,68

Complexing gelatin with surfactant gives rise to shifting isoelectric point of gelatin. Using photometric method it has been determined that for gelatin modified by SDS the isoelectric point is equal to 3,1; by CPC it is 6,8.

Thermodynamic (surface and interface tension) and rheological (yield shearing stress, viscosity) properties of the interface adsorbed layers, foam and emulsion films are extreme at pH near the isoelectric condition of complex (Table 2).

Table 2. Rheological (P_r^s, $P_v = P_r^s/d$), thermodynamic (σ) parameters and thickness (d) of the interface adsorption layers of gelatin and complexes with SDS at the interface with heptane at T = 295 K

pH	P_r^s, mN/m	d, nm	$P_v \cdot 10^{-3}$, mN/m^2	σ, mN/m
$C_G = 0{,}3\%$				
2,5	0	17	0	
3,5	0,02	31	0,6	
4,5	3,05	46	66	
6,2	0,34	46	7,4	
7,0	0,22	41	5,4	
8,0	0,17	26	6,5	
$C_G = 0{,}3\% + C_{SDS} = 2 \cdot 10^{-4}$ mol/dm^3				
2,5	0,08	≤ 10	≥ 8	28,0
3,0	2,50	81	31	16,0
4,1	0,96	26	37	21,5
6,1	0,25	23	11	26,0
7,0	0,20	8,0	25	

The gelatin-surfactant complexes are good stabilizers of the bilateral foam and emulsion films and the particular rheological parameters of films correlate with the rheological parameters of interface adsorption layers at the interface with air and organic liquid. This is caused by increasing role of structural-mechanical barrier as a result of formation of strengthened adsorbed layers.

4. ACKNOWLEDGMENT

This work was financed by the Russian Foundation for Fundamental Research.

A NEW CLASS OF SHEAR INDUCED STRUCTURE IN VISCOELASTIC MICELLAR SOLUTIONS

PETER FISCHER[1], ELIZABETH K. WHEELER[2], GERALD G. FULLER[2]

[1] ETH Zürich, ILW, Schmelzbergstrasse 9, 8092 Zürich, Switzerland
[2] Stanford University, Department of Chemical Engineering, Stanford, CA 94305-5025, USA

1. INTRODUCTION

In recent years much research has focused on wormlike micellar surfactant solutions. An appealing aspect of studying these systems is that their viscoelastic response can be described as a Maxwell fluid. However, in the nonlinear regime the rheological response of various systems is drastically different. For example, some solutions shear thicken, while others experience phenomena such as spurt effect (a), shear-induced phases (b), shear induced structures (c, d) as shown in Fig. 1.

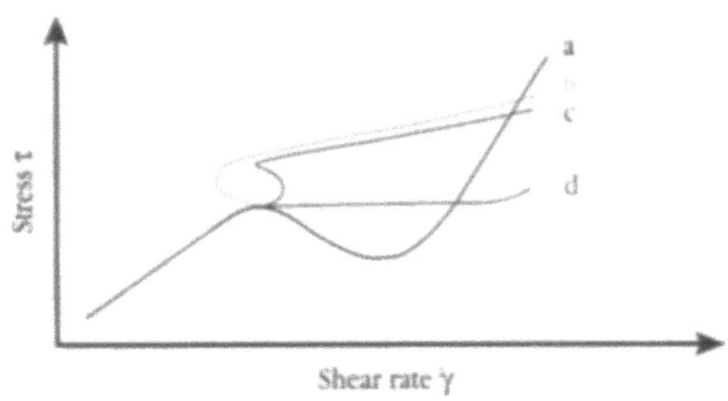

Figure 1: Behavior of shear stress for different viscoelastic surfactant solutions.

Both optical and mechanical techniques are used to investigate aqueous surfactant solutions of cetylpyridinium chloride (40mmol/L) and sodium salicylate (40mmol/L) in different flow geometries. Our studies show evidence of a unique transient shear induced structure (SIS) that results in a flow instability. This SIS is characterized by an initial increase in turbidity and, additionally, we see rings of alternating turbid and clear solution (Wheeler et al. (1998)). Based on the experimental results and a recent theoretical approach by Schmitt et al. (1995) we propose that a new class of transient SIS is observed here.

2. RESULTS

2.1 Plate-Plate Geometry

At rest and in the linear flow regime the solution was clear. As the flow entered the nonlinear regime, the solution became homogeneously turbid. Increasing the stress further resulted in rings of alternating turbid and clear solution as shown in Figure 2.

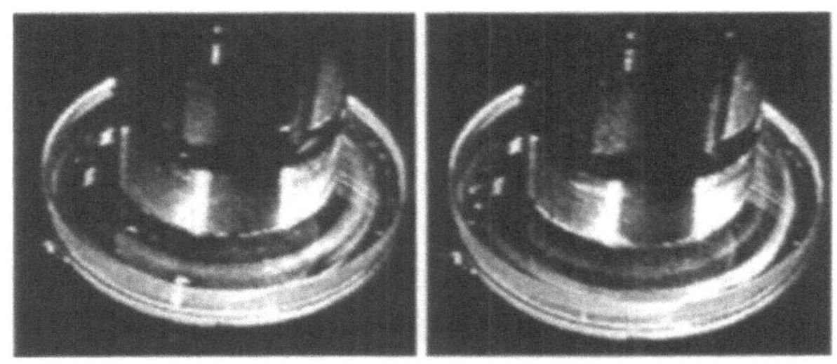

Figure 2: Alternating turbid and clear structures.

Not only did the rings' intensity vary radially but each ring also fluctuated between the turbid and clear state with a period of approximately one second.

2.2 Taylor-Couette Flow Cell

To investigate the structural transition we viewed the solution in a Taylor-Couette flow cell. The results are shown in Fig. 3 for a shear rate of 50 s^{-1} where bands of turbid and clear solution can be seen along the vorticity axis. The intensity along the vorticity direction at two instants in time that are separated by 1 second and are clearly out of phase (Fig.3). These bands bear a marked resemblance to those seen for Boger fluids that exhibit elastic instabilities above a critical Weissenberg number.

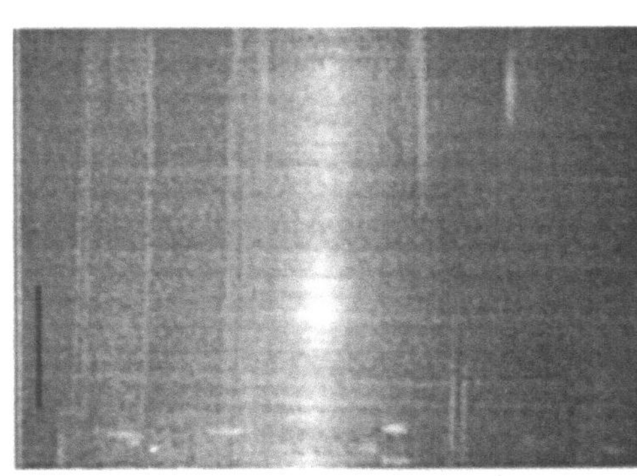

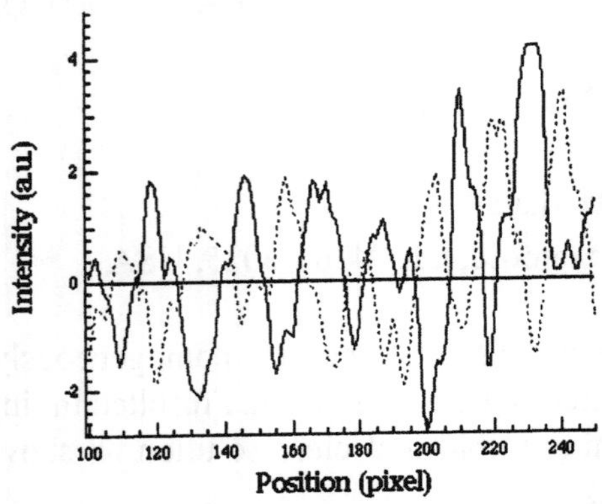

Figure 3: Band structure in Taylor-Couette flow cell. Intensity of bands plotted perpendicular to flow direction.

However, comparing critical parameters of the experimental solution to stability analyses shows that the solution should be stable, which is not the case. Therefore, it is not simply an elastic instability but a counteracting influence between SIS and an elastic flow instability.

2.3 Effects Of Temperature

To determine that this behavior is not unique to equimolar solutions of CPyCl/NaSal we varied the temperature.Further insight can be gained by looking at the shear stress, normalized by the plateau modulus, as a function of the Weissenberg number. It is interesting to note that data from different temperatures collapse onto a single curve as shown in Figure 4. The critical Wi at which we see SIS couple to an FI is 3.

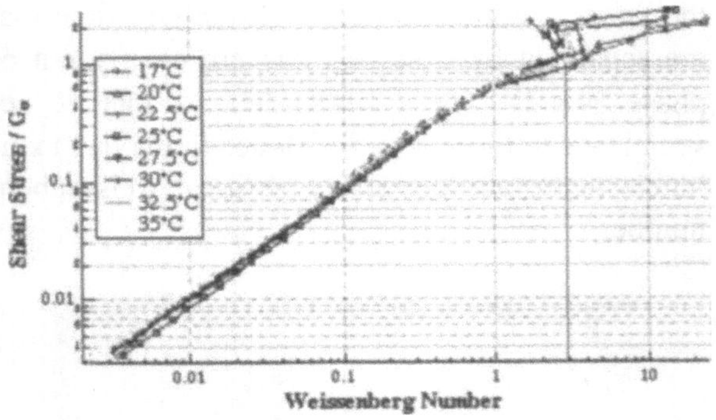

Figure 4: Viscosity as a function of Wi measured at different temperatures.

3. DISCUSSION

Recently Schmitt et al. (1995) presented a stability analysis of fluids where there is a coupling between flow and concentration ($\eta = \eta(c)$) which results in small, randomly generated perturbations in the shear velocity. Both shear rate and concentration are influencing the shear stress

$$\tau(\bar{v},c) = \tau(\bar{v}_0,c_0) + \left[\frac{\partial\tau}{\partial(\bar{v})}\right]\delta\bar{v} + \left[\frac{\partial\tau}{\partial c}\right]\delta c$$

where fluctuations in concentration δc and shear rate δv are introduced into the Navier-Stokes equations. Focusing on the diffusion coefficient, D, and a coupling term, C, a classification scheme for flow instabilities can be derived. The coupling is a function of the dependence of the viscosity on both concentration and the migration of molecules to regions of different shear rates. The relative viscosity of the SIS (C>0 less viscous, C<0 more viscous than the initial state) determines the type of instability. For C<0 concentration and flow inhomogeneities will develop in the vorticity direction, as seen in our observation. According to the observe band structure the diffusion coefficient D is negative and spinodal separation is superimposed on the instability. This situation is unstable and tends to break down. We postulate that while an SIP forms and is coupled with an FI, the destruction of this SIP is also coupled with flow. Within the coupling of flow and concentration there is a feedback circuit that allows an instability to continue to grow once it is present.

4. SUMMARY

The analysis of shear-induced transitions in micellar solutions is a complex problem that must take into account coupling between flow instabilities and structural changes. Inhomogeous shear flow layers of different concentration and viscosity field is generated by shear induced diffusion. For simple shear flow of viscoelastic surfactant solutions this effect is observed in bands of turbid and clear solution along the vorticity direction.

5. REFERENCES

Porte G, Berret J-F, Harden JL (1997) J. de Phys. II 7:459-472
Schmitt V, Marques CM, Lequeux, F (1995) Phys. Rev. E 52, 4009-4015.
Wheeler EK, Fischer P, Fuller GG (1998) JNNFM 75:193-209

MICELLAR STRUCTURE BREAK-UP AND RECOVERY IN TURBULENT FLOW OF DRAG-REDUCING SURFACTANT SOLUTIONS

K. GASLJEVIC, K. HOYER, E.F. MATTHYS

Dept. of Mechanical and Environmental Engg, University of California, Santa Barbara, CA 93106, USA

Some surfactant solutions in turbulent flow exhibit up to 80% wall friction reduction compared to water flow. This drag reduction is believed to result from the interactions between flow turbulence and large structures formed of cylindrical micelles. These shear-induced structures (SIS) form when the micelles are exposed to shear generated by the flow field. When the shear stress in the flow exceeds a critical level, the drag reduction disappears (i.e. "mechanical degradation" takes place), but only temporarily so. If the stress is subsequently reduced below the critical level, the drag reduction reappears. In this work, we studied the phenomenon of mechanical degradation as a tool for temporary local elimination of the drag-reducing effectiveness of surfactant solutions. Our fundamental objective here was a better understanding of the degradation phenomena under various flow conditions, and their possible connection with the micellar kinetics and relaxation phenomena observed in still fluids. Some of our findings are briefly summarized hereafter:

• Degradation was achieved under flow by placing an insert made of fine wire mesh in the measurement pipes. The fluid was then considered to be totally degraded when zero drag reduction was measured immediately downstream of the degrading device. It was seen that higher levels of stress beyond that necessary for total degradation did not result in any further lengthening in the recovery process.

• Interestingly, the best parameter for recovery quantification modeling appears to be time, and the recovery process is approximately a linear function of time (Fig. 1). In some cases a short interval is observed before the linear recovery takes place.

• The recovery time under flow for a fully-degraded fluid is nearly independent of the degradation device, the degradation work, or the flow velocity. It can be therefore be considered a property of the solution.

• For mesh inserts, the flow work per fluid volume (i.e. the pressure drop) needed for the total degradation of the fluid is independent of the flow velocity and size of the mesh.

• Fluid degradation appears to involve breakdown of micelles, rather than of shear-induced structures only. This is suggested by results showing that the time needed for SIS formation under flow in a fluid initially at rest (i.e. presumably with large micelles already in place) is much shorter than the recovery time after degradation by a mesh insert (Fig. 2).

• The recovery time is a strong function of temperature (Fig. 3). This dependence is similar to that found in kinetics studies of micellar rebuilding times after a sudden change of equilibrium conditions, another suggestion of micellar breakdown.

• The recovery after a sudden temperature drop (a common test in studies of micellar kinetics) under flow is faster than the recovery time after mechanical degradation by shear (Fig. 4). Since this surfactant exhibits reduced solubility at higher temperatures, the recovery after the temperature drop could presumably include a combination of solubility, micellar formation, and SIS formation times. This again supports the notion that mechanical degradation results in the breakdown of micelles rather than of the SIS only.

• Micellar recovery appears to be faster under flow than in a still fluid.

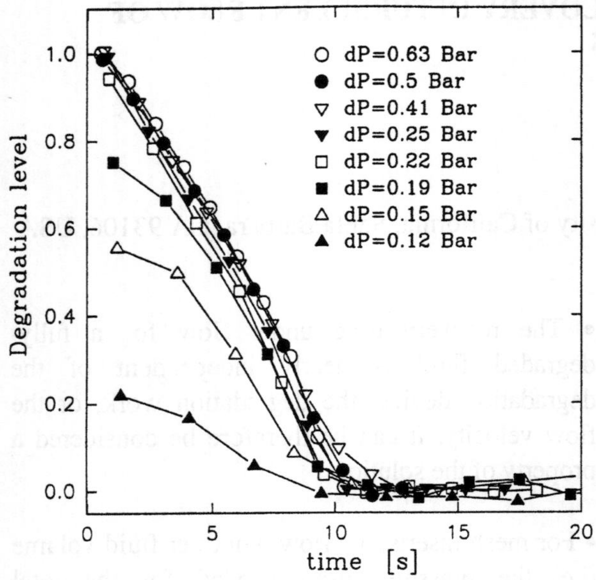

Fig.1: Recovery of a 1300 ppm SPE95285 (by Akzo Nobel) nonionic surfactant solution after degradation by a stack of wire meshes at various velocities. dP is the pressure drop across the stack.

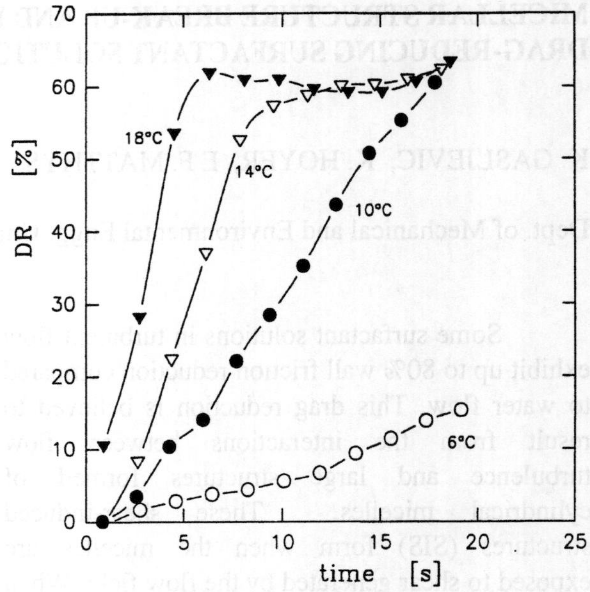

Fig. 3. Recovery of a 2200 ppm SPE95285 nonionic surfactant solution at various temperatures.

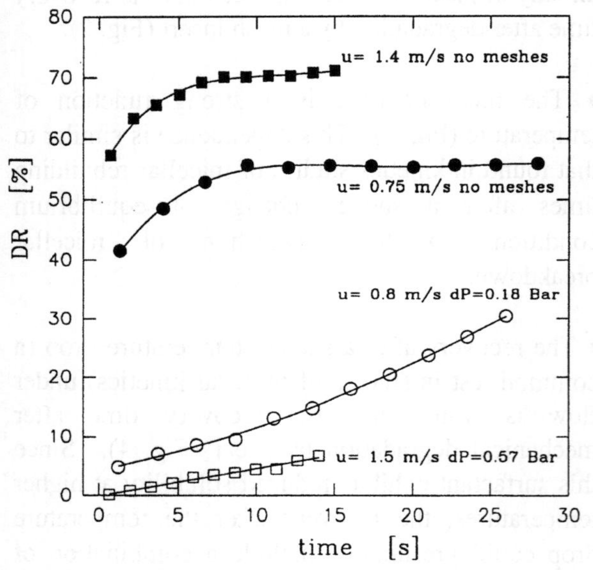

Fig. 2. Recovery after mesh degradation (open symbols) and SIS induction after rest (closed symbols) for an aged 1.8/2.5 mMol/l solution of a cationic surfactant (Ethoquad T13-W27) / NaSal solution.

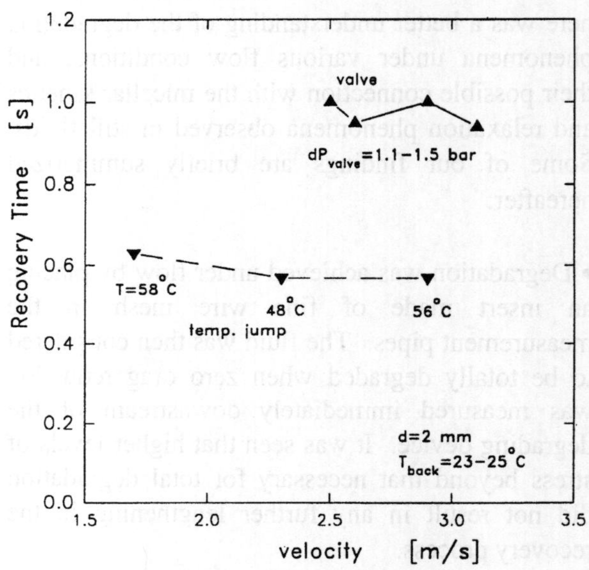

Fig. 4. Comparison of the recovery time under flow after a sudden temperature drop and after mechanical stress imposed by a valve. (Same solution as in Fig 3.)

COMPARISON OF THE EFFECTS OF METHYL AND CHLORO SUBSTITUTED SALICYLATE COUNTERIONS ON DRAG REDUCING AND RHEOLOGICAL BEHAVIORS AND MICROSTRUCTURES OF CATIONIC SURFACTANTS

Z. LIN, Y. ZHENG[1], Y. TALMON[2], H.T. DAVIS[1], L.E. SCRIVEN[1] AND J.L. ZAKIN

Department of Chemical Engineering, The Ohio State University,Columbus, OH 43210, U.S.A.
[1]Department of Chemical Engineering and Materials Science, University of Minnesota Minneapolis, MN 55455, U.S.A.
[2]Department of Chemical Engineering, Technion-Israel Institute of Technology, Haifa, 32000, Israel

1. INTRODUCTION

The chemical structure of the counterion affects the effectiveness of a cationic surfactant as a drag reducer, its rheological behavior and its microstructure. With the addition of certain negatively charged counterions, worm-like micelles can be formed. This worm-like micelle structure has been proposed as necessary for a surfactant solution to be drag reducing. Recently, however, we reported the first drag reducing system with a vesicle-dominated microstructure (Lin et al., 1997). The system was 5.0mM Arquad 16-50 (commercial CTAC) with 5.0mM 5-chlorosalicylate. This solution has a short swirl decay time but it has near zero first normal stress difference at 20 to $1000s^{-1}$ shear rates. In this paper, drag reduction, rheological properties, and microstructures from cryo-TEM imaging are described for Arquad 16-50 with 4-chlorosalicylate, 3-methylsalicylate, 4-methylsalicylate, and 5-methylsalicylate counterions, all at concentrations of 5.0mM Arquad 16-50/ 5.0mM counterion and they are compared with earlier results with 5-chlorosalicylate.

2. METHODS

Drag reduction, rheological and cryo-TEM measurements are described in detail in Lu (1997).

3. RESULTS AND DISCUSSION

3.1 Drag Reduction Results

All these systems are good drag reducers in the temperature range from 20-70 °C. The 4-methylsalicylate solution is effective at a temperature as low as 5°C. 4-chlorosalicylate and 3-methylsalicylate solutions are effective drag reducers up to 80°C. 5-chlorosalicylate is effective in the temperature range of 5-80°C.

3.2 Rheology Properties

3.2.1 Viscoelasticity

All the solutions have very short swirl decay times, thus showing strong viscoelastic behavior at room temperature. Rheological measurements show large first normal stress differences (N1) for the 3-, 4- and 5-methylsalicylate solutions, but low N1 for the 4-chlorosalicylate up to a shear rate of $200s^{-1}$ followed by a rapid increase. The 5-chlorosalicylate system shows near zero N1. Therefore, even though the first normal stress difference, N1, is often reported as a measure of viscoelastic behavior (Macosko, 1994), some viscoelastic surfactant systems have near-zero N1.

3.2.2 Shear Induced Structure

For all these systems it takes a finite time for the fluid to form highly elastic shear-induced micellar structures (SIS) (with accompanying shear thickening) after the start of shear flow from the quiescent state. SIS are suspected to be responsible for the drag reducing capacity of surfactant drag-reducing solutions (for example, Bewersdorff et al., 1986, 1989; Ohlendorf et al., 1986; Hu and

are shear thinning at shear rates above the SIS range.

3.2.3 Extensional Viscosity

All the systems show high ratios of apparent extensional viscosity to shear viscosity at room temperature. The mechanism of surfactant drag reduction is generally believed to be a suppression of small scale turbulent eddies (Lu, 1997). In the bursting and growth of these eddies, extensional motions dominate. High extensional viscosity increases the resistance of the fluid in the eddies and suppresses formation and growth of small scale eddies. High ratios (≥100) appear to correlate with good drag reduction (Lu, 1997).

3.3 Microstructure

Cryo-TEM images are used to reveal the microstructures of the drag reducing surfactant systems. It has often been stated in the literature and has been generally accepted that a network of thread-like micelles is necessary for a surfactant solution to be drag reducing.

Entangled thread-like micelles are observed in the 4-cholorosalicylate and 5-methylsalicylate solutions, while very long rigid thread-like micelles appear in the 4-methylsalicylate system. In the 3-methysalicylate system (Fig.1), however, a surprisingly different microstructure is observed. Vesicles dominate the microstructure and open vesicles are clearly seen.

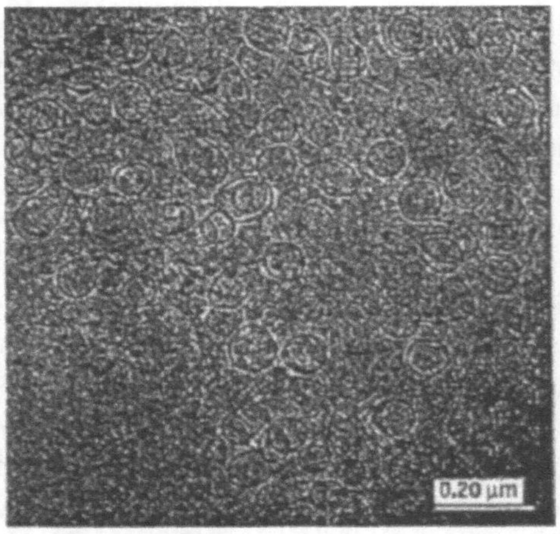

Fig.1. Cryo-TEM image of Arquad 16-50/ 3-methylsalicylate (5/5mM) solution

This is the first example that we have found with open vesicles in a surfactant drag reducing solution.

Coupled with the microstructures reported in the 5-chlorosalicylate system in which vesicles-dominated with a few thread-like micelles (Lin et al., 1997), we can see that thread-like micelles are not the only structure which can give viscoelasticity and drag reduction for surfactant solutions.

4. CONCLUSIONS

Cationic surfactant systems with counterions containing 4- or 5-chloro- and 3-, 4- or 5-methyl- substituent groups on the salicylate are all good drag reducers. All solutions show strong recoil, and thus are viscoelastic. However, the 5-Cl-salicylate has near zero N1, showing that a viscoelastic surfactant solution does not necessarily have large N1. All of these systems show shear-induced structures and high ratios of apparent extensional viscosity to shear viscosity. The microstructures can be quite different, however, especially the 3-methylsalicylate solution which has a vesicle-dominated microstructure, and in which open vesicles are clearly seen. This microstructure belies the previous proposals that a network of thread-like micelles is necessary for drag reduction to occur.

5. REFERENCES

Bewersdorff HW, Frings B, Linder P, Oberthür RC (1986) Rheol Acta 25, 642-464

Bewersdorff HW, Dohmann J, Langowski J, Linder P, Maack A, Oberthür RC(1989) Physica B, 157:508-511

Hu Y, Matthys EF (1996) Rheol Acta 35: 470-480

Lin Z, Lu B, Zheng Y, Digiulio D, Talmon Y, Davis HT, Scriven LE and Zakin JL, (1997), presented at AIChE Annual Meeting, Los Angeles, California

Lu B (1997), Ph.D. Dissertation, The Ohio State University, Columbus, OH

Macosko CW (1994), Rheology Principles, Measurements, and Applications, VCH Publishers, Inc., New York

Ohlendorf D, Interthal W, Hoffman H (1986) Rheol Acta 25: 468-486

INFLUENCE OF VISCOELASTICITY ON THE TURBULENT DRAG REDUCTION IN SURFACTANT SOLUTIONS

H. MIZUNUMA[1], T. KUMAGAI[1], K. UEDA[1]

[1]Tokyo Metropolitan University, 1-1, Minami-Ohsawa, Hachioji-shi, Tokyo, Japan.

1. INTRODUCTION

It is known that the turbulent drag reduction in surfactant solutions disappears over the critical point. To predict this critical point, its correlation with wall shear stress τ_c or average velocity V_c(1) was discussed. However, τ_c and V_c are not constant and increase with the concentration of the surfactant or the counter-ion. In this report, more general method is discussed to predict the critical point based on the viscoelastic characteristics of surfactant solutions.

2. EXPERIMENTAL SETUP

Three cationic surfactants were used: Ethoquad O/12, Arquad S-50 (Lion Co. Ltd.), and Habon-G(Hoechst Industry Co. Ltd.). As a counter-ion, equimolar sodium salicylate (NaSal) was used for Ethoquad O/12 and Arquad S-50. Habon-G was supplied as mixture with a counter-ion. In the case of Arquad S-50, the molar concentration ratio of the counter-ion to the surfactant was changed from 0 to 1. A capillary viscometer and a cone-and-plate rheogoniometer were used to measure the viscoelastic properties of the surfactant solutions.

The turbulent drag reduction was investigated in the pipes of 6.1mm and 10.0mm diameter. The pipe pressure loss and the flow rate were measured under the constant temperature of 20℃.

3. RESULTS AND DISCUSSION

Figure 1 shows the first normal stress difference $\Delta\sigma_1$ for each solution. In the case of Arquad S-50 134ppm, the influence of the counter-ion on $\Delta\sigma_1$ was measured as shown in Fig.2, where α means the molar concentration ratio of the counter-ion to the surfactant. The first normal stress difference increases with the surfactant concentration and α. The pipe friction factor λ was plotted in Fig.3 for Arquad S-50 134ppm where α was changed from 0 to 1. In Fig.3, we used a generalized Reynolds number Re* based on a power law model. When the drag reduction occurs, the friction factor approximately agrees with $\lambda=2.36\mathrm{Re}^{*\,-0.58}$ that was maximum drag reduction given by Virk(2) for dilute polymer solutions. If Re* is over a critical Reynolds number Rec^*, the drag reduction disappears. Table 1 shows the wall shear stress τ_c and the average velocity V_c at those critical points. With increase in the surfactant concentration and α, τ_c and V_c increase. Those critical values for the pipe of 6.1mm diameter are slightly higher than those of 10.0mm diameter. In the case of α=0 and 0.25, Arquad S-50 did not indicate the turbulent drag reduction.

Nextly, we introduce a general relationship between these critical points and the rheological characteristics of the solutions. At first, we obtain $\Delta\sigma_{1c}$ corresponding to τ_c , and calculate a critical Weissenberg number Wec as follows,

$$\mathrm{Wec} = \Delta\sigma_{1c} / \tau_c \, . \tag{1}$$

As shown in Fig.4, $\mathrm{Rec}^*/\mathrm{Ret}$ is plotted as a function of Wec, where Ret is a transitional Reynolds number for solvent. Although there is data scatter, the plotted results suggest the following relationship.

$$\mathrm{Rec}^*/\mathrm{Ret} = \mathrm{Wec}^{0.42} \tag{2}$$

From Re^*/Ret and We, the pipe friction is approximately given as follows.

$Re^*/Ret < 1$: $\lambda = 64/Re^*$ (Laminar flow).

$1 < Re^*/Ret < We^{0.42}$:

$\lambda \cong 2.36 Re^{*-0.58}$ (Drag-reduced turbulent flow).

$Re^*/Ret > We^{0.42}$:

$\lambda \cong 0.3164 Re^{*-0.25}$ (Normal turbulent flow).

4. CONCLUSION

At the critical points, the correlation was found between Wec and Rec^*/Ret. Although our results are confined to low concentrations and small pipe diameters, the results indicate that the viscoelasticity is one of the important factors to predict the turbulent drag reduction in surfactant solutions.

5. REFERENCES

(1)Gasljevic K., Matthy E.F.(1995) IMECE95:1-7.

(2)Virk P.S., Mickley H.S., Smith K.A. (1970) J. Applied Mech.:488-493.

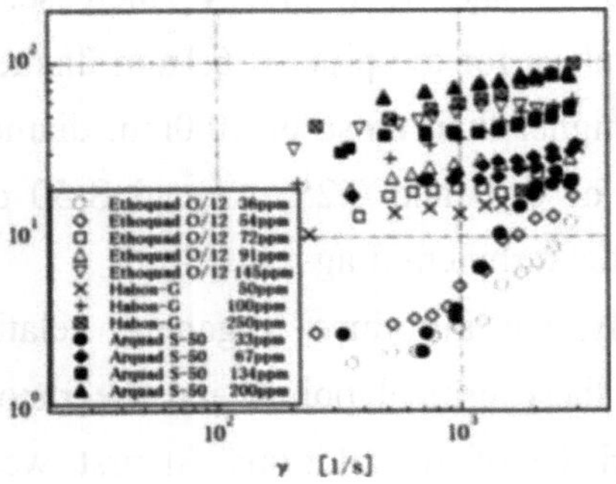

Fig.1 $\Delta\sigma_1$ versus shear rate γ.

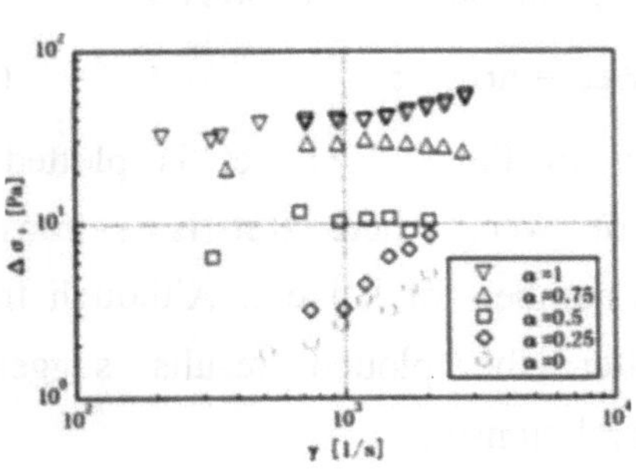

Fig.2 Influence of counter-ion concentration on $\Delta\sigma_1$. Arquad S-50 134ppm.

Table 1 Average velocity and shear stress at the critical points.

Surfac-tant	Conc. [ppm]	〈	V_c [m/s] D [mm] 6.1	V_c [m/s] D [mm] 10	τ_c [Pa] D [mm] 6.1	τ_c [Pa] D [mm] 10
Ethoquad O/12	36	1	0.50	0.49	0.89	0.63
	54	1	0.77	0.76	1.42	1.07
	72	1	0.93	0.90	2.00	1.46
	91	1	1.08	0.96	2.45	1.56
	145	1	1.60	1.13	4.18	1.95
Habon -G	50*	1	0.86	0.72	1.62	1.22
	100*	1	1.36	0.99	3.36	1.85
	250*	1	1.78	1.21	4.54	2.77
Arquad S-50	33	1	0.55	0.44	1.00	0.59
	67	1	0.71	0.70	2.49	1.95
	134	1	1.43	1.25	3.85	2.63
	200	1	1.94	1.60	6.14	4.19
	134	0	-	-	-	-
	134	0.25	-	-	-	-
	134	0.5	0.88	0.81	2.02	1.46
	134	0.75	0.96	0.92	2.29	1.61

* Concentration is for the mixture with counter-ion.

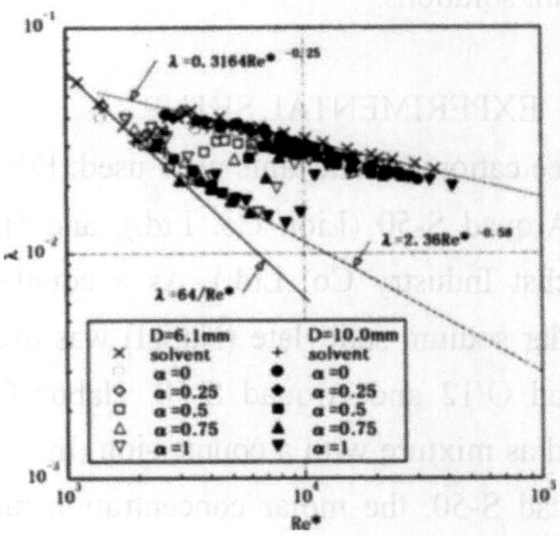

Fig.3 Influence of counter-ion concentration on λ. Arquad S-50 134ppm.

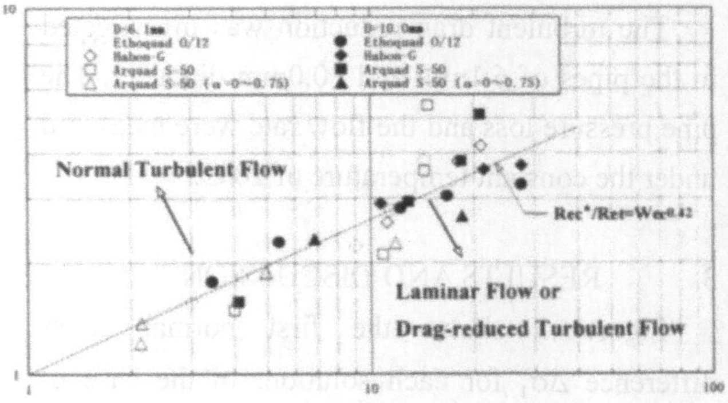

Fig.4 Rec^*/Ret versus Wec.

RELAXATION AND RETARDATION SPECTRA OF LAMELLAR LIQUID CRYSTALS IN A TOLUENE/NONIONIC SURFACTANT/WATER SYSTEM

J. MUÑOZ, M.C. ALFARO AND A.F. GUERRERO

Departamento de Ingeniería Química. Universidad de Sevilla, 41012 Sevilla

1. INTRODUCTION

Hydrophilic and lipophilic chemicals may be solubilized into either micelles or lyotropic liquid crystals (LC) forming monophasic ternary or more complex systems, which have interesting practical applications in detergency, emulsion technology, cosmetics, biology, monodispersed fine particles, as media for chemical reactions, etc. Lamellar liquid crystals (L_α), consist of surfactant bilayers, which may solubilize apolar chemicals among the hydrocarbon chains, separated by an aqueous medium that can incorporate hydrophilic substances. The whole microstructure is formed by the combination of clusters of flexible bilayers, each of them characterized by a director which defines the main orientation of the so-called monodomain. In other words, the L_α microstructure is formed by connected monodomains with random orientation so that from a macroscopic point of view the polidomain texture looks disoriented and with structural defects. Additionally, some vesicles may also be present.

The overall objective of this work was to analyse the influence of composition on the relaxation and retardation spectra shown by different systems in the Toluene/Triton N-101/Water phase diagram at 25°C.

2. METHODS

A polyoxyethylene glycol nonylphenyl ether, EO_a=10 (Triton N-101, a registered mark of Union Carbide Chemicals and Plastics) was purchased from Sigma and used without further purification. Toluene was supplied 98% pure by Panreac (Spain). Single distilled deionized water was used. All the samples had the same recent past mechanical and thermal history. The nomenclature used to denote the composition of the samples studied is as follows: XX/YY, where XX stands for the weight percentage of Triton-N101 and YY for the weight percentage of water.

Small amplitude oscillatory shear (SAOS) and creep compliance experiments were conducted on a CS rheometer Rheostress RS-100 from Haake with a cone and plate sensor system (60mm diameter, 4° angle).

LC's textures were observed under transmitted polarized light with a Carl Zeiss standard 142 microscope.

3. RESULTS AND DISCUSSION

3.1 Phase behaviour

A L_α region swelling from its location on the surfactant/water binary axis towards the water corner and two regions of highly stable dispersions; D_1 (characterized by small L_α portions embedded in an isotropic micellar phase) and D_α dispersions (formed by an aqueous micellar phase dispersed into a L_α continuous phase) were characterized.

3.2 Relaxation spectra

Mechanical spectra derived from SAOS experiments revealed that L_α LC's exhibited a typical response of highly structured materials. D_α dispersions showed qualitatively similar results but with lower values of G' and G''. D_1 dispersions presented more fluid-like viscoelastic properties.

Both moduli were fitted to the Generalized Maxwell model by an iterative linear regression method (Martínez, 1996).

In order to obtain a representation of the discrete relaxation spectrum, a normalization factor which takes into account the number of experimental data (Orbey and Dealy, 1991) was used. The normalized discrete relaxation spectra were fitted to an empirical equation (Madiedo, 1996):

$$H(\lambda) = \frac{\alpha\lambda^m + \beta\lambda^n}{1 + (\lambda/\lambda_t)^p}$$

Parameters m, n and p are related to characteristic slopes for the transition, plateau and terminal relaxation zones.

Figure 1 shows that H(λ) fitted the proposed model fairly well. H(λ) for L_α systems located well within the lamellar region are characterized by a wide plateau zone and by the onset of the transition zone.The influence of composition on H(λ) depended markedly on the surfactant/water mass ratio (SWMR). In fact, the highest H(λ) values and the appearance of a maximum in the plot of H(λ) vs. λ, that resulted in high positive values of parameter "n" (around +0.6), could be associated with systems containing ca.10% toluene and a relatively high SWMR.

20/60 dispersion belongs to D_1 region and H(λ) exhibited a less developed plateau zone ($n \approx -1$) and even the onset of the terminal zone, demonstrating that D_1 was more fluid-like than the remaining systems. The shape of H(λ) vs. λ of 25/65 D_α dispersion was rather similar to that shown by some L_α systems with a low SWMR (n = +0.12). However, the latter showed higher H(λ) values.

3.3 Retardation spectra

Retardation spectra, L(λ'), of L_α systems were derived from creep compliance by the Tikhonov regularization method using Rheolab 2 software (Madiedo, 1995). In order to check their validity, experimental J(t) values were recalculated using the Generalized Kelvin-Voigt model. On the whole, the shapes of the L(λ') were rather similar for the L_α systems (Figure 2). The smallest L(λ') values were found for systems with a SWMR embraced between 1,3 and 2.8. On the contrary, the highest values were obtained for L_α with a low SWMR. The fact that a binary system with a relatively high SWMR showed intermediate L(λ') values must be noted.

It was possible to estimate again the retardation spectra taking advantage of the approximate interrelations between linear viscoelastic functions. An excelent agreement with those obtained by the Tikhonov method was found.

The instantaneous compliance and the Newtonian viscosity along with the H(λ) and L(λ') values support the hypothesis that the more ordered L_α samples were restricted to a zone within the phase diagram defined by a relatively high SWMR and intermediate toluene concentration. Such zone may be characterized by optimum packing parameter values for the occurrence of bilayers giving rise to a more developed "entanglement" network among the monodomains forming the L_α structure. The nature of such "entanglements" is thought to be physical insofar they account for the density of contact zones among monodomains but also chemical when Van der Waals, steric and hydration forces (hydrogen bonds) are considered.

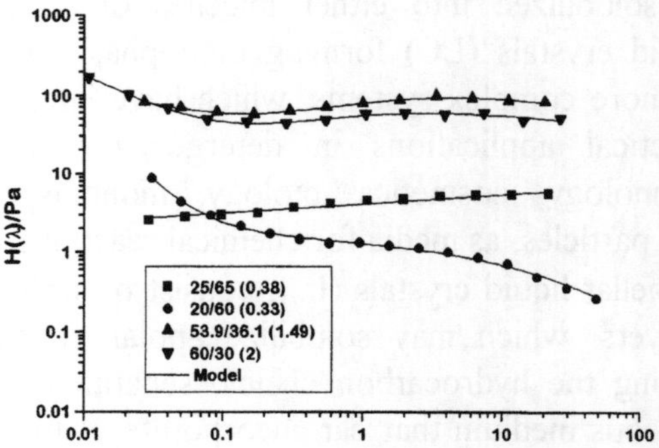

Fig 1. Relaxation spectra of two L_α samples, a D_1 dispersion and a D_α dispersion.

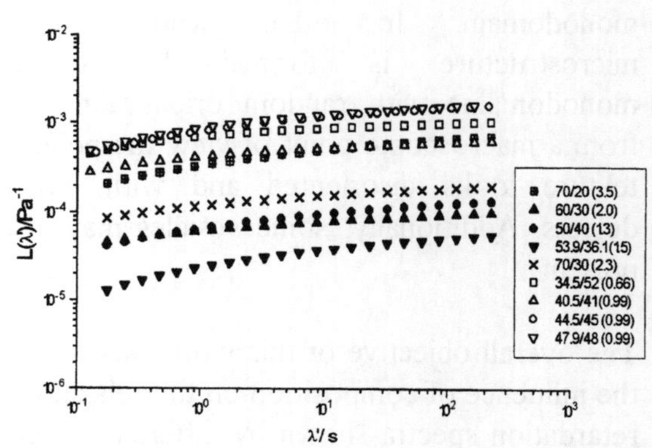

Fig. 2. Retardation spectra of several L_α systems

4. ACKNOWLEDGMENT

This contribution is part of a project (PB92-0664) sponsored by the CICYT, SPAIN. The authors gratefully acknowledge its financial support.

5. REFERENCES

Madiedo JM (1995) Rheolab© 2.0 for Windows

Madiedo JM (1996) Thesis University of Seville

Martinez F (1996) Thesis University of Seville

Orbey N, Dealy J (1991) J. Rheol. 35:1035-1049

NON LINEAR RHEOLOGY OF WORMLIKE MICELLES

A. PONTON, D. QUEMADA

Laboratoire de Biorhéologie et d'Hydrodynamique Physico-chimique (L.B.H.P.), CNRS ESA 7057, Univerité Paris VII, 2 Place Jassieu case 7056 - 75251 Paris Cedex 05

1. INTRODUCTION

The motivation of this study is a recent work of Porte et al. (1997) on shear induced inhomogeneous flows of solutions of giant micelles. The authors proposed a phenomenological analysis based on a effective potential that accounts for the free energy stored in a viscoelastic material under steady shear conditions. In such a scheme, mechanical flow instabilities have been described by a non-monotonic constitutive relation for the shear stress as a function of the shear rate. The model accounts for the existence of a stress plateau in which the set of stationnary states is highly degenerate in a shear rate controlled experiment. As a consequence, hysteresis is predicted but not observed experimentally.

In this paper, we present experimental results on rheological behavior of screened wormlike micelles of surfactant. We focus our interest on hysteresis behavior using shear rate controlled rheology.

2. EXPERIMENTAL

2.1 Materials

Our study was carried out on aqueous solutions of cetylpyridinium chloride (CPCl) and sodium salicylate with adding a salt (0.5 M Nacl). The phase diagram of this system has been derived by Berret et al. (1994) using X ray and birefringence measurements. The samples were chosen in the semi dilute range (isotropic phase) where the micelles are entangled. The total mass fraction of surfactant Φ was varied from 8% to 14%.

2.2. Rheology

The rheological measurements were performed with a stress controlled rheometer (Haake RS150) working in the constant shear rate mode and using a cone plate device (6 cm diameter, 4° cone angle).

The temperature was controlled with a thermostated bath and fixed to T=25°C. Shear rate measurements were carried out by increasing and decreasing lineary the shear rate $\dot{\gamma}$ for several duration times.

3. RESULTS

In Figure 1 is displayed the variation of the shear stress σ as a function of the shear rate $\dot{\gamma}$ at Φ = 10% and with a duration time of experiment of 60 min.

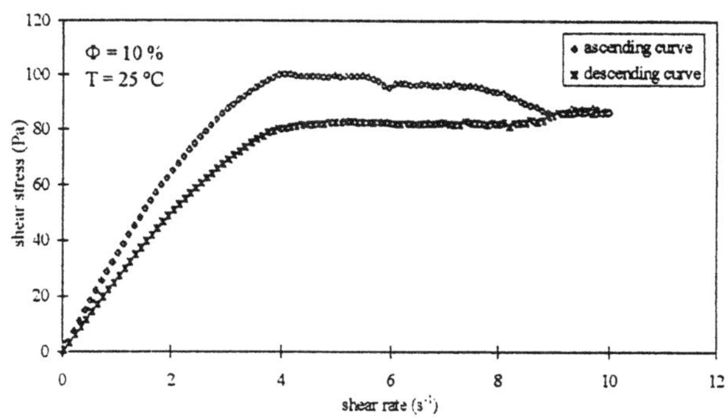

Figure 1: Flow curve for Φ = 10%.

Two domains can be distinguished on the ascending curve. The first corresponds approximatively to a Newtonian domain in which σ increases quasi linearly with the $\dot{\gamma}$ up to $\dot{\gamma}$ = 4 s^{-1}. As $\dot{\gamma}$ is increased, σ first decreases and tends to be nearly constant. In this domain, the flow is unstable so giving a shear banding structure. The evolution of $\sigma(\dot{\gamma})$ has been theoretically studied by Cates et al (1993) who proposed a constitutive scheme for reversibly breakable chain of surfactant solutions. It should be emphasized that no data points will be obtained in the second region from controlled stress experiments. The descending curve shows the same two regions but the plateau zone is clearly more pronounced than in the increasing curve. The difference in viscosity on the two curves suggests that an orientation of the micellar phase took place in the shear field when it was first subjected to shear stress. The orientation of the micelles is not defined before the first measurements are taken. Therefore its rheological properties vary distinctly.

As hysteresis is strongly time dependent, we have studied the time effect on the $\sigma(\dot{\gamma})$ curves for Φ = 10% .

The results are shown in Figure 2 where the $\sigma(\dot{\gamma})$ curves are reported for three duration times of experiment.

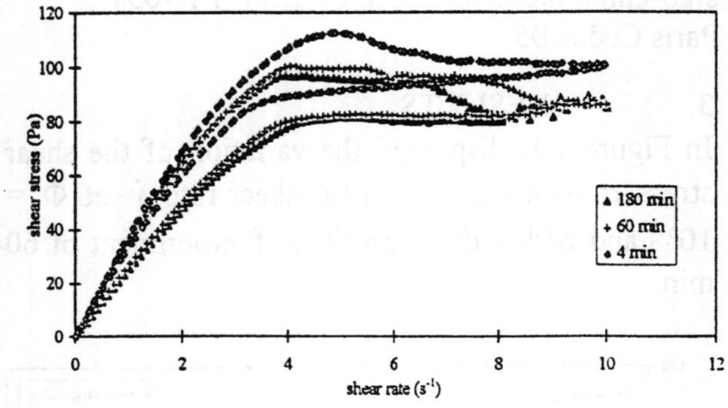

Figure 2 : Flow curves at different duration times of experiment for $\Phi = 10\%$.

Many points should be pointed. The ascending curve does not show a constant shear stress domain but rather a decrease of shear stress as a function of shear rate for the shortest time. In this case, the system is out of equilibrum and evolve during the experiment. The second and the third cycles are pratically superimposed. This effect is clearly seen on the descending curves. As a conquence, the system has reached a state close to the stationnary one.

In order to confirm this tendency, successive scanning of shear rate $\dot{\gamma}$ have been performed on the same sample. An example of results is shown in Figure 3 for $\Phi = 8\%$ and a duration time of experiment of 8 min.

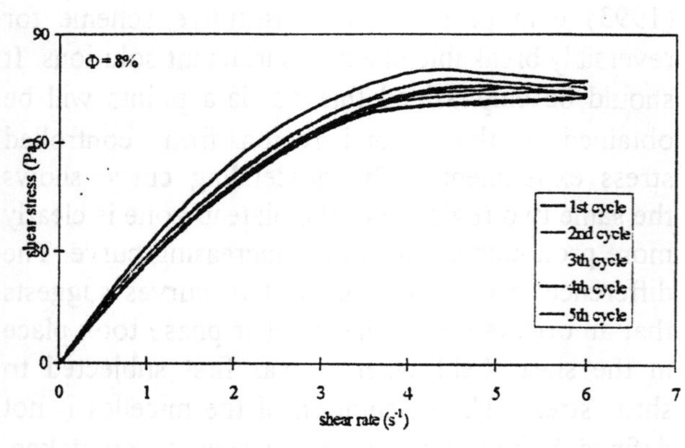

Figure 3 : Successive flow curves for $\Phi = 8\%$.

It is obvious that the last two cycles are superimposed. At first sight, this fact could be related to the existence of identical structure of the system for sucessive scannings of $\dot{\gamma}$.

The tendency to a limit cycle has also been observed for the other surfactant concentrations ($\Phi = 10\%, 12\%, 14\%$).

The same type of measurement have been carried out with a duration time of experiment of 4 min and a maximum shear rate of $10 s^{-1}$. As expected the difference between the ascending and descending curves for the first cycle has been emphasized and the limit cycle has been obtained after seven cycles. All the results seem to indicate that the shear history of the sample is an important parameter. We have studied the effect of preshearing on $\sigma(\dot{\gamma})$ curves by varing the strength and duration time of preshearing.

4. CONCLUSION

In this work, experimental results on unsteady flow behavior in a wormlike micelles of surfactant have been presented. The effect of duration of experiment has been investigated for several mass fractions of surfactant. Moreover successive scannings of shear rates on the same sample has led to a limit cycle corresponding to a limit structure for the system under flow. Finally the effect of preshear has been presented as a preliminary result. More experimental work are in progress to quantify these non linear phenomena.

6. REFERENCES

Berret JF, Roux DC, Porte G (1994) J Phys II France 4:1261-1279

Cates ME, McLeish TCB, Marrucci G (1993) Europhys Lett 4:451-456

Porte G, Berret JF, Harden JL (1997) J Phys II France 7:459-472

RHEOLOGY AND NMR MEASUREMENTS OF SODIUM DODECYL SULPHATE /DECANOL/WATER NEMATICS

T. Thiele[1], J.-F. Berret[1], S. Müller[2] and C. Schmidt[2]

[1] : Unité Mixte de Recherche CNRS / Université de Montpellier II n° 5581, Groupe de Dynamique des Phases Condensées F-34095 Montpellier Cedex 05 France

[2] : Institut für Makromolekulare Chemie, Universität Freiburg, Sonnenstrasse 5, D-79104 Freiburg, Germany

Lyotropic nematic mesophases in surfactant systems have been discovered nearly three decades ago. However, surfactant nematics have recently attracted much interest because of their remarkable analogies with liquid-crystalline polymers (LCP) when they are submitted to steady shear flow. In the wormlike micellar system made of cetylpyridinium chloride, hexanol and brine (CPCl/Hex), a tumbling instability has been found out [1]. Under stationary shear, tumbling results in oscillations in the shear stress response, related to the indefinite rotation of the director. The transient stress exhibits in addition the non-trivial scaling laws predicted by the Larson and Doi mesoscopic theory of textural materials [2]. These scaling laws are crucial since they suggest that, as far as the rheological history of the solution is controlled (say, presheared at $\dot{\gamma}_{presh}$), the transient stress response is then only a function of the deformation applied $\gamma(t) = \dot{\gamma}\, t$ and of the ratio $\dot{\gamma} / \dot{\gamma}_{presh}$. We have applied the same experimental approach to the thoroughly studied surfactant nematics, namely the ternary system sodium dodecyl sulfate / decanol / water (SDS/Dec). In this communication, we investigate the transient shear stress responses from nematic calamitic (N_c) and nematic discotic (N_d) mesophases and search for scaling laws. N_c and N_d mesophases are assumed to be colloidal suspensions of rod-like and disc-like particles, respectively. We also provide the preliminary rheo-NMR results that confirm that both SDS/Dec nematics are of flow alignment-type.

The phase diagram of SDS/decanol/water has been checked in the concentration ranges $\phi_{SDS} = 20$ - 30 % and $\phi_{Dec} = 0$ - 8 %. The phase boundaries were determined from their LC textures observed with optical microscopy, and were found in agreement with previous reports [3]. The three concentrations investigated here are all with a fixed decanol content, $\phi_{dec} = 4.5$ %. One has $\phi_{SDS} = 25.5$ % for the N_c-solution studied in rheology, and $\phi_{SDS} = 25.0$ % (N_c) and $\phi_{SDS} = 23.0$ % (N_d) in deuterated water for the solutions studied by means of rheo-NMR . Transient flow-curves of the SDS/Dec were recorded on a Rheometrics Fluid Spectrometer (RFS II) working in a Couette configuration with controlled shear rate (gap 1 mm). Flow reversal experiments were performed in the range 0.1 - 10 s^{-1}. Rheo-NMR measurements were carried out using a miniature cone-and-plate viscometer integrated into an NMR probe, and described in [4].

All nematic N_c and N_d SDS/Dec solutions are shear-thinning. The shear viscosity $\eta(\dot{\gamma})$ exhibits first a $\dot{\gamma}$-independent plateau up to ~ 10 s^{-1} (typical η_0 are between 0.1 and 1 Pas). Beyond this pseudo-Newtonian regime, $\eta(\dot{\gamma})$ decreases. More details on the rheology of SDS/Dec nematics can be found in [5]. In Fig. 1 are displayed the transient shear stress responses $\sigma(\gamma, \dot{\gamma})$ obtained from flow reversals at $\dot{\gamma}$ = 0.5 and 1 s^{-1}. For flow reversals, the sample is first presheared at fixed rate $\dot{\gamma}_{presh}$ until the steady state conditions are reached (~ 300 strain units). The direction of flow is then changed abruptly, the value of the strain rate remaining unchanged. The stress data are normalised to the stationary stress. As evidenced in Fig. 1, the superposition of both transients $\sigma(\gamma, \dot{\gamma})/\sigma(\gamma{=}\infty, \dot{\gamma})$ is observed. This is true at low deformations where the stress exhibits non periodic oscillations, as well as at large deformations where the stress relaxes to its long time stationary value (see inset). Stress maxima are at $\gamma = 0.5$, 3 and 15, minima are observed at $\gamma = 1$ and 7. We have also performed other types of experiments that still control the shear history. In step-change experiments, the solution is first presheared at $\dot{\gamma}_{presh}$, however in the same direction. In this configuration, no oscillations comparable to those of Fig. 1 appear in the transient response [5]. These later results suggest that SDS/decanol system in its N_c- and N_d-state belongs to the class of the flow-alignment nematics. Flow alignment is obtained when the nematic director has a fixed orientation with respect to the flow velocity in steady state conditions. The scaling laws and the long relaxation of the mechanical stress toward unity are indicative that mesoscopic textures play a crucial role in the flow properties.

In order to check the previous assumption of flow-alignment for SDS/Dec, rheo-NMR measurements were undertaken on solutions with deuterated solvent. Here, interest is focused on the line shape and splitting of the ^{2}H NMR spectra. The doublet structure is due to the quadrupole coupling between the deuterium and the electric field gradient arising from the surrounding electrons [4]. This splitting $\Delta\nu$ is related to the director orientation Θ between the nematic director and the magnetic field B_0 : $\Delta\nu = \frac{3}{4}\delta\left(3\cos^2\Theta - 1\right)$. Here δ is the quadrupolar coupling constant. In Fig. 2 are shown the ^{2}H NMR spectra of the two SDS/Dec nematic phases. N_c and N_d solutions have been investigated at rest, aligned by the field (top) and under shear in steady shear conditions at $\dot\gamma = 0.8\ s^{-1}$ (bottom). The spectra at rest can be interpreted as follows. As in many surfactant systems, the SDS/Dec particles, either rod-like or disc-like align in such a way that the long axis of the surfactant molecules is oriented perpendicular to B_0 (due to a negative anisotropy of the magnetic susceptibility). Thus the director of the cylindrical micelles aligns parallel to the magnetic field, whereas the director of the disc-like aggregates alignsperpendicular to B_0. This is demonstrated in Fig. 2. For the N_c-phase, maximum alignment gives rise to the maximum splitting observable for this solution. Under shear, and in the shear rate range of the rheology ($\dot\gamma > 0.1\ s^{-1}$), the splitting reduces to the half of the initial value, showing that the director is now aligned perpendicular to the magnetic field, but parallel to the velocity. The reverse transition is observed for the discotic phase, the splitting being increased by a factor two. In order to summarize Fig. 2, one can say that at rest the directors for N_c and N_d nematics are aligned by the external magnetic field, whereas under flow the shearing field overcomes this magnetic interaction and orients the nematic. In conclusion we have shown that the SDS/Dec nematics, both in N_c and N_d mesophases are nematics of the flow-alignment class. SDS/Dec behaves then differently from the well-studied CPCl/Hex tumbling nematic.

This work is funded by the European Union, under the TMR network « Rheology of Liquid-Crystalline Materials ».

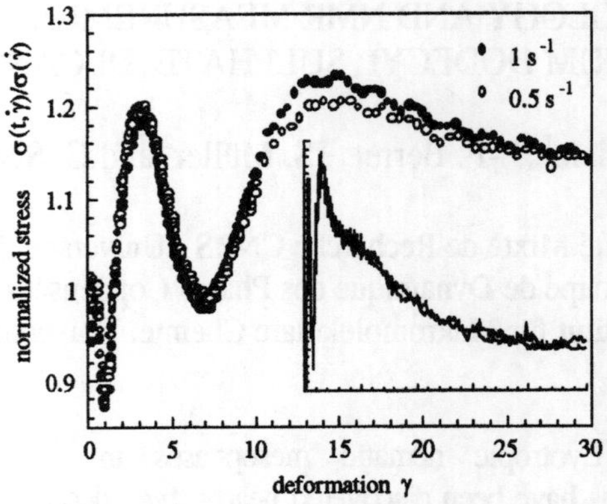

Figure 1 : Normalized shear stress after reversal of the flow, as a function of the applied strain for $\dot\gamma$ = 0.5 and 1 s^{-1}. The inset emphasizes the long time relaxation to the stationary state.

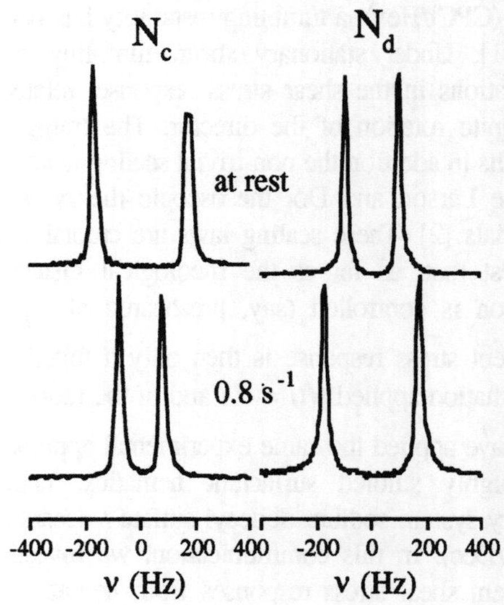

Figure 2 : ^{2}H NMR spectra of the two nematic phases as aligned by the magnetic field (top) and by the shearing field (bottom). Left : N_c-phase, right : N_d -phase.

REFERENCES

[1] J.-F. Berret *et al.*, Europhysics Letters **32** 137 (1995)
[2] R.G. Larson and M. Doi; J. Rheol. **35** 539 (1991)
[3] P.-O. Quist *et al.*, J. Chem. Phys. **96** 3875 (1992)
[4] H. Siebert et al., Rheol. Acta **36** 618 (1997)
[5] T. Thiele, J.-F. Berret, S. Müller and C. Schmidt submitted to J.Rheol (1998)

ON TENSILE TESTING OF CONCENTRATED SUSPENSIONS

A.N.Alexandrou[1], A.V.Bazilevsky[2], V.M.Entov[2,1], A.N.Rozhkov[2], A.Sharaf[1]

[1] Mechanical Engineering Dept., Worcester Polytechnic Institute, 100 Institute Road, Worcester MA 01609, USA
[2] Laboratory of Applied Continuum Mechanics, Institute for Problems in Mechanics of the Russian Academy of Sciences, pr.Vernadskogo, 101,117526, Moscow, Russia

Monitoring capillarity-driven necking and breakup of liquid bridges is used to study tensile rheology of suspensions. A specimen of the tested fluid is placed between two parallel disks and kept in the gap by capillary forces. The disks then are rapidly pulled apart and kept at a fixed distance. As a result, the fluid forms a liquid bridge which is unstable at sufficiently large aspect ratio and tends to break. The breakup process is driven by the lateral capillary pressure $p_q \propto \alpha/r$, which is maximum in the necking region. Here, α is the surface tension, r is the local radius of the bridge. According to simple theory developed for a uniform liquid filament squeezed by the lateral capillary pressure the neck radius obeys the differential equation

$$dr/dt = -\alpha/(2\eta_e) \qquad (1)$$

η_e being the effective elongational viscosity. For a Newtonian fluid with $\eta_e = 3\mu$ Eq.(1) implies

$$r = r_0 - \alpha/(6\mu t). \qquad (2)$$

For a nonlinear power-law fluid r(t) a linear function of $\alpha(t_{br}-t)^n/K$, t_{br} being the breakup time, K the consistency index and n the power law exponent.

These predictions are derived for long uniform filaments. If we assume that they also apply to the the later stages of the evolution of the neck radius versus time for non-uniform deformation in the liquid bridge it would be feasible to estimate the effective elongational viscosity by monitoring the bridge breakup and by using relation (1) to find the effective elongational viscosity:

$$\eta_e = 3\mu_e = \alpha/(2|dr/dt|) \quad (3)$$

Constant value of η_e indicates effectively Newtonian behavior; if it changes during the filament thinning, it is an indication of importance of non- Newtonian effects.

This basic initial idea was tested experimentally and proved to be both correct and incorrect. The experiments were performed at 18-19C with suspensions of crushed sand in glycerine. Two types of the sand powder were used having the surface area .850-.856 m^2/g and .086 -.088 m^2/g (particle diameters 1.5 μm and 15 μm); further they are referred to as suspensions 1 and 2. The concentration c ranged from 0 to 0.465. The suspensions were first tested in shear using rotational rheometer `Rheotest-2'. The data were approximated with the Bingham model.

The following conclusions are derived from the tensile experiments. First, there exists a stage of uniform filament-like neck thinning (Fig.1) that corresponds to the postulated pattern.

Fig.1

(At a later stage, a local neck develops on the suspension column and it breaks rapidly. Transition to localized necking is more pronounced in dense suspensions of coarse particles).

Secondly, the r(t) plots have sufficiently long linear segments (Fig.2) to be used to estimate the effective elongational viscosity using Eq.(3).

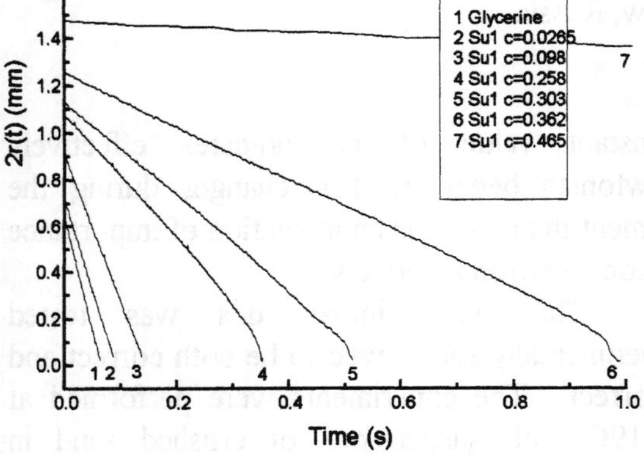

Fig.2

It is well correlated with the shear viscosity at high shear rates η over entire range of concentrations. Its dependence on the solid content can be approximated using Krieger-Dougherty formula

$$\mu_e (c)/ \mu_e =(1-c/c_m)^{-\beta} \quad (4)$$

The best-fit values c_m =0.53, β=2.00 for the suspension 1 and c_m =0.56, β=1.45 for the suspension 2. However, this effective viscosity is approximately twice as large as shear viscosity, and this discrepancy persists even for a Newtonian fluid (glycerine). The best fit linear correlation of these data is

$$\mu_e =(1.934 \pm 0.022)\, \eta \quad (5)$$

The discrepancy is due to the fact that the effective stress driving the neck thinning is less than capillary pressure in the neck by a factor S, further referred to as stress efficiency. Experiments with glycerine bridges and modeling were performed to prove this statement and derive respective correction factor.

In the experiments, breakup of bridges of glycerine was followed in detail for different initial lengths. In modeling, a simple quasi-one-dimensional theory of the liquid bridge evolution and breakup driven by capillary forces was used.

Comparison of the experimental observations and modeling results proves to be encouraging. The modeling shows that for a Newtonian fluid necking kinetic is fairly linear, but the thinning rate is less than that predicted by simple formula (2). This explains the iscrepancy between apparent and true viscosities. In modeling using experimentally observed initial bridge shapes as initial conditions the neck thinning and bridge shape are reproduced reasonably well. A slight increase in 'apparent viscosity ratio' with bridge length is observed both in modeling and experiments.

Thus in tensile testing of inelastic fluids (both Newtonian and nonlinearly viscous) the observed behavior is determined by an inteaction between the filament and the end drops. To be useful as an auxiliary rheometric device, the experiment should allow a kind of data processing to assess the fluid rheology.

To study this option, the modeling was performed for nonlinearly viscous power-law fluids. The thinning kinetics is markedly different for shear-thinning and shear-thickening fluids. Even relatively small deviation from the Newtonian behavior (n=1±.2) can be easily detected. Both n and K can be estimated using the r(t) dependence, if the efficiency factor S is introduced in the formula (3) of the elementary theory, and it is written as

$$\eta_e= \mu_e =S\alpha/(2|dr/dt|) \quad (6)$$

In a range of parameters studied the factor S varies between .2 and 0.45. Better estimates of rheological properties can be derived using more sophisticated techniques of data processing.

RHEOLOGY OF AN ASSOCIATIVELY THICKENED LATEX

T. AUBRY[1], J.M. CORPART[2], M. MOAN[1]

[1]Laboratoire de Mécanique et Matériaux, Université de Bretagne Occidentale, 6, avenue Victor Le Gorgeu, 29285 Brest Cedex, France.
[2]Elf Atochem, C.A.L., 95, rue Danton, 92300 Levallois Perret, France.

1. INTRODUCTION

Associative thickeners are long known to be good rheological modifiers for water-based paints (Hall et al., 1986).In spite of these industrially important applications, studies of the rheological behavior of associative polymers/solid particles systems are rather scarce (Pham et al. 1998; Fonnum et al. 1993; Schaller ans Sperry, 1992).In this work, the rheological properties of latex suspensions in the presence of telechelic hydrophobically modified polyacrylamides (HMPAM) have been studied.

2. MATERIALS AND METHODS

The materials used in this investigation were all synthetized at Elf Atochem.They are composed of a skeleton of polyacrylamide, end-capped by hydrophobic groups (C_{16}-H_{33}).The weight average molecular weight of this polymer is $9x10^4$g/mol.

The latex used in the study has been supplied by Elf Atochem; it is a styrene-acrylic copolymer.The average diameter of latex particles is 140 nm.The dispersing medium is a serum composed of water plus surfactants which stabilize the latex suspensions.

Rheological measurements have been performed at 25°C on a Contraves Low-Shear 30 viscometer in the dilute regime and on a Carri-Med CSL 50 controlled stress rheometer in the concentrated regime.The rheological properties were measured after 24h equilibration.

Adsorption of HMPAM on latex particles has been studied using a viscometric method described below.

3. RESULTS AND DISCUSSION

3.1. Adsorption properties

Adsorption of HMPAM on latex particles has been studied by measuring, if possible, the amount of HMPAM present in the supernatant after centrifugation of HMPAM/latex suspensions at different latex volume fractions and polymer concentrations.The results seem to show that hydrophobically modified PAM, contrary to non-modified PAM, do adsorb on latex particles; which points out the role played by hydrophobic groups in adsorption properties.Moreover, the results tend to confirm that multilayer adsorption of HMPAM on latex particles is a possible mechanism (Argillier et al., 1996).

3.2. Zero-shear viscosity

Figure 1 shows the zero-shear viscosity as a function of HMPAM concentrations at different latex volume fractions.

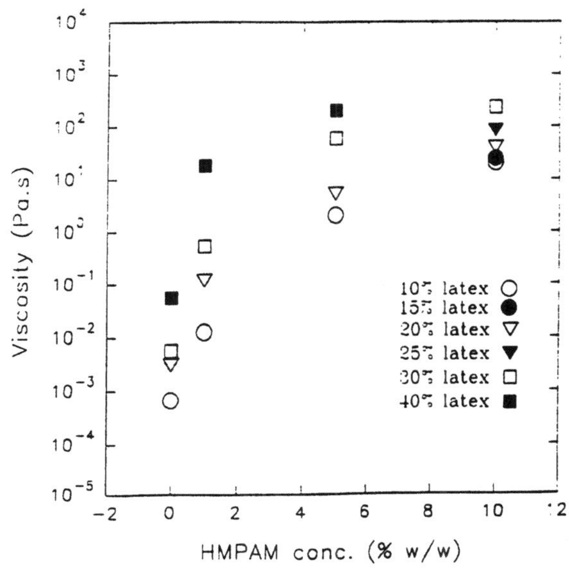

Figure 1 Zero-shear viscosity vs. HMPAM concentration at six latex volume fractions.

Figure 1 clearly shows that addition of HMPAM leads to a very significant thickening effect.At low polymer concentrations, this effect depends much on latex volume fraction; whereas, at high polymer concentrations, the thickening effect depends slightly on latex concentrations and may be attributed to the associative network (the viscosity level is nearly that of HMPAM alone in distilled water).

3.3. Non-Newtonian behavior

All samples tested exhibit non-Newtonian behavior, more precisely a shear-thinning behavior, above some shear rate.No shear-thickening is ever observed, contrary to what is obtained with HMPAM alone in distilled water.

3.4. Linear viscoelastic behavior

Figure 2 shows the storage modulus G' and the loss modulus G" as a function of frequency in the linear viscoelastic regime for a latex suspension at 20% latex volume fraction and at 10% w/w HMPAM concentration.The same behavior is observed for all suspensions above given latex volume fraction and polymer concentration.

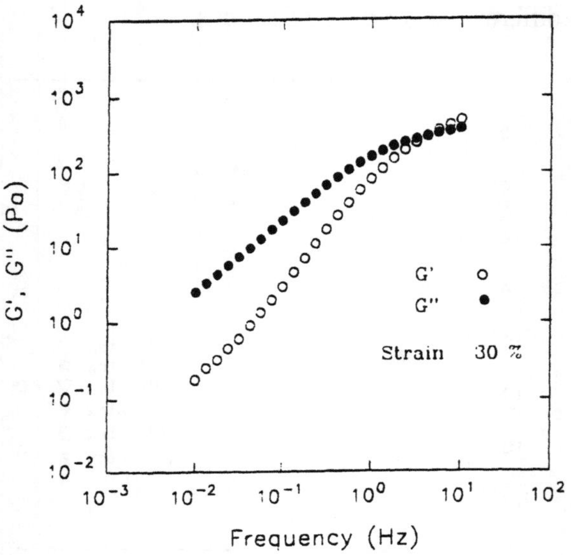

Figure 2 Storage modulus G' and loss modulus G" versus frequency at 20% latex volume fraction and 10% w/w HMPAM concentration.

These linear viscoelastic curves have mainly two characteristics:

- at a short time, that is about 0.1s (order of magnitude of the average lifetime of an associative junction), G' = G".
- at a long time, that is about 10s, the G' vs. frequency curve exhibits an inflexion point; which could be attributed to the relaxation of the interphase particle/dispersing medium.

4. CONCLUSION

The linear and non-linear rheological behavior of latex suspensions associatively thickened by hydrophobically modified polyacrylamide, at least above some latex volume fraction and polymer concentration, seems to agree with a microstructure composed of latex particles more or less linked with each other in an "associative continuum" (associated adsorbed and non-adsorbed polymer chains).

5. REFERENCES

Argillier J.F., Audibert A., Lecourtier J., Moan M., Rousseau L. (1996) Colloid Surf. A 113:247-257.

Fonnum G., Bakke J., Hansen F.K. (1993) Colloid Polym. Sci. 271:380-389.

Hall J.E., Hogdson P., Krivanek L., Malizia P. (1986) J. Coat. Technol. 58:65-73.

Pham Q.T., Russel W.B., Lau W.(1998) J. Rheol. 42:159-176.

Schaller E.J., Sperry P.R. (1992) Handbook of Coatings Additives 2:105-163.

MEASUREMENT OF SHEAR-INDUCED SELF-DIFFUSION IN CONCENTRATED SUSPENSIONS BY A NOVEL METHOD

V. BREEDVELD [1], A. TRIPATHI [2,3], A. ACRIVOS [2]

[1] Twente Institute of Mechanics, J.M. Burgerscenter, Rheology Group, University of Twente, P.O. Box 217, 7500 AE Enschede, The Netherlands.
[2] The Levich Institute, The City College of the City University of New York, New York 10031.
[3] Department of Mechanical Engineering, MIT, Cambridge, MA-02139

1. INTRODUCTION

Diffusion of particles in concentrated suspensions is traditionally associated with the Brownian motion of colloids due to interactions with solvent molecules or with inertial effects in case of turbulence. It has been shown experimentally however (Leighton & Acrivos (1987)), that another mechanism for diffusion exists: large, non-colloidal (~100 μm) particles show diffusion-like behavior even in low Reynolds numbers flow, where inertial effects are negligible. The origin of this so-called "shear-induced" diffusion is depicted in the figure below.

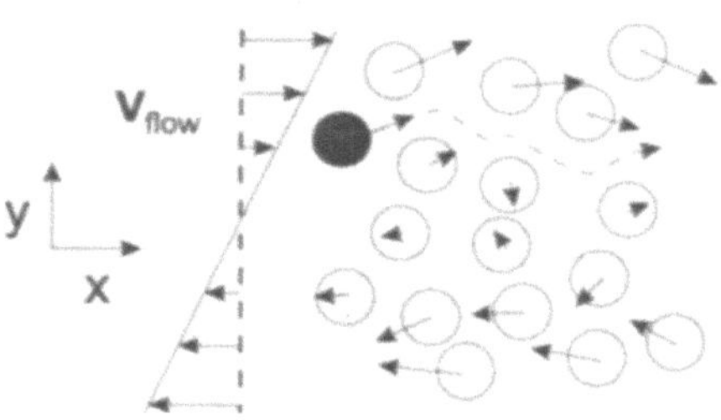

The figure represents a concentrated suspension in simple shear flow; on the left, the average flow field is drawn. Due to the interactions with its neighbors, the marked particle does not move along the original streamline. As a result, the marked particle exhibits a fluctuating motion when it tumbles around its neighbors whose positions have a random component. The displacements of the individual particles have been shown to be of diffusive nature and are usually referred to as shear-induced self-diffusion (e.g. Davis 1996).

The diffusion of individual particles can also lead to collective migration of particles if gradients are present in either the particle volume fraction or shear rate. The direction of such fluxes will always be down these gradients, since the frequency of interactions is higher on the side of high concentration or shear rate. Shear-induced gradient diffusion is responsible for the particles in a pipe flow to migrate towards the center, where the shear rate is lowest, or towards the outer wall in wide-gap Couette flow.

Various researchers have carried out experiments to study these phenomena, but still little is known about the processes by which shear-induced diffusion is driven. In order to investigate the fundamental aspects of shear-induced self-diffusion, a new experimental technique has been developed.

2. NOVEL EXPERIMENTAL TECHNIQUE

The experimental technique is based on the spatial correlation methods which are also applied in Particle Image Velocimetry (PIV). The experimental system consists of a concentrated suspension of density and refractive index matched particles. The suspension is seeded with black tracer particles of the same material, which can be detected optically. At a fixed position in the sheared suspension, a large ensemble of images (usually about 500) is taken with a CCD camera at a constant fixed time interval Δt and stored on a computer. Two consecutive images of such an ensemble are sketched.

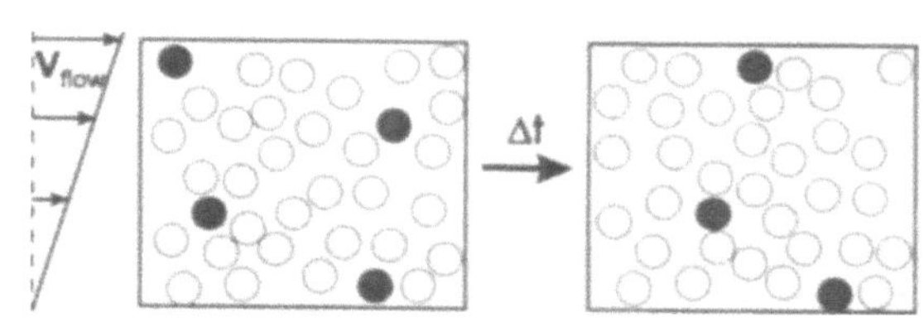

The positions of all tracer particles (black) in the images are then determined with image analysis software. All positions of the tracer particles (N = 3) in the second image are correlated with the tracer positions (M = 4) in the first image, by calculating the displacement vectors:

$$\Delta\mathbf{x}_{nm} = \left(x_n - x_n, y_n - y_n\right) \quad n = 1..N, m = 1..M \quad (1)$$

When these vectors are calculated for the entire ensemble of images, all correlation data can be combined and analyzed statistically to evaluate the motion of the particles. A characteristic example of the correlation data for an experimental situation is given below. The graph shows the experimental distribution of the displacement vectors (in pixels) in the Δx-Δz-plane. The experiment was carried out with the camera looking from the velocity gradient direction (y).

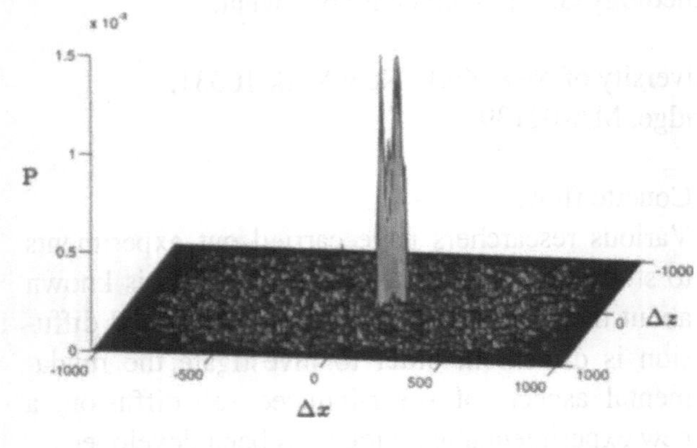

The peak contains information on the motion of the particles during the time-step Δt. The location of the center of the peak gives the average displacement of the particles due to the macroscopic flow (used in PIV). In addition, the width of the peak is a measure for the fluctuations in the particle displacements and is therefore the quantity of interest for investigating shear-induced diffusion.
It can be shown (Breedveld et al. (1998)) that in case of diffusive motion of the tracer particles, the peak in the correlation graph should have a Gaussian shape in the Δz-direction:

$$P = A \cdot \exp\left(-\frac{\Delta z^2}{2\sigma_z^2} \right) \qquad (2)$$

The width of the peak, σ_z, is expected to scale like (Eckstein et al (1977)):

$$\frac{\sigma_z^2}{a^2} = 2\hat{D}_{zz}\,\dot{\gamma}\,\Delta t \qquad (3)$$

with a the particle radius, $\dot{\gamma}$ the shear rate, Δt the correlation time and $\hat{D}_{zz}$ the dimensionless diffusion coefficient in the z-direction. When the system is viewed from the vorticity direction (z), similar equations can be derived, where z is simply replaced with y.

3. RESULTS

The technique was used to study the shear-induced self-diffusion in a narrow gap Couette device for spherical particles with a diameter of 90±15 μm. The volume fraction ranged from 20 to 50% and the dimensionless time $\dot{\gamma}\,\Delta t$ from 0.05 to 0.5. The equipment was designed so that the suspension could be studied from both the velocity gradient and vorticity direction. The figure below shows the dimensionless width σ^2/a^2 of the correlation peak as function of the dimensionless time $\dot{\gamma}\,\Delta t$ for a particle volume fraction of 30%.

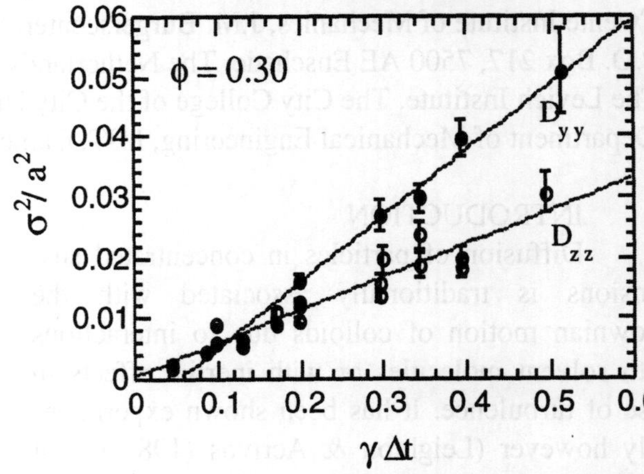

The linear scaling of eqn. (3) is clearly present in the experimental results. Therefore the motion of the particles is diffusive over the entire range of time scales. This is a striking result, because it is generally believed that the particle interactions in simple shear flow occur on the time scale $\dot{\gamma}\,\Delta t \sim O(1)$. The values of the diffusion coefficient were obtained from the slope of the linear fits in the figure above (see eqn. (3)). The results are in remarkable agreement with the results from previous experiments (e.g. Leighton & Acrivos (1987)) which were performed for $\dot{\gamma}\,\Delta t \sim O(10)$.

4. CONCLUSIONS

A novel experimental method has been developed to study shear-induced self-diffusion in concentrated suspensions in a simple shear flow over a range of time scales.
On the relatively short time scales of the experiment, the motion of the particles was shown to be of diffusive nature. The values for the diffusion coefficients were in remarkable agreement with previous experimental results on much longer time scales. This suggests that the process behind shear-induced self-diffusion has a typical time scale $\dot{\gamma}\,\Delta t$ which is significantly smaller than $O(1)$.

5. REFERENCES

Breedveld V, Van den Ende D, Tripathi A, Acrivos A (1998) J. Fluid Mech. in press.
Davis RH (1996) J. Fluid Mech. 310, 325-335.
Eckstein EC, Bailey DG, Shapiro AH (1977) J. Fluid Mech. 79, 191.
Leighton D, Acrivos A (1987) J. Fluid Mech. 177, 109-131.

RHEO-OPTICAL STUDY OF THIXOTROPY IN SYNTHETIC CLAY SUSPENSIONS

S. COCARD, J. F. TASSIN

Chimie et Physique des Matériaux Polymères, UMR 6515, Université du Maine, Avenue Olivier Messiaen, 72085 LE MANS Cedex 9, France

1. INTRODUCTION

Laponite, a synthetic Hectorite clay, manufactured by Laporte Industries (U. K.) is well known for its atypical rheological behavior when it is dispersed in water. A phase diagram of these anisotropic particles (small discs with a 30 nm diameter and a 1 nm thickness which carry some negative charges on their faces) was established by Mourchid et al (1995). They found a reversible sol-gel transition for these suspensions. Many authors, among them Ramsay et al (1986, 1993), Morvan et al (1994), Pignon et al (1997), Kröon et al (1996, 1998) have investigated with different methods the organization of the particles in the gel state. Here we present a study of these suspensions under shear flow using optical rheometry, emphasizing thixotropy in these materials.

2. MATERIALS AND METHODS

2.1. Samples preparation

Laponite is dispersed in deionised water at fixed ionic strength at 15000 rpm during 3 minutes at room temperature. At the end of this preparation phase, an opaque solution is obtained. Then, the suspension is let at rest during 24 hours, and gradually clears as the clay swells.

2.2 Rheo-optical experiments

These experiments were performed with a Rheometrics Optical Analyzer. Using this equipment which involves rapid modulation of the polarization, birefringence and orientation angle can be simulatneously measured. The present measurements were carried out using a concentric cylinder geometry, where light travels along the vorticity axis.

Experimental procedures

To characterize the thixotropy of the suspensions, we apply three types of experiments schematically depicted in Figure 1.

i) Steady state birefringence measurements at given shear rate.
ii) Studies of kinetics of recovery, where a low shear rate (phase II) is applied following a breakdown step at high shear rate(phase I).
iii) Relaxation experiments (phase III) after cessation of steady shear.

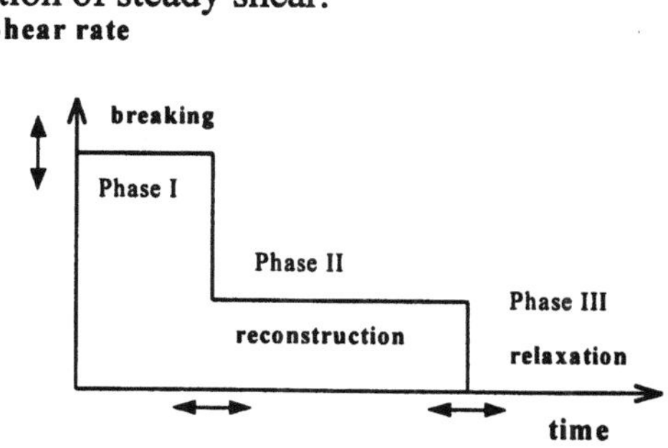

Figure 1 : Procedure applied to characterize the thixotropy of the Laponite suspensions.

3. RESULTS

3.1 The aging of dilute suspensions

Figure 2 gives some results about the time evolution of birefringence versus shear rate for a dilute Laponite suspension (1 % w/w). We can observe that few days after its preparation, the suspension has a Newtonian behavior, slightly shear thinning at high shear rates. After 40 days aging, we can observe a strong shear thinning behavior and after 100 days, the birefringence becomes independant of shear rate at low shear rates. Such a behavior is connected to the apparition of a yield stress for the suspension. It can be noted that essentially the low shear rate behavior is changing during the transition from the sol to the gel state. This implies that the entities that orient at high shear rates are the same whatever the state of the suspension. The increasing orientation at low shear rates can be understood by the formation of larger and larger aggregates during aging. These aggregates are progressively parceled in smaller ones as the shear rate increases. It is unlikely that single particles can be oriented at such moderate shear rates since Peclet number is on the order of $10^{-5}\,\dot{\gamma}$.

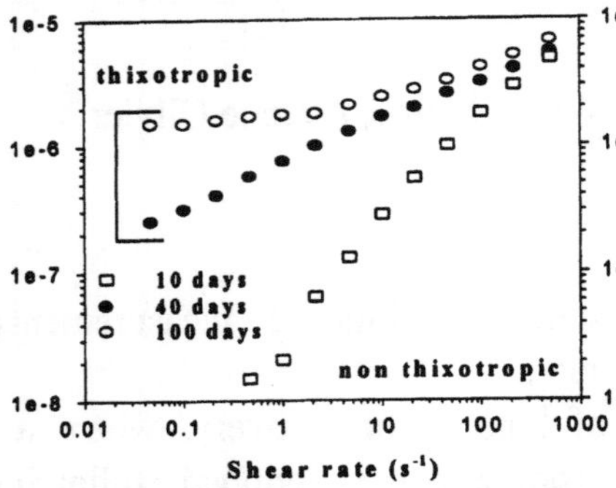

Figure 2 : Time evolution of birefringence for a 1% (W/W), I = 10^{-4}M Laponite suspension

3.2 Reconstruction experiments

Figure 3 gives some results of reconstruction experiments after a high shear rate. The preshearing is applied at 200 s^{-1} during 80 s. The birefringence response during this stage is, in this case, $\Delta n' = 2\ 10^{-5}$.

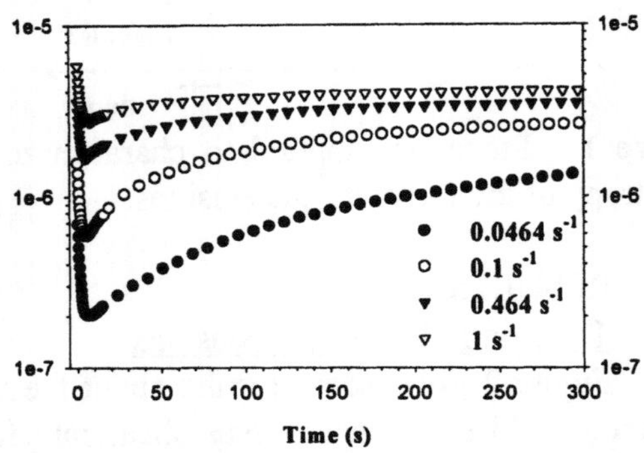

Figure 3 : Example of reconstruction experiments at different shear rates for a 1.6% (W/W), I = 10^{-4} M Laponite suspension

The shape of the curves can be explained by the competition between two phenomena. A strong decrease of birefringence is observed at short times due to the relaxation from steady state of small aggregates. Under low shear rates, larger and larger aggregates are formed which will yield an increase of birefringence but almost the same steady state value (see Figure 2).

It seems clear that the characteristic time during reconstruction decreases with shear rate. It can thus be though that shear enhances the formation of large structures, pointing towards a possible local nematic ordering of the particles.

3.4 Relaxation experiments

Figure 4 gives some results about normalized birefringence versus time at different shear rates.

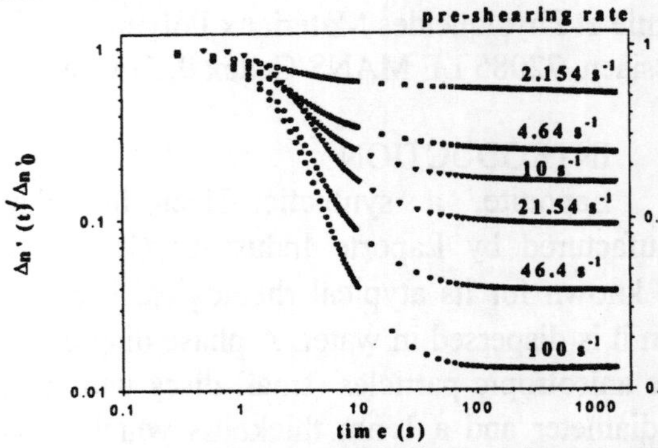

Figure 4 : Time evolution of normalized birefringence after different pre-shearing stages.

Two types of relaxation can be seen :

i) A short time relaxation the contribution of which increases with the preshearing rate. It corresponds to the relaxation of small objects.

ii) A long time relaxation, more evident at low shear rates which corresponds to the relaxation of large objects.

These two relaxation phenomena are always observed, and the weight of each relaxation depends on the pre-shearing rate applied. A short time relaxation, even after low shear rates, indicates that there are, in the structure, some small birefringent entities which don't contribute to the network.

4. REFERENCES

Kroon M, Wegdam G, Spirk R (1996) Phys. Rev. E 54:6541-6550

Kroon M, Vos WL, Wegdam G (1998) Phys. Rev. E 57:1962-1970

Morvan M, Espinat D, Lambard J, Zemb T (1994) Coll. Surf. A:Phys. Eng. 82:193-203

Mourchid A, Delville A, Lambard J, Lecolier E, Levitz P (1995) Langmuir 11:1942-1950

Pignon F, Magnin A, Piau JM, Cabane B, Linder P, Diat O (1997) Phys. Rev. E 56:3281-3289

Ramsay JDF, (1986) J. Coll. Int. Sci. 109:441-447

Ramsay JDF, Linder P (1993) J. Chem. Soc. Faraday Trans.89:4207-4214

A MODEL TO UNDERSTAND THE CONCRETE WORKABILITY

F. DECROIX, C. LANOS, C. CASANDJIAN, M. LAQUERBE

Laboratoire GTMa, INSA de Rennes, 20, avenue des Buttes de Coësmes 35 043 Rennes Cedex, France

1. INTRODUCTION

Concretes are mixes of aggregates (sand and/or gravel) and binding like cement and water or polymer resin. Concretes are generally considered as viscoplastic fluids whose static and dynamic behaviour parameters are very different because of a high sensibility to vibrations. So, the workability of the plastic paste, its ability to modelling processes in the fresh state, results from the right content of components associated with pouring vibration conditions. To understand its rheological behaviour, concrete can be seen as a concentrated multimodal suspension. This type of approach was recently developed by Sedran and al (1994), and by Hu (1995) respectively in adapting the model of Mooney and of Faris. Both are limited to the optimisation of the compacity in a dry aggregates mix. Another method consists in exploiting the principle of a double coating of grains, validated by Gamski and al (1973) and Lanos and al (1995). This principle proposes a distinction between just coated binding on the grains surface and "free" binding, which proportions significately depends on the vibration conditions and the kind of binding. We propose to adapt this principle and to build a model in order to understand the effect of vibration parameters on the workability. Then model and experimental results on polymer mortars are conpared.

2. THE ANALYTIC MODEL

2.1 Description

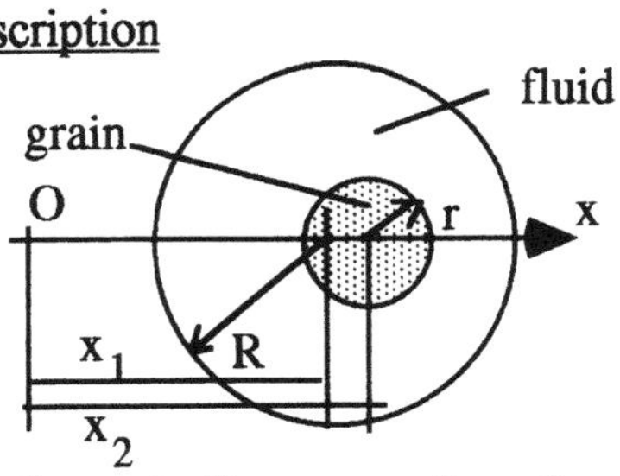

figure 1: Geometry of a cell

The local approach of the built model makes it possible to analyse the movement equation of a spherical grain (an aggregate of concrete) in a closed spherical cell, full of a fluid (the binding) considered as incompressible and newtonian (fig. 1). Thus, concrete is represented as the juxtaposition of different-sized cells.

2.2 Movement equation

The grain movement amplitude must be identified and parameterized by the vibration conditions of the cell (frequency, acceleration), its geometry and the rheological parameter values of the coating fluid. ρ_1 and ρ_2 are the unit masses respectively of the binding and the grain. The position (x_3) of the system's centre of gravity is given by :

$$m_1 = \frac{4}{3}\pi R^3 \rho_1 \qquad m_2 = \frac{4}{3}\pi r^3 (\rho_2 - \rho_1) \qquad (1)$$

$$m_1.x_1 + m_2.x_2 = (m_1 + m_2).x_3 \qquad (2)$$

The equation linking the relative movement $a = x_1 - x_2$ to the vibration conditions is built by neglecting the gravity effects. $\ddot{x}_3$ is the acceleration imposed by vibration to the centre of gravity. C(a) is the viscosity module whose expression was mentioned by Moore (1972).

$$\ddot{a} + \frac{(m_1 + m_2)}{m_1}\ddot{x}_3 + C(a)\dot{a}\frac{(m_1 + m_2)}{m_1 m_2} = 0 \qquad (3)$$

2.3 Answer of the model

The numerical resolution of (3) makes it possible to evaluate the influence of the sinusoidal vibration conditions (γ acceleration and f frequency) on the value of the grain movement amplitude in the cell, called A. α is a constant, μ is the fluid 's newtonian viscosity, V is the vibration speed ($V = \gamma/(2.\pi.f)$). E_p is the average coating thickness ($E_p = R - r$).

$$A = \alpha \frac{V.E_p^3}{r.\mu^{1,2}} \qquad (4)$$

3. RESULT ANALYSIS

3.1 Study

The local parameter A/r is characteristic of the global rheological behaviour of mix. Low value induces firm and not workable mix. The model allows to differently present the concept of double coating. Several points can be discussed :

•The juxtaposition of two cells induce the harmonisation of their movements. To avoid any imbrication or contact loss of the cells subjected to the same accelerations, their displacements must be identical. If this condition is not respected, stresses appear between neighbouring cells, what favours segregation in the vibrating mix.

•In the case of imbrication of small-sized cells in the saturing fluid of a big-sized cell, movement harmonisation imposes to the small-sized cell to have a displacement included between the one of the big-sized cell (x_1) and the one of the big-sized cell's grain (x_2). The stresses thus created between small cells are directly linked to the shearing of the saturing fluid of the big cell whose μ viscosity depends on the size distribution of small cells. The harmonisation of cell movements in a mix is also very influenced by the composition of the granular spectrum.

•The movement equation of cells placed near a wall is different from (3). In this case, the cell movement is set. The answer to this problem is similar to the one found, but turns out to be amplified.

3.2 Exploitations

The application of the model to the whole grain sizes, making up the granular spectrum of a concrete, can thus make it possible to identify pouring conditions reducing mix segregation risks. Another approach consists in optimising the granulometry of a concrete necessary for a good workability in given vibration conditions.

4. EXPERIMENTAL VALIDATION

Different mixes using various sizes of sand grains (0.2/5mm) and different contents of binding (polymer+fillers) are tested in order to valid the principle of the model. The following table presents the results obtained with polymer mortar. Two bindings (n°1 and n°2) with different polymer/fillers ratios and three different binding/sand proportions are used. One makes thus six mixes tested thanks to a slump test with two vibration speeds V. Thickness E_p is calculated by dividing the volume of binding by the surface of sand grains supposed to be spherical.

To validate the built model and thus to link up the rheology of the mix to A/r, an assumption for these mortars of an average grain radius r together with the analysis of slump times t obtained for the mixes tested with $V.E_p^3$ constant, can be done.

tableau 1: experimental results

binding	E_p µm	V m/s	$V.E_p^3$	t s
n°1	56.55	0.16	2.89	9.5
n°1	42.94	0.32	2.53	12
n°1	42.94	0.16	1.27	25
n°1	35.21	0.32	1.40	23
n°2	56.53	0.16	2.89	7
n°2	42.95	0.32	2.54	10
n°2	42.95	0.16	1.27	22.5
n°2	32.52	0.32	1.10	25

Results obtained indicate that two tests realised with the same binding and neighbouring values $V.E_p^3$ show neighbouring t times. Moreover, in similar test conditions, the use of a more supple binding reduces the values of t.

5. CONCLUSIONS

This work leads to a new method of firm concretes formulation taking into account rheological criteria and pouring condition. Parameters such as frequency and vibration acceleration, grain size, thickness and coating binding viscosity are taken into account in a simple model. Validation results on mortars are in accordance with the forecasts of the model even if it exploits perfect spheres and rigid cells. However, the extension of the model to the dynamic behaviour of the binding's fillers remains to be built. Moreover, it is suitable to associate to this model an evaluation of the coating binding's viscosity which depends on the granular spectrum of fillers and eventually of vibration conditions. Then, the optimisation of a concrete formulation or of a manufacturing process is conceivable.

REFERENCES

Gamski K., Degeimbre R., Petit D., (1973), "Contribution à l'étude des composites à liants résineux", Bulletin Technique n°3.

Hu Chong, (1995), "Rhéologie des bétons fluides", thèse E.N.P.C.

Lanos C., Laquerbe M., Casandjian C., (1996), RILEM Proc n°32, E & FN Spon, 343-353.

Moore D.F., (1972) "The friction and lubrification of elastomers", Pergamon Press,

Sedran T., De Larrard F., Angot D., (1994), Bul LCPC n°194, 59-93.

ORIENTATION AND INTERACTION BEHAVIOUR OF RODLIKE PARTICLES IN DILUTED, SEMI-DILUTED AND HIGLY-CONCENTRATED SUSPENSIONS

J.-C. EISCHEN, E. J. WINDHAB
ETH Zürich, ILW, LFO E19, ETH Zentrum, CH-8092 Zürich, Switzerland

1 INTRODUCTION

An essential part of foods contains rodlike particles like fibres (e.g. in cereals, doughs, spreads, sauces, and dressings). In the processing of such products the food industry deals with problems based on the non-Newtonian shear thinning and thixotropic behaviour caused by the orientation and interaction of the shaped particles in laminar flows (e.g. in extrusion, pipe transport, mixing and stirring, dispersing).

This behaviour is depends directly on the shape of the particles (aspect ratio), on their concentration and on the rheological properties of the fluid phase.

2 EXPERIMENTAL SETUP

2.1 The parallel band apparatus (PBA)

The PBA was originally designed by Taylor (1934) to study the formation of emulsions. This apparatus has been enhanced by an electronic regulating device allowing to "fix" a fibre (carbon fibre, plastic rod) at an arbitrary point in the linear shear field of a Newtonian fluid (silicon oil).

2.2 The tube apparatus (TA)

In a transparent glass tube, a fibre suspension (composed of polyamide fibres suspended in polyethylene glycol) is submitted to a laminar flow. Through this glass tube a laser light sheet is laid. A camera, fixed on a rail, can be moved with the speed of a chosen fibre, located in the laser light sheet. With this system, the investigation of fibre orientations in a Poiseuille flow field becomes possible.

When observing a fibre in the glass tube, the distortion due to optical refraction has to be taken into account.

2.3 The parallel plate shear rheometer (PPSR)

This apparatus, developed by Kraft (1996), creates a linear flow field between the two plates and can be used for studies on highly concentrated fibre suspensions. The used carbon fibres are embedded into a specially selected polymer melt, showing a Newtonian behaviour in the required shear rate / stress range. In order to obtain information on the orientation angles of the fibres after shearing, the samples are analyzed (after microtomic cutting) by microscopic imaging.

3 RESULTS

3.1 Diluted Suspensions

A single fibre, introduced into a linear shear field of a viscous fluid is subjected to different kinds of forces: pressure, shear and inertia. As described by Jeffrey (1922), these forces lead to a rotational movement of the fibre.

These experiments have been carried out in the PBA. The parameters of interest are the viscosity of the fluid, the difference in density of particle and fluid, and most important the dimensions of the rodlike particle. The measurement of the fibre movement (orientation angle related to the direction of the streamlines) is carried out by means of digital image processing. In order to allow a semi-automatic evaluation of the image data (e.g. capturing an image sequence from videotape to disk, "following" a fibre through an image sequence), a macro-program has been developed for the NIH-Image program, written by Rasband (1998).

The obtained results have been compared to the results predicted by Jeffrey (1922) and to numerical calculations performed by Nirschl and Denk (1995) and show a perfect agreement.

Fig. 1 shows the rotational movement of three fibres with different aspect ratios. It can be seen that the "longer" fibres lie more stable in the flow and that their flipping movement (compared to the period of their rotation) is faster than for the "shorter" fibres.

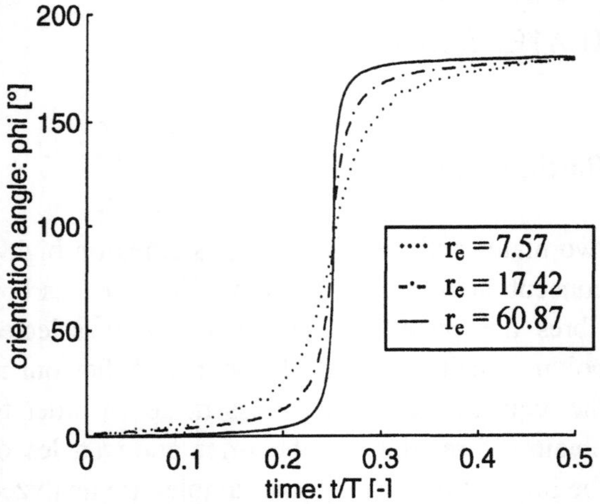

Fig. 1: Influence of the aspect ratio r_e

3.2 Semi-Diluted Suspensions

The orientation behaviour of fibres in semi-diluted suspensions is investigated in the PBA and in the TA. The aim is to describe the motion of a single fibre in the suspension flow. The results show the dependency of the fibre concentration on the rotational movement of the fibres. Further experiments describe the transient evolution of the orientation degree of fibres in a sheared suspension (starting from initially randomly orientated fibres).

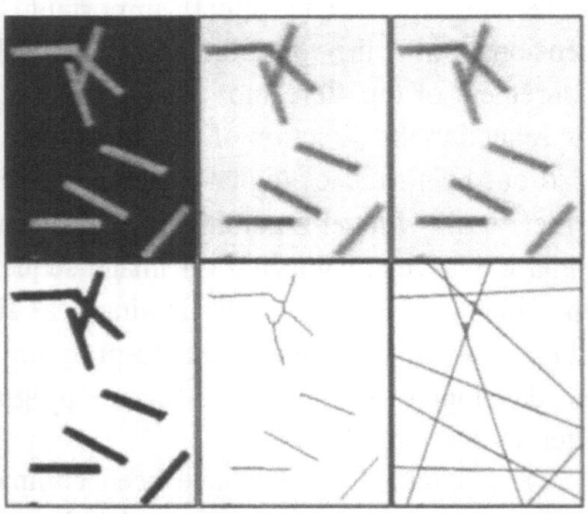

Fig. 2: Original image, inversion, de-interlacing thresholding, skeleton, found lines

In order to allow for an automated measurement of the orientation angles of the fibres in the obtained image sequences, a software ("FindLines") has been implemented on Apple Macintosh Computers. This software also allows performing some basic image transformations: inversion, de-interlacing, thresholding, skeletonizing (as shown in Fig. 2). The algorithm used to detect the angle of the fibres is the Hough transformation.

3.3 Highly-Concentrated Suspensions

The studies on highly concentrated fibre suspensions have been carried out the PPSR. To quantify the orientation degree of the fibres, a measure has been defined by Eischen (1997). The calculation of this orientation degree is based on a two-dimensional Fourier transformation of the digitized microscopical image (Fig. 3).

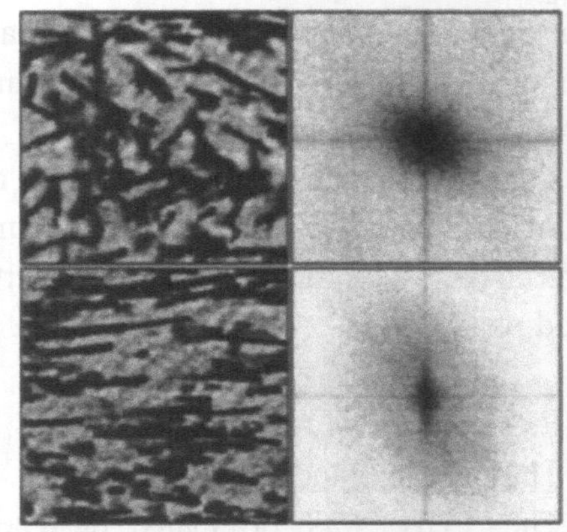

Fig. 3: Fibre orientation and Fourier space

Rheological measurements of the viscosity function of a suspension have been carried out and have been correlated with the orientation degree of the fibres.

4 REFERENCES

Eischen J-C (1997) Bildanalytische Auswertung des Orientierungsgrades von Fasern in hochkonzentrierten Fasersuspensionen, ETH Zürich.

Jeffrey GB (1922) Proceedings of the Royal Society of London, 102 A: 161-179.

Kraft M (1996) Untersuchungen zur scherinduzierten rheologischen Anisotropie von verschiedenen Polyethylen-Schmelzen, ETH Zürich.

Nirschl H, Denk V (1995) 6th International Symposium on Computational Fluid Dynamics, Lake Tahoe, Nevada.

Rasband W (1998) NIH-Image, U.S. National Institutes of Health.

Taylor GI (1934) Proceedings of the Royal Society of London 146 A: 501-523.

SUSPENSION OF EXTENSIBLE RODS MODEL OF EXTENSIONAL FLOWS OF POLYMER SOLUTIONS.

V.M.Entov

Laboratory of Applied Continuum Mechanics,Institute for Problems in Mechanics RAS, pr.Vernadskogo, 101, 117526, Moscow, Russia

A simple mechanistic model of a polymer solution in strong elongational flow is studied. A polymeric molecule is modeled by an uniform slender rod of incompressible rubber-like elastic material imbedded into a coaxial cylindrical cell filled with a viscous fluid (solvent). The objective is to estimate effect of strong hydrodynamic interaction between molecules in moderately concentrated solutions upon rheological behavior in elongational flow. It is assumed that the polymer solution can be treated as a regular lattice of identical individual cells aligned along the extension direction. So only dynamics of an individual cell is studied. Under these assumptions the cell deforms affinely with the bulk flow and the shear stress vanishes at the cell boundary.

The rod-fluid interaction within the cell of large aspect ratio is studied using the lubrication layer approximation. Then the rod extension dynamics is described by a nonlinear parabolic equation for the local stretch ratio $\lambda(x,t)$ as a function of time and Eulerian coordinate with a source term responsible for the externally imposed flow. This equation expresses balance between the squeezed solvent pressure drop and elastic force in the rod $F=K\Phi(\lambda)$ and is essentially the same as that derived by Hinch for an individual molecule in extensional flow. It should be solved subjected to boundary conditions which express vanishing of the elastic force at the unknown boundaries corresponding to the ends ($x=\pm X(t)$). The kinematic condition $L= \int dx/ \lambda(x,t)$ serves to find the moving boundary. Here, L is the initial length of the 'molecule', and the integral is taken over the extended length.

First, steady-state solutions for the extension with constant strain rate α are considered. Two steady-state solutions exist at small (subcritical) extension rate, one corresponding to a slightly extended molecule, and another to a highly extended one. It is argued that the second steady state is unstable and a highly extended rod should either collapse, or get involved into unbounded extension. The steady states cease to exist beyond a critical strain rate α_*. A "molecule" extends unboundedly in supercritical flow. At constant strain rate the extension asymptotically (after an adjustment period) tends to exponential elongation of the rod affinely with the bulk flow.

Then relaxation of an initially highly extended rod-molecule is considered. For an uniformly extended initial state relaxation proceeds non-uniformly, propagating from the rod ends to the center as a nonlinear wave of the type of self-similar heat wave. There is no relaxation of stress at the molecule center until the wave reaches the center point. Thus this solution implies slow relaxation in the case of high initial extension.

The model predictions expressed in terms of the microscopic variables, namely the elastic strain $\lambda(x,t)$ and force $F(x,t)$, distribution along the 'molecule' can be related to bulk variables, such as extension rate and effective tensile stress σ. In particular, the elastic component of the bulk stress proves to be proportional to the elastic force in the rod integrated over its (extended) length.

More general extension regimes are studied using numerical modeling of the rod dynamics equation. Those include constant-strain-rate extension with subsequent relaxation, step test with switching from a high initial to a lower extension rate, and extension followed by an oscillatory extension regime with zero mean extension rate. In all cases neo-Hookean local elastic behavior, $\Phi(\lambda)=\lambda-1/\lambda$, $K=const.$ was assumed.

Among the effects observed are those predicted by the asymptotic analysis, namely existence of a critical value of extension rate beyond which the rod extends unboundedly, exponential affine extension after some adjustment period, slow relaxation after very strong extension, and hystresis phenomena, such as unbounded extension at subcritical extension rate and asymmetric behavior of stress in oscillatory extension regime after preliminary strong extension. Simulations for constant rate extension show that an asymptotic regime eventually develops in which distributions of all properties (local stretch ratio, local elastic force) along the 'molecule' have almost constant shape, but an increasing 'amplitude':

$$\lambda(x,t)=\Lambda(t)\,\lambda_{\infty}(x/X(t)); \quad (1)$$
$$F(x,t)\cong K\Lambda(t)\lambda_{\infty}(x/X(t)). \quad (2)$$

Both $\Lambda(t)$ and $X(t)$ grow exponentially at large t: $\Lambda(t) \propto \Lambda_{\infty}\exp(\alpha t)$; $X(t)\propto X_{\infty}\exp(\alpha t)$. Therefore the bulk elastic stress σ, being proportional to the integrated elastic force, scales as $\Lambda(t)X(t) \propto X^2$, in exactly the same way as for an ideal elastic dumbbell, the microstructural unit of the Maxwell-Oldroyd model. In other words, in strong monotonic extension the elastic rod 'molecule' deforms coherently, and therefore effectively behaves as an individual elastic dumbbell molecule in dilute solution. However parameters of this 'equivalent elastic dumbbell' depend both on the polymer concentration and the 'molecule' strain history at the initial stage of deformation. The asymptotic stretch ratio distribution is different at different extension rates, and switching from an extension rate to another at a later stage of extension results in 'anomalous' asymptotic profiles. This is one more manifestation of permanent (non-decaying) effect of non-uniformity in initial distribution on accelerated extensional flow first discovered by McKinley and identified as a generic behavior by Pearson and Petrie.

This suggests a plausible explanation of some experimental observations, in particular, of unexpectedly successful application of the Maxwell-Oldroyd model (with appropriate correction for finite extensibility) to predominantly elongational strong flows of dilute, semi-dilute and moderately concentrated solutions, such as occur in constant-rate extension experiments of Sridhar et al. and McKinley et al., and in studies of capillary thinning of polymer solutions using the the Liquid Filament Microrheometer (Bazilevsky, Entov, Rozhkov). In both cases the molecule is involved into progressive extension with monotonically increasing stress, and effectively behaves as an individual elastic dumbbell.

Essential difference between the models having one degree of freedom (elastic dumbbell) and an infinite number degrees of freedom (suspension of elastic rods) becomes obvious after fast drop in extension rate (relaxation) and may be significant in other non-monotonic strain regimes. On the other hand, this study provides an additional evidence in favor of using the simple upper-convected Maxwell-Oldroyd (Oldroyd-B) model in fluid-dynamical studies of flows with prevailing elongation increasing in downstream direction (such as converging flows) even in the case of moderately concentrated polymer solutions.

RHEOLOGICAL INVESTIGATIONS OF SELF-ORGANIZING SORBITOL-NETWORKS IN AMORPHOUS POLYMER MATRICES

K. FUCHS, M. FAHRLÄNDER, R.-D. MAIER, Chr. FRIEDRICH

Freiburg Materials Research Center, Stefan-Meier-Str. 21, 79104 Freiburg, Germany

1. INTRODUCTION

1,3:2,4-dibenzylidene sorbitol (DBS) is known to be an efficient gelling agent for low molecular weight organic liquids, as well as to nucleate polyolefin crystallization. It was suggested that the DBS molecules self-organize by hydrogen bonding into a three-dimensional, nanofibrillar network. However, only few studies have attended to effects of DBS in amorphous, non-crystallizing polymers.

2. EXPERIMENTAL

We have investigated the rheological properties of three different polymers, i.e. polystyrene, polypropylene glycol, and atactic polypropylene, with small amounts (0.2 to 5 wt.%) of DBS. Temperature- and frequency-dependent measurements were performed on a Rheometrics Mechanical Spectrometer RMS-800 in parallel plate geometry. Differential scanning calorimetry (DSC) and transmission electron microscopy (TEM) were included to clear up the underlying processes and superstructures.

3. RESULTS

All investigated polymers show a strong increase of the dynamic moduli G' and G", if a critical DBS concentration is exceeded (see also *Fig. 1*). This reinforcement is not tied to the polarity or the chemical nature of the polymer, but it requires annealing of the quenched samples near a polymer dependent critical temperature. From DSC and rheological measurements, we assign this critical temperature to a recrystallization process. TEM images of films reveal that the DBS forms a network-like skeleton of arborescent bundles after annealing (see *Fig. 2*). Approximately 20 degrees above the recrystallization temperature, the DBS network melts and the mechanical properties resemble mainly the pure matrix. It was found that time-temperature superposition holds either above the melting point or in the annealed state at low shear rates. *Fig. 1* shows the frequency dependent master curves of a polypropylene glycol with various amounts of DBS.

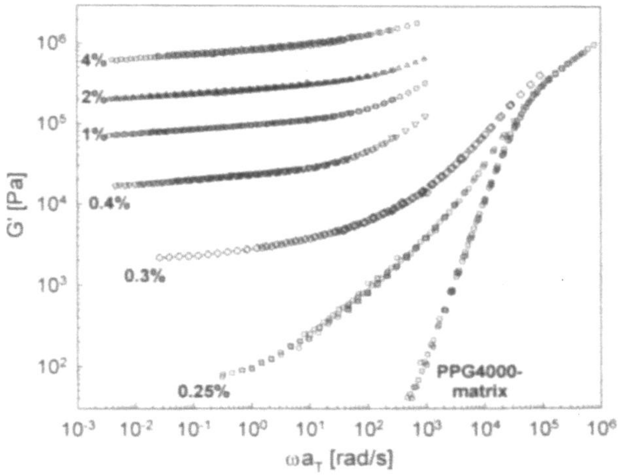

Fig. 1 Storage modulus of a polypropylene glycol (M_W = 4000 g/mol) with the indicated amount of DBS at the common reference temperature T_0 = 0 °C

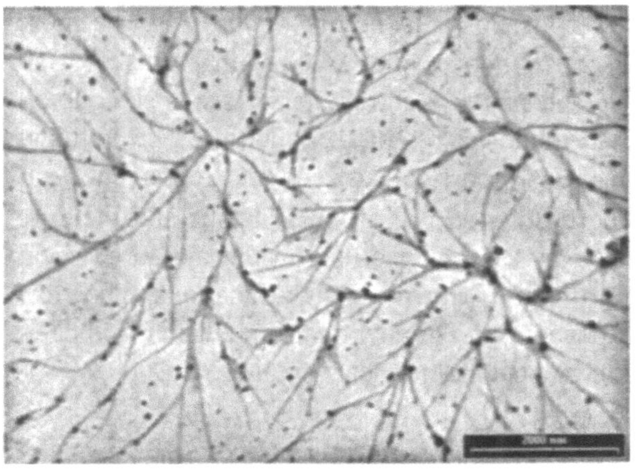

Fig. 2 TEM image of a solution casted film of an atactic polypropylene with 0.5 wt.% DBS after annealing

Approximately 0.3 wt.% DBS are sufficient to obtain a clear, gel-like state, which is characterized by a very small frequency dependence of the dynamic moduli and the absence of a terminal flow region. However, a completely frequency-independent equilibrium modulus G_∞, as it defines a permanent gel, is not found as well. According to this definition, the sorbitol compounds can not be designated as classical gels. An explanation might be the temporary nature of the network with physical junctions that can be released or created depending on external influences.

THE NORMAL-STRESS-BEHAVIOUR OF SUSPENSIONS WITH VISCOELASTIC MATRIX FLUIDS

S. GLEISSLE [1], W. GLEISSLE [1], G.H. MCKINLEY [2], H. BUGGISCH [1]

[1]Inst. für Mechanische Verfahrenstechnik und Mechanik, Universität Karlsruhe (TH), D-76128 Karlsruhe
[2]Department of Mechanical Engineering, Mass. Institute of Technology, Cambridge, Massachusetts 02139

1. INTRODUCTION

Normal stresses in viscoelastic suspensions may have similar amounts as the shear stresses. So the first and second normal stress difference of such materials are of high interest. Especially the second normal stress differences are unknown for all suspensions. To investigate N_2 in comparison to N_1 is of special interest.

2. EXPERIMENTAL

2.1 FLUIDS

Two model-suspension-groups were chosen. The first one consists of a silicone oil matrix with a zero shear viscosity of $1000\,Pa \cdot s$ whereas the second group was formulated with a lower viscous silicone oil ($400\,Pa \cdot s$). The more viscous silicone-oil is filled with particles of an average size of 4 and the less viscous silicone oil with monospheres of 0.5 microns. The volume-fractions C_v are considered up to 25 percent.

The flow curves of all suspensions and the pure matrices can be described in the measured shear rate regime with power-law-functions. The flow exponents are smaller than one which indicates that the material is shear thinning.

2.2 RHEOMETERS

To determine N_2 and N_1 the RMS800 is used. The first normal stress difference can be derived from a single cone and plate axial force measurement at a fixed shear rate.

In contrast to N_1, N_2 can only be backed out from axial force measuremensts by combining cone-and plate(CP) and parallel plate(PP) experiments. The upper limit of the shear rate regime is limited by the onset of edge fracture. To extend that shear rate range to about one decade for all suspensions a ring fixture around the cone or plate is added. This fixture influences the torque. Hence the measured shear stresses are shifted.

3. RESULTS

Our results confirme Tanaka, White (1980) and Ohl, Gleissle (1993): N_1 at a fixed shear stress decreases with increasing solid concentration:

$$(N_1(C_v))_\tau \downarrow \text{ with } C_v \uparrow \qquad (1)$$

N_1 can be described as power law function with respect to τ. The exponent n_1 is independent of the volume-fraction Eq.(2), Fig.1.

$$N_1 = C_{N_1}(C_v) \cdot \tau^{n_1} \quad \text{with} \quad n_1 \neq \mathrm{f}(C_v) \qquad (2)$$

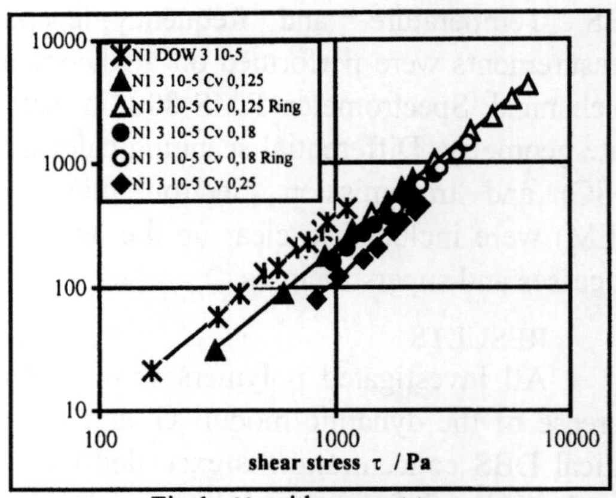

Fig.1: N_1 with respect to τ

First results of N_2 for viscoelastic suspensions are obtained. N_2 is negative for all suspensions and the pure silicone oils. In contrast to N_1 the magnitude of N_2 rises with increasing volume-fraction.

$$|N_2(C_v)|_\tau \uparrow \text{ with } C_v \uparrow \qquad (3)$$

$|N_2|$ can also be approximated with power law functions with the same exponent n_2 for all suspensions and the matrix fluid Eq.(4), Fig.2.

$$|N_2| = C_{N_2}(C_v) \cdot \tau^{n_2} \quad \text{with} \quad n_2 \neq \mathrm{f}(C_v) \qquad (4)$$

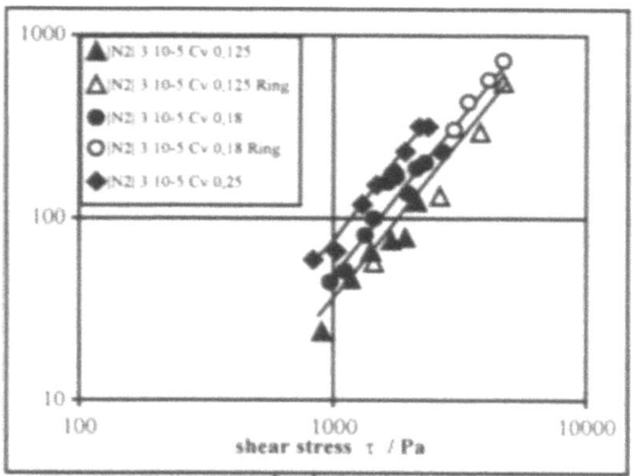

Fig.2: $|N_2|$ versus τ

N_2 for the pure silicone-oil with a viscosity of $400\,Pa\cdot s$ does not provide reliable data because the forces lie below the resolution of the transducer.

If one compares the first with the second normal stress difference within one filled system as shown in Fig.3, it turns out that the exponents of power-law-functions for N_1 and N_2 are the same Eq.(5).

$$n_1 \approx n_2 \overset{!}{=} n \qquad (5)$$

This observation leads to a constant ratio of $\frac{|N_2(C_v)|}{N_1(C_v)}$ for all shear stresses, which rises with increasing volume-fraction. This ratio is about 0.1 for the pure matrix and increases to a value of 0.7 for the lower viscous suspension-group and to a value of 0.5 for the more viscous system type. It was observed that $|N_2|$ rises by the same ratio as N_1 drops when the volume-fractions are increased. This observation leads to Eq.(6) which is found to be true for all suspensions in both suspension-groups as well as for the pure AK10^6 cSt. silicone oil.

$$N_1(C_v)\cdot|N_2(C_v)| = const. \overset{!}{=} G \quad \text{for } \tau = const. \qquad (6)$$

The constant G is standardized for a shear stress of 1 kPa (Eq.(7)). So that $C_{N_2}(C_v$ can be calculated from Eq.(8).

$$\left(N_1(C_v)\cdot|N_2(C_v)|\right)_{\tau=1kPa} \overset{!}{=} G_o \qquad (7)$$

$$C_{N_2}(C_v) = \frac{G_o}{C_{N_1}(C_v)} \qquad (8)$$

Eq.(7) is valid for the pure matrix fluid too, so that the constant G is independent of the particle properties and hence only a function of the matrix fluid.

With the above mentioned simple relations between the first and the second normal stress difference and the shear stress, it is possible to describe the normal stress behaviour of viscoelastic suspensions with the parameters $\left(C_{N_1}(C_v);n;G_o\right)$.

Fig.3 shows the measured data-points as well as the calculated functions for N_1 and $|N_2|$, Eqs.(2,4,8) and hence illustrate the usefulness of the relations described before.

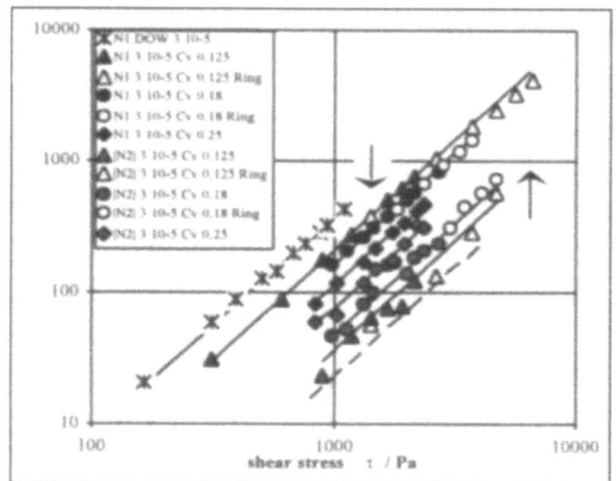

Fig.3: $|N_2|$, N_1 versus τ ,calc. power law functions

4. CONCLUSIONS

For the suspensions with viscoelastic matrix N_1 was measured and N_2 was quantified for the first time. It turns out that N_2 is negative for all materials whereas N_1 is positive. It was found that multiplying N_2 and N_1 at a fixed shear stress provides a constant which is independent of the volume-fraction. Now it is possible to predict the normal stress behaviour of viscoelastic suspensions with only three parameters: the constant n, and G_o and the normal stress parameter $C_{N_1}(C_v)$. This relationship permits calculating N_2 by measuring only N_1 and N_2 for one particular suspension or matrix fluid or assuming that the ratio of N_2 over N_1 for the matrix is approximately -0.1.

5. REFERENCES

Tanaka H, White J (1980) Pol.Eng.Sci.20: p949

Ohl N, Gleissle W (1993) J. Rheol.: p381

RHEOLOGICAL CHARACTERISTICS OF A LIGNITE SLURRY FUEL

T.B. GOUDOULAS, E.G. KASTRINAKIS, S.G. NYCHAS

Chemical Engineering Department, Aristotle University of Thessaloniki, GR-54006, Greece

1. INTRODUCTION

Work on lignite-water slurries has been carried out in the Laboratory of Chemical Engineering Fundamentals, with the goal of producing premium fuels.

Effects of coal type, loading and particle size on the rheological behavior of the slurries have been investigated by Turian et al. (1993) and Roh et al. (1995). Wildemuth and Williams (1985) reported a model for relative viscosity η_r, that invokes the stress dependence of the maximum packing fraction, φ_m, of coal suspensions and an expression for yield stress τ_y. Sengun and Probstein (1997) introduced a bimodal model for the viscosity, which results from the contribution of coarse and colloidal fractions. The dependence of viscoelastic functions on the applied stress and frequency on noncolloidal suspensions in non-Newtonian medium, was reported by Aral and Kalyon (1997).

2. EXPERIMENTAL

The preparation of a lignite-water slurry started with the slurryfication process via wet grinding in a ball mill. The lignite originated from the Ptolemais area in northern Greece. Surfactants and polyelectrolytes were used as additives in the slurryfication process, ensuring acceptable rheological behaviour and stability. Sodium lignosuplhonate was used as a dispersion agent and polyoxyethylene as a stabilizer. The suspending medium was a solution of the additives in deionized water. The total amount of additives in the slurries was less than 0.5% wt. The achieved lignite load in the slurries was up to ~50% wt. or in volume fraction, φ, equal to 0.47.

Extensive measurements with a controlled stress rheometer (CARRI-MED CLS 100) have demonstrated the effect of multimodality of the particle size distribution, PSD, in reducing the yield stress and the apparent non-Newtonian viscosity of the slurry. Dynamic experiments (oscillatory shear measurements) were performed to observe structural changes in samples. The range of the frequency sweep, was between 0.01 and 1.5 Hz and the applied stress in the range of 0.01 and 0.4 Pa.

3. RESULTS

The work is focused on optimizing the lignite content of the slurry in conjunction with its stability and rheological properties for a wide range of shear rates. The effect of solid volume fraction in the suspensions and various additives is shown in Fig.1.

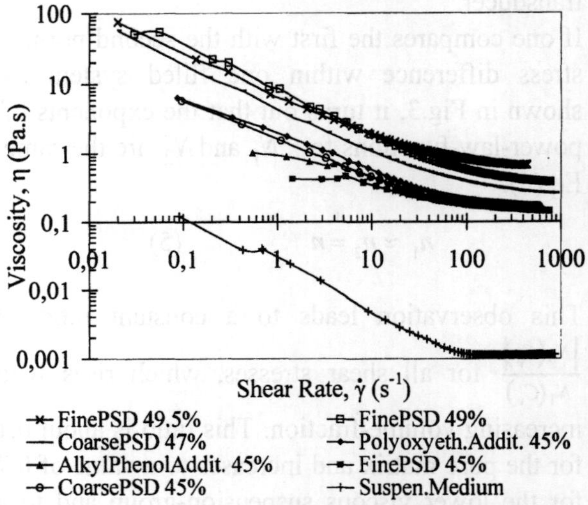

Fig.1 Effect of solids loading and additives in bimodal lignite-water slurries, at 20 °C. The terms Coarse and Fine refer to the upper limit of the PSD

The slurries have approximately the same bimodal PSD, with $d_{0.9}$=220 to 270μm and ~35% of the solids were in the range of 110 to 400μm, while ~55% were in the range of 5 to 80μm. All slurries contained 1%wt. on solids base additives. The effect of the various additives is pronounced at the low shear rates. The slurry with polyoxyethylene exhibits one order of magnitude higher viscosity than the slurry with lignosulphonate; on the other hand, at the high shear limit, they both show the same behavior. At this shear limit the solid fraction in the slurry, seems to dominate the rheological behavior. The effect of the solids loading can be observed in Fig.1, where the 49% wt. slurry exhibits a viscosity about one order of magnitude higher than the 45%wt. slurry.

A comparison of unimodal and bimodal slurries appears in Fig.2. The bimodal slurries for the same concentration of 45%wt, have a lower viscosity than the unimodal. The same Fig.2 shows also, that a slurry with a broad PSD (47%wt.) exhibits a lower viscosity in comparison to the unimodal slurries at the high shear limit.

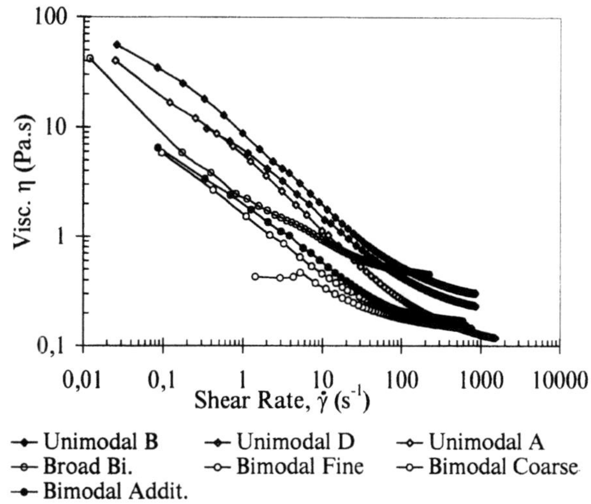

Fig.2 Effect of PSD in the lignite-water slurries, at 20 °C

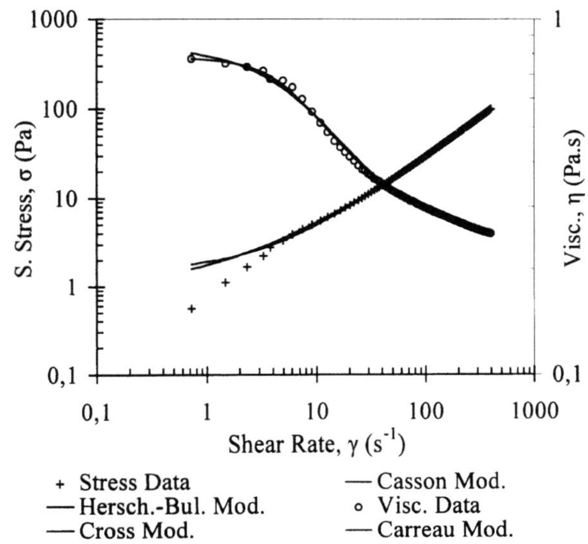

Fig.3 Fitting of viscosity and stress data for a bimodal slurry at 25 °C

Fig.3 shows an effort to compare present data with certain existing models. The observed agreement between models and experiment is good.

All slurries examined showed a thixotropic behavior. In the data reported here, the stress range was applied several times in order to reduce the degree of thixotropy. In Figures 1,2 and 3 the curves shown, are the down curves of the last applied shear stress loop.

Dynamic measurements performed on slurries showed that the linear viscoelastic region was limited and the viscoelastic behavior was typical of dense suspensions.

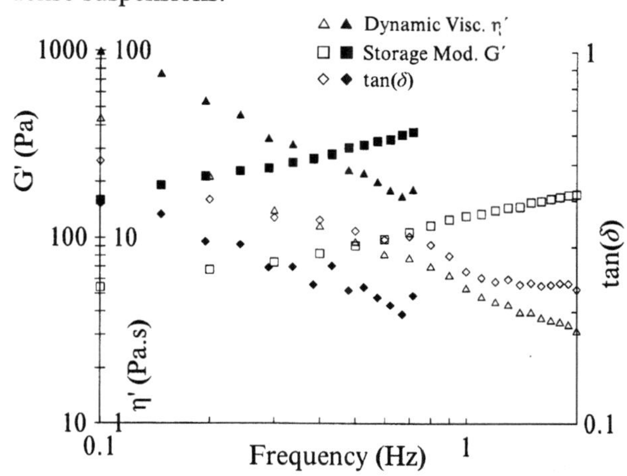

Fig.4 Effect of frequency on viscoelastic functions, G', η' and tan δ for various preshear conditions. Presheared at: open symbols 50Pa, filled symbols 100Pa

4. CONCLUSIONS

Concentrated, stable, suspensions of lignite in water of maximum concentration ~49%wt. were prepared as potential upgraded fuels; they were tested in the shear rate range of 0.01 to ~1600 s^{-1}. No effect of the additives on the apparent viscosity was observed at the high shear rates. A 3 to 4%wt. increase of particle loading, at the upper concentration range, causes an one order of magnitude increase of the viscosity. On the other hand, with a bimodal PSD a reduction of the viscosity by about 100% was achieved in comparison to a unimodal PSD.

5. REFERENCES

Aral B. K. and Kalyon D. M. (1997) J. Rheol. (41):599-620

Roh N.-S., Shin D.-H., Kim D.-C. Kim Jong-Duk (1995) Fuel 74:(1)1220-1225 (2)1313-1318

Sengun M. Z. and Probstein R. F. (1997) J. Rheol. 41:811-819

Turian M. R., Fakhreddine K. M., Avramidis S. K. Sung D.-J. (1993) Fuel 72:1313-1318

Wildemuth C. R. and Williams M. C. (1985) Rheol. Acta 24:75-91

CAPILLARY RHEOMETRY OF MICELLAR AQUEOUS SOLUTIONS

Sergio Hernández-Acosta[1], Alejandro González-Alvarez[1], Octavio Manero[2], Arturo F. Méndez-Sánchez[3], José Pérez-González[3], Lourdes de Vargas[3].

[1]Departamento de Ingeniería Química, Universidad de Guadalajara, MEXICO. [2]Intituto de Investigaciones en Materiales, Universidad Nacional Autónoma de México, MEXICO. [3]Departamento de Física, Escuela Superior de Física y Matemáticas, Instituto Politécnico Nacional, C. P. 07300, Apdo. Postal 75-685, México D. F., MEXICO. E-mail: lvega@esfm.ipn.mx

INTRODUCTION

Wormlike micellar systems show interesting rheological effects during shear flow, such as spurt flow and fluid structuration. Several authors have pointed out that for shear rates or shear stresses larger than a critical value, a nearly constant stress plateau where the flow becomes unstable is observed (Rehage and Hoffmann 1988, Spenley et al. 1993). It has been suggested that to maintain a steady flow along this region, the system may split into shear bands of low and high shear rates.

Most of the experimental studies with wormlike micelles have been performed in rotational rheometers such as cone and plate, where a homogeneous flow field is generated, and in Couette cells. Just a few experiments have been carried out with capillary rheometers where a non-homogeneous flow field is present (Ohlendorf et al. 1986, Callaghan et al., 1997).

In the present work, capillary flow experiments with Cetyltrimethylammonium p-toluene sulfonate (CTAT) at different concentrations in tridistilled water, were performed to determine the stable and unstable flow conditions in a non-homogeneous flow field. Results were compared to those obtained in a cone and plate rheometer. Important differences are highlighted between the results obtained from both rheometers in the characterization of these CTAT solutions. Finally, impressive spurt effects are shown for this system.

EXPERIMENTAL

The experiments were performed in a capillary rheometer (Pérez-González et al, 1992) at constant pressure, by using capillaries of different diameters and a length to diameter ratio equal to 400. The experiments were carried out at T = 30° C, above the Krafft temperature for this system. The solutions were prepared in concentrations of 1.0, 1.5 and 2.0 w/o in tridistilled water.The volumetric flow rate was measured by the catch and measure technique, while the experiments in homogeneous flow were carried out in a DVIII cone and plate Brookfield rheometer. Video images were obtained of the spurt instabilities in the capillary rheometer.

RESULTS AND DISCUSSION

The flow curves obtained with capillaries of different diameters and that of the cone and plate geometry for the 1.0 w/o solution are shown in figure 1. Region A belongs to a Newtonian behavior. Data from cone and plate and capillary almost superposed. At higher shear rates, in region B, the flow becomes a power-law. Data from different capillaries describes the same flow curve. At still higher shear rates, in regions C and D, the flow is unstable. Differences between these two regions consist in that in region C, both the pressure and the shear rate change continuously, while in region D only the shear rate is variable. It is interesting to note here that the driving pressure is kept constant along the experiment, which evidences the existence of flow instabilities. Slip is not observed in these data.

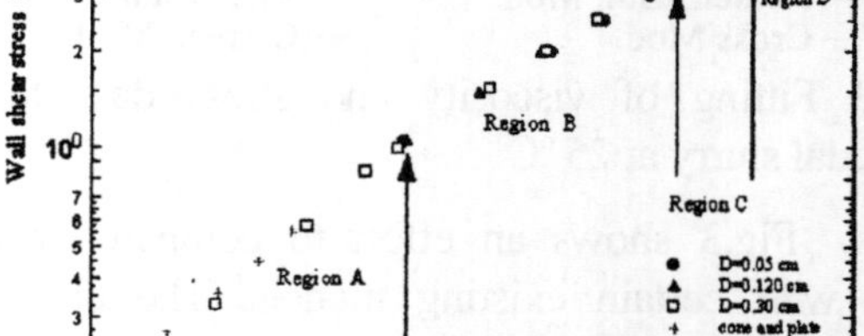

Figure 1.- Flow curves for the 1.0 w/o solution. T = 30° C and L/D = 400.

Flow curves for 1.5 w/o are shown in figure 2. Data presented here only exhibit the flow regions B, C and D, similar to those presented in figure 1, but Newtonian region A was not observed. Again, region B is described by a power law, however, regions C and D are now better distinguished than in figure 1, and show a clear tendency for the capillaries with smaller diameters to lie above those with larger diameters i.e., there appears to be adherence. This result is the opposite to the usual behavior in the presence of slip. On the other hand, data belonging to cone and plate lie close to those of the capillaries in regions B and C. Up to the end of region C, cone and plate data show the plateau often found in micellar systems. This result demonstrates the effect of residence time under flow, which has been interpreted as banded flow or fluid structuration.

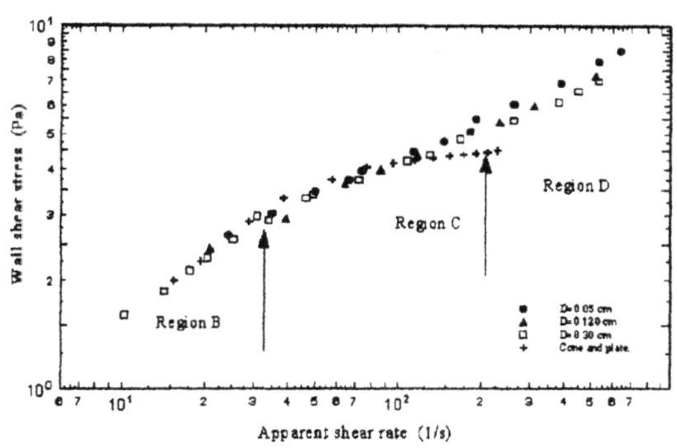

Figure 2.- Flow curves for the 1.5 w/o solution. T = 30° C and L/D = 400.

Figure 3 shows data corresponding to the 2.0 w/o solution. The plateau region extends over a much wider span of shear rates. For stresses lower than the critical stress (around 4 Pa) the behavior is non-Newtonian corresponding to the transition region toward the plateau. The banded flow region lies between those characteristic shear rates, where instabilities are detected. For shear rates higher than those correspondig to the plateau, there is a second non-Newtonian unstable region where the magnitude of the stress at a given shear rate again depends on the capillary diameter. The cone and plate data for this solution describe a wide plateau consistent with capillary data. At low shear rates, cone and plate data lie above those of the capillaries, but approach to the curve of the capillaries at the upper limit of the plateau. This more clearly defined plateau was accompanied by the larger variations in pressure and flow rate, i.e., spurt flow.

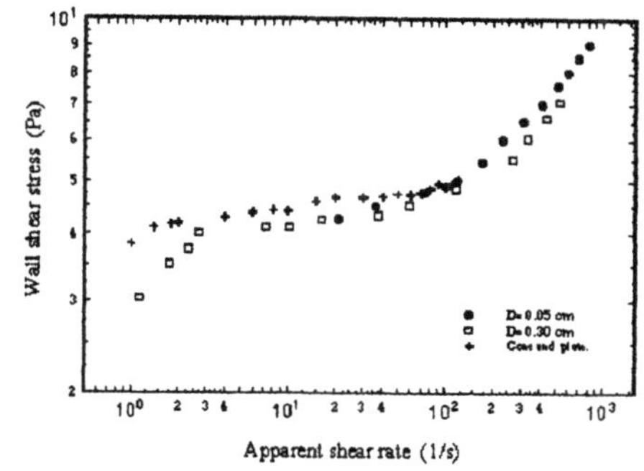

Figure 3.- Flow curves for the 2.0 w/o solution. T = 30° C and L/D = 400.

CONCLUSIONS

From the results presented in this work it can be concluded that there is a critical stress for the onset of instabilities and spurt flow, as well as a critical residence time of the fluid. This last result can be observed from the cone and plate rheometry where the fluid can be sheared for long times. The equivalent result is observed by using large enough capillaries, as observed in a complementary work (Hernández-Acosta et al. manuscript in preparation 1998).

No slip flow was observed during the experiments. On the contrary, what it is known as adherence was observed for the 1.5 and 2.0 w/o solutions in region D. This is a behavior not previously reported for the flow of this system through capillaries.

REFERENCES

Callaghan PT, Mair R.W. (1997) J. Rheolgy 41 901-924.

Ohlendorf D, Interthal W, Hoffmann H (1986) Rheol Acta 25:468-486.

Pérez-González J, de Vargas L, Tejero J (1992) Rheol Acta 31:83-93

Rehage H, Hoffmann H (1988) Jour. Phys. Chem. 92 4712-4719.

Spenley NA, Cates ME, MacLeish TBC (1993) Phys. Rev. Lett. 71 939-942.

INFLUENCE OF SURFACE MODIFIED PARTICLES ON THE RHEOLOGICAL PROPERTIES OF SUSPENSIONS

L. HEYMANN, N. AKSEL

Dept. of Fluid Mechanics, Institute of Mechanics, Chemnitz University of Technology, D-09107 Chemnitz, Germany

1. INTRODUCTION

The rheological properties of suspensions are influenced by a combination of several physico-chemical parameters. In contrast to the particle shape and solid volume concentration the influence of the physical and chemical properties of particle surfaces on the flow behavior of suspensions is less known. The aim of the present work was to investigate experimentally the influence of the particle porosity and surface modification on the plasto-viscoelastic properties of suspensions.

2. MATERIALS

2.1 Components

The continuous phase of the suspensions was a pure viscous silicone oil NM1-200 (Chemiewerk Nuenchritz GmbH, Germany) with a viscosity of 223 mPas at 20 °C. Because of the relatively low mean molecular weight (10^4 g/mol) no viscoelasticity was found. The disperse phase was a spherical, narrow-distributed LiChrospher® Si100 (Merck KGaA, Germany) with a mean diameter of approximately 5 µm. The particles consisting of an extremely porous silica matrix were applied in original and surface-modified form. In the latter case the particle surface was modified with a hydrophobic dispersion phase of octadecyl groups with a surface coverage of 2.6 μmol/m^2.

Suspensions of three solid volume concentrations were prepared. During the preparation of the suspensions nontypical viscosity levels were observed at the given solid volume concentrations. These observations could only be explained either by the formation of fluid bridges between particles at low fluid concentrations or by the influence of the inner particle porosity. The low surface tension of silicone oil (22 mN/m) implied the assumption that all particle surfaces were completely wetted. Additionally, some amount of the continuous phase diffused into the pores of the particles. Consequently, it contributed no longer to the macroscopic rheological properties of the suspensions. Hence, the particle porosity had to be taken into consideration.

2.2 Particle porosity

Due to the inner particle porosity the reduced density of the particles in water was lower than the density of the pure particle matrix. From the reduced solid and water densities an inner particle porosity of approximately 73.4 % was calculated. Assuming that the continuous phase contributes to the composition of the suspension until all particle pores are filled with fluid, the following relation between the inner porosity φ_i, the solid matrix volume content φ_s, and the effective solid volume content $\varphi_{s,eff}$ can be derived if the solid (V_s) and fluid volumes (V_f) are known:

$$\varphi_{s,eff} = \begin{cases} 1 & V_f < V_{f,i} \\ 1 - \dfrac{(1-\varphi_i)(\varphi_s^{-1}-1)-\varphi_i}{(1-2\varphi_i)+(1-\varphi_i)(\varphi_s^{-1}-1)} & V_f \geq V_{f,i} \end{cases}$$

$$\text{with } V_{f,i} = V_s(\varphi_i^{-1}-1)^{-1} \tag{1}$$

Using eq. (1) the solid volume concentration of the suspensions was recalculated after the suspensions had been prepared. The results of the correction are summarized in Table (1).

Table 1: Correction of the solid volume concentration

φ_s	$\varphi_{s,eff}$
0.10	0.138
0.15	0.256
0.20	0.446

3. DEVICE AND MEASUREMENTS

A universal dynamic spectrometer UDS 200 (Physica Messtechnik GmbH, Germany) with cone-and-plate- and parallel-plates-systems, respectively, was used to measure the yield stresses, flow curves and the complex modulus. All experiments were carried out at a temperature of 20°C in the controlled shear stress mode of the rheometer.

A logarithmic increase of the shear stress followed by a logarithmic decrease was applied to

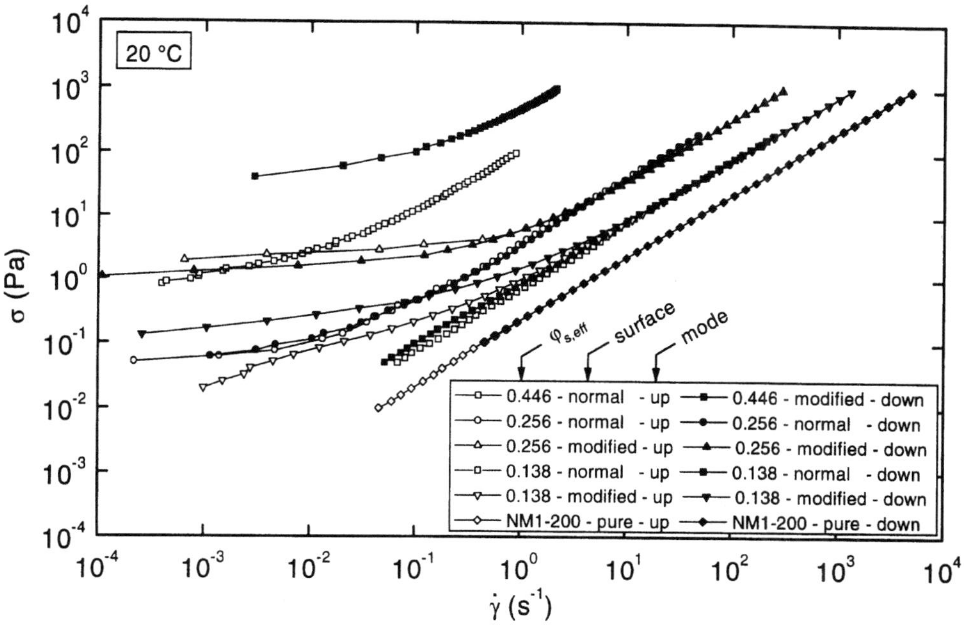

Fig. 1: Flow curves of normal and surface-modified particle suspensions as a function of effective solid volume concentration $\varphi_{s,eff}$

measure the flow curves of the suspensions and to prove whether thixotropy or rheopexy occured.

The measurement of the complex modulus was complicated due to the fact that no significant linear viscoelastic range in the amplitude sweep was found. However, the analysis of the primary strain signal as answer to a harmonic shear stress showed no significant deviation from a harmonic signal below a distinct minimum of storage modulus.

4. RESULTS

Fig. 1 shows the flow curves of the pure silicone oil and the suspensions investigated. It can be seen that the flow behavior, especially at low shear rates, is strongly influenced by the surface modification of the particles. From the flow curves the yield stresses were detected by fitting the Herschel-Bulkley model to the experimental data (see Fig. 2). The yield stresses of the unmodified suspensions depend only on the solid volume concentration. A significant increase of the yield stresses of more than one order of magnitude at the same effective solid volume concentration can be observed in the case of the surface-modified particle suspensions. This fact is due to the intensification of the interparticular interactions.

A slight thixotropy and rheopexy was found in the low shear range for the modified suspen-

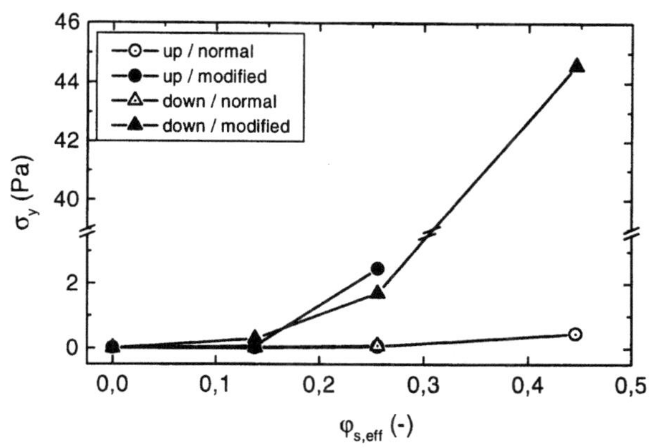

Fig. 2: Dependence of yield stresses on the effective solid volume concentration and surface modification (note the axis break)

sions with $\varphi_{s,eff} = 0.256$ and $\varphi_{s,eff} = 0.138$, respectively.

At high shear rates no significant difference between modified and unmodified suspensions could be observed. Furthermore, it is interesting to note that all suspensions have a Newtonian behavior in the high shear range.

The comparison of the complex moduli or the loss tangents of the various types of suspensions showed an increase of the elastic component for the surface-modified suspensions. This observation can be explained with the brush-like, elastic structure of the molecules of the surface modification which influences the interactions between the particles of the disperse phase.

RHEOLOGICAL EVALUATION AND AGGREGATION OF SUSPENSIONS OF LARGE GLASS SPHERES IN AQUEOUS MEDIA.

R L JONES[1], A D BURTON[1] AND R R HORSLEY[2]

[1] School of Applied Chemistry and [2] Executive Dean Engineering and Science,
The A J Parker Cooperative Research Centre for Hydrometallurgy,
Curtin University of Technology, GPO Box U 1987, Perth 6845, Australia.

1. INTRODUCTION

The rheology of suspensions is of critical importance in mineral processing operations where solid materials are often transported as suspensions in pipelines. Consequently a knowledge of suspension rheology is essential for efficient plant operations. The flow properties of the suspension determine the energy requirements for pumping, the minimum rate of flow which must be maintained to prevent sedimentation, and the dynamics of start-up processes involved after a period of no flow during which the solids may have settled (Condolis and Couratin, 1972). The rheology of dilute suspensions is relatively well understood but dilute suspensions are of limited technological interest. Concentrated suspensions, on the other hand, are of great importance, but are much less well understood, principally because they behave as non-Newtonian fluids and their rheology is influenced by many variables. This study investigates the rheology of concentrated suspensions of large (d_{50} = 70μ) glass spheres, looking at the dependence of plastic viscosity on volume fraction and temperature. Modifying the spheres was also investigated to determine the relative importance of factors which contribute to the rheology of coarse suspensions.

2. MATERIALS AND METHODS

Soda-lime glass spheres (Ballotini™ Impact Spheres, AH grade) were supplied by Potters Industries Pty. Ltd., Victoria, Australia. According to specifications at least 85% of the particles are true spheres, with the remainder being largely twins or clusters. Angular particles constitute no more than 3% by mass and this was confirmed by scanning electron microscopy. The spheres were used either as supplied, or after washing in 33% HCl, with repeated filtering and washing with deionised water. Suspensions were prepared using deionised water, and pH was adjusted using sodium hydroxide or nitric acid. Cation additions to acid washed spheres were made with magnesium, calcium and barium nitrates. Slurries were homogenised for one hour on a bottle roller prior to testing.

Rheological evaluation was carried out in a modified rotating cylinder viscometer designed to prevent settling as described by Overend *et al.* (1984), and temperatures were kept constant with a thermostatted water bath. Rheograms were determined in the directions of increasing and decreasing shear rate over time spans up to three hours. No evidence of hysteresis and hence time-dependent effects was observed. Apart from variable temperature runs, all determinations were carried out at 20°C. Maximum packing fractions were determined by centrifugation at 2600 rpm for a period of 30 min. and a value of 0.630 ± 0.001 was obtained for unwashed, washed and cation-treated materials.

3. RESULTS

3.1 Unwashed spheres

3.1.1 Temperature variation

Non-ideal plastic behaviour was found to be typical of concentrated suspensions of these spheres. Temperature variation runs for a 0.481 volume fraction suspension showed a decrease in plastic viscosity from 45.6 mPa s at 274.5 K to 25.3 mPa s at 313 K. An excellent linear fit to the data is obtained using the equation

$$\eta_s = (T/A)[1 - \phi/\phi_m]^{-s} \exp(Q/RT)$$

suggested by Chandler and Jones (1989), where η_s is the suspension viscosity, T is the absolute temperature, ϕ and ϕ_m are the volume fraction and maximum packing fraction, Q the flow activation energy for water and R the gas constant.

3.1.2 pH variation

Rheograms of suspensions at constant phase volume (ϕ = 0.534) and temperature (293 K), showed no significant change in rheology across the pH range 2.4 to 12.0.

3.1.3 Plastic viscosity *vs* volume fraction

These data are presented as phase volume *vs* viscosity in Fig. 1.

3.2 Acid washed spheres

3.2.1 Temperature and pH variation

Using acid washed spheres, similar results to the unwashed spheres were obtained.

3.2.2 Plastic viscosity *vs* volume fraction

The data from acid washed and calcium treated spheres are presented alongside those of the unwashed spheres in Fig. 1. It can be seen that data from acid washed spheres are well separated from the unwashed and acid washed-calcium treated spheres.

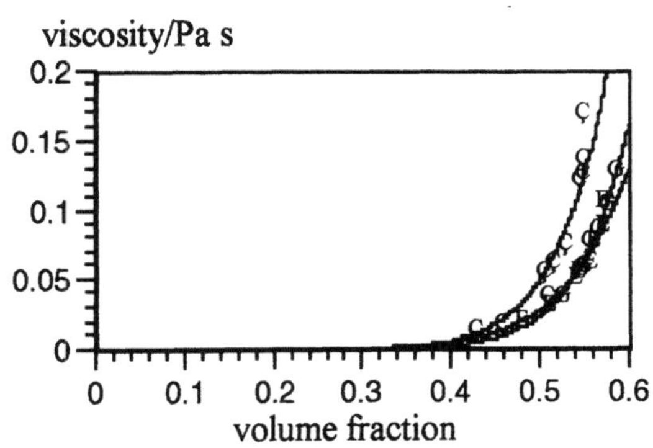

Figure 1. Viscosity *vs* volume fraction for unwashed (Ñ), acid washed (C) and Ca^{2+} treated (É) glass spheres

3.3 Addition of cations

The effect of varying concentrations of Mg^{2+}, Ca^{2+} and Ba^{2+} caused a discrete reduction in the high shear rate viscosity from 114 mPa s to 60 mPa s. The critical concentration at which this occurred depended on the cation, being greatest for magnesium and least for barium. Further increases in cation concentration caused no further reduction in viscosity. This reduction in viscosity was irreversible. When the ion concentration was diluted below the critical value, the high shear viscosity remained at the lower value.

4. DISCUSSION AND CONCLUSIONS

Generally the behaviour of the spheres tested follows expected trends and agrees with the various models put forward by Mooney (1951), Krieger and Dougherty (1954), Gillespie (1976), Chandler and Jones (1989). However, acid washing of the spheres unexpectedly induces aggregation, which can be reversed by the addition of alkali metal cations. The fact that pH changes cause no significant changes in rheology, and that particle size and shape are identical, leads to the conclusion that hydrodynamic forces are predominant in the rheology changes observed here. It seems that aggregation is induced by acid washing, and indeed this can be observed in dilute suspensions under optical microscopy. The relative effectiveness of the alkaline earth metals suggests the redispersion of aggregates is related to the surface adsorption of the ions.

6. ACKNOWLEDGMENTS

This research has been supported under the Australian Government's Cooperative Centre Programme. One of us (ADB) is grateful to the Centre for an Honours Scholarship. The support is gratefully acknowledged.

7. REFERENCES

Condolis E, Couratin P (1972) Minerals Science and Engineering 4(1): 25-37

Chandler HD, Jones RL (1989) Phil Mag A 59(3): 645-660

Gillespie T (1976) In: Kerker M (ed) Colloid and Interface Science Vol 4. Academic Press, New York. pp 465-471

Krieger IM, Dougherty TJ (1959) Trans Soc Rheol 3: 137-152

Mooney M (1970) J Colloid Sci 6: 162-170

Overend IJ, Horsley RR, Jones RL, Vinycomb RK (1984) In Mena B *et* al (eds*)* Advances in Rheology Vol 4. Univ Nacional Autonom de Mexico pp 583-616

ON THE TIME DEPENDENT RHEOLOGICAL BEHAVIOUR OF DISPERSIONS

J. KASCHTA, M. SCHMIDT, H. MÜNSTEDT

Institute of Polymer Materials, University Erlangen-Nürnberg,
Martensstr. 7, D-91058 Erlangen, Germany

1. INTRODUCTION

In the literature on concentrated dispersions a number of influences on the rheology of these systems is found as e.g. particle size, particle size distribution, specific surface area of the particles, chemical composition of the matrix, hydrodynamic interaction etc. (Ilic and Phan-Thien, 1994). It is also found in literature that concentrated dispersions exhibit structures, which may be shear induced (Gadala-Maria and Acrivos, 1980, Papenhuijzen 1972). In order to study the structure build-up and break-down in concentrated dispersions two model systems of PMMA-spheres in silicon oils and polyisobutylenes as matrix was chosen. In these systems the influence of the viscosity and of the flow behaviour (Newtonian or shear thinning) of the matrix can be studied at ambient temperatures without a change in the chemical structure of the chosen matrix. This paper describes the time dependences observed in these model systems.

2. METHODS

Poly(di-methyl-siloxanes) (AK-oils of Wacker AG) and polyisobutylenes (Glissopal of BASF AG) of different molecular weights were used as Newtonian matrices. Spherical particles (Plexidon 527 of Röhm GmbH) with a mean particle diameter of 37 µm were dispersed in the fluids in concentrations between 10 and 50 percent by volume. The particle size distribution obeys a log-normal distribution function. After mixing with a stirrer, the dispersions were degassed in vacuum. All measurements were performed at 30°C by a controlled stress rheometer (CS-melt, Bohlin Instruments). Most experiments were made using cone and plate geometry (cone angle 4°) but were checked for the influence of the chosen geometry in parallel plate geometry (1 mm gap), too.

3. RESULTS

In Fig. 1 the viscosity determined at a constant shear stress of 0.43 Pa is shown as a function of the measuring time for a dispersion of 40 vol.-% PMMA-spheres in silicon oil. The result is typical of all dispersions investigated and we therefore restrict ourselves to the system with silicon oil as a matrix. The insert in Fig. 1 shows the beginning of the experiment in more detail. It clearly demonstrates a pronounced time dependence of the viscosity which is not found in the unfilled matrix. The viscosity increases during the first 15 minutes of the experiment, passes through a maximum and decreases with increasing measuring time to a plateau.

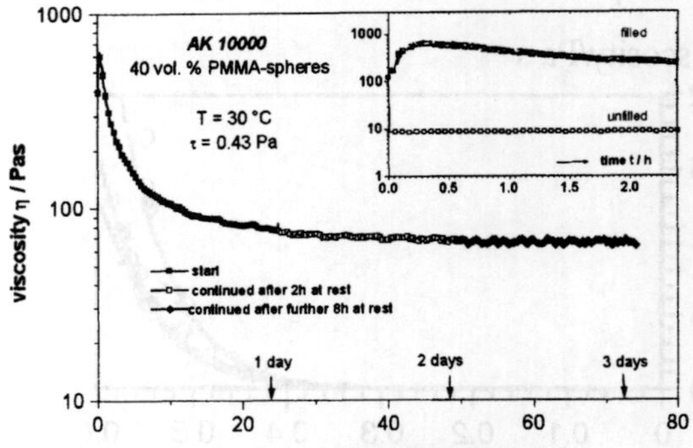

Fig. 1 Time-dependent viscosity of a dispersion with 40 vol. % PMMA-spheres (experimental details see text)

To answer the question whether the time dependence is caused by sedimentation of the particles the shearing was interruped for two hours after a measuring time of 25 h. When starting shearing after the period at rest again, the viscosity (open symbols) continues to decrease. Its starting value is the same as it was before the stop. A second stop of the experiment after 50 h with a rest time of 8 h gives a similar result. This intermittent shear experiment proves that sedimentation does not play an important role in the measurements. On the other hand it can be concluded, that a structural change has not occured during rest. It is worth to note that the equilibrium value of the viscosity was not reached before about 60 h.

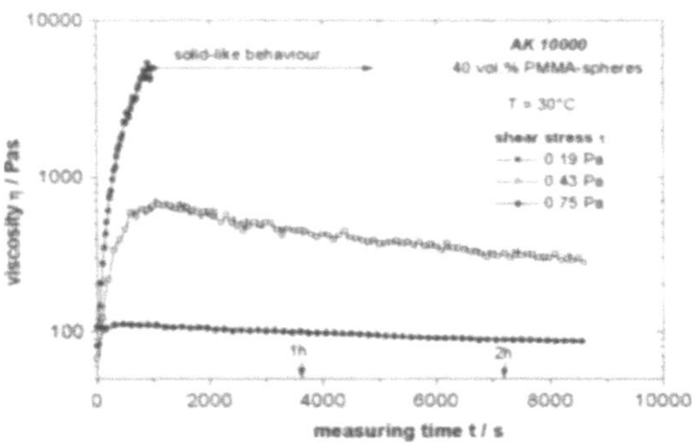

Fig. 2 Influence of shear stress on the time dependence of the viscosity for a dispersion with 40 vol.-% of PMMA-spheres

Fig. 2 shows the change of the time-dependence of the viscosity for three different constant shear stresses. At the lowest shear stress of 0.19 Pa a complete stop of the rotation of the cone was detected and the behaviour of the dispersion was solid-like. An increase of the shear stress leads to a less pronounced growth in viscosity at the beginning of the experiments followed by a decreasing viscosity with increasing measuring time. The time to reach stationary values becomes smaller with increasing stress level.

In Fig. 3 the influence of pre-shear time (at a constant shear stress) on the stress dependence of the viscosity is shown for two different volume concentrations. The viscosity curves exhibit a maximum in the low stress regime. This maximum is shifted to higher viscosities and to lower shear stress values if the applied pre-shear time is increased. The dependence of the viscosity on pre-shear time vanishes at higher shear stresses.

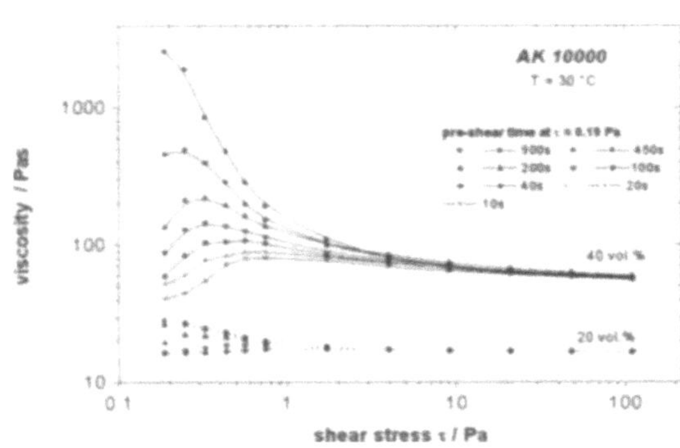

Fig. 3 Influence of pre-shearing at 0.19 Pa for different pre-shear times and two different volume concentrations of PMMA-spheres

The effect is less pronounced for lower volume concentrations (see. Fig. 3) and vanishes for concentrations below 20 vol.-% (measurements not shown here). For all matrix materials investigated it was found that the time dependence of the viscosity is more marked if the viscosity of the matrix is increased.

4. DISCUSSION

Our Computer simulations of the mean inter-particle distances (IPD) as a function of the volume concentration using a unit cell model give rise to the assumption of structures in the dispersions because the calculated IPD values are orders of magnitude larger than the distances which have to be considered when dealing with interactions of the particles itself or with interactions via matrix molecules adhered to the surface of the spheres (Castellani and Lomellini, 1993).

The results presented here can be interpreted in terms of a structure build-up and break-down. The equilibrium (time-independent) structure is reached when both processes, structure build-up and structure break-down, have the same kinetics. As the PMMA-particles do not have strong particle-particle interactions (e.g. no measurable surface charges) the structural rearrangements are found only at small shear stresses. At low shear stresses the structure build-up dominates at the beginning of the experiment. The break-down of the perfected structures becomes dominant at longer creep times because the particle network is weakened due to accumulated strain (Kaschta and Münstedt, 1998).

5. REFERENCES

Gadala-Maria F, Acrivos A (1980) J Rheol 24:799-814

Ilic V, Phan-Tien N (1994) Rheol Acta 33:283-291

Kaschta J, Münstedt H. (1998) In: Emri I, Knauss W (eds) Proc 2nd Conf Time-Dep Mat pp 98-99

Castellani L, Lomellini P (1993) Symposium Recent Developments in Structed Continua, Montreal

Papenhuijzen JMP (1972) Rheol Acta 11:73-88

RHEOLOGICAL CHARACTERISATION OF CLAY DISPERSIONS

N. KHERFELLAH[1] and K. BEKKOUR[2]

[1]Université des Sciences et de la Technologie Houari Boumédiène
Institut de Physique. Laboratoire de Rhéologie. BP 32 El Alia, Algiers (Algeria)
[2]Institut de Mécanique des fluides. UMR 7507 ULP-CNRS. 2,rue Boussingault
67000 Strasbourg (France)

1. INTRODUCTION

This work was devoted to an experimental study of steady and unsteady shear flow properties and time-dependent behaviour of clay dispersions frequently used in water-based drilling mud formulations.

2. MATERIALS AND METHODS

Montmorillonite clays are widespread thickening agents for industrial fluid formulations. Examples are : drilling fluids, paints, pesticides, cosmetic, pharmaceutical, , etc...A commercial bentonite (Na^+ - montmorillonite from Maghnia, West Algeria) has been used to prepare the clay dispersions without any further purification (clay contents 4 and 10% by weight). All measurements were performed at a controlled temperature of 20°C. To prevent changes in composition during measurements for water evaporation, a humidification chamber was placed around the cone-and-plate device.

3. RESULTS AND DISCUSSION

3.1 Measurement of the thixotropy

A Constant stress (Carri-Med CSL 100, cone and plate configuration) rheometer was used to carry out the measurements. Loading and unloading experiments, by increasing the applied stress from zero to a prefixed maximum value and thereafter decreasing it from this value to zero, were carried out systematically. The main feature revealed by these experiments is the existence of hysteresis loops which can be considered as a measure of thixotropy (Fig.1). Such behaviour is caused by progressive break-up of the structure under shear followed by gradual recovery. The area of the thixotropic loop increases when the bentonite content is increased which implies that the thixotropic properties are more important. Also, it can be observed that bentonite dispersions behave like a non-Newtonian fluid with a yield stress. which is given, as a first approximation, by the minimum value of the stress in these plots

3.2 Linear viscoelastic measurements

The linear viscoelastic properties of bentonite dispersions were first studied in terms of creep flow tests. A creep curve is obtained by applying at time t_0 a stress held constant and measuring the time evolution of the resultant deformation followed by a strain recovery curve obtained after removal of the constant stress. The elastic properties are defined by correlating the results with the classical viscoelastic models of Maxwell or Kelvin-Voigt which can be physically described by assemblies of dashpots and springs. From our experiments it was found that the viscoelastic behaviour of the bentonite dispersions could be represented by the generalised Kelvin-Voigt model which combines a Maxwell unit (Newtonian dashpot in series with purely elastic spring) in serial with one or more Kelvin-Voigt units (dashpots in parallel with springs). The creep compliance for the bentonite dispersions is then defined as :

$$f(t) = J_0 + \sum_{i=1}^{n} J_i(1 - e^{-t/\theta_i}) + \frac{1}{\eta_N} \quad (1)$$

where $f(t)$ is the over-all creep compliance at the time t during the test, J_0 the instantaneous elastic compliance; J_i the contribution to retarded elastic compliance of the ith component with a retardation time $\theta_i = (J_i\ \eta_i)$ and a viscosity η_i and η_N is the Newtonian viscosity. Fig. 2 shows an example of the creep curves obtained. The viscoelastic parameters were derived from the creep compliance-time curves by use of the method developed by Inokuchi to analyse the retarded elasticity region of each curve (Inokuchi, 1955). In addition to the creep properties discussed previously, the linear viscoelastic properties of bentonite dispersions were also studied in terms of complex dynamic properties using small amplitude oscillatory shear experiments. A sinusoidal stress $\tau = \tau_0 \sin \omega t$ is applied to the sample whereas its corresponding sinusoidal strain given by the following relationship is measured :

$$\gamma = \gamma_0 \sin(\omega t + \delta) \quad (2)$$

where, ω is the frequency of oscillation in rad/s, γ_0, the strain amplitude and τ_0 the stress amplitude, δ the phase angle between the stress and the strain (i.e., the loss angle) and t, the time.

The shear storage (G') and shear loss (G'') modules were calculated from the measured outputs, the strain amplitude γ_0 and the loss angle δ, by mean of the following equations:

$$G' = (\tau_0/\gamma_0)\cos\delta \quad (3)$$

$$G'' = (\tau_0/\gamma_0)\sin\delta \quad (4)$$

The effect of frequency was investigated with a constant deformation of 10%. As example, the modules G' and G'' are plotted in Fig. 3 as functions of frequency (concentration of 8%). It can be observed that both G' and G'' are frequency independent. This shows a pronounced elastic behaviour of the suspensions. Furthermore, the storage modules G' is larger than the loss modules G'', showing that the elastic behaviour of bentonite dispersions is dominant compared to the viscous behaviour. This result is in good accordance with those obtained by other authors (Khan et al., 1988).

An exhaustive study of clay suspensions may require the determination of their viscoelastic properties over a wide frequency scale. Unfortunately; due to sedimentation and evaporation processes, the experiments are limited in time. However, the high-frequency (short time) range may be investigated by a dynamic method such as oscillatory while the long time range (low-frequency) is accessible by a transient method such as creep because a periodic experiment at frequency ω is qualitatively equivalent to a creep test at time $1/\omega$. Mathematical methods (Barry, 1975) may then be used to unify data so as to describe the viscoelastic behaviour over the frequency range using only one type of response function. That is, it was possible to complete the dynamic response in the low-frequency range (long time) using data from the transient response in creep. Creep compliance curves analysed by the Inokuchi method were used to calculate storage and loss compliances (J' and J'') using the approximation method of Shwarzl (1969). One obtains

$$J'(\omega) = J_0 + \sum_{i=1}^{n} J_i / (1 + \omega^2 \tau_i^2) \quad (5)$$

$$J''(\omega) = \sum_{i=1}^{n} J_i \omega \tau_i / (1 + \omega^2 \tau) + \frac{t}{\eta_N} \quad (6)$$

Values for G' and G'' may be deduced using equations 7 and 8.

$$G' = \frac{J'}{J'^2 + J''^2} \quad (7) \qquad G'' = \frac{J''}{J'^2 + J''^2} \quad (8)$$

The results in terms of variation of the storage modulus G' with frequency are illustrated in Fig. 3. The high frequency values are derived from oscillatory measurements with the Carri-Med rheometer and the low frequency values are obtained by transforming creep data. One may observe that the transformed dynamic functions compare well with the functions obtained directly from oscillatory measurements.

3. CONCLUSION

Loading and unloading measurements have show that the bentonite suspensions exhibit pranced thixotropic properties. The viscoelastic properties were investigated in terms of creep flow and dynamic tests. It was found that the dispersions exhibit pronounced viscoelastic properties for concentrations higher than 6% by weight.

5. REFERENCES

BARRY B. W.(1975), 'Viscoelastic properties of concentrated emulsions', Advances in Colloid and Interface Science, **5**, 37-75.

INOKUCHI K. (1955) Bull. Chem. Society, Japan, **20**, 453.

KHAN S.A., SCHNEPPER C.A., ARMSTRONG R. C.(1988), Foam Rheology : III. Measurement of shear flow properties, J. Rheology, 32 (1), 69-92.

SCHWARZL F. R.(1969) Rheol. Acta, **8**, 6.

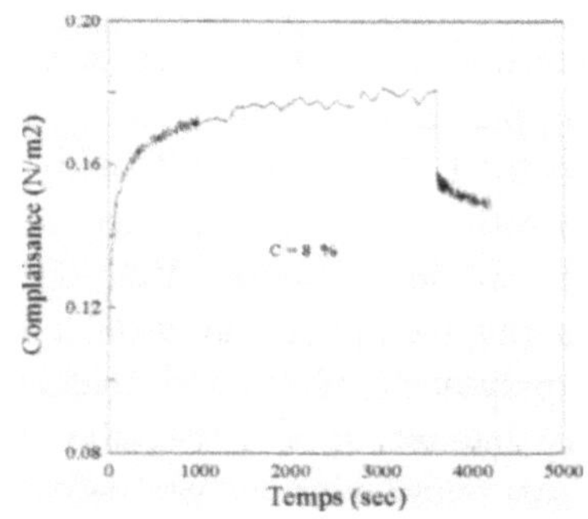

Figure 1. Thixotropic loops for two bentonite suspensions (5% and 8%)

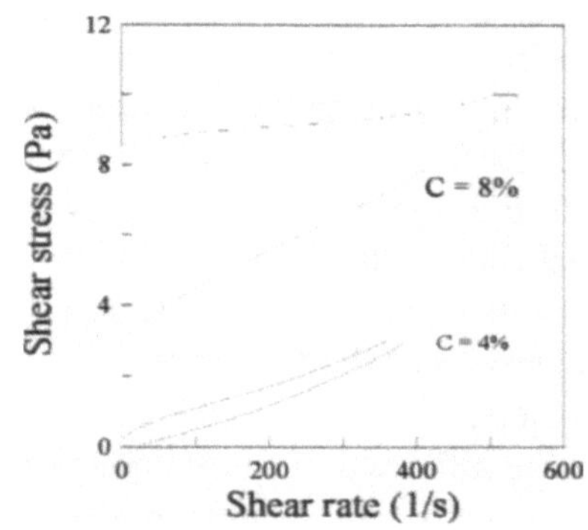

Figure 2. Creep compliance and recovery curve for a 5% bentonite suspension

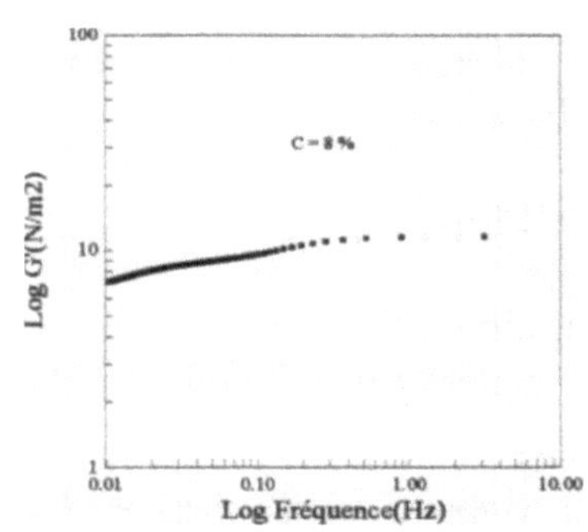

Figure 3. Variation of G' and G'' with frequency. Low frequency values (filled symbols) are derived by transforming creep data, and high frequency values (open symbols) are measured by oscillatory measurements with the Carri-Med rheometer.

RHEOLOGY OF MODIFED POLYSACCHA-RIDE POLYMERS IN REACTIVE PRINTING

V. KOKOL, S.ŠOSTAR

University of Maribor, Faculty of Mechanical Engineering, Textile Chemistry Institute
Smetanova 17, 2000 Maribor, Slovenia

1.0. INTRODUCTION

Alginate thickeners are the most widely used thickeners in reactive printing of viscose because they form printing pastes of high drawability at relatively low solid contenst. They wash off easily (except in combination with bifunctional reactive dyes) and do not increase fabric stiffness. But due to their unreliable supply, high price and unstable quality, the alternative polysaccharide guar thickeners have become very interesting. On the other hand, guar thickeners have many hydroxyl groups, so they have to be modified by a corresponding substitution level. Due to their different chemical constitution, the guar thickeners have different primary properties which are expressed, depending on the molecular weight and the density of the cluster structure, in different solid contents. In a solution the thickeners with a low solid content form large low-density clusters (which is ascribed to the high molecular mass of polymers), and the thickeners with a high solid content form small high-density clusters (which is ascribed to the low molecular mass of short polymers).

The objective of our research was to study and to compare the influence of the solid content on the viscoelastic properties of printing pastes containing alginate or different guar thickeners on the printing paste add-on, on dye penetration and on the depth of colour. In addition, the influence of the solid content of the guar thickener on the stiffness of printed viscose will be shown.

2.0. EXPERIMENTAL

2.1. Substrate

All experiments were made on viscose textiles with a mass per unit area of 126.1 g/m^2.

2.2. Thickeners

Sodium alginate, non-substituted guar gum, carboxymethyl guar gum of different substitution levels and depolymerised guar gum were used as polysaccharide thickeners (table 1).

2.3. Printing paste and printing process

Printing pastes were made according to the recipe: 60% thickener, 2.5% $NaHCO_3$, 1.5% reactive dye (bi-vinylsulphonic C.I. Reactive Black 5 dye) and 36% water. All printing pastes were adjusted to a constant viscosity η=7000±500 mPa s at shear rate γ=10s^{-1}. Stages of work: printing (the flat screen printing technique was used), drying, steaming and washing.

Thickener	Sign of abbreviation	Solid content [g/kg]
Sodium alginate low viscosity	AlgL DS=0	48.0
Nonsubstituted guar gum high viscosity	NSGH DS=0	12.0
Carboxymethyl guar gum	CMG DS=0.13	33.6
Carboxymethyl guar gum	CMG DS=0.26	39.6
Depolimerisired guar gum	DPG DS=0	34.5

Table 1: Thickener, substitution level (DS) and solid content

2.4. Methods

The rheological properties (rotation and oscillation measurements) of printing pastes were determined with a viscosimeter Rheometrics Fluids Spectrometer RFS II (Rheometrics GmbH, Germany).

In order to determine fabric handle, the stiffness of washed and climatized printed samples was measured according to DIN 53212. The measurements were performed using a fabric stiffness measuring apparatus Type 58963 (Frank GmbH, Germany). Dye penetration and the depth of colour were determined by remission measurements using the colourimetric apparatus Texflach DC 3881 (Datacolor, Germany).

3.0. RESULTS AND DISCUSSION

3.1. The influence of thickeners on viscosity and on viscoelastic properties of printing pastes

The viscosity (Fig.1) in the I. Newtonian and structurally viscous region depends on the structure of the polymer chain, the molecular mass, the degree of depolymerization and the thickener's solid content. The high viscosity guar thickener NSGH shows a strong decrease in the viscosity curve of the I. Newtonian and structurally viscous region of the increasing shear rate, while the viscosity curve of the low viscosity AlgL thickerner exhibits an explicitly Newtonian character in the given region. The comparison of the viscosity curves of medium viscosity thickeners (CMG DS=0.13, CMG DS=0.26 and DPG) shows a Newtonian character in the first Newtonian region and a strongly expressed pseudoplastic character in the structurally viscous region when compared to the standard AlgL thickener.The decrease of the viscosity curve depends on the strength of the crosslinking of macromolecules. Therefore, a strong decrease of the viscosity curve can be observed in the low solid content thickener (NSGH) which forms a loose crosslinking structure, and inversely.

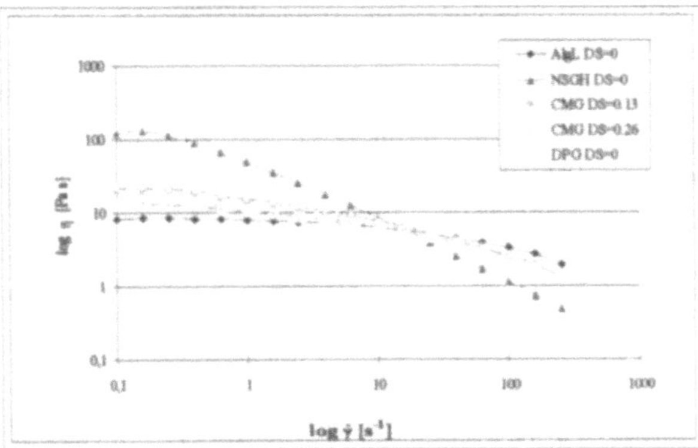

Figure 1: The influence of the thickener on the viscosity curve of printing pastes

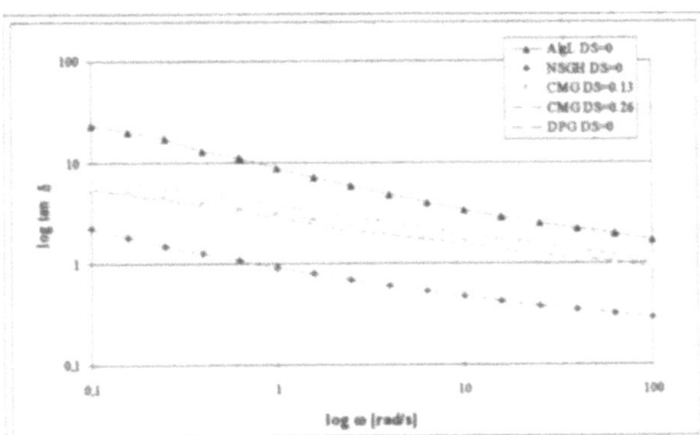

Figure 2: The influence of the thickener on the tan δ of the printing paste

From the flow curves (Fig.2) of printing pastes we can see, that the loss factor (tan δ) decreases with the increasing oscillation frequency (ω). Higher values of the loss factor (tan δ) indicate that the module of viscosity is higher than the elasticity module, and that the viscosity component, which is most strongly expressed in the printing paste containing the AlgL thickener, is the predominating component. Thus a good diffusion of the printing paste in the substrate and the resulting high quality of penetration (table 2) are characteristic for this type of printing pastes, which is ascribed to short polymers. The values of the loss factor (tan δ) of the polysaccharide thickener NSGH with a low solid content are substantially lower, which indicates a smaller viscous and a higher elastic component. The reason for this lies in the long polymer chains of high molecular mass. A high degree of drawability and surface printing with poor penetration (tabele 2) are typical of printing pastes of this kind. Among all medium viscosity thickeners the CMG=0.26 thickener has the highest loss factor (tan δ), and it offers the best penetration.

3.2. Penetration, depth of colour and fabric stiffness

Thickener	Paste add-on [g/m^2]	Penetration [%]	Fabric stiffness [mN cm^2]	K/S$_f$
AlgL DS=0	122.2	39.3	1.67	9.73
NSGH DS=0	34.2	15.0	5.60	2.93
CMG DS=0.13	88.2	10.0	15.76	12.21
CMG DS=0.26	89.8	9.7	12.19	10.83
DPG DS=0	85.0	11.3	16.56	11.10

Table 2: Paste add-on, penetration, fabric stiffness and depth of colour

The lowest K/S$_f$ values are observed with the NSG thickener, which is ascribed to the low solid content or the low printing paste add-on. The K/S$_f$ values are considerably higher in case of medium viscous thickeners with similar solid contents (CMG, DPG), which diffuse, due to shorter polymers, more easily in the substrate and cause, due to many free hydroxyl groups, greater stiffness. A higher degree of DS improves the better fabric stiffness.

4.0. CONCLUSION

With respect to the standard AlgL thickener, the alternative guar thickener is the CMG DS = 0.26 thickener.

5.0. REFERENCES

V.Kokol, R.Schneider, S.Šostar; (1997) Tekstilec 11/12: 346-350;

S.Šostar, V.Kokol, S.Jeler; (1996) Tekstil 8: 414-420;

V.Kokol, S.Šostar; (1998) Joint conference of Italian, Austrian, Slovenian rheologists.

The influence of coalescing aids on rheological and mechanical properties of final emulsions

Končnik Damjan, Golob Janvit, Zupančič Andreja

University of Ljubljana, Faculty of Chemistry and Chemistry Technology,
Aškrčeva 5 1000 Ljubljana, Slovenia, tel.:++386 61 176 0 434, fax.: ++0386 126 45 77

1. INTRODUCTON

Emulsion polymerization is a method of producing latexes with desired particle size, molecular weight, composition, morphology, film-forming properties etc. For this reason the emulsions were synthezised with different compositions of styrene and 2-ethylhexylacrilate. One of the most important characteristics of latexes from theoretical and practical point of view is viscosity. Rheological measurements often provide important interpretation for structure of latexes[Lovell, El-Aasser, 1997]. The accurate synthesis of emulsions is well described elsewhere [Končnik, 1998]. From these latexes emulsion 2 with weight ratio styrene/2-ethylhexylacrilate 54/46 and emulsion 3 with weight ratio styrene/2-ethylhexylacrilate 64/36 were taken. The emulsions contain 48 % of polymer. The minimum film formation temperature (MFT) of emulsion 2 is 11^{0}C and MFT of emulsion 3 is 34.4^{0}C. The average particle size of emulsion 2 is 0.21 mm and the average particle size of emulsion 3 is 0.24 mm.

2. EXPERIMENTAL

2.1. Materials

Two coalescing aids texanol and diethylenglycol butyl ether acetate (BDGAC) were taken in the range from 0 to 6% upon the emulsion. When a coalescing aid is added to an emulsion it will distribute itself between the aqueous phase and the polymer phase. A coalescing aid softens the polymer particles and after the film is formed it has a tendency to evaporate. The main task of coalescing aid is softening of polymer particles, but the softening of particles increases their viscosity and changes mechanical properties of films. The descries in the minimum film formation temperature of the samples depending on the added amount and type of coalescing aid.

2.2. Measurements

The rheological measurements were carried out with Haake RheoStress RS 150 rheometer by using cone and plate (6 cm/2^0) and double cone (6 cm/4^0) geometry at 20^{0}C. All the samples exhibited plastic flow behaviour. Time dependent effect on the rheological properties were not detected.

3. RESULTS AND DISCUSION

By rheological measurements the sample is subjected to the shear stress, which causes the deformation of polymer particles. The extent of deformation depends on the force and chemical structure of the polymer particles, in our case, particles containing more2-ethylheksylacrilate are softer and consequently more deformable than particles containing more styrene. The influence of shear stress is schematically represented in figure 1.

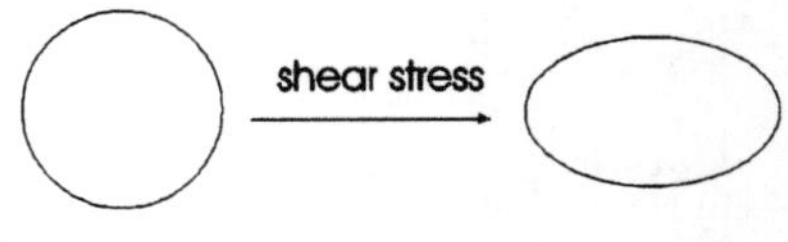

Figure 1. The influence of shear stress on polymer particle.

The picture 2 presents the flow curves for emulsions 2 by adding of texanol. The measurements show a significant increase of yield stress by adding of coalescing aid. The viscosity level of the samples also increases with adding amount of coalescing aid. The flow curves were fitted by the Casson model:

$$\tau^n = \tau_o^n + \left(\eta_\infty \dot{\gamma}\right)^n \qquad (1).$$

The yield stress is very important for applications of emulsions in water born paints. Figure 3 shows increase of the parameter with increasing concentration of texanol. The addition of BDGAC instead of texanol in the emulsions results in less pronounced changes of rheological behaviour with increasing coalescing aid concentration.

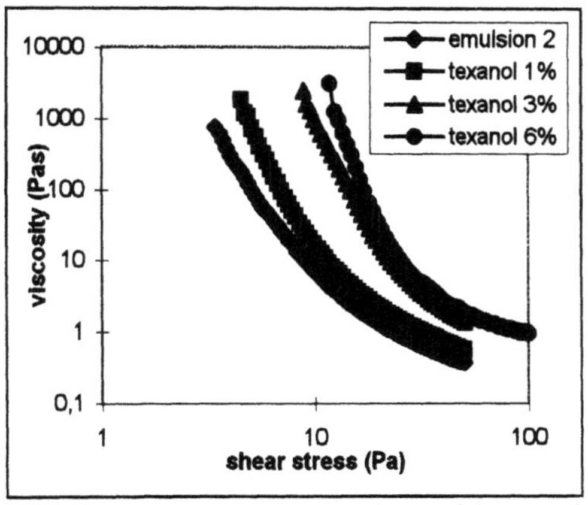

Figure 2. Flow curves of emulsion 2 by adding of texanol.

The effort of the addition of BDGAC is that lowers MFT more efficient than texanol. This is important from economical and technological point of view and gives rise to better mechanical properties of the final products. Already at presence of 1 % of BDGAC in the emulsion the drop of MFT for 10^0C was observed. The influence of coalescing aid concentration on MFT is shown in figure 4.

Under non-destructive conditions of oscillatory shear all the emulsions examined exhibited predominately elastic component of viscoelastic response. The phase lags ranges about 25^0 independently of coalescing aid concentration. Both moduli were almost insensible on the applied frequency.

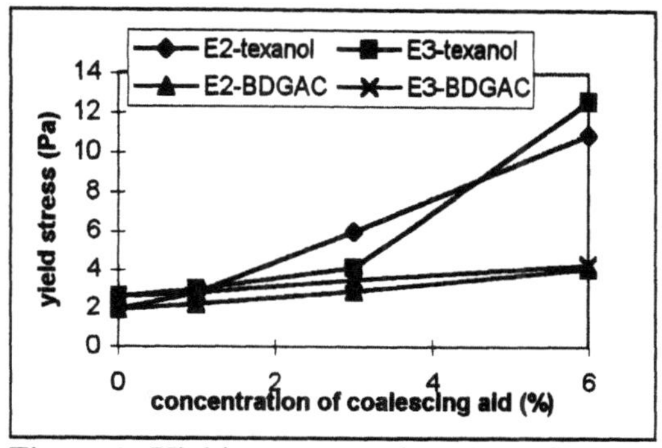

Figure 3. Yield stress as a function of added amount of coalescing aid.

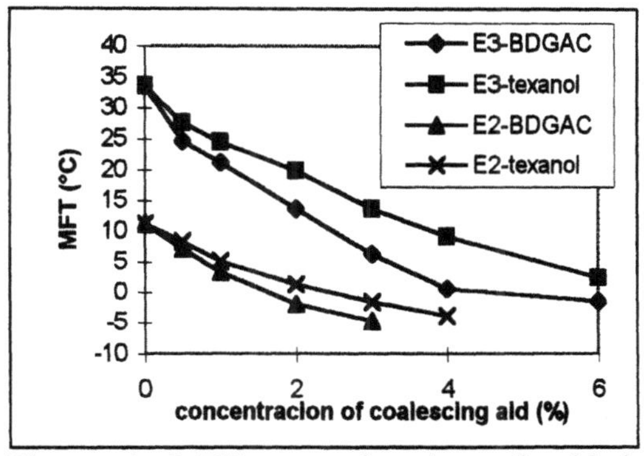

Figure 4. MFT as a function of added amount of coalescing aid.

4. CONCLUSON

On the basis of this study it could be suggested BDGAC lowers MFT more efficiently then texanol and causes less increase of viscosity, meanwhile addition of texanol causes strong changes in viscosity level and is slightly less efficient as a coalescing aid.

References:

1. Lovell, P.A. and El-Aasser, M.S., Emulsion Polymerisation and emulsion Polymers, Chichester, John Wiley and Sons, 1997 (437-441)
2. Končnik, D., Golob, J. and Šebenik, A., »Determination of the minimum film formation temperature and the hardness of the film as a function of latex blends«, submitted to Journal of Coatings Technology, 1998

The Rheology of Concentrated Suspension at a High Rate Deformation.

A. Krasheninnikov, V. Kornev

Moscow State Academy of Instrument Making and Information Sciences.
107846, Stromynka, 20, Russia

By concentrated suspensions we mean such suspensions which have space barriers at deformation, i.e. the particles run into one another. This effect may occur at concentration of over 60 vol. % if the dispersion phase particles is spherical and mono-dispersive. Viscosity of concentrated suspension is shown to greatly increase at high rate shift (the phenomenon of dilatation) when we study suspensions of mineral particles and plastisols. Unlike the Reynolds dilatation, which leads to irreversible changes in the system structure, instances of fully reversible dilatation have been discovered. If in found that by way of changing the nature of suspension components, its concentrations and thermotreatment we may come to altering the kind of viscosity anomaly of concentrated suspensions. These rheological transformations are most evident with plastisols, i.e. suspensions of polymer powders in peptizers. Transferring viscous-plastic currents into dilatantive ones and vice versa is easily achieved in such suspensions. Dilatations systems substantially grow in volume at high rate deformation, and becoming nearly solid they are subject to local cracking. The system gets back to normal after lifting the deformation. T.Gillespie's conception has been used to describe these rheological phenomenons. An equation is offered to connect the physical parameters of the system, which makes it possible to predict the complex interrelations between viscosity and shift rate.

Computer research of the rheological model has been undertaken with the aim of better understanding the technologies of dilatantive solidification of concentrated suspension at high rate deformation. A system has been devised to this effect including a viscous matrix and a dispersion filling made of some solid material with a high modulus of elasticity. The chief factors relevant to the mathematical model are interrelations of particles and the friction coefficient of the components. The evolution of structural modifications has been clearly described. The research of the model makes it possible to draw the following conclusions:

- the effect of suspension solidification depends upon the structural dislocation of particles in matrix volume;
- contact interrelations among the particles are destroyed after lefting the mechanical stress;
- destruction of contacts among the particles depends upon the time of "rest" of the suspension.

Gillespie T. (1966) J.Coll. Interface Sci 22:554-566

VISCOMETRIC AND PIPE FLOW OF INORGANIC WASTE SLURRIES

E. KRUSTEVA

Central Laboratory of Physico-Chemical Mechanics, Bulgarian Academy of Sciences,
Acad. G. Bonchev St., Bl.1, 1113 Sofia, Bulgaria, e-mail: krusteva@bgcict.acad.bg

1. INTRODUCTION

One of the most important fields for the practical application of rheological parameters is pipeline transport, hydraulic transport in particular, as is the case with drilling fluids in petroleum output, transport of residue mineral materials to tailings ponds, filling of abandoned mine galleries for the protection of geoenvironment, etc. (Altschul and Kiselev, 1975). The aqueous suspensions of mineral particles launch a number of challenges connected with their thermodynamic instability (Barnes et al., 1989; Shchukin et al., 1982). The colloidal forces of attraction or repulsion and the Brownian forces acting between the particles are the ever-present ones in the case of clay-containing suspensions. The viscous forces are mainly due to the dispersed phase and not to the medium ($\eta^{20}_{H_2O}$= 1.005 mPa.s only) and are not able to compensate the difference between the phase and the medium density. Hence, the ensuing difficulties in the correct measurement of the rheological characteristics and in the possibilities for transport due to rapid sedimentation and flocculation (Chára et al, 1974; Bournaski and Krusteva, 1997). In this context the rheological methods are indispensable for both formulating the optimal composition of the multicomponent aqueous slurries (possessing sufficient fluidity and stability at the possibly minimal water content) and for determining the necessary parameters for calculating the hydraulic parameters as kinematic viscosity, hydraulic losses, velocity profiles, pipe diameter, etc.

The present work considers the rheological properties of aqueous slurries based on different waste materials accumulated in many industrial regions with developed ore- and coal-mining in Bulgaria (blast furnace slag, fly ash, tailing materials, local materials from mine covers, etc.), which exhibit various types of non-Newtonian behaviour. Examples are given for the successfully applied hydraulic transport in the case of preventing the collapse in the vicinity of a railroad passing over abandoned mine galleries, liquidation of mines representing a health hazard, etc. (Krusteva, 1994).

2. EXPERIMENTAL

The numerous combinations of waste materials used for the formulations of the filling slurries were tested under laboratory conditions using a Rheotron Brabender rotational viscometer mainly in the automatic flow curve recording mode. Corrections taking under consideration the presence of coarser fractions were made further on during the *in situ* experiments.

2.1 Rheological Experiments

The flow curve tests were chosen as the most express method of obtaining information about both the rheological behaviour and the stability of the systems. The slurries were optimised during the rheological tests by addition of small amounts of plasticizers and highly dispersed admixtures. A typical example is shown in Fig.1. It is seen that the fast settling of the heavier particles is almost "visualised" by the flow curve.

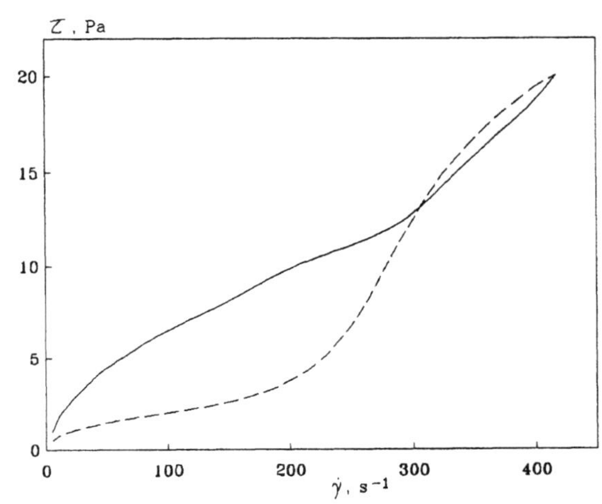

Fig. 1. Flow of a three-component system containing heavy particles (ρ = 4.2 g cm^{-3}) and ground blast furnace slag (ρ = 2.8 g cm^{-3}) in mass proportion 4:1 (w/s=0.35).

2.2. Combining Rheological and Field Experimental Data

All of the investigated slurries exhibited an yield-pseudoplastic behaviour. An approximation was made according to the Casson model

$$\tau^{1/2} = \tau_o^{\ 1/2} + (\eta_C \dot{\gamma})^{1/2} \qquad (1)$$

Assuming the dimensionless parameters $a = \tau_o / \tau_w$ and $\rho = r / R$, where τ_o is the yield stress, τ_w is the shear stress at the pipe wall, r and R - the current and the maximum radius, we obtain equations (2) for the plug region and for the rest of the cross-section:

$$\frac{2v\eta_C}{R\tau_w} = 1 - \frac{8}{3}a^{1/2} + 2a - \frac{1}{3}a^2 \qquad (2)$$

$$\frac{2v\eta_C}{R\tau_w} = (1-\rho^2) - \frac{8}{3}a^{1/2}(1-\rho^{3/2}) + 2(1-\rho)a$$

The average velocity is as follows:

$$\frac{4\bar{v}\eta_C}{R\tau_w} = 1 - \frac{16}{7}a^{1/2} + \frac{4}{3}a - \frac{1}{21}a^4. \qquad (3)$$

The dimensionless velocity distribution versus the dimensionless radius is plotted in Fig.2, the thick curve referring to the practically used composition with τ_o= 5.76 Pa and η_C=0.039 Pa.s (coal slime : fly ash : lime ash in the proportion of 5 : 5 : 1, w/s = 0.9).

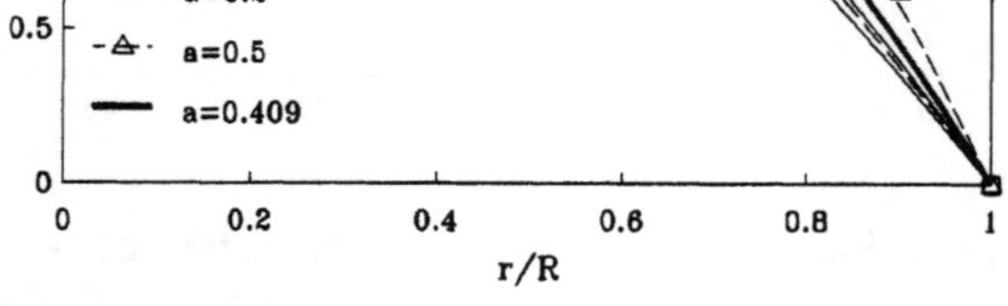

Fig. 2. Dimensionless velocity dependence on *r/R* and dimensionless shear stress

The value of τ_w was obtained substituting the parameters from the *in situ* experiment in the expression (4) (Kooijman and Van Zanten):

$$\frac{4\bar{v}\eta_C}{R\tau_o} = \left\{1.12 - \frac{1}{a}^{1/2}\right\}^2 \qquad (4)$$

3 CONCLUSIONS

A great variety of waste materials from different industrial regions in Bulgaria has been studied by rheological methods. The results were successfully applied in practice for filling of underground spaces with hydraulically transported waste slurries.

Acknowledgement: The present work was financially supported by the National Fund "Scientific Investigations" of the Bulgarian Ministry of Education and Science (project H3-417).

4 REFERENCES

Altschul AD, Kiselev PG (1975) Hydraulics and Aerodynamics. Stroiizdat, Moscow

Barnes HA, JF Hutton, K Walters (1989) An Introduction to Rheology. Elsevier, NY

Bournaski E, Krusteva E (1997) In: Proc. 9th Int. T&S Conf., Zeszyty Naukowe Akademii Polniezej we Wroclawiu, Nr 315, 427-436

Chára Z et al. (1994) In: Proc. 4th European Rheol. Conf., Sevilla, Progress and Trends in Rheology IV (Ed. C Gallegos) Steinkopff Verlag, 624-625

Kooijman JM, DC Van Zanten (1972) Chem. Eng. J., 4: 185-188

Krusteva E (1994) Les Cahiers de Rhéologie, Orléans, vol XIII Nr 1-2 141-149

Nguyen QD, DV Boger (1985) Rheol Acta 24: 427-437

Shchukin ED, AV Pertsov, EA Amelina (1982) Colloid Chemistry. Moscow (in Russian)

ZERO VISCOSITY OF A SUSPENSION SUBMITTED TO A DC ELECTRIC FIELD

E. LEMAIRE, L. LOBRY

LPMC, UMR 6622, Université de Nice, Parc Valrose, 06108 Nice cedex 2 France

1. INTRODUCTION

During the last ten years, many studies have been carried out on Electrorheological effect which is the enhancement of the viscosity of a suspension induced by an electric field (see for instance Halsey (1992)).In this paper, we show that the opposite effect, that is a decrease in viscosity, can be obtained with a DC field if the charge relaxation time, $\tau=\varepsilon/\sigma$, ε being the permittivities and σ the conductivities, is lower in the suspending liquid than in the particles ($\tau_1<\tau_2$).

2. MECHANISMS

In such a case ($\tau_1<\tau_2$), under the DC field, the free charges of the liquid migrate to the electrode of the opposite sign and, meeting the particles, they accumulate at their surface. This results in a charge distribution at the surface of the particles equivalent to that of dipoles antiparallel to the E field direction; this equilibrium is unstable and the particles begin to rotate in order to flip the direction of their polarisation, **P**. This phenomenon has been discovered by Quanincke (1896) and studied by Melcher (1974). Submitted to the field, the particles undergo a driving electric torque:

$$\mathbf{\Gamma^E}=\mathbf{P}\times\mathbf{E} \quad (1)$$

Cebers (1980) has calculated **P**, solving the relaxation equation for the retarding part of the polarisation:

$$\frac{D\mathbf{P}}{Dt}=\frac{\partial\mathbf{P}}{\partial t}+\boldsymbol{\omega}\times\mathbf{P}=\frac{1}{\tau_M}\left(\mathbf{P}-4\pi a^3(\chi^0-\chi^\infty)\mathbf{E}\right) \quad (2)$$

Where $\boldsymbol{\omega}$ is the angular velocity of the particles, τ_M, the Maxwell time and χ^0 and χ^∞, the polarisation coefficients at low and high frequency, respectively:

$$\tau_M=\frac{\varepsilon_2+2\varepsilon_1}{\sigma_2+2\sigma_1};\chi^0=\varepsilon_1\frac{\sigma_2-\sigma_1}{2\sigma_1+\sigma_2};\chi^\infty=\varepsilon_1\frac{\varepsilon_2-\varepsilon_1}{2\varepsilon_1+\varepsilon_2} \quad (3)$$

In stationary state, the angular velocity of the particles is given by the balance between the electric torque, $\mathbf{\Gamma^E}$, and the resistant viscous one, $\mathbf{\Gamma^H}$. The typical order of magnitude for the spin rate is of few hundreds rad s^{-1}. This is also a usual value for the spin rate of a particle in an ordinary shear rate so that Quincke rotation would be able to modify significantly the rheological behaviour of a suspension.

Indeed, when a suspension is sheared, if its particles are free to rotate, , that is they do not undergo any external torque, their angular velocity, ω_0, is equal to the vorticity which is given by a half of the shear rate, $\dot\gamma$:

$$\omega_0=\frac{\dot\gamma}{2} \quad (4)$$

Reciprocally, the control of the angular velocity of the particles permits to modify the effective shear rate inside the suspension. The viscosity can be increased when the rotation of particles is impeded or decreased when the rotation is favoured. Brener has proposed a theoretical expression for the dependence of the apparent viscosity on the particles' spin rate:

$$\eta=\eta_0(\phi)+\eta_r(\phi)$$

$$\eta_0(\phi)=\eta_0\frac{1+\frac{5}{2}\phi}{1-\phi};\ \eta_r(\phi)=\eta_0\frac{\frac{3}{2}\phi}{1-\phi}\frac{\frac{\dot\gamma}{2}-\omega}{\frac{\dot\gamma}{2}} \quad (5)$$

ϕ being the volume fraction of particles, η_o the viscosity of the suspending liquid $\eta_o(\phi)$ the shear viscosity of the suspension when the particles are free to rotate and $\eta_r(\phi)$, the rotational component of the viscosity which can be either positive (if $\omega<\dot\gamma/2$) or negative (if $\omega>\dot\gamma/2$).

In the following, the spin of the particles is induced by making use of Quincke rotation. Indeed, when the field is applied and the suspension is motionless, particles start rotating around themselves in any direction perpendicular to the field and so the average spin rate of the suspension is zero. But, when a velocity gradient (simple shear rate) is applied along the E field direction, the particles rotation axes will be favoured in the vorticity direction. This result is obtained by solving the equation $\Gamma^E=\Gamma^H$ where Γ^E is calculated by combining equations (1) and (2):

$$\Gamma^E=(\chi^\infty-\chi^0)\frac{\omega\tau_M}{1+(\omega\tau_M)^2} \qquad (6)$$

and Γ^H is given by:

$$\Gamma^H=8\pi\eta a^3(\omega-\omega_0) \qquad (7)$$

with $\omega_0=\dot{\gamma}/2$ in the vorticity direction and $\omega_0=0$ in the other directions.
The equation $\Gamma^E=\Gamma^H$ admits one solution for ω in the vorticity direction and two solutions with ω antiparallel to the vorticity. We (Lobry 1998) have shown that the rotation parallel to the vorticity direction is the only stable solution.

3. EXPERIMENTAL

We have measured the viscosity reduction induced by a DC field on a suspension made of 22% of PMMA particles dispersed in a mixture of transformer oil, another dielectric liquid (Ugilec, Elf Atochem) and AOT salts. The characteristics of the two components of the suspension are the following:

The fluid: $\varepsilon_1=3.4\varepsilon_0$, $\sigma_1=2\ 10^{-8}$S/m, $\eta_0=12$cP

The particles: $\varepsilon_2=2.6\varepsilon_0$, $\sigma_2<<\sigma_1$, $a=40\mu$m

The rheological measurements have been carried out with a controlled stress rheometer (Carrimed CSL100), in parallel plates geometry. The electrodes also serve as electrodes so that the E field is along the velocity gradient direction. The procedure for the rheological measurements is the following: a constant shear stress of 5 Pa is applied during 20 seconds so that all particles rotate in the vorticity direction. Then, the torque is decreased to zero, step by step, in order to measure the equilibrium value of the shear rate.

4. RESULTS AND DISCUSSSION

Figure 1 presents the variation of the viscosity of the suspension for different electric fields. During the measured shear rate decrease, we see that, for moderate DC field intensities (E=17kV/cm), the viscosity decreases to a minimum and then increases up to its value in the absence of the electric field. On the contrary, for higher fields, the viscosity decreases to zero, at

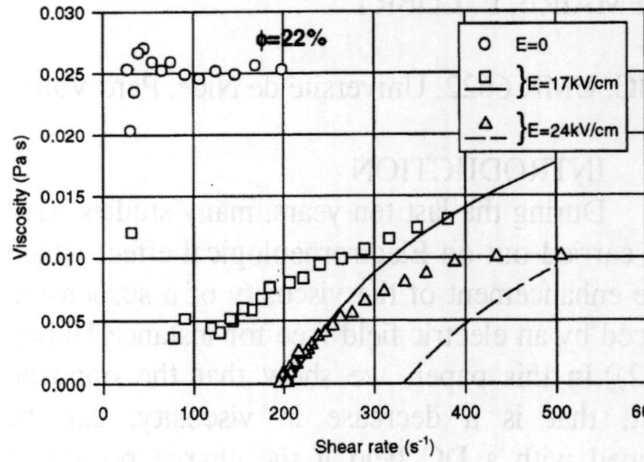

Figure 1

which point, without applying any torque to the rheometer, it keeps on rotating. As long as the DC field is on, the shear rate has a steady value $\dot{\gamma}_c$.The solid and dashed lines represent the theoretical predictions of (5) for E=17kV/cm and E=24kV/cm respectively. We have to admit that the agreement between the experimental results and the theoretical predictions is rather rough. This discrepancy between experimental results and theoretical predictions may have different origins: the calculation of the driving electric torque could overestimate its value, this is suggested by a previous experiment carried out by the authors (Lobry 1998). Furthermore, the hydrodynamic interactions between particles at this fairly large concentration (ϕ=22%) could modify the spin direction and the rotation rate of individual particles so that the average spin rate of the whole suspension is reduced.

6. REFERENCES

Brenner.J (1970) Fluid Mech 2: 137-176

Cebers.A (1980) Magn. Gidrodinamika.2:.81

T.C.Hasley.TC (1992) Science .258: 761-766

Lobry L, Lemaire E (1998) Proceedings of the International Workshop on Electrical Conduction, Convection and Breakdown in Fluids, Sevilla March 27 and 28th, 1998

Melcher JR (1974) IEEE Trans. Educ E-17: 100-110

Quincke.G (1896) Ann. Pys. Chem 59: 417-486

Lobry L, Lemaire E (1998) submitted to J Electrostatics

VISCOSITY OF AQUEOUS SUSPENSIONS OF BINARY AND TERNARY ALUMINA MIXTURES

R. LAPASIN, S. PRICL, M. STOLFA

Department of Chemical, Environmental and Raw Materials Engineering
University of Trieste, piazzale Europa 1, I 34127 Trieste, Italy

1. INTRODUCTION

At high solids concentration, the viscosity of suspensions is mainly governed by the proximity to maximum packing condition. Accordingly, a traditional way of lowering viscosity is to increase the maximum packing fraction by optimizing the particle size distribution, on condition that a good state of dispersion is achieved. So doing, high solids loadings can be achieved, being the suspension still processable.

Much attention was paid to binary model mixtures obtained by mixing monodisperse spherical particles of two different sizes, generally lying above the colloidal range, whereas very few experimental works have been performed on real systems, which are often characterized by wide particle size distributions and appreciable colloidal fractions.

In the present work we report the results of an experimental investigation undertaken to study the rheological properties of aqueous binary and ternary alumina suspensions, having the above mentioned particle characteristics. Concentrated aqueous alumina suspensions are very important materials in advanced ceramic processing such as tape casting, slip casting, etc. The preparation of aqueous alumina suspensions for tape casting requires not only a proper choice of dispersant type and concentration to ensure good dispersion, but also an appropriate binder in order to avoid significant interactions between the dispersant and the binder. So doing, the rheological properties of the final product are essentially governed by those of the alumina suspension without latex binder.

2. EXPERIMENTAL

The alumina powders (A16SG, CT1200 and CL2500, hereafter designated with A, C and D, respectively, and supplied by Alcoa, Germany) are characterized by different mean particle size (d_{50}: (A) 0.9 μm , (C) 1.4 μm, (D) 3.9 μm) and different specific surface area ((A) 8.23 m^2/cm^3, (C) 5.74 m^2/cm^3, (D) 2.65 m^2/cm^3). The dispersing agent Duramax D3021 (Rohm and Haas, USA) is a polyacrylic acid modified with substituted alkyl groups (MW 5000).

Binary and ternary systems were obtained by blending in different ratios the three reference suspensions prepared separately by dispersing only one powder in distilled water at volume fraction Φ = 0.5 and at a preset concentration of dispersant by ball milling in 0.5 litre polypropylene jars filled with 500 g of Si_3N_4 milling balls for 16 hours at a speed of 70 rpm.

Experimental tests were performed at 25± 0.1 °C, using a rotational rheometer Haake RV20 /RC20 with a measuring head CV100 equipped with coaxial cylinder sensor systems ZB15. The shear- and time-dependent properties were tested by applying shear rate ramps and stepwise procedures (constant shear rate sequences).

3. RESULTS AND DISCUSSION

The rheological behavior of the reference alumina suspensions (A, C and D) were examined at different dispersant concentrations in order to find its optimum concentration for each system. Such values can be linearly related to the specific surface area. All the systems examined exhibit plastic behavior, which can be suitably described by the generalized Casson model:

$$\tau^n = \tau_o^n + \eta_\infty^n \dot{\gamma}^n \qquad (1)$$

with relatively low apparent yield stress values. Appreciable time-dependent effects can be noticed only at dispersant concentrations below the optimum value.

Such shear- and time-dependent properties are observed also for binary systems prepared at constant and optimum dispersant concentration as well. Figure 1 reports a comparison between the experimental data of the A-D binary systems at constant dispersant concentration (0.2 wt%) and the flow curves calculated from the generalized Casson model.

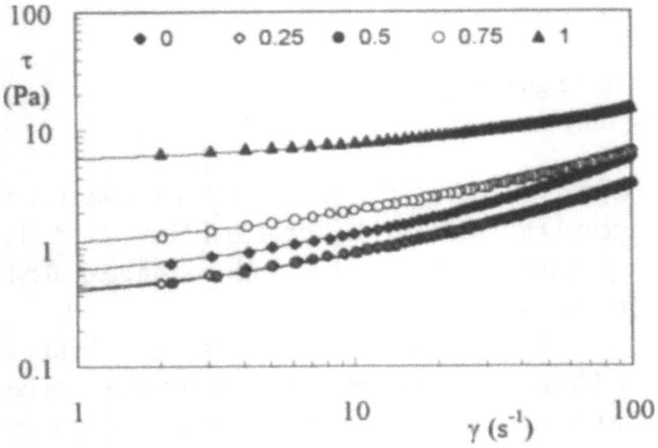

Fig. 1 – Flow curves of AD binary systems (at constant dispersant concentration and different χ_A (volume fraction of alumina A in the disperse phase)).

All the flow curves slightly diverge with decreasing shear rate, tending to an apparent yield stress, which approaches its minimum value for 1:1 mixture. The effects produced by changes in the size distribution (by powder mixing) are better examined in Fig. 2, showing the dependence of viscosity on the disperse phase composition at different shear rates.

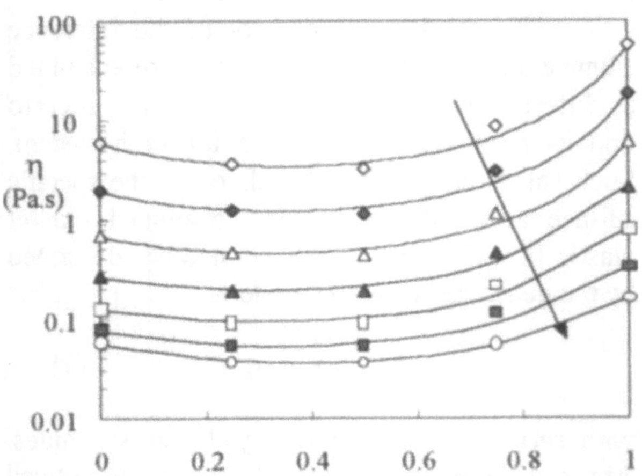

Fig. 2 – Viscosity vs disperse phase composition χ_A for AD binary systems at constant dispersant content (shear rates increasing in the arrow direction).

The minimum viscosity values are attained for χ_A ranging between 0.3 and 0.4. When the systems are prepared at optimum dispersant concentration, the beneficial effects produced by powder mixing on suspension fluidity can be detected only at sufficiently high shear rates, where a minimum appears (see Fig. 3).

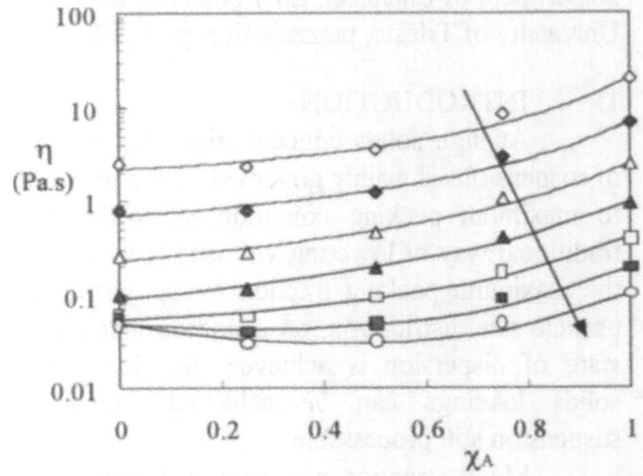

Fig. 3 – Viscosity vs disperse phase composition χ_A for AD binary systems at optimum dispersant content (shear rates increasing in the arrow direction).

Similar effects are observed for the other binary systems. Among the ternary systems having the same constant dispersant concentration, the lowest viscosity values were attained for χ_A = 0.166 and $\chi_C = \chi_D = 0.416$ in the whole shear rate explored, and, correspondingly, the minimum values are obtained for the Casson parameters.

Experimental viscosity data can be suitably described by the rheological model suggested for describing the dependence of the viscosity of multicomponent suspensions on the total volume fraction Φ, the composition of the disperse phase, expressed through the volume fractions χ_i and on the shear rate (Lapasin et al. (1998)).

The model parameters derived from fitting binary data can be used as the reference basis to predict the flow curves of the ternary systems with satisfactory approximation, by applying the same model.

4. REFERENCES

Lapasin R, Pricl S, Stolfa M (1998), Proc. Joint Conference of Italian, Austrian, Slovenian Rheologists, Trieste, in press.

VISCOELASTIC PROPERTIES OF WELAN SYSTEMS

R. LAPASIN, A. MARTORANA, M. GRASSI, S. PRICL

Department of Chemical, Environmental and Raw Materials Engineering
University of Trieste, piazzale Europa 1, I 34127 Trieste, Italy

1. INTRODUCTION

Polysaccharide systems may present a wide variety of structural conditions as a consequence of different chain conformations, inter- and intra-molecular interactions. Even moderate variations in temperature and solvent conditions may induce substantial changes in chain conformation and lead to polymer chain association through the formation of more or less extended junction zones. When a critical polymer concentration is exceeded, gelation can occur and very pronounced changes can be observed in the viscoelastic properties of the system. Such properties can be profitably exploited for many industrial applications, in particular in the food, pharmaceutical and ceramic fields.

The present paper reports the results of experimental studies carried out on welan systems under small and large deformation conditions. Welan gum is produced by *Alcaligenes spp.* ATCC 31555 and has the same basic backbone repeating units as in gellan gum, another microbial polysaccharide of widespread industrial use, but with a single α-L-rhamnose or α-L-mannose as side groups in the ratio of 2:1 per repeating unit. Even if the primary structure of these two bacterial polysaccharides differ, basically, only for the side chains, significant differences can be noted in their viscoelastic properties in aqueous media. Thus, whereas gellan can form stable gels in salt aqueous solutions, welan systems often behave as weak gels, so standing in between solutions and chemical gels, and their thermal stability represent an important characteristic for oilfield applications.

In the present work the dependence of the viscoelastic properties of welan systems on polymer concentration, temperature and solvent phase composition is examined and discussed in the light of fractional constitutive equations and in view of practical applications.

2. EXPERIMENTAL

Welan K1A96 was kindly supplied by Monsanto. The powdered polymer was dissolved at different concentrations C in distilled water and ethylene glycol by mechanical stirring at ambient temperature. All experiments were carried out using a Haake Rheo-Stress RS150 rheometer, equipped with cone-and-plate C60°/1 or plate-and-plate PP60 devices. The rheological properties were studied under small and large deformations, as well as in flow conditions, by applying different procedures: stress and frequency sweeps, creep tests, stress ramps and multistep sequences.

3. RESULTS AND DISCUSSION

The results obtained from stress sweeps show that the critical strain for linear viscoelastic regime decreases with increasing concentration. Figure 1 illustrates such peculiar features for some aqueous systems. Above a critical concentration, the transition from linear to strongly nonlinear properties is marked by an increase in the viscous component, that become more and more evident as C increases.

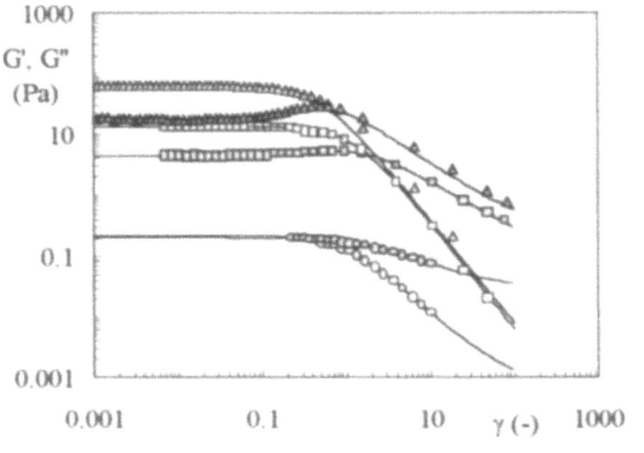

Fig. 1 – Strain dependence of G' (open symbols) and G'' (filled symbols) for three aqueous systems (0.1% (circles), 0.5 % (squares), 1% (triangles)) at 6.28 s^{-1}.

The curves reported in Fig. 1 are calculated from

the following dumping relation:

$$G = G_\infty + (G_0 - G_\infty)\frac{1+a\gamma}{1+b\gamma^n} \qquad (1)$$

with different parameters (a, b, n, G_0 and G_∞) for G' and G". Similar profiles and critical strains are observed at different frequencies for all aqueous systems. Critical strain values slightly depend on frequency in all the cases examined. In the case of ethylene glycol systems, no appreciable increase is noticed in G"- γ profiles with increasing γ and the critical strains are sensibly lower (below 0.01).

The mechanical spectra of aqueous welan systems at 25°C are reported in Fig. 2. When the polymer concentration exceed 0.1%, a net change is observed in the profiles of both moduli, so suggesting the appearance of a sol/gel transition.

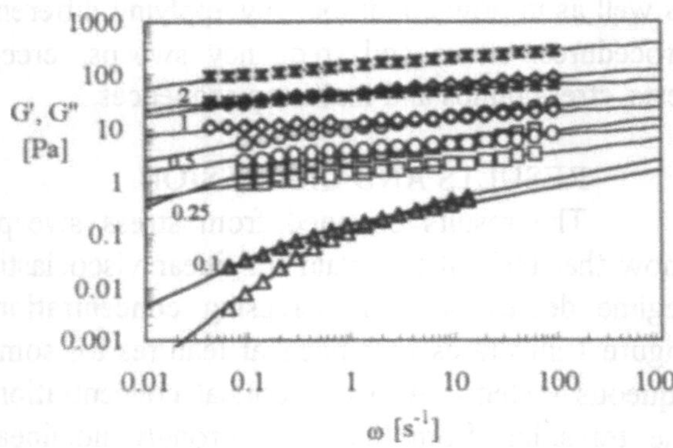

Fig. 2 – Frequency dependence of G' (open symbols) and G'' (filled symbols) for aqueous systems at 25°C, and curves calculated from the Friedrich model.

Above 0.1 %, G' and G'' scale as $C^{1.8\text{-}2.1}$ and $C^{1.5\text{-}1.75}$, respectively (both exponents increase with decreasing frequency). A C^2 dependence of the storage modulus is often observed for gelling systems, in accordance with theoretical models proposed for biopolymers with high functionality.

A satisfactory data fitting of mechanical spectra is provided by the fractional differential Maxwell model (Friedrich, (1991)):

$$\bar{\bar{\tau}} + \lambda^c D^c[\bar{\bar{\tau}}] = \Delta G \lambda^d D^d[\bar{\bar{\gamma}}] \qquad (2)$$

with c < d. The sol/gel transition is well described by the opposite trends of ΔG and the derivation orders, and the sharp increase in the relaxation time λ with increasing C (see Fig. 3).

No appreciable changes are noticed in the temperature interval explored (20 – 40°C) and for pH variations, with the only exception of strongly acid conditions. At pH = 1.36 the decrease in both moduli becomes appreciable (≅ 50% for G', ≅ 40% for G'') with slight changes of their profiles.

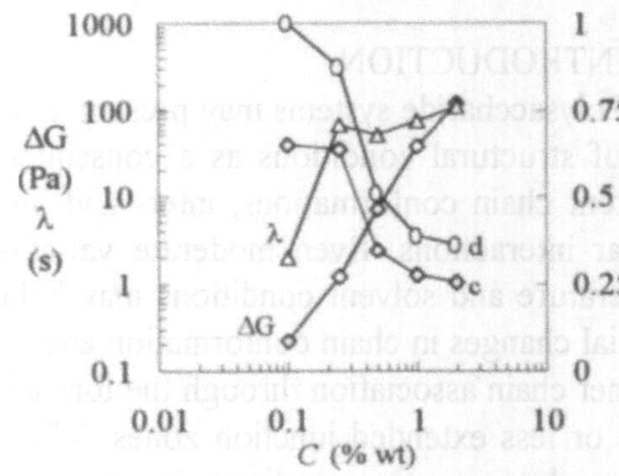

Fig. 3 – Parameters of the Friedrich model vs polymer concentration for aqueous systems at 25°C.

The ethylene glycol systems exhibit gel-like properties in the same concentration range (0.1-2 %) of aqueous systems, with less pronounced C dependences of both moduli and no evident transition. Even if the mechanical spectra of the two series tend to converge with increasing concentration, differences appear in the low frequency range and become more significant at large deformations. Moreover, the behavior of ethylene glycol systems is strongly affected by the shear history and all such features could be reasonably ascribed to their thermodynamical and structural state. As easily assessed by visual inspection, they are concentrated dispersions of swelled polymer particles, being the welan sample used in the present work scarcely soluble in ethylene glycol, and the swelling state depends on the maximum stresses previously applied. Striking differences in solubility of different welan samples in organic solvents and, hence, in their rheological properties were previously observed and discussed (Hember and Morris (1995)).

4. ACKNOWLEDGEMENTS

The authors would like to acknowledge MURST for its financial support.

5. REFERENCES

Friedrich Chr. (1991) Rheol Acta 30:151-158

Hember MWN, Morris ER (1995) Carbohydr Polym 27:23-36

RHEO-OPTICAL STUDY OF ASSOCIATIVE POLYMER SOLUTIONS

J.F LE MEINS , J.F TASSIN

Chimie et Physique des Matériaux Polymères , UMR 6515 Université du Maine,
Avenue Olivier Messiaen 72085 LE MANS.Cedex 09 FRANCE

1. INTRODUCTION

Associative polymers are materials which have found industrial application as thickeners and which raise stimulating fondamental problems. In order to get more insight in this field we have studied by rheo-optical methods two types of associative polymer: one Hydroxy Ethoxylated Urethane, (HEUR), wich is end-cap modified and one hydroxy Propyl Guar, (HPG), randomly modified along the chain backbone . These two polymers have been the subject of classical rheological studies (Annable et al. 1993), (Aubry and Moan. 1994).

2. MATERIALS AND METHODS.

2.1. Samples preparation.

We used a HEUR associative thickener with a molecular weight of 35000g/mol and end capped with n-hexadecanol. We will call it « POE35C16 ». The polymer with randomly distributed hydrophobic substituents along the chain was an HPG with a molecular weight of $1{,}8.10^6$g/mol. The hydrophobic substituents are a mixture of C22-C28 n-alkyl epoxides. Samples were dissolved in deionized water and were studied at the vicinity of C*.

2.2 Rheo-optical experiments.

These experiments were performed with a Rheometrics Optical Analyser. Using this equipment which involves rapid modulation of the polarization, birefringence and orientation angle can be simultaneously measured. The presents measurements were carried out using a concentric cylinder geometry, where light travels along the vorticity axis.

2.3 Experimental procedures.

To characterize our solutions we will focus on two types of experiments:
i)Relaxation after step shear.
ii)Steady state birefringence measurements.

3. RESULTS.

3.1. Step shear relaxation.

Figure 1 shows the relaxation of birefringence for the POE35C16. Two processes well separated in time are apparent. The short time correponds to lifetime of hydrophobic end cap, the long time has been attributed to relaxation of POE aggregates .

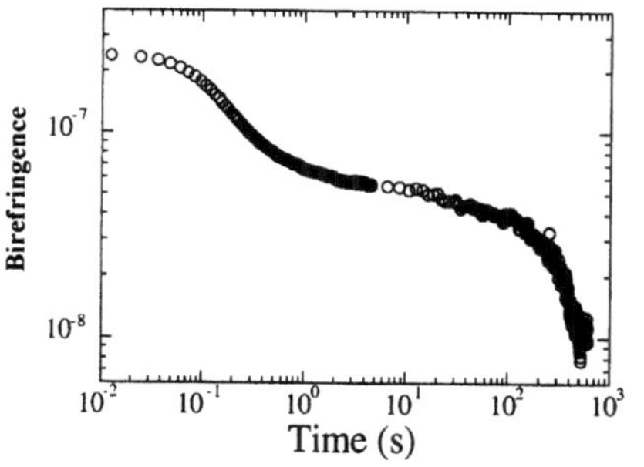

Figure1 Relaxation of birefringence after step shear for a 1% solution of Heur (POE35C16) after step shear of 50%

On the other hand the randomly side-chain modified polymer (HPG) shows a power law relaxation of birefringence (Figure 2) which indicates a large distribution of relaxation time. This evolution is not influenced by the temperature.

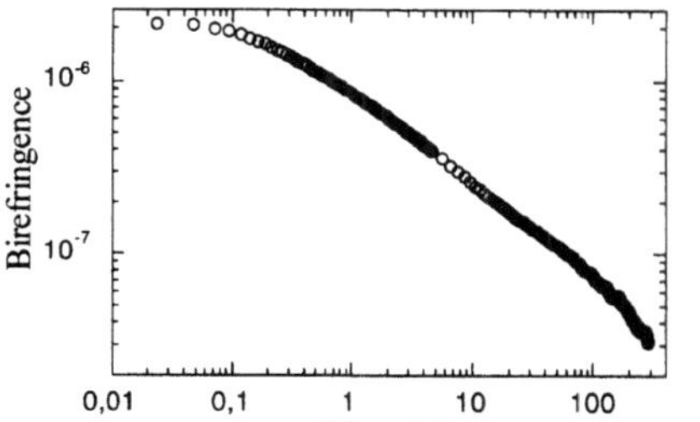

Figure2 Relaxation of birefringence after step shear of 400% for a 1,38% solution of HPG.

3.2 Steady state results.

Figure 3 shows the evolution of both birefringence and angle with shear rate. The behavior of the POE35C16 is quite different from what is usually observed for a polymer solution under flow. Indeed birefringence increases with shear rate has expected whereas the orientation angle also increases.

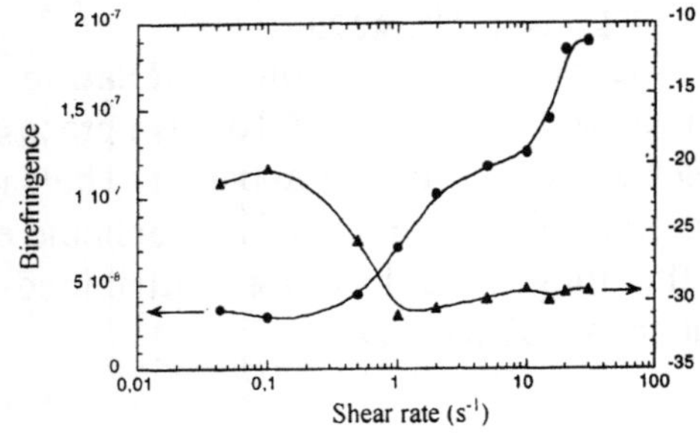

Figure3 Birefringence and orientation angle at steady state for POE35C16 at 2%.

In fact, birefringence at low shear rates is dominated by the response of POE aggregates, which explains the low value of the orientation angle at low shear rate. At high shear rates birefringence is controlled by the stretched chain segments.

Figure 4 shows the results obtained for the system randomly modified along the chain-backbone (HPG). A classical behavior of polymer solution with an increase of birefringence and a decrease of orientation angle is observed.

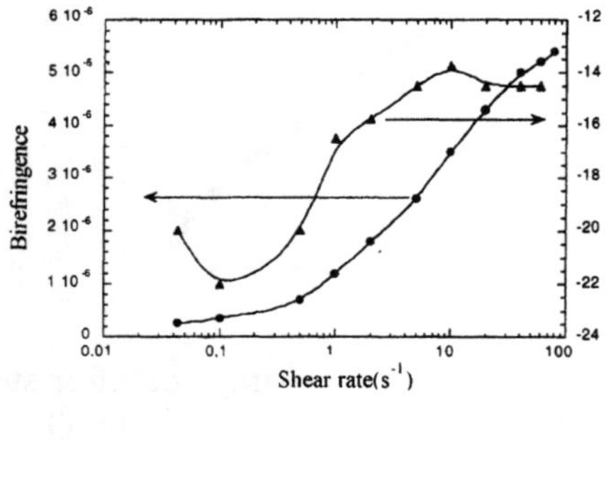

Figure 4 Birefringence and orientation angle at steady state for HPG at 1,38%.

4. CONCLUSION

The associative polymer with end-cap hydrophobic group (POE35C16) is a system in which the rheological properties are controlled by lifetime of hydrophobic end cap in micelles more than the concentration. The long time relaxation corresponds to relaxation of POE aggregates which cause the unusual behavior oberved under steady state. On the other hand the associative polymer randomly modified along the chain backbone (HPG) shows a very large distribution of relaxation time. At the opposite of the POE 35C16 this system shows a classical behavior of a concentrated polymer solution with a power law relaxation wich is reminiscent of the relaxation of highly branched structures.

5 REFERENCES

Annable T., Buscall R., Ettelaie R, Whittlestone D. (1993) J.Rheol. 37(4): 695-725

Aubry T, Moan M,(1994) J.Rheol. 38(6):1681-1691

MEASUREMENTS AND MODELING ON A COLLOIDAL SUSPENSION OF POLYMERICALLY STABILIZED SPHERES

P.A. NOMMENSEN, M.H.G. DUITS, D. VAN DEN ENDE, J. MELLEMA

Rheology Group (Twente Institute of Mechanics, JM Burgerscenter), Dept. of Applied Physics, University of Twente, P.O. box 217, 7500 AE, Enschede

1. INTRODUCTION

To understand the rheology of suspensions of sterically stabilized spheres it is in many cases possible to describe the particles as (effective) hard spheres. However when the stabilizing layer consists of polymer chains, such an approach must at some point break down. Our aim is to understand the influence of the properties of the polymer layer on the rheological behavior.

To this end we measured the high shear viscosity η_∞ and the elastic modulus G'. Theoretically we considered models focused on the interpenetration of the solvent in the polymer layer and on the deformation of the polymer layer.

2. PARTICLE SUSPENSION

The suspension used consisted of silica cores coated with endgrafted polydimethylsiloxane (PDMS) chains, suspended in heptane. The molecular weight of the PDMS-molecules M_n was $8{\cdot}10^4$ with a polydispersity M_w/M_n of 1.4. The synthesis and the characterization were reported by Nommensen et al. (1998).

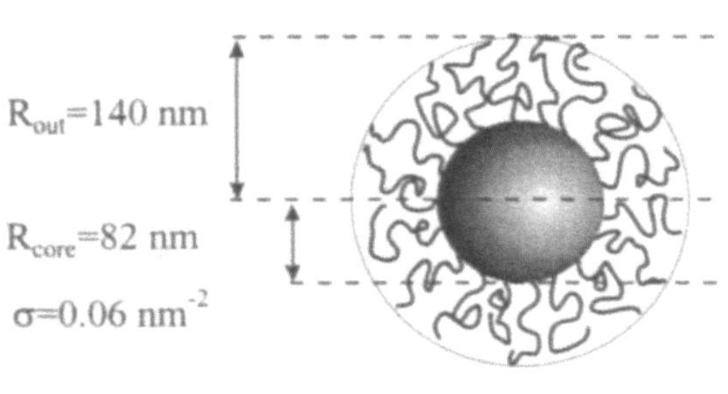

fig 1 Dimensions of the particles.

The radii R_{out} and R_{core} as measured with light scattering, indicate a layer thickness L of 58 nm. The grafting density σ was obtained with NMR. From σ the lateral distances between polymer molecules were obtained. They are much smaller than the layer thickness indicating a brushlike structure of the polymer layer.

3. HIGH SHEAR VISCOSITY η_∞

The polymer layer influences the hydrodynamic interaction between the particles in a specific way since the solvent can penetrate the layer. This effect can be seen clearly in the magnitude of η_∞ at high volume fractions.

Here the dissipation due to lubrication-forces is the dominant contribution to η_∞. When two particles are pushed together, the solvent in between the particles is squeezed out. The velocity profile of the squeezing flow determines the strength of the lubrication force, needed to push the particles together.

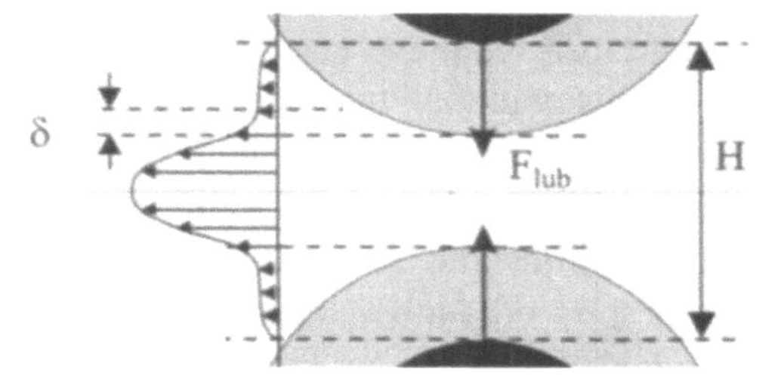

fig 2 The velocity profile of the sqeezing flow when 2 particles are pushed together.

By modeling the polymer layer (with thickness L) as a permeable medium with permeability δ^2, the velocity profile can be calculated from which η_{lub} can subsequently be obtained (Potanin and Russel (1995)).

limiting cases	qualitative behavior
$\delta > L$	polymers do not hinder the flow ⇒ "hard sphere" with radius R_{core}
$\delta < L$, H»2L	penetration of the flow can be neglected ⇒ "hard sphere" with radius R_{out}
$\delta < L$, $H \leq 2L$	flow penetrates the entire layer

By adding the solvent viscosity and the Einstein term to η_{lub}, we obtained a description for η_∞ which satisfies the expected behavior at low and high concentrations and gives a good interpolation at moderate concentrations. This can be seen by comparing the calculations to the experimental results for both hard sphere and polymerically stabilized suspensions.

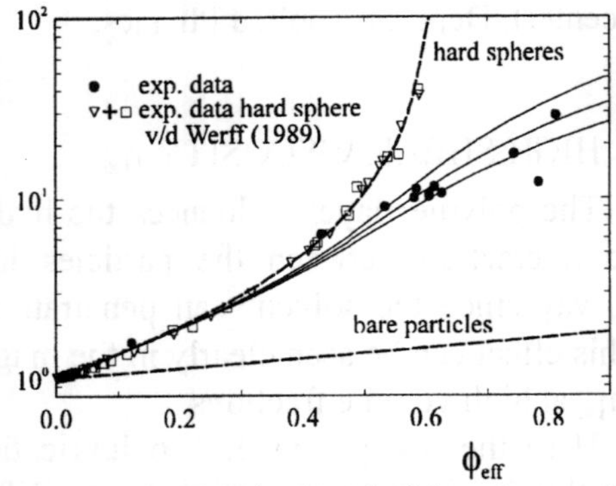

fig 3 Experimental (symbols) plus calculated (lines) high shear viscosity vs. effective volume fraction.

4. ELASTIC MODULUS G'

When two particles are pushed together such that H<2L their polymer layers deform which results in a repulsive elastic potential. The reflection of this potential into the volume fraction dependence of G' was modeled by Zwanzig and Mountain in 1965. Recently this model for G' was improved by Elliott and Russel (1998). The following table shows the most important differences of both models.

	Zwanzig & Mountain	Elliott & Russel
g(r): pair correlation function	assumed to be sharply peaked	expansion around equilibrium configurations, obtained from Monte Carlo simulations
interactions	interparticle potential	interparticle potential and Brownian motion
hydrodynamics	none	included using η_∞

The interparticle potential can be obtained by calculating the increase of the free energy of the polymer layers when 2 particle are pushed together. We obtained the free energy of the polymer layer following the variational approach of Li and Witten (1994) for starpolymers. Important parameters are R_{core}, M_n, σ and the solvent quality. The magnitude of G' was found to depend strongly on the pair correlation function g(r). Obviously g(r) in its turn is influenced by the interparticle potential. Monte Carlo simulations were used to obtain g(r).

When the results of the models are compared to the experimental data, it is seen that both models predict too high values but the volume fraction dependence is better described with the model of Elliott and Russel.

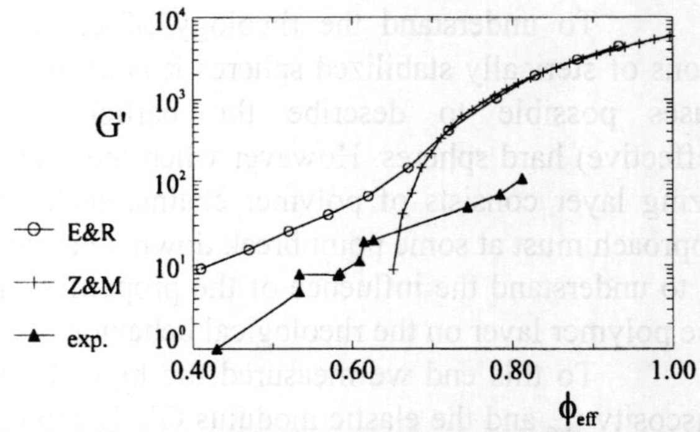

fig 4 Experimental and calculated G' as function of the effective volume fraction

5. CONCLUSIONS

The properties of the polymer layer in polymerically stabilized suspensions were used to describe the rheological behavior :

- The high shear viscosity can be described well by modeling the polymer layer as a permeable medium. The fitted value for δ is reasonably close to the expected value of $\sqrt{1/\sigma} = 4$ nm.
- Using the model of Elliott and Russel the volume fraction dependence of G' is described qualitatively well but the results of the calculation are one order of magnitude too large. Improvement of the Zwanzig & Mountain model is achieved at intermediate concentrations.

6. REFERENCES

Elliott SL, Russel WB (1998) J Rheology 42:361
Li H, Witten TA (1994) Macromolecules 27:449
Nommensen PA, Duits MHG, Lopulissa JS, Van den Ende D, Mellema J (1998) Prog Col & Polym Sci 110:00
Potanin AA, Russel WB (1995) Phys Rev E 52:750
Van der Werff JC, De Kruif CG (1989) J Rheology 33:421
Zwanzig R, Mountain RD (1965) J Chem Phys. 43:4464

PARTICLE-WALL INTERACTION IN LIQUIDS WITH DIFFERENT RHEOLOGY

A.N.PROKUNIN

Institute of Mechanics, Lomonosov Moscow State University, Michurinskii pr. 1,
119899 Moscow, Russia

In this paper, the study is performed for a Newtonian liquid with subsequent generalization for an power-law liquid and an viscoelastic liquid with low elasticity.

We study the particle-wall interaction by the example of a spherical solid particle of diameter d, which under action of its weight Psin(α) (α is the inclination angle measured from the horizontal) rolls down along an inclined plane or a pipe (d/D<<1, D is the pipe diameter) in a liquid at rest with a constant translational velocity u and rotates with a constant angular velocity Ω. For a Newtonian liquid, the torque and the force exerted on the spherical particle are determined by the parameters:

$$Re_p = \frac{ud\rho_f}{\eta},\ \omega = \frac{d\Omega}{2u},\ \varepsilon = \frac{2h}{d} \qquad (1)$$

Here, h is the clearance between the particle and the wall, ρ_f is the liquid density, and η is the viscosity.

When Re_p<<1, in the inertialess Stokes approximation the nondimensional force $B = \{B_\tau, B_N\}$ and torque B_M exerted on the particle for $\varepsilon \to 0$ are as follows (Goldman, et al. (1967), O'Neil and Stewartson (1967)):

$$B_\tau \approx -3{,}2\ln\varepsilon(4-\omega),\ B_M \approx -3{,}2\ln\varepsilon(4\omega-1) \qquad (2)$$

$B_N \equiv 0$, the subscripts τ and N correspond to the directions along and in normal to the wall. For the particle motion without rotation, far from the wall: $B_\tau = 24$.

A paradox follows from (2) "Goldman, et al. (1967)": for $\varepsilon \to 0$ the force and the torque tend to infinity and if h = 0 the particle motion down the wall is impossible. At the same time, for h = 0 the rolling down is also impossible because $B_N \equiv 0$. However, experiments show that such motion is realized.

To eliminate this paradox, Smart et al. (1993) assumed that the clearance h appears during particle contact with the wall due to the roughness of the contacting surfaces (h_p = h). In this case, the motion is affected by the friction force (for ω<0.98, it is the sliding friction "Johnson (1985)".

In our experiments (Re_p<<1) devoted to the study of the above-mentioned paradox, the working liquids were glycerin and silicone. Steel particles with the diameter d $\approx$ 0.4 cm were used, and the roughness was varied on the range 0.02-4 mkm. The roughness of the glass pipe was 0.02 mkm. The temperature variation (for glycerin, from 9 up to 30° C) demonstrated that the results (B_τ+B_M and $\omega(\alpha)$) are independent of the temperature, which is valid for inertialess motion of the liquid.

For h_p < 2 mkm, the experimental results for the same particles and different liquids with the same viscosity differ significantly. This fact cannot be explained by a 25% difference in the liquid densities since the flow is inertialess. This also cannot be explained by the friction between the particle and the wall since the friction depends only slightly on the liquid kind and, for h_p > 1 mkm, it coincided with the dry friction. Thus, for describing the particle motion along the wall only the mechanical parameters (1) (the parameters characterizing the surface roughness in a more detail can also be added) are not enough. We note also that for h_p = 4 mkm the experimental results for cilicon and glycerin were almost identical, i. e. for fairly large hp the list of the parameters (1) does not require the extension.

It follows from above that the theoretical description of "Smart, et al. (1993)" which used (2) with ε determined by the surface roughness is valid only for fairly large roughnesses. To attain agreement between the theory and the experiment for fairly small h, the scheme of "Smart, et al. (1993)" can be complemented by a certain lift force exerted on the particle and increasing with h. According to this scheme, for large angles (α) the particle travels without contact with the wall and, as alfa the clearance becomes equal to the roughness decreases, the particle "touches" the surface when height ($h = h_p$).

The lift force nature is not clear. The hypothesis of the possible manifestation of the liquid viscoelasticity in a thin clearance between the particle and the wall, formulated in "Goldman, et al (1967)", was not confirmed. Adding the polyacrilamid to the glycerin did not change the experimental results, though the relaxation time θ was as long as 0.1 sec. The lift force can appear because of the disjoining pressure "Deryagin (1986)" for maintaining the thermodynamic equilibrium in a thin liquid layer subjected to the simultaneous action of the particle and the wall. However, the thickness of such the layer is usually of the order of 0.01 mkm, so this effect cannot explain the appearance of the lift force for large clearances.

In the case of larqe Reynolds numbers ($Re_p = 10^2\text{-}10^5$), the contact motion of the spherical particle is accompanied by a noise "Prokunin and Williams (1996)". As follows from our experiments, the particle losses the contact with the wall when $\alpha \approx 74°$. For the contact motion, the generalized drag coefficient of the particle $C_p = (B_\tau + B_M)/Re_p$ is a function of Re_p and ω, while for $Re_p > 10^3$ it depends on Re_p only. The latter is attributed to a small particle slip ($\omega \approx 1$) relative to the wall. We note that the coefficient Cp is independent of e. This may be connected, for example, with the periodicity of the particle-wall contact, when the averaged clearance h is greater than the roughness height. The transversal force here is, likely, of the hydrodynamic nature. For $\alpha > 74°$, the particle loses contact with the wall and, with further increase in the angle α, the clearance more and more increases ($\omega \to 0$), and the drag coefficient C_p decreases and tends to the value measured far from the wall.

In the case of an power-law liquid (the aqueous solutions of the "milling yellow" Swanson W.M., Green R.L. (1969), a generalized Reynolds number $Re_p = (u^{2-n} d^n \rho_f)/k$ and the additional nondimensional parameter n appear in the system of governing parameters (1). The constants n and k are determined by the power-law $\sigma = k\gamma^n$. We note that, for small n, for the particles traveling in contact with the wall the drag coefficient C_p is a function of only Re_p on the whole range of the parameters considered ($n = 0.28$, $Re_p = 10\text{-}10^3$). In this experiment we did not measure ω. The angle of the particle detachment (deter-mined when the noise disappeared) is almost equal to that for the Newtonian liquid ($\alpha_m \approx 72\text{–}74°$).

REFERENCES

Deryagin VV (1986) Theory of stability of colloids and thin films, Moscow.

Goldman AI, Cox RG, Brenner H (1967) Chem Eng Sci 22: 637-651

Jonhson KL (1985) Contact mechanics, Cambridge

O'Neil ME, Stewartson K (1967) J Fluid Mech 27: 705-724

Prokunin AN, Wlliams MC (1996) Fluid Dynam 31: 567-572

Smart JR, Beimfohr S, Leighton OT (1993) Phys Fluids 5: 13-24

Swanson WM, Green R.L. (1969) J Colloid Sci 29: 161-164

RHEOLOGICAL STUDY OF THE PROCESS OF OIL/WATER EMULSIFICATION

S. RENOUF, H.Z. LI, L. CHOPLIN

Centre de Génie Chimique des Milieux Rhéologiquement Complexes (GEMICO), ENSIC-INPL
1 rue Grandville, 54001 Nancy Cedex, France

1. INTRODUCTION

An emulsion is generally a two-phase system composed of two immiscible liquids. One of the two liquids is dispersed in the other as droplets. The emulsification process under consideration is based on droplet break-up under intense stretching action in a high stress device, under dynamic conditions and short time scales. In batch process, macromixing performance is crucial for ensuring equilibrium droplet size distribution is reached. In situ rheological monitoring can be an interesting tool for studying the whole emulsification process, together with droplet size measurements and optical microscopy as well.

2. MATERIALS AND METHODS

The emulsification process is realised in a mixer-type geometry directly installed on a RFSII rheometer (Rheometric Scientific). The high stress device is an Ultra-Turrax disperser and the macromixing device is a helical-ribbon impeller, which allows to determine the torque evolution during the process (Figure 1). All experiments are carried out at 20°C.

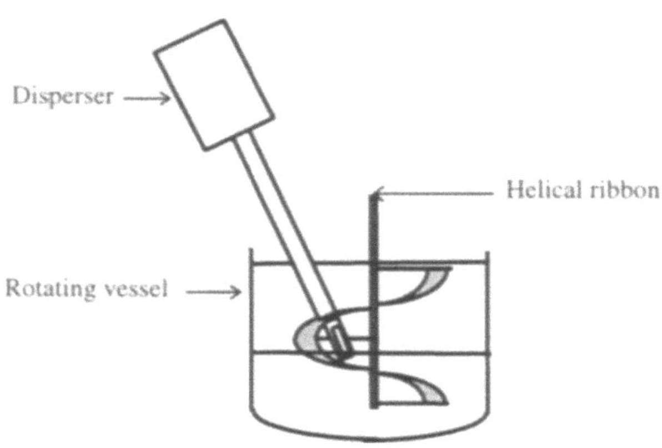

Figure 1 : Experimental set-up

Two systems were studied : in system 1, the oil phase is Cetiol SN (Sidobre Sinnova), in which the emulsifier Emulgade PL6850 (Sidobre Sinnova) is dissolved. In system 2, the oil phase is Carnation oil (Witco) and the emulsifier Tween 80 is dissolved in the aqueous phase. Both oils, without emulsifier have the same viscosity (20 mPa.s) at 20°C.

The two phases are first mixed with the macromixing device (N=50 rpm). Then the disperser is started and the torque response is monitored until an equilibrium value is reached. Sampling procedure for optical microscopy and droplet size distribution determination (Malvern Multisizer) is carried out throughout the process. At the end of the process, in situ rheological characterisation (steady and dynamic) is performed directly in the mixer-type geometry (Aït-Kadi and al., 1997).

3. EXPERIMENTAL RESULTS

Figure 2 shows typical torque evolution during the whole emulsification process.

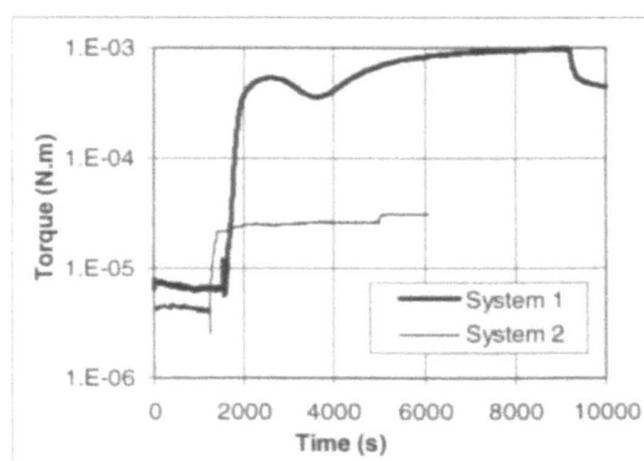

Figure 2 : Evolution of torque during emulsification

Whereas for system 2, a monotonic but rapid increase in torque is observed, the torque evolution in the case of system 1 is significantly different. It presents a minimum at a dimensionless critical time τ_c defined as :

$$\tau_c = \left(t_{\text{minimum}} - t_{\text{disperser starting}}\right) / \left(t_{\text{equilibrium}} - t_{\text{disperser starting}}\right)$$

However, this minimum does not appear for all oil concentrations.

In the case of system 1, droplet size distribution remains unimodal and the Sauter diameter has a value of 20 µm. As the torque goes through the minimum, a bimodal distribution appears. The Sauter diameter remains of about 20 µm.

Figure 3 is a typical photography of a 50µm-size aggregate in system 1 emulsion at the end of the emulsification process. Similar observations (Suzuki and al., 1984) have been reported for equivalent systems containing cetostearyl alcohol (this is the case of Emulgade PL6850) which are known to form complex structures (Junginger and al., 1984). In contrast, system 2 does not exhibit any complex structure. The Sauter diameter is of the order of 2 µm (Figure 4) and the droplet size distribution remains unimodal during the whole emulsification process.

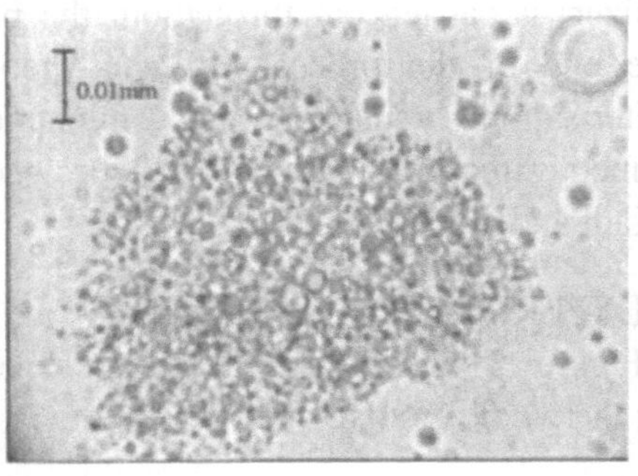

Figure 3 : System 1 at the end of the emulsification

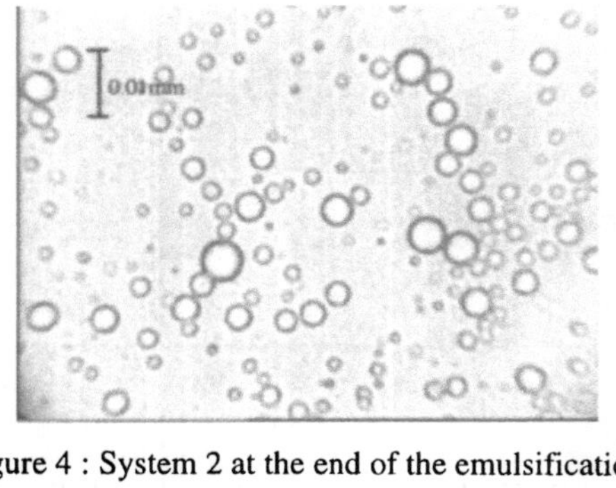

Figure 4 : System 2 at the end of the emulsification

In figure 5, we report the rheograms of several emulsions obtained directly in the mixer-type geometry of the rheometer, i.e. without any sampling procedure able to structure-damage the emulsions.

Also, small amplitude oscillatory strain experiments have been carried out directly in the mixer-type geometry of the rheometer. The torque information has been transformed into classical rheological functions, i.e. elastic (G') and loss (G'') moduli. We are then be able to obtain not only the mechanical spectra of the emulsions without sampling, but also to study their stability. For example, it is reported in Figure 6 the time evolution of the moduli of two particular emulsions. For a system 1 emulsion, significant evolution of the elastic modulus is seen, whereas the loss modulus remains constant over more than 100 hours. On the contrary, for a system 2 emulsion, some increase is observed for both moduli only during the first 20 hours after emulsification process has been stopped.

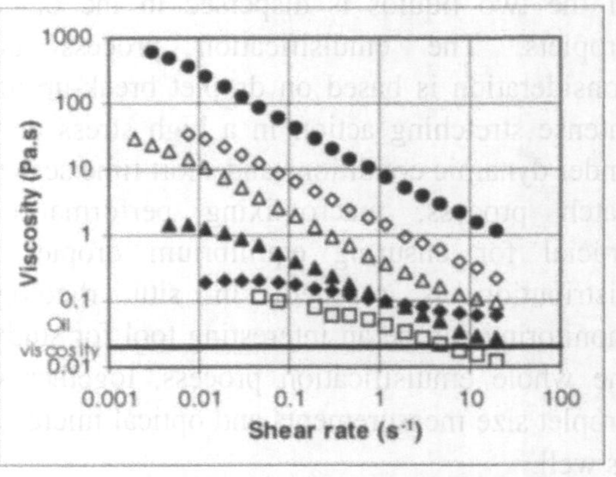

Figure 5: Viscous behaviour of emulsions

System 1 : ◆ Oil 45% - Emulsifier 1% ▲ Oil 15% - Emulsifier 1% ● Oil 45% - Emulsifier 3%

System 2 : □ Oil 20% - Emulsifier 3% △ Oil 45% - Emulsifier 10.75% ◇ Oil 70% - Emulsifier 10.75%

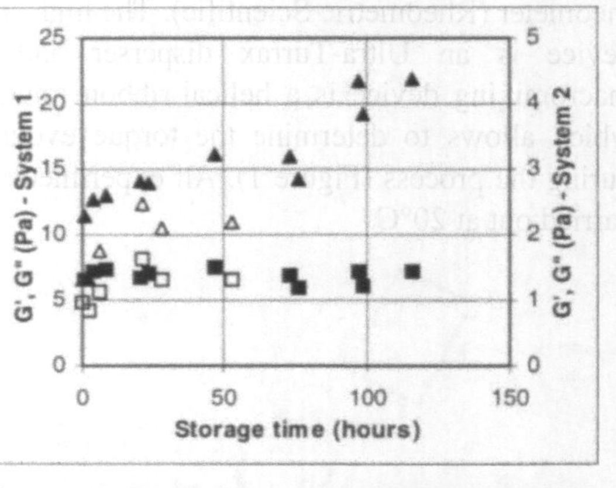

Figure 6: Evolution of G' and G'' after emulsification

System 1 : ▲ Storage modulus ■ Loss modulus

System 2 : △ Storage modulus □ Loss modulus

4. REFERENCES

Aït-Kadi A., Marchal P., Chrissemant A.S., Bousmina M., Choplin L. (1997), ASME, SG Advani and DA Siginer (eds.), FED-vol.243, MD-vol.78 , 159

Suzuki T., Tsutsumi H., Ishida A., (1984), J. Dispersion Science and Technology, 5(2), 119-141.

Junginger H, Akkermans A.A.M.D., Heering W., (1984), J. Soc. Cosmet. Chem., 35, 45-57.

SHEAR INDUCED REORIENTATIONS IN A DEFECTIVE LYOTROPIC LAMELLAR PHASE

W. RICHTERING[1], J. BERGHAUSEN[1], J. ZIPFEL[1,2], P. LINDNER[2]

[1]Institut für Makromolekulare Chemie, Universität Freiburg, Stefan-Meier Str. 31, D - 79104 Freiburg, Germany
[2]Institut Laue-Langevin, B.P. 156, F-38042 Grenoble, France

1. INTRODUCTION

Surfactants are known to form lyotropic liquid crystalline phases in concentrated aqueous solution. Usually samples of macroscopic size are not homogeneously aligned but a polydomain structure is observed. Shear flow has a profound influence on structure and macroscopic alignment of liquid crystals and consequently the rheological properties of such materials are rather complex. Recently, samples with lamellar morphology attracted much attention in order to find unifying principles in different types of materials as e.g. aqueous surfactant solutions and block copolymer melts.

In block copolymer melts different orientations of the lamellae were found as reviewed by Fredrickson and Bates (1996). Often the "parallel" layer orientation, i.e. with layer normal pointing along the velocity gradient direction, is found at low frequencies, and the "perpendicular" orientation, with the layer normal along the vorticity direction, at higher frequency deformation. Cates and Milner (1989) suggested that layer fluctuations close to the order-disorder transition increase the stability of the perpendicular orientation at higher frequencies. Recently, Maring and Wiesner (1997) reported on two transitions from *parallel to perpendicular to parallel orientation* which are not yet theoretically understood.

In a lyotropic surfactant lamellar phase, Diat et al. (1993) detected a parallel alignment of lamellae, whereas a transition from parallel to perpendicular orientation at high shear rates was observed in other systems by Mang et al. (1994), Penfold et al. (1997), Läuger et al. (1996) and Richtering (1997). A characteristic difference between lyotropic and polymer melt systems is the shear induced formation of multilamellar vesicles ("onions") which is frequently found in lyotropic systems, see e.g. Diat et al. (1993).

In the present work, we report on small angle neutron and light scattering (SANS, SALS) as well as birefringence experiments obtained from a *defective* lyotropic lamellar phase containing sodium dodecyl sulfate (SDS), decanol and water (Berger and Hiltrop, 1996). As will be shown below, a reorientation from "perpendicular" to "parallel" lamellae with increasing shear was found. This indicates that the *"perpendicular-to-parallel"* transition might be a general feature of complex fluids with lamellar morphology.

2. METHODS

The defective lamellar phase consisted of sodium dodecyl sulFate/decanol/D_2O with a D_2O mass fraction of 0.679. The decanol content x_c in the surfactant/co-surfactant mixture was $x_c = 0.301$. SANS experiments were performed on the instrument D11 of the Institut Laue-Langevin, Grenoble. The Couette type shear cell consisted of two concentric quartz cylinders with a gap of 1.0 mm. Two beam configurations were used. The radial position, where the beam direction is parallel to the velocity gradient direction and the tangential beam position, where the neutrons pass through the sample along the flow direction. For rheo-optical studies, a stress controlled Bohlin CS-10 rheometer equipped with a quartz glass cone/plate or plate/plate shear geometry was used. The incident beam of a He-Ne laser passes through the sample along the direction of the velocity gradient, see Schmidt et al. (1997) and Berghausen et al. (1997)

3. RESULTS AND DISCUSSION

SANS curves at rest displayed two characteristic features: i) a sharp "Bragg-peak" from the layer spacing at $q_{max} = 0.106$ Å^{-1}, and ii) a broad "defect-peak" corresponding to the distance between defects within the surfactant bilayers with $q_{max} = 0.066$ Å^{-1}.

Fig. 1 shows SANS patterns at three different shear rates for the radial (left) and tangential (right) beam position. Three different states were found. The lamellae were aligned parallel to the walls at low and high shear rates. In the radial beam position, the Bragg-peak was much weaker as compared to the defect-peak and the tangential beam showed a higher Bragg-peak intensity along the velocity gradient direction as compared to the vorticity direction.

At first glance, the behavior of this sample seems to be quite similar to what was observed by Diat et al. (1993). They found the parallel orientation at low and high shear rates as well. At intermediate shear rates they observed multilamellar vesicles ("onions"). However, in our case the behavior at intermediate shear rate is different.

The radial SANS pattern at intermediate shear rates shows a strong Bragg-peak *only* along the vorticity direction but not along the flow direction. If vesicles were present, the Bragg scattering from the layer spacing were observable on the entire azimuthal trace on the two dimensional multidetector as e.g.

described by Weigel et al. (1996). The radial SANS scattering pattern thus indicates the presence of perpendicularly aligned lamellae.

The tangential beam provides more detailed information. The Bragg-peak intensity along the vorticity direction is, by a factor of 2, higher as compared to the velocity direction. Thus the lamellae are predominately oriented perpendicularly at intermediate shear rates. The results from SANS in the tangential beam showed that the orientation at intermediate shear rates was not perfect. Birefringence and SANS experiments at different gap sizes indicate that the parallel orientation is preferred near the walls. With a further increase of shear, a reorientation of lamellae from *perpendicular to parallel orientation* is observed above 1000 s^{-1}.

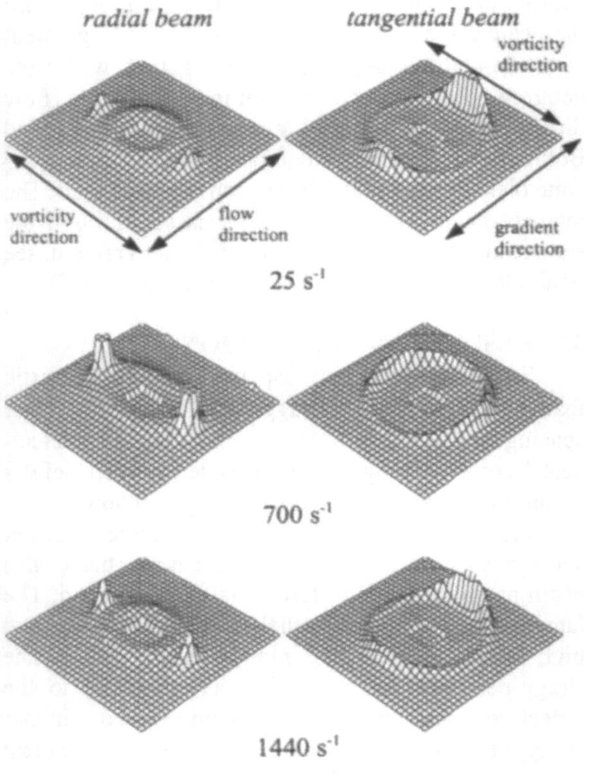

Fig.1: SANS patterns of the defective lamellar phase under shear for different rates of 25 s^{-1}, 700 s^{-1} and 1440 s^{-1} (top to bottom). For each beam direction the intensities are scaled to the same maximum.

The three different regions could be distinguished in flow birefringence as well. The birefringence curve showed low birefringence values at low and high shear rates supporting the homeotropic orientation with surfactant bilayers parallel to the walls of the shear cell. At intermediate shear rates, strong negative values were found, which indicate a perpendicular orientation of the lamellae (Berghausen et al (1998).

With depolarized SALS an intense streak perpendicular to the flow direction was observed at intermediate shear rates. This indicates the presence of an ordered texture with defect lines aligned in the flow direction. Furthermore this scattering pattern provides strong evidence against the presence of multilamellar vesicles. Bergenholtz and Wagner (1996) and Weigel et al. (1996) showed that multilamellar vesicles give rise to a four-lobe pattern in depolarized SALS. Thus SANS and SALS reveal that the intermediate region does not contain vesicles.

4. CONCLUSION

The defective lamellar phase investigated here showed three states of oriented lamellae. With increasing shear rate, *two reorientations from parallel to perpendicular to parallel* alignment were observed. Reducing the shear rate revealed that the transitions are reversible. Similar behavior was found with block copolymer melts where oscillatory shear was applied. Apparently, surfactant mesophases and polymer melts can show similar behavior under shear. This indicates that the *"perpendicular-to-parallel"* transition might be a general feature of complex fluids with lamellar morphology.

5. ACKNOWLEDGMENT

We thank the Deutsche Forschungsgemeinschaft for financial support.

6. REFERENCES

Berghausen J, Fuchs J, Richtering W Macromolecules (1997) 30:7574
Berghausen J, Zipfel J, Lindner P. Richtering W (1998) submitted
Bergenholtz J, Wanger NJ Langmuir (1996) 12:3122
Berger K, Hiltrop K Colloid Poly Sci (1996) 274: 269
Cates ME, Milner ST Phys Rev Lett (1989) 62:1856
Diat O, Roux D, Nallet F (1993) J Phys II 3: 1427
Fredrickson GH, Bates FS Ann Rev Mater Sci (1996) 26:501
Läuger J, Weigel R, Berger K, Hiltrop K, Richtering W J. Colloid Interface Sci (1996) 181:521
Mang JT, Kumar S, Hammouda B Europhys Lett (1994) 28:489
Maring D, Wiesner U Macromolecules (1997) 30:660
Penfold J, Staples E, Khan Lodi A, Tucker I, Tiddy GJT J Phys Chem B (1997) 101:66
Richtering W Prog Colloid Polym Sci (1997) 104:90
Schmidt J, Weigel R, Burchard W, Richtering W Macromol Symp (1997) 120:247
Weigel R, Läuger J, Richtering W, Lindner P J Phys II (1996) 6:529.

INFLUENCE OF THE TYPE OF EMULSIFIER ON THE RHEOKINETICS OF THE EMULSIFICATION PROCESS

M.C.SÁNCHEZ[1], M. BERJANO[1], A.GUERRERO[1], C.GALLEGOS[2]

[1]Departamento de Ingeniería Química. Universidad de Sevilla, 41012 Sevilla
[2] Departamento de Ingeniería Química. Universidad de Huelva,21819 Palos de la Frontera (Huelva)

1. INTRODUCTION

The consecutive disruption of oil droplets and their coalescence are the two critical steps during emulsification, both of which are favoured by an intense agitation. The role of a suitable surfactant is of paramount importance as it first lowers the interfacial tension favouring the disruption of droplets, and then prevents coalescence by increasing the repulsive potential between droplets.

The main objective of this work is to follow the evolution of the droplet size distribution and the rheological properties as a function of the emulsification time of o/w emulsions using different nonionic surfactants as emulsifier.

2. EXPERIMENTAL

The emulsification process was carried out in a batch mixer, using an anchor impeller driven from an IKA-Visc (Steufen, Germany) equipment, which allows to monitor emulsification by following the evolution of torque with time. Linear viscoelastic properties were performed by means of a Haake RS-100 rheometer (Karlsuhe, Germany). Droplet size distribution (DSD) were determined with a Mastersizer X analyzer (Malvern Instrument Ltd., UK).

Emulsions were prepared using 75%wt sunflower oil and 7%wt emulsifier. Different low-molecular weight emulsifiers were used : a poly(ethylene glycol) nonylphenyl ether (PEG-NP E), a poly(ethylene glycol) 4-tert-octylphenyl ether (PEG-tOPE); a sucrose laurate (SEL) and a polyoxyethylene sorbitan monolaurate, (POE-SL).

3. RESULTS AND DISCUSSION

3.1 Influence of the type of emulsifier

The emulsification was always preceded by a mixing stage and a resting period. Figure 1 shows the evolution of torque with the emulsification time for emulsions stabilized with different emulsifiers.

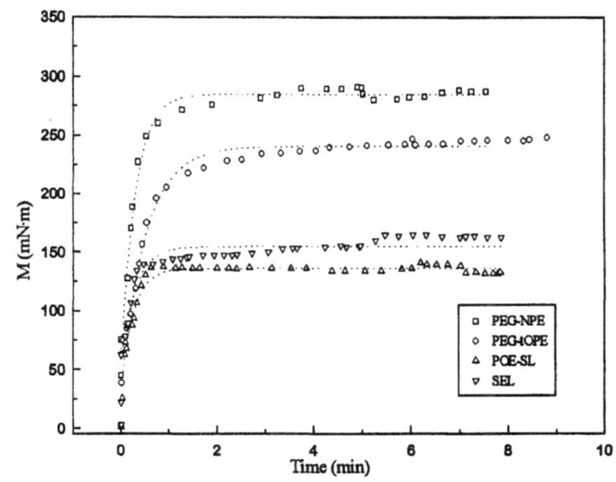

Figure1. Evolution of torque with emulsification time for emulsions, prepared at 250 rpm and 25°C, containing different emulsifiers.

A very rapid increase in the torque response is produced when the mixing process starts. In fact, the evolution of torque with time follows a first-order kinetic equation :

$$M = M_e\left[1 - \exp(-kt)\right] \qquad (1)$$

where M_e is the torque response for the final emulsion and k is the kinetic constant.

This kinetics may be explained in terms of the evolution of both the Reynolds and Capillary numbers. The emulsification process always starts under turbulent conditions. As emulsification proceeds the Re number decreases due to an increase in viscosity, whilst the Ca number decreases as a result of the length-scale reduction that is continuously taking place (Sánchez et al.,

1998). A dynamic equilibrium is finally obtained between coalescence and disruption of droplets.

DSD and linear viscoelastic results obtained for emulsions stabilized by different low-molecular-weight-nonionic surfactants are compared in Figures 2 and 3. The larger values obtained for droplet size and polidispersity correspond to the emulsions stabilized with (POE-SL) and SEL, both of which show a bimodal distribution with a smooth peak in the range of 1-2 µm and a well pronounced peak at much higher sizes. The emulsions stabilized by PEG-alkylphenyl ether, particularly PEG-NPE, showed much smaller values of the droplet size and much narrower unimodal DSD, giving rise to emulsions

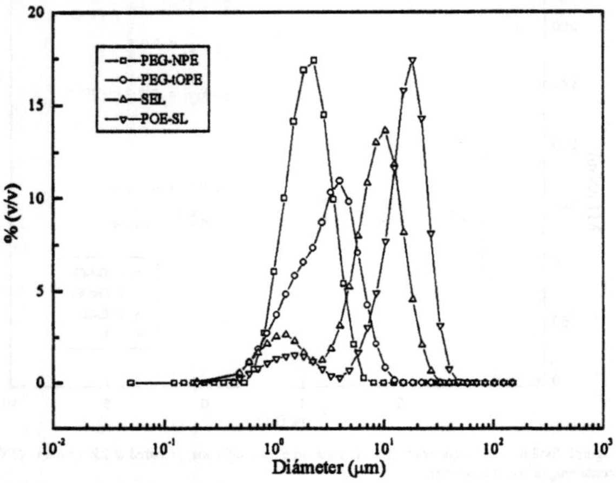

Figure 2. Droplet size distribution curves for emulsions, prepared at 250 rpm and 25°C, containing different emulsifiers.

with higher values of the linear viscoelasticity functions (Figure 3) and higher stability. An exception is the emulsion containing SEL which shows particularly high values for G' and G". The frequency-dependence obtained for this emulsion is also different, as may be seen in Figure 3. Thus, a very slow tendency to a crossover point between G' and G" at low frequency, is shown. This behaviour was also found in other emulsions containing sucrose esters and may be related to the tendency of sucrose ester to form gel-like structures in the continuous phase, in a wide range of concentrations and temperatures.(Franco et al., 1995a ; Guerrero et al., 1996; Madiedo, J.M, 1996). The preponderance of the elastic response over the viscous response is however higher for the emulsion containing PEG-NPE and the emulsion with PEG-tOPE (except al low frequency).Both emulsions show a well-defined plateau region. This behaviour may be attributed to a reversible flocculation by a depletion mechanism induced by the large excess of micellar emulsifier in the continuos phase, leading to weak gel structure (Dickinson, 1995).

The emulsions stabilized by POE-SL shows a fluid-like behaviour with a crossover point at intermediate frequency, a terminal region at low frequency and a small plateau region. This behaviour is characteristic of weakly flocculated emulsions in which the interactions among droplets are not strong (Tadros, 1993).

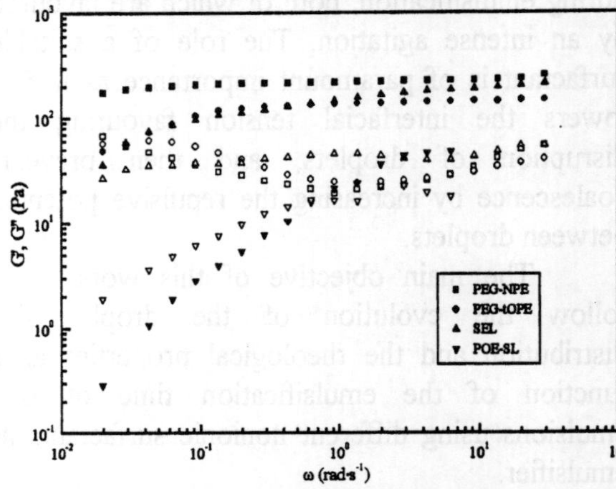

Figure 3. Evolution of storage and loss moduli for emulsions with frequency for emulsions, prepared at 250 rpm and 25°C, containing different emulsifiers.

4. REFERENCES

Dickinson E, y McClements, D. (1995) In: Advances in Food Colloids p: 272.

Franco JM, Guerrero A, Gallegos, C. (1995a) Rheol. Acta 34 : 513-524.

Guerrero A, Partal P, Berjano M, Gallegos C. (1996) Prog. Colloid Polym. Sci. 100 : 246-251.

Madiedo, J.M. (1996) Ph. D. Thesis. Univ. of Seville

Sánchez MC, Berjano M, Brito E, Guerrero A, Gallegos, C. (1998) accepted for publishing.

Tadros ThF. (1983) In : First World Congress on Emulsions 4 pp :237-266.

ACKNOWLEDGEMENTS

This work is part of a research project sponsored by the CICYT, Spain (research project PB92-0664), whose financial support is gratefully acknowledged.

RHEOLOGICAL CHARACTERIZATION OF SUNSCREEN EMULSIONS

A.Semenzato, M.Poropat, M.Marcazzan, A.Bettero, and *R.Lapasin.

Centro di Cosmetologia Chimica, Dipartimento di Scienze Farmaceutiche, Via Marzolo 5, 35131 PADOVA- Italy.
**DICAMP- Università di Trieste, Piazzale Europa 1 TRIESTE - Italy.*

INTRODUCTION

The addition of polymeric additives to sunscreen emulsions can improve the efficacy of the sunscreen emulsions. This effect could be related to the changes of the rheological properties of the emulsion, induced by the polymer, since a good distribution of the product on the skin strictly depends on the rheology of the formulation.
In this study we considered the structural effects induced by adding increasing percentage (1-5%) of an organic-modified silicone wax on the rheological features of an oil-in-water emulsions containing titanium dioxide, as physical sunscreen.
The rheology of the formulations was studied in parallel with their functionality, in order to optimize the performance of the products.
The rheological characterization of the emulsions was performed in continuous and oscillatory flow conditions using a stress controlled rheometer equipped with a plate-plate sensor; combined tests were also used for better understanding the destructive effects produced by the shear flow on the viscoelastic properties of the emulsions.

RESULTS AND DISCUSSION

Continuous flow tests were performed using the hysteresis cycles procedure (3 cycles 0-200 Pa in 6 minutes). The addition of the organic modified silicone wax progressively increased both the shear and the time dependent properties of the emulsions. The quantitative effects of the additive on the flow behaviour of the emulsions, calculated fitting the up curve of the first cycle with the Carreau equation, are reported in figure 1 where the trends of the viscosity at rest conditions (η_o), of the critical stress (τ_{cr}) and of the infinite viscosity (η_{inf}), in function of the polymer concentrations are reported.

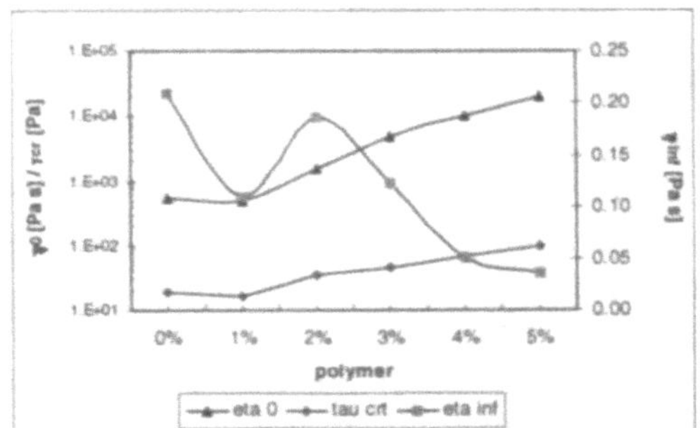

Figure 1. Parameters of the Carreau model.

Dynamic tests showed that all the emulsions could be considered weakly structured systems. In all the emulsions the elastic behaviour dominates always over the viscous component so that more energy is stored than dissipated by the material (Fig.2).

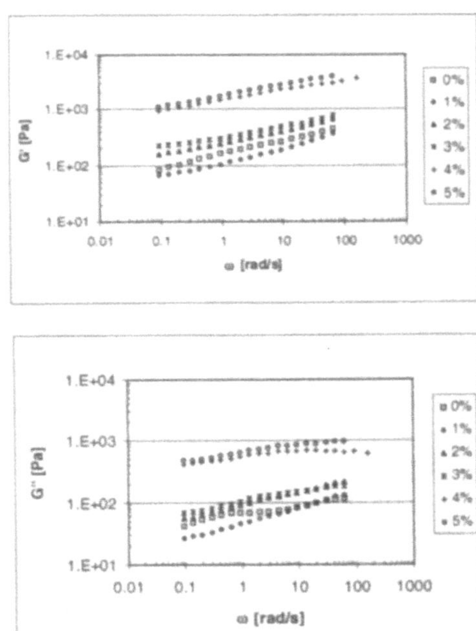

Figure 2. Mechanical spectra.

The absolute values of the storage and the loss modulus decreased, in respect to those of the base reference system, when 1% of the silicone wax was added. Higher concentrations of the additive provided a limited and gradual increase of the viscoelastic response up to 3%; this increase became dramatic in the samples prepared with 4 and 5% of polymer.

The mechanical spectra were successfully fitted using a generalized Maxwell model and the Friedrich-Braun model. Both these models describe systems with quite similar viscoelastic properties; the effects induced by the polymer result in a variation of the values of the responses.

A further investigation of the effects induced by the polymer on the emulsion rheology was performed using combined tests. This procedure, that consists in a sequence of oscillatory and continuous flow measurements, allows to simulate the in use conditions of the product and to evaluate the influence of the distribution properties on the efficacy of the emulsions. The obtained results (Fig. 3) show that the effects induced by the polymer to the reference base system depend on its concentration.

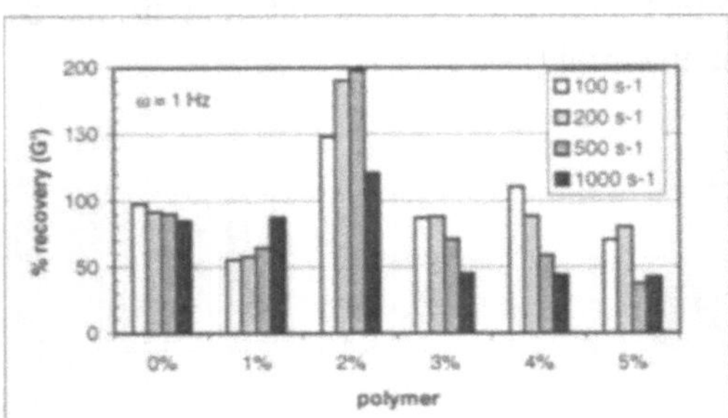

Figure 3. Recovery % of storage modulus, at the end of combined test procedures.

Low concentrations of the wax (1-2%) provides to the emulsion better structural rebuilding capacity than high concentrations (3-5%), even after high shear processing steps.

The structural features of the samples prepared with high concentration of the polymer did not guarantee an even distribution of the product, as confirmed by the optical microscope analysis (Fig. 4).

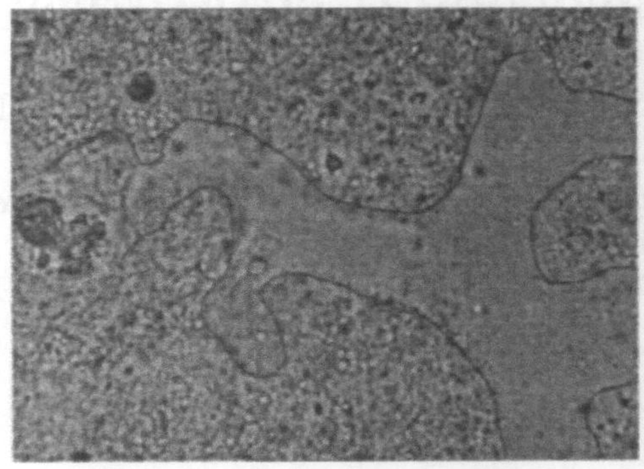

Figure 4. Photomicrograph of sample 5% after 5' at 500 sec^{-1}.

This polymer showed a moderate capacity to enhance the efficacy of titanium dioxide, at least in the oil-in-water emulsion considered (Fig. 5). The sun protection factors (SPF) were calculated by spreading each emulsion on a synthetic tape and measuring the UV transmittance in 5 different points of the tape, in order to evaluate the even distribution of the product.

The increment of the medium value of the relative standard deviations (calculated for each tape) seems to be consistent with the different spreading properties of the emulsions.

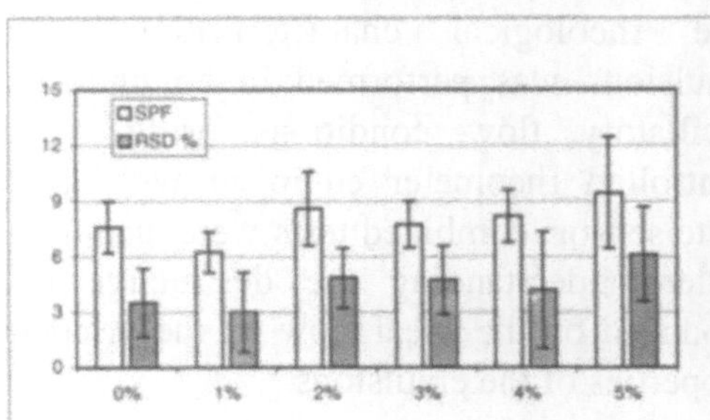

Figure 5. In vitro SPF values of emulsions.

These results clearly show that the rheological characterization of sunscreen emulsions can provide the basic elements to quantify the destructive effects of flow on the structure and the recovery capacity of the systems. The combined test procedures seem to be the more powerful measurements to correlate the spread and film forming properties with the efficacy of the emulsions.

RHEOLOGY OF TEMPERATURE SENSITIVE POLYMER DISPERSIONS

H. SENFF[1], W. RICHTERING[1], CH. NORHAUSEN[2], M. BALLAUFF[2]

[1]Institut für Makromolekulare Chemie, Universität Freiburg, Stefan-Meier Str. 31, D - 79104 Freiburg, Germany
[2]Polymer Institut, Universität Karlsruhe, Kaiserstr. 12, D-76128 Karlsruhe, Germany

1. INTRODUCTION

The rheology of concentrated polymer dispersions has been the subject of a number of studies. Recent contributions by Phan et al (1996) and Meeker et al. (1997) showed that the rheology of hard sphere suspensions is rather well understood by now. In many studies on systems with a more complicated interaction potential it is tried to describe the behavior within the frame work of an effective hard sphere. In sterically stabilized latices, the modification of the viscosity by an adsorbed polymer layer can be modeled in terms of an appropriate increase Δ of the radius a of the particle. Hence an effective volume fraction ϕ_{eff} may be defined through

$$\phi_{eff} = \phi_c \left(1 + \frac{\Delta}{a}\right)^3 \qquad (1)$$

where ϕ_c and a denote volume fraction and radius of the uncovered latex particles.

The effective volume fraction can be determined from the viscosity in the dilute regime by use of the expression derived by Batchelor

$$\frac{\eta_0}{\eta_s} = 1 + 2.5\phi_{eff} + 5.9\phi_{eff}^2 \qquad (2)$$

where η_s denotes the solvent viscosity. On the other hand, the thickness of the polymer layer can be directly determined by scattering methods and be compared to rheological data, see e.g. Weiss et al. (1998).

Gels composed of crosslinked poly(N-isopropyl acrylamide) chains (PNiPA) undergo a volume transition as a function of temperature. Recently Makino et al. (1994) and Okubo and Ahmad (1996) affixed crosslinked PNiPA chains to the surface of polymer latices and hence obtained thermosensitive colloidal particles.

In the present work, we report on rheological experiments from similar core-shell type latices the shell dimensions of which are sensitive to temperature. These systems will be compared to solutions of microgel particles of PNiPA.

2. METHODS

Core-shell particles were prepared by seeded emulsion polymerization. The polystyrene core was also synthesized by emulsion polymerization. Potassium peroxodisulfate was used as initiator in both steps. Purification was done by dialysis of the latex against 2.5mM KCl solution. The PNiPA chains were crosslinked with N,N'-methylene bisacrylamide. Details are given by Dingenouts et al. (1998). The microgel were polymerized in a similar fashion and dialyzed against water.

A stress controlled Bohlin CVO rheometer with cone/plate shear geometry was used.

3. RESULTS AND DISCUSSION

Core-shell particles. Small-angle X-ray scattering (SAXS) experiments from the core-shell particles revealed a core diameter of 80nm whereas the shell of crosslinked PNiPA has a thickness of Δ=32nm in the swollen state at 25°C which shrinks to 18nm during the continuous volume transition. The particles do not flocculate at high temperature despite the fact that the steric stabilization provided by the PNiPA-shell is greatly diminished under poor solvent conditions. Here the stability against flocculation is secured by the surface charges chemically affixed to the core-latex.

The temperature dependent swelling of the PNiPA shell gives rise to a strong temperature dependence of rheological properties. A sample with a solid content of 9.9% is a low viscous fluid at 30°C, but shows a yield stress at 10°C. The effective volume fraction of the core-shell particles was determined from viscometry in the dilute regime using eq. 2.

The zero shear viscosity is plotted vs. the effective volume fraction in Figure 1. All measurements below 30°C collapse to a master curve. For comparison, the results for hard sphere systems by Meeker et al. (1997) are shown as solid line. Up to effective volume fractions of ca. 0.5 a good agreement with the hard sphere behavior is found. At higher volume fractions deviations are observed which indicate a compression of the polymer shell at high solid content.

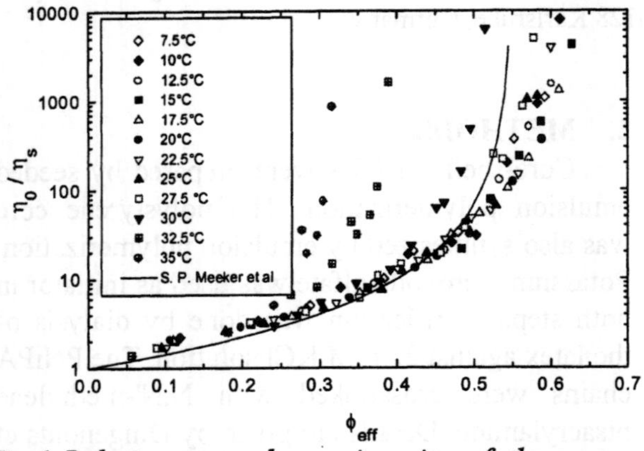

Fig.1:Relative zero shear viscosity of the core-shell particles vs. effective volume fraction.

At higher temperatures, the zero shear viscosity data no longer follow the master curve. As mentioned above, the particles do not flocculate at elevated temperature i.e. the latex is not purely sterically stabilized. This gives rise to deviations from hard sphere behavior when the PNiPA shell shrinks at higher temperature.

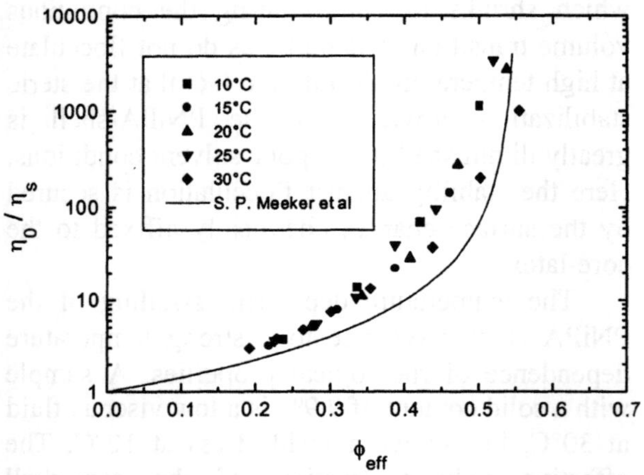

Fig.2:Relative zero shear viscosity of the microgel particles vs. effective volume fraction.

The behavior of the microgel particles is shown in Figure 2. Again the zero shear viscosity data at different temperature collapse to a single curve but deviations from the hard sphere behavior occur due to a low degree of crosslinking. When the crosslink density is increased, the hard sphere behavior is approached. In contrast to the core-shell particles, the microgels revealed the expected behavior of polymers under poor solvent conditions and flocculated at temperatures above 30°C.

The microgel particles form colloidal crystals at high effective volume fractions. This allows to determine the effective volume fraction for the different temperatures also by comparing the experimental phase diagram with the phase transitions from hard sphere suspensions. A very good agreement is observed. Thus the microgel shows similar behavior as compared to sterically stabilized colloidal suspensions.

4. CONCLUSION

The rheology of monodisperse temperature sensitive PNiPA microgel particles was studied. The effective volume fraction was determined by viscometry in the dilute regime and from the phase diagram i.e. the formation of colloidal crystals. A good agreement was found. The temperature dependent zero shear viscosity data collapsed to a master curve but due to the low degree of crosslinking deviations from the rheological behavior of hard spheres occurred at high concentration.

The core-shell particles showed similar behavior at low temperatures but deviations occurred above 30°C. The core-shell system revealed a high stability at elevated temperatures even under shear and these systems are suitable for technical applications.

5. ACKNOWLEDGMENT

We thank the Deutsche Forschungsgemeinschaft for financial support.

6. REFERENCES

Dingenouts N, Norhausen Ch, Ballauff M (1998) submitted

Makino K, Yamamoto S, Fujimoto K, Kawaguchi H, Oshima H J Colloid Interface Sci (1994) 166:251

Meeker SP, Poon WCK, Pusey PN Phys Rev E (1997) 55:5718

Okubo M, Ahmad H Colloid Polym Sci (1996) 274:112

Phan SE, Russel WB, Chen Z, Zhu J, Chaikin PM, Dunsmuir JH Ottewill RH Phys Rev E (1996) 54: 6633

Weiss A, Dingenouts N, Ballauff M, Senff H, Richtering W, (1998) submitted

A CONCEPT OF THE RHEOLOGICAL BEHAVIOR OF SUSPENSIONS

V. STOYANOV

Central Laboratory of Physico-Chemical Mechanics - Bulgarian Academy of Sciences,
Acad. G. Bonchev str., Block 1, 1113 Sofia, BULGARIA

1. INTRODUCTION

In the rheology there are substantial models, structural models, which imitate the inner structure of systems, and functional models, which describe the behavior of the systems. They depend on both the methods for analysis for solving the rheological problem and the degree of an abstraction. The functional models (the pure rheological model) are characterized as a double abstraction. First, the substance substratum is abstracted with calculation of the inner structure, then as a result of abstraction of structure the functional dependencies are calculated (based on Uemov AI, 1971).

In this paper, the structural model of the rheological behavior of the suspension is formulated. The particle orientation and migration, the wall effects and the value of the yield stress are included. On the strength of this model the suitable functional model can be deduced.

2. THEORY ANALYSIS

In rheology there are a large number of structural models. They are emphasized one phenomenon and the rest are neglected. The main restriction of these models is how to show the shear-induced structure of the suspension.

The greatest part of the structural models is based on structure building. In the rest of the systems the structure is a network formed (aggregates), which occupies the whole volume of sample. When these systems are sheared these structures are broken, which is register by value of apparent viscosity (e.g. Uriev NB, 1994). In that case the thixotropy is related with recovery of structure during the shearing.

The corresponding functional models contain structural parameters. Mostly it is used an effective volume fraction, which is always higher than real one. For example, in the model by Wildemuth CR and Williams MC (1984) the maximum packing fraction is used. The established structural parameters are related either to the shear stress or the shear rate.

That type of models is not executed to the suspensions in which coarse particles are disturbed. The final functional models do not explain the particle's size, shape and polydispersity, the particle orientation and migration, the phase separation influence on the apparent viscosity during shearing. For example the decrease of the viscosity of suspension, when the polydispersity size distribution of the particles is increased, is explained with the possibility that the fine particles are situated into the vicinity of the large ones, which is rather a geometrical condition.

3. STRUCTURAL MODEL

Let us consider the tube flow of the suspension (substantial model). The particle's sedimentation is neglected (abstraction). Since the flow is time-dependent process, the structure of the suspension is changed during the shear. This fact predetermines to present of the structure as a consecutive scheme. They show a momentary configuration of the suspension's components (abstraction of substratum) in concrete place during the tube. That configuration depends on both the Deborah and the Reynolds numbers. In other words it depends on both the pressure drop and the place location.

In an arbitrary section of the bulk flow (further a diametrical cross-section is examined) there are three regions. The liquid component is sheared and the solid particles are moved. There are transitional regions, which are surrounding the particles and aggregated (structural elements). In these ('hardening') layers the laminar flow of the liquid is disturbed (the hydrodynamics of the liquid in space around particles is different).

When particles are introduced into a liquid they usually assume a state of thermodynamics equilibrium. It is depends on the Brownian motion and the colloidal forces.

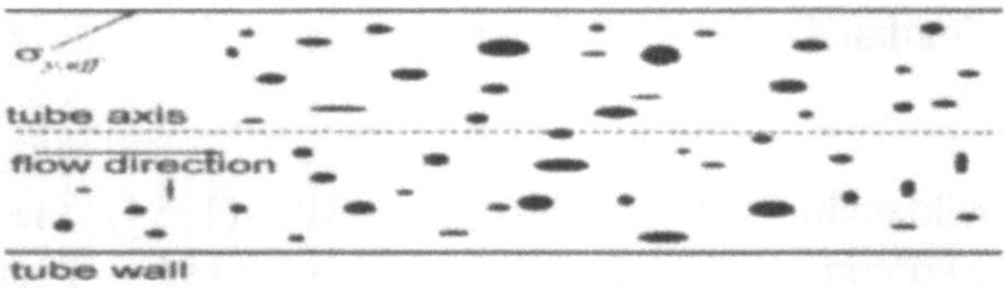

Figure 1

The suspension starts to flow when the pressure drop reaches a critical value (Fig.1). This state is determined by the maximal value of the shear stress, which is at the tube wall. It can be compared with real yield stress, corresponding to the transition from elastic to plastic deformation. As the Fig.1 shows, it is necessary to consider the adhesion between the suspension and the wall. By plastometers an effective yield stress is measured. The difference between effective and real yield stresses are depended on the presence of the colloidal particles, their concentration and homogeneity in vicinity at the wall (a model by Michaels AS and Bolger JC, 1962).

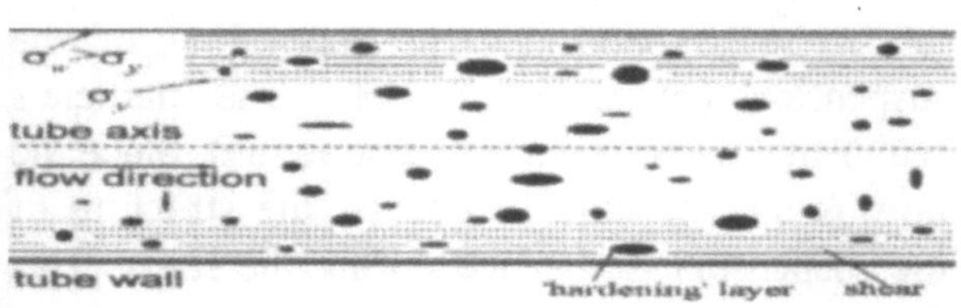

Figure 2

With the increasing the pressure drop (inner influence) there is a Bingham flow of the suspension (Fig.2). The formed plug is a cylinder, which radius is determined by yield stress on the its surface. The rest region of the suspension is sheared. Here the aggregates are broken; the particles are oriented and are moving to the regions with lower shear rates (Mondy L et al., 1994) (in considered case this is the tube axis).

For relative great values of the pressure drop the plug region may be neglected (Fig.3). The above mentioned phenomena dominate in the suspension. In the concentrated suspensions the effects of overlap of the "hardening" layers are existing. It is possible to put the particles into the 'hardening' layer, where the position of the particle is determined by another. In this state they are free of shearing and they are oriented randomly. Therefore the formed structural element is moving as a whole (Fig.3).

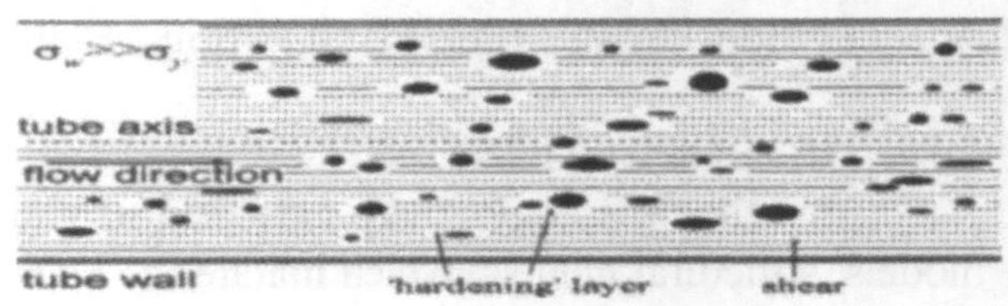

Figure 3

The shear-induced structure exists the three regions, but the correlation between them is changed. When the shear is increased, the volume of the 'hardening' layers is decreased. The particles are packed, too. This circumstance exerts an influence upon the value of apparent viscosity. During the shear the viscosity of the suspension decreases.

Only when the new critical value of the pressure drop is exceeded in the suspension the turbulent flow appears. This leads to the dilatant behavior of the suspension.

4. CONCLUSIONS

The present model allows to take account of the particle's orientation and migration, the wall effects and the value of the yield stress. The structure of the suspensions is considered as the location of the structural elements in bulk, which depends on the shear field. The next step is to formulate a functional model, which takes into account the different regions which exist in laminar flow of the suspensions.

5. REFERENCES

Michaels AS, Bolger JC (1962) Ind Eng Chem Fund 1: 24-33; 153-162

Mondy LA, Brenner H, Altobelli SA, Abbott JR, Graham A (1994) J Rheol 38: 444 -452

Uemov AI (1971) Logical basis of the methods of modeling, Misl, Moscow

Uriev NB (1994) Progress and Trends in Rheology. Proc IV-th Eur Congr. Rheology. pp 693-695

Wildemuth CR, Williams MC (1984) Rheol Acta 23: 627-635

ON THE VISCOSITY OF THE SUSPENSIONS

V. STOYANOV, R. KOTSILKOVA, Ya. IVANOV

Central Laboratory of Physico-Chemical Mechanics - Bulgarian Academy of Sciences,
Acad. G. Bonchev str., Block 1, 1113 Sofia, BULGARIA

1. INTRODUCTION

Gleißle and Baloch (1984) developed a method to describe the flow behavior of suspensions. They called it 'the concept of the shear stress equivalent intrinsic shear rate'(e.g. Gleißle, 1994).

The basis idea is that the 'viscosity' of suspension is higher than viscosity of the pure matrix liquid, because the shear rate is decreased. By means of an assumption for 'hardening' layer of Newtonian steady shear flow (Plattemodel - orig.), the empirical shift process under the constant value of the shear stress is used.

2. BACKGROUNDS

In viscometers the suspensions are investigated in simple shear flow. The Newton's postulate is used by suitably defined shear stress σ and shear rate $\dot{\gamma}$. The most used equation is

$$\sigma_w = \eta_{app} \cdot \dot{\gamma}_w, \tag{1}$$

where σ_w is the shear rate and $\dot{\gamma}_w$ is the shear rate both at the wall. For suspensions the η_{app} is not a coefficient, but a function of the shear rate.

We consider both the suspension and the matrix liquid flow by one and the same value of σ_w. The variation of the shear stress in the bulk of both systems is the same. The variation of the shear rate in bulk is changed, when we neglect the shear in the 'hardening' layers (particles - orig.). If we calculate the new form of the constitutive equation, the value of the apparent viscosity could be determined.

3. APPARENT VISCOSITY IN CASE OF POISEUILLE (CAPILLARY) FLOW

If the Newtonian liquid flows down in a straight circular tube of the radius R at a pressure gradient ($\mathrm{d}p/\mathrm{d}z$), the volume rate Q is given by the Poiseuille equation:

$$Q = \frac{\pi R^4}{8\eta_s}\left(-\frac{\mathrm{d}p}{\mathrm{d}z}\right), \tag{2}$$

where η_s is the viscosity of the liquid. The shear stress in the liquid varies linearly from maximal value σ_w at the capillary wall to zero at the center (capillary axis). The shear rate varies similarly.

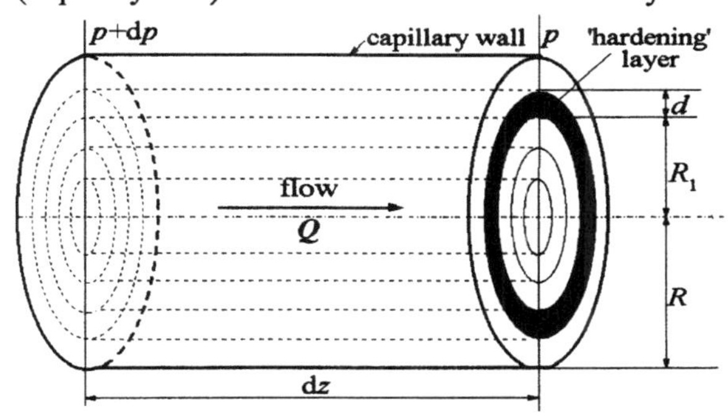

Figure 1. Scheme of Plattemodel of a suspension for Poiseuille flow

We consider the capillary flow of a suspension with the same liquid, flow geometry and pressure gradient ($\mathrm{d}p/\mathrm{d}z$). In Fig. 1 the solid particles are presented as one 'hardening' layer of thickness d, which is placed at R_1 distance from the tube axis. The shear in this layer is neglected, i.e. the layer is moved with a constant velocity. By integrating the velocity profile, the volume rate is:

$$Q = \frac{\pi}{8\eta_s}\left[R^4 - (R_1 + d)^4 - R_1^4\right]\left(-\frac{\mathrm{d}p}{\mathrm{d}z}\right). \tag{3}$$

From Eqs. (2) and (3), in the former η_{app} instead η_s, the apparent viscosity of the suspensions can be calculated as:

$$\eta_{app} = \eta_s \frac{R^4}{R^4 - (R_1 + d)^4 + R_1^4}. \tag{4}$$

The Eq. (4) allows to model both the particle migration and the wall adhesion.

Each particle changes the flow, which is concentrated in surrounding region around the particles (e.g. 'cell' model by Simha R,1952). We approximate that in these regions the energy is not dissipated. If they and the particles, with whole volume fraction ϕ_I, are pushed in one layer and it is located in the middle of the shear field, the R_1 and d depend both on the R and ϕ_I. The final result for η_{app} is:

$$\eta_{app} = \eta_s (1-\phi_I)^{-1}. \qquad (5)$$

4. APPARENT VISCOSITY IN CASE OF COUETTE CIRCULAR FLOW

We consider the Newtonian liquid flow in the gap between two concentric cylinders. The radii of the outer and the inner cylinders are R_O and R_I, respectively, and the angular velocity of the outer is Ω (the inner being stationary). The torque M, measured on the inner cylinder, is:

$$M = \frac{4\pi h \eta_s \Omega R_I^2 R_O^2}{R_O^2 - R_I^2}, \qquad (6)$$

where η_s is the viscosity of the liquid.

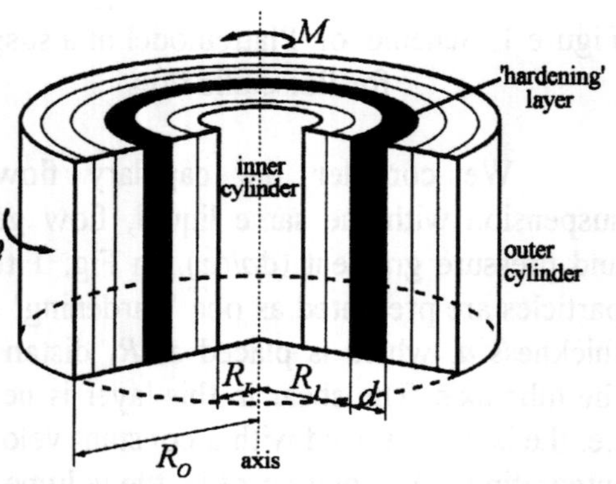

Figure 2. Scheme of Plattemodel of a suspension for Couette circular flow

We consider the Couette circular flow of the suspension with the same liquid, flow geometry and angular velocity of the outer cylinder Ω.In Fig. 2 the solid particles are presented as one 'hardening' layer of thickness d, which is placed at R_1 distance from the axis of the cylinders. The shear in this layer is neglected, i.e. the layer is moved with a constant velocity. By integrating the velocity profile, the torque on the inner cylinder is:

$$M = 4\pi h \eta_s \Omega (R_1 + d)\left[\frac{R_1^2 - R_I^2}{R_I^2 R_1} + \frac{R_O^2 - (R_1+d)^2}{R_O^2 (R_1+d)}\right]^{-1}. \qquad (7)$$

From Eqs. (6) and (7), in the former η_{app} instead η_s, the apparent viscosity of the suspensions is:

$$\eta_{app} = \eta_s \frac{(R_O^2 - R_I^2)(R_1+d)^2 R_1}{R_O^2 (R_1^2 - R_I^2)(R_1+d) + R_I^2 R_1 \left[R_O^2 - (R_1+d)^2\right]}. \qquad (8)$$

The Eq. (8) allows to model both the particle migration and the wall adhesion. The η_{app} depends on the radii of the cylinders, too. The Eqs. (4) and (8) show that the value of the apparent viscosity depends on the type of the viscometer used. The particle migration has higher influence on η_{app} in capillary flow than in the circular one.

If the particles and the regions, which are surrounded particles (the volume fraction, in which the energy is not dissipated, is ϕ_I) are pushed in one layer and it is located in the middle of the shear field, the R_1 and d depend on the R_I, R_O and ϕ_I. The final result for η_{app} is:

$$\eta_{app} = \eta_s \frac{\alpha_+ + \alpha_- \phi_I}{\alpha_+ + \kappa}(1-\phi_I)^{-1}, \qquad (9)$$

where β is the ratio of the radiuses of inner to outer cylinders, $\alpha_+ = 1+\beta^2$, $\alpha_- = 1-\beta^2$ and

$$\kappa = \sqrt{\frac{\alpha_+ + \alpha_- \phi_I}{\alpha_+ - \alpha_- \phi_I}} - 1.$$

5. CONCLUSIONS

The functional models show that the apparent viscosity of the suspensions is proportional to the viscosity of the basic fluid and it is reverse proportional to the volume fraction of regions, in which the laminar flow is disturbed. The particles orientation, its shape, size and polydispersity can be modeled by the ϕ_I value.

6. REFERENCES

Gleißle W, Baloch MK (1984) Advances in Rheology. Proc IXth Int Congr Rheology. pp 549-556

Gleißle W (1994) Progress and Trends in Rheology. Proc IVth Eur Congr Rheology. pp 645-647

Simha R (1952) J Appl Phys 23:1020-1024

SIMULATION OF THE DROPLET DEFORMATION AND THE VISCOSITY OF AN EMULSION OF NEWTONIAN DROPS IN A SIMPLE SHEAR FLOW

D. VAN DEN ENDE, E.M. TOOSE, J. MELLEMA

Twente Institute of Mechanics, J.M.Burgerscentrum, Rheology Group, Dept. of Applied Physics, University of Twente, P.O.Box 217, 7500 AE, Enschede, The Netherlands

1. INTRODUCTION

In many branches of industry (e.g. food industry) emulsifying devices play an important role. In these devices two immiscible fluids are mixed to obtain a distribution of droplets of one liquid immersed in the other. Changes of the size and deformability of the dispersed droplets have a significant effect on the properties of the product, hence there is a considerable interest in the dynamics of viscous (and viscoelastic) drops in a viscous fluid. Here we consider a concentrated two dimensional periodic suspension of Newtonian drops with radius a, viscosity η and constant surface tension σ, dispersed in a Newtonian fluid with viscosity μ. For this suspension we will present a numerical method to obtain the droplet deformation as a function the rate of shear $\dot{\gamma}$, the flow curve $\eta_{sus}(\dot{\gamma})$ and the linear viscoelastic behavior $\eta^*_{sus}(\omega)$, where ω is the angular frequency. For some limiting cases these results will be compared with analytically obtained results.

2. RHEOLOGICAL QUANTITIES

To obtain the macroscopic rheological properties like the shear viscosity or the first normal stress difference, in a simple shear flow, defined by $\underline{v} = y\dot{\gamma}\underline{e}_x$, the droplet contribution to the stress tensor has been calculated. For an emulsion of force-free Newtonian drops, this is given by (Batchelor, 1970):

$$\underline{\underline{T}}^p = \frac{1}{V}\sum_{p=1}^{N}\int_{\partial V_p}(a\sigma\kappa\,\underline{n}\,\underline{n} + (\eta-\mu)(\underline{v}\,\underline{n} + \underline{n}\,\underline{v}))\,d^2r$$

with ∂V_p the surface of drop p, $\underline{n}$ a unit vector normal to the drop surface, $\underline{v}$ the velocity, κ the local curvature of the surface and N is the number of drops in volume V. By solving the Stokes equations for creeping flow both inside and outside the drops $\underline{v}$ and κ are obtained. The boundary conditions are the continuity of velocity and tangential stress component on the drop-liquid interfaces and the jump in the normal stress component over the interfaces, which has to be balanced by the "interface-force" per unit area: $-\sigma\kappa\,\underline{n}$. From the stress contribution $\underline{\underline{T}}^p$ the viscosity of the suspension $\eta_{\text{sus}} = \mu + T^p_{xy}/\dot{\gamma}$ and the normal stress difference coefficient $\Psi_1 = (T^p_{xx} - T^p_{yy})/\dot{\gamma}^2$ can be calculated.

3. NUMERICAL METHOD

The numerical method to calculate the deformation of the individual dropls and the stress distribution over the droplet surface in a simple shear flow, is based on the boundary integral method (BIM), which is often used to study the deformation of a Newtonian drop in a creeping flow (Youngren and Acrivos, 1975). In order to use the method for a suspension of Newtonian drops, periodic Green's functions are used (Pozrikidis, 1996). It should be noted that the periodic dispersion considered, only consists of a finite number of drops. This implies that the periodic dispersion is only an approximation of a real dispersion which would contain an unlimited number of drops. The effect of this approximation can be investigated by increasing the number of drops in the periodic box while keeping the volume fraction constant. However, as the deformation of the drops is mostly governed by short range hydrodynamic interactions between the drops, it is expected that only a limited number of drops will be needed to yield reliable results (Brady, 1988). To calculate the drop deformation the displacement of the collocation points at the interface is calculated from time integration of the velocity at these points, which is obtained from the BIM-calculations. Only the displacements normal to the interface of the drops are calculated to garanty that the collocation points remain equally distributed over the interface. An extensive description of the numerical methods used has been given by Toose (1997).

4. SMALL DEFORMATION ANALYSIS

For small deformations and low concentrations the viscoelastic behavior of the suspension was obtained analytically by calculating the stream function $\Psi(x, y)$ defined by $v_x = \partial\Psi/\partial y$ and $v_y = -\partial\Psi/\partial x$ for the case of a single drop under small amplitude harmonically oscillating shear. The stream function obeys $\nabla^4\Psi = 0$, which is equivalent with the Stokes equations and the continuity equation. Once Ψ is known the complex viscosity in the low concentration limit can be calculated:

$$\eta^*_{\text{sus}} = \mu\left(1 + \phi\frac{(2\lambda+1) + 2i\Omega(\lambda^2-1)}{(\lambda+1) + i\Omega(\lambda+1)^2}\right)$$

where $\Omega = a\omega\mu/\sigma$ is a dimensionless frequency, $\lambda = \eta/\mu$ the viscosity ratio, $\phi = \pi a^2 N/V$ the area fraction and $i = \sqrt{-1}$. In the limit $\lambda \gg 1$ the equivalent of the Einstein correction on the viscosity of a hard sphere suspension has been obtained: $\eta = \mu(1+2\phi)$. At moderate concentrations we calculated the linear viscoelastic behavior using the cell model of Oosterbroek and Mellema (1981) adapted for the 2D situation. The cell radius b is a fitting parameter in this model and should be of the order $a/\sqrt{\phi}$.

5. RESULTS AND DISCUSSION

In figure 1 a snapshot is presented of a periodic suspension of 24 drops which have the same visocsity as the suspending fluid.

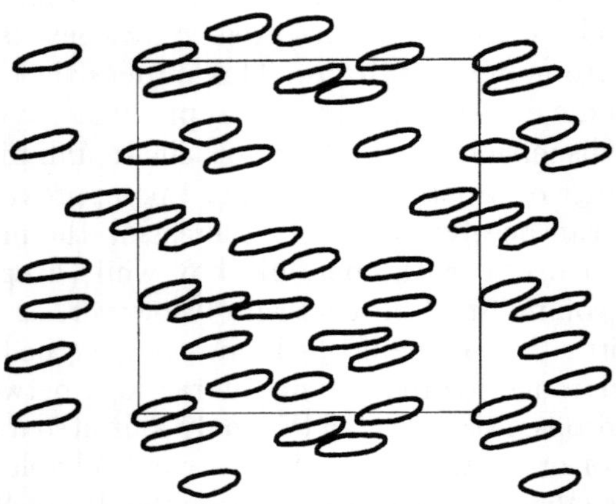

Figure 1: Snapshot of the periodic suspension for $C = 0.75$, $\lambda = 1.0$ and $\phi = 0.2$.

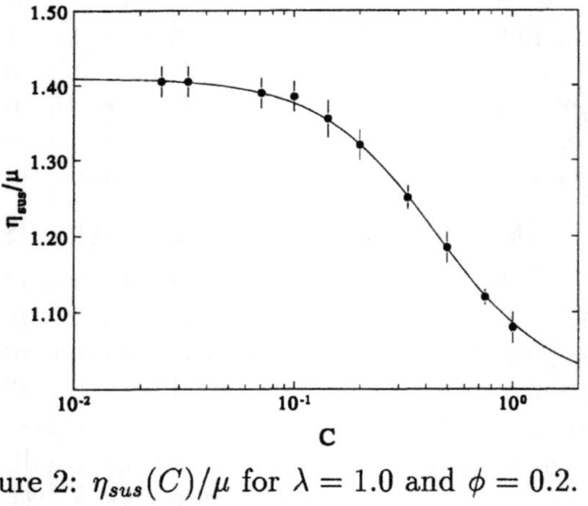

Figure 2: $\eta_{sus}(C)/\mu$ for $\lambda = 1.0$ and $\phi = 0.2$.

The area fraction ϕ of this suspension is 0.2 and the dimensionless shear rate $C = a\dot{\gamma}\mu/\sigma = 0.75$. At this shear rate the drops are highly deformed in the direction of the flow. In figure 2 the steady state viscosity of the same suspension is plotted against C. The observed shear thinning behavior is due to the increasing deformability of the drops with increasing rate of shear. The error bars result from variation of the number of drops in the periodic box at fixed concentration. To test the numerical code the linear viscoelastic behavior was calculated for ϕ is 0.01 and 0.2 and compared with the analytically obtained results. As shown in figure 3 there is good agreement between them not only for low concentrations but also at moderate concentrations where we used the cell model, with $b = 2a/\sqrt{\phi}$.

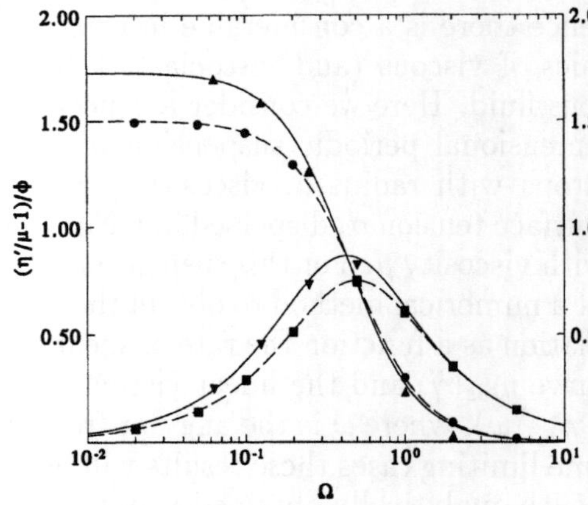

Figure 3: $(\eta^*_{sus}(\Omega)/\mu - 1)/\phi$ for $\phi = 0.01$ (- - -) and 0.20 (—) with $\lambda = 1$.

In this paper a boundary integral method has been presented to calculate the rheological properties of a two dimensional periodic suspension of viscous drops in a simple shear flow. For low concentrations the viscoelastic results have been compared with analytical results and showed good agreement with them. Moreover analytically not obtainable results for non linear behavior have been presented, showing the strength of the method. In the near future we will extend the code for full three dimensional calculations.

6. REFERENCES

Batchelor GK (1970) J Fluid Mech 41: 545-570

Brady JF (1988) Ann Rev Fluid Mech 20: 111-157

Oosterbroek M, Mellema J (1981) J Coll Int Sci 84: 14-26

Pozrikidis C (1996) J Eng Math 30: 131-150

Toose EM (1997) PhD-Thesis, University of Twente, The Netherlands

Youngren GK, Acrivos A (1975) J Fluid Mech 69: 377-403

FLOW-INDUCED DISORDER-ORDER TRANSITIONS IN CONCENTRATED STERICALLY STABILISED DISPERSIONS

J. VERMANT[1], J. MEWIS[1] AND G.G. FULLER[2]

[1]Chemical Engineering Department., K.U. Leuven, de Croylaan 46,B-3001 Leuven, Belgium
[2]Chemical Engineering Department., Stanford University, Stanford, CA-94305, USA

1. INTRODUCTION

In concentrated colloidal dispersions flow affects the microstructure. At relatively low stresses, the competition between the hydrodynamic and "elastic" (Brownian and interparticle) forces leads to a gradual distortion of the structure. At higher stresses the formation of ordered phases has been predicted by various numerical simulations, mainly for Brownian hard spheres. One structural feature for which clear experimental evidence is lacking is the formation of highly ordered phases, consisting of "strings" of particles aligned in the flow direction (Hess 1986, Phung et al. 1997) and packed in an hexagonal array (Rastogi and Wagner 1998). Some indications for such string-like ordering have been observed experimentally but the evidence remains rather weak (Ackerson and Pusey 1988, Clarke et al. 1995)

The existence of such structures under flow has been investigated here by means of rheological and rheo-optical techniques. Model colloidal dispersions, consisting of monodisperse particles with either chemically grafted or adsorbed polymeric stabiliser layers have been used

2. METHODS

Time resolved, 2D small angle light scattering (SALS) was selected as a suitable tool for detecting in-situ large ordered structures during flow. A stress rheometer (Rheometric Scientific) was used as a mechanical platform. A parallel plate geometry, with gap sizes down to 300 mm, was used to limit the optical path length because some samples were rather opaque. The suspending medium and the particles were not optically matched. As SALS is used as a diagnostic tool this is not a serious drawback. Effects of multiple scattering will tend to render the observed scattering patterns more isotropically, hence the observation of an anisotropic scattering pattern will reflect a forteriori the anisotropy of the underlying structure. The presence of multiple scattering does however limit to a certain extent the quantitative exploitation of the results. Linear conservative dichroism measurements have been used to complement the SALS data.

3. RESULTS

Figure 1 illustrates how, at relatively high volume fractions and under suitable flow conditions (σ=7.5 Pa), the characteristic ligh scattering patterns of string-like structures appear. This pattern was obtained on an aqueous suspension of poly(styrene-butylacrylate) latex particles with a core diameter (d_c) of 127 nm. The particles are stabilized by means of an adsorbed polymer layer with an hydrodynamic thickness (δ_h) of 10 nm. The interparticle potential is known from rheological measurements by Raynaud et al. (1996). The effective volume fraction ϕ_{eff}, which incorporates the core particle as well as the stabilizer layer, amounted to 0.75.

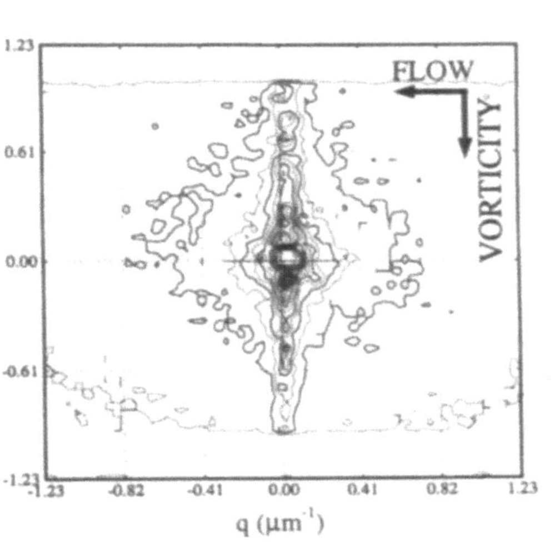

Fig.1.:Typical SALS pattern indicating string-like ordering (aqueous latex, d_c=127 nm, δ_h= 10 nm. ϕ_{eff} = 0.75, σ=7.5 Pa)

The pronounced streak in the scattering patterns only occured after a certain value of the stress was reached. This is shown more clearly in figure 2 where the intensity profile along the streaks is plotted as a function of the scattering vector for different stress levels. At low stresses the patterns were isotropic. Above a certain value of the stress, the intensity along the vorticity axis started to increase. Along the flow direction the intensity profiles remained essentially unaltered.

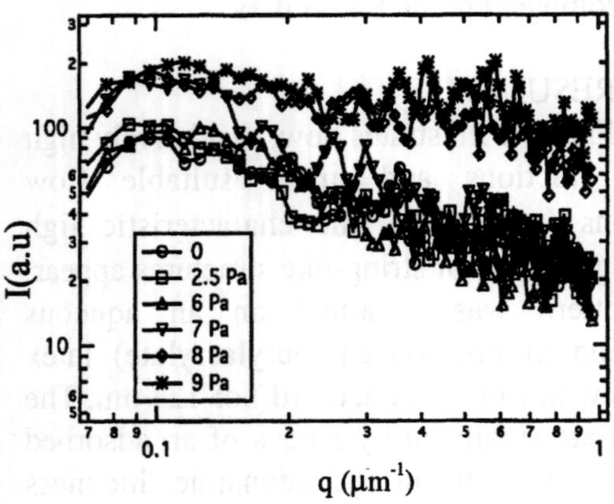

Fig.2.: Intensity as a function of the scattering vector for various stress levels.

The length scales probed in our SALS experiments are much larger than those corresponding to the diameter of individual strings. Hence the streak in the scattering patterns reflects the presence of at least domains with string-like order.

The increase in anisotropy in the optical properties could also be picked up by the linear conservative dichroism. The occurrence of the streak in the SALS patterns agreed with a distinct increase in dichroism. In some cases it coincides with the transition from shear thinning to a pseudo-Newtonian plateau (not shown here).

In order to investigate the effect of the interparticle potential, different model systems were used. Our data suggest that a relatively "soft" interparticle potential and a low medium viscosity favour the formation of large scale string phases. Upon further increasing the shear rate the string phases were observed to melt away. Hence flow induces consecutively a disorder-order and an order-disorder transition. Some of the systems investigated exhibited continuous shear thickening. In the latter case, the string phases always were observed to melt away.

4. CONCLUSIONS

The existence of phases, composed of strings of particles under shear flow, has been demonstrated clearly by using time-resolved Small Angle Light Scattering and linear conservative dichroism. The conditions under which string phases can be observed have been studied systematically. String phases could only be detected in concentrated systems (Φ_{eff}>0.50); a relatively "soft" interparticle potential and a low medium viscosity favour the formation of large scale string phases.

At relatively low stresses the structure is essentially fluid-like. As the shear stress increases string phases are formed. In some cases the strings were observed to melt away again.

5. ACKNOWLEDGEMENTS

The financial support of Elf-Aquitaine during part of this work is gratefully acknowledged. JV is a post-doctoral fellow of the Fund for Scientific Research - Flanders (Belgium).

6. REFERENCES

Ackerson BJ, Pusey, PN (1988) *Phys. Rev. Lett.*, **61**:1033

Ball RC, Melrose JR (1995) *Adv. Coll. Int. Sci.*, **59**:19

Clarke SM, Ottewill RH, Rennie AR (1995) *Adv. Colloid Int. Sci.*, **60**: 95

Hess S (1986) *Intern. J. Thermophys.*, **6**:657

Phung TN, Brady JF, Bossis G (1996) *J. Fluid Mech.* **313**: 181

Raynaud L, Ernst B, Vergé C, Mewis J (1996), *J. Coll. Int. Sci.*, **181**: 11

Rastogi SR, Wagner NJ (1998), *Computers Chem Eng.*, **19**, 693

THE EFFECT OF ADDITIVES ON FLOW BEHAVIOUR OF KAOLIN-WATER MIXTURES

P.VLASÁK, Z.CHÁRA, P.ŠTERN

Institute of Hydrodynamics, Acad. of Sci.Czech Rep., Pod Paťankou 5, 166 12 Prague 6, Czech Republic

Flow behaviour of highly concentrated fine-grained slurries depends not only on a concentration, density, particle size distribution and a shape of suspended particles, but it is also strongly affected by physico-chemical forces acting in the slurry and by an inner structure of the slurry. If the attractive forces acting in the slurry prevail, a process of coagulation and sedimentation is initiated. However, a simultaneous existence of the repulsive forces enables to stabilize the slurry and keeps individual particles separated. The effect of the electro-static repulsive forces on the stabilization process could quite well explain a mechanism of clay slurries liquefying, Šatava (1973).

The flow behaviour of the clay suspensions can often be modelled as the yield pseudoplastic rheological model (Bulkley-Herschel). On the contrary to e.g. fly-ash-water mixtures, the inner structure of a clay slurries prevents to achieve such high concentrations as for the ash slurries, Vlasák and Chára (1998). This fact is a result of the difference in the inner structure of the both mentioned kinds of slurries. The kaolin particles have an anizometric shape, mostly plates, infrequently rods. The coagulation process in the kaolin-water mixture gives to rise to voluminous aggregates with a loose structure where a large deal of water is fixed. Therefore, during the slurry flow a viscous friction in the carrier liquid can act only in a small-scale, more energy is consumed on the aggregates deformation. Nevertheless, if the aggregates are due to the presence of an additive destroyed on individual particles or smaller clusters of particles, water originally fixed in aggregates is liberated. Only a small part of the water remains bonded in a thin layer around the particles. During the slurry flow, the viscous friction can exert in much more scale - an apparent viscosity is decreasing and the slurry is liquefying.

The viscosity differences between coagulated and peptised slurries can achieve values of several orders depending on concentration and velocity gradient. To break the aggregates it is necessary to produce sufficient repulsive forces between the particles. One possibility is to replace originally bi-valent cations Ca and Mg by mono-valent cations (e.g. Na) since their ionic charge is working on a larger distance and hence the attractive Van der Walles forces do not exert and repulsive forces prevail, Vycudilík (1974).

The thickness of an electrical two-layer around a particle decreases if the concentration and valence of cations increases. The behaviour of the system is determined by a mutual effect of the attractive and the repulsive forces between the solid particles. A magnitude of the forces is determined by physical and chemical properties of the both phases. By addition of the appropriate cations in proper concentration the repulsive forces between particles prevail and the viscosity and the yield stress decreases.

To prove the possibility of liquefying of the kaolin slurries three kinds of the kaolin were tested. The kaolin differ by the size and also by a chemical composition (see Table 1). The sodium carbonate and soda water-glass were used as an additive (peptising agent) in different concentration. The flow behaviour of the kaolin slurry was measured by the overpressure capillary viscometer and by the rotational viscometer Haake Rotavisco RV 20. The rotational viscometer was used for velocity gradient range from 0 to 200 s^{-1} and for limited number of samples up to 1200 s^{-1}. With the capillary viscometer velocity gradient up to 10^4 s^{-1} was reached. The results of the both

Tab. 1 The kaolin properties

Kaolin		d_{50} μm	ρ kgm^{-3}	SiO_2 %	Al_2O_3 %
Sedlec	S1a	1,5	2507	47.84	35.83
Sedlec	MK	6	2505	50.56	34.56
H.Bříza	HB	3	2549	50.33	35.06

	Fe_2O_3	CaO	MgO	K_2O	Na_2O
S1a	0.92	0.50	0.37	1.11	0.06
MK	1.08	0.21	0.29	1.61	0.07
HB	0.67	0.07	0.20	1.38	0.09

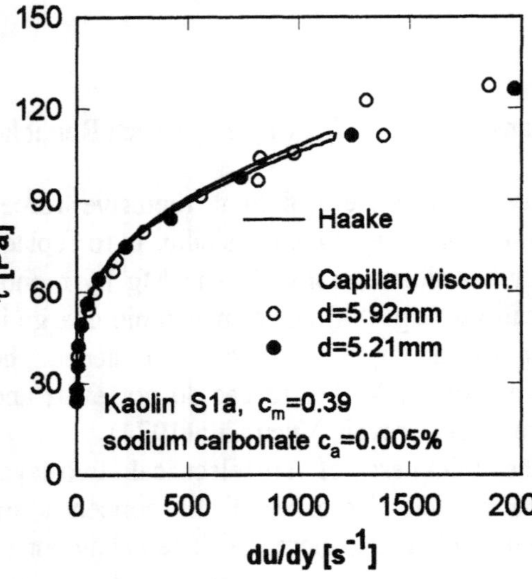

Fig. 1 Comparison of the results of capillary and rotational (Haake) viscometers. The effect of velocity gradient on shear stress.

viscometer measurements are in a good agreement (see Fig.1). In the Figure 1 the dependence of shear stress on velocity gradient is illustrated.

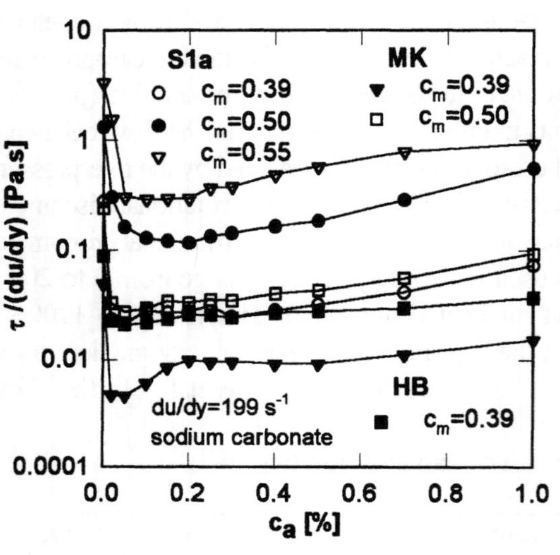

Fig.2 The effect of additive concentration, material and mass concentration

The effect of the both additives and their concentration for different slurry concentration was investigated. The significant decreasing of the apparent viscosity and the yield stress was observed due to the addition of the peptising agents. With increasing concentration of the agent,

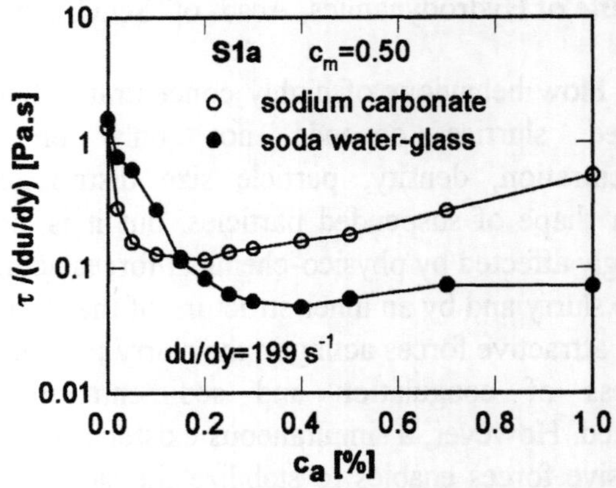

Fig. 3 Comparison of effect of different peptising agent

the apparent viscosity and the yield stress at first quickly decreases, after exceeding of the optimal value of the agent concentration, the both rheological quantities again increase and the recoagulation process resumes (see Fig. 2 and 3). A hysteresis of the flow curves depends on peptising agent concentration and on solid particles concentration.

CONCLUSIONS

It was demonstrated that the peptising agent can serve to decrease energy consumption and/or to increase operational solid concentration for kaolin or clay slurries. Probably a similar principle can be used also for some further fine-grained dense hydromixtures during the hydrotransport or industrial processing.

Acknowledgements

The support by Grand Agency of Academy of Sciences of the Czech Republic under Grand No. A 2060701 is gratefully acknowledged.

REFERENCES

Šatava V (1973) Sklář a keramik, XXIII., 195-199 (in Czech)

Vlasák P, Chára Z (1998) In: 9th International Freight Pipeline Symposium, IFPS, Monterey, Mexico, 6-3H

Vycudilík P (1974) Sklář a keramik, XXIV, 201-208 (in Czech)

ORDERING AND DYNAMICS OF MODEL SOFT SPHERES: MULTIARM STAR POLYMERS

D. VLASSOPOULOS[1], G. FYTAS[1], J. ROOVERS[2], G. FLEISCHER[3], T. PAKULA[4]

[1] FORTH, Institute of Electronic Structure and Laser, P.O. Box 1527, 71110 Heraklion, Crete, Greece
[2] Institute of Environmental Chemistry, National Research Council, Ottawa, Ontario K1A 0R6, Canada
[3] Fakultät für Physik und Geowissenschaften, Universität Leipzig, D-04103 Leipzig, Germany
[4] Max-Planck Institut für Polymerforschung, P.O. Box 3148, 55021 Mainz, Germany

1. INTRODUCTION

Polymer stars constitute a special case of tethered chains, in which linear macromolecules are attached by one end to a central core. In the simplest case of a single star in a good solvent, when the number of arms, f, is high, the segment density near the core is much higher than in the outside, as described by the model of Daoud and Cotton (1982) and confirmed in block copolymer micelles by small angle neutron scattering (SANS) (Förster et al, 1996); such a multiarm star polymer can be considered as a prototype spherical brush with soft colloidal character (core-shell structure). In good solvent above the overlap concentration, these stars order in a macrolattice due to the increased osmotic pressure, as shown by SANS and predicted by scaling arguments (Witten et al, 1986).

The model polybutadiene star polymers used in this work are shown in Table 1, along with their molecular characteristics (M_a, f); they have been prepared by coupling anionically synthesized narrow molecular weight distribution (<1.1) linear polybutadienes (arms) with small carbosilane dendritic coupling agents, constituting highly branched star centers of different generations depending on the desired functionality (Roovers et al, 1993).

The techniques used in this investigation are: dynamic light scattering, pulsed-field gradient NMR and shear rheology. High quality intermediate scattering functions for polybutadiene stars with f=64 and 128 in cyclohexane were obtained (Seghrouchni et al, 1998).

Figure 1 shows S(q,t) of a 12856 at different concentrations; S(q,t) is represented well by a double exponential decay. Its short-time behavior exhibits the global characteristics of cooperative diffusion. The second slow diffusive decay of S(q,t) relates to the self diffusion of the stars, due to their functionality (and thus size) polydispersity (Seghrouchni et al, 1998). At a given concentration, its intensity I_p increases with f and decreases with M_a, whereas I_p is a decreasing function of c. These findings conform to the self diffusion mechanism. In fact, the values of D are in good agreement with the self diffusion D_s measured by pulsed-field gradient (PFG)-NMR, shown in Figure 2. The strong slowing down of mass diffusion with c similar to that for the low-shear rate viscosity, is reminiscent of colloidal behavior. The largest stars 12856 and 12880 (qR=O(1)) exhibit an additional intermediate relaxation mode, which relates to the structural rearrangements of the ordered stars.

To examine the effects of this liquid-like ordering on the dynamics of the star melts, small amplitude oscillatory shear measurements were carried out in a wide temperature range (-103°C-60°C). Fig. 3 depicts master curves of elastic (G') and viscous (G") moduli with reference temperature -83°C. A striking effect of the increase of arm number is observed in the terminal relaxation region. Instead of a single terminal relaxation, the transition between the plateau range and the Newtonian flow range of the multiarm star melts takes place in two steps, indicating two well separated relaxation processes. From the relaxation times determined for all systems, it is clear that in the multiarm star melts the high frequency relaxation of the terminal relaxation regime can be attributed to the orientational relaxation of star arms (Milner and McLeish, 1997), depending (exponentially) on M_a and not on f. The slower terminal process, is only observed in systems with a high f, and its separation from the high frequency component is f-dependent; we attribute this process to a collective structural relaxation of the ordered stars (Vlassopoulos et al, 1997).

3. REFERENCES

Daoud M, Cotton JP (1982) J. Phys. (Paris) 43:531.

Förster S et al, (1996) Phys. Rev. Lett. **77**:95.

Milner ST, McLeish TCB (1997) Macromolecules 30:2159.

Roovers J et al (1993) Macromolecules 26:4324-

Seghrouchni R, Petekidis G, Vlassopoulos D, Fytas G, Semenov AN, Roovers J, Fleischer G (1998) Europhys. Lett. 42:271.

Vlassopoulos D, Pakula T, Fytas G, Roovers J, Karatasos K, Hadjichristidis N (1997) Europhys. Lett. 39:617.

Witten TA et al (1986) Europhys. Lett. 2:137.

Table 1
Molecular Characteristics of Multiarm Star Polymers

Code	f	M_a
PB	2	77,500
4S	4	42,000
6407	62	6,330
12807	124	6,800
12814	125	13,000
12828	114	26,000
12856	127	47,000
12880	122	72,000

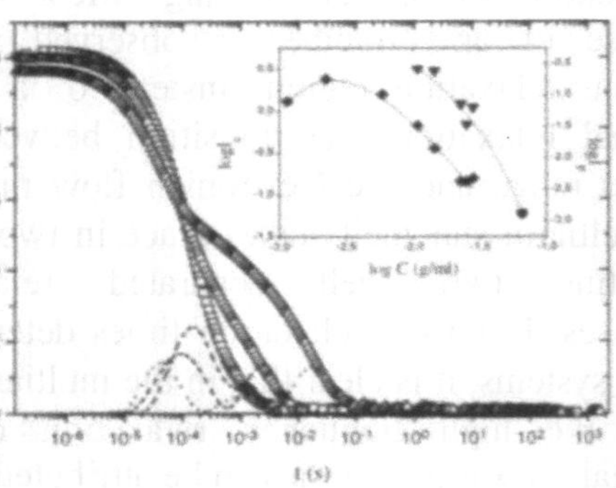

Figure 1: C(q,t) for 12856 in cyclohexane at q=0.035 nm^{-1} and three concentrations (1.38% wt, ❑; 1.84%, ❍; 2.89%, Δ), along with the respective distributions of relaxation times from ILT. Inset: Concentration dependence of the intensities of the cooperative (◆) and slow (▼) relaxation modes.

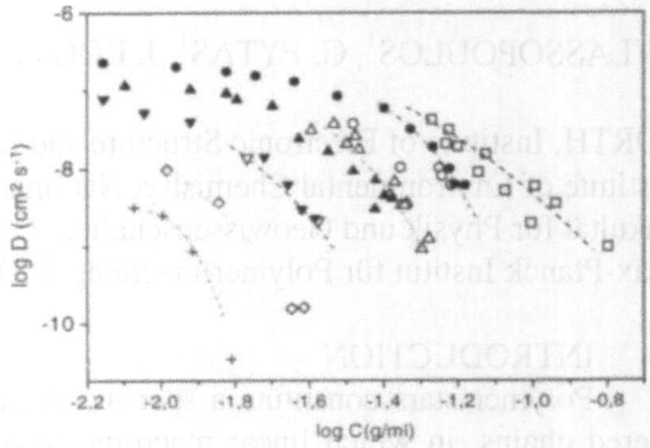

Firure 2: Compiled plot of the concentration dependence of the diffusion coefficients of the slow (polydispersity) mode, D_p for all multiarm star samples: 6407 (❑); 12807 (❍); 12814 (Δ) and 12828 (∇); 12856 (◊), 12880 (+). Also shown are the self-diffusion coefficients measured by PFG-NMR for 12807 (●); 12814 (▲); 12828 (▼). Lines are drawn to guide the eye.

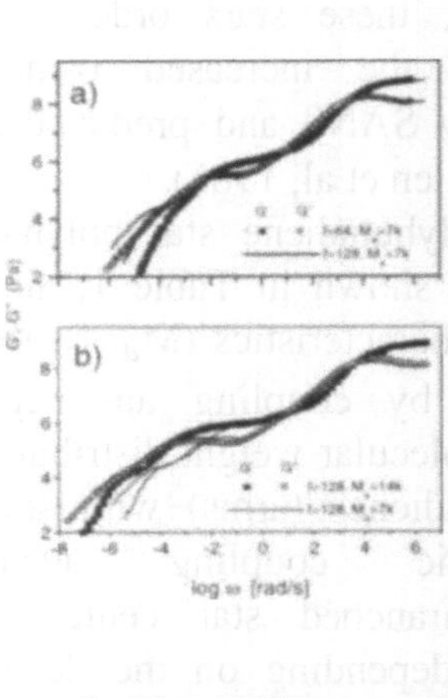

Figure 3: G' and G" mastercurves for 6407, 12807 and 12814 melts.

INFLUENCE OF PARTICLESHAPE AND -MORPHOLOGY ON THE RHEOLOGICAL BEHAVIOR OF HIGHLY CONCENTRATED SUSPENSIONS AND EMULSIONS

E. J. WINDHAB, J.-C. EISCHEN, P. BRAUN

Swiss Federal Institute of Technology Zürich (ETH); Institute of Food Science / Food Process Engineering, CH - 8092 Zürich; Switzerland

1. INTRODUCTION

Concentrated suspension and emulsion systems in principle show similar typical viscosity functions versus shear rate or shear stress with a lower Newtonian viscosity plateau η_0 or a yield value τ_0 in the low shear rate/shear stress domain. If a specific shear stress τ^* or the related shear rate $\dot{\gamma}(\tau^*)$ is exceeded, particle orientation and/or deformation is induced. As a consequence the viscosity is in general reduced (shear thinning and thixotropic behavior). A high orientation or deformation degree of the disperse structuring leads to a pure fluid rheology determined flow. At high shear stresses/shear rates a shear induced structure can either collapse or "superstructures"are generated, both providing dilatant and rheopectic flow behavior. The reversibility of disperse particle formation/ deformation also induces elastic properties.

2. RESULTS

To study the effect of particle size, particle shape/-morphology and particle deformation on the rheological properties of concentrated disperse systems systematically, model systems were composed based on a Newtonian continuous fluid phase (silicon oils). The disperse volume fraction, the particle shape and -morphology were varied. The rheological behavior, which was measured in steady and oscillatory shear with concentric cylinder gap geometries, is compared in terms of shear viscosity functions, relative viscosity and dynamic moduli.

2.1 Orientation effects in fibre suspensions

Figure 1 shows a characteristic shear viscosity dependency on shear stress $\eta(\tau)$ measured for a carbon fibre suspension (mean fibre length: 200 µm; fibre diameter 5 µm, volume fraction 0.25) which can be explained as follows, based on rheological measurements with different gap geometries and microstructure analysis:

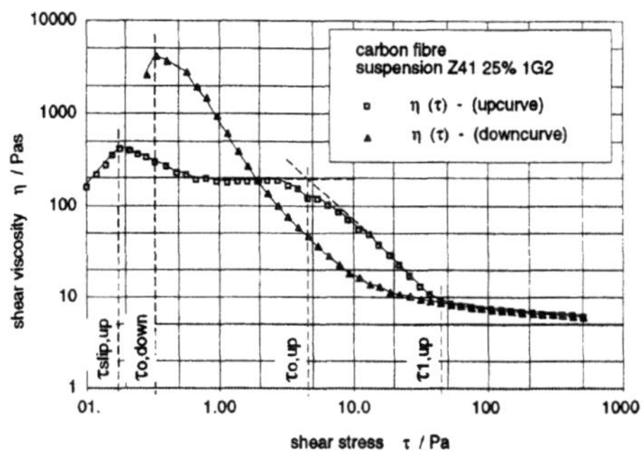

Figure 1: Shear viscosity functions from steady shear experiment (up- and down curve)$\eta(\tau)$ for carbon fibre suspension (ϕ = 0.25)with silicon oil AK1000

For $\tau < \tau_{slip}$, there is some minor elastic deformation. If τ_{slip} is exceeded, the fibre suspension flows plug-like on a wall slip layer of the pure fluid phase. An increase of the wall slip layer reduces the macroscopically measured viscosity to some extent until a viscosity plateau is reached, indicating that the pure fluid is increasingly sheared with further growing shear stress within a wall slip layer of constant thickness (5 microns calculated for fig.1). As soon as the up-curve yield value $\tau_{0,up}$ is exceeded, the carbon fibres orient in the shear flow field and approach a maximum orientation in the shear direction for $\tau \geq \tau_1$. At a volume fraction of 0.25 the oriented fibres are hindered from further rotation in the shear flow field, thus keeping an increased degree of orientation during the down curve measurement. The down curve crosses the up curve in the "up-curve wall slip domain". This indicates that there is a lower yield value $\tau_{0,down}$ for the oriented fibre system. Wall slippage seems to be negligible for the fibre suspension down curve structure with fibres oriented parallel to the shear gap walls.

2.2 Dispersion of agglomerate suspensions

Similar types of flow curves $\eta(\tau)$ like for fibre suspensions are found as well for concentrated suspensions of solid particle agglomerates/ aggregates which in general entrap some quantity of the fluid phase. If the particle agglomerates are deagglomerated in the shear flow field at high shear stresses, the efficient solid volume fraction feff is reduced due to a release of the immobilized entrapped fluid.

2.3 Deformation effects in emulsions

In the case of concentrated emulsions there is as well a strong dependency of viscosity on shear rates or shear stress. From the theory of diluted emulsions, this shear rate depen-dency is not predicted (Choi and Schowalter (1975)). However, the deformability of emulsion droplets allows for aligning them in the shear direction thus providing as well the type of viscosity functions like shown under 2.1 and 2.2. In all three cases demonstrated (2.1 to 2.3), two characteristic stresses τ_0 and τ_1 can be determined from steady shear experiments if a specific volume fraction of the disperse phase ϕ^* is exceeded. ϕ^* is a function of particle size distribution and interfacial properties. For $\tau < \tau_0$ wall slip based flow is mostly detected. For $\tau_0 \leq \tau < \tau_1$ flow induced microstructuring is found. If $\tau > \tau_1$ a "fully developed" shear induced structure is generated. This provides similarity of the micro-flow conditions for the continuous fluid phase "between" the disperse structuring units. Consequently, in this domain the relative viscosity $\eta_r(\phi)$ is shear rate or shear stress independent and can be modelled with a simple "disperse plate model (Gleissle (1984)). Equations of the general form

$$\eta_r(\phi) = (1 - \phi / \phi_{max})^{-q} \qquad (1)$$

allow for describing the relative viscosity if $\tau > \tau_1$. The validity of this equation can be expanded to the shear stress domain $\tau_0 \leq \tau < \tau_1$ if the so called efficient disperse volume fraction ϕ_{eff} or the maximum packing fraction ϕ_{max} are defined as shear rate or shear stress dependent. Such a concept is described by Quemada (1998). Due to the fact that most often there is no analytical structure information available, this concept leads to an approximation of the concentrations ϕ_{eff} and/or ϕ_{max}. The way suggested here is to use the measurable characteristic stresses τ_0 and τ_1 to describe the relative viscosity over the whole domain of shear rates or shear stresses (eq.2). Equation 3 describes the related flow function considering τ_0 and τ_1.

$$\eta_r(\phi) = \frac{\eta_{\tau > \tau_1}}{\eta_f} + \frac{\tau_1}{\eta_f \dot{\gamma}}[1 - (1 - \frac{\tau_0}{\tau_1})\exp\{-\dot{\gamma}/\dot{\gamma}_*\}] \qquad (2)$$

$$\tau(\dot{\gamma}) = \tau_0 + \eta_f \dot{\gamma} + [\tau_1 - \tau_0][1 - \exp\{-\dot{\gamma}/\dot{\gamma}_*\}] \qquad (3)$$

(η_f = pure fluid viscosity; $\dot{\gamma}^*$ = charact. shear rate)

To connect the structure-based modelling assuming ϕ_{eff} and/or ϕ_{max} as shear rate or shear stress dependent to the measured stress based (τ_0, τ_1) modelling, simple relationships for $\tau_0(\phi_{eff}/\phi_{max})$ and $\tau_1(\phi_{eff}/\phi_{max})$ were found from experiments (see figure 2).

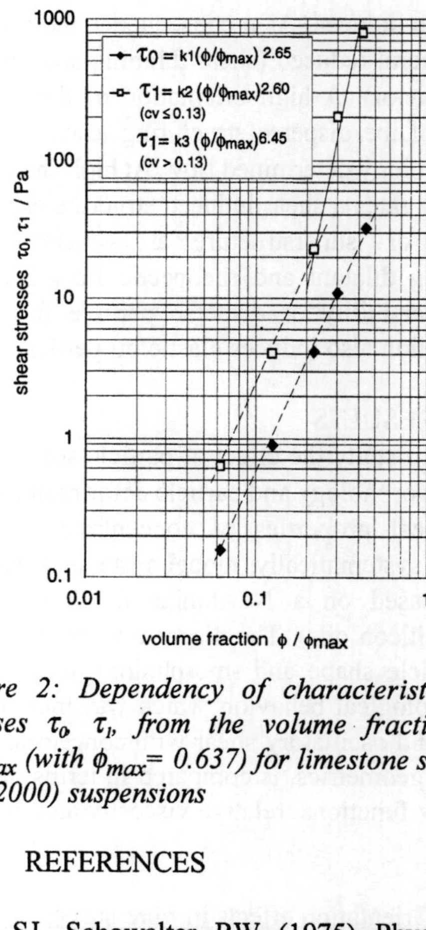

Figure 2: Dependency of characteristic shear stresses τ_0, τ_1, from the volume fraction ratio ϕ/ϕ_{max} (with ϕ_{max}= 0.637) for limestone silicon oil (AK 2000) suspensions

3. REFERENCES

Choi SJ, Schowalter RW (1975) Phys. Fluids 18:420-427

Gleissle W, Baloch MK (1984) Advances in Rheology, Proceedings IXth International Congress, Acapulco

Quemada D (1998) Eur. Phys. J. AP. 1:119-127

MULTI-STRUCTURAL THIXOTROPIC MODEL

J. LABANDA, J. LLORENS, C. MANS

Chemical Engineering Dept., University of Barcelona. c/ Martí i Franquès 1, 08028 Barcelona, Spain.

1. INTRODUCTION

Most colloidal solutions manifest thixotropy. When we handle these materials, the observed behaviour is nearly determined by its thixotropic character because the equilibrium viscosity is reached with difficulty.

Structural models have been proposed to explain thixotropic behaviour, Barnes (1997). These models deal with the rates of formation and breakdown of microstructures. Nevertheless, simple structural models cannot explain real thixotropic behaviour due to the simultaneous presence of different microstructures into the same material.

The purpose of this work is to advance in the development of a limitless structural model that can explain real thixotropic behaviour.

2. THEORY

We assume that several structures could be present simultaneously in a material. When some structures build-up other structures could break-down. This phenomenon can explain the real behaviour observed in materials that present positive and negative thixotropic behaviour at different shear rates.

The exposed model assigns a characteristic viscosity and kinetics for each assumed structure. Each i-th structure has a numerical scalar measure designed by λ_i. Completely built structure is represented by $\lambda_i = 1$ and completely broken-down structure as $\lambda_i = 0$. In order to normalize the model it is assumed that $\Sigma\lambda_i + \lambda_\infty = 1$. At very high shear rate, $\gamma \rightarrow \infty$, all structures are completely broken-down and $\lambda_\infty = 1$. In order to illustrate the model we depict an example of colloidal suspension with two microstructures in Fig.1.

Firstly, we assume that every structure contributes to the viscosity following a power-law viscosity

$$\eta(\dot{\gamma},\lambda_i) = \sum_i \eta_i(\dot{\gamma},\lambda_i) = \sum_i \eta_i' \cdot \dot{\gamma}^{m_i} \cdot \lambda_i \qquad (1)$$

Secondly, we presuppose a second order kinetic equation for the aggregation (or breakdown) with the shear rate.

$$\left.\frac{d\lambda_i}{dt}\right|_{\dot{\gamma}} = A_i \cdot \dot{\gamma}^{B_i} \cdot (\lambda_i - \lambda_{ie})^2 \qquad (2)$$

With this premises the model can deal with positive and negative thixotropy at different levels of the shear rate. The time-dependent viscosity when the material goes through a time dt and a shear rate $d\dot{\gamma}$ is :

$$d\eta = \sum_i m_i \frac{\eta_i}{\dot{\gamma}} d\dot{\gamma} + \sum_i \alpha_i \dot{\gamma}^{\beta} (\eta_i - \eta_{ei})^2 dt \qquad (3)$$

where α_i and β_i are kinetic constants and η_{ei} is the equilibrium viscosity.

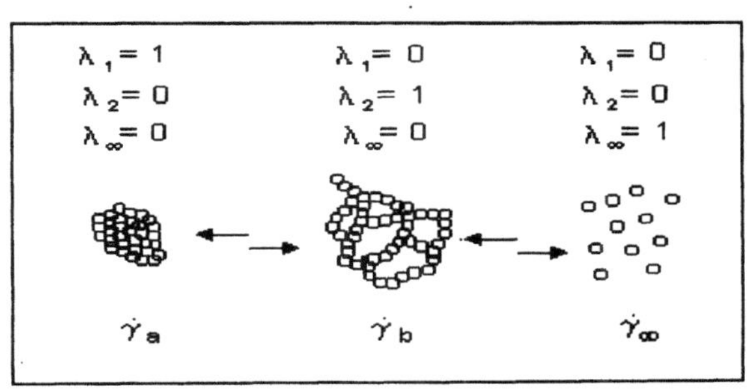

Fig. 1. Thixotropic model with two structures.

3. MATERIALS AND METHODS

The proposed model has been verified with polyelectrolite solutions of Carbopol (acrylic resins). Two Carbopol resins have been tested: Carbopol 941 (Mw = 1.250.000) and Carbopol 940 (Mw = 4.000.000).

The analyzed Carbopol solutions have been: 0.186% of Carbopol 941 at pH = 8.3, 0.153% Carbopol 940 at pH = 7.8, and a mixture of 0.1377% of Carbopol 940 and 0.0186% of Carbopol 941 at pH = 7.9.

The applied test has been: the sample has been keep at rest for t1 hours, then the shear rate has been increased for t2 hours, in a lineal form, from 0 s^{-1} to the maximum shear rate, $\dot{\gamma}_{max}$ s^{-1}, and followed by maintaining this maximum shear rate for a period of time t3 hours, finally the shear rate has been decreased for t4 hours, in a lineal form, from $\dot{\gamma}_{max}$ s^{-1} to 0 s^{-1}. Therefore, a particular test can be designated as t1-t2-t3-t4:$\dot{\gamma}_{max}$ (for example: 7-5-3-5:1,620).

The test have been carried out in the followed rheometers: Haake Rotovisco RV 100 with a coaxial cylinder sensor.

4. RESULTS

Many test have been done in the range of maximum shear rate of 200 to 1,620 s^{-1} with t1 = 7 hours, t2 = 2 and 5 hours, t3 = 0 and 3 hours, and t4 = 2 and 5 hours.

The solutions of Carbopol 941 show a clear positive thixotropy that can be modeled with a single structural parameter. The kinetic parameters have been adjusted and used in equation (3) to calculate the response for the shear stress for different shear rate *vs.* time inputs. The agreement between experimental and predicted shear stress is very satisfactory. The Fig. 2 shows the results obtained for the test 7-5-3-7:1,620

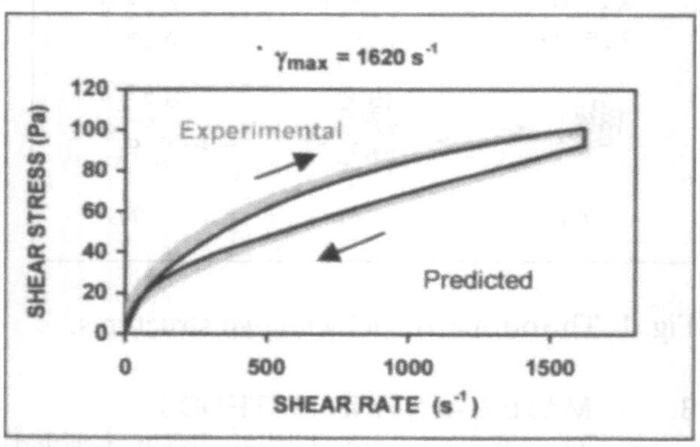

Fig. 2. Predicted and experimental shear stress for Carbopol 941 (0.186%) at pH = 8.3 in the test 7-5-3-7:1,620.

The solutions of Carbopol 940 show a clear negative thixotropy that can be modeled with a single structural parameter. The kinetic parameters have been adjusted and used in equation (3) to calculate the response for the shear stress for different shear rate *vs.* time inputs. The agreement between experimental and predicted shear stress is very satisfactory. The Fig. 3 shows the results obtained for the test 7-5-3-7:1,620

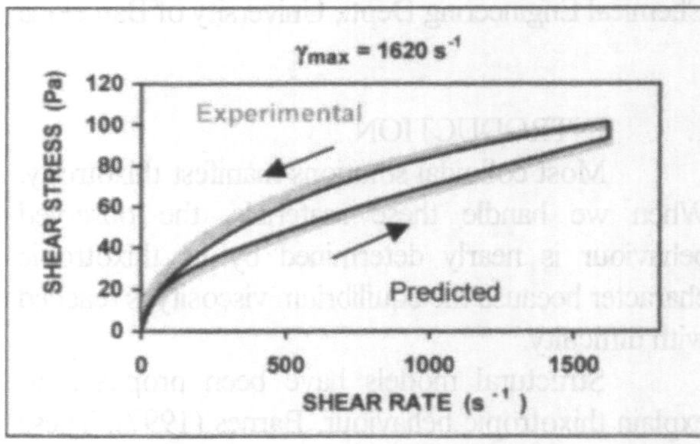

Fig. 3. Predicted and experimental shear stress for Carbopol 940 (0.153%) at pH = 7.8 in the test 7-5-3-7:1,620.

The mixture of 0.1377% of Carbopol 940 and 0.0186% of Carbopol 941 at pH = 7.9 show negative thixotropy that needs to be modeled with two structural parameters: one structural parameter for each polymer. We can observe a remarkable characteristic: the percentage of shear stress for every individual structure in front of the total shear stress is similar to the percentage of the components in the mixture (Fig. 4).

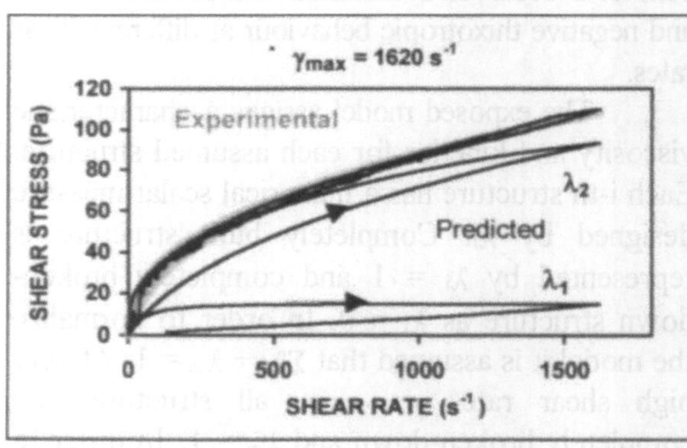

Fig. 4. Predicted and experimental shear stress for the mixture of 0.1377% of Carbopol 940 and 0.0186% of Carbopol 941 at pH = 7.9 in the test 7-5-5-0:1,620. Individual shear stress for each structure.

5. REFERNCES

Barnes HA (1997) J. Non-Newtonian Fluid Mech. 70:1-33

RHEOLOGICAL GEL POINT DETERMINATIONS IN SILICA AND TITANIUM BASED SOL-GEL SYSTEMS

E. RUDÉ, J. LLORENS, C. MANS

Chemical Engineering Dept., University of Barcelona, C/ Martí i Franquès 1-6, 08028 Barcelona, SPAIN

1. INTRODUCTION

This work deals with the characterization of the gel point (GP) from rheological data in the sol-gel transition. The main purposes of this work have been: (a) to develop a method for the GP determination of gelling systems with universal validity; (b) to analyze the composition dependence of the GP for two different kind of polymeric sols: silica and titanium based sol-gel systems.

2. MATERIALS AND METHODS

The studied sols have been prepared mixing different contents of the following starting materials: $Si(OEt)_4$, EtOH, HCl and H_2O for silica based sols and $Ti(OPr^i)_4$, CH_3COOH, Pr^iOH, HNO_3 and H_2O for titanium based sols. Three compositional parameters have been analyzed for each system: A, r, P and S, r, P for silica and titanium sols, respectively. These parameters are defined as follows: A= [EtOH] (%mol), r=[H_2O]/[alcoxide], P=[catalyst]/[alcoxide] and S=[AcOH]/[alcoxide]. The parameter ranges that have been studied are the following ones: $10 \leq A \leq 50$, $1.5 \leq r \leq 16.5$ and $0.03 \leq P \leq 0.07$ (silica sols) and $1.4 \leq S \leq 2.2$, $2.9 \leq r \leq 3.8$ and $0.06 \leq P \leq 0.12$ (titanium sols).

As regards the rheological monitoring of the sol-gel process, two different kind of rheological measurements have been used: *dynamic* tests and *creep-recovery* tests. Both tests have been performed in a *continuous* way (up to, through and after the gel point) on a controlled stress rheometer (HAAKE RS100).

3. RESULTS AND CONCLUSIONS

3.1. Gel point determination

Two methods have been developed to determine the gel point: one based on the evolution of the *dynamic* data and the other one based on the evolution of the *creep* data.

The first one, based on the evolution of the *dynamic* data, determines the GP by the system self-similarity proposed by Winter and Chambon (1986) for cross-linking systems. The self-similarity of a system is clearly depicted by the constancy of the loss factor (tan δ) in the whole frequency range (Holly et al., 1988). That means that the time evolution of the loss factor for all the frequencies tested will show a crossover point corresponding to the GP of the system. Figures 1 and 2 show the time evolution of the loss factor for the two systems studied. A quite different behavior is observed for both systems. From them it can be concluded that this method can be applied to silica sols but not to titanium ones. Therefore, a modification of the self-similarity method has been introduced to develop a new one, valid for both systems.

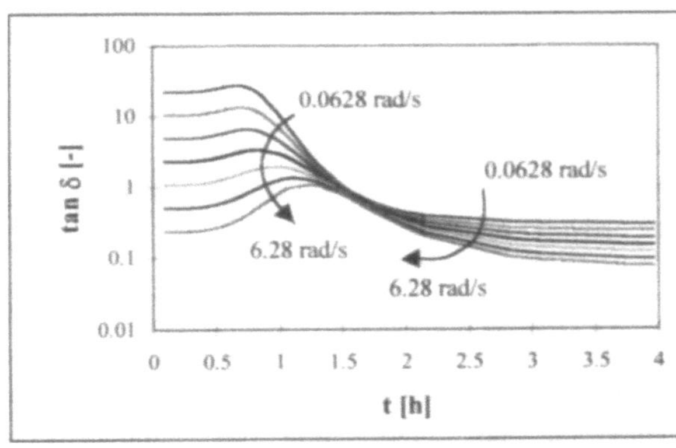

FIGURE 1. Typical time evolution of tanδ at different frequencies for a titanium sol

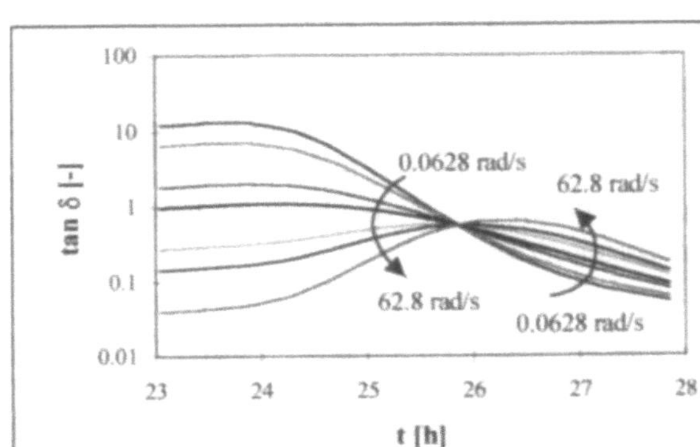

FIGURE 2. Typical time evolution of tanδ at different frequencies for a silica sol

The new method proposed lies the foundation in associating the GP to the occurrence of *"statistical loss factor self-similarity"* in the system. That means that the GP is characterized by a minimal $\log\left(s/\overline{\tan\delta}\right)$, where s is the standard deviation of $\tan\delta$ in a frequency sweep. This method allows to detect the GP in both systems and presents a universal validity.

The second method based on the evolution of the *creep* data associates the GP with the time corresponding to the inflection point of the zero-shear viscosity time evolution. This method has been validated with the first one. The results obtained with this last method lead to conclude that both methods are completely equivalent. Moreover, it has also been observed that both methods lead to gelation times lower than the macroscopic GP, determined by cessation of flow.

3.2 Influence of the sol composition on the GP

It is well known that the GP highly depends on the sol composition. The main conclusions achieved are the following:

It has been observed that the compositional parameters that have a greater influence on the GP are A, r and S, r for silica and titanium sols, respectively.

i) SILICA SOLS

The higher the molar percentage of ethanol (A), the higher the gelation time. The increase in the ethanol (solvent) concentration leads to a more diluted gel and, consequently, the entanglements among molecules are limited. The same results were obtained for Chang and Ring (1992) (see Fig. 3).

With regard to the influence of the water molar ratio, r, it has been found that the gelation time reaches a minimum at the equilibrium water concentration, which corresponds to r=4 (see Fig. 4). This minimum is due to the fact that the hydrolysis rate increases with r until it reaches the stoichiometric value, from which the hydrolysis rate begin to decrease because water is also a subproduct of the condensation reaction.

ii) TITANIUM SOLS

The addition of acetic acid produces an increase in the gelation time due to the reduction of the effective functionality of the alcoxide (see Fig. 5). The same effect is reached by a decrease in the hydrolysis ratio, r.

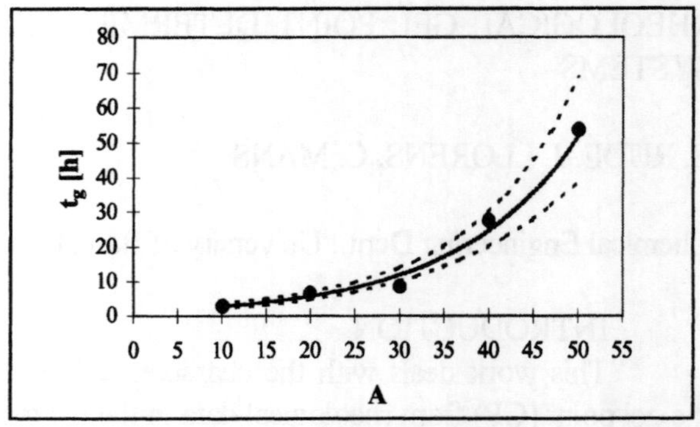

FIGURE 3. Variation of gelation time with ethanol molar percentage, A, for silica sols.

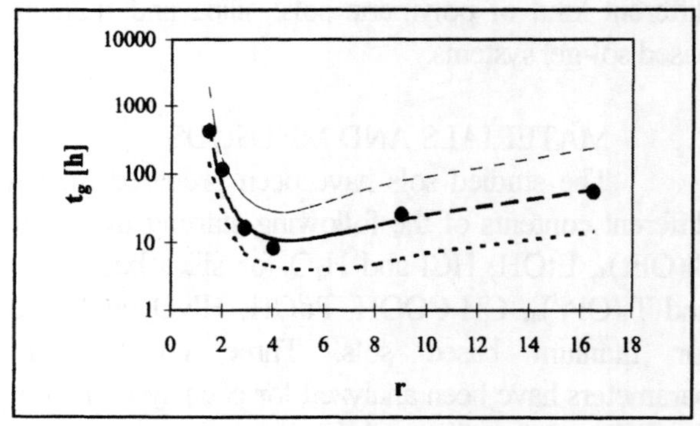

FIGURE 4. Variation of gelation time with the hydrolysis ratio, r, for silica sols.

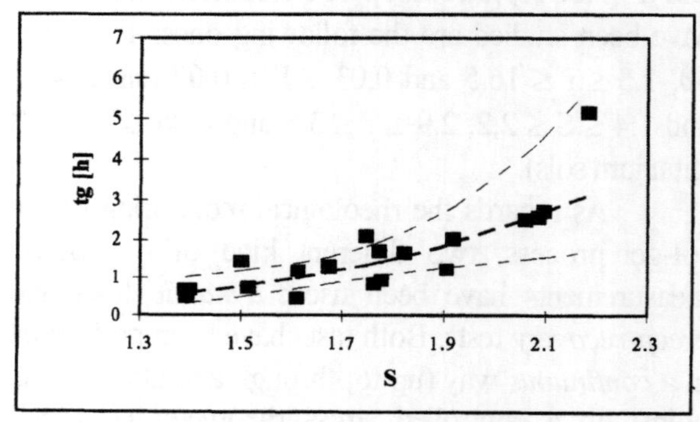

FIGURE 5. Variation of gelation time with acetic acid molar ratio, S, for titanium sols.

5. REFERENCES

Chang SY, Ring TA (1992) J Non-Cryst Solids 147&148:56-61

Holly EE, Venktaraman SK, Chambon F, Winter HH (1988) J Non-Newtonian Fluid Mech 13:17

Winter HH, Chambon F (1986) J Rheol 30:367

RELAXATION SPECTRUM EVOLUTION IN POLYMERIZATION REACTIONS

E. RUDÉ, J. LLORENS, C. MANS

Chemical Engineering Dept., University of Barcelona, C/ Martí i Franquès 1-6, 08028 Barcelona, SPAIN

1. INTRODUCTION

The evolution of the viscoelastic properties from liquid-like to solid-like behavior in a sol-to-gel transition has been measured starting immediately after mixing and proceeding up to, through and after the gel point. In this study the attention has been particularly focused on the time evolution of the *relaxation time spectrum*, $H(\lambda)$, since it is the function that characterizes the material in a most general way and its analysis yields information on the molecular weight distribution (MWD) and related properties (Llorens, 1993; Mead, 1996). Therefore, the main purpose of this work has been to characterize polymerization processes from the relaxation time spectra point of view.

2. MATERIALS

In the present work two kind of polymerization processes have been analyzed: the polymerization of titanium alkoxo-acetate sols and the polymerization of silica based sols.

The starting sols have been prepared mixing different proportions of $Ti(OPr^i)_4$, CH_3COOH, Pr^iOH, HNO_3, H_2O and $Si(OEt)_4$, EtOH, HCl, H_2O, for titanium and silica sols, respectively.

3. RHEOLOGICAL MONITORING

The polymerization reactions have been continuously monitored by dynamic measurements and creep-recovery tests -within the linear viscoelastic region- performed on a controlled stress rheometer (HAAKE RS100) at 30ºC for titanium sols and at 50ºC for silica sols. The creep-recovery tests allow the completion of the frequency sweep test, since it provides the rheological parameters for asymptotic frequency $\omega \to 0$, making it possible to get information about the terminal zones. Therefore, we get the whole rheological characterization of the system for all the reaction times explored during the process.

Direct experimental data of the viscous component η', the elastic component G' and the plateau modulus G_N^0 (information available from dynamic measurements), as well as the zero-shear viscosity (η_0) and the steady-state compliance (J_e^0) (information available from creep-recovery tests) up to, through and after the gel point have been obtained.

4. RELAXATION TIME SPECTRUM DETERMINATION

The relaxation time spectrum cannot be directly measured; rather it must be calculated from an experimentally determined material function. In this work the relaxation spectrum has indeed been calculated from the following expression:

$$G'(\omega) + \eta'(\omega) = \int_{-\infty}^{+\infty} \frac{(\omega\lambda)^2 + \lambda}{1 + (\omega\lambda)^2} H(\lambda)\, d(\ln \lambda) \qquad (1)$$

It is well known that inversion of such integral equation results in an extremely unstable problem (ill-posed problem). In order to solve the integral equation we have used the *maximum entropy method* with relaxation times equally spaced in log time and constrained with the integral moments given by the zero shear viscosity, η_0:

$$\eta_0 = \int_{-\infty}^{+\infty} \lambda\, H(\lambda)\, d(\ln \lambda) \qquad (2)$$

the steady-state recoverable compliance, $J^o{}_e$:

$$J_e^0 = \frac{\int_{-\infty}^{+\infty} \lambda^2\, H(\lambda)\, d(\ln \lambda)}{\left(\int_{-\infty}^{+\infty} \lambda\, H(\lambda)\, d(\ln \lambda)\right)^2} \qquad (3)$$

and the plateau modulus, G_N^0.

$$G_N^0 = \int_{-\infty}^{+\infty} H(\lambda)\, d(\ln \lambda) \qquad (4)$$

5. RESULTS AND CONCLUSIONS

In order to monitor the sol-gel process through the time evolution of the relaxation spectrum, several reaction times have been strategically chosen along the whole process to calculate the corresponding spectrum. To compare the spectra at different reaction times, the spectra must be "normalized". That means that each spectrum must be divided by the corresponding plateau modulus (see Eq. 4). Figure 1 shows the time evolution of the spectra for a particular silica sol. The ln $[H(\lambda)/G_N^0]$ is represented in front of the $\ln(t_r)$ and $\ln(\lambda)$, where t_r is the relative reaction time (t/t_g) and λ is the relaxation time.

On the other hand, Figure 2 shows the time evolution of the plateau modulus corresponding to the same silica sol.

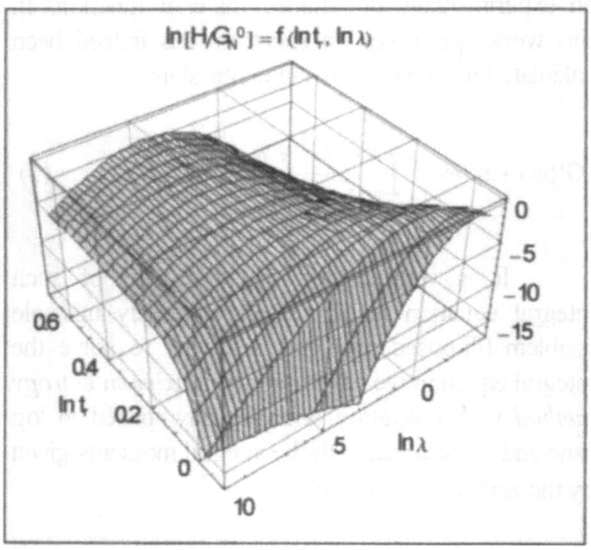

FIGURE 1. Time evolution of relaxation time spectra for a silica sol with 30% mol of EtOH, $4H_2O$:TEOS and 0.05HCl:TEOS.

Both figures are necessary to characterize the gelation process through the relaxation time spectra. From these graphics some deductions can be drawn about the way the system evolves from a structural and molecular point of view.

Initially, short relaxation time structures predominate in the system. As the gelation process is taking place, higher molecular species emerge in the reacting mixture, while the smallest ones disappear or considerably decrease.

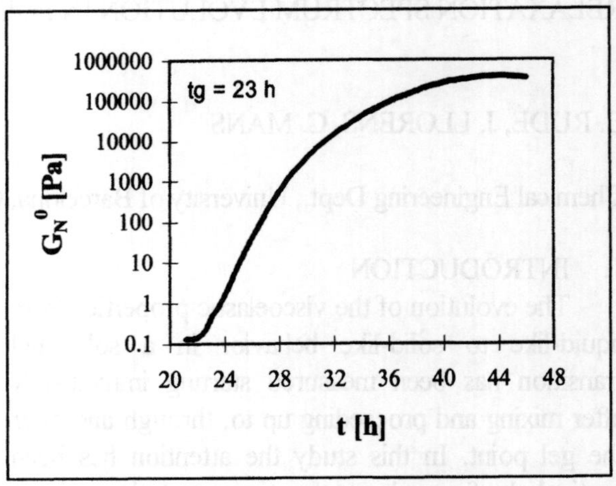

FIGURE 2. Time evolution of the plateau modulus for a silica sol with 30% mol of EtOH, $4H_2O$:TEOS and 0.05HCl:TEOS.

The gel point $[\ln(t_r)=0]$ can be easily identified by an *abrupt* change in relaxation time spectra. It also can be characterized by a linear spectrum in a wide relaxation time window, with a slope equal to the relaxation exponent Δ. This is indeed the exponent that characterizes the power law behavior of the dynamic moduli at the gel point.

The results have also shown that titanium based sols evolve to gel in a more gradual and smooth way than silica sols do, which gel much more abruptly.

Moreover, the quality of the relaxation spectra have been evaluated from their ability to reconstruct the experimental dynamic data (η'and G'). This also validates the mathematical procedure used in the present work to inverse the integral equation.

6. REFERENCES

Llorens J (1993) J Non Cryst Solids 162:188-196

Mead DW (1996) J Rheol 40:633-661

THE MODELLING RESEARCH OF THE OPTOTELECOMMUNICATION CABLE AND INNERDUCT DURING ITS INSTALLATION

R. Sikora, T. Klepka

Department of Polymer Processing, Technical University of Lublin, 36 Nadbystrzycka St., 20-618 Lublin, Poland

1. INTRODUCTION

It is possible to transmit high - quality signals such as data, acoustic signals and interactive pictures using the telecommunication techniques. In opposition to copper wires, modern glass optic fibres have the best transmission properties, but are extremely fragile. This is the reason for which thin fibres must be protected from moisture and stress to preserve their mechanical integrity, as well as to mitigate against signal distortion.

Optotelecommunication cables are provided with several plastic protection layers. In many cases the outer layer of the cable is made of polyethylene to obtain adequate strength properties. Cables are installed inside plastic pipes (called innerducts), which are also made of polyethylene and placed under the ground. The innerduct protects cables from mechanical failure and aggressive reaction of natural environment for a long term. In practice the cable installation inside innerduct for a long distance (max. to about 2 km) is a very difficult technological problem.

High resistance occurring in these circumstances is related to the friction between the cable and its innerduct. The assessment of such a complicated trybological effect is connected with polymer pair analysis and possibilities to reduce the value of friction resistance.

2. METHODS

There are many methods of trybological properties measurements developed by different research institutes. The test machine, which was constructed in our Department makes it possible to measure, precisely and reproducibly the friction resistance between the outer surface of the cable and inner surface of the innerduct. The front view of the test machine is shown in fig. 1. This construction enables simulation of the installation process in laboratory conditions. We can change some input factors such as: pipe wrap angle, speed of cable movement, load on its end, according to special requirements.

Fig. 1. Front view of the test machine to measure friction resistance

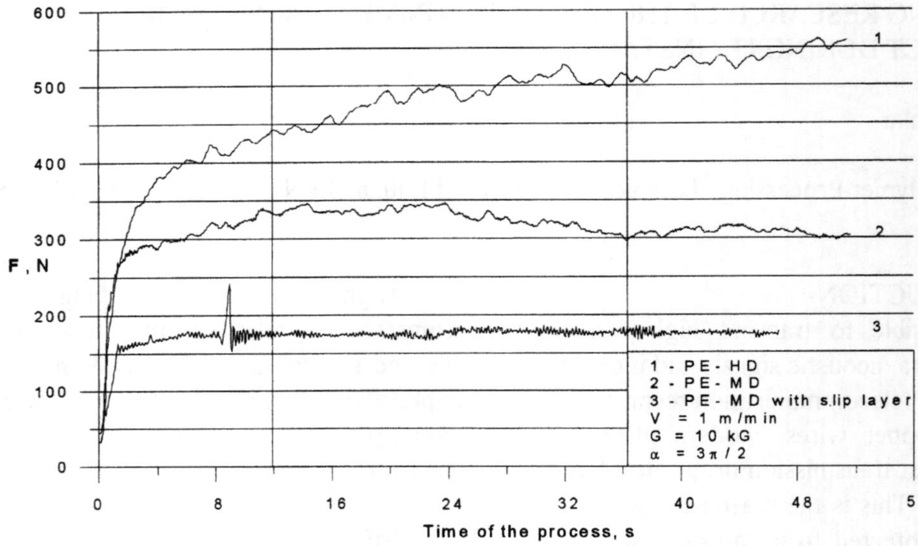

Fig. 2. Diagram of the example variation of the friction resistance force for different kinds of polyethylene.

Maximal values of the friction resistance force measured during laboratory tests enable to calculate a sliding coefficient of friction value for the analysed pair of polymer. The test cable is installed in the innerduct and then raped around the drum with the total angle 5π/2 rad (α). In addition innerducts are immobilised in the holders. One end of the cable was attached to the variable speed winch. The winch shall initiate the pulling of the cable with an approximate rate of 0,1 to 7 m/min. The second end of the cable was loaded with the weight (G) from 5 to 20 kg. Each test was finished, when the cable moved on the distance not less than about 0.7 m. Measurements of maximum friction resistance force (F) were made by means of force load-cell and inspection testing set. The value of sliding coefficient of friction (μ) was calculated according to the requirements of the Bellcore standards with an equivalent mathematical relation:

$$\mu = \frac{\ln \frac{F}{G}}{\alpha}$$

3. RESULTS

Tests were done using 11 mm outside diameter cable with the outer layer made of PE-HD. The example results of resistance force variation for different types of polyethylene innerducts are presented in fig. 2. Polyethylene pipes that are used in telecommunication industry have good mechanical and trybological properties, which depend on their material. Pipes with a special slip layer (with small thickness of 0,3 mm, achieved by co-extrusion or spraying) give the best results. This layer guarantees adequate trybological properties and effective resistance decrease, which occurs during cable installation in the innerduct.

4. FINAL NOTES

From presented research results of the friction resistance during installation of the optotelecommunication cable inside the PE pipes we get to know, that it is a very difficult and complicated process, especially when high mechanical and exploitation properties are required. In our Department some optimisation research is conducted as well, related to the influence of the innerducts material on the possibility of resistance decrease in the course of the cable installation inside them.

5. REFERENCES

Sikora R., Klepka T. (1995) Duct pipes for optotelecommunication,XIVInternational Scientific Conference Advanced Materials and Technologies AMT"95

GR-356-CORE (1995) Generic requirements for optical cable innerduct, Bellcore, USA.

Patent (1984) nr 4 565 351.

Patent (1992) nr 35 29 541.

AUTHOR INDEX

A

B

C

D

E

F

M

N

O

P

Q

R

S

MIX
Papier aus verantwortungsvollen Quellen
Paper from responsible sources
FSC® C105338

FSC
www.fsc.org

If you have any concerns about our products,
you can contact us on
ProductSafety@springernature.com

In case Publisher is established outside the EU,
the EU authorized representative is:
Springer Nature Customer Service Center GmbH
Europaplatz 3, 69115 Heidelberg, Germany

Printed by Libri Plureos GmbH
in Hamburg, Germany